Modular Theory in Operator Algebras

The first edition of this book appeared in 1981 as a direct continuation of *Lectures of von Neumann Algebras* (Ş. V. Strătilă and L. Zsidó) and, until 2003, was the only comprehensive monograph on the subject. Addressing students of mathematics and physics and researchers interested in operator algebras, noncommutative geometry and free probability, this revised edition covers the fundamentals and some latest developments in the field of operator algebras. The intent is to make modular theory accessible, with complete proofs, to readers having elementary training in operator algebras

The text provides detailed discussion of normal weights, conditional expectations and U. Haagerup's operator-valued weights, groups of automorphisms and their spectral theory, duality theory for noncommutative groups, and crossed products of von Neumann algebras by actions of groups and duals of groups, which enables the extension of M. Takesaki duality for noncommutative groups of automorphisms. It also contains detailed discussion on the group-measure space construction of factors and information about ITPFI factors and Krieger factors.

The core of the book is the continuous decomposition of A. Connes and M. Takesaki and discrete decomposition of A. Connes for type III factors.

It also explores new results, such as the A. Ocneanu's result on the actions of amenable groups on the hyperfinite factor, H. Kosaki's extension of the V. Jones index to arbitrary factors and F. Rădulescu's examples of non-hyperfinite factors of type III_λ, $\lambda \in (0,1)$ and of type III_1.

Şerban Strătilă is Professor at the Institute of Mathematics, Romanian Academy, and at the Department of Mathematics, University of Bucharest, Romania. His current research areas include operator algebras and representation theory. He received the 1975 Simion Stoilow Prize for Mathematics from the Romanian Academy. He has published *Lectures on von Neumann Algebras, 2nd edition* (2019) with Cambridge University Press.

CAMBRIDGE–IISc SERIES

Cambridge–IISc Series aims to publish the best research and scholarly work in different areas of science and technology with emphasis on cutting-edge research.

The books aim at a wide audience including students, researchers, academicians and professionals and are being published under three categories: research monographs, centenary lectures and lecture notes.

The editorial board has been constituted with experts from a range of disciplines in diverse fields of engineering, science and technology from the Indian Institute of Science, Bangalore.

Titles in print in this series:

- *Continuum Mechanics: Foundations and Applications of Mechanics* by C. S. Jog
- *Fluid Mechanics: Foundations and Applications of Mechanics* by C. S. Jog
- *Noncommutative Mathematics for Quantum Systems* by Uwe Franz and Adam Skalski
- *Mechanics, Waves and Thermodynamics* by Sudhir Ranjan Jain
- *Finite Elements: Theory and Algorithms* by Sashikumaar Ganesan and Lutz Tobiska
- *Ordinary Differential Equations: Principles and Applications* by A. K. Nandakumaran, P. S. Datti and Raju K. George
- *Lectures on von Neumann Algebras, 2nd Edition* by Serban Valentin Strătilă and László Zsidó
- *Biomaterials Science and Tissue Engineering: Principles and Methods* by Bikramjit Basu
- *Knowledge Driven Development: Bridging Waterfall and Agile Methodologies* by Manoj Kumar Lal
- *Partial Differential Equations: Classical Theory with a Modern Touch* by A. K. Nandakumaran and P. S. Datti

Cambridge–IISc Series

Modular Theory in Operator Algebras

Second Edition

Şerban Valentin Strătilă

CAMBRIDGE
UNIVERSITY PRESS

CAMBRIDGE
UNIVERSITY PRESS

University Printing House, Cambridge CB2 8BS, United Kingdom

One Liberty Plaza, 20th Floor, New York, NY 10006, USA

477 Williamstown Road, Port Melbourne, VIC 3207, Australia

314 to 321, 3rd Floor, Plot No.3, Splendor Forum, Jasola District Centre, New Delhi 110025, India

79 Anson Road, #06 04/06, Singapore 079906

Cambridge University Press is part of the University of Cambridge.

It furthers the University's mission by disseminating knowledge in the pursuit of education, learning and research at the highest international levels of excellence.

www.cambridge.org
Information on this title: www.cambridge.org/9781108489607

First edition published by Editura Academiei and Abacus Press, 1981
Second edition published by Cambridge University Press, 2020

Printed in India by Nutech Print Services, New Delhi 110020

A catalogue record for this publication is available from the British Library

ISBN 978-1-108-48960-7 Hardback

To the memory of my wife Sanda Strătilă,
the writer Alexandra Stănescu (1944–2011)

Contents

Preface to the Second Edition

The first edition of this book appeared in 1981 as a direct continuation of *Lectures of von Neumann Algebras* by Ş.V. Strătilă and L. Zsidó and, until 2003, was the only comprehensive monograph on the subject.

The book *Lectures on von Neumann Algebras*, 2nd Edition, Cambridge University Press, 2019, will be always referred to as [L]. It is assumed that the reader is familiar with the material contained in this book, including the terminology and notation.

The present book contains the continuous decomposition and the discrete decomposition for factors of type III and all the necessary results such as the extensive theory of normal weights including the U. Haagerup characterization of normality, the A. Connes theorem of Radon–Nikodym type and the Pedersen–Takesaki construction, the conditional expectations and the operator-valued weights, a detailed consideration of groups of automorphisms and their spectral theory, and the theory of crossed products. In order to include our extension of the Takesaki duality theorem to noncommutative groups of automorphisms, we considered a simultaneous generalization of groups and duals of groups, namely the Kac algebras, and their actions on von Neumann algebras, as well as the corresponding crossed products. Instead of Kac algebras we can also consider quantum groups, the arguments being exactly the same as for Kac algebras.

In this second edition we added information and references for several results, which appeared after 1981, untill now, including A. Ocneanu's theorem concerning actions of amenable groups and its extensions, H. Kosaki's extension of the index to arbitrary factors and F. Rădulescu's examples of non-hyperfinite factors of type III_λ, $\lambda \in (0, 1)$ and of type III_1 relying on D.-V. Voiculescu's theory of free probability.

For subjects such as the equivalence of injectivity and hyperfiniteness (originally due to A. Connes) we indicate references for more recent shorter proofs. This guarantees the uniqueness of the hyperfinite factor of type II_∞ and of type III_λ factors, $\lambda \in (0, 1)$. Another important result which could not be considered in detail, due to its length, is the proof of U. Haagerup of the A. Connes conjecture that the hyperfinite factor of type III_1 is unique. For the same reason we restricted ourselves to only a few results concerning Krieger factors and ITPFI factors. However, we considered in detail the "group-measure space construction" which produced many examples of factors, among them the first example of a type III factor by Murray and von Neumann, the Pukanszky non-hyperfinite factor of type III and also the famous Powers factors which are actually the only hyperfinite factors of type III_λ, $\lambda \in (0, 1)$.

I am grateful to Gadadhar Misra and V. Sunder for proposing this second edition, to Gadadhar Misra for his considerable help with the Latex conversion of the book, to Ms Rajitha Reddy for her excellent typing, to Florin Rădulescu for his help for Ocneanu's theorem and his own theorems, and to Alexandru Negrescu for his help in typing the new material for the second edition.

I am also grateful to Cambridge University Press and to Ms Taranpreet Kaur for editing this second edition.

Şerban Strătilă Bucharest, January 20, 2020

Preface to the First Edition

The discovery of the modular operator and the modular automorphism group associated with a normal semifinite faithful weight has led to a powerful theory – the modular theory – which is nowadays essential to the consideration of many problems concerning operator algebras. This theory has been developed in close association with the effort to understand the structure and produce examples and refined classifications of factors. Thus, the crossed product construction, which gave rise to the first non-trivial examples of factors, has been shown to play a fundamental role in the structure theory as well, by reducing the study of the purely infinite algebras to the study of the more familiar semifinite algebras and their automorphisms. Moreover, several algebraic invariants, previously defined only in some special cases, have been introduced via the modular theory for arbitrary factors and the corresponding classification has been proved to be almost complete for approximately finite dimensional factors.

The present book is a unified exposition of the technical tools of the modular theory and of its applications to the structure and classification of factors. It is based on several works recently published in periodicals or just circulated as preprints. The main sources used in writing this book are the works of W. B. Arveson, A. Connes, U. Haagerup, M. Landstad, G. K. Pedersen, M. Takesaki, and J. Tomiyama. The general treatment of crossed products follows an article by Ş. Strătilă, D. Voiculescu and L. Zsidó. Due to the wealth and variety of results recently obtained, it has not been possible to include here a detailed exposition of the classification of injective factors and their automorphisms; these topics and several others are just mentioned in the Notes sections, together with appropriate references.

The reader is assumed to have a good knowledge of the general theory of von Neumann algebras, including the standard forms. Actually, the present book can be viewed as a sequel to a previous book, Ş. Strătilă and L. Zsidó – *Lectures on von Neumann algebras*, Editura Academiei & Abacus Press, 1979, which is often quoted here and referred to as [L]. There is also an Appendix which contains some supplementary results on positive self-adjoint operators and introduces the terminology connected with W^*-algebras.

The list of references in the present book contains only those items which have been used, quoted or consulted. A more extensive bibliography is contained in [L] (and in the Preprint Series, INCREST, Bucharest) and the new preprints are periodically recorded in C^*–News (issued by CPT/CNRS, Marseille).

I am very indebted to Zoia Ceauşescu, Alain Connes, Sanda Strătilă and Dan Voiculescu for the moral support they offered me in writing this book. I am grateful to my colleagues Constantin Apostol, Grigore Arsene, Zoia Ceauşescu, Radu Gologan, Adrian Ocneanu, Cornel Pasnicu, Mihai Pimsner, Sorin Popa and Dan Timotin for several useful discussions and a critical reading of various parts of the manuscript. During his short visit in Romania, Alain Connes kindly informed me of the most recent developments of the theory. Thanks are also due to the National Institute for Scientific and Technical Creation, for the technical assistance.

It is a pleasure for me to acknowledge the most efficient and understanding cooperation of the Publishing House of the Romanian Academy (Editura Academiei) and Abacus Press, especially Mrs Sorana Gorjan, who edited this book, and Dr Simon Wassermann of Glasgow University, whose comments on the original translation were most helpful.

Şerban Strătilă Bucureşti, Romania, October 1979

CHAPTER I

Normal Weights

1 Characterizations of Normality

In this section, we prove the Theorem of Haagerup asserting that every normal weight on a W^*-algebra is the pointwise least upper bound of the normal positive forms it majorizes.

1.1. Let $\mathscr{A}$ be a C^*-algebra. A *weight* on $\mathscr{A}$ is a mapping $\varphi : \mathscr{A}^+ \to [0, +\infty]$ with the properties

$$\varphi(x + y) = \varphi(x) + \varphi(y), \varphi(\lambda x) = \lambda\varphi(x) \quad (x, y \in \mathscr{A}^+, \lambda \in \mathbb{R}^+).$$

The set

$$\mathfrak{F}_\varphi = \{x \in \mathscr{A}^+; \varphi(x) < +\infty\}$$

is a face of $\mathscr{A}^+$, the set

$$\mathfrak{N}_\varphi = \{x \in \mathscr{A}; \varphi(x^*x) < +\infty\}$$

is a left ideal of $\mathscr{A}$, and the set

$$\mathfrak{M}_\varphi = \mathfrak{N}_\varphi^*\mathfrak{N}_\varphi = \text{lin}\,\mathfrak{F}_\varphi$$

is a facial subalgebra of $\mathscr{A}$ with $\mathfrak{M}_\varphi \cap \mathscr{A}^+ = \mathfrak{F}_\varphi([L], 3.21)$, hence φ can be extended uniquely to a positive linear form, still denoted by φ, on the *-algebra $\mathfrak{M}_\varphi$.

A family $\mathscr{F}$ of weights on $\mathscr{A}$ is called *sufficient* if

$$x \in \mathscr{A} \text{ and } \varphi(a^*x^*xa) = 0 \text{ for all } \varphi \in \mathscr{F}, a \in \mathfrak{N}_\varphi \Rightarrow x = 0$$

and is called *separating* if

$$x \in \mathscr{A} \text{ and } \varphi(x^*x) = 0 \text{ for all } \varphi \in \mathscr{F} \Rightarrow x = 0.$$

In particular, the weight φ is called *faithful* if

$$x \in \mathscr{A} \text{ and } \varphi(x^*x) = 0 \Rightarrow x = 0.$$

1.2. Let φ be a weight on the C^*-algebra $\mathscr{A}$. The formula

$$(a|b)_\varphi = \varphi(b^*a) \quad (a, b \in \mathfrak{N}_\varphi)$$

defines a prescalar product on $\mathfrak{N}_\varphi$ with the properties:

$$(xa|xa)_\varphi \leq \|x\|^2 (a|a)_\varphi \quad (x \in \mathscr{A}, a \in \mathfrak{N}_\varphi),$$
$$(xa|b)_\varphi = (a|x^*b)_\varphi \quad (x \in \mathscr{A}, \quad a, b \in \mathfrak{N}_\varphi).$$

Let $\mathscr{H}_\varphi$ be the Hilbert space associated with $\mathfrak{N}_\varphi$ with the scalar product $(\cdot|\cdot)_\varphi$. It follows that there exists a *-representation $\pi_\varphi : \mathscr{A} \to \mathscr{B}(\mathscr{H}_\varphi)$, uniquely determined, such that

$$(\pi_\varphi(x)a_\varphi|b_\varphi)_\varphi = \varphi(b^*xa) \quad (x \in \mathscr{A}, a, b \in \mathfrak{N}_\varphi), \tag{1}$$

where $\mathfrak{N}_\varphi \ni a \mapsto a_\varphi \in \mathscr{H}_\varphi$ denotes the canonical mapping. The *-representation π_φ is called the *GNS representation* or the *standard representation* associated with φ. We remark that

$$\varphi(x^*) = \overline{\varphi(x)} \quad (x \in \mathfrak{M}_\varphi), \tag{2}$$
$$|\varphi(b^*a)|^2 \leq \varphi(a^*a)\varphi(b^*b) \quad (a, b \in \mathfrak{N}_\varphi). \tag{3}$$

1.3. Let $\mathscr{M}$ be a W^*-algebra. A weight φ on $\mathscr{M}$ is called *normal* if

$$\varphi(\sup_i x_i) = \sup_i \varphi(x_i)$$

for every norm-bounded increasing net $\{x_i\}_i \subset \mathscr{M}^+$, and *lower w-semicontinuous* if the convex sets

$$\{x \in \mathscr{M}^+; \varphi(x) \leq \lambda\} \quad (\lambda \in \mathbb{R}^+)$$

are w-closed. An important result concerning weights on W^*-algebras is the following characterization of normality:

Theorem (U. Haagerup). *Let φ be a weight on the W^*-algebra $\mathscr{M}$. The following statements are equivalent:*

(i) *φ is normal;*
(ii) *φ is lower w-semicontinuous;*
(iii) *$\varphi(x) = \sup\{f(x); f \in \mathscr{M}_*^+, f \leq \varphi\}$ for all $x \in \mathscr{M}^+$.*

Later (2.10, 5.8) we shall see that φ is normal if and only if it is a sum of normal positive forms, in accordance with the definition used in ([L], 10.14).

In Sections 1.4–1.7, we present some general results that will be used in the proof of the theorem; Sections 1.6–1.12 contain the main steps of the proof.

1.4 Proposition. *If $x, x_1, \ldots, x_n \in \mathscr{B}(\mathscr{H})$ and $x^*x = x_1^*x_1 + \ldots + x_n^*x_n$, then there exist $z_1, \ldots, z_n \in \mathscr{R}\{x, x_1, \ldots, x_n\}$ such that $z_1^*z_1 + \ldots + z_n^*z_n = \mathbf{s}(xx^*)$ and $x_k = z_k x$ for all $1 \leq k \leq n$.*

Proof. The equations $z_k(x\xi) = x_k\xi (\xi \in \mathscr{H})$ and $z_k\eta = 0 (\eta \in \mathscr{H} \bigoplus \overline{x\mathscr{H}})$ define operators $z_k \in \mathscr{B}(\mathscr{H})$, $\|z_k\| \leq 1$, with $x_k = z_k x$ and $z_k(\mathscr{H} \ominus \overline{x\mathscr{H}}) = 0$. Using the double commutant theorem

([L], 3.2) it is easy to check that $z_k \in \mathscr{R}\{x, x_k\}$. Also the relation $\sum_k z_k^* z_k = \mathbf{s}(xx^*)$ follows, since the positive operator $(\sum_k z_k^* z_k)^{1/2}$ vanishes on $\mathscr{H} \ominus \overline{x\mathscr{H}}$ and $x^*(\sum_k z_k^* z_k)x = x^*x$.

In particular if $x, y \in \mathscr{B}(\mathscr{H})$ and $y^*y \leq x^*x$, there exists $z \in \mathscr{R}\{x, y\}$ such that $z^*z \leq \mathbf{s}(xx^*)$ and $y = zx$. □

1.5. For each $\alpha > 0$ we shall consider the function

$$f_\alpha : (-\alpha^{-1}, +\infty) \to \mathbb{R}$$

defined by $f_\alpha(t) = t(1 + \alpha t)^{-1} = a^{-1}(1 - (1 + \alpha t)^{-1})$. These functions have the following properties:

$$f_\alpha(t) \leq \min\{t, \alpha^{-1}\} \qquad (t \in (-\alpha^{-1}, +\infty)) \tag{1}$$

$$\alpha \leq \beta \Rightarrow f_\alpha(t) \geq f_\beta(t) \qquad (t \in (-\alpha^{-1}, +\infty)) \tag{2}$$

$$\alpha \leq \beta \Rightarrow \alpha f_\alpha(t) \leq \beta f_\beta(t) \qquad (t \in (0, +\infty)) \tag{3}$$

$$f_\alpha(f_\beta(t)) = f_{\alpha+\beta}(t) \qquad (t \in (-(\alpha + \beta)^{-1}, +\infty)) \tag{4}$$

$$\lim_{\alpha \to 0} f_\alpha(t) = t \textit{ uniformly on compact subsets of } \mathbb{R} \tag{5}$$

$$\lim_{\alpha \to \infty} \alpha f_\alpha(t) = 1 \textit{ uniformly on compact subsets of } \mathbb{R}^+. \tag{6}$$

A continuous function $f : I \to \mathbb{R}$ is called *operator monotone* on the interval $I \subset \mathbb{R}$ if for every $x, y \in \mathscr{B}(\mathscr{H}), x = x^*, y = y^*$, with $Sp(x) \subset I, Sp(y) \subset I$, we have $x \leq y \Rightarrow f(x) \leq f(y)$. For instance, it is easy to see that

$$\text{the functions } f_\alpha \text{ are operator monotone } (\alpha \in \mathbb{R}^+). \tag{7}$$

Also, we recall that (see Pedersen, 1973–1977 or Strătilă & Zsidó, 1977, 1979)

$$\text{the functions } t \to t^\gamma \text{ are operator monotone } (0 < \gamma < 1). \tag{8}$$

On the other hand, using (A.2) we see that

$$\text{the functions } f_\alpha \text{ are operator continuous } (\alpha \in \mathbb{R}^+). \tag{9}$$

1.6. Let $\mathscr{X}$ be a locally convex Hausdorff real vector space with a partial ordering defined by a convex cone $\mathscr{X}^+ \subset \mathscr{X}$ such that $\mathscr{X}^+ \cap (-\mathscr{X}^+) = \{0\}$ and $\mathscr{X} = \overline{(\mathscr{X}^+ - \mathscr{X}^+)}$. The dual cone $\mathscr{X}_+^* = \{f \in \mathscr{X}^*; f(x) \geq 0 \text{ for all } x \in \mathscr{X}^+\}$ defines a partial ordering on $\mathscr{X}^*$. A subset $\mathscr{E}$ of $\mathscr{X}^+$ is called *hereditary* if

$$x \in \mathscr{E}, y \in \mathscr{X}^+, x - y \in \mathscr{X}^+ \Rightarrow y \in \mathscr{E}.$$

For $\mathscr{E} \subset \mathscr{X}^+$ and $\mathscr{F} \subset \mathscr{X}_+^*$, we define $\mathscr{E}^\wedge$ and $\mathscr{F}^\wedge$ by

$$\mathscr{E}^\wedge = \{f \in \mathscr{X}_+^*; f(x) \leq 1 \text{ for all } x \in \mathscr{E}\},$$
$$\mathscr{F}^\wedge = \{x \in \mathscr{X}^+; f(x) \leq 1 \text{ for all } f \in \mathscr{F}\}.$$

Proposition. *For $\mathscr{X}$ as above the following statements are equivalent:*

(i) $\mathscr{E} = \overline{(\mathscr{E} - \mathscr{X}^+)} \cap \mathscr{X}^+$ *for every closed hereditary convex subset* $\mathscr{E}$ *of* $\mathscr{X}^+$;
(ii) $\mathscr{E} = \mathscr{E}^{\wedge\wedge}$ *for every closed hereditary convex subset* $\mathscr{E}$ *of* $\mathscr{X}^+$;
(iii) *every subadditive, positively homogeneous, increasing and lower semicontinuous function* $\varphi : \mathscr{X}^+ \to [0, +\infty]$ *has the property*

$$\varphi(x) = \sup\{f(x); f \in \mathscr{X}^*_+, f \leq \varphi\} \quad (x \in \mathscr{X}^+).$$

Proof. We shall denote by $\mathscr{S}^0$ the polar of a subset $\mathscr{S}$ of $\mathscr{X}$ or $\mathscr{X}^*$.

(i) $\Rightarrow$ (ii). The sets $\mathscr{F} = \mathscr{E}^\wedge$ and $\mathscr{F}' = -(\mathscr{E} - \mathscr{X}^+)^0 = \{f \in \mathscr{X}^*; f(x) \leq 1$ for all $x \in \mathscr{E} - \mathscr{X}^+\}$ are equal. Indeed, it is clear that $\mathscr{F} \subset \mathscr{F}'$. Let $f \in \mathscr{F}'$ and $x \in \mathscr{X}^+$. Since $0 \in \mathscr{E}$, we have $f(-\lambda x) \leq 1$ for all $\lambda \geq 0$, whence $f(x) \geq 0$. Thus $\mathscr{F}' \subset \mathscr{X}^*_+$ and so $\mathscr{F}' \subset \mathscr{F}$.

By the bipolar theorem it follows that $\overline{(\mathscr{E} - \mathscr{X}^+)} = (\mathscr{E} - \mathscr{X}^+)^{00} = (-\mathscr{F})^0 = \{x \in \mathscr{X}; f(x) \leq 1$ for all $f \in \mathscr{F}\}$ and, using (i), we get $\mathscr{E} = \overline{(\mathscr{E} - \mathscr{X}^+)} \cap \mathscr{X}^+ = \{x \in \mathscr{X}^+; f(x) \leq 1$ for all $f \in \mathscr{F}\} = \mathscr{E}^{\wedge\wedge}$.

(ii) $\Rightarrow$ (iii). If φ satisfies the conditions required in (iii), then the set $\mathscr{E} = \{x \in \mathscr{X}^+; \varphi(x) \leq 1\}$ is closed, hereditary and convex. Also, $\mathscr{F} = \mathscr{E}^\wedge = \{f \in \mathscr{X}^*_+; f(x) \leq \varphi(x)$ for all $x \in \mathscr{X}^+\}$ and, by (ii), $\{x \in \mathscr{X}^+; \varphi(x) \leq 1\} = \mathscr{E} = \mathscr{F}^\wedge = \{x \in \mathscr{X}^+; \sup_{f \in \mathscr{F}} f(x) \leq 1\}$. It follows that $\varphi(x) = \sup\{f(x); f \in \mathscr{F}\}$, for all $x \in \mathscr{X}^+$.

(iii) $\Rightarrow$ (i). Let $\mathscr{E} \subset \mathscr{X}^+$ be closed, hereditary and convex. Define $\varphi(x) = \inf\{\lambda > 0; x \in \lambda\mathscr{E}\}$ if $x \in \bigcup_{\lambda>0} \lambda\mathscr{E}$ and $\varphi(x) = +\infty$ otherwise. Then φ satisfies the hypotheses in (iii) and therefore $\varphi(x) = \sup\{f(x); f \in \mathscr{F}\} (x \in \mathscr{X}^+)$, where $\mathscr{F} = \{f \in \mathscr{X}^*_+; f(x) \leq \varphi(x)$ for all $x \in \mathscr{X}^+\}$. It follows that $\mathscr{E} - \mathscr{X}^+ \subset \{x \in \mathscr{X}; f(x) \leq 1$ for all $f \in \mathscr{F}\}$ and, since the latter set is closed, we get $\overline{(\mathscr{E} - \mathscr{X}^+)} \cap \mathscr{X}^+ \subset \{x \in \mathscr{X}^+; f(x) \leq 1$ for all $f \in \mathscr{F}\} \subset \mathscr{E}$, hence $\overline{(\mathscr{E} - \mathscr{X}^+)} \cap \mathscr{X}^+ = \mathscr{E}$. □

1.7 Proposition. *Let $\mathscr{M}$ be a W^*-algebra and $\mathscr{E} \subset \mathscr{M}^+$ a w-closed hereditary convex set. Then $\mathscr{E} = \overline{(\mathscr{E} - \mathscr{M}^+)}^w \cap \mathscr{M}^+$.*

Proof. We shall use the properties of the functions f_α from 1.5. For $x \in \mathscr{M}_h$ let $\alpha_x = \sup\{\alpha > 0; -\alpha^{-1} \leq x\}$. Consider the set

$$\mathscr{S} = \{x \in \mathscr{M}_h; f_\alpha(x) \in \mathscr{E} - \mathscr{M}^+ \text{ for all } \alpha \in (0, \alpha_x)\},$$

and let $\mathscr{M}_\lambda;. = \{x \in \mathscr{M}; \|x\| \leq \lambda\}$.

We first show that for every $\lambda > 0$ the set $\mathscr{S} \cap \mathscr{M}_\lambda$ is s-closed.

Indeed, let $x \in \overline{\mathscr{S} \cap \mathscr{M}_\lambda}^s$. There is a net $\{x_i\}_{i \in I} \subset \mathscr{S}$ such that $\|x_i\| \leq \lambda$ and $x_i \xrightarrow{s} x$. Then $\alpha_{x_i} \geq 1/\lambda$, hence $f_\alpha(x_i) \in \mathscr{E} - \mathscr{M}^+$ for every $\alpha \in (0, 1/2\lambda)$ and every $i \in I$. Let $\alpha \in (0, 1/2\lambda)$ be fixed. There is a net $\{y_i\}_{i \in I} \subset \mathscr{E}$ such that

$$f_\alpha(x_i) \leq y_i \quad (i \in I).$$

Since f_α is operator monotone,

$$f_{2\alpha}(x_i) = f_\alpha(f_\alpha(x_i)) \leq f_\alpha(y_i) \quad (i \in I).$$

Since $f_{2\alpha}$ is operator continuous on $[-\lambda, +\lambda]$,

$$f_{2\alpha}(x_i) \xrightarrow{s} f_{2\alpha}(x).$$

Since $0 \leq f_\alpha(y_i) \leq \alpha^{-1}$ and $\mathcal{M}_1$ is w-compact, we may assume that there is $y \in \mathcal{M}$ such that

$$f_\alpha(y_i) \xrightarrow{w} y.$$

Since $0 \leq f_\alpha(y_i) \leq y_i \in \mathcal{E}$ and $\mathcal{E}$ is hereditary, $f_\alpha(y_i) \in \mathcal{E}$ and, since $\mathcal{E}$ is w-closed, it follows that $y \in \mathcal{E}$. Then

$$y - f_{2\alpha}(x) = w\text{-}\lim_i (f_\alpha(y_i) - f_{2\alpha}(x_i)) \geq 0,$$

hence $f_{2\alpha}(x) \in \mathcal{E} - \mathcal{M}^+$. We have thus proved that

$$f_\alpha(x) \in \mathcal{E} - \mathcal{M}^+ \text{ for every } \alpha \in (0, 1/\lambda).$$

Consider now $\alpha \in [1/\lambda, \alpha_x)$ and $\beta \in (0, 1/\lambda)$. Then $f_\alpha(x) \leq f_\beta(x)$, hence $f_\alpha(x) \in (\mathcal{E} - \mathcal{M}^+) - \mathcal{M}^+ = \mathcal{E} - \mathcal{M}^+$. We conclude that $x \in \mathcal{S} \cap \mathcal{M}_\lambda$. □

We now show that $\mathcal{S}$ is convex.

Indeed, it is sufficient to show that each $\mathcal{S} \cap \mathcal{M}_\lambda$ is convex, and this will follow from the equality

$$\mathcal{S} \cap \mathcal{M}_\lambda = \overline{((\mathcal{E} - \mathcal{M}^+) \cap \mathcal{M}_\mu)}^s \cap \mathcal{M}_\lambda \quad \text{for } \mu > \lambda.$$

If $x \in \mathcal{S} \cap \mathcal{M}_\lambda$, then $f_\alpha(x) \in \mathcal{E} - \mathcal{M}^+$ for $\alpha \in (0, \alpha_x)$ and $f_\alpha(x) \in \mathcal{M}_\mu$ for small $\alpha > 0$, hence

$$x = s\text{-}\lim_{\alpha \to 0} f_\alpha(x) \in \overline{((\mathcal{E} - \mathcal{M}^+) \cap \mathcal{M}_\mu)}^s \cap \mathcal{M}_\lambda.$$

Conversely, since $\mathcal{E}$ is hereditary and $f_\alpha(x) \leq x$ for all $\alpha \in (0, \alpha_x)$, we have $\mathcal{E} - \mathcal{M}^+ \subset \mathcal{S}$, hence $(\mathcal{E} - \mathcal{M}^+) \cap \mathcal{M}_\mu \subset \mathcal{S} \cap \mathcal{M}_\mu$. Using the first part of the proof we get $\overline{((\mathcal{E} - \mathcal{M}^+) \cap \mathcal{M}_\mu)}^s \subset \mathcal{S} \cap \mathcal{M}_\mu$, and the desired inclusion follows.

Using the Krein-Šmulian theorem ([L], C.1.1; Dunford & Schwartz, 1958, 1963, V.5.7), from the earlier it follows that $\mathcal{S}$ is w-closed. We have seen that $\mathcal{E} - \mathcal{M}^+ \subset \mathcal{S}$. Since $x = w\text{-}\lim_{\alpha \to 0} f_\alpha(x)$, we obtain $\mathcal{S} \subset \overline{(\mathcal{E} - \mathcal{M}^+)}^w$. Consequently, $\mathcal{S} = \overline{(\mathcal{E} - \mathcal{M}^+)}^w$.

Finally, let $x \in \overline{(\mathcal{E} - \mathcal{M}^+)}^w \cap \mathcal{M}^+ = \mathcal{S} \cap \mathcal{M}^+$. For every $\alpha > 0$ we have $f_\alpha(x) \in (\mathcal{E} - \mathcal{M}^+) \cap \mathcal{M}^+$, hence $f_\alpha(x) \in \mathcal{E}$, as $\mathcal{E}$ is hereditary. It follows that $x = w\text{-}\lim_{\alpha \to 0} f_\alpha(x) \in \mathcal{E}$.

From Propositions 1.6 and 1.7, we obtain the equivalence (ii) ⇒ (iii) in Theorem 1.3, as the implication (iii) ⇒ (ii) is obvious.

In Sections 1.8–1.12, we assume that φ is a fixed normal weight on the W^*-algebra $\mathcal{M}$.

1.8 Lemma. *There exists a linear mapping $\Phi : \mathfrak{M}_\varphi \to \pi_\varphi(\mathcal{M})'_*$, uniquely determined, such that*

$$\Phi(b^* a)(T') = (T' a_\varphi | b_\varphi)_\varphi \quad (T' \in \pi_\varphi(\mathcal{M})', a, b \in \mathfrak{N}_\varphi). \tag{1}$$

Moreover, for every $x \in \mathfrak{M}_\varphi \cap \mathcal{M}_h$ *we have*

$$\|\Phi(x)\| = \inf\{\varphi(y) + \varphi(z); y, z \in \mathfrak{M}_\varphi \cap \mathcal{M}^+, x = y - z\}. \tag{2}$$

Proof. The uniqueness of Φ follows from the relation $\mathfrak{M}_\varphi = \mathfrak{N}_\varphi^* \mathfrak{N}_\varphi$.

If $a, b, c \in \mathfrak{N}_\varphi, c^* = c$ and $c^*c = a^*a + b^*b$, then, by Proposition 1.4, there exist $x, y \in \mathcal{M}$ such that $a = xc, b = yc$ and $x^*x + y^*y = \mathbf{s}(cc^*) = \mathbf{s}(c)$, and for every $T' \in \pi_\varphi(\mathcal{M})'$ we have

$$\begin{aligned}(T'c_\varphi|c_\varphi)_\varphi &= (T'\pi_\varphi(x^*x + y^*y)c_\varphi|c_\varphi)_\varphi \\ &= (T'\pi_\varphi(x)c_\varphi|\pi_\varphi(x)c_\varphi)_\varphi + (T'\pi_\varphi(y)c_\varphi|\pi_\varphi(y)c_\varphi)_\varphi \\ &= (T'a_\varphi|a_\varphi)_\varphi + (T'b_\varphi|b_\varphi)_\varphi.\end{aligned}$$

It follows that the mapping

$$\Phi_0 : \mathfrak{M}_\varphi \cap \mathcal{M}^+ \ni a^*a \mapsto \omega_{a_\varphi}|\pi_\varphi(\mathcal{M})' \in \pi_\varphi(\mathcal{M})'_*.$$

is well defined and additive. Clearly, Φ_0 is positively homogeneous. Since $\mathfrak{M}_\varphi = \operatorname{lin}(\mathfrak{M}_\varphi \cap \mathcal{M}^+)$, Φ_0 has a unique linear extension Φ to $\mathfrak{M}_\varphi$ and (1) follows using the polarization relation ([L], p. 75).

The function ρ defined on $\mathfrak{M}_\varphi \cap \mathcal{M}_h$ by the right-hand side of (2) is a semi-norm on $\mathfrak{M}_\varphi \cap \mathcal{M}_h$. If $x \in \mathfrak{M}_\varphi \cap \mathcal{M}^+$, then $\|\Phi(x)\| = \Phi(x)(1) = ((x^{1/2})_\varphi|(x^{1/2})_\varphi)_\varphi = \varphi(x) = \rho(x)$. Consequently, for $x = y - z$, with $y, z \in \mathfrak{M}_\varphi \cap \mathcal{M}^+$, we have $\|\Phi(x)\| \le \|\Phi(y)\| + \|\Phi(z)\| = \varphi(y) + \varphi(z)$. Hence $\|\Phi(x)\| \le \rho(x)$ for all $x \in \mathfrak{M}_\varphi \cap \mathcal{M}_h$.

Let $x_0 \in \mathfrak{M}_\varphi \cap \mathcal{M}_h$. By the Hahn–Banach theorem, there exists a real linear form f on $\mathfrak{M}_\varphi \cap \mathcal{M}_h$ such that $f(x_0) = \rho(x_0)$ and $|f(x)| \le \rho(x)$ for every $x \in \mathfrak{M}_\varphi \cap \mathcal{M}_h$. Then f can be extended to a complex linear form, still denoted by f, on $\mathfrak{M}_\varphi$. Since $-\varphi(x) \le f(x) \le \varphi(x)$ for any $x \in \mathfrak{M}_\varphi \cap \mathcal{M}^+$, we may consider $\varphi + f$ and $\varphi - f$ as weights on $\mathcal{M}$. Consequently, using the Schwarz inequality 1.2.(3), for $a, b \in \mathfrak{N}_\varphi$ we obtain

$$\begin{aligned}f(b^*a) &\le 2^{-1}[|(\varphi + f)(b^*a)| + |(\varphi - f)(b^*a)|] \\ &\le 2^{-1}[(\varphi + f)(a^*a)^{1/2}(\varphi + f)(b^*b)^{1/2} + (\varphi - f)(a^*a)^{1/2}(\varphi - f)(b^*b)^{1/2}] \\ &\le 2^{-1}[(\varphi + f)(a^*a) + (\varphi - f)(a^*a)]^{1/2}[(\varphi + f)(b^*b) + (\varphi - f)(b^*b)]^{1/2} \\ &= \varphi(a^*a)^{1/2}\varphi(b^*b)^{1/2} = \|a_\varphi\|_\varphi \|b_\varphi\|_\varphi.\end{aligned}$$

Thus, there exists an operator $T' \in \mathcal{B}(\mathcal{H}_\varphi), \|T'\| \le 1$, such that $f(b^*a) = (T'a_\varphi/b_\varphi)_\varphi$ for all $a, b \in \mathfrak{N}_\varphi$. Moreover $T' \in \pi_\varphi(\mathcal{M})'$, since for every $x \in \mathcal{M}$ and every $a, b \in \mathfrak{N}_\varphi$ we have

$$(T'\pi_\varphi(x)a_\varphi|b_\varphi)_\varphi = f(b^*xa) = (\pi_\varphi(x)T'a_\varphi|b_\varphi)_\varphi.$$

It follows that $\rho(x_0) = |f(x_0)| = |\Phi(x_0)(T')| \le \|\Phi(x_0)\|\|T'\| \le \|\Phi(x_0)\|$. □

1.9 Lemma. *Let* $\{x_n\}$ *be a norm-bounded sequence in* $\mathfrak{M}_\varphi \cap \mathcal{M}^+$ *such that the sequence* $\{\Phi(x_n)\}$ *is norm-convergent in* $\pi_\varphi(\mathcal{M})'_*$. *Then*

$$x_n \overset{s}{\to} x \in \mathcal{M} \Rightarrow x \in \mathfrak{M}_\varphi \cap \mathcal{M}^+, \tag{1}$$

$$x_n \overset{s}{\to} 0 \Rightarrow \|\Phi(x_n)\| \to 0. \tag{2}$$

Proof. Let $\varepsilon > 0$ and $\psi = \lim_n \Phi(x_n) \in \pi_\varphi(\mathcal{M})'_*$. Without loss of generality we may assume that $\|\Phi(x_n) - \psi\| < \varepsilon/2^n$, so that $\|\Phi(x_{n+1} - x_n)\| < \varepsilon/2^{n-1}$ for all $n \in \mathbb{N}$. By Lemma 1.8, there exist sequences $\{y_n\}$ and $\{z_n\}$ in $\mathfrak{M}_\varphi \cap \mathcal{M}^+$ such that

$$x_{n+1} - x_n = y_n - z_n \text{ and } \varphi(y_n) + \varphi(z_n) < \varepsilon/2^{n-1} \quad (n \in \mathbb{N}).$$

We shall again use the functions f_α from Section 1.5.

(1) Since $x_{n+1} \le x_1 + \sum_{k=1}^n y_k$ and $x_{n+1} \xrightarrow{s} x$, we obtain

$$f_\alpha(x) = s\text{-}\lim_n f_\alpha(x_{n+1}) \le \sup_n f_\alpha\left(x_1 + \sum_{k=1}^n y_k\right)$$

and then, using the normality of φ,

$$\begin{aligned}\varphi(f_\alpha(x)) &\le \sup_n \varphi\left(f_\alpha\left(x_1 + \sum_{k=1}^n y_k\right)\right) \le \sup_n \varphi\left(x_1 + \sum_{k=1}^n y_k\right)\\ &\le \varphi(x_1) + \sum_{k=1}^\infty \varphi(y_k) \le \varphi(x_1) + \sum_{k=1}^\infty \varepsilon/2^{k-1} = \varphi(x_1) + 2\varepsilon.\end{aligned}$$

Since $f_\alpha(x) \uparrow x$, again using the normality of φ we get

$$\varphi(x) = \sup_{\alpha>0} \varphi(f_\alpha(x)) \le \varphi(x_1) + 2\varepsilon < +\infty,$$

hence $x \in \mathfrak{M}_\varphi \cap \mathcal{M}^+$.

(2) Since $-\sup_n \|x_n\| \le x_1 - x_{n+1} \le \sum_{k=1}^n z_k$, for $\alpha > (\sup_n \|x_n\|)^{-1}$ we obtain

$$f_\alpha(x_1 - x_{n+1}) \le \sup_n f_n\left(\sum_{k=1}^n z_k\right).$$

Since $x_1 - x_{n+1} \xrightarrow{s} x_1$, it follows that

$$f_\alpha(x_1) = s\text{-}\lim_n f_\alpha(x_1 - x_{n+1}) \le \sup_n f_\alpha\left(\sum_{k=1}^n z_k\right).$$

Using the normality of φ we infer that

$$\begin{aligned}\varphi(x_1) &= \sup_{\alpha>0} \varphi(f_\alpha(x_1)) \le \sup_{\alpha>0}\sup_n \varphi\left(f_\alpha\left(\sum_{k=1}^n z_k\right)\right)\\ &\le \sup_n \varphi\left(\sum_{k=1}^n z_k\right) \le \sum_{k=1}^\infty \varepsilon/2^{k-1} = 2\varepsilon.\end{aligned}$$

Consequently, $\|\psi\| \le \|\psi - \Phi(x_1)\| + \|\Phi(x_1)\| \le \varepsilon/2 + 2\varepsilon = 3\varepsilon/2$. We conclude that $\psi = 0$. □

1.10. Let $\mathscr{G}_\varphi = \{(x, x_\varphi); x \in \mathfrak{N}_\varphi\} \subset \mathscr{M} \times \mathscr{H}_\varphi$. Since every Hilbert space is a reflexive Banach space, $\mathscr{M} \times \mathscr{H}_\varphi$ is the dual of the Banach space $\mathscr{M}_* \times \mathscr{H}_\varphi$. For $\lambda, \mu > 0$, let $\mathscr{M}_\lambda = \{x \in \mathscr{M}; \|x\| \leq \lambda\}$ and $(\mathscr{H}_\varphi)_\mu = \{\xi \in \mathscr{H}_\varphi; \|\xi\| \leq \mu\}$.

Lemma. *If $\mathscr{M}$ is countably decomposable, then $\mathscr{G}_\varphi \cap (\mathscr{M}_\lambda \times (\mathscr{H}_\varphi)_\mu)$ is $\sigma(\mathscr{M} \times \mathscr{H}_\varphi, \mathscr{M}_* \times \mathscr{H}_\varphi)$-compact, for ever $\mu > 0$.*

Proof. Since $\mathscr{G}_\varphi \cap (\mathscr{M}_\lambda \times (\mathscr{H}_\varphi)_\mu)$ is convex and bounded, it is sufficient to prove that it is closed with respect to the product topology τ on $\mathscr{M} \times (\mathscr{H}_\varphi)$ of the s^*-topology on $\mathscr{M}$ and the norm-topology on $\mathscr{H}_\varphi$. Since $\mathscr{M}$ is countably decomposable, $\mathscr{M}_\lambda$ is s^*-metrizable ([L], E.5.7; Strătilă & Zsidó, 1977, 1979, 8.12). □

If $(x, \xi) \in \mathscr{M} \times \mathscr{H}_\varphi$ is τ-adherent to $\mathscr{G}_\varphi \cap (\mathscr{M}_\lambda \times (\mathscr{H}_\varphi)_\mu)$, then there exists a sequence $\{x_n\}$ in $\mathscr{M}_\lambda$ such that $x_n \xrightarrow{s^*} x$, $\|(x_n)_\varphi\|_\varphi \leq \mu$ and $\|(x_n)_\varphi - \xi\| \to 0$. Then $x_n^* x_n \xrightarrow{s^*} x^* x$ and $\Phi(x_n^* x_n) = \omega_{(x_n)_\varphi}$ is norm convergent to ω_ξ, whence $x \in \mathfrak{N}_\varphi$ by Lemma 1.8.(1). On the other hand, $(x_n - x)^*(x_n - x) \xrightarrow{s} 0$ and

$$\Phi((x_n - x)^*(x_n - x)) = \omega_{(x_n)_\varphi - x_\varphi} \to \omega_{\xi - x_\varphi}$$

so that $\omega_{\xi - x_\varphi} = 0$ by Lemma 1.8.(2). Thus $\xi = x_\varphi$ and $(x, \xi) \in \mathscr{G}_\varphi$.

1.11. If $\mathscr{M}$ is not countably decomposable, we consider the set $\mathscr{P}_0$ of all countably decomposable projections of $\mathscr{M}$ and put

$$\mathscr{M}_0 = \bigcup_{p \in \mathscr{P}_0} p\mathscr{M}p.$$

It is easy to check that $\mathscr{M}_0$ is a self-adjoint ideal in $\mathscr{M}$.

Lemma. *Let $\mathscr{E}$ be a hereditary convex subset of $\mathscr{M}_0 \cap \mathscr{M}^+$. Then $\mathscr{E}$ is w-closed in $\mathscr{M}_0$ if and only if $\mathscr{E} \cap p\mathscr{M}p$ is w-closed for every $p \in \mathscr{P}_0$.*

Proof. Assume that $\mathscr{E} \cap p\mathscr{M}p$ is w-closed for every $p \in \mathscr{P}_0$. The set $\mathscr{F} = \{x \in \mathscr{M}; x^*x \in \mathscr{E}\}$ is convex and $a\mathscr{F} \subset \mathscr{F}$ for all $a \in \mathscr{M}_1$.

We first show that $p\mathscr{F}$, or equivalently $\mathscr{F}^*p$, is w-closed for any $p \in \mathscr{P}_0$. Using the Krein-Šmulian theorem and the fact that any s-closed convex set is also w-closed, it is sufficient to show that $\mathscr{F}^*p \cap \mathscr{M}_\lambda$ is s-closed for every $\lambda > 0$. Let $x \in \mathscr{M}$ be such that x^* is s-adherent to $\mathscr{F}^*p \cap \mathscr{M}_\lambda$. Since $\mathscr{M}p \cap \mathscr{M}_\lambda$ is s-metrizable, there exists a sequence $\{x_n\}$ in $p\mathscr{F}$, $\|x_n\| \leq \lambda$, with $x_n^* \xrightarrow{s} x^*$. There exists a projection $q \in \mathscr{P}_0$ such that $x_n \in q\mathscr{M}q$ for all $n \in \mathbb{N}$. Thus

$$x_n \in \mathscr{F} \cap q\mathscr{M}q = \{y \in q\mathscr{M}q; y^*y \in \mathscr{E} \cap q\mathscr{M}q\} \quad (n \in \mathbb{N}).$$

By assumption, $\mathscr{E} \cap q\mathscr{M}q$ is w-closed, hence $\mathscr{F} \cap q\mathscr{M}q$ is s-closed. It follows that $\mathscr{F} \cap q\mathscr{M}q$ is w-closed and, consequently, $x \in \mathscr{F} \cap q\mathscr{M}q$. Since $px = x$ and $\|x\| \leq \lambda$, we get $x^* \in \mathscr{F}^*p \cap \mathscr{M}_\lambda$. Hence $p\mathscr{F}$ is w-closed. □

Let $x \in \mathcal{M}_0$ be w-adherent to $\mathcal{E}$. There exists a net $\{x_i\}_{i\in I} \subset \mathcal{E}$ such that $x_i \xrightarrow{s} x$. Then $p = \mathbf{l}(x) \in \mathcal{P}_0$ and $px_i^{1/2} \xrightarrow{s} px^{1/2} = x^{1/2}$. By the earlier paragraph, we know that $p\mathcal{F}$ is s-closed, hence $x^{1/2} \in p\mathcal{F} \subset \mathcal{F}$, that is, $x \in \mathcal{E}$. Hence $\mathcal{E}$ is w-closed in $\mathcal{M}_0$.

The converse is obvious.

1.12. *Proof of Theorem* 1.3. As we have already seen (1.7), (ii) $\Leftrightarrow$ (iii). The implication (ii) $\Rightarrow$ (i) is obvious. To show that (i) $\Rightarrow$ (ii), we have to prove that the set

$$\mathcal{E} = \{x \in \mathcal{M}^+; \varphi(x) \leq 1\}$$

is w-closed. Clearly, $\mathcal{E}$ is hereditary and convex.

Assume first that $\mathcal{M}$ is countably decomposable. As in the last part of the proof of Lemma 1.11, it is sufficient to show that the set $\mathcal{F} = \{x \in \mathcal{M}; \varphi(x^*x) \leq 1\}$ is w-closed. Since $\mathcal{F} \cap \mathcal{M}_\lambda$ is the image of $\mathcal{G}_\varphi \cap (\mathcal{M}_\lambda \times (\mathcal{H}_\varphi)_1)$ by the canonical projection mapping $(x, \xi) \mapsto x$, from Lemma 1.10 it follows that $\mathcal{F} \cap \mathcal{M}_\lambda$ is w-compact for every $\lambda > 0$. Since $\mathcal{F}$ is convex, we infer that $\mathcal{F}$ is w-closed.

Consider now the general case. By the earlier argument and by Lemma 1.11, it follows that $\mathcal{E} \cap \mathcal{M}_0$ is w-closed in $\mathcal{M}_0$. Let $x \in \mathcal{M}^+$ be w-adherent to $\mathcal{E}$. There exists a net $\{x_i\}_{i\in I} \subset \mathcal{E}$ such that $x_1 \xrightarrow{s} x$. Also, there exists an increasing net $\{p_k\}_{k\in K} \subset \mathcal{P}_0$ with $p_k \uparrow 1$. Since $\mathcal{M}_0$ is a two-sided ideal in $\mathcal{M}$, for every $k \in K$ we have

$$\mathcal{E} \cap \mathcal{M}_0 \ni x_i^{1/2} p_k x_i^{1/2} \xrightarrow{w}_{i\in I} x^{1/2} p_k x^{1/2} \in \mathcal{M}_0,$$

hence $x^{1/2}p_k x^{1/2} \in \mathcal{E} \cap \mathcal{M}_0$. Since $x^{1/2}p_k x^{1/2} \uparrow x$, using the normality of φ we infer that $\varphi(x) = \sup_{k\in K} \varphi(x^{1/2}p_k x^{1/2}) \leq 1$, that is, $x \in \mathcal{E}$.

1.13. We recall that a positive form φ on the W^*-algebra $\mathcal{M}$ is normal if and only if it is completely additive on projections ([L], 5.6, 5.11). This statement cannot be extended to weights, as the following example shows.

Let $\ell^\infty(\mathbb{N})$ be the W^*-algebra of all bounded complex sequences. The weight φ defined on $\ell^\infty(\mathbb{N})$ by $\varphi(\{a_n\}) = \sum_n a_n$ if the set $\{n \in \mathbb{N}; a_n \neq 0\}$ is finite, and $\varphi(\{a_n\}) = +\infty$ otherwise, is completely additive on projections, but is not normal.

1.14 Proposition. *Let φ be a normal weight on the W^*-algebra $\mathcal{M}$ and $a, b \in \mathfrak{N}_\varphi$. Then the mapping*

$$\varphi(b^* \cdot a) : \mathcal{M} \ni x \mapsto \varphi(b^*xa) \in \mathbb{C}$$

is a w-continuous linear form on $\mathcal{M}$.

Proof. Since $a, b \in \mathfrak{N}_\varphi$, for any $x \in \mathcal{M}$ we have $b^*xa \in \mathcal{R}_\varphi^*\mathfrak{N}_\varphi = \mathfrak{M}_\varphi$, hence $\varphi(b^* \cdot a)$ is well defined. If $x_i \uparrow x$ in $\mathcal{M}^+$, then $a^*x_ia \uparrow a^*xa$ in $\mathcal{M}^+$, and hence $\varphi(a^*x_ia) \uparrow \varphi(a^*xa)$, since φ is normal. It follows that $\varphi(a^* \cdot a)$ is w-continuous ([L], 5.11) and the general case is obtained using a polarization relation ([L], 3.21). □

1.15. Notes. The main result (Thm. 1.3) of this section is due to Haagerup (1975a).

For our exposition we have used Haagerup (1975a) and [L].

2 The Standard Representation

In this section, we prove that every normal semifinite weight is the supremum of an upward directed family of normal positive forms; also, we review and complete the results in ([L], Chapter 10) concerning the associated standard representation.

2.1. Let φ be a normal weight on the W^*-algebra $\mathscr{M}$.

Using ([L], 2.22) and the normality of φ it is easy to see that

$$x \in \mathscr{M}^+, \varphi(x) = 0 \Rightarrow \varphi(\mathbf{s}(x)) = 0. \tag{1}$$

If $e, f \in \mathscr{M}$ are projections and $\varphi(e) = \varphi(f) = 0$, then $\varphi(e \vee f) = \varphi(\mathbf{s}(e + f)) = 0$. Thus the family $\mathscr{E} = \{e \in \text{Proj}(\mathscr{M}); \varphi(e) = 0\}$ is upward directed. Let $e_0 = \sup \mathscr{E}$. By the normality of φ, it follows that $\varphi(e_0) = 0$, so that e_0 is the greatest projection in $\mathscr{M}$ annihilated by φ. The projection $\mathbf{s}(\varphi) = 1 - e_0$ is called the *support* of φ. Using (1), we obtain

$$\varphi(x^*x) = 0 \Leftrightarrow x\mathbf{s}(\varphi) = 0 \quad (x \in \mathscr{M}). \tag{2}$$

In particular, φ is faithful (1.1) if and only if $\mathbf{s}(\varphi) = 1$. Also

$$\varphi(x) = \varphi(\mathbf{s}(\varphi)x\mathbf{s}(\varphi)) \quad (x \in \mathscr{M}^+). \tag{3}$$

On the other hand, the w-closure $\overline{\mathfrak{N}_\varphi}^w$ of $\mathfrak{N}_\varphi$ is a w-closed left ideal of $\mathscr{M}$, hence $\overline{\mathscr{R}_\varphi}^w = \mathscr{M}e$ for some projection $e \in \mathscr{M}$ and $\overline{\mathfrak{M}_\varphi}^w = e\mathscr{M}e$ ([L], 3.20, 3.21). The weight φ is called *semifinite* if $e = 1$, that is, if $\mathfrak{N}_\varphi$, or equivalently, $\mathfrak{M}_\varphi$, is w-dense in $\mathscr{M}$. In this case, there exists an increasing net $\{u_i\}_{i \in I}$ in $\mathfrak{F}_\varphi = \mathfrak{M}_\varphi \cap \mathscr{M}^+$ such that $u_i \uparrow 1$ ([L], 3.20, 3.21).

We abbreviate the words "normal semifinite faithful" to *n.s.f.* Recall that on every W^*-algebra there exists an n.s.f. weight, while the countably decomposable W^*-algebras are characterized by the existence of a normal faithful positive form ([L], 10.14, E.5.6).

2.2 Theorem. *Let φ be a normal weight on the W^*-algebra $\mathscr{M}$. Then the associated GNS representation $\pi_\varphi : \mathscr{M} \to \mathscr{B}(\mathscr{H}_\varphi)$ is normal and nondegenerate. If φ is semifinite, then*

$$((\mathfrak{M}_\varphi)^n)_\varphi \text{ is dense in } \mathscr{H}_\varphi \quad (n \in \mathbb{N}). \tag{1}$$

*If φ is an n.s.f. weight, then π_φ is a *-isomorphism of $\mathscr{M}$ onto the von Neumann algebra $\pi_\varphi(\mathscr{M}) \subset \mathscr{B}(\mathscr{H}_\varphi)$.*

Proof. Clearly, $\pi_\varphi(1) = 1$, hence π_φ is nondegenerate. To show that π_φ is normal, that is, w-continuous, we have to check that $\omega \circ \pi_\varphi \in \mathscr{M}_*$ for every $\omega \in \mathscr{B}(\mathscr{H}_\varphi)_*$. Since the vector forms are total in $\mathscr{B}(\mathscr{H}_\varphi)_*$ ([L], 1.3) and $\mathfrak{N}_\varphi$ is dense in $\mathscr{H}_\varphi$, it is sufficient to do this only for $\omega = \omega_{a_\varphi, b_\varphi}$ with $a, b \in \mathfrak{N}_\varphi$. In this case, we have $\omega_{a_\varphi, b_\varphi} \circ \pi_\varphi = \varphi(b^* \cdot a) \in \mathscr{M}_*$, by Proposition 1.14. Since π_φ is normal and nondegenerate, $\pi_\varphi(\mathscr{M}) \subset \mathscr{B}(\mathscr{H}_\varphi)$ is a von Neumann algebra ([L], 3.12).

If φ is semifinite, then there exists an increasing net $\{u_i\}_{i \in I}$ in $\mathfrak{F}_\varphi = \mathfrak{M}_\varphi \cap \mathscr{M}^+$ with $u_i \uparrow 1$. For $a \in \mathfrak{N}_\varphi$, we have

$$\|a_\varphi - (u_i a)_\varphi\|_\varphi^2 = \varphi((a - u_i a)^*(a - u_i a)) \leq 2[\varphi(a^*a) - \varphi(a^*u_i a)] \to 0. \tag{2}$$

Since $u_i \in \mathfrak{F}_\varphi \subset \mathfrak{N}_\varphi^*$ and $a \in \mathfrak{N}_\varphi$, we have $u_i a \in \mathfrak{N}_\varphi^* \mathfrak{N}_\varphi = \mathfrak{M}_\varphi$ and from (2) it follows that $(\mathfrak{M}_\varphi)_\varphi$ is dense in $\mathfrak{N}_\varphi$, hence also in $\mathscr{H}_\varphi$. Statement (1) follows now using (2) repeatedly.

Assume that φ is an n.s.f. weight. If $x \in \mathscr{M}$ and $\pi_\varphi(x) = 0$, then $\varphi((xa)^*(xa)) = \|\pi_\varphi(x)a_\varphi\|_\varphi^2 = 0$ for all $a \in \mathfrak{N}_\varphi$. Since φ is faithful it follows that $x\mathfrak{N}_\varphi = 0$ and hence $x = 0$, as φ is semifinite. Consequently, π_φ is a *-isomorphism.

In the next three sections, we study the majorization relation between weights in terms of the associated GNS representation. □

2.3 Proposition. *Let φ, ψ be weights on the C^*-algebra $\mathscr{A}$ such that $\psi \le \varphi$, that is, $\psi(x) \le \varphi(x)$ for all $x \in \mathscr{A}^+$. There exists a unique operator $T' \in \pi_\varphi(\mathscr{A})', 0 \le T' \le 1$, such that*

$$\psi(b^*a) = (T'a_\varphi | T'b_\varphi)_\varphi \quad (a, b \in \mathfrak{N}_\varphi). \tag{1}$$

Proof. Since $\psi \le \varphi$ we have $\mathfrak{N}_\varphi \subset \mathfrak{N}_\psi$ and, for every $a \in \mathfrak{N}_\varphi$, $\|a_\psi\|_\psi^2 = \psi(a^*a) \le \varphi(a^*a) = \|a_\varphi\|_\varphi^2$. It follows that there exists a unique linear operator $S' : \mathscr{H}_\varphi \to \mathscr{H}_\psi$, $\|S'\| = 1$, such that $S'a_\varphi = a_\psi$ for all $a \in \mathfrak{N}_\varphi$. Then $T' = (S'^*S')^{1/2} \in \mathscr{B}(\mathscr{H}_\varphi), 0 \le T' \le 1$. For every $a, b \in \mathfrak{N}_\varphi$ and every $x \in \mathscr{A}$ we have

$$\psi(b^*a) = (a_\psi | b_\psi)_\psi = (S'a_\varphi | S'b_\varphi)_\psi = (T'^2 a_\varphi | b_\varphi)_\varphi = (T'a_\varphi | T'b_\varphi)_\varphi$$
$$(S'^*S'\pi_\varphi(x)a_\varphi | b_\varphi)_\varphi = \psi(b^*xa) = \psi((x^*b)^*a) = (\pi_\varphi(x)S'^*S'a_\varphi | b_\varphi)_\varphi,$$

hence $T'^2 = S'^*S' \in \pi_\varphi(\mathscr{A})'$. Since $\pi_\varphi(\mathscr{A})'$ is a C^*-algebra, we infer that $T' = (T'^2)^{1/2} \in \pi_\varphi(\mathscr{A})'$. □

From (1) it follows that the numbers $(T'^2 a_\varphi | b_\varphi)_\varphi (a, b \in \mathfrak{N}_\varphi)$ are uniquely determined by ψ and φ, and this implies the uniqueness of T'.

If φ and ψ are finite and $\mathscr{A}$ is unital, then from (1) it follows that

$$\psi(x) = (\pi_\varphi(x)T'1_\varphi | T'1_\varphi)_\varphi \quad (x \in \mathscr{A}). \tag{2}$$

2.4 Corollary. *Let φ be a weight on the C^*-algebra $\mathscr{A}$ and denote by $\mathscr{T}_\varphi'$ the set of all $T' \in \pi_\varphi(\mathscr{A})'$ such that there exists some $\lambda_{T'} > 0$ with the property that $\|T'a_\varphi\|_\varphi \le \lambda_{T'}\|a\|$ for all $a \in \mathfrak{N}_\varphi$. Then $\mathscr{T}_\varphi'$ is a left ideal of the W^*-algebra $\pi_\varphi(\mathscr{A})'$ and:*

(1) for every positive form f on $\mathscr{A}$ with $f \le \varphi$ there exist a unique $T' \in \mathscr{T}_\varphi', 0 \le T' \le 1$, and a unique $\eta \in \overline{\pi_\varphi(\mathfrak{N}_\varphi^*)\mathscr{H}_\varphi}$ such that

$$f(b^*a) = (T'a_\varphi | T'b_\varphi)_\varphi \quad (a, b \in \mathfrak{N}_\varphi);$$
$$f(x) = (\omega_\eta \circ \pi_\varphi)(x) \quad (x \in \mathfrak{M}_\varphi).$$

(2) for every $T' \in \mathscr{T}_\varphi', 0 \le T' \le 1$, there exist a positive form $f \le \varphi$ on $\mathscr{A}$ and a unique $\eta \in \overline{\pi_\varphi(\mathfrak{N}_\varphi^*)\mathscr{H}_\varphi}$ such that

$$f(b^*a) = (T'a_\varphi | T'b_\varphi)_\varphi \quad (a, b \in \mathfrak{N}_\varphi);$$
$$T'a_\varphi = \pi_\varphi(a)\eta \quad (a \in \mathfrak{N}_\varphi).$$

Proof. It is clear that $\mathcal{T}_\varphi'$ is a left ideal of $\pi_\varphi(\mathcal{A})'$. Also, if $f \leq \varphi$ is a positive form on $\mathcal{A}$, we infer from 2.3.(1) that $\|T'a_\varphi\|_\varphi \leq \|f\|^{1/2}\|a\|(a \in \mathfrak{N}_\varphi)$, hence $T' \in \mathcal{T}_\varphi'$.

Let $\{u_i\}_{i\in I}$ be a right approximate unit for the left ideal $\mathfrak{N}_\varphi$ of $\mathcal{A}$ ([L], 3.20). For $T' \in \mathcal{T}_\varphi'$ and $a \in \mathfrak{N}_\varphi$ it follows that

$$\pi_\varphi(a)T'(u_i)_\varphi = T'(au_i)_\varphi \to T'a_\varphi.$$

Thus, if $a_k \in \mathfrak{N}_\varphi, \xi_k \in \mathcal{H}_\varphi (1 \leq k \leq n)$, and $\zeta = \sum_{k=1}^n \pi_\varphi(a_k^*)\xi_k \in \pi_k(\mathfrak{N}_\varphi^*)\mathcal{H}_\varphi$, then

$$\left|\sum_{k=1}^n (\xi_k|T'(a_k)_\varphi)_\varphi\right| = \lim_i |(\zeta|T'(u_i)_\varphi)_\varphi| \leq \lambda_{T'}\|\zeta\|_\varphi.$$

It follows that the mapping $\zeta \mapsto \sum_{k=1}^n (\xi_k|T'(a_k)_\varphi)_\varphi$ defines a bounded linear form on $\pi_\varphi(\mathfrak{N}_\varphi^*)\mathcal{H}_\varphi$ and, consequently, there exists a unique vector $\eta \in \overline{\pi_\varphi(\mathfrak{N}_\varphi^*)\mathcal{H}_\varphi}$ such that

$$(\xi|T'a_\varphi)_\varphi = (\xi|\pi_\varphi(a)\eta)_\varphi \quad (a \in \mathfrak{N}_\varphi, \xi \in \mathcal{H}_\varphi),$$

that is, $T'a_\varphi = \pi_\varphi(a)\eta$ for all $a \in \mathfrak{N}_\varphi$. In particular, $f = \omega_\eta \circ \pi_\varphi$ is a positive form on $\mathcal{A}$ and $f(b^*a) = (T'a_\varphi|T'b_\varphi)_\varphi$ for all $a, b \in \mathfrak{N}_\varphi$. □

2.5 Corollary. *Let φ be a normal weight on the W^*-algebra $\mathcal{M}$ and f_0 a positive form on $\mathcal{M}$. If $f_0 \leq \varphi$, then there exists a normal positive form f on $\mathcal{M}$ such that $f \leq \varphi$ and $f|\mathfrak{M}_\varphi = f_0|\mathfrak{M}_\varphi$.*

Proof. By Corollary 2.4.(1), there exists a vector $\eta \in \mathcal{H}_\varphi$ such that $f_0(x) = (\omega_\eta \circ \pi_\varphi)(x)$ for every $x \in \mathfrak{M}_\varphi$. Since φ is normal, π_φ is normal (2.2), so we can take $f = \omega_\eta \circ \pi_\varphi$. □

2.6. By Haagerup's theorem (1.3), every normal weight φ on the W^*-algebra $\mathcal{M}$ is the pointwise supremum of the family $\mathfrak{F}_\varphi = \{f \in \mathcal{M}_*^+; f \leq \varphi\}$; thus φ is also the pointwise supremum of the family $\{f \in \mathcal{M}_*^+;$ there exists $\varepsilon > 0$ such that $(1+\varepsilon)f \leq \varphi\}$. The next result shows, in particular, *that every normal semifinite weight on a W^*-algebra is the pointwise supremum of an upward directed family of normal positive forms.*

Theorem (F. Combes). *Let φ be a normal semifinite weight on the W^*-algebra $\mathcal{M}$. Then the family*

$$\{f \in \mathcal{M}_*^+; \textit{ there exists } \varepsilon > 0 \textit{ such that } (1+\varepsilon)f \leq \varphi\}$$

is upward directed.

We first prove two general results of independent interest.

2.7 Proposition. *Let $\mathfrak{N}$ be a left ideal of the W^*-algebra $\mathcal{M}$. Then $\mathcal{F} = (\mathfrak{N}^*\mathfrak{N}) \cap \mathcal{M}^+$ is a face of $\mathcal{M}^+, \mathfrak{N} = \{x \in \mathcal{M}; x^*x \in \mathcal{F}\}$ and $\mathfrak{N}^*\mathfrak{N} = lin\, \mathcal{F}$. In particular, $\mathfrak{N}^*\mathfrak{N}$ is a facial subalgebra of $\mathcal{M}$.*

Proof. Clearly, $\mathfrak{N} \subset \{x\mathcal{M}; x^*x \in \mathcal{F}\}$ and, by the polarization relation ([L], 3.21), $\mathfrak{N}^*\mathfrak{N} = \operatorname{lin} \mathcal{F}$.

Let $x \in \mathcal{M}$ be such that $x^*x \leq b \in \mathcal{F}$. Since b is self-adjoint, using again the polarization relation we can find $x_k, y_k \in \mathfrak{N}, (1 \leq k \leq n)$ such that

$$x^*x \leq b = \sum_{k=1}^n x_k^*x_k - \sum_{k=1}^n y_k^*y_k \leq \sum_{k=1}^n x_k^*x_k = a$$

By Proposition 1.4, there exist $z, z_k \in \mathcal{M}(1 \leq k \leq n)$ such that $x = za^{1/2}, x_k = z_k a^{1/2}$ and $\sum_{k=1}^{n} z_k^* z_k =$ s(a). It follows that

$$x = za^{1/2} = z\left(\sum_{k=1}^{n} z_k^* z_k\right) a^{1/2} = \sum_{k=1}^{n} z z_k^* x_k \in \mathfrak{N}.$$

Hence $\mathcal{F}$ is a face of $\mathcal{M}^+$ and $\{x \in \mathcal{M}; x^* x \in \mathcal{F}\} \subset \mathfrak{N}$. □

2.8 Proposition. *Let $\mathcal{A}$ be a C^*-algebra and $\mathfrak{M}$ a facial subalgebra of $\mathcal{A}$. Then the set $\{x \in \mathfrak{M} \cap \mathcal{A}^+; \|x\| < 1\}$ is upward directed.*

Proof. Let $x, y \in \mathfrak{M} \cap \mathcal{A}^+$, $\|x\| < 1$, $\|y\| < 1$, and let $u = x(1-x)^{-1}, v = y(1-y)^{-1}, z = (u+v)(1+u+v)^{-1}$. Then $u, v, z \in \mathcal{A}^+$, $\|z\| < 1$ and $x = u(1+u)^{-1}, y = v(1+v)^{-1}$. Since the function f_1 defined in Section 1.5 is operator monotone, we have $x \leq z, y \leq z$. Since $\mathfrak{M}$ is a facial subalgebra in $\mathcal{A}$, $\mathfrak{M} \cap \mathcal{A}^+$ is a face of $\mathcal{A}^+$. We have $u \leq \|(1-x)^{-1}\| x \in \mathfrak{M} \cap \mathcal{A}^+$, so that $u \in \mathfrak{M} \cap \mathcal{A}^+$ and, similarly, $v \in \mathfrak{M} \cap \mathcal{A}^+$ and $z \in \mathfrak{M} \cap \mathcal{A}^+$. □

2.9. *Proof of Theorem 2.6.* We have to show that for every $f_1, f_2 \in \mathfrak{F}_\varphi$ and every $\varepsilon > 0$ there exists $f \in \mathfrak{F}_\varphi$ such that $(1-\varepsilon)f_1 \leq f$ and $(1-\varepsilon)f_2 \leq f$.

Let $f_1, f_2 \in \mathfrak{F}_\varphi$ and $\varepsilon > 0$. By Corollary 2.4, the set $\mathcal{T}_\varphi'$ of all the operators $T' \in \pi_\varphi(\mathcal{M})'$ such that there exists some $\lambda_{T'} > 0$ with the property

$$\|T' a_\varphi\|_\varphi \leq \lambda_{T'} \|a\| \quad (a \in \mathfrak{N}_\varphi)$$

is a left ideal of the von Neumann algebra $\pi_\varphi(\mathcal{M})'$; there exist $T_1', T_2' \in \mathcal{T}_\varphi'$ such that $0 \leq T_j' \leq 1$ and

$$f_j(b^* a) = (T_j' a_\varphi | T_j' b_\varphi)_\varphi \quad (a, b \in \mathfrak{N}_\varphi, j = 1, 2). \tag{1}$$

By Proposition 2.7, it follows that $(\mathcal{T}_\varphi')^* \mathcal{T}_\varphi'$ is a facial subalgebra of $\pi_\varphi(\mathcal{M})'$. Using Proposition 2.8, we obtain an element $X' \in (\mathcal{T}_\varphi')^* \mathcal{T}_\varphi'$ such that $(1-\varepsilon) T_j'^* T_j' \leq X' \leq 1, (j = 1, 2)$. Let $T' = (X')^{1/2}$. Using again Proposition 2.7, we see that

$$T' \in \mathcal{T}_\varphi', 0 \leq T' \leq 1 \text{ and } (1-\varepsilon) T_j'^* T_j' \leq T'^* T' \quad (j = 1, 2). \tag{2}$$

By Corollaries 2.4 and 2.5, there exists a normal positive form $f \leq \varphi$, that is, $f \in \mathfrak{F}_\varphi$, such that

$$f(b^* a) = (T' a_\varphi | T' b_\varphi)_\varphi \quad (a, b \in \mathfrak{N}_\varphi). \tag{3}$$

From (1), (2), and (3), it follows that

$$(1-\varepsilon) f_j(x) \leq f(x) \quad (x \in \mathfrak{M}_\varphi, j = 1, 2). \tag{4}$$

Since φ is semifinite, $\mathfrak{M}_\varphi$ is w-dense in $\mathcal{M}$. As f, f_1, f_2 are w-continuous, it follows that inequalities (4) remain valid for every $x \in \mathcal{M}$, that is, $(1-\varepsilon) f_1 \leq f, (j = 1, 2)$.

2.10 Corollary. *Let φ be a normal semifinite weight on the W^*-algebra $\mathscr{M}$. For each w-continuous function $\mathbb{R} \ni t \mapsto x(t) \in \mathscr{M}^+$ such that $\int_{-\infty}^{\infty} \|x(t)\| \mathrm{d}t < +\infty$ we have*

$$\varphi\left(\int_{-\infty}^{+\infty} x(t)\mathrm{d}t\right) = \int_{-\infty}^{\infty} \varphi(x(t))\mathrm{d}t.$$

Proof. Since $\mathscr{M} = (\mathscr{M}_*)^*$ ([L], 1.10; A.16), the properties of the function $t \mapsto x(t)$ show that there exists a unique element $x = \int x(t)\mathrm{d}t \in \mathscr{M}^+$ such that $f(x) = \int f(x(t))\mathrm{d}t$ for all $f \in \mathscr{M}_*$.

By Theorem 2.6, there exists an increasing net $\{f_i\}_{i\in I} \subset \mathscr{M}_*^+$ such that $\varphi = \sup_i f_i$. For each $i \in I$, we have $f_i(x) = \int f_i(x(t))\mathrm{d}t$. Since $f_i(x(t)) \uparrow_i \varphi(x(t))(t \in \mathbb{R})$, using the classical Beppo-Levi theorem we obtain $\varphi(x) = \sup_i f_i(x) = \sup_i \int f_i(x(t))\mathrm{d}t = \int \sup_i f_i(x(t))\mathrm{d}t = \int \varphi(x(t))\mathrm{d}t$. □

2.11. Using the theorems of Haagerup (1.3) and Combes (2.6), we can now extend, without any modification, the statement and the proof of the standard respresentation theorem ([L], 10.14) for weights that are normal in the sense defined in Section 1.3:

Theorem. *Let φ be an n.s.f. weight on the W^*-algebra $\mathscr{M}$. Then $\mathfrak{N}_\varphi \cap \mathfrak{N}_\varphi^*$, endowed with the $*$-algebra structure inherited from $\mathscr{M}$ and the scalar product of $\mathscr{H}_\varphi$, is a left Hilbert algebra $\mathfrak{A}_\varphi \subset \mathscr{H}_\varphi$ such that $\mathfrak{A}_\varphi = \mathfrak{A}_\varphi''$, $\pi_\varphi(\mathscr{M}) = \mathfrak{L}(\mathfrak{A}_\varphi)$ and*

$$\varphi(a) = \begin{cases} \|\xi\|^2 \textit{ if there exists } \xi \in \mathfrak{A}_\varphi \text{ with } \pi_\varphi(a)^{1/2} = L_\xi \\ +\infty, \textit{otherwise.} \quad (a \in \mathscr{M}^+). \end{cases}$$

Indeed, from Theorems 1.3 and 2.6, it follows that every normal semifinite weight is the supremum of an upward directed family of normal positive forms and it is exactly this definition of normality which is used in the proof of ([L], Thm. 10.14).

Consequently, every n.s.f. weight on a W^*-algebra is the natural weight associated with a left Hilbert algebra ([L], 10.16). Since these natural weights are sums of normal positive forms ([L], 10.18), any n.s.f. weight has the same property. In Section 5.8, we shall give a simpler proof of this result.

2.12. We shall use the notation and results of ([L], Chapter 10) for left Hilbert algebras and the objects associated with them.

Let φ be an n.s.f. weight on the W^*-algebra $\mathscr{M}$ and $\mathfrak{A}_\varphi \subset \mathscr{H}_\varphi$ the associated left Hilbert algebra (2.11). In this section, we recall some of the notation and results just used, together with some new results.

Since the weight φ is faithful, the left ideal $\mathfrak{N}_\varphi \subset \mathscr{M}$ will be also considered as a linear subspace of $\mathscr{H}_\varphi$ via the mapping $\mathfrak{N}_\varphi \ni x \mapsto x_\varphi \in \mathscr{H}_\varphi$.

The closed antilinear operator $S = S_\varphi$ in $\mathscr{H}_\varphi$ is the closure of the preclosed antilinear operator

$$S_\varphi^0 : \mathfrak{A}_\varphi \ni x_\varphi \mapsto (x_\varphi)^* = (x^*)_\varphi \in \mathscr{H}_\varphi.$$

For each $\eta \in \mathscr{H}_\varphi$, one defines a linear operator R_η^0 on $\mathscr{H}_\varphi$ affiliated to $\pi_\varphi(\mathscr{M})'$, with domain $D(R_\eta^0) = \mathfrak{A}_\varphi = \mathfrak{N}_\varphi \cap \mathfrak{N}_\varphi^*$ and

$$R_\eta^0 x_\varphi = \pi_\varphi(x)\eta \quad (x \in \mathfrak{N}_\varphi \cap \mathfrak{N}_\varphi^*).$$

If $\eta \in D(S^*)$, then R_η^0 is preclosed and its closure $R_\eta = \overline{R_\eta^0}$ satisfies $R_\eta^* \supset R_{S^*\eta}$ *). The right Hilbert algebra $\mathfrak{A}'_\varphi \subset \mathscr{H}_\varphi$ is the set

$$\mathfrak{A}'_\varphi = \{\eta \in D(S^*); R_\eta \in \mathscr{B}(\mathscr{H}_\varphi)\}$$

with the scalar product of $\mathscr{H}_\varphi$ and with the operations

$$\eta^b = S^*\eta, \eta_1\eta_2 = R_{\eta_2}\eta_1 \quad (\eta, \eta_1, \eta_2 \in \mathfrak{A}'_\varphi).$$

We have

$$\pi_\varphi(\mathscr{M})' = \mathfrak{R}(\mathfrak{A}'_\varphi) = \overline{\{R_\eta; \eta \in \mathfrak{A}'_\varphi\}}^{so}.$$

If R_η^0 $(\eta \in \mathscr{H}_\varphi)$ is bounded, then

$$R_\eta x_\varphi = \pi_\varphi(x)\eta \quad (x \in \mathfrak{N}_\varphi). \tag{1}$$

Indeed, if $\{y_i\}_{i \in I} \subset \mathfrak{N}_\varphi$ is a norm-bounded net with $y_i \xrightarrow{s^*} 1$, then for every $x \in \mathfrak{N}_\varphi$ we have $y_i^* x \in \mathfrak{N}_\varphi \cap \mathfrak{N}_\varphi^*$ and, by the definition of R_η^0,

$$\begin{aligned} R_\eta x_\varphi &= \lim_i \pi_\varphi(y_i^*) R_\eta x_\varphi = \lim_i R_\eta \pi_\varphi(y_i^*) x_\varphi \\ &= \lim_i R_\eta^0 (y_i^* x)_\varphi = \lim_i \pi_\varphi(y_i^* x)\eta = \pi_\varphi(x)\eta. \end{aligned}$$

Similarly, for each $\xi \in \mathscr{H}_\varphi$ one defines a linear operator L_ξ^0 on $\mathscr{H}_\varphi$, affiliated to $\pi_\varphi(\mathscr{M})$, with domain $D(L_\xi^0) = \mathfrak{A}'_\varphi$ and

$$L_\xi^0 \eta = R_\eta \xi \quad (n \in \mathfrak{A}'_\varphi).$$

If $\xi \in D(S)$, then L_ξ^0 is preclosed and its closure $L_\xi = \overline{L_\xi^0}$ satisfies the relation: $L_\xi^* \supset L_{S\xi}$ *). By Theorem 2.11, we have

$$\begin{aligned} \mathfrak{A}_\varphi = \mathfrak{A}''_\varphi &= \{\xi \in D(S); L_\xi \in \mathscr{B}(\mathscr{H}_\varphi)\} \\ \pi_\varphi(\mathscr{M}) = \mathfrak{L}(\mathfrak{A}_\varphi) &= \overline{\{L_\xi; \xi \in \mathfrak{A}_\varphi\}}^{so}. \end{aligned}$$

For $\xi \in \mathscr{H}_\varphi$, we have

$$\begin{gathered} L_\xi^0 \text{ is bounded} \Leftrightarrow \text{there exists } x \in \mathfrak{N}_\varphi \text{ with } \xi = x_\varphi; \\ \text{in this case } L_\xi = \pi_\varphi(x). \end{gathered} \tag{2}$$

---

*The equality is not always true (cf. [L], C.10.1 and an example by M. Pimsner).

Indeed, let $x \in \mathfrak{N}_\varphi$. By (1), for every $\eta \in \mathfrak{A}'_\varphi$ we have

$$L^0_{x_\varphi}\eta = R_\eta x_\varphi = \pi_\varphi(x)\eta.$$

Conversely, assume that L^0_ξ is bounded. Then $L_\xi \in \pi_\varphi(\mathscr{M})$, hence there exists $x \in \mathscr{M}$ with $\pi_\varphi(x) = L_\xi$. Let $x = v|x|$ be the polar decomposition of x in $\mathscr{M}$. It is easy to check that

$$L_{\pi_\varphi(v^*)\xi} = \pi_\varphi(v^*)L_\xi = \pi_\varphi(v^*x) = \pi_\varphi(|x|) = |L_\xi| \geq 0.$$

Using ([L], 10.8) it follows that $\pi_\varphi(v^*)\xi \in \mathfrak{A}''_\varphi$ and then, using Theorem 2.11 we get $\varphi(x^*x) = \|\pi_\varphi(v^*)\xi\|^2_\varphi = \|\xi\|^2_\varphi < +\infty$, hence $x \in \mathfrak{N}_\varphi$. Since $L_{x_\varphi} = \pi_\varphi(x) = L_\xi$, we conclude that $\xi = x_\varphi$.

Also, note that $x \in \mathfrak{N}_\varphi, x_\varphi \in D(S_\varphi) \Rightarrow x_\varphi \in \mathfrak{A}_\varphi$, that is,

$$\mathfrak{N}_\varphi \cap D(S_\varphi) = \mathfrak{N}_\varphi \cap \mathfrak{N}^*_\varphi. \tag{3}$$

Indeed, the inclusion $\mathfrak{N}_\varphi \cap \mathfrak{N}^*_\varphi = \mathfrak{A}_\varphi \subset \mathfrak{N}_\varphi \cap D(S_\varphi)$ is obvious. Conversely, if $x \in \mathfrak{N}_\varphi$ and $x_\varphi \in D(S_\varphi)$, it follows from (2) that $x_\varphi = \mathfrak{A}''_\varphi = \mathfrak{A}_\varphi$ hence $x \in \mathfrak{N}_\varphi \cap \mathfrak{N}^*_\varphi$.

From the polar decomposition $S_\varphi = J_\varphi \Delta^{1/2}_\varphi$ of S_φ, one obtains the modular operator $\Delta_\varphi = S^*_\varphi S_\varphi$ and the canonical conjugation $J_\varphi = J^*_\varphi = J^{-1}_\varphi$, associated with $\mathfrak{A}_\varphi \subset \mathscr{H}_\varphi$. Since J_φ is antilinear and $J_\varphi \Delta_\varphi J_\varphi = \Delta^{-1}_\varphi$, it follows that

$$J_\varphi f(\Delta_\varphi) J_\varphi = \bar{f}(\Delta^{-1}_\varphi)$$

for every Borel function f. In particular, $J_\varphi \Delta^{it}_\varphi = \Delta^{it}_\varphi J_\varphi (t \in \mathbb{R})$, and $S_\varphi = J_\varphi \Delta^{1/2}_\varphi = \Delta^{-1/2}_\varphi J_\varphi, S^*_\varphi = J_\varphi \Delta^{-1/2}_\varphi = \Delta^{1/2}_\varphi J_\varphi$.

By Tomita's fundamental theorem ([L], 10.12), we have $\Delta^{it}_\varphi \mathfrak{A}_\varphi = \mathfrak{A}_\varphi, \Delta^{it}_\varphi \mathfrak{A}'_\varphi = \mathfrak{A}'_\varphi (t \in \mathbb{R})$, $J_\varphi \mathfrak{A}_\varphi = \mathfrak{A}'_\varphi$ and

$$\begin{aligned} L_{\Delta^{it}_\varphi \xi} &= \Delta^{it}_\varphi L_\xi \Delta^{-it}_\varphi, R_{J_\varphi \xi} = J_\varphi L_\xi J_\varphi \quad (\xi \in \mathfrak{A}_\varphi), \\ R_{\Delta^{it}_\varphi \eta} &= \Delta^{it}_\varphi R_\eta \Delta^{-it}_\varphi, L_{J_\varphi \eta} = J_\varphi R_\eta J_\varphi \quad (\eta \in \mathfrak{A}'_\varphi). \end{aligned} \tag{4}$$

Using the definition of the operators L^0, R^0 the validity of (4) can be extended to arbitrary vectors $\xi, \eta \in \mathscr{H}_\varphi$, replacing the operators L, R by L^0, R^0, respectively. If the operators L^0_ξ, R^0_η are preclosed, these identities can be extended to their closures, that is, they remain valid in the earlier form.

From Tomita's fundamental theorem it follows that the mapping $j_\varphi : x \mapsto J_\varphi \pi_\varphi(x^*) J_\varphi$ defines a $*$-antiisomorphism j_φ of $\mathscr{M}$ onto $\pi_\varphi(\mathscr{M})'$, which coincides with π_φ on the center of $\mathscr{M}$ ([L], 10.13). We note that

$$J_\varphi \pi_\varphi(x) J_\varphi y_\varphi = \pi_\varphi(y) J_\varphi x_\varphi \quad (x, y \in \mathfrak{N}_\varphi). \tag{5}$$

Indeed, using the extension of (4) as well as (1) and (2); we obtain $J_\varphi \pi_\varphi(x) J_\varphi y_\varphi = J_\varphi L_{x_\varphi} J_\varphi y_\varphi = R_{J_\varphi x_\varphi} y_\varphi = \pi_\varphi(y) J_\varphi x_\varphi$.

Using the isometric character of J_φ it is easy to check that the n.s.f. weight φ' on $\pi_\varphi(\mathcal{M})'$ defined by

$$\varphi'(j_\varphi(x)) = \varphi(x) \quad (x \in \mathcal{M}^+) \tag{6}$$

is just the natural weight on $\pi_\varphi(\mathcal{M})' = \mathfrak{R}(\mathfrak{A}'_\varphi)$ associated with the right Hilbert algebra $\mathfrak{A}'_\varphi \subset \mathcal{H}_\varphi$ ([L], p. 331), that is,

$$\varphi'(R^*_\zeta R_\eta) = (\eta|\zeta)_\varphi \tag{7}$$

for every $\eta, \zeta \in \mathcal{H}_\varphi$, such that R_η, R_ζ are bounded, in particular for every $\eta, \zeta \in \mathfrak{A}'_\varphi$. We note that the standard representation of $\pi_\varphi(\mathcal{M})'$ associated with the n.s.f. weight φ' is unitarily equivalent to the identity representation $\pi_\varphi(\mathcal{M})' \subset \mathcal{B}(\mathcal{H}_\varphi)$ and we have

$$S_{\varphi'} = S^*_\varphi, \quad \Delta_{\varphi'} = \Delta^{-1}_\varphi, \quad J_{\varphi'} = J_\varphi. \tag{8}$$

On the other hand, it follows from Tomita's fundamental theorem that the relation

$$\pi_\varphi(\sigma^\varphi_t(x)) = \Delta^{it}_\varphi \pi_\varphi \Delta^{-it}_\varphi \quad (x \in \mathcal{M}, t \in \mathbb{R})$$

defines an s^*-continuous group $\{\sigma^\varphi_t\}_{t\in\mathbb{R}}$ of $*$-automorphisms of $\mathcal{M}$ which act identically on the center of $\mathcal{M}$ ([L], 10.13). With an argument similar to that used in proving (5), we obtain

$$R_\eta(\sigma^\varphi_t(x))_\varphi = \pi_\varphi(\sigma^\varphi_t(x))\eta = R_\eta \Delta^{it}_\varphi x_\varphi$$

for every $x \in \mathfrak{N}_\varphi$ and every $\eta \in \mathfrak{A}'_\varphi$. Letting $R_\eta \overset{so}{\to} 1$, we conclude

$$(\sigma^\varphi_t(x))_\varphi = \Delta^{it}_\varphi x_\varphi \quad (x \in \mathfrak{N}_\varphi, t \in \mathbb{R}). \tag{9}$$

Since $\Delta_{\varphi'} = \Delta^{-1}_\varphi$ for the weight φ' on $\pi_\varphi(\mathcal{M})'$ we have

$$\sigma^{\varphi'}_t(j_\varphi(x)) = j_\varphi(\sigma^\varphi_{-t}(x)) \quad (x \in \mathcal{M}, t \in \mathbb{R}). \tag{10}$$

The weight φ is invariant with respect to the modular automorphism group $\{\sigma^\varphi_t\}_{t\in\mathbb{R}}$, that is, $\varphi \circ \sigma^\varphi_t = \varphi (t \in \mathbb{R})$, and satisfies the *KMS* condition with respect to $\{\sigma^\varphi_t\}_{t\in\mathbb{R}}$ in any two elements $x, y \in \mathfrak{N}_\varphi \cap \mathfrak{N}^*_\varphi$, that is, there exists a function $f = f_{x,y}$ defined, continuous and bounded on the strip $\{\alpha \in \mathbb{C}; 0 \le \operatorname{Re} \alpha \le 1\}$, analytic in the interior of this strip and such that

$$f(it) = \varphi(x\sigma^\varphi_t(y)), \quad f(1+it) = \varphi(\sigma^\varphi_t(y)x) \quad (t \in \mathbb{R}).$$

These properties characterize the modular automorphism group associated with φ. More exactly, in ([L], 10.17) one actually proves the following uniqueness statement:

$$\begin{aligned}&\textit{if } \{\sigma_t\}_{t\in\mathbb{R}} \textit{ is a group of } * \textit{-automorphisms of } \mathscr{M} \textit{ with the properties}:\\ &(a)\ \varphi\circ\sigma_t=\varphi \textit{ for all } t\in\mathbb{R};\\ &(b)\textit{ there exists a } * \textit{-subalgebra } \mathscr{X}\subset\mathfrak{A}_\varphi \textit{ such that } \overline{S_\varphi|\mathscr{X}}=S_\varphi \\ &\textit{and } \varphi \textit{ satisfies the KMS condition with respect to } \{\sigma_t\}_{t\in\mathbb{R}}\\ &\textit{in any two elements of } \mathscr{X};\\ &\textit{then } \sigma_t=\sigma_t^\varphi \textit{ for } t\in\mathbb{R}.\end{aligned} \tag{11}$$

If the weight φ is finite, that is, if $\varphi\in\mathscr{M}_*^+$, then

$$x_i\xrightarrow{s}x \text{ in } \mathscr{M}\Rightarrow \|(x_i)_\varphi-x_\varphi\|_\varphi\to 0. \tag{12}$$

Thus, in this case, we can replace $\mathscr{X}$ in condition (b) of (11) by any w-dense $*$-subalgebra of $\mathscr{M}$.

If φ is not necessarily finite, we still have the following convergence result

$$\mathfrak{N}_\varphi\ni x_i\xrightarrow{w}x\in\mathscr{M},(x_i)_\varphi\xrightarrow{\text{weakly}}\xi\in\mathscr{H}_\varphi\Rightarrow x\in\mathfrak{N}_\varphi,\xi=x_\varphi. \tag{13}$$

Indeed, for $\eta\in\mathfrak{A}'_\varphi$ and $\zeta\in\mathscr{H}_\varphi$, we have $(L^0_\xi\eta|\zeta)_\varphi=(R_\eta\xi|\zeta)_\varphi=\lim_i(R_\eta(x_i)_\varphi|\zeta)_\varphi=\lim_i(\pi_\varphi(x_i)\eta|\zeta)=(\pi_\varphi(x)\eta|\zeta)_\varphi$ and using (2) we conclude $x\in\mathfrak{N}_\varphi$ and $\xi=x_\varphi$.

Using (13) and the w-compactness of the closed unit ball of $\mathscr{M}$, we obtain also the following result:

$$\begin{aligned}&\textit{if } \{x_i\}\subset\mathfrak{N}_\varphi \textit{ is a norm-bounded net and if } \{(x_i)_\varphi\} \textit{ is weakly}\\ &\textit{convergent to some } \xi\in\mathscr{H}_\varphi, \textit{ then there exists } x\in\mathfrak{N}_\varphi \textit{ such that}\\ &x_i\xrightarrow{s}x \textit{ and } x_\varphi=\xi.\end{aligned} \tag{14}$$

An important technical tool in the standard representation associated with φ is the Tomita algebra ([L], 10.20, 10.21)

$$\mathfrak{T}_\varphi\subset\mathfrak{A}_\varphi\cap\mathfrak{A}'_\varphi\cap\bigcap_{\alpha\in\mathbb{C}}D(\Delta_\varphi^\alpha).$$

Recall that $\mathfrak{T}_\varphi$ is a left Hilbert subalgebra of $\mathfrak{A}_\varphi$, equivalent to $\mathfrak{A}_\varphi$, and $J_\varphi\mathfrak{T}_\varphi=\mathfrak{T}_\varphi,\Delta_\varphi^\alpha\mathfrak{T}_\varphi=\mathfrak{T}_\varphi\ \overline{\Delta_\varphi^\alpha|\mathfrak{T}_\varphi}=\Delta_\varphi^\alpha,(\alpha\in\mathbb{C})$. The identities

$$\Delta_\varphi^\alpha(\xi\eta)=(\Delta_\varphi^\alpha\xi)(\Delta_\varphi^\alpha\eta),J_\varphi(\xi\eta)=(J_\varphi\eta)(J_\varphi\xi)\quad(\xi,\eta\in\mathfrak{T}_\varphi)$$

are straightforward consequences of (9) and (5).

The arguments in ([L], 10.21) prove that for $\xi\in\mathscr{H}_\varphi$ we have

$$\xi\in\mathfrak{T}_\varphi\Leftrightarrow\xi\in\bigcap_{\alpha\in\mathbb{C}}D(\Delta_\varphi^\alpha)\textit{ and } \Delta_\varphi^n\xi\in\mathfrak{N}_\varphi\subset\mathscr{H}_\varphi \textit{ for all } n\in\mathbb{Z}. \tag{15}$$

Using this criterion and arguing as in ([L], p. 302), it is easy to obtain the following approximation result:

$$\text{for every } x \in \mathfrak{N}_\varphi \text{ there exists a sequence } \{x_n\} \subset \mathfrak{T}_\varphi \text{ such that} \\ \|x_n\| \leq \|x\|, x_n \xrightarrow{s} x \text{ and } \|(x_n)_\varphi - x_\varphi\|_\varphi \to 0; \tag{16}$$

in fact we can take

$$x_n = \sqrt{n/\pi} \int_{-\infty}^{+\infty} \mathrm{e}^{-nt^2} \sigma_t^\varphi(x) \mathrm{d}t.$$

If $x \in \mathfrak{N}_\varphi \cap \mathfrak{N}_\varphi^*$, then the approximation is stronger, namely we have also ([L], Cor. 2/10.21) $x_n^* \xrightarrow{s} x^*$ and $\|(x_n^*)_\varphi - (x^*)_\varphi\|_\varphi \to 0$.

In the next sections, we define the translation of a weight by certain elements (2.13), consider analytic elements with respect to a weight (2.14–2.16), give some useful reformulations of the *KMS* condition (2.17–2.10), study the centralizer of a weight (2.21, 2.22), and use the standard representation in order to introduce a natural topology on the group of all *-automorphisms (2.23–2.26).

2.13 Proposition. *Let φ be an n.s.f. weight on the W^*-algebra $\mathscr{M}$. If $a \in \mathfrak{T}_\varphi^2$, then the linear form*

$$x \mapsto \varphi(xa) \quad (\textit{resp.}\ x \mapsto \varphi(ax))$$

is defined and w-continuous on the set

$$\{x \in \mathscr{M}; x\mathfrak{T}_\varphi \subset \mathfrak{A}_\varphi\} \quad (\textit{resp.}\ \{x \in \mathscr{M}; \mathfrak{T}_\varphi x \subset \mathfrak{A}_\varphi\})$$

which contains $\mathfrak{T}_\varphi$ and can therefore be extended to a w-continuous linear form on $\mathscr{M}$, denoted by $\varphi(\cdot a)$ (resp. $\varphi(a\cdot)$).

The sets $\{\varphi(\cdot a); a \in \mathfrak{T}_\varphi^2\}$ and $\{\varphi(a\cdot); a \in \mathfrak{T}_\varphi^2\}$ are norm-dense linear subspaces of $\mathscr{M}$.

Proof. Let $a = bc^*$ with $b, c \in \mathfrak{T}_\varphi$ and $x \in \mathscr{M}$ with $x\mathfrak{T}_\varphi \subset \mathfrak{A}_\varphi$. Then $xb \in \mathfrak{A}_\varphi \subset \mathfrak{N}_\varphi^*, c^* \in \mathfrak{N}_\varphi$, hence $xa \in \mathfrak{M}_\varphi$ and we have

$$\varphi(xa) = (\pi_\varphi(x)b_\varphi | \Delta_\varphi c_\varphi)_\varphi. \tag{1}$$

Indeed,

$$\begin{aligned} \varphi(xbc^*) &= ((c^*)_\varphi | (b^*x^*)_\varphi)_\varphi = (S_\varphi c_\varphi | S_\varphi (xb)_\varphi)_\varphi \\ &= (J_\varphi \Delta_\varphi^{1/2} c_\varphi | J_\varphi \Delta_\varphi^{1/2} (xb)_\varphi)_\varphi = (\Delta_\varphi^{1/2}(xb)_\varphi | \Delta_\varphi^{1/2} c_\varphi)_\varphi \\ &= ((xb)_\varphi | \Delta_\varphi c_\varphi)_\varphi = (\pi_\varphi(x)b_\varphi | \Delta_\varphi c_\varphi)_\varphi. \end{aligned}$$

Similarly,

$$\varphi(ax) = (\pi_\varphi(x)\Delta_\varphi b_\varphi | c_\varphi)_\varphi. \tag{2}$$

This proves the first part of the proposition. Moreover, (1) and (2) give the explicit form of the extensions $\varphi(\cdot a)$ and $\varphi(a\cdot)$.

Assume that the linear subspace $\{\varphi(\cdot a); a \in \mathfrak{T}^2_\varphi\}$ is not norm-dense in $\mathcal{M}_*$. Then, by the Hahn–Banach theorem, there exists $x \in \mathcal{M}, x \neq 0$, such that $\varphi(xa) = 0$ for all $a \in \mathfrak{T}^2_\varphi$. Using (1) we infer that $(\pi_\varphi(x)\xi|\eta)_\varphi = 0$ for all $\xi, \eta \in \mathfrak{T}_\varphi$, hence $\pi_\varphi(x) = 0$, contradicting $x \neq 0$. □

Note that, conversely,

$$\text{if } 0 \leq a \in \mathfrak{N}_\varphi \text{ and } \mathfrak{N}^*_\varphi \ni x \mapsto \varphi(xa) \text{ is } w\text{-continuous, then } a \in \mathfrak{M}_\varphi. \tag{3}$$

Indeed, there exists a sequence $\{e_n\}$ of spectral projections of a such that $ae_n \geq n^{-1}e_n$ and $e_n \uparrow \mathbf{s}(a)$. We have $e_n \leq n^2 ae_n a \in \mathfrak{M}_\varphi$, so that $\mathfrak{N}^*_\varphi \ni e_n \uparrow 1$ and hence, by assumption, $\sup_n \varphi(e_n a) < +\infty$. On the other hand, we have $e_n a = a^{1/2}e_n a^{1/2} \uparrow a$, hence $\varphi(a) = \sup_n \varphi(e_n a) < +\infty$, that is a $a \in \mathfrak{M}_\varphi$.

2.14. Let φ be an n.s.f. weight on the W^*-algebra $\mathcal{M}$ and $a \in \mathcal{M}$.

The element a is called *analytic in the vertical strip* $\{\alpha \in \mathbb{C}; -\varepsilon_1 \leq \operatorname{Re} \alpha \leq \varepsilon_2\}, (0 \leq \varepsilon_1, \varepsilon_2 < +\infty)$, if there exists an $\mathcal{M}$-valued function F, defined and w-continuous on this strip and analytic in the interior of the strip, such that

$$F(\mathrm{i}t) = \sigma^\varphi_t(a) \quad (t \in \mathbb{R}).$$

In this case, for each $\alpha \in \mathbb{C}, -\varepsilon_1 \leq \operatorname{Re} \alpha \leq \varepsilon_2$, we let

$$\sigma^\varphi_{-\mathrm{i}\alpha}(a) = F(\alpha).$$

For $\alpha \in \mathbb{C}$, we shall write $a \in D(\sigma^\varphi_\alpha)$ if the element a is analytic in some vertical strip containing $\mathrm{i}\alpha$. The following statements are easily verified:

$$a \in D(\sigma^\varphi_\alpha) \Rightarrow a^* \in D(\sigma^\varphi_{\bar{\alpha}}), \sigma^\varphi_{\bar{\alpha}}(a^*) = \sigma^\varphi_\alpha(a)^*; \tag{1}$$

$$a, b \in D(\sigma^\varphi_\alpha) \Rightarrow ab \in D(\sigma^\varphi_\alpha), \sigma^\varphi_\alpha(ab) = \sigma^\varphi_\alpha(a)\sigma^\varphi_\alpha(b). \tag{2}$$

Using ([L], 9.21), from the relation $\pi_\varphi(\sigma^\varphi_t(a))\Delta^{\mathrm{i}t}_\varphi\xi = \Delta^{\mathrm{i}t}_\varphi\pi_\varphi(a)\xi$ we infer that

$$a \in D(\sigma^\varphi_{-\mathrm{i}\alpha}), \xi \in D(\Delta^\alpha_\varphi) \Rightarrow \pi_\varphi(a)\xi \in D(\Delta^\alpha_\varphi), \Delta^\alpha_\varphi\pi_\varphi(a)\xi = \pi_\varphi(\sigma^\varphi_{-\mathrm{i}\alpha}(a))\Delta^\alpha_\varphi\xi \tag{3}$$

or, using ([L], 9.24) and replacing α by $-\mathrm{i}\alpha$,

$$a \in D(\sigma^\varphi_\alpha) \Rightarrow \pi_\varphi(\sigma^\varphi_\alpha(a)) = \overline{\Delta^{\mathrm{i}\alpha}_\varphi\pi_\varphi(a)\Delta^{-\mathrm{i}\alpha}_\varphi|D(\Delta^{-\mathrm{i}\alpha}_\varphi)}. \tag{4}$$

It follows that if the element a is analytic in the strip $\{\alpha \in \mathbb{C}; 0 \leq \operatorname{Re} \alpha \leq \varepsilon\}$ then the function $\alpha \mapsto \sigma^\varphi_{-\mathrm{i}\alpha}(a)$ is norm-continuous and norm-bounded on this strip.

Also, we have

$$a \in D(\sigma^\varphi_\beta), \sigma^\varphi_\beta(a) \in D(\sigma^\varphi_\alpha) \Rightarrow a \in D(\sigma^\varphi_{\alpha+\beta}), \sigma^\varphi_{\alpha+\beta}(a) = \sigma^\varphi_\alpha(\sigma^\varphi_\beta(a)); \tag{5}$$

$$a \in D(\sigma^\varphi_\alpha) \Rightarrow \sigma^\varphi_\alpha(a) \in D(\sigma^\varphi_{-\alpha}), \sigma^\varphi_{-\alpha}(\sigma^\varphi_\alpha(a)) = a. \tag{6}$$

Proposition. *Let φ be an n.s.f. weight on the W^*-algebra $\mathscr{M}$, $a \in \mathscr{M}$ and $\lambda \in (0, +\infty)$. The following statements are equivalent:*

(i) $\varphi(ax^*xa^*) \leq \lambda^2 \varphi(x^*x)$ *for every* $x \in \mathscr{M}$;
(ii) $x \in \mathfrak{N}_\varphi \Rightarrow xa^* \in \mathfrak{N}_\varphi$ *and* $\|(xa^*)_\varphi\|_\varphi \leq \lambda \|x_\varphi\|_\varphi$;
(iii) $a \in D(\sigma^\varphi_{-\mathrm{i}/2})$ and $\|\sigma^\varphi_{-\mathrm{i}/2}(a)\| \leq \lambda$.

If $a \in D(\sigma^\varphi_{-\mathrm{i}/2})$, *then*

$$(xa^*)_\varphi = J_\varphi \pi_\varphi(\sigma^\varphi_{-\mathrm{i}/2}(a)) J_\varphi x_\varphi \quad (x \in \mathfrak{N}_\varphi) \tag{7}$$

and if moreover $\sigma^\varphi_t(aa^*) = aa^* (t \in \mathbb{R})$, *then*

$$\varphi(\sigma^\varphi_{-\mathrm{i}/2}(a)^* x^* x \sigma^\varphi_{-\mathrm{i}/2}(a)) = \varphi(aa^*x^*x) \quad (x \in \mathfrak{N}_\varphi). \tag{8}$$

Proof. It is clear that (i) $\Leftrightarrow$ (ii).

(ii) $\Rightarrow$ (iii). From (ii) it follows that there exists $T \in \mathscr{B}(\mathscr{H}_\varphi)$, $\|T\| \leq \lambda$, such that $Tx_\varphi = (xa^*)_\varphi$ $(x \in \mathfrak{N}_\varphi)$. For $x \in \mathfrak{A}_\varphi$, we have

$$Tx_\varphi = (xa^*)_\varphi = S_\varphi(ax^*)_\varphi = S_\varphi \pi_\varphi(a) S_\varphi x_\varphi = J_\varphi \Delta^{1/2}_\varphi \pi_\varphi(a) \Delta^{-1/2}_\varphi J_\varphi x_\varphi$$

so the operator $\Delta^{1/2}_\varphi \pi_\varphi(a) \Delta^{-1/2}_\varphi | \mathfrak{A}'_\varphi$ is bounded with norm $\leq \lambda$. Since $\overline{\Delta^{-1/2}_\varphi | \mathfrak{A}'_\varphi} = \Delta^{-1/2}_\varphi$, using ([L], 9.24) we infer that $a \in D(\sigma^\varphi_{-\mathrm{i}/2})$ and $\|\sigma^\varphi_{-\mathrm{i}/2}(a)\| \leq \lambda$.

(iii) $\Rightarrow$ (i). Let $x \in \mathfrak{A}_\varphi = \mathfrak{N}_\varphi \cap \mathfrak{N}^*_\varphi$. Then $(x^*)_\varphi \in D(\Delta^{1/2}_\varphi)$ and, using (3) with $\alpha = 1/2$, we get

$$\Delta^{1/2}_\varphi \pi_\varphi(a)(x^*)_\varphi = \pi_\varphi(\sigma^\varphi_{-\mathrm{i}/2}(a)) \Delta^{1/2}_\varphi (x^*)_\varphi.$$
$$S_\varphi(ax^*)_\varphi = J_\varphi \pi_\varphi(\sigma^\varphi_{-\mathrm{i}/2}(a)) J_\varphi S_\varphi (x^*)_\varphi,$$

hence

$$(xa^*)_\varphi = J_\varphi \pi_\varphi(\sigma^\varphi_{-\mathrm{i}/2}(a)) J_\varphi x_\varphi \quad (x \in \mathfrak{A}_\varphi). \tag{9}$$

It follows that

$$\varphi(aza^*) \leq \|\sigma^\varphi_{-\mathrm{i}/2}(a)\| \varphi(z) \quad (z \in \mathscr{M}^+),$$

and this proves (i). Indeed, if $\varphi(z) = +\infty$ the inequality is obvious and if $\varphi(z) < +\infty$ then $x = z^{1/2} \in \mathfrak{A}_\varphi$ and we use (9).

Consider now $x \in \mathfrak{N}_\varphi$. There is a sequence $\{x_n\} \subset \mathfrak{A}_\varphi$ such that $\|(x_n)_\varphi - x_\varphi\|_\varphi \to 0$. From (ii) it follows that $xa^* \in \mathfrak{N}_\varphi$ and $\|(x_n a^*)_\varphi - (xa^*)_\varphi\|_\varphi \to 0$. Thus, (7) follows from (9) in the limit.

Finally, we prove (8). Let $b = \sigma^\varphi_{-\mathrm{i}/2}(a)$. Using (1) and (6) it follows that $b^* \in D(\sigma^\varphi_{-\mathrm{i}/2})$ and $\sigma^\varphi_{-\mathrm{i}/2}(b^*) = a^*$. On the other hand, since $\sigma^\varphi_t(aa^*) = aa^* (t \in \mathbb{R})$, it is obvious that $\sigma^\varphi_{-\mathrm{i}/2}(aa^*) = aa^*$. Using

(7) we obtain

$$\begin{aligned}\varphi(b^*x^*xb) &= \|(xb)_\varphi\|_\varphi^2 = \|J_\varphi\pi_\varphi(a^*)J_\varphi x_\varphi\|_\varphi^2\\ &= (x_\varphi | J_\varphi\pi_\varphi(aa^*)J_\varphi x_\varphi)_\varphi = (x_\varphi|(xaa^*)_\varphi)_\varphi = \varphi(aa^*x^*x).\end{aligned}$$ □

2.15. An element $a \in \mathscr{M}$ such that $a \in D(\sigma_\varphi^\alpha)$ for all $\alpha \in \mathbb{C}$ is called an *entire analytic element*. We put

$$\mathscr{M}_\infty^\varphi = \{a \in \mathscr{M}; a \text{ is an entire analytic element}\}.$$

Using ([L], 10.20, 9.24) we see that

$$a \in \mathfrak{T}_\varphi \Rightarrow a \in \mathscr{M}_\infty^\varphi \textit{ and } \sigma_\alpha^\varphi(a) \in \mathfrak{T}_\varphi \text{ for all } \alpha \in \mathbb{C}. \tag{1}$$

From Section 2.14, it follows that $\mathscr{M}_\infty^\varphi$ is an s^*-dense $*$-subalgebra of $\mathscr{M}$. Moreover, the sets $\mathfrak{N}_\varphi, \mathfrak{A}_\varphi, \mathfrak{M}_\varphi$ are all invariant under left or right multiplications by elements of $\mathscr{M}_\infty^\varphi$. Note also that

$$a \in \mathfrak{T}_\varphi \Rightarrow \Delta_\varphi^{i\alpha} a_\varphi = (\sigma_\alpha^\varphi(a))_\varphi \textit{ and } L_{\Delta_\varphi^{i\alpha} a_\varphi} = \pi_\varphi(\sigma_\alpha^\varphi(a)). \tag{2}$$

Indeed, for $\xi \in \mathfrak{T}_\varphi$ we have

$$\begin{aligned}L_{\Delta_\varphi^{i\alpha}a_\varphi}\xi &= R_\xi \Delta_\varphi^{i\alpha} a_\varphi = \Delta_\varphi^{i\alpha} R_{\Delta_\varphi^{-i\alpha}\xi} a_\varphi = \Delta_\varphi^{i\alpha} L_{a_\varphi} \Delta_\varphi^{-i\alpha}\xi\\ &= \Delta_\varphi^{i\alpha}\pi_\varphi(a)\Delta_\varphi^{-i\alpha}\xi = \pi_\varphi(\sigma_\alpha^\varphi(a))\xi = L_{(\sigma_\alpha^\varphi(a))_\varphi}\xi.\end{aligned}$$

In particular,

$$R_{\Delta_\varphi^{i\bar{\alpha}} J_\varphi a_\varphi} = R_{J_\varphi \Delta_\varphi^{i\alpha} a_\varphi} = J_\varphi L_{\Delta_\varphi^{i\alpha} a_\varphi} J_\varphi = J_\varphi \pi_\varphi(\sigma_\alpha^\varphi(a)) J_\varphi. \tag{3}$$

2.16 Proposition. *Let φ be an n.s.f. weight on the W^*-algebra $\mathscr{M}$, $\{x_k\}_{k\in K} \subset \mathfrak{A}_\varphi$ a net such that $x_k \xrightarrow{s^*} 1$, $\sup_k \|x_k\| \le 1$ and*

$$a_k = \sqrt{1/\pi}\int_{-\infty}^{+\infty} \mathrm{e}^{-t^2}\sigma_t^\varphi(x_k)\mathrm{d}t \quad (k \in K). \tag{1}$$

Then $\{a_k\}_{k\in K} \subset \mathfrak{T}_\varphi \subset \mathscr{M}_\infty^\varphi$ and for every $\alpha \in \mathbb{C}, k \in K$, we have

$$\sigma_\alpha^\varphi(a_k) \xrightarrow{s^*} 1, \tag{2}$$

$$\|\sigma_\alpha^\varphi(a_k)\| \le \exp((\operatorname{Im}\alpha)^2). \tag{3}$$

Proof. Arguing as in ([L], p. 347) we see that $a_k \in \mathfrak{T}_\varphi$ and

$$\sigma_\alpha^\varphi(a_k) = \sqrt{1/\pi}\int_{-\infty}^{+\infty} \mathrm{e}^{-(t-\alpha)^2}\sigma_t^\varphi(x_k)\mathrm{d}t.$$

Let $r = \mathrm{Re}\,\alpha, s = \mathrm{Im}\,\alpha$. Then $(t-\alpha)^2 = -s^2 + (t-r)^2 - 2is(t-r)$, so

$$\|\sigma_\alpha^\varphi(a_k)\| \le \mathrm{e}^{s^2}\sqrt{1/\pi}\int_{-\infty}^{+\infty} \mathrm{e}^{-(t-r)^2}\|\sigma_t(x_k)\|\mathrm{d}t \le \mathrm{e}^{s^2}\sqrt{1/\pi}\int_{-\infty}^{+\infty} \mathrm{e}^{-t^2}\mathrm{d}t = \mathrm{e}^{s^2}.$$

Let $\mathcal{M} \subset \mathcal{B}(\mathcal{H})$ be realized as a von Neumann algebra. For $\xi \in \mathcal{H}$, we have

$$\begin{aligned}\|\xi - \sigma_\alpha^\varphi(a_k)\xi\| &\le \mathrm{e}^{s^2}\int_{-\infty}^{+\infty} \mathrm{e}^{-(t-r)^2}\|\sigma_t^\varphi(1-x_k)\xi\|\mathrm{d}t\\ &= \mathrm{e}^{s^2}\int_{-\infty}^{+\infty} \mathrm{e}^{-t^2}\|\sigma_t^\varphi(1-x_k)\xi\|dt \to 0,\end{aligned}$$

since $\lim_k \|\sigma_t^\varphi(1-x_k)\xi\| = 0\,(t \in \mathbb{R})$, using the Lebesgue dominated convergence theorem. Consequently, $\sigma_\alpha^\varphi(a_k) \overset{so}{\to} 1$ and, similarly, $\sigma_\alpha^\varphi(a_k)^* \overset{so}{\to} 1$. □

2.17 Proposition. *Let φ be an n.s.f. weight on the W^*-algebra $\mathcal{M}, x, y \in \mathcal{M}$ and $\alpha \in \mathbb{C}$. If $x \in \mathfrak{N}_\varphi^* \cap D(\sigma_{\alpha-\mathrm{i}}^\varphi), \sigma_{\alpha-\mathrm{i}}^\varphi(x) \in \mathfrak{N}_\varphi$ and $y \in \mathfrak{N}_\varphi \cap D(\sigma_\alpha^\varphi), \sigma_\alpha^\varphi(y) \in \mathfrak{N}_\varphi^*$, then*

$$\varphi(xy) = \varphi(\sigma_\alpha^\varphi(y)\sigma_{\alpha-\mathrm{i}}^\varphi(x)). \tag{1}$$

Proof. By Proposition 2.16, there exists a net $\{a_k\} \subset \mathfrak{T}_\varphi$ such that $\sigma_\beta^\varphi(a_k) \overset{s^*}{\to} 1$ for all $\beta \in \mathbb{C}$. Using Properties 2.14.(1), 2.14.(4), 2.14.(6) and 2.15.(3), we obtain

$$\begin{aligned}\pi_\varphi(a_k)y_\varphi &= (a_k y)_\varphi = S_\varphi(y^* a_k^*)_\varphi = S_\varphi \pi_\varphi(y^*)S_\varphi(a_k)_\varphi\\ &= S_\varphi\pi_\varphi(\sigma_{-\bar\alpha}^\varphi(\sigma_{\bar\alpha}^\varphi(y^*)))S_\varphi(a_k)_\varphi = S_\varphi\Delta_\varphi^{-\mathrm{i}\bar\alpha}\pi_\varphi(\sigma_{\bar\alpha}^\varphi(y^*))\Delta_\varphi^{\mathrm{i}\bar\alpha}S_\varphi(a_k)_\varphi\\ &= J_\varphi\Delta_\varphi^{-\mathrm{i}\bar\alpha+(1/2)}\pi_\varphi(\sigma_\alpha^\varphi(y)^*)\Delta_\varphi^{\mathrm{i}\bar\alpha-(1/2)}J_\varphi(a_k)_\varphi\\ &= J_\varphi\Delta_\varphi^{-\mathrm{i}\bar\alpha+(1/2)}R_{\Delta_\varphi^{\mathrm{i}\bar\alpha-(1/2)}J_\varphi(a_k)_\varphi}(\sigma_\alpha^\varphi(y)^*)_\varphi\\ &= J_\varphi\Delta_\varphi^{-\mathrm{i}\bar\alpha+(1/2)}J_\varphi\pi_\varphi(\sigma_{\alpha-(\mathrm{i}/2)}^\varphi(a_k))J_\varphi(\sigma_\alpha^\varphi(y)^*)_\varphi\end{aligned}$$

and taking the limit over k it follows that

$$y_\varphi = J_\varphi\Delta_\varphi^{-\mathrm{i}\bar\alpha+(1/2)}(\sigma_\alpha^\varphi(y)^*)_\varphi.$$

Similarly, we obtain

$$(x^*)_\varphi = J_\varphi\Delta_\varphi^{-\mathrm{i}\alpha-(1/2)}(\sigma_{\alpha-i}^\varphi(x))_\varphi.$$

Consequently,

$$\begin{aligned}\varphi(xy) &= (y_\varphi|(x^*)_\varphi)_\varphi\\ &= (J_\varphi\Delta_\varphi^{-\mathrm{i}\bar\alpha+(1/2)}(\sigma_\alpha^\varphi(y)^*)_\varphi|\,J_\varphi\Delta_\varphi^{-\mathrm{i}\alpha-(1/2)}(\sigma_{\alpha-\mathrm{i}}^\varphi(x))_\varphi)_\varphi\\ &= (\sigma_{\alpha-\mathrm{i}}^\varphi(x)_\varphi|(\sigma_\alpha^\varphi(y)^*)_\varphi)_\varphi = \varphi(\sigma_\alpha^\varphi(y)\sigma_{\alpha-\mathrm{i}}^\varphi(x)).\end{aligned}$$

In particular, for $\alpha = 0, \alpha = \mathrm{i}$, and $\alpha = \mathrm{i}/2$, we have

$$\varphi(xy) = \varphi(y\sigma^{\varphi}_{-\mathrm{i}}(x)) = \varphi(\sigma^{\varphi}_{\mathrm{i}}(y)x) = \varphi(\sigma^{\varphi}_{1/2}(y)\sigma^{\varphi}_{-1/2}(x)), \tag{2}$$

whenever $x, y \in \mathfrak{T}_{\varphi}$. These identities replace for weights the relation $\varphi(xy) = \varphi(yx)$, which is valid only for traces. □

2.18. Another similar result, which (formally) follows from 2.17.(1), is contained in the next statement:

$$a \in \mathscr{M}^{\varphi}_{\infty}, z \in \mathfrak{M}_{\varphi} \Rightarrow \varphi(z\sigma^{\varphi}_{\alpha}(a)) = \varphi(\sigma^{\varphi}_{\alpha+\mathrm{i}}(a)z) \text{ for all } \alpha \in \mathbb{C}. \tag{1}$$

We give a direct proof here. Since $z \in \mathfrak{M}_{\varphi} = \mathfrak{N}^*_{\varphi}\mathfrak{N}_{\varphi}$, we may assume $z = y^*x$ with $x, y \in \mathfrak{N}_{\varphi}$. Using Proposition 2.14, we get

$$\begin{aligned}
\varphi(y^*x\sigma^{\varphi}_{\alpha}(a)) &= ((x\sigma^{\varphi}_{\alpha}(a))_{\varphi}|y_{\varphi})_{\varphi} = (J_{\varphi}\pi_{\varphi}(\sigma^{\varphi}_{-\mathrm{i}/2}(\sigma^{\varphi}_{\alpha}(a)^*)J_{\varphi}x_{\varphi}|y_{\varphi})_{\varphi} \\
&= (J_{\varphi}\pi_{\varphi}(\sigma^{\varphi}_{\alpha+(\mathrm{i}/2)}(a))^*J_{\varphi}x_{\varphi}|y_{\varphi})_{\varphi} = (x_{\varphi}|J_{\varphi}\pi_{\varphi}(\sigma^{\varphi}_{\alpha+(\mathrm{i}/2)}(a))J_{\varphi}y_{\varphi})_{\varphi} \\
&= (x_{\varphi}|J_{\varphi}\pi_{\varphi}(\sigma^{\varphi}_{-\mathrm{i}/2}(\sigma^{\varphi}_{\alpha+\mathrm{i}}(a)))J_{\varphi}y_{\varphi})_{\varphi} = ((y\sigma^{\varphi}_{\alpha+\mathrm{i}}(a)^*)_{\varphi}|x_{\varphi})_{\varphi} \\
&= \varphi(\sigma^{\varphi}_{\alpha+\mathrm{i}}(a)y^*x).
\end{aligned}$$

As $\alpha \mapsto (x_{\varphi}|J_{\varphi}\pi_{\varphi}(\sigma^{\varphi}_{\alpha+(\mathrm{i}/2)}(a))J_{\varphi}y_{\varphi})_{\varphi}$ is an entire analytic function, bounded on horizontal strips, it follows that *the functions*

$$\alpha \mapsto \varphi(z\sigma^{\varphi}_{\alpha}(a)) \text{ and } \alpha \mapsto \varphi(\sigma^{\varphi}_{\alpha}(a)z)$$

are entire analytic and bounded on horizontal strips, for all $a \in \mathscr{M}^{\varphi}_{\infty}$ *and all* $z \in \mathfrak{M}_{\varphi}$.

On the other hand, if $a \in \mathscr{M}^{\varphi}_{\infty}$, and $z \in \mathfrak{M}_{\varphi} \cap \mathscr{M}^+$, then for all $\alpha \in \mathbb{C}$ we have

$$\begin{aligned}
\varphi(\sigma^{\varphi}_{\alpha}(a)z\sigma^{\varphi}_{\alpha}(a)^*) &\leq \|\sigma^{\varphi}_{\alpha-(\mathrm{i}/2)}(a)\|^2\varphi(z), \\
\varphi(\sigma^{\varphi}_{\alpha}(a)^*z\sigma^{\varphi}_{\alpha}(a)) &\leq \|\sigma^{\varphi}_{\alpha+(\mathrm{i}/2)}(a)\|^2\varphi(z).
\end{aligned} \tag{2}$$

Indeed, $z = x^*x$ with $x = z^{1/2} \in \mathfrak{N}_{\varphi}$ and using Proposition 2.14, we get

$$\begin{aligned}
\varphi(\sigma^{\varphi}_{\alpha}(a)z\sigma^{\varphi}_{\alpha}(a)^*) &= \|(x\sigma^{\varphi}_{\alpha}(a)^*)_{\varphi}\|^2_{\varphi} \\
&\leq \|J_{\varphi}\pi_{\varphi}(\sigma^{\varphi}_{-\mathrm{i}/2}(\sigma^{\varphi}_{\alpha}(a)))J_{\varphi}\|^2\|x_{\varphi}\|^2_{\varphi} = \|\sigma^{\varphi}_{\alpha-(\mathrm{i}/2)}(a)\|^2\varphi(z)
\end{aligned}$$

and the other inequality is verified in a similar way.

We note that the right-hand side of (2) depends just on $|\mathrm{Im}\,\alpha|$, as the $\sigma^{\varphi}_t (t \in \mathbb{R})$ being $*$-automorphisms, are isometric.

Properties (1) and (2) characterize the entire analytic functions of the type $\alpha \mapsto \sigma^{\varphi}_{\alpha}(a)$ with $a \in \mathscr{M}^{\varphi}_{\infty}$ as we shall see in Theorem 2.19.

The identity (1) is meaningful also for certain other a and z. Indeed, since $\mathfrak{M}_{\varphi}$ is contained and w-dense in the set $\{z \in \mathscr{M}; z\mathfrak{T}_{\varphi} \subset \mathfrak{A}_{\varphi}, \mathfrak{T}_{\varphi}z \subset \mathfrak{A}_{\varphi}\}$, using Proposition 2.13, we infer from (1), taking the limit, the following statement:

$$\begin{aligned}&\textit{if } a \in \mathfrak{T}_\varphi^2 \textit{ and } z \in \mathcal{M}; z\mathfrak{T}_\varphi \subset \mathfrak{A}_\varphi, \mathfrak{T}_\varphi z \subset \mathfrak{A}_\varphi, \textit{then}\\ &\varphi(z\sigma_\alpha^\varphi(a)) = \varphi(\sigma_{\alpha+\mathrm{i}}^\varphi(a)z) \text{ for all } \alpha \in \mathbb{C}.\end{aligned} \tag{3}$$

Under the same conditions as in statement (3), the functions

$$\alpha \mapsto \varphi(z\sigma_\alpha^\varphi(a)) \textit{ and } \alpha \mapsto \varphi(\sigma_\alpha^\varphi(a)z)$$

are entire analytic and bounded on horizontal strips. Indeed, if $a = bc^*$ with $b, c \in \mathfrak{T}_\varphi$, then, using 2.13.(1) and 2.13.(2), we get

$$\begin{aligned}\varphi(z\sigma_\alpha^\varphi(a)) &= \varphi(z\sigma_\alpha^\varphi(b)\sigma_{\bar\alpha}^\varphi(c)^*) = (\pi_\varphi(z)\Delta_\varphi^{\mathrm{i}\alpha} b_\varphi | \Delta^{\mathrm{i}\bar\alpha+1} c_\varphi)_\varphi,\\ \varphi(\sigma_\alpha^\varphi(a)z) &= \varphi(\sigma_\alpha^\varphi(b)\sigma_{\bar\alpha}^\varphi(c)^* z) = (\pi_\varphi(z)\Delta_\varphi^{\mathrm{i}\alpha+1} b_\varphi | \Delta^{\mathrm{i}\bar\alpha} c_\varphi)_\varphi.\end{aligned}$$

2.19 Theorem (A. Connes). *Let φ be an n.s.f. weight on the W^*-algebra $\mathcal{M}$ and consider a function $F : \mathbb{C} \to \mathcal{M}$ such that*

(a) *for every $z \in \mathfrak{M}_\varphi \cap \mathcal{M}^+$ we have $F(\alpha)z \in \mathfrak{M}_\varphi, zF(\alpha) \in \mathfrak{M}_\varphi$ the functions $\alpha \mapsto \varphi(F(\alpha)z)$ and $\alpha \mapsto \varphi(zF(\alpha))$ are entire analytic and*

$$\varphi(zF(\alpha)) = \varphi(F(\alpha + \mathrm{i})z) \textit{ for all } \alpha \in \mathbb{C};$$

(b) *for every $\varepsilon > 0$ there exists $\delta > 0$ such that if $\alpha \in \mathbb{C}, |\mathrm{Im}\ \alpha| \le \delta$, and $z \in \mathfrak{M}_\varphi \cap \mathcal{M}^+$, then $\varphi(F(\alpha)zF(\alpha)^*) \le \varepsilon\varphi(z)$ and $\varphi(F(\alpha)^* zF(\alpha)) \in \varepsilon\varphi(z)$.*

Then $F(0) \in \mathcal{M}_\infty^\varphi$ and $F(\alpha) = \sigma_\alpha^\varphi(F(0))$ for all $\alpha \in \mathbb{C}$.

Proof. We have to show that F is an entire analytic function and $F(t) = \sigma_\alpha^\varphi(F(0))$ for all $t \in \mathbb{R}$.

By assumption, $\mathfrak{N}_\varphi F(\alpha) \subset \mathfrak{N}_\varphi, \mathfrak{N}_\varphi F(\alpha)^* \subset \mathfrak{N}_\varphi, \mathfrak{M}_\varphi F(\alpha) \subset \mathfrak{M}_\varphi, F(\alpha)\mathfrak{M}_\varphi \subset \mathfrak{M}_\varphi$ and, for $x \in \mathfrak{N}_\varphi$ and $\alpha \in \mathbb{C}, |\mathrm{Im}\ \alpha| \le \delta$, we have

$$\|(xF(\alpha))_\varphi\|_\varphi \le \varepsilon^{1/2}\|x_\varphi\|_\varphi, \|(xF(\alpha)^*)_\varphi\|_\varphi \le \varepsilon^{1/2}\|x_\varphi\|_\varphi. \tag{1}$$

Let $a, b \in \mathfrak{T}_\varphi$. For every $\gamma \in \mathbb{C}$, we have $\sigma_\gamma^\varphi(ab) = \sigma_\gamma^\varphi(a)\sigma_\gamma^\varphi(b) \in \mathfrak{T}_\varphi\mathfrak{T}_\varphi \subset \mathfrak{N}_\varphi^*\mathfrak{N}_\varphi = \mathfrak{M}_\varphi$. We define a function G of two complex variables by

$$G(\alpha, \beta) = \varphi(F(\alpha)\sigma_\beta^\varphi(ab)) = ((\sigma_\beta^\varphi(b))_\varphi | (\sigma_\beta^\varphi(a)^* F(\alpha)^*)_\varphi)_\varphi. \tag{2}$$

By assumption and by the first equation in (2) it follows that $\alpha \mapsto G(\alpha, \beta)$ is an entire analytic function and

$$G(\alpha + \mathrm{i}, \beta) = \varphi(\sigma_\beta^\varphi(ab)F(\alpha)). \tag{3}$$

Using (1) and the second equation in (2) it follows that $\beta \mapsto G(\alpha, \beta)$ is also an entire analytic function. Consequently, G is an entire analytic function in both variables, by the Hartogs theorem (Hormander, 1966, p. 2.2.8). On the other hand, using (3) and 2.18.(3), we obtain

$$G(\alpha + \mathrm{i}, \beta + \mathrm{i}) = \varphi(\sigma_{\beta+\mathrm{i}}^\varphi(ab)F(\alpha)) = \varphi(F(\alpha)\sigma_\beta^\varphi(ab)) = G(\alpha, \beta).$$

Thus, $\alpha \mapsto g(\alpha) = G(\alpha, \alpha)$ is an entire analytic function and $g(\alpha + \mathrm{i}) = g(\alpha)$. Using (1) with $\delta = 1$ and (2), for $\alpha \in \mathbb{C}$ with $|\mathrm{Im}\,\alpha| \leq 1$ we get

$$\begin{aligned}|g(\alpha)| &= |((\sigma_\alpha^\varphi(b))_\varphi | (\sigma_\alpha^\varphi(a)^* F(\alpha)^*)_\varphi)_\varphi \\ &\leq \varepsilon^{1/2} \|(\sigma_\alpha^\varphi(b))_\varphi\|_\varphi \|(\sigma_\alpha^\varphi(a)^*)_\varphi\|_\varphi = \varepsilon^{1/2} \|\Delta_\varphi^{\mathrm{i}\alpha} b_\varphi\|_\varphi \|\Delta_\varphi^{\mathrm{i}\alpha+(1/2)} a_\varphi\|_\varphi.\end{aligned}$$

Therefore, the entire analytic function g is bounded, and hence constant, by the Liouville theorem. In particular, for $t \in \mathbb{R}$ we have $\varphi(\sigma_{-t}^\varphi(F(t))ab) = \varphi(F(t)\sigma_t^\varphi(ab)) = g(t) = g(0) = \varphi(F(0)ab)$. Since $\mathfrak{T}_\varphi$ is dense in $\mathscr{H}_\varphi$ it follows that $F(t) = \sigma_t^\varphi(F(0))$ for all $t \in \mathbb{R}$.

In order to prove that F is an entire analytic function, it is sufficient to show that F is bounded on each compact subset of $\mathbb{C}$, as the set $\{\varphi(\cdot z); z \in \mathfrak{T}_\varphi^2\}$ is norm-dense in $\mathscr{M}_*$ (2.13) and the functions $\alpha \mapsto \varphi(F(\alpha)z)(z \in \mathfrak{T}_\varphi^2)$ are, by assumption, entire analytic (use the Montel theorem and [L], Lemma 9.24). □

According to (1), the boundedness of F on compact subsets of $\mathbb{C}$ will follow once we establish the following identity (compare with 2.14.(7)):

$$J_\varphi \pi_\varphi(F(\alpha)) J_\varphi a_\varphi = (aF(\alpha + (\mathrm{i}/2))^*)_\varphi \quad (a \in \mathfrak{T}_\varphi, \alpha \in \mathbb{C}). \tag{4}$$

Since the assumptions are stable under translations $\alpha \mapsto \alpha + \alpha_0$, it is sufficient to prove (4) only for $\alpha = 0$.

To this end, consider $a, b \in \mathfrak{T}_\varphi$. For $\beta \in \mathbb{C}$ let

$$\begin{aligned}f_1(\beta) &= (\pi_\varphi(F(0))\Delta_\varphi^{-\mathrm{i}\beta}(a^*)_\varphi | \Delta_\varphi^{-\mathrm{i}\bar{\beta}} b_\varphi)_\varphi, \\ f_2(\beta) &= (J_\varphi \Delta_\varphi^{-1/2} b_\varphi | (aF(\beta)^*)_\varphi)_\varphi.\end{aligned}$$

The function f_1 is obviously entire analytic and the function f_2 is entire analytic by the assumption (b). For $t \in \mathbb{R}$, it is easy to check that $f_1(t) = f_2(t)$, since $F(t) = \sigma_t^\varphi F(0)$. Hence $f_1 = f_2$. In particular, $f_1(\mathrm{i}/2) = f_2(\mathrm{i}/2)$, that is,

$$(J_\varphi \pi_\varphi(F(0)) J_\varphi a_\varphi | J_\varphi \Delta_\varphi^{-1/2} b_\varphi)_\varphi = (aF(\mathrm{i}/2)^*)_\varphi | J_\varphi \Delta_\varphi^{-1/2} b_\varphi)_\varphi.$$

Since $b \in \mathfrak{T}_\varphi$ was arbitrary, we obtain (4) for $\alpha = 0$. □

2.20. The results presented in Section 2.18 involve several variants of the *KMS* condition.

For instance, if $a \in \mathscr{M}_\infty^\varphi$ and $z \in \mathfrak{M}_\varphi$, then from 2.18.(1) it follows that the equation

$$f(\alpha) = \varphi(z\sigma_{-\mathrm{i}\alpha}^\varphi(a)) \quad (\alpha \in \mathbb{C})$$

defines an entire analytic function f, bounded on vertical strips, such that

$$f(\mathrm{i}t) = \varphi(z\sigma_t^\varphi(a)), f(1 + \mathrm{i}t) = \varphi(\sigma_t^\varphi(a)z) \quad (t \in \mathbb{R}).$$

Also, if $a \in \mathfrak{T}_\varphi^2$ and $z \in \mathscr{M}, z\mathfrak{T}_\varphi \subset \mathfrak{A}_\varphi, \mathfrak{T}_\varphi z \subset \mathfrak{A}_\varphi$, then the same conclusion is obtained from 2.18.(3).

We record here one more variant of the *KMS* condition, where the similarity to, as well as the contrast with, the trace property is very striking.

Proposition. *Let φ be an n.s.f. weight on the W^*-algebra $\mathscr{M}$ and let $x \in \mathfrak{N}_\varphi$. There exists a bounded regular positive Borel measure μ on $(0, +\infty)$ such that*

$$\varphi(x^*\sigma_t^\varphi(x)) = \int_0^\infty \lambda^{it}\mathrm{d}\mu(\lambda), \quad \varphi(xx^*) = \int_0^\infty \lambda\,\mathrm{d}\mu(\lambda) \quad (t \in \mathbb{R}).$$

Proof. Let $\{\mathrm{e}_\lambda = \chi_{[0,\lambda)}(\Delta_\varphi)\}_{\lambda>0}$ be the spectral scale of Δ_φ ([L], E.9.10). Since $x \in \mathfrak{N}_\varphi$, we obtain a bounded regular positive Borel measure μ on $(0, +\infty)$ setting

$$\mathrm{d}\mu(\lambda) = \mathrm{d}(\mathrm{e}_\lambda x_\varphi | x_\varphi)_\varphi,$$

that is, μ is "the spectral measure associated with Δ_φ and $x_\varphi \in \mathscr{H}_\varphi$." According to ([L], E.9.11) we get

$$\varphi(x^*\sigma_t^\varphi(x)) = (\Delta_\varphi^{it} x_\varphi | x_\varphi)_\varphi = \int_0^\infty \lambda^{it}\mathrm{d}\mu(\lambda)$$

and if $x_\varphi \in \mathfrak{N}_\varphi \cap \mathfrak{N}_\varphi^*$,

$$\varphi(xx^*) = \|(x^*)_\varphi\|_\varphi^2 = \|S_\varphi x_\varphi\|_\varphi^2 = \| J_\varphi \Delta_\varphi^{1/2} x_\varphi\|_\varphi^2 = \|\Delta_\varphi^{1/2} x_\varphi\|_\varphi^2 = \int_0^\infty \lambda \mathrm{d}\mu(\lambda).$$

The proof is completed by the remark that $x \in D(S_\varphi) = D(\Delta_\varphi^{1/2}) \Leftrightarrow \int_0^\infty \lambda\,\mathrm{d}\mu(\lambda) < +\infty$ and $\mathfrak{N}_\varphi \cap D(S_\varphi) = \mathfrak{N}_\varphi \cap \mathfrak{N}_\varphi^*$g (2.12.(3)). □

2.21. Let φ be an n.s.f. weight on the W^*-algebra $\mathscr{M}$. The centralizer of φ is the W^*-subalgebra of $\mathscr{M}$ defined by

$$\mathscr{M}^\varphi = \{a \in \mathscr{M}; \sigma_t^\varphi(a) = a \text{ for all } t \in \mathbb{R}\}.$$

Since $\pi_\varphi(\sigma_t^\varphi(a)) = \Delta_\varphi^{it}\pi_\varphi(a)\Delta_\varphi^{-it}(t \in \mathbb{R})$, it follows that $a \in \mathscr{M}^\varphi$ if and only if $\pi_\varphi(a)$ commutes with Δ_φ ([L], E.9.20, E.9.23).

Clearly,

$$a \in \mathscr{M}^\varphi \Rightarrow a \in \mathscr{M}_\infty^\varphi \textit{ and } \sigma_\alpha^\varphi(a) = a \text{ for all } \alpha \in \mathbb{C}.$$

Also, it follows from statement (7) of 2.14 that

$$a \in \mathscr{M}^\varphi, x \in \mathfrak{N}_\varphi \Rightarrow xa^* \in \mathfrak{N}_\varphi \text{ and } (xa^*)_\varphi = J_\varphi \pi_\varphi(a) J_\varphi x_\varphi. \quad (1)$$

The Pedersen–Takesaki theorem ([L], 10.27) shows that if $a \in \mathscr{M}$, then

$$a \in \mathscr{M}^\varphi \Leftrightarrow a\mathfrak{M}_\varphi \subset \mathfrak{M}_\varphi, \mathfrak{M}_\varphi a \subset \mathfrak{M}_\varphi \textit{ and } \varphi(ax) = \varphi(xa) \textit{ for } x \in \mathfrak{M}_\varphi. \quad (2)$$

The implication ($\Rightarrow$) follows obviously from 2.15 and 2.18.(1). Conversely, we have $a\mathfrak{T}_\varphi \subset \mathfrak{A}_\varphi$, $\mathfrak{T}_\varphi a \subset \mathfrak{A}_\varphi$, hence (2.20, 2.18(3)) for every $x \in \mathfrak{T}_\varphi^2$ the function $f_x(\alpha) = \varphi(a\sigma_{-i\alpha}^\varphi(x))$ is an entire analytic function, bounded on vertical strips and such that

$$f_x(1+\mathrm{i}t) = \varphi(\sigma_t^\varphi(x)a) = \varphi(a\sigma_t^\varphi(x)) = f_x(\mathrm{i}t) \quad (t \in \mathbb{R}).$$

By the Liouville theorem, it follows that f_x is constant and hence

$$\varphi(\sigma_t^\varphi(a)x) = \varphi(a\sigma_{-t}^\varphi(x)) = f_x(-\mathrm{i}t) = f_x(0) = \varphi(ax) \quad (t \in \mathbb{R}).$$

Since $x \in \mathfrak{T}_\varphi^2$ was arbitrary, using Proposition 2.13 we infer that $\sigma_t^\varphi(a) = a (t \in \mathbb{R})$, so that $a \in \mathscr{M}^\varphi$.

Let $v \in \mathscr{M}$ be a partial isometry such that $vv^* \in \mathscr{M}^\varphi$. We define a normal semifinite weight φ_v on $\mathscr{M}$ by

$$\varphi_v(x) = \varphi(vxv^*) \quad (x \in \mathscr{M}^+).$$

It is easy to check that $\mathbf{s}(\varphi_v) = v^*v$.

In particular, for every projection $e \in \mathscr{M}^\varphi$, we have defined a subweight φ_e on $\mathscr{M}$ with $\mathbf{s}(\varphi_e) = e$.

Proposition. *Let φ be an n.s.f. weight on the W^*-algebra $\mathscr{M}$ and $v \in \mathscr{M}$ a partial isometry such that $e = v^*v \in \mathscr{M}^\varphi$ and $f = vv^* \in \mathscr{M}^\varphi$. Then: $\varphi_v = \varphi_e \Leftrightarrow v \in \mathscr{M}^\varphi$.*

Proof. We assume first that $v \in \mathscr{M}^\varphi$. By (2) we have $\mathfrak{N}_\varphi v \subset \mathfrak{N}_\varphi$ and $\mathfrak{N}_\varphi v^* \subset \mathfrak{N}_\varphi$. Since $\mathfrak{N}_{\varphi_v} = \{x \in \mathscr{M}; xv^* \in \mathfrak{N}_\varphi\}$ and $\mathfrak{N}_{\varphi_e} = \{x \in \mathscr{M}; xe \in \mathfrak{N}_\varphi\}$, and since $e = v^*v, v^* = ev^*$, it follows that $\mathfrak{N}_{\varphi_v} = \mathfrak{N}_{\varphi_e}$ and hence $\mathfrak{M}_{\varphi_v} = \mathfrak{M}_{\varphi_e}$. For $x \in \mathfrak{M}_{\varphi_v} = \mathfrak{M}_{\varphi_e}$ we obtain $vxv^* \in \mathfrak{M}_\varphi, exe \in \mathfrak{M}_\varphi$ and, since $v = ve \in \mathscr{M}^\varphi$, $\varphi(vxv^*) = \varphi(vexv^*) = \varphi(exv^*v) = \varphi(exe)$. We conclude that $\varphi_v = \varphi_e$.

Conversely, assume that $\varphi_v = \varphi_e$. For every $x \in \mathscr{M}^+$ we have $\varphi(vxv^*) = \varphi(exe)$. Replacing x by v^*xv here, we get $\varphi(v^*xv) = \varphi(fxf)$. If $x \in \mathfrak{N}_\varphi$, then $xf \in \mathfrak{N}_\varphi$, as $f \in \mathscr{M}^\varphi$; consequently, $\varphi(v^*x^*xv) = \varphi(fx^*xf) < +\infty$, that is, $xv \in \mathfrak{N}_\varphi$. Thus, $\mathfrak{N}_\varphi v \subset \mathfrak{N}_\varphi$ and, similarly, $\mathfrak{N}_\varphi v^* \subset \mathfrak{N}_\varphi$. It follows that $v\mathfrak{M}_\varphi \subset \mathfrak{M}_\varphi$ and $\mathfrak{M}_\varphi v \subset \mathfrak{M}_\varphi$. Then, for $x \in \mathfrak{M}_\varphi$ we have

$$\varphi(vx) = \varphi(fvx) = \varphi(vxf) = \varphi(vxvv^*) = \varphi(exve) = \varphi(xve) = \varphi(xv).$$

Using (2) we conclude that $v \in \mathscr{M}^\varphi$. □

In particular, for a unitary element $u \in \mathscr{M}$ we have $\varphi_u = \varphi$ if and only if $u \in \mathscr{M}^\varphi$.

2.22. If φ is a normal, semifinite, but not necessarily faithful, weight on the W^*-algebra $\mathscr{M}$, then we shall denote by $\{\sigma_t^\varphi\}_{t\in\mathbb{R}}$ the modular automorphism group associated with the n.s.f. weight on the W^*-algebra $\mathbf{s}(\varphi)\mathscr{M}\mathbf{s}(\varphi)$ defined by the restriction of φ; also, we shall denote by $\mathscr{M}_\infty^\varphi$, and $\mathscr{M}^\varphi$ the *-subalgebra of all entire analytic elements and the centralizer of $\varphi|\mathbf{s}(\varphi)\mathscr{M}\mathbf{s}(\varphi)$ respectively.

The results obtained for n.s.f. weights, in particular the characterization of the modular automorphism group with the aid of the *KMS* condition, Proposition 2.20, and so on, can easily be extended to normal semifinite weights.

For instance, if φ is a normal semifinite weight on $\mathscr{M}$ and $v \in \mathscr{M}$ is a partial isometry with $vv^* \in \mathscr{M}^\varphi$, then we get a new normal semifinite weight φ_v on $\mathscr{M}$, with $\mathbf{s}(\varphi_v) = v^*v$ and, using the *KMS* condition it follows that

$$\sigma_t^{\varphi_v}(x) = v^*\sigma_t^\varphi(vxv^*)v \quad (x \in v^*v\mathscr{M}v^*v, t \in \mathbb{R}), \tag{1}$$

hence

$$\mathcal{M}^{\varphi_v} = v^* \mathcal{M}^{\varphi} v. \tag{2}$$

In particular, if $e \in \mathcal{M}^{\varphi}$ is a projection, then

$$\sigma_t^{\varphi_e}(x) = \sigma_t^{\varphi}(x) \quad (x \in e\mathcal{M}e, t \in \mathbb{R}), \tag{3}$$

$$\mathcal{M}^{\varphi_e} = e\mathcal{M}^{\varphi}e. \tag{4}$$

We mention also that if φ is an n.s.f. weight on $\mathcal{M}$ and σ is a *-automorphism of $\mathcal{M}$, then, using the *KMS* condition we get

$$\sigma_t^{\varphi \circ \sigma} = \sigma^{-1} \circ \sigma_t^{\varphi} \circ \sigma \quad (t \in \mathbb{R}), \tag{5}$$

$$\mathcal{M}^{\varphi \circ \sigma} = \sigma^{-1}(\mathcal{M}^{\varphi}). \tag{6}$$

2.23. Recall ([L), 10.23) that a von Neumann algebra $\mathcal{M} \subset \mathcal{B}(\mathcal{H})$ is called hyperstandard if there exists a conjugation $J : \mathcal{H} \to \mathcal{H}$ and a self-polar convex cone $\mathfrak{P} \subset \mathcal{H}$ with the following properties: (a) the mapping $x \mapsto Jx^*J$ is a *-antiisomorphism of $\mathcal{M}$ onto $\mathcal{M}'$, which acts identically on the center; (b) $J\xi = \xi$ for every $\xi \in \mathfrak{P}$; c) $[x(JxJ)]\mathfrak{P} \subset \mathfrak{P}$ for every $x \in \mathcal{M}$.

For any n.s.f. weight φ on the W^*-algebra $\mathcal{M}$, the von Neumann algebra $\pi_{\varphi}(\mathcal{M}) \subset \mathcal{B}(\mathcal{H}_{\varphi})$ with the conjugation J_{φ} and the self-polar convex cone

$$\mathfrak{P}_{\varphi} = \overline{\{\pi_{\varphi}(x)J_{\varphi}x_{\varphi}; x \in \mathfrak{T}_{\varphi}\}} = \overline{\{\pi_{\varphi}(x)J_{\varphi}x_{\varphi}; x \in \mathfrak{A}_{\varphi}\}}$$

is a hyperstandard form of $\mathcal{M}$ ([L], 10.23). Using statement (16) of 2.12 and arguing as in ([L], 10.23), one shows that

$$\mathfrak{P}_{\varphi} = \overline{\{\pi_{\varphi}(x)J_{\varphi}x_{\varphi};\ x \in \mathfrak{N}_{\varphi}\}}. \tag{1}$$

Also, any two hyperstandard forms of $\mathcal{M}$ are spatially isomorphic by a unique unitary operator that preserves the self-polar convex cones and hence also the conjugations ([L], 10.26, 10.23).

Let $\mathcal{M}$ be a W^*-algebra and $(\mathcal{M}, \mathcal{H}, J, \mathfrak{P})$ a hyperstandard form of $\mathcal{M}$. As usual, we denote by $Aut(\mathcal{M})$ the group of *-automorphisms of $\mathcal{M}$ and by $U(\mathcal{H})$ the group of unitary operators on $\mathcal{H}$. Also, for $u \in U(\mathcal{H})$ with $u\mathcal{M}u^* = \mathcal{M}$, we denote by $Ad(u) \in Aut(\mathcal{M})$ the *-automorphism defined by $[Ad(u)](x) = uxu^*(x \in \mathcal{M})$.

It follows that there exists an injective group homomorphism

$$Aut(\mathcal{M}) \ni \sigma \mapsto u_{\sigma} \in U(\mathcal{H}),$$

uniquely determined, such that $\sigma = Ad(u_{\sigma})$ and $u_{\sigma}(\mathfrak{P}) = \mathfrak{P}(\sigma \in Aut(\mathcal{M}))$. Clearly.

$$\{u_{\sigma}; \sigma \in Aut(\mathcal{M})\} = \{u \in U(\mathcal{H}); u\mathcal{M}u^* = \mathcal{M}, u(\mathfrak{P}) = \mathfrak{P}\}.$$

The mapping $\sigma \mapsto u_{\sigma}$ is called the *canonical implementation* of $Aut(\mathcal{M})$.

On the linear space $\mathscr{B}_w(\mathscr{M})$ of all w-continuous linear mappings $T : \mathscr{M} \to \mathscr{M}$, we may consider several locally convex topologies, for instance:

the *n-topology*, defined by the seminorms $T \mapsto \|Tx\|$,
the *p-topology*, defined by the seminorms $T \mapsto \|\varphi(Tx)\|$,
the *u-topology*, defined by the seminorms $T \mapsto \|\varphi \circ T\|$,

where $x \in \mathscr{M}$ and $\varphi \in \mathscr{M}_*$. Clearly, the u-topology and the n-topology are stronger than the p-topology.

By restriction, we obtain the topologies n, p, and u on $Aut(\mathscr{M}) \subset \mathscr{B}_w(\mathscr{M})$.

On the other hand, on $U(\mathscr{H})$ the topologies wo, so, so^*, w, s, s^*, and also the Mackey topologies τ_{w_0}, τ_w all coincide and, with this topology, $U(\mathscr{H})$ is a topological group (on the unit ball of $\mathscr{B}(\mathscr{H})$, the Mackey topology τ_w coincides with the s^*-topology by the theorem of Akemann (Aarnes, 1968; Akemann, 1967).

If $\mathscr{H}$ is separable, then $U(\mathscr{H})$ with the two-sided uniform structure associated with this topology is a complete separable metric space, that is, a polish group.

Theorem (U. Haagerup). *The canonical implementation $\sigma \mapsto u_\sigma$ of $Aut(\mathscr{M})$ is an isomorphism of topological groups between $Aut(\mathscr{M})$ with the u-topology and a closed subgroup of $U(\mathscr{H})$.*

Proof. Let $\sigma \in Aut(\mathscr{M})$ and let $\{\sigma_i\}$ be a net in $Aut(\mathscr{M})$; also let $u = u_\sigma \in U(\mathscr{H})$ and $u_i = u_{\sigma_i} \in U(\mathscr{H})$. We have $\sigma_i \overset{u}{\to} \sigma$ in $Aut(\mathscr{M})$ if and only if $\|\varphi \circ \sigma_i - \varphi \circ \sigma\| \to 0$ for every $\varphi \in \mathscr{M}_*^+$, that is, ([L], 10.25) if and only if $\|\omega_\xi \circ \sigma_i - \omega_\xi \circ \sigma\| \to 0$ for every $\xi \in \mathfrak{P}$, that is, if and only if $\|\omega_{u_i^*\xi} - \omega_{u^*\xi}\| \to 0$ for every $\xi \in \mathfrak{P}$. According to ([L], Proposition 10.24), this means that $\|u_i^*\xi - u^*\xi\| \to 0$ for every $\xi \in \mathfrak{P}$ and, since $\mathscr{H}$ is the linear span of $\mathfrak{P}$ ([L], 10.23), it follows that $u_i \overset{wo}{\to} u$.

Thus, the canonical implementation is a homeomorphism of $Aut(\mathscr{M})$, with the u-topology, onto the set $\{u \in U(\mathscr{M}) : u\mathscr{M}u^* = \mathscr{M}, u(\mathfrak{P}) = \mathfrak{P}\}$, which is clearly a closed subgroup of $U(\mathscr{H})$. In particular, $Aut(\mathscr{M})$ with the u-topology is a topological group.

In view of the above Theorem, we shall consider the u-topology as the *natural topology* on $Aut(\mathscr{M})$. In general, the u-topology does not coincide either with the p-topology, or with the n-topology. For instance, if $\mathscr{M}$ is the von Neumann algebra $\mathscr{L}^\infty([0,1]) \subset \mathscr{B}(\mathscr{L}^2([0,1]))$, then the u-topology is not comparable with the n-topology and hence both of them are strictly stronger than the p-topology (Haagerup, 1975b, p. 3.14).

Note that *if the W^*-algebra $\mathscr{M}$ has a separable predual (i.e., if $\mathscr{H}$ is separable), then $Aut(\mathscr{M})$ with the u-topology is a polish group.*

Finally, we note that if $\{\sigma_n\}_{n\geq 0} \subset Aut(\mathscr{M})$ is a sequence such that $\sigma_n \overset{p}{\to} \sigma_0$, then, for every compact set $K \subset U(\mathscr{M})$, we have

$$\sigma_n(u) \overset{s^*}{\to} \sigma_0(u) \text{ uniformly for } u \in K.$$

Clearly, it is sufficient to show that $\sigma_n(u) \overset{w}{\to} \sigma_0(u)$ uniformly for $u \in K$. Let $\varphi \in \mathscr{M}_*$ and $\varepsilon > 0$ be fixed. The set $\{\varphi \circ \sigma_n; n \geq 0\} \subset \mathscr{M}_*$ is norm-compact and hence also $\sigma(\mathscr{M}_*, \mathscr{M})$-compact. Since the topology on $U(\mathscr{M})$ is equal, in particular, to the topology induced by the Mackey topology τ_w on $\mathscr{M}$, it follows that the function $u \mapsto \sup\{|\varphi(\sigma_n(u))|; n \geq 0\}$ is continuous on $U(\mathscr{M})$. Thus, there exist $u_1, \ldots, u_m \in K$ such that $\inf\{|\varphi(\sigma_n(u - u_j))|; 1 \leq j \leq m\} < \varepsilon/3$ for all $u \in K$ and all $n \geq 0$. On the other hand, as $\sigma_n \overset{p}{\to} \sigma_0$, there exists $n_0 \geq 1$ such that $|\varphi(\sigma_n(u_j)) - \varphi(\sigma_0(u_j))| < \varepsilon/3$ for all $1 \leq j \leq m$ and all $n \geq n_0$. It follows that $|\varphi(\sigma_n(u)) - \varphi(\sigma_0(u))| < \varepsilon$ for all $u \in K$ and all $n \geq n_0$. □

2.24. Let σ be an *action* of the locally compact group G on the W^*-algebra $\mathcal{M}$, that is a group homomorphism

$$\sigma : G \to Aut(\mathcal{M}).$$

In Section 13.5, we prove that the homomorphism σ is p-continuous if and only if it is u-continuous; in this case we say that σ is a *continuous action*.

The following result is an obvious consequence of Theorem 2.23:

Corollary. *Let $\sigma : G \to Aut(\mathcal{M})$ be a continuous action of the locally compact group G on the W^*-algebra $\mathcal{M}$, let $(\mathcal{M}, \mathcal{H}, J, \mathfrak{P})$ be a hyperstandard form of $\mathcal{M}$ and denote by $u(g) = u_{\sigma_g} (g \in G)$ the canonical implementation. Then*

$$G \ni g \mapsto u(g) \in U(\mathcal{H})$$

is an so-continuous unitary representation and $\sigma_g = Ad(u(g)), (g \in G)$.

2.25. We now show that in certain cases the u-topology coincides with the p-topology.

Let φ be an n.s.f. weight on the W^*-algebra $\mathcal{M}$ and let

$$Aut_\varphi(\mathcal{M}) = \{\sigma \in Aut(\mathcal{M}); \varphi \circ \sigma = \varphi\}.$$

If $\sigma \in Aut_\varphi(\mathcal{M})$, then the canonical implementation of σ in the hyperstandard form $(\pi_\varphi(\mathcal{M}), \mathcal{H}_\varphi, J_\varphi, \mathfrak{P}_\varphi)$ has an explicit form, namely

$$u_\sigma a_\varphi = (\sigma(a))_\varphi \quad (a \in \mathfrak{N}_\varphi). \tag{1}$$

Indeed, for $a \in \mathfrak{N}_\varphi$ we have $\|(\sigma(a))_\varphi\|_\varphi^2 = \varphi(\sigma(a^*a)) = \varphi(a^*a) = \|a_\varphi\|_\varphi^2$ and hence (1) defines a unitary operator $u_\sigma \in U(\mathcal{H}_\varphi)$. For every $x \in \mathcal{M}, a \in \mathfrak{N}_\varphi$, we have $u_\sigma \pi_\varphi(x) u_\sigma^* a_\varphi = u_\sigma \pi_\varphi(x)(\sigma^{-1}(a))_\varphi = u_\sigma(x\sigma^{-1}(a))_\varphi = \pi_\varphi(\sigma(x))a_\varphi$. On the other hand, for $a \in \mathfrak{A}_\varphi$ we have $u_\sigma S_\varphi a_\varphi = u_\sigma(a^*)_\varphi = (\sigma(a^*))_\varphi = (\sigma(a)^*)_\varphi = S_\varphi(\sigma(a))_\varphi = S_\varphi u_\sigma a_\varphi$, hence u_σ commutes with S_φ. Consequently, u_σ commutes with J_φ and with Δ_φ. Moreover, we have $u_\sigma \pi_\varphi(a) J_\varphi a_\varphi = \pi_\varphi(\sigma(a)) u_\sigma J_\varphi a_\varphi = \pi_\varphi(\sigma(a)) J_\varphi u_\sigma a_\varphi = \pi_\varphi(\sigma(a)) J_\varphi(\sigma(a))_\varphi$ hence $u_\sigma(\mathfrak{P}_\varphi) \subset \mathfrak{P}_\varphi$ by 2.23.(1). We conclude that u_σ is the canonical implementation of σ.

Corollary. *Let φ be an n.s.f. weight on the W^*-algebra $\mathcal{M}$. Then the p-topology coincides with the u-topology on the u-closed subgroup $Aut_\varphi(\mathcal{M})$ of $Aut(\mathcal{M})$.*

Proof. Let $\sigma_i \xrightarrow{p} \sigma$ in $Aut_\varphi(\mathcal{M})$. Let $\xi = a_\varphi \in \mathfrak{N}_\varphi$ and $\eta = R_{\eta_1}^* \eta_2$ with $\eta_1, \eta_2 \in \mathfrak{A}'_\varphi$. Using (1) we obtain

$$\begin{aligned}(u_{\sigma_t}\xi|\eta)_\varphi &= ((\sigma_t(a))_\varphi|R_{\eta_1}^*\eta_2)_\varphi = (R_{\eta_1}(\sigma_i(a))_\varphi|\eta_2)_\varphi \\ &= (\pi_\varphi(\sigma_i(a))\eta_1|\eta_2)_\varphi \to (\pi_\varphi(\sigma(a))\eta_1|\eta_2)_\varphi \\ &= (R_{\eta_1}(\sigma(a))_\varphi|\eta_2)_\varphi = ((\sigma(a))_\varphi|R_{\eta_1}^*\eta_2)_\varphi = (u_\sigma\xi|\eta)_\varphi.\end{aligned}$$

It follows that $u_{\sigma_t} \xrightarrow{wo} u$ and hence, by Theorem 2.23, $\sigma_i \xrightarrow{u} \sigma$ in $Aut_\varphi(\mathcal{M})$. □

In particular, *if $\mathcal{M}$ is a factor of type I, or a factor of type II_1, then the p-topology coincides with the u-topology on Aut($\mathcal{M}$)*, because in these cases every *-automorphism preserves the trace on $\mathcal{M}$.

2.26. Finally, we consider the particular case of a von Neumann algebra, $\mathcal{M} \subset \mathcal{B}(\mathcal{H})$ with a cyclic and separating vector $\xi \in \mathcal{H}$. Then $\varphi = \varphi_\xi|\mathcal{M}$ is a faithful normal positive on $\mathcal{M}$ and we can identify $\mathcal{H}_\varphi$ with $\mathcal{H}$ via the mapping $x_\varphi \mapsto x\xi (x \in \mathcal{M})$; let $S = S_\varphi, J = J_\varphi, \Delta = \Delta_\varphi$. We have the convex cones ([L], 10.8)

$$\mathfrak{P}_S = \overline{\{x\xi : x \in \mathcal{M}^+\}}, \quad \mathfrak{P}_{S^*} = \overline{\{x'\xi : x' \in \mathcal{M}'^+\}}$$

polar to one another, and the self-polar convex cone ([L], 10.23)

$$\mathfrak{P} = \Delta^{1/4}\mathfrak{P}_S = \Delta^{-1/4}\mathfrak{P}_{S^*} = \overline{\{xJxJ\xi : x \in \mathcal{M}\}}.$$

Proposition. $\mathfrak{P} = \overline{\{xJxJ\xi; x \in \mathcal{M}^+\}}$.

Proof. It is sufficient to show that for every $x \in \mathcal{M}^+$ there exists a sequence $\{x_n\} \subset \mathcal{M}^+$ such that $x_nJx_nJ\xi \in \Delta^{1/4}x\xi$. Moreover, we may assume that x is invertible. In this case, both ξ and $x^{1/2}\xi$ are cyclic and separating vectors for $\mathcal{M}' \subset \mathcal{B}(\mathcal{H})$. By the uniqueness of the hyperstandard form of a von Neumann algebra, there exists a unitary operator $u \in (\mathcal{M}')' = \mathcal{M}$ such that $ux^{1/2}\xi \in \mathfrak{P}$.

Thus, $y = ux^{1/2} \in \mathcal{M}$ and $y\xi \in \mathfrak{P}$. If

$$y_n = \sqrt{n/\pi} \int_{-\infty}^{+\infty} e^{-nt^2} \sigma_t^\varphi(y) \mathrm{d}t \quad (n \in \mathbb{N}),$$

then (2.12.(16)) $\mathcal{M}_\infty^\varphi \ni y_n \overset{s^*}{\mapsto} y$ and ([L], 10.23.(6))

$$y_n\xi = \sqrt{n/\pi} \int_{-\infty}^{+\infty} e^{-nt^2} \Delta^{\mathrm{i}t} y\xi \, \mathrm{d}t \in \mathfrak{P} \quad (n \in \mathbb{N}).$$

Let $x_n = \sigma_{\mathrm{i}/4}^\varphi(y_n) = \overline{\Delta^{-1/4}y_n\Delta^{1/4}} \in \mathcal{M} (n \in \mathbb{N})$. Then $\Delta^{1/4}x_n\xi = y_n\xi \in \mathfrak{P}$, that is, $x_n\xi \in \Delta^{-1/4}\mathfrak{P} \subset \mathfrak{P}_S$, and hence $x_n \geq 0$. In particular $x_n = x_n^* = \sigma_{\mathrm{i}/4}^\varphi(y_n)^* = \sigma_{-\mathrm{i}/4}^\varphi(y_n^*)$, by 2.14.(1), and using this identity it is easy to check that

$$x_nJx_n\xi = \Delta^{1/4}y_n^*y_n\xi \to \Delta^{1/4}y^*y\xi = \Delta^{1/4}x\xi. \qquad \square$$

2.27. Notes. Theorem 2.6 and the technical results (2.3, 2.4) used for its proof are due to Combes (1968). Proposition 2.15 is due to Connes (1973b). Proposition 2.16 and the various forms of the *KMS* condition (2.17, 2.18, 2.20) are due to Pedersen and Takesaki (1973) and Connes (1980). Theorem 2.19 is due to Connes (1973a). The canonical implementation of *-automorphisms (2.23–2.25) is explicitly stated by Haagerup (1975b), having been obtained also by Araki (1974) and Connes (1974c). Proposition 2.26 is due to Connes (1974c).

For our exposition we have used Combes (1968, 1971a, 1971b); Connes (1973a, 1972, 1973b, 1980); Connes and Takesaki (1977); Van Daele (1973); Digernes (1975); Haagerup (1975b); Pedersen (1973–1977); Strătilă, Voiculescu and Zsidó (1976, 1977); [L]; Takesaki (1969–1970); Takesaki (1970).

In the framework of quantum field theory, Bisognano and Wichmann (1975) explicitly computed the modular operator and the canonical conjugation for some von Neumann algebras associated with a hermitian scalar field. Recent developments concerning geometric aspects of standard representations are due to Haagerup and Skau (see Skau, 1980). Other related references are Haagerup (1979c) and Woronowicz (1979).

3 The Balanced Weight

This section contains a detailed study, with applications, of the balanced weight, which is the main technical idea in the proof of the Connes cocycle theorem.

3.1. Let $\mathscr{M}$ be a W^*-algebra, φ an n.s.f. weight on $\mathscr{M}$ and ψ a normal semifinite (but not necessarily faithful) weight on $\mathscr{M}$.

We define the *balanced weight* $\theta = \theta(\varphi, \psi)$ on the W^*-algebra $\mathscr{N} = Mat_2(\mathscr{M})$ of all 2×2 matrices over $\mathscr{M}$, by

$$\theta\begin{pmatrix} x_{11} & x_{12} \\ x_{21} & x_{22} \end{pmatrix} = \varphi(x_{11}) + \psi(x_{12}) \qquad \left(\begin{pmatrix} x_{11} & x_{12} \\ x_{21} & x_{22} \end{pmatrix} \in \mathscr{N}^+\right).$$

It is easy to check that θ is a normal weight on $\mathscr{N}$. We have

$$\begin{pmatrix} x_{11} & x_{12} \\ x_{21} & x_{22} \end{pmatrix} \in \mathfrak{N}_\theta \Leftrightarrow x_{11}, x_{12} \in \mathfrak{N}_\varphi \; and \; x_{11}, x_{12} \in \mathfrak{N}_\psi,$$

hence

$$\mathfrak{N}_\theta = \begin{pmatrix} \mathfrak{N}_\varphi & \mathfrak{N}_\psi \\ \mathfrak{N}_\varphi & \mathfrak{N}_\psi \end{pmatrix}, \quad \mathfrak{N}_\theta^* = \begin{pmatrix} \mathfrak{N}_\varphi^* & \mathfrak{N}_\varphi^* \\ \mathfrak{N}_\psi^* & \mathfrak{N}_\psi^* \end{pmatrix}; \tag{1}$$

$$\mathfrak{N}_\theta \cap \mathfrak{N}_\theta^* = \begin{pmatrix} \mathfrak{N}_\varphi \cap \mathfrak{N}_\varphi^* & \mathfrak{N}_\psi \cap \mathfrak{N}_\varphi^* \\ \mathfrak{N}_\varphi \cap \mathfrak{N}_\psi^* & \mathfrak{N}_\psi \cap \mathfrak{N}_\psi^* \end{pmatrix}; \tag{2}$$

$$\mathfrak{M}_\theta = \mathfrak{N}_\theta^*\mathfrak{N}_\theta = \begin{pmatrix} \mathfrak{N}_\varphi^*\mathfrak{N}_\varphi & \mathfrak{N}_\varphi^*\mathfrak{N}_\psi \\ \mathfrak{N}_\psi^*\mathfrak{N}_\varphi & \mathfrak{N}_\psi^*\mathfrak{N}_\psi \end{pmatrix} = \begin{pmatrix} \mathfrak{M}_\varphi & \mathfrak{N}_\varphi^*\mathfrak{N}_\psi \\ \mathfrak{N}_\psi^*\mathfrak{N}_\varphi & \mathfrak{M}_\psi \end{pmatrix}. \tag{3}$$

In particular, θ is semifinite. Also

$$\mathbf{s}(\theta) = \begin{pmatrix} 1 & 0 \\ 0 & \mathbf{s}(\psi) \end{pmatrix}. \tag{4}$$

In the proof of the Connes cocycle theorem ([L], 10.28), we have already considered the case when both φ and ψ are faithful. Taking into account the remarks made in Section 2.22, we can extend the Connes theorem to the more general case considered here.

Using Proposition 2.21, we obtain

$$\begin{pmatrix} 1 & 0 \\ 0 & \mathbf{s}(\psi) \end{pmatrix}, \begin{pmatrix} 1 & 0 \\ 0 & -\mathbf{s}(\psi) \end{pmatrix} \in \mathcal{N}^\theta \text{ and hence } \begin{pmatrix} 1 & 0 \\ 0 & 0 \end{pmatrix}, \begin{pmatrix} 0 & 0 \\ 0 & \mathbf{s}(\psi) \end{pmatrix} \in \mathcal{N}^\theta.$$

Using, as in ([L], 10.28), the *KMS* condition, we infer that

$$\sigma_t^\theta\left(\begin{pmatrix} x & 0 \\ 0 & 0 \end{pmatrix}\right) = \begin{pmatrix} \sigma_t^\varphi(x) & 0 \\ 0 & 0 \end{pmatrix} \quad (x \in \mathcal{M}), \tag{5}$$

$$\sigma_t^\theta\left(\begin{pmatrix} 0 & 0 \\ 0 & y \end{pmatrix}\right) = \begin{pmatrix} 0 & 0 \\ 0 & \sigma_t^\psi(y) \end{pmatrix} \quad (x \in \mathbf{s}(\psi)\mathcal{M}\mathbf{s}(\psi)). \tag{6}$$

Since $\begin{pmatrix} 0 & 0 \\ \mathbf{s}(\psi) & 0 \end{pmatrix} \in \mathbf{s}(\theta)\mathcal{N}\mathbf{s}(\theta)$ and since

$$\begin{pmatrix} 1 & 0 \\ 0 & 0 \end{pmatrix}\sigma_t^\theta\left(\begin{pmatrix} 0 & 0 \\ \mathbf{s}(\psi) & 0 \end{pmatrix}\right) = \sigma_t^\theta\left(\begin{pmatrix} 1 & 0 \\ 0 & 0 \end{pmatrix}\begin{pmatrix} 0 & 0 \\ \mathbf{s}(\psi) & 0 \end{pmatrix}\right) = 0,$$

$$\sigma_t^\theta\left(\begin{pmatrix} 0 & 0 \\ \mathbf{s}(\psi) & 0 \end{pmatrix}\right)\begin{pmatrix} 0 & 0 \\ 0 & \mathbf{s}(\psi) \end{pmatrix} = \sigma_t^\theta\left(\begin{pmatrix} 0 & 0 \\ \mathbf{s}(\psi) & 0 \end{pmatrix}\begin{pmatrix} 0 & 0 \\ 0 & \mathbf{s}(\psi) \end{pmatrix}\right) = 0,$$

it follows that there exists $u_t \in \mathcal{M}$ such that

$$\sigma_t^\theta\left(\begin{pmatrix} 0 & 0 \\ \mathbf{s}(\psi) & 0 \end{pmatrix}\right) = \begin{pmatrix} 0 & 0 \\ u_t & 0 \end{pmatrix}.$$

Thus, we get an s^*-continuous mapping $\mathbb{R} \ni t \mapsto u_t \in \mathcal{M}$ such that

$$u_t u_t^* = \mathbf{s}(\psi) = u_0, \quad u_t^* u_t = \sigma_t^\varphi(\mathbf{s}(\psi)), \tag{7}$$

$$u_{t+s} = u_t \sigma_t^\varphi(u_s), \tag{8}$$

$$u_{-t} = \sigma_{-t}^\varphi(u_t^*), \tag{9}$$

$$\sigma_t^\psi(x) = u_t \sigma_t^\varphi(x) u_t^* \quad (x \in \mathbf{s}(\psi)\mathcal{M}\mathbf{s}(\psi)). \tag{10}$$

The arguments for checking (7), (8), and (10) are similar to those given in ([L], 10.28) and (9) is an easy consequence of (7) and (8).

For $X = [x_{ij}] \in \mathcal{N}$, we have $X \in \mathbf{s}(\theta)\mathcal{N}\mathbf{s}(\theta)$ if and only if $x_{12} = x_{12}\mathbf{s}(\psi), x_{21} = \mathbf{s}(\psi)x_{21}, x_{22} = \mathbf{s}(\psi)x_{22} = x_{22}\mathbf{s}(\psi)$; in this case

$$\sigma_t^\theta\left(\begin{pmatrix} x_{11} & x_{12} \\ x_{21} & x_{22} \end{pmatrix}\right) = \begin{pmatrix} \sigma_t^\varphi(x_{11}) & \sigma_t^\varphi(x_{12})u_t^* \\ u_t\sigma_t^\varphi(x_{21}) & \sigma_t^\psi(x_{22}) \end{pmatrix}. \tag{11}$$

On the other hand, the weight θ satisfies the *KMS* condition with respect to $\{\sigma_t^\theta\}_{t\in\mathbb{R}}$. In order to avoid notational complications, we assume that $\mathbf{s}(\psi) = 1$. Then, for any $X, Y \in \mathfrak{N}_\theta \cap \mathfrak{N}_\theta^*$, there exists a function F defined, continuous and bounded on the strip $\{\alpha \in \mathbb{C}; 0 \le \operatorname{Re}\alpha \le 1\}$, analytic in the interior of this strip, and such that $F(it) = \theta(X\sigma_t^\theta(Y)), F(1+it) = \theta(\sigma_t^\theta(Y)X)$ for all $t \in \mathbb{R}$.

In particular, if $X = \begin{pmatrix} 0 & x \\ 0 & 0 \end{pmatrix}, Y = \begin{pmatrix} 0 & 0 \\ y & 0 \end{pmatrix}$ with $x \in \mathscr{R}_\psi \cap \mathfrak{N}_\varphi^*, y \in \mathfrak{N}_\varphi \cap \mathscr{R}_\psi^*$, then $F(it) = \varphi(xu_t\sigma_t^\varphi(y)), F(1+it) = \psi(\sigma_t^\psi(y)u_tx)$ for all $t \in \mathbb{R}$.

We have thus proved the existence part of the following

Theorem (A. Connes). *Let $\mathscr{M}$ be a W^*-algebra, φ an n.s.f. weight on $\mathscr{M}$ and ψ a normal semifinite weight on $\mathscr{M}$. There exists an s^*-continuous mapping $\mathbb{R} \ni t \mapsto u_t \in \mathscr{M}$, uniquely determined, such that*

$$u_tu_t^* = \mathbf{s}(\psi) = u_0, \quad u_t^*u_t = \sigma_t^\varphi(\mathbf{s}(\psi)) \tag{1}$$

$$u_{t+s} = u_t\sigma_t^\varphi(u_s) \tag{2}$$

$$u_{-t} = \sigma_{-t}^\varphi(u_t^*) \tag{3}$$

$$\sigma_t^\psi(x) = u_t\sigma_t^\varphi(x)u_t^* \quad (x \in \mathbf{s}(\psi)\mathscr{M}\mathbf{s}(\psi)), \tag{4}$$

(5) *for every $x = x\mathbf{s}(\psi) \in \mathfrak{N}_\psi \cap \mathfrak{N}_\varphi^*$ and every $y = \mathbf{s}(\psi) \in \mathfrak{N}_\varphi \cap \mathfrak{N}_\psi^*$, there exists a function F defined, continuous and bounded on the strip $\{\alpha \in \mathbb{C}; 0 \le \operatorname{Re} \alpha \le 1\}$, analytic in the interior of this strip, such that $F(it) = \varphi(xu_t\sigma_t^\varphi(y)), F(1+it) = \psi(\sigma_t^\psi(y)u_tx)$ for all $t \in \mathbb{R}$.*

To prove the uniqueness part of the theorem, we consider an s^*-continuous mapping $\mathbb{R} \ni t \mapsto v_t \in \mathscr{M}$ with the same properties and we assume that $\mathbf{s}(\psi) = 1$. For each $t \in \mathbb{R}$, we define a mapping $\sigma_t : \mathscr{N} \to \mathscr{N}$ by putting

$$\sigma_t\left(\begin{pmatrix} x_{11} & x_{12} \\ x_{21} & x_{22} \end{pmatrix}\right) = \begin{pmatrix} \sigma_t^\varphi(x_{11}) & v_t^*\sigma_t^\psi(x_{12}) \\ v_t\sigma_t^\varphi(x_{21}) & \sigma_t^\psi(x_{22}) \end{pmatrix}.$$

Conditions (1)–(4) insure that $\{\sigma_t\}_{t\in\mathbb{R}}$ is an s^*-continuous one-parameter group of $*$-automorphisms of $\mathscr{N}$, which preserve the weight θ.

For $X = [x_{ij}] \in \mathfrak{N}_\theta \cap \mathfrak{N}_\theta^*$ and $Y = [y_{ij}] \in \mathfrak{N}_\theta \cap \mathfrak{N}_\theta^*$ we have

$$\theta(X\sigma_t(Y)) = \varphi(x_{11}\sigma_t^\varphi(y_{11})) + \varphi(x_{12}u_t\sigma_t^\varphi(y_{211})) + \psi(x_{21}u_t^*\sigma_t^\psi(y_{12})) + \psi(x_{22}\sigma_t^\psi(y_{22})),$$
$$\theta(\sigma_t(Y)X) = \varphi(\sigma_t^\varphi(y_{11})x_{11}) + \psi(\sigma_t^\psi(y_{21})u_tx_{12}) + \varphi(\sigma_t^\varphi(y_{12})u_t^*x_{21}) + \psi(\sigma_t^\psi(y_{22})x_{22}).$$

Using the *KMS* condition satisfied by φ and ψ, we find two function F_{11}, F_{22} defined, continuous and bounded on the strip $\{\alpha \in \mathbb{C}; 0 \le \operatorname{Re} \alpha \le 1\}$, analytic in the interior of this strip, such that

$$F_{11}(it) = \varphi(x_{11}\sigma_t^\varphi(y_{11})), F_{11}(1+it) = (\sigma_t(y_{11})x_{11}),$$
$$F_{22}(it) = \psi(x_{22}\sigma_t^\psi(y_{22})), F_{22}(1+it) = \psi(\sigma_t^\psi(y_{22})x_{22}).$$

By condition (5), there exist two functions F_{12}, G_{21} defined, continuous and bounded on the strip $\{\alpha \in \mathbb{C}; 0 \le \operatorname{Re} \alpha \le 1\}$, analytic in the interior of this strip, such that

$$F_{12}(it) = \varphi(x_{12}u_t\sigma_t^\varphi(y_{21})), F_{12}(1+it) = \psi(\sigma_t^\psi(y_{21})u_tx_{12}),$$
$$G_{21}(it) = \varphi(x_{21}^*u_t\sigma_t^\varphi(y_{12}^*)), G_{21}(1+it) = \psi(\sigma_t^\psi(y_{12}^*)u_tx_{21}^*).$$

Then the function $F_{21}(\alpha) = \overline{G_{21}(-\overline{\alpha} + 1)}$ satisfies the equalities

$$F_{21}(\mathrm{i}t) = \overline{G_{21}(\mathrm{i}t + 1)} = \overline{\psi(\sigma_t^\psi(y_{21}^*)u_t x_{21}^*)} = \psi(x_{21}u_t^*\sigma_t^\psi(y_{12})),$$
$$F_{21}(1 + \mathrm{i}t) = \overline{G_{21}(\mathrm{i}t)} = \overline{\varphi(x_{21}^* u_t \sigma_t^\varphi(y_{12}^*))} = \varphi(\sigma_t^\varphi(y_{12})u_t^* x_{21}).$$

Putting $F = F_{11} + F_{12} + F_{21} + F_{22}$, it follows that

$$F(\mathrm{i}t) = \theta(X\sigma_t(Y)), F(1 + \mathrm{i}t) = \theta(\sigma_t(Y)X) \quad (t \in \mathbb{R}).$$

We have proved that θ satisfies the *KMS* condition with respect to $\{\sigma_t\}_{t\in\mathbb{R}}$. Consequently, $\sigma_t = \sigma_t^\theta$ and hence $v_t = u_t$ for all $t \in \mathbb{R}$.

The mapping $\mathbb{R} \ni t \mapsto u_t \in \mathscr{M}$, uniquely determined by the above theorem, is called the *Connes cocycle* associated with the normal semifinite weight ψ with respect to the n.s.f. weight φ, and is denoted by $[D\psi : D\varphi]$, that is,

$$[D\psi : D\varphi]_t = u_t \quad (t \in \mathbb{R}).$$

3.2. Let $\mathscr{M}$ be a W^*-algebra. Let $U(\mathscr{M}) = \{u \in \mathscr{M}; u^*u = uu^* = 1\}$, $Int(\mathscr{M}) = \{Ad(u); u \in U(\mathscr{M})\} \subset Aut(\mathscr{M})$. Then $Int(\mathscr{M})$ is a normal subgroup of $Aut(\mathscr{M})$:

$$\sigma \cdot Ad(u) \cdot \sigma^{-1} = Ad(\sigma(u)) \quad (\sigma \in Aut(\mathscr{M}), u \in U(\mathscr{M})).$$

Let $Out(\mathscr{M}) = Aut(\mathscr{M})/Int(\mathscr{M})$ and let $\mathfrak{a}_{\mathscr{M}} : Aut(\mathscr{M}) \to Out(\mathscr{M})$ be the canonical quotient mapping.

An obvious and important consequence of Theorem 3.1 is that the mapping $\delta_{\mathscr{M}} : \mathbb{R} \to Out(\mathscr{M})$ defined by $\delta_{\mathscr{M}}(t) = \mathfrak{a}_{\mathscr{M}}(\sigma_t^\varphi)$, $(t \in \mathbb{R})$, is a group homomorphism, independent of the choice of the n.s.f. weight φ on $\mathscr{M}$. The mapping $\delta_{\mathscr{M}} : \mathbb{R} \to Out(\mathscr{M})$ is called the *modular homomorphism* of $\mathscr{M}$.

Using 2.22.(5), we see that $\delta_{\mathscr{M}}(\mathbb{R})$ is contained in the center of the group $Out(\mathscr{M})$.

3.3. Let $\varphi_1, \varphi_2, \ldots, \varphi_n$ be n.s.f. weights on the W^*-algebra $\mathscr{M}$.

The equation

$$\theta([x_{ij}]) = \sum_{k=1}^{n} \varphi_k(x_{kk}) \quad ([x_{ij}] \in Mat_n(\mathscr{M}))$$

defines an n.s.f. weight θ on the W^*-algebra $\mathscr{N} = Mat_n(\mathscr{M})$ of $n \times n$ matrices over $\mathscr{M}$.

For each $i,j \in \{1, \ldots, n\}$ we denote by $e_{ij} \in Mat_n(\mathbb{C})$ the matrix $e_{ij} = [\delta_{hi}\delta_{kj}]_{1<h,k<n}$. Then $\{e_{ij}\}_{1\le i,j\le n}$ is a system of matrix units in $Mat_n(\mathbb{C})$, that is

$$e_{ij}^* = e_{ji}, e_{ij}e_{hk} = \delta_{jh}e_{ik} \quad (1 \le i,j,h,k \le n),$$

and the mapping

$$Mat_n(\mathscr{M}) \ni [x_{ij}] \mapsto \sum_{ij} x_{ij} \otimes e_{ij} \in \mathscr{M} \otimes Mat_n(\mathbb{C})$$

is a $*$-isomorphism.

Using, as in Section 3.1, Proposition 2.21, we see that all elements of the form $\sum_{ij} \pm 1 \otimes e_{kk}$ are contained in the centralizer of θ, hence

$$1 \otimes e_{kk} \in \mathcal{N}^{\theta} \quad (1 \leq k \leq n).$$

Proposition. *For every $i, j, k \in \{1, \ldots, n\}$ and every $x \in \mathcal{M}$ we have*

$$\sigma_t^{\theta}(x \otimes e_{kk}) = \sigma_t^{\varphi_k}(x) \otimes e_{kk} \quad (t \in \mathbb{R}). \tag{1}$$

$$\sigma_t^{\theta}(1 \otimes e_{ij}) = [D\varphi_i : D\varphi_j]_t \otimes e_{ij} \quad (t \in \mathbb{R}). \tag{2}$$

Proof. Equation (1) is easily proved using the *KMS* condition. To prove (2) we may assume, for instance, that $i = 2$ and $j = 1$. By the preceding remarks, the projection $e = 1 \otimes (e_{11} + e_{22})$ belongs to $\mathcal{N}^{\theta}$, so that we may consider the n.s.f. weight θ_e on the W^*-algebra $e\mathcal{N}e$ (see 2.21). It is clear that $e\mathcal{N}e$ can be identified with the W^*-algebra $Mat_2(\mathcal{M})$ in such a way that θ_e corresponds to the balanced weight $\theta(\varphi_1, \varphi_2)$. Using this identification, with 2.22.(3) and the definition of the Connes cocycle we obtain

$$\sigma_t^{\theta}(1 \otimes e_{21}) = \sigma_t^{\theta_e}(1 \otimes e_{21}) = \sigma_t^{\theta(\varphi_1, \varphi_2)}\left(\begin{pmatrix} 0 & 0 \\ 1 & 0 \end{pmatrix}\right)$$
$$= \begin{pmatrix} 0 & 0 \\ D\varphi_2 : D\varphi_1]_t & 0 \end{pmatrix} = [D\varphi_2 : D\varphi_1]_t \otimes e_{21}.$$

□

3.4 Corollary. *Let φ_1, φ_2 be n.s.f. weights on the W^*-algebra $\mathcal{M}$. Then*

$$[D\varphi_1 : D\varphi_2]_t = [D\varphi_2 : D\varphi_1]_t^{-1} \quad (t \in \mathbb{R}).$$

Proof. This follows from Proposition 3.3, with $n = 2$, as $e_{12} = e_{21}^*$. □

3.5 Corollary. *Let $\varphi_1, \varphi_2, \varphi_3$ be n.s.f. weights on the W^*-algebra $\mathcal{M}$. Then the following "chain rule" holds:*

$$[D\varphi_1 : D\varphi_3]_t = [D\varphi_1 : D\varphi_2]_t [D\varphi_2 : D\varphi_3]_t \quad (t \in \mathbb{R}).$$

Proof. This follows from Proposition 3.3, with $n = 3$, as $e_{13} = e_{12}e_{23}$. □

3.6 Corollary. *Let φ be an n.s.f. weight on the W^*-algebra $\mathcal{M}$ and let ψ_1, ψ_2 be normal semifinite weights on $\mathcal{M}$. Then*

$$[D\psi_1 : D\varphi] = [D\psi_2 : D\varphi] \Leftrightarrow \psi_1 = \psi_2.$$

Proof. We remark first that $\mathbf{s}(\psi_1) = [D\psi_1 : D\varphi]_0 = [D\psi_2 : D\varphi]_0 = \mathbf{s}(\psi_2)$. Thus, using a slight extension of Corollary 3.5, we get

$$[D\psi_2 : D\psi_1]_t [D\psi_1 : D\varphi]_t = [D\psi_2 : D\varphi]_t \quad (t \in \mathbb{R}).$$

If $[D\psi_1 : D\varphi] = [D\psi_2 : D\varphi]$, it follows that $[D\psi_2 : D\psi_1] = 1$ for all $t \in \mathbb{R}$.

Consequently, it is enough to show that if φ and ψ are n.s.f. weights on $\mathscr{M}$ and $[D\psi : D\varphi] = 1$, then $\psi = \varphi$. Let $\theta = \theta(\varphi, \psi)$ be the balanced weight on $\mathscr{N} = Mat_2(\mathscr{M})$. From our assumption, it follows that $\sigma_t^\theta\left(\begin{pmatrix} 0 & 1 \\ 1 & 0 \end{pmatrix}\right) = \begin{pmatrix} 0 & 1 \\ 1 & 0 \end{pmatrix}$, $(t \in \mathbb{R})$, that is, $\begin{pmatrix} 0 & 1 \\ 1 & 0 \end{pmatrix} \in \mathscr{N}^\theta$. Using Proposition 2.12, we infer that

$$\varphi(x) = \theta\left(\begin{pmatrix} x & 0 \\ 0 & 0 \end{pmatrix}\right) = \theta\left(\begin{pmatrix} 0 & 1 \\ 1 & 0 \end{pmatrix}\begin{pmatrix} x & 0 \\ 0 & 0 \end{pmatrix}\begin{pmatrix} 0 & 1 \\ 1 & 0 \end{pmatrix}\right)$$
$$= \theta\left(\begin{pmatrix} 0 & 0 \\ 0 & x \end{pmatrix}\right) = \psi(x) \quad x \in \mathscr{M}^+$$

hence $\psi = \varphi$. □

3.7 Corollary. *Let φ be an n.s.f. weight on the W^*-algebra $\mathscr{M}$ and $v \in \mathscr{M}$ a partial isometry with $vv^* \in \mathscr{M}^\varphi$. Then*

$$[D\varphi_v : D\varphi]_t = v^*\sigma_t^\varphi(v), \quad \sigma_t^\varphi(v) = v[D\varphi_v : D\varphi]_t \quad (t \in \mathbb{R}).$$

Proof. Consider the balanced weights $\theta(\varphi, \varphi)$ and $\theta(\varphi, \varphi_v)$ on $Mat_2(\mathscr{M})$ and $u = \begin{pmatrix} 1 & 0 \\ 0 & v \end{pmatrix} \in Mat_2(\mathscr{M})$. Then uu^* belongs to the centralizer of $\theta(\varphi, \varphi)$ and $\theta(\varphi, \varphi_v) = \theta(\varphi, \varphi)_u$. Using 2.21.(1) and the definition of the Connes cocycle, we obtain

$$\begin{pmatrix} 0 & 0 \\ [D\varphi_v : D\varphi]_t & 0 \end{pmatrix} = \sigma_t^{\theta(\varphi,\varphi_v)}\left(\begin{pmatrix} 0 & 0 \\ \mathbf{s}(\varphi_v) & 0 \end{pmatrix}\right) = \sigma_t^{\theta(\varphi,\varphi)_u}\left(\begin{pmatrix} 0 & 0 \\ \mathbf{s}(\varphi_v) & 0 \end{pmatrix}\right)$$
$$= \begin{pmatrix} 1 & 0 \\ 0 & v^* \end{pmatrix}\sigma_t^{\theta(\varphi,\varphi)}\left(\begin{pmatrix} 1 & 0 \\ 0 & v \end{pmatrix}\begin{pmatrix} 0 & 0 \\ v^*v & 0 \end{pmatrix}\begin{pmatrix} 1 & 0 \\ 0 & v^* \end{pmatrix}\right)\begin{pmatrix} 1 & 0 \\ 0 & v \end{pmatrix}$$
$$= \begin{pmatrix} 1 & 0 \\ 0 & v^* \end{pmatrix}\sigma_t^{\theta(\varphi,\varphi)}\left(\begin{pmatrix} 0 & 0 \\ v & 0 \end{pmatrix}\right)\begin{pmatrix} 1 & 0 \\ 0 & v \end{pmatrix}$$
$$= \begin{pmatrix} 1 & 0 \\ 0 & v^* \end{pmatrix}\begin{pmatrix} 0 & 0 \\ \sigma_t^\varphi(v) & 0 \end{pmatrix}\begin{pmatrix} 1 & 0 \\ 0 & v \end{pmatrix} = \begin{pmatrix} 0 & 0 \\ v^*\sigma_t^\varphi(v) & 0 \end{pmatrix}.$$

This proves the first identity in the statement. Since $vv^* \in \mathscr{M}^\varphi$, we also have $v[D\varphi_v : D\varphi]_t = vv^*\sigma_t^\varphi(v) = \sigma_t^\varphi(vv^*v) = \sigma_t^\varphi(v)$. □

3.8 Corollary. *Let φ be an n.s.f. weight on the W^*-algebra $\mathscr{M}$, ψ a normal semifinite weight on $\mathscr{M}$ and $\sigma \in Aut(\mathscr{M})$. Then*

$$[D(\psi \circ \sigma) : D(\varphi \circ \sigma)]_t = \sigma^{-1}([D\psi : D\varphi]_t) \quad (t \in \mathbb{R}).$$

Proof. This follows by checking conditions (1)–(5) of Theorem 3.1 and using the uniqueness assertion of that theorem. □

3.9. Let φ_1, φ_2 be normal positive forms on the W^*-algebras $\mathscr{M}_1, \mathscr{M}_2$, respectively. Consider the W^*-algebras $\mathscr{M}_1, \mathscr{M}_2$ realized as standard von Neumann algebras $\mathscr{M}_1 \subset \mathscr{B}(\mathscr{H}_1), \mathscr{M}_2 \subset \mathscr{B}(\mathscr{H}_2)$. By

([L], 10.25), there exist $\xi_1 \in \mathcal{H}_1, \xi_2 \in \mathcal{H}_2$ such that $\varphi_1 = \omega_{\xi_1}, \varphi_2 = \omega_{\xi_2}$. Then $\omega_{\xi_1 \otimes \xi_2}$ is the unique normal positive form φ on the W^*-algebra $\mathcal{M}_1 \bar{\otimes} \mathcal{M}_2$ such that $\varphi(x_1 \bar{\otimes} x_2) = \varphi_1(x_1)\varphi_2(x_2)$ for all $x_1 \in \mathcal{M}_1, x_2 \in \mathcal{M}_2$. Accordingly, this form is denoted by $\varphi_1 \bar{\otimes} \varphi_2$ and is called the *tensor product* of φ_1 and φ_2.

It is easy to check that

$$\mathbf{s}(\varphi_1 \bar{\otimes} \varphi_2) = \mathbf{s}(\varphi_1) \bar{\otimes} \mathbf{s}(\varphi_2). \tag{1}$$

Also, using the *KMS* condition (see 2.12.(11) and 2.12.(12)), we get

$$\sigma_t^{\varphi_1 \bar{\otimes} \varphi_2} = \sigma_t^{\varphi_1} \bar{\otimes} \sigma_t^{\varphi_2} \quad (t \in \mathbb{R}). \tag{2}$$

Consider now two other normal positive forms ψ_1, ψ_2 on $\mathcal{M}_1, \mathcal{M}_2$, respectively, such that $\mathbf{s}(\psi_1) \leq \mathbf{s}(\varphi_1), \mathbf{s}(\psi_2) \leq \mathbf{s}(\varphi_2)$. Then

$$[D(\psi_1 \bar{\otimes} \psi_2) : D(\varphi_1 \bar{\otimes} \varphi_2)]_t = [D\psi_1 : D\varphi_1]_t \bar{\otimes} [D\psi_2; D\varphi_2]_t. \tag{3}$$

Indeed, using the chain rule (3.5) we may assume that $\psi_1 = \varphi_1$. For the computation of the left-hand side of (3), we consider the balanced weight $\theta = \theta(\varphi_1 \bar{\otimes} \varphi_2, \varphi_1 \bar{\otimes} \psi_2)$ on the W^*-algebra $Mat_2(\mathcal{M}_1 \bar{\otimes} \mathcal{M}_2) = \mathcal{M}_1 \bar{\otimes} \mathcal{M}_2 \otimes Mat_2(\mathbb{C}) = \mathcal{M}_1 \bar{\otimes} Mat_2(\mathcal{M}_2)$. It is easy to see that $\theta = \varphi_1 \bar{\otimes} \theta_2$ with $\theta_2 = \theta(\varphi_2, \psi_2)$ so that, by (2)

$$\sigma_t^\theta = \sigma_t^{\varphi_1} \bar{\otimes} \sigma_t^{\varphi_2} \quad (t \in \mathbb{R}).$$

Now (3) follows by applying the two sides of this identity to the element "$e_{21} = 1 \otimes e_{21}$."

3.10. Let φ, ψ be n.s.f. weights on the W^*-algebra $\mathcal{M}$, let $\theta = \theta(\varphi, \psi)$, the balanced weight on $Mat_2(\mathcal{M})$ and let $u_t = [D\psi : D\varphi]_t$ $(t \in \mathbb{R})$, the corresponding Connes cocycle. For $x \in \mathcal{M}, t \in \mathbb{R}$, put

$$\sigma_t^{\psi,\varphi}(x) = u_t\sigma_t^\varphi(x) = \sigma_t^\psi(x)u_t.$$

By 3.1.(11), for $[x_{ij}] \in Mat_2(\mathcal{M})$, we have

$$\sigma_t^\theta\left(\begin{pmatrix} x_{11} & x_{12} \\ x_{21} & x_{22} \end{pmatrix}\right) = \begin{pmatrix} \sigma_t^\varphi(x_{11}) & \sigma_t^{\varphi,\psi}(x_{12}) \\ \sigma_t^{\psi,\varphi}(x_{21}) & \sigma_t^\psi(x_{22}) \end{pmatrix} \tag{1}$$

It is easy to see that

$$\{\sigma_t^{\psi,\varphi}\}_{t\in\mathbb{R}} \textit{ is an } s^*\textit{-continuous one-parameter group of isometries of } \mathcal{M} \textit{ onto } \mathcal{M}. \tag{2}$$

Indeed, for $s, t \in \mathbb{R}$, and $x \in \mathcal{M}$, we have

$$\sigma_s^{\psi,\varphi}(\sigma_t^{\psi,\varphi}(x)) = u_s\sigma_s^\varphi(u_t\sigma_t^\varphi(x)) = u_s\sigma_s^\varphi(u_t)\sigma_s^\varphi(\sigma_t^\varphi(x)) = u_{s+t}\sigma_{s+t}^\varphi(x) = \sigma_{s+t}^{\psi,\varphi}(x).$$

Also, for $x \in \mathscr{M}$ and $t \in \mathbb{R}$, we have

$$\sigma_t^{\psi,\varphi}(x^*) = (\sigma_t^{\varphi,\psi}(x))^*, \tag{3}$$

and, if τ is a third n.s.f. weight on $\mathscr{M}$, then for $x, y \in \mathscr{M}$, and $t \in \mathbb{R}$, we have

$$\sigma_t^{\psi,\varphi}(xy) = \sigma_t^{\psi,\tau}(x)\sigma_t^{\tau,\varphi}(y), \tag{4}$$

using the chain rule (3.5).

The one-parameter group $\{\sigma_t^{\psi,\varphi}\}_{t\in\mathbb{R}}$ satisfies also the following *KMS* condition:

for every $x \in \mathfrak{N}_\psi \cap \mathfrak{N}_\varphi^*$ and every $y \in \mathfrak{N}_\varphi \cap \mathfrak{N}_\psi^*$
there exists a function F defined, continuous and bounded on the strip (5)
$\{\alpha \in \mathbb{C}; 0 \leq \operatorname{Re} \alpha \leq 1\}$, analytic in the interior of this strip and such that
$F(it) = \varphi(x\sigma_t^{\psi,\varphi}(y)), F(1+it) = \psi(\sigma_t^{\psi,\varphi}(y)x)$ *for all* $t \in \mathbb{R}$.

Properties (2) and (5), together with (4) with $\tau = \varphi$ and with $\tau = \psi$, determine uniquely the group $\{\sigma_t^{\psi,\varphi}\}_{t\in\mathbb{R}}$. The proof is similar to the proof of Theorem 3.1.

3.11. Let φ, ψ be n.s.f. weights on the W^*-algebra $\mathscr{M}$ and let $\theta = \theta(\varphi, \psi)$, the balanced weight on the W^*-algebra $\mathscr{N} = Mat_2(\mathscr{M})$. Using 3.1.(1) and the fact that, for $X = [x_{ij}] \in \mathscr{N}$, we have

$$\|X_\theta\|_\theta^2 = \|(x_{11})_\varphi\|_\varphi^2 + \|(x_{11})_\psi\|_\psi^2 + \|(x_{21})_\varphi\|_\varphi^2 + \|(x_{22})_\psi\|_\psi^2,$$

it follows that the mapping

$$\begin{pmatrix} x_{11} & x_{12} \\ x_{21} & x_{22} \end{pmatrix} \mapsto \begin{pmatrix} (x_{11})_\varphi \\ (x_{12})_\psi \\ (x_{21})_\varphi \\ (x_{22})_\psi \end{pmatrix}$$

determines an identification of Hilbert spaces:

$$\mathscr{H}_\theta = \mathscr{H}_\varphi \oplus \mathscr{H}_\psi \oplus \mathscr{H}_\varphi \oplus \mathscr{H}_\psi. \tag{1}$$

Since $\mathfrak{M}_\theta$ is densely imbedded in $\mathscr{H}_\theta$, from 3.1.(3), we infer that

$$\mathfrak{N}_\psi^* \mathfrak{N}_\varphi \textit{ is densely imbedded in } \mathscr{H}_\varphi. \tag{2}$$

Consider the antilinear operator $S_{\psi,\varphi}^0$ from $\mathscr{H}_\varphi$ to $\mathscr{H}_\psi$ defined by

$$S_{\psi,\varphi}^0(x_\varphi) = (x^*)_\psi \text{ for } x_\varphi \in D(S_{\psi,\varphi}^0) = \mathfrak{N}_\varphi \cap \mathfrak{N}_\psi^* \subset \mathscr{H}_\varphi.$$

It is easy to check that

$$(S_{\psi,\varphi}^0)^{-1} = S_{\varphi,\psi}^0.$$

On the other hand, we have the preclosed antilinear operator S_θ^0 in $\mathscr{H}_\theta$ defined by the n.s.f. weight θ as in Section 2.12. With respect to the decomposition (1) of $\mathscr{H}_\theta$ into a direct sum, we have

$$D(S_\theta^0) = D(S_\varphi^0) \oplus D(S_{\varphi,\psi}^0) \oplus D(S_{\psi,\varphi}^0) \oplus D(S_\psi^0)$$

and

$$S^0 = \begin{pmatrix} S_\varphi^0 & 0 & 0 & 0 \\ 0 & 0 & S_{\psi,\varphi}^0 & 0 \\ 0 & S_{\varphi,\psi}^0 & 0 & 0 \\ 0 & 0 & 0 & S_\psi^0 \end{pmatrix}.$$

Since S_θ^0 is preclosed, it follows that

$$S_{\psi,\varphi}^0 \textit{ is preclosed.}$$

Indeed, if $D(S_{\psi,\varphi}^0) \ni \xi_n \to 0$ and $S_{\psi,\varphi}^0 \xi_n \to \eta$, then $D(S_\theta^0) \ni (0,0,\xi_2,0) \to (0,0,0,0)$ and $S_\theta^0(0,0,\xi_n,0) = (0, S_{\psi,\varphi}^0 \xi_n, 0,0) \to (0,\eta,0,0)$, hence $\eta = 0$.

Let $S_{\psi,\varphi}$ be the closure of $S_{\psi,\varphi}^0$. With a similar argument we get

$$S = \begin{pmatrix} S_\varphi & 0 & 0 & 0 \\ 0 & 0 & S_{\psi,\varphi} & 0 \\ 0 & S_{\varphi,\psi} & 0 & 0 \\ 0 & 0 & 0 & S_\psi \end{pmatrix}. \tag{3}$$

In particular,

$$D(S_\theta) = D(S_\varphi) \oplus D(S_{\varphi,\psi}) \oplus D(S_{\psi,\varphi}) \oplus D(S_\psi) \tag{4}$$

$$S_{\psi,\varphi}^{-1} = S_{\varphi,\psi}. \tag{5}$$

Now, $\Delta_{\psi,\varphi} = S_{\psi,\varphi}^* S_{\psi,\varphi}$ is a linear positive self-adjoint operator in $\mathscr{H}_\varphi$ and the polar decomposition of $S_{\psi,\varphi}$ is

$$S_{\psi,\varphi} = J_{\psi,\varphi} \Delta_{\psi,\varphi}^{1/2}$$

with $J_{\psi,\varphi} : \mathscr{H}_\varphi \to \mathscr{H}_\psi$ a surjective isometric antilinear operator. Since

$$S_{\varphi,\psi} = S_{\psi,\varphi}^{-1} = (J_{\psi,\varphi} \Delta_{\psi,\varphi}^{1/2})^{-1} = J_{\psi,\varphi}^{-1} (J_{\psi,\varphi} \Delta_{\psi,\varphi}^{-1/2} J_{\psi,\varphi}^{-1}),$$

it follows from the uniqueness of the polar decomposition that

$$J_{\psi,\varphi}^* = J_{\psi,\varphi}^{-1} = J_{\varphi,\psi} \tag{6}$$

$$\Delta_{\psi,\varphi}^{-1} = J_{\varphi,\psi} \Delta_{\varphi,\psi} J_{\psi,\varphi}. \tag{7}$$

It is easy to check that

$$S_\theta^* = \begin{pmatrix} S_\varphi^* & 0 & 0 & 0 \\ 0 & 0 & S_{\varphi,\psi}^* & 0 \\ 0 & S_{\psi,\varphi}^* & 0 & 0 \\ 0 & 0 & 0 & S_\psi^* \end{pmatrix}. \tag{8}$$

$$\Delta_\theta = \begin{pmatrix} \Delta_\varphi & 0 & 0 & 0 \\ 0 & \Delta_{\varphi,\psi} & 0 & 0 \\ 0 & 0 & \Delta_{\psi,\varphi} & 0 \\ 0 & 0 & 0 & \Delta_\psi \end{pmatrix}, \quad J_\theta = \begin{pmatrix} J_\varphi & 0 & 0 & 0 \\ 0 & 0 & J_{\psi,\varphi} & 0 \\ 0 & J_{\varphi,\psi} & 0 & 0 \\ 0 & 0 & 0 & J_\psi \end{pmatrix}. \tag{9}$$

For $X = [x_{ij}] \in \mathscr{N}$, the operator $\pi_\theta(X) \in \mathscr{B}(\mathscr{H}_\theta)$ is defined by $\pi_\theta(X)Y_\theta = (XY)_\theta$ for all $Y = [y_{ij}] \in \mathscr{R}_\theta$. It follows that

$$\pi_\theta\left(\begin{pmatrix} x_{11} & x_{12} \\ x_{21} & x_{22} \end{pmatrix}\right) = \begin{pmatrix} \pi_\varphi(x_{11}) & 0 & \pi_\varphi(x_{12}) & 0 \\ 0 & \pi_\psi(x_{11}) & 0 & \pi_\psi(x_{12}) \\ \pi_\varphi(x_{21}) & 0 & \pi_\varphi(x_{22}) & 0 \\ 0 & \pi_\psi(x_{21}) & 0 & \pi_\psi(x_{22}) \end{pmatrix}. \tag{10}$$

In particular, for $X = \begin{pmatrix} 0 & 0 \\ x & 0 \end{pmatrix}$ we have $\sigma_t^\theta(X) = \begin{pmatrix} 0 & 0 \\ \sigma_t^{\psi,\varphi}(x) & 0 \end{pmatrix}$ and

$$\pi_\theta(X) = \begin{pmatrix} 0 & 0 & 0 & 0 \\ 0 & 0 & 0 & 0 \\ \pi_\varphi(x) & 0 & 0 & 0 \\ 0 & \pi_\psi(x) & 0 & 0 \end{pmatrix},$$

so that, as $\pi_\theta(\sigma_t^\theta(X)) = \Delta_\theta^{\mathrm{i}t}\pi_\theta(X)\Delta_\theta^{(-\mathrm{i}t)}$, we get

$$\pi_\varphi(\sigma_t^{\psi,\varphi}(x)) = \Delta_{\psi,\varphi}^{\mathrm{i}t}\pi_\varphi(x)\Delta_\varphi^{-\mathrm{i}t}, \tag{11}$$

$$\pi_\psi(\sigma_t^{\psi,\varphi}(x)) = \Delta_\psi^{\mathrm{i}t}\pi_\psi(x)\Delta_{\varphi,\psi}^{-\mathrm{i}t}. \tag{12}$$

3.12. Let φ, ψ be n.s.f. weights on the W^*-algebra $\mathscr{M}$. For the s^*-continuous one-parameter group $\{\sigma_t^{\psi,\varphi}\}_{t\in\mathbb{R}}$ of isometries of $\mathscr{M}$, we can develop the same theory of analytic extensions as in the case $\psi = \varphi$, considered in Section 2.14. Thus, for $a, b \in \mathscr{M}$ and $\alpha \in \mathbb{C}$ with $\mathrm{Re}\ \alpha > 0$, we shall write

$$a \in D(\sigma_{-\mathrm{i}\alpha}^{\psi,\varphi}) \text{ and } \sigma_{-\mathrm{i}\alpha}^{\psi,\varphi}(a) = b$$

if there exists an $\mathscr{M}$-valued function F, defined and w-continuous on the strip $\{\beta \in \mathbb{C}; 0 \leq \mathrm{Re}\beta \leq \mathrm{Re}\ \alpha\}$, analytic in the interior of this strip, such that

$$F(\mathrm{i}t) = \sigma_t^{\psi,\varphi}(a) \text{ and } F(\alpha) = b.$$

It follows from 3.11.(11) and 3.11.(12) by analytic continuation that for every $a \in D(\sigma_\alpha^{\psi,\varphi})$ we have

$$\pi_\varphi(\sigma_\alpha^{\psi,\varphi}(a)) = \Delta_{\psi,\varphi}^{\mathrm{i}\alpha}\pi_\varphi(a)\Delta_\varphi^{-\mathrm{i}\alpha}|D(\Delta_\varphi^{-\mathrm{i}\alpha}); \tag{1}$$

$$\pi_\psi(\sigma_\alpha^{\psi,\varphi}(a)) = \Delta_\psi^{\mathrm{i}\alpha}\pi_\psi(a)\Delta_{\varphi,\psi}^{-\mathrm{i}\alpha}|D(\Delta_{\varphi,\psi}^{-\mathrm{i}\alpha}). \tag{2}$$

Thus if $a \in D(\sigma_{-\mathrm{i}\alpha}^{\psi,\varphi})$, the function $\beta \mapsto \sigma_{-\mathrm{i}\beta}^{\psi,\varphi}(a)$ is norm-continuous and norm-bounded on the strip $\{\beta \in \mathbb{C}; 0 \le \operatorname{Re}\beta \le \operatorname{Re}\alpha\}$.

On the other hand, using 3.10.(1) we see that

$$a \in D(\sigma_\alpha^{\psi,\varphi}) \textit{ and } \sigma_\alpha^{\psi,\varphi}(a) = b \Leftrightarrow \tag{3}$$

$$\begin{pmatrix} 0 & 0 \\ a & 0 \end{pmatrix} \in D(\sigma_\alpha^{\theta(\varphi,\psi)}) \textit{ and } \sigma_\alpha^{\theta(\varphi,\psi)}\left(\begin{pmatrix} 0 & 0 \\ a & 0 \end{pmatrix}\right) = \begin{pmatrix} 0 & 0 \\ b & 0 \end{pmatrix} \tag{4}$$

so that, in view of the results of Section 2.14, we can obtain several properties of $\sigma_\alpha^{\psi,\varphi}$ in a very simple way.

For instance, statement 2.14.(1), applied to the balanced weight, gives

$$a \in D(\sigma_\alpha^{\psi,\varphi}) \Rightarrow a^* \in D(\sigma_{\overline{\alpha}}^{\varphi,\psi}) \textit{ and } \sigma_{\overline{\alpha}}^{\varphi,\psi}(a^*) = \sigma_\alpha^{\psi,\varphi}(a)^*. \tag{5}$$

Note that the same result can be obtained from 3.10.(3) by analytic continuation.

Also, by applying Proposition 2.14 to the balanced weight $\theta(\varphi,\psi)$ and the element $\begin{pmatrix} 0 & 0 \\ a & 0 \end{pmatrix}$, and using the results in Section 3.11, we obtain the following:

Proposition. *Let φ, ψ be n.s.f. weights on the W^*-algebra $\mathcal{M}$, $a \in \mathcal{M}$ and $\lambda \in (0,+\infty)$. The following statements are equivalent:*

(i) $\psi(ax^*xa^*) \le \lambda^2\varphi(x^*x)$ for all $x \in \mathcal{M}$;
(ii) $x \in \mathfrak{N}_\varphi \Rightarrow xa^* \in \mathfrak{N}_\psi$ and $\|(xa^*)_\psi\|_\psi \le \lambda\|x_\varphi\|_\varphi$;
(iii) $a \in D(\sigma_{-\mathrm{i}/2}^{\psi,\varphi})$ and $\|\sigma_{-\mathrm{i}/2}^{\psi,\varphi}(a)\| \le \lambda$.

If $a \in D(\sigma_{-\mathrm{i}/2}^{\psi,\varphi})$, then

$$(xa^*)_\psi = J_{\psi,\varphi}\pi_\varphi(\sigma_{-\mathrm{i}/2}^{\psi,\varphi}(a))J_\varphi x_\varphi \quad (x \in \mathfrak{N}_\varphi), \tag{6}$$

and if moreover $aa^* \in \mathcal{M}^\psi$, then

$$\varphi(\sigma_{-\mathrm{i}/2}^{\psi,\varphi}(a)^*y^*y\sigma_{-\mathrm{i}/2}^{\psi,\varphi}(a)) = \psi(aa^*y^*y) \quad (y \in \mathfrak{N}_\varphi). \tag{7}$$

3.13. In particular, taking $\lambda = 1$ and $a = 1$ in Proposition 3.12, we obtain:

Corollary. *Let φ, ψ be n.s.f. weights on the W^*-algebra $\mathcal{M}$. The following statements are equivalent:*

(i) $\psi \le \varphi$;
(ii) *there exists an $\mathcal{M}$-valued function F, defined and w-continuous on the strip $\{\alpha \in \mathbb{C}; 0 \le \operatorname{Re}\alpha \le 1/2\}$, analytic in the interior of this strip, such that $F(\mathrm{i}t) = [D\psi : D\varphi]_t$ $(t \in \mathbb{R})$, and $\|F(1/2)\| \le 1$.*

If $\psi \le \varphi$ there exists an element $a(= F(1/2)^*)$ in $\mathcal{M}$ with $\|a\| \le 1$, such that

$$\psi(x^*x) = \varphi(ax^*xa^*) \quad (x \in \mathfrak{N}_\psi).$$

Indeed, since $\sigma_t^{\psi,\varphi}(1) = [D\psi : D\varphi]_t, (t \in \mathbb{R})$, we can take $F(\alpha) = \sigma_{-i\alpha}^{\psi,\varphi}(1), (\alpha \in \mathbb{C}, 0 \le \text{Re } \alpha \le 1/2)$.

We remark that it follows from 3.1.(8), by analytic continuation, that $F(it + \alpha) = [D\psi : D\varphi]_t \sigma_t^\varphi(F(\alpha))$, hence $\|F(it)\| \le 1, \|F(it + (1/2))\| \le 1$ for all $t \in \mathbb{R}$ and the Phragmen–Lindelöf theorem implies that $\|F(\alpha)\| \le 1$ for all $\alpha \in \mathbb{C}, 0 \le \text{Re } \alpha \le 1/2$.

3.14. Since $F(1/2) = \sigma_{-i/2}^{\psi,\varphi}(1)$, it follows from 3.12.(4) that $F(1/2)^* = \sigma_{i/2}^{\varphi,\psi}(1)$ and hence (see 2.14.(6))

$$F(1/2)^* \in D(\sigma_{-i/2}^{\varphi,\psi}) \text{ and } \quad \sigma_{-i/2}^{\varphi,\psi}(F(1/2)^*) = 1.$$

Corollary. *Let φ, ψ be n.s.f. weights on the W^*-algebra $\mathcal{M}$. The following statements are equivalent:*

(i) $\psi(x^*x) = \varphi(x^*x)$ *for every* $x \in \mathfrak{N}_\varphi$;
(ii) *there exists an $\mathcal{M}$-valued function F, defined and w-continuous on the strip $\{\alpha \in \mathbb{C}; 0 \le \text{Re } \alpha \le 1/2\}$, analytic in the interior of this strip, such that $F(it) = [D\psi : D\varphi]_t (t \in \mathbb{R})$, and $F(1/2)^*F(1/2) = 1$.*

Proof. We shall use the notation of Section 3.13.

(ii) $\Rightarrow$ (i). We have $a = F(1/2)^* \in D(\sigma_{-i/2}^{\varphi,\psi}), \sigma_{-i/2}^{\varphi,\psi}(a) = 1$ and, by (ii), $aa^* = 1$. Thus, (i) follows from 3.12.(6) with ψ instead of φ, and φ instead of ψ.

(i) $\Rightarrow$ (ii). From (i) it follows that $\psi \le \varphi$, so that the function F of Corollary 3.13 does exist and $\|F(1/2)\| \le 1$. Let

$$x = \begin{pmatrix} x_{11} & x_{12} \\ x_{21} & x_{22} \end{pmatrix} \in \mathfrak{A}_\theta = \mathfrak{N}_\theta \cap \mathfrak{N}_\theta^*,$$

and $e_{11} = \begin{pmatrix} 1 & 0 \\ 0 & 0 \end{pmatrix}, e_{21} = \begin{pmatrix} 0 & 0 \\ 1 & 0 \end{pmatrix} \in \mathcal{N} = Mat_2(\mathcal{M})$. Using Proposition 2.14 and 2.21.(1), we get

$$\begin{aligned}
&\| J_\theta \pi_\theta \left(\begin{pmatrix} 0 & 0 \\ F(1/2) & 0 \end{pmatrix} \right) J_\theta x_\theta \|_\theta^2 = \| J_\theta \pi_\theta (\sigma_{-i/2}^\theta (e_{21})) J_\theta x_\theta \|_\theta^2 \\
&= \|(xe_{21}^*)_\theta\|_\theta^2 = \theta(e_{21}x^*e_{21}) \\
&= \psi(x_{11}^*x_{11} + x_{21}^*x_{21}) = \varphi(x_{11}^*x_{11} + x_{21}^*x_{21}) \\
&= \theta(e_{11}x^*xe_{11}) = \|(xe_{11})_\theta\|_\theta^2 = \| J_\theta \pi_\theta(e_{11}) J_\theta x_\theta \|_\theta^2.
\end{aligned}$$

Taking the limit it follows that

$$\|\pi_\theta \left(\begin{pmatrix} 0 & 0 \\ F(1/2) & 0 \end{pmatrix} \right) \xi\|_\theta = \|\pi_\theta(e_{11})\xi\|_\theta$$

for all $\xi \in \mathcal{H}_\theta$. Consequently, $\begin{pmatrix} 0 & 0 \\ F(1/2) & 0 \end{pmatrix}$ is a partial isometry with initial support equal to e_{11}, hence $F(1/2)^*F(1/2) = 1$. □

3.15. The next result is both an extension of the Pedersen–Takesaki theorem concerning the centralizer (2.21.(2)), and an algebraic characterization of the operator $\sigma_{-\mathrm{i}}^{\psi,\varphi}$.

Theorem. *Let φ, ψ be n.s.f. weights on the W^*-algebra $\mathcal{M}$ and $a, b \in \mathcal{M}$. The following statements are equivalent:*

(i) $a \in D(\sigma_{-\mathrm{i}}^{\psi,\varphi})$ *and* $\sigma_{-\mathrm{i}}^{\psi,\varphi}(a) = b$;
(ii) $a\mathfrak{N}_{\varphi}^* \subset \mathfrak{N}_{\psi}^*$, $\mathfrak{N}_{\psi} b \subset \mathfrak{N}_{\varphi}$ *and* $\psi(ax) = \varphi(xb)$ *for all* $x \in \mathfrak{N}_{\varphi}^*\mathfrak{N}_{\psi}$.

Proof. Taking into account remark 3.12.(3), 3.1.(1), and the definition of the balanced weight, we see that it is sufficient to consider only the case $\psi = \varphi$.

(i) $\Rightarrow$ (ii). If $a \in D(\sigma_{-\mathrm{i}}^{\varphi})$ and $\sigma_{-\mathrm{i}}^{\varphi}(a) = b$, then (2.14) $a \in D(\sigma_{-\mathrm{i}/2}^{\varphi})$, $b \in D(\sigma_{\mathrm{i}/2}^{\varphi})$ and $\sigma_{-\mathrm{i}/2}^{\varphi}(a) = \sigma_{\mathrm{i}/2}^{\varphi}(b)$. Since $a \in D(\sigma_{-\mathrm{i}/2}^{\varphi})$ and $b^* \in D(\sigma_{-\mathrm{i}/2}^{\varphi})$, it follows that (2.14) $\mathfrak{N}_{\varphi} a^* \subset \mathfrak{N}_{\varphi}$, that is, $a\mathfrak{N}_{\varphi}^* \subset \mathfrak{N}_{\varphi}^*$ and $\mathfrak{N}_{\varphi} b \subset \mathfrak{N}_{\varphi}$. Moreover, for every $x \in \mathfrak{N}_{\varphi}$ we have

$$
\begin{aligned}
\varphi(ax^*x) &= (x_{\varphi} | (xa^*)_{\varphi})_{\varphi} \\
&= (x_{\varphi} | J_{\varphi}\pi_{\varphi}(\sigma_{-\mathrm{i}/2}^{\varphi}(a))J_{\varphi}x_{\varphi})_{\varphi} = (x_{\varphi} | J_{\varphi}\pi_{\varphi}(\sigma_{\mathrm{i}/2}^{\varphi}(b))J_{\varphi}x_{\varphi})_{\varphi} \\
&= (J_{\varphi}\pi_{\varphi}(\sigma_{-\mathrm{i}/2}^{\varphi}(b^*))J_{\varphi}x_{\varphi} | x_{\varphi})_{\varphi} = ((xb)_{\varphi} | x_{\varphi})_{\varphi} = \varphi(x^*xb).
\end{aligned}
$$

(ii) $\Rightarrow$ (i). Let $x, y \in \mathfrak{T}_{\varphi}$. Assuming (ii), it follows from 2.13.(1), 2.13.(2), and 2.15.(2) that

$$
\begin{aligned}
(\pi_{\varphi}(a)\Delta_{\varphi}^{-1}x_{\varphi} | \Delta_{\varphi}y_{\varphi})_{\varphi} &= (\pi_{\varphi}(a)(\sigma_{i}^{\varphi}(x))_{\varphi} | \Delta_{\varphi}y_{\varphi})_{\varphi} \\
&= \varphi(a\sigma_{\mathrm{i}}^{\varphi}(x)y^*) = \varphi(\sigma_{\mathrm{i}}^{\varphi}(x)y^*b) \\
&= (\pi_{\varphi}(b)\Delta_{\varphi}(\sigma_{\mathrm{i}}^{\varphi}(x))_{\varphi} | y_{\varphi})_{\varphi} = (\pi_{\varphi}(b)x_{\varphi} | y_{\varphi})_{\varphi}.
\end{aligned}
$$

Consequently, $\pi_{\varphi}(a)\Delta_{\varphi}^{-1}x_{\varphi} \in D(\Delta_{\varphi})$ and $\Delta_{\varphi}\pi_{\varphi}(a)\Delta_{\varphi}^{-1}x_{\varphi} = \pi_{\varphi}(b)x_{\varphi}$ for all $x_{\varphi} \in \mathfrak{T}_{\varphi}$. By ([L], 9.24) it follows that

$$
\alpha \mapsto F(\alpha) = \Delta_{\varphi}^{\alpha}\pi_{\varphi}(a)\Delta_{\varphi}^{-\alpha} \in \mathcal{B}(\mathcal{H}_{\varphi})
$$

is a function defined and w-continuous on the strip $\{\alpha \in \mathbb{C}; 0 \leq \operatorname{Re}\alpha \leq 1\}$, analytic in the interior of this strip, such that $F(\mathrm{i}t) = \pi_{\varphi}(\sigma_t^{\varphi}(a)) \in \pi_{\varphi}(\mathcal{M})$, $F(1 + \mathrm{i}t) = \pi_{\varphi}(\sigma_t^{\varphi}(b)) \in \pi_{\varphi}(\mathcal{M})$ for all $t \in \mathbb{R}$. We conclude that $F(\alpha) \in \pi_{\varphi}(\mathcal{M})$ for all $\alpha \in \mathbb{C}$ with $0 \leq \operatorname{Re}\alpha \leq 1$, and hence $a \in D(\sigma_{-\mathrm{i}}^{\varphi})$ and $\sigma_{-\mathrm{i}}^{\varphi}(a) = b$. □

In particular, for $v \in U(\mathcal{M})$ and $\lambda \in (0, +\infty)$ we have

$$
\sigma_t^{\psi,\varphi}(v) = \lambda^{\mathrm{i}t}v, (t \in \mathbb{R}) \Leftrightarrow \psi = \lambda\varphi_v \tag{1}
$$

as both sides are equivalent to $\sigma_{-\mathrm{i}}^{\psi,\varphi}(v) = \lambda v$.

3.16. Let φ, ψ be n.s.f. weights on the W^*-algebra $\mathcal{M}$. We consider the standard representations $\pi_{\varphi} : \mathcal{M} \to \mathcal{B}(\mathcal{H}_{\varphi})$, $\pi_{\psi} : \mathcal{M} \to \mathcal{B}(\mathcal{H}_{\psi})$ the self-polar convex cones $\mathfrak{P}_{\varphi} \subset \mathcal{H}_{\varphi}$, $\mathfrak{P}_{\psi} \subset \mathcal{H}_{\psi}$, and the canonical conjugations $J_{\varphi} : \mathcal{H}_{\varphi} \to \mathcal{H}_{\varphi}$, $J_{\psi} : \mathcal{H}_{\psi} \to \mathcal{H}_{\psi}$ (2.23). Let $\theta = \theta(\varphi, \psi)$ be the balanced weight.

By ([L], 10.26) there exists a unitary operator $V_{\psi,\varphi} : \mathscr{H}_\varphi \to \mathscr{H}_\psi$, uniquely determined, such that

$$\pi_\psi(x) = V_{\psi,\varphi} \circ \pi_\varphi(x) \circ V^*_{\psi,\varphi} \quad (x \in \mathscr{M}) \tag{1}$$

$$V_{\psi,\varphi}(\mathfrak{P}_\varphi) = \mathfrak{P}_\psi. \tag{2}$$

Since the self-polar convex cone determines the canonical conjugation ([L], 10.23), we have also

$$V_{\psi,\varphi} \circ J_\varphi = J_\psi \circ V_{\psi,\varphi}. \tag{3}$$

From the uniqueness assertion, we infer that

$$V_{\varphi,\psi} = V^*_{\psi,\varphi}. \tag{4}$$

Proposition. *We have*

$$J_{\psi,\varphi} = V_{\psi,\varphi} \circ J_\varphi = J_\psi \circ V_{\psi,\varphi}. \tag{5}$$

Proof. By ([L], 10.14, 10.1, 10.4) we know that $\overline{S_\theta|\mathfrak{M}_\theta} = S_\theta$. Using 3.1.(3) and 3.11.(3), we deduce that $\overline{S_{\psi,\varphi}|\mathfrak{N}^*_\psi \mathfrak{N}_\varphi} = S_{\psi,\varphi}$. Thus, if we show that

$$J_\varphi V_{\psi,\varphi} S_{\psi,\varphi}|\mathfrak{N}^*_\psi \mathfrak{N}_\varphi \geq 0, \tag{6}$$

the uniqueness of the polar decomposition of $S_{\psi,\varphi}$, will imply the equalities $J_{\psi,\varphi} = (J_\varphi V_{\varphi,\psi})^* = V_{\psi,\varphi} J_\varphi$.

For $x \in \mathfrak{N}_\varphi, y \in \mathfrak{N}_\psi$ we have

$$\begin{aligned}
(J_\varphi V_{\varphi,\psi} S_{\psi,\varphi}(y^*x)_\varphi|(y^*x)_\varphi)_\varphi &= (J_\varphi V_{\varphi,\psi}(x^*y)_\psi|(y^*x)_\varphi)_\varphi \\
= (V_{\varphi,\psi} J_\psi \pi_\psi(x^*) y_\psi | \pi_\varphi(y^*) x_\varphi)_\varphi &= (\pi_\varphi(y) V_{\varphi,\psi} J_\psi \pi_\psi(x^*) y_\psi | x_\varphi)_\varphi \\
= (V_{\varphi,\psi} \pi_\psi(y) J_\psi \pi_\psi(x^*) J_\psi J_\psi y_\psi | x_\varphi)_\varphi &= (V_{\varphi,\psi} J_\psi \pi_\psi(x^*) J_\psi \pi_\psi(y) J_\psi y_\psi | x_\varphi)_\varphi \\
= (J_\varphi V_{\varphi,\psi} \pi_\psi(x^*) J_\psi \pi_\psi(y) J_\psi y_\psi | x_\varphi)_\varphi &= (J_\varphi \pi_\varphi(x)^* V_{\varphi,\psi} J_\psi \pi_\psi(y) J_\psi y_\psi | x_\varphi)_\varphi \\
= (J_\varphi \pi_\varphi(x)^* J_\varphi V_{\varphi,\psi} \pi_\psi(y) J_\psi y_\psi | x_\varphi)_\varphi &= (V_{\varphi,\psi} \pi_\psi(y) J_\psi y_\psi \,|\, J_\varphi \pi_\varphi(x) J_\varphi x_\varphi)_\varphi.
\end{aligned}$$

By 2.23.(1) we have $\pi_\psi(y) J_\psi y_\psi \in \mathfrak{P}_\psi$ and $\pi_\varphi(x) J_\varphi x_\varphi \in \mathfrak{P}_\varphi$. Since $V_{\varphi,\psi}(\mathfrak{P}_\psi) = \mathfrak{P}_\varphi$ and $J_\varphi(\mathfrak{P}_\varphi) = \mathfrak{P}_\varphi$, it follows that both sides of the last scalar product belong to the self-polar cone $\mathfrak{P}_\varphi$ and therefore this scalar product is positive.

We have thus proved (6) and hence the proposition. □

We record also the following identities that are easy consequences of the proposition:

$$J_\psi = V_{\psi,\varphi} \circ J_\varphi \circ V_{\varphi,\psi}, \tag{7}$$

$$V_{\psi,\varphi} = J_{\psi,\varphi} \circ J_\varphi = J_\psi \circ J_{\psi,\varphi}, \tag{8}$$

and, if τ is a third n.s.f. weight on $\mathscr{M}$,

$$V_{\psi,\varphi} = V_{\psi,\tau} \circ V_{\tau,\varphi}. \tag{9}$$

3.17. Let φ, ψ be n.s.f. weights on the W^*-algebra $\mathcal{M}$. We shall identify the Hilbert space $\mathcal{H}_\psi$ with the Hilbert space $\mathcal{H}_\varphi$ via the unitary operator $V_{\varphi,\psi} : \mathcal{H}_\psi \to \mathcal{H}_\varphi$ (3.16). Then we have $J_\psi = J_\varphi$ and $\pi_\psi(x) = \pi_\varphi(x)(x \in \mathcal{M})$. Also, there is a $*$-antiisomorphism

$$J : \mathcal{M} \to \pi_\varphi(\mathcal{M})' = \pi_\psi(\mathcal{M})'$$

by means of which we define the weights φ' and ψ' (see 2.12.(6), 2.12.(10)). With these identifications it is easy to check that

$$[D\psi' : D\varphi']_t = j([D\psi| : D\varphi]_{-t}) \quad (t \in \mathbb{R}).$$

3.18. Finally, we consider the particular case of cyclic and separating vectors.

Let $\mathcal{M} \subset \mathcal{B}(\mathcal{H})$ be a von Neumann algebra with two cyclic and separating vectors $\xi_1, \xi_2 \in \mathcal{H}$ and $\varphi_1 = \omega_{\xi_1}|\mathcal{M}, \varphi_2 = \omega_{\xi_2}|\mathcal{M}$ the corresponding faithful normal positive forms on $\mathcal{M}$. For $k \in \{1,2\}$, identify the Hilbert space $\mathcal{H}_{\varphi_k}$ with $\mathcal{H}$ via the mapping $x_{\varphi_k} \mapsto x\xi_k, (x \in \mathcal{M})$, and for every $i,j \in \{1,2\}$ let $S_{ij} = S_{\varphi_i,\varphi_j}, J_{ij} = J_{\varphi_i,\varphi_j}, \Delta_{ij} = \Delta_{\varphi_i,\varphi_j}$, and $V_{ij} = V_{\varphi_i,\varphi_j} \in \mathcal{M}'$.

In order to identify the balanced normal positive form $\varphi = \theta(\varphi_1, \varphi_2)$ on $Mat_2(\mathcal{M}) = \mathcal{M} \otimes Mat_2(\mathbb{C})$ as a vector form, we consider the four-dimensional Hilbert space $\mathcal{H}_4$ with the orthonormal basis $\{\varepsilon_{ij}; i,j = 1,2\}$, the factor $\mathcal{F} \subset \mathcal{B}(\mathcal{H}_4)$ linearly spanned by the operators $e_{ij} \in \mathcal{B}(\mathcal{H}_4)$,

$$e_{ij}\varepsilon_{hk} = \delta_{jh}\varepsilon_{ik} \quad (i,j,h,k \in \{1,2\})$$

and its commutant $\mathcal{F}' \subset \mathcal{B}(\mathcal{H}_4)$ linearly spanned by the operators $e'_{ij} \in \mathcal{B}(\mathcal{H}_4)$,

$$e'_{ij}\varepsilon_{hk} = \delta_{ki}\varepsilon_{hj} \quad (i,j,h,k \in \{1,2\}).$$

Then $Mat_2(\mathcal{M}) = \mathcal{M} \otimes \mathcal{F}$ acts on the Hilbert space $\tilde{\mathcal{H}}_4 = \mathcal{H} \bar{\otimes} \mathcal{H}_4$ and its commutant is $Mat_2(\mathcal{M}') = \mathcal{M}' \otimes \mathcal{F}'$. Moreover, the vector

$$\xi_\varphi = \xi_1 \otimes \varepsilon_{11} + \xi_2 \otimes \varepsilon_{22} \in \tilde{\mathcal{H}}_4 \tag{1}$$

is cyclic and separating for $\mathcal{M} \otimes \mathcal{F} \subset \mathcal{B}(\tilde{\mathcal{H}}^4)$ and the balanced form $\varphi = \theta(\varphi_1, \varphi_2)$ is the vector form $\varphi = \omega_{\xi_\varphi}|\mathcal{M} \otimes \mathcal{F}$. As earlier, we identify the Hilbert space $\mathcal{H}_\varphi$ with $\tilde{\mathcal{H}}_4$ and we consider the operators $S_\varphi, J_\varphi, \Delta_\varphi$ on $\mathcal{H}_\varphi$.

Consider also the isometries

$$u_{ij} : \mathcal{H} \ni \xi \mapsto \xi \otimes \varepsilon_{ij} \in \tilde{\mathcal{H}}_4 \quad (i,j \in \{1,2\}).$$

Then $\sum_{ij} u_{ij}u_{ij}^* = 1 \in \mathcal{B}(\tilde{\mathcal{H}}_4)$ and $u_{ij}^*u_{hk} = \delta_{ih}\delta_{jk} \in \mathcal{B}(\mathcal{H})$ for $i,j,h,k \in \{1,2\}$. As in Section 3.11, we see that

$$S_\varphi = \sum_{ij} u_{ij}S_{ji}u_{ji}^*, \qquad S_\varphi^* = \sum_{ij} u_{ij}S_{ij}^*u_{ji}^*, \tag{2}$$

$$J_\varphi = \sum_{ij} u_{ij}J_{ji}u_{ji}^*, \qquad \Delta_\varphi = \sum_{ij} u_{ij}\Delta_{ij}u_{ij}^*, \tag{3}$$

Using the above expression for J_φ, we get

$$J_\varphi(1 \otimes e_{kk})J_\varphi = 1 \otimes e'_{kk} \quad (k \in \{1,2\}) \tag{4}$$

and, using also 3.16.(8), it is easy to check that

$$J_\varphi(1 \otimes e_{21})J_\varphi = V_{21} \otimes e'_{12}. \tag{5}$$

Proposition. *Let $(\mathscr{M}, \mathscr{H}, J, \mathfrak{P})$ be a hyperstandard von Neumann algebra with two cyclic and separating vectors $\xi_1, \xi_2 \in \mathfrak{P}$ and $\varphi_1 = \omega_{\xi_1}|\mathscr{M}, \varphi_2 = \omega_{\xi_2}|\mathscr{M}$ the corresponding faithful normal positive forms on $\mathscr{M}$. The following statements are equivalent:*

(i) *$\xi_2 \leq \xi_1$ with respect to $\mathfrak{P}$;*
(ii) *there exists on $\mathscr{M}$-valued function F, defined and w-continuous on the strip $\{\alpha \in \mathbb{C}; 0 \leq \operatorname{Re} \alpha \leq 1/4\}$, analytic in the interior of this strip, such that $F(it) = [D\varphi_2 : D\varphi_1]_t$ for every $t \in \mathbb{R}$ and $\|F(1/4)\| \leq 1$;*
(iii) *$\|\Delta_{\xi_2}^{1/4} x\xi_2\| \leq \|\Delta_{\xi_1}^{1/4} x\xi_1\|$ for every $x \in \mathscr{M}$.*

Proof. We shall use the notation introduced in this section. Since $\xi_1, \xi_2 \in \mathfrak{P}$ we have ([L], 10.24) $\mathfrak{P}_{\xi_1} = \mathfrak{P}_{\xi_2} = \mathfrak{P}$, hence $J_{11} = J_{22} = J$ and $V_{21} = 1$; using Proposition 3.16, we obtain also $J_{12} = J_{21} = J$.

(i) $\Rightarrow$ (ii). Let $x \in \mathscr{M}$. Then $JxJx\xi_1 \in \mathfrak{P}$ and, assuming (i), $\xi_1 - \xi_2 \in \mathfrak{P}$, so that $(\xi_1 - \xi_2 | JxJx\xi_1) \geq 0$ and hence

$$\begin{aligned}\|\Delta_{21}^{1/4} x\xi_1\|^2 &= (\Delta_{21}^{1/2} x\xi_1 | x\xi_1) = (JS_{21} x\xi_1 | x\xi_1) = (Jx^*\xi_2 | x\xi_1)\\ &= (\xi_2 | JxJx\xi_1) \leq (\xi_1 | JxJx\xi_1) = \|\Delta_{11}^{1/4} x\xi_1\|^2.\end{aligned}$$

It follows that

$$\xi \in D(\Delta_{11}^{1/4}) \Rightarrow \xi \in D(\Delta_{21}^{1/4}) \text{ and } \|\Delta_{21}^{1/4}\xi\| \leq \|\Delta_{11}^{1/4}\xi\|.$$

Using ([L], 9.24) we infer that the function $\alpha \mapsto F(\alpha) = \Delta_{21}^{\alpha}\Delta_{11}^{-\alpha}$ is defined and w-continuous on the strip $\{\alpha \in \mathbb{C}; 0 \leq \operatorname{Re} \alpha \leq 1/4\}$, analytic in the interior of this strip and, clearly, $\|F(1/4)\| \leq 1$. Also, for $t \in \mathbb{R}$ and $\xi \in \mathscr{H}$ we have

$$\begin{aligned}(\Delta_{21}^{it}\xi) \otimes \varepsilon_{21} &= \Delta_\varphi^{it}(\xi \otimes \varepsilon_{21}) = \Delta_\varphi^{it}(1 \otimes e_{21})\Delta_\varphi^{-it}\Delta_\varphi^{it}(\xi \otimes \varepsilon_{11})\\ &= \sigma_t^\varphi(1 \otimes e_{21})((\Delta_{11}^{it}\xi) \otimes \varepsilon_{11}) = ([D\varphi_2 : D\varphi_1]_t \otimes e_{21})((\Delta_{11}^{it}\xi) \otimes \varepsilon_{11})\\ &= ([D\varphi_2 : D\varphi_1]_t \Delta_{11}^{it}\xi) \otimes \varepsilon_{21}\end{aligned}$$

hence $F(it) = \Delta_{21}^{it}\Delta_{11}^{-it} = [D\varphi_2 : D\varphi_1]_t$.

(ii) $\Rightarrow$ (iii). Since $\sigma_t^\varphi(1 \otimes e_{21}) = [D\varphi_2 : D\varphi_1] \otimes e_{21}, (t \in \mathbb{R})$, assuming (ii) it follows that

$$b = 1 \otimes e_{21} \in D(\sigma_{-i/4}^\varphi) \text{ and } \sigma_{-i/4}^\varphi(b) = F(1/4) \otimes e_{21}. \tag{6}$$

For every $y \in \mathscr{M} \otimes \mathscr{F}$ we have $y^*\xi_\varphi \in D(\Delta_{1/4}^\varphi)$ and, using 2.14.(3), we get

$$\sigma_{-i/4}^\varphi(b)\Delta_\varphi^{1/4} y^*\xi_\varphi = \Delta_\varphi^{1/4} by^*\xi_\varphi, \tag{7}$$

that is $\sigma^{\varphi}_{-i/4}(b)J_\varphi\Delta^{1/4}_\varphi y\xi_\varphi = \Delta^{\varphi}_{1/4}by^*\xi_\varphi$. Replacing here y by by^* we obtain $\sigma^{\varphi}_{-i/4}(b)J_\varphi\Delta^{1/4}_\varphi by^*\xi_\varphi = \Delta^{\varphi}_{1/4}byb^*\xi_\varphi$ and by (7) we conclude that

$$\sigma^{\varphi}_{-i/4}(b)J_\varphi\sigma^{\varphi}_{-i/4}(b)J_\varphi\Delta^{1/4}_\varphi\xi_\varphi = \Delta^{\varphi}_{1/4}byb^*\xi_\varphi. \tag{8}$$

In particular, using (1), (3), (6), and (8) for $y = x \otimes e_{11}$ with $x \in \mathscr{M}$, we obtain

$$(\Delta^{1/4}_{22}x\xi_2) \otimes \varepsilon_{22} = (F(1/4)JF(1/4)\Delta^{1/4}_{11}x\xi_1) \otimes \varepsilon_{22}.$$

Since $\|F(1/4)\| \le 1$, the desired conclusion (iii) follows.

(iii) $\Rightarrow$ (i). Assuming (iii), for every $x \in \mathscr{M}^+$ we have

$$(xJxJ\xi_2|\xi_2) = \|\Delta^{1/4}_{22}x\xi_2\|^2 \le \|\Delta^{1/4}_{11}x\xi_1\|^2 = (xJxJ\xi_1|\xi_1). \tag{9}$$

Since $y = xJxJ \ge 0$, by the Schwarz inequality we get $|(y\xi_1|\xi_2)|^2 \le (y\xi_1|\xi_1)(y\xi_2|\xi_2) \le (y\xi_1|\xi_1)^2$. We have $\xi_2 \in \mathfrak{P}$ and also $y\xi_1 \in \mathfrak{P}$, so that $(y\xi_1|\xi_2) \ge 0$ and hence $(y\xi_1|\xi_2) \le (y\xi_1|\xi_1)$, that is,

$$(\xi_1 - \xi_2|\,JxJx\xi_1) \ge 0 \quad (x \in \mathscr{M}^+).$$

By Proposition 2.26, it follows that $(\xi_1 - \xi_2|\xi) \ge 0$ for every $\xi \in \mathfrak{P}$, hence $\xi_1 - \xi_2 \in \mathfrak{P}$ as $\mathfrak{P}$ is a self-polar cone. Thus, $\xi_2 \le \xi_1$ with respect to $\mathfrak{P}$. □

3.19. Notes. The results included in this section are essentially due to Connes (1973a, 1973b, 1974c). Proposition 3.15 is explicitly stated in Haagerup (1979a). The implication (i) $\Rightarrow$ (iii) for normal forms has been obtained earlier by Sakai, S. (see [L], 5.21), with $a \in \mathscr{M}, 0 \le a \le 1$. Further remarks are due to Kosaki (1992).

For our exposition, we have used Combes and Delaroche (1975); Connes (1973a, 1973b, 1974c); Connes and Takesaki (1977); Digernes (1975); Elliot (1975); and Haagerup (1979a).

4 The Pedersen–Takesaki Construction

In this section, we present the canonical construction and properties of the weight φ_A associated with a given weight φ and a given positive self-adjoint operator A affiliated to the centralizer of φ. We also consider several commutation properties of weights.

4.1. Let φ be a normal semifinite weight on the W^*-algebra $\mathscr{M}$ and $a \in \mathscr{M}^\varphi, a \ge 0$. A weight φ_a on $\mathscr{M}$ is defined by

$$\varphi_a(x) = \varphi(a^{1/2}xa^{1/2}) \quad (x \in \mathscr{M}^+).$$

It is clear that φ_a is normal. Since $a \in \mathscr{M}^\varphi$ we have (2.21.(2)) $\mathfrak{N}_\varphi \subset \mathfrak{N}_{\varphi_a}, \mathfrak{M}_\varphi \subset \mathfrak{M}_{\varphi_a}$, and

$$\varphi_a(x) = \varphi(ax) = \varphi(xa) \quad (x \in \mathfrak{M}_\varphi).$$

In particular, φ_a is semifinite. Also, $\mathbf{s}(\varphi_a) = \mathbf{s}(a)$.

We remark that the notation φ_a introduced here agrees with the notation φ_v introduced in Section 2.21, since the only positive partial isometries are projections.

4.2. We begin the study of the weight φ_a by considering the case when φ is an n.s.f. weight on $\mathscr{M}$ and $a \in \mathscr{M}^\varphi, a \geq 0$ is invertible. In this case, it is clear that

$$\mathfrak{M}_{\varphi_a} = \mathfrak{M}_\varphi, \quad \mathfrak{N}_{\varphi_a} = \mathfrak{N}_\varphi. \tag{1}$$

Recall (2.21) that the condition $a \in \mathscr{M}^\varphi$ means that $\pi_\varphi(a)$ commutes with Δ_φ.

Let $x, y \in \mathfrak{N}_\varphi = \mathfrak{N}_{\varphi_a}$. By 2.21.(1)

$$(x|y)_{\varphi_a} = \varphi_a(y^*x) = \varphi(ay^*x) = (x_\varphi|(ya)_\varphi)_\varphi = (x_\varphi|\, J_\varphi\pi_\varphi(a)J_\varphi y_\varphi)_\varphi.$$

As $\|a^{-1}\|^{-1} \leq J_\varphi\pi_\varphi(a)J_\varphi \leq \|a\|$, it follows that the scalar products $(\cdot|\cdot)_{\varphi_a}$ and $(\cdot|\cdot)_\varphi$ are equivalent and hence the Hilbert spaces $\mathscr{H}_{\varphi_a}$ and $\mathscr{H}_\varphi$ are the same set. We put

$$\mathscr{H} = \mathscr{H}_{\varphi_a} = \mathscr{H}_\varphi, \tag{2}$$

but distinguish the two scalar products. It is clear that

$$S_{\varphi_a} = S_\varphi, \quad \pi_{\varphi_a} = \pi_\varphi. \tag{3}$$

On the other hand, it is easy to check that if T is a closed linear or antilinear operator in $\mathscr{H}$, and T^* denotes the adjoint of T with respect to the scalar product $(\cdot|\cdot)_\varphi$, then the adjoint of T with respect to the scalar product $(\cdot|\cdot)_{\varphi_a}$ is $J_\varphi\pi_\varphi(a)^{-1}J_\varphi T^* J_\varphi\pi_\varphi(a)J_\varphi$. Consequently,

$$\begin{aligned}
S^*_{\varphi_a} &= J_\varphi\pi_\varphi(a)^{-1}J_\varphi S^*_\varphi J_\varphi\pi_\varphi(a)J_\varphi, \text{ and hence}\\
\Delta_{\varphi_a} &= S^*_{\varphi_a}S_{\varphi_a} = J_\varphi\pi_\varphi(a)^{-1}J_\varphi\Delta_\varphi^{1/2}J_\varphi\, J_\varphi\pi_\varphi(a)J_\varphi\, J_\varphi\Delta_\varphi^{1/2}\\
&= J_\varphi\pi_\varphi(a)^{-1}J_\varphi\Delta_\varphi^{1/2}\pi_\varphi(a)\Delta_\varphi^{1/2} = (J_\varphi\pi_\varphi(a)^{-1}J_\varphi)\pi_\varphi(a)\Delta_\varphi.
\end{aligned}$$

The three positive self-adjoint operators appearing in the last term of the earlier equation mutually commute so that, for $t \in \mathbb{R}$, we get

$$\Delta^{it}_{\varphi_a} = (J_\varphi\pi_\varphi(a)^{-1}J_\varphi)^{it}\pi_\varphi(a)^{it}\Delta^{it}_\varphi = J_\varphi\pi_\varphi(a)^{it}J_\varphi\pi_\varphi(a)^{it}\Delta^{it}_\varphi$$

as J_φ is antilinear. Thus, for $x \in \mathscr{M}, t \in \mathbb{R}$, we have

$$\begin{aligned}
\pi_{\varphi_a}(\sigma_t^{\varphi_a}(x)) &= \Delta^{it}_{\varphi_a}\pi_{\varphi_a}(x)\Delta^{-it}_{\varphi_a}\\
&= (J_\varphi\pi_\varphi(a)^{it}J_\varphi)\pi_\varphi(a)^{it}\Delta^{it}_\varphi\pi_\varphi(x)\Delta^{-it}_\varphi\pi_\varphi(a)^{-it}(J_\varphi\pi_\varphi(a)^{-it}J_\varphi)\\
&= (J_\varphi\pi_\varphi(a)^{it}J_\varphi)\pi_\varphi(a^{it}\sigma^\varphi_t(x)a^{-it})(J_\varphi\pi_\varphi(a)^{-it}J_\varphi).
\end{aligned}$$

Since $J_\varphi\pi_\varphi(a)^{it}J_\varphi \in \pi_\varphi(\mathscr{M})'$, we conclude that

$$\sigma_t^{\varphi_a}(x) = a^{it}\sigma^\varphi_t(x)a^{-it} \quad (x \in \mathscr{M}, t \in \mathbb{R}). \tag{4}$$

Note that this conclusion can be also obtained by using the *KMS* condition.

4.3. Let φ be a normal semifinite weight on the W^*-algebra $\mathcal{M}$ and $a, b \in \mathcal{M}^\varphi, a, b \geq 0$. We have

$$\varphi_{a+b} = \varphi_a + \varphi_b. \tag{1}$$

Indeed, if a and b are invertible, then $\mathfrak{M}_{\varphi_a} = \mathfrak{M}_{\varphi_b} = \mathfrak{M}_{\varphi_{a+b}} = \mathfrak{M}_\varphi$ (4.2.(1)) and for every $x \in \mathfrak{M}_\varphi$ we have (4.1) $\varphi_{a+b}(x) = \varphi((a+b)x) = \varphi(ax) + \varphi(bx) = \varphi_a(x) + \varphi_b(x)$; so (1) is obvious in this case.

In the general case, consider the elements

$$u = w\text{-}\lim_{\varepsilon\to 0} a^{1/2}(a+b+\varepsilon)^{-1/2}, \quad v = w\text{-}\lim_{\varepsilon\to 0} b^{1/2}(a+b+\varepsilon)^{-1/2}.$$

Then we have $u, v \in \mathcal{M}^\varphi$ and (compare with 1.4) $a^{1/2} = u(a+b)^{1/2}, b^{1/2} = v(a+b)^{1/2}, u^*u + v^*v = s(a+b)$. If $x \in \mathfrak{M}_{\varphi_{a+b}} \cap \mathcal{M}^+$, then $y = (a+b)^{1/2}x(a+b)^{1/2} \in \mathfrak{M}_\varphi$ and, using 2.21.(2), we deduce further that $uyu^* \in \mathfrak{M}_\varphi$ and $vyv^* \in \mathfrak{M}_\varphi$, that is,

$$+\infty > \varphi(uyu^*) + \varphi(vyv^*) = \varphi(a^{1/2}xa^{1/2}) + \varphi(b^{1/2}xb^{1/2}).$$

Thus, again by 2.21.(2), we get

$$\varphi_a(x) + \varphi_b(x) = \varphi(y(u^*u + v^*v)) = \varphi(y) = \varphi_{a+b}(x).$$

Conversely let $x \in \mathfrak{M}_{\varphi_a+\varphi_b} \cap \mathcal{M}^+$. Then $a^{1/2}xa^{1/2} \in \mathfrak{M}_\varphi \cap \mathcal{M}^+, b^{1/2}xb^{1/2} \in \mathfrak{M}_\varphi \cap \mathcal{M}^+$ and by a further application of 2.21.(2), we infer that $u^*a^{1/2}xa^{1/2}u \in \mathfrak{M}_\varphi \cap \mathcal{M}^+, v^*b^{1/2}xb^{1/2}v \in \mathfrak{M}_\varphi \cap \mathcal{M}^+$. As

$$\begin{aligned}(a+b)^{1/2}x(a+b)^{1/2} &= w\text{-}\lim_{\varepsilon\to 0}(a+b+\varepsilon)^{-1/2}(a+b)x(a+b)(a+b+\varepsilon)^{-1/2}\\ &\leq w\text{-}\lim_{\varepsilon\to 0} 2(a+b+\varepsilon)^{-1/2}(axa+bxb)(a+b+\varepsilon)^{-1/2}\\ &= 2(u^*a^{1/2}xa^{1/2}u + v^*b^{1/2}xb^{1/2}v),\end{aligned}$$

it follows that $x \in \mathfrak{M}_{\varphi_{a+b}} \cap \mathcal{M}^+$. Thus, (1) is completely proved.

From (1) we infer that

$$a \leq b \Rightarrow \varphi_a \leq \varphi_b. \tag{2}$$

Finally, if $a \in \mathcal{M}^\varphi, a \geq 0$, and $\{a_i\}_{i\in I} \subset \mathcal{M}^\varphi$ is a net of positive elements, then

$$a_i \uparrow a \Rightarrow \varphi_{a_i} \uparrow \varphi_a. \tag{3}$$

Indeed, let $x \in \mathcal{M}^+$. From (2) it follows that $\sup_i \varphi_{a_i}(x) \leq \varphi_a(x)$. On the other hand, since φ is normal (1.3) and $a_i^{1/2}xa_i^{1/2} \xrightarrow{w} a^{1/2}xa^{1/2}$, we have $\varphi_a(x) = \varphi(a^{1/2}xa^{1/2}) \leq \liminf_i \varphi(a_i^{1/2}xa_i^{1/2}) \leq \sup_i \varphi_{a_i}(a)$.

4.4. Let φ be a normal semifinite weight on the W^*-algebra $\mathcal{M}$ and A a positive self-adjoint operator affiliated to $\mathcal{M}^\varphi$ (A.16).

We shall consider the bounded positive operators

$$A_\varepsilon = A(1+\varepsilon A)^{-1} \in \mathcal{M}^\varphi \quad (\varepsilon > 0);$$

Recall (A.5) that $A_\varepsilon \uparrow A$ for $\varepsilon \downarrow 0$. Also, for each $n \in \mathbb{N}, n \geq 1$, let

$$e_n = \chi_{[1/n,n]}(A) \in \mathcal{M}^\varphi;$$

we recall ([L], 9.9) that Ae_n is a bounded positive operator, invertible in $e_n\mathcal{M}^\varphi e_n$, and $e_n \uparrow \mathbf{s}(A)$.

In view of 4.3.(2), a normal weight φ_A on $\mathcal{M}$ is defined by

$$\varphi_A(x) = \sup_{\varepsilon\to 0} \varphi_{A_\varepsilon}(x) = \lim_{\varepsilon\to 0} \varphi_{A_\varepsilon}(x) \quad (x \in \mathcal{M}^+).$$

If A is bounded, then by 4.3.(3), we see that the weight φ_A defined here coincides with the weight defined in Section 4.1.

If $x \in \mathcal{M}^+$ and $\mathbf{s}(x)\mathbf{s}(A) = 0$, then clearly $\varphi_A(x) = 0$. On the other hand, if $x \in e_n(\mathfrak{M}_\varphi \cap \mathcal{M}^+)e_n \subset \mathfrak{M}_\varphi \cap \mathcal{M}^+ \subset \mathfrak{M}_{\varphi_{Ae_n}} \cap \mathcal{M}^+$, then $\varphi_A(x) = \lim_{\varepsilon\to 0} \varphi(A_\varepsilon^{1/2} e_n x e_n A_\varepsilon^{1/2}) = \lim_{\varepsilon\to 0} \varphi((Ae_n)_\varepsilon^{1/2} x (Ae_n)_\varepsilon^{1/2}) = \varphi_{Ae_n}(x) < +\infty$. It follows that the weight φ_A is finite and $\mathbf{s}(\varphi_A) \leq \mathbf{s}(A)$.

Actually, we have

$$\mathbf{s}(\varphi_A) = \mathbf{s}(A). \tag{1}$$

Indeed, if $x \in \mathcal{M}^+$ and $\varphi_A(x) = 0$, then for every $\varepsilon > 0$ we have $\varphi(A_\varepsilon^{1/2} x A_\varepsilon^{1/2}) = 0$, $A_\varepsilon^{1/2} x A_\varepsilon^{1/2}\mathbf{s}(\varphi) = 0$, $xA_\varepsilon = 0$, $x\mathbf{s}(A_\varepsilon) = 0$, and hence $x\mathbf{s}(A) = 0$.

We remark that for every $x \in \mathfrak{N}_\varphi$ we have

$$\varphi_A(x^*x) = \|\pi_\varphi(A)^{1/2} J_\varphi x_\varphi\|_\varphi^2 \leq +\infty. \tag{2}$$

Indeed, using 2.21.(1) and (A.5), we obtain

$$\begin{aligned}\varphi_A(x^*x) &= \lim_{\varepsilon\to 0} \varphi(A_\varepsilon^{1/2} x^* x A_\varepsilon^{1/2}) = \lim_{\varepsilon\to 0} \varphi(A_\varepsilon x^* x) = \lim_{\varepsilon\to 0} (x_\varphi | (xA_\varepsilon)_\varphi)_\varphi \\ &= \lim_{\varepsilon\to 0} (x_\varphi | J_\varphi \pi_\varphi(A_\varepsilon) J_\varphi x_\varphi)_\varphi = \|\pi_\varphi(A)^{1/2} J_\varphi x_\varphi\|_\varphi^2.\end{aligned}$$

4.5 Proposition. *Let φ be a normal semifinite weight on the W^*-algebra $\mathcal{M}$ and let A, B be positive self-adjoint operators affiliated to $\mathcal{M}^\varphi$. Then: $A \leq B \Leftrightarrow \varphi_A \leq \varphi_B$.*

Proof. If $A \leq B$, then (A.4) for every $\varepsilon > 0$ we have $A_\varepsilon \leq B_\varepsilon$ so $\varphi_{A_\varepsilon} \leq \varphi_{B_\varepsilon}$ by 4.3.(2), and hence $\varphi_A(x) = \lim_\varepsilon \varphi_{A_\varepsilon}(x) \leq \lim_\varepsilon \varphi_{B_\varepsilon}(x) = \varphi_B(x)$ for all $x \in \mathcal{M}^+$, that is, $\varphi_A \leq \varphi_B$.

Conversely, assume that $\varphi_A \leq \varphi_B$. Let $\varepsilon > 0$ and $f_n = \chi_{[0,n]}(B) \in \mathcal{M}^\varphi$ for each $n \in \mathbb{N}$. Then for every $x \in \mathfrak{N}_\varphi$ we have (2.21.(1)):

$$\begin{aligned}(\pi_\varphi(f_n A_\varepsilon f_n) J_\varphi x_\varphi | J_\varphi x_\varphi)_\varphi &= \varphi_{A_\varepsilon}(f_n x^* x f_n) \leq \varphi_B(f_n x^* x f_n) \\ &= (\pi_\varphi(Bf_n) J_\varphi x_\varphi | J_\varphi x_\varphi)_\varphi\end{aligned}$$

so that $f_n A_\varepsilon f_n \le B f_n$. Since $f_n \uparrow \mathbf{s}(B) = \mathbf{s}(\varphi_B) \ge \mathbf{s}(\varphi_A) = \mathbf{s}(A)$, it follows that $A_\varepsilon \le B$. Since $A_\varepsilon \uparrow A$, we conclude $A \le B$. □

4.6 Proposition. *Let φ be a normal semifinite weight on the W^*-algebra $\mathcal{M}$ and let $A, \{A_i\}_{i\in I}$ be positive self-adjoint operators affiliated to $\mathcal{M}^\varphi$. Then: $A_i \uparrow A \Leftrightarrow \varphi_{A_i}(x) \uparrow \varphi_A(x)$ for all $x \in \mathcal{M}^+$.*

Proof. Assume that $A_i \uparrow A$ and let $x \in \mathcal{M}^+$. From Proposition 4.5, it follows that $\sup_i \varphi_{A_i}(x) \le \varphi_A(x)$. On the other hand, since $A_i \uparrow A$, we have $(A_i)_\varepsilon \uparrow A_\varepsilon$, for all $\varepsilon > 0$ (A.5) and using 4.3.(3) we deduce $\sup_i \varphi_{(A_i)_\varepsilon}(x) = \varphi_{A_\varepsilon}(x)$ for all $\varepsilon > 0$. Consequently, $\varphi_A(x) = \sup_\varepsilon \varphi_{A_\varepsilon}(x) = \sup_\varepsilon \sup_i \varphi_{(A_i)_\varepsilon}(x) \le \sup_i \varphi_{A_i}(x)$.

Conversely, if $\varphi_{A_i} \uparrow \varphi_A$, then, by Proposition 4.5, $\{A_i\}_{i\in I}$ is an increasing net bounded above by A. By (4.5) there exists a positive self-adjoint operator B such that $A_i \uparrow B$. It follows that B is affiliated to $\mathcal{M}^\varphi$ and, by the first part of the proof, $\varphi_{A_i} \uparrow \varphi_B$. Consequently, $\varphi_B = \varphi_A$. Using Proposition 4.5 again we obtain $A \le B$ and $B \le A$, that is, (A.4) $A = B$. □

In particular, with the notation of Section 4.4, we have the following equivalent definition of φ_A:

$$\varphi_A(x) = \sup_n \varphi_{Ae_n}(x) = \lim_n \varphi_{Ae_n}(x) \quad (x \in \mathcal{M}^+). \tag{1}$$

4.7 Proposition. *Let φ be a normal semifinite weight on the W^*-algebra $\mathcal{M}$ and A a positive self-adjoint operator affiliated to $\mathcal{M}^\varphi$. Then*

$$\sigma_t^{\varphi_A}(x) = A^{it}\sigma_t^\varphi(x)A^{-it} \quad (x \in \mathbf{s}(A)\mathcal{M}\mathbf{s}(A), t \in \mathbb{R}).$$

Proof. Assume the W^*-algebra $\mathcal{M}$ realized as a von Neumann algebra $\mathcal{M} \subset \mathcal{B}(\mathcal{H})$. Then $A|\mathbf{s}(A)\mathcal{H}$ is a nonsingular positive self-adjoint operator on the Hilbert space $\mathbf{s}(A)\mathcal{H}$, so that for each $t \in \mathbb{R}$ we can form the unitary operator $A^{it} = (A|\mathbf{s}(A))^{it}$ on $\mathbf{s}(A)\mathcal{H}$. Of course, the operator A^{it} can also be regarded as a partial isometry in $\mathcal{M}$. □

We shall use the notation introduced in Section 4.4.

Let $n \in \mathbb{N}$ be fixed. Then φ_{e_n} is an n.s.f. weight on the W^*-algebra $e_n\mathcal{M}e_n$ and Ae_n is a bounded invertible positive operator in $e_n\mathcal{M}^\varphi e_n$, which is the centralizer of the weight φ_{e_n}. Using 4.6.(1), we see that $(\varphi_A)_{e_n} = (\varphi_{e_n})_{Ae_n}$ as n.s.f. weights on $e_n\mathcal{M}e_n$. Thus, taking into account 2.22.(3) and 4.2.(4), for $x \in e_n\mathcal{M}e_n$ we obtain

$$\begin{aligned}\sigma_t^{\varphi_A}(x) &= \sigma^{(\varphi_A)_{e_n}}(x) = \sigma^{(\varphi_{e_n})_{Ae_n}}(x) = (Ae_n)^{it}\sigma_t^{\varphi_{e_n}}(x)(Ae_n)^{-it}\\ &= A^{it}e_n\sigma_t^\varphi(x)e_nA^{-it} = A^{it}\sigma_t^\varphi(x)A^{-it}.\end{aligned}$$

Since $e_n \uparrow \mathbf{s}(A)$, the set $\bigcup_{n\in\mathbb{N}} e_n\mathcal{M}e_n$ is w-dense in $\mathbf{s}(A)\mathcal{M}\mathbf{s}(A)$. Thus, the assertion of the proposition follows.

4.8 Corollary. *Let φ be a normal semifinite weight on the W^*-algebra $\mathcal{M}$ and A a positive self-adjoint operator affiliated to $\mathcal{M}$. Then*

$$[D\varphi_A : D\varphi]_t = A^{it} \quad (t \in \mathbb{R}).$$

Proof. Consider the balanced weights $\theta = \theta(\varphi, \varphi)$ and $\tau = \theta(\varphi, \varphi_A)$ on $Mat_2(\mathcal{M})$. Then $\tau = \theta_B$ with $B = \begin{pmatrix} 1 & 0 \\ 0 & A \end{pmatrix}$. By Proposition 4.7, we have $\sigma_t^\tau(X) = B^{\mathrm{i}t}\sigma_t^\theta(X)B^{-\mathrm{i}t}$ for all $X \in \mathbf{s}(B)[Mat_2(\mathcal{M})]\mathbf{s}(B)$. In particular, for $X = \begin{pmatrix} 0 & 0 \\ \mathbf{s}(A) & 0 \end{pmatrix}$ we obtain the desired result. □

From the above corollary and from Corollary 3.6, it follows again (see 4.5) that $\varphi_A = \varphi_B \Leftrightarrow A = B$.

4.9 Corollary. *Let φ be an n.s.f. weight on the W^*-algebra $\mathcal{M}$ and let A, B be two commuting positive self-adjoint operators affiliated to $\mathcal{M}^\varphi$. Then the weights $\varphi_{\overline{AB}}, (\varphi_A)_B, (\varphi_B)_A$ are defined and equal:*

$$(\varphi_A)_B = \varphi_{\overline{AB}} = (\varphi_B)_A$$

Proof. Since A and B commute, the closure $\overline{AB}$ of AB is a positive self-adjoint operator and $(AB)^{\mathrm{i}t} = A^{\mathrm{i}t}B^{\mathrm{i}t}(t \in \mathbb{R})$ (A.6). It is easy to check that $\overline{AB}$ (resp. A, B) is affiliated to the centralizer of φ (resp. φ_B, φ_A), hence the weights $\varphi_{\overline{AB}}, (\varphi_A)_B, (\varphi_B)_A$ are defined. Using Corollaries 3.5 and 4.8, we see that the Connes cocycles of these weights with respect to φ coincide and hence, by Corollary 3.6, these weights are equal. □

4.10 Theorem (Pedersen & Takesaki, 1973). *Let φ be an n.s.f. weight on the W^*-algebra $\mathcal{M}$ and ψ a normal semifinite weight on M. The following conditions are equivalent:*

(i) $\psi \circ \sigma_t^\varphi = \psi$ *for all* $t \in \mathbb{R}$;
(ii) $[D\psi : D\varphi]_t \in \mathcal{M}^\psi$ *for all* $t \in \mathbb{R}$;
(iii) $[D\psi : D\varphi]_t \in \mathcal{M}^\varphi$ *for all* $t \in \mathbb{R}$;
(iv) $\{[D\psi : D\varphi]_t\}_{t\in\mathbb{R}}$ *is an s-continuous group of unitary elements of* $\mathbf{s}(\psi)\mathcal{M}\mathbf{s}(\psi)$;
(v) *there exists a positive self-adjoint operator A affiliated to* $\mathcal{M}^\varphi$ *such that* $\psi = \varphi_A$.

If moreover ψ is faithful, then also the following statement is equivalent to those earlier:

(vi) $\varphi \circ \sigma_t^\psi = \varphi$ *for all* $t \in \mathbb{R}$.

Proof. Let $u_t = [D\psi : D\varphi]_t (t \in \mathbb{R})$.

(i) ⇒ (ii). From (i) it follows that $\mathbf{s}(\psi)$ and $\mathcal{M}^\psi$ are σ^φ-invariant. In particular, for every $t \in \mathbb{R}$ we have

$$\sigma_{-t}^\varphi(u_t)\sigma_{-t}^\varphi(u_t)^* = \mathbf{s}(\psi) = \sigma_{-t}^\varphi(u_t)^*\sigma_{-t}^\varphi(u_t).$$

Then, for every $x \in (\mathbf{s}(\psi)\mathcal{M}\mathbf{s}(\psi))^+$ and every $t \in \mathbb{R}$ we obtain

$$\begin{aligned}\psi(\mathbf{s}(\psi)x\mathbf{s}(\psi)) &= \psi(x) = \psi(\sigma_t^\psi(x)) = \psi(u_t\sigma_t^\varphi(x)u_t^*) \\ &= \psi(\sigma_t^\varphi(\sigma_{-t}^\varphi(u_t)x\sigma_{-t}^\varphi(u_t^*))) = \psi(\sigma_{-t}^\varphi(u_t)x\sigma_{-t}^\varphi(u_t)^*).\end{aligned}$$

By Proposition 2.21, it follows that $\sigma_{-t}^\varphi(u_t) \in \mathcal{M}^\psi$, and hence $u_t = \sigma_t^\varphi(\sigma_{-t}^\varphi(u_t)) \in \mathcal{M}^\psi$.

(ii) ⇒ (iv). From (ii) it follows that $u_t = \sigma_s^\psi(u_t) = u_s\sigma_s^\varphi(u_t)u_s^* = u_{s+t}u_s^*$ and therefore $u_{s+t} = u_tu_s$ for all $s, t \in \mathbb{R}$.

(iii) ⇒ (iv). If $u_t \in \mathcal{M}^\varphi$, then $u_{s+t} = u_s\sigma_s^\varphi(u_t) = u_su_t$. Conversely, if $u_su_t = u_{s+t} = u_s\sigma_s^\varphi(u_t)$, then $u_t = \sigma_s^\varphi(u_t)$, hence $u_t \in \mathcal{M}^\varphi$.

(iv) ⇒ (v). If the condition (iv) holds then, by Stone's theorem ([L], 9.20), there exists a positive self-adjoint operator A affiliated to $\mathcal{M}^\varphi$, with $\mathbf{s}(A) = \mathbf{s}(\psi)$, such that $[D\psi : D\varphi]_t = A^{it} = [D\varphi_A : D\varphi]_t (t \in \mathbb{R})$. By Corollary 3.6, it follows that $\psi = \varphi_A$.

(v) ⇒ (i). This follows obviously from the definition of φ_A.

If ψ is faithful, then (vi) is equivalent to the other conditions owing to the symmetry between (ii) and (iii). □

If the above equivalent conditions are satisfied, we shall say that the normal semifinite weight ψ *commutes* with the n.s.f. weight φ (see also [L], Cor. 10.28). Thus, the weights of the form φ_A are exactly the weights commuting with φ.

Clearly, if ψ commutes with φ, then $\mathbf{s}(\psi) \in \mathcal{M}^\varphi$.

4.11 Corollary. *Let φ be an n.s.f. weight on the W^*-algebra $\mathcal{M}$ with center $\mathcal{Z}(\mathcal{M})$ and ψ a normal semifinite weight on $\mathcal{M}$. The following conditions are equivalent:*

(i) *ψ satisfies the* KMS *condition with respect to $\{\sigma_t^\varphi\}_{t\in\mathbb{R}}$;*
(ii) *$\mathbf{s}(\psi) \in \mathcal{Z}(\mathcal{M})$ and $\sigma_t^\psi = \sigma_t^\varphi|\mathcal{M}\mathbf{s}(\psi)$ for all $t \in \mathbb{R}$;*
(iii) *there exists a positive self-adjoint operator A affiliated to $\mathcal{Z}(\mathcal{M})$ such that $\psi = \varphi_A$.*

Proof. Let $q = \mathbf{s}(\psi)$ and $p = 1 - q$.

(i) ⇒ (ii). Since $\psi \circ \sigma_t^\varphi = \psi$, we have $p, q \in \mathcal{M}^\varphi$. Let $x, y \in \mathfrak{N}_\psi \cap \mathfrak{N}_\psi^*$. Then $\psi((xp)^*(xp)) = \psi(px^*xp) = 0$ and $\psi((py)^*(py)) = \psi(y^*py) \le \psi(y^*y) < +\infty$, hence $xp, py \in \mathfrak{N}_\psi \cap \mathfrak{N}_\psi^*$. By assumption, there exists a function f, defined, continuous and bounded on the strip $\{\alpha \in \mathbb{C}; 0 \le Re\,\alpha \le 1\}$, analytic in the interior of this strip, such that $f(it) = \psi(\sigma_t^\varphi(xp)py) = \psi(\sigma_t^\varphi(x)py)$ and $f(1+it) = \psi(py\sigma_t^\varphi(xp)) = 0$ for all $t \in \mathbb{R}$. It follows that f is identically zero, in particular $\psi(\sigma_t^\varphi(x)py) = 0\ (t \in \mathbb{R})$.

Consequently, $\psi(xpx^*) = 0$ for every $x \in \mathfrak{N}_\psi \cap \mathfrak{N}_\psi^*$, so that $xpx^* = pxpx^*p$ for all $x \in \mathcal{M}$. In particular, for $x = v \in \mathcal{M}$, unitary, we have $vpv^* \le p, vpv^* = p$ and therefore $p \in \mathcal{Z}(\mathcal{M})$ and $\mathbf{s}(\psi) = q = 1 - p \in \mathcal{Z}(\mathcal{M})$.

Now, the restriction of ψ to $\mathcal{M}q$ is an n.s.f. weight on $\mathcal{M}q$ which satisfies the *KMS* condition with respect to $\{\sigma_t^\varphi|\mathcal{M}q\}_{t\in\mathbb{R}}$, so that $\sigma_t^\varphi|\mathcal{M}q = \sigma_t^\psi (t \in \mathbb{R})$, by 2.12.(11).

(ii) ⇒ (iii). From (ii) it follows that $\psi \circ \sigma_t^\varphi = \psi, (t \in \mathbb{R})$. By Theorem 4.10, we have $\psi = \varphi_A$ for some positive self-adjoint operator A affiliated to $\mathcal{M}^\varphi$. Then for every $x \in \mathcal{M}q$ and every $t \in \mathbb{R}$, we have $\sigma_t^\varphi(x) = \sigma_t^\psi(x) = A^{it}\sigma_t^\varphi(x)A^{-it}$, hence A is affiliated to $\mathcal{Z}(\mathcal{M})q \subset \mathcal{Z}(\mathcal{M})$.

Finally, the implication (iii) ⇒ (ii) follows from Proposition 4.7, as $\mathbf{s}(\psi) = \mathbf{s}(A) \in \mathcal{Z}(\mathcal{M})$, while the implication (ii) ⇒ (i) is obvious. □

The equivalent conditions in the above corollary hold, for instance, if φ, ψ, are n.s.f. traces on $\mathcal{M}$.

4.12. From ([L], 10.29), we know that a *W^*-algebra $\mathcal{M}$ is semifinite if and only if there exist an n.s.f weight φ on $\mathcal{M}$ and an s-continuous one-parameter group $\{u_t\}_{t\in\mathbb{R}}$ of unitary operators in $\mathcal{M}$ such that $\sigma_t^\varphi(x) = u_txu_t^*(x \in \mathcal{M}, t \in \mathbb{R})$.* The preceding results allow a simple proof of this theorem.

If μ is an n.s.f. trace on $\mathcal{M}$ and φ is any n.s.f. weight on $\mathcal{M}$, then $u_t = [D_\varphi : D\mu]_t (t \in \mathbb{R})$ satisfies the required condition because any trace commutes with any weight. Conversely, if this condition is satisfied and A is the unique positive self-adjoint operator such that $A^{-it} = u_t (t \in \mathbb{R})$, then $\mu = \varphi_A$ is an n.s.f. trace on $\mathcal{M}$ because $\sigma_t^\mu(x) = x(x \in \mathcal{M}, t \in \mathbb{R})$.

4.13 Corollary. *Let φ be an n.s.f weight on the W^*-algebra $\mathcal{M}$ of type III and $\beta \in \mathbb{R}, \beta \ne 1$. Then no n.s.f. weight on $\mathcal{M}$ satisfies the* KMS *condition with respect to $\{\sigma_{\beta t}^\varphi\}_{t\in\mathbb{R}}$.*

Proof. Assume to the contrary; then there exists an n.s.f. weight ψ on $\mathcal{M}$ such that $\sigma^{\varphi}_{\beta t} = \sigma^{\psi}_{t}$ ($t \in \mathbb{R}$), in particular ψ commutes with φ. Consequently, there exists a positive self-adjoint operator A affiliated to $\mathcal{M}^{\varphi}$ such that $\psi = \varphi_A$. It follows that $\sigma^{\varphi}_{\beta t}(x) = A^{it}\sigma^{\varphi}_{t}(x)A^{-it}(x \in \mathcal{M}, t \in \mathbb{R})$. Putting $\alpha = (\beta - 1)^{-1}$, we obtain $\sigma^{\varphi}_{t}(x) = (A^{\alpha})^{it}x(A^{\alpha})^{-it}(x \in \mathcal{M}, t \in \mathbb{R})$, contradicting the fact that $\mathcal{M}$ is not semifinite (4.12). □

4.14. Let φ, ψ be n.s.f. weights on the W^*-algebra $\mathcal{M}$ and $\sigma \in Aut(\mathcal{M})$. By 2.22.(5), we get

$$\varphi \circ \sigma = \varphi \Rightarrow \sigma^{\varphi}_{t} \circ \sigma = \sigma \circ \sigma^{\varphi}_{t} \text{ for all } t \in \mathbb{R}. \tag{1}$$

In particular,

$$\psi \text{ commutes with } \varphi \Rightarrow \sigma^{\varphi}_{t} \circ \sigma^{\psi}_{s} = \sigma^{\psi}_{s} \circ \sigma^{\varphi}_{t} \text{ for all } s, t \in \mathbb{R}. \tag{2}$$

In general, the converse of (1) is not true. For instance, there exist measurable but not measure-preserving transformations on measure spaces. Even with the supplementary assumption that σ acts identically on the center $\mathcal{Z}(\mathcal{M})$ of $\mathcal{M}$, the converse of (1) need not hold, as there exist $*$-automorphisms of type II_{∞}-factors which do not preserve the trace.

Also, the converse of statement (2) is not valid in general, as we shall see in the next section in an important example. Following this example, we shall say that the weights φ, ψ *anticommute* if they do not commute but the corresponding modular automorphism groups commute.

However, there are certain special cases in which the converses of (1) and (2) are true. These cases are considered in Sections 4.17–4.20.

4.15. Consider the Hilbert space $\mathcal{H} = \mathcal{L}^2(\mathbb{R})$ and the operators $u_t, v_s \in \mathcal{B}(\mathcal{H})$ defined by

$$(u_t\xi)(r) = \xi(r+t), (v_s\xi)(r) = e^{isr}\xi(r) \quad (\xi \in \mathcal{H}, r, s, t \in \mathbb{R}),$$

Then $\{u_t\}_{t\in\mathbb{R}}$ and $\{v_s\}_{s\in\mathbb{R}}$ are *so*-continuous one-parameter groups of unitary operators on $\mathcal{H}$, which satisfy the following *anticommutation relations*:

$$v_s u_t = e^{-its}u_t v_s \quad (s, t \in \mathbb{R}). \tag{1}$$

By Stone's theorem ([L], 9.20), there exist positive self-adjoint operators A, B in $\mathcal{H}$, uniquely determined, such that $u_t = A^{it}(t \in \mathbb{R})$, and $v_s = B^{is}(s \in \mathbb{R})$. Thus $B^{is}A^{it} = e^{-its}A^{it}B^{is}(s, t \in \mathbb{R})$, and using the definition of the operator A ([L], 9.20) we infer that

$$B^{-is}AB^{is} = e^{s}A \quad (s \in \mathbb{R}). \tag{2}$$

Consider the W^*-algebra $\mathcal{M} = \mathcal{B}(\mathcal{H})$ with the canonical trace tr and the n.s.f. weights $\varphi = tr_A, \psi = tr_B$ on $\mathcal{M}$, Then, by Proposition 4.7, we have $\sigma^{\varphi}_{t} = Ad(u_t), \sigma^{\psi}_{s} = Ad(v_s)(s, t \in \mathbb{R})$, and it follows from (1) that

$$\sigma^{\varphi}_{t} \circ \sigma^{\psi}_{s} = \sigma^{\psi}_{s} \circ \sigma^{\varphi}_{t} \quad (s, t \in \mathbb{R}). \tag{3}$$

However, φ and ψ do not commute, more precisely we have

$$\varphi \circ \sigma_s^{\psi} = e^s \varphi \quad (s \in \mathbb{R}). \tag{4}$$

Indeed, let $x \in \mathscr{M}^+$ and $A_\varepsilon = A(1+\varepsilon A)^{-1} (\varepsilon > 0)$. From (2), it follows that $B^{-is} A_\varepsilon B^{is} = (e^s A)_\varepsilon$ $(s \in \mathbb{R}, \varepsilon > 0)$, and therefore

$$\begin{aligned}
\varphi(\sigma_s^{\psi}(x)) &= tr_A(B^{is} x B^{-is}) = \lim_{\varepsilon \to 0} tr(A_\varepsilon^{1/2} B^{is} x B^{-is} A_\varepsilon^{1/2}) \\
&= \lim_{\varepsilon \to 0} tr(x^{1/2} B^{-is} A_\varepsilon B^{is} x^{1/2}) = \lim_{\varepsilon \to 0} tr(x^{1/2} (e^s A)_\varepsilon x^{1/2}) \\
&= \lim_{\varepsilon \to 0} tr((e^s A)_\varepsilon^{1/2} x (e^s A)_\varepsilon^{1/2}) = tr_{e^s A}(x) = e^s \varphi(x).
\end{aligned}$$

4.16. Let φ be an n.s.f. weight on the W^*-algebra $\mathscr{M}, \sigma \in Aut(\mathscr{M})$ and A a positive self-adjoint operator affiliated to $\mathscr{M}$.

The operator A is affiliated to $\mathscr{M}^{\varphi \circ \sigma}$ if and only if the operator $\sigma(A)$ is affiliated to $\mathscr{M}^{\varphi}$ (see 2.22.(6) and [L], 9.25) and in this case we have

$$(\varphi \circ \sigma)_A = \varphi_{\sigma(A)} \circ \sigma. \tag{1}$$

Using Corollary 4.8, it follows that if A is affiliated to $\mathscr{M}^{\varphi}$, then

$$\varphi \circ \sigma = \varphi, \ \varphi_A \circ \sigma = \varphi_A \Rightarrow \sigma(A) = A. \tag{2}$$

Note that the set of all nonsingular positive self-adjoint operators affiliated to the center $\mathscr{Z}(\mathscr{M})$ of $\mathscr{M}$ is a group with respect to the operation $(A, B) \mapsto \overline{AB}$.

Proposition. *Let φ be an n.s.f. weight on the W^*-algebra $\mathscr{M}$ and $\sigma : G \to Aut(\mathscr{M})$ an action of the group G on $\mathscr{M}$. We assume that each *-automorphism $\sigma_g (g \in G)$, acts identically on $\mathscr{Z}(\mathscr{M})$. Then the following statements are equivalent:*

(i) $\sigma_g \circ \sigma_t^{\varphi} = \sigma_t^{\varphi} \circ \sigma_g$ *for every $g \in G$ and every $t \in \mathbb{R}$;*

(ii) *there exists a homomorphism $g \mapsto A_g$ of the group G into the group of all nonsingular positive self-adjoint operators affiliated to $\mathscr{Z}(\mathscr{M})$ such that $\varphi \circ \sigma_g = \varphi_{A_g}$ for every $g \in G$.*

If moreover $G = \mathbb{R}$ and the action $\sigma : \mathbb{R} \to Aut(\mathscr{M})$ is continuous, then the following statement is equivalent to (i) and (ii).

(iii) *there exists a nonsingular positive self-adjoint operator A affiliated to $\mathscr{Z}(\mathscr{M})$ such that $\varphi \circ \sigma_s = \varphi_{A^s}$ for every $s \in \mathbb{R}$.*

Proof. It is clear that (iii) $\Rightarrow$ (ii), and (ii) $\to$ (i) follows by using Corollary 4.11 and 2.22.(5). With the same arguments, from (i) it follows that for each $g \in G$ there exists a unique A_g such that $\varphi \circ \sigma_g = \varphi_{A_g}$. Since each σ_g acts identically on $\mathscr{Z}(\mathscr{M})$, for $g, h \in G$ we obtain

$$\varphi_{A_{gh}} = (\varphi \circ \sigma_g) \circ \sigma_h = \varphi_{A_g} \circ \sigma_h = (\varphi \circ \sigma_h)_{A_g} = (\varphi_{A_h})_{A_g} = \varphi_{\overline{A_g A_h}}$$

by Corollary 4.9. Hence (i) $\Rightarrow$ (ii).

We now show that (ii) ⇒ (iii). By assumption, there exists a homomorphism $\mathbb{R} \ni s \mapsto A_s$ such that $\varphi \circ \sigma_s = \varphi_{A_s} (s \in \mathbb{R})$. Let $A = A_1$. Since $A_{s+t} = \overline{A_s A_t}$, it follows that $A_s = A^s$ for every rational number $s \in \mathbb{R}$.

Let $e_n = \chi_{[1/n,n]}(A)$ and $x \in \mathfrak{N}_\varphi$. The function

$$\mathbb{R} \ni s \mapsto \varphi(A^s e_n x^* x) = \|\pi_\varphi(A^s e_n)^{1/2} x_\varphi\|_\varphi^2 \tag{3}$$

is continuous and bounded. On the other hand, we have

$$\begin{aligned}\varphi(\sigma_s(e_n x^* x)) &= \lim_{\varepsilon \to 0} \varphi((A_s)_\varepsilon e_n x^* x)\\ &= \lim_{\varepsilon \to 0} \|\pi_\varphi(A_s)_\varepsilon^{1/2} \pi_\varphi(e_n) x_\varphi\|_\varphi^2 = \|\pi_\varphi(A_s e_n)^{1/2} x_\varphi\|_\varphi^2.\end{aligned}$$

Since the weight φ is normal, it follows that the function

$$\mathbb{R} \ni s \mapsto \varphi(\sigma_s(e_n x^* x)) = \|\pi_\varphi(A_s e_n)^{1/2} x_\varphi\|_\varphi^2 \in [0, +\infty] \tag{4}$$

is lower w-semicontinuous. As the functions (3) and (4) coincide for every rational $s \in \mathbb{R}$ and every $x \in \mathfrak{N}_\varphi$, we infer that

$$A_s e_n \leq A^s e_n \quad (s \in \mathbb{R}).$$

Consequently, $s \mapsto (A_s e_n)(A^s e_n)^{-1}$ is a one-parameter group of positive operators with norm ≤ 1 on $e_n \mathcal{H}$. Since the only such group is the trivial one, it follows that $A_s e_n = A^s e_n$, and, since $e_n \uparrow 1$, we conclude that $A_s = A^s$ for all $s \in \mathbb{R}$. □

Let $\sigma : G \to Aut(\mathcal{M})$ be an action of G on $\mathcal{M}$ which is trivial on $\mathcal{Z}(\mathcal{M})$ and satisfies the equivalent conditions (i) and (ii) of the above proposition. Then $G_0 = \{g \in G; \varphi \circ \sigma_g = \varphi\}$ is the kernel of the homomorphism $g \mapsto A_g$ and hence a normal subgroup of G. Since $\{A_g\}_{g \in G}$ is an abelian group, it follows that the quotient group G/G_0 is abelian.

4.17 Corollary. *Let φ be a faithful normal state on the W^*-algebra $\mathcal{M}$ and $\sigma : G \to Aut(\mathcal{M})$ an action of the group G on $\mathcal{M}$ which is trivial on $\mathcal{Z}(\mathcal{M})$. Then the following statements are equivalent:*

(i) $\varphi \circ \sigma_g = \varphi$ *for all* $g \in G$;
(ii) $\sigma_t^\varphi \circ \sigma_g = \sigma_g \circ \sigma_t^\varphi$ *for all* $g \in G$ *and all* $t \in \mathbb{R}$.

Proof. By 4.14.(1), we know that (i) ⇒ (ii). If (ii) holds, then we can write $\varphi \circ \sigma_g = \varphi_{A_g} (g \in G)$, as in Proposition 4.16. Let $g \in G, \varepsilon > 0$ and let $e_\varepsilon = \chi_{[1+\varepsilon,+\infty)}(A_g)$. Then for every $k \in \mathbb{N}$ we get $\|\varphi\| \geq \varphi(\sigma_g^k(e_\varepsilon)) = \varphi(A_g^k e_\varepsilon) \geq (1+\varepsilon)^k \varphi(e_\varepsilon)$, hence $e_\varepsilon = 0$. It follows that $A_g \leq 1$, so $1 \leq A_g^{-1} = A_{g^{-1}} \leq 1$ and hence $A_g = 1$ for all $g \in G$. □

4.18 Corollary. *Let φ, ψ be n.s.f. weights on the W^*-algebra $\mathcal{M}$. If φ is finite, or if $\varphi \leq \psi$, then the following are equivalent:*

(i) φ *commutes with* ψ;
(ii) $\sigma_t^\varphi \circ \sigma_s^\psi = \sigma_s^\psi \circ \sigma_t^\varphi$ *for all* $s, t \in \mathbb{R}$.

Proof. By 4.14.(2), we know that (i) ⇒ (ii). Also, if φ is finite, then (ii) ⇒ (i) by Corollary 4.17.

Assume that (ii) holds and $\varphi \leq \psi$. By Proposition 4.16, there exists a nonsingular positive self-adjoint operator A affiliated to the center $\mathscr{Z}(\mathscr{M})$ such that $\varphi \circ \sigma_s^{\psi} = \varphi_{A^s}(s \in \mathbb{R})$. Let $e_\varepsilon = \chi_{[1+\varepsilon,+\infty)}(A)$. For $x \in \mathfrak{M}_\psi \cap \mathscr{M}^+$ and $s \in \mathbb{R}, s \geq 0$, we have $+\infty > \psi(x) \geq \psi(e_\varepsilon x) = \psi(e_s\sigma_s^{\psi}(x)) \geq \varphi(e_s\sigma_s^{\psi}(x)) = \varphi(A^s e_\varepsilon x) \geq (1+\varepsilon)^s\varphi(e_\varepsilon x)$. Consequently, $\varphi(e_\varepsilon x) = 0$. Since φ is normal and semifinite, it follows that $\varphi(e_\varepsilon) = 0$; since φ is faithful we conclude $e_\varepsilon = 0$. Thus, $A \leq 1$. Similarly, using $f = \chi_{[0,1-\varepsilon]}(A)$, we get $A \geq 1$. Hence $A = 1$ and φ commutes with ψ. □

4.19. Let $\mathscr{M}$ be a W^*-algebra and $\varphi \in \mathscr{M}_*$. Put

$$\mathfrak{L}_\varphi = \{x \in \mathscr{M}; \varphi(ax) = 0 \text{ for all } a \in \mathscr{M}\},$$
$$\mathfrak{R}_\varphi = \{x \in \mathscr{M}; \varphi(xa) = 0 \text{ for all } a \in \mathscr{M}\}.$$

Then $\mathfrak{L}_\varphi$ (resp. $\mathfrak{R}_\varphi$) is a w-closed left (resp. right) ideal of $\mathscr{M}$, hence ([L], 3.20) there exists a unique projection $e \in \mathscr{M}$ (resp. $f \in \mathscr{M}$) such that $\mathfrak{L}_\varphi = \mathscr{M}e$ (resp. $\mathfrak{R}_\varphi = f\mathscr{M}$). The projection $\mathbf{r}(\varphi) = 1 - e$ (resp. $\mathbf{l}(\varphi) = 1 - f$) is called the *right support* of φ (resp. the *left support* of φ) in $\mathscr{M}$. Thus,

$$\mathfrak{L}_\varphi = \mathscr{M}(1 - \mathbf{r}(\varphi)), \mathfrak{R}_\varphi = (1 - \mathbf{l}(\varphi))\mathscr{M}. \tag{1}$$

Since $1 - \mathbf{r}(\varphi) \in \mathfrak{L}_\varphi$ and $1 - \mathbf{l}(\varphi) \in \mathfrak{R}_\varphi$, we get

$$\varphi = \varphi(\cdot\mathbf{r}(\varphi)) = \varphi(\mathbf{l}(\varphi)\cdot). \tag{2}$$

Also, using the Hahn–Banach theorem, we infer from (1) that

$$\begin{aligned} &\{\varphi(a\cdot); a \in \mathscr{M}\} \text{ is norm-dense in } \mathscr{M}_* \cdot \mathbf{r}(\varphi), \\ &\{\varphi(\cdot a); a \in \mathscr{M}\} \text{ is norm-dense in } \mathbf{l}(\varphi) \cdot \mathscr{M}_*. \end{aligned} \tag{3}$$

Since $\mathfrak{L}_\varphi = (\mathfrak{R}_{\varphi^*})^*$, we have $\mathbf{r}(\varphi) = \mathbf{l}(\varphi^*)$. In particular, if $\varphi = \varphi^*$, then $\mathbf{r}(\varphi) = \mathbf{l}(\varphi)$ is called the *support* of φ and is denoted by $\mathbf{s}(\varphi)$. If φ is positive, then the Schwarz inequality implies $\{x \in \mathscr{M}; \varphi(x^*x) = 0\} = \mathfrak{L}_\varphi = \mathscr{M}(1 - \mathbf{s}(\varphi))$, hence $\mathbf{s}(\varphi)$ is the usual (2.1) support of φ.

Consider now $\varphi, \psi \in \mathscr{M}_*^+$. It is easy to check that if φ is faithful, then the left and right supports of the forms $\varphi + \mathrm{i}\psi$ and $\varphi - \mathrm{i}\psi$ are all equal to 1, hence their absolute values $|\varphi + \mathrm{i}\psi|$ and $|\varphi - \mathrm{i}\psi|$ are faithful normal positive forms and the partial isometries from the corresponding polar decompositions are unitary elements ([L], 5.16, E.5.10).

Proposition. *Let φ, ψ be normal positive forms on the W^*-algebra $\mathscr{M}$ and $\mathbf{s}(\varphi) = 1$. Then the following conditions are equivalent:*

(i) *ψ commutes with φ;*
(ii) *$|\varphi + \mathrm{i}\psi| = |\varphi - \mathrm{i}\psi|$;*
(iii) *[assume also $\mathbf{s}(\psi) = 1$] $\sigma_t^{\varphi} \circ \sigma_s^{\psi} = \sigma_s^{\psi} \circ \sigma_t^{\varphi}$ for all $s, t \in \mathbb{R}$.*

Proof. The equivalence (i) ⇔ (iii) follows from 4.18.

(i) ⇒ (ii). By assumption (4.10), we have $\psi = \varphi_A$ for some positive self-adjoint operator A affiliated to $\mathscr{M}^{\varphi}$. It follows that

$$(\varphi + \mathrm{i}\psi)(\cdot(1 + \mathrm{i}A)^{-1}) = \varphi = (\varphi - \mathrm{i}\psi)(\cdot(1 - \mathrm{i}A)^{-1}),$$

hence $u = (1 - \mathrm{i}A)(1 + \mathrm{i}A)^{-1} \in \mathscr{M}^{\varphi}$ is a unitary operator and we have $(\varphi + \mathrm{i}\psi)(\cdot u) = \varphi - \mathrm{i}\psi$.

If $\varphi + i\psi = \omega(\cdot v)$ is the polar decomposition of $\varphi + i\psi$, then $\varphi - i\psi = \omega(\cdot uv)$ is the polar decomposition of $\varphi - i\psi$, and hence $|\varphi + i\psi| = \omega = |\varphi - i\psi|$.

(ii) ⇒ (i). Let $\varphi + i\psi = \omega(\cdot v)$ be the polar decomposition of $\varphi + i\psi$, with $\omega = |\varphi + i\psi|$ and $v \in \mathcal{M}$ unitary. By assumption, we have $\omega = |\varphi - i\psi| = |(\varphi + i\psi)^*| = \omega(v^* \cdot v)$. It follows that $\varphi + i\psi = (\varphi + i\psi)(v^* \cdot v)$, that is, $\varphi = \varphi(v^* \cdot v)$ and $\psi = \psi(v^* \cdot v)$. By Proposition 2.21, we infer that $v \in \mathcal{M}^\varphi$. On the other hand, we have $\varphi + i\psi = \omega(\cdot v) = \omega(\cdot v^2 v^*) = (\varphi - i\psi)(\cdot v^2)$, that is, $\varphi(\cdot(v^2 - 1)) = i\psi(\cdot(v^2 + 1))$. Since $v \in \mathcal{M}^\varphi$, it follows that $\psi(\cdot(v^2 + 1))$ is σ_t^φ-invariant ($t \in \mathbb{R}$).

For fixed $t \in \mathbb{R}$ let $e = \mathbf{s}(\psi - \psi \circ \sigma_t^\varphi)$. By the above arguments, we have $(v^2 + 1)e = 0$, hence $e = -v^2 e$. Consequently, $\varphi(e) - i\psi(e) = -\varphi(v^2 e) + i\psi(v^2 e) = -(\varphi - i\psi)(ev^2) = -(\varphi + i\psi)(e) = -\varphi(e) - i\psi(e)$, so that $\varphi(e) = 0$ and $e = 0$. Thus, $\psi \circ \sigma_t^\varphi = \psi (t \in \mathbb{R})$, that is, ψ commutes with φ. □

4.20 Proposition. *Let φ be an n.s.f. weight on the W^*-algebra $\mathcal{M}$ and $\sigma : G \to Aut(\mathcal{M})$ a continuous action of the compact group G on $\mathcal{M}$ which is trivial on the center $\mathcal{Z}(\mathcal{M})$ of $\mathcal{M}$. The following conditions are equivalent:*

(i) *$\varphi \circ \sigma_g = \varphi$ for all $g \in G$;*
(ii) *$\sigma_t^\varphi \circ \sigma_g = \sigma_g \circ \sigma_t^\varphi$ for all $g \in G$ and all $t \in \mathbb{R}$.*

Proof. By 4.14.(1), we know that (i) ⇒ (ii). If (ii) holds, then by Proposition 4.16, we can write

$$\varphi \circ \sigma_g = \varphi_{A_g} \quad (g \in G).$$

Let $g \in G$ be fixed. For the proof, we may assume that G is topologically generated by the element g.

If G is discrete, G is finite so that there exists $n \in \mathbb{N}, n \geq 1$, such that σ_g^n is the identity automorphism. It follows that $A_g^n = 1$, hence $A_g = 1$ and $\varphi \circ \sigma_g = \varphi$.

If G is not discrete, then there exists a sequence $n_k \to +\infty$ of positive integers such that $\{g^{n_k}\}_k$ converges to the neutral element of G. Let $\varepsilon > 0$, $e_\varepsilon = \chi_{[1+\varepsilon,+\infty)}(A_g)$ and $x \in \mathfrak{M}_\varphi \cap \mathcal{M}^+$. Since σ is a continuous action, the sequence $\{\sigma_g^{-n_k}(e_\varepsilon x)\}_k$ is w-convergent to $e_\varepsilon x$, hence

$$\begin{aligned}\varphi(e_\varepsilon x) &\leq \liminf_{k\to\infty} \varphi(\sigma_g^{-n_k}(e_\varepsilon x)) = \liminf_{k\to\infty} \varphi(A_g^{-n_k} e_\varepsilon x)\\ &\leq \liminf_{k\to\infty} (1+\varepsilon)^{-n_k} \varphi(e_\varepsilon x) = 0,\end{aligned}$$

so that $\varphi(e_\varepsilon x) = 0$ and $e_\varepsilon x = 0$. Since $\mathfrak{M}_\varphi$ is s-dense in $\mathcal{M}$ we get $e_\varepsilon = 0$. Consequently, $A_g \leq 1$. Similarly we obtain $A_g \geq 1$, hence $A_g = 1$ and $\varphi \circ \sigma_g = \varphi$. □

4.21 Proposition. *Let $\mathcal{M}$ be a semifinite factor and $\sigma : G \to Aut(\mathcal{M})$ an action of the group G on $\mathcal{M}$. If there exists some nonzero σ-invariant normal state φ on $\mathcal{M}$, then the trace μ on $\mathcal{M}$ is σ-invariant.*

Proof. By Theorem 4.10, there exists a positive self-adjoint operator A affiliated to $\mathcal{M}$ such that $\varphi = \mu_A$ and by Proposition 4.16, there exists a group homomorphism $G \ni g \mapsto \lambda_g \in \mathbb{R}^+$ such that $\mu \circ \sigma_g = \lambda_g \mu (g \in G)$. Since $\varphi = \mu_A$, is σ-invariant, it follows that $\sigma_g(A) = \lambda_g A$ $(g \in G)$.

Let $g \in G$ be fixed and assume that $\lambda_g < 1$. For each $\varepsilon > 0$ and each $k \in N$, we have

$$\varphi(\chi_{(0,\varepsilon)}(A)) = \varphi(\sigma_g^k(\chi_{(0,\varepsilon)}(A))) = \varphi(\chi_{(0,\varepsilon)}(\sigma_g^k(A))) = \varphi(\chi_{(0,\varepsilon)}(\lambda_g^k(A))).$$

Since $\lambda_g < 1$, we have $\lim_{k\to\infty} \chi_{(0,+\infty)}(\lambda_g^k t) = \chi_{(0,+\infty)}(t)$ for every $t \in \mathbb{R}$. It follows that $\varphi(\chi_{(0,\varepsilon)}(A)) = \varphi(\chi_{(0,+\infty)}(A))$, hence $A = A\mathrm{s}(\varphi) \le \varepsilon$. As $\varepsilon > 0$ was arbitrary, we obtain $A = 0$, contradicting the fact that $\varphi \ne 0$. Consequently, $\lambda_g \ge 1$ and, similarly, $\lambda_g \le 1$, that is, $\lambda_g = 1$. □

4.22. Even though the commutation of the modular automorphism groups does not insure the commutation of the n.s.f. weights, it does imply that the sum of the two weights is still semifinite.

Proposition. *Let φ, ψ be n.s.f. weights on the W^*-algebra $\mathcal{M}$ such that $\sigma_t^\varphi \circ \sigma_s^\psi = \sigma_s^\psi \circ \sigma_t^\varphi$ for all $s, t \in \mathbb{R}$. Then the normal faithful weight $\varphi + \psi$ is semifinite.*

Proof. By Proposition 4.16, there exists a nonsingular positive self-adjoint operator A affiliated to the center of $\mathcal{M}$ such that

$$\psi \circ \sigma_t^\varphi = \psi_{A^t} \quad (t \in \mathbb{R}).$$

Putting $e_m = \chi_{(0,m]}(A)$ $(m \in \mathbb{N})$, we have $e_m \uparrow \mathbf{s}(A) = 1$.

Let $x \in \mathcal{M}$ and $n \in \mathbb{N}$. Consider the elements

$$x_n = \frac{n}{\pi} \int_{-\infty}^{+\infty} \int_{-\infty}^{+\infty} e^{-n(t^2+s^2)} \sigma_t^\varphi(\sigma_s^\psi(x))\, \mathrm{d}t\, \mathrm{d}s \in \mathcal{M}.$$

It is easy to check that $x_n \xrightarrow{s} x$, $x_n \in \mathcal{M}_\infty^\varphi \cap \mathcal{M}_\infty^\psi$ and

$$\sigma_\alpha^\varphi(x_n) = \frac{n}{\pi} \int_{-\infty}^{+\infty} \int_{-\infty}^{+\infty} e^{-n((t-\alpha)^2+s^2)} \sigma_t^\varphi(\sigma_s^\psi(x))\, \mathrm{d}t\, \mathrm{d}s \quad (\alpha \in \mathbb{C}),$$

$$\sigma_\alpha^\psi(x_n) = \frac{n}{\pi} \int_{-\infty}^{+\infty} \int_{-\infty}^{+\infty} e^{-n(t^2+(s-\alpha)^2)} \sigma_t^\varphi(\sigma_s^\psi(x))\, \mathrm{d}t\, \mathrm{d}s \quad (\alpha \in \mathbb{C}).$$

If $x \in \mathfrak{M}_\psi \cap \mathcal{M}^+$, then

$$\begin{aligned}
\psi(x_n e_m) &= \frac{n}{\pi} \int_{-\infty}^{+\infty} \int_{-\infty}^{+\infty} e^{-nt^2} e^{-ns^2} \psi(\sigma_s^\psi(\sigma_t^\varphi(x) e_m))\, \mathrm{d}t\, \mathrm{d}s \\
&= \sqrt{\frac{n}{\pi}} \int_{-\infty}^{+\infty} e^{-nt^2} \psi(\sigma_t^\varphi(x e_m))\mathrm{d}t = \sqrt{\frac{n}{\pi}} \int_{-\infty}^{+\infty} e^{-nt^2} \psi(A^t e_m x)\mathrm{d}t \\
&\le m^t \sqrt{\frac{n}{\pi}} \int_{-\infty}^{+\infty} e^{-nt^2} \psi(x)\mathrm{d}t = m^t \psi(x) < +\infty,
\end{aligned}$$

hence $x_n e_m \in \mathcal{M}_\infty^\varphi \cap \mathcal{M}_\infty^\psi \cap \mathfrak{M}_\psi$. Since $x_n e_m \xrightarrow{s} x$, it follows that

$$\mathcal{M}_\infty^\varphi \cap \mathcal{M}_\infty^\psi \cap \mathfrak{M}_\psi \text{ is s-dense in } \mathcal{M}.$$

Similarly,

$$\mathcal{M}_\infty^\varphi \cap \mathcal{M}_\infty^\psi \cap \mathfrak{M}_\varphi \text{ is s-dense in } \mathcal{M}.$$

Consequently, the product of these two sets is w-dense in $\mathscr{M}$. On the other hand, as we noted in Section 2.15, this product is contained in $\mathfrak{M}_\varphi \cap \mathfrak{M}_\psi \subset \mathfrak{M}_{\varphi+\psi}$, hence $\varphi + \psi$ is semifinite. □

4.23. Let tr be the canonical trace on $\mathscr{B}(\mathscr{H}), A$ a positive self-adjoint operator in $\mathscr{H}$ and $\varphi = tr_A$. For $\xi, \eta \in \mathscr{H}$, we shall use the notation $\xi \otimes \bar{\eta}$ for the operator

$$\mathscr{H} \ni \zeta \mapsto (\zeta|\eta)\xi \in \mathscr{H}.$$

It is clear that

$$(\xi \otimes \bar{\eta})^* = \eta \otimes \bar{\xi}, (\xi \otimes \bar{\eta})(\xi' \otimes \bar{\eta}') = (\eta|\xi')\xi \otimes \bar{\eta}' \tag{1}$$

and, for $x \in \mathscr{B}(\mathscr{H})$,

$$x(\xi \otimes \bar{\eta}) = x\xi \otimes \bar{\eta}, (\xi \otimes \bar{\eta})x = \xi \otimes \overline{x^*\eta}. \tag{2}$$

On the other hand, we have

$$\xi \otimes \bar{\eta} \in \mathfrak{N}_\varphi \Leftrightarrow \eta \in D(A^{1/2}), \tag{3}$$

$$\xi \otimes \bar{\eta} \in \mathfrak{M}_\varphi \Leftrightarrow \xi \otimes \bar{\eta} \in \mathfrak{N}_\varphi \cap \mathfrak{N}_\varphi^* \Leftrightarrow \xi, \eta \in D(A^{1/2}) \tag{4}$$

$$\xi, \eta \in D(A^{1/2}) \Rightarrow \varphi(\xi \otimes \bar{\eta}) = (A^{1/2}\xi|A^{1/2}\eta). \tag{5}$$

Indeed, let $x = \xi \otimes \bar{\eta}$. Then $x^*x = \|\xi\|^2(\eta \otimes \bar{\eta})$ and so, by (2),

$$A_\varepsilon^{1/2}x^*xA_\varepsilon^{1/2} = \|\xi\|^2(A_\varepsilon^{1/2}\eta \otimes \overline{A_\varepsilon^{1/2}\eta})$$

is a multiple of the orthogonal projection onto the linear subspace spanned by the vector $A^{1/2}\eta$. It follows that

$$\begin{aligned}\varphi(x^*x) &= \sup_{\varepsilon>0} tr(A_\varepsilon^{1/2}x^*xA_\varepsilon^{1/2}) = \|\xi\|^2 \sup_{\varepsilon>0} tr(A_\varepsilon^{1/2}\eta \otimes \overline{A_\varepsilon^{1/2}\eta}) \\ &= \|\xi\|^2 \sup_{\varepsilon>0} \|A_\varepsilon^{1/2}\eta\|^2 = \|\xi\|^2\|A^{1/2}\eta\|^2.\end{aligned}$$

From this we obtain (3) and then, by polarization, (4) and (5).

4.24. Consider again the canonical trace tr on $\mathscr{B}(\mathscr{H})$ and two positive self-adjoint operators A and B on $\mathscr{H}$. We recall (A.11) that the weak sum $A\hat{+}B$ is defined if and only if

$$D = D(A^{1/2}) \cap D(B^{1/2}) \text{ is dense in } \mathscr{H} \tag{1}$$

and, in this case, $A\hat{+}B$ is determined by

$$\|A^{1/2}\xi\|^2 + \|B^{1/2}\xi\|^2 = \|(A\hat{+}B)^{1/2}\xi\|^2, \xi \in D = D((A\hat{+}B)^{1/2}). \tag{2}$$

On the other hand, if the normal weight $tr_A + tr_B$ is semifinite, then (4.10) there exists a unique positive self-adjoint operator C on $\mathscr{H}$ such that $tr_A + tr_B = tr_C$. If A and B are bounded, then $C = A + B$ by 4.3.(1). In the general case, we have the following result:

Proposition. *The normal weight $tr_A + tr_B$ is semifinite if and only if the weak sum $A \hat{+} B$ is defined. In this case, we have*

$$tr_A + tr_B = tr_{A \hat{+} B}.$$

Proof. Assume first that the weight $tr_A + tr_B$ is semifinite and write $tr_A + tr_B = tr_C$ as above. Since $tr_A \leq tr_C, tr_B \leq tr_C$, we have (4.5) $A \leq C, B \leq C$. Thus, $D = D(A^{1/2}) \cap D(B^{1/2}) \supset D(C^{1/2})$ is dense in $\mathscr{H}$ that is $A \hat{+} B$ is defined. If $\xi \in D$, then (4.23.(5))

$$\|C^{1/2}\xi\|^2 = tr_C(\xi \otimes \xi) = tr_A(\xi \otimes \xi) + tr_B(\xi \otimes \xi) = \|A^{1/2}\xi\|^2 + \|B^{1/2}\xi\|^2,$$

hence $C = A \hat{+} B$.

Conversely, assume that $A \hat{+} B$ is defined and consider the increasing sequence $\{A_\varepsilon + B_\varepsilon\}_{\varepsilon > 0}$ of bounded positive operators. It is clear that

$$\lim_\varepsilon ((A_\varepsilon + B_\varepsilon)\xi | \xi) < +\infty \Leftrightarrow \xi \in D = D(A^{1/2}) \cap D(B^{1/2}).$$

Using (A.5) it follows that there exists a unique positive self-adjoint operator C in $\mathscr{H}$ such that $D(C^{1/2}) = D$ and $A_\varepsilon + B_\varepsilon \uparrow C$. By Proposition 4.6, we have $tr_{A_\varepsilon + B_\varepsilon} \uparrow tr_C$. Since $tr_{A_\varepsilon + B_\varepsilon} = tr_{A_\varepsilon} + tr_{B_\varepsilon}$, we conclude that $tr_A + tr_B = tr_C$ is semifinite. □

4.25. Notes. The construction of the weight φ_A, the definition of commutation for weights and almost all the results contained in this section are due to Pedersen and Takesaki (1973). Corollary 4.17 is due to Takesaki (1970) and Proposition 4.19 is due to Herman and Takesaki (1970). Several simplifications of the proofs given in Pedersen and Takesaki (1973) were made possible by the use of the Connes cocycle theorem.

For our exposition we have used Elliott (1975); Pedersen and Takesaki (1973); Takesaki (1969–1970) and Zsidó (1978).

Further results related to Proposition 4.21 are contained in Pedersen and Takesaki (1973) and Størmer (1971a).

5 The Converse of the Connes Theorem

In this section, we state and prove a converse to Theorem 3.1 and give some applications.

5.1. Let $\mathscr{M}$ be a W^*-algebra, G a locally compact group and $\sigma : G \to Aut(\mathscr{M})$ a continuous action of G on $\mathscr{M}$. We denote by $e \in G$ the neutral element of G.

A *σ-cocycle* (of degree 1) is an s^*-continuous function $w : G \to \mathscr{M}$ with the properties:

$$w(gh) = w(g)\sigma_g(w(h)), w(g^{-1}) = \sigma_g^{-1}(w(g)^*) \quad (g, h \in G).$$

In this case $w(g)$ are partial isometries and

$$w(g)w(g)^* = w(e), \qquad w(g)^*w(g) = \sigma_g(w(e)) \quad (g, h \in G).$$

The set of σ-cocycles is denoted by $Z_\sigma(G; \mathscr{M})$. A detailed study of $Z_\sigma(G; \mathscr{M})$ is contained in Section 20.

If φ is an n.s.f. weight on $\mathscr{M}$, the modular automorphism group $\{\sigma_t^\varphi\}_{t\in\mathbb{R}}$ is a continuous action $\sigma^\varphi : \mathbb{R} \to Aut(\mathscr{M})$ and, by the Connes theorem (3.1), for every normal semifinite weight ψ on $\mathscr{M}$ the Connes cocycle $[D\psi : D\varphi]$ is a σ^φ-cocycle.

Conversely, we have the following result:

Theorem (A. Connes). *Let φ be an n.s.f. weight on the W^*-algebra $\mathscr{M}$. For every σ^φ-cocycle $w \in Z_{\sigma^\varphi}(\mathbb{R}; \mathscr{M})$ there exists a unique normal semifinite weight ψ on $\mathscr{M}$ such that $[D\psi : D\varphi] = w$.*

The proof is contained in Sections 5.2–5.7.

We shall consider the W^*-algebra $\mathscr{M}$ realized as a von Neumann algebra $\mathscr{M} \subset \mathscr{B}(\mathscr{H})$. Also, we denote by $\mathscr{F}_\infty$ the discrete factor $\mathscr{B}(\mathscr{L}^2(\mathbb{R}))$ and identify the W^*-algebras $\mathscr{M} \bar{\otimes} \mathscr{F}_\infty$ and $Mat_\infty(\mathscr{M})$ as in ([L], 3.17). Thus, every element $x \in \mathscr{M} \bar{\otimes} \mathscr{F}_\infty$ is determined by a certain matrix $[x_{ij}]$ with elements in $\mathscr{M}$. Let $e_n \in \mathscr{M} \bar{\otimes} \mathscr{F}_\infty$ be the element corresponding to the matrix $[x_{ij}]$ with $x_{ij} = 0$ if $i \neq j$ or if $i > n$, and $x_{11} = x_{22} = \cdots = x_{nn} = 1$.

We shall successively construct weights Φ, Φ', Ψ, Ψ' on $\mathscr{M} \bar{\otimes} \mathscr{F}_\infty$ and weights ψ', ψ on $\mathscr{M}$.

5.2. We first define the n.s.f. weight Φ on $\mathscr{M} \bar{\otimes} \mathscr{F}_\infty$ by

$$\Phi(x) = \sum_k \varphi(x_{kk}) \quad (x = [x_{ij}] \in (\mathscr{M} \bar{\otimes} \mathscr{F}_\infty)^+).$$

Then we have

$$\sigma_t^\Phi = \sigma_t^\varphi \bar{\otimes} id \quad (t \in \mathbb{R}). \tag{1}$$

Indeed, using 2.21.(2), it is easy to check that the projection e_n belongs to the centralizer of Φ and using 2.22.(3) and Proposition 3.3 we obtain

$$\sigma_t^\Phi(x) = (\sigma_t^\varphi \bar{\otimes} id)(x) \quad (x \in e_n(\mathscr{M} \bar{\otimes} \mathscr{F}_\infty)e_n).$$

Since $e_n \uparrow 1$, this proves (1).

Actually, Φ is nothing but the tensor product $\varphi \bar{\otimes} tr$ where tr denotes the canonical trace on $\mathscr{F}_\infty$ (see 8.2).

5.3. Let $\{u_t\}_{t\in\mathbb{R}} \subset \mathscr{B}(\mathscr{L}^2(\mathbb{R}))$ be the *so*-continuous unitary group defined in Section 4.15. Using Stone's theorem ([L], 9.20) and the Pedersen–Takesaki construction (4.7), we obtain a new n.s.f. weight Φ' on $\mathscr{M} \bar{\otimes} \mathscr{F}_\infty$ such that

$$\sigma_t^{\Phi'} = Ad(1 \otimes u_t) \circ \sigma_t^\Phi = \sigma_t^\varphi \bar{\otimes} Ad(u_t) \quad (t \in \mathbb{R}).$$

5.4. The von Neumann algebra $\mathscr{M} \bar{\otimes} \mathscr{F}_\infty$ acts on the Hilbert space $\mathscr{H} \bar{\otimes} \mathscr{L}^2(\mathbb{R})$ which can be identified with $\mathscr{L}^2(\mathbb{R}, \mathscr{H})$ via the mapping

$$\mathscr{H} \bar{\otimes} \mathscr{L}^2(\mathbb{R}) \ni \xi \otimes f \mapsto \{t \mapsto f(t)\xi\} \in \mathscr{L}^2(\mathbb{R}, \mathscr{H}).$$

Using the given cocycle $w \in Z_{\sigma^\varphi}(\mathbb{R}; \mathcal{M})$, we define a partial isometry W on this Hilbert space by

$$(W\zeta)(t) = w(t)\zeta(t) \quad (\zeta \in \mathcal{L}^2(\mathbb{R}, \mathcal{H}), t \in \mathbb{R}). \tag{1}$$

It is easy to check that W commutes with the commutant $\mathcal{M}' \bar{\otimes} \mathbb{C}$ of $\mathcal{M} \bar{\otimes} \mathcal{F}_\infty$ and hence $W \in \mathcal{M} \bar{\otimes} \mathcal{F}_\infty$ by the von Neumann double commutant theorem. Also, we have

$$W\sigma_t^{\Phi'}(W^*) = w(t) \otimes 1 \quad (t \in \mathbb{R}). \tag{2}$$

Indeed, W is defined in (1) by the function $s \mapsto w(s)$ and, similarly, $\sigma_t^{\Phi'}(W^*)$ is defined by the function $s \mapsto \sigma_t^\varphi(w(s-t)^*)$, so that $W\sigma_t^{\Phi'}(W^*)$ is defined by the constant function $s \mapsto w(s)\sigma_t^\varphi(w(s-t)^*) = w(t)$ and hence it is equal to $w(t) \otimes 1$.

We define the normal semifinite weight $\Psi = \Phi'(W^* \cdot W)$ on $\mathcal{M} \bar{\otimes} \mathcal{F}_\infty$. Using (2) and 5.3.(1) and Corollary 3.7, we obtain

$$\sigma_t^\Psi = (Ad(w(t)) \circ \sigma_t^\varphi) \bar{\otimes} Ad(u_t) \quad (t \in \mathbb{R}). \tag{3}$$

5.5. Using the Pedersen–Takesaki construction again, we obtain from Ψ a new normal semifinite weight Ψ' on $\mathcal{M} \bar{\otimes} \mathcal{F}_\infty$ such that

$$\sigma_t^{\Psi'} = (Ad(w(t)) \circ \sigma_t^\varphi) \bar{\otimes} id \quad (t \in \mathbb{R}). \tag{1}$$

5.6. Let p be a minimal projection of $\mathcal{F}_\infty$. Using the mapping $x \mapsto x\bar{\otimes}p$ we identify $\mathcal{M}$ with $(1 \bar{\otimes} p)(\mathcal{M} \bar{\otimes} \mathcal{F}_\infty)(1 \bar{\otimes} p)$. It is clear that $1 \bar{\otimes} p$ belongs to the centralizer of Ψ', hence $\psi' = \Psi'_{(1 \bar{\otimes} p)}$ is a normal semifinite weight on $\mathcal{M} = (1 \bar{\otimes} p)(\mathcal{M} \bar{\otimes} \mathcal{F}_\infty)(1 \bar{\otimes} p)$. For $x \in \mathbf{s}(\psi')\mathcal{M}\mathbf{s}(\psi')$, we have (2.22.(3))

$$\sigma_t^{\psi'}(x) = \sigma_t^{\Psi'}(x \bar{\otimes} p) = w(t)\sigma_t^\varphi(x)w(t)^* \bar{\otimes} p = w(t)\sigma_t^\varphi(x)w(t)^*,$$

hence

$$\sigma_t^{\psi'} = Ad(w(t)) \circ \sigma_t^\varphi \quad (t \in \mathbb{R}). \tag{1}$$

5.7. Let $w'(t) = [D\psi' : D\varphi]_t$ and $a(t) = w'(t)^*w(t), (t \in \mathbb{R})$. Since $Ad(w(t)) \circ \sigma_t^\varphi = \sigma_t^{\psi'} = Ad(w'(t)) \circ \sigma_t^\varphi$, it follows that $\{a(t)\}_{t\in\mathbb{R}}$ is a *so*-continuous group of unitary operators in the center of the von Neumann algebra $\mathbf{s}(\psi')\mathcal{M}\mathbf{s}(\psi')$. By Stone's theorem, there exists a positive self-adjoint operator A in $\mathcal{H}$, affiliated to the center of $\mathcal{M}$, such that $a(t) = A^{it}, (t \in \mathbb{R})$.

Let $\psi = \psi'_A$. Then ψ is a normal semifinite weight on $\mathcal{M}$ and, for every $t \in \mathbb{R}$, we have $[D\psi : D\varphi]_t = [D\psi'_A : D\psi']_t[D\psi' : D\varphi]_t = a(t)w'(t) = w'(t)a(t) = w(t)$.

Thus, the existence assertion of Theorem 5.1 is proved. The uniqueness part has been already considered in Corollary 3.6.

5.8 Corollary (Pedersen & Takesaki, 1973). *For every normal weight φ on a W^*-algebra $\mathcal{M}$ there exists a family $\{\varphi_{\mathrm{i}}\}_{\mathrm{i}\in I}$ of normal positive forms on $\mathcal{M}$ such that*

$$\varphi(x) = \sum_{\mathrm{i}\in I} \varphi_{\mathrm{i}}(x) \quad (x \in \mathcal{M}^+).$$

Proof. We shall say that a weight φ has property S if φ is a sum of normal positive forms.

If $\{\omega_i\}_{i\in I}$ is a maximal family of normal positive forms on $\mathscr{M}$ with mutually orthogonal supports, then $\omega = \sum_{i\in I}\omega_i$ is an n.s.f. weight on $\mathscr{M}$ with property S.

Let ψ be an arbitrary normal semifinite weight on $\mathscr{M}$ and $w(t) = [D\psi : D\omega]_t (t \in \mathbb{R})$. By the proof of Theorem 5.1, the weight ψ is obtained from ω by a repeated application of the following operations:

$$\varphi \mapsto \Phi = \varphi \bar{\otimes}\, tr, (see\ 5.2); \tag{2}$$

$$\varphi \mapsto \varphi_e\ (e \in \mathscr{M}^{\varphi}); \tag{3}$$

$$\varphi \mapsto \varphi \circ \sigma, (\sigma \in Aut(\mathscr{M})); \tag{4}$$

$$\varphi \mapsto \varphi_A, (see\ 4.4). \tag{5}$$

The point is that all these operations preserve property S. This is obvious for (1)–(3) and also for (4) if A is bounded. In the general case, there exists a sequence $\{e_n\}$ of mutually orthogonal spectral projections of A such that Ae_n are bounded and $\varphi_A = \sum_n \varphi_{Ae_n}$. Thus every normal semifinite weight has property S.

Finally, let φ be an arbitrary normal weight and denote by e the unique projection in $\mathscr{M}$ such that $\overline{\mathfrak{N}_{\varphi}}^{w} = \mathscr{M}e$. Then φ_e is a normal semifinite weight on $e\mathscr{M}e$, hence φ_e has property S. On the other hand, φ_{1-e} takes only the values 0 and $+\infty$ so that it is obvious that φ_{1-e} has property S. Finally, it is easy to check that $\varphi = \varphi_e + \varphi_{1-e}$. □

5.9 Corollary. *Let φ be a normal weight on the von Neumann algebra $\mathscr{M} \subset \mathscr{B}(\mathscr{H})$. Then there exists a family $\{\xi_i\}_{i\in I} \subset \mathscr{H}$ such that*

$$\varphi(x) = \sum_{i\in I}(x\xi_i|\xi_i) \quad (x \in \mathscr{M}^+).$$

Proof. Every positive normal form on $\mathscr{M}$ has the stated property ([L], 8.17), so this result follows from the preceding corollary. □

In particular it follows that the function

$$\mathscr{M} \ni x \mapsto \varphi(x^*x)^{1/2} = \left(\sum_{i\in I}\|x\xi_i\|^2\right)^{1/2} \in [0, +\infty]$$

is subadditive and lower w-semicontinuous (see also 2.12).

5.10. Recall the notation (3.2) $\mathfrak{o}_{\mathscr{M}} : Aut(\mathscr{M}) \to Out(\mathscr{M})$ for the canonical quotient mapping and $\delta_{\mathscr{M}} : \mathbb{R} \to Out(\mathscr{M})$ for the modular homomorphism of the W^*-algebra $\mathscr{M}$.

Another consequence of Theorem 5.1 is the following

Corollary. *Let $\mathscr{M}$ be a W^*-algebra with separable predual and $\sigma : \mathbb{R} \to Aut(\mathscr{M})$ a continuous action of $\mathbb{R}$ on $\mathscr{M}$. Then σ is the modular automorphism group of some n.s.f. weight on $\mathscr{M}$ if and only if $\mathfrak{o}_{\mathscr{M}}(\sigma_t) = \delta_{\mathscr{M}}(t)$ for all $t \in \mathbb{R}$.*

We give here only a sketch of the proof (Connes (1973a)). Let φ be any n.s.f. weight on $\mathscr{M}$. Since $\mathfrak{o}_{\mathscr{M}}(\sigma_t) = \delta_{\mathscr{M}}(t) = \mathfrak{o}_{\mathscr{M}}(\sigma_t^{\varphi}), (t \in \mathbb{R})$, there is a mapping $v : \mathbb{R} \to U(\mathscr{M})$ such that $\sigma_t = Ad(v(t)) \circ \sigma_t^{\varphi}$, that is,

$$\sigma_t(x) = v(t)\sigma_t^{\varphi}(x)v(t)^* \quad (x \in \mathcal{M}, t \in \mathbb{R}).$$

Moreover, it is possible to choose a Borel mapping $v : \mathbb{R} \to U(\mathcal{M})$ with this property (Kallman, 1971; Moore, 1976). Then, for $s, t \in \mathbb{R}$ and $x \in \mathcal{M}$, we have

$$v(s)\sigma_s^{\varphi}(v(t))\sigma_{s+t}^{\varphi}(x)\sigma_s^{\varphi}(v(t)^*)v(s)^* = \sigma_{s+t}(x) = v(s+t)\sigma_{s+t}^{\varphi}(x)v(s+t)^*$$

so that we obtain a Borel function

$$a : \mathbb{R}^2 \ni (s,t) \mapsto a(s,t) = v(s)\sigma_s^{\varphi}(v(t))v(s+t)^* \in U(\mathcal{Z}(\mathcal{M})).$$

Since $a(s,t) \in \mathcal{Z}(\mathcal{M})$, we have $\sigma_r^{\varphi}(a(s,t)) = a(s,t)$ and an easy computation based on this remark gives

$$a(r,s)a(r+s,t) = a(s,t)a(r,s+t) \quad (r,s,t) \in \mathbb{R}.$$

Now, by arguments of Borel cohomology (Kadison, 1965; Moore, 1976) it follows that there exists a Borel mapping $b : \mathbb{R} \to U(\mathcal{Z}(\mathcal{M}))$ such that

$$a(s,t) = b(s+t)b(s)^{-1}b(t)^{-1} \quad (s,t \in \mathbb{R}).$$

Let $w(t) = b(t)v(t)(t \in \mathbb{R})$. Then $w : \mathbb{R} \to U(\mathcal{M})$ is a Borel mapping and $w(s+t) = w(s)\sigma_s^{\varphi}(w(t))(s,t \in \mathbb{R})$, so that (see Connes, 1973a) this mapping is continuous, that is, $w \in Z_{\sigma^{\varphi}}(\mathbb{R};\mathcal{M})$. Also, we have $\sigma_t = Ad(w(t)) \circ \sigma_t^{\varphi}(t \in \mathbb{R})$.

By Theorem 5.1, there exists an n.s.f. weight ψ on $\mathcal{M}$ such that $[D\psi : D\varphi] = w$ and it follows that $\sigma_t = \sigma_t^{\psi}(t \in \mathbb{R})$.

5.11. Notes. Theorem 5.1 and Corollary 5.10 are due to Connes (1973a). Corollary 5.8 is due to Pedersen and Takesaki (1973) and the proof given here is that of Elliott (1975).

For our exposition, we have used Connes (1973a) and Elliott (1975).

6 Equality and Majorization of Weights

In this section, we present an important criterion that insures the equality of two weights, study various order relations between weights, and give some examples.

6.1 Proposition. *Let φ be an n.s.f. weight on the W^*-algebra $\mathcal{M}$, $a \in \mathcal{M}^{\varphi}$, $a \geq 0$, and ψ a normal semifinite weight on $\mathcal{M}$. If there exists a *-subalgebra $\mathcal{B} \subset \mathfrak{N}_{\varphi}$, σ^{φ}-invariant and w-dense in $\mathcal{M}$ such that*

$$\psi(y^*y) = \varphi_a(y^*y) \quad (y \in \mathcal{B}),$$

then $\psi \leq \varphi_a$.

Proof. By the Kaplansky density theorem ([L], 3.10), there exists a net $\{b_j\}_{j\in J} \subset \mathscr{B}$ such that $0 \le b_j \xrightarrow{s} 1$ and $\sup_j \|b_j\| \le 1$.

Then

$$0 \le a_j = \sqrt{\frac{1}{\pi}} \int_{-\infty}^{+\infty} e^{-t^2} \sigma_t^{\varphi}(b_j)\, \mathrm{d}t \in \mathfrak{T}_{\varphi} \subset \mathscr{M}_{\infty}^{\varphi}$$

and $\sigma_{\alpha}^{\varphi}(a_j) \xrightarrow{s} 1$ for all $a \in \mathbb{C}$, by Proposition 2.16.

Let $y \in \mathscr{B}$. By Proposition 1.14, $\varphi_a(y^* \cdot y)$ and $\psi(y^* \cdot y)$ are normal positive forms on $\mathscr{M}$ and

$$\psi(y^* \cdot y) = \varphi_a(y^* \cdot y).$$

In particular, this equality is valid for

$$y_{s,t;j,k} = \sigma_s^{\varphi}(b_j) + \mathrm{i}^k \sigma_t^{\varphi}(b_j) \in \mathscr{B} \quad (s, t \in \mathbb{R}, j \in J, k = 0, 1, 2, 3).$$

Since for every $x \in \mathscr{M}$ we have

$$a_j x a_j = \frac{1}{\pi} \sum_{k=0}^{3} \mathrm{i}^k \int_{-\infty}^{+\infty} \int_{-\infty}^{+\infty} e^{-(s^2+t^2)} y_{s,t;j,k}^* x y_{s,t;j,k}\, \mathrm{d}s\, \mathrm{d}t,$$

by Corollary 2.10 we infer that

$$\psi(a_j x a_j) = \varphi_a(a_j x a_j) \quad (x \in \mathscr{M}^+).$$

Let $x \in \mathfrak{N}_{\varphi}$. Since ψ is normal, $a_j x^* x a_j \xrightarrow{w} x^* x$ and $\sigma_{-\mathrm{i}/2}^{\varphi}(a_j) \xrightarrow{s} 1$, using Proposition 2.14 we obtain

$$\begin{aligned}
\psi(x^* x) &\le \liminf_j \psi(a_j x^* x a_j) \le \liminf_j \varphi_a(a_j x^* x a_j) \\
&= \liminf_j \varphi(a^{1/2} a_j x^* x a_j a^{1/2}) = \liminf_j \|(x a_j a^{1/2})_{\varphi}\|_{\varphi}^2 \\
&= \liminf_j \| J_{\varphi} \pi_{\varphi}(a^{1/2} \sigma_{-\mathrm{i}/2}^{\varphi}(a_j)) J_{\varphi} x_{\varphi} \|_{\varphi}^2 \\
&= \| J_{\varphi} \pi_{\varphi}(a^{1/2}) J_{\varphi} x_{\varphi} \|_{\varphi}^2 = \|(x a^{1/2})_{\varphi}\|_{\varphi}^2 = \varphi(a^{1/2} x^* x a^{1/2}) \\
&= \varphi_a(x^* x).
\end{aligned}$$

In particular, it follows that $\mathbf{s}(\psi) \le \mathbf{s}(a)$.

Consider now $x \in \mathfrak{N}_{\varphi_a}$, that is $x a^{1/2} \in \mathfrak{N}_{\varphi}$. There exists a sequence $\{e_n\} \subset \mathscr{M}^{\varphi}$, of spectral projections of a such that $e_n \uparrow 1$ and $a e_n \ge n^{-1} e_{\alpha}, (n \in \mathbb{N})$. Since $a \in \mathscr{M}^{\varphi}$ and $a e_n$ is invertible in $e_n \mathscr{M}^{\varphi} e_n$, we see again by Proposition 2.14 that $x e_n \in \mathfrak{N}_{\varphi}$. Consequently, we have

$$\begin{aligned}
\psi(x^* x) = \psi(\mathbf{s}(a) x^* x \mathbf{s}(a)) &\le \liminf_n \psi(e_n x^* x e_n) \\
&\le \liminf_n \varphi_a(e_n x^* x e_n) \le \varphi_a(x^* x);
\end{aligned}$$

the last inequality is obtained by applying Proposition 2.14 once more.

Hence $\psi(z) \le \varphi_a(z)$ for every $z \in \mathfrak{M}_{\varphi_a} \cap \mathscr{M}^+$, that is, $\psi \le \varphi_a$. □

6.2 Theorem. *Let φ be an n.s.f weight on the W^*-algebra $\mathscr{M}$, $a \in \mathscr{M}^\varphi, a \geq 0$, and ψ a normal semifinite weight on $\mathscr{M}$. If ψ commutes with φ and there exists a $*$-subalgebra $\mathscr{B} \subset \mathfrak{N}_\varphi, \sigma^\varphi$-invariant and w-dense in $\mathscr{M}$ such that*

$$\psi(y^*y) = \varphi_a(y^*y) \quad (y \in \mathscr{B}),$$

then $\psi = \varphi_a$.

Proof. Since ψ commutes with φ, there exists a positive self-adjoint operator A affiliated to $\mathscr{M}$ such that $\psi = \varphi_A$ (4.10).

By Proposition 6.1, it follows that $\varphi_A = \psi \leq \varphi_a$ and then, using Proposition 4.5, we deduce that $A \leq a$, so that $A \in \mathscr{M}^\varphi$ is bounded. Again by Proposition 6.1, with φ_a instead of ψ and A instead of a, we get $\varphi_a \leq \varphi_A = \psi$. Hence $\psi = \varphi_a$. □

The particular case $a = 1$ of the preceding theorem is known as "the Pedersen–Takesaki theorem on the equality of weights."

6.3. By the Radon–Nikodym type theorem of Sakai ([L], 5.21), if φ, ψ are normal positive forms on the W^*-algebra $\mathscr{M}$ and $\psi \leq \varphi$, there exists $a \in \mathscr{M}$ (actually, $0 \leq a \leq 1$) such that $\psi = \varphi(a \cdot a^*)$. The extension of this result to weights is contained in Corollary 3.13.

We remark that Proposition 2.14 gives necessary and sufficient conditions for the inequality $\varphi(a \cdot a^*) \leq \varphi$ to hold.

6.4. We now consider another form of Radon–Nikodym type theorems, which has been pointed out by S. Sakai (see [L], C.5.5).

We begin with the case of normal positive forms. Let $\mathscr{M}$ be a W^*-algebra, $\varphi, \psi \in \mathscr{M}_*^+$ such that $\psi \leq \varphi$, and $\lambda \in \mathbb{C}$ with $\lambda + \bar{\lambda} = 1$. The set $\mathscr{X} = \{\varphi(\lambda a \cdot + \bar{\lambda} \cdot a); a \in \mathscr{M}, a = a^*, \|a\| \leq 1\} \subset \mathscr{M}^h$ is convex and $\sigma(\mathscr{M}_*, \mathscr{M})$-compact. If $\psi \notin \mathscr{X}$, then by the Hahn–Banach theorem there exists $b \in \mathscr{M}, b = b^*$ and $t \in \mathbb{R}$ such that $\psi(b) > t$ while $\varphi(\lambda ab + \bar{\lambda}ba) \leq t$ for every $a\mathscr{M}, a = a^*, \|a\| \leq 1$. Let $b = v|b|$ be the polar decomposition. Then $v \in \mathscr{M}, v = v^*, \|v\| \leq 1$ and $|b| = vb = bv$, hence $t < \psi(b) \leq \psi(|b|) \leq \varphi(|b|) = \varphi(\lambda vb + \bar{\lambda}bv) \leq t$, a contradiction. Thus, there exists $a_1 \in \mathscr{M}$, with $a_1 = a_1^*, \|a_1\| \leq 1$, such that $\psi(x) = \varphi(\lambda a_1 x + \bar{\lambda} x a_1), (x \in \mathscr{M})$. Since $\varphi \geq 0$, it follows that $\psi(a_1^-) = 0$, hence $a = a_1^+ \in \mathscr{M}, 0 \leq a \leq 1$ and

$$\psi(x) = \varphi(\lambda ax + \bar{\lambda}xa)(x \in \mathscr{M}). \tag{1}$$

Proposition. *Let φ be an n.s.f. weight on the W^*-algebra $\mathscr{M}$ and $\psi \in \mathscr{M}_*^+, \psi \leq \varphi$. There exists a unique $a \in \mathfrak{M}_\varphi, 0 \leq a \leq 1$, such that*

$$\psi(x) = \varphi(ax + xa)/2 \quad (x \in \mathfrak{M}_\varphi).$$

Proof. By Corollary 2.4, there exists a vector $\eta \in \mathscr{H}_\varphi$ and an operator $T' \in \pi_\varphi(\mathscr{M})', 0 \leq T' \leq 1$, such that

$$\psi(x) = (\pi_\varphi(x)\eta|\eta)_\varphi \quad (x \in \mathscr{M}_\varphi), \tag{2}$$

$$T'x_\varphi = \pi_\varphi(x)\eta \qquad (x \in \mathfrak{A}_\varphi). \tag{3}$$

Let $\zeta = T'\eta$ and $\xi = 2(\Delta_\varphi + 1)^{-1}\zeta$. From (3), it follows that $\eta \in \mathfrak{A}'_\varphi, S^*_\varphi\eta = \eta$ and $R_\eta = T'$. Hence $\zeta \in \mathfrak{A}'_\varphi, S^*_\varphi\zeta = \zeta$ and $R_\zeta = T'^2$. Using ([L], 10.11), we infer that $\xi \in \mathfrak{A}_\varphi$, and, since $S^*_\varphi\zeta = \zeta$ and $J_\varphi\Delta_\varphi J_\varphi^{-1} = \Delta_\varphi^{-1}$, we have $S_\varphi\xi = \xi$. Using ([L], Cor. 9,23), we get

$$\xi = S_\varphi\xi = 2J_\varphi\Delta_\varphi^{1/2}(\Delta_\varphi + 1)^{-1}\zeta = \int_{-\infty}^{+\infty} 2(e^{\pi t} + e^{-\pi t})^{-1}\Delta_\varphi^{\mathrm{i}t} J_\varphi\zeta \, \mathrm{d}t.$$

Since (2.12.(4))

$$L_{\Delta_\varphi^{\mathrm{i}t} J_\varphi\zeta} = \Delta_\varphi^{\mathrm{i}t} J_\varphi R_\zeta J_\varphi \Delta_\varphi^{-\mathrm{i}t} = \Delta_\varphi^{\mathrm{i}t} J_\varphi T'^2 J_\varphi \Delta_\varphi^{-\mathrm{i}t},$$

it follows that

$$L_\xi = \int_{-\infty}^{+\infty} 2(e^{\pi t} + e^{-\pi t})^{-1}\Delta_\varphi^{\mathrm{i}t} J_\varphi T'^2 J_\varphi \Delta_\varphi^{-\mathrm{i}t} \, \mathrm{d}t.$$

Consequently, $a = L_\xi \in \mathscr{M}, 0 \le a \le 1$. Also, since $J_\varphi T' J_\varphi = J_\varphi R_\eta J_\varphi = L_{J_\varphi\eta}$, we get (2.11) $\varphi(a) = \varphi(J_\varphi T'^2 J_\varphi) = \| J_\varphi\eta\|_\varphi^2 < +\infty$, hence $a \in \mathfrak{M}_\varphi$. For $x \in \mathfrak{M}_\varphi$, we have

$$\begin{aligned} 2\psi(x) &= 2(\pi_\varphi(x)\eta|\eta)_\varphi = 2(T'x_\varphi|\eta)_\varphi = 2(x_\varphi|T'\eta)_\varphi \\ &= 2(x_\varphi|\zeta)_\varphi = (x_\varphi|(\Delta_\varphi + 1)\xi)_\varphi = (x_\varphi|\xi)_\varphi + (x_\varphi|S^*_\varphi S_\varphi\xi)_\varphi \\ &= (x_\varphi|\xi)_\varphi + (\xi|S_\varphi x_\varphi)_\varphi = (x_\varphi|a_\varphi)_\varphi + (a_\varphi|(x^*)_\varphi)_\varphi \\ &= \varphi(ax + xa). \end{aligned}$$

To prove the uniqueness assertion, assume that $a \in \mathfrak{M}_\varphi, 0 \le a \le 1$, and $\varphi(ax + xa) = 0$ for every $x \in \mathfrak{M}_\varphi$. Then $\xi = a_\varphi \in \mathfrak{A}_\varphi, S_\varphi\xi = \xi$, and for every $\eta \in \mathfrak{T}_\varphi$ we have

$$(\eta| -\xi)_\varphi = (\xi|S_\varphi\eta)_\varphi = (S_\varphi\xi|S_\varphi\eta)_\varphi = (\Delta_\varphi\eta|\xi)_\varphi.$$

Since $\Delta_\varphi = \Delta_\varphi^* = [\Delta_\varphi|\mathfrak{T}_\varphi]^*$, it follows that $\xi \in D(\Delta_\varphi)$ and $\Delta_\varphi\xi = -\xi$, which is possible only if $\xi = 0$, as Δ_φ is positive. Hence $a = 0$. □

6.5. In order to treat the general case of two weights, we recall ([L], 10.22) that for any n.s.f. weight φ on the W^*-algebra $\mathscr{M}$, the set

$$\mathfrak{G}_\varphi = \mathfrak{T}_\varphi \cap \mathcal{S}(\Delta_\varphi) \cap \mathcal{S}(\Delta_\varphi^{-1})$$

is a left Hilbert subalgebra of $\mathfrak{A}_\varphi$, equivalent with $\mathfrak{A}_\varphi$, and $J_\varphi\mathfrak{G}_\varphi = \mathfrak{G}_\varphi, \Delta_\varphi^\alpha\mathfrak{G}_\varphi = \mathfrak{G}_\varphi$ and $\overline{\Delta_\varphi^\alpha|\mathfrak{G}_\varphi} = \Delta_\varphi^\alpha$. Note that

$$\mathfrak{G}_\varphi \subset D(\ln \Delta_\varphi).$$

Also, the following approximation result holds:

$$\begin{array}{l} \textit{for every } \xi \in D(\Delta_\varphi^{1/2}) \cap D(\Delta_\varphi^{-1/2}) \textit{ there exists a} \\ \textit{sequence } \{\xi_n\} \subset \mathfrak{G}_\varphi \textit{ such that } \Delta_\varphi^\alpha \xi_n \to \Delta_\varphi^\alpha \xi \\ \textit{for each } \alpha \in \mathbb{C} \textit{ with } -1/2 \le \operatorname{Re} \alpha \le 1/2 \end{array} \tag{1}$$

Indeed, since $\Delta_\varphi^{-1/2}\mathfrak{G}_\varphi = \mathfrak{G}_\varphi$ is dense in $\mathscr{H}_\varphi$, there exists a sequence $\{\zeta_n\} \subset \mathfrak{G}_\varphi$ such $\Delta_\varphi^{-1/2}\zeta_n \to \Delta_\varphi^{1/2}\xi + \Delta_\varphi^{-1/2}\xi$. If $-1/2 \le \operatorname{Re}\alpha \le 1/2$, then the operator $\Delta_\varphi^{\alpha+(1/2)}(1+\Delta_\varphi)^{-1}$ is bounded and sends the vector $\Delta_\varphi^{1/2}\xi + \Delta_\varphi^{-1/2}\xi$ into the vector $\Delta_\varphi^\alpha\xi$. Thus (1) follows with $\xi_n = (1+\Delta_\varphi)^{-1}\zeta_n$.

Proposition. *Let φ be an n.s.f. weight on the W^*-algebra $\mathscr{M}$ and ψ any weight on $\mathscr{M}$ with $\psi \le \varphi$. There exists a unique element $a \in \mathscr{M}, 0 \le a \le 1$, such that $\mathfrak{G}_\varphi a \subset \mathfrak{N}_\varphi$ and*

$$x \in \mathfrak{G}_\varphi^2 \Rightarrow ax + xa \in \mathfrak{M}_\varphi \text{ and } \psi(x) = \varphi(ax+xa)/2.$$

Proof. By Proposition 2.3, there exists an operator $T' \in \pi_\varphi(\mathscr{M})', 0 \le T' \le 1$, such that

$$\psi(x^*x)(T'x_\varphi|T'x_\varphi)_\varphi \quad (x \in \mathfrak{N}_\varphi).$$

Since $J_\varphi T'^2 J_\varphi \in \pi_\varphi(\mathscr{M})$, there exists $b \in \mathscr{M}, 0 \le b \le 1$, such that $\pi_\varphi(b) = J_\varphi T'^2 J_\varphi$. Define an element $a \in \mathscr{M}, 0 \le a \le 1$, by

$$a = \int_{-\infty}^{+\infty} 2(e^{\pi t} + e^{-\pi t})^{-1} \Delta_\varphi^{\mathrm{i}t} \Delta_\varphi^{-\mathrm{i}t}\, \mathrm{d}t.$$

Let $x \in \mathfrak{G}_\varphi$. Then $J_\varphi x_\varphi \in \mathfrak{G}_\varphi \subset D(\Delta_\varphi^{-1/2}) \cap D(\ln \Delta_\varphi)$. Using Proposition A.13, it follows that

$$(ax^*)_\varphi = \pi_\varphi(a)\Delta_\varphi^{-1/2} J_\varphi x_\varphi \in D(\Delta_\varphi^{1/2}) = D(S_\varphi),$$

hence (2.12.(1)) $xa \in \mathfrak{N}_\varphi$ and

$$\begin{aligned} (xa)_\varphi &= S_\varphi \pi_\varphi(a) S_\varphi x_\varphi = J_\varphi \Delta_\varphi^{1/2} \pi_\varphi(a) \Delta_\varphi^{-1/2} J_\varphi x_\varphi \\ &= J_\varphi \pi_\varphi(b) J_\varphi x_\varphi - 2\mathrm{i}\, PV \int_{-\infty}^{+\infty} (e^{\pi t} + e^{-\pi t})^{-1} J_\varphi \Delta_\varphi^{\mathrm{i}t} \pi_\varphi(b) \Delta_\varphi^{-\mathrm{i}t} J_\varphi x_\varphi \, \mathrm{d}t. \end{aligned}$$

Since $\pi_\varphi(b) \ge 0$, we get

$$\operatorname{Re}((xa)_\varphi | x_\varphi)_\varphi = (J_\varphi \pi_\varphi(b) J_\varphi x_\varphi | x_\varphi)_\varphi = (T'^2 x_\varphi | x_\varphi)_\varphi,$$

that is $\varphi(ax^*x + x^*xa)/2 = \psi(x^*x)$. The desired conclusion follows now by polarization.

To prove the uniqueness part we consider an element $a = a^* \in \mathscr{M}$, such that $\mathfrak{G}_\varphi a \subset \mathfrak{N}_\varphi$ and $\varphi(ay^*x + y^*xa) = 0$ for every $x, y \in \mathfrak{G}_\varphi$.

Let $X = J_\varphi \pi_\varphi(a) J_\varphi$. Then, for $x, y \in \mathfrak{G}_\varphi$, we get

$$\begin{aligned}0 &= (x_\varphi | (ya)_\varphi)_\varphi + ((xa)_\varphi | y_\varphi)_\varphi \\ &= (x_\varphi | S_\varphi \pi_\varphi(a) S_\varphi y_\varphi)_\varphi + (S_\varphi \pi_\varphi(a) S_\varphi x_\varphi | y_\varphi)_\varphi \\ &= (X\Delta_\varphi^{-1/2} x_\varphi | \Delta_\varphi^{1/2} y_\varphi)_\varphi + (X\Delta_\varphi^{1/2} x_\varphi | \Delta_\varphi^{-1/2} y_\varphi)_\varphi.\end{aligned}$$

Using the approximation result (1), for every $\xi, \eta \in D(\Delta_\varphi^{1/2}) \cap D(\Delta_\varphi^{-1/2})$ we obtain $(X\Delta_\varphi^{-1/2}\xi | \Delta_\varphi^{1/2}\eta)_\varphi + (X\Delta_\varphi^{1/2}\xi | \Delta_\varphi^{-1/2}\eta)_\varphi = 0$ and by ([L], 9.23) it follows that $X = 0$, that is, $a = 0$. □

We now consider some examples.

6.6. Let K be an infinite dimensional separable Hilbert space and consider the von Neumann algebras $\mathscr{M} = \mathscr{B}(\mathscr{K}) \otimes 1_\mathscr{K} \otimes 1_\mathscr{K}$ and $\mathscr{N} = \mathscr{B}(\mathscr{K}) \bar{\otimes} \mathscr{B}(\mathscr{K}) \otimes 1_\mathscr{K}$ acting on the Hilbert space $\mathscr{H} = \mathscr{K} \bar{\otimes} \mathscr{K} \bar{\otimes} \mathscr{K}$. By ([L], 8.15) we know that each of the von Neumann algebras $\mathscr{M}, \mathscr{M}', \mathscr{N}, \mathscr{N}'$ has a separating vector. By the Dixmier–Marechal theorem (Dixmier & Marechal, 1971; [L], C.3.5), the set of separating vectors of any von Neumann algebra is either empty or is a dense G_δ set. We conclude that there exists a vector in $\mathscr{H}$ which is simultaneously cyclic and separating for both $\mathscr{M}$ and $\mathscr{N}$.

Thus, there exist von Neumann algebras $\mathscr{M} \subset \mathscr{N} \subset \mathscr{B}(\mathscr{H})$, $\mathscr{M} \neq \mathscr{N}$, having a common cyclic and separating vector $\xi_0 \in \mathscr{H}$ [*)].

Let $A = \Delta(\mathscr{M}, \xi_0)$ and $B = \Delta(\mathscr{N}, \xi_0)$ be the associated modular operators ([L], 10.6). Then A and B are nonsingular positive self-adjoint operators in $\mathscr{H}$ such that

$$D(A^{1/2}) \subset D(B^{1/2}) \text{ and } \|B^{1/2}\xi\| = \|A^{1/2}\xi\| \text{ for all } \xi \in D(A^{1/2}) \tag{1}$$

but

$$B \neq A. \tag{2}$$

Indeed, for the corresponding S operators we have $S(\mathscr{M}, \xi_0) \subset S(\mathscr{N}, \xi_0)$ and, consequently, for $\xi \in D(A^{1/2})$ we get

$$\begin{aligned}\|A^{1/2}\xi\| &= \| J(\mathscr{M}, \xi_0) S(\mathscr{M}, \xi_0)\xi\| = \|S(\mathscr{M}, \xi_0)\xi)\| = \|S(\mathscr{N}, \xi_0)\xi\| \\ &= \| J(\mathscr{N}, \xi_0) S(\mathscr{N}, \xi_0)\xi)\| = \|B^{1/2}\xi\|,\end{aligned}$$

hence assertion (1) holds. Also (2) holds, for if $A = B$ then $J(\mathscr{M}, \xi_0) = J(\mathscr{N}, \xi_0) = J$ and, since $\mathscr{M} \subset \mathscr{N}$, we obtain $\mathscr{N}' \subset \mathscr{M}'$ and also $\mathscr{M}' = J\mathscr{M}J \subset J\mathscr{N}J = \mathscr{N}'$, that is $\mathscr{M} = \mathscr{N}$, a contradiction.

Consider now the canonical trace tr on $\mathscr{B}(\mathscr{H})$ and the n.s.f. weights $\varphi = tr_A, \psi = tr_B$ on $\mathscr{B}(\mathscr{H})$ (cf. 4.4). From assertion (1), it follows that $B \leq A$ and hence (4.5)

$$\psi \leq \varphi. \tag{3}$$

while from assertion (2) we obtain (4.5, 4.8)

$$\psi \neq \varphi. \tag{4}$$

[*)] Using the T-theorem ([L]; C.6.1), it is easy to see that this situation can occur only if $\mathscr{M}, \mathscr{N}, \mathscr{M}', \mathscr{N}'$ are all infinite.

Also, we have

$$\psi(x^*x) = \varphi(x^*x) \text{ for every } x \in \mathfrak{N}_\varphi. \tag{5}$$

Indeed, from (1) it follows that the operator $B^{1/2}A^{-1/2}$ is defined and isometric on $D(A^{-1/2})$ so, by ([L], Proposition 9.24), the function $\alpha \mapsto F(\alpha) = \overline{B^\alpha A^{-\alpha}}$ satisfies condition 3.14.(ii).

This example shows that *the equivalent conditions of Corollary 3.14 do not insure that* $\psi = \varphi$, *and that in Theorem 6.2 we cannot omit the condition that* ψ *commutes with* φ. *Also, the example shows that Proposition 6.1 is no longer valid if the operator a is not assumed to be bounded.*

6.7. Consider again the canonical trace tr on $\mathscr{B}(\mathscr{H})$, a nonsingular positive self-adjoint operator A in $\mathscr{H}$ and the n.s.f. weight $\varphi = tr_A$ on $\mathscr{B}(\mathscr{H})$.

Let $a \in \mathfrak{N}_\varphi$ and $x_0 = \xi \otimes \bar{\eta}$ with $\xi \in D(A^{1/2})$ and $\eta \notin D(A^{1/2})$. Then (4.23) $x_0 \in \mathfrak{N}_\varphi^*$ hence $x_0 a \in \mathfrak{M}_\varphi$, but $ax_0 = a\xi \otimes \bar{\eta} \notin \mathfrak{M}_\varphi$ hence $ax_0 + x_0 a \notin \mathfrak{M}_\varphi$.

It follows that *the assertion of Proposition 6.4 cannot be extended to all* $x \in \mathscr{M}$.

In what follows we show that *nor can the assertion of Proposition 6.5 be extended to all* $x \in \mathfrak{M}_\varphi$.

By (A.14) there exist be $b \in \mathscr{B}(\mathscr{H}), 0 \leq b \leq 1$, a nonsingular positive self-adjoint operator A in $\mathscr{H}$ and a vector $\zeta \in D(A^{-1/2})$ such that, putting

$$a = \int_{-\infty}^{+\infty} 2(e^{\pi t} + e^{-\pi t})^{-1} A^{it} b A^{-it}\, dt \in \mathscr{B}(\mathscr{H}), \tag{1}$$

we have $0 \leq a \leq 1$ and $aA^{-1/2}\zeta \notin D(A^{1/2})$. Then, for $\eta = A^{-1/2}\zeta$, we have

$$\eta \in D(A^{1/2}), a\eta \notin D(A^{1/2}).$$

On the other hand, for an arbitrary vector $\xi_0 \in D(A^{-1/2}) \cap D(\ln A)$, the vector $\xi = A^{-1/2}\xi_0$ has the properties (A.13):

$$\xi \in D(A^{1/2}), a\xi \in D(A^{1/2}).$$

Consider the n.s.f. weight $\varphi = tr_A$ on $\mathscr{B}(\mathscr{H})$ and the operator $x_0 = \xi \otimes \bar{\eta}$. Then (4.23) $x_0 \in \mathfrak{M}_\varphi$ and $ax_0 = a\xi \otimes \bar{\eta} \in \mathfrak{M}_\varphi$ but $x_0 a = \xi \otimes \overline{a\eta} \notin \mathfrak{M}_\varphi$, hence $ax_0 + x_0 a \notin \mathfrak{M}_\varphi$.

Using the same arguments as in ([L], 10.16.(1), (2), (3)) one shows that a weight ψ on $\mathscr{B}(\mathscr{H})$ is defined by

$$\psi(y) = \begin{cases} \| J_\varphi \pi_\varphi(b^{1/2}) J_\varphi(y^{1/2})_\varphi \|_\varphi^2, & \text{if } y \in \mathfrak{M}_\varphi \\ +\infty, & \text{in the contrary case} \end{cases} (y \in \mathscr{B}(\mathscr{H})^+).$$

Since $b \leq 1$, we have $\psi \leq \varphi$. The element a defined by (1) is exactly the element given by the proof of Proposition 6.5, that is, the unique element $a \in \mathscr{M}, 0 \leq a \leq 1$, such that $\mathfrak{G}_\varphi a \subset \mathfrak{N}_\varphi$ and

$$x \in \mathfrak{G}_\varphi^2 \Rightarrow ax + xa \in \mathfrak{M}_\varphi \text{ and } \psi(x) = \varphi(ax + xa)/2,$$

but $x_0 \in \mathfrak{M}_\varphi$ and $ax_0 + x_0 a \notin \mathfrak{M}_\varphi$.

6.8. We now give an example connected with the Radon–Nikodym type theorem of Sakai ([L], 5.21), namely we show that *on* $\mathscr{B}(\mathscr{H})$ *there exist an n.s.f. weight* φ *and a normal positive form* $\psi \leq \varphi$ *such that* $\psi \neq \varphi(a \cdot a)$ *for all* $a \in \mathscr{M}^+$.

Indeed, let $\{\xi_n\}_{n\in\mathbb{N}}$ be an orthonormal basis of the separable infinite dimensional Hilbert space $\mathscr{H}$. Let A be the positive self-adjoint operator in $\mathscr{H}$, diagonalizable with respect to $\{\xi_n\}_{n\in\mathbb{N}}$, such that

$$A\xi_n = n\xi_n \quad (n \in \mathbb{N}),$$

and let $e \in \mathscr{B}(\mathscr{H})$ be the orthogonal projection onto the linear subspace spanned by the vector $\xi = \sum_n n^{-1}\xi_n \in \mathscr{H}$.

Define $\varphi = tr_A$ and $\psi = tr_e$. Then φ is an n.s.f. weight on $\mathscr{B}(\mathscr{H})$, ψ is a normal positive form on $\mathscr{B}(\mathscr{H})$, and $\psi \leq \varphi$ as $e \leq 1 \leq A$.

Assume that there exists $a \in \mathscr{B}(\mathscr{H})^+$ such that $\psi = \varphi(a \cdot a)$.

Let $\eta \in \mathscr{H}$, $\|\eta\| = 1$ and $f = \eta \otimes \bar{\eta}$. Then

$$\begin{aligned}\lim_{\varepsilon\to 0}(aA_\varepsilon a\eta|\eta) &= \lim_{\varepsilon\to 0} tr(faA_\varepsilon af) = \lim_{\varepsilon\to 0} tr(A_\varepsilon^{1/2}afaA_\varepsilon^{1/2}) = \varphi(afa)\\ &= \psi(f) = tr(efe) = tr(fef) = (e\eta|\eta),\end{aligned}$$

hence $aA_\varepsilon a \xrightarrow{wo} e$. Since $A_\varepsilon \uparrow A$, it follows that $aA_\varepsilon a \uparrow e$. As $A \geq 1$, we have $A_\varepsilon = A(1+\varepsilon A)^{-1} \geq (1+\varepsilon)^{-1}$, hence $a^2 \leq e$. Since e is a minimal projection, we conclude $a = \lambda e$ with $0 \leq \lambda \leq 1$, so that

$$\|\xi\|^2 = (e\xi|\xi) \geq (aA_\varepsilon a\xi|\xi) = \lambda^2(A_\varepsilon\xi|\xi) = \lambda^2\sum_n n^{-1}(1+\varepsilon n)^{-1}$$

tends to $+\infty$ when $\varepsilon \to 0$, a contradiction.

6.9. Let us denote by $W_{nsf}(\mathscr{M})$ the set of all n.s.f. weights on the W^*-algebra $\mathscr{M}$.

For $\lambda \in [0, +\infty]$ and $\varphi_1, \varphi_2 \in W_{nsf}(\mathscr{M})$ write

$$\varphi_2 \leq \varphi_1(\lambda)$$

if there exists an $\mathscr{M}$-valued function F, defined and w-continuous on the strip $\{\alpha \in \mathbb{C}; 0 \leq \operatorname{Re}\alpha \leq \lambda\}$, analytic in the interior of this strip, such that $F(it) = [D\varphi_2 : D\varphi_1]$ for every $t \in \mathbb{R}$ and $\|F(\alpha)\| \leq 1$ for every $\alpha \in \mathbb{C}, 0 \leq \operatorname{Re}\alpha \leq \lambda$.

Proposition. *Let* $\mathscr{M}$ *be a* W^**-algebra. For each* $\lambda \in [0, +\infty]$, *the relation "*$\varphi_2 \leq \varphi_1(\lambda)$*" is an order relation on* $W_{nsf}(\mathscr{M})$.

Proof. Let $S(\lambda) = \{\alpha \in \mathbb{C}; 0 \leq \operatorname{Re}\alpha \leq \lambda\}$ and $D(\lambda) = \{\alpha \in \mathbb{C}; -\lambda \leq \operatorname{Re}\alpha \leq \lambda\}$.

It is obvious that $\varphi \leq \varphi(\lambda)$ for all $\varphi \in W_{nsf}(\mathscr{M})$.

Let $\varphi_1, \varphi_2, \varphi_3 \in W_{nsf}(\mathscr{M})$ be such that $\varphi_3 \leq \varphi_2(\lambda), \varphi_2 \leq \varphi_1(\lambda)$ and denote by $F_{32} : S(\lambda) \to \mathscr{M}, F_{21} : S(\lambda) \to \mathscr{M}$ the corresponding functions. Using the chain rule (3.5), it is easy to check that the properties of the function $F_{31} : S(\lambda) \to \mathscr{M}$ defined by $F_{31}(\alpha) = F_{32}(\alpha)F_{21}(\alpha), (\alpha \in S(\lambda))$, yield the relation $\varphi_3 \leq \varphi_1(\lambda)$.

Let $\varphi_1, \varphi_2 \in W_{nsf}(\mathscr{M})$ be such that $\varphi_2 \leq \varphi_1(\lambda)$ and $\varphi_1 \leq \varphi_2(\lambda)$, and denote by $F_{21} : S(\lambda) \to \mathscr{M}, F_{12} : S(\lambda) \to \mathscr{M}$ the corresponding functions. Let ψ be an arbitrary normal state on $\mathscr{M}$ and $f_{21} = \psi \circ F_{21}, f_{12} = \psi \circ F_{12}$. By Corollary 3.4, we have $F_{21}(it) = F_{12}(it)^*$, hence $f_{21}(it) = \overline{f_{12}(it)} = \overline{f_{12}(-\overline{it})}$

for all $t \in \mathbb{R}$. It follows that the function f_{21} can be extended to an analytic function, still denoted by f_{21}, defined on $D(\lambda)$, and such that $|f_{21}(\alpha)| \leq 1 (\alpha \in D(\lambda))$. Since $f_{21}(0) = 1$, by the maximum modulus principle we get $f_{21}(\alpha) = 1 (\alpha \in D(\lambda))$. In particular, $\psi(F_{21}(it)) = f_{21}(it) = 1$ for all $t \in \mathbb{R}$. Since ψ was an arbitrary normal state on $\mathcal{M}$ we infer that $[D\varphi_2 : D\varphi_1]_t = F_{21}(it) = 1 (t \in \mathbb{R})$, that is, $\varphi_2 = \varphi_1$ by Corollary 3.6. □

By Corollary 3.13, the ordering $\varphi_2 \leq \varphi_1(1/2)$ is the usual pointwise ordering $\varphi_2(x) \leq \varphi_1(x)$, $(x \in \mathcal{M})^+$. On the other hand, the ordering $\varphi_2 \leq \varphi_1(1/4)$ for faithful normal positive forms is studied in Proposition 3.18.

It is clear that if $\varphi_2 \leq \varphi_1(\lambda)$ and $\mu < \lambda$, then $\varphi_2 \leq \varphi_1(\mu)$.

In particular,

$$\varphi_2 \leq \varphi_1(\infty) \Rightarrow \varphi_2 \leq \varphi_1. \tag{1}$$

6.10. With the help of the ordering corresponding to $\lambda = \infty$, we can define a metric d on the set $W_{nsf}(\mathcal{M})$ by putting, for $\varphi_1, \varphi_2 \in W_{nsf}(\mathcal{M})$,

$$d(\varphi_1, \varphi_2) = \inf\{\delta > 0, \varphi_2 \leq e^{\delta}\varphi_1(\infty), \varphi_1 \leq e^{\delta}\varphi_2(\infty)\}.$$

By Proposition 6.9, d is indeed a metric on $W_{nsf}(\mathcal{M})$.

By remark 6.9.(1), we see that if $d(\varphi_1, \varphi_2) \leq \delta$, then

$$e^{-\delta}\varphi_1(x) \leq \varphi_2(x) \leq e^{\delta}\varphi_1(x) \quad (x \in \mathcal{M}^+).$$

It follows that for every $x \in \mathcal{M}^+$

$$\textit{the function } W_{nsf}(\mathcal{M}) \ni \varphi \mapsto \varphi(x) \textit{ is d-continuous.} \tag{1}$$

Also, if φ_1, φ_2 are faithful normal positive forms on $\mathcal{M}$ and $d(\varphi_1, \varphi_2) \leq \delta$, then

$$\begin{aligned} \|\varphi_2 - \varphi_1\| &\leq 2\sup\{|\varphi_2(x) - \varphi_1(x)|; x \in \mathcal{M}^+, \|x\| \leq 1\} \\ &\leq 2\max\{e^{\delta} - 1, 1 - e^{-\delta}\} \leq 4\varepsilon, \end{aligned}$$

hence

$$\|\varphi_1 - \varphi_2\| \leq 4d(\varphi_1, \varphi_2). \tag{2}$$

Let $\varphi_1, \varphi_2 \in W_{nsf}(\mathcal{M})$ and assume that $d(\varphi_1, \varphi_2) \leq \delta < +\infty$. Then there exist $\mathcal{M}$-valued functions G_{21}, G_{12} defined and w-continuous on $\{\alpha \in \mathbb{C}; \mathrm{Re}\,\alpha \geq 0\}$, analytic in $\{\alpha \in \mathbb{C}; \mathrm{Re}\,\alpha > 0\}$, such that $\|G_{21}(\alpha)\| \leq 1$, $\|G_{12}(\alpha)\| \leq 1$ and

$$\begin{aligned} G_{21}(it) &= [D\varphi_2 : D(e^{\delta}\varphi_1)]_t = e^{-it\delta}[D\varphi_2 : D\varphi_1]_t, \\ G_{12}(it) &= [D\varphi_1 : D(e^{\delta}\varphi_2)]_t = e^{-it\delta}[D\varphi_1 : D\varphi_2]_t. \end{aligned}$$

Taking into account Corollary 6.4, we see that the function $F_{21} : \mathbb{C} \to \mathcal{M}$ defined by

$$F_{21}(\alpha) = \begin{cases} e^{\delta\alpha}G_{21}(\alpha), & \textit{if } \mathrm{Re}\,\alpha \geq 0 \\ e^{-\delta\alpha}G_{12}(-\bar{\alpha})^*, & \textit{if } \mathrm{Re}\,\alpha \leq 0 \end{cases}$$

is entire analytic and we have

$$\|F_{21}(\alpha)\| \le e^{\delta|\text{Re}\,\alpha|} \quad (\alpha \in \mathbb{C}), \tag{3}$$

$$F_{21}(it) = [D\varphi_2 : D\varphi_1]_t \quad (t \in \mathbb{R}). \tag{4}$$

Thus, *the relation* $d(\varphi_1, \varphi_2) \le \delta$ *is equivalent to the existence of an entire analytic* $\mathcal{M}$*-valued function* F_{21} *satisfying (3) and (4).*

To continue the study of the metric d we need the following

Lemma. *For every* $\varepsilon > 0$ *and every* $r > 0$*, there exists* $\delta = \delta(\varepsilon, r) > 0$ *such that if* F *is a Banach space valued entire analytic function with the property* $\|F(\alpha)\| \le e^{\delta|\text{Re}\,\alpha|}$, $(\alpha \in \mathbb{C})$*, then*

$$\|F(\alpha) - F(0)\| \le \varepsilon \text{ for all } \alpha \in \mathbb{C} \text{ with } |\alpha| < r.$$

Proof. Let $\mathscr{A}$ be the linear space of all entire analytic complex valued functions equipped with the compact-open topology and denote by $\mathscr{K}(\delta)$ the set of those $f \in \mathscr{A}$ such that $|f(\alpha)| \le e^{\delta|\text{Re}\,\alpha|}$ for all $\alpha \in \mathbb{C}$. Then each $\mathscr{K}(\delta)$ is a compact subset of $\mathscr{A}$ and $\bigcap_{\delta>0} \mathscr{K}(\delta) = \{\alpha \in \mathbb{C}; |\alpha| \le 1\} = \mathscr{K}(0)$. On the other hand, $\mathscr{D}(\varepsilon, r) = \{f \in \mathscr{A}; |f(\alpha) - f(0)| < \varepsilon \text{ for all } \alpha \in \mathbb{C} \text{ with } |\alpha| \le r\}$ is an open subset of $\mathscr{A}$ and $\mathscr{D}(\varepsilon, r) \supset \mathscr{K}(0)$. It follows that there exists $\delta = \delta(\varepsilon, r) > 0$ such that $\mathscr{D}(\varepsilon, r) \supset \mathscr{K}(\delta)$ and it is easy to check that this $\delta = \delta(\varepsilon, r)$ satisfies the requirements of the lemma. □

Thus, if $d(\varphi_1, \varphi_2) \le \delta = \delta(\varepsilon, r)$, then

$$\|F_{21}(\alpha) - 1\| \le \varepsilon \text{ for all } \alpha \in \mathbb{C} \text{ with } |\alpha| \le r. \tag{5}$$

Moreover, for any other $\varphi_0 \in W_{nsf}(\mathcal{M})$ such that $d(\varphi_1, \varphi_0) = \gamma < +\infty$ we have $F_{20}(\alpha) = F_{21}(\alpha)F_{10}(\alpha)(\alpha \in \mathbb{C})$, and using (3) and (5) we obtain

$$\|F_{20}(\alpha) - F_{10}(\alpha)\| \le \varepsilon e^{\gamma|Re\,\alpha|} \text{ for all } \alpha \in \mathbb{C} \text{ with } |\alpha| \le r. \tag{6}$$

Proposition. *The metric space* $(W_{nsf}(\mathcal{M}), d)$ *is complete.*

Proof. Let $\{\varphi_n\}_{n\ge0}$ be a d-Cauchy sequence in $W_{nsf}(\mathcal{M})$. By (6) it follows that the sequence $\{F_{n0}\}_{n\ge0}$ converges uniformly on compact subsets of $\mathbb{C}$ to a function $F_{\infty0} : \mathbb{C} \to \mathcal{M}$. Consequently, the function $F_{\infty0}$ is entire analytic and the mapping $\mathbb{R} \ni t \mapsto F_{\infty0}(it) \in U(\mathcal{M})$ is a σ^{φ_0}-cocycle. By Theorem 5.1, there exists $\varphi_\infty \in W_{nsf}(\mathcal{M})$ such that $F_{\infty0}(it) = [D\varphi_\infty : D\varphi_0]_t$ for every $t \in \mathbb{R}$.

Let $\varepsilon > 0$. Since $\{\varphi_n\}$ is a d-Cauchy sequence there exists $m \in \mathbb{N}$ such that $\|F_{nm}(\alpha)\| \le e^{\varepsilon|Re\,\alpha|}$ $(\alpha \in \mathbb{C})$ for every $n \ge m$.

It follows that

$$\begin{aligned}\|F_{\infty m}(\alpha)\| &= \|F_{\infty0}(\alpha)F_{0m}(\alpha)\| = \lim_n \|F_{n0}(\alpha)F_{0m}(\alpha)\| \\ &= \lim_n \|F_{nm}(\alpha)\| \le e^{\varepsilon|Re\alpha|}\end{aligned}$$

for all $\alpha \in \mathbb{C}$, that is $d(\varphi_\infty, \varphi_m) \le \varepsilon$. Hence $d(\varphi_\infty, \varphi_m) \to 0$. □

6.11. Notes. Theorem 6.2 is a refinement, given in Zsidó (1978), of the Pedersen–Takesaki theorem on the equality of weights (Pedersen & Takesaki, 1973). Propositions 6.4, 6.5 and the examples in

Section 6.7 are due to Van Daele (1976). The examples in Sections 6.6 and 6.8 are from Connes (1973b) and Pedersen and Takesaki (1973). The material contained in Sections 6.9 and 6.10 is due to Connes and Takesaki (1977).

For our exposition, we have used Connes (1973b); Connes and Takesaki (1977); Van Daele (1976); Pedersen and Takesaki (1973) and Zsidó (1978).

7 The Spatial Derivative

In this section, we introduce a positive self-adjoint operator called the spatial derivative of a weight φ on a von Neumann algebra $\mathcal{M} \subset \mathcal{B}(\mathcal{H})$ with respect to a weight φ' on its commutant $\mathcal{M}' \subset \mathcal{B}(\mathcal{H})$, as a generalization of the modular operator. By way of applications we give several continuity properties.

7.1. Let ψ be an n.s.f. weight on the von Neuman algebra $\mathcal{N} \subset \mathcal{B}(\mathcal{H})$.

A vector $\eta \in \mathcal{H}$ is called *ψ-bounded* if there exists a constant $0 < \lambda < +\infty$ such that $\|b\eta\| \leq \lambda\|b_\psi\|_\psi$ for every $b \in \mathfrak{N}_\psi$. Let

$$D(\mathcal{H}, \psi) = \{\eta \in \mathcal{H}; \eta \text{ is } \psi\text{-bounded}\}.$$

We show that

$$D(\mathcal{H}, \psi) \textit{ is a dense linear subspace of } \mathcal{H}. \tag{1}$$

Indeed, by Corollary 5.9 there is a family $\{\eta_k\}_{k\in K} \subset \mathcal{H}$ such that $\psi(y) = \sum_k (y\eta_k|\eta_k)$ for all $y \in \mathcal{N}^+$. For each $k \in K$, we have $\eta_k \in D(\mathcal{H}, \psi)$ as $\|b\eta_k\|^2 = (b^*b\eta_k|\eta_k) \leq \psi(b^*b) = \|b_\psi\|_\psi^2 (b \in \mathfrak{N}_\psi)$. On the other hand, it is clear that $D(\mathcal{H}, \psi)$ is an $\mathcal{N}$-invariant linear subspace of $\mathcal{H}$, so that the orthogonal projection $f \in \mathcal{B}(\mathcal{H})$ onto the closure of $D(\mathcal{H}, \psi)$ belongs to $\mathcal{N}$. Since $f\eta_k = \eta_k (k \in K)$, it follows that $\psi(1-f) = 0$, hence $f = 1$ since ψ is faithful.

Each vector $\eta \in D(\mathcal{H}, \psi)$ defines a bounded operator $R_\eta^\psi : \mathcal{H}_\psi \to \mathcal{H}$, uniquely determined, such that $R_\eta^\psi b_\psi = b\eta, (b \in \mathfrak{N}_\psi)$, and its adjoint $(R_\eta^\psi)^* : \mathcal{H} \to \mathcal{H}_\psi$, is uniquely determined by $((R_\eta^\psi)^*\xi|b_\psi)_\psi = (\xi|b\eta)$ for every $b \in \mathfrak{N}_\psi, \xi \in \mathcal{H}$.

It is easy to check that

$$\eta \in D(\mathcal{H}, \psi) \textit{ and } y' \in \mathcal{N}' \Rightarrow y'\eta \in D(\mathcal{H}, \psi) \textit{ and } R_{y'\eta}^\psi = y'R_\eta^\psi \tag{2}$$

and that the operators $R_\eta^\psi : \mathcal{H}_\psi \to \mathcal{H}$ intertwine the standard representation π_ψ of $\mathcal{N}$ and the identity representation of $\mathcal{N}$:

$$\eta \in D(\mathcal{H}, \psi) \textit{ and } y \in \mathcal{N} \Rightarrow yR_\eta^\psi = R_\eta^\psi \pi_\psi(y). \tag{3}$$

Also,

$$\eta \in D(\mathcal{H}, \psi) \Rightarrow \eta \in \overline{R_\eta^\psi(\mathcal{H}_\psi)} = \mathbf{s}(R_\eta^\psi(R_\eta^\psi)^*)\mathcal{H} \tag{4}$$

since if $\mathfrak{N}_\psi \ni b_k \overset{so}{\to} 1$ then $R_\eta^\psi(b_k)_\psi = b_k\eta \to \eta$.

Let $\mathcal{S}(\mathcal{H},\psi)$ be the linear subspace spanned in $\mathcal{B}(\mathcal{H})$ by the operators of form $R^\psi_\eta(R^\psi_\zeta)^*$ with $\eta,\zeta \in D(\mathcal{H},\psi)$. Then

$$\mathcal{S}(\mathcal{H},\psi) \textit{ is an } s^*\textit{-dense two-sided ideal in } \mathcal{N}'. \tag{5}$$

Indeed, the fact that $\mathcal{S}(\mathcal{H},\psi)$ is a two-sided ideal in $\mathcal{N}'$ follows from (2) and (3). Let $y' \in \mathcal{N}', y' \neq 0$. From (1), it follows that there exists $\eta \in D(\mathcal{H},\psi)$ with $y'\eta \neq 0$. Using (4) and (2), we get $0 \neq R^\psi_{y'\eta}(R^\psi_{y'\eta})^* = y'R^\psi_\eta(R^\psi_\eta)^*y'^* \leq \|R^\psi_\eta\|^2 y'y'^*$. Thus, for every positive element $b' = y'y'^* \neq 0$ of $\mathcal{N}'$, there exists a positive element $c' = R^\psi_{y'\eta}(R^\psi_{y'\eta})^* \neq 0$ of the two-sided ideal $\mathcal{S}(\mathcal{H},\psi)$ such that $c' \leq b'$. By ([L], 3.20) it follows that $\mathcal{S}(\mathcal{H},\psi)$ is s^*-dense in $\mathcal{N}'$.

Also,

$$\textit{every positive element in } \mathcal{S}(\mathcal{H},\psi) \textit{ is of the form } \sum_{k=1} R^\psi_{\eta_k}(R^\psi_{\eta_k})^* \textit{ with } \eta_k \in D(\mathcal{H},\psi) \textit{ and } \eta \in \mathbb{N}. \tag{6}$$

Indeed, let $b' = \sum_{k=1}^n R^\psi_{\eta_k}(R^\psi_{\zeta_k})^* \in J(\mathcal{H},\psi), b' \geq 0$. Then $b' = (b' + b'^*)/2$, so

$$0 \leq 2^{-1}\sum_{k=1}^n (R^\psi_{\eta_k}(R^\psi_{\zeta_k})^* + R^\psi_{\zeta_k}(R^\psi_{\eta_k})^*) = b' \leq c' = 2^{-1}\sum_{k=1}^n R^\psi_{\eta_k+\zeta_k}(R^\psi_{\eta_k+\zeta_k})^*.$$

Since $0 \leq b' \leq c'$, it follows (1.4) that there exists $y' \in \mathcal{N}'$ such that $b' = y'c'y'^*$. Consequently,

$$b' = 2^{-1}\sum_{k=1}^n R^\psi_{y'(\eta_k+\zeta_k)}(R^\psi_{y'(\eta_k+\zeta_k)})^*.$$

Finally, we show that

$$\textit{there exists a family } \{\eta_k\}_k \subset D(\mathcal{H},\psi) \textit{ such that } s^*\text{-}\sum_k R^\psi_{\eta_k}(R^\psi_{\eta_k})^* = 1. \tag{7}$$

Indeed, since the two-sided ideal $\mathcal{S}(\mathcal{H},\psi)$ is s^*-dense in $\mathcal{N}'$, there exists a series of positive elements in $\mathcal{S}(\mathcal{H},\psi)$, which is s^*-convergent to 1. Thus (7) follows using (6).

7.2. In the particular case when $\mathcal{N} = \pi_\psi(\mathcal{N}) \subset \mathcal{B}(\mathcal{H}_\psi)$, the operators R^ψ_η are just the operators $R_\eta \in \pi_\psi(\mathcal{N})'$ considered in Section 2.12. If ψ' is the natural weight on $\pi_\psi(\mathcal{N})'$, then (2.12.(7))

$$\psi'(R^*_\zeta R_\eta) = (\eta|\zeta)_\psi \quad (\eta,\zeta \in D(\mathcal{H}_\psi,\psi)).$$

In the general case considered in Section 7.1, the following similar result holds:

$$\eta,\zeta \in D(\mathcal{H},\psi) \Rightarrow (R^\psi_\zeta)^*R^\psi_\eta \in \mathfrak{M}_{\psi'} \textit{ and } \psi'((R^\psi_\zeta)^*R^\psi_\eta) = (\eta|\zeta). \tag{1}$$

Indeed, due to the polarization relation ([L], 3.21) it is sufficient to consider only the case $\zeta = \eta$. Then let $(R^\psi_\eta)^* = V(R^\psi_\eta(R^\psi_\eta)^*)^{1/2}$ be the polar decomposition of $(R^\psi_\eta)^*$. Using 7.1.(4) and 7.1.(3), we see that the partial isometry $V : \mathscr{H} \to \mathscr{H}_\psi$ has the properties

$$V^* V\eta = \eta \text{ and } \pi_\psi(y)V = Vy \text{ for every } y \in \mathscr{N}.$$

Since $V\eta \in \mathscr{H}_\psi$ and $\pi_\psi(b)V\eta = Vb\eta = VR_\eta^\psi b_\psi (b \in \mathfrak{N}_\psi)$, it follows that $V\eta \in D(\mathscr{H}_\psi, \psi)$ and $R_{V\eta} = VR_\eta^\psi$. Consequently, $R_\eta^\psi = V^* R_{V\eta}, (R_\eta^\psi)^* R_\eta^\psi = R_{V\eta}^* R_{V\eta}$ and $\psi'((R_\eta^\psi)^* R_\eta^\psi) = \psi'(R_{V\eta}^* R_{V\eta}) = \|V\eta\|_\psi^2 = \|\eta\|^2$.

7.3. Consider now a von Neumann algebra $\mathscr{M} \subset \mathscr{B}(\mathscr{H})$ with commutant $\mathscr{N} = \mathscr{M}' \subset \mathscr{B}(\mathscr{H})$; let φ be a normal semifinite weight on $\mathscr{M}$ and ψ an n.s.f. weight on $\mathscr{N}$.

We show that the function

$$q : D(\mathscr{H}, \psi) \ni \eta \mapsto q(\eta) = \varphi(R_\eta^\psi (R_\eta^\psi)^*) \in [0, +\infty]$$

is lower semicontinuous and has the following properties:

$$D(q) = \{\eta \in D(\mathscr{H}, \psi); q(\eta) < +\infty\} \text{ is dense in } \mathscr{H}; \tag{1}$$

$$q(\lambda\eta) = |\lambda|^2 q(\eta) \text{ for all } \eta \in D(\mathscr{H}, \psi) \text{ and } \lambda \in \mathbb{C}; \tag{2}$$

$$q(\eta + \zeta) + q(\eta - \zeta) = 2q(\eta) + 2q(\zeta) \text{ for all } \eta, \zeta \in D(\mathscr{H}, \psi). \tag{3}$$

Indeed, by Corollary 5.9, there exists a family $\{\xi_i\}_{i \in I} \subset \mathscr{H}$ such that $\varphi(x) = \sum_i (x\xi_i | \xi_i) (x \in \mathscr{M}^+)$. Then, for $\eta \in D(\mathscr{H}, \psi)$, we get

$$\begin{aligned} q(\eta) &= \sum_i \|(R_\eta^\psi)^* \xi_i\|_\psi^2 = \sum_i \sup_{b \in \mathfrak{N}_\psi \|b_\psi\|_\psi \le 1} |((R_\eta^\psi)^* \xi_i | b_\psi)_\psi| \\ &= \sum_i \sup_{b \in \mathfrak{N}_\psi \|b_\psi\|_\psi \le 1} |(b^* \xi_i | \eta)|, \end{aligned}$$

hence q is lower semicontinuous.

Since φ is semifinite, $\mathfrak{N}_\varphi^*$ is s-dense in $\mathscr{M}$ so that the set $\{x\eta; x \in \mathfrak{N}_\varphi^*, \eta \in D(\mathscr{H}, \psi)\}$ is total in $\mathscr{H}$ and contained in $D(\mathscr{H}, \psi)$. For $x \in \mathfrak{N}_\varphi^*$ and $\eta \in D(\mathscr{H}, \psi)$ we have $q(x\eta) = \varphi(R_{x\eta}^\psi (R_{x\eta}^\psi)^*) = \varphi(xR_\eta^\psi (R_\eta^\psi)^* x^*) \le \|R_\eta^\psi\|^2 \varphi(xx^*) < +\infty$, so that $x\eta \in D(q)$. This proves (1), while (2) and (3) are obvious.

By (A.10) it follows that there exists a greatest positive self-adjoint operator $\Delta(\varphi/\psi)$ in $\mathscr{H}$ such that

$$D(\Delta(\varphi/\psi)^{1/2}) \supset D(q), \tag{4}$$

$$\eta \in D(\mathscr{H}, \psi) \Rightarrow \|\Delta(\varphi/\psi)^{1/2}\eta\|^2 = \varphi(R_\eta^\psi (R_\eta^\psi)^*), \tag{5}$$

and we have

$$\overline{\Delta(\varphi/\psi)^{1/2} | D(q)} = \Delta(\varphi/\psi)^{1/2}. \tag{6}$$

The operator $\Delta(\varphi/\psi)$ is called the *spatial derivative* of the normal semifinite weight φ on $\mathscr{M} \subset \mathscr{B}(\mathscr{H})$ with respect to the n.s.f. weight ψ on $\mathscr{N} = \mathscr{M}' \subset \mathscr{B}(\mathscr{H})$.

Note that if $\varphi \in \mathcal{M}_*^+$, then $D(q) = D(\mathcal{H}, \psi)$ and hence

$$\overline{\Delta(\varphi/\psi)^{1/2}|D(\mathcal{H}, \psi)} = \Delta(\varphi/\psi)^{1/2}. \tag{7}$$

Using (5) and 7.1(2), it is easy to check that

$$\mathbf{s}(\Delta(\varphi/\psi)) \leq \mathbf{s}(\varphi), \tag{8}$$

but we shall see that in fact equality holds (7.4).

Let us compute the spatial derivative in a very simple case, namely $\mathcal{M} = \mathcal{B}(\mathcal{H})$, so that $\mathcal{N} = \mathcal{M}' = \mathbb{C} \cdot 1_{\mathcal{H}}$. On $\mathcal{N}$ we consider the weight $\psi : \lambda \cdot 1_{\mathcal{H}} \mapsto \lambda$. An arbitrary normal semifinite weight φ on $\mathcal{M}$ is of the form $\varphi = tr_A$ with A a positive self-adjoint operator in $\mathcal{H}$ (4.10). It is clear that $\mathcal{H}_\psi = \mathbb{C}$ and $\|\lambda \cdot 1_{\mathcal{H}}\|_\psi = |\lambda|$ $(\lambda \cdot 1_{\mathcal{H}} \in \mathcal{N})$. Also, $D(\mathcal{H}, \psi) = \mathcal{H}$ and for $\eta \in \mathcal{H}$ we have $R_\eta^\psi : \mathbb{C} \ni \lambda \mapsto \lambda\eta \in \mathcal{H}$, hence $(R_\eta^\psi)^* : \mathcal{H} \ni \xi \mapsto (\xi|\eta) \in \mathbb{C}$. Thus, $R_\eta^\psi(R_\eta^\psi)^* = \eta \otimes \bar{\eta}$ and $q(\eta) = \varphi(R_\eta^\psi(R_\eta^\psi)^*) = \varphi(\eta \otimes \bar{\eta}) = \|A^{1/2}\eta\|^2$ (see 4.23), hence $\Delta(\varphi/\psi) = A$. Denoting the weight ψ by 1, we can write the conclusion as follows:

$$\Delta(tr_A/1) = A \quad (\mathcal{M} = \mathcal{B}(\mathcal{H}), \mathcal{N} = \mathcal{M}' = \mathbb{C} \cdot 1_{\mathcal{H}}). \tag{9}$$

7.4. The main result concerning the spatial derivative is the following

Theorem (A. Connes). *Let $\mathcal{M} \subset \mathcal{B}(\mathcal{H})$ be a von Neumann algebra with commutant $\mathcal{N} = \mathcal{M}' \subset \mathcal{B}(\mathcal{H})$, φ an n.s.f. weight on $\mathcal{M}$ and ψ an n.s.f. weight on $\mathcal{N}$. Then*

$$\mathbf{s}(\Delta(\varphi/\psi)) = 1 \textit{ and } \Delta(\varphi/\psi)^{-1} = \Delta(\psi/\varphi); \tag{1}$$

$$\sigma_t^\varphi(x) = \Delta(\varphi/\psi)^{it} x \Delta(\varphi/\psi)^{-it} \quad (x \in \mathcal{M}, t \in \mathbb{R}); \tag{2}$$

$$\sigma_t^\psi(y) = \Delta(\varphi/\psi)^{-it} y \Delta(\varphi/\psi)^{it} \quad (y \in \mathcal{N}, t \in \mathbb{R}); \tag{3}$$

and for every normal semifinite φ_1 on $\mathcal{M}$ we have

$$\mathbf{s}(\Delta(\varphi_1/\psi)) = \mathbf{s}(\varphi_1); \tag{4}$$

$$\Delta(\varphi_1/\psi)^{it} = [D\varphi_1 : D\varphi]_t \Delta(\varphi/\psi)^{it} \quad (t \in \mathbb{R}). \tag{5}$$

The proof is given in Sections 7.5–7.10.

7.5. For the W^*-algebra $\mathcal{N}$, we have the identity representation $id : \mathcal{N} \to \mathcal{B}(\mathcal{H})$ and the standard representation $\pi_\psi : \mathcal{N} \to \mathcal{B}(\mathcal{H}_\psi)$ associated with ψ. We consider the direct sum of these representations

$$\pi : \mathcal{N} \to \mathcal{B}(\mathcal{H} \oplus \mathcal{H}_\psi)$$

and the von Neumann algebra

$$\pi(\mathcal{N})' \subset \mathcal{B}(\mathcal{H} \oplus \mathcal{H}_\psi).$$

Every operator $T \in \mathscr{B}(\mathscr{H} \oplus \mathscr{H}_\psi)$ is a matrix

$$T = \begin{pmatrix} T_{11} & T_{12} \\ T_{21} & T_{22} \end{pmatrix},$$

with $T_{11} \in \mathscr{B}(\mathscr{H}), T_{12} \in \mathscr{B}(\mathscr{H}_\psi, \mathscr{H}), T_{21} \in \mathscr{B}(\mathscr{H}, \mathscr{H}_\psi), T_{22} \in \mathscr{B}(\mathscr{H}_\psi)$ such that for any vector $\zeta \in \mathscr{H} \oplus \mathscr{H}_\psi$,

$$\zeta = \begin{pmatrix} \zeta_1 \\ \zeta_2 \end{pmatrix},$$

with $\zeta_1 \in \mathscr{H}, \zeta_2 \in \mathscr{H}_\psi$, we have

$$T\zeta = \begin{pmatrix} T_{11}\zeta_1 + T_{12}\zeta_2 \\ T_{21}\zeta_1 + T_{22}\zeta_2 \end{pmatrix}.$$

Define

$$\begin{aligned} \mathscr{I}(\pi_\psi, id) &= \{X \in \mathscr{B}(\mathscr{H}_\psi, \mathscr{H}); yX = X\pi_\psi(y) \text{ for all } y \in \mathscr{N}\} \\ \mathscr{I}(id, \pi_\psi) &= \{X \in \mathscr{B}(\mathscr{H}, \mathscr{H}_\psi); Xy = \pi_\psi(y)X \text{ for all } y \in \mathscr{N}\}. \end{aligned}$$

It is easy to check that

$$T \in \pi(\mathscr{N})' \Leftrightarrow T_{11} \in \mathscr{N}' = \mathscr{M}, T_{12} \in \mathscr{I}(\pi_\psi, id), T_{21} \in \mathscr{I}(\pi_\psi, id), T_{22} \in \pi_\psi(\mathscr{N})'$$

Let ψ' be the natural weight on $\pi_\psi(\mathscr{N})'$ (2.12.(6)). Then the weight ψ' on $\pi_\psi(\mathscr{N})'$ and the weight φ on $\mathscr{M}$ define an n.s.f. weight $\theta = \theta(\varphi, \psi')$ on $\pi(\mathscr{N})'$

$$\theta(T) = \varphi(T_{11}) + \psi'(T_{22}) \quad (T \in \pi(\mathscr{N})'^{+}).$$

As for the balanced weight (3.1, 3.10) one shows that

$$\sigma_t^\theta \begin{pmatrix} T_{11} & 0 \\ 0 & T_{22} \end{pmatrix} = \begin{pmatrix} \sigma_t^\varphi(T_{11}) & 0 \\ 0 & \sigma_t^{\psi'}(T_{22}) \end{pmatrix} \quad (T_{11} \in \mathscr{M}, T_{22} \in \pi_\psi(\mathscr{N})')$$

and that there exists a group of isometrics $\{S_t^\varphi\}_{t\in\mathbb{R}}$ on the Banach space $\mathscr{I}(\pi_\psi, id)$, uniquely determined, such that

$$\sigma_t^\theta \begin{pmatrix} 0 & T_{12} \\ 0 & 0 \end{pmatrix} = \begin{pmatrix} 0 & S_t^\varphi(T_{12}) \\ 0 & 0 \end{pmatrix} \quad (T_{12} \in \mathscr{I}(\pi_\psi, id)).$$

Recall (7.1) that $R_\eta^\psi \in \mathscr{I}(\pi_\psi, id)$ for all $\eta \in D(\mathscr{H}, \psi)$.

7.6. With the assumptions of Theorem 7.4 and the notation of Section 7.5, we have the following

Lemma. *There exists an so-continuous unitary representation*

$$\mathbb{R} \ni t \mapsto u_t^\varphi \in \mathscr{B}(\mathscr{H}),$$

uniquely determined, such that

$$\sigma_t^{\varphi}(x) = u_t^{\varphi} x u_{-t}^{\varphi} \quad (x \in \mathcal{M}) \tag{1}$$

$$\sigma_t^{\psi}(x) = u_{-t}^{\varphi} y u_t^{\varphi} \quad (y \in \mathcal{N}) \tag{2}$$

$$S_t^{\varphi}(R_\eta^{\psi}) = R_{u_t^{\varphi}\eta}^{\psi} \quad (\eta \in D(\mathcal{H}, \psi)). \tag{3}$$

Proof. The uniqueness of u_t^{φ} follows obviously from (3). The proof of the existence statement is divided into three steps.

(I) If the lemma is valid for a certain n.s.f. weight φ_0 on $\mathcal{M}$, then it is valid for any other n.s.f. weight φ on $\mathcal{M}$.

Indeed, it is easy to check that the mapping

$$\mathbb{R} \ni t \mapsto u_t^{\varphi} = [D\varphi : D\varphi_0]_t u_t^{\varphi_0} \in \mathcal{B}(\mathcal{H})$$

is an *so*-continuous unitary representation with properties (1) and (2). Also (3) follows if we note that

$$R_{u_t^{\varphi}\eta}^{\psi} = [D\varphi : D\varphi_0]_t R_{u_t^{\varphi_0}\eta}^{\psi} \quad (\eta \in D(\mathcal{H}, \psi))$$

and that for $\theta_0 = \theta(\varphi_0, \psi')$ we have

$$[D\theta : D\theta_0]_t = \begin{pmatrix} [D\varphi : D\varphi_0]_t & 0 \\ 0 & 1 \end{pmatrix} \quad (t \in \mathbb{R}).$$

(II) If the lemma is valid for some particular realization of $\mathcal{N}$ (i.e., of $\mathcal{M}$) as a von Neumann algebra, then it remains valid for any other realization.

Indeed, if we have two different realizations of $\mathcal{N}$, then the corresponding von Neumann algebras are *-isomorphic. Since any *-isomorphism between von Neumann algebras is the composition of an amplification, an injective induction and a spatial isomorphism ([L], E.8.8), we may consider separately these three cases.

The case of a spatial isomorphism is trivial.

The case of an amplification: we assume the lemma is valid for the von Neumann algebras $\mathcal{N} \subset \mathcal{B}(\mathcal{H}), \mathcal{M} = \mathcal{N}' \subset \mathcal{B}(\mathcal{H})$; we have to prove it for the von Neumann algebras $\mathcal{N} \otimes 1 \subset \mathcal{B}(\mathcal{H})\overline{\otimes}\mathcal{B}(\ell^2(I)), \mathcal{M}\overline{\otimes}\mathcal{B}(\ell^2(I)) = (\mathcal{N} \otimes 1)' \subset \mathcal{B}(\mathcal{H})\overline{\otimes}\mathcal{B}(\ell^2(I))$, where I is an arbitrary set.

A vector $\eta = \{\eta_i\}_{i \in I}$ in $\mathcal{H}\overline{\otimes}\ell^2(I) = \ell(I, \mathcal{H})$ is ψ-bounded whenever each $\eta_i \in \mathcal{H}(i \in I)$ is ψ-bounded and $\sum_i \|R_{\eta_i}^{\psi}\|^2 < +\infty$. In this case, the operator $R_\eta^{\psi} : \mathcal{H}_\psi \to \mathcal{H}\overline{\otimes}\ell^2(I)$ acts as follows:

$$(R_\eta^{\psi}\zeta)_i = R_{\eta_i}^{\psi}\zeta \quad (\zeta \in \mathcal{H}).$$

If φ is any n.s.f. weight on $\mathcal{M}$ and *tr* is the canonical trace on $\mathcal{B}(\ell^2(I))$, then $\Phi = \varphi\overline{\otimes}tr$ (see 5.2 or 8.2) is an n.s.f. weight on $\mathcal{M}\overline{\otimes}\mathcal{B}(\ell^2(I))$ and $\sigma_t^{\Phi} = \sigma_t^{\varphi} \otimes t (t \in \mathbb{R})$.

It is now easy to check that if $\{u_t^{\varphi}\}_t$ satisfies the requirements of the lemma for $(\mathcal{N}, \psi; \mathcal{M}, \varphi)$, then $\{U_t^{\Phi} = u_t^{\varphi}\overline{\otimes}1\}_t$ satisfies those requirements for $(\mathcal{N} \otimes 1, \psi; \mathcal{M}\overline{\otimes}\mathcal{B}(\ell^2(I)), \Phi)$.

Taking into account step (I) of the proof, it follows that the lemma is true for the von Neumann algebras $\mathcal{N} \otimes 1, \mathcal{M}\overline{\otimes}\mathcal{B}(\ell^2(I))$.

The case of an injective induction: we assume the lemma is valid for the von Neumann algebras $\mathcal{N} \subset \mathcal{B}(\mathcal{H}), \mathcal{M} = \mathcal{N}' \subset \mathcal{B}(\mathcal{H})$ and we have to prove it for the von Neumann algebras $\mathcal{N}e \subset \mathcal{B}(e\mathcal{H}), e\mathcal{M}e = (\mathcal{N}e)' \subset \mathcal{B}(e\mathcal{H})$, where $e \in \mathcal{M}$ is a projection with the central support $\mathbf{z}(e) = 1$.

It is easy to see that $D(e\mathcal{H}, \psi) = eD(\mathcal{H}, \psi)$ and that for $\eta \in D(e\mathcal{H}, \psi) \subset D(\mathcal{H}, \psi)$ the operator $R^\psi_\eta : \mathcal{H}_\psi \to e\mathcal{H}$ is the corestriction of the operator $R^\psi_\eta : \mathcal{H}_\psi \to \mathcal{H}$.

There exists an n.s.f. weight φ on $\mathcal{M}$ such that $e \in \mathcal{M}^\varphi$. Indeed, if $\mathcal{M}$ is semifinite, we can take φ equal to any n.s.f. trace on $\mathcal{M}$; if $\mathcal{M}$ is of type III, then $\mathcal{M} \approx Mat_2(e\mathcal{M}e)$ and we can take φ equal to the balanced weight corresponding to an arbitrary n.s.f. weight on $e\mathcal{M}e$ with respect to itself. With this choice of φ, the restriction of the weight φ_e (2.22) to $e\mathcal{M}e$ is an n.s.f. weight on $e\mathcal{M}e$; its modular automorphism group is the restriction of the modular automorphism group of φ.

If $\{u^\varphi_t\}_t$ satisfies the requirements of the lemma for $(\mathcal{N}, \psi; \mathcal{M}, \varphi)$, then $u^\varphi_t e = eu^\varphi_t$ as $e \in \mathcal{M}^\varphi$ and it is easy to check that $\{u^\varphi_t e\}_t$ satisfies the requirements of the lemma for $(\mathcal{N}e, \psi; e\mathcal{M}e, \varphi_e)$.

Taking into account step (I) of the proof, it follows that the lemma is true for the von Neumann algebras $\mathcal{N}e, e\mathcal{M}e$.

(III) By (I) and (II), it is sufficient to prove the lemma assuming $\mathcal{N} = \pi_\psi(\mathcal{N}) \subset \mathcal{B}(\mathcal{H}_\psi), \mathcal{M} = \pi_\psi(\mathcal{N})'$, and $\varphi = \psi'$. In this case, the mapping

$$\mathbb{R} \ni t \mapsto u^\varphi_t = \Delta^{-it}_\psi \in \mathcal{B}(\mathcal{H}_\psi)$$

is an *so*-continuous unitary representation satisfying (1) and (2). Also, the weight $\theta = \theta(\varphi, \psi')$ is just the balanced weight, so that $S^\varphi_t = \sigma^\varphi_t$ by 3.10.(1), and consequently

$$S^\varphi_t(R_\eta) = \sigma^\varphi_t(R_\eta) = \Delta^{-it}_\psi R_\eta \Delta^{it}_\psi = R_{\Delta^{-it}_\psi \eta} = R_{u^\varphi_t \eta},$$

which proves (3). □

7.7. By Stone's theorem ([L], 9.20), there exists a unique positive self-adjoint operator A_φ in $\mathcal{H}$ with $\mathbf{s}(A_\varphi) = 1$ such that $u^\varphi_t = A^{it}_\varphi (t \in \mathbb{R})$. In this section, we show that $\Delta(\varphi/\psi) = A_\varphi$ thus proving statements (2) and (3) of Theorem 7.4, as well as showing that $\mathbf{s}(\Delta(\varphi/\psi)) = 1$.

By Proposition 2.20, we obtain for every $T \in \pi(\mathcal{N})'$ with $\theta(T^*T) < +\infty$, a bounded regular positive Borel measure μ on $(0, +\infty)$ such that

$$\theta(T^*\sigma^\theta_t(T)) = \int_0^\infty \lambda^{it}\, \mathrm{d}\mu(\lambda), \theta(TT^*) = \int_0^\infty \lambda\, \mathrm{d}\mu(\lambda) \leq +\infty.$$

Let $\eta \in D(\mathcal{H}, \psi)$ and $T \in \pi(\mathcal{N})'$ with $T_{12} = R^\psi_\eta$ and $T_{11} = T_{21} = T_{22} = 0$. Then, using 7.6.(3) and 7.2.(1), we get

$$\int_0^\infty \lambda^{it}\, \mathrm{d}\mu(\lambda) = \psi'((R^\psi_\eta)^* S^\varphi_t(R^\psi_\eta)) = \psi'((R^\psi_\eta)^* R^\psi_{u^\varphi_t \eta}) = (u^\varphi_t \eta | \eta) = (A^{it}_\varphi \eta | \eta)$$

hence μ is the spectral measure associated with A_φ and $\eta \in \mathscr{H}$. Consequently,

$$\|A_\varphi^{1/2}\eta\|^2 = \int_0^\infty \lambda \, \mathrm{d}\mu(\lambda) = \varphi(R_\eta^\psi (R_\eta^\psi)^*) = \|\Delta(\varphi|\psi)^{1/2}\eta\|^2 \quad (\eta \in D(\mathscr{H}, \psi)). \tag{1}$$

On the other hand, we infer from 7.6.(1) and 7.6.(2) that the $*$-automorphisms $\mathscr{B}(\mathscr{H}) \ni z \mapsto u_t^\varphi z u_{-t}^\varphi \in \mathscr{B}(\mathscr{H})(t \in \mathbb{R})$ preserve $\mathscr{M}, \mathscr{N}, \varphi, \psi$, hence $u_t^\varphi \Delta(\varphi/\psi) u_{-t}^\varphi = \Delta(\varphi/\psi)(t \in \mathbb{R})$, that is,

$$\text{the operators } \Delta(\varphi/\psi) \text{ and } A_\varphi \text{ commute.} \tag{2}$$

Now, the conclusions (1) and (2) imply that $\Delta(\varphi/\psi) = A_\varphi$ (see, for instance, Theorem 6.2).

7.8. We prove that for two n.s.f. weights ψ_1, ψ_2 on $\mathscr{N}$

$$\Delta(\varphi/\psi_2)^{-it}\Delta(\varphi/\psi_1)^{it} = [D\psi_2 : D\psi_1]_t \quad (t \in \mathbb{R}). \tag{1}$$

To this end consider the balanced weight $\psi = \theta(\psi_1, \psi_2)$ on the von Neumann algebra $Mat_2(\mathscr{N}) \subset \mathscr{B}(\mathscr{H} \oplus \mathscr{H})$ with commutant $\widetilde{\mathscr{M}} \subset \mathscr{B}(\mathscr{H} \oplus \mathscr{H})$ isomorphic to $\mathscr{M}$ by amplification, and the n.s.f. weight $\tilde{\varphi}$ on $\widetilde{\mathscr{M}}$ corresponding to φ under this amplification. Also, as in 3.11.(1), identify the Hilbert spaces $\mathscr{H}_\psi = \mathscr{H}_{\psi_1} \oplus \mathscr{H}_{\psi_2} \oplus \mathscr{H}_{\psi_1} \oplus \mathscr{H}_{\psi_2}$.

It is easy to check that the vector $\eta = \eta_1 \oplus \eta_2 \in \mathscr{H} \oplus \mathscr{H}$ is ψ-bounded if and only if $\eta_1 \in D(\mathscr{H}, \psi_1), \eta_2 \in D(\mathscr{H}, \psi_2)$ and in this case

$$R_\eta^\psi = \begin{pmatrix} R_{\eta_1}^{\psi_1} & R_{\eta_2}^{\psi_2} & 0 & 0 \\ 0 & 0 & R_{\eta_1}^{\psi_1} & R_{\eta_2}^{\psi_2} \end{pmatrix} : \mathscr{H}_\psi \to \mathscr{H} \oplus \mathscr{H},$$

hence

$$\tilde{\varphi}(R_\eta^\psi (R_\eta^\psi)^*) = \varphi(R_{\eta_1}^{\psi_1}(R_{\eta_1}^{\psi_1})^*) + \varphi(R_{\eta_2}^{\psi_2}(R_{\eta_2}^{\psi_2})^*).$$

Consequently,

$$\Delta(\tilde{\varphi}/\psi) = \Delta(\varphi/\psi_1) \oplus \Delta(\varphi/\psi_2).$$

Using 7.4.(3) which has already been proved, we get

$$\begin{aligned} \begin{pmatrix} 0 & 0 \\ [D\psi_2 : D\psi_1]_t & 0 \end{pmatrix} &= \sigma_t^\psi \begin{pmatrix} 0 & 0 \\ 1 & 0 \end{pmatrix} \\ &= \begin{pmatrix} \Delta(\varphi/\psi_1)^{-it} & 0 \\ 0 & \Delta(\varphi/\psi_2)^{-it} \end{pmatrix} \begin{pmatrix} 0 & 0 \\ 1 & 0 \end{pmatrix} \begin{pmatrix} \Delta(\varphi/\psi_1)^{it} & 0 \\ 0 & \Delta(\varphi/\psi_2)^{it} \end{pmatrix} \\ &= \begin{pmatrix} 0 & 0 \\ \Delta(\varphi/\psi_2)^{-it}\Delta(\varphi/\psi_1)^{it} & 0 \end{pmatrix}. \end{aligned}$$

7.9. We now prove that $\Delta(\psi, \varphi) = \Delta(\varphi/\psi)^{-1}$.

Taking into account the argument in 7.6.(I) and 7.8.(1), we see that it is sufficient to prove the earlier identity only for a particular pair of weights φ, ψ. Then by the arguments of 7.6.(II), it follows that it is sufficient for the proof to assume that $\mathcal{N} = \pi_\psi(\mathcal{N}) \subset \mathcal{B}(\mathcal{H}_\psi)$, $\mathcal{M} = \pi_\psi(\mathcal{N})' \subset \mathcal{B}(\mathcal{H}_\psi)$, and $\varphi = \psi'$. In this last case, equality is obvious.

7.10. To complete the proof of Theorem 7.4, we still have to check 7.4.(4) and 7.4.(5). If the weight φ_1 is faithful, this has already been done in Sections 7.7 and 7.6.(I). Since, by 7.3.(8), $\mathbf{s}(\Delta(\varphi_1/\psi)) \leq \mathbf{s}(\varphi_1)$, the general case follows by considering the induction of $\mathcal{N}$ by $\mathbf{s}(\varphi_1)$, as the restriction of φ_1 to $\mathbf{s}(\varphi_1)\mathcal{M}\mathbf{s}(\varphi_1)$ is faithful.

7.11 Corollary. *Let $\mathcal{M} \subset \mathcal{B}(\mathcal{H})$ be a von Neumann algebra with commutant $\mathcal{N} = \mathcal{M}' \subset \mathcal{B}(\mathcal{H})$, φ an n.s.f. weight on $\mathcal{M}$ and ψ an n.s.f. weight on $\mathcal{N}$. For every $\xi \in D(\mathcal{H}, \varphi)$ and every $\eta \in D(\mathcal{H}, \psi)$, we have*

$$|(\xi|\eta)|^2 \leq \varphi(R^\psi_\eta(R^\psi_\eta)^*)\psi(R^\varphi_\xi(R^\varphi_\xi)^*) \leq +\infty.$$

Proof. Let $\Delta = \Delta(\varphi/\psi)$, hence $\Delta^{-1} = \Delta(\psi/\varphi)$. Using Definition 7.3.(5), we see that if either $\xi \notin D(\Delta^{-1/2})$ or $\eta \notin D(\Delta^{1/2})$ then the right-hand side of the inequality is equal to $+\infty$, while if $\xi \in D(\Delta^{-1/2})$ and $\eta \in D(\Delta^{1/2})$ the inequality follows from the estimate $|(\xi|\eta)|^2 = |(\Delta^{-1/2}\xi|\Delta^{1/2}\eta)|^2 \leq \|\Delta^{-1/2}\xi\|^2\|\Delta^{1/2}\eta\|^2$. □

7.12 Corollary (Haagerup). *Let $\mathcal{M} \subset \mathcal{B}(\mathcal{H})$ be a von Neumann algebra with commutant $\mathcal{N} = \mathcal{M}' \subset \mathcal{B}(\mathcal{H})$ and $\mathbb{R} \ni t \mapsto u_t \in \mathcal{B}(\mathcal{H})$ an so-continuous unitary representation. Then the following conditions are equivalent:*

(i) *there exists $\varphi \in W_{nsf}(\mathcal{M})$ such that $\sigma^\varphi_t = Ad(u_t)_t$, $(t \in \mathbb{R})$;*
(ii) *there exists $\psi \in W_{nsf}(\mathcal{N})$ such that $\sigma^\psi_t = Ad(u^*_t)_t$, $(t \in \mathbb{R})$.*

Proof. By symmetry it is sufficient to prove (i) ⇒ (ii). Let ψ_0 be an arbitrary n.s.f. weight on $\mathcal{N}$. Using (i) it follows that

$$\mathbb{R} \ni t \mapsto w_t = u^*_t\Delta(\varphi/\psi_0)^{it} \in \mathcal{M}' = \mathcal{N}$$

is a σ^{ψ_0}-cocycle hence, by Theorem 5.1, there exists an n.s.f. weight ψ on $\mathcal{N}$ such that $[D\psi : D\psi_0]_t = w_t (t \in \mathbb{R})$. Then

$$\sigma^\psi_t(y) = w_t\sigma^{\psi_0}_t(y)w^*_t = u^*_t y u_t \quad (y \in \mathcal{N}, t \in \mathbb{R}).$$ □

7.13. Some computation rules for the spatial derivative are contained in the following

Proposition. *Let $\mathcal{M} \subset \mathcal{B}(\mathcal{H})$ be a von Neumann algebra with commutant $\mathcal{N} = \mathcal{M}' \subset \mathcal{B}(\mathcal{H})$, $\varphi, \varphi_1, \varphi_2$ normal semifinite weights on $\mathcal{M}$, ψ an n.s.f. weight on $\mathcal{N}$ and $a \in \mathcal{M}$ an invertible element. Then*

$$\varphi_1 \leq \varphi_2 \Leftrightarrow \Delta(\varphi_1/\psi) \leq \Delta(\varphi_2/\psi) \tag{1}$$

$$\Delta(\varphi(a \cdot a^*)/\psi) = a^*\Delta(\varphi/\psi)a \tag{2}$$

and, if $\varphi_1, \varphi_2 \in \mathcal{M}^+_*$,

$$\Delta(\varphi_1 + \varphi_2/\psi) = \Delta(\varphi_1/\psi)\hat{+}\Delta(\varphi_2/\psi). \tag{3}$$

Proof. (1) Let q_1, q_2 be the functions associated as in Section 7.3 with the weights φ_1, φ_2, respectively. If $\varphi_1 \leq \varphi_2$, then we obviously have $D(q_2) \subset D(q_1)$ and $q_1(\eta) \leq q_2(\eta)$ for any $\eta \in D(q_2)$, so it follows that $\Delta(\varphi_1/\psi) \leq \Delta(\varphi_2/\psi)$.

Conversely, if $\Delta(\varphi_1/\psi) \leq \Delta(\varphi_2/\psi)$, then by 7.1.(6) we infer that $\varphi_1(x) \leq \varphi_2(x)$ for every positive element $x \in \mathcal{S}(\mathcal{H}, \psi)$. Since $\mathcal{S}(\mathcal{H}, \psi)$ is an s^*-dense two-sided ideal of $\mathcal{M} = \mathcal{N}'$ (7.1.(5)), any element $x \in \mathcal{M}^+$ is the s^*-limit of an increasing net of positive elements in $\mathcal{S}(\mathcal{H}, \psi)$ ([L], 3.21). Thus, the inequality $\varphi_1(x) \leq \varphi_2(x)$ remains valid for all $x \in \mathcal{M}^+$, since φ_1, φ_2 are normal.

(2) By 7.1.(2), for every $\eta \in D(\mathcal{H}, \psi)$ we have $a\eta \in D(\mathcal{H}, \psi)$ and

$$\|(a^*\Delta(\varphi/\psi)a)^{1/2}\eta\|^2 = \|\Delta(\varphi/\psi)^{1/2}a\eta\|^2 = \varphi(R^{\psi}_{a\eta}(R^{\psi}_{a\eta})^*) = \varphi(aR^{\psi}_{\eta}(R^{\psi}_{\eta})^*a^*).$$

Since a is invertible, it follows from the definition of $\Delta(\varphi/\psi)$ that $a^*\Delta(\varphi/\psi)a$ is the greatest positive self-adjoint operator which satisfies the above equation; hence it coincides with the operator $\Delta(\varphi(a \cdot a^*)/\psi)$.

(3) This follows from the definition of the weak sum (A.11), using 7.3.(7). □

7.14. The operators of form $\Delta(\varphi/\psi)$ are characterized by the following

Theorem (A. Connes). *Let $\mathcal{M} \subset \mathcal{B}(\mathcal{H})$ be a von Neumann algebra with commutant $\mathcal{N} = \mathcal{M}' \subset \mathcal{B}(\mathcal{H})$ and ψ an n.s.f. weight on $\mathcal{N}$. For a positive self-adjoint operator A in $\mathcal{H}$, the following statements are equivalent:*

(i) *there exists a normal semifinite weight φ on $\mathcal{M}$ such that $A = \Delta(\varphi/\psi)$;*
(ii) *$A^{it}y = \sigma^{\psi}_{-t}(y)A^{it}$ for all $y \in \mathcal{N}$ and all $t \in \mathbb{R}$;*
(iii) *$\overline{A^{1/2}|D(A^{1/2}) \cap D(\mathcal{H}, \psi)} = A^{1/2}$ and, for any $\eta_1, \ldots, \eta_n \in D(\mathcal{H}, \psi)$, the number $\sum_{k=1}^n \|A^{1/2}\eta_k\|^2$ depends only on $\sum_{k=1}^n R^{\psi}_{\eta_k}(R^{\psi}_{\eta_k})^*$.*

Proof. The implications (i) ⇒ (ii) and (i) ⇒ (iii) are obvious.

(ii) ⇒ (i). If φ_0 is any n.s.f. weight on $\mathcal{M}$, then by (ii) the mapping $\mathbb{R} \ni t \mapsto A^{it}\Delta(\varphi_0/\psi)^{-it}$ is a σ^{φ_0}-cocycle; thus by Theorem 5.1, there exists a unique normal semifinite weight φ on $\mathcal{M}$ such that $[D\varphi : D\varphi_0]_t = A^{it}\Delta(\varphi_0/\psi)^{-it}(t \in \mathbb{R})$. By 7.4.(5), we get $\Delta(\varphi/\psi)^{it} = A^{it}(t \in \mathbb{R})$, whence $A = \Delta(\varphi/\psi)$.

(iii) ⇒ (i). Assuming (iii), we infer from 7.1.(6), 7.1.(5) that there exists a unique weight φ_0 on $\mathcal{S}(\mathcal{H}, \psi)$ such that

$$\varphi_0(a) = \sum_{k=1}^n \|A^{1/2}\eta_k\|^2 \text{ for } 0 \leq a = \sum_{k=1}^n R^{\psi}_{\eta_k}(R^{\psi}_{\eta_k})^* \in \mathcal{S}(\mathcal{H}, \psi). \tag{1}$$

If $\{z_i\} \subset \mathcal{M}$ is a net such that $z_i \xrightarrow{s} 1$, then

$$\varphi_0(a) \leq \liminf_i \varphi_0(z_i a z_i^*), \tag{2}$$

since the mapping $\mathcal{H} \ni \zeta \mapsto \|A^{1/2}\zeta\|^2$ is lower semicontinuous and $\varphi_0(z_i a z_i^*) = \sum_{k=1}^n \|A^{1/2}z_i\eta_k\|^2$.

For $x \in \mathcal{M}^+$, let

$$\varphi(x) = \sup\{\varphi_0(a); a \in \mathcal{S}(\mathcal{H}, \psi), 0 \leq a \leq x\}.$$

We thus obtain a function $\varphi : \mathscr{M}^+ \to [0, +\infty]$ with the properties

$$\varphi(\lambda x) = \lambda\varphi(x), \quad \varphi(x + y) \geq \varphi(x) + \varphi(y) \quad (x, y \in \mathscr{M}^+, \lambda \geq 0). \tag{3}$$

Assume that $x_i \uparrow x$ in $\mathscr{M}^+$. By Proposition 1.4, there exist $z_i \in \mathscr{M}$, with $\|z_i\| \leq 1$, such that $x_i = z_i x z_i^*$ and $z_i \xrightarrow{s} 1$. If $a \in \mathscr{S}(\mathscr{H}, \psi)$ and $0 \leq a \leq x$, then $0 \leq z_i a z_i^* \leq z_i x z_i^* = x_i$ and

$$\varphi_0(a) \leq \liminf_i \varphi_0(z_i a z_i^*) \leq \sup_i \varphi(x_i).$$

It follows that $\varphi(x) \leq \sup \varphi(x_i)$, and hence φ is normal. Since $\mathscr{S}(\mathscr{H}, \psi)$ is an s^*-dense two-sided ideal in $\mathscr{M}$, the elements of $\mathscr{M}^+$ can be approximated by increasing positive nets in $\mathscr{S}(\mathscr{H}, \psi)$. Since φ is normal and superadditive (cf. (3)) and since $\varphi|\mathscr{S}(\mathscr{H}, \psi) = \varphi_0$ is additive, it follows that φ is a normal weight on $\mathscr{M}$ such that

$$\varphi(R_\eta^\psi (R_\eta^\psi)^*) = \|A^{1/2}\eta\|^2 \quad (\eta \in D(\mathscr{H}, \psi)). \tag{4}$$

Now, by 7.1.(7) we see that φ is semifinite. Thus (4) is equivalent to

$$\|\Delta(\varphi/\psi)^{1/2}\eta\| = \|A^{1/2}\eta\| \quad (\eta \in D(\mathscr{H}, \psi)).$$

Since each of the operators $\Delta(\varphi/\psi)^{1/2}$ and $A^{1/2}$ is equal to the closure of its restriction to the intersection of $D(\mathscr{H}, \psi)$ with its domain of definition, it follows that $A = \Delta(\varphi/\psi)$. □

7.15. The following corollary characterizes the operators of the form $\Delta(\varphi/\psi)$ with $\varphi \in \mathscr{M}_*^+$, the assumptions being as in Theorem 7.14.

Corollary. *For a positive self-adjoint operator A in $\mathscr{H}$, the following statements are equivalent:*

(i) *there exists $\varphi \in \mathscr{M}_*^+$ such that $A = \Delta(\varphi/\psi)$;*
(ii) *$A^{it}y = \sigma_{-t}^\psi(y)A^{it}$ for all $y \in \mathscr{N}, t \in \mathbb{R}$, and there exists a family $\{\eta_k\}$ in $D(\mathscr{H}, \psi)$ such that*

$$\sum_k R_{\eta_k}^\psi (R_{\eta_k}^\psi)^* = 1 \text{ and } \sum_k \|A^{1/2}\eta_k\|^2 < +\infty.$$

(iii) *$D(\mathscr{H}, \psi) \subset D(A^{1/2}), \overline{A^{1/2}|D(\mathscr{H}, \psi)} = A^{1/2}$ and there exists a constant $0 < \lambda < +\infty$ such that for any $\eta_1, \zeta_1, \ldots, \eta_n, \zeta_n \in D(\mathscr{H}, \psi)$ we have*

$$\left|\sum_{k=1}^n (A^{1/2}\eta_k | A^{1/2}\zeta_k)\right| \leq \lambda \left\|\sum_{k=1}^n R_{\eta_k}^\psi (R_{\zeta_k}^\psi)^*\right\|.$$

Proof. (i) ⇔ (iii). If $A = \Delta(\varphi, \psi)$ with $\varphi \in \mathscr{M}_*^+$, then (iii) is obviously satisfied with $\lambda = \|\varphi\|$. Conversely, if (iii) holds, then 7.14.(iii) holds also, and so there exists a normal semifinite weight φ on $\mathscr{M}$ with $A = \Delta(\varphi/\psi)$. By the construction of φ and by (iii), it follows that $|\varphi(a)| \leq \lambda\|a\|$ for every element $0 \leq a \in \mathscr{S}(\mathscr{H}, \psi)$, hence $\varphi \in \mathscr{M}_*^+$ and $\|\varphi\| = \varphi(1) \leq \lambda$.

(i) and (iii) ⇒ (ii). This follows from 7.3.(7) and 7.1.(7).

(ii) ⇒ (i). By Theorem 7.14, there is a normal semifinite weight φ on $\mathscr{M}$ such that $\Delta(\varphi/\psi) = A$ and we have

$$\varphi(1) = \sum_k \varphi(R^{\psi}_{\eta_k}(R^{\psi}_{\eta_k})^*) = \sum_k \|A^{1/2}\eta_k\|^2 < +\infty. \qquad \square$$

7.16. Let ψ be an n.s.f. weight on the von Neumann algebra $\mathscr{N} \subset \mathscr{B}(\mathscr{H})$. A closed linear operator T in $\mathscr{H}$, with polar decomposition $T = u|T|$, is called *homogeneous of degree $s \in \mathbb{R}$ with respect to ψ*, if $u \in \mathscr{N}'$ and

$$|T|^{it}y = \sigma^{\psi}_{st}(y)|T|^{it} \quad (y \in \mathscr{N}, t \in \mathbb{R}).$$

Thus, the operators of form $A = \Delta(\varphi/\psi)$ are characterized by Theorem 7.14 as the only positive self-adjoint operators homogeneous of degree $s = -1$ with respect to ψ. For such an operator, it is possible to define "the integral with respect to ψ" by choosing any family $\{\eta_k\} \subset D(\mathscr{H}, \psi)$ with $\sum_k R^{\psi}_{\eta_k}(R^{\psi}_{\eta_k})^* = 1$ (7.1.(7)) and putting

$$\int A \,\mathrm{d}\psi = \sum_k \|A^{1/2}\eta_k\|^2 = \varphi(1). \tag{1}$$

A is called "ψ-integrable" if $\varphi(1) < +\infty$, that is if A satisfies the equivalent conditions of Corollary 7.15.

Exploiting these ideas, it is possible to develop a "noncommutative integration theory" by defining spaces $\mathscr{L}^p(\mathscr{H}, \psi)$, where $\mathscr{L}^p(\mathscr{H}, \psi)$ is the set of closed linear operators T on $\mathscr{H}$ which are homogeneous of degree $s = -1/p$ with respect to ψ and such that $|T|^p$ is ψ-integrable. As an example, we show that

$$T \in \mathscr{L}^2(\mathscr{H}, \psi) \Rightarrow T^* \in \mathscr{L}^2(\mathscr{H}, \psi) \textit{ and } \int TT^* \,\mathrm{d}\psi = \int T^*T \,\mathrm{d}\psi. \tag{2}$$

Indeed, let $T = uA = Bu$ be the polar decompositions of T with $u^*u = \mathbf{s}(A), uu^* = \mathbf{s}(B), A = |T|, B = |T^*| = uAu^*$ ([L], 9.30). If $\{\eta_k\}$ is a family in $D(\mathscr{H}, \psi)$ with $\sum_k R^{\psi}_{\eta_k}(R^{\psi}_{\eta_k})^* = \mathbf{s}(A)$, then $\{\zeta_k = u\eta_k\}$ is a family in $D(\mathscr{H}, \psi)$ with $\sum_k R_{\zeta_k}(R_{\zeta_k})^* = \mathbf{s}(B)$ and

$$\int TT^* \,\mathrm{d}\psi = \int B^2 \,\mathrm{d}\psi = \sum_k \|B\zeta_k\|^2 = \sum_k \|A\eta_k\|^2 = \int A^2 \,\mathrm{d}\psi = \int T^*T \,\mathrm{d}\psi.$$

We also record the following form of (1):

$$\int \Delta(\varphi/\psi) \,\mathrm{d}\psi = \varphi(1). \tag{3}$$

7.17. Another important application of Theorem 7.14 is contained in the following

Proposition. *Let φ, $\{\varphi_k\}_{k\in K}$ be normal semifinite weights on the von Neumann algebra $\mathscr{M} \subset \mathscr{B}(\mathscr{H})$ with commutant $\mathscr{N} = \mathscr{M}' \subset \mathscr{B}(\mathscr{H})$ and ψ an n.s.f. weight on $\mathscr{N}$. If $\varphi_k(x) \uparrow \varphi(x)$ for all $x \in \mathscr{M}^+$,* *then*

$$\sigma_t^{\varphi_k}(x) \xrightarrow{s^*} \sigma_t^{\varphi}(x) \text{ for all } x \in \bigcup_k \mathbf{s}(\varphi_k)\mathscr{M}\mathbf{s}(\varphi_k) \text{ and all } t \in \mathbb{R}; \tag{1}$$

$$[D\varphi_k : D\tau] \xrightarrow{s^*} [D\varphi : D\tau]_t \text{ for all } \tau \in W_{nsf}(\mathscr{M}) \text{ and all } t \in \mathbb{R}; \tag{2}$$

$$\Delta(\varphi_k/\psi) \uparrow \Delta(\varphi/\psi). \tag{3}$$

Moreover, the convergence in (1) and (2) is uniform for $|t| \leq t_0$.

Proof. Let $\Delta_k = \Delta(\varphi_k/\psi)$, $\Delta = \Delta(\varphi/\psi)$. Using 7.13.(1) and (A.5), we see that there exists a unique positive self-adjoint operator A in $\mathscr{H}$ such that $\Delta_k \uparrow A \leq \Delta$, hence (A.3)

$$\Delta_k^{it} \xrightarrow{s^*} A^{it} \text{ uniformly for } |t| \leq t_0.$$

Since Δ_k are homogeneous of degree $s = -1$ with respect to ψ, it follows that A enjoys the same property and hence, by Theorem 7.14, there exists a unique normal semifinite weight μ on $\mathscr{M}$ such that $A = \Delta(\mu/\psi)$. Since $A \leq \Delta$, it follows (7.13.(1)) that $\mu \leq \varphi$ and since $\Delta_k \leq A$ we have $\varphi_k \leq \mu$. Hence $\mu = \varphi$.

We have thus proved that (3) holds and that $\Delta_k^{it} \xrightarrow{s^*} A^{it}$ uniformly for $|t| \leq t_0$. Now (1) follows on applying Theorem 7.4, and (2) is obtained by applying (1) to the balanced weights $\theta(\varphi_k, \tau) \uparrow \theta(\varphi, \tau)$ on $Mat_2(\mathscr{M})$ and to the element $e_{21} \in Mat_2(\mathscr{M})$. □

7.18. A similar result holds for norm convergence of normal positive forms:

Proposition. *Let* $\mathscr{M} \subset \mathscr{B}(\mathscr{H})$ *be a von Neumann algebra with commutant* $\mathscr{N} = \mathscr{M}' \subset \mathscr{B}(\mathscr{H})$, φ, $\{\varphi_n\}_{n\in\mathbb{N}}$ *faithful normal positive forms on* $\mathscr{M}$ *and* ψ *an n.s.f. weight on* $\mathscr{N}$. *If* $\|\varphi_n - \varphi\| \to 0$, *then*

$$\sigma_t^{\varphi_n}(x) \xrightarrow{s^*} \sigma_t^{\varphi}(x) \text{ for all } x \in \mathscr{M} \text{ and all } t \in \mathbb{R}; \tag{1}$$

$$[D\varphi_n : D\tau]_t \xrightarrow{s^*} [D\varphi : D\tau]_t \text{ for all } \tau \in W_{nsf}(\mathscr{M}) \text{ and all } t \in \mathbb{R}; \tag{2}$$

$$\Delta(\varphi_n/\psi)^{it} \xrightarrow{s^*} \Delta(\varphi/\psi)^{it} \text{ for all } t \in \mathbb{R}. \tag{3}$$

Moreover, in each case convergence is uniform for $|t| \leq t_0$.

Proof. It is sufficient to prove (1). Indeed, (2) for $\tau \in \mathscr{M}_*^+$ follows from (1) by considering the balanced forms $\|\theta(\varphi_n, \tau) - \theta(\varphi, \tau)\| \to 0$ and then, for an arbitrary $\tau \in W_{nsf}(\mathscr{M})$, we can use the chain rule (3.5). Also, (3) follows from (2) using 7.4.(5):

$$\Delta(\varphi_n/\psi)^{it} = [D\varphi_n : D\varphi]_t \Delta(\varphi/\psi)^{it} \xrightarrow{s^*} \Delta(\varphi/\psi)^{it}.$$

To prove (1) we may assume that $\varphi = \omega_{\xi_0}$ where $\xi_0 \in \mathscr{H}$ is a cyclic and separating vector for $\mathscr{M}$. Then ([L], Thm. 10.25) for each $n \in \mathbb{N}$ there exists a unique vector $\xi_n \in \mathfrak{P}_{\xi_0}$, cyclic and separating for $\mathcal{M}$, such that $\varphi_n = \omega_{\xi_n}$. By ([L], Proposition 10.24), we have

$$\|\xi_n - \xi\| \le \|\varphi_n - \varphi\|^{1/2}.$$

Let $\Delta_0 = \Delta_{\xi_0}, \Delta_n = \Delta_{\xi_n}$ be the corresponding modular operators, and $J = J_{\xi_0} = J_{\xi_n}$ ([L], Lemma 2/10.24) the canonical conjugation. For every $x \in \mathcal{M}$, we have

$$\begin{aligned}\|\Delta_n^{1/2} x\xi_n - \Delta_0^{1/2} x\xi_0\| &= \|J\Delta_n^{1/2} x\xi_n - J\Delta_0^{1/2} x\xi_0\| = \|x^*\xi_n - x^*\xi_0\| \\ &\le \|x\|\|\varphi_n - \varphi\|^{1/2},\end{aligned}$$

hence $\|(\Delta_n^{1/2} + 1)x\xi_n - (\Delta_0^{1/2} + 1)x\xi_0\| \le 2\|x\|\|\varphi_n - \varphi\|^{1/2}$ and

$$\begin{aligned}&\|[(\Delta_n^{1/2} + 1)^{-1} - (\Delta_0^{1/2} + 1)^{-1}](\Delta_0^{1/2} + 1)x\xi_0\| \\ &\le \|(\Delta_n^{1/2} + 1)^{-1}[(\Delta_0^{1/2} + 1)x\xi_0 - (\Delta_n^{1/2} + 1)x\xi_n] + (x\xi_n - x\xi_0\| \\ &\le 3\|x\|\|\varphi_n - \varphi\|^{1/2}.\end{aligned}$$

Since $\Delta_0^{1/2} = \overline{\Delta_0^{1/2}|\mathcal{M}\xi_0}$, the linear subspace $(\Delta_0^{1/2} + 1)\mathcal{M}\xi_0$ is dense in $\mathcal{H}$ ([L], E.9.1) and since $\|(\Delta_n^{1/2}+1)^{-1}\| \le 1$, it follows that $(\Delta_n^{1/2}+1)^{-1} \xrightarrow{s} (\Delta_0^{1/2}+1)^{-1}$ and hence (A.3) $\Delta_n^{it} \xrightarrow{s} \Delta_0^{it}$ uniformly for $|t| \le t_0$. Consequently, for $x \in \mathcal{M}$ we have

$$\sigma_t^{\varphi_n}(x) = \Delta_n^{it} x \Delta_t^{-it} \xrightarrow{s^*} \Delta_0^{it} x \Delta_0^{-it} = \sigma_t^{\varphi}(x),$$

uniformly for $|t| \le t_0$. □

7.19. A related result is contained in the next proposition.

Proposition. *For each $t \in \mathbb{R}$, there exists a constant $0 < C_t < +\infty$ such that for every W^*-algebra $\mathcal{M}$ and every pair φ, ψ of normal states on $\mathcal{M}$ with $\mathbf{s}(\psi) \le \mathbf{s}(\varphi)$ we*

$$|1 - \varphi([D\psi : D\varphi]_t)| \le C_t\|\psi - \varphi\|.$$

Proof. If the statement is not true, then there exist W^*-algebras $\mathcal{M}_n$ and normal states φ_n, ψ_n on $\mathcal{M}_n$ with $\mathbf{s}(\varphi_n) = 1$ such that

$$\|\varphi_n - \psi_n\| \le 2^{-n}|1 - \varphi_n([D\psi_n : D\varphi_n]_t)| \quad (n \in \mathbb{N}).$$

We have

$$\sum_n \|\varphi_n - \psi_n\| < +\infty \tag{1}$$

and, by repeating some pairs φ_n, ψ_n if necessary, we may further assume that

$$\sum_n |1 - \varphi_n([D\psi_n : D\varphi_n]_t)| = +\infty. \tag{2}$$

We may assume $\mathscr{M}_n$ realized as a von Neumann algebra $\mathscr{M}_n \subset \mathscr{B}(\mathscr{H}_n)$ with cyclic and separating vector $\xi_n \in \mathscr{H}_n$ such that $\varphi_n = \omega_{\xi_n}$. By ([L], 10.24, 10.25) there exists $\eta_n \in \mathscr{H}_n$ such that $\psi_n = \omega_{\eta_n}, (\eta_n|\xi_n) > 0$ and

$$\|\eta_n - \xi_n\|^2 \leq \|\varphi_n - \psi_n\| \quad (n \in \mathbb{N}). \tag{3}$$

The infinite tensor product W^*-algebra $\mathscr{M} = \otimes_n(\mathscr{M}_n, \varphi_n)$ (see A.17) can be realized as an infinite tensor product von Neumann algebra acting on the Hilbert space $\mathscr{H} = \otimes_n(\mathscr{H}_n, \xi_n)$, and on this von Neumann algebra we have vector states

$$\Phi = \otimes_n \varphi_n \textit{ and } \Phi_k = \otimes_{n=1}^k \psi_n \otimes \otimes_{m=k+1}^\infty \varphi_m$$

defined, respectively, by the vectors

$$\xi = \otimes_n \xi_n \in \mathscr{H} \textit{ and } \eta^{(k)} = \otimes_{n=1}^k \eta_n \otimes \otimes_{m=k+1}^\infty \xi_m \in \mathscr{H}.$$

From (1) and (3), it follows that the sequence $\{\eta^{(k)}\}$ is norm-convergent in $\mathscr{H}$, so that the sequence $\{\Phi_k\}$ is norm-convergent in $\mathscr{M}_*$. Taking into account the results of Section 3.9 and Proposition 7.18, it follows that the sequence

$$[D\Phi_k : D\Phi]_t = [D\psi_1 : D\varphi_1) \otimes \ldots \otimes [D\psi_k : D\varphi_k]_t \otimes 1$$

is s-convergent in $\mathscr{M}$. In particular, the sequence

$$\prod_{n=1}^k ([D\psi_n : D\varphi_n]_t \xi_n | \xi_n) = ([D\Phi_k : D\Phi]_t \xi | \xi)$$

is convergent, and hence

$$\sum_n |1 - ([D\psi_n : D\varphi_n]_t \xi_n | \xi_n)| < +\infty.$$

contradicting (2). □

7.20. Let φ be a normal positive form on the W^*-algebra $\mathscr{M}$. Besides the notation $\|x\|_\varphi = \|x_\varphi\|_\varphi = \varphi(x^*x)^{1/2}$, we shall also write $\|x\|_\varphi^\sharp = \varphi(x^*x + xx^*)^{1/2}, (x \in \mathscr{M})$. For $a \in \mathscr{M}$, we shall consider the commutator $[a, \varphi] = \varphi(\cdot a) - \varphi(a \cdot)$.

Corollary. *There exists an absolute constant* $0 < C < +\infty$ *such that if* $\mathscr{M}$ *is a* W^**-algebra,* φ *a normal state on* $\mathscr{M}$ *and* $a \in \mathscr{M}$*, with* $\|a\| \leq 1$*, we have*

$$\|\sigma_t^\varphi(a) - a\|_\varphi^* \leq C(1 + |t|)\|[a, \varphi]\|^{1/2} \quad (t \in \mathbb{R}).$$

Proof. We divide the proof into several steps.

(I) First, let $a = v \in \mathscr{M}$ be unitary. Then, using Proposition 7.19 for φ and $\psi = \varphi(v \cdot v^*)$, and the identity $[D\psi : D\varphi]_t = v^*\sigma_t^\varphi(v)$ (cf. 3.7), we obtain

$$\|\sigma_t^\varphi(v) - v\|_\varphi^\sharp \leq 2C_t^{1/2}\|[v, \varphi]\|^{1/2} \quad (t \in \mathbb{R}).$$

(II) If $0 \le a \le 1/2$, then $\|[(1-a^2)^{1/2}, \varphi]\| \le \frac{2}{3}\|[a,\varphi]\|$.
Indeed, by induction over n it is easy to check that

$$\|[a^n, \varphi]\| \le n\|a\|^{n-1}\|[a,\varphi]\| \le n2^{-n+1}\|[a,\varphi)\|$$

and the desired inequality follows using the expansion

$$(1-a^2)^{1/2} = \sum_{n=0}^{\infty}(-1)^n \frac{2^{-1}(2^{-1}-1)\dots(2^{-1}-n+1)}{n!}a^{2n}.$$

(III) If $a = a^*$ and $\|a\| \le 1$, then

$$\|\sigma_t^{\varphi}(a) - a\|_{\varphi}^{*} \le 8C_t^{1/2}\|[a,\varphi]\|^{1/2} \quad (t \in \mathbb{R}).$$

Indeed, let $\|[a,\varphi]\| = \varepsilon$. We have $0 \le (1+a)/4 \le 1/2$. Putting $v = (1+a)/4 + \mathrm{i}(1-((1+a)/4)^2)^{1/2}$, it follows from (II) that $\|[v,\varphi]\| \le \varepsilon/2$ and $\|[v^*,\varphi]\| \le \varepsilon/2$; and, since v is unitary, we get from (I) that $\|\sigma_t^{\varphi}(v) - v\|_{\varphi}^{\sharp} \le 2C_t^{1/2}\varepsilon^{1/2}$, $\|\sigma_t^{\varphi}(v^*) - v^*\|_{\varphi}^{\sharp} \le 2C_t^{1/2}\varepsilon^{1/2}$. Thus, the desired conclusion follows, as $(1+a)/2 = v + v^*$.

(IV) Consider now $a \in \mathscr{M}$, $\|a\| \le 1$. By applying (III) for $(a+a^*)/2$ and $(a-a^*)/2\mathrm{i}$, we get

$$\|\sigma_t^{\varphi}(a) - a\|_{\varphi}^{\sharp} \le 16C_t^{1/2}\|[a,\varphi]\|^{1/2} \quad (t \in \mathbb{R}).$$

For each $t \in \mathbb{R}$, we define $K(t)$ as the least upper bound of the numbers $\lambda > 0$ such that $\|\sigma_t^{\varphi}(a) - a\|_{\varphi}^{\sharp} \le \lambda\|[a,\varphi]\|^{1/2}$ for every $\mathscr{M}, \varphi, a$ as in the statement of the corollary. Then the function $\mathbb{R} \ni t \mapsto K(t)$ is lower semicontinuous, $K(0) = 0$, $K(-t) = K(t)$ and $K(t+s) \le K(t) + K(s)$, $(t, s \in \mathbb{R})$. With these conditions, it is easy to show that $C = \sup\{K(t)/(1+|t|) : t \in \mathbb{R}\} < +\infty$ and this proves the corollary. □

In particular, we obtain, as an immediate consequence, the implication ($\Leftarrow$) of 2.21.(2) for $\varphi \in \mathscr{M}_*^+$. Moreover, if $\{a_i\}_i \subset \mathscr{M}$ is a norm-bounded net, then

$$\|[a_i,\varphi]\| \to 0 \Rightarrow \|\sigma^{\varphi}(a_i) - a_i\|_{\varphi}^{*} \to 0 \tag{1}$$

uniformly for $|t| \le t_0$.

7.21. Notes. The results in this section are due to Connes (1980). An important part of the motivation of Connes (1980) was the work of Haagerup (1979a), which included Corollary 7.12 and several arguments used in the proof of Lemma 7.6. The proof of Proposition 7.18 is due to Araki (1974).

For our exposition, we have used Araki (1974); Connes (1980) and Haagerup (1979a).

Noncommutative integration theory, initiated and developed in the semifinite case by Dixmier (1953) and Segal (1953) has been extended to the general case by Haagerup (1979b). Introducing the spaces $\mathscr{L}^p(\mathscr{H},\varphi)$ (7.16), Connes (1980) proposed the problem of establishing the properties of these spaces and their connection with the earlier theory of Haagerup (1979b); this has been done by Hilsum (1981). Another general approach is contained in the recent work of Connes (1979a).

8 Tensor Products

In this section, we introduce the tensor product of weights, starting with the tensor product of left Hilbert algebras, and study its properties.

8.1 Proposition. *Let* $\mathfrak{A}_k \subset \mathscr{H}_k$ *be a left Hilbert algebra with associated operators* $S_k, S_k^*, J_k, \Delta_k (k = 1, 2)$. *Then*

$$\mathfrak{A} = \mathfrak{A}_1 \otimes \mathfrak{A}_2 \subset \mathscr{H}_1 \bar{\otimes} \mathscr{H}_2 = \mathscr{H},$$

equipped with the tensor product involutive algebra structure and with the scalar product of $\mathscr{H}$ *is a left Hilbert algebra with associated operators*

$$S = S_1 \bar{\otimes} S_2, S^* = S_1^* \bar{\otimes} S_2^*, J = J_1 \bar{\otimes} J_2, \Delta = \Delta_1 \bar{\otimes} \Delta_2; \tag{1}$$

we have

$$L_{\xi_1 \otimes \xi_2} = L_{\xi_1} \bar{\otimes} L_{\xi_2} \quad (\xi_1 \in D(S_1), \xi_2 \in D(S_2)) \tag{2}$$

$$R_{\eta_1 \otimes \eta_2} = R_{\eta_1} \bar{\otimes} R_{\eta_2} \quad (\eta_1 \in D(S_1^*), \eta_2 \in D(S_2^*)) \tag{3}$$

$$\mathfrak{L}(\mathfrak{A}) = \mathfrak{L}(\mathfrak{A}_1) \bar{\otimes} \mathfrak{L}(\mathfrak{A}_2), \mathfrak{R}(\mathfrak{A}') = \mathfrak{R}(\mathfrak{A}_1') \bar{\otimes} \mathfrak{R}(\mathfrak{A}_2'). \tag{4}$$

Proof. We begin by checking the axioms of a left Hilbert algebra ([L], 10.1. (i)–(iv)) for $\mathfrak{A}$. In order to avoid notational complications, we shall write the sums which define the elements of $\mathfrak{A} = \mathfrak{A}_1 \otimes \mathfrak{A}_2$ without specifying summation indices. Let $\xi_1, \eta_1, \zeta_1 \in \mathfrak{A}_1$ and $\xi_2, \eta_2, \zeta_2 \in \mathfrak{A}_2$.

(i) We have $(\sum \xi_1 \otimes \xi_2)(\sum \eta_1 \otimes \eta_2) = \sum \xi_1\eta_1 \otimes \xi_2\eta_2 = \sum L_{\xi_1}\eta_1 \otimes L_{\xi_2}\eta_2 = (\sum L_{\xi_1} \otimes L_{\xi_2})(\sum \eta_1 \otimes \eta_2)$, hence the mapping $(\sum \eta_1 \otimes \eta_2) \mapsto (\sum \xi_1 \otimes \xi_2)(\sum \eta_1 \otimes \eta_2)$ is continuous and

$$L_{\xi_1 \otimes \xi_2} = L_{\xi_1} \bar{\otimes} L_{\xi_2} \quad (\xi_1 \in \mathfrak{A}_1, \xi_2 \in \mathfrak{A}_2).$$

(ii) We have

$$\begin{aligned} \left(\left(\sum \xi_1 \otimes \xi_2\right)\left(\sum \eta_1 \otimes \eta_2\right) \Big| \left(\sum \zeta_1 \otimes \zeta_2\right)\right) &= \sum (\xi_1\eta_1|\zeta_1)(\xi_2\eta_2|\zeta_2) \\ = \sum (\eta_1|\xi_1^*\zeta_1)(\eta_2|\xi_2^*\zeta_2) &= \left(\left(\sum \eta_1 \otimes \eta_2\right) \Big| \left(\sum \xi_1 \otimes \xi_2\right)^* \left(\sum \zeta_1 \otimes \zeta_2\right)\right) \end{aligned}$$

(iii) It is clear that $\mathfrak{A}^2$ is dense in $\mathfrak{A}$ and hence

$$\begin{aligned} \mathfrak{L}(\mathfrak{A}) &= \overline{\{L_{\xi_1 \otimes \xi_2}; \xi_1 \in \mathfrak{A}_1, \xi_2 \in \mathfrak{A}_2\}}^{so} \\ &= \overline{\{L_{\xi_1} \bar{\otimes} L_{\xi_2}; \xi_1 \in \mathfrak{A}_1, \xi_2 \in \mathfrak{A}_2\}}^{so} = \mathfrak{L}(\mathfrak{A}_1) \bar{\otimes} \mathfrak{L}(\mathfrak{A}_2). \end{aligned}$$

(iv) We have $(\sum \xi_1\eta_1 \otimes \xi_2\eta_2)^* = \sum (\xi_1\eta_1)^* \otimes (\xi_2\eta_2)^* = \sum S_1(\xi_1\eta_1) \otimes S_2(\xi_2\eta_2) = (S_1 \otimes S_2)(\sum \xi_1\eta_1 \otimes \xi_2\eta_2)$. Since S_1 and S_2 are preclosed, it follows that $S_1 \otimes S_2$ is preclosed ([L], 9.33) and $S = S_1 \bar{\otimes} S_2$. □

Thus, $\mathfrak{A}$ is indeed a left Hilbert algebra, $\mathfrak{L}(\mathfrak{A}) = \mathfrak{L}(\mathfrak{A}_1) \bar{\otimes} \mathfrak{L}(\mathfrak{A}_2)$ and $S = S_1 \bar{\otimes} S_2$. From ([L], 9.34) it follows that $S^* = S_1^* \bar{\otimes} S_2^*$. Then

$$J\Delta^{1/2} = S = S_1 \bar{\otimes} S_2 = J_1\Delta_1^{1/2} \bar{\otimes} J_2\Delta_2^{1/2} = (J_1 \bar{\otimes} J_2)(\Delta_1^{1/2} \bar{\otimes} \Delta_2^{1/2}),$$

the last equality being an easy exercise. Since $\Delta_1^{1/2} \bar{\otimes} \Delta_2^{1/2}$ is a positive self-adjoint operator ([L], 9.34), it follows by the uniqueness of the polar decomposition that $J = J_1 \bar{\otimes} J_2$ and $\Delta^{1/2} = \Delta_1^{1/2} \bar{\otimes} \Delta_2^{1/2}$; hence $\Delta = \Delta_1 \bar{\otimes} \Delta_2$. The other assertions in the statement are now easily verified.

Using Stone's theorem ([L], 9.20) we obtain

$$\Delta^{it} = \Delta_1^{it} \bar{\otimes} \Delta_2^{it} \quad (t \in \mathbb{R}) \tag{5}$$

and then, by analytic continuation,

$$\Delta^{\alpha} = \Delta_1^{\alpha} \bar{\otimes} \Delta_2^{\alpha} \quad (\alpha \in \mathbb{C}). \tag{6}$$

We shall denote by $\mathfrak{A}_1 \bar{\otimes} \mathfrak{A}_2$ the maximal left Hilbert algebra $\mathfrak{A}''$ associated with $\mathfrak{A}$.

From (3), it follows that $\mathfrak{A}_1' \otimes \mathfrak{A}_2' \subset \mathfrak{A}'$. On the other hand, we have

$$\begin{aligned}\overline{S^*|\mathfrak{A}_1' \otimes \mathfrak{A}_2'} &= \overline{S_1^* \otimes S_2^*|\mathfrak{A}_1' \otimes \mathfrak{A}_2'} = \overline{(S_1^*|\mathfrak{A}_1') \otimes (S_2^*|\mathfrak{A}_2')} \\ &= \overline{(S_1^*|\mathfrak{A}_1')} \bar{\otimes} \overline{(S_2^*|\mathfrak{A}_2')} = S_1^* \bar{\otimes} S_2^* = S^*\end{aligned}$$

and consequently ([L], Lemma 3/10.5)

$$\mathfrak{A}_1' \bar{\otimes} \mathfrak{A}_2' = \mathfrak{A}', \tag{7}$$

that is, $\mathfrak{A}'$ is the maximal right Hilbert algebra associated with the right Hilbert algebra $\mathfrak{A}_1' \otimes \mathfrak{A}_2' \subset \mathscr{H}$. Similarly, we get

$$\mathfrak{A}_1'' \bar{\otimes} \mathfrak{A}_2'' = \mathfrak{A}_1 \bar{\otimes} \mathfrak{A}_2. \tag{8}$$

By ([L], 10.4.(2)) we have $\mathfrak{L}(\mathfrak{A})' = \mathfrak{R}(\mathfrak{A}')$, so that from (4) we once again obtain the result ([L], Thm. 10.7)

$$(\mathfrak{L}(\mathfrak{A}_1) \bar{\otimes} \mathfrak{L}(\mathfrak{A}_2))' = \mathfrak{L}(\mathfrak{A}_1)' \bar{\otimes} \mathfrak{L}(\mathfrak{A}_2)'.$$

8.2 Theorem. *Let φ_k be a normal semifinite weight on the W*-algebra $\mathscr{M}_k$, $(k = 1, 2)$. There exists a unique normal semifinite weight $\varphi = \varphi_1 \bar{\otimes} \varphi_2$ on the W*-algebra $\mathscr{M} = \mathscr{M}_1 \bar{\otimes} \mathscr{M}_2$ such that*

$$x_1 \in \mathfrak{M}_{\varphi_1}, x_2 \in \mathfrak{M}_{\varphi_2} \Rightarrow x_1 \bar{\otimes} x_2 \in \mathfrak{M}_{\varphi} \text{ and } \varphi(x_1 \bar{\otimes} x_2) = \varphi_1(x_1)\varphi_2(x_2) \tag{1}$$

$$\mathbf{s}(\varphi) = \mathbf{s}(\varphi_1) \bar{\otimes} \mathbf{s}(\varphi_2) \text{ and } \sigma_t^{\varphi} = \sigma_t^{\varphi_1} \bar{\otimes} \sigma_t^{\varphi_2} \quad (t \in \mathbb{R}). \tag{2}$$

Proof. Assume first that φ_1 and φ_2 are n.s.f. weights. To prove the existence assertion, we consider the standard representations $\pi_k : \mathscr{M}_k \to \mathscr{B}(\mathscr{H}_{\varphi_k})$ associated with φ_k and the maximal left Hilbert algebras $\mathfrak{A}_{\varphi_k} \subset \mathscr{H}_{\varphi_k}, (k = 1, 2)$. Let $\mathfrak{A} = \mathfrak{A}_{\varphi_1} \bar{\otimes} \mathfrak{A}_{\varphi_2} \subset \mathscr{H}_{\varphi_1} \bar{\otimes} \mathscr{H}_{\varphi_2} = \mathscr{H}$ be the tensor product left Hilbert algebra (8.1). Then $\pi = \pi_1 \bar{\otimes} \pi_2$ is a *-isomorphism of the W^*-algebra $\mathscr{M}$ onto the von Neumann algebra $\mathfrak{L}(\mathfrak{A})$. If $\varphi_{\mathfrak{A}}$ denotes the natural weight on $\mathfrak{L}(\mathfrak{A})$, then $\varphi = \varphi_{\mathfrak{A}} \cdot \pi$ is an n.s.f. weight on $\mathscr{M}$ and $\pi : \mathscr{M} \to \mathscr{B}(\mathscr{H})$ can be identified with the standard representation associated with φ.

Let $0 \leq x_1 \in \mathfrak{M}_{\varphi_1}; 0 \leq x_2 \in \mathfrak{M}_{\varphi_2}$. Then $x_1^{1/2} \in \mathfrak{A}_{\varphi_1}, x_2^{1/2} \in \mathfrak{A}_{\varphi_2}$ and

$$\pi(x_1 \bar{\otimes} x_2)^{1/2} = L_{(x_1^{1/2})_{\varphi_1} \bar{\otimes} (x_2^{1/2})_{\varphi_2}}.$$

Hence

$$\begin{aligned}\varphi(x_1 \bar{\otimes} x_2) &= \varphi_{\mathfrak{A}}(\pi(x_1 \bar{\otimes} x_2)) = \|(x_1^{1/2})_{\varphi_1} \bar{\otimes} (x_2^{1/2})_{\varphi_2}\|^2 \\ &= \|(x_1^{1/2})_{\varphi_1}\|_{\varphi_1}^2 \|(x_2^{1/2})_{\varphi_2}\|_{\varphi_2}^2 = \varphi_1(x_1)\varphi_2(x_2).\end{aligned}$$

On the other hand, using (8.1.(5)) we obtain

$$\begin{aligned}\pi(\sigma_t^{\varphi}(x_1 \bar{\otimes} x_2)) &= \Delta_{\varphi}^{it} \pi(x_1 \bar{\otimes} x_2) \Delta_{\varphi}^{-it} \\ &= (\Delta_{\varphi_1}^{it} \pi_1(x_1) \Delta_{\varphi_1}^{-it}) \bar{\otimes} (\Delta_{\varphi_2}^{it} \pi_2(x_2) \Delta_{\varphi_2}^{-it}) \\ &= \pi_1(\sigma_t^{\varphi_1}(x_1)) \bar{\otimes} \pi_2(\sigma_t^{\varphi_2}(x_2)) = \pi((\sigma_t^{\varphi_1} \bar{\otimes} \sigma_t^{\varphi_2})(x_1 \bar{\otimes} x_2)).\end{aligned}$$

Thus, φ satisfies conditions (1) and (2). If ψ is another n.s.f. weight on $\mathscr{M}$ satisfying (1) and (2), ψ commutes with φ and coincides with φ on the σ^{φ}-invariant and w-dense *-subalgebra $\mathfrak{M}_{\varphi_1} \otimes \mathfrak{M}_{\varphi_2}$ of $\mathscr{M}$, so that $\psi = \varphi$ by the Pedersen–Takesaki theorem (6.2).

If φ_1 and φ_2 are not necessarily faithful, we define $\varphi = \varphi_1 \bar{\otimes} \varphi_2$ as an n.s.f. weight on the W^*-algebra $e\mathscr{M}e$, where $e = \mathbf{s}(\varphi_1) \bar{\otimes} \mathbf{s}(\varphi_2)$, and then consider φ as a weight on $\mathscr{M}$, that is, $\varphi(x) = \varphi(exe)(x \in \mathscr{M}^+)$.

If φ_1 and φ_2 are normal positive forms on the W^*-algebras $\mathscr{M}_1$ and $\mathscr{M}_2$, respectively, the normal positive form $\varphi_1 \bar{\otimes} \varphi_2$ on $\mathscr{M}_1 \bar{\otimes} \mathscr{M}_2$ is determined uniquely by condition 8.2.(1) alone (see 3.9).

Let φ be an n.s.f. weight on the W^*-algebra $\mathscr{M}$ and let tr be the canonical trace on a factor $\mathscr{F}_n$ of type I_n. Then $\mathscr{M} \bar{\otimes} \mathscr{F}_n$ can be identified with $Mat_n(\mathscr{M})$ such that (see 5.2)

$$(\varphi \bar{\otimes} tr)(x) = \sum_k \varphi(x_{kk}) \quad (x = [x_{ij}] \in Mat_n(\mathscr{M})^+).$$ □

8.3. Recall (2.6) that for any normal semifinite weight φ on the W^*-algebra $\mathscr{M}$ there exists an increasing net $\{\varphi_{\rm i}\}_{{\rm i}\in I}$ of normal positive forms on $\mathscr{M}$ such that $\varphi_{\rm i} \uparrow \varphi$, that is, $\varphi_{\rm i}(x) \uparrow \varphi(x)$ for all $x \in \mathscr{M}^+$.

Proposition. *Let φ, ψ be normal semifinite weights and $\{\varphi_{\rm i}\}_{{\rm i}\in I}, \{\psi_j\}_{j\in J}$ increasing nets of normal positive forms on the W^*-algebra $\mathscr{M}, \mathscr{N}$, respectively. If $\varphi_{\rm i} \uparrow \varphi$ and $\psi_{\rm j} \uparrow \psi$, then $\varphi_{\rm i} \bar{\otimes} \psi_{\rm j} \uparrow \varphi \bar{\otimes} \psi$.*

Proof. Indeed, $\{\varphi_{\rm i} \otimes \psi_{\rm i}\}_{{\rm i}\in I, {\rm j}\in J}$ is an increasing net of normal positive forms -on $\mathscr{M} \bar{\otimes} \mathscr{N}$ and $\varphi_{\rm i} \bar{\otimes} \psi_{\rm j} \leq \varphi \bar{\otimes} \psi$ for all ${\rm i} \in I$, ${\rm j} \in J$. Consequently, we define a normal weight ω on $\mathscr{M} \bar{\otimes} \mathscr{N}$ by

$$\omega(z) = \sup_{\rm ij}(\varphi_{\rm i} \bar{\otimes} \psi_{\rm j})(z) = \lim_{\rm ij}(\varphi_{\rm i} \bar{\otimes} \psi_{\rm j})(z) \quad (z \in (\mathscr{M} \bar{\otimes} \mathscr{N})^+).$$

For $a \in \mathfrak{M}_\varphi \cap \mathcal{M}^+, b \in \mathfrak{M}_\psi \cap \mathcal{N}^+$ we have

$$\omega(a \bar{\otimes} b) = \sup_{ij} \varphi_i(a)\psi_j(b) = \sup_i \varphi_i(a) \sup_j \psi_j(b) = \varphi(a)\psi(b).$$

On the other hand, it is easy to see that $\mathbf{s}(\varphi_i) \uparrow \mathbf{s}(\varphi), \mathbf{s}(\psi_j) \uparrow \mathbf{s}(\psi)$, and $\mathbf{s}(\varphi_i \bar{\otimes} \psi_j) \uparrow \mathbf{s}(\omega)$, whence $\mathbf{s}(\omega) = \mathbf{s}(\varphi) \bar{\otimes} \mathbf{s}(\psi)$. Moreover, by Proposition 7.17, for $t \in \mathbb{R}, x \in \bigcup_{i \in I} \mathbf{s}(\varphi_i)\mathcal{M}\mathbf{s}(\varphi_i), y \in \bigcup_{j \in J} \mathbf{s}(\psi_j)\mathcal{N}\mathbf{s}(\psi_j)$ we have

$$\sigma_t^{\varphi_i}(x) \xrightarrow{s} \sigma_t^{\varphi}(x), \sigma_t^{\psi_j}(y) \xrightarrow{s} \sigma_t^{\psi}(y), \sigma_t^{\varphi_i \bar{\otimes} \psi_j}(x \bar{\otimes} y) \xrightarrow{s} \sigma_t^{\omega}(x \bar{\otimes} y),$$

hence $\sigma_t^{\omega}(x \bar{\otimes} y) = \sigma_t^{\varphi}(x) \bar{\otimes} \sigma_t^{\psi}(y)$. Since $\mathbf{s}(\varphi_i) \uparrow \mathbf{s}(\varphi), \mathbf{s}(\psi_j) \uparrow \mathbf{s}(\psi)$, this equality holds for every $x \in \mathbf{s}(\varphi)\mathcal{N}\mathbf{s}(\varphi), y \in \mathbf{s}(\psi)\mathcal{M}\mathbf{s}(\psi)$. Thus, the weight ω satisfies 8.2.(1) and 8.2.(2), which determine $\varphi \bar{\otimes} \psi$; hence $\omega = \varphi \bar{\otimes} \psi$, that is, $\varphi_i \bar{\otimes} \psi_j \uparrow \varphi \bar{\otimes} \psi$. □

8.4. As a first application we obtain the distributive law for the tensor product with respect to addition:

Corollary. *Let φ_1, φ_2 be normal semifinite weights on the W^*-algebra $\mathcal{M}$ and ψ a normal semifinite weight on the W^*-algebra $\mathcal{N}$. If $\varphi_1 + \varphi_2$ is semifinite, then $(\varphi_1 + \varphi_2) \bar{\otimes} \psi = \varphi_1 \bar{\otimes} \psi + \varphi_2 \bar{\otimes} \psi$.*

Proof. Let $\{\psi_k\}$ be an increasing net of normal positive forms on $\mathcal{N}$ such that $\psi_k \uparrow \psi$.

Assume that φ_1, φ_2 are normal positive forms. Since the distributive law is obvious for normal positive forms, we have by Proposition 8.3, $(\varphi_1 + \varphi_2) \bar{\otimes} \psi = \sup_k (\varphi_1 \bar{\otimes} \varphi_2) \bar{\otimes} \psi_k = \sup_k \varphi_1 \bar{\otimes} \psi_k + \sup_k \varphi_2 \bar{\otimes} \psi_k = \varphi_1 \bar{\otimes} \psi + \varphi_2 \bar{\otimes} \psi$.

In the general case, let $\{\varphi_{1i}\}, \{\psi_{2j}\}$ be increasing nets of normal positive forms on $\mathcal{M}$ such that $\varphi_{1i} \uparrow \varphi_1, \psi_{2j} \uparrow \psi_2$. It is then obvious that $\varphi_{1i} + \varphi_{2j} \uparrow \varphi_1 + \varphi_2$. We have, again by Proposition 8.3 and the first part of the proof, $(\varphi_1 + \varphi_2) \bar{\otimes} \psi = \sup_{ij} (\varphi_{1i} + \varphi_{2j}) \bar{\otimes} \psi = \sup_{ij} (\varphi_{1i} \bar{\otimes} \psi + \varphi_{2j} \bar{\otimes} \psi) = \sup_i \varphi_{1i} \bar{\otimes} \psi + \sup_j \varphi_{2j} \bar{\otimes} \psi = \varphi_1 \bar{\otimes} \psi + \varphi_2 \bar{\otimes} \psi$. □

8.5. Another application concerns the relation between the tensor product and the balanced weight:

Corollary. *Let φ_1, φ_2 be normal semifinite weights on the W^*-algebra $\mathcal{M}$ and ψ a normal semifinite weight on the W^*-algebra $\mathcal{N}$. Then $\theta(\varphi_1 \bar{\otimes} \psi, \varphi_2 \bar{\otimes} \psi) = \theta(\varphi_1, \varphi_2) \bar{\otimes} \psi$ as weights on the W^*-algebra $Mat_2(\mathcal{M} \bar{\otimes} \mathcal{N}) \approx (\mathcal{M} \bar{\otimes} \mathcal{N}) \otimes Mat_2(\mathbb{C}) \approx Mat_2(\mathcal{M}) \bar{\otimes} \mathcal{N}$.*

Proof. Let $\{\varphi_{1i}\}, \{\varphi_{2j}\}, \{\psi_k\}$ be increasing nets of normal positive forms such that $\varphi_{1i} \uparrow \varphi_1, \varphi_{2j} \uparrow \varphi_2, \psi_k \uparrow \psi$. For the balanced weights is obvious that $\theta(\varphi_{1i}, \varphi_{2j}) \uparrow \theta(\varphi_1, \varphi_2)$; using Proposition 8.3, we get $\theta(\varphi_{1i}, \varphi_{2j}) \bar{\otimes} \psi_k \uparrow \theta(\varphi_1, \varphi_2) \bar{\otimes} \psi$.

Also, we have $\varphi_{1i} \bar{\otimes} \psi_k \uparrow \varphi_1 \bar{\otimes} \psi, \varphi_{2j} \bar{\otimes} \psi_k \uparrow \varphi_2 \bar{\otimes} \psi$, hence $\theta(\varphi_{1i} \bar{\otimes} \psi_k, \varphi_{2j} \bar{\otimes} \psi_k) \uparrow \theta(\varphi_1 \bar{\otimes} \psi, \varphi_2 \bar{\otimes} \psi)$. Since $\theta(\varphi_{1i} \bar{\otimes} \psi_k, \varphi_{2j} \bar{\otimes} \psi_k) = \theta(\varphi_{1i}, \varphi_{2j}) \bar{\otimes} \psi_k$ for all i, j, k (see 3.9), it follows that $\theta(\varphi_1 \bar{\otimes} \psi, \varphi_2 \bar{\otimes} \psi) = \theta(\varphi_1, \varphi_2) \bar{\otimes} \psi$. □

8.6. If $\mathbf{s}(\varphi_2) \le \mathbf{s}(\varphi_1)$, then, arguing as in Section 3.9, we infer from Corollary 8.5 that

$$[D(\varphi_2 \bar{\otimes} \psi) : D(\varphi_1 \bar{\otimes} \psi)]_t = [D\varphi_2 : D\varphi_1]_t \bar{\otimes} \mathbf{s}(\psi) \quad (t \in \mathbb{R}).$$

Using the chain rule (3.5), we obtain the following general result:

Corollary. *Let* φ_1, φ_2 *be normal semifinite weights on* $\mathcal{M}$ *with* $\mathbf{s}(\varphi_2) \le \mathbf{s}(\varphi_1)$ *and* ψ_1, ψ_2 *normal semifinite weights on* $\mathcal{N}$ *with* $\mathbf{s}(\psi_2) \le \mathbf{s}(\psi_1)$. *Then*

$$[D(\varphi_2 \bar{\otimes} \psi_2) : D(\varphi_1 \bar{\otimes} \psi_1)]_t = [D\varphi_2 : D\varphi_1] \bar{\otimes} [D\psi_2 : D\psi_1]_t \quad (t \in \mathbb{R}).$$

8.7. In particular, using Corollary 4.8, we obtain:

Corollary. *Let* φ, ψ *be normal semifinite weights on the* W^**-algebras* $\mathcal{M}, \mathcal{N}$*, respectively, and* A, B *positive self-adjoint operators affiliated to* $\mathcal{M}^{\varphi}, \mathcal{N}^{\psi}$*, respectively. Then* $A \bar{\otimes} B$ *is a positive self-adjoint operator affiliated to* $(\mathcal{M} \bar{\otimes} \mathcal{N})^{\varphi \bar{\otimes} \psi}$ *and*

$$(\varphi \bar{\otimes} \psi)_{A \bar{\otimes} B} = \varphi_A \bar{\otimes} \psi_B.$$

8.8. Let $\varphi, \varphi_1, \varphi_2, \{\varphi_i\}$ be normal semifinite weights on the W^*-algebra $\mathcal{M}$ and $\psi, \psi_1, \psi_2, \{\psi_j\}$ normal semifinite weights on the W^*-algebra $\mathcal{N}$.

By Corollary 8.6 and Corollary 3.13,

$$\varphi_1 \le \varphi_2, \psi_1 \le \psi_2 \Rightarrow \varphi_1 \bar{\otimes} \psi_1 \le \varphi_2 \bar{\otimes} \psi_2. \tag{1}$$

Also, arguing as in the proof of Proposition 8.3,

$$\varphi_{\mathrm{i}} \uparrow \varphi, \psi_{\mathrm{j}} \uparrow \psi \Rightarrow \varphi_{\mathrm{i}} \bar{\otimes} \psi_{\mathrm{j}} \uparrow \varphi \bar{\otimes} \psi, \tag{2}$$

and, by Corollary 8.4,

$$\left(\sum_{\mathrm{i}} \varphi_{\mathrm{i}}\right) \bar{\otimes} \left(\sum_{\mathrm{j}} \psi_{\mathrm{j}}\right) = \sum_{\mathrm{ij}} \varphi_{\mathrm{i}} \bar{\otimes} \psi_{\mathrm{j}}. \tag{3}$$

8.9. If φ, ψ are n.s.f. weights on the W^*-algebras $\mathcal{M}, \mathcal{N}$, respectively, and $\sigma \in Aut(\mathcal{M}), \tau \in Aut(\mathcal{N})$, by the definition (8.2) of the tensor product weight and 2.22.(5), we get

$$(\varphi \bar{\otimes} \psi) \circ (\sigma \bar{\otimes} \tau) = (\varphi \circ \sigma) \bar{\otimes} (\psi \circ \tau).$$

More generally, arguing as in the proofs of Corollaries 8.4 and 8.5, we obtain the following

Corollary. *Let* $\Phi : \mathcal{M}_1 \to \mathcal{M}, \Psi : \mathcal{N}_1 \to \mathcal{N}$ *be normal completely positive linear mappings between* W^**-algebras and* φ, ψ *normal semifinite weights on* $\mathcal{M}, \mathcal{N}$ *respectively. If the weights* $\varphi \circ \Phi, \psi \circ \Psi$ *are semifinite, then*

$$(\varphi \bar{\otimes} \psi) \circ (\Phi \bar{\otimes} \Psi) = (\varphi \circ \Phi) \bar{\otimes} (\psi \circ \Psi).$$

Recall (Sakai, 1971; Strătilă & Zsidó, 1977–2005, p. 8.8) that the algebraic tensor product $\Phi \otimes \Psi$ of two normal completely positive linear mappings extends to a normal completely positive linear mapping $\Phi \bar{\otimes} \Psi : \mathcal{M}_1 \bar{\otimes} \mathcal{N}_1 \to \mathcal{M} \bar{\otimes} \mathcal{N}$.

8.10. While the equality of normal forms is equivalent to their equality on w-dense subsets, the equality of normal weights is, as we have seen (6.2, 6.6), a more delicate problem. The next result concerns the equality of certain tensor product weights.

Proposition. *Let φ be an n.s.f. weight on the W*-algebra $\mathscr{M}$ and f a normal positive form on the factor $\mathscr{F}$ of type I. If ψ is a normal semifinite weight on $\mathscr{M} \bar{\otimes} \mathscr{F}$ such that $\psi \circ (\sigma_t^\varphi \otimes 1_{\mathscr{F}}) = \psi$, $(t \in \mathbb{R})$, and there exists σ^φ-invariant w-dense *-subalgebra $\mathscr{B}$ of $\mathfrak{M}_\varphi$ such that*

$$\psi(x^*x \bar{\otimes} y^*y) = \varphi(x^*x)f(y^*y) \quad (x \in \mathscr{B}, y \in \mathscr{F}),$$

then $\psi = \varphi \bar{\otimes} f$.

Proof. Let tr be the canonical trace on $\mathscr{F}$. By ([L], E.7.8) there exists $a \in \mathscr{F}$, with $a \geq 0$, such that $f = tr_a$. Then $\varphi \bar{\otimes} tr$ is an n.s.f. weight on $\mathscr{M} \bar{\otimes} \mathscr{F}$, $\sigma_t^{\varphi \bar{\otimes} tr} = \sigma_t^\varphi \otimes id_{\mathscr{F}}$, $(t \in \mathbb{R})$, $\varphi \bar{\otimes} f = (\varphi \bar{\otimes} tr)_{1 \bar{\otimes} a}$ and $\mathscr{B} \bar{\otimes} \mathfrak{M}_{tr}$ is a $\sigma^{\varphi \bar{\otimes} tr}$-invariant w-dense *-subalgebra of $\mathscr{M}_{\varphi \bar{\otimes} tr}$ such that

$$\psi(z^*z) = (\varphi \bar{\otimes} tr)_{1 \bar{\otimes} a}(z^*z) \quad (z \in \mathscr{B} \bar{\otimes} \mathfrak{M}_{tr}).$$

Since ψ is $\sigma^{\varphi \bar{\otimes} tr}$-invariant, we conclude by Theorem 6.2 that $\psi = (\varphi \bar{\otimes} tr)_{1 \bar{\otimes} a} = \varphi \bar{\otimes} f$. □

Notes. Proposition 8.1 is due to Tomita and Takesaki (1970). The definition of the tensor product weight appears in Takesaki (1970) and Connes (1973a). The other results in this section are from Digernes (1975); Katayama (1974); Strătilă (1977); Zsidó (1978).

For our exposition we have used Connes (1973a); Strătilă (1979); Takesaki (1970) and Zsidó (1978).

CHAPTER II

Conditional Expectations and Operator-Valued Weights

9 Conditional Expectations

In this section, we introduce a special kind of positive linear mapping, called a conditional expectation, and give some applications to tensor products.

9.1. Let $\mathscr{A}$ be a C^*-algebra and $\mathscr{B} \subset \mathscr{A}$ a C^*-subalgebra.

A linear mapping $\Phi : \mathscr{A} \to \mathscr{B}$ is called *a projection* if $\Phi(b) = b$ for every $b \in \mathscr{B}$. In this case $\Phi \circ \Phi = \Phi$ and $\|\Phi\| \geq 1$.

A linear mapping $\Phi : \mathscr{A} \to \mathscr{B}$ is called *$\mathscr{B}$-linear* if $\Phi(ab) = \Phi(a)b$ and $\Phi(ba) = b\Phi(a)$ for every $a \in \mathscr{A}, b \in \mathscr{B}$.

A $\mathscr{B}$-linear projection $\Phi : \mathscr{A} \to \mathscr{B}$ which is also a positive mapping, that is, $\Phi(\mathscr{A}^+) \subset \mathscr{B}^+$, is called *a conditional expectation.*

Theorem (J. Tomiyama). *Every projection of norm 1 of the C^*-algebra $\mathscr{A}$ onto the C^*-subalgebra $\mathscr{B} \subset \mathscr{A}$ is a conditional expectation.*

Proof. Consider first a norm 1 projection $\Phi : \mathscr{M} \to \mathscr{N}$ of a W^*-algebra $\mathscr{M}$ onto a W^*-subalgebra $\mathscr{N} \subset \mathscr{M}$. Then $\mathscr{N}$ is a W^*-algebra and its unit element is a projection $e_{\mathscr{N}} \in \mathscr{M}$.

Let e be a projection in $\mathscr{M}$, put $f = 1 - e$, and assume that either e or f is in $\mathscr{N}$. For any $x \in \mathscr{N}$, both ex and fx are then in $\mathscr{N}$ and for $x, y \in \mathscr{M}$ we have

$$\|ex + fy\|^2 = \|(ex + fy)^*(ex + fy)\| = \|x^*ex + y^*fy\| \leq \|ex\|^2 + \|fy\|^2.$$

So for $\lambda \in \mathbb{R}$,

$$(\lambda + 1)^2\|f\Phi(ex)\|^2 = \|f\Phi(ex + \lambda f\Phi(ex))\|^2 \leq \|ex + \lambda f\Phi(ex)\|^2$$
$$\leq \|ex\|^2 + \|\lambda f\Phi(ex)\|^2 = \|ex\|^2 + \lambda^2\|f\Phi(ex)\|^2.$$

As λ is arbitrary, we have $(1 - e)\Phi(ex) = f\Phi(ex) = 0$, that is, $\Phi(ex) = e\,\Phi(ex)$. Interchanging e and f, we have $e\,\Phi(x - ex) = e\,\Phi(fx) = 0$, or $e\,\Phi(x) = e\,\Phi(ex)$. Thus

$$e\,\Phi(x) = \Phi(ex) \quad (x \in \mathscr{M}). \tag{1}$$

Putting $x = 1, e_{\mathcal{N}} = \Phi(e_{\mathcal{N}}) = e_{\mathcal{N}}\Phi(1) = \Phi(1)$. Let ψ be any positive form on $\mathcal{N}$ and $\varphi = \psi \circ \Phi$. Since $\|\varphi\| \leq \|\psi\| = \psi(e_{\mathcal{N}}) = \varphi(1) \leq \|\varphi\|$, by ([L], 5.4.) it follows that φ is positive. Hence Φ is positive and so self-adjoint. Taking adjoints in (1) we have

$$\Phi(x)e = \Phi(xe) \quad (x \in \mathcal{M}). \tag{2}$$

Since the W^*-algebra $\mathcal{N}$ is the closed linear span of its projections ([L], 2.23), by (1) and (2) it follows that $\Phi(yx) = y\Phi(x)$ and $\Phi(xy) = \Phi(x)y$ for $x \in \mathcal{M}, y \in \mathcal{N}$.

In the general case, when $\Phi : \mathcal{A} \to \mathcal{B}$ is a projection of norm 1 of the C^*-algebra $\mathcal{A}$ onto its C^*-subalgebra $\mathcal{B}$, we consider the second transpose $\Psi = \Phi^{tt}$, mapping of the second dual W^*-algebra $\mathcal{M} = \mathcal{A}^{**}$ onto the second dual W^*-algebra $\mathcal{N} = \mathcal{B}^{**}$ (A.15, A.16). Since $\mathcal{A}$ is w-densely imbedded in $\mathcal{A}^{**}$ and $\mathcal{B}$ is w-densely imbedded in $\mathcal{B}^{**}$, we may identify $\mathcal{N}$ with a W^*-subalgebra of $\mathcal{M}$ and then $\Psi : \mathcal{M} \to \mathcal{N}$ is obviously a projection of norm 1. By the first part of the proof, Ψ is a conditional expectation, which implies that its restriction $\Phi = \Psi|\mathcal{A}$ is also a conditional expectation. □

9.2. Let $\mathcal{A}, \mathcal{B}$ be C^*-algebras. A linear mapping $\Phi : \mathcal{A} \to \mathcal{B}$ is called a *Schwarz mapping* if $\Phi(a)^*\Phi(a) \leq \Phi(a^*a)$ for all $a \in \mathcal{A}$.

Note that if $\Phi : \mathcal{A} \to \mathcal{B}$ is a Schwarz mapping and if $a \in \mathcal{A}$ satisfies $\Phi(a)^*\Phi(a) = \Phi(a^*a)$, then for $x \in \mathcal{A}$ we have

$$\Phi(x^*a) = \Phi(x)^*\Phi(a), \quad \Phi(a^*x) = \Phi(a)^*\Phi(x).$$

Indeed, for $x \in \mathcal{A}, t \in \mathbb{R}$ we have

$$\begin{aligned} t(\Phi(a)^*\Phi(x) + \Phi(x)^*\Phi(a)) &= \Phi(ta+x)^*\Phi(ta+x) - t^2\Phi(a)^*\Phi(a) - \Phi(x)^*\Phi(x) \\ &\leq \Phi((ta+x)^*(ta+x)) - t^2\Phi(a^*a) - \Phi(x)^*\Phi(x) \\ &= t\Phi(a^*x + x^*a) + (\Phi(x^*x) - \Phi(x)^*\Phi(x)). \end{aligned}$$

Dividing this inequality by $t \gtrless 0$ and letting $|t| \to +\infty$, we get

$$\Phi(a)^*\Phi(x) + \Phi(x)^*\Phi(a) = \Phi(a^*x) + \Phi(x^*a).$$

Replacing a by $-ia$ here and then multiplying by i, we obtain

$$\Phi(a)^*\Phi(x) - \Phi(x)^*\Phi(a) = \Phi(a^*x) - \Phi(x^*a);$$

our assertion follows from the last two equations.

In particular, it follows that a projection $\Phi : \mathcal{A} \to \mathcal{B} \subset \mathcal{A}$ which is also a Schwarz mapping, is a conditional expectation.

Proposition. *Every conditional expectation $\Phi : \mathcal{A} \to \mathcal{B} \subset \mathcal{A}$ is a Schwarz mapping and a projection of norm 1. If $\mathcal{A}$ is unital, then $\mathcal{B}$ is also unital and $\Phi(1_{\mathcal{A}}) = 1_{\mathcal{B}}$.*

Proof. Indeed, for any $a \in \mathcal{M}$ we have $0 \leq \Phi((\Phi(a)-a)^*(\Phi(a)-a)) = -\Phi(a)^*\Phi(a) + \Phi(a^*a)$ and, since $a^*a \leq \|a\|^2 \cdot 1_{\mathcal{A}^{**}}$, we obtain $\|\Phi(a)\|^2 = \|\Phi(a)^*\Phi(a)\| \leq \|\Phi(a^*a)\| \leq \|a\|^2$, hence $\|\Phi\| = 1$.

Also, by the preceding remark, $\Phi(1_{\mathscr{A}})\Phi(a) = \Phi(a) = \Phi(a)\Phi(1_{\mathscr{A}})$, so that $\Phi(1_{\mathscr{A}})$ is the unit element of $\mathscr{B}$. □

A positive linear mapping $\Phi : \mathscr{A} \to \mathscr{B}$ between C^*-algebras is called *faithful* if for $a \in \mathscr{A}, \Phi(a^*a) = 0 \Rightarrow a = 0$.

If $\Phi : \mathscr{M} \to \mathscr{N}$ is a Schwarz mapping between W^*-algebras, it is easy to check that Φ is w-continuous if and only if Φ is s^*-continuous; recall that in this case Φ is called *normal*.

9.3. Recall ([L], C.5.2; Araki & Masuda, 1982; Strătilă & Zsidó, 1977, 2005) that a linear mapping $\Phi : \mathscr{A} \to \mathscr{B}$ between the C^*-algebras $\mathscr{A}$ and $\mathscr{B}$ is said to be *completely positive* if, for each $n \in \mathbb{N}$, the natural extension $\Phi_n : Mat_n(\mathscr{A}) \to Mat_n(\mathscr{B})$ is a positive mapping.

If $\mathscr{B} = \mathscr{B}(\mathscr{H})$, then $Mat_n(\mathscr{B}) = \mathscr{B}(\tilde{\mathscr{H}}_n)$ where $\tilde{\mathscr{H}}_n$ is the Hilbert space direct sum of n copies of $\mathscr{H}$. An element $X \in \mathscr{B}(\tilde{\mathscr{H}}_n)$ is positive if and only if $(X\xi|\xi) > 0$ for any $\xi = [\xi_1, \dots, \xi_n] \in \tilde{\mathscr{H}}_n$. On the other hand, it is easy to check that every positive element of the C^*-algebra $Mat_n(\mathscr{A})$ is a finite sum of matrices $[a_{ij}] \in Mat_n(\mathscr{A})$ with $a_{ij} = a_i^* a_j (1 \leq i,j \leq n)$, where $a_1, \dots, a_n \in \mathscr{A}$. Consequently, a linear mapping $\Phi : \mathscr{A} \to \mathscr{B}(\mathscr{H})$ is completely positive if and only if $\sum_{ij}(\Phi(a_i^* a_j)\xi_j|\xi_i) \geq 0$ for n-tuples $a_1, \dots a_n \in \mathscr{A}$, and $\xi_1, \dots, \xi_n \in \mathscr{H} (n = 1, 2, \dots)$.

It is easy to see that if $\mathscr{H} = \oplus_{i\in I}\mathscr{H}_i$ and $\Phi(\mathscr{A})\mathscr{H}_i \subset \mathscr{H}_i (i \in I)$, then Φ is completely positive if and only if each of the mappings $\Phi_i : \mathscr{A} \ni a \mapsto \Phi(a)|\mathscr{H}_i \in \mathscr{B}(\mathscr{H}_i)$ is completely positive.

Every C^*-algebra $\mathscr{B}$ may be regarded as a concrete C^*-algebra $\mathscr{B} \subset \mathscr{B}(\mathscr{H})$ such that $\overline{\mathscr{B}\mathscr{H}} = \mathscr{H}$; in this case, there exist a direct sum decomposition $\mathscr{H} = \oplus_{i\in I}\mathscr{H}_i$ and vectors $\xi_i \in \mathscr{H}_i$ such that $\overline{\mathscr{B}\xi_i} = \mathscr{H}_i (i \in I)$. Thus, in proving that a linear mapping $\Phi : \mathscr{A} \to \mathscr{B}$ is completely positive we may assume, without loss of generality, that $\mathscr{B} \subset \mathscr{B}(\mathscr{H})$ and $\overline{\mathscr{B}\xi} = \mathscr{H}$ for some $\xi \in \mathscr{H}$.

Proposition. *Every conditional expectation is completely positive.*

Proof. Let $\Phi : \mathscr{A} \to \mathscr{B} \subset \mathscr{A}$ be a conditional expectation and assume that $\mathscr{B} \subset \mathscr{B}(\mathscr{H})$ with $\overline{\mathscr{B}\xi} = \mathscr{H}$ for some $\xi \in \mathscr{H}$. Let $a_k \in \mathscr{A}, x_k \in \mathscr{A}$ and $\xi_k = \Phi(x_k)\xi (1 \leq k \leq n)$. We have

$$\sum_{ij}(\Phi(a_i^* a_j)\xi_j|\xi_i) = \sum_{ij}(\Phi(x_i)^*\Phi(a_i^* a_j)\Phi(x_j)\xi|\xi)$$

$$= \sum_{ij}(\Phi(\Phi(x_i)^* a_i^* a_j \Phi(x_j))\xi|\xi) = \left(\Phi\left(\sum_{ij}\Phi(x_i)^* a_i^* a_j \Phi(x_j)\right)\xi|\xi\right) \geq 0,$$

as $\sum_{ij}\Phi(x_i)^* a_i^* a_j \Phi(x_j) \geq 0$ and Φ is positive. Since $\overline{\mathscr{B}\xi} = \mathscr{H}$, the above inequality holds for arbitrary vectors $\xi_1, \dots, \xi_n \in \mathscr{H}$, hence Φ is completely positive. □

9.4. If $\Phi : \mathscr{A} \to \mathscr{B}(\mathscr{H})$ is a completely positive linear mapping, the Stinespring theorem ([L], C.5.2; Arveson (1969); Strătilă & Zsidó, 1977, 2005) shows that there exist a Hilbert space $\mathscr{K}$, a $*$-representation $\pi : \mathscr{A} \to \mathscr{B}(\mathscr{K})$ and a bounded linear mapping $V : \mathscr{H} \to \mathscr{K}$ such that $\|V\| = \|\Phi\|^{1/2}$, $\mathscr{K}$ is the closed linear space generated by $\pi(\mathscr{A})V\mathscr{H}$, and

$$\Phi(a) = V^*\pi(a)V \quad (a \in \mathscr{A}).$$

The triple $\{\pi, V, \mathscr{K}\}$ is uniquely determined by these conditions and is called the *Stinespring dilation* of Φ.

By means of the Stinespring dilation, we obtain

$$\Phi(a)^*\Phi(a) \leq \|\Phi\|\Phi(a^*a) \quad (a \in \mathscr{A}); \tag{1}$$

so if $\|\Phi\| = 1, \Phi$ is a Schwarz mapping. If $\mathscr{A}$ is unital, then $\Phi(1) = V^*V$ and $\|\Phi\| = \|\Phi(1)\|$.

Again, using the Stinespring dilation, we see that if $\Phi_1 : \mathscr{A}_1 \to \mathscr{B}_1$ and $\Phi_2 : \mathscr{A}_2 \to \mathscr{B}_2$ are (faithful) completely positive linear mappings, there exists a unique (faithful) completely positive linear mapping $\Phi_1 \otimes_{C^*} \Phi_2 : \mathscr{A}_1 \otimes_{C^*} \mathscr{A}_1 \to \mathscr{B}_1 \otimes_{C^*} \mathscr{B}_2$ such that

$$(\Phi_1 \otimes_{C^*} \Phi_2)(a_1 \otimes a_2) = \Phi_1(a_1) \otimes \Phi_2(a_2)$$

for $a_1 \in \mathscr{A}_1, a_2 \in \mathscr{A}_2$, and moreover

$$\|\Phi_1 \otimes_{C^*} \Phi_2\| = \|\Phi_1\|\|\Phi_2\|. \tag{2}$$

It is easy to check that *if* Φ_1 *and* Φ_2 *are conditional expectations, then* $\Phi_1 \otimes_{C^*} \Phi_2$ *is also a conditional expectation.*

If $\Phi : \mathscr{M} \to \mathscr{B}(\mathscr{H})$ is a normal completely positive linear mapping of the W^*-algebra $\mathscr{M}$, then the Stinespring dilation π is also a normal $*$-representation.

It follows that if $\Phi_1 : \mathscr{M}_1 \to \mathscr{N}_1$ and $\Phi_2 : \mathscr{M}_2 \to \mathscr{N}_2$ are (faithful) normal completely positive linear mappings between W^*-algebras, then there exists a unique (faithful) normal completely positive linear mapping $\Phi_1 \bar{\otimes} \Phi_2 : \mathscr{M}_1 \bar{\otimes} \mathscr{M}_2 \to \mathscr{N}_1 \bar{\otimes} \mathscr{N}_2$ which extends the algebraic tensor product mapping $\Phi_1 \otimes \Phi_2 : \mathscr{M}_1 \otimes \mathscr{M}_2 \to \mathscr{N}_1 \otimes \mathscr{N}_2$. Note that $\mathscr{M}_1 \otimes_{C^*} \mathscr{M}_2$ is a w-dense C^*-subalgebra of $\mathscr{M}_1 \bar{\otimes} \mathscr{M}_2$ and the restriction of $\Phi_1 \bar{\otimes} \Phi_2$ to $\mathscr{M}_1 \otimes_{C^*} \mathscr{M}_2$ is just $\Phi_1 \otimes_{C^*} \Phi_2$, so that

$$\|\Phi_1 \bar{\otimes} \Phi_2\| = \|\Phi_1\|\|\Phi_2\|. \tag{3}$$

If Φ_1 *and* Φ_2 *are normal conditional expectations,* $\Phi_1 \bar{\otimes} \Phi_2$ *is also a normal conditional expectation.*

The extension of the algebraic tensor product mapping $\Phi_1 \otimes \Phi_2$ to the W^*-algebra $\mathscr{M}_1 \bar{\otimes} \mathscr{M}_2$ is possible even if Φ_1, Φ_2 are not normal:

Proposition (J. Tomiyama). *Let* $\Phi_1 : \mathscr{M}_1 \to \mathscr{N}_1, \Phi_2 : \mathscr{M}_2 \to \mathscr{N}_2$ *be completely positive linear mappings between* W^**-algebras. There exists a completely positive linear mapping* $\Phi : \mathscr{M}_1 \bar{\otimes} \mathscr{M}_2 \to \mathscr{N}_1 \bar{\otimes} \mathscr{N}_2$, $\|\Phi\| = \|\Phi_1\|\|\Phi_2\|$, *such that*

$$\Phi(x_1 \bar{\otimes} x_2) = \Phi_1(x_1) \bar{\otimes} \Phi_2(x_2) \quad (x_1 \in \mathscr{M}_1, x_2 \in \mathscr{M}_2).$$

If Φ_1 *and* Φ_2 *are conditional expectations, we can choose* Φ *to be a conditional expectation.*

Before we give the proof (in Section 9.7) some preparation is needed.

9.5 Lemma. *Let* $\Phi : \mathscr{M} \to \mathscr{N}$ *be a completely positive linear mapping between the* W^**-algebras* $\mathscr{M}, \mathscr{N}$*. There exists a completely positive linear mapping* $\Psi : \mathscr{M} \to \mathscr{N}$ *with* $\Psi(1) = 1$ *such that*

$$\Phi(x) = \Phi(1)^{1/2}\Psi(x)\Phi(1)^{1/2} \quad (x \in \mathscr{M}).$$

Proof. Let $b = \Phi(1) \in \mathcal{N}^+, f = \mathbf{s}(b)$ and let φ be any state of $\mathcal{M}$. The equation

$$\Psi_n(x) = (b + n^{-1})^{-1/2}\Phi(x)(b + n^{-1})^{-1/2} + \varphi(x)(1 - f) \quad (x \in \mathcal{M})$$

defines completely positive linear mappings $\Psi_n : \mathcal{M} \to \mathcal{N} (n \in \mathbb{N})$.

Let $a \in \mathcal{M}, 0 \le a \le 1$. Since $0 \le \Phi(a) \le b$, there exists ([L], E.2.6) $y \in \mathcal{N}, \|y\| \le 1$, such that $\Phi(a)^{1/2} = yb^{1/2}$. As $b(b + n^{-1})^{-1} \xrightarrow{s^*} f$, it follows that $\Phi(a)^{1/2}(b + n^{-1})^{-1/2} \xrightarrow{s^*} yf$, and the sequence $\{\Psi_n(a)\}$ is s^*-convergent to the element $\Psi(a) = fy^*yf + \varphi(a)(1 - f)$. We have $b^{1/2}\Psi(a)b^{1/2} = b^{1/2}y^*yb^{1/2} = \Phi(a)$. If $a = 1$, then $y = 1$ and $\varphi(a) = 1$, hence $\Psi(1) = 1$.

Since every element $x \in \mathcal{M}$ is a linear combination of elements $0 \le a \le 1$, it follows that the mappings Ψ_n are pointwise convergent, with respect to the s^*-topology on $\mathcal{N}$, to a completely positive linear mapping $\Psi : \mathcal{M} \to \mathcal{N}$ with the required properties. □

9.6. Let I be a directed set. The space of all bounded nets $\{f(i)\}_{i \in I}$ of complex numbers is just the C^*-algebra $\mathcal{B}(I)$ of bounded functions on I. For each $i \in I$, the equation

$$\Lambda_i(f) = f(i) \quad (f \in \mathcal{B}(I))$$

defines a positive form $\Lambda_i \in \mathcal{B}(I)^*$ such that $\|\Lambda_i\| = 1$. Since the closed unit ball of $\mathcal{B}(I)^*$ is $\sigma(\mathcal{B}(I)^*, \mathcal{B}(I))$-compact, it follows that the intersection

$$\bigcap_{i \in I} \overline{\{\Lambda_j : j \ge i\}} \subset \mathcal{B}(I)^*$$

is not empty. An arbitrary element Λ in this intersection will be called a *Banach limit with respect to I*. For $f \in \mathcal{B}(I)$, we shall write

$$\Lambda(f) = LIM_i f(i).$$

The properties of the Banach limits follow immediately from the fact that Λ is a positive form of norm 1 on $\mathcal{B}(I)$. Moreover, we have $LIM_i f(i) = \lim_i f(i)$ whenever $\lim_i f(i)$ exists.

Consider now a norm-bounded net $\{x_i\}_{i \in I} \subset \mathcal{B}(\mathcal{H})$. The mapping $\mathcal{H} \times \mathcal{H} \ni (\xi, \eta) \mapsto LIM_i(x_i\xi|\eta)$ is a bounded sesquilinear form on $\mathcal{H}$, so there exists a unique operator

$$x = LIM_i x_i \in \mathcal{B}(\mathcal{H}),$$

such that $(x\xi|\eta) = LIM_i(x_i\xi|\eta)$ for $\xi, \eta \in \mathcal{H}$.

It is easy to check the following properties:

$$LIM_i(x_i + y_i) = LIM_i x_i + LIM_i y_i, \tag{1}$$

$$LIM_i(x_i^*) = (LIM_i x_i)^*, \tag{2}$$

$$LIM_i(ax_ib) = a(LIM_i x_i)b, \tag{3}$$

$$x_i \xrightarrow{wo} x \Rightarrow LIM_i x_i = x. \tag{4}$$

Also, using the Hahn–Banach theorem we see that

$$\text{if } \mathcal{X} : \mathcal{B}(\mathcal{H}) \text{ is a wo-closed convex set and } x_i \in \mathcal{X} \text{ whenever } i \geq i_0, \text{ then } LIM_i x_i \in \mathcal{X}. \tag{5}$$

9.7. *Proof of Proposition 9.4.* We divide the proof into three steps.

(I) Let $\Phi : \mathcal{M} \to \mathcal{N}$ be a completely positive linear mapping between the W^*-algebras $\mathcal{M}$, $\mathcal{N}$ and $\iota : \mathcal{F} \to \mathcal{F}$ the identity mapping on a factor $\mathcal{F}$ of type I. Let $\{e_k\}_{k \in K}$ be a family of minimal projections in $\mathcal{F}$ with $\sum_k e_k = 1$, put $I = \{J \subset K; J\ \textit{finite}\}$, $e_J = \sum_{k \in J} e_k$ for $J \in I$, and let LIM be any Banach limit with respect to I.

Since $e_J \mathcal{F} e_J$ is finite dimensional, we have ([L], 3.17)

$$(1 \bar{\otimes} e_J)(\mathcal{M} \bar{\otimes} \mathcal{F})(1 \bar{\otimes} e_J) = \mathcal{M} \bar{\otimes} e_J \mathcal{F} e_J = \mathcal{M} \otimes e_J \mathcal{F} e_J$$

and so we can define a completely positive linear mapping

$$\widetilde{\Phi}_J : \mathcal{M} \bar{\otimes} \mathcal{F} \to \mathcal{N} \bar{\otimes} \mathcal{F}, \|\widetilde{\Phi}_J\| \leq \|\Phi\|, by$$
$$\widetilde{\Phi}_J(x) = (\Phi \otimes \iota_J)((1 \bar{\otimes} e_J)x(1 \bar{\otimes} e_J)) \quad (x \in \mathcal{M} \bar{\otimes} \mathcal{F}).$$

Now, using the results of Section 9.6, we obtain a completely positive linear mapping $\widetilde{\Phi} : \mathcal{M} \bar{\otimes} \mathcal{F} \to \mathcal{N} \bar{\otimes} \mathcal{F}, \|\widetilde{\Phi}\| \leq \|\Phi\|$, by putting

$$\widetilde{\Phi}(x) = LIM_J \widetilde{\Phi}_J(x) \quad (x \in \mathcal{M} \bar{\otimes} \mathcal{F}).$$

Since $e_J \uparrow 1$ in $\mathcal{F}$, for $a \in \mathcal{M}, b \in \mathcal{F}$ we have

$$\widetilde{\Phi}(a \bar{\otimes} b) = LIM_J(\Phi(a) \bar{\otimes} e_J b e_J) = w\text{-}\lim_J(\Phi(a) \bar{\otimes} e_J b e_J) = \Phi(a) \bar{\otimes} b;$$

thus $\widetilde{\Phi}$ extends the algebraic tensor product mapping $\Phi \otimes \iota$.

If Φ is a conditional expectation, then $\|\widetilde{\Phi}\| = \|\Phi\| = 1$ and, as $1 \bar{\otimes} e_J \uparrow 1 \bar{\otimes} 1$, for $y \in \mathcal{N} \bar{\otimes} \mathcal{F} \subset \mathcal{M} \bar{\otimes} \mathcal{F}$ we obtain

$$\widetilde{\Phi}(y) = w\text{-}\lim_J(\Phi \otimes \iota_J)((1 \bar{\otimes} e_J)y(1 \bar{\otimes} e_J)) = w\text{-}\lim_J(1 \bar{\otimes} e_J)y(1 \bar{\otimes} e_J) = y;$$

it follows by Theorem 9.1, that $\widetilde{\Phi}$ is a conditional expectation.

(II) We assume that $\Phi_1(1) = 1, \Phi_2(1) = 1$ and assume $\mathcal{M}_1 \subset \mathcal{B}(\mathcal{H}_1)$, $\mathcal{M}_2 \subset \mathcal{B}(\mathcal{H}_2)$, $\mathcal{N}_1 \subset \mathcal{B}(\mathcal{K}_1)$, $\mathcal{N}_2 \subset \mathcal{B}(\mathcal{K}_2)$ realized as von Neumann algebras. By (I) there exist completely positive linear mappings

$$\widetilde{\Phi}_1 : \mathcal{M}_1 \bar{\otimes} \mathcal{B}(\mathcal{H}_2) \to \mathcal{N}_1 \bar{\otimes} \mathcal{B}(\mathcal{H}_2) \text{ and } \widetilde{\Phi}_2 : \mathcal{B}(\mathcal{K}_1) \bar{\otimes} \mathcal{M}_2 \to \mathcal{B}(\mathcal{K}_1) \bar{\otimes} \mathcal{N}_2$$

which extend the algebraic tensor product mappings $\Phi_1 \otimes \iota_2$ and $\iota_1 \otimes \Phi_2$, respectively. Let us show that

$$x \in \mathcal{M}_1 \bar{\otimes} \mathcal{M}_2 \Rightarrow \widetilde{\Phi}_1(x) \in \mathcal{N}_1 \bar{\otimes} \mathcal{M}_2.$$

Indeed, let $x_1' \in \mathscr{M}_2'$. Since $\Phi_1(1) = 1$, we have

$$\widetilde{\Phi}_1(1 \bar{\otimes} x_2')^*\widetilde{\Phi}_1(1 \bar{\otimes} x_2') = \widetilde{\Phi}_1((1 \bar{\otimes} x_2')^*(1 \bar{\otimes} x_2')).$$

Using 9.4. (1) and the first remark in Section 9.2, we obtain

$$\begin{aligned}\widetilde{\Phi}_1(x)(1 \bar{\otimes} x_2') &= \widetilde{\Phi}_1(x)\widetilde{\Phi}_1(1 \bar{\otimes} x_2') = \widetilde{\Phi}_1(x(1 \bar{\otimes} x_2')) \\ &= \widetilde{\Phi}_1((1 \bar{\otimes} x_2')x) = \widetilde{\Phi}_1(1 \bar{\otimes} x_2')\widetilde{\Phi}_1(x) = (1 \bar{\otimes} x_2')\widetilde{\Phi}_1(x).\end{aligned}$$

It follows that $\widetilde{\Phi}_1(x) \in \mathscr{N}_1 \bar{\otimes} \mathscr{M}_2$, as asserted.

Similarly,

$$y \in \mathscr{N}_1 \bar{\otimes} \mathscr{M}_2 \Rightarrow \widetilde{\Phi}_2(y) \in \mathscr{N}_1 \bar{\otimes} \mathscr{N}_2.$$

Thus, $\Phi = \widetilde{\Phi}_2 \circ \widetilde{\Phi}_1 : \mathscr{M}_1 \bar{\otimes} \mathscr{M}_2 \to \mathscr{N}_1 \bar{\otimes} \mathscr{N}_2$ is a completely positive linear mapping which extends the algebraic tensor product mapping $\Phi_1 \otimes \Phi_2$. If Φ_1 and Φ_2 are conditional expectations, $\widetilde{\Phi}_1$ and $\widetilde{\Phi}_2$ are also conditional expectations, and so Φ is a conditional expectation.

(III) The general case of Proposition 9.4 now follows easily using Lemma 9.5. □

9.8. Let $\mathscr{M}$ and $\mathscr{N}$ be W^*-algebras and $\varphi \in \mathscr{M}_*$. Since $\mathscr{N} = (\mathscr{N}_*)^*$, for every $x \in \mathscr{M} \bar{\otimes} \mathscr{N}$, there exists a unique element $E_{\mathscr{N}}^{\varphi}(x) \in \mathscr{N}$ such that

$$\psi(E_{\mathscr{N}}^{\varphi}(x)) = (\varphi \bar{\otimes} \psi)(x) \quad (\psi \in \mathscr{N}_*).$$

It is easy to check that $E_{\mathscr{N}}^{\varphi} : \mathscr{M} \bar{\otimes} \mathscr{N} \to \mathscr{N}$ is a w-continuous linear mapping, $\|E_{\mathscr{N}}^{\varphi}\| = \|\varphi\|$, and

$$E_{\mathscr{N}}^{\varphi}(a \bar{\otimes} b) = \varphi(a)b \quad (a \in \mathscr{M}, b \in \mathscr{N}). \tag{1}$$

Using the w-continuity of $E_{\mathscr{N}}^{\varphi}$ and the w-density of $\mathscr{M} \otimes \mathscr{N}$ in $\mathscr{M} \bar{\otimes} \mathscr{N}$, from (1) we infer that

$$E_{\mathscr{N}}^{\varphi}((1 \bar{\otimes} b)x(1 \bar{\otimes} c)) = bE_{\mathscr{N}}^{\varphi}(x)c \quad (x \in \mathscr{M} \bar{\otimes} \mathscr{N}, b, c \in \mathscr{N}). \tag{2}$$

If $\varphi \in \mathscr{M}_*$ is positive, $E_{\mathscr{N}}^{\varphi}$ is also positive. Identifying $\mathscr{N}$ with $1 \otimes \mathscr{N}$ by amplification, it follows that *if φ is a normal state on $\mathscr{M}$, then $E_{\mathscr{N}}^{\varphi}$ is a normal conditional expectational of $\mathscr{M} \otimes \mathscr{N}$ onto $\mathscr{N}$*. The family of conditional expectations $\{E_{\mathscr{N}}^{\varphi}; \varphi$ normal state on $\mathscr{M}\}$ is separating in the following sense:

for every $0 \neq x \in \mathscr{M} \bar{\otimes} \mathscr{N}$ there exists a normal state (3)
φ on $\mathscr{M}$ such that $E_{\mathscr{N}}^{\varphi}(x) \neq 0$.

From (1) it follows that

$$E_{\mathscr{N}}^{\varphi} = \varphi \bar{\otimes} \iota_{\mathscr{N}} \tag{4}$$

in the sense defined in Section 9.4, where $\iota_{\mathscr{N}}$ denotes the identity mapping on $\mathscr{N}$. In view of their similarity to "partial integration" procedures, the mappings $E_{\mathscr{N}}^{\varphi}$ are also called *Fubini mappings*.

If $\varphi \in \mathcal{M}_*^+$, then for every normal semifinite weight ψ on $\mathcal{N}$ we have

$$\psi(E_{\mathcal{N}}^{\varphi}(x)) = (\varphi \bar{\otimes} \psi)(x) \quad (x \in (\mathcal{M} \bar{\otimes} \mathcal{N})^+). \tag{5}$$

This can be easily verified from the definition of $E_{\mathcal{N}}^{\varphi}$, using Corollary 5.8 and Proposition 8.3.

The next result is equivalent to the commutation theorem for tensor products ([L], 10.7):

Proposition. *Let $\mathcal{M} \subset \mathcal{B}(\mathcal{H}), \mathcal{N} \subset \mathcal{B}(\mathcal{H})$ be von Neumann algebras, and let $\mathcal{S} \subset \mathcal{B}(\mathcal{H} \bar{\otimes} \mathcal{K})$ be a von Neumann algebra such that $\mathcal{M} \bar{\otimes} \mathcal{N} \subset \mathcal{S}$ and $E_{\mathcal{B}(\mathcal{K})}^{\varphi}(\mathcal{S}) \subset \mathcal{N}, E_{\mathcal{B}(\mathcal{H})}^{\psi}(\mathcal{S}) \subset \mathcal{M}$ for every $\varphi \in \mathcal{B}(\mathcal{H})_*, \psi \in \mathcal{B}(\mathcal{K})_*$. Then $\mathcal{S} = \mathcal{M} \bar{\otimes} \mathcal{N}$.*

Proof. Let $x \in \mathcal{S}$ and $a' \in \mathcal{M}'$. For $\varphi \in \mathcal{B}(\mathcal{H})_*, \psi \in \mathcal{B}(\mathcal{K})_*$, we have

$$\begin{aligned}(\varphi \bar{\otimes} \psi)((a' \bar{\otimes} 1)x) &= \varphi(E_{\mathcal{B}(\mathcal{H})}^{\psi}((a' \bar{\otimes} 1)x) = \varphi(a' E_{\mathcal{B}(\mathcal{H})}^{\psi}(x)) \\ &= \varphi(E_{\mathcal{B}(\mathcal{H})}^{\psi}(x)a') = \varphi(E_{\mathcal{B}(\mathcal{H})}^{\psi}(x(a' \bar{\otimes} 1))) = (\varphi \bar{\otimes} \psi)(x(a' \bar{\otimes} 1)),\end{aligned}$$

hence x commutes with $a' \bar{\otimes} 1$. Similarly, for $b' \in \mathcal{N}', x$ commutes with $1 \bar{\otimes} b'$. Thus, $x \in (\mathcal{M}' \bar{\otimes} \mathcal{N}')' = \mathcal{M} \bar{\otimes} \mathcal{N}$. □

9.9 Corollary. *Let $\sigma : G \to Aut(\mathcal{M}), \tau : \mathcal{H} \to Aut(\mathcal{N})$ be actions of the groups G, H on the W^*-algebras $\mathcal{M}, \mathcal{N}$, respectively, and let $\sigma \bar{\otimes} \tau : G \times H \to Aut(\mathcal{M} \bar{\otimes} \mathcal{N})$ be the tensor product action, that is, $(\sigma \bar{\otimes} \tau)_{g,h} = \sigma_g \bar{\otimes} \tau_h$ ($g \in G, h \in H$). Then $(\mathcal{M} \bar{\otimes} \mathcal{N})^{\sigma \bar{\otimes} \tau} = \mathcal{M}^{\sigma} \bar{\otimes} \mathcal{N}^{\tau}$.*

Proof. Clearly, $\mathcal{M}^{\sigma} \bar{\otimes} \mathcal{N}^{\tau} \subset (\mathcal{M} \bar{\otimes} \mathcal{N})^{\sigma \bar{\otimes} \tau}$. Now let $x \in (\mathcal{M} \bar{\otimes} \mathcal{N})^{\sigma \bar{\otimes} \tau}, \varphi \in \mathcal{M}_*$ and $h \in H$, and denote by $e \in G$ the neutral element of G. For $\psi \in \mathcal{N}_*$, we have

$$\begin{aligned}\psi(\tau_h(E_{\mathcal{N}}^{\varphi}(x))) &= (\varphi \bar{\otimes} \psi \circ \tau_h)(x) = (\varphi \bar{\otimes} \psi)((\sigma_e \bar{\otimes} \tau_h)(x)) \\ &= (\varphi \bar{\otimes} \psi)(x) = \psi(E_{\mathcal{N}}^{\varphi}(x)),\end{aligned}$$

hence $\tau_h(E_{\mathcal{N}}^{\varphi}(x)) = E_{\mathcal{N}}^{\varphi}(x)$. Thus, $E_{\mathcal{N}}^{\varphi}(x) \in \mathcal{N}^{\tau}$. Similarly, $E_{\mathcal{M}}^{\psi}(x) \in \mathcal{M}^{\sigma}$ for $\psi \in \mathcal{N}_*$. By Proposition 9.8, it follows that $x \in \mathcal{M}^{\sigma} \bar{\otimes} \mathcal{N}^{\tau}$. □

We obtain the next two results from Proposition 9.8, in a similar manner.

9.10 Corollary. *Let $\mathcal{A}, \mathcal{B}$ be maximal abelian *-subalgebras of the W^*-algebras $\mathcal{M}, \mathcal{N}$, respectively. Then $\mathcal{A} \bar{\otimes} \mathcal{B}$ is a maximal abelian *-subalgebra of the W^*-algebra $\mathcal{M} \bar{\otimes} \mathcal{N}$.*

9.11 Corollary. *Let $\mathcal{M}_1 \subset \mathcal{B}(\mathcal{H}), \mathcal{M}_2 \subset \mathcal{B}(\mathcal{H}), \mathcal{N}_1 \subset \mathcal{B}(\mathcal{K}), \mathcal{N}_2 \subset \mathcal{B}(\mathcal{K})$ be von Neumann algebras. Then*

$$\mathcal{M}_1 \bar{\otimes} \mathcal{N}_1 \cap \mathcal{M}_2 \bar{\otimes} \mathcal{N}_2 = (\mathcal{M}_1 \cap \mathcal{M}_2) \bar{\otimes} (\mathcal{N}_1 \cap \mathcal{N}_2).$$

9.12. The next result is a characterization of tensor product W^*-algebras:

Theorem. *Let $\mathcal{R}$ be a W^*-algebra and $\mathcal{M}, \mathcal{N} \subset \mathcal{R}$ W^*-subalgebras with the properties:*

(i) *$\mathcal{R}$ is the W^*-algebra generated by $\mathcal{M}$ and $\mathcal{N}$;*
(ii) *$ab = ba$ for every $a \in \mathcal{M}, b \in \mathcal{N}$;*

(iii) *there exists a family* $\{E_i\}_{i\in I}$ *of* w*-continuous* $\mathcal{N}$*-linear mappings* $E_i : \mathcal{R} \to \mathcal{N}$*, such that*

$$E_i(a) \in \mathbb{C} \cdot 1 \text{ for } a \in \mathcal{M}, i \in I, \tag{1}$$

$$x \in \mathcal{R}, E_i(x^*x) = 0 \text{ for all } i \in I \Rightarrow x = 0. \tag{2}$$

*Then there exists a *-isomorphism* $\Phi : \mathcal{M} \bar{\otimes} \mathcal{N} \to \mathcal{R}$ *such that*

$$\Phi(a \bar{\otimes} b) = ab \quad (a \in \mathcal{M}, b \in \mathcal{N}).$$

Proof. From the assumptions, it follows that the family $\{E_i(a\cdot); a \in \mathcal{M}, i \in I\}$ consists of w-continuous $\mathcal{N}$-linear mappings satisfying (1), and the condition

$$x \in \mathcal{R}, E_i(ax) = 0 \text{ for } a \in \mathcal{M}, i \in I \Rightarrow x = 0.$$

Thus, we may assume that the family $\{E_i\}_{i\in I}$ is separating, that is,

$$x \in \mathcal{R}, E_i(x) = 0 \text{ for all } i \in I \Rightarrow x = 0.$$

In this case, the set $\{\psi \circ E_i; i \in I, \psi \in \mathcal{N}_*^+\}$ is total in $\mathcal{R}_*$.

From condition (1), it follows that for each $i \in I$ there exists $\varphi_i \in \mathcal{M}_*$ such that $E_i(a) = \varphi_i(a) \cdot 1 (a \in \mathcal{M})$. Since the family $\{E_i\}_{i\in I}$ is separating, the set $\{\varphi_i; i \in I\}$ is total in $\mathcal{M}_*$ and hence the set $\{\varphi_i \bar{\otimes} \psi; i \in I, \psi \in \mathcal{N}_*^+\}$ is total in $(\mathcal{M} \bar{\otimes} \mathcal{N})_*$. □

Let $\mathcal{F} \subset \mathcal{M}_*$ be the linear subspace generated by $\{\varphi_i; i \in I\}$. For $\varphi = \sum_i \lambda_i\varphi_i \in \mathcal{F}$ let $E^\varphi = \sum_i \lambda_i E_i$. Then the sets $\{\varphi \bar{\otimes} \psi; \varphi \in \mathcal{F}, \psi \in \mathcal{N}_*^+\} \subset (\mathcal{M} \bar{\otimes} \mathcal{N})_*$ and $\{\psi \circ E^\varphi; \varphi \in \mathcal{F}, \psi \in \mathcal{N}_*^+\} \subset \mathcal{R}_*$ are dense linear subspaces and, for every $\varphi \in \mathcal{F}, \psi \in \mathcal{N}_*^+$, we have

$$\begin{aligned} \|\psi \circ E^\varphi\| &\geq \|(\psi \circ E^\varphi)|\mathcal{M}\| = \sup\{|(\psi \circ E^\varphi)(a)|; a \in \mathcal{M}, \|a\| \leq 1\} \qquad (3)\\ &= \sup\{|\psi(1)\varphi(a)|; a \in \mathcal{M}, \|a\| \leq 1\} = \|\varphi\|\|\psi\| = \|\varphi \bar{\otimes} \psi\|; \end{aligned}$$

moreover, if $a_k \in \mathcal{M}, b_k \in \mathcal{N} (1 \leq k \leq n)$, then

$$(\psi \circ E^\varphi)\left(\sum_k a_k b_k\right) = (\varphi \bar{\otimes} \psi)\left(\sum_k a_k \otimes b_k\right). \tag{4}$$

Using (ii) and (4), we see that the equation

$$\Phi_0\left(\sum_k a_k \bar{\otimes} b_k\right) = \sum_k a_k b_k$$

defines a *-isomorphism of the *-algebra $\mathcal{M} \bar{\otimes} \mathcal{N}$ onto the *-subalgebra of $\mathcal{R}$ generated by $\mathcal{M} \cup \mathcal{N}$. Furthermore, using (3) and (4) we have $\|\Phi_0\| \leq 1$. Since Φ_0 is bounded and $(\psi \circ E^\varphi) \circ \Phi_0 = \varphi \bar{\otimes} \psi (\varphi \in \mathcal{F}, \psi \in \mathcal{N}_*^+)$, it follows that for any $\theta \in \mathcal{R}_*$ the linear form $\theta \circ \Phi$ on $\mathcal{M} \otimes \mathcal{N}$ is w-continuous. It follows that Φ_0 is w-continuous and so can be extended to a normal *-homomorphism Φ of $\mathcal{M} \bar{\otimes} \mathcal{N}$ onto the whole of $\mathcal{R}$, by assumption (i). If $x \in \mathcal{M} \bar{\otimes} \mathcal{N}$ and

$\Phi(x) = 0$, then $(\varphi \bar{\otimes} \psi)(x) = (\psi \circ E^{\varphi})(\Phi(x)) = 0$ for all $\varphi \in \mathcal{F}, \psi \in \mathcal{N}_*^+$, so that $x = 0$. We conclude that $\Phi : \mathcal{M} \otimes \mathcal{N} \to \mathcal{R}$ is the required *-isomorphism. □

9.13. *If, in Theorem 9.12, $\mathcal{R}$ is a factor, the conclusion remains valid assuming (i), (ii) and the weaker condition*

(iii_0) there exists a nonzero w-continuous $\mathcal{N}$-linear mapping $E : \mathcal{R} \to \mathcal{N}$.

Indeed, let $I = \mathcal{M} \times \mathcal{M}$ and, for $i = (a_1, a_2) \in I$, define

$$E_i(x) = E(a_1 x a_2) \quad (x \in \mathcal{R}).$$

We thus get a family $\{E_i\}_{i \in I}$ of w-continuous $\mathcal{N}$-linear mappings $E_i : \mathcal{R} \to \mathcal{N}$. If $a \in \mathcal{M}, E(a) \in \mathcal{N}$ commutes with the elements of $\mathcal{M}$ and for any $b \in \mathcal{N}$, we have $E(a)b = E(ab) = E(ba) = bE(a)$, hence $E(a)$ belongs to the center of the factor $\mathcal{R}$. It follows that the family $\{E_i\}_{i \in I}$ satisfies condition 9.12.(1). On the other hand, it is easy to check that the set $\mathcal{J} = \{x \in \mathcal{R}; E_i(x) = 0 \text{ for all } i \in I\}$ is a w-closed two-sided ideal of $\mathcal{R}$. Since $E \neq 0$ we have $\mathcal{J} \neq \mathcal{R}$ and hence $\mathcal{J} = \{0\}$, as $\mathcal{R}$ is a factor. Thus, the family $\{E_i\}_{i \in I}$ also satisfies condition 9.12.(2). Hence, our assertion follows from Theorem 9.12.

9.14. For a subset $\mathcal{N}$ of the W^*-algebra $\mathcal{M}$, we shall denote by

$$\mathcal{N}' \cap \mathcal{M} = \{x \in \mathcal{M}; xy = yx \text{ for all } y \in \mathcal{N}\}$$

the relative commutant of $\mathcal{N}$ in $\mathcal{M}$. If $\mathcal{N} = \mathcal{N}^*, \mathcal{N}' \cap \mathcal{M}$ is a W^*-subalgebra of $\mathcal{M}$. We record here an obvious consequence of Theorem 9.12:

Corollary. *Let $\mathcal{M}$ be a W^*-algebra and $\mathcal{N} \subset \mathcal{M}$ a W^*-subfactor of $\mathcal{M}, 1_{\mathcal{N}} = 1_{\mathcal{M}}$, such that $\mathcal{M}$ is generated, as a W^*-algebra, by $\mathcal{N}$ and $\mathcal{N}' \cap \mathcal{M}$, and such that there exists a faithful normal conditional expectation $E : \mathcal{M} \to \mathcal{N}$. Then there exists a *-isomorphism $\Phi : \mathcal{N} \bar{\otimes} (\mathcal{N}' \cap \mathcal{M}) \to \mathcal{M}$ such that*

$$\Phi(a \bar{\otimes} b) = ab \quad (a \in \mathcal{N}, b \in \mathcal{N}' \cap \mathcal{M}).$$

9.15. Let $\mathcal{M}$ be a W^*-algebra and $\mathcal{F} \subset \mathcal{M}$ a unital W^*-subalgebra which is a factor of type I.

Since $\mathcal{F}$ is *-isomorphic with some $\mathcal{B}(\mathcal{H})$, there exists a system of matrix units $\{e_{ij}\}_{i,j \in I}$ in $\mathcal{F}$, that is,

$$e_{ij} = e_{ji}^*, e_{ij} e_{hk} = \delta_{jh} e_{ik} \quad (i, j, h, k \in I),$$

the $e_{ii} (i \in I)$ are mutually orthogonal minimal projections in $\mathcal{F}$, and

$$\sum_{i \in J} e_{ii} = 1.$$

Let $1 \in I$ be a fixed index. Every element $a \in \mathcal{F}$ determines a scalar matrix $[a_{ij}]$ satisfying

$$e_{1i} a e_{j1} = a_{ij} e_{11} \quad (i, j \in I),$$

and we have

$$a = \sum_{ij} e_{ii} a e_{jj} = \sum_{ij} e_{i1} e_{1i} a e_{j1} e_{1j} = \sum_{ij} a_{ij} e_{ij}.$$

On the other hand, w-continuous linear mappings $E_{ij} : \mathcal{M} \to \mathcal{F}' \cap \mathcal{M}$ are defined by

$$E_{ij}(x) = \sum_k e_{ki} x e_{jk} \quad (x \in \mathcal{M}, i, j \in I)$$

with the properties:

$$E_{ij}(xb) = E_{ij}(x)b,\ E_{ij}(bx) = bE_{ij}(x) \quad (x \in \mathcal{M}, b \in \mathcal{F}' \cap \mathcal{M}), \tag{1}$$

$$E_{ij}(a) = a_{ij} \cdot 1 \quad (a \in \mathcal{F}), \tag{2}$$

$$\sum_{ij} e_{ij} E_{ij}(x) = x \quad (x \in \mathcal{M}). \tag{3}$$

It follows that $\mathcal{M}$ is the W^*-algebra generated by $\mathcal{F}$ and $\mathcal{F}' \cap \mathcal{M}$. Moreover, by Theorem 9.12 we have a canonical $*$-isomorphism

$$\Phi : \mathcal{F} \bar{\otimes} (\mathcal{F}' \cap \mathcal{M}) \to \mathcal{M} \tag{4}$$

such that $\Phi(a \bar{\otimes} b) = ab$ for every $a \in \mathcal{F}, b \in \mathcal{F}' \cap \mathcal{M}$.

Note that the Fubini mappings $\mathcal{F} \bar{\otimes} (\mathcal{F}' \cap \mathcal{M}) \to \mathcal{F}$ give rise to a separating family of normal conditional expectations of $\mathcal{M}$ onto $\mathcal{F}$ (see 9.8.(3)).

Also, the mappings $E_{ii} : \mathcal{M} \to \mathcal{F}' \cap \mathcal{M} (i \in I)$ are normal conditional expectations.

If $\mathcal{F}$ is of finite type $I_n (n \in \mathbb{N})$ then $I = \{1, \dots, n\}$, and the mapping

$$E = \frac{1}{n} \sum_{i=1}^{n} E_{ii} : \mathcal{M} \to \mathcal{F}' \cap \mathcal{M} \tag{5}$$

is a faithful normal conditional expectation. Identifying $\mathcal{M}$ with $\mathcal{F} \otimes (\mathcal{F}' \cap \mathcal{M})$ via Φ, it is easy to check that $E = \mu \bar{\otimes} \iota$, where $\mu = \frac{1}{n} tr$ is the normalized trace on $\mathcal{F}$ and ι is the identity mapping on $\mathcal{F}' \cap \mathcal{M}$. Also, for every finite trace τ on $\mathcal{M}$ we have

$$\tau \circ E = \tau. \tag{6}$$

9.16. As an application, we obtain the following factorization result:

Corollary. *Let $\sigma : G \to Aut(\mathcal{M})$ be an action of the group G on the W^*-algebra $\mathcal{M}$ such that the centralizer $\mathcal{M}^\sigma$ is properly infinite and let $\mathcal{F}$ be the countably decomposable infinite factor of type I. Then there exists a $*$-isomorphism*

$$(\mathcal{M}, \sigma) \approx (\mathcal{M} \bar{\otimes} \mathcal{F}, \sigma \bar{\otimes} \iota_{\mathcal{F}}).$$

Proof. By assumption, we may consider $\mathcal{F}$ as a unital W^*-subalgebra of $\mathcal{M}^\sigma \subset \mathcal{M}$. By the preceding Section, we may identify $\mathcal{M}$ with $\mathcal{F} \bar{\otimes} (\mathcal{F}' \cap \mathcal{M})$ via the *-isomorphism Φ (9.15.(4)). Since $\mathcal{F} \subset \mathcal{M}^\sigma$, we have also $\sigma_g(\mathcal{F}'\cap\mathcal{M}) = \mathcal{F}'\cap\mathcal{M} (g \in G)$, and hence $(\mathcal{M}, \sigma) \approx (\mathcal{F} \bar{\otimes} (\mathcal{F}'\cap\mathcal{M}), \iota_{\mathcal{F}} \bar{\otimes} (\sigma|\mathcal{F}'\cap\mathcal{M}))$. Since $(\mathcal{F}, \iota_{\mathcal{F}}) \approx (\mathcal{F} \bar{\otimes} \mathcal{F}, \iota_{\mathcal{F}} \bar{\otimes} \iota_{\mathcal{F}})$, it follows that $(\mathcal{M}, \sigma) \approx (\mathcal{M} \bar{\otimes} \mathcal{F}, \sigma \bar{\otimes} \iota_{\mathcal{F}})$. □

9.17. A related factorization result concerning weights on W^*-algebras is the following

Proposition. *Let φ be an n.s.f. weight on the W^*-algebra $\mathcal{M}$ such that the centralizer $\mathcal{M}^\varphi$ contains a unital W^*-subfactor $\mathcal{F}$ of type I. Then there exists an n.s.f. weight ψ on $\mathcal{F}' \cap \mathcal{M}$ such that $\sigma_t^\psi = \sigma_t^\varphi|\mathcal{F}' \cap \mathcal{M} (t \in \mathbb{R})$, and there exists a *-isomorphism*

$$(\mathcal{M}, \varphi) \approx (\mathcal{F} \bar{\otimes} (\mathcal{F}' \cap \mathcal{M}), tr \bar{\otimes} \psi).$$

Proof. By Section 9.15, we may identify $\mathcal{M}$ with $\mathcal{F} \bar{\otimes} (\mathcal{F}'\cap\mathcal{M})$ via the *-isomorphism Φ (9.15.(4)). Also, we shall use the notation $\{e_{ij}\} \subset \mathcal{F}$ and $E_{ij} : \mathcal{M} \to \mathcal{F}' \cap \mathcal{M}$ of Section 9.15.

The mapping $\psi : (\mathcal{F}' \cap \mathcal{M})^+ \ni b \mapsto \varphi(e_{11}b) \in [0, +\infty]$ is a normal weight on $\mathcal{F}' \cap \mathcal{M}$. If $b \in (\mathcal{F}' \cap \mathcal{M})^+$ and $\psi(b) = 0$, then $e_{11}b = 0$ because φ is faithful. It follows that $e_{ii}b = e_{i1}e_{11}be_{1i} = 0$ for all $i \in I$ and hence $b = \sum_i e_{ii}b = 0$. Thus ψ is faithful. Since φ is semifinite, there exists a net $\{x_s\} \subset \mathfrak{M}_\varphi \cap \mathcal{M}^+$ such that $x_s \xrightarrow{w} 1$. Then $b_s = E_{11}(x_s) \in (\mathcal{F}' \cap \mathcal{M})^+, b_s \xrightarrow{w} 1$ and $\psi(b_s) = \varphi(e_{11}E_{11}(x_s)e_{11}) = \varphi(e_{11}x_se_{11}) < +\infty$, as $e_{11} \in \mathcal{F} \subset \mathcal{M}^\varphi$ (2.21.(2)). Hence ψ is an n.s.f. weight on $\mathcal{F}' \cap \mathcal{M}$.

For each $b \in (\mathcal{F}'\cap\mathcal{M})^+$, the mapping $\tau_b : \mathcal{F}^+ \ni a \mapsto \varphi(ab) \in [0, +\infty]$ is a normal weight on $\mathcal{F}$. Since $\mathcal{F} \subset \mathcal{M}^\varphi$, by 2.21.(2) we see that τ_b is actually a normal trace on $\mathcal{F}$, and hence $\tau_b = \lambda(b) \cdot tr$, with $0 \leq \lambda(b) = \tau_b(e_{11}) = \varphi(e_{11}b) = \psi(b) \leq +\infty$. Consequently, for $a \in \mathcal{F}^+, b \in (\mathcal{F}' \cap \mathcal{M})^+$ and $t \in \mathbb{R}$, we have

$$\varphi(ab) = \psi(b)tr(a) = (tr \bar{\otimes} \psi)(ab)$$

and

$$\begin{aligned}(tr \bar{\otimes} \psi)(\sigma_t^\varphi(ab)) &= (tr \bar{\otimes} \psi)(a\sigma_t^\varphi(b)) = tr(a)\psi(\sigma_t^\varphi(b)) \\ &= tr(a)\varphi(e_{11}\sigma_t^\varphi(b)) = tr(a)\varphi(\sigma_t^\varphi(e_{11}b)) \\ &= tr(a)\varphi(e_{11}b) = tr(a)\psi(b) \\ &= (tr \bar{\otimes} \psi)(ab).\end{aligned}$$

Using the Pedersen–Takesaki theorem on the equality of weights (6.2), we conclude that $\varphi = tr \bar{\otimes} \psi$.

Since $\sigma_t^\varphi = \sigma_t^{tr\bar{\otimes}\psi} = \iota_{\mathcal{F}} \bar{\otimes} \sigma_t^\psi$ (8.2), it follows that $\sigma_t^\psi = \sigma_t^\varphi|\mathcal{F}' \cap \mathcal{M}$ for $t \in \mathbb{R}$. □

9.18. A normal semifinite weight φ on the W^*-algebra $\mathcal{M}$ is called of *infinite multiplicity* if the centralizer $\mathcal{M}^\varphi$ is properly infinite.

Corollary. *If φ is a normal semifinite weight of infinite multiplicity on the W^*-algebra $\mathcal{M}$, then there exists a *-isomorphism:*

$$(\mathcal{M}, \varphi) \approx (\mathcal{M} \bar{\otimes} \mathcal{F}, \varphi \bar{\otimes} tr)$$

where $\mathcal{F}$ is the countably decomposable infinite type I factor.

Proof. Since $\mathcal{M}^\varphi$ is properly infinite, we have $\mathcal{M}^\varphi \supset \mathcal{A} \bar{\otimes} \mathcal{B}$ with $\mathcal{A} \approx \mathcal{B} \approx \mathcal{F}$. By Proposition 9.17, there exists a normal semifinite weight ψ on $\mathcal{A}' \cap \mathcal{M}$ such that $(\mathcal{M}, \varphi) \approx (\mathcal{A} \bar{\otimes} (\mathcal{A}' \cap \mathcal{M}), tr \bar{\otimes} \psi)$ and $(\mathcal{A}' \cap \mathcal{M})^\psi \supset \mathcal{B}$. Thus, there exists a W^*-algebra $\mathcal{N}$ and a normal semifinite weight ψ of infinite multiplicity on $\mathcal{N}$ such that $(\mathcal{M}, \varphi) \approx (\mathcal{F} \bar{\otimes} \mathcal{N}, tr \bar{\otimes} \psi)$. With the same argument, we find a W^*-algebra $\mathcal{P}$ and a normal semifinite weight θ on $\mathcal{P}$ such that $(\mathcal{N}, \psi) \approx (\mathcal{F} \bar{\otimes} \mathcal{P}, tr \bar{\otimes} \theta)$. Since $(\mathcal{F} \bar{\otimes} \mathcal{F}, tr \bar{\otimes} tr) \approx (\mathcal{F}, tr)$, it follows that $(\mathcal{M}, \varphi) \approx (\mathcal{F} \bar{\otimes} \mathcal{F} \bar{\otimes} \mathcal{P}, tr \bar{\otimes} tr \bar{\otimes} \theta) \approx (\mathcal{F} \bar{\otimes} \mathcal{P}, tr \bar{\otimes} \theta)$ and therefore $(\mathcal{M}, \varphi) \approx (\mathcal{F} \bar{\otimes} \mathcal{M}, tr \bar{\otimes} \varphi)$. □

9.19. Notes. Conditional expectations in a noncommutative setting were introduced by Dixmier (1949, 1952, 1953) and Umegaki (1954, 1956, 1959, 1962). The main results (9.1, 9.4, 9.6) concerning projections of norm one are due to Tomiyama (1957–1959, 1959, 1969, 1971, 1972, 1981). Theorem 9.12 and its consequences (9.13, 9.14) appeared in Nakamura (1954); Takesaki (1958); Tomiyama (1971). The simple proof of Theorem 9.1 appeared also in a course by E.C. Lance. Thanks are due to Simon Wassermann for mentioning this fact and for further simplifications of our proof.

For our exposition, we have used Connes and Takesaki (1977a); Palmer (1974); Størmer (1974a); Strătilă and Zsidó (1977, 2005); and Tomiyama (1971).

Theorem 9.1 allows an easy proof of Sakai's characterization of von Neumann algebras as W^*-algebras (A.16). Further properties of conditional expectations are contained in Tomiyama (1971).

10 Existence and Uniqueness of Conditional Expectations

In this section, we give several criteria for the existence and uniqueness of conditional expectations, and some applications to the type theory of W^*-algebras.

10.1 Theorem (M. Takesaki). *Let $\mathcal{N}$ be a unital W^*-subalgebra of the W^*-algebra $\mathcal{M}$ and φ an n.s.f. weight on $\mathcal{M}$. The following conditions are equivalent:*

(i) the faithful normal weight $\varphi|\mathcal{N}^+$ is semifinite and $\sigma_t^\varphi(\mathcal{N}) = \mathcal{N}$ for every $t \in \mathbb{R}$;

(ii) there exists a faithful normal conditional expectation $E : \mathcal{M} \to \mathcal{N}$ such that

$$\varphi(x) = \varphi(E(x)) \quad (x \in \mathcal{M}^+). \tag{1}$$

Condition (1) determines uniquely the faithful normal conditional expectation $E : \mathcal{M} \to \mathcal{N}$.

The proof is given in Sections 10.2–10.3.

10.2. *Assume that the weight $\varphi|\mathcal{N}^+$ is semifinite.* Thus, we have an n.s.f. weight φ on $\mathcal{M}$ and an n.s.f. weight $\psi = \varphi|\mathcal{N}^+$ on $\mathcal{N}$. In this section, we study the relationship between the standard representations associated with φ and ψ.

It is obvious that $\mathfrak{N}_\psi = \mathfrak{N}_\varphi \cap \mathcal{N}$ and that the scalar products $(\cdot|\cdot)_\psi, (\cdot|\cdot)_\varphi$ coincide on $\mathfrak{N}_\psi$, hence $\mathcal{H}_\psi$ can be identified with a closed linear subspace of $\mathcal{H}_\varphi$ such that

$$b_\psi = b_\varphi \quad (b \in \mathfrak{N}_\psi). \tag{1}$$

Let P be the orthogonal projection of $\mathcal{H}_\varphi$ onto $\mathcal{H}_\psi$.

For $y \in \mathcal{N}$, $b \in \mathfrak{N}_\psi$, we have $\pi_\varphi(y)b_\psi = \pi_\varphi(y)b_\varphi = (yb)_\varphi = (yb)_\psi = \pi_\psi(y)b_\psi$, hence $\mathscr{H}_\psi$ is $\pi_\varphi(\mathcal{N})$-invariant, that is, $P \in \pi_\varphi(\mathcal{N})'$, and

$$\pi_\varphi(y)|\mathscr{H}_\psi = \pi_\psi(y) \quad (y \in \mathcal{N}). \tag{2}$$

Also, $\mathfrak{A}_\psi = \mathfrak{A}_\varphi \cap \mathcal{N}$, so that

$$\mathfrak{A}_\psi \subset \mathfrak{A}_\varphi \cap \mathscr{H}_\psi, \tag{3}$$

and $\mathfrak{A}_\psi$ is a *-subalgebra of $\mathfrak{A}_\varphi$ with the same scalar product. It follows that

$$D(S_\psi) \subset D(S_\varphi) \cap \mathscr{H}_\psi \text{ and } S_\varphi\xi = S_\psi\xi \text{ for } \xi \in D(S_\psi). \tag{4}$$

Assume moreover that $\sigma_t^\varphi(\mathcal{N}) = \mathcal{N} (t \in \mathbb{R})$. Using the *KMS* condition (2.12.(11)), we see that $\sigma_t^\varphi(y) = \sigma_t^\psi(y)$ for $y \in \mathcal{N}, t \in \mathbb{R}$.

For $b \in \mathfrak{N}_\psi$, we have (2.12.(9)) $\Delta_\varphi^{it} b_\psi = \Delta_\varphi^{it} b_\varphi = (\sigma_t^\varphi(b))_\varphi = (\sigma_t^\psi(b))_\psi = \Delta_\psi^{it} b_\psi (t \in \mathbb{R})$, hence $\mathscr{H}_\psi$ is Δ_φ^{it}-invariant, that is, $\Delta_\varphi^{it}P = P\Delta_\varphi^{it}$, and

$$\Delta_\varphi^{it}|\mathscr{H}_\psi = \Delta_\psi^{it} \quad (t \in \mathbb{R}). \tag{5}$$

Thus, Δ_φ commutes with P and, using ([L], 9.21), we infer from (5) that for any $\alpha \in \mathbb{C}$

$$D(\Delta_\psi^\alpha) = D(\Delta_\varphi^\alpha) \cap \mathscr{H}_\psi \text{ and } \Delta_\varphi^\alpha\xi = \Delta_\psi^\alpha\xi \text{ for } \xi \in D(\Delta_\psi^\alpha). \tag{6}$$

For $\alpha = 1/2$, we get from (6) that

$$D(S_\psi) = D(S_\varphi) \cap \mathscr{H}_\psi; \tag{7}$$

and, using also (4), we obtain $J_\psi\Delta_\psi^{1/2}\xi = S_\psi\xi = S_\varphi\xi = J_\varphi\Delta_\varphi^{1/2}\xi = J_\varphi\Delta_\psi^{1/2}\xi$ for $\xi \in D(\Delta_\psi^{1/2})$. It follows that $\mathscr{H}_\psi$ is J_φ-invariant, that is, $J_\varphi P = PJ_\varphi$, and

$$J_\varphi\xi = J_\psi\xi \quad (\xi \in \mathscr{H}_\psi). \tag{8}$$

Also, by (6) for $\alpha = -1/2$ and (8), we get

$$D(S_\psi^*) = D(S_\varphi^*) \cap \mathscr{H}_\psi \text{ and } S_\varphi^*\eta = S_\psi^*\eta \text{ for } \eta \in D(S_\psi^*). \tag{9}$$

If $\eta \in \mathfrak{A}'_\varphi \cap \mathscr{H}_\psi$, then $\eta \in D(S_\varphi^*) \cap \mathscr{H}_\psi = D(S_\psi^*)$ and for $b \in \mathfrak{A}_\psi \subset \mathfrak{A}_\varphi$ we have $\pi_\varphi(b)\eta = \pi_\psi(b)\eta = R_\eta^\varphi b_\varphi = R_\eta^\varphi b_\psi$. Consequently,

$$\mathfrak{A}'_\varphi \cap \mathscr{H}_\psi \subset \mathfrak{A}'_\psi \tag{10}$$

and, for $\eta \in \mathfrak{A}'_\varphi \cap \mathscr{H}_\psi$ we have $R_\eta^\psi = R_\eta^\varphi|\mathscr{H}_\psi$ and also $R_{S_\varphi^*\eta}^\psi = R_{S_\varphi^*\eta}^\varphi|\mathscr{H}_\psi = (R_\eta^\varphi)^*|\mathscr{H}_\psi$. Since $\mathscr{H}_\psi$ is invariant with respect to R_η^φ and $(R_\eta^\varphi)^*$, it follows that

$$PR_\eta^\varphi = R_\eta^\varphi P = R_\eta^\psi P \quad (\eta \in \mathfrak{A}'_\varphi \cap \mathscr{H}_\psi). \tag{11}$$

Applying J_φ to (10) and using (8), (3) and Tomita's fundamental theorem (2.12), we get

$$\mathfrak{A}_\psi = \mathfrak{A}_\varphi \cap \mathscr{H}_\psi. \tag{12}$$

Then, applying J_φ to (12), we obtain

$$\mathfrak{A}'_\psi = \mathfrak{A}'_\varphi \cap \mathscr{H}_\psi. \tag{13}$$

If $a \in \mathfrak{A}_\varphi$, then $a_\varphi \in D(S_\varphi) = D(\Delta_\varphi^{1/2})$, and hence $Pa_\varphi \in D(\Delta_\varphi^{1/2}) = D(S_\varphi)$; for $\eta \in \mathfrak{A}'_\psi = \mathfrak{A}'_\varphi \cap \mathscr{H}_\psi$, we have $L^\psi_{Pa_\varphi}\eta = R^\psi_\eta Pa_\varphi = PR^\varphi_\eta a_\varphi = P\pi_\varphi(a)\eta$, so that

$$\mathfrak{A}_\psi = P\mathfrak{A}_\varphi \text{ and } L^\psi_{Pa_\varphi} = P\pi_\varphi(a)P \text{ for } a \in \mathfrak{A}_\varphi. \tag{14}$$

Thus, $\pi_\psi(\mathscr{N}) = \mathfrak{L}(\mathfrak{A}_\psi) = P\pi_\varphi(\mathscr{M})P$ and so there exists a w-continuous positive linear mapping $E : \mathscr{M} \to \mathscr{N}$, uniquely determined, such that

$$\pi_\psi(E(x)) = P\pi_\varphi(x)P \quad (x \in \mathscr{M}). \tag{15}$$

Using (2), for $x \in \mathscr{M}, y \in \mathscr{N}$ and $\xi \in \mathscr{H}_\psi$ we get $\pi_\psi(E(xy))\xi = P\pi_\varphi(xy)P\xi = P\pi_\varphi(x)\pi_\varphi(y)\xi = P\pi_\varphi(x)P\pi_\psi(y)\xi = \pi_\psi(E(x))\pi_\psi(y)\xi = \pi_\psi(E(x)y)\xi$, hence $E(xy) = E(x)y$.

If $x \in \mathscr{M}$ and $E(x^*x) = 0$, for $b \in \mathfrak{N}_\psi$ we have $\varphi((xb)^*(xb)) = \|\pi_\varphi(x)b_\varphi\|^2_\varphi = (P\pi_\varphi(x)Pb_\varphi|b_\varphi)_\varphi = (\pi_\psi(E(x))b_\psi|b_\psi)_\psi = 0$, so that $xb = 0$. Since $\mathfrak{N}_\psi$ is w-dense in $\mathscr{N}$, it follows that $x = 0$.

Thus, $E : \mathscr{M} \to \mathscr{N}$ is a faithful normal conditional expectation. Moreover, we have

$$E \circ \sigma^\varphi_t = \sigma^\psi_t \circ E \quad (t \in \mathbb{R}). \tag{16}$$

Indeed, using (5), for $x \in \mathscr{M}$ and $\xi \in \mathscr{H}_\psi$, we get

$$\begin{aligned}\pi_\psi(E(\sigma^\varphi_t(x)))\xi &= P\pi_\varphi(\sigma^\varphi_t(x))P\xi \\ &= P\Delta^{it}_\varphi\pi_\varphi(x)\Delta^{-it}_\varphi P\xi = \Delta^{it}_\psi P\pi_\varphi(x)P\Delta^{-it}_\psi\xi \\ &= \Delta^{it}_\psi\pi_\psi(E(x))\Delta^{-it}_\psi\xi = \pi_\psi(\sigma^\psi_t(E(x)))\xi.\end{aligned}$$

It follows that $\psi \circ E$ is a σ^φ-invariant n.s.f. weight on $\mathscr{M}$, that is, $\psi \circ E$ commutes with φ. On the other hand, $\mathscr{B} = \mathfrak{N}^*_\psi\mathscr{M}\mathfrak{N}_\psi \subset \mathfrak{N}^*_\varphi\mathfrak{N}_\varphi = \mathfrak{M}_\varphi$ is a σ^φ-invariant and w-dense $*$-subalgebra of $\mathscr{M}$ and for every $x \in \mathscr{M}, b \in \mathfrak{N}_\psi$ we have $(\psi \circ E)(b^*xb) = \psi(b^*E(x)b) = (\pi_\psi(E(x)b_\psi|b_\psi)_\psi = (P\pi_\varphi(x)Pb_\varphi|b_\varphi)_\varphi = (\pi_\varphi(x)b_\varphi|b_\varphi)_\varphi = \varphi(b^*xb)$. By the Pedersen–Takesaki theorem on the equality of weights (6.2), it follows that $\psi \circ E = \varphi$, that is, $\varphi(E(x)) = \varphi(x)$ for $x \in \mathscr{M}^+$.

Thus, *we have proved the implication (i) ⇒ (ii) of Theorem 10.1.*

Let us note that by analogy with (14) we have

$$\mathfrak{A}'_\psi = P\mathfrak{A}'_\varphi \text{ and } R^\psi_{P\eta} = PR^\varphi_\eta P \text{ for } \eta \in \mathfrak{A}'_\varphi. \tag{17}$$

Using (2.12.(15) or [L], 10.21), we obtain

$$\mathfrak{T}_\psi = \mathfrak{T}_\varphi \cap \mathscr{H}_\psi = P\mathfrak{T}_\varphi. \tag{18}$$

Also, since the inclusion $\mathcal{N} \hookrightarrow \mathcal{M}$ is an isometric normal *-homomorphism, it follows that the mapping $\rho : \mathfrak{R}(\mathfrak{A}'_\psi) \to \mathfrak{R}(\mathfrak{A}'_\varphi)$ defined by $\rho(J_\psi \pi_\psi(y) J_\psi) = J_\varphi \pi_\varphi(y) J_\varphi (y \in \mathcal{N})$ is an isometric normal *-homomorphism. In particular, for $\eta \in \mathfrak{A}'_\psi$, we have $\|R^\psi_\eta\| = \|R^\varphi_\eta\|$ and $R^\psi_\eta \xrightarrow{s} 1 \Leftrightarrow R^\varphi_\eta \xrightarrow{s} 1$.

10.3. *Assume now that there exists a normal conditional expectation $E : \mathcal{M} \to \mathcal{N}$ such that $\varphi \circ E = \varphi$*; then it follows that E is also faithful.

Let $\psi = \varphi|\mathcal{N}^+$. If $x_i \in \mathfrak{M}_\varphi \cap \mathcal{M}^+$ and $x_i \xrightarrow{w} 1$, then $\psi(E(x_i)) = \varphi(x_i) < +\infty$, hence $E(x_i) \in \mathfrak{M}_\psi \cap \mathcal{N}^+$ and $E(x_i) \xrightarrow{w} 1$. Consequently, *ψ is an n.s.f. weight on $\mathcal{N}$*.

For $x \in \mathcal{M}$ and $b \in \mathfrak{N}_\psi$, we have $(\pi_\psi(E(x)) b_\psi | b_\psi)_\psi = \psi(b^* E(x) b) = \varphi(E(b^* x b)) = \varphi(b^* x b)$. Thus, *the normal conditional expectation $E : \mathcal{M} \to \mathcal{N}$ is uniquely determined by the condition $\varphi \circ E = \varphi$*.

In what follows we shall use the notation introduced in the first part of Section 10.2. For $a \in \mathfrak{N}_\varphi$, we have $\psi(E(a)^* E(a)) \leq \psi(E(a^* a)) = \varphi(a^* a)$, hence $E(a) \in \mathfrak{N}_\psi$. Since for every $b \in \mathfrak{N}_\psi$, we have $(a_\varphi | b_\varphi)_\varphi = \varphi(b^* a) = \psi(E(b^* a)) = \psi(b^* E(a)) = ((E(a))_\psi | b_\psi)_\psi$, it follows that

$$a \in \mathfrak{N}_\varphi \Rightarrow E(a) \in \mathfrak{N}_\psi \text{ and } (E(a))_\psi = P a_\varphi. \tag{1}$$

Since $E : \mathcal{M} \to \mathcal{N}$ is a self-adjoint mapping, we further obtain

$$P\mathfrak{A}_\varphi = \mathfrak{A}_\psi \subset \mathfrak{A}_\varphi, \text{ and } S_\varphi P a_\varphi = P S_\varphi a_\varphi \text{ for } a \in \mathfrak{A}_\varphi. \tag{2}$$

As $S_\varphi = \overline{S_\varphi | \mathfrak{A}_\varphi}$, it follows that $PS_\varphi \subset S_\varphi P$, that is, $(1-2P)S_\varphi(1-2P) = S_\varphi$, where $1-2P$ is a unitary operator. Then $(1-2P)S^*_\varphi(1-2P) = S^*_\varphi, (1-2P)\Delta_\varphi(1-2P) = \Delta_\varphi$, so that P commutes with Δ_φ, that is,

$$\Delta^{it}_\varphi P = P \Delta^{it}_\varphi \quad (t \in \mathbb{R}). \tag{3}$$

From (2) and (3), we infer that

$$\Delta^{it}_\varphi \mathfrak{A}_\psi = \Delta^{it}_\varphi P \mathfrak{A}_\varphi = P \Delta^{it}_\varphi \mathfrak{A}_\varphi = P\mathfrak{A}_\varphi = \mathfrak{A}_\varphi. \tag{4}$$

Let $b \in \mathfrak{A}_\varphi$ and $t \in \mathbb{R}$. By (4), there exists an element $y \in \mathfrak{A}_\psi$ such that $\Delta^{it}_\varphi b_\psi = y_\psi$. Then $(\sigma^\varphi_t(b))_\varphi = \Delta^{it}_\varphi b_\varphi = \Delta^{it}_\varphi b_\psi = y_\psi = y_\varphi$, hence $\sigma^\varphi_t(b) = y$. Thus, for every $t \in \mathbb{R}$, we have $\sigma^\varphi_t(\mathfrak{A}_\psi) \subset \mathfrak{A}_\psi$ and so $\sigma^\varphi_t(\mathcal{N}) = \mathcal{N}$, as $\mathfrak{A}_\psi$ is w-dense in $\mathcal{N}$.

We have proved the implication (ii) $\Rightarrow$ (i) in Theorem 10.1.

10.4. Let $\mathcal{N}$ be a unital W^*-subalgebra of the W^*-algebra $\mathcal{M}$. For every n.s.f. weight φ on $\mathcal{M}$ such that $\varphi|\mathcal{N}^+$ is semifinite and $\sigma^\varphi_t(\mathcal{N}) = \mathcal{N} (t \in \mathbb{R})$, we shall denote by

$$E^{\mathcal{N}}_\varphi : \mathcal{M} \to \mathcal{N}$$

the unique faithful normal conditional expectation such that $\varphi \circ E^{\mathcal{N}}_\varphi = \varphi$.

Corollary. *In the above situation, for $\sigma \in Aut(\mathcal{M})$ we have*

$$E^{\sigma^{-1}(\mathcal{N})}_{\varphi \circ \sigma} = \sigma^{-1} \circ E^{\mathcal{N}}_\varphi \circ \sigma.$$

10.5 Corollary. *Let $E : \mathcal{M} \to \mathcal{N}$ be a faithful normal conditional expectation of the W^*-algebra $\mathcal{M}$ onto its W^*-subalgebra $\mathcal{N}$. If ψ is any n.s.f. weight on $\mathcal{N}$, then $\varphi = \psi \circ E$ is an n.s.f. weight on $\mathcal{M}$ and*

$$\sigma_t^{\psi}(E(x)) = \sigma_t^{\varphi}(E(x)) = E(\sigma_t^{\varphi}(x)) \quad (x \in \mathcal{M}, t \in \mathbb{R}). \tag{1}$$

If ψ_1, ψ_2 are n.s.f. weights on $\mathcal{N}$ and $\varphi_1 = \psi_1 \circ E, \varphi_2 = \psi_2 \circ E$, then

$$[D\varphi_2 : D\varphi_1]_t = [D\psi_2 : D\psi_1]_t \in \mathcal{N} \qquad (t \in \mathbb{R}). \tag{2}$$

Proof. Indeed, φ is an n.s.f. weight on $\mathcal{M}, \psi = \varphi|\mathcal{N}^+$ is semifinite, and from Theorem 10.1 it follows that $\mathcal{N}$ is σ^{φ}-invariant and $E = E_{\varphi}^{\mathcal{N}}$. Now (1) follows using the *KMS* condition and Corollary 10.4.

Then $E \otimes \iota$ is a faithful normal conditional expectation of $Mat_2(\mathcal{M}) = \mathcal{M} \bar{\otimes} Mat_2(\mathbb{C})$ onto $Mat_2(\mathcal{N}) = \mathcal{N} \otimes Mat_2(\mathbb{C})$ (see 9.4) and for the balanced weights (see 3.1) we have $\theta(\varphi_1, \varphi_2) = \theta(\psi_2, \psi_2) \circ (E \bar{\otimes} \iota)$, so that (2) follows from the first equation in (1) applied to the balanced weights and to the element $\begin{pmatrix} 0 & 0 \\ 1 & 0 \end{pmatrix} \in Mat_2(\mathcal{N})$.

(1) and (2) can also be expressed in the form:

$$\sigma_t^{\psi_2,\psi_1}(E(x)) = \sigma_t^{\varphi_2,\varphi_1}(E(x)) = E(\sigma_t^{\varphi_2,\varphi_1}(x)) \quad (x \in \mathcal{M}, t \in \mathbb{R}). \tag{3}$$

Let ψ and φ be as in the statement of the corollary and ω an arbitrary n.s.f. weight on $\mathcal{M}$. An easy application of the corollary and Theorem 5.1 shows that

$$\textit{if } [D\omega : D\varphi]_t \in \mathcal{N} \textit{ for all } t \in \mathbb{R}, \textit{ then } \omega|\mathcal{N}^+ \textit{ is semifinite.} \tag{4}$$

Indeed, since $\sigma_t^{\psi} = \sigma_t^{\varphi}|\mathcal{N}$, there exists an n.s.f. weight τ on $\mathcal{N}$ such that $[D(\tau \circ E) : D\varphi]_t = [D\tau : D\psi]_t = [D\omega : D\varphi]_t (t \in \mathbb{R})$, hence $\omega = \tau \circ E$, and $\omega|\mathcal{N}^+ = \tau$ is semifinite. □

10.6. From Theorem 10.1, it follows that if τ is an n.s.f. trace on the W^*-algebra $\mathcal{M}$ and $\mathcal{N} \subset \mathcal{M}$ is a unital W^*-subalgebra such that $\tau|\mathcal{N}^+$ is semifinite, there exists a unique faithful normal conditional expectation $E_{\tau}^{\mathcal{N}} : \mathcal{M} \to \mathcal{N}$ such that $\tau \circ E_{\tau}^{\mathcal{N}} = \tau$. In particular:

Corollary. *Let τ be a faithful normal finite trace on the W^*-algebra $\mathcal{M}$. For every unital W^*-subalgebra $\mathcal{N} \subset \mathcal{M}$, there exists a unique faithful normal conditional expectation $E_{\tau}^{\mathcal{N}} : \mathcal{M} \to \mathcal{N}$ such that $\tau \circ E_{\tau}^{\mathcal{N}} = \tau$.*

If $\mathcal{M}$ is a countably decomposable finite W^*-algebra and $\mathcal{N} = \mathcal{Z}(\mathcal{M})$ is its center, then all the conditional expectations $E_{\tau}^{\mathcal{Z}(\mathcal{M})}$ coincide with the canonical central trace $\natural : \mathcal{M} \to \mathcal{Z}(\mathcal{M})$ ([L], 7.11). Also, if $\mathcal{F} \subset \mathcal{M}$ is a finite type I factor and $\mathcal{N} = \mathcal{F}' \cap \mathcal{M}$, then all the $E_{\tau}^{\mathcal{N}}$ coincide with the conditional expectation defined in 9.15.(5).

10.7 Corollary. *Let $\mathcal{M}$ be a W^*-algebra, φ a faithful normal state on $\mathcal{M}$ and $\mathcal{N} \subset \mathcal{M}$ a σ^{φ}-invariant unital W^*-subfactor such that $\mathcal{M}$ generated as a W^*-algebra by $\mathcal{N}$ and $\mathcal{N}' \cap \mathcal{M}$. There exists a *-isomorphism $\Phi : \mathcal{N} \bar{\otimes} (\mathcal{N}' \cap \mathcal{M}) \to \mathcal{M}$ such that $\Phi(a \bar{\otimes} b) = ab (a \in \mathcal{N}, b \in \mathcal{N}' \cap \mathcal{M})$, and $\varphi \circ \Phi = (\varphi|\mathcal{N}) \bar{\otimes} (\varphi|\mathcal{N}' \cap \mathcal{M})$, that is,*

$$\varphi(ab) = \varphi(a)\varphi(b) \quad (a \in \mathcal{N}, b \in \mathcal{N}' \cap \mathcal{M}).$$

Proof. By Theorem 10.1, there exists a unique faithful normal conditional expectation $E : \mathcal{M} \to \mathcal{N}$ with $\varphi \circ E = \varphi$. Thus, the existence of Φ follows from Corollary 9.14. For $b \in \mathcal{N}' \cap \mathcal{M}$, the element $E(b)$ belongs to the center of the factor $\mathcal{N}$, and hence $\varphi(b) = \varphi(E(b)) = E(b)$; if $a \in \mathcal{N}$, then $\varphi(ab) = \varphi(E(ab)) = \varphi(aE(b)) = \varphi(a)E(b) = \varphi(a)\varphi(b)$. □

10.8. Let $\sigma : G \to Aut(\mathcal{M})$ be an action of the group G on the W^*-algebra $\mathcal{M}$. We shall say that $\mathcal{M}$ is *σ-finite* if for every $x \in \mathcal{M}, x \neq 0$, there exists a σ-invariant normal state φ on $\mathcal{M}$ such that $\varphi(x) \neq 0$.

Corollary (Kovács & Szücs, 1966). *Let $\sigma : G \to Aut(\mathcal{M})$ be an action of the group G on the W^*-algebra $\mathcal{M}$. The following statements are equivalent:*

(i) *$\mathcal{M}$ is σ-finite;*
(ii) *there exists a σ-invariant faithful normal conditional expectation $E : \mathcal{N} \to \mathcal{M}^\sigma$.*

Proof. The implication (ii) ⇒ (i) is obvious. Conversely, if $\mathcal{M}$ is σ-finite, there exists a family $\{\varphi_i\}_{i\in I}$ of σ-invariant normal states on $\mathcal{M}$ with $\sum_i \mathbf{s}(\varphi_1) = 1$, and $\varphi = \sum_i \varphi_i$ is a σ-invariant n.s.f. weight on $\mathcal{M}$. Then each $\sigma_t^\varphi (t \in \mathbb{R})$ commutes with each $\sigma_g (g \in G)$, in particular $\mathcal{M}^\sigma$ is σ^φ-invariant. Since $\mathbf{s}(\varphi_i) \in \mathcal{M}^\sigma$ and $\varphi(\mathbf{s}(\varphi_i)) = \varphi_i(\mathbf{s}(\varphi_i)) < +\infty$, it follows that $\varphi|(\mathcal{M}^\sigma)^+$ is semifinite.

By Theorem 10.1, there exists a unique faithful normal conditional expectation $E : \mathcal{M} \to \mathcal{M}^\sigma$ with $\varphi \circ E = \varphi$. For $g \in G$, we have $\varphi \circ \sigma_g = \varphi$ and $\sigma_g^{-1}(\mathcal{M}^\sigma) = \mathcal{M}^\sigma$, hence $E \circ \sigma_g = \sigma_g \circ E = E$, by Corollary 10.4. □

10.9 Corollary (F. Combes). *Let φ be an n.s.f. weight on the W^*-algebra $\mathcal{M}$. The following statements are equivalent:*

(i) *$\varphi|(\mathcal{M}^\varphi)^+$ is an n.s.f. trace on $\mathcal{M}^\varphi$;*
(ii) *there exists a faithful normal conditional expectation $E : \mathcal{M} \to \mathcal{M}^\varphi$ such that $\varphi = \varphi \circ E$;*
(iii) *there exists a σ-invariant faithful normal conditional expectation $E : \mathcal{M} \to \mathcal{M}^\varphi$ such that $\varphi = \varphi \circ E$;*
(iv) *there exists a family $\{\varphi_i\}_{i\in I} \subset \mathcal{M}_*^+$ with $\sum_i \mathbf{s}(\varphi_i) = 1$ and $\varphi = \sum_i \varphi_i$;*
(v) *$\mathcal{M}$ is σ-finite.*

Proof. (i) ⇒ (ii), by Theorem 10.1.

(ii) ⇒ (iii), by Corollary 10.4.

(iii) ⇒ (i). By Theorem 10.1, it follows, assuming (iii), that $\varphi|(\mathcal{M}^\varphi)^+$ is an n.s.f. weight on $\mathcal{M}^\varphi$ and the trace property follows from 2.21.(2).

(i) ⇒ (iv). Since $\psi = \varphi|(\mathcal{M}^\varphi)^+$ is an n.s.f. trace on $\mathcal{M}^\varphi$, by ([L], E.7.11) we know that there exists a family $\{\psi_i\}_{i\in I}$ of normal positive forms on $\mathcal{M}^\varphi$ with mutually orthogonal supports such that $\psi = \sum_i \psi_i$. If E is the conditional expectation given by (iii) (cf. (i) ⇒ (iii)), then $\varphi_i = \psi_i \circ E \in \mathcal{M}_*^+ (i \in I)$, $\sum_i \mathbf{s}(\varphi_i) = \sum \mathbf{s}(\psi_i) = 1$, and $\varphi = \psi \circ E = \sum_i \varphi_i$.

(iv) ⇒ (v). Since $\varphi(\cdot \mathbf{s}(\varphi_i)) = \varphi(\mathbf{s}(\varphi_i)\cdot) = \varphi_i$, it follows from 2.21.(2) that $\mathbf{s}(\varphi_i) \in \mathcal{M}^\varphi$ and from 2.22.(3) that φ_i is σ^φ-invariant $(i \in I)$; hence $\mathcal{M}$ is σ-finite.

(v) ⇒ (iii), by Corollary 10.8. □

If the equivalent conditions of Corollary 10.9 are satisfied, then the n.s.f. weight φ is called *strictly semifinite* and we shall abbreviate this by saying that φ is an *n.ss.f. weight.*

With obvious modifications, Corollary 10.9 can be extended to weights which are not necessarily faithful and we get the notion of normal strictly semifinite weight.

Note that any normal semifinite trace and any normal positive form are strictly semifinite.

If there exists $t_0 \in \mathbb{R}, t_0 > 0$, such that $\sigma^\varphi_{t_0} = \iota_{\mathscr{M}}$, the identity mapping on $\mathscr{M}$, then the weight φ is strictly semifinite, as the mapping

$$E : \mathscr{M} \ni x \mapsto \frac{1}{t_0}\int_0^{t_0} \sigma^\varphi(x)\,\mathrm{d}t \in \mathscr{M}^\sigma$$

is then a faithful normal conditional expectation with $\varphi \circ E = \varphi$.

Finally, we note that the tensor product of two normal strictly semifinite weights is again strictly semifinite (8.8.(3)).

10.10. *On every W^*-algebra $\mathscr{M}$, there exists an n.ss.f. weight.* Indeed, there exists a family $\{\varphi_i\}_{i\in I}$ of normal states on $\mathscr{M}$ with supports $e_i = \mathbf{s}(\varphi_i)$ mutually orthogonal and $\sum_i e_i = 1$. Then $\varphi = \sum_i \varphi_i$ is an n.ss.f. weight on $\mathscr{M}$.

Let $\xi_i = (e_i)_\varphi \in \mathscr{H}_\varphi$. Then the set $\{\xi_i\}_{i\in I} \in \mathscr{H}_\varphi$ is cyclic and separating for $\pi_\varphi(\mathscr{M})$ and, for $i \neq j$, we have $(\xi_i|\xi_j)_\varphi = ((e_i)_\varphi|(e_j)_\varphi)_\varphi = \varphi(e_je_i) = 0$.

Since every *-isomorphism between two standard von Neumann algebras is spatial ([L], 10.15), it follows that *if a von Neumann algebra $\mathscr{M} \subset \mathscr{B}(\mathscr{H})$ is standard, then there exists a family $\{\xi_i\}_{i\in I} \subset \mathscr{H}$ of mutually orthogonal vectors, cyclic and separating for $\mathscr{M}$*. The converse of this assertion is also true (Rousseau, Van Daele & Van Heeswijck, 1977).

10.11 Proposition. *Let φ be an n.ss.f. weight on the W^*-algebra $\mathscr{M}$. If $\sigma : \mathbb{R} \to Aut(\mathscr{M})$ is an action of $\mathbb{R}$ on $\mathscr{M}$ such that $\sigma_t(\mathfrak{M}_\varphi) = \mathfrak{M}_\varphi$, $(t \in \mathbb{R})$, and φ satisfies the KMS condition with respect to $\{\sigma_t\}_{t\in\mathbb{R}}$ in any two elements of $\mathfrak{M}_\varphi$, then φ is σ-invariant and hence $\sigma_t = \sigma^\varphi_t(t \in \mathbb{R})$.*

Proof. Since $\psi = \varphi|(\mathscr{M}^\varphi)^+$ is an n.s.f. trace on $\mathscr{M}^\varphi$, the w-dense two-sided ideal of $\mathfrak{M}_\psi$ of $\mathscr{M}^\varphi$ has an increasing approximate unit $u_k \uparrow 1$. Let $x \in \mathfrak{M}_\varphi \cap \mathscr{M}^+$ and let f_k be a complex function, defined, continuous and bounded on the strip $\{\alpha \in \mathbb{C}; 0 \leq Re\ \alpha \leq 1\}$, analytic in the interior of this strip, such that $f_k(it) = \varphi(\sigma_t(x)u_k)$ and $f_k(1 + it) = \varphi(u_k\sigma_t(x))(t \in \mathbb{R})$. Since $u_k \in \mathscr{M}^\varphi$, using 2.21.(2) it follows that $f_k(it) = f_k(1 + it)(t \in \mathbb{R})$, and hence f_k is constant, that is, $\varphi(\sigma_t(x)u_k) = \varphi(xu_k)(t \in \mathbb{R})$. Let $E : \mathscr{M} \to \mathscr{M}^\varphi$, be the faithful normal conditional expectation with $\varphi = \psi \circ E$. We have

$$\begin{aligned}
&\psi(E(\sigma_t(x))^{1/2}u_kE(\sigma_t(x))^{1/2}) = \psi(u_k^{1/2}E(\sigma_t(x))u_k^{1/2}) \\
&= \psi(E(u_k^{1/2}\sigma_t(x)u_k^{1/2})) = \varphi(u_k^{1/2}\sigma_t(x)u_k^{1/2}) = \varphi(\sigma_t(x)u_k) = \varphi(xu_k) \\
&= \varphi(u_k^{1/2}xu_k^{1/2}) = \psi(E(u_k^{1/2}xu_k^{1/2})) = \psi(u_k^{1/2}E(x)u_k^{1/2}) = \psi(E(x)^{1/2}u_kE(x)^{1/2}).
\end{aligned}$$

Since $u_k \uparrow 1$ we get $\varphi(\sigma_t(x)) = \psi(E(\sigma_t(x)) = \psi(E(x)) = \varphi(x)$, for every $y \in \mathfrak{M}_\varphi \cap \mathscr{M}^+$, and because of the assumption $\sigma_t(\mathfrak{M}_\varphi) = \mathfrak{M}_\varphi$ it follows that $\varphi \circ \sigma_t = \varphi$. □

The above result holds, in particular, for normal positive forms. In this case, the proof is simpler because we can take $u_k = 1$ and the assumption $\sigma_t(\mathfrak{M}_\varphi) = \mathfrak{M}_\varphi(t \in \mathbb{R})$ is automatically satisfied.

10.12. Let G be a locally compact group. For a function f defined on G and an element $g \in G$, we define the function ${}_gf$ on G by $({}_gf)(h) = f(g^{-1}h)(h \in G)$.

A *left invariant mean* on $\mathscr{L}^\infty(G)$ is a positive linear form m on the C^*-algebra $\mathscr{L}^\infty(G)$ with m(1) = 1 and

$$\mathfrak{m}({}_gf) = \mathfrak{m}(f) \quad (f \in \mathscr{L}^\infty(G), g \in G).$$

The locally compact group G is called *amenable* if there exists a left invariant mean on $\mathscr{L}^\infty(G)$. There are many other equivalent definitions of amenability for which we refer to (Day, 1957; Greenleaf, 1969).

Clearly, every compact group is amenable, the invariant mean being given by the normalized Haar measure of G. Also, every commutative discrete group is amenable (Day, 1957; Greenleaf, 1969).

If the group G is discrete, then every left invariant mean $\mathfrak{m}$ on $\ell^\infty(G)$ is also right invariant (Greenleaf, 1969), that is,

$$\mathfrak{m}(f_g) = \mathfrak{m}(f) \quad (f \in \ell^\infty(G), g \in G),$$

where $f_g(h) = f(hg)$ $(h \in G)$.

We record the following criterion concerning the existence of (not necessarily normal) conditional expectations:

Proposition. *Let $\sigma : G \to Aut(\mathscr{M})$ be an action of the discrete group G on the W^*-algebra $\mathscr{M}$. If G is amenable, then there exists a σ-invariant conditional expectation $E : \mathscr{M} \to \mathscr{M}^\sigma$.*

Proof. Let $\mathfrak{m}$ be an invariant mean on $\ell^\infty(G)$. For each $x \in \mathscr{M}$, the mapping $\mathscr{M}_* \ni \varphi \mapsto \mathfrak{m}(g \mapsto \varphi(\sigma_g(x)) \in \mathbb{C}$ is a bounded linear form on $\mathscr{M}_*$, with norm $\leq \|x\|$, hence there exists a unique element $E(x) \in \mathscr{M}$, $\|E(x)\| \leq \|x\|$, such that $\varphi(E(x)) = \mathfrak{m}(g \mapsto \varphi(\sigma_g(x))$ for every $\varphi \in \mathscr{M}_*$. Since $\mathfrak{m}$ is invariant, it follows that $E(x) \in \mathscr{M}^\sigma$ and $E(\sigma_g(x)) = E(x)$ for all $x \in \mathscr{M}, g \in G$. If $x \in \mathscr{M}$, then clearly $E(x) = x$. Thus, $E : \mathscr{M} \to \mathscr{M}^\sigma$ is a σ-invariant projection of norm 1, and hence a conditional expectation, by the Tomiyama theorem (9.1). □

10.13. Even if the group G is not necessarily amenable, the following result is known:

Theorem (J. T. Schwartz). *Let $\sigma : G \to Aut(\mathscr{M})$ be an action of the discrete group G on the W^*-algebra $\mathscr{M}$ with the property*

$$\overline{co}^w(\{\sigma_g(x); g \in G\}) \cap \mathscr{M}^\sigma \neq \emptyset \quad (x \in \mathscr{M}). \tag{1}$$

For every $x_0 \in \mathscr{M}$ and every $a_0 \in \overline{co}^w(\{\sigma_g(x_0); g \in G\}) \cap \mathscr{M}^\sigma$, there exists a conditional expectation $E : \mathscr{M} \to \mathscr{M}^\sigma$ such that $E(x_0) = a_0$.

Proof. Let $\mathscr{E}$ be the set of all bounded linear mappings $E : \mathscr{M} \to \mathscr{M}$ with the following properties: (a) $E(x_0) = a_0$; (b) $E(x) \in \overline{co}^w(\{\sigma_g(x); g \in G\})$ for $x \in \mathscr{M}$; (c) $E(y) = y$ for $y \in \mathscr{M}^\sigma$.

Let Λ be the set of all finitely supported functions $\lambda : G \to [0, 1]$ such that $\sum_{g\in G} \lambda(g) = 1$. For $\lambda \in \Lambda$ and $x \in \mathscr{M}$, we shall write $\lambda[x] = \sum_{g\in G} \lambda(g)\sigma_g(x)$.

Since $a_0 \in \overline{co}^w(\{\sigma_g(x); g \in G\})$, there exists a net $\{\lambda_i\}_{i\in I} \subset \Lambda$ such that $\lambda_i[x_0] \xrightarrow{w} a_0$. If LIM is a Banach limit with respect to I and $E_0(x) = LIM_i \lambda_i[x](x \in \mathscr{M})$, then $E_0 \in \mathscr{E}$, and hence $\mathscr{E}$ is not empty.

We define a preorder relation "$\leq$" on $\mathscr{E}$ by writing $E_1 \leq E_2$ for $E_1, E_2 \in \mathscr{E}$ if and only if

$$\overline{co}^w(\{\sigma_g(E_1(x)); g \in G\}) \supseteq \overline{co}^w(\{\sigma_g(E_2(x)); g \in G\})$$

for every $x \in \mathcal{M}$. If $\{E_k\}_{k\in K}$ is an increasing net in $\mathcal{E}$ and LIM is a Banach limit with respect to K, then the equation $E(x) = LIM_k E_k(x), (x \in \mathcal{M})$, defines an upper bound $E \in \mathcal{E}$ of $\{E_k\}_{k\in K}$. Thus, $\mathcal{E}$ is inductively ordered.

Let E be a maximal element of $\mathcal{E}$. We show that $E(\mathcal{M}) = \mathcal{M}^\sigma$. If this is not the case, there exists $x_1 \in \mathcal{M}$ such that $E(x_1) \notin \mathcal{M}^\sigma$. By assumption (1), there exists $a_1 \in \overline{co}^w(\{\sigma_g(x_1); g \in G\}) \cap \mathcal{M}^\sigma$. Let $\{\lambda_j\}_{j\in J} \subset \Lambda$ be a net such that $\lambda_j[x_1] \overset{w}{\to} a_1$ and consider a Banach limit LIM with respect to J. Then the equation $E_0(x) = LIM\lambda_j[E(x)](x \in \mathcal{M})$ defines an element $E_0 \in \mathcal{E}, E_0 \geq E$ such that $E_0 \neq E$, as $E(x_1) \in \overline{co}^w(\{\sigma_g(E(x_1)); g \in G\})$ but $\overline{co}^w(\{\sigma_g(E_0(x_1)); g \in G\}) = \{a_1\} \not\ni E(x_1)$, and this contradicts the maximality of E.

It follows that the maximal element E of $\mathcal{E}$ is a projection of norm 1 of $\mathcal{M}$ onto $\mathcal{M}^\sigma$, and hence a conditional expectation, by the Tomiyama theorem (9.1). □

If there exists a σ-invariant normal conditional expectation $E : \mathcal{M} \to \mathcal{M}^\sigma$, then from (1) it follows that

$$\overline{co}^w(\{\sigma_g(x); g \in G\}) \cap \mathcal{M}^\sigma = \{E(x)\} \quad (x \in \mathcal{M}), \tag{2}$$

so that E is uniquely determined and faithful.

To see that E is faithful, choose a family $\{\psi_i\}_i$ of normal states on $\mathcal{M}$ with $\sum_i \mathbf{s}(\psi_i) = 1$. Then each $\psi_i \circ E$ is a σ-invariant normal state on $\mathcal{M}$, hence $\mathbf{s}(\psi_i \circ E) \in \mathcal{M}^\sigma$. It follows that $\mathbf{s}(\psi_i \circ E) = \mathbf{s}(\psi_i)$, $(i \in I)$, hence $\mathcal{M}$ is σ-finite and E is faithful (10.8).

Note that if the W^*-algebra $\mathcal{M}$ is σ-finite, then using the Ryll-Nardzewski fixed point theorem ([L], A.3) one can show that (1), and hence (2) hold (Abdalla & Szücs, 1974; Kovács & Szücs, 1966).

10.14 Proposition. *Let $\mathcal{N}$ be a unital W^*-subalgebra of the W^*-algebra $\mathcal{M}$ and τ a faithful normal trace on $\mathcal{M}$. If $x \in \mathfrak{N}_\tau$, then*

$$\overline{co}^w(\{vxv^*; v \in U(\mathcal{N})\}) \subset \mathfrak{N}_\tau, \tag{1}$$
$$\overline{co}^w(\{vxv^*; v \in U(\mathcal{N})\}) \cap (\mathcal{N}' \cap \mathcal{M}) \neq \emptyset. \tag{2}$$

Proof. If $a = vxv^*$, then $\tau(a^*a) = \tau(x^*x) < +\infty$. Since the function $\mathcal{M} \ni a \mapsto \tau(a^*a)^{1/2}$ is a lower w-semicontinuous seminorm (5.9), statement (1) follows.

The set $\mathcal{K} = \overline{co}^w(\{vxv^*; v \in U(\mathcal{N})\})$ is norm bounded (by $\|x\|$) and w-closed, thus $\mathcal{K}$ is w-compact and the lower w-semicontinuous function $a \mapsto \tau(a^*a)$ attains its greatest lower bound on $\mathcal{K}$, in some $a_0 \in \mathcal{K}$; let $\lambda = \tau(a_0^*a_0)$. Clearly, $va_0v^* \in \mathcal{K}$ and $\tau((va_0v^*)^*(va_0v^*)) = \tau(a_0^*a_0) = \lambda$ for every $v \in U(\mathcal{N})$. To prove (2), it is therefore sufficient to show that the set $\mathcal{K}_0 = \{a \in \mathcal{K}; \tau(a^*a) = \lambda\}$ reduces to the singleton $\{a_0\}$.

If $a, b \in \mathcal{K}_0$, then $(a+b)/2 \in \mathcal{K}$ and hence $\lambda \leq \tau((a+b)^*(a+b))/4 = \tau(a^*a)/2 + \tau(b^*b)/2 - \tau((a-b)^*(a-b))/4 = \lambda - \tau((a-b)^*(a-b))/4$, so that $\tau((a-b)^*(a-b)) = 0$ and $a = b$, as τ is faithful. □

Statement (2) is a fixed point theorem. If $\mathcal{N}$ is abelian, then (2) follows by the Markov–Kakutani theorem ([L], A.1) without any restriction on $x \in \mathcal{M}$. Also, if $\mathcal{N} = \mathcal{M}$, then $\mathcal{N}' \cap \mathcal{M} = \mathcal{Z}(\mathcal{M})$ and (2) holds for all $x \in \mathcal{M}$ by a theorem of Dixmier ([L], C.4.4).

Note that $\mathcal{N}' \cap \mathcal{M} = \mathcal{M}^G$, where $G = \{Ad(v); v \in U(\mathcal{N})\} \subset Aut(\mathcal{M})$. Thus, by Theorem 10.13 we obtain existence criteria for conditional expectations $E : \mathcal{M} \to \mathcal{N}' \cap \mathcal{M}$.

10.15. We shall say that a von Neumann algebra $\mathcal{P} \subset \mathcal{B}(\mathcal{H})$ has the *property P of J. T. Schwartz* if, for every $x \in \mathcal{B}(\mathcal{H})$,

$$\overline{co}^w(\{vxv^*; v \in U(\mathcal{P}')\}) \cap \mathcal{P} \neq \emptyset. \tag{1}$$

In this case, there exists by Theorem 10.13 a conditional expectation $E : \mathcal{B}(\mathcal{H}) \to \mathcal{P}$.

In particular, by the Markov–Kakutani fixed point Theorem (see [L] Appendix), it follows that every von Neumann algebra $\mathcal{P} \subset \mathcal{B}(\mathcal{H})$ with an abelian commutant $\mathcal{P}' \subset \mathcal{B}(\mathcal{H})$ has property P. Consequently,

for every von Neumann algebra $\mathcal{P} \subset \mathcal{B}(\mathcal{H})$ with abelian commutant there exists a conditional expectation $E : \mathcal{B}(\mathcal{H}) \to \mathcal{P}$. (2)

If $\mathcal{A}$ is a maximal abelian $*$-subalgebra of the W^*-algebra $\mathcal{M}$, then $\mathcal{A}' \cap \mathcal{M} = \mathcal{A}$, and again using the Markov–Kakutani theorem, we see that $\overline{co}^w(\{vxv^*; v \in U(\mathcal{A})\}) \cap \mathcal{A} \neq \emptyset (x \in \mathcal{M})$. Thus, by Theorem 10.13, it follows that

for every maximal abelian $*$-subalgebra $\mathcal{A}$ of the W^*-algebra $\mathcal{M}$ there exists a conditional expectation $E : \mathcal{M} \to \mathcal{A}$. (3)

Also, by the remark made in Section 10.13, we see that

if there exists a normal conditional expectation $E : \mathcal{M} \to \mathcal{A}$, then E is uniquely determined and faithful. (4)

10.16. Note that *for every C^*-subalgebra $\mathcal{A}$ of an abelian W^*-algebra $\mathcal{M}$ there exists a $*$-homomorphism $\pi : \mathcal{M} \to \mathcal{A}$ such that $\pi(a) = a$ for $a \in \mathcal{A}$* (Gleason, 1958; Hasumi, 1958; Strătilă & Zsidó, 1977–1979; 9.27). Combined with 10.15.(3), this result shows that *for every abelian C^*-subalgebra $\mathcal{A}$ of an arbitrary W^*-algebra $\mathcal{M}$ there exists a conditional expectation $E : \mathcal{M} \to \mathcal{A}$.*

We record the following result concerning the existence of normal conditional expectations:

Proposition. *Let $\mathcal{M}$ be a W^*-algebra and $\mathcal{C}$ any unital W^*-subalgebra of the center $\mathcal{Z}(\mathcal{M})$ of $\mathcal{M}$. For every $x_0 \in \mathcal{M}, x_0 \neq 0$, there exists a normal conditional expectation $E : \mathcal{M} \to \mathcal{C}$ such that $E(x_0) \neq 0$.*

Proof. Let φ be a normal state on $\mathcal{M}$ with $\varphi(x_0) \neq 0$.

Let $x \in \mathcal{M}^+$. For every $z \in \mathcal{C}^+$ we have $\varphi(xz) \leq \|x\|\varphi(z)$ so, by the Radon–Nikodym theorem ([L], 5.21), there exists a unique $E(x) \in \mathcal{C}, 0 \leq E(x) \leq \|x\|$, such that $\varphi(xz) = \varphi(E(x)z)(z \in \mathcal{C})$, and $E(x) = E(x)\mathbf{s}(\varphi|\mathcal{C})$.

It is easy to check that the mapping $\mathcal{M}^+ \ni x \mapsto E(x) \in \mathcal{C}$ can be uniquely completed to a normal conditional expectation $E : \mathcal{M} \to \mathcal{C}$ such that $\varphi(xz) = \varphi(E(x)z)$ for all $x \in \mathcal{M}, z \in \mathcal{C}$. In particular, $\varphi(E(x_0)) = \varphi(x_0) \neq 0$. □

10.17. The uniqueness of the normal conditional expectation onto a maximal commutative *-subalgebra follows also from the next result:

Proposition. *Let $\mathcal{N}$ be a unital W*-subalgebra of the W*-algebra $\mathcal{M}$ such that $\mathcal{N}' \cap \mathcal{M} \subset \mathcal{N}$. If there exists a normal conditional expectation $E : \mathcal{M} \to \mathcal{N}$, then this is uniquely determined and faithful. Moreover, for $u \in U(\mathcal{M})$ we have*

$$u\mathcal{N}u^* = \mathcal{N} \Leftrightarrow E(uxu^*) = uE(x)u^* \text{ for all } x \in \mathcal{M}. \tag{1}$$

Proof. As for weights (2.1), we can define the support $\mathbf{s}(E)$ of E as the complement of the greatest projection $p \in \mathcal{M}$ with $E(p) = 0$. For every $v \in U(\mathcal{N})$ we have $E(vpv^*) = vE(p)v^* = 0$, so that $vpv^* = p$ and hence $p \in \mathcal{N}' \cap \mathcal{M} \subset \mathcal{N}$. Consequently, $p = E(p) = 0, \mathbf{s}(E) = 1$ and E is faithful.

Let $E_1 : \mathcal{M} \to \mathcal{N}$ be another faithful normal conditional expectation, ψ an n.s.f. weight on $\mathcal{N}, \varphi = \psi \circ E, \varphi_1 = \psi \circ E_1$ n.s.f. weights on $\mathcal{M}$ and $u_t = [D\varphi_1 : D\varphi]_t (t \in \mathbb{R})$.

For any $y \in \mathcal{N}$ we have (10.5.(1)) $\sigma_t^\varphi(y) = \sigma_t^\psi(y) = \sigma_t^{\varphi_1}(y) = u_t\sigma_t^\varphi(y)u_t^*$, hence $u_t \in \mathcal{N}' \cap \mathcal{M} \subset \mathcal{N} (t \in \mathbb{R})$. Then, for $x \in \mathcal{M}^+$ and $t \in \mathbb{R}$ we obtain $\varphi_1(\sigma_t^\varphi(x)) = \psi(E_1(u_t^*\sigma_t^{\varphi_1}(x)u_t)) = \psi(u_t^*E_1(\sigma_t^{\varphi_1}(x))u_t) = \psi(E_1(\sigma_t^{\varphi_1}(x))) = \varphi_1(x)$, so that φ_1 commutes with φ, that is, (4.10) there exists a positive self-adjoint operator A affiliated to $\mathcal{M}^\varphi$ such that $\varphi_1 = \varphi_A$. We have $A^{it} = u_t \in \mathcal{N}' \cap \mathcal{M} = \mathcal{Z}(\mathcal{N})$ and $\psi = \psi_A$, hence $A = 1$ and $\varphi_1 = \varphi$.

We have proved that $\psi \circ E = \psi \circ E_1$ for every n.s.f. weight ψ on $\mathcal{N}$. It follows that $E = E_1$.

If $u \in U(\mathcal{M})$ and $u\mathcal{N}u^* = \mathcal{N}$, then it is easy to check that the mapping $E_1 : \mathcal{M} \to \mathcal{N}$ defined by $E_1(x) = u^*E(uxu^*)u (x \in \mathcal{M})$, is a faithful normal conditional expectation. By the first part of the proof, we infer that $E_1 = E$ and statement (1) follows. □

For any conditional expectation $E : \mathcal{M} \to \mathcal{N} \subset \mathcal{M}$ define the *normalizer* $\mathcal{N}(E)$ of E by

$$\mathcal{N}(E) = \{u \in U(\mathcal{M}); E(uxu^*) = uE(x)u^* \text{ for } x \in \mathcal{M}\}.$$

If E is normal and $\mathcal{N}' \cap \mathcal{M} \subset \mathcal{M}$, the above proposition shows that

$$\mathcal{N}(E) = \{u \in U(\mathcal{M}); u\mathcal{N}u^* = \mathcal{N}\}.$$

10.18 Corollary. *Let $\mathcal{N}$ be a factor and $\mathcal{A}$ an abelian W*-algebra. If $\mathcal{M}$ is a subfactor of $\mathcal{N} \bar{\otimes} \mathcal{A}$ and $\mathcal{M} \supset \mathcal{N} \otimes 1$, then $\mathcal{M} = \mathcal{N} \otimes 1$.*

Proof. Consider $\mathcal{N}$ realized as a von Neumann algebra $\mathcal{N} \subset \mathcal{B}(\mathcal{H})$ and $\mathcal{A}$ realized as a maximal abelian von Neumann algebra $\mathcal{A} \subset \mathcal{B}(\mathcal{H})$. Then $\mathcal{N} \otimes 1 \subset \mathcal{M} \subset \mathcal{N} \bar{\otimes} \mathcal{A} \subset \mathcal{B}(\mathcal{H} \bar{\otimes} \mathcal{K})$ and, by Corollary 9.11,

$$(\mathcal{N} \otimes 1)' \cap (\mathcal{N} \bar{\otimes} \mathcal{A}) = (\mathcal{N}' \cap \mathcal{N}) \bar{\otimes} (\mathcal{B}(\mathcal{K}) \cap \mathcal{A}) = 1 \otimes \mathcal{A},$$

since $\mathcal{N}$ is a factor. Consequently,

$$(\mathcal{N} \otimes 1)' \cap \mathcal{M} = (1 \otimes \mathcal{A}) \cap \mathcal{M} = \mathbb{C} \cdot 1 \subset \mathcal{N} \otimes 1,$$

as $\mathcal{M}$ is also a factor and $1 \otimes \mathcal{A}$ is the center of $\mathcal{N} \bar{\otimes} \mathcal{A}$.

Thus, by Proposition 10.17, it follows that all the Fubini mappings (9.8) $E^{\omega}_{\mathscr{N}}|\mathscr{M} : \mathscr{M} \to \mathscr{N}(\omega \in \mathscr{A}^+_*, \omega(1) = 1)$, are the same faithful normal "conditional expectation" $E : \mathscr{M} \to \mathscr{N}$.

Let $x \in \mathscr{M}$. For $\psi \in \mathscr{N}_*$ and $\omega \in \mathscr{A}^+_*, \omega(1) = 1$, we have

$$(\psi \bar{\otimes} \omega)(E(x) \bar{\otimes} 1) = \psi(E(x)) = \psi(E^{\omega}_{\mathscr{N}}(x)) = (\psi \bar{\otimes} \omega)(x),$$

so that $x = E(x) \bar{\otimes} 1 \in \mathscr{N} \otimes 1$. □

10.19. If $\mathscr{A} \subset \mathscr{B}(\mathscr{H})$ is a maximal abelian von Neumann algebra such that every nonzero projection of $\mathscr{A}$ dominates a minimal projection of $\mathscr{A}$, then there exists a system of matrix units (9.15) $\{e_{ij}\}_{i,j \in I}$ in $\mathscr{B}(\mathscr{H})$ such that $\mathscr{A}$ is the von Neumann algebra generated by $\{e_{ii}; i \in I\}$ and it is easy to see that the mapping

$$\mathscr{B}(\mathscr{H}) \ni a \mapsto \sum_{i \in I} e_{ii} a e_{ii} = \sum_{i \in I} a_{ii} e_{ii} \in \mathscr{A}$$

is a faithful normal conditional expectation.

On the other hand, let $\mathscr{H} = \mathscr{L}^2([0,1])$ with respect to Lebesgue measure and let $\mathscr{A} \subset \mathscr{B}(\mathscr{H})$ be von Neumann algebra $\mathscr{L}^\infty([0,1])$ of multiplication operators. Since $1 \in \mathscr{H}$ is a cyclic vector for $\mathscr{A}$, it follows that $\mathscr{A}$ is maximal abelian in $\mathscr{B}(\mathscr{H})$ (see [L], E.3.9, E.3.10; Strătilă & Zsidó, 1977, 2005, Cor. 3/8.13, Prop. 3/9.37), but it is obvious that $\mathscr{A}$ has no minimal projections.

Consider now a general maximal abelian von Neumann algebra $\mathscr{A} \subset \mathscr{B}(\mathscr{H})$ without minimal projections and let $E : \mathscr{B}(\mathscr{H}) \to \mathscr{A}$ be a conditional expectation.

Let γ be a character of the abelian C^*-algebra $\mathscr{A}$. Then γ is a singular form ([A.16]) on the W^*-algebra $\mathscr{A}$. Indeed, let $0 \neq p \in \mathscr{A}$ be a projection. Since $\mathscr{A}$ has no minimal projections, there exists a projection $0 \neq q \leq p$ in A with $r = p - q \neq 0$. If $\gamma(p) \neq 0$ and $\gamma(q) \neq 0$, then $\gamma(p) = \gamma(q) = 1$, hence $\gamma(r) = 0$. Thus, every nonzero projection of $\mathscr{A}$ dominates a nonzero projection of $\mathscr{A}$ annihilated by γ, that is, γ is singular.

Then $\gamma \circ E$ is an extension to $\mathscr{B}(\mathscr{H})$ of the singular form γ on $\mathscr{A}$ and hence (Tomiyama, 1971, Lemma 4.3; Strătilă & Zsidó, 1977, 2005, p. 8.5) $\gamma \circ E$ is a singular form on $\mathscr{B}(\mathscr{H})$. It follows that $\gamma(E(e)) = 0$ for every minimal projection $e \in \mathscr{B}(\mathscr{H})$.

Since γ was an arbitrary character of $\mathscr{A}$ we have $E(e) = 0$ for every minimal projection $e \in \mathscr{B}(\mathscr{H})$. Hence, E is not normal, for the normality of E would imply that $E(x) = 0$ for all $x \in \mathscr{B}(\mathscr{H})$, a contradiction.

In conclusion, *there exists a maximal abelian von Neumann algebra $\mathscr{A} \subset \mathscr{B}(\mathscr{H})$ without minimal projections; for any such an algebra $\mathscr{A}$ there is no normal conditional expectation of $\mathscr{B}(\mathscr{H})$ onto $\mathscr{A}$.*

10.20. As an application of the above remarks we establish a characterization of finiteness for W^*-algebras:

Proposition. *A W^*-algebra $\mathscr{M}$ is finite if and only if for every maximal abelian *-subalgebra $\mathscr{A}$ of $\mathscr{M}$ there exists a normal conditional expectation of $\mathscr{M}$ onto $\mathscr{A}$.*

Proof. If $\mathscr{M}$ is finite, the desired conclusion follows from Corollary 10.6. Assume that $\mathscr{M}$ is properly infinite. Then $\mathscr{M} \approx \mathscr{M} \bar{\otimes} \mathscr{B}(\mathscr{H})$ with $\mathscr{H}$ a separable infinite dimensional Hilbert space. Let $\mathscr{C}$ be a maximal abelian *-subalgebra of $\mathscr{M}$ and $\mathscr{A} \subset \mathscr{B}(\mathscr{H})$ a maximal abelian von Neumann algebra. By Corollary 9.10, $\mathscr{C} \bar{\otimes} \mathscr{A}$ is a maximal abelian *-subalgebra of $\mathscr{M} \bar{\otimes} \mathscr{B}(\mathscr{H})$.

Suppose that there exists a normal conditional expectation E of $\mathscr{M} \bar{\otimes} \mathscr{B}(\mathscr{H})$ onto $\mathscr{C} \bar{\otimes} \mathscr{A}$ and let φ be a normal state on $\mathscr{M}$. Then the mapping

$$\mathscr{B}(\mathscr{H}) \ni x \mapsto (\varphi \bar{\otimes} \iota)(E(1 \bar{\otimes} x)) \in \mathscr{A}$$

is a normal conditional expectation of $\mathscr{B}(\mathscr{H})$ onto $\mathscr{A}$, contradicting the conclusion of Section 10.19. □

10.21. The existence of a normal conditional expectation $E : \mathscr{M} \to \mathscr{N}$ transfers certain properties of $\mathscr{M}$ to $\mathscr{N}$.

Recall ([L], E.6.11; Strătilă & Zsidó, 1977, 2005, 9.40) that the supremum of all the minimal projections (or atoms) of a W^*-algebra $\mathscr{M}$ is a central projection of $\mathscr{M}$ and that the W^*-algebra $\mathscr{M}$ is called *atomic* if this central projection is equal to 1.

Proposition. *Let $E : \mathscr{M} \to \mathscr{N}$ be a normal conditional expectation of the W^*-algebra $\mathscr{M}$ onto its unital W^*-subalgebra $\mathscr{N}$. Then*

$$\mathscr{M} \text{ is semifinite} \Rightarrow \mathscr{N} \text{ is semifinite} \tag{1}$$

$$\mathscr{M} \text{ is discrete} \Rightarrow \mathscr{N} \text{ is discrete} \tag{2}$$

$$\mathscr{M} \text{ is atomic} \Rightarrow \mathscr{N} \text{ is atomic} \tag{3}$$

Proof. (1) Assume that $\mathscr{M}$ is semifinite and $\mathscr{N}$ is of type III. Let τ be an n.s.f. trace on $\mathscr{M}$. Since $E \neq 0$ is normal, there exists a projection $e \in \mathscr{M}$ with $\tau(e) < +\infty$ and $E(e) \neq 0$. Since $E(e) \in \mathscr{N}^+$, there exists a projection $0 \neq f \in \mathscr{N}$ and $\lambda \in (0, +\infty)$ such that $\lambda f \leq E(e)$.

We shall obtain a contradiction by showing that $f\mathscr{N}f$ is finite with the help of Sakai's topological criterion of finiteness ([L], 7.23). So, let $\{y_i\}_i \subset f\mathscr{N}f$ be a net such that $\|y_i\| \leq 1$ and $y_i \overset{s}{\to} 0$. Since $e \in \mathscr{M}$ is a finite projection and $y_i e \overset{s}{\to} 0$, we have $ey_i^* \overset{s}{\to} 0$ ([L], 7.23). By the last remark of Section 9.2, we infer that $E(e)y_i^* \overset{s}{\to} 0$ and therefore

$$y_i^* = (fE(e)f + (1 - f))^{-1} fE(e)fy_i^* \overset{s}{\to} 0.$$

Hence ([L], 7.23) $f\mathscr{N}f$ is finite, a contradiction.

(2) If $\mathscr{M}$ is discrete, then by (1), $\mathscr{N}$ is semifinite and hence ([L], E.4.14) there exists a family $\{\mathscr{N}_i\}_{i \in I}$ of finite W^*-algebras and a family $\{\mathscr{H}\}_{i \in I}$ of Hilbert spaces such that $\mathscr{N} = \bigoplus_{i \in I} \mathscr{N}_i \bar{\otimes} \mathscr{B}(\mathscr{H}_i)$. Taking into account the existence of Fubini mappings (9.8), we see that in order to prove (2) we may assume $\mathscr{N}$ is finite.

Moreover, if $\mathscr{N}$ is not discrete, we may assume that $\mathscr{N}$ is of type II_1. Let τ be an n.s.f. trace on $\mathscr{M}$ and μ a nonzero normal finite trace on $\mathscr{N}$.

Then $\varphi = \mu \circ E$ is a nonzero normal positive form on $\mathscr{M}$ and, by Theorem 4.10, there exists a nonzero positive self-adjoint operator A affiliated to $\mathscr{M}$ such that $\varphi = \tau_A$. For $y \in \mathscr{N}$ and $x \in \mathscr{M}$, we have $\varphi(xy) = \mu(E(x)y) = \mu(yE(x)) = \varphi(yx)$, and hence (by 2.21.(2) and 4.7) $y = \sigma_t^{\varphi}(y) = A^{it}yA^{-it}(t \in \mathbb{R})$. Thus, A is affiliated to $\mathscr{N}' \cap \mathscr{M}$. There exists a nonzero spectral projection e of A and an element $0 \leq a \in e\mathscr{M}e$ such that Ae is bounded and $Aea = e$. Then $e \in \mathscr{N}' \cap \mathscr{M}$ and $\tau(e) = \tau(Aea) = \varphi(a) < +\infty$, hence e is finite in $\mathscr{M}$. The W^*-algebra $e\mathscr{M}e$ is finite and discrete so that every W^*-subalgebra of $e\mathscr{M}e$ is discrete (see [L], 7.16, 7.17). However, the induced algebra $\mathscr{N}e = e\mathscr{N}e \subset e\mathscr{M}e$ is of type ([L], E.6.10), a contradiction.

(3) If $\mathcal{M}$ is atomic and $\mathcal{N}$ contains no minimal projections, then the desired contradiction can be obtained arguing as in Section 10.19, and replacing the characters by pure states of $\mathcal{N}$. □

Since there exist conditional expectations $\mathcal{M} \bar{\otimes} \mathcal{N} \to \mathcal{M} \otimes 1$ (9.8), from the above proposition we get the conclusions listed in ([L], C.7.4) concerning the type of the tensor product. Recall in particular that

$$\textit{if } \mathcal{M} \textit{ is of type III, then } \mathcal{M} \bar{\otimes} \mathcal{N} \textit{ is of type III;} \tag{4}$$

$$\textit{if } \mathcal{M} \textit{ is continuous, then } \mathcal{M} \bar{\otimes} \mathcal{N} \textit{ is continuous.} \tag{5}$$

10.22. A W^*-algebra $\mathcal{M}$ is called *injective* if, whenever $\mathcal{M}$ is imbedded as a C^*-subalgebra of a C^*-algebra $\mathcal{A}$, there exists a conditional expectation $E : \mathcal{A} \to \mathcal{M}$. Of course, we could similarly define the notions of an injective C^*-algebra, but it appears that the injective C^*-algebras so defined are almost W^*-algebras (more precisely, they are monotone complete AW^*-algebras, see Strătilă & Zsidó, 1977, 2005, §9).

It is easy to check that injectivity is stable under *-isomorphisms.

Theorem. *Let $\mathcal{M} \subset \mathcal{B}(\mathcal{H})$ be a von Neumann algebra. The following statements are equivalent:*

(i) *$\mathcal{M}$ is injective;*
(ii) *there exists a conditional expectation $\mathcal{B}(\mathcal{H}) \to \mathcal{M}$;*
(iii) *there exists a conditional expectation $\mathcal{B}(\mathcal{H}) \to \mathcal{M}'$;*
(iv) *$\mathcal{M}'$ is injective.*

Proof. Taking into account the structure of *-isomorphisms between von Neumann algebras ([L], E.8.8), the existence of Fubini mappings (9.8) and using Proposition 9.4, it is easy to check that properties (ii) and (iii) are stable under *-isomorphisms. We shall therefore assume that $\mathcal{M} \subset \mathcal{B}(\mathcal{H})$ is a standard von Neumann algebra ([L], 10.15). Let $J : \mathcal{H} \to \mathcal{H}$ be a conjugation such that $J\mathcal{M}J = \mathcal{M}'$.

(ii) ⇔ (iii). If $E : \mathcal{B}(\mathcal{H}) \to \mathcal{M}$ is a conditional expectation, then the mapping $E' : \mathcal{B}(\mathcal{H}) \to \mathcal{M}'$ defined by $E'(x) = JE(JxJ)J(x \in \mathcal{B}(\mathcal{H}))$ is a conditional expectation.

(i) ⇔ (ii). It is clear that (i) ⇒ (ii). Conversely, let $\mathcal{A}$ be any C^*-algebra containing $\mathcal{M}$ as a C^*-subalgebra. Then $\mathcal{M}^{**}$ is a W^*-subalgebra of $\mathcal{A}^{**}$ and there exist a central projection $p \in \mathcal{M}^{**}$ and a *-isomorphism $\pi : \mathcal{M}^{**}p \to \mathcal{M}$ (A.16). Since $\mathcal{M}$ has property (ii), which is stable to *-isomorphisms, there exists a conditional expectation $E_0 : \mathcal{A}^{**}p \to \mathcal{M}^{**}p$. The mapping $E : \mathcal{A} \to \mathcal{M}$ defined by $E(x) = \pi(E_0(xp))$ $(x \in \mathcal{A} \subset \mathcal{A}^{**})$, is then a conditional expectation.

(iii) ⇔ (iv) follows from (i) ⇔ (ii). □

There are many other important characterizations of injective W^*-algebras (see 10.31).

In what follows we give some examples of injective W^*-algebras.

10.23 Proposition. *Every discrete W^*-algebra is injective.*

Proof. By ([L], 6.5) we can realize $\mathcal{M}$ as a von Neumann algebra $\mathcal{M} \subset \mathcal{B}(\mathcal{H})$ with an abelian commutant, so that there exists a conditional expectation $E : \mathcal{B}(\mathcal{H}) \to \mathcal{M}$ (10.15.(2)) and $\mathcal{M}$ is injective by Theorem 10.22. □

10.24 Proposition. *Let $\mathcal{M}, \mathcal{N}$ be W^*-algebras. Then*

$$\mathcal{M} \bar{\otimes} \mathcal{N} \text{ is injective} \Leftrightarrow \mathcal{M} \text{ and } \mathcal{N} \text{ are injective}; \tag{1}$$

$$\mathcal{M} \bar{\otimes} \mathcal{N} \text{ is injective} \Leftrightarrow \mathcal{M}' \text{ and } \mathcal{N}' \text{ are injective}. \tag{2}$$

Proof. Use Theorem 10.22, Proposition 9.4, and the Fubini mappings. □

10.25 Proposition. *Let $\mathcal{M}$ be a W^*-algebra and $\{\mathcal{M}_i\}_{i\in I}$ an upward directed family of W^*-subalgebras of $\mathcal{M}$ such that $\mathcal{M}$ is generated by $\bigcup_{i\in I}\mathcal{M}_i$. If each $\mathcal{M}_i$ is injective, then $\mathcal{M}$ is also injective.*

Proof. Let $\mathcal{M} \subset \mathcal{B}(\mathcal{H})$ be realized as a von Neumann algebra and choose a Banach limit LIM with respect to I (9.6). Since the $\mathcal{M}_i'$ are injective (10.22), there exist conditional expectations $E_i' : \mathcal{B}(\mathcal{H}) \to \mathcal{M}_i' (i \in I)$. The mapping $E' : \mathcal{B}(\mathcal{H}) \to \mathcal{M}' = \bigcap_{i\in I}\mathcal{M}_i'$ defined by $E'(x) = LIM_i E_i'(x) (x \in \mathcal{B}(\mathcal{H}))$ is a conditional expectation, so that $\mathcal{M}'$ is injective (10.22). □

10.26. A W^*-algebra $\mathcal{M}$ is called *approximately finite dimensional* if it is generated by an upward directed family of finite dimensional W^*-subalgebras.

By Propositions 10.23 and 10.25, it follows that every approximately finite dimensional W^*-algebra is injective. Conversely, Connes (1976a) has proved that every injective factor with separable predual is approximately finite dimensional. Simpler proofs were obtained by Haagerup (1985) and Popa (1986).

A W^*-algebra $\mathcal{M}$ is called *semidiscrete* if the identity mapping on $\mathcal{M}$ can be approximated, with respect to the p-topology (2.23), by normal, finite rank, completely positive linear contractions (Effors & Lance, 1977). It is known that $\mathcal{M}$ is semidiscrete if and only if it is injective (Choi & Effros, 1976a; Connes, 1976a; Wassermann, 1977).

On the other hand, it is easy to check directly that every approximately finite dimensional von Neumann algebra $\mathcal{M} \subset \mathcal{B}(\mathcal{H})$ has property P of J .T. Schwartz, so that $\mathcal{M}$ is injective (10.15.(1)).

*If $\mathcal{M}$ is a W^*-algebra with separable predual, then the assertions: $\mathcal{M}$ is injective, $\mathcal{M}$ is semidiscrete, $\mathcal{M} \subset \mathcal{B}(\mathcal{H})$ has property P, are all equivalent to saying that $\mathcal{M}$ is generated by an increasing sequence of finite dimensional *-subalgebras.*

Another property equivalent to injectivity (in the case of separable predual) is amenability, which we shall consider in the next sections.

10.27. We first prove another existence criterion for conditional expectations.

Theorem. *Let $\mathcal{A}$ be a unital C^*-algebra and $\mathcal{M}$ a countably decomposable finite W^*-algebra contained as a unital C^*-subalgebra in $\mathcal{A}$. The following statements are equivalent:*

(i) *there exists a conditional expectation $E : \mathcal{A} \to \mathcal{M}$;*
(ii) *there exists an $\mathcal{M}$-linear projection $P : \mathcal{A} \to \mathcal{M}$;*
(iii) *there exists a bounded linear form φ on $\mathcal{A}$ such that $\varphi|\mathcal{M}$ is positive and faithful and $\varphi(x\cdot) = \varphi(\cdot x)$ for every $x \in \mathcal{M}$;*
(iv) *there exists a positive linear form φ on $\mathcal{A}$ such that $\varphi|\mathcal{M}$ is faithful and $\varphi(x\cdot) = \varphi(\cdot x)$ for every $x \in \mathcal{M}$.*

Proof. (i) ⇒ (ii) is obvious.

(ii) ⇒ (iii). If τ is any faithful finite trace on $\mathcal{M}$, then $\varphi = \tau \circ P$ is as required in (iii).

(iii) $\Rightarrow$ (iv). Replacing the bounded linear form φ given in (iii) by $\varphi + \varphi^*$, we may assume that φ is self-adjoint. Then let $\varphi = \varphi^+ - \varphi^-$ be the Jordan decomposition of φ ([L], 5.17). For every $u \in U(\mathscr{M})$, we have $\varphi = \varphi(u \cdot u^*) = \varphi^+(u \cdot u^*) - \varphi^-(u \cdot u^*)$, so that $\varphi^+ = \varphi^+(u \cdot u^*)$, by the uniqueness of the Jordan decomposition. Since $\varphi^+|\mathscr{M} \geq \varphi|\mathscr{M}$, it follows that $\varphi^+|\mathscr{M}$ is faithful. Hence φ^+ is the required form.

(iv) $\Rightarrow$ (i). Recall first that for every bounded linear form ψ on a W^*-algebra $\mathscr{N}$ there exist unique normal and singular linear forms ψ_{nor} and ψ_{sing} such that $\psi = \psi_{nor} + \psi_{sing}$; if ψ is positive and faithful, then ψ_{nor} is also positive and faithful (A.16; Strătilă & Zsidó, 1977, 2005, p. 8.4).

We define a linear mapping $F : \mathscr{A} \to \mathscr{M}_*$ by $F(a) = (\varphi(a \cdot)|\mathscr{M})_{nor} (a \in \mathscr{A})$. Then $\tau = F(1) = (\varphi|\mathscr{M})_{nor}$ is a faithful normal finite trace on $\mathscr{M}$.

Let $a \in \mathscr{A}$ with $a \geq 0$. For $x \in \mathscr{M}$, we have $\varphi(ax^*x) = \varphi(a|x|^2) = \varphi(|x|a|x|) \geq 0$, hence $F(a) \geq 0$. Since $0 \leq a \leq \|a\| \cdot 1$, we obtain $0 \leq F(a) \leq \|a\|\tau$. By the Radon–Nikodym type theorem ([L], 5.21), there exists a unique element $E(a) \in \mathscr{M}^+, \|E(a)\| \leq \|a\|$, such that $F(a) = \tau(E(a)^{1/2} \cdot E(a)^{1/2}) = \tau(E(a) \cdot)$.

We thus get a positive linear mapping $E : \mathscr{A} \to \mathscr{M}$, uniquely determined, such that $F(a) = \tau(E(a) \cdot)(a \in \mathscr{A})$. It is easy to check that E is a projection of $\mathscr{A}$ onto $\mathscr{M}$ and that $\|E\| = \|E(1)\| = \|1\| = 1$, as E is positive. By Theorem 9.1, it follows that E is a conditional expectation. □

10.28 Corollary. *A von Neumann algebra $\mathscr{M} \subset \mathscr{B}(\mathscr{H})$ is injective if and only if there exists an $\mathscr{M}$-linear projection $P : \mathscr{B}(\mathscr{H}) \to \mathscr{M}$.*

Proof. If $\mathscr{M}$ is finite and countably decomposable, then the corollary follows obviously from Theorem 10.27.

Assume that $\mathscr{M}$ is semifinite and there exists an $\mathscr{M}$-linear projection $P : \mathscr{B}(\mathscr{H}) \to \mathscr{M}$. There exists an increasing net $\{e_i\}_{i \in I}$ of countably decomposable finite projections in $\mathscr{M}$ with $e_i \uparrow 1$ (see [L], 4.20, 7.2). For each $i \in I$, P defines by restriction an $e_i\mathscr{M}e_i$-linear projection $P_i : \mathscr{B}(e_i\mathscr{H}) = e_i\mathscr{B}(\mathscr{H})e_i \to e_i\mathscr{M}e_i$. Hence each $e_i\mathscr{M}e_i$ is injective so that $\mathscr{M}$ is also injective, by Proposition 10.25.

Assume that $\mathscr{M}$ is properly infinite and that there exists an $\mathscr{M}$-linear projection $P : \mathscr{B}(\mathscr{H}) \to \mathscr{M}$. In this case, we need the Connes–Takesaki continuous decomposition theorem, which will be proved later (23.6). According to this theorem, we may assume that there exists a semifinite von Neumann subalgebra $\mathscr{N} \subset \mathscr{M}$ together with an *so*-continuous unitary representation $\mathbb{R} \ni t \mapsto u(t) \in \mathscr{M}$ such that $u(t)\mathscr{N}u(t)^* = \mathscr{N} (t \in \mathbb{R})$, and $\mathscr{M} = (\mathscr{N} \cup u(\mathbb{R}))''$. Moreover, $\sigma : \mathbb{R} \ni t \mapsto Ad(u(t))|\mathscr{N} \in Aut(\mathscr{N})$ is a continuous action of $\mathbb{R}$ on $\mathscr{N}$ which defines a dual action $\hat{\sigma}$ of $\mathbb{R}$ on $\mathscr{M}$ such that $\mathscr{N} = \mathscr{M}^{\hat{\sigma}}$.

Since the discrete abelian group $\mathbb{R}$ is amenable, by Proposition 10.12 there exists a conditional expectation $Q : \mathscr{M} \to \mathscr{M}^{\hat{\sigma}} = \mathscr{N}$. Then $Q \circ P : \mathscr{B}(\mathscr{H}) \to \mathscr{N}$ is an $\mathscr{N}$-linear projection. Since $\mathscr{N}$ is semifinite, by the first part of the proof it follows that $\mathscr{N}$ is injective and so there exists a conditional expectation (10.22) $F' : \mathscr{B}(\mathscr{M}) \to \mathscr{N}'$.

Since $\mathscr{M} = (\mathscr{N} \cup u(\mathbb{R}))''$, we have $\mathscr{M}' = \mathscr{N}' \cap u(\mathbb{R})'$, that is, $\mathscr{M}' = (\mathscr{N}')^{\sigma'}$ where $\sigma' : \mathbb{R} \ni t \mapsto Ad(u(t))|\mathscr{N}' \in Aut(\mathscr{N}')$ is an action of the discrete abelian group $\mathbb{R}$ on $\mathscr{N}'$. We obtain again by Proposition 10.12 a conditional expectation $E' : \mathscr{N}' \to (\mathscr{N}')^{\sigma'} = \mathscr{M}'$. Then $E' \circ F' : \mathscr{B}(\mathscr{H}) \to \mathscr{M}'$ is a conditional expectation and so $\mathscr{M}$ is injective (10.22).

In the general case, there exists a central projection $p \in \mathscr{M}$ such that $\mathscr{M}p$ is semifinite and $\mathscr{M}(1-p)$ is of type III, hence properly infinite. If there exists an $\mathscr{M}$-linear projection $P : \mathscr{B}(\mathscr{H}) \to \mathscr{M}$, then the von Neumann algebras $\mathscr{M}p \subset \mathscr{B}(p\mathscr{H}), \mathscr{M}(1-p) \subset \mathscr{B}((1-p)\mathscr{H})$ enjoy the same property so that they are injective as well as their direct sum $\mathscr{M}$ (10.24.(2)). □

10.29. Let $\mathscr{M}$ be a W^*-algebra. *A normal dual Banach $\mathscr{M}$-bimodule* is a Banach $\mathscr{M}$-bimodule $\mathscr{X}^{*)}$ such that the Banach space $\mathscr{X}$ is the dual of a certain Banach space $\mathscr{X}_*$ and, for $f \in \mathscr{X}_*, a_0 \in \mathscr{M}, x_0 \in$

$\mathscr{X}$, the mappings

$$\mathscr{M} \ni a \mapsto f(a \cdot x_0) \in \mathbb{C}, \quad \mathscr{M} \ni a \mapsto f(x_0 \cdot a) \in \mathbb{C}$$

are w-continuous and the mappings

$$\mathscr{X} \ni x \mapsto f(a_0 \cdot x) \in \mathbb{C}, \quad \mathscr{X} \ni x \mapsto f(x \cdot a_0) \in \mathbb{C}$$

are $\sigma(\mathscr{X}, \mathscr{X}_*)$-continuous.

A *derivation* of $\mathscr{M}$ into $\mathscr{X}$ is a bounded linear mapping $\delta : \mathscr{M} \to \mathscr{X}$ such that $\delta(ab) = \delta(a)\cdot b + a \cdot \delta(b) (a, b \in \mathscr{M})$. Every element $x \in \mathscr{X}$ determines an *inner derivation* $\delta_x : \mathscr{M} \ni a \mapsto x \cdot a - a \cdot x \in \mathscr{X}$.

The W^*-algebra $\mathscr{M}$ is called *amenable* if every derivation of $\mathscr{M}$ into some normal dual Banach $\mathscr{M}$-bimodule $\mathscr{X}$ is inner.

Proposition. *If $\mathscr{M}$ is an amenable unital W^*-subalgebra of the W^*-algebra $\mathscr{B}$, then there exists an $(\mathscr{M}' \cap \mathscr{B})$-linear projection $P : \mathscr{B} \to \mathscr{M}' \cap \mathscr{B}$.*

Proof. Starting from the inclusion $\mathscr{M} \subset \mathscr{B}$, we shall construct a normal dual Banach $\mathscr{M}$-bimodule $\mathscr{X}$.

Let $\mathscr{Y}_* = \mathscr{B} \hat{\otimes} \mathscr{B}_*$ be the Banach space completion of the algebraic tensor product $\mathscr{B} \otimes \mathscr{B}_*$ with respect to the greatest cross-norm (Schatten, 1950) and let $\mathscr{Y} = (\mathscr{Y}_*)^*$. It is known (Schatten, 1950) that $\mathscr{Y}$ can be identified isometrically with the Banach space $\mathscr{B}(\mathscr{B})$ of all bounded linear mappings $\mathscr{B} \to \mathscr{B}$ in such a way that

$$S(b \otimes \psi) = \psi(S(b)) \quad (S \in \mathscr{B}(\mathscr{B}), b \in \mathscr{B}, \psi \in \mathscr{B}_*).$$

If we define the left and right actions of $\mathscr{M}$ on $\mathscr{Y} = \mathscr{B}(\mathscr{B})$ by

$$(a \cdot S)(b) = aS(b), (S \cdot a)(b) = S(b)a \quad (a \in \mathscr{M}, S \in \mathscr{B}(\mathscr{B}), b \in \mathscr{B}),$$

then $\mathscr{Y}$ becomes a normal dual Banach $\mathscr{M}$-bimodule.

Now let $\mathscr{X}$ be the set of those $S \in \mathscr{Y}$ with the property that for every $a' \in \mathscr{M}' \cap \mathscr{B}, b \in \mathscr{B}, \psi \in \mathscr{B}_*$ we have

$$S(a' \otimes \psi) = 0, \text{ that is, } S|(\mathscr{M}' \cap \mathscr{B}) = 0; \tag{1}$$
$$S(a'b \otimes \psi - b \otimes \psi(a' \cdot)) = 0, \text{ that is, } S(a'b) = a'S(b); \tag{2}$$
$$S(ba' \otimes \psi - b \otimes \psi(\cdot a')) = 0, \text{ that is, } S(ba') = S(b)a'. \tag{3}$$

Then $\mathscr{X}$ is a $\sigma(\mathscr{Y}, \mathscr{Y}_*)$-closed sub-$\mathscr{M}$-bimodule of $\mathscr{Y}$ and $\mathscr{X}$ becomes a normal dual Banach $\mathscr{M}$-bimodule with the induced structure.

Let $\iota_{\mathscr{B}} \in \mathscr{B}(\mathscr{B}) = \mathscr{Y}$ be the identity mapping on $\mathscr{B}$. It is easy to check that the inner derivation of $\mathscr{M}$ into $\mathscr{Y}$ determined by $\iota_{\mathscr{B}} \in \mathscr{Y}$ takes values in $\mathscr{X}$, thus defining a derivation $\delta : \mathscr{M} \to \mathscr{X}$.

---

*That is, there are bounded bilinear mappings

$$\mathscr{M} \times \mathscr{X} \ni (a, x) \mapsto a \cdot x \in \mathscr{X} \text{ and } \mathscr{X} \times \mathscr{M} \ni (x, a) \mapsto x \cdot a \in \mathscr{X}$$

called the left and right actions of $\mathscr{M}$ on $\mathscr{X}$, such that, for every $a, b \in \mathscr{M}, x \in \mathscr{X}$, we have $1 \cdot x = x \cdot 1 = x, a \cdot (b \cdot x) = (ab) \cdot x, (x \cdot a) \cdot b = x \cdot (ab), a \cdot (x \cdot b) = (a \cdot x) \cdot b$.

Since $\mathscr{M}$ is amenable, there exists an element $S \in \mathscr{X}$ such that δ is the inner derivation determined by S.

Let $P = \iota_{\mathscr{B}} - S \in \mathscr{Y} = \mathscr{B}(\mathscr{B})$. Then $P \cdot a - a \cdot P = 0$ for all $a \in \mathscr{M}$, hence $P : \mathscr{B} \to \mathscr{M}' \cap \mathscr{B}$ is a bounded linear mapping. Since $S \in \mathscr{X}$, it follows from (1) that P is a projection of $\mathscr{B}$ onto $\mathscr{M}' \cap \mathscr{B}$ while from (2) and (3) we deduce that P is $(\mathscr{M}' \cap \mathscr{B})$-linear. □

10.30. Taking $\mathscr{B} = \mathscr{B}(\mathscr{H})$ in Proposition 10.29 and using Corollary 10.28 and Theorem 10.22, we obtain the following

Corollary. *Every amenable W^*-algebra is injective.*

Conversely, if $\mathscr{M}$ is an injective W^*-algebra with separable predual, then $\mathscr{M}$ is generated by an increasing sequence of finite dimensional *-subalgebras (as mentioned in 10.26) and by Johnson, Kadison and Ringrose (1972) it follows that $\mathscr{M}$ is amenable.

The equivalence of the notions of amenability, semidiscreteness and injectivity in the case of separable predual, to the structural property of approximate finite dimensionality is one of the deepest and most important results in the theory of operator algebras to date.

10.31. Notes. Theorem 10.1 and Corollary 10.7 are due to Takesaki (1972a). The particular case of Theorem 10.1 contained in Corollary 10.6 is due to Dixmier (1949) and Umegaki (1956). Corollary 10.8 is due to Kovács and Szücs (1966); actually, they proved a better result, namely that $\overline{co}^w(\{\sigma_g(x); g \in G\}) \cap \mathscr{M}^\sigma = \{E(x)\}$ whenever $\mathscr{M}$ is σ-finite (this result can be obtained using the Ryll-Nardzewski fixed point theorem, see Abdalla & Szücs, 1974). The strictly semifinite weights were introduced by Combes (1971a, 1971b) who proved Corollary 10.9 and Proposition 10.11. Theorem 10.13 is due to Schwartz (1963); in the same article, Schwartz introduced property P (see also Choda & Echigo, 1963, 1964). Proposition 10.14 is due to Dixmier (1957, 1969), Proposition 10.16 appears in Halpern (1969) and Strătilă, Voiculescu, and Zsidó (1976, 1977), Proposition 10.17 is due to Connes [1973a] and Corollary 10.18 is due to Connes and Takesaki (1977). Proposition 10.20 is the answer to a problem posed by Kadison, given by Takesaki (1972a) and Tomiyama (1972). Proposition 10.21 is due to Sakai (1957) and Tomiyama (1957–1959, 1971). For the injectivity property for W^*-algebras or properties equivalent to injectivity, as well as for the results contained in Sections 10.22–10.26, we refer to Arveson (1969); Choda and Echigo (1963, 1964); Choi and Effros (1977a, 1976a, 1977b); Connes (1973a); Effors (1972); Effors and Lance (1977); Hakeda and Tomiyama (1967); Lance (1973); Schwartz (1963); and Wassermann (1977). Theorem 10.27 and Corollary 10.28 are due to Connes (1977a, 1978a) and Bunce and Paschke (1978). For the definition of an amenable W^*-algebra (10.29), we refer to Johnson (1972); Johnson, Kadison, and Ringrose (1972). Corollary 10.30 is due to Connes (1978a) and its proof, based on Proposition 10.29, is due to Bunce and Paschke (1978).

For our exposition, we have used Bunce and Paschke (1978); Choi and Effros (1977a); Combes (1971a, 1971b); Diximer (1969); Takesaki (1972a) and Tomiyama (1971).

Further results related to the Kovács–Szücs theorem are contained in Guichardet (1974); Størmer (1970, 1971a, 1972a, 1973, 1972b, 1974b).

The property of a W^*-algebra of being approximately finite dimensional was considered for the first time by Murray and von Neumann (1936, 1937, 1943, p. IV) in the case of finite factors. They obtained several characterizations of this property and showed that all approximately finite dimensional factors of type II_1 are *-isomorphic [*)] (see Dixmier, 1957, 1969). The approximately

[*)] We consider only W^*-algebras with separable preduals.

finite dimensional factor of type II_1, usually called the hyperfinite II_1 factor, will be denoted by $\mathcal{R}$.

For properly infinite W^*-algebras, several characterizations of this property are given in Elliott and Woods (1976) and interesting stability results appeared in Golodec (1971); Connes et al. (1973).

Connes (1976a) proved the fundamental result that all injective factors of type II_1 are $*$-isomorphic to $\mathcal{R}$. In particular, it follows that any subfactor of $\mathcal{R}$ is either finite dimensional, or $*$-isomorphic to $\mathcal{R}$. Also, it follows that all injective factors of type II_∞, are $*$-isomorphic to $\mathcal{R}_\infty = \mathcal{R} \bar{\otimes} \mathcal{F}_\infty$, where $\mathcal{F}_\infty$ is the only factor of type I_∞ (see Kadison, 1971). Recently, Connes, Feldman, and Weiss (1981) proved that for any two maximal abelian $*$-subalgebras $\mathcal{A}_1$, $\mathcal{A}_2$ of $\mathcal{R}$ whose normalizers generate $\mathcal{R}$ there exists $\sigma \in Aut(\mathcal{R})$ such that $\sigma(\mathcal{A}_1) = \mathcal{A}_2$.

Among the most important examples of injective W^*-algebras, we have the crossed products of abelian W^*-algebras by continuous actions of amenable locally compact groups, the von Neumann algebras generated by *so*-continuous unitary representations of connected locally compact groups and the von Neumann algebras generated by any $*$-representation of a nuclear C^*-algebra (Choi & Effros, 1976a, 1977b; Connes, 1976a; Connes, 1977b).

Note that if G is a discrete *ICC*-group (22.6), then the factor $\mathcal{L}(G)$ is injective, that is, $\mathcal{L}(G) \approx \mathcal{R}$, if and only if G is amenable (see Connes, 1976a; Sakai, 1971; Schwartz, 1963; Tomiyama, 1971). Since the free group F_k on k generators, $(k \geq 2)$, is not amenable, it follows that $\mathcal{L}(F_k)$ is not $*$-isomorphic to $\mathcal{R}(\approx \mathcal{L}(S(\infty)$, see 22.6).

11 Operator-Valued Weights

In this section, we introduce the extended positive part of a W^*-algebra and operator-valued weights, together with their main properties.

11.1. Let $\mathcal{M}$ be a W^*-algebra. The *extended positive part* $\bar{\mathcal{M}}^+$ of $\mathcal{M}$ is the set of all functions $\mathfrak{m} : \mathcal{M}_*^+ \to [0, \infty]$ such that

$$\mathfrak{m}(\varphi + \psi) = \mathfrak{m}(\varphi) + \mathfrak{m}(\psi) \quad (\varphi, \psi \in \mathcal{M}_*^+); \tag{1}$$

$$\mathfrak{m}(\lambda\varphi) = \lambda\mathfrak{m}(\varphi) \quad (\varphi \in \mathcal{M}_*^+, \lambda \geq 0); \tag{2}$$

$$\mathfrak{m} \text{ is lower semicontinuous.} \tag{3}$$

If $\mathfrak{m}, \mathfrak{n} \in \bar{\mathcal{M}}^+, x \in \mathcal{M}$ and $\lambda > 0$, then one defines the elements $\mathfrak{m} + \mathfrak{n}, \lambda\mathfrak{m}$ and $x^*\mathfrak{m}x$ of $\bar{\mathcal{M}}^+$ by

$$(\mathfrak{m} + \mathfrak{n})(\varphi) = \mathfrak{m}(\varphi) + \mathfrak{n}(\varphi) \quad (\varphi \in \mathcal{M}_*^+); \tag{4}$$

$$(\lambda\mathfrak{m})(\varphi) = \lambda\mathfrak{m}(\varphi) \quad (\varphi \in \mathcal{M}_*^+); \tag{5}$$

$$(x^*\mathfrak{m}x)(\varphi) = \mathfrak{m}(\varphi(x^* \cdot x)) \quad (\varphi \in \mathcal{M}_*^+). \tag{6}$$

If z is a positive element in the center $\mathcal{Z}(\mathcal{M})$ of $\mathcal{M}$, then we write $z\mathfrak{m}$ or $\mathfrak{m}z$ instead of $z^{1/2}\mathfrak{m}z^{1/2}$.

For $\mathfrak{m}_1, \mathfrak{m}_2 \in \bar{\mathcal{M}}^+$, write $\mathfrak{m}_1 \le \mathfrak{m}_2$ if $\mathfrak{m}_1(\varphi) \le \mathfrak{m}_2(\varphi)$ for all $\varphi \in \mathcal{M}_*^+$. If $\{\mathfrak{m}_i\}_{i\in I} \subset \bar{\mathcal{M}}^+$ is an increasing net, then one defines an element $\mathfrak{m} = \sup_{i\in I} \mathfrak{m}_i \in \bar{\mathcal{M}}^+$ by $\mathfrak{m}(\varphi) = \sup_{i\in I} \mathfrak{m}_i(\varphi)(\varphi \in \mathcal{M}_*^+)$; in this case we also write $\mathfrak{m}_i \uparrow \mathfrak{m}$. In particular, for an arbitrary family $\{\mathfrak{m}_i\}_{i\in I} \subset \bar{\mathcal{M}}^+$, an element $\mathfrak{m} = \sum_{i\in I} \mathfrak{m}_i \in \bar{\mathcal{M}}^+$ is defined by $\mathfrak{m}(\varphi) = \sum_{i\in I} \mathfrak{m}_i(\varphi)(\varphi \in \mathcal{M}_*^+)$.

11.2. Let $\mathcal{M} \subset \mathcal{B}(\mathcal{H})$ be a von Neumann algebra, $e \in \mathcal{M}$ a projection and A a positive self-adjoint operator in the Hilbert space $e\mathcal{H}$, affiliated to the von Neumann algebra $e\mathcal{M}e \subset \mathcal{B}(e\mathcal{H})$. Let

$$e_n = \chi_{[n-1,n)}(A) \le e \quad (n = 1, 2, \ldots).$$

For each n, Ae_n is a bounded positive operator in $e\mathcal{M}e$ and $\mathbf{s}(Ae_n) \le e_n$. Define an element $\mathfrak{m}_A \in \bar{\mathcal{M}}^+$ by

$$\mathfrak{m}_A(\varphi) = \sum_{n=1}^{\infty} \varphi(Ae_n) + \infty \cdot \varphi(1-e) \quad (\varphi \in \mathcal{M}_*^+).$$

If $e = 1$ and $A = a \in \mathcal{M}^+$, then $\mathfrak{m}_a(\varphi) = \varphi(a), (\varphi \in \mathcal{M}_*^+)$. The mapping $\mathcal{M}^+ \ni a \mapsto \mathfrak{m}_a \in \bar{\mathcal{M}}^+$ is injective and preserves the operations introduced in Section 11.1, so that we can identify $\mathfrak{m}_a$ with a and write $\mathcal{M}^+ \subset \bar{\mathcal{M}}^+$.

If $e = 1 - f$ and $A = 0$, then the element $\mathfrak{m}_A$ will be denoted by $\infty \cdot f$; for $\varphi \in \mathcal{M}_*^+$ we have $(\infty \cdot f)(\varphi) = \infty \cdot \varphi(f)$, that is, this value is 0 if $\varphi(f) = 0$ and $+\infty$ if $\varphi(f) > 0$.

Thus, in the general case considered above we have

$$\mathfrak{m}_A = \sum_{n=1}^{\infty} Ae_n + \infty \cdot (1-e), \tag{1}$$

so that $\mathfrak{m}_A$ is a sum of elements in $\mathcal{M}^+$ with mutually orthogonal supports and a symbol ∞ on a projection which is orthogonal on all these supports.

If $\xi \in \mathcal{H}$, then $\omega_\xi \in \mathcal{M}_*^+$ and, by ([L], 9.9), we obtain

$$\mathfrak{m}_A(\omega_\xi) = \begin{cases} \|A^{1/2}\xi\|^2 \text{ if } \xi \in D(A^{1/2}) \subset e\mathcal{H} \\ +\infty, \text{ in the contrary case.} \end{cases} \tag{2}$$

In particular, it follows that the element $\mathfrak{m}_A$ determines the projection e and the operator A uniquely: if $f \in \mathcal{M}$ is another projection and B is a positive self-adjoint operator in $f\mathcal{H}$ affiliated to $f\mathcal{M}f$, then

$$\mathfrak{m}_A = \mathfrak{m}_B \Leftrightarrow e = f \textit{ and } A = B. \tag{3}$$

It is easy to check that

$$\mathbf{s}(A) = e \Leftrightarrow \mathfrak{m}_A(\varphi) > 0 \textit{ for all } 0 \ne \varphi \in \mathcal{M}_*^+; \tag{4}$$

in this case we say that $\mathfrak{m} = \mathfrak{m}_A \in \bar{\mathcal{M}}^+$ is *faithful* or *nonsingular*. Also,

$$e = 1 \Leftrightarrow \{\varphi \in \mathcal{M}_*^+; \mathfrak{m}_A(\varphi) < +\infty\} \text{ is dense in } \mathcal{M}_*^+; \tag{5}$$

in this case we say that $\mathfrak{m} = \mathfrak{m}_A \in \bar{\mathcal{M}}^+$ is *semifinite*. Moreover,

$$e = 1 \text{ and } A \text{ is bounded } \Leftrightarrow \mathfrak{m}_A(\varphi) < +\infty \text{ for all } \varphi \in \mathcal{M}_*^+. \tag{6}$$

If $\{e_\lambda\}_{\lambda\in(0,+\infty)}$ is the spectral scale of A (see [L), E.9.10), then $e_\lambda \uparrow e$ for $\lambda \uparrow +\infty$ and

$$\mathfrak{m}_A(\varphi) = \int_0^\infty \lambda \, \mathrm{d}\varphi(e_\lambda) + \infty \cdot \varphi(1-e) \quad (\varphi \in \mathcal{M}_*^+). \tag{7}$$

Indeed, this identity is easily verified for $\varphi = \omega_\xi (\xi \in \mathcal{H})$, using (2) and ([L], E.9.10), and then extended for any $\varphi \in \mathcal{M}_*^+$, since $\varphi = \sum_n \omega_{\xi_n}$ with $\sum_n \|\xi_n\|^2 < +\infty$ (see [L], E.7.9, 8.17).

11.3. The example considered in the preceding section is general, as is shown by the following

Proposition. *Let $\mathcal{M} \subset \mathcal{B}(\mathcal{H})$ be a von Neumann algebra and $\mathfrak{m} \in \bar{\mathcal{M}}^+$. There exists a projection $e \in \mathcal{M}$ and a positive self-adjoint operator A in $e\mathcal{H}$ affiliated to $e\mathcal{M}e$, uniquely determined, such that $\mathfrak{m} = \mathfrak{m}_A$.*

Proof. The uniqueness follows from 11.2.(3).

Since $\mathfrak{m} \in \bar{\mathcal{M}}^+$, the function $q : \mathcal{H} \ni \xi \mapsto q(\xi) = \mathfrak{m}(\omega_\xi) \in [0, +\infty]$ is lower semicontinuous and has the following properties:

$$q(\lambda\xi) = |\lambda|^2 q(\xi) \quad (\xi \in \mathcal{H}, \lambda \in \mathbb{C});$$
$$q(\xi + \eta) + q(\xi - \eta) = 2q(\xi) + 2q(\eta) \quad (\xi, \eta \in \mathcal{H});$$
$$q(u'\xi) = q(\xi) \quad (\xi \in \mathcal{H}, u' \in \mathcal{M}' \text{ unitary}).$$

It follows that the set $D(q) = \{\xi \in \mathcal{H}; q(\xi) < +\infty\}$ is a linear subspace of $\mathcal{H}$, stable under $\mathcal{M}'$, so that there exists a unique projection $e \in \mathcal{M}$ such that $\bar{D(q)} = e\mathcal{H}$.

By (A.10) there exists a unique positive self-adjoint operator A in $e\mathcal{H}$ such that $\|A^{1/2}\xi\|^2 = q(\xi) = \mathfrak{m}(\omega_\xi)$ for all $\xi \in D(A^{1/2}) = D(q)$. Using 11.2.(2), it follows that $\mathfrak{m}(\omega_\xi) = \mathfrak{m}_A(\omega_\xi)(\xi \in \mathcal{H})$, and since any $\varphi \in \mathcal{M}_*^+$ is a sum of vector forms we conclude that $\mathfrak{m} = \mathfrak{m}_A$. □

By this proposition and the results in Section 11.2, it follows that for each element m in the extended positive part of a W^*-algebra $\mathcal{M}$ there exists a sequence $\{a_n\}$ of positive elements in $\mathcal{M}$ with mutually orthogonal supports and a projection $e \in \mathcal{M}, e \geq \sum_n \mathbf{s}(a_n)$, such that

$$\mathfrak{m} = \sum_n a_n + \infty \cdot (1-e). \tag{1}$$

Also, by 11.2.(7), each element $\mathfrak{m} \in \bar{\mathcal{M}}^+$ has a unique "spectral decomposition":

$$\mathfrak{m} = \int_0^\infty \lambda \, \mathrm{d}e_\lambda + \infty \cdot (1-e). \tag{2}$$

Finally, by 11.2.(6), for $\mathfrak{m} \in \bar{\mathcal{M}}^+$ we have

$$\mathfrak{m}(\varphi) < +\infty \text{ for all } \varphi \in \mathcal{M}_*^+ \Leftrightarrow m = a \in \bar{\mathcal{M}}^+. \tag{3}$$

Let $\mathcal{Z}$ be an abelian W^*-algebra. In ([L], 7.20) we defined an extension $\widetilde{\mathcal{Z}}$ of $\mathcal{Z}$. It is easy to check that $\widetilde{\mathcal{Z}}^+$ consists of those elements $\mathfrak{m}$ of the extended positive part $\bar{\mathcal{Z}}^+$ such that the set $\{\varphi \in \mathcal{Z}_*^+; m(\varphi) < +\infty\}$ is dense in $\mathcal{Z}_*^+$. Also, each element $\mathfrak{m} \in \bar{\mathcal{Z}}^+$ is of the form

$$\mathfrak{m} = z + \infty \cdot p \textit{ with } z \in \widetilde{\mathcal{Z}}^+, p \in \mathcal{Z}, pz = 0. \tag{4}$$

Let $\mathcal{M} \subset \mathcal{B}(\mathcal{H})$ be a von Neumann algebra. We say that an element $\mathfrak{m} \in \overline{\mathcal{B}(\mathcal{H})}^+$ is *affiliated* to $\mathcal{M}$ if $u^*\mathfrak{m}u = \mathfrak{m}$ for all unitaries $u \in \mathcal{M}'$. Let $e \in \mathcal{B}(\mathcal{H})$ and A be the projection and the positive self-adjoint operator in $e\mathcal{H}$, respectively, such that $\mathfrak{m} = \mathfrak{m}_A$. Then

$$\mathfrak{m} = \mathfrak{m}_A \textit{ is affiliated to } \mathcal{M} \Leftrightarrow e \in \mathcal{M} \textit{ and A is affiliated to } e\mathcal{M}e. \tag{5}$$

Thus, there is a natural bijective correspondence between $\bar{\mathcal{M}}^+$ and the elements in $\overline{\mathcal{B}(\mathcal{H})}^+$ affiliated to $\mathcal{M}$.

11.4 Proposition. *Let $\mathcal{M}$ be a W^*-algebra. Every normal weight φ on $\mathcal{M}$ has a unique extension to $\bar{\mathcal{M}}^+$, still denoted by φ, such that*

$$\varphi(\mathfrak{m} + \mathfrak{n}) = \varphi(\mathfrak{m}) + \varphi(\mathfrak{n}) \quad (\mathfrak{m}, \mathfrak{n} \in \bar{\mathcal{M}}^+) \tag{1}$$

$$\varphi(\lambda\mathfrak{m}) = \lambda\varphi(\mathfrak{m}) \quad (\mathfrak{m} \in \bar{\mathcal{M}}^+, \lambda > 0) \tag{2}$$

$$\mathfrak{m}_i, \mathfrak{m} \in \bar{\mathcal{M}}^+, \mathfrak{m}_i \uparrow \mathfrak{m} \Rightarrow \varphi(\mathfrak{m}_i) \uparrow \varphi(\mathfrak{m}). \tag{3}$$

If $\varphi \in \mathcal{M}_*^+$, then

$$\varphi(\mathfrak{m}) = \mathfrak{m}(\varphi) \quad (\mathfrak{m} \in \bar{\mathcal{M}}^+). \tag{4}$$

Proof. If $\varphi \in \mathcal{M}_*^+$, define the extension of φ to $\bar{\mathcal{M}}^+$ by (4); (1), (2), (3) then follow.

By Corollary 5.8, every normal weight φ on $\mathcal{M}$ can be written $\varphi = \sum_i \varphi_i$ with $\varphi_i \in \mathcal{M}_*^+$, so that we can define $\varphi(\mathfrak{m}) = \sum_i \varphi_i(\mathfrak{m})(\mathfrak{m} \in \bar{\mathcal{M}}^+)$, and (1), (2), (3) again follow easily. □

For any $m \in \bar{\mathcal{M}}^+$, there exists an increasing sequence $\{a_n\} \subset \bar{\mathcal{M}}^+$ such that $a_n \uparrow \mathfrak{m}$ (see 11.3.(1)); condition (3) implies that $\varphi(a_n) \uparrow \varphi(\mathfrak{m})$. This proves the uniqueness of the extension of φ.

11.5. Let $\mathcal{M}$ be a W^*-algebra and $\mathcal{N} \subset \mathcal{M}$ a W^*-subalgebra. An *operator-valued weight* on $\mathcal{M}$ with values in $\mathcal{N}$, or an $\mathcal{N}$-valued weight on $\mathcal{M}$, is a mapping $E : \mathcal{M}^+ \to \bar{\mathcal{N}}^+$ with the properties:

$$E(a + b) = E(a) + E(b) \quad (a, b \in \mathcal{M}^+) \tag{1}$$

$$E(\lambda a) = \lambda E(a) \quad (a \in \mathcal{M}^+, \lambda > 0) \tag{2}$$

$$E(y^*ay) = y^*E(a)y \quad (a \in \mathcal{M}^+, y \in \mathcal{N}). \tag{3}$$

Any weight φ on $\mathcal{M}$ can be regarded as an operator-valued weight $\varphi : \mathcal{M}^+ \to \overline{(\mathbb{C} \cdot 1_{\mathcal{M}})}^+$.

Every conditional expectation (9.1) $E : \mathcal{M} \to \mathcal{N}$ is an operator-valued weight such that $E(1_{\mathcal{M}}) = 1_{\mathcal{N}}$. Conversely, every operator-valued weight $E : \mathcal{M}^+ \to \bar{\mathcal{N}}^+$ such that $E(1_{\mathcal{M}}) = 1_{\mathcal{N}}$ can be extended by linearity to a conditional expectation $E : \mathcal{M} \to \mathcal{N}$.

For an operator-valued weight $E : \mathcal{M}^+ \to \bar{\mathcal{N}}^+$ consider the sets

$$\mathfrak{F}_E = \{a \in \mathcal{M}^+; E(a) \in \mathcal{N}^+\}, \mathfrak{N}_E = \{x \in \mathcal{M}; E(x^*x) \in \mathcal{N}^+\}, \mathfrak{M}_E = \mathfrak{N}_E^*\mathfrak{N}_E.$$

$\mathfrak{F}_E$ is a face of $\mathcal{M}^+$, $\mathfrak{N}_E$ is a left ideal of $\mathcal{M}$ and $\mathfrak{M}_E$ is a facial subalgebra of $\mathcal{M}$; $\mathfrak{M}_E = lin\, \mathfrak{F}_E$, $\mathfrak{M}_E \cap \mathcal{M}^+ = \mathfrak{F}_E$. Consequently, E can be extended uniquely up to a linear mapping $E : \mathfrak{M}_E \to \mathcal{N}$. Also, it is easy to see that $\mathfrak{N}_E$ and $\mathfrak{M}_E$ are $\mathcal{N}$-bimodules, that is,

$$\mathcal{N} \cdot \mathfrak{N}_E \cdot \mathcal{N} \subset \mathfrak{N}_E \quad \mathcal{N} \cdot \mathfrak{M}_E \cdot \mathcal{N} \subset \mathfrak{M}_E \tag{4}$$

and

$$E(y_1 x y_2) = y_1 E(x) y_2 \quad (x \in \mathfrak{M}_E, y_1, y_2 \in \mathcal{N}). \tag{5}$$

In particular, it follows from (5) that

$$E(\mathfrak{M}_E) \textit{ is a two sided ideal in } \mathcal{N}. \tag{6}$$

The face $\mathfrak{F}_E$ contains an increasing right approximate unit $\{u_i\}_{i\in I}$ for the left ideal $\mathfrak{N}_E$, which is s^*-convergent to the unique projection $e \in \mathcal{M}$ such that $\bar{\mathfrak{N}}_E^w = \mathcal{M}e$ and $\bar{\mathfrak{M}}_E^w = e\mathcal{M}e$. If $\mathcal{N}$ contains the unit element of $\mathcal{M}$, then from (4) it follows that $e \in \mathcal{N}' \cap \mathcal{M}$. We say that E *is semifinite* if $e = 1$. On the other hand, we say that E is *faithful* if for $x \in \mathcal{M}$, $E(x^*x) = 0 \Rightarrow x = 0$.

Note that

$$\textit{if } E \textit{ is semifinite and faithful, then } \overline{E(\mathfrak{M}_E)}^w = \mathcal{N}. \tag{7}$$

Indeed, by (6) there exists a unique central projection q in $\mathcal{N}$ such that $\overline{E(\mathfrak{M}_E)}^w = \mathcal{N}q$. Assume that $q \neq 1$. Since E is semifinite, there exists $x \in \mathfrak{N}_E$ with $x(1-q) \neq 0$, hence $0 \neq (1-q)x^*x(1-q) \in \mathfrak{M}_E$. Since E is faithful, it follows that $0 \neq E((1-q)x^*x(1-q)) = (1-q)E(x^*x)(1-q) \in \mathcal{N}(1-q)$, contradicting $E(\mathfrak{M}_E) \subset \mathcal{N}q$.

We say that E is *normal* if $E(a_i) \uparrow E(a)$ whenever $a_i, a \in \mathcal{M}^+$ and $a_i \uparrow a$. In this case, there exists a unique projection $\mathbf{s}(E) \in \mathcal{N}' \cap \mathcal{M}$, called the *support* of E, such that for $x \in \mathcal{M}$ we have $E(x^*x) = 0$ if and only if $x\mathbf{s}(E) = 0$. Moreover, E is faithful if and only if $\mathbf{s}(E) = 1$.

If E is normal, then, by Proposition 11.4, for every $\psi \in \mathcal{N}_*^+$ we get a normal weight $\psi \circ E$ on $\mathcal{M}$. Again by Proposition 11.4, for each $\mathfrak{m} \in \bar{\mathcal{M}}^+$ we can define an element $E(\mathfrak{m}) \in \bar{\mathcal{N}}^+$ such that $E(\mathfrak{m})(\psi) = (\psi \circ E)(\mathfrak{m})$, $(\psi \in \mathcal{N}_*^+)$. We thus get a unique extension of E to a normal, additive and positively homogeneous mapping $E : \bar{\mathcal{M}}^+ \to \bar{\mathcal{N}}^+$.

Note that

$$\textit{if } E \textit{ is normal semifinite and faithful, then } E(\bar{\mathcal{M}}^+) = \bar{\mathcal{N}}^+. \tag{8}$$

Indeed, consider first $b \in \mathcal{N}^+ \cap E(\mathfrak{M}_E)$. Then b is the image of a hermitian element of $\mathfrak{M}_E$, and there exist $a, a' \in \mathfrak{M}_E \cap \mathcal{M}^+$ such that $b = E(a) - E(a') \leq E(a)$. By ((L], E.2.6) there exists $y \in \mathcal{N}$ such that $b = y^*E(a)y = E(y^*ay)$, hence $b \in E(\mathcal{M}^+)$.

Consider now $b \in \mathcal{N}^+$ and let $\{v_i\}$ be an increasing approximate unit for the w-dense two sided ideal $E(\mathfrak{M}_E)$ of $\mathcal{N}$ (see (6), (7)). Since $b \geq b^{1/2}v_i b^{1/2} \uparrow b$, it follows that there exists a family

$\{b_j\} \subset E(\mathfrak{M}_E) \cap \mathcal{N}^+$ such that $b = \sum_j b_j$. For each j there exists $a_j \in \mathcal{M}^+$ with $E(a_j) = b_j$. Hence $a = \sum_j a_j \in \bar{\mathcal{M}}^+$ and $E(a) = b$.

Finally, any element $b \in \bar{\mathcal{N}}^+$ is a sum of elements in $\mathcal{N}^+$ (11.3.(1)), and the same argument as above shows that $b \in E(\bar{\mathcal{M}}^+)$.

The set of all normal semifinite faithful $\mathcal{N}$-valued weights on $\mathcal{M}$ will be denoted by $P(\mathcal{M}, \mathcal{N})$. An $\mathcal{N}$-valued weight E on $\mathcal{M}$ such that $E(x^*x) = E(xx^*)$ for all $x \in \mathcal{M}$ will be called an *operator-valued trace*. In this case, $\mathfrak{N}_E$ and $\mathfrak{M}_E$ are two sided ideals of $\mathcal{M}$, $E(ab) = E(ba)$ for $a, b \in \mathfrak{N}_E$ and $E(xa) = E(ax)$ for $a \in \mathfrak{M}_E, x \in \mathcal{M}$.

11.6 Proposition. *Let $\mathcal{M}$ be a W*-algebra, $Q \subset \mathcal{N}$ W*-subalgebras of $\mathcal{M}$ and $E \in P(\mathcal{M}, \mathcal{N})$, $F \in P(\mathcal{N}, Q)$. Then $F \circ E \in P(\mathcal{M}, Q)$.*

Proof. Clearly, $F \circ E$ is a normal and faithful operator-valued weight. Let $x \in \mathfrak{N}_E$ and let $\{v_j\}_{j\in J} \subset \mathfrak{N}_F \subset \mathcal{N}$ be such that $0 \le v_j \uparrow 1$. Then $(F \circ E)(v_j^* x^* x v_j) = F(v_j^* E(x^*x) v_j) \le \|E(x^*x)\| F(v_j^* v_j) < +\infty$, so that $xv_j \in \mathfrak{N}_{F\circ E}$ and $xv_j \xrightarrow{w} x$. Thus, $\mathfrak{N}_{F\circ E}$ is w-dense in $\mathfrak{N}_E$ which is w-dense in $\mathcal{M}$. Consequently, $F \circ E$ is also semifinite. □

It is easy to check that

$$(F \circ E)(x) = F(E(x)) \text{ for all } x \in lin(\mathfrak{M}_E \cap \mathfrak{M}_{F\circ E} \cap \mathcal{M}^+). \tag{1}$$

In particular, if $E \in P(\mathcal{M}, \mathcal{N})$ and ψ is an n.s.f. weight on $\mathcal{N}$, then $\psi \circ E$ is an n.s.f. weight on $\mathcal{M}$ and

$$(\psi \circ E)(x) = \psi(E(x)) \text{ for all } x \in lin(\mathfrak{M}_E \cap \mathfrak{M}_{\psi\circ E} \cap \mathcal{M}^+). \tag{2}$$

11.7. Conversely,

Proposition. *Let $E : \mathcal{M}^+ \to \bar{\mathcal{N}}^+$ be a normal operator-valued weight. If there exists a faithful normal weight ψ on $\mathcal{N}$ such that the normal weight $\varphi = \psi \circ E$ on $\mathcal{M}$ is faithful and semifinite, then E is faithful and semifinite.*

Proof. Indeed, for $x \in \mathcal{M}$ we have $E(x^*x) = 0 \Rightarrow \varphi(x^*x) = 0 \Rightarrow x = 0$; hence E is faithful. Let $a \in \mathfrak{M}_\varphi \cap \mathcal{M}^+$ and let $\{b_n\}$ be a family of elements in $\mathcal{N}^+$ with mutually orthogonal supports $f_n = \mathbf{s}_{\mathcal{N}}(b_n)$ and $f \ge \sum_n f_n$ such that $E(a) = \sum_n b_n + \infty \cdot (1 - f)$ (see 11.3. (1)). Then $\psi(E(a)) = \varphi(a) < +\infty$, hence $f = 1$. Putting $e_k = 1 - \sum_{n=k+1}^{\infty} f_n \in \mathcal{N}$, we have $e_k \uparrow 1$, hence $e_k x e_k \xrightarrow{w} x$; also $e_k x e_k \in \mathfrak{N}_E$, as $E(e_k x e_k) = e_k E(x) e_k = \sum_{n=1}^{k} b_n \in \mathcal{N}^+$. Consequently $\mathfrak{N}_E$ is w-dense in $\mathfrak{N}_\varphi$, which is w-dense in $\mathcal{M}$, that is, E is semifinite. □

11.8. Let $E : \mathcal{M}^+ \to \bar{\mathcal{N}}^+$ be an n.s.f. operator-valued weight, $\mathcal{F}_2 = Mat_2(\mathbb{C})$ a factor of type I_2 and $\iota : \mathcal{F}_2 \to \mathcal{F}_2$ the identity mapping on $\mathcal{F}_2$. Consider the tensor product linear mapping

$$\widetilde{E} = E \otimes \iota : \mathfrak{M}_E \otimes \mathcal{F}_2 \to \mathcal{N} \otimes \mathcal{F}_2.$$

Using the polarization relation it is easy to check that $\mathfrak{M}_E \otimes \mathcal{F}_2 = (\mathfrak{N}_E \otimes \mathcal{F}_2)^*(\mathfrak{N}_E \otimes \mathcal{F}_2)$ is a *-subalgebra of $\mathcal{M} \otimes \mathcal{F}_2$, linearly spanned by its positive elements. Also, $\mathfrak{M}_E \otimes \mathcal{F}_2$ is an

$(\mathcal{N} \otimes \mathcal{F}_2)$-bimodule and we have

$$\widetilde{E}(y_1 x y_2) = y_1 \widetilde{E}(x) y_2 \quad (x \in \mathfrak{M}_E \otimes \mathcal{F}_2, y_1, y_2 \in \mathcal{N} \otimes \mathcal{F}_2). \tag{1}$$

Note that

$$x \in \mathfrak{M}_E \otimes \mathcal{F}_2, x \geq 0 \Rightarrow \widetilde{E}(x) \geq 0. \tag{2}$$

Indeed, let $\psi \in \mathcal{N}_*^+, \theta = \psi \bar{\otimes} tr \in (\mathcal{N} \otimes \mathcal{F}_2)_*^+$ and $\widetilde{\theta} = (\psi \circ E) \bar{\otimes} tr$ be a normal semifinite weight on $\mathcal{M} \otimes \mathcal{F}_2$. We have $\mathfrak{M}_E \otimes \mathcal{F} \subset \mathfrak{M}_{\widetilde{\theta}}$ and $\widetilde{\theta}(x) = \theta(\widetilde{E}(x))$ for all $x \in \mathfrak{M}_E \otimes \mathcal{F}_2$. For every $y \in \mathcal{N} \otimes \mathcal{F}_2$, we have $(\pi_\theta(\widetilde{E}(x)) y\xi_\theta | y\xi_\theta)_\theta = \theta(y^* \widetilde{E}(x) y) = \theta(\widetilde{E}(y^* x y)) = \widetilde{\theta}(y^* x y) \geq 0$, hence $\pi_\theta(\widetilde{E}(x)) \geq 0$. Since $\psi \in \mathcal{N}_*^+$ was arbitrary, it follows that $\widetilde{E}(x) \geq 0$.

Lemma. *Let $E : \mathcal{M}^+ \to \bar{\mathcal{N}}^+$ be an n.s.f. operator-valued weight, let φ and ψ be n.s.f. weights on $\mathcal{N}$ and let $\widetilde{\varphi} = \varphi \circ E, \widetilde{\psi} = \psi \circ E$. Then*

$$x \in (\mathfrak{N}_{\widetilde{\varphi}} \cap \mathfrak{N}_E)^* (\mathfrak{N}_{\widetilde{\psi}} \cap \mathfrak{N}_E) \Rightarrow E(x) \in \mathfrak{N}_\varphi^* \mathfrak{N}_\psi.$$

Proof. We may assume that $x = a^* b$ with $a \in \mathfrak{N}_{\widetilde{\varphi}} \cap \mathfrak{N}_E, b \in \mathfrak{N}_{\widetilde{\psi}} \cap \mathfrak{N}_E$, Then $0 \leq \widetilde{x} = \begin{pmatrix} a & b \\ 0 & 0 \end{pmatrix}^* \begin{pmatrix} a & b \\ 0 & 0 \end{pmatrix} = \begin{pmatrix} a^* a & x \\ x^* & b^* b \end{pmatrix} \in \mathfrak{M}_E \otimes \mathcal{F}_2$ and hence, by (2),

$$\widetilde{E}(\widetilde{x}) = \begin{pmatrix} E(a^* a) & E(x) \\ E(x^*) & E(b^* b) \end{pmatrix} \geq 0.$$

Let $\theta = \theta(\varphi, \psi)$ be the balanced weight on $\mathcal{N} \otimes \mathcal{F}_2$. Then $\theta(\widetilde{E}(\widetilde{x})) = \varphi(E(a^* a)) + \psi(E(b^* b)) = \widetilde{\varphi}(a^* a) + \widetilde{\psi}(b^* b) < +\infty$, hence $\widetilde{E}(\widetilde{x}) \in \mathfrak{M}_\theta$ and consequently (see 3.1.(3)) $E(x) \in \mathfrak{N}_\varphi^* \mathfrak{N}_\psi$. □

11.9 Theorem (U. Haagerup). *Let $E : \mathcal{M}^+ \to \bar{\mathcal{N}}^+$ be an n.s.f. operator-valued weight and φ, ψ n.s.f. weights on $\mathcal{N}$. Then*

$$\sigma_t^{\psi \circ E, \varphi \circ E}(y) = \sigma_t^{\psi, \varphi}(y) \quad (y \in \mathcal{N}, t \in \mathbb{R}), \tag{1}$$

that is,

$$\sigma_t^{\varphi \circ E}(y) = \sigma_t^{\varphi}(y) \quad (y \in \mathcal{N}, t \in \mathbb{R}) \tag{2}$$

$$[D(\psi \circ E) : D(\varphi \circ E)]_t = [D\psi : D\varphi]_t \quad (t \in \mathbb{R}). \tag{3}$$

In particular, it follows from this statement that $\mathcal{N}$ is $\sigma^{\varphi \circ E}$-invariant and $[D(\psi \circ E) : D(\varphi \circ E)]_t \in \mathcal{N} (t \in \mathbb{R})$.

The proof of the theorem will be given in Section 11.12. In Sections 11.10 and 11.11, we consider properties of s^*-continuous one-parameter groups of isometries on W^*-algebras which are necessary for the proof.

Note that if $E(1_{\mathcal{M}}) = 1_{\mathcal{N}}$, then the theorem follows from Corollary 10.5.

11.10. Let $\{\sigma_t\}_{t\in\mathbb{R}}$ be an s^*-continuous one-parameter group of isometries on the W^*-algebra $\mathcal{M}$. Define the analytic extensions $\sigma_\alpha(\alpha \in \mathbb{C})$ as in Sections 3.12 and 2.14.

We say that an element $a \in \mathcal{M}$ is of *exponential type* with respect to $\{\sigma_t\}$ if $a \in D(\sigma_\alpha)$ for all $\alpha \in \mathbb{C}$ and there exist two constants $\gamma, \delta > 0$ such that $\|\sigma_\alpha(a)\| \leq \gamma exp(-\delta\|\mathrm{Im}\ \alpha|)(\alpha \in \mathbb{C})$. The set of all elements in $\mathcal{M}$ of exponential type will be denoted by $\mathcal{M}^{\alpha}_{exp}$. Note that for $a \in \mathcal{M}$ we have

$$a \in \mathcal{M}^{\sigma}_{exp} \Leftrightarrow a \in D((\sigma_{-\mathrm{i}})^n)\ \textit{for all}\ n \in \mathbb{Z}\ \textit{and there exist}\ y, \delta > 0 \qquad (1)$$
$$\textit{such that}\ \|(\sigma_{-\mathrm{i}})^n(a)\| \leq \gamma \exp(\delta|n|) \quad (n \in \mathbb{Z}).$$

Indeed, the implication $(\Rightarrow)$ is obvious. Note that $(\sigma_{-\mathrm{i}})^n = \sigma_{-n\mathrm{i}}(n \in \mathbb{Z})$, and $\|\sigma_{t+\mathrm{i}s}(a)\| = \|\sigma_{\mathrm{i}s}(a)\| = \sup_{t\in\mathbb{R}} \|\sigma_{t+\mathrm{i}s}(a)\|(s, t \in \mathbb{R})$, since all the $\sigma_t(t \in \mathbb{R})$ are isometries. Thus the converse implication $(\Leftarrow)$ follows using the "three lines theorem" (Dunford & Schwartz, 1958, 1963, VI.10.3).

Lemma. *$\mathcal{M}^{\sigma}_{exp}$ is w-dense in $\mathcal{M}$.*

Proof. For each $\lambda > 0$ consider the entire analytic function $F_\lambda(\alpha) = (1 - \cos\lambda\alpha)/\pi\lambda\alpha^2(\alpha \in \mathbb{C})$. It is well known that

$$\int_{-\infty}^{+\infty} F_\lambda(t)\,\mathrm{d}t = 1 \qquad (2)$$

and that for any bounded continuous function $f : \mathbb{R} \to \mathbb{R}$ we have

$$\lim_{\lambda\to+\infty} \int_{-\infty}^{+\infty} F_\lambda(t) f(t)\,\mathrm{d}t = f(0). \qquad (3)$$

We shall show that

$$\int_{-\infty}^{+\infty} |F_\lambda(t + \mathrm{i}s)|\,\mathrm{d}t \leq \exp(\lambda|s|); \quad s \in \mathbb{R}. \qquad (4)$$

To this end we consider the functions f_λ defined by

$$f_\lambda(r) = \begin{cases} \lambda^{-1/2} \text{ if } |r| \leq \lambda/2 \\ 0 \text{ if } |r| > \lambda/2 \end{cases}$$

and their Fourier–Laplace transforms

$$\hat{f}_\lambda(\alpha) = (2\pi)^{-1/2} \int_{-\infty}^{+\infty} f_\lambda(r)\mathrm{e}^{-\mathrm{i}\alpha r}\,\mathrm{d}r = (2/\pi\lambda x)^{-1/2} \sin(\lambda r/2).$$

Since

$$\hat{f}_\lambda(\alpha)^2 = F_\lambda(\alpha) \text{ and } \hat{f}_\lambda(t + \mathrm{i}s) = (2\pi)^{-1/2} \int_{-\infty}^{+\infty} f_\lambda(r)\mathrm{e}^{sr}\mathrm{e}^{-\mathrm{i}tr}\,\mathrm{d}r,$$

we get using the Plancherel formula

$$\int_{-\infty}^{+\infty} |F_\lambda(t+\mathrm{i}s)|\,\mathrm{d}t = \int_{+\infty}^{-\infty} |\hat{f}_\lambda(t+\mathrm{i}s)|^2\,\mathrm{d}t = \int_{-\infty}^{+\infty} |f_\lambda(r)\mathrm{e}^{sr}|^2\,\mathrm{d}r$$
$$= \lambda^{-1}\int_{-\lambda/2}^{+\lambda/2} \mathrm{e}^{2sr}\,\mathrm{d}r \le \mathrm{e}^{\lambda|s|}.$$

Consider now $a \in \mathcal{M}$ and $a_\lambda = \int_{-\infty}^{+\infty} F_\lambda(r)\sigma_r(a)\,\mathrm{d}r\,(\lambda > 0)$. We shall show that $a_\lambda \in \mathcal{M}^\sigma_{exp}$ and $a_\lambda \xrightarrow{w} a$ when $\lambda \to +\infty$.

Indeed, it is easy to check that a_λ is an entire analytic element and $\sigma_\alpha(a_\lambda) = \int_{-\infty}^{+\infty} F_\lambda(r-\alpha)\sigma_r(a)\,\mathrm{d}r$ $(\alpha \in \mathbb{C})$. Moreover, using (4) we obtain $\|\sigma_\alpha(a_\lambda)\| \le \|a\| \exp(\lambda|\mathrm{Im}\,\alpha|)$, hence $a_\lambda \in \mathcal{M}^\sigma_{exp}$. On the other hand, using (2) and (3), we get, for every $\varphi \in \mathcal{M}_*$,

$$\varphi(a_\lambda - a) = \int_{-\infty}^{+\infty} F_\lambda(r)\varphi(\sigma_r(a) - a)\,\mathrm{d}r \to 0$$

hence $a_\lambda \xrightarrow{w} a$ when $\lambda \to +\infty$. □

11.11 Proposition. *Let $\mathcal{M}$ be a W^*-algebra, $\mathcal{N} \subset \mathcal{M}$ a W^*-subalgebra and $\{\sigma_t\}_{t\in\mathbb{R}}, \{\tau_t\}_{t\in\mathbb{R}}$ s^*-continuous one-parameter groups of isometries $\mathcal{M}, \mathcal{N}$ respectively. If*

$$b \in D(\tau_{-\mathrm{i}}) \subset \mathcal{N} \Rightarrow b \in D(\sigma_{-\mathrm{i}}) \text{ and } \sigma_{-\mathrm{i}}(b) = \tau_{-\mathrm{i}}(b)$$

then

$$\sigma_t(y) = \tau_t(y) \quad (y \in \mathcal{N}, t \in \mathbb{R}).$$

Proof. By assumption and by 11.10. (1) it follows that $\mathcal{N}^\tau_{exp} \subset \mathcal{M}^\sigma_{exp}$. For $b \in \mathcal{N}^\tau_{exp}$ we have $\sigma_{-n\mathrm{i}}(b) = \tau_{n\mathrm{i}}(b)(n \in \mathbb{N})$, and there exist $\gamma, \delta > 0$ such that $\|\sigma_n(b) - \tau_n(b)\| \le \gamma \exp(\delta|\mathrm{Im}\,\alpha|)(\alpha \in \mathbb{C})$. By a theorem of F. Carlson (Polya & Szegö, 1964, part 3, chapter 6, problem 328) it follows that $\sigma_\alpha(b) = \tau_\alpha(b)(\alpha \in \mathbb{C})$. Thus, $\sigma_t(b) = \tau_t(b)$ for all $b \in \mathcal{N}^\tau_{exp}, t \in \mathbb{R}$. Since the isometrics σ_t and τ_t are automatically w-continuous and since, by Lemma 11.10, $\mathcal{N}^\tau_{exp}$ is w-dense in $\mathcal{N}$, we conclude that $\sigma_t(y) = \tau_t(y)$ for $y \in \mathcal{N}, t \in \mathbb{R}$. □

11.12. *Proof of Theorem* 11.9. Let $\widetilde{\varphi} = \varphi \circ E, \widetilde{\psi} = \psi \circ E$ be n.s.f. weights on $\mathcal{M}$. By Proposition 11.11, we have to show that, if

$$a \in D(\sigma^{\psi,\varphi}_{-\mathrm{i}}) \subset \mathcal{N} \text{ and } b = \sigma^{\psi,\varphi}_{-\mathrm{i}}(a) \in \mathcal{N}, \tag{1}$$

then

$$a \in D(\sigma^{\widetilde{\psi},\widetilde{\varphi}}_{-\mathrm{i}}) \text{ and } b = \sigma^{\widetilde{\psi},\widetilde{\varphi}}_{-\mathrm{i}}(a). \tag{2}$$

By Theorem 3.15, this amounts to showing that

$$a\mathfrak{N}^*_{\widetilde{\varphi}} \subset \mathfrak{N}^*_{\widetilde{\psi}}, \mathfrak{N}_{\widetilde{\psi}} b \subset \mathfrak{N}_{\widetilde{\varphi}} \tag{3}$$
$$\widetilde{\psi}(ax) = \widetilde{\varphi}(xb) \text{ for all } x \in \mathfrak{N}^*_{\widetilde{\varphi}}\mathfrak{N}_{\widetilde{\psi}}. \tag{4}$$

From (1) it follows that $a \in D(\sigma_{-i/2}^{\psi,\varphi})$ and $b^* \in D(\sigma_{-i/2}^{\varphi,\psi})$ (see 3.12. (4)). Using Proposition 3.12 we infer that there exists $\lambda > 0$ such that $\psi(aya^*) \leq \lambda^2\varphi(y)$ and $\varphi(b^*yb) \leq \lambda^2\psi(y)$ for all $y \in \mathcal{N}^+$. These inequalities remain valid for any $y \in \bar{\mathcal{N}}^+$ and hence we get $\widetilde{\psi}(axa^*) \leq \lambda^2\widetilde{\varphi}(x)$ and $\widetilde{\varphi}(b^*xb) \leq \lambda^2\widetilde{\psi}(x)$ for all $x \in \mathcal{M}^+$. By Proposition 3.12 again, we obtain the required inclusions (3) as well as the inequalities

$$\|(xa^*)_{\widetilde{\psi}}\|_{\widetilde{\psi}} \leq \lambda\|x_{\widetilde{\varphi}}\|_{\widetilde{\varphi}} \quad (x \in \mathfrak{N}_{\widetilde{\varphi}}) \tag{5}$$

$$\|(xb)_{\widetilde{\varphi}}\|_{\widetilde{\varphi}} \leq \lambda\|x_{\widetilde{\psi}}\|_{\widetilde{\psi}} \quad (x \in \mathfrak{N}_{\widetilde{\psi}}). \tag{6}$$

We now prove (4) in the particular case

$$x = y^*z \text{ with } y \in \mathfrak{N}_{\widetilde{\varphi}} \cap \mathfrak{N}_E, z \in \mathfrak{N}_{\widetilde{\psi}} \cap \mathfrak{N}_E.$$

Using the first inclusion in (3) and the fact that $\mathfrak{N}_E$ is a right $\mathcal{N}$-module, we obtain $ax = (ya^*)^*z \in (\mathfrak{N}_{\widetilde{\psi}} \cap \mathfrak{N}_E)^*(\mathfrak{N}_{\widetilde{\psi}} \cap \mathfrak{N}_E) \subset lin(\mathfrak{M}_E \cap \mathfrak{M}_{\widetilde{\psi}} \cap \mathcal{M}^+)$ and, similarly, $xb \in lin(\mathfrak{M}_E \cap \mathfrak{M}_{\widetilde{\varphi}} \cap \mathcal{M}^+)$. We have $x \in \mathfrak{M}_E$ and, by Lemma 11.7, $E(x) \in \mathfrak{N}_{\varphi}^*\mathfrak{N}_{\psi}$. Using 11.6.(2), assumption (1) and Theorem 3.15, we get $\widetilde{\psi}(ax) = \psi(E(ax)) = \psi(aE(x)) = \varphi(E(x)b) = \varphi(E(xb)) = \widetilde{\varphi}(xb)$.

Finally, we consider the general case:

$$x = y^*z \; with \; y \in \mathfrak{N}_{\widetilde{\varphi}}, z \in \mathfrak{N}_{\widetilde{\psi}}.$$

Since $\varphi(E(y^*y)) < +\infty$ and φ is faithful, from 11.3 and 11.2.(7) we infer that $E(y^*y)$ has a spectral decomposition of the form $E(y^*y) = \int_0^\infty t\, \mathrm{d}e_t$. For any $t \geq 0$, we have

$$ye_t \in \mathfrak{N}_{\widetilde{\varphi}} \cap \mathfrak{N}_E \tag{7}$$

and for $t \to +\infty$ we get

$$\begin{aligned} \|(ye_t - y)_{\widetilde{\varphi}}\|_{\widetilde{\varphi}}^2 &= \varphi(E((ye_t - y)^*(ye_t - y)) = \\ &= \varphi((1 - e_t)E(y^*y)(1 - e_t)) = \varphi\left(\int_t^\infty s\, \mathrm{d}e_s\right) \to 0 \end{aligned} \tag{8}$$

so that, using (5), we further deduce

$$(ye_ta^*)_{\widetilde{\psi}} \to (ya^*)_{\widetilde{\psi}} \; in \; \mathcal{H}_{\widetilde{\psi}}. \tag{9}$$

Similarly, we have a spectral decomposition $E(z^*z) = \int_0^\infty t\, \mathrm{d}f_t$ such that, for any $t \geq 0$

$$zf_t \in \mathfrak{N}_{\widetilde{\psi}} \cap \mathfrak{N}_E \tag{10}$$

and, for $t \to +\infty$,

$$(zf_t)_{\widetilde{\psi}} \to z_{\widetilde{\psi}} \; in \; \mathcal{H}_{\widetilde{\psi}}, \tag{11}$$

$$(zf_tb)_{\widetilde{\varphi}} \to (zb)_{\widetilde{\varphi}} \; in \; \mathcal{H}_{\widetilde{\varphi}}. \tag{12}$$

Using (7)–(12) and the particular case of statement (4) proved above, we conclude that

$$\begin{aligned}\widetilde{\psi}(ax) = (z_{\widetilde{\psi}}|(ya^*)_{\widetilde{\psi}})_{\widetilde{\psi}} &= \lim_{t\to\infty}((zf_t)_{\widetilde{\psi}}|(ye_ta^*)_{\widetilde{\psi}})_{\widetilde{\psi}} \\ &= \lim_{t\to\infty}\widetilde{\psi}(a(ye_t)^*(zf_t)) = \lim_{t\to\infty}\widetilde{\varphi}((ye_t)^*(zf_t)b) \\ &= \lim_{t\to\infty}((zf_tb)_{\widetilde{\varphi}}|(ye_t)_{\widetilde{\varphi}})_{\widetilde{\varphi}} = ((zb)_{\widetilde{\varphi}}|y_{\widetilde{\varphi}})_{\widetilde{\varphi}} = \widetilde{\varphi}(xb),\end{aligned}$$

and this completes the proof of Theorem 11.9. □

11.13 Corollary. *Let $E_1, E_2 : \mathscr{M}^+ \to \bar{\mathscr{N}}^+$ be n.s.f. operator-valued weights. If there exists an n.s.f. weight φ on $\mathscr{N}$ such that $\varphi \circ E_1 = \varphi \circ E_2$, then $E_1 = E_2$.*

Proof. Let ψ be another n.s.f. weight on $\mathscr{N}$. By Theorem 11.9, we have $[D(\psi \circ E_1) : D(\varphi \circ E_1)]_t = [D\psi : D\varphi]_t = [D(\psi \circ E_2) : D(\varphi \circ E_2)]_t (t \in \mathbb{R})$, so that, using Corollary 3.6, it follows from the assumption $\varphi \circ E_1 = \varphi \circ E_2$ that $\psi \circ E_1 = \psi \circ E_2$.

Now let $\omega \in \mathscr{N}_*^+$ and $e = \mathbf{s}_{\mathscr{N}}(\omega)$. There exists an n.s.f. weight ψ on $\mathscr{N}$ such that $\omega(y) = \psi(eye)$ for all $y \in \mathscr{N}^+$ and this identity extends to all $y \in \bar{\mathscr{N}}^+$. Thus, for $x \in \mathscr{M}^+$ we have $E_1(x)(\omega) = \omega(E_1(x)) = \psi(E_1(exe)) = \psi(E_2(exe)) = \omega(E_2(x)) = E_2(x)(\omega)$. Since $\omega \in \mathscr{N}_*^+$ was arbitrary, it follows that $E_1(x) = E_2(x) (x \in \mathscr{M}^+)$. □

Note that if φ_1, φ_2 are n.s.f. weights on $\mathscr{N}$, then

$$[D(\varphi_1 \circ E_1) : D(\varphi_2 \circ E_2)]_t = [D\varphi_1 : D\varphi_2]_t (t \in \mathbb{R}) \Rightarrow E_1 = E_2. \tag{1}$$

Indeed, using 11.9.(3), we have by assumption $[D(\varphi_1 \circ E_1) : D(\varphi_2 \circ E_1)]_t = [D(\varphi_1 \circ E_2) : D(\varphi_2 \circ E_2)]_t (t \in \mathbb{R})$, so that $\varphi_1 \circ E_1 = \varphi_1 \circ E_2$ by Corollary 3.6 and hence $E_1 = E_2$.

Arguing as in the second part of the proof of the above corollary we also see that

$$E_1 \le E_2 \Leftrightarrow \varphi \circ E_1 \le \varphi \circ E_2 \text{ for all } \varphi \in W_{nsf}(\mathscr{N}). \tag{2}$$

11.14. Every normal positive linear mapping between W^*-algebras $\Phi : \mathscr{M} \to \mathscr{N}$ can be extended to a normal, additive and positively homogeneous mapping $\Phi : \bar{\mathscr{M}}^+ \to \bar{\mathscr{N}}^+$ by

$$\Phi(\mathfrak{m})(\psi) = \mathfrak{m}(\psi \circ \Phi) \quad (\psi \in \mathscr{N}_*^+, \mathfrak{m} \in \bar{\mathscr{M}}^+).$$

In particular, every *-automorphism σ of $\mathscr{M}$ can be extended to $\bar{\mathscr{M}}^+$. Note that

$$(\bar{\mathscr{M}}^\sigma)^+ = \{\mathfrak{m} \in \bar{\mathscr{M}}^+; \sigma(\mathfrak{m}) = \mathfrak{m}\}. \tag{1}$$

Indeed, it is clear that $(\bar{\mathscr{M}}^\sigma)^+ \subset \{\mathfrak{m} \in \bar{\mathscr{M}}^+; \sigma(\mathfrak{m}) = \mathfrak{m}\}$. Conversely, let $\mathfrak{m} \in \bar{\mathscr{M}}^+$, with $\sigma(\mathfrak{m}) = \mathfrak{m}$, have spectral decomposition

$$\mathfrak{m} = \int_0^\infty \lambda \, \mathrm{d}e_\lambda + \infty \cdot (1 - e); \ \textit{then } \sigma(\mathfrak{m}) = \int_0^\infty \lambda \, \mathrm{d}\sigma(e_\lambda) + \infty \cdot \sigma(1 - e).$$

Since $\sigma(\mathfrak{m}) = \mathfrak{m}$, it follows that $e_\lambda, e \in \mathscr{M}^\sigma$, hence $\mathfrak{m} \in (\bar{\mathscr{M}}^\sigma)^+$.

Corollary. *Let $E : \mathcal{M}^+ \to \bar{\mathcal{N}}^+$ be an n.s.f. operator-valued weight and φ an n.s.f. weight on $\mathcal{N}$. Then*

$$E(\sigma_t^{\varphi \circ E}(x)) = \sigma_t^{\varphi}(E(x)) \quad (x \in \mathcal{M}^+, t \in \mathbb{R}). \tag{2}$$

In particular,

$$\sigma_t^{\varphi \circ E}(\mathfrak{M}_E) = \mathfrak{M}_E \quad (t \in \mathbb{R}). \tag{3}$$

Proof. For each $t \in \mathbb{R}$ consider the mapping

$$E_t = \sigma_{-t}^{\varphi} \circ E \circ \sigma_t^{\varphi \circ E} : \mathcal{M}^+ \to \bar{\mathcal{N}}^+.$$

It is easy to check that E_t is an n.s.f. operator-valued weight; condition 11.5.(3) follows using Theorem 11.9. Since $\varphi \circ E_t = \varphi \circ E \circ \sigma_t^{\varphi \circ E} = \varphi \circ E$, using Corollary 11.13 we infer that $E_t = E$, the desired conclusion. □

11.15. Let $E : \mathcal{M}^+ \to \mathcal{N}^+$ be an n.s.f. operator-valued weight and φ, ψ n.s.f. weights on $\mathcal{N}$.

By 11.9.(2) we have $\sigma_t^{\varphi \circ E}(\mathcal{N}) = \sigma_t^{\varphi}(\mathcal{N}) = \mathcal{N}$, so that

$$\sigma_t^{\varphi \circ E}(\mathcal{N}' \cap \mathcal{M}) = \mathcal{N}' \cap \mathcal{M} \quad (t \in \mathbb{R}). \tag{1}$$

Since $[D(\psi \circ E) : D(\varphi \circ E)]_t = [D\psi : D\varphi]_t \in \mathcal{N}$, for any $z \in \mathcal{N}' \cap \mathcal{M}$ we have $\sigma_t^{\varphi \circ E}(z) \in \mathcal{N}' \cap \mathcal{M}$, hence

$$\sigma_t^{\psi \circ E}(z) = [D(\psi \circ E) : D(\varphi \circ E)]_t \sigma_t^{\varphi \circ E}(z) [D(\psi \circ E) : D(\varphi \circ E)]_t^* = \sigma_t^{\varphi \circ E}(z).$$

Thus, $\sigma_t^{\varphi \circ E} | \mathcal{N}' \cap \mathcal{M}$ does not depend on the n.s.f. weight φ on $\mathcal{N}$, so that we can define

$$\sigma_t^E = \sigma_t^{\varphi \circ E} | \mathcal{N}' \cap \mathcal{M} \quad (t \in \mathbb{R}), \tag{2}$$

by choosing an arbitrary n.s.f. weight φ on $\mathcal{N}$.

The one-parameter group $\{\sigma_t^E\}_{t \in \mathbb{R}} \subset Aut(\mathcal{N}' \cap \mathcal{M})$ is called *the modular automorphism group associated with the operator-valued weight $E : \mathcal{M}^+ \to \bar{\mathcal{N}}^+$.*

Consider now two n.s.f. operator-valued weights $E, F : \mathcal{M}^+ \to \bar{\mathcal{N}}^+$ and two n.s.f. weights φ, ψ on $\mathcal{N}$.

We define an n.s.f. operator-valued weight $\Theta = \Theta(E, F)$,

$$\Theta : Mat_2(\mathcal{M})^+ \to \overline{(\mathcal{N} \otimes 1)}^+$$

by (compare with 3.1)

$$\Theta\left(\begin{pmatrix} x_{11} & x_{12} \\ x_{21} & x_{22} \end{pmatrix}\right) = \begin{pmatrix} E(x_{11}) + F(x_{22}) & 0 \\ 0 & E(x_{11} + F(x_{22}) \end{pmatrix} \quad ([x_{ij}] \in Mat_2(\mathcal{M})^+)$$

Consider also the weight $\widetilde{\varphi} = \varphi \otimes 1$ on $\mathcal{N} \otimes 1$ and the balanced weight $\theta = \theta(\varphi \circ E, \varphi \circ F)$ on $Mat_2(\mathcal{M})$. It is easy to check that $\widetilde{\varphi} \circ \Theta = \theta$ and $(\mathcal{N} \otimes 1)' \cap Mat_2(\mathcal{M}) = Mat_2(\mathcal{N}' \cap \mathcal{M})$. By applying (1) in this case we get

$$\begin{pmatrix} 0 & 0 \\ [D(\varphi \circ F) : D(\varphi \circ E)]_t & 0 \end{pmatrix} = \sigma_t^{\widetilde{\varphi} \circ \Theta}\left(\begin{pmatrix} 0 & 0 \\ 1 & 0 \end{pmatrix}\right) \in Mat_2(\mathcal{N}' \cap \mathcal{M}),$$

hence

$$[D(\varphi \circ F) : D(\varphi \circ E)]_t \in \mathcal{N}' \cap \mathcal{M} \quad (t \in \mathbb{R}). \tag{3}$$

Using (3), 3.4, 3.5, and 11.9.(3), we obtain

$$\begin{aligned}
&[D(\psi \circ F) : D(\psi \circ E)]_t \\
&= [D(\psi \circ F) : D(\varphi \circ F)]_t [D(\varphi \circ F) : D(\varphi \circ E)]_t [D(\varphi \circ E) : D(\psi \circ E)]_t \\
&= [D\psi : D\varphi]_t [D(\varphi \circ F) : D(\varphi \circ E)]_t [D\varphi : D\psi]_t \\
&= [D(\varphi \circ F) : D(\varphi \circ E)]_t.
\end{aligned}$$

Thus, $[D(\varphi \circ F) : D(\varphi \circ E)]_t$ does not depend on the n.s.f. weight φ on $\mathcal{N}$, so that we can define

$$[DF : DE]_t = [D(\varphi \circ F) : D(\varphi \circ E)]_t \in \mathcal{N}' \cap \mathcal{M} \quad (t \in \mathbb{R}), \tag{4}$$

with an arbitrary n.s.f. weight φ on $\mathcal{N}$.

The function $t \mapsto [DF : DE]_t$ is called the *σ^E-cocycle associated with F*. Indeed, using the results in Sections 3.1–3.5, it is easy to check that

$$\sigma_t^F = Ad([DF : DE]_t) \circ \sigma_t^E, \tag{5}$$

$$[DF : DE]_{s+t} = [DF : DE]_s \sigma_s^E([DF : DE]_t), \tag{6}$$

$$[DE : DF]_t = [DF : DE]_t^*. \tag{7}$$

Also, the "chain rule" holds.

Using 11.13.(2) and Corollary 3.13, we obtain

$$F \leq E \Leftrightarrow \left[\begin{array}{l} \textit{there exists a w-continuous function} \\ f : \{\alpha \in \mathbb{C}; 0 \leq \operatorname{Re} \alpha \leq 1\} \to \mathcal{M}, \textit{ analytic in} \\ \{\alpha \in \mathbb{C}; 0 < \operatorname{Re} \alpha < 1\} \textit{ such that} \\ f(it) = [DF : DE]_t, (t \in \mathbb{R}), \textit{ and } \|f(1/2)\| \leq 1. \end{array}\right. \tag{8}$$

Note that all the above results apply in particular for normal faithful conditional expectations.

11.16. Notes. Operator-valued weights appeared in the works of Connes and Takesaki (1977) and Landstad (1979), but their systematic study is due to Haagerup (1979). The modular automorphism group associated with an operator-valued weight was first considered in the case of conditional expectations by Combes and Delaroche (1975).

For our exposition, we have used Haagerup (1979).

12 Existence and Uniqueness of Operator-Valued Weights

In this section, we give several criteria for the existence and uniqueness of operator-valued weights, together with some applications.

12.1. The main existence criterion for n.s.f. operator-valued weights is the following

Theorem (U. Haagerup). *Let φ be an n.s.f. weight on the W^*-algebra $\mathscr{M}$ and ψ an n.s.f. weight on the W^*-algebra $\mathscr{N}$ of $\mathscr{M}$. If*

$$\sigma_t^{\varphi}(y) = \sigma_t^{\psi}(y) \quad (y \in \mathscr{N}, t \in \mathbb{R}),$$

then there exists a unique n.s.f. operator-valued weight $E : \mathscr{M}^+ \to \mathscr{N}^+$ such that $\varphi = \psi \circ E$.

The proof of this theorem is contained in Sections 12.2–12.5. After some preparation (12.2), we prove the theorem in the particular case when φ and ψ are n.s.f. traces in Section 12.3. In Section 12.4, we present a brief review of some results concerning the crossed product of a W^*-algebra by the modular automorphism group associated with an n.s.f. weight; the exposition here is essentially self-contained although these results will be considered later in greater generality. Finally, using these results, we complete the proof of the theorem in Section 12.5.

12.2. Let τ be an n.s.f. trace on the semifinite W^*-algebra $\mathscr{M}$. Each semifinite element $A \in \bar{\mathscr{M}}^+$ (see 11.2.(5)) defines a normal semifinite weight τ_A on $\mathscr{M}$ as in Sections 4.1, 4.4. By Theorem 4.10, the mapping $A \mapsto \tau_A$ establishes a bijective correspondence between the semifinite elements $A \in \bar{\mathscr{M}}^+$ and the normal semifinite weights on $\mathscr{M}$.

Note that for $a, b \in \mathscr{M}^+$ we have

$$\tau_a(b) = \tau_b(a). \tag{1}$$

Indeed, $\tau_a(b) = \tau(a^{1/2}ba^{1/2}) = \tau((b^{1/2}a^{1/2})^*(b^{1/2}a^{1/2})) = \tau((b^{1/2}a^{1/2})(b^{1/2}a^{1/2})^*) = \tau(b^{1/2}ab^{1/2}) = \tau_b(a)$.

Consider now $A \in \bar{\mathscr{M}}^+$. Using 4.3.(1), 4.3.(3), and Proposition 11.4, we see that the equation

$$\tau_A(b) = \tau_b(A) \quad (b \in \mathscr{M}^+)$$

defines a normal weight τ_A on $\mathscr{M}$, which extends to $\bar{\mathscr{M}}^+$. By the definition of τ_A and by Proposition 11.4, it follows that

$$\tau_{A+B} = \tau_A + \tau_B, \tau_{\lambda A} = \lambda \tau_A \quad (A, B \in \bar{\mathscr{M}}^+, \lambda > 0). \tag{2}$$

Also, it is easy to check that

$$\mathscr{M}^+ \ni a_n \uparrow A \in \bar{\mathscr{M}}^+, \mathscr{M}^+ \ni b_m \uparrow B \in \bar{\mathscr{M}}^+ \Rightarrow \tau_{a_n}(b_m) \uparrow \tau_A(B). \tag{3}$$

Using (1) and (3), we infer that

$$\tau_A(B) = \tau_B(A) \quad (A, B \in \bar{\mathscr{M}}^+), \tag{4}$$

and from (4) and 11.4. (3) it follows that

$$A_i \uparrow A, B_j \uparrow B \text{ in } \bar{\mathcal{M}}^+ \Rightarrow \tau_{A_i}(B_j) \uparrow \tau_A(B); \tag{5}$$

indeed, $\tau_A(B) = \sup_j \tau_A(B_j) = \sup_j \tau_{B_j}(A) = \sup_j \sup_i \tau_{B_j}(A_i) = \sup_{ij} \tau_{A_i}(B_j)$. Note also that

$$\tau_{xAx^*}(B) = \tau_A(x^*Bx) \quad (A, B \in \bar{\mathcal{M}}^+, x \in \mathcal{M}); \tag{6}$$

indeed, if $A = a \in \mathcal{M}^+, B = b \in \mathcal{M}^+$, then $\tau_{xax^*}(b) = \tau_b(xax^*) = \tau(b^{1/2}xax^*b^{1/2}) = \tau(a^{1/2}x^*bxa^{1/2}) = \tau_a(x^*bx)$, and the general case is obtained using (3).

It is easy to check that the normal weight τ_A is semifinite if and only if the element $A \in \bar{\mathcal{M}}^+$ is semifinite and that in this case the above definition of τ_A agrees with the definition given in Section 4.4.

Also, the normal weight τ_A is faithful if and only if the clement $A \in \bar{\mathcal{M}}^+$ is faithful (11.2.(4)).

Since any element of $\mathcal{M}_*^+$ is of the form τ_A with $A \in \bar{\mathcal{M}}^+$ (by Theorem 4.10) and since every normal weight on $\mathcal{M}$ is a sum of elements in $\mathcal{M}_*^+$ (by Corollary 5.8), it follows that every normal weight on $\mathcal{M}$ is of the form τ_A with $A \in \bar{\mathcal{M}}^+$.

If $A, B \in \bar{\mathcal{M}}^+$, then again by Theorem 4.10 we get: $\tau_A = \tau_B \Leftrightarrow \tau_A(X) = \tau_B(X)$ for all $X \in \bar{\mathcal{M}}^+ \Leftrightarrow \tau_X(A) = \tau_X(B)$ for all $X \in \bar{\mathcal{M}}^+ \Leftrightarrow \varphi(A) = \varphi(B)$ for all $\varphi \in \mathcal{M}_*^+ \Leftrightarrow A = B$. Consequently,

the mapping $A \mapsto \tau_A$ establishes a bijective correspondence between the sets $\mathcal{M}^+$ and $W_n(\bar{\mathcal{M}})$. (7)

Similarly, for $A, A_i, B \in \bar{\mathcal{M}}^+$, we obtain

$$A \leq B \Leftrightarrow \tau_A \leq \tau_B, \tag{8}$$

$$A_i \uparrow A \Leftrightarrow \tau_{A_i} \uparrow \tau_A. \tag{9}$$

12.3. We now prove Theorem 12.1 in the case when φ and ψ are n.s.f. traces on $\mathcal{M}$ and $\mathcal{N}$, respectively.

For each $a \in \mathcal{M}^+$, $\varphi_a|\mathcal{N}^+$ is a normal weight on $\mathcal{N}$ and hence, by 12.2.(7), there exists a unique element $E(a) \in \bar{\mathcal{N}}^+$ such that

$$\varphi_a|\mathcal{N}^+ = \psi_{E(a)}. \tag{1}$$

We thus define a mapping $E : \mathcal{M}^+ \to \mathcal{N}^+$. Using the results of Section 12.2, it is easy to check that E is a normal operator-valued weight and that $\varphi = \psi \circ E$. Moreover, using Proposition 11.7 it follows that E is faithful and semifinite.

If $F : \mathcal{M}^+ \to \bar{\mathcal{N}}^+$ is another operator-valued weight such that $\varphi = \psi \circ F$, then for any $a \in \mathcal{M}^+, b \in \mathcal{N}^+$ we get $\varphi_a(b) = \varphi_b(a) = \varphi(b^{1/2}ab^{1/2}) = \psi(b^{1/2}F(a)b^{1/2}) = \psi_b(F(a)) = \psi_{F(a)}(b)$, hence $F(a) = E(a)$. Actually, the uniqueness of E follows also from Corollary 11.13.

12.4. Let φ be an n.s.f. weight on the von Neumann algebra $\mathcal{M} \subset \mathcal{B}(\mathcal{H})$. The modular automorphism group $\{\sigma_t^\varphi\}_{t\in\mathbb{R}}$ determines a continuous action $\sigma = \sigma^\varphi$ of the group $\mathbb{R}$ on $\mathcal{M}$. Consider the Hilbert space $\mathcal{L}^2(\mathbb{R}, \mathcal{H}) = \mathcal{H} \bar{\otimes} \mathcal{L}^2(\mathbb{R})$.

The crossed product of $\mathscr{M}$ by the action σ of $\mathbb{R}$ is the von Neumann algebra $\mathscr{R}(\mathscr{M},\sigma) \subset \mathscr{B}(\mathscr{L}^2(\mathbb{R},\mathscr{H}))$ generated by the operators $I(x) = I_\varphi(x)(x \in \mathscr{M})$,

$$(I(x)\xi)(r) = \sigma_{-r}(x)\xi(r) \quad (\xi \in \mathscr{L}^2(\mathbb{R},\mathscr{H}), r \in \mathbb{R}),$$

and by the unitary operators $\boldsymbol{\lambda}(t)(t \in \mathbb{R})$,

$$(\boldsymbol{\lambda}(t)\xi)(r) = \xi(r-t) \quad (\xi \in \mathscr{L}^2(\mathbb{R},\mathscr{H}), r \in \mathbb{R}).$$

The mapping $I : \mathscr{M} \to \mathscr{R}(\mathscr{M},\sigma)$ is an injective normal *-homomorphism. We shall identify $\mathscr{M}$ with $I(\mathscr{M}) \subset \mathscr{R}(\mathscr{M},\sigma)$. With this identification, we have

$$\sigma_t(x) = \boldsymbol{\lambda}(t)x\boldsymbol{\lambda}(t^*) \quad (x \in \mathscr{M}, t \in \mathbb{R}). \tag{1}$$

The unitary operators $\mathbf{m}(s)(s \in \mathbb{R})$,

$$\mathbf{m}(s)\xi(r) = e^{-isr}\xi(r) \quad (\xi \in \mathscr{L}^2(\mathbb{R},\mathscr{H}), r \in \mathbb{R}),$$

define a dual action $\theta = \theta^\varphi = \hat{\sigma}^\varphi$ of $\mathbb{R}$ on $\mathscr{R}(\mathscr{M},\sigma)$:

$$\theta_s(X) = \mathbf{m}(s)X\mathbf{m}(s)^* \quad (X \in \mathscr{R}(\mathscr{M},\sigma), s \in \mathbb{R}),$$

which is characterized by the equalities

$$\theta_s(x) = x \quad (x \in \mathscr{M}, s \in \mathbb{R}), \tag{2}$$

$$\theta_s(\boldsymbol{\lambda}(t)) = e^{-ist}\boldsymbol{\lambda}(t) \quad (s,t \in \mathbb{R}). \tag{3}$$

Moreover, we shall see later (Proposition 19.3) that the centralizer of the dual action coincides with $\mathscr{M}$:

$$\mathscr{R}(\mathscr{M},\sigma)^\theta = \mathscr{M}. \tag{4}$$

We shall show that

$$E = E_\varphi = \int_{-\infty}^{+\infty} \theta_s \,\mathrm{d}s : \mathscr{R}(\mathscr{M},\sigma)^+ \to \bar{\mathscr{M}}^+$$

is an n.s.f. $\mathscr{M}$-valued weight on $\mathscr{R}(\mathscr{M},\sigma)$ and

$$E(\theta_s(X)) = E(X) \quad (X \in \mathscr{R}(\mathscr{M},\sigma)^+, s \in \mathbb{R}), \tag{5}$$

$$E(\boldsymbol{\lambda}(t)X\boldsymbol{\lambda}(t)^*) = \boldsymbol{\lambda}(t)E(X)\boldsymbol{\lambda}(t)^* = \sigma_t(E(X)) \quad (X \in \mathscr{R}(\mathscr{M},\sigma), t \in \mathbb{R}). \tag{6}$$

More precisely, E is defined by

$$E(X)(\omega) = \int_{-\infty}^{+\infty} \omega(\theta_s(X))\,\mathrm{d}s \quad (X \in \mathscr{R}(\mathscr{M},\sigma)^+, \omega \in \mathscr{R}(\mathscr{M},\sigma)_*^+)$$

as a mapping $E : \mathscr{R}(\mathscr{M},\sigma)^+ \to \overline{\mathscr{R}(\mathscr{M},\sigma)}^+$. Let $X \in \mathscr{R}(\mathscr{M},\sigma)^+$. Using the translation invariance of the Lebesgue measure, we obtain $\theta_s(E(X)) = E(X)(s \in \mathbb{R})$, and hence (see (4) and 11.14.(1)) $E(X) \in (\overline{\mathscr{R}(\mathscr{M},\sigma)}^\theta)^+ = \overline{\mathscr{M}}^+$. It is now easy to check that E is a normal faithful $\mathscr{M}$-valued weight on $\mathscr{R}(\mathscr{M},\sigma)$ which satisfies (5) and (6).

To show that E is semifinite, we consider a continuous function $f : \mathbb{R} \to \mathbb{C}$ with compact support and the operator

$$\boldsymbol{\lambda}(f) = \int_{-\infty}^{+\infty} f(r)\boldsymbol{\lambda}(r)\,\mathrm{d}r \in \mathscr{R}(\mathscr{M},\sigma).$$

We have $\boldsymbol{\lambda}(f)^*\boldsymbol{\lambda}(f) = \boldsymbol{\lambda}(f^* * f)$, where $f^* * f$ is the convolution of the function $f^*(r) = \overline{f(-r)}$ with the function f. Also, we consider the functions $g_n(s) = \exp(s^2/2n^2)(s \in \mathbb{R}, n \in \mathbb{N})$, and their Fourier transforms $h_n = \hat{g}_n, h_n(t) = \int_{-\infty}^{+\infty} g_n(s)e^{-ist}\,\mathrm{d}t = n\sqrt{2}\exp(n^2t^2/2)(t \in \mathbb{R}, n \in \mathbb{N})$. Note that $g_n(s) \uparrow 1$ uniformly on compact sets, as $n \uparrow \infty, h_n(t) \geq 0$ and $\int_{-\infty}^{+\infty} h_n(t)\,\mathrm{d}t = 2\pi$.

Since $\|\boldsymbol{\lambda}((f^* * f)h_n)\| \leq \|(f^* * f)h_n\|_1 \leq \|f^* * f\|_\infty \|h_n\|_1 \leq 2\pi\|f\|_2^2$, it follows by Fatou's lemma and Fubini's theorem that

$$\begin{aligned}
E(\boldsymbol{\lambda}(f)^*\boldsymbol{\lambda}(f)) &= \sup_n \int_{-\infty}^{+\infty} \theta_s(\boldsymbol{\lambda}(f^* * f))g_n(s)\,\mathrm{d}s \\
&= \sup_n \int_{-\infty}^{+\infty} g_n(s) \int_{-\infty}^{+\infty} (f^* * f)(t)e^{-\mathrm{i}st}\boldsymbol{\lambda}(t)\,\mathrm{d}t\,\mathrm{d}s \\
&= \sup_n \int_{-\infty}^{+\infty} (f^* * f)(t)h_n(t)\boldsymbol{\lambda}(t)\,\mathrm{d}t \\
&= \sup_n \boldsymbol{\lambda}((f^* * f)h_n) \in \mathscr{M}^+,
\end{aligned}$$

that is, $\boldsymbol{\lambda}(f) \in \mathfrak{N}_E$.

Since $\mathfrak{N}_E$ is a right $\mathscr{M}$-module, it follows that $\boldsymbol{\lambda}(f)x \in \mathfrak{N}_E$ for any $x \in \mathscr{M}$ and any continuous function f with compact support. Since $\mathscr{R}(\mathscr{M},\sigma)$ is generated by $\mathscr{M}$ and $\{\boldsymbol{\lambda}(t); t \in \mathbb{R}\}$, it follows that $\mathfrak{N}_E$ is dense in $\mathscr{R}(\mathscr{M},\sigma)$, that is, E is semifinite.

Thus, we can define an n.s.f. weight $\hat{\varphi} = \varphi \circ E$ on $\mathscr{R}(\mathscr{M},\sigma)$ called the dual weight of φ. By (5) it follows that the dual weight is invariant with respect to the dual action:

$$\hat{\varphi} \circ \theta_s = \hat{\varphi} \quad (s \in \mathbb{R}). \tag{7}$$

By Theorem 11.9, we have

$$\sigma_t^{\hat{\varphi}}(x) = \sigma_t^{\varphi}(x) = \boldsymbol{\lambda}(t)x\boldsymbol{\lambda}(t)^* \quad (x \in \mathscr{M}, t \in \mathbb{R}). \tag{8}$$

On the other hand, by (6), for $X \in \mathscr{R}(\mathscr{M},\sigma)^+$ and $t \in \mathbb{R}$ we have

$$\hat{\varphi}(\boldsymbol{\lambda}(s)X\boldsymbol{\lambda}(s)^*) = \varphi(\sigma_s^{\varphi}(E(X))) = \varphi(E(X)) = \hat{\varphi}(X),$$

so that, using Corollary 3.7 we deduce that

$$\sigma_t^{\hat{\varphi}}(\boldsymbol{\lambda}(s)) = \boldsymbol{\lambda}(s) = \boldsymbol{\lambda}(t)\boldsymbol{\lambda}(s)\boldsymbol{\lambda}(t)^* \quad (s, t \in \mathbb{R}). \tag{9}$$

Since $\mathscr{R}(\mathscr{M},\sigma)$ is generated by $\mathscr{M}$ and $\{\boldsymbol{\lambda}(s); s \in \mathbb{R}\}$, from (8) and (9) it follows that $\sigma_t^{\hat{\varphi}} = Ad(\boldsymbol{\lambda}(t))(t\in\mathbb{R})$.

There exists a unique positive self-adjoint operator A affiliated to $\mathscr{R}(\mathscr{M},\sigma)$ such that

$$\boldsymbol{\lambda}(t) = A^{-it} \quad (t \in \mathbb{R}). \tag{10}$$

Then the n.s.f. weight $\tau = \tau_\varphi = \hat{\varphi}_A$, constructed as in Section 4.4, has a trivial modular automorphism group (see Corollary 4.8) and hence is an n.s.f. trace on $\mathscr{R}(\mathscr{M},\sigma)$ (see, for instance, 2.18.(1)). Note that

$$[D\hat{\varphi} : D\tau]_t = \boldsymbol{\lambda}(t) \quad (t \in \mathbb{R}). \tag{11}$$

From (3) and (10), it follows that

$$\theta_s(A) = e^s A \quad (s \in \mathbb{R}), \tag{12}$$

and using (7) and (12), it is easy to check that

$$\tau \circ \theta_s = e^{-s}\tau \quad (s \in \mathbb{R}), \tag{13}$$

12.5. *Proof of Theorem* 12.1. We may consider $\mathscr{N} \subset \mathscr{M} \subset \mathscr{B}(\mathscr{H})$ realized as von Neumann algebras. Since $\sigma_t^\psi = \sigma_t^\varphi|\mathscr{N}(t \in \mathbb{R})$, it follows that $\mathscr{R}(\mathscr{N},\sigma^\psi)$ is the von Neumann subalgebra of $\mathscr{R}(\mathscr{M},\sigma^\varphi)$ generated by $\mathscr{N}$ and $\{\boldsymbol{\lambda}(t); t \in \mathbb{R}\}$, and

$$\theta_s^\psi = \theta_s^\varphi|\mathscr{R}(\mathscr{N},\sigma^\psi) \quad (s \in \mathbb{R}). \tag{1}$$

Consequently,

$$E_\psi = E_\varphi|\mathscr{R}(\mathscr{N},\sigma^\psi). \tag{2}$$

Consider the n.s.f. traces τ_φ and τ_ψ defined on $\mathscr{R}(\mathscr{M},\sigma^\varphi)$ and $\mathscr{R}(\mathscr{N},\sigma^\psi)$ as in Section 12.4. By the result of Section 12.3, there exists a unique n.s.f. operator-valued weight

$$F : \mathscr{R}(\mathscr{M},\sigma^\varphi)^+ \to \overline{\mathscr{R}(\mathscr{N},\sigma^\psi)}^+$$

such that

$$\tau_\psi \circ F = \tau_\varphi. \tag{3}$$

By (1) it follows that for each $s \in \mathbb{R}$

$$F_s = \theta_{-s}^\psi \circ F \circ \theta_\varphi^s : \mathscr{R}(\mathscr{M},\sigma^\varphi)^+ \to \overline{\mathscr{R}(\mathscr{N},\sigma^\psi)}^+$$

is an n.s.f. operator-valued weight. Since $\tau_\varphi \circ \theta_s^\varphi = e^{-s}\tau_\varphi$ and $\tau_\psi \circ \theta_s^\psi = e^{-s}\tau_\psi$, using equality (3) we obtain $\tau_\psi \circ F_s = \tau_\varphi$, hence $F_s = F$, that is, $F \circ \theta_s^\varphi = \theta_s^\psi \circ F(s \in \mathbb{R})$. It follows that

$$f \circ E_\varphi = E_\psi \circ F. \tag{4}$$

From 11.5.(8), we infer that

$$F(\mathscr{M}^+) \subset F(E_\varphi(\overline{\mathscr{R}(\mathscr{M}, \sigma^\varphi)}^+)) = E_\psi(F(\overline{\mathscr{R}(\mathscr{M}, \sigma^\varphi)}^+)) \subset \bar{\mathscr{N}}^+$$

hence $E = F|\mathscr{M}^+ : \mathscr{M}^+ \to \bar{\mathscr{N}}^+$ is a normal $\mathscr{N}$-valued weight on $\mathscr{M}$.

Recall (12.4.(11)) that for the dual weights $\hat{\varphi} = \varphi \circ E_\varphi$ and $\hat{\psi} = \psi \circ E_\psi$ we have $[D\hat{\varphi} : D\tau_\varphi]_t = \boldsymbol{\lambda}(t) = [D\hat{\psi} : D\tau_\psi]_t (t \in \mathbb{R})$. Using (3) and Theorem 11.9, we obtain $(D(\hat{\psi} \circ F) : D\tau_\varphi]_t = [D(\hat{\psi} \circ F) : D(\tau_\psi \circ F)]_t = [D\hat{\psi} : D\tau_\psi]_t = [D\hat{\varphi} : D\tau_\varphi]_t (t \in \mathbb{R})$. By Corollary 3.6, we deduce that $\hat{\varphi} = \hat{\psi} \circ F$, that is, $\varphi \circ E_\varphi = \psi \circ E_\psi \circ F = \psi \circ E \circ E_\varphi$. Since $E_\varphi(\overline{\mathscr{R}(\mathscr{M}, \sigma^\varphi)}^+) = \bar{\mathscr{M}}^+$ (see 11.5.(8)), we conclude that $\varphi = \psi \circ E$. According to Proposition 11.7, it follows that the normal operator-valued weight E is semifinite and faithful.

The uniqueness of E follows from Corollary 11.13. □

12.6 Corollary. *Let $\mathscr{N} \subset \mathscr{M} \subset \mathscr{B}(\mathscr{H})$ be von Neumann algebras. Then $P(\mathscr{M}, \mathscr{N}) \neq \emptyset$ if and only if there exist n.s.f. weights φ and ψ on $\mathscr{M}$ and $\mathscr{N}$, respectively, such that $\sigma_t^\psi = \sigma_t^\varphi|\mathscr{N} (t \in \mathbb{R})$.*

Proof. Follows from Theorems 11.9 and 12.1. □

12.7 Corollary. *Let $\mathscr{N} \subset \mathscr{M} \subset \mathscr{B}(\mathscr{H})$ be von Neumann algebras and let*

$$W_{nsf}(\mathscr{N}) \ni \varphi \mapsto \widetilde{\varphi} \in W_{nsf}(\mathscr{M})$$

be a mapping with the following properties

$$\sigma_t^\varphi = \sigma_t^{\widetilde{\varphi}}|\mathscr{N}\ (t \in \mathbb{R}, \varphi \in W_{nsf}(\mathscr{N}))$$
$$[D\widetilde{\psi} : D\widetilde{\varphi}]_t = [D\psi : D\varphi]_t \quad (t \in \mathbb{R}, \varphi, \psi \in W_{nsf}(\mathscr{N})).$$

Then there exists a unique n.s.f. operator-valued weight $E : \mathscr{M}^+ \to \mathscr{N}^+$ such that $\widetilde{\varphi} = \varphi \circ E$ for all $\varphi \in W_{nsf}(\mathscr{N})$.

Proof. Let $\varphi \in W_{nsf}(\mathscr{N})$ be fixed. By Theorem 12.1, there exists a unique n.s.f. operator-valued weight $E : \mathscr{M}^+ \to \bar{\mathscr{N}}^+$ such that $\widetilde{\varphi} = \varphi \circ E$. By our assumption and Theorem 11.9, for any $\psi \in W_{nsf}(\mathscr{N})$ we have $[D\widetilde{\psi} : D\widetilde{\varphi}]_t = [D\psi : D\varphi]_t = [D(\psi \circ E) : D(\varphi \circ E)]_t = [D(\psi \circ E) : D\widetilde{\varphi}]_t (t \in \mathbb{R})$; hence $\widetilde{\psi} = \psi \circ E$ by Corollary 3.6. □

12.8 Corollary. *Let $\mathscr{N}_1 \subset \mathscr{M}_1 \subset \mathscr{B}(\mathscr{H}_1), \mathscr{N}_2 \subset \mathscr{M}_2 \subset \mathscr{B}(\mathscr{H}_2)$ be von Neumann algebras and $E_1 \in P(\mathscr{M}_1, \mathscr{N}_1), E_2 \in P(\mathscr{M}_2, \mathscr{N}_2)$. There exists a unique element $E = E_1 \bar{\otimes} E_2 \in P(\mathscr{M}_1 \bar{\otimes} \mathscr{M}_2, \mathscr{N}_1 \bar{\otimes} \mathscr{N}_2)$ such that*

$$(\psi_1 \bar{\otimes} \psi_2) \circ E = (\psi_1 \circ E_1) \bar{\otimes} (\psi_2 \circ E_2) \textit{ for all } \psi_1 \in W_{nsf}(\mathscr{N}_1), \psi_2 \in W_{nsf}(\mathscr{N}_2) \tag{1}$$

Proof. Let $\varphi_1 \in W_{nsf}(\mathscr{N}_1), \varphi_2 \in W_{nsf}(\mathscr{N}_2)$ be fixed. According to the definition of the tensor product of n.s.f. weights (8.2) and to Theorem 11.9, we have

$$\sigma_t^{(\varphi_1 \circ E_1) \bar{\otimes} (\varphi_2 \circ E_2)}|\mathscr{N}_1 \bar{\otimes} \mathscr{N}_2 = \sigma_t^{\varphi_1 \bar{\otimes} \varphi_2} \quad (t \in \mathbb{R}).$$

By Theorem 12.1, there exists a unique element $E \in P(\mathscr{M}_1 \bar{\otimes} \mathscr{M}_2, \mathscr{N}_1 \bar{\otimes} \mathscr{N}_2)$ such that $(\varphi_1 \bar{\otimes} \varphi_2) \circ E = (\varphi_1 \circ E_1) \bar{\otimes} (\varphi_2 \circ E_2)$.

For other n.s.f. weights ψ_1, ψ_2, (1) now follows again using Theorem 11.9, Corollary 8.6, and Corollary 3.6. □

The operator-valued weight $E_1 \bar{\bar{\otimes}} E_2$ is called the *tensor product* of E_1 and E_2. It is easy to check that if $a_1 \in \mathfrak{M}_{E_1}, a_2 \in \mathfrak{M}_{E_2}$ then $a_1 \bar{\otimes} a_2 \in \mathfrak{M}_{E_1 \bar{\otimes} E_2}$ and

$$(E_1 \bar{\otimes} E_2)(a_1 \bar{\otimes} a_2) = E_1(a_1) \bar{\otimes} E_2(a_2),$$

but the uniqueness of $E_1 \bar{\otimes} E_2$ requires (1) for some ψ_1, ψ_2.

From Section 11.15, it is clear that

$$\sigma_t^{E_1 \bar{\otimes} E_2} = \sigma_t^{E_1} \bar{\otimes} \sigma_t^{E_2} \quad (t \in \mathbb{R}). \tag{2}$$

If E_1 and E_2 are normal faithful conditional expectations, then, by Corollary 8.7, the present definition of $E_1 \bar{\otimes} E_2$ agrees with that considered in Section 9.4.

12.9. Let $\mathscr{N}_0 \subset \mathscr{M}_0 \subset \mathscr{B}(\mathscr{H})$ be von Neumann algebras, $\mathscr{F}$ a type I factor and $\mathscr{N} = \mathscr{N}_0 \bar{\otimes} \mathscr{F} \subset \mathscr{M}_0 \bar{\otimes} \mathscr{F} = \mathscr{M}$. Consider an n.s.f. operator-valued weight $E : \mathscr{M}^+ \to \hat{\mathscr{N}}^+$.

Let $x \in \mathscr{M}_0^+$. For $u \in \mathscr{F}$ unitary, we have $1 \bar{\otimes} u \in \mathscr{N}$ and

$$(1 \bar{\otimes} u)E(x \bar{\otimes} 1)(1 \bar{\otimes} u)^* = E((1 \bar{\otimes} u)(x \bar{\otimes} 1)(1 \bar{\otimes} u)^*) = E(x \bar{\otimes} 1),$$

hence $E(x \bar{\otimes} 1) \in \overline{(\mathscr{N}_0 \bar{\otimes} 1)}^+$. Thus, there exists $E_0(x) \in \overline{\mathscr{N}}_0^+$ such that

$$E(x \bar{\otimes} 1) = E_0(x) \bar{\otimes} 1. \tag{1}$$

It is easy to see that $E_0 : \mathscr{M}_0^+ \to \bar{\mathscr{N}}_0^+$ is a normal faithful operator-valued weight.

Let ψ be an n.s.f. weight on $\mathscr{N}_0$ and tr the canonical trace on $\mathscr{F}$. For $u \in \mathscr{F}$ we have $1 \bar{\otimes} u \in \mathscr{N}$ and hence (11.9.(2)) $\sigma_t^{(\psi \bar{\otimes} tr) \circ E}(1 \bar{\otimes} u) = \sigma_t^{\psi \bar{\otimes} tr}(1 \bar{\otimes} u) = 1 \bar{\otimes} u$. Thus, $1 \bar{\otimes} \mathscr{F}$ is contained in the centralizer of the n.s.f. weight $(\psi \bar{\otimes} tr) \circ E$ on $\mathscr{M} = \mathscr{M}_0 \bar{\otimes} \mathscr{F}$. By Proposition 9.17, there exists an n.s.f. weight φ on $\mathscr{M}_0$ such that $(\psi \bar{\otimes} tr) \circ E = \varphi \bar{\otimes} tr$. Then, for any $x \in \mathscr{M}_0^+$ and any minimal projection $e \in \mathscr{F}$, we have $\varphi(x) = (\varphi \bar{\otimes} tr)(x \bar{\otimes} e) = (\psi \bar{\otimes} tr)(E(x \bar{\otimes} e)) = (\psi \bar{\otimes} tr)(E_0(x) \bar{\otimes} e) = \psi(E_0(x))$, hence $\varphi = \psi \circ E_0$.

Consequently,

$$(\psi \bar{\otimes} tr) \circ E = (\psi \circ E_0) \bar{\otimes} tr. \tag{2}$$

From (2) and Proposition 11.7, it follows that E_0 is semifinite. Then, by (2) and Corollary 12.8, we get

$$E = E_0 \bar{\otimes} \iota_{\mathscr{F}}, \tag{3}$$

where $\iota_{\mathscr{F}}$ is the identity mapping on $\mathscr{F}$.

Thus, every $E \in P(\mathscr{M}_0 \bar{\otimes} \mathscr{F}, \mathscr{N}_0 \bar{\otimes} \mathscr{F})$ is of the form (3), with $E_0 \in P(\mathscr{M}_0, \mathscr{N}_0)$.

12.10 Corollary. *Let φ be an n.s.f. weight on the W^*-algebra $\mathscr{M}$. The centralizer $\mathscr{M}^\varphi$ of φ is semifinite if and only if there exists a σ^φ-invariant n.s.f. $\mathscr{M}^\varphi$-valued weight on $\mathscr{M}$.*

Proof. Assume that $\mathscr{M}^\varphi$ is semifinite and let τ be an n.s.f. trace on $\mathscr{M}$. Since $\sigma_t^\varphi|\mathscr{M} = \iota = \sigma_t^\tau (t \in \mathbb{R})$, it follows by Theorem 12.1 that there exists a unique n.s.f. operator-valued weight $E : \mathscr{M}^+ \to (\bar{\mathscr{M}}^\varphi)^+$ such that $\varphi = \tau \circ E$. Since $\tau \circ (E \circ \sigma_t^\varphi) = \varphi \circ \sigma_t^\varphi = \varphi = \tau \circ E$, it follows that $E \circ \sigma_t^\varphi = E$, that is, E is σ^φ-invariant.

Conversely, let $E : \mathscr{M}^+ \to (\bar{\mathscr{M}}^\varphi)^+$ be a σ^φ-invariant n.s.f. operator-valued weight and let ψ be an n.s.f. weight on $\mathscr{M}^\varphi$. Then the n.s.f. weights φ and $\psi \circ E$ on $\mathscr{M}$ commute and so, by Theorem 4.10, there exists a semifinite element $A \in (\bar{\mathscr{M}}^\varphi)^+$ such that $\psi \circ E = \varphi_A$. Using Theorem 11.9 and Corollary 4.8, we obtain $\sigma_t^\psi = \sigma_t^{\psi \circ E} = \sigma_t^{\varphi_A} = \mathrm{Ad}(A^{it}) (t \in \mathbb{R})$. Consequently, $\tau = \psi_{A^{-1}}$ is an n.s.f. trace on the W^*-algebra $\mathscr{M}$. □

12.11. Another consequence of Theorem 12.1, as well as Theorems 7.4, 7.14, and 11.9, is the following

Corollary. *Let $\mathscr{N} \subset \mathscr{M} \subset \mathscr{B}(\mathscr{H})$ be von Neumann algebras. There is a bijection*

$$P(\mathscr{M}, \mathscr{N}) \ni E \mapsto E' \in P(\mathscr{N}', \mathscr{M}'),$$

uniquely determined, with the property

$$\Delta(\psi/\varphi' \circ E') = \Delta(\psi \circ E/\varphi') \quad (\psi \in W_{nsf}(\mathscr{N}), \varphi' \in W_{nsf}(\mathscr{M}')), \tag{1}$$

where $\Delta(\cdot/\cdot)$ stands for the spatial derivative (7.3).

For any $E, E_1, E_2 \in P(\mathscr{M}, \mathscr{N})$ and any $t \in \mathbb{R}$ we have

$$\sigma_t^{E'} = \sigma_{-t}^E, \tag{2}$$
$$[DE_1' : DE_2']_t = [DE_1 : DE_2]_{-t} \tag{3}$$
$$E_1 \le E_2 \Leftrightarrow E_2' \le E_1'. \tag{4}$$

Proof. Let $\psi \in W_{nsf}(\mathscr{N})$ and $\varphi' \in W_{nsf}(\mathscr{M}')$ be fixed.

Let $E \in P(\mathscr{M}, \mathscr{N})$. Then $\varphi = \psi \circ E \in W_{nsf}(\mathscr{M})$. Writing $u_t = \Delta(\varphi/\varphi')^{it}$, by Theorem 7.4 we have $\sigma_t^\varphi = Ad(u_t)|\mathscr{M}$ and $\sigma_t^{\varphi'} = Ad(u_t^*)|\mathscr{M}'(t \in \mathbb{R})$. By Theorem 11.9, we get $\sigma_t^\psi = \sigma_t^\varphi|\mathscr{N} = Ad(u_t)|\mathscr{N}(t \in \mathbb{R})$. Using 7.4.(1) and Theorem 7.14, we infer the existence of a unique weight $\psi' \in W_{nsf}(\mathscr{N}')$ such that $u_t = \Delta(\psi/\psi')^{it}(t \in \mathbb{R})$. Then $\sigma_t^{\psi'} = Ad(u_t^*)|\mathscr{N}'$, and hence $\sigma_t^{\varphi'} = \sigma_t^{\psi'}|\mathscr{M}'(t \in \mathbb{R})$. By Theorem 12.1, there exists a unique operator-valued weight $E' \in P(\mathscr{N}', \mathscr{M}')$ such that $\psi' = \varphi' \circ E'$ and we have

$$\Delta(\psi/\varphi' \circ E') = \Delta(\psi \circ E/\varphi'). \tag{5}$$

Equation (5) determines the weight $\varphi' \circ E'$ and hence the operator-valued weight E', uniquely (see 7.13.(1) and 11.13). □

Similarly, one can construct a mapping $P(\mathscr{N}', \mathscr{M}') \ni E' \mapsto E \in P(\mathscr{M}, \mathscr{N})$, satisfying (5). It follows that the mappings $E \mapsto E'$ and $E' \mapsto E$ are reciprocal bijections.

If $\bar{\psi} \in W_{nsf}(\mathcal{N}), \bar{\varphi}' \in W_{nsf}(\mathcal{M}')$ are other weights, then, using Theorem 7.4.(5) and Theorem 11.9.(3), we deduce from (5) that $\Delta(\bar{\psi}/\bar{\varphi}' \circ E') = \Delta(\bar{\psi} \circ E/\bar{\varphi}')$, proving (1).

For $z \in \mathcal{N}' \cap \mathcal{M} = (\mathcal{M}')' \cap \mathcal{N}'$, we have by definition (11.15.(2)) $\sigma_t^{E'}(z) = \sigma_t^{\varphi' \circ E'}(z) = \sigma_t^{\psi'}(z) = u_t^* z u_t = u_{-t}^* z u_{-t}^* = \sigma_{-t}^{\varphi}(z) = \sigma_{-t}^{\psi \circ E}(z) = \sigma_{-t}^{E}(z)$. This proves (2).

Consider now $E_1, E_2 \in P(\mathcal{M}, \mathcal{N})$ and let $\varphi_1, u_t^1, \psi_1', E_1'$ and $\varphi_2, u_t^2, \psi_2', E_2'$ be associated, as in the first part of the proof, with E_1 and E_2, respectively. By Theorem 7.4.(5), we have $[D\psi_1' : D\psi_2']_t = u_{-t}^1 u_t^2 = [D\varphi_1 : D\varphi_2]_{-t} (t \in \mathbb{R})$. Since $\psi_j' = \varphi' \circ E_j', \varphi_j = \psi \circ E_j (j = 1, 2)$, using Definition 11.15.(4) we obtain (3).

Finally, (4) is an immediate consequence of (3) and 11.15.(8).

12.12. In particular, we have the following

Corollary. *Let $\mathcal{M} \subset \mathcal{B}(\mathcal{H})$ be von Neumann algebra. There is a uniquely determined bijection*

$$W_{nsf}(\mathcal{M}') \ni \psi \mapsto E_\psi \in P(\mathcal{B}(\mathcal{H}), \mathcal{M})$$

such that

$$\varphi \circ E_\psi = tr_{\Delta(\varphi/\psi)} \textit{ for any } \varphi \in W_{nsf}(\mathcal{M}). \tag{1}$$

Moreover, we have

$$E_\psi(\eta \bar{\otimes} \bar{\eta}) = R_\eta^\psi (R_\eta^\psi)^* \textit{ for all } \eta \in D(\mathcal{H}, \psi). \tag{2}$$

Proof. The notation is as in Sections 4.23 and 7.1.

We apply Corollary 12.11, replacing $\mathcal{M}, \mathcal{N}, E, \psi, \mathcal{M}', \mathcal{N}', E', \varphi'$ in 12.11 by $\mathcal{M}', \mathbb{C} \cdot 1_{\mathcal{H}}, \psi, t, \mathcal{M}, \mathcal{B}(\mathcal{H}), E_\psi, \varphi$, respectively, where $\mathcal{M} \subset \mathcal{B}(\mathcal{H})$, $\psi \in W_{nsf}(\mathcal{M}')$, $E_\psi \in (\mathcal{B}(\mathcal{H}), \mathcal{M})$ are as in the statement of 12.12, φ is an n.s.f. weight on $\mathcal{M}$, and the positive real number $t > 0$ is regarded as the weight $\mathbb{C} \cdot 1_{\mathcal{M}} \ni \lambda \mapsto t\lambda \in \mathbb{C}$. We thus deduce the existence of the required bijection $\psi \mapsto E_\psi$, uniquely determined by the condition

$$\Delta(t/\varphi \circ E_\psi) = \Delta(t\psi/\varphi) \textit{ for all } \varphi \in W_{nsf}(\mathcal{M}) \textit{ and all } t > 0.$$

According to 7.4.(1), 7.13.(2), and 7.3.(6), the above condition is equivalent to the condition

$$\Delta(\varphi \circ E_\psi/1) = \Delta(\varphi/\psi) = \Delta(tr_{\Delta(\varphi/\psi)}/1) \textit{ for all } \varphi \in W_{nsf}(\mathcal{M}).$$

Using 7.13.(1), we get

$$\varphi \circ E_\psi = tr_{\Delta(\varphi/\psi)} \textit{ for all } \varphi \in W_{nsf}(\mathcal{M}).$$

Thus, for any $\eta \in D(\mathcal{H}, \psi)$ and any $\varphi \in W_{nsf}(\mathcal{M})$ we have (see 4.23.(4) and 7.3.(2))

$$\varphi(E_\psi(\eta \bar{\otimes} \bar{\eta})) = tr_{\Delta(\varphi/\psi)}(\eta \bar{\otimes} \bar{\eta}) = \|\Delta(\varphi/\psi)^{1/2}\eta\|^2 = \varphi(R_\eta^\psi(R_\eta^\psi)^*)$$

and hence $E_\psi(\eta \bar{\otimes} \bar{\eta}) = R_\eta^\psi (R_\eta^\psi)^* (\eta \in D(\mathcal{H}, \psi))$. □

12.12. bis As an application of Corollary 12.11, we get an extension of the V. Jones' theory of index ([L], C.7.7) to arbitrary factors, due to Kosaki (1986).

Let $\mathcal{N} \subset \mathcal{M} \subset \mathcal{B}(\mathcal{H})$ be arbitrary factors and assume that there is a normal conditional expectation $E : \mathcal{M} \to \mathcal{N}$ (otherwise, $\mathcal{N}$ is too small in $\mathcal{M}$, so that the index should be $+\infty$). By Corollary 12.11, we get an operator-valued weight $E' \in P(\mathcal{N}', \mathcal{M}')$. For any unitary $u' \in \mathcal{M}'$ we have

$$u'E'(1)u'^* = E'(u' \cdot 1 \cdot u'^*) = E'(1)$$

and, since $\mathcal{M}'$ is a factor, this means that $E'(1)$ is a scalar, possibly $+\infty$. By definition,

$$Index\, E = E'(1)$$

and, when E is self-understood, we denote

$$[\mathcal{M} : \mathcal{N}] = Index\, E.$$

It is shown that the number $E'(1)$ does not depend on $\mathcal{H}$. If $\mathcal{M}$ is a factor of type II_1 and $E : \mathcal{M} \to \mathcal{N}$ is the unique conditional expectation with respect to the normalized trace, then this index coincides with the V. Jones' index.

When $Index\, E < +\infty$, the operator-valued weight $E' \in P(\mathcal{N}', \mathcal{M}')$ is a scalar muultiple of a conditional expectation, namely, with $\tau = (E'(1))^{-1}$, we have

$$(\tau E')(x') = (\tau E')(x' \cdot 1) = x'(\tau E')(1) = x', \quad (x' \in \mathcal{M}').$$

By imitating the arguments of V. Jones, H. Kosaki obtained the same restrictions for the values of the index as V. Jones in the case of finite factors.

12.13. Another consequence of Theorem 12.1 is an extension of Theorem 5.1 for operator-valued weights:

Corollary. *Let $\mathcal{N} \subset \mathcal{M} \subset \mathcal{B}(\mathcal{H})$ be von Neumann algebras and $E \in P(\mathcal{M}, \mathcal{N}) \neq \emptyset$. Then the mapping*

$$F \mapsto \{[DF : DE]_t\}_{t\in\mathbb{R}}$$

establishes a bijection between $P(\mathcal{M}, \mathcal{N})$ and the set of all unitary σ^E-cocycles $\{w_t\}_{t\in\mathbb{R}} \subset \mathcal{N}' \cap \mathcal{M}$.

Proof. Let $\{w_t\}_{t\in\mathbb{R}} \subset \mathcal{N}' \cap \mathcal{M}$ be a unitary σ^E-cocyle and $\psi \in W_{nsf}(\mathcal{N})$. Since $\sigma_t^E = \sigma_t^{\psi\circ E}|\mathcal{N}' \cap \mathcal{M}$ (11.15.(2)), $\{w_t\}_{t\in\mathbb{R}}$ is a unitary $\sigma^{\psi\circ E}$-cocycle so that, by Theorem 5.1, there exists $\varphi \in W_{nsf}(\mathcal{M})$ such that

$$[D\varphi : D(\psi \circ E)]_t = w_t (t \in \mathbb{R}). \tag{1}$$

Since $w_t \in \mathcal{N}' \cap \mathcal{M}$, for $y \in \mathcal{N}$ we have (11.9.(2)) $\sigma_t^\varphi(y) = w_t\sigma_t^{\psi\circ E}(y)w_t^* = w_t\sigma_t^\psi(y)w_t^* = \sigma_t^\psi(y)(t \in \mathbb{R})$. By Theorem 12.1, there exists an operator-valued weight $F \in P(\mathcal{M}, \mathcal{N})$ such that

$\varphi = \psi \circ F$. Then, from (1) and 11.15.(4) it follows that $[DF : DE]_t = w_t (t \in \mathbb{R})$. Thus, the mapping considered in the statement is surjective.

The injectivity of this mapping follows immediately from 3.6 and 11.13. □

12.14. Let $\mathcal{M}$ be a semifinite von Neumann algebra and $\mathcal{Z}$ a von Neumann subalgebra of the center $\mathcal{Z}(\mathcal{M})$ of $\mathcal{M}$.

Let us fix an n.s.f. trace τ on $\mathcal{M}$ and an n.s.f. weight ν on $\mathcal{Z}$. By Theorem 12.1 (actually, by the particular case considered in Section 12.3), there exists a unique n.s.f. operator-valued weight

$$\natural = \natural(\mathcal{Z}, \tau, \nu) : \mathcal{M}^+ \ni x \mapsto x^\natural \in \mathcal{Z}^+$$

such that $\tau = \nu \circ \natural$, that is,

$$\tau(x) = \nu(x^\natural) \quad (x \in \mathcal{M}^+). \tag{1}$$

Since $\mathcal{Z}$ is contained in $\mathcal{Z}(\mathcal{M})$, we have

$$(ax)^\natural = ax^\natural \quad (x \in \mathcal{M}^+, a \in \mathcal{Z}^+). \tag{2}$$

For every $x \in \mathcal{M}$ and every $a \in \mathcal{Z}^+$, we have $\nu(a(x^*x)^\natural) = \nu((ax^*x)^\natural) = \tau(ax^*x) = \tau(axx^*) = \nu((axx^*)^\natural) = \nu(a(xx^*)^\natural)$. Since ν is faithful and $a \in \mathcal{Z}^+$ was arbitrary, it follows that

$$(x^*x)^\natural = (xx^*)^\natural \quad (x \in \mathcal{M}) \tag{3}$$

that is, $\natural$ is actually an operator-valued trace. Thus, $\mathfrak{M}_\natural$ and $\mathfrak{N}_\natural$ are w-dense two sided ideals of $\mathcal{M}$ and, consequently, any nonzero element of $\mathcal{M}^+$ dominates a nonzero element of $\mathfrak{M}_\natural \cap \mathcal{M}^+$.

In what follows we assume that $\mathcal{Z} = \mathcal{Z}(\mathcal{M})$.

Then, for two projections $e, f \in \mathcal{M}$ we have

$$e \prec f \Leftrightarrow e^\natural \leq f^\natural; \tag{4}$$

this follows easily, using (3), the faithfulness of $\natural$ and the comparison theorem ([L], 4.6).

Consider now an arbitrary normal operator-valued trace $E : \mathcal{M}^+ \to \bar{\mathcal{Z}}^+$. Then $\nu \circ E$ is a normal trace on $\mathcal{M}$ and hence (see 12.2.(7) and [L], E.7.14, C.10.4) there exists $A \in \bar{\mathcal{Z}}^+$ such that $\nu \circ E = \tau_A$. Thus, for $x \in \mathcal{M}^+$ and $a \in \mathcal{Z}^+$ we have $\nu(aE(x)) = \nu(E(ax)) = \tau(Aax) = \nu((Aax)^\natural) = \nu(aAx^\natural)$. Consequently,

$$E(x) = Ax^\natural \quad (x \in \mathcal{M}^+). \tag{5}$$

If $\mathcal{M}$ is finite and countably decomposable and if $\tau(1) = 1, \nu(1) = 1$, then $\natural : \mathcal{M} \to \mathcal{Z}(\mathcal{M})$ is just the canonical central trace ([L], 7.11).

If $\mathcal{M}$ is properly infinite, then $\nu(z) = +\infty$ for all $0 \neq z \in \mathcal{Z}^+$, hence

$$z^\natural = \infty \circ \mathbf{s}(z) \quad (z \in \mathcal{Z}^+). \tag{6}$$

Proposition. *Let $\mathcal{M}$ be a type II_∞ von Neumann algebra with center $\mathcal{Z}$, τ an n.s.f. trace on $\mathcal{M}$, ν an n.s.f. weight on $\mathcal{Z}$ and $\natural = \natural(\mathcal{Z}, \tau, \nu) : \mathcal{M}^+ \to \mathcal{Z}^+$. Then*

$$\{e^\natural; e \in \mathcal{M}, \textit{projection}\} = \bar{\mathcal{Z}}^+.$$

Proof. Let $0 \neq A \in \bar{\mathcal{Z}}^+$ and let $f \in \mathcal{M}$ be a projection such that $A \leq f^\natural$. We first show that there exists a projection $e \in \mathcal{M}, 0 \neq e \leq f$, such that $e^\natural \leq A$.

If $A = \infty \cdot p$ with p a central projection, then, according to (6), we can take $e = fp$. Otherwise we may assume that A is bounded (11.3.(1)). Since $\natural$ is semifinite, there exists $x \in \mathcal{M}^+, 0 \neq x \leq f$, with $\|x^\natural\| \leq 1$. Then ([L], 2.21) there exists a spectral projection $g \neq 0$ of xA and a positive integer $n > 1$ such that $g \leq nxA$; note that $g \leq \mathbf{s}(xA) \leq f$. Since $\mathcal{M}$ is a continuous von Neumann algebra, it follows by ([L], 4.11, E.4.10) that there exist mutually orthogonal and equivalent projections $e_1, \ldots, e_n \in \mathcal{M}$ such that $g = e_1 + \ldots + e_n$. Then, for $e = e_1$, we have $0 \neq e \leq f$ and $e^\natural = n^{-1}f \leq n^{-1}(nxA)^\natural = x^\natural A \leq A$.

Consider now an arbitrary clement $0 \neq A \in \bar{\mathcal{Z}}^+$. There exists a projection $e \in \mathcal{M}$, which is maximal with the property $e^\natural \leq A$.

If $A = \infty \cdot \mathbf{s}(A)$, then we have $e = \mathbf{s}(A)$ and hence $e^\natural = A$.

Assume that $A \neq \infty \cdot \mathbf{s}(A)$ and $e^\natural \neq A$. Then there is a central projection $p \leq \mathbf{s}(A)$ such that Ap and $e^\natural p$ be bounded. Also, there is a central projection $q \leq p$ such that $Aq - e^\natural q \leq (1-e)^\natural q$ and $(1-e)^\natural(p-q) \leq A(p-q) - e^\natural(p-q)$. We have $q \neq 0$ since $q = 0$ would imply $\infty \cdot p = p^\natural = e^\natural p + (1-e)^\natural p \leq e^\natural p + Ap - e^\natural p = Ap$, contradicting the fact that Ap is bounded. By the first part of the proof, there exists a projection $h \in \mathcal{M}, 0 \neq h \leq (1-e)q$ such that $h^\natural \leq Aq - e^\natural q$. Then $e + h \in \mathcal{M}$ is a projection, $e + h \geq e, e + h \neq e$ and $(e+h)^\natural = e^\natural + h^\natural = e^\natural(1-q) + e^\natural q + h^\natural \leq A(1-q) + Aq = A$, contradicting the maximality of e. Hence $e^\natural = A$. □

Corollary. *Let $\mathcal{M}$ be a type II_∞ von Neumann algebra with center $\mathcal{Z}$, τ an n.s.f. trace on $\mathcal{M}$ and ω a normal weight on $\mathcal{Z}$. There exists a projection $e \in \mathcal{M}$ such that $\omega(z) = \tau(ez)(z \in \mathcal{Z}^+)$.*

Proof. Let ν be an n.s.f. weight on $\mathcal{Z}$ and $\natural = \natural(\mathcal{Z}, \tau, \nu)$. By 12.2.(7) there exists $A \in \bar{\mathcal{Z}}^+$ such that $\omega = \nu_A$ and by the preceding proposition there exists a projection $e \in \mathcal{M}$ such that $e^\natural = A$. Then, for any $z \in \mathcal{Z}^+$, we have $\omega(z) = \nu(Az) = \nu(e^\natural z) = \nu((ez)^\natural) = \tau(ez)$. □

12.15. Let $\mathcal{N} \subset \mathcal{M} \subset \mathcal{B}(\mathcal{H})$ be von Neumann algebras and $E : \mathcal{M}^+ \to \bar{\mathcal{N}}^+$ a normal operator-valued weight. For $x \in \mathcal{N}' \cap \mathcal{M}$ and unitary $v \in \mathcal{N}$, we have $v^*E(x)v = E(v^*xv) = E(x)$, hence $E(x) \in \overline{\mathcal{Z}(\mathcal{N})}^+$. Consequently, putting $\mathcal{N}^c = \mathcal{N}' \cap \mathcal{M}$, we obtain a normal operator-valued weight

$$E^c = E|(\mathcal{N}^c)^+; (\mathcal{N}^c)^+ \to \overline{\mathcal{Z}(\mathcal{N})}^+.$$

Clearly, if E is faithful, then E^c is also faithful.

Theorem (U. Haagerup). *Let $\mathcal{N} \subset \mathcal{M} \subset \mathcal{B}(\mathcal{H})$ be von Neumann algebras. The following statements are equivalent:*

(i) *there exists $E \in P(\mathcal{M}, \mathcal{N})$ such that $E^c \in P(\mathcal{N}^c, \mathcal{Z}(\mathcal{N}))$;*
(ii) *$P(\mathcal{M}, \mathcal{N}) \neq \emptyset$ and $E \in P(\mathcal{M}, \mathcal{N}) \Rightarrow E^c \in P(\mathcal{N}^c, \mathcal{Z}(\mathcal{N}))$;*
(iii) *there exists a separating family of bounded normal $\mathcal{N}$-valued weights on $\mathcal{M}$;*
(iv) *there exists a separating family of normal conditional expectations of $\mathcal{M}$ onto $\mathcal{N}$.*

If these conditions are satisfied, then the mapping

$$P(\mathcal{M}, \mathcal{N}) \ni E \mapsto E^c \in P(\mathcal{N}^c, \mathcal{Z}(\mathcal{N}))$$

is a bijection and, for any $E, F \in P(\mathcal{M}, \mathcal{N})$ *and any* $t \in \mathbb{R}$, *we have*

$$\sigma_t^{E^c} = \sigma_t^E, \tag{1}$$

$$[DF^c : DE^c]_t = [DF : DE]_t. \tag{2}$$

Proof. (1) We first assume that $\mathcal{N}$ is countably decomposable. Consider a fixed faithful normal state φ on $\mathcal{N}$ and put $\omega = \varphi|\mathcal{Z}(\mathcal{N})$.

(i) $\Leftrightarrow$ (ii). Let $E, F \in P(\mathcal{M}, \mathcal{N})$. Assume that E^c is semifinite, that is, $E^c \in P(\mathcal{N}^c, \mathcal{Z}(\mathcal{N}))$. It is clear that

$$(\varphi \circ E)|(\mathcal{N}^c)^+ = \omega \circ E^c \text{ is semifinite.} \tag{3}$$

Since (11.15.(4)) $[D(\varphi \circ F) : D(\varphi \circ E)]_t = [DF : DE]_t \in \mathcal{N}^c (t \in \mathbb{R})$, it follows also that

$$\omega \circ F^c = (\varphi \circ F)|(\mathcal{N}^c)^+ \text{ is semifinite} \tag{4}$$

and hence, by Proposition 11.7, F^c is semifinite, that is, $F^c \in P(\mathcal{N}^c, \mathcal{Z}(\mathcal{N}))$.

(i) $\Leftrightarrow$ (iii). Let $E \in P(\mathcal{M}, \mathcal{N})$ and assume that E^c is semifinite. Then there exists a net $\{v_i\}_{i \in I} \subset \mathcal{N}^c \cap \mathfrak{M}_F$ such that $v_i \uparrow 1$. It follows that the mappings

$$F_i : \mathcal{M}^+ \ni x \mapsto E(v_i^* x v_i) \in \mathcal{N}^+ \quad (i \in I)$$

constitute a separating family of bounded normal operator-valued weights.

(iii) $\Rightarrow$ (iv). Let $x_0 \in \mathcal{M}^+, x_0 \neq 0$. Assuming (iii), there exists a bounded normal operator-valued weight $F_0 : \mathcal{M}^+ \to \mathcal{N}^+$ such that $F_0(x_0) \neq 0$. Since $1 \in \mathcal{N}^c$, we have $F_0(1) \in \mathcal{Z}(\mathcal{N})$ and we may assume that $F_0(1) \leq 1$. Let $\{F_i\}_{i \in I}$ be a maximal family of bounded normal $\mathcal{N}$-valued weights on $\mathcal{M}$ which contains F_0, such that $F_i(1) \leq 1$ and the supports $\mathbf{s}(F_i(1)) \in \mathcal{Z}(\mathcal{N})$ are mutually orthogonal. In view of (iii) we get $\sum_{i \in I} \mathbf{s}(F_i(1)) = 1$. Consequently, $F = \sum_{i \in I} F_i$ is a bounded normal $\mathcal{N}$-valued weight on $\mathcal{M}, F(x_0) \neq 0$ and $a = F(1) \in \mathcal{Z}(\mathcal{N})^+, 0 \leq a \leq 1, \mathbf{s}(a) = 1$. Then $a^{-1} \in \overline{\mathcal{Z}(\mathcal{N})}^+$ is a semifinite element and the equation

$$E(x) = a^{-1} F(x); x \in \mathcal{M}^+$$

defines a normal conditional expectation E of $\mathcal{M}$ onto $\mathcal{N}$ such that $E(x_0) \neq 0$.

(iv) $\Rightarrow$ (i). Let $\{E_i\}_{i \in I}$ be a maximal family of bounded normal $\mathcal{N}$-valued weights on $\mathcal{M}$ with mutually orthogonal supports $0 \neq \mathbf{s}(E_i) \in \mathcal{N}^c$ and let $e = 1 - \sum_{i \in I} \mathbf{s}(E_i) \in \mathcal{N}^c$. If $e \neq 0$, then by (iv) there exists a normal conditional expectation $F_0 : \mathcal{M} \to \mathcal{N}$ with $F_0(e) \neq 0$. Since $e \in \mathcal{N}^c$ it follows that the mapping $E_0 : \mathcal{M} \ni x \mapsto F_0(exe) \in \mathcal{N}$ is a bounded normal operator-valued weight with $0 \neq \mathbf{s}(E_0) \leq e$, contradicting the maximality of the family $\{E_i\}_{i \in I}$. Hence $e = 0$, that is,

$\sum_{i\in I} \mathbf{s}(E_i) = 1$. Thus, the equation

$$E(x) = \sum_{i\in I} E_i(x) \quad (x \in \mathcal{M}^+)$$

defines a normal faithful operator-valued weight $E : \mathcal{M}^+ \to \bar{\mathcal{N}}^+$. Since $\mathbf{s}(E_i) \in \mathcal{N}^c$, $E(\mathbf{s}(E_i)) = E_i(1)$ is bounded, and $\sum_{i\in I} \mathbf{s}(E_i) = 1$, it follows that E^c and E are semifinite.

Now consider $E, F \in P(\mathcal{M}, \mathcal{N})$ and assume that E^c, F^c are semifinite. Using (3), 11.15.(2), 10.1, and 10.5, it follows that

$$\sigma_t^{E^c} = \sigma_t^{\omega\circ E^c} | \mathcal{N}^c = \sigma_t^{\varphi\circ E} | \mathcal{N}^c = \sigma_t^E \quad (t \in \mathbb{R}).$$

Consider also the operator-valued weight $\Theta = \Theta(E, F) \in P(Mat_2(\mathcal{M}), \mathcal{N} \bar{\otimes} 1)$ defined in Section 11.15. It is easy to check that

$$\Theta^c = \Theta|(\mathcal{N} \bar{\otimes} 1)' \cap Mat_2(\mathcal{M}) = \Theta(E^c, F^c) \in P(Mat_2(\mathcal{N}^c), \mathcal{Z}(\mathcal{N}) \bar{\otimes} 1),$$

that is, Θ^c is semifinite. It follows that $\sigma_t^{\Theta^c} = \sigma_t^{\Theta} (t \in \mathbb{R})$, and hence $[DF^c : DE^c]_t = [DF : DE]_t (t \in \mathbb{R})$. We have thus proved statements (1) and (2).

The injectivity of the mapping $E \mapsto E^c$ follows from (2); its surjectivity can be easily proved using (2) and Corollary 12.13.

(II) Assume now that $\mathcal{N}$ is uniform of type γ ([L], 8.5). Then there exists a countably decomposable von Neumann algebra $\mathcal{N}_0$ and a type I factor $\mathcal{F}$ such that $\mathcal{N} = \mathcal{N}_0 \bar{\otimes} \mathcal{F}$. Let $\mathcal{M} = \mathcal{M}_0 \bar{\otimes} \mathcal{F}$ be the corresponding factorization of $\mathcal{M}$ with $\mathcal{M}_0 \supset \mathcal{N}_0$ (see 9.15). By 12.9, every $E \in P(\mathcal{M}, \mathcal{N})$ is of the form $E = E_0 \bar{\otimes} \iota_{\mathcal{F}}$ with $E_0 \in P(\mathcal{M}_0, \mathcal{N}_0)$. It is easy to check that E^c is semifinite if and only if E_0^c is semifinite and, in this case,

$$\sigma_t^E = \sigma_t^{E^c} \bar{\otimes} \iota_{\mathcal{F}}, \sigma_t^{E^c} = \sigma_t^{E_0^c} \bar{\otimes} \iota_{\mathcal{F}} (t \in \mathbb{R}).$$

Using these facts, it is easy to see that the proof of the theorem in this case reduces to the case considered in the first part of the proof.

(III) In the general case, there exist a family Γ of distinct infinite cardinals and a family $\{q_\gamma\}_{\gamma\in\Gamma}$ of projections in $\mathcal{Z}(\mathcal{N})$ such that $\sum_\gamma q_\gamma = 1$ and each $\mathcal{N} q_\gamma$ is uniform of type γ ([L], 8.5). If $E \in P(\mathcal{M}, \mathcal{N})$ and $E_\gamma = E|q_\gamma \mathcal{M} q_\gamma$, $E_\gamma^c = E|q_\gamma \mathcal{N}^c$, $(\gamma \in \Gamma)$, then E^c is semifinite if and only if each E_γ^c is semifinite and, in this case,

$$\sigma_t^{E^c} = \sigma_t^E (t \in \mathbb{R}) \Leftrightarrow \sigma_t^{E_\gamma^c} = \sigma_t^{E_\gamma} (t \in \mathbb{R}, \gamma \in \Gamma).$$

Thus, the general case reduces to the case (II). □

12.16. Let $\mathcal{N} \subset \mathcal{M} \subset \mathcal{B}(\mathcal{H})$ be von Neumann algebras. *The equivalent conditions in Theorem 12.15 are satisfied in each of the following particular cases:*

(1) *$\mathcal{N}$ is a direct sum of type I factors;*
(2) *$\mathcal{M} = \mathcal{N} \bar{\otimes} \mathcal{R}$, where $\mathcal{N}$ is identified with $\mathcal{N} \otimes 1 \subset \mathcal{M}$;*
(3) *$\mathcal{N} = \mathcal{Z}(\mathcal{M})$.*

Indeed, in all these cases there are separating families of normal conditional expectations of $\mathcal{M}$ onto $\mathcal{N}$ (see 10.23, 9.8.(3), 10.16).

In particular, *in all these cases we have* $P(\mathcal{M},\mathcal{N}) \neq \emptyset$. According to Corollary 12.11, *we have* $P(\mathcal{M},\mathcal{N}) \neq \emptyset$ *also in the case when*

$$\mathcal{M} \textit{ is a direct sum of type I factors.} \tag{4}$$

Finally, by Theorem 12.1 it follows that $P(\mathcal{M},\mathcal{N}) \neq \emptyset$ whenever

$$\mathcal{M} \textit{ and } \mathcal{N} \textit{ are semifinite.} \tag{5}$$

12.17. In this section we consider some results which show that the equivalent conditions in Theorem 12.15 are not always satisfied.

We begin with two general remarks. Let $\mathcal{N} \subset \mathcal{M} \subset \mathcal{B}(\mathcal{H})$ be von Neumann algebras.

$$\begin{gathered}\textit{If } \mathcal{M} \textit{ and } \mathcal{N} \textit{ are semifinite, then, for every } E \in P(\mathcal{M},\mathcal{N}),\\ \textit{the modular automorphism group } \{\sigma_t^E\}_{t\in\mathbb{R}} \subset Aut(\mathcal{N}' \cap \mathcal{M})\\ \textit{is implemented by a unitary representation } \mathbb{R} \mapsto U(\mathcal{N}' \cap \mathcal{M}).\end{gathered} \tag{1}$$

Indeed, let φ and ψ be n.s.f. traces on $\mathcal{M}$ and $\mathcal{N}$, respectively, and let $E \in P(\mathcal{M}, \mathcal{N})$ be uniquely determined such that $\varphi = \psi \circ E$ (see 12.3). Then, by 11.15.(2), $\sigma_t^E = \sigma_t^{\psi\circ E}|\mathcal{N}'\cap\mathcal{M} = \sigma_t^\varphi = \iota (t \in \mathbb{R})$, and hence, for any $F \in P(\mathcal{M}, \mathcal{N})$, we have $w_t = [DF : DE]_t \in \mathcal{N}' \cap \mathcal{M}$ and $\sigma_t^F = Ad(w_t)(t \in \mathbb{R})$ (see 11.15.(4), 11.15.(5)).

$$\begin{gathered}\textit{If } \mathcal{M} \textit{ and } \mathcal{N} \textit{ satisfy the equivalent conditions in } 12.15, \textit{ then}\\ \textit{for every } E \in P(\mathcal{M}, \mathcal{N}) \textit{ there exists } \varphi \in W_{nsf}(\mathcal{N}' \cap \mathcal{M})\\ \textit{such that } \sigma_t^E = \sigma_t^\varphi (t \in \mathbb{R}).\end{gathered} \tag{2}$$

Indeed, let ν be an n.s.f. weight on $\mathcal{Z}(\mathcal{N})$ and $\varphi = \nu \circ E^c$. Then for every $x \in \mathcal{N}' \cap \mathcal{M}$ we have (12.15.(1)): $\sigma_t^E(x) = \sigma_t^{E^c}(x) = \sigma_t^{\psi\circ E^c}(x) = \sigma_t^\varphi(x)(t \in \mathbb{R})$.

Haagerup (1977) showed that *there exists an approximately finite dimensional type* II_∞ *factor* $\mathcal{R} \subset \mathcal{B}(\mathcal{H})$, *with* $\mathcal{H}$ *a separable Hilbert space, and an abelian von Neumann algebra* $\mathcal{A} \subset \mathcal{R}$ *such that the relative commutent* $\mathcal{A}^c = \mathcal{A}' \cap \mathcal{R}$ *is of type* III.

Then $\mathcal{R}$ and $\mathcal{A}$ *are semifinite, hence* $P(\mathcal{R}, \mathcal{A}) \neq \emptyset$, *but they do not satisfy the equivalent conditions in Theorem 12.15*, as is easily seen using (1), (2) and ([L], 10.29).

From this example of Haagerup, it also follows that *there exists an n.s.f. weight* φ *on the von Neumann algebra* $\mathcal{M} = \mathcal{R}$ *such, that the centralizer* $\mathcal{M}^\varphi$ *is of type* III. Indeed, since $\mathcal{A} \subset \mathcal{B}(\mathcal{H})$ is abelian and $\mathcal{H}$ is separable, there exists a positive operator $a \in \mathcal{A}, 0 \leq a \leq 1, \mathrm{s}(a) = 1$, such that $\mathcal{A}$ is the von Neumann algebra generated by a, that is, $\mathcal{A} = \{a\}''$ (Strătilă & Zsidó, 1977–1979, Prop. 8.14). Let τ be an n.s.f. trace on $\mathcal{R}$ and $\varphi = \tau_a$ an n.s.f. weight on $\mathcal{M} = \mathcal{R}$. Since $\sigma_t^\varphi = Ad(a^{it})(t \in \mathbb{R})$, it follows that $\mathcal{M}^\varphi = \mathcal{R} \cap \{a\}' = \mathcal{A}' \cap \mathcal{R}$ is of type III. By Corollary 12.10, we see that *in this case there is no* σ^φ*-invariant n.s.f.* $\mathcal{M}^\varphi$*-valued weight on* $\mathcal{M}$.

Another instance when $P(\mathcal{M}, \mathcal{N}) \neq \emptyset$ but $\mathcal{M}$ and $\mathcal{N}$ do not satisfy the equivalent conditions in Theorem 12.15 is the case of the continuous decomposition $\mathcal{M} = \mathcal{R}(\mathcal{N}, \theta)$ of a properly infinite

W^*-algebra $\mathscr{M}$ (see 23.7). In this case there are no nonzero normal conditional expectations of $\mathscr{M}$ onto $\mathscr{N}$ (Haagerup, U. [1979a]).

12.18. Let $\mathscr{M}, \mathscr{N}$ be W^*-algebras and φ a normal weight on $\mathscr{M}$. We shall identify $\mathscr{N}$ with $1_{\mathscr{M}} \bar{\otimes} \mathscr{N} \subset \mathscr{M} \bar{\otimes} \mathscr{N}$. As in Section 9.8, we define a normal operator-valued weight $E^{\varphi}_{\mathscr{N}}$: $(\mathscr{M} \bar{\otimes} \mathscr{N})^+ \to \bar{\mathscr{N}}^+$ by

$$E^{\varphi}_{\mathscr{N}}(x)(\psi) = (\varphi \bar{\otimes} \psi)(x) \quad (x \in (\mathscr{M} \bar{\otimes} \mathscr{N})^+, \psi \in \mathscr{N}^+_*); \tag{1}$$

$E^{\varphi}_{\mathscr{N}}$ is called the *Fubini mapping associated with* φ. It is easy to check that $E^{\varphi}_{\mathscr{N}}$ is semifinite (resp. faithful) if and only if φ is semifinite (resp. faithful). Using Combes' theorem (2.6), (1) has the extension

$$\psi \circ E^{\varphi}_{\mathscr{N}} = \varphi \bar{\otimes} \psi, \tag{2}$$

valid for any normal weight ψ on $\mathscr{N}$.

If φ is an n.s.f. weight on $\mathscr{M}$, then equation (2) means that

$$E^{\varphi}_{\mathscr{N}} = \varphi \bar{\otimes} \iota_{\mathscr{N}}, \tag{3}$$

where the tensor product is defined as in Corollary 12.8.

12.19. Notes. The results contained in this section are due to Haagerup (1979a). Proposition 12.12 contains an improvement due to Connes (1980) and Proposition 12.14 is a classical result.

For our exposition, we have used Connes (1980) and Haagerup (1979a).

CHAPTER III

Groups of Automorphisms

13 Groups of Isometries on Banach Spaces

In this section, we describe the general framework for the spectral analysis of groups of isometrics on Banach spaces.

13.1. Let $\mathscr{X}$ be a Banach space and $\mathscr{X}_* \subset \mathscr{X}^*$ a closed linear subspace. Besides the norm topologies, we shall also consider the weak topologies $w = \sigma(\mathscr{X}, \mathscr{X}_*)$ on $\mathscr{X}$ and $w_* = \sigma(\mathscr{X}_*, \mathscr{X})$ on $\mathscr{X}_*$.

Consider the following conditions on the pair $(\mathscr{X}, \mathscr{X}_*)$:

$$\|x\| = \sup\{|\rho(x)|; \rho \in \mathscr{X}_*, \|\rho\| \leq 1\} \text{ for every } x \in \mathscr{X}; \tag{$1_{\mathscr{X}}$}$$

$$\text{if } \mathscr{K} \subset \mathscr{X} \text{ is } w\text{-compact, then } \overline{\text{co}}^{w}(\mathscr{K}) \subset \mathscr{X} \text{ is also } w\text{-compact;} \tag{$2_{\mathscr{X}}$}$$

$$\text{if} \mathscr{K} \subset \mathscr{X}_* \text{ is } w_*\text{-compact, then } \overline{\text{co}}^{w_*}(\mathscr{K}) \subset \mathscr{X}_* \text{ is also } w_*\text{-compact.} \tag{$3_{\mathscr{X}}$}$$

Lemma 1. *Let $(\mathscr{X}, \mathscr{X}_*)$ be a pair-satisfying conditions $(1_{\mathscr{X}}), (2_{\mathscr{X}})$ and μ a bounded regular Borel measure on a separable locally compact Hausdorff space S. For every w-continuous norm-bounded function $x(\cdot) : S \to \mathscr{X}$, there exists a unique element $x \in \mathscr{X}$ such that*

$$\rho(x) = \int_S \rho(x(s))\, \mathrm{d}\mu(s) \quad (\rho \in \mathscr{X}_*). \tag{1}$$

Proof. The equation

$$f(\rho) = \int_S \rho(x(s))\, \mathrm{d}\mu(s) \quad (\rho \in \mathscr{X}_*)$$

defines a linear form f on $\mathscr{X}_*$. To prove the lemma it is sufficient to show that f is $\sigma(\mathscr{X}_*, \mathscr{X})$-continuous or, equivalently, that f is continuous with respect to the Mackey topology $\tau(\mathscr{X}_*, \mathscr{X})$. Thus, we have to show that there exist an absolutely convex w-compact set $\mathscr{L} \subset \mathscr{X}$ and $c > 0$ such that

$$|f(\rho)| \leq c \sup\{|\rho(x)|; x \in \mathscr{L}\} \quad (\rho \in \mathscr{X}_*). \tag{2}$$

We first assume that $C = supp\ \mu$ is compact. Then we have

$$|f(\rho)| \leq \|\mu\| \sup\{|\rho(x(s))|; s \in C\} \quad (\rho \in \mathscr{X}_*). \tag{3}$$

Since C is compact and $x(\cdot)$ is w-continuous, the set $x(C) \subset \mathscr{X}$ is w-compact. Then the set $\mathscr{K} = \{\lambda \cdot x(s); s \in C, |\lambda| = 1\}$ is w-compact, and hence $\mathscr{L} = \overline{co}^w(\mathscr{K})$ is absolutely convex and w-compact by condition $(2_{\mathscr{X}})$. Since $x(C) \subset \mathscr{L}$, inequality (2) follows from (3).

In the general case, there exists an increasing sequence $\{C_n\}_{n\geq 1}$ of compact subsets of S such that $|\mu|(S\backslash C_n) \to 0$. By the first part or the proof, there exist $x_n \in \mathscr{X}$ such that

$$\rho(x_n) = \int_{C_n} \rho(x(s))\, \mathrm{d}\mu(s) \quad (\rho \in \mathscr{X}_*, n \geq 1).$$

Then, for every $\rho \in \mathscr{X}_*$, we have

$$|\rho(x_n) - f(x)| \leq \|\rho\| \sup\{\|x(s)\|; s \in S\} |\mu|(S\backslash C_n) \to 0 \tag{4}$$

and using condition $(1_{\mathscr{X}})$ it follows that $\{x_n\}_{n\geq 1}$ is a Cauchy sequence in $\mathscr{X}$. If $x \in \mathscr{X}$ is the limit of this sequence, it then follows from (4) that $\rho(x) = f(\rho)$ for all $\rho \in \mathscr{X}_*$.

The uniqueness of the element x satisfying (1) follows obviously using $(1_{\mathscr{X}})$. □

The unique element $x \in \mathscr{X}$ satisfying (1) will be denoted by

$$x = \int_S x(s)\, \mathrm{d}\mu(s).$$

Consider now two pairs $(\mathscr{X}, \mathscr{X}_*)$ and $(\mathscr{Y}, \mathscr{Y}_*)$ satisfying conditions $(1_{\mathscr{X}}), (2_{\mathscr{X}})$ and $(1_{\mathscr{Y}}), (2_{\mathscr{Y}})$. Let $\mathscr{B}(\mathscr{X}, \mathscr{Y})$ be the Banach space or all bounded linear operators $\mathscr{X} \to \mathscr{Y}$ and $\mathscr{B}_w(\mathscr{X}, \mathscr{Y})$ the linear space or all w-continuous linear operators $\mathscr{X} \to \mathscr{Y}$. Using the Banach–Steinhauss theorem, it is easy to check that $\mathscr{B}_w(\mathscr{X}, \mathscr{Y}) \subset \mathscr{B}(\mathscr{X}, \mathscr{Y})$ is a norm-closed linear subspace. In particular, $\mathscr{B}_w(\mathscr{X}, \mathscr{Y})$ is a Banach space. For $\rho \in \mathscr{Y}_*$ and $x \in \mathscr{X}$ define a bounded linear form $\rho(\cdot x)$ on $\mathscr{B}_w(\mathscr{X}, \mathscr{Y})$ by

$$\rho(\cdot x)(T) = \rho(Tx) \quad (T \in \mathscr{B}_w(\mathscr{X}, \mathscr{Y}))$$

and define the norm-closed linear subspace $\mathscr{B}_w(\mathscr{X}, \mathscr{Y})_*$ by

$$\mathscr{B}_w(\mathscr{X}, \mathscr{Y})_* = \overline{\text{lin}}\{\rho(\cdot x); \rho \in \mathscr{Y}_*, x \in \mathscr{X}\} \subset \mathscr{B}_w(\mathscr{X}, \mathscr{Y})^*.$$

Lemma 2. *In the above situation, if the pair $(\mathscr{X}, \mathscr{X}_*)$ also satisfies condition $(3_{\mathscr{X}})$, then the pair $(\mathscr{B}_w(\mathscr{X}, \mathscr{Y}), \mathscr{B}_w(\mathscr{X}, \mathscr{Y})_*)$ satisfies conditions $(1_{\mathscr{B}_w(\mathscr{X},\mathscr{Y})})$, $(2_{\mathscr{B}_w(\mathscr{X},\mathscr{Y})})$.*

Proof. Let $T \in \mathscr{B}_w(\mathscr{X}, \mathscr{Y})$. By $(1_{\mathscr{Y}})$ we have $\|Tx\| = \sup\{|\rho(Tx)|; \rho \in \mathscr{Y}_*, \|\rho\| \leq 1\}$ for all $x \in \mathscr{X}$ and hence $\|T\| = \sup\{|\rho(Tx)|; \rho \in \mathscr{Y}_*, x \in \mathscr{X}, \|\rho\| \leq 1, \|x\| \leq 1\} \leq \sup\{|\varphi(T)|; \varphi \in \mathscr{B}_w(\mathscr{X}, \mathscr{Y})_*, \|\varphi\| \leq 1\| \leq \|T\|$. This proves that condition $(1_{\mathscr{B}_w(\mathscr{X},\mathscr{Y})})$ is satisfied.

Note that each separately (w, w_*)-continuous bilinear form $\mathscr{X} \times \mathscr{Y}_* \ni (x, \rho) \mapsto \langle x, \rho \rangle \in \mathbb{C}$ defines a unique element $T \in \mathscr{B}_w(\mathscr{X}, \mathscr{Y})$ such that $\rho(Tx) = \langle x, \rho \rangle$ for $x \in \mathscr{X}$ and $\rho \in \mathscr{Y}_*$.

Now let $\mathscr{K} \subset \mathscr{B}_w(\mathscr{X}, \mathscr{Y})$ be a w-compact set and denote by $\mathfrak{T}$ the convex set of all regular Borel probability measures on $\mathscr{K}$. For each $\mu \in \mathfrak{T}$, we can define a bilinear form $\langle ., . \rangle_\mu$ on $\mathscr{X} \times \mathscr{Y}_*$ by

$$\langle x, \rho \rangle_\mu = \int_{\mathscr{K}} \rho(Tx)\, \mathrm{d}\mu(T) \quad (x \in \mathscr{X}, \rho \in \mathscr{Y}_*).$$

We show that $\langle ., . \rangle_\mu$ is separately (w, w_*)-continuous. Let $\rho \in \mathscr{Y}_*$ be fixed. The mapping $\mathscr{B}_w(\mathscr{X}, \mathscr{Y}) \ni T \mapsto \rho \circ T \in \mathscr{X}_*$ is continuous with respect to the w-topology on $\mathscr{B}_w(\mathscr{X}, \mathscr{Y})$ and the w_*-topology on $\mathscr{X}_*$, so that the set $\{\rho \circ T; T \in \mathscr{K}\} \subset \mathscr{X}_*$ is w_*-compact. According to condition $(3_{\mathscr{X}})$, it follows that the w_*-closed absolutely convex envelope $\mathscr{L} \subset \mathscr{X}$ of $\{\rho \circ T; T \in \mathscr{K}\}$ is w_*-compact. We have

$$|\langle x, \rho \rangle_\mu| = \int_{\mathscr{K}} |\rho(Tx)| d|\mu|(T) \leq \|\mu\| \sup\{|\varphi(x)|; \varphi \in \mathscr{L}\}.$$

Thus, the mapping $\mathscr{X} \ni x \mapsto \langle x, \rho \rangle_\mu$ is continuous with respect to the Mackey topology $\tau(\mathscr{X}, \mathscr{X}_*)$ and hence also with respect to the weak topology $w = \sigma(\mathscr{X}, \mathscr{X}_*)$. Similarly, for each fixed $x \in \mathscr{X}$, the mapping $\mathscr{Y}_* \ni \rho \mapsto \langle x, \rho \rangle_\mu$ is w_*-continuous.

Consequently, for every measure $\mu \in \mathfrak{T}$ there exists a unique element $T_\mu \in \mathscr{B}_w(\mathscr{X}, \mathscr{Y})$ such that

$$\rho(T_\mu x) = \int_{\mathscr{K}} \rho(Tx)\, \mathrm{d}\mu(T) \quad (\rho \in \mathscr{Y}_*, x \in \mathscr{X}). \tag{5}$$

Since these equalities are valid in particular for the Dirac measures on $\mathscr{K}$, it follows that $\{T_\mu; \mu \in \mathfrak{T}\}$ is a convex set containing $\mathscr{K}$. Thus, in order to check condition $(2_{\mathscr{B}_w(\mathscr{X},\mathscr{Y})})$, it is sufficient to show that the set $\{T_\mu; \mu \in \mathfrak{T}\} \subset \mathscr{B}_w(\mathscr{X}, \mathscr{Y})$ is w-compact. By the Alaoglu theorem, $\mathfrak{T}$ is a $\sigma(\mathscr{M}(\mathscr{K}), \mathscr{C}(\mathscr{K}))$-compact subset of $\mathscr{M}(\mathscr{K}) = \mathscr{C}(\mathscr{K})^*$, hence it is enough to show that the mapping $\mathfrak{T} \ni \mu \mapsto T_\mu \in \mathscr{B}_w(\mathscr{X}, \mathscr{Y})$ is continuous with respect to the corresponding topologies. Thus, we have to show that for each $F \in \mathscr{B}_w(\mathscr{X}, \mathscr{Y})_*$ the mapping $\mathfrak{T} \ni \mu \mapsto F(T_\mu)$ is $\sigma(\mathscr{M}(\mathscr{K}), \mathscr{C}(\mathscr{K}))$-continuous. Indeed, the function $\mathscr{K} \ni T \mapsto F(T)$ belongs to $\mathscr{C}(\mathscr{K})$ and therefore the mapping $\mathfrak{T} \ni \mu \mapsto \int_{\mathscr{K}} F(T)\, \mathrm{d}\mu(T)$ is $\sigma(\mathscr{M}(\mathscr{K}), \mathscr{C}(\mathscr{K}))$-continuous, while from (5) it follows that $F(T_\mu) = \int_{\mathscr{K}} F(T)\, \mathrm{d}\mu(T)$. □

Note that

$$x = \int_S x(s)\, \mathrm{d}\mu(s) \in \mathscr{X}, T \in \mathscr{B}_w(\mathscr{X}, \mathscr{Y}) \Rightarrow Tx = \int_S Tx(s)\, \mathrm{d}\mu(s) \in \mathscr{Y}. \tag{6}$$

13.2. Let $(\mathscr{X}, \mathscr{X}_*)$ be a pair-satisfying conditions $(1_{\mathscr{X}}), (2_{\mathscr{X}})$ and let G be a separable locally compact group with neutral element $e \in G$. A *continuous representation* of G on $\mathscr{X}$ is a mapping $U : G \to \mathscr{B}_\omega(\mathscr{X})$ such that $U_e = \iota_{\mathscr{X}}, U_{st} = U_s U_t$ and $\|U_t\| = 1$ for $s, t \in G$, and the functions

$$G \ni t \mapsto \rho(U_t x) \in \mathbb{C} \quad (x \in \mathscr{X}, \rho \in \mathscr{X}_*)$$

are continuous.

Sometimes it is necessary to impose stronger continuity conditions, such as the norm-continuity of the functions

$$G \ni t \mapsto U_t x \in \mathscr{X} \quad (x \in \mathscr{X}) \tag{C_U}$$

or the norm-continuity of the functions

$$G \ni t \mapsto \rho \circ U_t \in \mathscr{X}_* \quad (x \in \mathscr{X}_*). \tag{C_U^*}$$

Let $\mathscr{M}(G)$ be the convolution Banach algebra of bounded regular Borel measures on G; the $*$-subalgebra of those measures that are absolutely continuous with respect to the Haar measure can be identified with $\mathscr{L}^1(G)$.

Let $\mu \in \mathscr{M}(G)$. For each $x \in \mathscr{X}$, the function $G \ni t \mapsto U_t x \in \mathscr{X}$ is norm-bounded and w-continuous, so that Lemma 1/13.1 assures us that there is a well-defined element $U_\mu x$ such that $U_\mu x = \int_G U_t x \,\mathrm{d}\mu(t)$. We thus obtain an element $U_\mu \in \mathscr{B}(\mathscr{X})$ with $\|U_\mu\| \leq \|\mu\|$.

The mapping $\mathscr{M}(G) \ni \mu \mapsto U_\mu \in \mathscr{B}(\mathscr{X})$ is a Banach algebra homomorphism, in particular we have $U_{\mu * \nu} = U_\mu U_\nu (\mu, \nu \in \mathscr{M}(G))$. Indeed, for any $x \in \mathscr{X}$ we have

$$\begin{aligned} U_\mu U_\nu x &= \int U_s U_\nu x \,\mathrm{d}\mu(s) = \int U_s \left(\int U_t x \,\mathrm{d}\nu(t) \right) \mathrm{d}\mu(s) \\ &= \int \int U_{st} x \,\mathrm{d}\mu(s)\,\mathrm{d}\nu(t) = \int U_r x \,\mathrm{d}(\mu * \nu)(r) = U_{\mu * \nu} x. \end{aligned}$$

For the Dirac measures δ_t, we obviously have $U_{\delta_t} = U_t (t \in G)$.

On the other hand, the set $\{U_f x; f \in \mathscr{L}^1(G), x \in \mathscr{X}\}$ is w-dense in X. More precisely, we have the following result:

Lemma. *Let $\{V_i\}_{i \in I}$ be a fundamental system of neighborhoods of the neutral element of G. For each $i \in I$, let f_i be a positive continuous function on G with compact support supp $f_i \subset V_i$ and $\int f_i(t) dt = 1$. Then $U_{f_i} x \underset{i \in I}{\overset{w}{\rightarrow}} x$ for all $x \in \mathscr{X}$.*

Proof. Let $x \in \mathscr{X}, \rho \in \mathscr{X}_*$, and $\varepsilon > 0$. Since U is a continuous representation, there exists $i \in I$ such that $|\rho(U_t x - x)| < \varepsilon$ for $t \in V_i$. Then $|\rho(U_{f_i} x - x)| = |\int_{V_i} \rho(U_t x - x) f_i(t)\,\mathrm{d}t| \leq \varepsilon$. □

Note that the strong continuity condition (C_U) implies that $\|U_{f_i} x - x\| \underset{i \in I}{\rightarrow} 0$ for all $x \in \mathscr{X}$.

If the pair $(\mathscr{X}, \mathscr{X}_*)$ also satisfies condition $(3_{\mathscr{X}})$, then $U_\mu \in \mathscr{B}_w(\mathscr{X})$ for $\mu \in \mathscr{M}(G)$.

Indeed, for any $\rho \in \mathscr{X}_*$, the function $G \ni t \mapsto \rho \circ U_t \in \mathscr{X}_*$ is norm-bounded and w_*-continuous. Condition $(3_{\mathscr{X}})$ shows that we can apply Lemma 1/13.1 to the pair $(\mathscr{X}_*, \mathscr{X})$ and obtain an element $\rho' \in \mathscr{X}_*$ such that $\rho'(x) = \int \rho(U_t x)\,\mathrm{d}\mu(t) = \rho(U_\mu x)$ for $x \in \mathscr{X}$. It follows that $\rho \circ U_\mu = \rho' \in \mathscr{X}_*$ for each $\rho \in \mathscr{X}_*$, that is, $U_\mu \in \mathscr{B}_w(\mathscr{X})$.

13.3 Lemma. *Consider two pairs $(\mathscr{X}, \mathscr{X}_*)$ and $(\mathscr{Y}, \mathscr{Y}_*)$ satisfying conditions $(1_{\mathscr{X}}), (2_{\mathscr{X}}), (3_{\mathscr{X}})$ and $(1_{\mathscr{Y}}), (2_{\mathscr{Y}})$. Consider also two continuous representations $U : G \rightarrow \mathscr{B}_w(\mathscr{X})$ and $V : G \mapsto \mathscr{B}_w(\mathscr{Y})$ of the separable locally compact group G on $\mathscr{X}$ and $\mathscr{Y}$, respectively, and assume that at least one of the strong continuity conditions (C_U) or (C_V^*) holds. Then the equation*

$$\mathfrak{S}_s T = V_s T U_s^{-1} \quad (T \in \mathscr{B}_w(\mathscr{X}, \mathscr{Y}), s \in G)$$

defines a continuous representation $\mathfrak{S} : G \mapsto \mathscr{B}_w(\mathscr{B}_w(\mathscr{X}, \mathscr{Y}))$ and for every $\mu \in \mathscr{M}(G)$ we have $\mathfrak{S}_\mu \in \mathscr{B}_w(\mathscr{B}_w(\mathscr{X}, \mathscr{Y}))$.

Proof. By Lemma 2/13.1, the pair $(\mathscr{B}_w(\mathscr{X},\mathscr{Y}),\mathscr{B}_w(\mathscr{X},\mathscr{Y})_*)$ satisfies conditions $(1_{\mathscr{B}_w(\mathscr{X},\mathscr{Y})})$, $(2_{\mathscr{B}_w(\mathscr{X},\mathscr{Y})})$. It is clear that $\mathfrak{S}_e$ is the identity mapping, $\mathfrak{S}_{st} = \mathfrak{S}_s\mathfrak{S}_t$, $\|\mathfrak{S}_s\| \le \|V_s\|\|U_s^{-1}\| = 1$ and $\mathfrak{S}_s \in \mathscr{B}_w(\mathscr{B}_w(\mathscr{X},\mathscr{Y}))$ for $s,t \in G$. To check the continuity of $\mathfrak{S}$, we must show that for each $T \in \mathscr{B}_w(\mathscr{X},\mathscr{Y})$ and each $F \in \mathscr{B}_w(\mathscr{X},\mathscr{Y})_*$ the function $G \ni s \mapsto F(\mathfrak{S}_s T)$ is continuous. Clearly, we may assume that $F = \rho(\cdot x)$ with $\rho \in \mathscr{Y}_*$ and $x \in \mathscr{X}$. Then $F(\mathfrak{S}_s T) = \rho(V_s T U_s^{-1}x) = (\rho \circ V_s)(TU_{s^{-1}}x)$. Using, for instance, condition (C_V^*), for $s \to s_0$ we obtain

$$\begin{aligned}&\|F(\mathfrak{S}_s T) - F(\mathfrak{S}_{s_0} T)\|\\ &\quad\le \|\rho\circ V_s - \rho\circ V_{s_0}\|\|T\|\|x\| + |(\rho\circ V_{s_0})(TU_{s^{-1}}x - TU_{s_o^{-1}}x)| \to 0.\end{aligned}$$

Consider now $\mu \in \mathscr{M}(G)$. In order to check that $\mathfrak{S}_\mu \in \mathscr{B}_w(\mathscr{B}_w(\mathscr{X},\mathscr{Y}))$, we have to show that $F\circ\mathfrak{S}_\mu \in \mathscr{B}_w(\mathscr{X},\mathscr{Y})_*$ for any $F \in \mathscr{B}_w(\mathscr{X},\mathscr{Y})_*$, but it is sufficient to consider only $F = \rho(\cdot x)$ with $\rho \in \mathscr{Y}_*$ and $x \in \mathscr{X}$. Then

$$(F\circ\mathfrak{S}_\mu)(T) = \int (\rho\circ V_s)(TU_s^{-1}x)\,\mathrm{d}\mu(s).$$

Assume first that *supp* $\mu = C$ is compact and let $\varepsilon > 0$. Using, for instance, the strong continuity condition (C_V^*), it follows that the set $\{\rho\circ V_s; s \in C\} \subset \mathscr{Y}_*$ is norm-compact, so that there exists a finite set $\{\rho_1,\dots,\rho_n\} \subset \mathscr{Y}_*$ such that the union of the sets $E_k = \{s \in G; \|\rho\circ V_s - \rho_k\| \le \varepsilon\}$, $(k = 1,\dots,n)$, contains $\{\rho\circ V_s; s \in C\}$. Then $S_1 = E_1$ and $S_k = E_k\backslash(E_1 \cup \cdots \cup E_{k-1})(k = 2,\dots,n)$ are mutually disjoint Borel sets and $\bigcup_k S_k = \bigcup_k E_k \supset \{\rho\circ V_s; s \in C\}$. With $x_k = \int_{S_k} U_s^{-1}x\,\mathrm{d}\mu(s) \in \mathscr{X}$, $(k = 1,\dots,n)$, it is easy to see that

$$\left\|F\circ\mathfrak{S}_\mu - \sum_{k=1}^{n} \rho_k(\cdot x_k)\right\| \le \varepsilon\|x\|\|\mu\|.$$

Since $x > 0$ was arbitrary, it follows that $F\circ\mathfrak{S}_\mu \in \mathscr{B}_w(\mathscr{X},\mathscr{Y})_*$.

In the general case, there exists an increasing sequence $\{C_n\}_{n\ge1}$ of compact sets in G with $|\mu|(G\backslash C_n) \to 0$. By the preceding paragraph, there exist $F_n \in \mathscr{B}_w(\mathscr{X},\mathscr{Y})_*$ such that $F_n(T) = \int_{C_n} F(\mathfrak{S}_s T)\,\mathrm{d}\mu(s)(n \ge 1)$. It follows that the sequence $\{F_n\}_{n\ge1}$ is norm-convergent to $F\circ\mathfrak{S}_\mu$, and so $F\circ\mathfrak{S}_\mu \in \mathscr{B}_w(\mathscr{X},\mathscr{Y})_*$. □

13.4. Let $\mathscr{X}$ be a Banach space and $\mathscr{X}_* = \mathscr{X}^*$ Then the pair $(\mathscr{X},\mathscr{X}^*)$ satisfies condition $(1_{\mathscr{X}})$ by the Hahn–Banach theorem, and condition $(2_{\mathscr{X}})$ by a theorem of Krein and Smulian (Dunford & Schwartz, 1958, 1963, V.6.4); condition $(3_{\mathscr{X}})$ is an obvious consequence of the Alaoglu theorem.

In this case, any continuous representation $U : G \to \mathscr{B}_w(\mathscr{X})$ also satisfies the strong continuity condition (C_U). More precisely, we have the following result:

Lemma. *Let $U : G \to \mathscr{B}_w(\mathscr{X})$ be a homomorphism of the locally compact group G into the group of all bounded linear bijections on the Banach space $\mathscr{X}$. The following statements are equivalent:*

(i) *the mappings $G \ni t \mapsto \rho(U_t x) \in \mathbb{C}$ are continuous for all $x \in \mathscr{X}, \rho \in \mathscr{X}^*$;*
(ii) *the mappings $G \ni\mapsto U_t x \in \mathscr{X}$ are norm-continuous for all $x \in \mathscr{X}$;*
(iii) *the mapping $G \times \mathscr{X} \ni (t,x) \mapsto U_t x \in \mathscr{X}$ is norm-continuous.*

Proof. It is clear that (iii) $\Rightarrow$ (ii) $\Rightarrow$ (i).

(i) $\Rightarrow$ (ii). Using the Banach–Steinhauss theorem, from (i) we infer that $\lambda_K = \sup\{\|U_t\|; t \in K\} < +\infty$ for every compact set $K \subset G$. The set $\mathscr{S}$ of those elements $x \in \mathscr{X}$ such that the mapping $G \ni t \mapsto U_t x \in \mathscr{X}$ is norm-continuous is a norm-dense linear subspace of $\mathscr{X}$ and so also $\sigma(\mathscr{X}, \mathscr{X}_*)$-closed. On the other hand, for $x \in \mathscr{X}, f \in \mathscr{L}^1(G)$ with support contained in a compact neighborhood K of 0 in G and $t \in G, t \to e$, we have (see Hewitt & Ross, 1963, 1970, 20.4):

$$\begin{aligned}\|U_t(U_f x) - U_f x\| &= \left\|U_t \int f(s) U_s x \,\mathrm{d}s - \int f(s) U_s x \,\mathrm{d}s\right\| \\ &= \left\|\int f(s) U_{ts} x \,\mathrm{d}s - \int f(s) U_s x \,\mathrm{d}s\right\| \\ &= \left\|\int f(t^{-1}s) U_s x \,\mathrm{d}s - \int f(s) U_s x \,\mathrm{d}s\right\| \\ &\le \lambda_K \|x\| \int |f(t^{-1}s) - f(s)| \,\mathrm{d}s \to 0.\end{aligned}$$

Thus, $\mathscr{S}$ contains the set $\{U_f x; x \in \mathscr{X}, f \in \mathscr{L}^1(G), \mathit{supp}\, f\ \mathit{compact}\}$ which, by (i) and Lemma 13.2, is $\sigma(\mathscr{X}, \mathscr{X}^*)$-dense in $\mathscr{X}$. Hence $\mathscr{S} = \mathscr{X}$.

(ii) $\Rightarrow$ (iii). Let $K \subset G$ be a compact set. From (ii) it follows that $\lambda_K = \sup\{\|U_t\|; t \in K\} < +\infty$. If $K \ni t_n \mapsto t \in G$ and $x_n \to x$ in $\mathscr{X}$, then, again by assumption (ii), we get

$$\begin{aligned}\|U_{t_n} x_n - U_t x\| &\le \|U_{t_n}(x_n - x)\| + \|U_{t_n} x - U_t x\| \\ &\le \lambda_K \|x_n - x\| + \|U_{t_n} x - U_t x\| \to 0.\end{aligned}$$

□

13.5. Let $\mathscr{M}$ be a W^*-algebra with predual $\mathscr{M}_* \subset \mathscr{M}^*$. Then $\mathscr{M} = (\mathscr{M}_*)^*$, condition $(1_{\mathscr{M}})$ is clearly satisfied for the pair $(\mathscr{M}, \mathscr{M}_*)$, $(2_{\mathscr{M}})$ follows from the Alaoglu theorem and $(3_{\mathscr{M}})$ follows from the Krein–Šmulian theorem (Dunford & Schwartz, 1958, 1963, V.6.4). The topology $w = \sigma(\mathscr{M}, \mathscr{M}_*)$ is just the usual w-topology on $\mathscr{M}$. Recall ([L], C.5.1) that on $\mathscr{M}$ we can consider also the topologies s, s^*, as well as the Mackey topology τ_w associated with the w-topology.

A representation of the locally compact group G by $*$-automorphisms of $\mathscr{M}$, or an action (2.24) of G on $\mathscr{M}$, is a group homomorphism $\sigma : G \to \mathit{Aut}(\mathscr{M})$. The weak continuity condition (13.2) for such a representation is just the continuity of the mapping σ with respect to the p-topology (2.23) on $\mathit{Aut}(\mathscr{M})$, while the strong continuity condition (C^*_σ) amounts to continuity with respect to the u-topology (2.23) on $\mathit{Aut}(\mathscr{M})$. Actually, these two conditions are equivalent, as the following shows.

Proposition. *Let $\sigma : G \to \mathit{Aut}(\mathscr{M})$ be an action of the locally compact group G on the W^*-algebra $\mathscr{M}$. The following statements are equivalent:*

(i) *the mappings $G \ni t \mapsto \sigma_t(x) \in \mathscr{M}$ are w-continuous $(x \in \mathscr{M})$;*
(ii) *the mappings $G \ni t \mapsto \sigma_t(x) \in \mathscr{M}$ are s-continuous $(x \in \mathscr{M})$;*
(iii) *the mappings $G \ni t \mapsto \sigma_t(x) \in \mathscr{M}$ are s^*-continuous $(x \in \mathscr{M})$;*
(iv) *the mappings $G \ni t \mapsto \sigma_t(x) \in \mathscr{M}$ are τ_w-continuous $(x \in \mathscr{M})$;*
(v) *for every $\sigma(\mathscr{M}_*, \mathscr{M})$-compact subset $\mathscr{L}$ of $\mathscr{M}_*$, the mapping $G \times \mathscr{L} \ni (t, \varphi) \mapsto \varphi \circ \sigma_t \in \mathscr{M}_*$ is continuous with respect to the $\sigma(\mathscr{M}_*, \mathscr{M})$-topology on $\mathscr{L}$ and $\mathscr{M}_*$;*
(vi) *the mappings $G \ni t \mapsto \varphi \circ \sigma_t \in \mathscr{M}_*$ are norm-continuous $(\varphi \in \mathscr{M}_*)$;*
(vii) *the mapping $G \times \mathscr{M}_* \ni (t, \varphi) \mapsto \varphi \circ \sigma_t \in \mathscr{M}_*$ is norm-continuous.*

Proof. It is clear that (iv) ⇒ (iii) ⇒ (ii) ⇒ (i) and it is easy to check that (iv) ⇒ (v) ⇒ (i). Since $(\mathcal{M}_*)^* = \mathcal{M}$, it follows from Lemma 13.4 that (i)⇔ (vi) ⇔ (vii).

By a result due to Akemann ((1967); Strătilă & Zsidó, 1977, 2005, Cor. 8.17) the restriction of the Mackey topology τ_w to the closed unit ball of $\mathcal{M}$ coincides with the restriction of the s^*-topology; since $\|\sigma_t(x)\| = \|x\|$, it follows that (iii) ⇔ (iv).

For $x \in \mathcal{M}, \varphi \in \mathcal{M}_*, t \in G$, we have $\varphi((\sigma_t(x) - x)^*(\sigma_t(x) - x)) = \varphi(\sigma_t(x^*x)) - \varphi(\sigma_t(x)^*x) - \varphi(x^*\sigma_t(x)) + \varphi(x^*x)$. If $t \in G$ converges to $e \in G$, it follows from (i) that the right hand side of this equation converges to 0. Thus, (i) ⇒ (ii) and (i) ⇒ (iii).

Recall (2.24) that if σ satisfies the equivalent conditions of the above proposition, we say that σ is a continuous action of G on $\mathcal{M}$. □

13.6. Notes. The exposition in this section is based on the article of Arveson (1974).

14 Spectra and Spectral Subspaces

In this section we introduce the spectral subspaces associated to a continuous representation of a locally compact abelian group on a Banach space, together with their main properties and some applications.

14.1. Let G be a locally compact abelian group with dual group $\hat{G}$. For $t \in G$ and $\gamma \in \hat{G}$ we shall denote by $\langle t, \gamma \rangle$ the value of the character γ at t. We shall denote the group operation by addition; in particular, 0 will denote the neutral element of G or of $\hat{G}$.

For $\mu \in \mathcal{M}(G), f \in \mathcal{L}^1(G)$, we define the Fourier transforms $\hat{\mu}, \hat{f}$ by

$$\hat{\mu}(\gamma) = \int \langle t, \gamma \rangle \, \mathrm{d}\mu(t), \ \hat{f}(\gamma) = \int f(t)\langle t, \gamma \rangle \, \mathrm{d}t \quad (\gamma \in \hat{G}).$$

For $f \in \mathcal{L}^1(G)$, we write $Z(f) = \{\gamma \in \hat{G}; \hat{f}(\gamma) = 0\}$ and for an ideal $\mathcal{J}$ of $\mathcal{L}^1(G)$ define its "hull" to be the closed set $Z(\mathcal{J}) = \cap\{Z(f); f \in \mathcal{J}\} \subset \hat{G}$. Clearly, $Z(\bar{\mathcal{J}}) = Z(\mathcal{J})$. Recall the following important result:

The Maximal Tauberian Theorem. *Let $f \in \mathcal{L}^1(G)$ and let $\mathcal{J} \subset \mathcal{L}^1(G)$ be a closed ideal such that $Z(\mathcal{J}) \subset Z(f)$. If the intersection of the boundaries of the sets $Z(\mathcal{J})$ and $Z(f)$ does not contain any nonempty perfect set, then $f \in \mathcal{J}$.*

In particular, for a closed ideal $\mathcal{J} \subset \mathcal{L}^1(G)$ we have

$$f \in \mathcal{L}^1(G),\ Z(\mathcal{J}) \subset int\, Z(f) \Rightarrow f \in \mathcal{J}. \tag{1}$$

$$Z(\mathcal{J}) = \emptyset \Rightarrow \mathcal{J} = \mathcal{L}^1(G). \tag{2}$$

$$Z(\mathcal{J}) = \{\gamma\} \Rightarrow \mathcal{J} = \{f \in \mathcal{L}^1(G); \hat{f}(\gamma) = 0\}. \tag{3}$$

Let $F \subset \hat{G}$ be a closed set. Then the set of all closed ideals $\mathcal{J} \subset \mathcal{L}^1(G)$ with $Z(\mathcal{J}) = F$ has a greatest element

$$\mathcal{K}(F) = \{f \in \mathcal{L}^1(G); F \subset Z(f)\}$$

and a smallest element $\mathscr{I}(F)$ which is the closure of each of the following ideals:

$$\mathscr{I}_0(F) = \{f \in \mathscr{L}^1(G); F \subset int\, Z(f)\}$$
$$\mathscr{I}_{00}(F) = \{f \in \mathscr{L}^1(G);\ supp\, \hat{f}\, compact, F \subset int\, Z(f)\}.$$

In particular, the ideal $\{f \in \mathscr{L}^1(G);\ supp\, \hat{f}\, compact\}$ is dense in $\mathscr{L}^1(G)$.

Note that *every dense ideal* $\mathscr{I}$ *of* $\mathscr{L}^1(G)$ *contains an approximate unit of* $\mathscr{L}^1(G)$ *with elements of norm* ≤ 1. Indeed, if $\{f_i\}_{i\in I}$ is an approximate unit of $\mathscr{L}^1(G)$ with $\|f_i\|_i = 1$, then, for each $i \in I$ and each $n \geq 1$, there exists an element $h_{i,n} \in \mathscr{I}$ such that $\|f_i - h_{i,n}\|_1 < 1/n$ and the family $\{h_{i,n}/\|h_{i,n}\|_1\}_{i,n} \subset \mathscr{I}$ is an approximate unit of $\mathscr{L}^1(G)$.

Consequently, there exists an approximate unit $\{k_i\}_{i\in I}$ of $\mathscr{L}^1(G)$ with $\|k_i\|_1 \leq 1$ and *supp* $\hat{k}_i$ compact ($i \in I$). In particular,

$$\text{for every } f \in \mathscr{L}^1(G) \text{ and every } \varepsilon > 0 \text{ there exists } k \in \mathscr{L}^1(G) \text{ with } \|k\|_1 \leq 1 \text{ and supp } \hat{k} \text{ compact such that } \|f - f * k\|_1 < \varepsilon. \tag{4}$$

Moreover,

$$\text{for every compact set } C \subset \hat{G} \text{ and every open set } D \subset \hat{G} \text{ with } C \subset D \text{ there exists a function } f \in \mathscr{L}^1(G) \text{ with supp } \hat{f} \subset D \text{ such that } \hat{f}(\gamma) = 1 \text{ for any } \gamma \in C. \tag{5}$$

For all the above results, and for various other results in harmonic analysis which we shall use in the sequel, we refer to Hewitt and Ross (1963, 1970); Rudin (1962); Strătilă (1967, 1968).

14.2. Let $(\mathscr{X}, \mathscr{X}_*)$ be a pair consisting of a Banach space $\mathscr{X}$ and a closed linear subspace $\mathscr{X}_* \subset \mathscr{X}^*$ which satisfies conditions $(1_\mathscr{X})$ and $(2_\mathscr{X})$ of Section 13.1. Let $U : G \to \mathscr{B}_w(\mathscr{X})$ be a continuous representation of the locally compact abelian group G on $\mathscr{X}$ such that $U_\mu \in \mathscr{B}_w(\mathscr{X})$ for all $\mu \in \mathscr{M}(G)$; this last condition is implied, for instance, by condition $(3_\mathscr{X})$.

For each element $x \in \mathscr{X}$, we consider the closed ideal

$$\mathscr{I}_x^U = \{f \in \mathscr{L}^1(G); U_f x = 0\} \subset \mathscr{L}^1(G)$$

and define *the spectrum of x with respect to U* by

$$Sp_U(x) = Z(\mathscr{I}_x^U) = \{\gamma \in \hat{G}; f \in \mathscr{L}^1(G), U_f x = 0 \Rightarrow \hat{f}(\gamma) = 0\}.$$

Proposition. *For* $x, y \in \mathscr{X}; \mu, \mu_1, \mu_2 \in \mathscr{M}(G); t \in G; \gamma \in \hat{G}; 0 \neq \lambda \in \mathbb{C}$ *we have*

$$Sp_U(\lambda x) = Sp_U(x) \tag{1}$$
$$Sp_U(x + y) \subset Sp_U(x) \cup Sp_U(y) \tag{2}$$
$$Sp_U(U_\mu x) \subset Sp_U(x) \cap supp\, \hat{\mu} \tag{3}$$
$$Sp_U(U_t x) = Sp_U(x) \tag{4}$$
$$Sp_U(x) = \emptyset \Leftrightarrow x = 0 \tag{5}$$

$$Sp_U(x) = \{\gamma\} \Leftrightarrow x \neq 0 \text{ and } U_s x = (s, \gamma)x \text{ for } s \in G \tag{6}$$

$$Sp_U(x) = \{0\} \Leftrightarrow U_s x = x \neq 0 \text{ for } s \in G \tag{7}$$

$$Sp_U(x) \subset \text{int } \{\omega \in \hat{G}; \hat{\mu}_1(\omega) = \hat{\mu}_2(\omega)\} \Rightarrow U_{\mu_1} x = U_{\mu_2} x \tag{8}$$

$$Sp_U(x) \subset \text{int } \{\omega \in \hat{G}; \hat{\mu}(\omega) = 1\} \Rightarrow U_\mu x = x. \tag{9}$$

Proof. (1) follows from $\mathcal{J}_x^U = \mathcal{J}_{\lambda x}^U$ and (2) from $\mathcal{J}_x^U \cap \mathcal{J}_y^U \subset \mathcal{J}_{x+y}^U$.

Let us prove (3). If $\gamma \notin Sp_U(x)$, there exists $f \in \mathcal{L}^1(G)$ with $U_f x = 0$ and $\hat{f}(\gamma) \neq 0$, so that $U_f U_\mu x = U_\mu U_f x = 0$ and $\hat{f}(\gamma) \neq 0$, hence $\gamma \notin Sp_U(U_\mu x)$. If $\gamma \neq$ *supp* $\hat{\mu}$, there exists $f \in \mathcal{L}^1(G)$ with $\hat{f}(\gamma) = 1$ and *supp* $\hat{f} \cap$ *supp* $\hat{\mu} = \emptyset$ (see 14.1.(5)); we have $(f * \mu)\hat{} = \hat{f}\hat{\mu} = 0$, hence $f * \mu = 0$ and $U_f U_\mu x = U_{f*\mu} x = 0$, but $\hat{f}(\gamma) \neq 0$, so that $\gamma \notin Sp_U(U_\mu x)$.

Equation (4) follows from (3) since for the Dirac measure δ_t we have $\hat{\delta}_t(\gamma) = \langle t, \gamma\rangle, (\gamma \in \hat{G})$.

Let us prove (5). If $x \neq 0$, then by Lemma 13.2 there exists $f \in \mathcal{L}^1(G)$ with $U_f x \neq 0$, hence $\mathcal{J}_x^U \neq \mathcal{L}^1(G)$ and using 14.1.(2) we see that $Sp_U(x) \neq \emptyset$. If $x = 0$, then $\mathcal{J}_x^U = \mathcal{L}^1(G)$ and therefore (14.1.(5)) $Sp_U(x) = \emptyset$.

Let us prove (6). If $Sp_U(x) = \{\gamma\}$, then (14.1.(3)) $\mathcal{J}_x^U = \{f \in \mathcal{L}^1(G); \hat{f}(\gamma) = 0\}$, that is, for $f \in \mathcal{L}^1(G)$ we have $U_f x = 0 \Leftrightarrow \hat{f}(\gamma) = 0$. Then, for $\mu \in \mathcal{M}(G)$ we get

$$U_\mu x = 0 \Leftrightarrow \hat{\mu}(\gamma) = 0.$$

Indeed, $U_\mu x = 0 \Rightarrow U_f U_\mu x = 0 \Rightarrow U_{f*\mu} x = 0 \Rightarrow (f * \mu)\hat{}(\gamma) = 0 \Rightarrow \hat{f}(\gamma)\hat{\mu}(\gamma) = 0$ for any $f \in \mathcal{L}^1(G)$ and therefore (14.1.(5)) $\hat{\mu}(\gamma) = 0$; conversely, if $\hat{\mu}(\gamma) = 0$, then we obtain similarly $U_f U_\mu x = 0$ for any $f \in \mathcal{L}^1(G)$, and, using Lemma 13.2, we conclude that $U_\mu x = 0$. Since $\mu - \hat{\mu}(\gamma)\delta_0 \in \mathcal{M}(G)$ and $(\mu - \hat{\mu}(\gamma)\delta_0)\hat{}(\gamma) = 0$, it follows that $U_\mu x = \hat{\mu}(\gamma)x$ for all $\mu \in \mathcal{M}(G)$. Taking in particular $\mu = \delta_s$ we obtain $U_s x = \langle s, \gamma\rangle x$ $(s \in G)$. Conversely, if $U_s x = \langle s, \gamma\rangle x$ $(s \in G)$, then $U_f x = \hat{f}(\gamma)x$ for any $f \in \mathcal{L}^1(G)$ and if $x \neq 0$ it follows that $\mathcal{J}_x^U = \{f \in \mathcal{L}^1(G); \hat{f}(\gamma) = 0\}$, hence $Sp_U(x) = \{\gamma\}$.

Equation (7) follows obviously from (6).

Let us prove (8). For $\mu = \mu_1 - \mu_2$ we have $Sp_U(x) \subset$ *int* $\{\omega \in \hat{G}; \hat{\mu}(\omega) = 0\}$ by assumption, hence $Sp_U(x) \cap$ *supp* $\hat{\mu} = \emptyset$. Using (3) and (5) it follows that $U_\mu x = 0$, hence $U_{\mu_1} x = U_{\mu_2} x$.

Finally, (9) follows from (8) since $\hat{\delta}_0(\omega) = 1$ $(\omega \in \hat{G})$. □

In this proof, we have also shown that

$$Sp_U(x) = \{\gamma \in \hat{G}; \mu \in \mathcal{M}(G), U_\mu x = 0 \Rightarrow \hat{\mu}(\gamma) = 0\}. \tag{10}$$

Note that $Sp_U(x)$ is "the support of the vector distribution $\hat{f} \mapsto U_f x$," that is, $\hat{G} \backslash Sp_U(x)$ is the greatest open set $D \subset \hat{G}$ with the property: $f \in \mathcal{L}^1(G)$, *supp* $\hat{f} \subset D \Rightarrow U_f x = 0$.

14.3. We continue with the notation of the previous section.

For each set $E \subset \hat{G}$ define *the spectral subspace* $\mathcal{X}(U; E)$ *of* $\mathcal{X}$ *associated with* U *and* E to be the w-closure of the set

$$\mathcal{X}_0(U; E) = \{x \in \mathcal{X}; Sp_U(x) \subset E\}.$$

Clearly,

$$E_1 \subset E_2 \Rightarrow \mathcal{X}(U;E_1) \subset \mathcal{X}(U;E_2). \tag{1}$$

Proposition. *For every subset $E \subset \hat{G}$, the spectral subspace $\mathcal{X}(U;E)$ is a w-closed U-invariant linear subspace of $\mathcal{X}$ and is equal to the w-closure of the set*

$$\mathcal{X}_{00}(U;E) \subset \{x \in \mathcal{X}; Sp_U(x) \subset E \ compact\}.$$

For every closed subset $F \subset \hat{G}$ we have

$$\mathcal{X}(U;F) = \mathcal{X}_0(U;F) = \{x \in X; \mathcal{I}_0(F) \subset \mathcal{I}_x^U\} = \{x \in X; \mathcal{I}_{00}(F) \subset \mathcal{I}_x^U\}. \tag{2}$$

For every open subset $D \subset \hat{G}, \mathcal{X}(U;D)$ is the w-closure of the set

$$\{U_f x; x \in \mathcal{X}, f \in \mathcal{L}^1(G), \ supp \hat{f} \subset D \ compact\}.$$

If $\{F_i\}$ is any family of closed subsets of $\hat{G}$, then

$$\mathcal{X}(U;\bigcap_i F_i) = \bigcap_i \mathcal{X}(U;F_i) \tag{3}$$

and if $\{D_i\}$ is any family of open subsets of $\hat{G}$, then

$$\mathcal{X}(U;\bigcup_i D_i) = \bigvee_i \mathcal{X}(U;D_i). \tag{4}$$

Proof. By Proposition 14.2, it follows that $\mathcal{X}_0(U;E)$ is a U-invariant linear subspace of $\mathcal{X}$, so that its w-closure $\mathcal{X}(U;E)$ is a w-closed U-invariant linear subspace of $\mathcal{X}$.

By Lemma 13.2 and Proposition 14.2.(3), the set of elements in $\mathcal{X}(U;E)$ of the form $U_f x$ is w-dense in $\mathcal{X}(U;E)$ and hence, using 14.1.(4) and 14.2.(3), it follows that $\mathcal{X}_{00}(U;E)$ is w-dense in $\mathcal{X}(U;E)$. Consequently,

$$\mathcal{X}(U;E) = \overline{\mathcal{X}_0(U;E)}^w = \overline{\mathcal{X}_{00}(U;E)}^w. \tag{5}$$

Consider now a closed set $F \subset \hat{G}$ and $x \in \mathcal{X}_0(U;F)$. If $f \in \mathcal{I}_0(F)$, that is, $F \cap \ supp \hat{f} = \emptyset$, then (14.2.(3), 14.2.(5)) $U_f x = 0$, that is, $f \in \mathcal{I}_x^U$. It follows that $\mathcal{X}_0(U;F) \subset \{x \in \mathcal{X}; \mathcal{I}_0(F) \subset \mathcal{I}_x^U\} \subset \{x \in \mathcal{X}; \mathcal{I}_{00}(F) \subset \mathcal{I}_x^U\}$. Conversely, if $\mathcal{I}_{00}(F) \subset \mathcal{I}_x^U$, then $Sp_U(x) = Z(\mathcal{I}_x^U) \subset Z(\mathcal{I}_{00}(F)) = F$ (14.1), hence $x \in \mathcal{X}_0(U;F)$. Thus, the above inclusions are actually equalities and it follows that

$$\mathcal{X}_0(U;F) = \bigcap \{Ker \ U_f; f \in \mathcal{I}_0(F)\} \tag{6}$$

is w-closed, since the U_f's are w-continuous.

For any set $D \subset \hat{G}$ we have (14.2.(3)) $\{U_f x; x \in \mathcal{X}, supp\, \hat{f} \subset D\ compact\} \subset \mathcal{X}_0(U;D)$. Assume that D is open and let $x \in \mathcal{X}_{00}(U;D)$. Then $K = Sp_U(x) \subset D$ is compact and hence (14.1.(5)) there exists $f \in \mathcal{L}^1(G)$ with $supp\, \hat{f} \subset D$ compact and $K \subset int\, \{\gamma \in \hat{G}; \hat{f}(\gamma) = 1\}$. It follows (14.1.(9)) that $x = U_f x$ with $f \in \mathcal{L}^1(G)$, $supp\, \hat{f} \subset D$, and using (5) we obtain

$$\mathcal{X}(U;D) = \overline{\{U_f x; x \in \mathcal{X}, f \in \mathcal{L}^1(G),\ supp\, \hat{f} \subset D\ compact\}}^{w}. \tag{7}$$

Equation (3) follows easily using (2) and (6).

To prove (4) we note first that $\mathcal{X}(U;D_i) \subset \mathcal{X}(U;\bigcup_i D_i)$, so that the w-closed linear subspace $\bigvee_i \mathcal{X}(U;D_i)$ generated by $\bigcup_i \mathcal{X}(U;D_i)$ is contained in $\mathcal{X}(U;\bigcup_i D_i)$. Conversely, let $x \in \mathcal{X}_{00}(U;\bigcup_i D_i)$. Then $K = Sp_U(x) \subset \bigcup_i D_i$ is compact and there exist $i_1, \ldots, i_n$ such that $K \subset \bigcup_{k=1}^n D_{i_k}$. Using 14.1.(5), we find functions $f_1, \ldots, f_n \in \mathcal{L}^1(G)$ such that $supp \hat{f}_k \subset D_{i_k} (k = 1, \ldots, n)$, and $K \subset int\, \{\gamma \in \hat{G}; \sum_{k=1}^n \hat{f}_k(\gamma) = 1\}$. Then (14.2) $x = \sum_{k=1}^n U_{f_k} x$ with $U_{f_k} x \in \mathcal{X}(U;D_{i_k})$, and hence $x \in \bigvee_i \mathcal{X}(U,D_i)$. □

Let $F \subset \hat{G}$ be a closed set. From (3) and (1), it follows that if $\{D_i\}$ is any family of open sets in $\hat{G}$ such that $F = \bigcap_i D_i = \bigcap_i \bar{D}_i$ then

$$\mathcal{X}(U;F) = \bigcap_i \mathcal{X}(U;D_i) = \bigcap_i \mathcal{X}(U;D_i). \tag{8}$$

Thus, for every fundamental system $\{N_i\}_{i \in I}$ of open and relatively compact neighborhoods of 0 in G with $\bigcap_i \overline{N}_i = \{0\}$, we have

$$\mathcal{X}(U;F) = \bigcap_i \mathcal{X}(U;F + N_i) = \bigcap_i \mathcal{X}(U;F + \overline{N}_i). \tag{9}$$

In particular,

$$\mathcal{X}(U;F) = \bigcap \{\mathcal{X}(U;D); D \supset F \text{ open}\}. \tag{10}$$

On the other hand, by (5), for every set $E \subset \hat{G}$ we have

$$\mathcal{X}(U;E) = \bigvee \{\mathcal{X}(U;K); K \subset E \text{ compact}\}. \tag{11}$$

Equations (10) and (11) prove the "regularity" of the family of spectral subspaces associated with U. We have also (14.2.(5))

$$\mathcal{X}(U;\emptyset) = \{0\}, \mathcal{X}(U;\hat{G}) = X. \tag{12}$$

For $\gamma \in \hat{G}$ the spectral subspace (14.2.(6))

$$\mathcal{X}(U;\{\gamma\}) = \{x \in \mathcal{X}; U_t x = \langle t, \gamma \rangle x \text{ for all } t \in G\} \tag{13}$$

will be called *the eigenspace of U corresponding to the eigenvalue* $\gamma \in \hat{G}$. In particular,

$$\mathscr{X}^U = \mathscr{X}(U; \{0\}) = \{x \in \mathscr{X}; U_t x = x \text{ for all } t \in G\} \tag{14}$$

will be called *the centralizer of U*.

14.4. We shall need the following extension of 14.3.(13):

Lemma. *For each $\gamma \in \hat{G}$, each compact set $K \subset G$, and $\varepsilon > 0$, there exists a compact neighborhood W of 0 in $\hat{G}$ such that*

$$\|U_t x - \langle t, \gamma \rangle x\| \leq \varepsilon \|x\| \textit{ for } x \in \mathscr{X}(U; \gamma + W) \quad (t \in K).$$

Proof. Let W_0 be a compact neighborhood of γ in $\hat{G}$ and $f \in \mathscr{L}^1(G)$ be such that $\hat{f}(W_0) = \{1\}$.
For each $t \in K$, we define a function $f_t \in \mathscr{L}^1(G)$ by

$$f_t(s) = f(s - t) - \langle t, \gamma \rangle f(s) \quad (s \in G).$$

Then $\hat{f}_t(\gamma) = 0$ and hence (Rudin, 1962, 2.6.3) there exist a function $k_t \in \mathscr{L}^1(G)$ and a neighborhood W_t of γ in $\hat{G}$ such that $\hat{k}_t(W_t) = \{1\}$ and $\|f_t * k_t\|_1 < \varepsilon$. Since the mapping $G \ni s \mapsto f_s \in \mathscr{L}^1(G)$ is norm-continuous, there exists a neighborhood N_t of t in G such that $\|f_s * k_t\|_1 < \varepsilon$ for all $s \in N_t$.
Since K is compact, there exist $t_1, \dots, t_n \in K$ such that $K \subset \bigcup_{i=1}^n N_{t_i}$. Let W be a compact neighborhood of γ in $\hat{G}$ such that $W \subset int(W_0 \cap W_{t_1} \cap \dots \cap W_{t_n})$. For each $t \in K$, there exists a function $h_t \in \{k_{t_1}, \dots, k_{t_n}\} \subset \mathscr{L}^1(G)$ such that $W \subset int\{\omega \in \hat{G}; \hat{h}_t(\omega) = 1\}$ and $\|f_t * h_t\}_1 < \varepsilon$.
Let $x \in \mathscr{X}(U; W)$ and $t \in K$. Then $(\delta_t - \langle t, \gamma \rangle \delta_0)\hat{}$ coincides with $(f_t * h_t)\hat{}$ on a neighborhood of $Sp_U(x)$ and hence (14.2.(8)) $\|U_t x - \langle t, \gamma \rangle x\| = \|U_{f_t * h_t} x\| \leq \varepsilon \|x\|$. □

14.5. In the setting of Section 14.2, the set $Ker\ U = \{f \in \mathscr{L}^1(G); U_f = 0\}$ is a closed ideal of $\mathscr{L}^1(G)$. The *spectrum of the representation U* is defined as being the closed set

$$Sp\ U = Z(Ker\ U) = \{\gamma \in \hat{G}; f \in \mathscr{L}^1(G), U_f = 0 \Rightarrow \hat{f}(\gamma) = 0\}.$$

Using 14.1.(5), it is easy to check that for every w-total set $\mathscr{X}_0$ in $\mathscr{X}$ we have

$$Sp\ U = \overline{\bigcup \{Sp_U(x); x \in \mathscr{X}_0\}}. \tag{1}$$

It is also easy to check that $Sp\ U$ is "the support of the vector distribution $\hat{f} \mapsto U_f$," that is, $\hat{G} \setminus Sp\ U$ is the greatest open set D in $\hat{G}$ with the property: $f \in \mathscr{L}^1(G)$, $supp\,\hat{f} \subset D \Rightarrow U_f = 0$.

Proposition. *For $\gamma \in \hat{G}$ the following statements are equivalent:*

(i) $\gamma \in Sp\ U$;
(ii) $\mathscr{X}(U; \gamma + W) \neq \{0\}$ *for every neighborhood W of 0 in* $\hat{G}$;
(iii) *there exists a net* $\{x_i\} \subset \mathscr{X}$, $\|x_i\| = 1$, *such that*

$$\lim_i \|U_t x_i - \langle t, \gamma \rangle x_i\| = 0$$

uniformly for t in compact subsets of G;

(iv) $|\hat{\mu}(\gamma)| \le \|U_\mu\|$ *for all* $\mu \in \mathscr{M}(G)$;
(v) $|\hat{f}(\gamma)| \le \|U_f\|$ *for all* $f \in \mathscr{L}^1(G)$.

Proof. (i) $\Rightarrow$ (ii). Let W be a neighborhood of γ in $\hat{G}$. By 14.1.(5), there exists $f \in \mathscr{L}^1(G)$ with $\hat{f}(\gamma) = 1$ and $supp\,\hat{f} \subset W$. Since $\gamma \in Sp\ U$, we have $U_f \neq 0$ so that $0 \neq U_f x \in \mathscr{X}(U; W)$ for some $x \in X$.

(ii) $\Rightarrow$ (iii). Using Lemma 14.4, we infer, assuming (ii), that for each compact set $K \subset G$ and $\varepsilon > 0$ there exists an element $x_{K,\varepsilon} \in \mathscr{X}$, $\|x_{K,\varepsilon}\| = 1$, such that $\|U_t x_{K,\varepsilon} - \langle t, \gamma\rangle x_{K,\varepsilon}\| < \varepsilon$ for all $t \in K$. Then $\{x_{K,\varepsilon}\}_{K,\varepsilon}$ is the required net.

(iii) $\Rightarrow$ (iv). Let $\mu \in \mathscr{M}(G)$ and $\varepsilon > 0$. There exists a compact set $K \subset G$ with $|\mu|(G\backslash K) < \varepsilon/4$. By assumption (iii) there exists $x \in \mathscr{X}$, $\|x\| \le 1$, such that $\|U_t x - \langle t, \gamma\rangle x\| < \varepsilon/2|\mu|(K)$ for all $t \in K$. Then

$$\begin{aligned}\|U_\mu x - \hat{\mu}(\gamma)x\| &\le \int_G \|U_t x - \langle t,\gamma\rangle x\|\, \mathrm{d}|\mu| \\ &\le \int_K \|U_t x - \langle t,\gamma\rangle x\|\, \mathrm{d}|\mu| + 2|\mu|(G\backslash K) \le \frac{\varepsilon}{2} + \frac{\varepsilon}{2} = \varepsilon.\end{aligned}$$

It follows that $|\hat{\mu}(\gamma)| = \|\hat{\mu}(\gamma)x\| \le \|U_\mu x\| + \varepsilon \le \|U_\mu\| + \varepsilon$. Since $\varepsilon > 0$ was arbitrary, we get $|\hat{\mu}(\gamma)x| \le \|U_\mu\|$.

(iv) $\Rightarrow$ (v). Obvious.

(v) $\Rightarrow$ (i). $f \in \mathscr{L}^1(G)$ and $U_f = 0$, then $\hat{f}(\gamma) = 0$ by (v). Hence $\gamma \in Sp\ U$. □

Corollary. *Let $\mathscr{A}$ be the Banach algebra defined as the norm-closure of the subalgebra $\{U_f; f \in \mathscr{L}^1(G)\}$ of $\mathscr{B}(\mathscr{X})$ and denote by $\Omega_{\mathscr{A}}$ the Gelfand spectrum of $\mathscr{A}$. Then the mapping*

$$\Omega_{\mathscr{A}} \ni \omega \mapsto \omega \circ U \in \Omega_{\mathscr{L}^1(G)} = \hat{G}$$

is a homeomorphism of $\Omega_{\mathscr{A}}$ onto Sp U.

Proof. If ω is a continuous character of $\mathscr{A}$, then the mapping $\gamma_\omega : \mathscr{L}^1(G) \ni f \mapsto \omega(U_f)$ is a continuous character of $\mathscr{L}^1(G)$; hence $\gamma_\omega \in \hat{G}$ and $|f(\gamma_\omega)| = |\omega(U_f)| \le \|U_f\|$ for every $f \in \mathscr{L}^1(G)$. The above proposition shows that $\gamma_\omega \in Sp\ U$. The mapping $\omega \mapsto \gamma_\omega = \omega \circ U$ is clearly continuous. □

If $\gamma \in Sp\ U$, the above proposition shows that the mapping $U_f \mapsto \hat{f}(\gamma)$ can be extended to a continuous character ω_γ of $\mathscr{A}$, which is uniquely determined by the condition $\omega_\gamma(U_f) = \hat{f}(\gamma)$ ($f \in \mathscr{L}^1(G)$). The mapping $\gamma \mapsto \omega_\gamma$ is continuous and its inverse is just the above-defined mapping $\omega \mapsto \gamma_\omega$.

14.6. With the help of the set *Sp U*, we can describe the usual spectrum $Sp(U_t)$ of the operator $U_t \in \mathscr{B}(\mathscr{X})$:

Proposition. $Sp(U_t) = \overline{\{\langle t,\gamma\rangle; \gamma \in Sp\ U\}}$ $(t \in G)$.

Proof. If $\gamma \in Sp\ U$, then, by Proposition 14.5, there exists a net $\{x_i\} \subset \mathcal{X}$, $\|x_i\| = 1$, such that $\|(U_t - \langle t, \gamma\rangle)x_i\| \to 0$, hence $\langle t, \gamma\rangle \in Sp(U_t)$ for $t \in G$.

Since $\|U_t\| = \|U_t^{-1}\| = 1$, we have $Sp(U_t) \subset \mathbb{T} = \{z \in \mathbb{C}; |z| = 1\}$. Assume that there exists $\lambda \in \mathbb{T}$, $\lambda \notin \overline{\{\langle t, \gamma\rangle; \gamma \in Sp\ U\}}$ and consider two open neighborhoods $V \subset W$ of the set $\overline{\{\langle t, \gamma\rangle; \gamma \in Sp\ U\}}$ with $\lambda \notin W$. There exists a C^∞-function φ on $\mathbb{T}$ which is identically equal to 1 on V and $supp\ \varphi \subset W$. Then $\mathbb{T} \ni z \mapsto f(z) = \varphi(z)(z - \lambda)^{-1}$ is a C^∞-function on $\mathbb{T}$, equal to $(z-\lambda)^{-1}$ on V and $supp\ f \subset W$. The Fourier series associated with f is absolutely convergent, and so we can write

$$f(z) = \sum_{n\in\mathbb{Z}} a_n z^n \text{ with } \sum_{n\in\mathbb{Z}} |a_n| < +\infty.$$

We consider the operator

$$T = \sum_{n\in\mathbb{Z}} a_n U_{nt} = U_{\sum_n a_n \delta_{nt}} \in \mathcal{B}(\mathcal{X}).$$

Then

$$T(U_t - \lambda) = (U_t - \lambda)T = U_\mu \text{ with } \mu = (\delta_t - \lambda\delta_0) * \sum_n a_n \delta_{nt},$$

and we have

$$\hat{\mu}(\gamma) = (\langle t, \gamma\rangle - \lambda)f(\langle t, \gamma\rangle) = \varphi(\langle t, \gamma\rangle) \quad (\gamma \in \hat{G}).$$

Since φ is equal to 1 on V, it follows that $Sp\ U \subset int\{\gamma \in G; \hat{\mu}(\gamma) = 1\}$, and so U_μ = the identity mapping on $\mathcal{X}$. Thus, $U_t - \lambda$ is invertible in $\mathcal{B}(\mathcal{X})$, that is, $\lambda \notin Sp(U_t)$. □

From the above proposition, it is easy to obtain the following expression for the spectral radius of $U_t - I$:

$$\|U_t - I\|_{Sp} = \sup\{|1 - \langle t, \gamma\rangle|; y \in Sp\ U\}.$$

On the other hand, it is easy to check that a closed set $K \subset \hat{G}$ is compact if and only if $\lim_{t\to 0} \sup\{|1 - \langle t, \gamma\rangle|; \gamma \in K\} = 0$. Consequently, we have

$$\lim_{t\to 0} \|U_t - I\|_{Sp} = 0 \Leftrightarrow Sp\ U \text{ is compact.}$$

Corollary. *The representation $U : G \to \mathcal{B}_w(\mathcal{X})$ is norm-continuous if and only if $Sp\ U$ is compact.*

Proof. By the above remarks, if $\lim_{t\to 0} \|U_t - I\| = 0$, then $Sp\ U$ is compact.

Conversely, if $Sp\ U$ is compact, then there exists a function $f \in \mathcal{L}^1(G)$ with $Sp\ U \subset \{\gamma \in \hat{G}; \hat{f}(\gamma) = 1\}$, hence $U_f = I$ and

$$\lim_{t\to 0} \|U_t - I\| = \lim_{t\to 0} \|U_{\delta_t * f} - U_f\| \le \lim_{t\to 0} \|\delta_t * f - f\|_1 = 0.$$

□

14.7. Consider now two pairs $(\mathcal{X}, \mathcal{X}_*)$ and $(\mathcal{Y}, \mathcal{Y}_*)$ satisfying conditions $(1_{\mathcal{X}})$, $(2_{\mathcal{X}}), (3_{\mathcal{X}})$ and $(1_{\mathcal{Y}}), (2_{\mathcal{Y}}), (3_{\mathcal{Y}})$ of Section 13.1. By Lemma 2/13.1, the pair $(\mathscr{B}_w(\mathcal{X}, \mathcal{Y}), \mathscr{B}_w(\mathcal{X}, \mathcal{Y})_*)$ satisfies conditions $(1_{\mathscr{B}_w(\mathcal{X},\mathcal{Y})}), (2_{\mathscr{B}_w(\mathcal{X},\mathcal{Y})})$.

Consider also a separable locally compact abelian group G and two continuous representations $U : G \to \mathscr{B}_w(\mathcal{X}), V : G \to \mathscr{B}_w(\mathcal{Y})$ of G on $\mathcal{X}$ and $\mathcal{Y}$, respectively. We assume that V satisfies the strong continuity condition (C_V^*). By Lemma 13.3, we then obtain a continuous representation $\mathfrak{S} : G \to \mathscr{B}_w(\mathscr{B}_w(\mathcal{X}, \mathcal{Y}))$ defined by $\mathfrak{S}_s T = V_s T U_s^{-1} (T \in \mathscr{B}_w(\mathcal{X}, \mathcal{Y}), s \in G)$, such that $\mathfrak{S}_\mu \in \mathscr{B}_w(\mathscr{B}_w(\mathcal{X}, \mathcal{Y}))$ for all $\mu \in \mathscr{M}(G)$.

We shall write $\mathscr{B} = \mathscr{B}_w(\mathcal{X}, \mathcal{Y})$.

Theorem (W. B. Arveson). *In the above situation, for $T \in \mathscr{B}, Q \subset \hat{G}$ a closed set and $\{W_i\}_{i\in I}$ a family of neighborhoods of 0 in $\hat{G}$ such that $Q = \bigcap_i \overline{Q + W_i}$, the following statements are equivalent*

(i) *$T \in \mathscr{B}(\mathfrak{S}; Q)$;*
(ii) *$T\mathcal{X}(U; F) \subset \mathcal{Y}(V; \overline{Q + F})$ for any closed set $F \subset \hat{G}$;*
(iii) *$T\mathcal{X}(U; E) \subset \mathcal{Y}(V; Q + E)$ for any set $E \subset \hat{G}$;*
(iv) *$T\mathcal{X}(U; D) \subset \mathcal{Y}(V; Q + D)$ for any open set $D \subset \hat{G}$;*
(v) *$T\mathcal{X}(U; K) \subset \mathcal{Y}(V; Q + K)$ for any compact set $K \subset \hat{G}$;*
(vi) *$T\mathcal{X}(U; \gamma + W_i) \subset \mathcal{Y}(V; \gamma + W_i + Q)$ for any $\gamma \in \hat{G}$ and any $i \in I$.*

Proof. (i) $\Rightarrow$ (ii). Let $x_0 \in \mathcal{X}(U; F)$. To show that $Tx_0 \in \mathcal{Y}(V; \overline{Q + F})$, it is sufficient to show that for every neighborhood W_0 of 0 in $\hat{G}$ we have $Tx_0 \in \mathcal{Y}(V; Q + F + W_0)$ (see 14.3.(8)). Let W be an open and relatively compact neighborhood of 0 in $\hat{G}$ such that $W + W \subset W_0$.

Since $x_0 \in \mathcal{X}(U; F) \subset \mathcal{X}(U; F + W)$, it follows using 14.3.(7) that x_0 is the w-limit of a net of elements of the form $U_g x$ with $x \in \mathcal{X}$ and $g \in \mathscr{L}^1(G)$, *supp* $\hat{g} \subset F + W$ compact. Since $T \in \mathscr{B}_w(\mathcal{X}, \mathcal{Y})$, it is sufficient to show that $TU_g x \in \mathcal{Y}(V; Q+F+W_0)$ for any $x \in X$ and any $g \in \mathscr{L}^1(G)$ with *supp* $\hat{g} \subset F + W$ compact.

Similarly, T is the w-limit of a net of elements of form $\mathfrak{S}_f S$ with $S \in \mathscr{B}_w(\mathcal{X}, \mathcal{Y})$ and $f \in \mathscr{L}^1(G)$, *supp* $\hat{f} \subset Q + W$ compact; hence it is sufficient to show that

$$(\mathfrak{S}_f S)U_g x \in \mathcal{Y}(V; Q + F + W_0)$$

for every $S \in \mathscr{B}_w(\mathcal{X}, \mathcal{Y}), x \in \mathcal{X}, f \in \mathscr{L}^1(G)$ with *supp* $\hat{f} \subset Q + W$ compact and $g \in \mathscr{L}^1(G)$ with *supp* $\hat{g} \subset F + W$ compact.

It is easy to see that

$$\begin{aligned}(\mathfrak{S}_f S)U_g x &= \left(\int f(t) V_t S U_{-t} \, dt \right) \int g(s) U_s x \, ds \\ &= \int \int f(t) g(s) V_t S U_{s-t} x \, ds \, dt = \int \left(\int f(t) g(s + t) V_t S U_s x \, dt \right) ds;.\end{aligned}$$

thus it is sufficient to show that for every $s \in G$ we have

$$V_{fg_s}(SU_s x) = \int f(t) g(s + t) V_t S U_s x \, dt \in \mathcal{Y}(V; Q + F + W_0), \tag{1}$$

where $g_s(t) = g(s+t), (t \in G)$. The functions $\hat{f}$ and $\hat{g}_s$ are continuous and compactly supported, so they belong to $\mathscr{L}^2(G)$. By the Plancherel theorem, it follows that $f, g_s \in \mathscr{L}^2(G)$, hence $fg_s \in \mathscr{L}^1(G)$, and $supp\,(fg_s)\hat{} = supp\,\hat{f} * \hat{g}_s \subset supp\,\hat{f} + supp\,\hat{g} \subset Q + W + F + W \subset Q + F + W_0$, thus proving (1).

(ii) ⇒ (iii). By 14.3.(5), it is sufficient to show that $T\mathscr{X}_{00}(U;E) \subset \mathscr{Y}(V;Q+E)$. Let $x \in \mathscr{X}_{00}(U;E)$. Then $F = Sp_U(x) \subset E$ is compact; hence $\overline{Q+F} = Q+F$, and $x \in \mathscr{X}(U;F)$. By assumption (ii) it follows that $Tx \in \mathscr{Y}(V;Q+F) \subset \mathscr{Y}(V;Q+E)$.

(iii) ⇒ (iv). Obvious.

(iv) ⇒ (v). Let $\{N_j\}_{j\in J}$ be a fundamental system of open and relatively compact neighborhoods of 0 in $\hat{G}$. Assuming (iv) it follows that $T\mathscr{X}(U;K) \subset T\mathscr{X}(U;K+N_j) \subset \mathscr{Y}(V;Q+K+N_j)$ for all $j \in J$, and using 14.3.(9) we get $T\mathscr{X}(U;K) \subset \mathscr{Y}(V;Q+K)$, as $Q+K$ is closed.

(v) ⇒ (vi). This follows easily using 14.3.(5).

(vi) ⇒ (i). To prove (i) it is sufficient (14.3.(3)) to show that $T \in \mathscr{B}(\mathfrak{S};\overline{Q+W_i})$ for every $i \in I$. Accordingly, consider a neighborhood W of 0 in $\hat{G}$ such that

$$T\mathscr{X}(U;\gamma+W) \subset \mathscr{Y}(W;\gamma+Q+W) \quad (\gamma \in \hat{G}) \tag{2}$$

and let us show that $T \in \mathscr{B}(\mathfrak{S};\overline{Q+W})$. To this end it is sufficient (14.3.(2)) to show that

$$\mathfrak{S}_h T = 0 \textit{ for all } h \in \mathscr{I}_{00}(\overline{Q+W}). \tag{3}$$

Let $f \in \mathscr{L}^1(G)$ with $\hat{f}(0) \neq 0$ and $supp\,\hat{f} \subset W$ compact. For $\gamma \in \hat{G}$ we denote by $\bar{\gamma}f$ the $\mathscr{L}^1$-function $G \ni t \mapsto \overline{\langle t,\gamma\rangle} f(t)$. Then $supp(\bar{\gamma}f)\hat{} \subset \gamma + W$, hence $U_{\bar{\gamma}f} \in \mathscr{X}(U;\gamma+W)$ for all $x \in \mathscr{X}$. Using (2) we get $TU_{\bar{\gamma}f}x \in \mathscr{Y}(V;\gamma+Q+W)$ $(x \in \mathscr{X})$. If $h \in \mathscr{I}_{00}(\overline{Q+W})$, then $\bar{\gamma}h \in \mathscr{I}_{00}(\gamma+\overline{Q+W})$ and hence (14.3.(2)) $V_{\bar{\gamma}h}TU_{\bar{\gamma}f}x = 0 (x \in \mathscr{X}, \gamma \in \hat{G})$. Consequently,

$$\begin{aligned} 0 &= \int\int \overline{\langle s+t,\gamma\rangle} f(s)h(t)V_tTU_sx \, \mathrm{d}s \, \mathrm{d}t \\ &= \int \overline{\langle t,\gamma\rangle} \left(\int f(s)h(t-s)V_{t-s}TU_sx \, \mathrm{d}s \right) \mathrm{d}t. \end{aligned} \tag{4}$$

For every $\rho \in \mathscr{Y}_*$, the formula $k_\rho(t) = \int f(s)h(t-s)\rho(V_{t-s}TU_sx) \, \mathrm{d}s \ (t \in G)$ determines a function $k_\rho \in \mathscr{L}^1(G)$; from (4) it follows that $\hat{k}_\rho = 0$, so that $k_\rho = 0$ as an element of $\mathscr{L}^1(G)$. Since the functions $\hat{f}$ and $\hat{h}$ are compactly supported, and so belong to $\mathscr{L}^2(G)$, we have $f, h \in \mathscr{L}^2(G)$, while the function $s \mapsto \rho(V_{t-s}TU_sx)$ is bounded and continuous. Thus, the function k_ρ is continuous, and it follows that $k_\rho(0) = 0$, that is, $\int f(-s)h(s)\rho(V_sTU_{-s}x) \, \mathrm{d}s = 0$. Replacing the function f here by its translates we get

$$\int f(r-s)h(s)\rho((\mathfrak{S}_sT)x) \, \mathrm{d}s = 0 \quad (r \in G). \tag{5}$$

Furthermore, the formula $g_\rho(s) = h(s)\rho((\mathfrak{S}_sT)x)$ $(s \in G)$ determines a function $g_\rho \in \mathscr{L}^1(G)$; from (5) it follows that $f * g_\rho = 0$, and $0 = (f * g_\rho)\hat{}(0) = \hat{f}(0)\hat{g}_\rho(0)$. Since $\hat{f}(0) \neq 0$, we obtain $\hat{g}_\rho(0) = 0$, that is, $\rho((\mathfrak{S}_hT)x) = \int h(s)\rho((\mathfrak{S}_sT)x) \, \mathrm{d}s = 0$. Since $\rho \in \mathscr{Y}_*$ and $x \in \mathscr{X}$ were arbitrary, we obtain the required conclusion (3). □

Corollary 1. *For any sets* $P \subset \hat{G}, E \subset \hat{G}$ *we have* $\mathscr{B}(\mathfrak{S}; P)\mathscr{X}(U; E) \subset \mathscr{Y}(V; P + E)$.

Proof. Indeed, if $T \in \mathscr{B}_0(\mathfrak{S}; P)$, then $Q = Sp_{\mathfrak{S}}(T) \subset P$ is closed and $T \in \mathscr{B}(\mathfrak{S}; Q)$, hence $T\mathscr{X}(U; E) \subset \mathscr{Y}(V; Q + E) \subset \mathscr{Y}(V; P + E)$ by the previous theorem. Thus, Corollary 1 follows using 14.3.(5). □

Corollary 2. *An operator* $T \in \mathscr{B}_w(\mathscr{X}, \mathscr{Y})$ *intertwines representations U and V, that is,*

$$V_s T = T U_s \quad (s \in G) \tag{6}$$

if and only if the inclusion $T\mathscr{X}(U; E) \subset \mathscr{Y}(V; E)$ *holds for any* $E \subset \hat{G}$ *which is closed (or open, or compact, or of the form* $\gamma + W_i$*).*

Proof. Equation (6) means that $\mathfrak{S}_s T = T$ $(s \in G)$, that is, (14.4.(13)) $T \in \mathscr{B}(\mathfrak{S}; \{0\})$ and the corollary follows from the theorem by taking $Q = \{0\}$. □

In particular, taking $(\mathscr{X}, \mathscr{X}_*) = (\mathscr{Y}, \mathscr{Y}_*)$ and $T = I$ in Corollary 2, we see that the spectral subspaces associated with a continuous representation of G determine the representation uniquely:

Corollary 3. *Given two continuous representations* $U : G \to \mathscr{B}_w(\mathscr{X})$ *and* $V : G \to \mathscr{B}_w(\mathscr{X})$ *of G on* $\mathscr{X}$, *we have* $U = V$ *if and only if the inclusion* $\mathscr{X}(U; E) \subset \mathscr{X}(V; E)$ *holds for every set* $E \subset \hat{G}$ *which is closed (or open, or compact, or of the form* $\gamma + W_i$*).*

In Corollary 4, we assume that $\hat{G}$ is ordered by a closed semigroup $S \subset \hat{G}$ with $S \cap (-S) = \{0\}$, $S \cup (-S) = \hat{G}$ and that 0 is adherent to the interior $\mathring{S}$ of S.

Corollary 4. *Let* $T \in \mathscr{B}_w(\mathscr{X}, \mathscr{Y})$ *and* $\gamma \in \hat{G}$. *The following statements are equivalent:*

(i) $T \in \mathscr{B}(\mathfrak{S}; \gamma + S)$;
(ii) $T\mathscr{X}(U; \omega + S) \subset \mathscr{Y}(V; \gamma + \omega + S)$ *for* $\omega \in \hat{G}$;
(iii) $T\mathscr{X}(U; \omega + \mathring{S}) \subset \mathscr{Y}(V; \gamma + \omega + \mathring{S})$ *for* $\omega \in \hat{G}$.

Proof. (i) ⇒ (iii). Since $(\gamma + S) + (\omega + \mathring{S}) \subset \gamma + \omega + \mathring{S}$, this implication follows from Corollary 1.

(iii) ⇒ (ii). Let $\{\gamma_i\}$ be a net in $\mathring{S}$ which converges to 0. Then, for every $\lambda \in \hat{G}$ we have $\lambda + S = \bigcap_i (\lambda - \gamma_i + \mathring{S})$. Thus, using assumption (iii) we get $T\mathscr{X}(U; \omega + S) \subset T\mathscr{X}(U; \omega - \gamma_i + \mathring{S}) \subset \mathscr{Y}(V; \gamma + \omega - \gamma_i + \mathring{S})$ and hence $T\mathscr{X}(U; \omega + S) \subset \bigcap_i \mathscr{Y}(V; \gamma + \omega - \gamma_i + \mathring{S}) = \mathscr{Y}(V; \gamma + \omega + S)$.

(ii) ⇒ (i). Let $Q = \gamma + S$. For each $\lambda \in \mathring{S}$, we consider the set $W_\lambda = (-\lambda + S) \cap (\lambda - S)$. Since $W_\lambda \supset (-\lambda + \mathring{S}) \cap (\lambda - \mathring{S})$, W_λ is a neighborhood of 0 in $\hat{G}$ and we have $\bigcap \{\overline{Q + W_\lambda}; \lambda \in \mathring{S}\} = \bigcap \{\gamma - \lambda + S; \lambda \in \mathring{S}\} = Q$. Using assumption (ii), for every $\omega \in \hat{G}$ we obtain $T\mathscr{X}(U; \omega + W_\lambda) \subset T\mathscr{X}(U; \omega - \lambda + S) \subset \mathscr{Y}(V; \gamma + \omega - \lambda + S) = \mathscr{Y}(V; \omega + W_\lambda + Q)$ and using the implication (vi) ⇒ (i) from the previous theorem we get the required conclusion (i). □

14.8. In the context of Section 14.7, we also note the following result:

Proposition. *Let* $x \in \mathscr{X}$ *and* $T \in \mathscr{B}_w(\mathscr{X}, \mathscr{Y})$. *If T intertwines the representations U and V, that is,* $V_s T = T U_s$ $(s \in G)$, *then* $Sp_V(Tx) \subset Sp_U(x)$, *and if moreover T is injective, then* $Sp_V(Tx) = Sp_U(x)$.

Proof. For $f \in \mathscr{L}^1(G)$, we get $V_f Tx = TU_f x$, so it follows that $\mathscr{I}_x^U \subset \mathscr{I}_{Tx}^V$, and indeed $\mathscr{I}_x^U = \mathscr{I}_{Tx}^V$ if T is injective. Thus, the proposition follows by the definition of the spectrum of an element (14.2). □

14.9. Let $U : G \to \mathscr{B}_w(\mathscr{X})$ be a continuous representation. Consider another separable locally compact abelian group H and a continuous homomorphism $\varphi : H \to G$. Then $U \circ \varphi : H \to \mathscr{B}_w(\mathscr{X})$ is a continuous representation and the mapping $\gamma \to \gamma \circ \varphi$ defines the dual homomorphism $\hat{\varphi} : \hat{G} \to \hat{H}$.

Proposition. *In the above situation we have*

$$Sp_{U\circ\varphi}(x) = \overline{\hat{\varphi}(Sp_U(x))} \quad (x \in \mathscr{X}). \tag{1}$$

$$\mathscr{X}(U\circ\varphi; E) \subset \mathscr{X}(U; \hat{\varphi}^{-1}(E)) \quad (E \subset \hat{H}). \tag{2}$$

$$\mathscr{X}(U\circ\varphi; F) \subset \mathscr{X}(U; \hat{\varphi}^{-1}(F)) \quad (F \subset \hat{H},\ closed). \tag{3}$$

$$Sp\ U\circ\varphi = \overline{\hat{\varphi}(Sp\ U)}. \tag{4}$$

Proof. For every $\nu \in \mathscr{M}(H)$, we denote by $\varphi(\nu) \in \mathscr{M}(G)$ the image of the measure ν by the continuous homomorphism φ. It is easy to check that

$$(U\circ\varphi)_\nu = U_{\varphi(\nu)} \quad (\nu \in \mathscr{M}(H)) \tag{5}$$

$$\widehat{\varphi(\nu)}(\gamma) = \hat{\nu}(\hat{\varphi}(\gamma)) \quad (\nu \in \mathscr{M}(H), \gamma \in \hat{G}). \tag{6}$$

Let us prove (1). Let $\gamma \in Sp_U(x)$ and $\hat{\varphi}(\gamma) \in \hat{H}$. If $\nu \in \mathscr{M}(H)$ and $(U\circ\varphi)_\nu x = 0$, then $U_{\varphi(\nu)}x = 0$, hence $\hat{\nu}(\hat{\varphi}(\gamma)) = \widehat{\varphi(\nu)}(\gamma) = 0$, since $\gamma \in Sp_U(x)$ (see 14.3.(10)). Consequently, $\hat{\varphi}(\gamma) \in Sp_{U\circ\varphi}(x)$. We have thus proved that $\overline{\hat{\varphi}(Sp_U(x))} \subset Sp_{U\circ\varphi}(x)$. Conversely, assume that $\omega \in \hat{H}$ and $\omega \notin \overline{\hat{\varphi}(Sp_U(x))}$. There exists a function $h \in \mathscr{L}^1(H) \subset \mathscr{M}(H)$ with $\hat{h}(\omega) \neq 0$ and $supp\ \hat{h} \cap \overline{\hat{\varphi}(Sp_U(x))} = \emptyset$. Then $\mu = \varphi(h) \in \mathscr{M}(G)$ and $\hat{\mu} = \hat{h}\circ\hat{\varphi}$ vanishes on a neighborhood of $Sp_U(x)$, so that $(U\circ\varphi)_h x = U_\mu x = 0$. Since $\hat{h}(\omega) \neq 0$, it follows that $\omega \notin Sp_{U\circ\varphi}(x)$. We have thus proved that $Sp_{U\circ\varphi}(x) \subset \overline{\hat{\varphi}(Sp_U(x))}$.

The proof of (4) is quite similar to the above proof.

Finally, (2) and (3) follow obviously from (1) using also 14.3.(5). □

In what follows we consider some examples.

14.10. Let $\mathscr{H}$ be a Hilbert space and G a separable locally compact abelian group. Then the pair $(\mathscr{H}, \mathscr{H}^*)$ satisfies conditions $(1_\mathscr{H}), (2_\mathscr{H}), (3_\mathscr{H})$ of Section 13.1 and every *so*-continuous unitary representation $u : G \to \mathscr{B}(\mathscr{H})$ is a continuous representation of G on $\mathscr{H}$ according to the definition given in Section 13.2; moreover, u satisfies the strong continuity conditions (C_u) and (C_u^*).

In this case, as is easily verified, the mapping $\mathscr{L}^1(G) \ni f \mapsto u_f \in \mathscr{B}(\mathscr{H})$ is a *-representation of the involutive Banach algebra $\mathscr{L}^1(G)$, where the involution $f \mapsto f^*$ is defined by $f^*(t) = \overline{f(-t)}$ $(t \in G)$:

$$u_{f^*} = u_f^* \quad (f \in \mathscr{L}^1(G)).$$

On the other hand, for each $\xi \in \mathscr{H}$, the function $G \ni t \mapsto (u_t\xi|\xi)$ is a positive definite function on G and hence, by Bochner's theorem, there exists a positive measure $\nu_\xi \in \mathscr{M}(\hat{G})$ such that

$$(u_t\xi|\xi) = \int_G \langle t, \gamma\rangle\, \mathrm{d}\nu_\xi(\gamma) \quad (t \in G).$$

Then, for $f \in \mathscr{L}^1(G)$, we obtain

$$\|u_f\xi\|^2 = (u_{f^* * f}\xi|\xi) = \int_G |\hat{f}(\gamma)|^2 \, d\nu_\xi(\gamma) \le \|\hat{f}\|_\infty^2 \|\xi\|^2.$$

The set $\mathscr{A}(G) = \{\hat{f}; f \in \mathscr{L}^1(G)\}$ is a *-subalgebra of $\mathscr{C}_0(\hat{G})$ which separates the points of G and hence is norm-dense in $\mathscr{C}_0(\hat{G})$, by the Stone–Weierstrass theorem. The above inequality shows that the mapping $\mathscr{A}(G) \ni \hat{f} \mapsto u_f$ can be extended to a *-representation $\pi_u : \mathscr{C}_0(\hat{G}) \to \mathscr{B}(\mathscr{H})$ of the C^*-algebra $\mathscr{C}_0(\hat{G})$, uniquely determined, such that

$$\pi_u(\hat{f}) = u_f \quad (\hat{f} \in \mathscr{L}^1(G)).$$

Furthermore (see, e.g., Strătilă & Zsidó, 1977, 2005, 7.14) the *-representation π_u can be uniquely extended to a *-representation, still denoted by π_u, of the C^*-algebra $Baire(\hat{G})$ of all bounded Baire functions on $\hat{G}$, such that $\pi_u(\varphi_n) \overset{so}{\to} \pi_u(\varphi)$ for every norm-bounded sequence $\{\varphi_n\} \subset Baire(\hat{G})$ which is pointwise convergent to $\varphi \in Baire(\hat{G})$, and we have

$$(\pi_u(\varphi)\xi|\xi) = \int_{\hat{G}} \varphi(\gamma) \, d\nu_\xi(\gamma) \quad (\varphi \in Baire(\hat{G}), \xi \in \mathscr{H}).$$

Then, putting $p_u(E) = \pi_u(\chi_E)$ for every Baire set $E \subset \hat{G}$, we obtain a $\mathscr{B}(\mathscr{H})$-valued spectral measure $p_u(\cdot)$ on $\hat{G}$, uniquely determined, such that

$$(u_t\xi|\xi) = \int_{\hat{G}} \langle t, \gamma \rangle \, d(p_u(\gamma)\xi|\xi) \quad (t \in G, \xi \in \mathscr{H}).$$

p_u is called the *Stone spectral measure associated with u*. This spectral measure has regularity properties similar to 14.3.(10) and 14.3.(11).

We are now in a position to state the next result:

Proposition. *Let $u : G \to \mathscr{B}(\mathscr{H})$ be an so-continuous unitary representation of G on $\mathscr{H}$. For every Baire set $E \subset \hat{G}$ we have*

$$\mathscr{H}(u; E) = p_u(E)\mathscr{H}.$$

Sp u is the support of the Stone spectral measure $p_u(\cdot)$.

Proof. For every $f \in \mathscr{L}^1(G)$ and every Baire set $E \subset \hat{G}$ we have

$$u_f p_u(E) = p_u(E) u_f = \int_E \hat{f}(\gamma) \, dp_u(\gamma). \tag{1}$$

Let $D \subset \hat{G}$ be an open set. If $f \in \mathscr{L}^1(G)$ and $supp\,\hat{f} \subset D$, then $u_f\mathscr{H} \subset p_u(D)\mathscr{H}$, by (1). Hence (14.3.(7)).

$$\mathscr{H}(u; D) \subset p_u(D)\mathscr{H}. \tag{2}$$

Let $F \subset \hat{G}$ be a closed set. For every $f \in \mathcal{J}_0(F)$, we have $u_f p_u(F) = 0$ again by (1). Hence (14.3.(2))

$$p_u(F)\mathcal{H} \subset \mathcal{H}(u; F).$$

On the other hand, using (2), 14.3.(10), and the regularity of the Stone spectral measure, we obtain

$$\mathcal{H}(u; F) = \bigcap\{\mathcal{H}(u; D); D \supset F \text{ open}\} \subset \bigcap\{p_u(D)\mathcal{H}; D \supset F \text{ open}\} = p_u(F)\mathcal{H}.$$

Hence

$$\mathcal{H}(u; F) = p_u(F)\mathcal{H}. \tag{3}$$

Consider now any Bairc set $E \subset \hat{G}$. Using (3), 14.3.(11), and the regularity of the Stone spectral measure, we obtain

$$\begin{aligned}\mathcal{H}(u; E) &= \bigvee\{\mathcal{H}(u; K); K \subset E \text{ compact}\}\\ &= \bigvee\{p_u(K)\mathcal{H}; K \subset E \text{ compact}\} = p_u(E)\mathcal{H}.\end{aligned}$$

Since, for an open set $D \subset \hat{G}$, we have $\mathcal{H}(u; D) = 0$ if and only if $p_u(D)\mathcal{H} = 0$, it follows that $Sp\ u$ is just the support of $p_u(\cdot)$. □

Recall (Strătilă & Zsidó, 1977, 2005, 7.14) that if $\hat{G}$ has a countable basis of open sets, then any Borel set in $\hat{G}$ is also a Baire set.

If $G = \mathbb{R}$, then we can identify $\hat{G}$ with $\mathbb{R}$ so that for $t \in \mathbb{R} = G$ and $s \in \mathbb{R} = \hat{G}$ we have

$$\langle t, s\rangle = e^{\mathrm{i}tx}.$$

In this case, the Stone theorem shows that there exists a unique self-adjoint operator A in $\mathcal{H}$ such that $u_t = exp(itA)(t \in \mathbb{R})$; A is called the infinitesimal generator of the *so*-continuous unitary representation u. For any Borel set $E \subset \mathbb{R} = \hat{G}$ we have

$$p_u(E) = \chi_E(A). \tag{4}$$

Also,

$$Sp\ u = Sp\ A \tag{5}$$

$$\hat{f}(A) = \int_{\mathbb{R}} f(t)e^{\mathrm{i}tA}\,\mathrm{d}t \qquad (f \in \mathcal{L}^1(\mathbb{R})). \tag{6}$$

If $(\mathcal{X}, \mathcal{X}_*) = (\mathcal{Y}, \mathcal{Y}_*) = (\mathcal{H}, \mathcal{H}^*)$, then $\mathcal{B}_w(\mathcal{X}, \mathcal{Y}) = \mathcal{B}(\mathcal{H})$ and the w-topology on $\mathcal{B}_w(\mathcal{X}, \mathcal{Y})$ is just the w-topology on the W^*-algebra $\mathcal{B}(\mathcal{H})$, defined by the predual $\mathcal{B}(\mathcal{H})_* = \mathcal{B}_w(\mathcal{X}, \mathcal{Y})_*$ of $\mathcal{B}(\mathcal{H})$.

Given an *so*-continuous unitary representation $u : G \to \mathcal{B}(\mathcal{H})$, the representation $\mathfrak{S}(u, u)$ of G on $\mathcal{B}(\mathcal{H})$ is just the continuous action $\sigma : G \to Aut(\mathcal{B}(\mathcal{H}))$ defined by

$$\sigma_t(x) = u_t x u_t^* \quad (x \in \mathcal{B}(\mathcal{H}), t \in G).$$

Assume that $\hat{G}$ is ordered by a closed semigroup $S \subset \hat{G}$ with $S \cap (-S) = \{0\}, S \cup (-S) = \hat{G}$ and that 0 is adherent to the interior $\mathring{S}$ of S. Then, applying Corollary 4/14.7, we obtain

Corollary. *Let $u: G \to \mathscr{B}(\mathscr{H})$ be a so-continuous unitary representation and $\sigma: G \to Aut(\mathscr{B}(\mathscr{H}))$ the continuous action defined by $\sigma_t = Ad(u_t)(t \in G)$. For $x \in \mathscr{B} = \mathscr{B}(\mathscr{H})$ and $\gamma \in \hat{G}$, the following statements are equivalent:*

(i) $x \in \mathscr{B}(\sigma; \gamma + S)$;
(ii) $xp_u(\omega + S)\mathscr{H} \subset p_u(\gamma + \omega + S)\mathscr{H}$ *for all* $\omega \in \hat{G}$; *and*
(iii) $xp_u(\omega + \mathring{S})\mathscr{H} \subset p_u(\gamma + \omega + \mathring{S})\mathscr{H}$ *for all* $\omega \in \hat{G}$.

In particular, for $G = \mathbb{R}$, we can take $S = [0, +\infty) \subset \hat{G}$ or $S = (-\infty, 0] \subset \hat{G}$.

14.11. Let $\mathscr{X}$ be a Banach space and $D \in \mathscr{B}(\mathscr{X})$ be such that

$$\| \exp(\mathrm{i}tD)\| = 1 \text{ for all } t \in \mathbb{R}. \tag{1}$$

Since $\| \exp(\mathrm{i}tD) - I\| \leq \exp(t\|D\|) - 1$, it follows that the formula

$$U_t = \exp(\mathrm{i}tD) \quad (t \in \mathbb{R}) \tag{2}$$

defines a norm-continuous representation $U : \mathbb{R} \to \mathscr{B}(\mathscr{H})$ which is also a continuous representation in the sense defined in Section 13.2, with respect to the pair $(\mathscr{X}, \mathscr{X}_*)$.

From (1), it follows that the usual spectrum $Sp(D) = Sp_{\mathscr{B}(\mathscr{H})}(D)$ of the operator $D \in \mathscr{B}(\mathscr{H})$ is real, and, clearly, $Sp(D) \subset [-\|D\|, \|D\|]$. We have

$$Sp(\exp(\mathrm{i}tD)) = \{e^{\mathrm{i}ts}; s \in Sp(D)\} \quad (t \in \mathbb{R}). \tag{3}$$

On the other hand, since the representation U is norm-continuous, the set $Sp\ U \in \mathbb{R}$ is compact, by Corollary 14.5. Using Proposition 14.5, it follows that

$$Sp(U_t) = \{e^{\mathrm{i}ts}; s \in Sp\ U\} \quad (t \in \mathbb{R}). \tag{4}$$

Using (2), (3), and (4), we conclude that

$$Sp\ U = Sp(D) \subset [-\|D\|, \|D\|]. \tag{5}$$

It is a classical result that for any norm-continuous representation $U : \mathbb{R} \to \mathscr{B}(\mathscr{X})$ of $\mathbb{R}$ by isometries on $\mathscr{X}$, there exists a unique operator $D \in \mathscr{B}(\mathscr{X})$ which satisfies conditions (1) and (2). This result can easily be obtained using the compactness of $Sp\ U$. Indeed, one can take $D = U_f$ for any compactly supported C^∞-function f such that $Sp\ U \subset int\{s \in \mathbb{R}; \hat{f}(s) = s\}$. Moreover, $D =$ norm-$\lim_{t\to 0} \frac{1}{t}(U_t - I)$.

14.12. Let $(\mathscr{X}, \mathscr{X}_*)$ be a pair-satisfying conditions $(1_{\mathscr{X}}), (2_{\mathscr{X}}), (3_{\mathscr{X}})$ of Section 13.1 and let $T \in \mathscr{B}_w(\mathscr{X})$ be an isometry on $\mathscr{X}$. Then we obtain a representation $U : \mathbb{Z} \to \mathscr{B}_w(\mathscr{X})$ defined by

$$U_n = T^n \quad (n \in \mathbb{Z}). \tag{1}$$

The dual group $\hat{\mathbb{Z}}$ of $\mathbb{Z}$ can be identified with the one-dimensional torus $\mathbb{T} = \{\lambda \in \mathbb{C}; |\lambda| = 1\}$ in such a way that for $n \in \mathbb{Z}$ and $\lambda \in \mathbb{T}$ if we have $\langle n, \lambda \rangle = \lambda^n$.

Since $Sp\, U \in \mathbb{T}$ is compact, using Proposition 14.5 we get $Sp(T) = Sp(U_1) = \{\lambda; \lambda \in Sp\, U\}$, that is,

$$Sp\, U = Sp(T) \tag{2}$$

where $Sp(T)$ is the usual spectrum $Sp_{\mathscr{B}(\mathscr{X})}(T)$ of the operator $T \in \mathscr{B}(\mathscr{X})$.

For $\lambda \in \mathbb{C}$, $|\lambda| = 1$ we have

$$T = \lambda I \Leftrightarrow Sp(T) = \{\lambda\}. \tag{3}$$

Indeed, the implication ($\Rightarrow$) is obvious. Conversely, if $Sp(T) = \{\lambda\}$, then $Sp\, U = \{\lambda\}$, hence $\mathscr{X}(U; \{\lambda\}) = \mathscr{X}$. Consequently (14.3.(13)), for every $x \in \mathscr{X}$ we have $T^n x = U_n x = \lambda^n x$ $(n \in \mathbb{Z})$, hence $T = \lambda I$.

14.13. Recall that if a, b are two commuting elements of a unital Banach algebra $\mathscr{A}$, then

$$Sp_{\mathscr{A}}(a+b) \subset \{\lambda + \mu; \lambda \in Sp_{\mathscr{A}}(a), \mu \in Sp_{\mathscr{A}}(b)\},$$
$$Sp_{\mathscr{A}}(ab) \subset \{\lambda\mu; \lambda \in Sp_{\mathscr{A}}(a), \mu \in Sp_{\mathscr{A}}(b)\}.$$

Now, for arbitrary elements $a, b \in \mathscr{A}$, the mappings $L_a : \mathscr{A} \ni x \mapsto ax \in \mathscr{A}, R_b : \mathscr{A} \ni x \mapsto xb \in \mathscr{A}$ define two commuting elements L_a, R_b of the Banach algebra $\mathscr{B}(\mathscr{A})$ of all bounded linear operators on $\mathscr{A}$, and it is clear that $Sp_{\mathscr{B}(\mathscr{A})}(L_a) \subset Sp_{\mathscr{A}}(a), Sp_{\mathscr{B}(\mathscr{A})}(R_b) \subset Sp_{\mathscr{A}}(b)$. Consequently,

$$Sp_{\mathscr{B}(\mathscr{A})}(L_a + R_b) \subset \{\lambda + \mu; \lambda \in Sp_{\mathscr{A}}(a), \mu \in Sp_{\mathscr{A}}(b)\}, \tag{1}$$
$$Sp_{\mathscr{B}(\mathscr{A})}(L_a R_b) \subset \{\lambda\mu; \lambda \in Sp_{\mathscr{A}}(a), \mu \in Sp_{\mathscr{A}}(b)\}. \tag{2}$$

Proposition. *Let $\mathscr{M}$ be a W*-factor, and let $u \in U(\mathscr{M}), \sigma = Ad(u) \in Aut(\mathscr{M})$. Then*

$$Sp_{\mathscr{B}(\mathscr{M})}(\sigma) = \{\lambda\mu^{-1}; \lambda, \mu \in Sp_{\mathscr{M}}(u)\}.$$

Proof. By the above remarks, it is clear that $Sp_{\mathscr{B}(\mathscr{M})}(\sigma) \subset \{\lambda\mu^{-1}; \lambda, \mu \in Sp_{\mathscr{M}}(u)\}$.

Now, let $\lambda, \mu \in Sp_{\mathscr{M}}(u)$ and $\varepsilon > 0$. There exist two nonzero spectral projections $e, f \in \mathscr{M}$ of u such that $\|ue - \lambda e\| \leq \varepsilon/2$ and $\|fu - \mu f\| \leq \varepsilon/2$. Since $\mathscr{M}$ is a factor, there exists a nonzero partial isometry $v \in \mathscr{M}$ such that $vv^* \leq e$ and $v^*v \leq f$. We have

$$\begin{aligned}\|\sigma(v) - \lambda\mu^{-1}v\| &= \|uv - \lambda\mu^{-1}vu\| = \|\lambda^{-1}uv - \mu^{-1}vu\| \\ &\leq \|\lambda^{-1}uv - v\| + \|v - \mu^{-1}vu\| \\ &= \|uv - \lambda v\| + \|vu - \mu v\| \\ &= \|(ue - \lambda e)v\| + \|v(fu - \mu f)\| \leq \varepsilon.\end{aligned}$$

Hence $Sp_{\mathscr{B}(\mathscr{M})}(\sigma) = \{\lambda\mu^{-1}; \lambda, \mu \in Sp_{\mathscr{M}}(u)\}$. □

14.14. Notes. The spectral theory of the action of a locally compact abelian group on a Banach space appeared in the works of Borchers, Colojoară and Foiaş, Godement, Lyubic, Macaev and Feldman, Wightman, and others, but it was Arveson (1974) who, as well as giving a systematic, self-contained exposition of this subject with several new results, showed the relevance of the theory for the study of operator algebras. The main result of this section, Theorem 14.7, is due to Arveson (1974). Propositions 14.4, 14.5, 14.9, and 14.13 are due to Connes (1973a) (see also Olesen, 1976).

Four our exposition, we have used Arveson (1974); Combes and Delaroche (1973–1974); Connes (1973a), and Olesen (1976).

15 Continuous Actions on W^*-algebras

In this section, we apply the spectral theory so far developed to continuous actions of locally compact abelian groups by *-automorphisms of W^*-algebras. The main results concern the inner implementation of *-automorphisms and derivations.

15.1. Let $u : G \to Aut(\mathscr{M})$ be a continuous action (13.5) of the separable locally compact abelian group G on the W^*-algebra $\mathscr{M}$.

Since σ is u-continuous, we infer (see 13.2) that every $\varphi \in \mathscr{M}_*$ belongs to the norm-closure of the set $\{\varphi\circ\sigma_f; f \in \mathscr{L}^1(G)\}$. It follows that

$$\begin{aligned}&\textit{for every norm-bounded approximate unit } \{f_i\}_{i\in I}\\ &\textit{of the Banach algebra } \mathscr{L}^1(G) \textit{ we have}\\ &\qquad \|\varphi\circ\sigma_{f_i} - \varphi\| \to 0 \quad (\varphi \in \mathscr{M}_*).\end{aligned} \tag{1}$$

Indeed, let $\varphi \in \mathscr{M}_*$, $\|\varphi\| \leq 1$, and $\varepsilon > 0$. We assume that $\|f_i\| \leq 1$ for each $i \in I$. There exists $f \in \mathscr{L}^1(G)$ with $\|\varphi\circ\sigma_f - \varphi\| < \varepsilon/3$ and then $i_\varepsilon \in I$ such that $\|f * f_i - f\|_1 < \varepsilon/3$ for every $i \geq i_\varepsilon$. Thus, for $i \geq i_\varepsilon$, we get $\|\varphi\circ\sigma_{f_i} - \varphi\| \leq \|\varphi\circ\sigma_{f_i} - \varphi\circ\sigma_f\circ\sigma_{f_i}\| + \|\varphi\circ\sigma_f\circ\sigma_{f_i} - \varphi\circ\sigma_f\| + \|\varphi\circ\sigma_f - \varphi\| \leq \|f_i\|_1\|\varphi - \varphi\circ\sigma_f\| + \|\varphi\|\|f * f_i - f\|_1 + \|\varphi\circ\sigma_f - \varphi\| < \varepsilon$.

Also (13.2), every $x \in \mathscr{M}$ belongs to the w-closure of the set $\{\sigma_f x; f \in \mathscr{L}^1(G)\}$. Using (1) it follows that

$$\begin{aligned}&\textit{for every norm-bounded approximate unit } \{f\}_{i\in I}\\ &\textit{of the Banach algebra } \mathscr{L}^1(G) \textit{ we have}\\ &\qquad \sigma_{f_i} x \xrightarrow{w} x \quad (x \in \mathscr{M}).\end{aligned} \tag{2}$$

Indeed, let $x \in \mathscr{M}$, $\|x\| \leq 1$, $\varphi \in \mathscr{M}_*$, $\|\varphi\| \leq 1$, and $\varepsilon > 0$. There exists $f \in \mathscr{L}^1(G)$ with $|\varphi(x - \sigma_f x)| < \varepsilon/4$ and there exists $i_\varepsilon \in I$ such that $\|f_i * f - f\| < \varepsilon/4$ and $\|\varphi - \varphi\circ\sigma_{f_i}\| < (\varepsilon/4)\|x - \sigma_f x\|$ for $i \geq i_\varepsilon$. Thus, for $i \geq i_\varepsilon$ we get $|\varphi(\sigma_{f_i} x - x)| \leq |(\varphi\circ\sigma_{f_i})(x - \sigma_f x)| + |\varphi(\sigma_{f_i}\sigma_f x - \sigma_f x)| + |\varphi(\sigma_f x - x)| \leq 2|\varphi(x - \sigma_f x)| + |\varphi\circ\sigma_{f_i} - \varphi\|\|x - \sigma_f x\| + \|f_i * f - f\|_1\|\varphi\|\|x\| < \varepsilon$.

In particular, for every $x \in \mathscr{M}$

$$x \in \overline{\{\sigma_k x; k \in \mathscr{L}^1(G), \|k\|_1 \leq 1, \textit{ supp } \hat{k} \text{ compact}\}}^{s^*}. \tag{3}$$

Indeed, $\{\sigma_k x; k \in \mathscr{L}^1(G), \|k\|_1 \leq 1, supp\ \hat{k}$ compact$\}$ is a convex set and, by (2) and 14.1.(4), x is w-adherent and hence also s^*-adherent to this set.

We shall also use the following result:

Lemma. *Let $\{D_i\}_{i\in I}$ be an open covering of $\hat{G}$. There exist a directed set Λ and a family $\{f_{i,\lambda}\}_{i\in I,\lambda\in\Lambda} \subset \mathscr{L}^1(G)$ such that*

(a) *for each $\lambda \in \Lambda$ the set $\{i \in I; f_{i,\lambda} \neq 0\}$ is finite and*

(b) *for each $i \in I$ and each $\lambda \in \Lambda$ we have $supp\,\hat{f}_{i,\lambda} \subset D_i$ and, moreover, for every $x \in \mathscr{M}$ we have*

$$\sum_{i\in I} \sigma_{f_{i,\lambda}} x \underset{\lambda\in\Lambda}{\overset{w}{\to}} x.$$

Proof. Indeed, the linear subspace of $\mathscr{L}^1(G)$ spanned by the set $\{f \in \mathscr{L}^1(G)$; there exists $i \in I$ with $supp\,\hat{f} \subset D_1\}$ is a norm-dense ideal of $\mathscr{L}^1(G)$ and hence (14.1) it contains a norm-bounded approximate unit of $\mathscr{L}^1(G)$, so that the lemma follows using statement (2). □

15.2. Let $\sigma : G \to Aut(\mathscr{M})$ be a continuous action of G on $\mathscr{M}$. In this section, we study the relationship between the spectral subspaces associated with σ and the *-operation on $\mathscr{M}$.

Let $f \in \mathscr{L}^1(G)$ and let $\bar{f} \in \mathscr{L}^1(G)$ be its complex conjugate. It is easy to see that for any $x \in \mathscr{M}$ and any $\gamma \in \hat{G}$ we have

$$(\sigma_f(x))^* = \sigma_{\bar{f}}(x^*), (\bar{f})^\wedge(\gamma) = \bar{\hat{f}}(-\gamma). \tag{1}$$

It follows that

$$Sp_\sigma(x^*) = -Sp_\sigma(x). \tag{2}$$

Consequently, for any set $E \subset \hat{G}$ we have

$$\mathscr{M}(\sigma; E)^* = \mathscr{M}(\sigma; -E) \tag{3}$$

and $Sp\ \sigma$ is a symmetric subset of $\hat{G}$, that is,

$$Sp\ \sigma = -Sp\ \sigma. \tag{4}$$

Let $\sigma' : G \to Aut(\mathscr{M})$ be another continuous action of G on $\mathscr{M}$. We consider the pair $(\mathscr{B}, \mathscr{B}_*)$ consisting of the Banach space $\mathscr{B}$ of all w-continuous linear mappings $\mathscr{M} \to \mathscr{M}$ and the norm-closed linear subspace $\mathscr{B}_* \subset \mathscr{B}^*$ generated by $\{\varphi(\cdot x); \varphi \in \mathscr{M}_*, x \in \mathscr{M}\}$ (cf. 13.1), together with the continuous representation $\mathfrak{S}$ of G on $\mathscr{B}$ defined by (cf. 13.3)

$$\mathfrak{S}_s T = \sigma'_s \circ T \circ \sigma_{-s} \quad (T \in \mathscr{B}, s \in G).$$

Note that there is a natural *-operation $T \mapsto T^*$ on the Banach space $\mathscr{B}$, namely $T^*(x) = T(x^*)^* (T \in \mathscr{B}, x \in \mathscr{M})$, and it is easy to check that $\mathfrak{S}$ is a *-representation, that is, $\mathfrak{S}_s(T^*) = (\mathfrak{S}_s T)^* (T \in \mathscr{B},$

$s \in G$). The same argument as above shows that

$$Sp_{\mathfrak{S}}(T^*) = -Sp_{\mathfrak{S}}(T) \quad (T \in \mathscr{B}), \tag{5}$$

so that the spectrum of every self-adjoint element of $\mathscr{B}$ is symmetric with respect to $0 \in \hat{G}$.

Assume that $\hat{G}$ is ordered by a closed semigroup $S \subset \hat{G}$ with $S \cap (-S) = \{0\}, S \cup (-S) = \hat{G}$ and that 0 is adherent to the interior $\mathring{S}$ of S. Using (5) and Corollary 4/14.7, we get the following result:

Proposition. *Let $\sigma : G \to Aut(\mathscr{M}), \sigma' : G \to Aut(\mathscr{M})$ be continuous actions of G on $\mathscr{M}$ and $\theta \in Aut(\mathscr{M})$. The following statements are equivalent:*

(i) $\sigma'_t \circ \theta = \theta \circ \sigma_t$ *for all $t \in G$;*
(ii) $\theta(\mathscr{M}(\sigma; \gamma + S)) \subset \mathscr{M}(\sigma'; \gamma + S)$ *for all $\gamma \in \hat{G}$; and*
(iii) $\theta(\mathscr{M}(\sigma; \gamma + \mathring{S})) \subset \mathscr{M}(\sigma'; \gamma + \mathring{S})$ *for all $\gamma \in \hat{G}$.*

Proof. By Corollary 4/14.7, statements (ii) and (iii) are both equivalent to the fact that the $*$-automorphism $\theta \in Aut(\mathscr{M}) \subset \mathscr{B}$ belongs to the spectral subspace $\mathscr{B}(\mathfrak{S}; S)$. Since θ is a self-adjoint element of $\mathscr{B}$, the spectrum of θ is symmetric and so $Sp_{\mathfrak{S}}(\theta) \subset S \cap (-S) = \{0\}$, that is, (14.3.(14)) $\mathfrak{S}_t(\theta) = \theta$ and $\sigma'_t \circ \theta = \theta \circ \sigma_t$ for all $t \in G$. □

Clearly, the previous proposition still holds for any self-adjoint element $\theta \in \mathscr{B}$ and can be extended to actions on different W^*-algebras.

If $\theta =$ the identity mapping on $\mathscr{M}$, we get spectral conditions that are equivalent to the equalities $\sigma_t = \sigma'_t$ $(t \in G)$.

The previous proposition is particularly useful in the case $G = \mathbb{R}$, when we can take $S = [0, +\infty) \subset \hat{G}$ or $S = (-\infty, 0] \subset \hat{G}$.

15.3. Let $\sigma : G \to Aut(\mathscr{M})$ be a continuous action of G on $\mathscr{M}$. We now study the relationship between the spectral subspaces associated with σ and the multiplication operation on $\mathscr{M}$.

Consider again the pair $(\mathscr{B}, \mathscr{B}_*)$ as in Section 15.2 and the continuous representation $\mathfrak{S}$ of G on $\mathscr{B}$ defined by $\mathfrak{S}_s T = \sigma_s \circ T \circ \sigma_{-s} (T \in \mathscr{B}, s \in G)$. We define an operator $L : \mathscr{M} \to \mathscr{M}$ by putting

$$L_a(x) = ax \quad (a \in \mathscr{M}, x \in \mathscr{M}).$$

The operator L is injective, w-continuous, and intertwines the representations σ and $\mathfrak{S}$, that is,

$$\mathfrak{S}_s \circ L = L \circ \sigma_s \quad (s \in G)$$

since for $a, x \in \mathscr{M}$ we have $(\mathfrak{S}_s(L_a))(x) = \sigma_s(a\sigma_{-s}(x)) = \sigma_s(a)x = (L_{\sigma_s(a)})(x)$. By Proposition 14.8 it follows that

$$Sp_{\mathfrak{S}}(L_a) = Sp_\sigma(a) \quad (a \in \mathscr{M}).$$

Thus, using Theorem 14.7, we obtain the following result:

Proposition. *Let $\sigma : G \to Aut(\mathscr{M})$ be a continuous action of G on $\mathscr{M}$, $a \in \mathscr{M}$, $Q \subset \hat{G}$ a closed set, and $\{W_i\}_{i \in I}$ a fundamental system of compact neighborhoods of 0 in $\hat{G}$. The following statements are equivalent:*

(i) $a \in \mathscr{M}(\sigma; Q)$;
(ii) $a\mathscr{M}(\sigma; F) \subset \mathscr{M}(\sigma; Q + F)$ *for any closed set* $F \subset \hat{G}$;
(iii) $a\mathscr{M}(\sigma; E) \subset \mathscr{M}(\sigma; Q + E)$ *for any set* $E \subset \hat{G}$;
(iv) $a\mathscr{M}(\sigma; D) \subset \mathscr{M}(\sigma; Q + D)$ *for any open set* $D \subset \hat{G}$;
(v) $a\mathscr{M}(\sigma; K) \subset \mathscr{M}(\sigma; Q + K)$ *for any compact set* $K \subset \hat{G}$; *and*
(vi) $a\mathscr{M}(\sigma; \gamma + W_i) \subset \mathscr{M}(\sigma; \gamma + W_i + Q)$ *for any* $\gamma \in \hat{G}$ *and any* $i \in I$.

In particular, for every $a, b \in \mathscr{M}$ we have

$$Sp_\sigma(ab) \subset Sp_\sigma(a) + Sp_\sigma(b) \tag{1}$$

and for every $E_1, E_2 \subset \hat{G}$ we have

$$\mathscr{M}(\sigma; E_1)\mathscr{M}(\sigma; E_2) \subset \mathscr{M}(\sigma; E_1 + E_2). \tag{2}$$

We note also the following obvious identity:

$$\sigma_\mu(axb) = a\sigma_\mu(x)b \quad (a, b \in \mathscr{M}^\sigma, x \in \mathscr{M}, \mu \in \mathscr{M}(G)). \tag{3}$$

15.4. Let $\sigma : G \to Aut(\mathscr{M})$ be a continuous action of G on $\mathscr{M}$.

If $0 \neq e \in \mathscr{M}^\sigma$ is a projection, then σ obviously defines a continuous action $\sigma^e : G \to Aut(e\mathscr{M}e)$ of G on the reduced W^*-algebra $e\mathscr{M}e : \sigma^e_t = \sigma_t|e\mathscr{M}e$ $(t \in G)$.

It is easy to check that for every $x \in e\mathscr{M}e$ we have

$$Sp_{\sigma^e}(x) = Sp_\sigma(x). \tag{1}$$

For every set $E \subset \hat{G}$ we shall write $\mathscr{M}(\sigma^e; E)$ instead of $e\mathscr{M}e(\sigma^e; E)$. From (1) it follows that

$$\mathscr{M}(\sigma^e; E) = \mathscr{M}(\sigma; E) \cap e\mathscr{M}e. \tag{2}$$

In particular,

$$(e\mathscr{M}e)^{\sigma^e} = \mathscr{M}^\sigma \cap e\mathscr{M}e. \tag{3}$$

If e_1 and e_2 are both projections in $\mathscr{M}^\sigma$, then

$$0 \neq e_1 \leq e_2 \Rightarrow Sp\, \sigma^{e_1} \subset Sp\, \sigma^{e_2}. \tag{4}$$

Let $0 \neq e \in \mathscr{M}^\sigma$ be a projection and denote by $\bar{e} \in \mathscr{Z}(\mathscr{M}^\sigma)$ the central support of e in $\mathscr{M}^\sigma$. Then we have

$$Sp\, \sigma^{\bar{e}} = Sp\, \sigma^e. \tag{5}$$

Indeed, let $E \subset \hat{G}$ be any set and suppose that there exists $0 \neq x \in \mathscr{M}(\sigma^{\bar{e}}; E) = \mathscr{M}(\sigma; E) \cap \bar{e}\mathscr{M}\bar{e}$. Since $\bar{e} = \vee\{ueu^*; u \in U(\mathscr{M}^\sigma)\}$, we can find $u, v \in U(\mathscr{M}^\sigma)$ such that $ueu^*xvev^* \neq 0$. Then $0 \neq eu^*xve \in e\mathscr{M}e$ and $eu^*xve \in \mathscr{M}(\sigma; E)$ as $e, u, v \in \mathscr{M}^\sigma, x \in \mathscr{M}(\sigma; E)$ (15.3.(2)), so that $\mathscr{M}(\sigma^e; E) = \mathscr{M}(\sigma; E) \cap e\mathscr{M}e \neq \{0\}$. Thus, (5) follows using Proposition 14.5.

15.5. Let $\sigma : G \to Aut(\mathscr{M})$ be a continuous action of G on the von Neuman algebra $\mathscr{M} \subset \mathscr{B}(\mathscr{H})$.

Let $E \subset \hat{G}$ be an arbitrary set. We consider the w-closed left ideal $\mathscr{T}(\sigma; E) = \{x \in \mathscr{M}; x\mathscr{M}(\sigma; E) = 0\}$ of $\mathscr{M}$ and denote by $q(\sigma; E)$ the greatest projection in $\mathscr{T}(\sigma; E)$, that is,

$$\mathscr{T}(\sigma; E) = \mathscr{M}q(\sigma; E).$$

From this definition, it follows that

$$1 - q(\sigma; E) = \vee\{\mathbf{l}(x); x \in \mathscr{M}(\sigma; E)\} \tag{1}$$

or, equivalently,

$$(1 - q(\sigma; E))\mathscr{H} = \overline{\mathscr{M}(\sigma; E)\mathscr{H}}. \tag{2}$$

The spectral subspace $\mathscr{T}(\sigma; E)$ is σ-invariant (14.2.(4)) and also invariant under left or right multiplications by elements in $\mathscr{M}^\sigma$ (15.3.(2), 14.3.(14)). It follows that $\mathscr{T}(\sigma; E)$ enjoys the same properties and hence

$$q(\sigma; E) \in \mathscr{Z}(\mathscr{M}^\sigma). \tag{3}$$

If $\theta \in Aut(\mathscr{M})$ and $\theta \circ \sigma_t = \sigma_t \circ \theta$ $(t \in G)$, then, by Corollary 2/14.7, $\theta(\mathscr{M}(\sigma; E)) = \mathscr{M}(\sigma; E)$, and hence

$$\theta \in Aut(\mathscr{M}), \theta \circ \sigma_t = \sigma_t \circ \theta \Rightarrow \theta(q(\sigma; E)) = q(\sigma; E). \tag{4}$$

15.6. We now restrict ourselves to the case $G = \mathbb{R}$, and consider a continuous action $\sigma : \mathbb{R} \to Aut(\mathscr{M})$ of $\mathbb{R}$ on the von Neumann algebra $\mathscr{M} \subset \mathscr{B}(\mathscr{H})$ identifying $\hat{G}$ with $\mathbb{R}$ so that $\langle t, s\rangle = e^{ist}(t \in G = \mathbb{R}, s \in \hat{G} = \mathbb{R})$.

By the construction in Section 15.5, for each $t \in \hat{G} = \mathbb{R}$ we have a projection

$$q_t^\sigma = q(\sigma; (t, +\infty)) \in \mathscr{Z}(\mathscr{M}^\sigma) \tag{1}$$

such that

$$1 - q_t^\alpha = \vee\{\mathbf{l}(x); x \in \mathscr{M}(\sigma; (t, +\infty))\} = \overline{\mathscr{M}(\sigma; (t, +\infty))\mathscr{H}}. \tag{2}$$

We also consider the projection

$$q_\infty^\sigma = \bigvee_{t \in \mathbb{R}} q_t^\sigma \in \mathscr{Z}(\mathscr{M}^\sigma); \tag{3}$$

clearly

$$1 - q_\infty^\sigma = \bigwedge_{t \in \mathbb{R}} \overline{\mathscr{M}(\sigma; (t, +\infty))\mathscr{H}}. \tag{4}$$

Lemma. *The mapping* $\mathbb{R} \ni t \mapsto q_t^\sigma \in Proj(\mathscr{Z}(\mathscr{M}^\sigma))$ *has the following properties:*

$$t_1 \le t_2 \Rightarrow q_{t_1}^\sigma \le q_{t_2}^\sigma, \tag{a}$$

$$t_n \ge t, t_n \to t \Rightarrow q_{t_n}^\sigma \xrightarrow{s} q_t^\sigma, \tag{b}$$

$$t_n \to +\infty \Rightarrow q_{t_n}^\sigma \xrightarrow{s} q_\infty^\sigma, \tag{c}$$

$$t < 0 \Rightarrow q_t^\sigma = 0. \tag{d}$$

If there exists $\varepsilon > 0$ *such that* $Sp\, \sigma \subset [-\varepsilon, \varepsilon]$, *then*

$$t > \varepsilon \Rightarrow q_t^\sigma = 1. \tag{e}$$

Proof. For the proof put $\mathscr{M}(t) = \mathscr{M}(\sigma; (t, +\infty)), \mathcal{Q}(t) = \mathcal{Q}(\sigma; (t, +\infty))$ the left annihilator of $\mathscr{M}(t)$ in $\mathscr{M}_*$, $q_t = q_t^\sigma$ and $q_\infty = q_\infty^\sigma$. We have $\mathscr{T}(t) = \mathscr{M} q_t$ and $q_\infty = \vee_t q_t$.

Statement (a) follows from the fact that the mapping $t \mapsto \mathscr{M}(t)$ is decreasing and hence the mapping $t \mapsto \mathscr{T}(t)$ is increasing. Statement (c) is now obvious as $q_\infty = \vee_t q_t$.

Let $t_n \ge t, t_n \to t$. Since the mapping $t \mapsto q_t$ is increasing, in order to show that $q_{t_n} \xrightarrow{s} q_t$ we may assume that the sequence $\{t_n\}$ is decreasing. Then (14.3.(4)) $\mathscr{M}(t) = \vee_n \mathscr{M}(t_n)$, hence $\mathscr{T}(t) = \bigcap_n \mathscr{T}(t_n)$ and $q_t = \wedge_n q_{t_n} = s\text{-}\lim_n q_{t_n}$. We have thus proved statement (b).

If $t < 0$, then $\mathscr{M}(t) \supset \mathscr{M}^\sigma \ni 1$, hence $\mathscr{T}(t) = 0$ and $q_t = 0$.

If $Sp\, \sigma \subset [-\varepsilon, \varepsilon]$, then for $t > \varepsilon$ we have $\mathscr{M}(t) = \{0\}$, hence $\mathscr{T}(t) = \mathscr{M}$ and $q_t = 1$. □

15.7. Continuing with the notation of the previous section, we assume that

$$q_\infty^\sigma = 1.$$

Then Lemma 15.6 shows that the mapping $\mathbb{R} \ni t \mapsto q_t^\sigma \in \mathscr{Z}(\mathscr{M}^\sigma)$ defines a spectral resolution of the identity on the Hilbert space $\mathscr{H}$ (see Dunford & Schwartz, 1958, 1963, XI.5) and hence a self-adjoint operator A in $\mathscr{H}$ affiliated to the von Neumann algebra $\mathscr{Z}(\mathscr{M}^\sigma)$,

$$A = \int_{-\infty}^{+\infty} t \, \mathrm{d}q_t^\sigma, \tag{1}$$

and an *so*-continuous unitary representation $u : \mathbb{R} \to \mathscr{Z}(\mathscr{M}^\sigma) \subset \mathscr{M}$,

$$u_s = \exp(\mathrm{i}sA) \quad (s \in \mathbb{R}). \tag{2}$$

Since $q_t^\sigma = 0$ for $t < 0$, the self-adjoint operator A is positive. Let $p_u(\cdot)$ be the Stone spectral measure associated with u (14.10). Using Proposition 14.10 and 14.10.(4), 14.10.(5), we get

$$\mathscr{H}(u; (t, +\infty)) = p_u((t, +\infty))\mathscr{H} = \chi_{(t,+\infty)}(A)\mathscr{H} = (q_\infty^\sigma - q_t^\sigma)\mathscr{H} \quad (t \in \mathbb{R}).$$

Since, by assumption, $q_\infty^\sigma = 1$, we obtain using 15.6.(2)

$$\mathscr{H}(u; (t, +\infty)) = \overline{\mathscr{M}(\sigma; (t, +\infty))\mathscr{H}} \quad (t \in \mathbb{R}). \tag{3}$$

Since (15.3.(2)) $\mathscr{M}(\sigma;(s,+\infty))\mathscr{M}(\sigma;(t,+\infty)) \subset \mathscr{M}(\sigma;(s+t,+\infty))$, we infer from (3) that

$$\mathscr{M}(\sigma;(s,+\infty))\mathscr{H}(u;(t,+\infty)) \subset \mathscr{H}(u;(s+t,+\infty)) \quad (s,t \in \mathbb{R}). \tag{4}$$

Using Corollary 14.10, we further deduce from (4) that

$$\mathscr{M}(\sigma;(s,+\infty)) \subset \mathscr{M}(Ad(u);(s,+\infty)) \quad (s \in \mathbb{R}). \tag{5}$$

Finally, with the help of Proposition 15.2 we conclude from (5) that

$$\sigma_t = Ad(u_t) \quad (t \in \mathbb{R}). \tag{6}$$

Recall that the *so*-continuous unitary representation $u : \mathbb{R} \to \mathscr{M}$ defined above has positive spectrum,

$$Sp\, u = Sp(A) \subset [0,+\infty),$$

that is, its infinitesimal generator A is positive.

If $Sp\,\sigma \subset [-\varepsilon,\varepsilon]$, then (15.6.(e)) $q_t^\sigma = 1$ for every $t > \varepsilon$, in particular $q_\infty^\sigma = 1$, and from the definition (1) of the operator A it follows that

$$0 \le A \le \varepsilon. \tag{7}$$

15.8. As a conclusion to the above considerations we have the following remarkable result:

Theorem (Borchers & Arveson). *Let $\sigma : \mathbb{R} \to Aut(\mathscr{M})$ be a continuous action of $\mathbb{R}$ on the von Neuman algebra $\mathscr{M} \subset \mathscr{B}(\mathscr{H})$. The following statements are equivalent:*

(i) *there exists an so-continuous unitary representation $u : \mathbb{R} \to \mathscr{B}(\mathscr{H})$ with positive spectrum such that $\sigma_t = Ad(u_t)(t \in \mathbb{R})$;*
(ii) *there exists an so-continuous unitary representation $u : \mathbb{R} \to \mathscr{M}$ with positive spectrum such that $\sigma_t = Ad(u_t)(t \in \mathbb{R})$;*
(iii) $q_\infty^\sigma = 1$; *and*
(iv) $\bigcap_{t\in\mathbb{R}} \overline{\mathscr{M}(\sigma;(t,+\infty))\mathscr{H}} = \{0\}$.

Proof. (i) ⇒ (iv). Since $Sp\,u \subset [0,+\infty)$ we have $\mathscr{H} = \mathscr{H}(u;[0,+\infty))$. Let $p_u(\cdot)$ be the Stone spectral measure associated with u (14.10). Since $\sigma = Ad(u)$, using Corollary 1/14.7 and Proposition 14.10, we obtain, for every $t \in \mathbb{R}$, $\mathscr{M}(\sigma;(t,+\infty))\mathscr{H} = \mathscr{M}(\sigma;(t,+\infty))\mathscr{H}(u;[0,+\infty)) \subset \mathscr{H}(u;(t,+\infty)) = p_u((t,+\infty))\mathscr{H}$ hence

$$\bigcap_{t\in\mathbb{R}} \overline{\mathscr{M}(\sigma;(t,+\infty))\mathscr{H}} \subset \bigcap_{t\in\mathbb{R}} p_u((t,+\infty))\mathscr{H} = \left(\bigwedge_{t\in\mathbb{R}} p_u((t,+\infty))\right)\mathscr{H} = \{0\}.$$

(iv) ⇒ (iii). Follows from 15.6.(4).
(iii) ⇒ (iii). This was proved in Section 15.7.
(ii) ⇒ (i). Obvious. □

Clearly, the theorem still remains valid if we replace the condition concerning the positivity of the spectrum by the requirement that the spectrum be left or right bounded.

15.9. Let $\sigma : \mathbb{R} \to Aut(\mathcal{M})$ be a continuous action of $\mathbb{R}$ on the von Neumann algebra $\mathcal{M} \subset \mathcal{B}(\mathcal{H})$ and assume that $q^{\sigma}_{\infty} = 1$. Then the *so*-continuous unitary representation $u : \mathbb{R} \to \mathcal{M}$ constructed in Section 15.7 has the following *minimality property*:

If $v : \mathbb{R} \to \mathcal{B}(\mathcal{H})$ *is any so-continuous unitary representation with positive spectrum such that* $\sigma_s = Ad(v_s)(s \in \mathbb{R})$, then

$$\mathcal{H}(u; (t, +\infty)) \subset: \mathcal{H}(v; (t, +\infty)) \quad (t \in \mathbb{R}),$$

that is, for the corresponding infinitesimal generators A, B *we have*

$$\chi_{(t,+\infty)}(A) \leq \chi_{(t,+\infty)}(B) \quad (t \in \mathbb{R}).$$

Indeed, since $\mathcal{H} = \mathcal{H}(v; [0, +\infty))$ and $\sigma = Ad(v)$, using Corollary 1/14.7 we obtain $\mathcal{M}(\sigma; (t, +\infty))\mathcal{H} = \mathcal{M}(\sigma; (t, +\infty))\mathcal{H}(v; [0, +\infty)) \subset \mathcal{H}(v; (t, +\infty))$, so that the desired conclusion follows using 15.7.(3).

15.10. The arguments presented in Section 15.7 also give the following:

Proposition. *Let* $\sigma : \mathbb{R} \to Aut(\mathcal{M})$ *be a continuous action of* $\mathbb{R}$ *on the* W^**-algebra* $\mathcal{M}$. *If there exists* $\varepsilon > 0$ *such that* $Sp\, \sigma \subset [-\varepsilon, \varepsilon]$, *then there exists* $a = a^* \in \mathcal{Z}(\mathcal{M}^{\sigma})$, $\|a\| \leq \varepsilon/2$, *such that* $\sigma_t = Ad(exp(ita))$
$(t \in \mathbb{R})$.

Proof. By the last remark in Section 15.7, there exists $A \in \mathcal{Z}(\mathcal{M}^{\sigma}), 0 \leq A \leq \varepsilon$, such that $\sigma_t = Ad(\exp(itA))(t \in \mathbb{R})$, and we can take $a = A - \frac{\varepsilon}{2}$. □

15.11. Let $\sigma : G \to Aut(\mathcal{M})$ and $\tau : G \to Aut(\mathcal{M})$ be two continuous actions of G on $\mathcal{M}$. We shall say that the actions σ and τ are outer conjugate, and we shall write $\sigma \sim \tau$, if there exists a unitary σ-cocycle $w \in Z_{\alpha}(G; U(\mathcal{M}))$ (see 5.1) such that $\tau_t = Ad(w_t) \circ \sigma_t (t \in \mathbb{R})$. It is easy to check that "$\sim$" is an equivalence relation.

In this case, we can define the *balanced action* $\theta = \theta(\sigma, w)$ of G on the W^*-algebra $\mathcal{P} = Mat_2(\mathcal{M})$ by putting

$$\theta_t \begin{pmatrix} x_{11} & x_{12} \\ x_{21} & x_{22} \end{pmatrix} = \begin{pmatrix} \sigma_t(x_{11}) & \sigma_t(x_{12})w_t^* \\ w_t\sigma_t(x_{21}) & w_t\sigma_t(x_{22})w_t^* \end{pmatrix} \quad ([x_{ij}] \in \mathcal{P}). \tag{1}$$

Since $w \in Z_{\sigma}(G, U(\mathcal{M}))$, putting $v_t = \begin{pmatrix} 1 & 0 \\ 0 & w_t \end{pmatrix}$ $(t \in G)$, we define a unitary $(\sigma \bar{\otimes} 1)$-cocycle $v \in Z_{\sigma \bar{\otimes} s}(G, U(\mathcal{P}))$ such that $\theta_t = Ad(v_t) \circ (\sigma \bar{\otimes} \iota)$ $(t \in G)$. Thus, $\theta : G \to Aut(\mathcal{P})$ is indeed a continuous action of G on $\mathcal{P}$. It is clear that for $t \in G$ and $x \in \mathcal{M}$ we have

$$\theta_t \begin{pmatrix} x & 0 \\ 0 & 0 \end{pmatrix} = \begin{pmatrix} \sigma_t(x) & 0 \\ 0 & 0 \end{pmatrix}, \quad \theta_t \begin{pmatrix} 0 & 0 \\ 0 & x \end{pmatrix} = \begin{pmatrix} 0 & 0 \\ 0 & \tau_t(x) \end{pmatrix}. \tag{2}$$

The projections $e = \begin{pmatrix} 1 & 0 \\ 0 & 0 \end{pmatrix}, f = \begin{pmatrix} 0 & 0 \\ 0 & 1 \end{pmatrix}$ are equivalent in $\mathscr{P}$ and we have $e, f \in \mathscr{P}^{\theta}$. Moreover, from (2) it follows that

$$(\mathscr{M}, \sigma) \approx (e\mathscr{P}e, \theta^e), \quad (\mathscr{M}, \tau) \approx (f\mathscr{P}f, \theta^f), \tag{3}$$

In particular, for every $x \in \mathscr{M}$ we have

$$Sp_{\sigma}(x) = Sp_{\theta} \begin{pmatrix} x & 0 \\ 0 & 0 \end{pmatrix}, Sp_{\tau}(x) = Sp_{\theta} \begin{pmatrix} 0 & 0 \\ 0 & x \end{pmatrix}. \tag{4}$$

Note that, conversely, if there exists a continuous action $\theta : G \to Aut(Mat_2(\mathscr{M}))$ which satisfies (2), then the actions σ and τ are outer conjugate.

15.12. Using the results of Sections 15.10 and 15.11, we now prove a result which we shall need later.

Proposition. *Let $\sigma : \mathbb{R} \to Aut(\mathscr{M})$ be a continuous action of $\mathbb{R}$ on the W^*-algebra $\mathscr{M}$ and $\varepsilon > 0$ such that*

$$Sp\, \sigma \cap \{[-2\varepsilon, -\varepsilon] \cup [\varepsilon, 2\varepsilon]\} = \emptyset. \tag{1}$$

Then there exists $a = a^* \in \mathscr{Z}(\mathscr{M}^{\sigma})$, $\|a\| \leq \varepsilon/2$, such that for the continuous action $\tau \in \mathbb{R} \to Aut(\mathscr{M})$ defined by $\tau_t = Ad(exp(ita)) \circ \sigma_t$ $(t \in \mathbb{R})$, we have

$$Sp\, \tau \cap (-\varepsilon, \varepsilon) = \{0\}. \tag{2}$$

Proof. From (1) and statements 15.2.(3), 15.3.(2), it follows that the spectral subspace $\mathscr{N} = \mathscr{M}(\sigma; [-\varepsilon, \varepsilon])$ is a σ-invariant unital W^*-subalgebra of $\mathscr{M}$. Let $\sigma'_t = \sigma_t|\mathscr{N}$ $(t \in \mathbb{R})$. Then $\sigma' : \mathbb{R} \to Aut(\mathscr{N})$ is a continuous action of $\mathbb{R}$ on $\mathscr{N}$, $\mathscr{N}^{\sigma'} = \mathscr{M}^{\sigma}$ and $Sp\, \sigma' \subset [-\varepsilon, \varepsilon]$. By Proposition 15.10, there exists $a = a^* \in \mathscr{Z}(\mathscr{N}^{\sigma'}) = \mathscr{Z}(\mathscr{M}^{\sigma})$, $\|a\| \leq \varepsilon/2$, such that $\sigma'_t = Ad(\exp(ita))$ $(t \in \mathbb{R})$. Since $a \in \mathscr{M}^{\sigma}$, the equation $\tau_t(x) = \exp(ita)\sigma_t(x)\exp(-ita)$ $(x \in \mathscr{M}, t \in \mathbb{R})$, defines a continuous action $\tau : \mathbb{R} \to Aut(\mathscr{M})$.

By (1) and Lemma 15.1, we see that the set $\mathscr{N} \cup \mathscr{M}(\sigma; \mathbb{R}\setminus[-2\varepsilon, 2\varepsilon])$ is w-total in $\mathscr{M}$. Thus, in order to prove (2), it is sufficient (14.5.(1)) to show that

$$x \in \mathscr{N} \cup \mathscr{M}(\sigma; \mathbb{R}\setminus[-2\varepsilon, 2\varepsilon]) \Rightarrow Sp_{\tau}(x) \cap (-\varepsilon, \varepsilon) = \{0\}.$$

Since $\mathscr{N} \subset \mathscr{M}^{\tau}$, for $x \in \mathscr{N}$ we obviously have $Sp_{\tau}(x) = \{0\}$.

Consider now the balanced action $\theta = \theta(\sigma, w)$ of $\mathbb{R}$ on $\mathscr{P} = Mat_2(\mathscr{M})$ constructed as in Section 15.11 with $w_t = \exp(ita)$ $(t \in \mathbb{R})$. Then we have

$$\theta_t \begin{pmatrix} 0 & 0 \\ 1 & 0 \end{pmatrix} = \begin{pmatrix} 0 & 0 \\ \exp(ita) & 0 \end{pmatrix} \quad (t \in \mathbb{R})$$

and hence for every $f \in \mathscr{L}^1(\mathbb{R})$ we get (see 14.10.(6))

$$\theta_f \begin{pmatrix} 0 & 0 \\ 1 & 0 \end{pmatrix} = \begin{pmatrix} 0 & 0 \\ \hat{f}(a) & 0 \end{pmatrix}.$$

It follows that

$$Sp_\theta \begin{pmatrix} 0 & 0 \\ 1 & 0 \end{pmatrix} = Sp(a) \subset [-\varepsilon/2, \varepsilon/2]. \tag{3}$$

Let $x \in \mathscr{M}(\sigma; \mathbb{R}, [-2\varepsilon, 2\varepsilon])$. Since $\begin{pmatrix} 0 & 0 \\ 0 & x \end{pmatrix} = \begin{pmatrix} 0 & 0 \\ 1 & 0 \end{pmatrix}\begin{pmatrix} x & 0 \\ 0 & 0 \end{pmatrix}\begin{pmatrix} 0 & 1 \\ 0 & 0 \end{pmatrix}$, using 15.3.(2), 15.2.(2), 15.11.(4), and (3) above, we get $Sp_\tau(x) \subset [-\varepsilon/2, \varepsilon/2] + \{\mathbb{R} \setminus [-2\varepsilon, 2\varepsilon]\} + [-\varepsilon/2, \varepsilon/2]$; hence $Sp_\tau(x) \cap (-\varepsilon, \varepsilon) = \{0\}$. □

15.13. We now give another application of the technique developed in Section 15.7.

Let $\mathscr{M}$ be a W^*-algebra. A linear mapping $\delta : \mathscr{M} \to \mathscr{M}_*$ such that $\delta(xy) = x\delta(y) + \delta(x)y\ (x, y \in \mathscr{M})$ is called a *derivation* on $\mathscr{M}$. Every derivation on a W^*-algebra is bounded (Sakai, 1971, 4.1.3).

Let δ be a derivation on $\mathscr{M}$. Then the mappings $\delta_1, \delta_2 : \mathscr{M} \to \mathscr{M}$ defined by

$$\delta_1(x) = (\delta(x) - \delta(x^*)^*)/2, \delta_2(x) = (\delta(x) + \delta(x^*)^*)/2i$$

are *antihermitian derivations*, that is, $\delta_k(x^*) = -\delta_k(x)^* (k = 1, 2)$ and $\delta = \delta_1 + i\delta_2$.

Every element $a \in \mathscr{M}$ determines an *inner derivation* δ_a on $\mathscr{M}$ defined by $\delta_a(x) = ax - xa (x \in \mathscr{M})$.

Theorem (Kadison & Sakai). *Every derivation on a W^*-algebra is inner.*

Proof. Without loss of generality, we consider an antihermitian derivation δ on the W^*-algebra $\mathscr{M}$. It is then easy to check that the elements $\sigma_t = \exp(it\delta) (t \in \mathbb{R})$ of the Banach algebra $\mathscr{B}(\mathscr{M})$ are $*$-automorphisms of $\mathscr{M}$ and that the mapping $\sigma : \mathbb{R} \ni t \mapsto \sigma_t \in Aut(\mathscr{M})$ is a norm-continuous action of $\mathbb{R}$ on $\mathscr{M}$. In Section 14.11, we proved that $Sp\,\sigma = Sp_{\mathscr{B}(\mathscr{M})}(\delta) \subset [-\|\delta\|, \|\delta\|]$.

As we saw in Section 15.7, $a = \int t\, \mathrm{d}q_t^\sigma$ is a positive element of $\mathscr{M}$, $\|a\| \le \|\delta\|$ and $\sigma_t = Ad(\exp(ita))\ (t \in \mathbb{R})$, that is,

$$(\exp(it\delta))(x) = \exp(ita)x\exp(-ita) \quad (x \in \mathscr{M}, t \in \mathbb{R}).$$

Taking the derivative here with respect to t and then putting $t = 0$, it follows that $\delta(x) = ax - xa$ $(x \in \mathscr{M})$, that is, $\delta = \delta_a$. □

15.14. *For an antihermitian derivation δ of $\mathscr{M}_*$, the element $a \in \mathscr{M}$ constructed in the proof of Theorem 15.13 is the smallest positive element $b \in \mathscr{M}$ such that $\delta = \delta_b$; moreover, for every central projection e in $\mathscr{M}$ we have*

$$\|ae\| = \|\delta | e\mathscr{M}e\|.$$

Indeed, let $e \in \mathscr{M}$ be a central projection. We have $\delta(e) = \delta_a(e) = 0$. It follows that $\delta(exe) = e\delta(x)e\ (x \in \mathscr{M})$, and $\sigma_t(e) = e (t \in \mathbb{R})$, that is, $e \in \mathscr{M}^\sigma$. Consequently, the restriction of δ to $e\mathscr{M}e$ is a derivation δ^e on the W^*-algebra $e\mathscr{M}e$ and $\exp(it\delta^e) = \sigma_t^e (t \in \mathbb{R})$. By the proof of Theorem 15.13, we have $\sigma_t^e = Ad(\exp(ita'))(t \in \mathbb{R})$, where the positive operator $a' \in e\mathcal{M}e$ with $\|a'\| \leq \|\delta^e\|$ is defined by $a' = \int t dq_t^{\sigma^e}$. It is easy to check that $q_t^{\sigma^e} = eq_t^\sigma = q_t^\sigma e \ (t \in \mathbb{R})$, hence $a' = ae$ and therefore

$$\|ae\| \leq \|\delta|e\mathcal{M}e\|.$$

If $c = ae - \|ae\|/2$, then $\|c\| \leq \|ae\|/2$ and $\delta_c = \delta_{ae} = \delta^e$, hence $\|c\| > \|\delta^e\|/2$. Consequently

$$\|ae\| > \|\delta|e\mathcal{M}e\|. \tag{1}$$

Consider now $0 \leq b \in \mathcal{M}$ with $\delta = \delta_b$. Then the mapping $v : \mathbb{R} \to e\mathcal{M}e$ defined by $v_t = \exp(itbe) \ (t \in \mathbb{R})$ is an *so*-continuous unitary representation with positive spectrum and $\sigma_t^e = Ad(v_t)(t \in \mathbb{R})$. By the minimality property established in Section 15.9, we have

$$\chi_{(t,+\infty)}(ae) \leq \chi_{(t,+\infty)}(be) \quad (t \in \mathbb{R}).$$

In particular, taking $t = \|be\|$, we obtain

$$\|ae\| \leq \|be\| \quad (e \in Proj(\mathcal{Z}(\mathcal{M}))). \tag{2}$$

Assume that $(a - b)^+ \neq 0$. Then there exist $\lambda > 0$ and a spectral projection $e \neq 0$ of $(a - b)$ such that

$$(a - b)e \geq \lambda e. \tag{3}$$

Since $\delta_a = \delta_b$, we have $a - b \in \mathcal{Z}(\mathcal{M})$, hence $(a - b)^+ \in \mathcal{Z}(\mathcal{M})$ and $e \in Proj(\mathcal{Z}(\mathcal{M}))$. Now from (3) we deduce that $\|be\| < \|be\| + \lambda = \|be + \lambda\| \leq \|ae\|$, contradicting (2). Thus, $(a - b)^+ = 0$ and $a \leq b$.

15.15. Using Theorem 15.13, we can also obtain a similar result for automorphisms of W^*-algebras. To this end it is necessary to consider, when possible, the logarithm of an automorphism and to show that it is a derivation.

We shall denote by ln the principal branch of the logarithm defined on the domain $\mathbb{C}\setminus(-\infty, 0]$. Then

$$\ln : \mathbb{C}\setminus(-\infty, 0] \to \{\lambda \in \mathbb{C}; |Im\ \lambda| < \pi\}$$

is an analytic function, namely the inverse function of

$$\exp : \{\lambda \in \mathbb{C}; |Im\ \lambda| < \pi\} \to \mathbb{C}\setminus(-\infty, 0].$$

Let $\mathcal{A}$ be a unital Banach algebra and $\sigma \in \mathcal{A}$ such that

$$Sp_{\mathcal{A}}(\sigma) \subset \{\lambda \in \mathbb{C}; Re\ \lambda > 0\}. \tag{1}$$

Then σ is an invertible element and so defines a bounded linear mapping

$$Ad(\sigma) : \mathcal{A} \ni \alpha \mapsto \sigma\alpha\sigma^{-1} \in \mathcal{A}.$$

Since $Sp_{\mathscr{A}}(\sigma) \subset \mathbb{C}\setminus(-\infty, 0]$, it is meaningful to consider $\delta = \ln(\sigma) \in \mathscr{A}$. We have $\sigma = \exp(\delta)$ and, by (1),

$$Sp_{\mathscr{A}}(\delta) \subset \{\lambda \in \mathbb{C}; |Im\ \lambda| < \pi/2\}. \tag{2}$$

$\delta \in \mathscr{A}$ defines a bounded linear mapping

$$ad(\delta) : \mathscr{A} \ni \alpha \mapsto \delta\alpha - \alpha\delta \in \mathscr{A}.$$

Then $Ad(\sigma)$ and $ad(\delta)$ are elements of the Banach algebra $\mathscr{B}(\mathscr{A})$ and, using 14.13.(1), 14.13.(2), we deduce from (1), (2) that

$$Sp_{\mathscr{B}(\mathscr{A})}(Ad(\sigma)) \subset \{\lambda\mu^{-1}; \lambda, \mu \in Sp_{\mathscr{A}}(\sigma)\} \subset \mathbb{C}\setminus(-\infty, 0] \tag{3}$$

$$Sp_{\mathscr{B}(\mathscr{A})}(ad(\delta)) \subset \{\lambda - \mu; \lambda, \mu \in Sp_{\mathscr{A}}(\delta)\} \subset \{\lambda \in \mathbb{C}; |Im\ \lambda| < \pi\}. \tag{4}$$

Since $\sigma = \exp(\delta)$, it is easy to check that

$$Ad(\sigma) = \exp(ad(\delta)); \tag{5}$$

and using (3) and (4) we deduce that

$$ad(\delta) = \ln(Ad(\sigma)). \tag{6}$$

Now *let $\mathscr{M}$ be a unital Banach algebra and $\mathscr{A} = \mathscr{B}(\mathscr{M})$. If $\sigma \in \mathscr{A}$ is an automorphism of $\mathscr{M}$ satisfying (1), then $\delta = \ln(\sigma) \in \mathscr{A}$ is a derivation on $\mathscr{M}$.*

Indeed, consider the mapping $L : \mathscr{M} \to \mathscr{A}$ defined by $L_x(y) = xy$ $(x, y \in \mathscr{M})$. Then L is a bounded injective Banach algebra homomorphism and we have $(Ad(\sigma))(L_x) = L_{\sigma(x)}$ $(x \in \mathscr{M})$, that is, $Ad(\sigma) \circ L = L \circ \sigma$. It follows that $p(Ad(\sigma)) \circ L = L \circ p(\sigma)$ for every polynomial p and, using Runge's theorem (Hormander, 1966, 1.3.2), we also obtain $\ln(Ad(\sigma)) \circ L = L \circ \ln(\sigma)$. By (6), $ad(\delta) \circ L = L \circ \delta$. Since $ad(\delta)$ is a derivation of $\mathscr{A}$ and L is an injective homomorphism, we conclude that δ is a derivation on $\mathscr{M}$.

If δ is an inner derivation, that is, $\delta = \delta_a = ad(a)$ with $a \in \mathscr{M}$, then $\sigma = \exp(\delta) = \exp(ad(a)) = Ad(\exp(a))$.

Using Theorem 3.13, we thus obtain the following:

Corollary. *Let σ be an automorphism of the W^*-algebra $\mathscr{M}$ such that $Sp_{\mathscr{B}(\mathscr{M})}(\sigma) \subset \{\lambda \in \mathbb{C}; Re\ \lambda > 0\}$. There exists an invertible element $v \in \mathscr{M}$ such that $\sigma = Ad(v)$, that is,*

$$\sigma(x) = vxv^{-1} \quad (x \in \mathscr{M}).$$

*If, moreover, σ is a *-automorphism, then there exists a unitary element $u \in \mathscr{M}$ such that $\sigma = Ad(u)$.*

Proof. The first statement has already been proved. Assume that $\sigma = Ad(v)$ is a *-automorphism and let $v = ua$ be the polar decomposition of v. Then $u \in \mathscr{M}$ is a unitary element and $a \in \mathscr{M}$ is an invertible positive element. For every $x \in \mathscr{M}$, we have $uax^*a^{-1}u^* = \sigma(x^*) = \sigma(x)^* = ua^{-1}x^*au^*$. It follows that $a^2 \in \mathscr{Z}(\mathscr{M})$, so that $a \in \mathscr{Z}(\mathscr{M})$ and $\sigma = Ad(u)$. □

We note that if $\sigma \in Aut(\mathcal{M})$ and $\|\sigma - \iota_{\mathcal{M}}\| < 1$, then the requirement $Sp_{\mathcal{B}(\mathcal{M})}(\sigma) \subset \{\lambda; Re\,\lambda > 0\}$ is satisfied and hence $\sigma \in Int(\mathcal{M})$.

Actually, Kadison and Ringrose (1967) proved that if $\sigma \in Aut(\mathcal{M})$ and $\|\sigma - \iota_{\mathcal{M}}\| < 2$, then $\sigma \in Int(\mathcal{M})$ (see also Borchers, 1973, Thm. 5.4.A). Since for any *-automorphism σ we have $\|\sigma - \iota_{\mathcal{M}}\| \leq 2$, it follows that $\|\sigma - \iota_{\mathcal{M}}\| = 2$ whenever σ is outer. A simple proof of the implication $\|\sigma - \iota_{\mathcal{M}}\| < \sqrt{3} \Rightarrow \sigma \in Int(\mathcal{M})$ can be found in Dixmier (1957, 1969).

15.16. Let $\sigma : G \to Aut(\mathcal{M})$ be a continuous action of the locally compact group G on the W^*-algebra $\mathcal{M}$ such that $\sigma_t \in Int(\mathcal{M})$ for all $t \in G$. Without the assumption that the predual $\mathcal{M}_*$ of $\mathcal{M}$ is separable, this property of σ does *not* imply the existence of an s-continuous unitary representation $u : G \to \mathcal{M}$ such that $\tau_t = Ad(u_t)(t \in G)$ (see Connes, 1973a, 1.5.8.c). In the case when $\mathcal{M}_*$ is separable several positive results are known (Dixmier, 1957, 1969; Hansen, 1977; Kadison, 1965; Kallman, 1971; Moore, 1976). We note here the following result:

Theorem (Kallman & Moore). *Let $\sigma : \mathbb{R} \to Aut(\mathcal{M})$ be a continuous action of $\mathbb{R}$ on the W^*-algebra $\mathcal{M}$. If the predual $\mathcal{M}_*$ is separable and $\sigma_t \in Int(\mathcal{M})$ for all $t \in \mathbb{R}$, then there exists an s-continuous unitary representation $u : \mathbb{R} \to \mathcal{M}$ such that $\sigma_t = Ad(u_t)$ for all $t \in \mathbb{R}$.*

For the proof of the full theorem, we refer to Kallman (1971) and Moore (1976). In this section, we present some arguments, due to Hansen (1977), which lead to a simple and elementary proof of the theorem when $\mathcal{M}$ is a factor.

Let $\sigma : \mathbb{R} \to Aut(\mathcal{M})$ be a continuous action of $\mathbb{R}$ on the W^*-algebra $\mathcal{M}$ and $\{v_t\}_{t\in\mathbb{R}} \subset U(\mathcal{M})$ such that $\sigma_t = Ad(v_t)(t \in \mathbb{R})$. Then

$$v_s v_t = v_t v_s (s, t \in \mathbb{R}). \tag{1}$$

Since $Ad(v_s v_t) = \sigma_{s+t} = Ad(v_t v_s)$, there exists a mapping $c : \mathbb{R} \times \mathbb{R} \to U(\mathcal{Z}(\mathcal{M}))$ such that $\sigma_t(v_s) = v_t v_s v_t^* = c(t,s)v_s$ for all $s, t \in \mathbb{R}$. It is clear that the mapping c is separately w-continuous, $c(0,s) = c(t,t) = 1, c(s,t) = c(-t,s) = c(t,s)^*$ and $c(t+t',s) = c(t,s)c(t',s)$. It follows that $c(2t,t) = c(t,t)^2 = 1$, hence $c(q2^{-n}t,t) = c(2^{-n}t,t)^q = 1$ for any $t \in \mathbb{R}, n \in \mathbb{N}, q \in \mathbb{Z}$ and the continuity of c allows us to conclude that $c(s,t) = 1$ for all $s, t \in \mathbb{R}$.

Assume now that $\mathcal{M}$ is a *factor*. Then the elements $v_t \in U(\mathcal{M})$ with $\sigma_t = Ad(v_t)$ are uniquely determined modulo the normal subgroup $\mathbb{T} = \{\lambda \cdot 1_{\mathcal{M}}; \lambda \in \mathbb{C}, |\lambda| = 1\}$ of $U(\mathcal{M})$.

Consider the topological group $U(\mathcal{M})$ with the w-topology and the group $Int(\mathcal{M})$ with the p-topology (2.23). Then the mapping $Ad : U(\mathcal{M}) \ni u \mapsto Ad(u) \in Int(\mathcal{M})$ is the composition of the canonical quotient mapping $k : U(\mathcal{M}) \to U(\mathcal{M})/\mathbb{T}$ with an injective homomorphism

$$\alpha : U(\mathcal{M})/\mathbb{T} \to Int(\mathcal{M})$$

of the quotient topological group $U(\mathcal{M})/\mathbb{T}$ into $Int(\mathcal{M})$.

Then the equation $\nu_t = k(v_t)(t \in \mathbb{R})$ defines a group homomorphism

$$\nu : \mathbb{R} \to U(\mathcal{M})/\mathbb{T},$$

uniquely determined, such that $\alpha \circ \nu = \sigma$.

Finally assume also that the factor $\mathcal{M}$ has *separable predual*. Then $U(\mathcal{M})$ and $U(\mathcal{M})/\mathbb{T}$ are polish groups. Using the measurable selection theorem of von Neumann (Dixmier, 1957, 1969) and the continuity of the action σ, we find a Lebesgue measurable mapping $v : \mathbb{R} \to U(\mathcal{M})$ such that $Ad(v_t) = \sigma_t(t \in \mathbb{R})$. It follows that the mapping ν is a Lebesgue measurable homomorphism and hence ν is a *continuous* homomorphism.

To prove the theorem, it is sufficient to show that there exists a continuous homomorphism $u : \mathbb{R} \to U(\mathcal{M})$ such that $k(u_t) = v_t$ for $t \in \mathbb{R}$.

By statement (1), the set $G = \{(u,t) \in U(\mathcal{M})\times\mathbb{R}; k(u) = \nu_t\}$ is an abelian subgroup of $U(\mathcal{M})\times\mathbb{R}$. With the topology inherited from $U(\mathcal{M}) \times \mathbb{R}$, G is a topological group. The mapping

$$j : G \ni (u,t) \mapsto t \in \mathbb{R}$$

is an open, continuous and surjective homomorphism, whose kernel is isomorphic and homeomorphic to $\mathbb{T}$. Since $\mathbb{T}$ and $\mathbb{R}$ are locally compact groups, it follows (Hewitt & Ross, 1963, 1970, 5.25) that G is also locally compact. On the other hand, the mapping

$$i : \mathbb{T} \ni \lambda \mapsto (\lambda, 0) \in G$$

is a homeomorphism of $\mathbb{T}$ onto a compact subgroup of G. The character $\lambda \mapsto \lambda$ on $\mathbb{T}$ can be extended (Hewitt & Ross, 1963, 1970, 24.12; Rudin, 1962, 2.1.4) to a continuous character γ on G. Thus, there exists a continuous homomorhism $\gamma : G \to \mathbb{T}$ such that $\gamma \circ i = \iota_{\mathbb{T}}$. Thus, the short exact sequence

$$0 \to \mathbb{T} \xrightarrow{i} G \xrightarrow{j} \mathbb{R} \to 0$$

is split and therefore there exists a continuous homomorphism $\delta : \mathbb{R} \to G$ with $j \circ \delta = \iota_{\mathbb{R}}$, that is, a continuous homomorphism $u : \mathbb{R} \to U(\mathcal{M})$ with $k(u_t) = v_t$ $(t \in \mathbb{R})$.

15.17. In this final section, we prove a general result which we shall use later in some particular situations.

Proposition. *Let $\sigma : G \to Aut(\mathcal{M})$ be a continuous action of the separable locally compact abelian group G on the countably decomposable W*-algebra $\mathcal{M}$. Let $x \in \mathcal{M}$, $e = \mathbf{r}(x)$, $e' = \mathbf{l}(x)$ and let $x = u|x|$ the polar decomposition of x. We assume that there exist projections $f, f' \in \mathcal{M}^\sigma$, $f \geq e, f' \geq e'$, such that*

$$(Sp_\sigma(x) - Sp_\sigma(x)) \cap (Sp\ \sigma^f \cup Sp\ \sigma^{f'}) = \{0\}. \tag{1}$$

Then

$$Sp_\sigma(u) = Sp_\sigma(x), |x| \in \mathcal{M}^\sigma, e \in \mathcal{M}^\sigma, e' \in \mathcal{M}^\sigma. \tag{2}$$

If, moreover, $\mathcal{M}$ is properly infinite, then there exists a partial isometry $v \in \mathcal{M}$ such that

$$Sp_\sigma(v) \subset Sp_\sigma(x) \tag{3}$$

$$v^*v = \textit{the central support of } e \text{ in } \mathcal{M}^\sigma \tag{4}$$

$$v^*v = \textit{the central support of } e' \text{ in } \mathcal{M}^\sigma \tag{5}$$

$$v\mathcal{M}^\sigma v^* \subset \mathcal{M}^\sigma, v^*\mathcal{M}^\sigma v \subset \mathcal{M}^\sigma \tag{6}$$

$$a = v^*x \in \mathcal{M}^\sigma \tag{7}$$

$$x = va. \tag{8}$$

Proof. Let $E = Sp_\sigma(x)$. We have $x \in \mathcal{M}(\sigma; E), x^* \in \mathcal{M}(\sigma; -E)$ hence $x^*x \in \mathcal{M}(\sigma; E - E) \cap f\mathcal{M}f \subset \mathcal{M}^\sigma$, by (1). Thus, $|x| \in \mathcal{M}^\sigma$ and $Sp_\sigma(x) \subset Sp_\sigma(u)$, as $x = u|x|$. There exists a sequence $\{a_n\} \subset \mathcal{M}^\sigma$ such that $|x|a_n \xrightarrow{w} \mathbf{s}(|x|) = e$. It follows that $e \in \mathcal{M}^\sigma$ and $\mathcal{M}(\sigma; E) \ni xa_n = u|x|a_n \xrightarrow{w} u$, hence $Sp_\sigma(u) \subset Sp_\sigma(x)$. Consequently, $Sp_\sigma(u) = Sp_\sigma(x) = E$ and $e' = uu^* \in \mathcal{M}(\sigma; E - E) \cap f'\mathcal{M}f' \subset \mathcal{M}^\sigma$, by (1). We have thus proved (2).

Let $c = \mathbf{z}_{\mathcal{M}^\sigma}(e), c' = \mathbf{z}_{\mathcal{M}^\sigma}(e') \in \mathcal{Z}(\mathcal{M}^\sigma)$ be the central supports of the projections $e, e' \in \mathcal{M}^\sigma$.

There exist projections $e_0, e_1 \in \mathcal{Z}(e\mathcal{M}^\sigma e)$, uniquely determined, such that $e = e_0 + e_1, e_0\mathcal{M}^\sigma e_0$ is properly infinite, and $e_1\mathcal{M}^\sigma e_1$ is finite. Also, there exist projections $e'_0, e'_1 \in \mathcal{Z}(e'\mathcal{M}^\sigma e')$, uniquely determined, such that $e' = e'_0 + e'_1, e'_0\mathcal{M}^\sigma e'_0$ is properly infinite, and $e'_1\mathcal{M}^\sigma e'_1$ is finite.

Then there exist projections $c_0, c_1 \in \mathcal{Z}(\mathcal{M}^\sigma)$ with $c_0 + c_1 = c, c_0 e = e_0, c_1 e = e_1$ an $c'_0, c'_1 \in \mathcal{Z}(\mathcal{M}^\sigma)$ with $c'_0 + c'_1 = c', c'_0 e' = e'_0, c'_1 e' = e'_1$.

For $y \in e\mathcal{M}^\sigma e$, we have $uyu^* \in e'\mathcal{M}e'$ and $Sp_\sigma(uyu^*) \subset (E - E) \cap Sp\sigma^{e'} = \{0\}$ hence $uyu^* \in e'\mathcal{M}^\sigma e'$. Thus, the mappings

$$e\mathcal{M}^\sigma e \ni y \mapsto uyu^* \in e'\mathcal{M}^\sigma e', e'\mathcal{M}^\sigma e' \ni z \mapsto u^*zu \in e\mathcal{M}^\sigma e$$

are reciprocal *-isomorphisms. By the uniqueness of the decompositions $e = e_0 + e_1, e' = e'_0 + e'_1$, it follows that

$$ue_0u^* = e'_0, \quad ue_1u^* = e'_1, \quad u^*e'_0u = e_0, u^*e'_1u = e_1.$$

By construction, e_1 is a finite projection in $\mathcal{M}^\sigma$ with central support c_1. Since $\mathcal{M}^\sigma$ is properly infinite, we can write

$$c_1 = \sum_{n=1}^{\infty} e_n \text{ with } e_n \in Proj(\mathcal{M}^\sigma), e_n \sim e_1 \text{ in } \mathcal{M}^\sigma,$$

so there exist $w_n \in \mathcal{M}^\sigma$ with $w_n^*w_n = e_1, w_nw_n^* = e_n$ and $w_1 = e_1$.

Similarly, we can write

$$c'_1 = \sum_{n=1}^{\infty} e'_n \text{ with } e'_n \in Proj(\mathcal{M}^\sigma), e'_n \sim e'_1 \text{ in } \mathcal{M}^\sigma$$

and there exist $w'_n \in \mathcal{M}^\sigma$ with $w'^*_n w'_n = e'_1, w'_nw'^*_n = e'_n$ and $w'_1 = e'_1$.

Then $u_n = w'_n u w^*_n \in \mathcal{M}(\sigma; E), u^*_n u_n = e_n, u_n u^*_n = e'_n$, hence

$$v_1 = \sum_{n=1}^{\infty} u_n \in \mathcal{M}(\sigma; E), v_1^*v_1 = c_1, v_1v_1^* = c'_1.$$

On the other hand, e_0 is a properly infinite projection in $\mathcal{M}^\sigma$ with central support equal to c_0, hence $e_0 \sim c_0$ in $\mathcal{M}^\sigma$, that is, there exists $w \in \mathcal{M}^\sigma$ with $w^*w = e_0, ww^* = c_0$. Similarly, there exists $w' \in \mathcal{M}^\sigma$ with $w'^*w' = e'_0, w'w'^* = c'_0$. Then

$$v_0 = w'uw^* \in \mathcal{M}(\sigma; E), v_0^*v_0 = c_0, v_0v_0^* = c'_0.$$

Thus, $v = v_0 + v_1 \in \mathcal{M}(\sigma; E), v^*v = c, vv^* = c'$ and this proves (3), (4), (5).

From (1) and 15.4.(5) it follows that

$$(E - E) \cap (Sp\ \sigma^c \cup Sp\ \sigma^{c'}) = \{0\}.$$

Thus, if $y \in \mathscr{M}^\sigma$, then $vyv^* \in \mathscr{M}(\sigma; E-E) \cap c'\mathscr{M}c' = \mathscr{M}^\sigma$ and $v^*yv \in \mathscr{M}(\sigma; E-E) \cap c\mathscr{M}c = \mathscr{M}^\sigma$, proving (6).

Also, $a = v^*x \in \mathscr{M}(\sigma; E-E) \cap c\mathscr{M}c = \mathscr{M}^\sigma$ and, as $vv^* = c' \geq e' = \mathbf{l}(x)$, we have $va = vv^*x = x$, proving (7) and (8). □

Corollary. *Let* $\sigma : \mathbb{R} \to Aut(\mathscr{M})$ *be a continuous action of* $\mathbb{R}$ *on the factor* $\mathscr{M}$ *such that* $\mathscr{M}^\sigma$ *is properly infinite. Let* $\varepsilon > 0$ *and let* $p_1, p_2 \in \mathscr{Z}(\mathscr{M}^\sigma)$ *be nonzero projections such that 0 be an isolated point in* $Sp\ \sigma^{p_1} \cup Sp\ \sigma^{p_2}$*. Then there exists a partial isometry* $0 \neq v \in \mathscr{M}$ *with* $p_1 \geq v^*v \in \mathscr{Z}(\mathscr{M}^\sigma), p_2 \geq vv^* \in \mathscr{Z}(\mathscr{M}^\sigma)$*, and* $Sp_\sigma(v) - Sp_\sigma(v) \subset [-\varepsilon, \varepsilon]$.

Proof. We may assume that $(Sp\ \sigma^{p_1} \cup Sp\ \sigma^{p_2}) \cap [-\varepsilon, \varepsilon] = \{0\}$. Since $\mathscr{M}$ is a factor, there exists $y \in \mathscr{M}$ with $p_2yp_1 \neq 0$. By Lemma 15.1, there exists $f \in \mathscr{L}^1(\mathbb{R})$ with $supp\hat{f} - supp\hat{f} \subset [-\varepsilon, \varepsilon]$ and $x = p_2\sigma_f(y)p_1 = \sigma_f(p_2yp_1) \neq 0$. We have $p_2x = xp_1 = x$ and $Sp_\sigma(x) - Sp_\sigma(x) \subset [-\varepsilon, \varepsilon]$. From the proposition, we infer that $e_1 = \mathbf{r}(x) \in \mathscr{M}^\sigma, e_2 = \mathbf{l}(x) \in \mathscr{M}^\sigma$. Let $q_1 = \mathbf{z}_{\mathscr{M}^\sigma}(e_1) \leq p_1, q_2 = \mathbf{z}_{\mathscr{M}^\sigma}(e_2) \leq p_2$. Again by the proposition, there exists $0 \neq v \in \mathscr{M}$ such that $v^*v = q_1, vv^* = q_2$, and $Sp_\sigma(v) \subset Sp_\sigma(x)$, hence $Sp_\sigma(v) - Sp_\sigma(v) \subset [-\varepsilon, \varepsilon]$. □

15.18. Notes. The proofs of Borchers' theorem (15.8, (i) ⇒ (ii); Borchers, 1969) and of the Kadison–Sakai theorem (15.13; Kadison, 1966; Sakai, 1966) given in this section are due to Arveson (1974). Theorem 15.15 appears in Sakai (1966) and the proof is based on a device due to Zeller–Meier (1967). The results contained in Sections 15.10–15.12 and 15.17 are due to Connes (1973a).

For our exposition, we have used Arveson (1974), Combes and Delaroche (1975), Connes (1973a), Hansen (1977); and Olesen (1974).

Different proofs and extensions of Borchers' theorem are contained in Combes and Delaroche (1975), Kadison (1969), Olesen (1976), and Pedersen (1978). The following references also contain applications of the theory of spectral subspaces to groups of *-automorphisms on C^*algebras: Akemann et al. (1976), Bisognano and Wichmann (1975), Borchers (1973), Olesen (1974, 1975), Olesen and Pedersen (1974), and Pedersen (1973–1977).

16 The Connes Invariant $\Gamma(\sigma)$

In this section, we introduce an outer conjugacy invariant for continuous actions.

16.1. Let $\sigma : G \to Aut(\mathscr{M})$ be a continuous action of the separable commutative locally compact group G on the W^*-algebra $\mathscr{M}$. We define a closed set $\Gamma(\sigma) \subset \hat{G}$ by

$$\Gamma(\sigma) = \bigcap \{Sp\ \sigma^e; 0 \neq e \in Proj(\mathscr{M}^\sigma)\}. \quad (1)$$

According to 15.4.(5), we also have

$$\Gamma(\sigma) = \bigcap \{Sp\ \sigma^e; 0 \neq e \in Proj(\mathscr{Z}(\mathscr{M}^\sigma))\}. \quad (2)$$

In particular,

$$\textit{if } \mathscr{M}^\sigma \textit{ is a factor, then } \Gamma(\sigma) = Sp\ \sigma. \tag{3}$$

Clearly, for every nonzero projection $e \in \mathscr{M}^\sigma$ we have

$$\Gamma(\sigma) \subset \Gamma(\sigma^e). \tag{4}$$

Proposition. *$\Gamma(\sigma)$ is a closed subgroup of $\hat{G}$ and for every nonzero projection $e \in \mathscr{M}^\sigma$ we have*

$$\Gamma(\sigma) + Sp\ \sigma^e = Sp\ \sigma^e. \tag{5}$$

Proof. We remark first that $0 \in \Gamma(\sigma)$, since for every nonzero projection $e \in \mathscr{M}^\sigma$ we have $Sp\ \sigma^e \supset Sp_{\sigma^e}(e) = \{0\}$.

Therefore, $\Gamma(\sigma) + Sp\ \sigma \supset Sp\ \sigma$. Conversely, let $\gamma_1 \in Sp\ \sigma$ and $\gamma_2 \in \Gamma(\sigma)$. In order to prove that $\gamma_1 + \gamma_2 \in Sp\ \sigma$, we have to show (14.5) that $\mathscr{M}(\sigma; V) \neq \{0\}$ for every neighborhood V of $\gamma_1 + \gamma_2$. Let V_1 and V_2 be neighborhoods of γ_1 and γ_2, respectively, such that $V_1 + V_2 \subset V$. Since $\gamma_1 \in Sp\ \sigma$, there exists $0 \neq x_1 \in \mathscr{M}(\sigma; V_1)$. Then $e = \bigvee_{t\in G} \mathbf{r}(\sigma_t(x_1))$ is a nonzero projection in $\mathscr{M}^\sigma$. Since $\gamma_2 \in \Gamma(\sigma) \subset Sp\ \sigma^e$, it follows (15.4.(2)) that there exists $0 \neq x_2 \in \mathscr{M}(\sigma; V_2) \cap e\mathscr{M}e$. Since $ex_2 = x_2 \neq 0$, there exists $t \in G$ with $x = \sigma_t(x_1)x_2 \neq 0$ and (14.2.(4), 15.3.(1)) $x \in \mathscr{M}(\sigma; V_1 + V_2) \subset \mathscr{M}(\sigma; V)$. Hence $\Gamma(\sigma) + Sp\ \sigma = Sp\ \sigma$.

Using this conclusion and (4) we see that $\Gamma(\sigma) + Sp\ \sigma^e = Sp\ \sigma^e$ for every $0 \neq e \in Proj(\mathscr{M}^\sigma)$.

From (1) and (5) we infer that $\Gamma(\sigma) + \Gamma(\sigma) = \Gamma(\sigma)$. On the other hand, from (1) and 15.2.(4) it follows that $\Gamma(\sigma) = -\Gamma(\sigma)$. Hence $\Gamma(\sigma)$ is a closed subgroup of $\hat{G}$. □

16.2. Let $\sigma : G \to Aut(\mathscr{M})$ be a continuous action of G on $\mathscr{M}$.

Proposition. *Let $e_1, e_2 \in \mathscr{M}^\sigma$ be two projections with $\mathbf{z}_{\mathscr{M}}(e_1)\mathbf{z}_{\mathscr{M}}(e_2) \neq 0$ and V a compact neighborhood of 0 in $\hat{G}$. There exist two nonzero projections $f_1, f_2 \in \mathscr{M}^\sigma, f_1 \leq e_1, f_2 \leq e_2$, such that*

$$Sp\ \sigma^{f_1} \subset V + Sp\ \sigma^{f_2},\ Sp\ \sigma^{f_2} \subset V + Sp\ \sigma^{f_1}. \tag{1}$$

Proof. Since $\mathbf{z}_{\mathscr{M}}(e_1)\mathbf{z}_{\mathscr{M}}(e_2) \neq 0$, there exists $0 \neq y \in \mathscr{M}$ such hat $e_1 y = y = ye_2$ ([L], 4.5). By Lemma 15.1, there exists $h \in \mathscr{L}^1(G)$ with $supp\ \hat{h} - supp\ \hat{h} \subset V$ and $x = \sigma_h(y) \neq 0$. Then (15.3.(3)) $e_1 x = e_1\sigma_h(y) = \sigma_h(e_1 y) = \sigma_h(y) = x = \cdots = xe_2$ and (14.2.(3)) $Sp_\sigma(x) - Sp_\sigma(x) \subset V$. We define

$$f_1 = \bigvee_{t\in G} \mathbf{l}(\sigma_t(x)), \quad f_2 = \bigvee_{t\in G} \mathbf{r}(\sigma_t(x)).$$

Then $f_1, f_2 \in \mathscr{M}^\sigma, 0 \neq f_1 \leq e_1, 0 \neq f_2 \leq e_2$.

Let $\gamma_1 \in Sp\ \sigma^{f_1}$. Since V is compact, the set $V + Sp\ \sigma^{f_2}$ is closed. Thus, in order to prove that $\gamma_1 \in V + Sp\ \sigma^{f_2}$, it is sufficient to show that $(V_1 - V) \cap Sp\ \sigma^{f_2} \neq \emptyset$ for any compact neighborhood V_1 of γ_1. Since $\gamma_1 \in Sp\ \sigma^{f_1}$, there exists $0 \neq x_1 \in \mathscr{M}(\sigma; V_1) \cap f_1\mathscr{M}f_1$. Since $x_1 f_1 = x_1 \neq 0$, there exists $t \in G$ with $x_1\sigma_t(x) \neq 0$. Since $\sigma_1(x)^* x_1^* f_1 = \sigma_1(x)^* x_1^* \neq 0$, there exists $s \in G$ such that $\sigma_1(x)^* x_1^* \sigma_s(x) \neq 0$. Therefore, $x_2 = \sigma_s(x^*)x_1\sigma_t(x) \neq 0$. We obviously have $x_2 f_2 = x_2 = f_2 x_2$ and

(15.2.(2), 15.3(1)) $Sp_\sigma(x_2) \subset Sp_\sigma(x_1) + Sp_\sigma(x) - Sp_\sigma(x) \subset V_1 - V$, hence $(V_1 - V) \cap Sp\, \sigma^{f_2} \neq \emptyset$. We have thus proved the first inclusion in (1). The second one is proved similarly. □

Corollary 1. *If $e_1, e_2 \in Proj(\mathscr{M}^\sigma)$ and $\mathbf{z}_{\mathscr{M}}(e_1) = \mathbf{z}_{\mathscr{M}}(e_2) \neq 0$, then*

$$\Gamma(\sigma^{e_1}) = \Gamma(\sigma^{e_2}). \tag{2}$$

In particular, if $e \in Proj(\mathscr{M}^\sigma)$ and $\mathbf{z}_{\mathscr{M}}(e) = 1$, then

$$\Gamma(\sigma^e) = \Gamma(\sigma). \tag{3}$$

Proof. It is clear that if $e \in \mathscr{M}^\sigma$, then $\mathbf{z}_{\mathscr{M}}(e) \in \mathscr{M}^\sigma$. Thus, it is sufficient to prove (2) only for $e_2 = \mathbf{z}_{\mathscr{M}}(e_1)$, that is, it is sufficient to prove (3).

We have (16.1.(5)) $\Gamma(\sigma) \subset \Gamma(\sigma^e)$. To prove the opposite inclusion we note that

$$\Gamma(\sigma) = \bigcap \{V + Sp\, \sigma^f\} \tag{4}$$

with $0 \neq f \in Proj(\mathscr{M}^\sigma)$ and V compact neighborhood of 0 in $\hat{G}$. The above proposition shows that for any f and any V there exists $0 \neq e' \in Proj(\mathscr{M}^\sigma)$ with $e' \leq e$ and $Sp\, \sigma^{e'} \subset V + Sp\, \sigma^f$, hence $\Gamma(\sigma^e) \subset Sp\, \sigma^{e'} \subset V + Sp\, \sigma^f$. Therefore, $\Gamma(\sigma^e) \subset \Gamma(\sigma)$. □

Corollary 2. *If $\mathscr{M}$ is a factor, then the family*

$$\mathfrak{F}(\sigma) = \{(V + Sp\, \sigma^e; 0 \neq e \in Proj\, \mathscr{M}^\sigma), V \text{ a compact neighborhood of } 0 \in \hat{G}\}$$

is the basis of a filter with intersection equal to $\Gamma(\sigma)$.

Proof. We have already noticed (4) that the intersection of the family $\mathfrak{F}(\sigma)$ is equal to $\Gamma(\sigma)$. Now, let $e_1, e_2 \in \mathscr{M}^\sigma$ be two nonzero projections and V_1, V_2 two compact neighborhoods of 0 in $\hat{G}$. There exists a compact neighborhood V of 0 in $\hat{G}$ such that $V \subset V_1$ and $V + V \subset V_2$. By the previous proposition, there exist two nonzero projections $f_1, f_2 \in \mathscr{M}^\sigma, f_1 \leq e_1, f_2 \leq e_2$, which satisfy (1). Since $V \subset V_1$ and $f_1 \leq e_1$, we have (15.4.(4)) $V + Sp\, \sigma^{f_1} \subset V_1 + Sp\, \sigma^{e_1}$. Since $V + V \subset V_2$ and $f_2 \leq e_2$, it follows from (1) that $V + Sp\, \sigma^{f_1} \subset V_2 + Sp\, \sigma^{e_2}$. □

16.3. Let $\sigma : G \to Aut(\mathscr{M})$ and $\tau : G \to Aut(\mathscr{M})$ be two continuous actions of G on the W^*-algebra $\mathscr{M}$. In Section 15.11, we defined the outer conjugacy relation $\sigma \sim \tau$.

Proposition. *If $\sigma \sim \tau$, then $\Gamma(\sigma) = \Gamma(\tau)$.*

Proof. Let $\mathscr{P} = Mat_2(\mathscr{M})$ and $e = \begin{pmatrix} 1 & 0 \\ 0 & 0 \end{pmatrix}, f = \begin{pmatrix} 0 & 0 \\ 0 & 1 \end{pmatrix} \in \mathscr{P}$. Then e and f are equivalent projections in $\mathscr{P}$ and, as proved in Section 15.11, there exists a continuous action $\theta : G \to Aut(\mathscr{P})$ such that $e, f \in \mathscr{P}^\theta$ and $(\mathscr{M}, \sigma) \approx (e\mathscr{P}e, \theta^e), (\mathscr{M}, \tau) \approx (f\mathscr{P}f, \theta^f)$. Using Corollary 1/16.2, we conclude that $\Gamma(\sigma) = \Gamma(\theta^e) = \Gamma(\theta^f) = \Gamma(\tau)$. □

By the above proposition and by Corollary 2/16.2, it follows that *if $\mathscr{M}$ is a factor and $\sigma \sim \tau$, then for every $F \in \mathfrak{F}(\sigma)$ there exists $F' \in \mathfrak{F}(\tau)$ with $F' \subset F$.*

16.4. Let $\sigma : G \to Aut(\mathscr{M})$ be a continuous action of G on the W^*-algebra $\mathscr{M}$. Put

$$Ker\,\sigma = \{t \in G; \sigma_t = \iota_{\mathscr{M}}\},$$
$$Int\,\sigma = \{t \in G; \text{ there exists } u \in U(\mathscr{M}^\sigma) \text{ with } \sigma_t = Ad(u)\}.$$

Note that $Ker\,\sigma \subset Int\,\sigma$ and

$$u \in U(\mathscr{M}^\sigma), \sigma_t = Ad(u) \Rightarrow u \in U(\mathscr{Z}(\mathscr{M}^\sigma)). \tag{1}$$

For each subset $E \subset \hat{G}$, the set

$$E^\perp = \{t \in G; \langle t, \gamma\rangle = 1 \text{ for all } \gamma \in E\}$$

is a closed subgroup of G and $E^{\perp\perp}$ is the smallest closed subgroup of $\hat{G}$ containing E (Rudin, 1962, 2.1.3).

Using 14.13.(1) and Proposition 14.6, we get

$$Ker\,\sigma = (Sp\,\sigma)^\perp \subset \Gamma(\sigma)^\perp, \tag{2}$$

since, by definition, $\Gamma(\sigma) \subset Sp\,\sigma$. It follows that

$$Sp\,\sigma = \Gamma(\sigma) \Leftrightarrow Ker\,\sigma = \Gamma(\sigma)^\perp. \tag{3}$$

Recall (16.1) that these conditions are satisfied if $\mathscr{M}^\sigma$ is a factor and that they imply that $Sp\,\sigma$ is a closed subgroup of $\hat{G}$. Conversely,

Proposition. *If $\Gamma(\sigma)$ is discrete, then*

$$Sp\,\sigma = \Gamma(\sigma) \Rightarrow \mathscr{Z}(\mathscr{M}^\sigma) = \mathscr{Z}(\mathscr{M})^\sigma. \tag{4}$$

In particular, if $\mathscr{M}$ is a factor and $\Gamma(\sigma)$ is discrete, then

$$Sp\,\sigma = \Gamma(\sigma) \Leftrightarrow \mathscr{M}^\sigma \textit{ is a factor}. \tag{5}$$

Proof. Let $e, f \in \mathscr{M}^\sigma$ be projections with $\mathbf{z}_{\mathscr{M}}(e)\mathbf{z}_{\mathscr{M}}(f) \neq 0$. Then ([L],4.5) there exists $y \in \mathscr{M}$ with $eyf \neq 0$. Let $\gamma \in Sp_\sigma(eyf)$. Since $Sp\,\sigma = \Gamma(\sigma)$ is discrete, there exists (14.1.(5)) $h \in \mathscr{L}^1(G)$ such that $\hat{h}(\gamma) \neq 0$ and $\hat{h}(\omega) = 0$ for all $\omega \in Sp\,\sigma\setminus\{\gamma\}$. Then $x = \sigma_h(eyf) \neq 0, Sp_\sigma(x) = \{y\}$, and (15.3.(3)) $ex = x = xf$. Since $Sp_\sigma(x) = \{\gamma\}$ is a singleton, using 15.2.(2), 15.3.(1), and 14.3.(14) we obtain $x^*x \in \mathscr{M}^\sigma, xx^* \in \mathscr{M}^\sigma$, hence $\mathbf{r}(x) = \mathbf{s}(x^*x) \in \mathscr{M}^\sigma, \mathbf{l}(x) = \mathbf{s}(xx^*) \in \mathscr{M}^\sigma$; note that $\mathbf{l}(x) \leq e, \mathbf{r}(x) \leq f$. Since $\gamma \in Sp\,\sigma = \Gamma(\sigma), -\gamma \in \Gamma(\sigma)$ (16.1), and since $0 \neq \mathbf{r}(x) \in Proj(\mathscr{M}^\sigma)$, there exists $0 \neq z \in \mathscr{M}(\sigma; \{-\gamma\}) \cap \mathbf{r}(x)\mathscr{M}\mathbf{r}(x)$. Then $0 \neq a = xz \in \mathscr{M}^\sigma$ and $eaf = a \neq 0$. Hence ([L], 4.5) $\mathbf{z}_{\mathscr{M}^\sigma}(e)\mathbf{z}_{\mathscr{M}^\sigma}(f) \neq 0$.

It is clear that $\mathscr{Z}(\mathscr{M}^\sigma) \subset \mathscr{Z}(\mathscr{M}^\sigma)$. Conversely, $e \in \mathscr{Z}(\mathscr{M}^\sigma)$ be a projection and let $f = 1 - e$. Since $ef = 0$, by the above arguments we infer that $\mathbf{z}_{\mathscr{M}}(e)\mathbf{z}_{\mathscr{M}}(f) = 0$, hence $e = \mathbf{z}_{\mathscr{M}}(e) \in \mathscr{Z}(\mathscr{M})$. □

This proposition can be applied, in particular, if the group G is compact, since then $\hat{G}$ is discrete (Rudin, 1962, 1.2.5).

16.5. The main duality result is the following:

Theorem. *Let $\sigma : G \to Aut(\mathscr{M})$ be a continuous action of G on the W^*-algebra $\mathscr{M}$. Then*

$$Int\,\sigma \subset \Gamma(\sigma)^\perp. \tag{1}$$

If $\mathscr{M}$ is a factor and the set $Sp\,\sigma/\Gamma(\sigma)$ is compact in $\hat{G}/\Gamma(\sigma)$, then

$$Int\,\sigma = \Gamma(\sigma)^\perp. \tag{2}$$

Proof. Let $t \in Int\,\sigma$ and let $u \in U(\mathscr{Z}(\mathscr{M}^\sigma))$ be such that $\sigma_t = Ad(u)$. Let $\varepsilon > 0$. There exist $\lambda_0 \in \mathbb{C}$, with $|\lambda_0| = 1$, and a spectral projection $e \in \mathscr{Z}(\mathscr{M}^\sigma)$ of u, such that $\|ue - \lambda_0 e\| < \varepsilon$. Then the spectrum of ue in $e\mathscr{M}e$ is contained in $\{\lambda \in \mathbb{C}; |\lambda - \lambda_0| < \varepsilon\}$. Using 14.13.(2), it follows that the spectrum of the automorphism $\sigma_t^e = Ad(ue) \in Aut(e\mathscr{M}e)$ as an element of $\mathscr{B}(e\mathscr{M}e)$ is contained in the set $\{\lambda \in \mathbb{C}; |\lambda - 1| < 2\varepsilon\}$. On the other hand, by Proposition 14.6, the spectrum of σ_t^e in $\mathscr{B}(e\mathscr{M}e)$ is equal to the closure of the set $\{\langle t, \gamma\rangle; \gamma \in Sp\,\sigma^e\}$. Consequently, $|\langle t, \gamma\rangle - 1| < 2\varepsilon$ for every $\gamma \in Sp\,\sigma^e$ in particular for every $\gamma \in \Gamma(\sigma)$. Since $\varepsilon > 0$ was arbitrary, it follows that $t \in \Gamma(\sigma)^\perp$.

We assume now that $\mathscr{M}$ is a factor and that $Sp\,\sigma/\Gamma(\sigma)$ is a compact subset of $\hat{G}/\Gamma(\sigma)$.

Let $k : \hat{G} \to \hat{G}/\Gamma(\sigma)$ be the canonical quotient mapping. Since, by assumption, $k(Sp\,\sigma)$ is compact, using Corollary 2/16.2 we see that the family $\{k(F); F \in \mathfrak{F}(\sigma)\}$ is the basis of a filter on $\hat{G}/\Gamma(\sigma)$ with intersection equal to $\{k(0)\}$. By Proposition 16.1, we have $k^{-1}(k(F)) = F$ for every $F \in \mathfrak{F}(\sigma)$.

Let $t \in \Gamma(\sigma)^\perp$ and $0 < \varepsilon < 1$. The set $D_e = \{y \in \hat{G}; Re\,\langle t, \gamma\rangle > 1 - \varepsilon\}$ is an open neighborhood of $k(0)$ in $\hat{G}/\Gamma(\sigma)$. Consequently, there exists $F \in \mathfrak{F}(\sigma)$ such that $k(F) \subset k(D_\varepsilon)$. Since $t \in \Gamma(\sigma)^\perp$, we have $k^{-1}(k(D_\varepsilon)) = D_\varepsilon$ and therefore $F \subset D_\varepsilon$. We conclude that there exists a nonzero projection $e \in \mathscr{M}^\sigma$ such that $Sp\,\sigma^e \subset D_\varepsilon$.

Using Proposition 14.6 it follows that

$$Sp\,(\sigma_t^e) \subset \{\lambda \in \mathbb{C}; |\lambda| = 1, Re\,\lambda \geq 1 - \varepsilon\}, \tag{3}$$

where, we recall, $Sp(\sigma_t^e)$ is the spectrum of σ_t^e in $\mathscr{B}(e\mathscr{M}e)$. By Corollary 15.15 we infer that $\sigma_t^e \in Int(e\mathscr{M}e)$. Since $\mathscr{M}$ is a factor, using Proposition 17.1 we deduce that $\sigma_t \in Int(\mathscr{M})$, that is, there exists $u \in U(\mathscr{M})$ such that $\sigma_t = Ad(u)$. We still have to show that $u \in \mathscr{M}^\sigma$.

Note that $ueu^* = \sigma_t(e) = e, ue = eu$, and so $\sigma_t^e = Ad(ue)$. Since $e\mathscr{M}e$ is a factor, it follows from Proposition 14.13 that

$$Sp(\sigma_t^e) = \{\lambda\mu^{-1}; \lambda, \mu \in Sp_{\mathscr{M}}(ue)\}. \tag{4}$$

Comparing (3) and (4) and using the spectral theorem, we deduce by some elementary computations that

$$\inf\{\|ue - \lambda e\|; \lambda \in \mathbb{C}, |\lambda| = 1\} \leq \sqrt{2\varepsilon}. \tag{5}$$

Let $f \in \mathcal{M}^\sigma$ be an arbitrary nonzero projection. Since $\Gamma(\sigma^f) = \Gamma(\sigma)$, it follows from what has just been shown that there exists a nonzero projection $e \in \mathcal{M}^\sigma, e \leq f$, such that (3) holds. As above, we see that $ue = eu$ and $\sigma_t^e = Ad(ue)$; and we deduce inequality (5).

Using a standard argument it follows that for each $\varepsilon > 0$ there exist a family $\{e_i\}$ of mutually orthogonal nonzero projections in $\mathcal{M}^\sigma$ with $\sum_i e_i = 1$ and a family $\{\lambda_i\}$ of unimodular complex numbers such that $\|ue_i - \lambda_i e_i\| < \varepsilon$, hence $\|u - \sum_i \lambda_i e_i\} < \varepsilon$. We conclude that $u \in \mathcal{M}^\sigma$. Hence $t \in Int\ \sigma$.

A different proof of (1) is given in Section 21.6. □

16.6. The second main result concerning the invariant Γ is the following:

Theorem. *Let $\sigma : G \to Aut(\mathcal{M})$ be a continuous action of G on the factor $\mathcal{M}$. For every nonzero projection $e \in \mathcal{M}^\sigma$, there exists a continuous action $\tau : G \to Aut(\mathcal{M})$ such that $\tau \sim \sigma$ and $Sp\ \tau \subset Sp\ \sigma^e$.*

In particular,

$$\Gamma(\sigma) = \bigcap_{\tau\sim\sigma} Sp\ \tau. \tag{1}$$

By Proposition 14.6 and by the definition (16.1) of $\Gamma(\sigma)$, the inclusion

$$\Gamma(\sigma) \subset \bigcap_{\tau\sim\sigma} Sp\ \tau \tag{2}$$

is valid in general, for any W^*-algebra. If $\mathcal{M}$ is a factor, then (1) follows obviously from the first assertion of the theorem and the definition of $\Gamma(\sigma)$.

Note that (1) is trivially true if $\mathcal{M}^\sigma$ is a factor (16.1.(3)). However, (1) is not true for all W^*-algebras, as can easily be seen by considering a direct sum $(\mathcal{M}, \sigma) = (\mathcal{M}_1, \sigma_1) \oplus (\mathcal{M}_2, \sigma_2)$.

The proof of Theorem 16.6 will be given in Section 16.11, while in Sections 16.7–16.10 we present some auxiliary results which are of independent interest. In fact, Propositions 16.7, 16.8, and 16.9 have stronger conclusions, but in special situations.

16.7 Proposition. *Let $\sigma : G \to Aut(\mathcal{M})$ be a continuous action of G on the W^*-algebra $\mathcal{M}$. For every projection $e \in \mathcal{M}^\sigma$ which is equivalent to 1 in $\mathcal{M}$ there exists a continuous action $\tau : G \to Aut(\mathcal{M})$ such that $\tau \sim \sigma$ and $Sp\ \tau = Sp\ \sigma^e$.*

Proof. Let $u \in \mathcal{M}$ with $u^*u = 1, uu^* = e$ and put $w_t = u^*\sigma_t(u), (t \in \mathbb{R})$. It is easy to check that the mapping $t \mapsto w_t$ is a unitary σ-cocycle $w \in Z_\sigma(G; U(\mathcal{M}))$ which defines a continuous action $\tau \sim \sigma$ where $\tau_t(x) = w_t\sigma_t(x)w_t^* = u^*\sigma_t(uxu^*)u\ (x \in \mathcal{M}, t \in \mathbb{R})$. Thus, the mapping $x \mapsto uxu^*$ defines a $*$-automorphism $(\mathcal{M}, \tau) \approx (e\mathcal{M}e, \sigma^e)$ and $Sp\ \tau = Sp\ \sigma^e$. □

This proposition already implies Theorem 16.6 for countably decomposable type III factors.

16.8 Lemma. *Let $\sigma : G \to Aut(\mathcal{M})$ and $\tau : G \to Aut(\mathcal{M})$ be two continuous actions of G on the W^*-algebra $\mathcal{M}$. If there exists a projection $e \in \mathcal{M}^\sigma \cap \mathcal{M}^\tau$ with $\mathbf{z}_{\mathcal{M}}(e) = 1$ such that $\sigma^e \sim \tau^e$, then $\sigma \sim \tau$.*

Proof. By assumption, there exists an s-continuous mapping $t \mapsto v_t$ of G into $U(e\mathcal{M}e)$ such that $v_{s+t} = v_s\sigma_s(v_t)$ and $\tau_t\sigma_t^{-1}(x) = v_txv_t^*$ for $s, t \in G, x \in e\mathcal{M}e$. By Proposition 17.1, for each $t \in G$ there exists a unique $w_t \in U(\mathcal{M})$ such that $w_te = v_t = ew_t$ and $\tau_t\sigma_t^{-1} = Ad(w_t)\ (t \in G)$. Since $\tau_{s+t}\sigma_{s+t}^{-1}(x) =$

$\tau_s\tau_t\sigma_t^{-1}\sigma_s^{-1}(x) = \tau_s(w_t\sigma_s^{-1}((x)w_t^*) = \tau_s\sigma_s^{-1}(\sigma_s(w_t)x\sigma_s(w_t^*)) = w_s\sigma_s(w_t)x\sigma_s(w_t^*)w_t^*(x \in \mathscr{M})$, and $w_s\sigma_s(w_t)e = v_s\sigma_s(v_t) = v_{s+t} = ew_s\sigma_s(w_t)$, by uniqueness it follows that $w_s\sigma_s(w_t) = w_{s+t}(s, t \in G)$. Also, the mapping $t \mapsto w_t$ is s-continuous (17.1.(2)). Hence $\tau \sim \sigma$. □

Proposition. *Let $\sigma : G \to Aut(\mathscr{M})$ be a continuous action of G on the W^*-algebra $\mathscr{M}$. If $e \in \mathscr{M}^\sigma$ is a projection such that $1 - e$ is the sum of a family of mutually orthogonal projections in $\mathscr{M}$, each equivalent to e, then there exists a continuous action $\tau : G \to Aut(\mathscr{M})$ such that $\tau \sim \sigma$ and $Sp\ \tau = Sp\ \sigma^e$.*

Proof. By assumption it follows that there exist a type I factor $\mathscr{F}$, a minimal projection p in $\mathscr{F}$ and a *-isomorphism π of $\mathscr{M}$ onto $(e\mathscr{M}e) \bar{\otimes} \mathscr{F}$ such that $\pi(x) = x \bar{\otimes} p$ for $x \in e\mathscr{M}e$. We then define a continuous action $\tau : G \to Aut(\mathscr{M})$ by putting $\tau_t = \pi^{-1}\circ(\sigma_t^e \bar{\otimes} \iota_{\mathscr{F}})\circ\pi\ (t \in G)$. We have $\tau_t^e = \sigma_t^e\ (t \in G)$, hence $\tau^e \sim \sigma^e$ and so $\tau \sim \sigma$ by the previous lemma. Thus, $Sp\ \tau = Sp(\sigma^e \bar{\otimes} \iota_{\mathscr{F}}) = Sp\ \sigma^e$ (see 16.16.(1)). □

16.9 Lemma. *Let G be a closed subgroup of the locally compact abelian group G' and $\mathscr{M}$ a W^*-algebra. Every s-continuous unitary representation $u : G \to U(\mathscr{M})$ can be extended to an s-continuous unitary representation $u' : G' \to U(\mathscr{M})$.*

Proof. Without loss of generality, we may assume that the W^*-algebra $\mathscr{M}$ is generated by $u(G)$. Then $\mathscr{M}$ is abelian and can be written as a direct sum of countably decomposable W^*-algebras ([L], 7.2), so we can also assume that $\mathscr{M}$ is countably decomposable. Then ([L], 10.15) we can realize $\mathscr{M}$ as a von Neumann algebra $\mathscr{M} \subset \mathscr{B}(\mathscr{H})$ with a cyclic and separating vector $\xi \in \mathscr{H}$. As we have seen in Section 14.10, there exists a unique *-homomorphism $\pi_u : \mathscr{C}_0(\hat{G}) \to \mathscr{M}$ such that

$$\pi_u(\hat{f}) = u_f \quad (f \in \mathscr{L}^1(G)), \tag{1}$$

and there exists a unique positive measure $\mu \in \mathscr{M}(\hat{G})$ such that

$$(\pi_u(\varphi)\xi|\xi) = \int \varphi(\gamma)\ d\mu(\gamma) \quad (\varphi \in \mathscr{C}_0(\hat{G})). \tag{2}$$

$\mathscr{C}_0(\hat{G})$ is a w-dense C^*s ubalgebra of the W^*-algebra $\mathscr{L}^\infty(\hat{G}, \mu), \pi_u(\mathscr{C}_0(\hat{G}))$ is a w-dense C^*-subalgebra of the W^*-algebra $\mathscr{M}$ and, using (2), it is easy to check that the *-homomorphism π_u is normal, that is, w-continuous. Therefore, π_u can be extended to a *-isomorphism $\pi_u : \mathscr{L}^\infty(\hat{G}, \mu) \to \mathscr{M}$. Using (1) and approximate units in $\mathscr{L}^1(G)$, it is easy to check that

$$(\pi_u^{-1}(u_t))(\gamma) = \langle t, \gamma\rangle \quad (t \in G, \gamma \in \hat{G}).$$

Thus, we can assume that $\mathscr{M} = \mathscr{L}^\infty(\hat{G}, \mu)$ and $u_t(\gamma) = \langle t, \gamma\rangle(t \in G, \gamma \in \hat{G})$. Furthermore, by decomposing $\mathscr{M}$ into a direct sum, we may assume that the measure μ has compact support (see, e.g., Dinculeanu, 1967, Prop. 41, §15).

Let $p : \hat{G}' \to \hat{G}$ be the dual homomorphism of the inclusion $G \to G'$. There exists a compact set $K' \subset \hat{G}'$ such that $p(K') = supp\ \mu$. Let μ' be an extreme point of the weakly compact convex set of all positive measures ν' on K' such that $p(\nu') = \mu$. Then (Mokobodsky, 1962) the mapping

$$\Phi : \mathscr{M} = \mathscr{L}^\infty(\hat{G}, \mu) \ni \varphi \mapsto \varphi\circ p \in \mathscr{L}^\infty(\hat{G}', \mu')$$

is a *-isomorphism. For $t' \in G'$, we define $v'_{t'} \in \mathscr{L}^{\infty}(\hat{G}', \mu')$ by $v'_{t'}(\gamma') = \langle t', \gamma' \rangle (\gamma' \in \hat{G}')$, and $u'_t = \Phi^{-1}(v'_{t'})$. Then $u' : G' \to U(\mathscr{M})$ is an s-continuous unitary representation and for $t \in G, \gamma' \in \hat{G}'$ we have $(\Phi(u'_t))(\gamma') = v'_t(\gamma') = \langle t, \gamma' \rangle = u_t(p(\gamma')) = (\Phi(u_t))(\gamma')$, hence $u'_t = u_t$. □

16.10 Proposition. *Let $\sigma : G \to Aut(\mathscr{M})$ be a continuous action of G on the factor $\mathscr{M}$. If $\mathscr{Z}(\mathscr{M}^{\sigma})$ contains a minimal projection, then there exists a continuous action $\tau : G \to Aut(\mathscr{M})$ such that $\tau \sim \sigma$ and $Sp\, \tau = \Gamma(\tau) = \Gamma(\sigma)$.*

Proof. Let $e \in \mathscr{Z}(\mathscr{M}^{\sigma})$ be a minimal projection. Then $\mathscr{M}^{\sigma^e} = e\mathscr{M}^{\sigma}e$ is a factor, hence (16.1.(3)) $Sp\, \sigma^e = \Gamma(\sigma^e)$ and therefore (16.4.(3)) $Ker\, \sigma^e = \Gamma(\sigma^e)^{\perp}$. Since $\mathscr{M}$ is a factor, we have $\Gamma(\sigma^e) = \Gamma(\sigma)$ (16.2.(3)). Thus, for every $t \in \Gamma(\sigma)^{\perp}$ we have $\sigma^e_t = \iota \in Int(e\mathscr{M}e)$. Using Proposition 17.1, it follows that for every $t \in \Gamma(\sigma)^{\perp}$ there exists a unique $u_t \in U(\mathscr{M})$ such that $u_t e = eu_t = e$ and $\sigma_t = Ad(u_t)$; since, for any $s \in G$, the element $\sigma_s(u_t)$ satisfies the same conditions as u_t, we have $u_t \in \mathscr{M}^{\sigma}$. Also, by the uniqueness of u_t and 17.1.(2) we see that the mapping $u : \Gamma(\sigma)^{\perp} \ni t \mapsto u_t \in U(\mathscr{M}^{\sigma})$ is an s-continuous unitary representation. By Lemma 16.9, there exists an s-continuous unitary representation $v : G \to U(\mathscr{M}^{\sigma})$ such that $v_t = u_t$ for $t \in \Gamma(\sigma)^{\perp}$. Then v is a unitary σ-cocycle; v defines a continuous action $\tau : G \to Aut(\mathscr{M}), \tau \sim \sigma$, by the formula $\tau_t = Ad(v^*_t) \circ \sigma_t$ $(t \in G)$. If $t \in \Gamma(\tau)^{\perp} = \Gamma(\sigma)^{\perp}$, then $\tau_t = Ad(v^*_t u_t) = \iota$, that is, $t \in Ker\, \tau$. Consequently, $\Gamma(\tau)^{\perp} = Ker\, \tau$ and hence (16.4.(3)) $Sp\, \tau = \Gamma(\tau) = \Gamma(\sigma)$. □

16.11. *Proof of Theorem 16.6.* If $\mathscr{M}^{\sigma}$ has a minimal projection, then its central support in $\mathscr{M}^{\sigma}$ is a minimal projection in $\mathscr{Z}(\mathscr{M}^{\sigma})$. Taking into account Propositions 16.10 and 16.7, we see that, in order to prove the theorem, we may assume that *$\mathscr{M}^{\sigma}$ has no minimal projections* and that *the nonzero projection $e \in \mathscr{M}^{\sigma}$ is not equivalent to 1 in $\mathscr{M}$*. In this case, we shall show that there exists a nonzero projection $f \in \mathscr{M}^{\sigma}, f \leq e$, such that $1 - f$ is the sum of a family of mutually orthogonal projections in $\mathscr{M}$, each equivalent to f. According to Proposition 16.8, it will follow that there exists $\tau \sim \sigma$ with $Sp\, \tau = Sp\, \sigma^f \subset Sp\, \sigma^e$.

Assume first that $\mathscr{M}$ is properly infinite. Since, by assumption, e is not equivalent to 1, we can take in this case $f = e$.

Assume now that $\mathscr{M}$ is finite and let μ be the n.s.f. trace on $\mathscr{M}$ with $\mu(1) = 1$. By assumption, for every nonzero projection $p \in \mathscr{M}^{\sigma}$ and every $\varepsilon > 0$ there exists a nonzero projection $q \in \mathscr{M}^{\sigma}$ with $\mu(q) < \varepsilon$. Let $n \in \mathbb{N}$ be such that $1/n \leq \mu(e)$ and let $\{f_i\}$ be a maximal family of mutually orthogonal nonzero projections in $\mathscr{M}^{\sigma}$, majorized by e and such that $\sum_i \mu(f_i) \leq 1/n$. Then $\sum_i \mu(f_i) = 1/n$. Let $f = \sum_i f_i$. Then $0 \neq f \in \mathscr{M}^{\sigma}, f \leq e$, and $\mu(f) = 1/n$, so that there exist $(n-1)$ mutually orthogonal projections in $\mathscr{M}$, each equivalent to f, with sum equal to $1 - f$. □

16.12 Proposition. *Let $\sigma : G \to Aut(\mathscr{M})$ be a continuous action of G on the factor $\mathscr{M}$. If $\hat{G}/\Gamma(\sigma)$ is compact, then there exists a continuous action $\tau : G \to Aut(\mathscr{M})$ such that $\tau \sim \sigma$ and $Sp\, \tau = \Gamma(\tau) = \Gamma(\sigma)$.*

Proof. Since $\hat{G}/\Gamma(\sigma)$ is compact, its dual $\Gamma(\sigma)^{\perp}$ is a discrete, hence closed, subgroup of G. Moreover, by Theorem 16.5, we have $\Gamma(\sigma)^{\perp} = Int\, \sigma$. Let $k : Int\, \sigma \to Int\, \sigma / Ker\, \sigma$ be the canonical quotient mapping. □

Let U be the subgroup of $U(\mathscr{Z}(\mathscr{M}^{\sigma}))$ consisting of those $u \in U(\mathscr{Z}(\mathscr{M}^{\sigma}))$ with the property that there exists $t = t(u) \in Int\, \sigma$ such that $\sigma_t = Ad(u)$. For each $u \in U$ we put $j(u) = k(t(u)) \in Int\, \sigma / Ker\, \sigma$. Then the mapping $j : U \to Int\, \sigma / Ker\, \sigma$ is a well-defined surjective homomorphism.

Let $\mathbb{T} = \{\lambda \in \mathbb{C}; |\lambda| = 1\}$ and denote by $i : \mathbb{T} \to U$ the injective homomorphism defined by $i(\lambda) = \lambda \cdot 1_{\mathcal{M}} (\lambda \in \mathbb{T})$. Since $\mathcal{M}$ is a factor we have a short exact sequence

$$0 \to \mathbb{T} \overset{i}{\to} U \overset{i}{\to} Int\,\sigma / Ker\sigma \to 0,$$

which is split as $\mathbb{T}$ a divisible group (Hewitt & Ross, 1963, 1970, Thm. A.7). Hence there exists a homomorphism $h : Int\,\sigma/Ker\,\sigma \to U$ such that $j \circ h$ = the identity mapping. Then $u = h \circ k : \Gamma(\sigma)^{\perp} = Int\,\sigma \to U \subset U(\mathcal{Z}(\mathcal{M}^{\sigma}))$ is a unitary representation of the discrete group $\Gamma(\sigma)^{\perp}$ in $\mathcal{Z}(\mathcal{M}^{\sigma})$ such that $\sigma_t = Ad(u_t)$ for all $t \in \Gamma(\sigma)^{\perp}$. By Lemma 16.9, u can be extended to an s-continuous unitary representation $u : G \to U(\mathcal{Z}(\mathcal{M}^{\sigma}))$. Then u is a unitary σ-cocycle and the equation $\tau_t = \mathrm{Ad}(u_t^*) \circ \sigma_t (t \in G)$ defines a continuous action $\tau \sim \sigma$ such that $\Gamma(\sigma)^{\perp} \subset Ker\,\tau$. It follows (16.3, 16.4.(2)) that $Ker\,\tau = \Gamma(\tau)^{\perp}$ and hence (16.4.(3)) $Sp\,\tau = \Gamma(\tau) = \Gamma(\sigma)$.

In general, all the sets in the family

$$\{Ker\,\tau; \tau \sim \sigma\} \tag{1}$$

are contained in $\Gamma(\sigma)^{\perp}$ (16.3, 16.4.(2)). *If $\mathcal{M}$ is a factor and either $\hat{G}/\Gamma(\sigma)$ is compact or $\mathcal{Z}(\mathcal{M}^{\sigma})$ has a minimal projection, then,* by the above proposition and by Proposition 16.10, *family (1) has a greatest element, equal to* $\Gamma(\sigma)^{\perp}$ (16.4.(3)).

16.13. We now give two applications of Theorem 16.5 to *-automorphisms of W^*-algebras.

Let $\mathcal{M}$ be a W^*-algebra and $\sigma \in Aut(\mathcal{M})$. Then the mapping $n \mapsto \sigma^n$ defines an action $\sigma : \mathbb{Z} \to Aut(\mathcal{M})$. As usual, we identify the dual $\hat{\mathbb{Z}}$ of $\mathbb{Z}$ with $\mathbb{T}$ in such a way that $\langle n, \lambda \rangle = \lambda^n (n \in \mathbb{Z}, \lambda \in \mathbb{T})$. By Proposition 14.6, the spectrum of $\sigma \in Aut(\mathcal{M})$ in $\mathcal{B}(\mathcal{M})$ is equal to the spectrum $Sp\,\sigma \subset \mathbb{T}$ of the action $\sigma : \mathbb{Z} \to Aut(\mathcal{M})$.

If $\mathcal{M}$ is a factor, then Theorem 16.5 can be applied to the action σ since $\mathbb{T} = \mathbb{Z}$ is compact.

Proposition 1. *Let $\mathcal{M}$ be a factor and $\sigma \in Aut(\mathcal{M})$. If $\lambda(Sp\,\sigma) \neq Sp\,\sigma$ for every $\lambda \in \mathbb{T}, \lambda \neq 1$, then $\sigma \in Int(\mathcal{M})$.*

Proof. By assumption and Proposition 16.1, it follows that $\Gamma(\sigma) = \{1\}$, hence $\Gamma(\sigma)^{\perp} = \mathbb{Z} \ni 1$. By Theorem 16.5, we conclude that $\sigma \in Int(\mathcal{M})$. □

Proposition 2. *Let $\mathcal{M}$ be a factor and $\sigma \in Aut(\mathcal{M})$. If $Sp\,\sigma \neq \mathbb{T}$, then there exists $n \in \mathbb{Z}, n \neq 0$, such that $\sigma^n \in Int(\mathcal{M})$.*

Proof. Since $Sp\,\sigma \neq \mathbb{T}$ we have $\Gamma(\sigma) \neq \mathbb{T}$ and hence $\Gamma(\sigma)^{\perp} \neq \{0\}$. If $n \in \Gamma(\sigma)^{\perp}, n \neq 0$, then $\sigma^n \in Int(\mathcal{M})$ by Theorem 16.5. □

The requirement that $\mathcal{M}$ be a factor comes from Theorem 16.5, but *Proposition 2 is valid for any W^*-algebra*, as shown by Borchers (1974) with rather similar methods.

16.14. We now give *an example of two continuous actions $\sigma : G \to Aut(\mathcal{M}), \tau : G \to Aut(\mathcal{M})$ such that $\tau_t \sigma_t^{-1} \in Int(\mathcal{M})$ for every $t \in G$ but σ and τ are not outer conjugate.*

Let $\mathcal{H} = \ell^2(\mathbb{Z}), \mathcal{M} = \mathcal{B}(\mathcal{H}), G = \mathbb{Z} \times \mathbb{Z}$, and $\sigma : G \ni t \mapsto \iota_{\mathcal{M}} \in Aut(\mathcal{M})$ be the trivial action. There exists $\lambda \in \mathbb{T}$ such that the set $\{\lambda^n; n \in \mathbb{Z}\}$ is dense in $\mathbb{T}$. We define the unitary operators $u, v \in \mathcal{B}(\mathcal{H})$ by

$$(u\xi)(n) = \xi(n-1), (v\xi)(n) = \lambda^n \xi(n) \quad (\xi \in \ell^2(\mathbb{Z}), n \in \mathbb{Z}),$$

and for $t = (p, q) \in G$ we define $\tau_t = Ad(u^p v^q)$. Since $vu = \lambda uv$, it follows that $\tau : G \ni t \mapsto \tau_t \in Aut(\mathscr{M})$ is an action.

By construction, we have $\tau_t \sigma_t^{-1} \in Int(\mathscr{M})$ for all $t \in G$.

On the other hand, it is clear that $Ker\, \tau = \{0\}$. Thus, if we can show that $\mathscr{M}^\tau = \mathbb{C}$, it will follow from 16.1.(1) that $\Gamma(\tau) = Sp\, \tau$, so that (16.4.(3)) $\Gamma(\tau)^\perp = Ker\, \tau = \{0\}$ and $\Gamma(\tau) = \mathbb{T}$, while $\Gamma(\sigma) = \{0\}$, so that (16.3) the actions σ and τ cannot be outer conjugate.

Let us show that $\mathscr{M}^\tau = \mathbb{C}$. Since v is the multiplication operator by the character $\lambda \in \ell^\infty(\mathbb{Z})$ and since the set $\{\lambda^n; n \in \mathbb{Z}\}$ is dense in $\mathbb{T}$, the commutant of v is just the von Neumann algebra $\ell^\infty(\mathbb{Z}) \subset \mathscr{B}(\mathscr{H})$. Since the only translation invariant functions are the constants, the commutant $\{u, v\}' \subset \mathscr{B}(\mathscr{H}) = \mathscr{M}$ reduces to scalar operators, that is, $\mathscr{M}^\tau = \mathbb{C}$.

16.15. Let $\mathscr{M}$ be a W^*-algebra and $\sigma, \tau \in Aut(\mathscr{M})$. Consider also the corresponding actions $\sigma : \mathbb{Z} \ni n \mapsto \sigma^n \in Aut(\mathscr{M}), \tau : \mathbb{Z} \ni n \mapsto \tau^n \in Aut(\mathscr{M})$. Then

$$\tau \sim \sigma \Leftrightarrow \tau \equiv \sigma (\text{mod } Int(\mathscr{M})). \tag{1}$$

Indeed, if there exists $u \in U(\mathscr{M})$ such that $\tau = Ad(u) \circ \sigma$, then equation

$$u_0 = 1 \text{ and } u_n = u\sigma(u) \ldots \sigma^{n-1}(u), u_{-n} = \sigma^{-1}(u^*) \ldots \sigma^{-n}(u^*), (n \geq 1) \tag{2}$$

defines a unitary σ-cocycle with $\tau^n = Ad(u^n) \circ \sigma^n$ $(n \in \mathbb{Z})$, and $u_1 = u$. Conversely, any unitary σ-cocycle is defined as in (2) by a unitary element of $\mathscr{M}$.

16.16. Let $\mathscr{M}, \mathscr{F}$ be W^*-algebras, $\sigma : G \to Aut(\mathscr{M})$ a continuous action of G on $\mathscr{M}$, and $\iota : G \to Aut(\mathscr{F})$ the trivial action of G on $\mathscr{F}$. Since for every $f \in \mathscr{L}^1(G)$ we have $(\sigma \bar{\otimes} \iota)_f = \sigma_f \bar{\otimes} \iota$, it follows (14.5) that

$$Sp(\sigma \bar{\otimes} \iota) = Sp\, \sigma. \tag{1}$$

Also, using Corollary 9.9, we obtain $\mathscr{Z}((\mathscr{M} \bar{\otimes} \mathscr{F})^{\sigma \bar{\otimes} \iota}) = \mathscr{Z}(\mathscr{M}^\sigma \bar{\otimes} \mathscr{F}) = \mathscr{Z}(\mathscr{M}^\sigma) \bar{\otimes} \mathscr{Z}(\mathscr{F})$. By statement 16.1.(2) it follows that *if $\mathscr{F}$ is a factor, then*

$$\Gamma(\sigma \bar{\otimes} \iota) = \Gamma(\sigma). \tag{2}$$

16.17. The next group of results is concerned with periodic automorphisms, that is, with actions of finite cyclic groups.

Let $\mathscr{M}$ be a W^*-algebra. A $*$-automorphism $\sigma \in Aut(\mathscr{M})$ is called *minimal periodic* if there exists $n \in \mathbb{Z}, n \geq 2$, such that

$$\sigma^n = 1 \text{ and } \Gamma(\sigma) \supset \{\lambda \in \mathbb{C}; \lambda^n = 1\}; \tag{1}$$

the number n is called the *minimal period* of σ.

In this case, for every nonzero projection $e \in \mathscr{M}^\sigma$ we have

$$Sp\, \sigma^e = \Gamma(\sigma^e) = \{\lambda \in \mathbb{C}; \lambda^n = 1\} = \{1, \gamma, \ldots, \gamma^{n-1}\}. \tag{2}$$

where $\gamma = \exp(2\pi i/n)$.

Indeed, if $\lambda \in Sp\ \sigma$, then, by Proposition 14.5, there exists a net $\{x_i\} \subset \mathcal{M}, \|x_i\| = 1$, such that $\|\sigma^k(x_i) - \lambda^k x_i\| \to 0$ for all $k \in \mathbb{Z}$; in particular, for $k = n$ we have $|1 - \lambda^n| = \|x_i - \lambda^n x_i\| \to 0$, that is, $\lambda^n = 1$. Therefore, $Sp\ \sigma \subset \{\lambda \in \mathbb{C}; \lambda^n = 1\} \subset \Gamma(\sigma) \subset Sp\ \sigma$, proving (2) for $e = 1$. For an arbitrary projection $e \in \mathcal{M}^\sigma$, the *-automorphism σ^e also satisfies (1) since (16.1.(4)) $\Gamma(\sigma^e) \supset \Gamma(\sigma)$; hence (2) is valid in general.

From (2) and Proposition 16.4 it follows that

$$\mathcal{Z}(e\mathcal{M}^\sigma e) = e\mathcal{Z}(\mathcal{M})^\sigma \quad (e \in Proj(\mathcal{M}^\sigma)). \tag{3}$$

In particular, for central supports we have

$$\mathbf{z}_{\mathcal{M}^\sigma}(e) = \mathbf{z}_{\mathcal{M}}(e) = z(e) \quad (e \in Proj(\mathcal{M}^\sigma)). \tag{4}$$

Let $x \in \mathcal{M}$ and write

$$\hat{x}(k) = \frac{1}{n} \sum_{j=0}^{n-1} \gamma^{-kj} \sigma^j(x) \quad (k = 0, 1, \ldots, n-1). \tag{5}$$

Then $\|\hat{x}(k)\| \leq \|x\|, \hat{x}(k)^* = \hat{x}^*(n-k)\, (k = 0, 1, \ldots, n-1)$, and

$$x = \sum_{j=0}^{n-1} \hat{x}(k) \text{ and } \hat{x}(k) \in \mathcal{M}(\sigma; \{\gamma^k\}) \quad (k = 0, 1, \ldots, n-1). \tag{6}$$

Proposition. *Let $\sigma \in Aut(\mathcal{M})$ be a minimal periodic *-automorphism on the properly infinite W^*-algebra $\mathcal{M}$. Then the centralizer $\mathcal{M}^\sigma$ is also properly infinite.*

Proof. For every nonzero projection $p \in \mathcal{Z}(\mathcal{M}^\sigma) = \mathcal{Z}(\mathcal{M})^\sigma, \sigma^P$ is also a minimal periodic *-automorphism on the properly infinite W^*-algebra $\mathcal{M}p$. Hence it is sufficient just to prove that $\mathcal{M}^\sigma$ is infinite.

By assumption, there exists $n \in \mathbb{Z}, n \geq 2$, such that (2) holds.

Let $e, f \in \mathcal{M}^\sigma$ be nonzero projections such that $\mathbf{z}(e)\,\mathbf{z}(f) \neq 0$. Then ([L], 4.5) there exists $x \in \mathcal{M}$ with $exf \neq 0$. By (6) there exist $k \in \{0, 1, \ldots, n-1\}$ and $x_k \in \mathcal{M}(\sigma; \{\gamma^k\})$ such that $ex_kf \neq 0$. Let $r = \mathbf{r}(ex_kf) \leq f$ and $j \in \{0, 1, \ldots, n-1\}$ with $k + j = 1 (\mathrm{mod}\, n)$. Since $0 \neq r \in Proj(\mathcal{M}^\sigma)$, there exists $x_j \in \mathcal{M}(\sigma; \{\gamma^j\})$ such that $0 \neq x_j r x_j r$. Then $y = ex_kfx_j \in \mathcal{M}(\sigma; \{\gamma\})$ and $eyf = y \neq 0$. If $y = v|y|$ is the polar decomposition of y, then $v \in \mathcal{M}(\sigma; \{\gamma\})$ is a nonzero partial isometry such that $v^*v \leq f, vv^* \leq e$.

Now let u be a maximal partial isometry such that

$$u \in \mathcal{M}(\sigma; \{\gamma\}).$$

If $\mathbf{z}(1-uu^*)\mathbf{z}(1-u^*u) \neq 0$, the preceding argument shows that there exists a nonzero partial isometry $v \in \mathcal{M}(\sigma; \{\gamma\})$ with $v^*v \leq 1 - u^*u, vv^* \leq 1 - uu^*$, and $w = u + v \in \mathcal{M}(\sigma; \{\gamma\})$ is a partial isometry, contradicting the maximality of u. Consequently, $\mathbf{z}(1-uu^*)\mathbf{z}(1-u^*u) = 0$, and there exists a nonzero central projection p such that either $u^*up = p$ or $uu^*p = p$. Replacing $(\mathcal{M}, \sigma)$ by $(\mathcal{M}p, \sigma^p)$ we may assume that

$$\text{either } u^*u = 1 \text{ or } uu^* = 1.$$

If $u^*u = 1$ but $uu^* \neq 1$, then u^n is a nonunitary isometry in $\mathcal{M}(\sigma; \{\gamma^n\}) = \mathcal{M}^\sigma$, hence $\mathcal{M}^\sigma$ is infinite. If $uu^* = 1$ but $u^*u \neq 1$, then $(u^*)^n$ is a nonunitary isometry in $\mathcal{M}^\sigma$, and $\mathcal{M}^\sigma$ is again infinite.

Finally, consider the case $u^*u = uu^* = 1$ and assume, to the contrary, that $\mathcal{M}^\sigma$ is finite. Then ([L], 7.23) the *-operation is s-continuous on the unit ball of $\mathcal{M}^\sigma$. We shall show that the same property is valid for $\mathcal{M}$; it will follow that $\mathcal{M}$ is finite ([L], 7.23), a contradiction.

So, consider $x \in \mathcal{M}$, $\|x\| \leq 1$, and a net $\{x_i\}_{i\in I} \subset \mathcal{M}$, $\|x_i\| \leq 1$, such that $x_i \overset{s}{\underset{i\in I}{\to}} x$. For each $k \in \{0, 1, \dots, n-1\}$, the elements $\hat{x}(k)u^{-k}, \hat{x}_i(k)u^{-k}$ belong to the closed unit ball of $\mathcal{M}^\sigma$, and $\hat{x}_i(k)u^{-k} \overset{s}{\underset{i\in I}{\to}} \hat{x}(k)u^{-k}$ since σ is s-continuous. By our last assumption $\mathcal{M}^\sigma$ is finite, so $(\hat{x}_i(k)u^{-k})^* \overset{s}{\underset{i\in I}{\to}} (\hat{x}(k)u^{-k})^*$, that is, $u^k\hat{x}_i^*(n-k) \overset{s}{\underset{i\in I}{\to}} u^k\hat{x}^*(n-k)$ and hence $\hat{x}_i^*(n-k) \overset{s}{\underset{i\in I}{\to}} \hat{x}^*(n-k)$. It follows that $x_i^* \overset{s}{\underset{i\in I}{\to}} x^*$. □

Corollary. *Let $\sigma \in Aut(\mathcal{M})$ be a minimal *-automorphism of the countably decomposable W^*-algebra $\mathcal{M}$ with minimal period $n \geq 2$. Let $e, f \in Proj(\mathcal{M}^\sigma)$, $v \in U(\mathcal{M})$, and $\lambda \in Sp\, \sigma$. Then*

$$e \sim f \text{ in } \mathcal{M} \Leftrightarrow \text{there exists } w \in \mathcal{M}(\sigma; \{\lambda\}) \text{ with } w^*w = e, ww^* = f; \tag{7}$$

$$v\sigma(v)\dots\sigma^{n-1}(v) = 1 \Leftrightarrow \text{there exists } u \in U(\mathcal{M}) \text{ with } v = u^*\sigma(u). \tag{8}$$

In particular,

$$e \sim f \text{ in } \mathcal{M} \Rightarrow e \sim f \text{ in } \mathcal{M}^\sigma; \tag{9}$$

$$\text{there exists } u \in U(\mathcal{M}) \text{ with } u^n = 1 \text{ and } \sigma(u) = \lambda u; \tag{10}$$

$$\text{there exists a nonzero projection } p \in \mathcal{M} \text{ with } \sigma(p)p = 0. \tag{11}$$

Proof. We first prove (9). In view of (3) we can consider separately the cases e finite and e properly infinite in $\mathcal{M}$.

If e is finite, then f and $e \vee f$ are also finite, so that we may may assume that $\mathcal{M}$ is finite. Let $\natural : \mathcal{M} \to \mathcal{Z}(\mathcal{M})$ be the canonical central trace on $\mathcal{M}$. Using (3) it is easy to check that $\natural|\mathcal{M}^\sigma : \mathcal{M}^\sigma \to \mathcal{Z}(\mathcal{M}^\sigma)$ is the canonical central trace on $\mathcal{M}$. Since $e \sim f$ in $\mathcal{M}$, we have $e^\natural = f^\natural$, hence $e \sim f$ in $\mathcal{M}^\sigma$ also ([L], 7.11, 7.12).

If e is properly infinite in $\mathcal{M}$, then so is f and, by the previous proposition, e and f are also properly infinite in $\mathcal{M}^\sigma$. Since $e \sim f$ in $\mathcal{M}$, we have $\mathbf{z}(e) = \mathbf{z}(f)$. Therefore ([L], 4.13), $e \sim f$ in $\mathcal{M}^\sigma$, since $\mathcal{M}$ is assumed countably decomposable.

We now prove (8). If $v = u^*\sigma(u)$ with $u \in U(\mathcal{M})$, then $v\sigma(v)\dots\sigma^{n-1}(v) = u^*\sigma(u)\sigma(u^*)\sigma^2(u)\dots\sigma^{n-1}(u)^*\sigma^n(u) = u^*\sigma^n(u) = u^*u = 1$. Conversely, assume that $v\sigma(v)\dots\sigma^{n-1}(v) = 1$. Let $\mathcal{P} = Mat_2(\mathcal{M}) = \mathcal{M} \otimes Mat_2(\mathbb{C})$, $V = \begin{pmatrix} 1 & 0 \\ 0 & v \end{pmatrix} \in U(\mathcal{P})$, and $\mathfrak{S} = Ad(V)\circ(\sigma \bar{\otimes} \iota) \in Aut(\mathcal{P})$. Since $\sigma^n = \iota$ and $v\sigma(v)\dots\sigma^{n-1}(v) = 1$, we have $\mathfrak{S}^n = \iota$. Also, using the results of Sections 16.3, 16.15, and 16.16, we get $\Gamma(\mathfrak{S}) = \Gamma(\sigma \bar{\otimes} \iota) = \Gamma(\sigma) = \{\omega \in \mathbb{C}; \omega^n = 1\}$. Thus, $\mathfrak{S}$ is a minimal periodic

*-automorphism. On the other hand, the projections $\begin{pmatrix} 1 & 0 \\ 0 & 0 \end{pmatrix}$ and $\begin{pmatrix} 0 & 0 \\ 0 & 1 \end{pmatrix}$ belong to $\mathscr{P}^{\mathfrak{G}}$ and are equivalent in $\mathscr{P}$ so that, by (9), they are also equivalent in $\mathscr{P}^{\mathfrak{G}}$. A partial isometry implementing this equivalence is necessarily of the form $\begin{pmatrix} 0 & u \\ 0 & 0 \end{pmatrix}$ with $u \in U(\mathscr{M})$; the fact that $\begin{pmatrix} 0 & u \\ 0 & 0 \end{pmatrix}$ is $\mathfrak{G}$-invariant means that $u = \sigma(u)v^*$, that is, $v = u^*\sigma(u)$.

In particular, for $v = \lambda \cdot 1_{\mathscr{M}}$ with $\lambda^n = 1$ there exists $u_1 \in U(\mathscr{M})$ such that $\sigma(u_1) = \lambda u_1$. Then $u_1^n \in \mathscr{M}^\sigma$, and there exists $u_0 \in U(\mathscr{M}^\sigma)$ a Borel function of u_1^n such that $u_0^{-n} = u_1^n$. Since u_1^n commutes with u_1 and u_0 is a Borel function of u_1^n, it follows that u_0 commutes with u_1. Then $u = u_0u_1 \in U(\mathscr{M}), \sigma(u) = \lambda u$, and $u^n = 1$, proving (10).

It is now easy to prove (7). If $e \sim f$ in $\mathscr{M}$, then, by (9), there exists $x \in \mathscr{M}^\sigma$ with $x^*x = e$ and $xx^* = f$. Since σ^e is also minimal periodic with minimal period n, using (10) we obtain $u \in \mathscr{M}$ with $u^*u = uu^* = e$ and $\sigma(u) = \lambda u$. Then $w = xu \in \mathscr{M}, \sigma(w) = \lambda w$, and $w^*w = e, ww^* = f$.

Finally, we prove (11). Let $\gamma = \exp(2\pi i/n)$. By (10) there exists $u \in U(\mathscr{M})$ with $u^n = 1$, and $\sigma(u) = \gamma u$. Since $u^n = 1$, the spectral decomposition of u is of the form $u = \sum_{k=0}^{n-1} \gamma^k p_k$ with $p_0, \dots, p_{n-1}$ mutually orthogonal projections in $\mathscr{M}$ and $\sum_{k=0}^{n-1} p_k = 1$. Then $\sum_{k=0}^{n-1} \gamma^k \sigma(p_k) = \sigma(u) = \gamma u = \sum_{k=0}^{n-1} \gamma^{k+1} p_k$. Thus, if $p_k \neq 0, \sigma(p_k) = p_{k-1}$ and hence $\sigma(p_k)p_k = 0$. □

16.18. Notes. The results in this section are due to Connes (1973a, 1977a, 1976b).

For our exposition, we have used Combes and Delaroche (1973–1974); Combes and Delaroche (1975) and Connes (1977a). For the proofs of Proposition 16.17 and Corollary 17.24 (which are not explicit in the literature) the author has benefited from several useful discussions with Apostol and Digernes.

We record the following related references: Ikunishi and Nakagami (1976); Kishimoto and Takai (1978); Olesen (1975); Olesen and Pedersen (1978, 1980); Olesen et al. (1977); Pedersen (1973–1977); Pedersen (1980).

17 Outer Automorphisms

In this section, we derive a canonical decomposition of a *-automorphism into an inner part and a properly outer part. We give an important characterization of properly outer *-automorphisms and some applications.

17.1 Proposition. *Let $\mathscr{M}$ be a W*-algebra, $\sigma \in Aut(\mathscr{M})$ and let $e \in \mathscr{M}^\sigma$ be a projection with central support $p = \mathbf{z}_{\mathscr{M}}(e)$. If there exists $u \in U(e\mathscr{M}e)$ such that $\sigma^e = Ad(u)$, then there exists a unique $v \in U(\mathscr{M}p)$ such that $\sigma^p = Ad(v)$ and $ve = u = ev$.*

If $a \in \mathscr{M}p, ae = u = ea$, and $ax = \sigma(x)a$ for all $x \in \mathscr{M}p$, then $a = v$.

Proof. Since $e \in \mathscr{M}^\sigma$, we also have $\mathbf{z}_{\mathscr{M}}(e) \in \mathscr{M}^\sigma$. Thus, without loss of generality, we may assume that $p = \mathbf{z}_{\mathscr{M}}(e) = 1$.

Let $\mathscr{M} \subset \mathscr{B}(\mathscr{H})$ be realized as a von Neumann algebra. We recall ([L], 3.9) that $\mathbf{z}_{\mathscr{M}}(e)\mathscr{H} = \overline{\mathscr{M}e\mathscr{H}}$. For $x_1, \dots, x_n \in \mathscr{M}$ and $\xi_1, \dots, \xi_n \in e\mathscr{H}$ we have

$$\left\| \sum_k \sigma(x_k)u\xi_k \right\|^2 = \sum_{i,j}(u^*\sigma(ex_j^*x_ie)u\xi_i|\xi_j)$$

$$= \sum_{i,j}(ex_j^*x_ie\xi_i|\xi_j) = \left\| \sum_k x_k\xi_k \right\|^2 .$$

Since $\mathbf{z}_{\mathscr{M}}(e) = 1$, it follows that there exists a unique unitary operator $v \in \mathscr{B}(\mathscr{H})$ such that

$$vxe\xi = \sigma(x)ue\xi \quad (x \in \mathscr{M}, \xi \in \mathscr{H}). \tag{1}$$

It is easy to check that $v \in U(\mathscr{M}')' = \mathscr{M}$, and $v|e\mathscr{H} = u$. For every $x, y \in \mathscr{M}$ and $\xi \in \mathscr{H}$ we have $vxye\xi = \sigma(xy)ue\xi = \sigma(x)\sigma(y)ue\xi = \sigma(x)vye\xi$, hence $vx = \sigma(x)v$, that is, $\sigma = Ad(v)$.

If $a \in \mathscr{M}, ae = u = ea$, and $ax = \sigma(x)a$ for all $x \in \mathscr{M}$, then $v^*a \in \mathscr{Z}(\mathscr{M})$ and $v^*ae = e$; hence $v^*a = 1$ and $a = v$. □

With the same assumptions as in the proposition, let $\{\sigma_i\}_{i\in I} \subset Aut(\mathscr{M})$ and $\{u_i\}_{i\in I} \subset U(e\mathscr{M}e)$ be two nets with $\sigma_i^e = Ad(u_i), (i \in I)$. For each $i \in I$, let $v_t \in U(\mathscr{M}p)$ be the unique element such that $\sigma_i^p = Ad(v_i)$ and $v_ie = u_i = ev_i$. Then

$$\sigma_i(x) \xrightarrow{w} \sigma(x) \, for \, all \, x \in \mathscr{M} \, and \, u_i \xrightarrow{w} u \Rightarrow v_i \xrightarrow{w} v. \tag{2}$$

Indeed, arguing as above we may assume that $p = 1$ and $\mathscr{M} \subset \mathscr{B}(\mathscr{H})$. For $x \in \mathscr{M}$ and $\xi, \eta \in \mathscr{M}$ we have, by (1),

$$(v_ixe\xi|\eta) = (\sigma_i(x)u_ie\xi|\eta) \to (\sigma(x)ue\xi|\eta) = (vxe\xi|\eta),$$

hence $v_i \xrightarrow{w} v$.

17.2. Let $\mathscr{M}$ be a W^*-algebra and $\sigma \in Aut(\mathscr{M})$. Let $\{p_i\}_{i\in I} \subset \mathscr{Z}(\mathscr{M})$ be a maximal family of mutually orthogonal projections with the property $\sigma^{p_i} \in Int(\mathscr{M}p_i)\,(i \in I)$, and put $p(\sigma) = \sum_{i\in I} p_i \in \mathscr{Z}(\mathscr{M})$. Then it is clear that $\sigma^{p(\sigma)} \in Int(\mathscr{M}p(\sigma))$ and from Proposition 17.1 it follows that $p(\sigma)$ *is the greatest projection* $e \in \mathscr{M}^\sigma$ *such that* $\sigma^e \in Int(e\mathscr{M}e)$.

We call $p(\sigma)$ the *inner part* of σ. If $p(\sigma) = 0$, then the *-automorphism σ is called *properly outer*. It is clear that $p(\sigma^{1-p(\sigma)}) = 0$. We call $1 - p(\sigma)$ the *properly outer part* of σ. Thus:

Corollary. *Let* $\mathscr{M}$ *be a* W^**-algebra and* $\sigma \in Aut(\mathscr{M})$*. There exist two central projections* $p, q \in \mathscr{Z}(\mathscr{M})$ *with* $p+q = 1$*, such that* σ^p *is inner and* σ^q *is properly outer.* p *and* q *are uniquely determined by these conditions.*

Since $p(\sigma)$ is a central projection, it is easy to see that

$$p(Ad(u)\circ\sigma) = p(\sigma) = p(\sigma\circ Ad(u)) \quad (\sigma \in Aut(\mathscr{M}), u \in U(\mathscr{M})). \tag{1}$$

It follows that

$$\textit{if there exists } 0 \neq e \in Proj(\mathcal{M}) \textit{ and } u \in U(\mathcal{M}) \qquad (2)$$
$$\textit{such that } \sigma(x) = uxu^* \textit{ for all } x \in e\mathcal{M}e, \textit{ then } p(\sigma) \geq \mathbf{z}_{\mathcal{M}}(e) \neq 0$$

Indeed, by assumption it follows that $(Ad(u^*)\circ\sigma)(x) = x$ for all $x \in e\mathcal{M}e$, hence $p(\sigma) = p(Ad(u^*)\circ\sigma) \geq e$.

17.3. Let $\mathcal{M}$ be a W^*-algebra and $G \subset Aut(\mathcal{M})$ a subgroup. Let

$$[G] = \{\sigma \in Aut(\mathcal{M}) : \bigvee_{g\in G} p(g^{-1}\sigma) = 1\}.$$

Let $\sigma \in [G]$ and denote by $\mathfrak{F}(\sigma, G)$ the set of all families of the form $\{p_i, u_i, g_i\}_{i\in I}$ such that

$$\{p_i; i \in I\} \text{ are mutually orthogonal nonzero projections in } \mathcal{Z}(\mathcal{M}); \qquad (1)$$
$$u_i \in \mathcal{M}, u_i^* u_i = u_i u_i^* = p_i; \qquad (2)$$
$$g_i \in G, (g_i^{-1}\sigma)(p_i) = p_i \text{ and } (g_i^{-1}\sigma)(x) = u_i x u_i^* \text{ for all } x \in \mathcal{M}p_i. \qquad (3)$$

The set $\mathfrak{F}(\sigma, G)$ is inductively ordered by inclusion and hence there exists a maximal element $\{p_i, u_i, g_i\}_{i\in I} \in \mathfrak{F}(\sigma, G)$. Since $\sigma \in [G]$, it follows that $\sum_{i\in I} p_i = 1$. For $x \in \mathcal{M}$ we have

$$x = \sum_{i\in I} xp_i \text{ and } \sigma(x) = \sum_{i\in I} g_i(u_i x u_i^*). \qquad (4)$$

Thus, the maximal element $\{p_i, u_i, g_i\}_{i\in I}$ of $\mathfrak{F}(\sigma, G)$ determines completely the *-automorphism $\sigma \in [G]$.

It is now easy to check that $[G]$ is a subgroup of $Aut(\mathcal{M}), G \subset [G]$ and $[[G]] = [G]$. The group $[G]$ is called *the full group associated with G*.

17.4 Proposition. *Let $\mathcal{M}$ be a W^*-algebra and $\sigma \in Aut(\mathcal{M})$. Then*

$$a \in \mathcal{M}, ax = \sigma(x)a \textit{ for all } x \in \mathcal{M} \Rightarrow a = ap(\sigma). \qquad (1)$$

In particular, σ is properly outer if and only if

$$a \in \mathcal{M}, ax = \sigma(x)a \textit{ for all } x \in \mathcal{M} \Rightarrow a = 0, \qquad (2)$$

and σ is outer, that is, $\sigma \notin Int(\mathcal{M})$, if and only if

$$a \in \mathcal{M}, ax = \sigma(x)a \textit{ for all } x \in \mathcal{M} \Rightarrow \mathbf{z}(a) \neq 1. \qquad (3)$$

Proof. Let $a \in \mathcal{M}$ be such that $ax = \sigma(x)a$ for all $x \in \mathcal{M}$. If $x \in \mathcal{M}$ is unitary, then $x^*a^* = a^*\sigma(x)^*, a^*a = a^*\sigma(x)^*\sigma(x)a = x^*a^*ax, aa^* = axx^*a^* = \sigma(x)aa^*\sigma(x)^*$, hence $a^*a \in \mathcal{Z}(\mathcal{M}), aa^* \in \mathcal{Z}(\mathcal{M}), \mathbf{r}(a) = \mathbf{s}(a^*a) \in \mathcal{Z}(\mathcal{M}), \mathbf{l}(a) = \mathbf{s}(aa^*) \in \mathcal{Z}(\mathcal{M})$. In particular, $|a| \in \mathcal{Z}(\mathcal{M})$ and $\mathbf{r}(a) = \mathbf{l}(a) = \mathbf{z}(a) = p \in \mathcal{Z}(\mathcal{M})$. Also, $pa = ap = \sigma(p)a$, hence $(p - \sigma(p))p = 0$ and $\sigma(p) = p$, that is, $p \in \mathcal{Z}(\mathcal{M})^\sigma$.

Let $a = u|a|$ be the polar decomposition of a. Then $u^*u = uu^* = p$ and for $x \in \mathcal{M}p$ we have $ux|a| = u|a|x = ax = \sigma(x)a = \sigma(x)u|a|$, hence $ux = uxp = \sigma(x)up = \sigma(x)pu = \sigma(xp)u = \sigma(x)u$, that is, $\sigma^p = Ad(u)$. Consequently, $p \leq p(\sigma)$ and $a = ap = ap(\sigma)$.

If σ is properly outer, that is, $p(\sigma) = 0$, the (2) follows from (1). Conversely, if $p = p(\sigma) \neq 0$ and $u \in \mathcal{M}, u^*u = uu^* = p, \sigma^p = Ad(u)$, it follows from (2) that $u = 0$, a contradiction.

If $\sigma \notin Int(\mathcal{M})$, then $p(\sigma) \neq 1$ and (3) follows from (1). Conversely, if $\sigma \in Int(\mathcal{M})$ and $u \in U(\mathcal{M}), \sigma = Ad(u)$, it follows from (3) that $1 = \mathbf{z}(u) \neq 1$, a contradiction. □

The main assertion (1) is obviously equivalent to

$$p(\sigma) = \vee\{\mathbf{z}(a); \quad a \in \mathcal{M}, ax = \sigma(x)a \text{ for all } x \in \mathcal{M}\}. \tag{4}$$

An element $a \in \mathcal{M}$ such that $ax = \sigma(x)a$ for all $x \in \mathcal{M}$ will be called a *σ-dependent element.*

The above proof shows that if *$a \in \mathcal{M}$ is σ-dependent and $a = u|a|$ is its polar decomposition,* then

$$|a| = |a^*| \in \mathcal{Z}(\mathcal{M})^\sigma, \mathbf{l}(a) = \mathbf{r}(a) = \mathbf{z}(a) \in \mathcal{Z}(\mathcal{M})^\sigma, \tag{5}$$

$$u^*u = uu^* = \mathbf{z}(a) \text{ and } \sigma(x) = uxu^* \quad (x \in \mathcal{M}\mathbf{z}(a)). \tag{6}$$

It is easy to check that, for $a \in \mathcal{M}$, we have

$$a \text{ is } \sigma\text{-dependent} \Leftrightarrow a^* \text{ is } \sigma^{-1}\text{-dependent}. \tag{7}$$

Note that all the above statements remain valid if we replace the conditions of the form $ax = \sigma(x)a$ by $xa = a\sigma(x)$, the only change appearing in (6), where u must be replaced by u^*.

17.5. Let $\mathcal{Z}$ be an abelian W^*-algebra and $\sigma \in Aut(\mathcal{Z})$. In this case, $p(\sigma)$ is the greatest projection $p \in \mathcal{Z}$ such that $\sigma^p = \iota =$ the identity mapping on $\mathcal{Z}p$. We shall say that σ *acts freely* on $\mathcal{Z}$ if for every nonzero projection $p \in \mathcal{Z}$ there exists a nonzero projection $q \in \mathcal{Z}, q \leq p$, such that $q\sigma(q) = 0$. The next proposition shows, in particular, that σ acts freely on $\mathcal{Z}$ if and only if $p(\sigma) = 0$.

Proposition. *Let $\mathcal{M}$ be a W^*-algebra and $\sigma \in Aut(\mathcal{M})$. The following statements are equivalent:*

(i) *$p(\sigma|\mathcal{Z}(\mathcal{M})) = 0$;*
(ii) *σ acts freely on $\mathcal{Z}(\mathcal{M})$;*
(iii) *$a \in \mathcal{M}$ and $az = \sigma(z)a$ for all $z \in \mathcal{Z}(\mathcal{M}) \Rightarrow a = 0$.*

In particular, if σ acts freely on $\mathcal{Z}(\mathcal{M})$, then σ is properly outer.

Proof. (i) ⇒ (ii). Let $p \in \mathcal{Z}(\mathcal{M})$ be a nonzero projection. We first show that there exists a projection $r \in \mathcal{Z}(\mathcal{M})$ with $\sigma(r) \neq r$. Otherwise, for every projection $z \in \mathcal{Z}(\mathcal{M})$ we have $pz = \sigma(pz) = \sigma(p)\sigma(z) = p\sigma(z) = \sigma(z)p$ and the identity $pz = \sigma(z)p$ holds for any $z \in \mathcal{Z}(\mathcal{M})$; this, by (i) and 17.4.(2), implies that $p = 0$, a contradiction. Thus, there exists a projection $r \in \mathcal{Z}(\mathcal{M})$ with $r \leq p$ and $\sigma(r) \neq r$. If $q = r - r\sigma(r) \neq 0$, then we have $0 \neq q \leq r \leq p$ and $q\sigma(q) = 0$. If $r = r\sigma(r)$, then $r \leq \sigma(r), \sigma^{-1}(r) \leq r$, hence $q' = r - \sigma^{-1}(r) \neq 0, q' \leq r \leq p$ and $q'\sigma(q') = 0$.

(ii) ⇒ (iii). Let $a \in \mathcal{M}$ be such that $az = \sigma(z)a$ for all $z \in \mathcal{Z}(\mathcal{M})$, and suppose that $p = \mathbf{z}(a) \in \mathcal{Z}(\mathcal{M})$. If $a \neq 0$, then, by (ii), there exists a nonzero projection $q \in \mathcal{Z}(\mathcal{M}), q \leq p$, with $q\sigma(q) = 0$; it follows that $aq = \sigma(q)a = 0$, hence $0 \neq q = qp = 0$, a contradiction. Thus, $a = 0$.

(iii) ⇒ (i). Obvious. □

17.6 Proposition. *Let $\mathcal{M}, \mathcal{N}$ be W^*-algebras, $\sigma \in Aut(\mathcal{M}), \tau \in Aut(\mathcal{N})$. Then*

$$p(\sigma \bar{\otimes} \tau) = p(\sigma) \bar{\otimes} p(\tau). \tag{1}$$

In particular,

$$\sigma \bar{\otimes} \tau \text{ is properly outer} \Leftrightarrow \text{ either } \sigma \text{ or } \tau \text{ is properly outer.} \tag{2}$$

$$\sigma \bar{\otimes} \tau \in Int(\mathcal{M} \bar{\otimes} \mathcal{N}) \Leftrightarrow \sigma \in Int(\mathcal{M}) \text{ and } \tau \in Int(\mathcal{N}). \tag{3}$$

Proof. It is clear that if $\sigma = Ad(u)$ and $\tau = Ad(v)$ with $u \in U(\mathcal{M})$ and $v \in U(\mathcal{N})$, then $\sigma \bar{\otimes} \tau = Ad(u \bar{\otimes} v)$ with $u \bar{\otimes} v \in U(\mathcal{M} \bar{\otimes} \mathcal{N})$. Hence $p(\sigma \bar{\otimes} \tau) \geq p(\sigma) \bar{\otimes} p(\tau)$.

Conversely, let $a \in \mathcal{M} \bar{\otimes} \mathcal{N}$ be such that $az = (\sigma \bar{\otimes} \tau)(z)a$ for all $z \in \mathcal{M} \bar{\otimes} \mathcal{N}$. Let $\psi \in \mathcal{N}_*$ and $x \in \mathcal{M}$. We have (9.8)

$$E^{\psi}_{\mathcal{M}}(a)x = E^{\psi}_{\mathcal{M}}(a(x \bar{\otimes} 1)) = E^{\psi}_{\mathcal{M}}((\sigma(x) \bar{\otimes} 1)a) = \sigma(x)E^{\psi}_{\mathcal{M}}(a).$$

Using Proposition 17.4.(1), we infer that

$$0 = E^{\psi}_{\mathcal{M}}(a) - E^{\psi}_{\mathcal{M}}(a)p(\sigma) = E^{\psi}_{\mathcal{M}}(a((1 - p(\sigma)) \bar{\otimes} 1)).$$

Since $\psi \in \mathcal{N}_*$, was arbitrary, it follows (9.8.(3)) that $a = a(p(\sigma) \bar{\otimes} 1)$. Similarly, we get $a = a(1 \bar{\otimes} p(\tau))$. Hence $a = a(p(\sigma) \bar{\otimes} p(\tau))$. Using 17.4.(4), we infer that $p(\sigma \bar{\otimes} \tau) \leq p(\sigma) \bar{\otimes} p(\tau)$. We have thus proved (1).

Equations (2) and (3) follow immediately from (1). □

17.7. Let $\sigma : G \to Aut(\mathcal{M})$ be an action of the discrete group G on the W^*-algebra $\mathcal{M}$. We shall say that the action σ is *properly outer* if for every $t \in G, t \neq$ the neutral element of G, the *-automorphism $\sigma_t \in Aut(\mathcal{M})$ is properly outer. We recall that the action is called *ergodic* if $\mathcal{M}^{\sigma} = \mathbb{C} \cdot 1_{\mathcal{M}}$.

Every *-automorphism $\sigma \in Aut(\mathcal{M})$ defines an action $\sigma : \mathbb{Z} \ni n \mapsto \sigma^n \in Aut(\mathcal{M})$. Clearly, the action σ is ergodic if and only if the *-automorphism σ is ergodic. We shall say that the *-automorphism is *aperiodic* if the action σ is properly outer, that is, if $p(\sigma^n) = 0$ for all $n \in \mathbb{Z}, n \neq 0$.

Proposition. *Let $\mathcal{M}$ be a W^*-algebra without minimal projections and $\sigma \in Aut(\mathcal{M})$. If σ is ergodic, then σ is aperiodic.*

Proof. The set of all states $\{\varphi \in \mathcal{M}_* : \varphi \geq 0, \varphi(1) = 1\}$ on $\mathcal{M}$ is a $\sigma(\mathcal{M}^*, \mathcal{M})$-compact convex subset of $\mathcal{M}^*$, invariant under the transformations $\varphi \mapsto \varphi \circ \sigma^n, (n \in \mathbb{Z})$. By the Markov–Kakutani fixed point theorem ([L], A.1) it follows that there exists a state φ on $\mathcal{M}$ such that $\varphi \circ \sigma = \varphi$.

Let $n \in \mathbb{Z}, n > 0$, be such that $p = p(\sigma^n) \neq 0$. There exists $u \in \mathcal{M}, u^*u = uu^* = p$ such that $\sigma^n(px) = upxu^*(x \in \mathcal{M})$. Then $v = u + (1 - p) \in U(\mathcal{M})$ and $\sigma^n(px) = vpxv^*(x \in \mathcal{M})$. Let $\mathcal{A}$ be a maximal abelian *-subalgebra of $\mathcal{M}$ containing v. For every $a \in \mathcal{A}$, we have $\sigma^n(pa) = pa$.

It is easy to check that every minimal projection of $\mathcal{A}$ is also a minimal projection of $\mathcal{M}$ so that, by assumption, $\mathcal{A}$ has no minimal projections. Thus there exists a nonzero projection $q \in \mathcal{A}, q \leq p$, such that $\varphi(q) \leq 1/n$. Since $\sigma^n(q) = q$, we have $a = q + \sigma(q) + ... + \sigma^{n-1}(q) \in \mathcal{M}^{\sigma}$. Since σ is ergodic, there exists $\lambda \in \mathbb{C}$ with $a = \lambda \cdot 1_{\mathcal{M}}$. We have $a \geq q$ hence $\lambda \geq 1$. On the other hand, $\lambda = \varphi(a) = \varphi(q) + \varphi(\sigma(q)) + \ldots + \varphi(\sigma^{n-1}(q)) < n/n = 1$, a contradiction.

Hence $p(\sigma^n) = 0$ for all $n \in \mathbb{Z}, n \neq 0$. □

17.8. *Let $\mathscr{M}$ be a W*-algebra and $G \hookrightarrow Aut(\mathscr{M})$ an ergodic action of the discrete group G on $\mathscr{M}$.* Then

$$\sigma \in Aut(\mathscr{M}), \sigma g = g\sigma \text{ for all } g \in G \Rightarrow \text{ either } p(\sigma) = 0 \text{ or } p(\sigma) = 1, \tag{1}$$

since it follows from the assumption that $p(\sigma)$ is G-invariant.

Assume moreover that G is *commutative* and let $\iota \in G$ be the neutral element of G. Then from (1) it follows that

$$g \notin Int(\mathscr{M}) \text{ for all } g \in G,\ g \neq \iota \Rightarrow p(g) = 0 \text{ for all } g \in G, g \neq \iota. \tag{2}$$

Also,

$$\sigma \in [G], \sigma g = g\sigma \text{ for all } g \in G \Rightarrow \text{ there exists } g \in G \text{ with } g^{-1}\sigma \in Int(\mathscr{M}). \tag{3}$$

Indeed, we have $(g^{-1}\sigma)h = h(g^{-1}\sigma)$ for all $g, h \in G$ so, by (1), either $p(g^{-1}\sigma) = 0$ for all $g \in G$, or $p(g^{-1}\sigma) = 1$ for some $g \in G$. Since $\sigma \in [G]$, the desired conclusion follows.

Finally, if the W*-algebra $\mathscr{M}$ *is commutative*, it follows from (3) that

$$\sigma \in [G], \sigma g = g\sigma \text{ for all } g \in G \Rightarrow \sigma \in G, \tag{4}$$

that is, G is "maximal commutative" in $[G]$.

17.9. The next theorem is a remarkable noncommutative extension of the equivalence (i) ⇔ (ii) of Proposition 17.5.

Theorem (A. Connes). *Let $\mathscr{M}$ be a countably decomposable W*-algebra and $\sigma \in Aut(\mathscr{M})$. Then $p(\sigma)$ is the smallest central projection p in $\mathscr{M}$ with the following property:*

$$\text{for every nonzero projection } e \in \mathscr{M}, e \leq 1 - p, \text{ and every } \varepsilon > 0 \text{ there exists a nonzero projection } f \in \mathscr{M}, f \leq e, \text{ such that } \|f\sigma(f)\| < \varepsilon. \tag{1}$$

In particular, σ is properly outer if and only if

$$\text{for every nonzero projection } e \in \mathscr{M} \text{ and every } \varepsilon > 0 \text{ there exists a nonzero projection } f \in \mathscr{M}, f \leq e, \text{ such that } \|f\sigma(f)\| < \varepsilon. \tag{2}$$

Also, if $\sigma \notin Int(\mathscr{M})$, then

$$\text{for every } \varepsilon > 0 \text{ there exists a nonzero projection } f \in \mathscr{M} \text{ such that } \|f\sigma(f)\| < \varepsilon. \tag{3}$$

The proof is contained in Sections 17.10–17.16, which are also of independent interest.

17.10. In this section, we show that *if a central projection $p \in \mathcal{Z}(\mathcal{M})$ satisfies condition 17.9.(1), then $p \geq p(\sigma)$.*

If $p \ngeq p(\sigma)$, then $0 \neq q = p(\sigma) - p(\sigma)p \leq 1 - p$ and there exists $u \in \mathcal{M}, u^*u = uu^* = q$ such that $\sigma(x) = uxu^*$ for all $x \in \mathcal{M}q$. There exists $\lambda \in \mathbb{C}, |\lambda| = 1$ and a nonzero spectral projection $e \in \mathcal{M}q$ of u such that $\|ue - \lambda e\| \leq 1/4$. Since $e \in \mathcal{M}^\sigma$, it follows that $\sigma^e = Ad(ue) \in Aut(e\mathcal{M}e)$ and $\|\sigma^e - \iota\| \leq 1/2$. Then for every projection $f \in \mathcal{M}, f \leq e$, we have $\|\sigma(f) - f\| \leq 1/2$. For $f \neq 0$ we get

$$\begin{aligned}2\|f\sigma(f)\| &\geq \|f\sigma(f) - \sigma(f)f\| = \|(\sigma(f) + f) - (\sigma(f) - f)^2\| \\ &\geq \|\sigma(f) + f\| - \|(\sigma(f) - f)^2\| > \|f\| - \|\sigma(f) - f\|^2 \geq 1 - \frac{1}{4} = \frac{3}{4},\end{aligned}$$

that is, $\|f\sigma(f)\| \geq 3/8$, contradicting 17.9.(1).

17.11. In order to complete the proof of Theorem 17.9, we still have to show that $p(\sigma)$ satisfies condition 17.9.(1). To this end we shall first prove (17.11–17.14) that *every outer *-automorphism $\sigma \in Aut(\mathcal{M})$ satisfies condition* 17.9.(3).

If σ does not act identically on $\mathcal{Z}(\mathcal{M})$, that is, if $p(\sigma|\mathcal{Z}(\mathcal{M})) \neq 1$, then 17.9.(3) follows obviously from Proposition 17.5.

Therefore, we shall assume that *σ acts identically on $\mathcal{Z}(\mathcal{M})$*. Then a standard maximality argument shows that for each $n \in \mathbb{Z}, n > 1$, there exists a greatest central projection $p_n(\sigma) \in \mathcal{Z}(\mathcal{M})$ such that the following statement concerning $0 \neq p \in Proj(\mathcal{Z}(\mathcal{M}))$ and $1 \leq k \in \mathbb{Z}$:

$$\textit{there exists } u \in \mathcal{M}^\sigma, u^*u = uu^* = p, \textit{ such that } \sigma^k(x) = uxu^* \textit{ for all } x \in \mathcal{M}p \tag{1}$$

is true for $p = p_n(\sigma)$ when $k = n$, but false for every $p \leq p_n(\sigma)$ when $k < n$.

The projections $\{p_n(\sigma)\}_{n>1}$ are mutually orthogonal and

$$\sum_{n\geq 1} p_n(\sigma) = \bigvee_{n\geq 1} p(\sigma^n). \tag{2}$$

More precisely, we shall show that

$$p(\sigma^n) \leq \sum_{k=1}^{n^2} p_k(\sigma) \quad (n \geq 1). \tag{3}$$

Indeed, let $p = p(\sigma^n)$ and let $v \in \mathcal{M}$ and $v^*v = vv^* = p$, such that $\sigma^n(x) = vxv^*, (x \in \mathcal{M}p)$. For $x \in \mathcal{M}p$ we have $\sigma(v)x\sigma(v)^* = \sigma(v\sigma^{-1}(x)v^*) = \sigma(\sigma^n(\sigma^{-1}(x))) = \sigma^n(x) = vxv^*$, hence $v^*\sigma(v) = z \in \mathcal{Z}(\mathcal{M}p)$. Since $v = \sigma^n(v) = z^n v$, it follows that $z^n = p$. Therefore, $\sigma(v^n) = \sigma(v)^n = z^n v^n = v^n$ and $\sigma^{n^2}(x) = (v^n)x(v^n)^*$ for all $x \in \mathcal{M}p$. This proves (3) and hence also (2).

Note that $\sigma \in Int(\mathcal{M}) \Leftrightarrow p_1(\sigma) = 1$. So, in order to prove that $\sigma \notin Int(\mathcal{M}) \Rightarrow$ 17.9.(3), we distinguish two cases: either

$$\textit{there exists } n \in \mathbb{Z}, n \geq 2, \textit{ such that } p_n(\sigma) = 1 \tag{a}$$

or $p_n(\sigma) = 0$ for all $n \geq 1$, that is, by (3),

$$\sigma \in Aut(\mathscr{M}) \text{ is aperiodic.} \tag{b}$$

17.12. In each of the cases 17.11.(a) and 17.11.(b), we now compute the invariant $\Gamma(\sigma) \subset \hat{\mathbb{Z}} = \mathbb{T} = \{\gamma \in \mathbb{C}; |\gamma| = 1\}$ of the action $\sigma : \mathbb{Z} \ni n \mapsto \gamma^n \in Aut(\mathscr{M})$. If $\mathscr{M}$ is a factor, it is easy to see, using Theorem 16.5, that

$$\textit{in case } 17.11.(a) \textit{ we have } \Gamma(\sigma) = (n\mathbb{Z})^{\perp} = \{\gamma \in \mathbb{C}; \gamma^n = 1\}, \tag{1}$$

$$\textit{in case } 17.11.(b) \textit{ we have } \Gamma(\sigma) = \{0\}^{\perp} = \{\gamma \in \mathbb{C}; |\gamma| = 1\}. \tag{2}$$

Statements (1) and (2) are true in general, without the assumption that $\mathscr{M}$ is a factor. However, in proving the general case we shall use another characterization of invariant $\Gamma(a)$ and some elementary results concerning crossed products by discrete groups, which are given in Sections 21.1 and 22.6.

By Theorem 21.1, $\Gamma(\sigma)$ is the kernel of the restriction of the dual action $\hat{\sigma}$ of $\mathbb{T} = \hat{\mathbb{Z}}$ to the centre $\mathscr{Z}(\mathscr{R}(\mathscr{M}, \sigma))$ of the crossed product $\mathscr{R}(\mathscr{M}, \sigma)$. An arbitrary element $X \in \mathscr{R}(\mathscr{M}, \sigma)$ is of the form (22.1)

$$X = \sum_{k \in \mathbb{Z}} \pi_\sigma(a(k))(1 \bar{\otimes} \boldsymbol{\lambda}(k)) \tag{3}$$

with $a(k) \in \mathscr{M}$ $(k \in \mathbb{Z})$, and for every $\gamma \in \mathbb{T}$ we have (19.3)

$$\hat{\sigma}_\gamma(X) = \sum_{k \in \mathbb{Z}} \gamma^k \pi_\sigma(a(k))(1 \bar{\otimes} \boldsymbol{\lambda}(k)). \tag{4}$$

By Theorem 22.6, we have $X \in \mathscr{Z}(\mathscr{R}(\mathscr{M}, \sigma))$ if and only if for every $k \in \mathbb{Z}$.

$$a(k) \in \mathscr{M}^\sigma \text{ and } a(k)\sigma^k(x) = xa(k) \textit{ for } x \in \mathscr{M}. \tag{5}$$

In case 17.11.(b), we have $p(\sigma^k) = 0$ for all $k \neq 0$ so, using (5) and 17.4.(2), it follows that $a(k) = 0$ for all $k \neq 0$ and $a(0) \in \mathscr{Z}(\mathscr{M})^\sigma$. Thus, in this case we have $\mathscr{Z}(\mathscr{R}(\mathscr{M}, \sigma)) = \pi_\sigma(\mathscr{Z}(\mathscr{M})^\sigma)$ and so the restriction of the dual action to $\mathscr{Z}(\mathscr{R}(\mathscr{M}, \sigma))$ is just the trivial action, that is its kernel is the whole dual group $\hat{\mathbb{Z}} = \mathbb{T}$; hence $\Gamma(\sigma) = \{\gamma \in \mathbb{C}; |\gamma| = 1\}$.

Consider now case 17.11.(a). For each $k \in \mathbb{Z}, a(k)^* \in \mathscr{M}$ is a σ^k-dependent element, so that (17.4.(5), 17.4.(6)) the partial isometry appearing in the polar decomposition of $a(k)^*$ is σ-invariant and implements the restriction of σ^k to the (central) support of $a(k)^*$. It follows from condition 17.11.(a) that

$$a(k) \neq 0 \Rightarrow k \in n\mathbb{Z}.$$

From 17.11.(a) it also follows that there exists a unitary element $u \in \mathscr{M}^\sigma$ such that $\sigma^n = Ad(u)$. Then $\mathscr{Z}(\mathscr{R}(\mathscr{M}, \sigma))$ consists of all elements of the form

$$X = \sum_{m \in \mathbb{Z}} \pi_\sigma(z_m u^{-m})(1 \bar{\otimes} \boldsymbol{\lambda}(nm))$$

with $z_m \in \mathscr{Z}(\mathscr{M})$ $(m \in \mathbb{Z})$, and for every $\gamma \in \mathbb{T}$ we have

$$\hat{\sigma}_\gamma(X) = \sum_{m\in\mathbb{Z}} \gamma^{nm}\pi_\sigma(z_m u^{-m})(1 \bar{\otimes} \boldsymbol{\lambda}(nm)).$$

Consequently, $\hat{\sigma}_\gamma(X) = X$ for $X \in \mathscr{Z}(\mathscr{R}(\mathscr{M},\sigma))$ if and only if $\gamma^n = 1$, that is, $\Gamma(\tau) = \{\gamma \in \mathbb{C}; \gamma^n = 1\}$.

17.13. We show that 17.9.(3) holds true in case 17.11.(b). In this case, by 16.1.(1) and 17.12.(2), we have $Sp\ \sigma = \Gamma(\sigma) = \{\gamma \in \mathbb{C}; |\gamma| = 1\}$, so that it is sufficient to prove the following:

Lemma. *Let $\sigma \in Aut(\mathscr{M})$ be such that $-1 \in Sp\ \sigma$. Then, for every $\varepsilon > 0$ there exists a nonzero projection $f \in \mathscr{M}$ such that $\|f\sigma(f)\| \leq \varepsilon$.*

Proof. Since $-1 \in Sp\ \sigma$, there exists, by Proposition 14.5. $x \in \mathscr{M}$, $\|x\| = 1$, such that $\|\sigma(x) + x\| \leq \varepsilon/4 = \delta$. Write $x = b + ic$ with $b = b^* \in \mathscr{M}$, $c = c^* \in \mathscr{M}$. Then $\|b\| + \|c\| \geq 1$, $\|\sigma(b) + b\| \leq \delta$ and $\|\sigma(c) + c\| \leq \delta$. Therefore, we may assume that $\|b\| \geq 1/2$ and then, putting $a = \pm b/\|b\|$, we have

$$a = a^* \in \mathscr{M}, \|a\| = 1, 1 \in Sp(a), \|\sigma(a) + a\| \leq 2\delta.$$

Since $1 \in Sp(a)$, we have $f = \chi_{[1-\delta,1]}(a) \neq 0$ and $af \geq (1-\delta)f$.

Consider $\mathscr{M} \subset \mathscr{B}(\mathscr{H})$ realized as a standard von Neumann algebra; then ([L], 10.15) there exists a unitary operator $v \in \mathscr{B}(\mathscr{H})$ such that $\sigma = Ad(v)|\mathscr{M}$. For every $\xi \in f\mathscr{H}$, $\|\xi\| = 1$, we have

$$\begin{aligned}\|a\xi - \xi\|^2 &= \|(af - f)\xi\|^2 = \omega_\xi((af-f)^2) = \omega_\xi(a^2 f - 2af + f)\\ &\leq \omega_\xi(af - 2af + f) = \omega_\xi(f - af) \leq \omega_\xi(f - (1-\delta)f) = \delta.\end{aligned}$$

Then $\sigma(f)\mathscr{H} = vfv^*\mathscr{H} = vf\mathscr{H}$ and, for $\eta = v\xi \in \sigma(f)\mathscr{H}$, we have

$$\begin{aligned}\|\sigma(a)\eta - \eta\| &= \|vav^*\eta - \eta\| = \|va\xi - v\xi\| = \|a\xi - \xi\| \leq \delta,\\ \|a\eta + \eta\| &= \|(a\eta + \sigma(a)\eta) + (\eta - \sigma(a)\eta)\|\\ &\leq \|a + \sigma(a)\|\|\eta\| + \|\eta - \sigma(a)\eta\| \leq 3\delta.\end{aligned}$$

Consequently, for $\xi \in f\mathscr{H}$, $\|\xi\| = 1$, and $\eta \in \sigma(f)\mathscr{H}$, $\|\eta\| = 1$, we get

$$|(\xi|\eta) - (a\xi|\eta)| \leq \delta,\ |(\xi|a\eta) + (\xi|\eta)| \leq 3\delta,$$

and hence $|(\xi|\eta)| \leq 4\delta = \varepsilon$. Therefore $\|fa(f)\| \leq \varepsilon$. □

17.14. Let us show that statement 17.9.(3) holds also in case 17.11.(a). In this case we have (17.11, 17.12)

$$\Gamma(\sigma) = \{\gamma \in \mathbb{C}; \gamma^n = 1\} \textit{ and } \sigma^n = Ad(u) \textit{ with } u \in U(\mathscr{M}^\sigma).$$

Since $u \in U(\mathscr{M}^\sigma)$, there exists $v \in U(\mathscr{M}^\sigma)$ such that $v^n = u$. Then $\tau = Ad(v^*)\circ\sigma \in Aut(\mathscr{M})$ and, since $v \in \mathscr{M}^\sigma$ and $v^n = u$, we have $\tau^n = Ad(v^*)^n\circ\sigma^n = \iota$. Note that the corresponding actions σ and τ of $\mathbb{Z}$ on $\mathscr{M}$ are outer conjugate (16.15) and hence, by Proposition 16.3, $\Gamma(\sigma) = \Gamma(\tau)$.

Let $\varepsilon > 0$. There exist $\lambda \in \mathbb{C}$, $|\lambda| = 1$, and a nonzero spectral projection $e \in \mathcal{M}^\sigma \cap \mathcal{M}^\tau$ of v such that $\|ve - \lambda e\| \leq \varepsilon/2$. Then

$$\|\sigma^e - \tau^e\| \leq \varepsilon. \tag{1}$$

On the other hand, it is clear that $(\tau^e)^n = \iota$ and (16.1.(4)) $\Gamma(\tau^e) \supset \Gamma(\tau) = \Gamma(\sigma) = \{\gamma \in \mathbb{C}; \gamma^n = 1\}$, so that τ^e is minimal periodic. By Corollary 16.17.(11), there exists a nonzero projection $f \in \mathcal{M}$, $f \leq e$, such that $f\tau(f) = 0$.

Finally, using (1), we obtain $\|f\sigma(f)\| \leq \varepsilon$.

17.15. Before completing the proof of Theorem 17.9 we shall review some elementary facts about the "relative position" of two projections.

Let $\mathcal{H}$ be a Hilbert space, e and f two projections in $\mathcal{B}(\mathcal{H})$, $\mathcal{M} = \mathcal{R}\{e,f\} \subset \mathcal{B}(\mathcal{H})$ the von Neumann algebra generated by e,f, and let

$$s(e,f) = |e - f| \in \mathcal{M}, c(e,f) = |e \vee f - e - f| \in \mathcal{M}.$$

It is clear that

$$e \text{ and } f \text{ are abelian projections in } \mathcal{M} \tag{1}$$

since $e\mathcal{M}e = \mathcal{R}\{e, efe\}$ and $f\mathcal{M}f = \mathcal{R}\{f, fef\}$.

Also,

$$\begin{gathered} p = e \wedge f + (1-e) \wedge f + e \wedge (1-f) + (1-e) \wedge (1-f) \\ \text{is the greatest projection } q \in \mathcal{Z}(\mathcal{M}) \text{ such that } \mathcal{M}q \text{ is abelian and} \\ \mathcal{M}(1-p) \text{ is of type } I_2 \end{gathered} \tag{2}$$

Indeed, p is clearly a projection, p is central since it commutes with e and f, and $\mathcal{M}p = \mathcal{R}\{ep, fp\}$ is abelian as $epfp = e \wedge f = fpep$. If $q \in \mathcal{M}$ is a central projection and $\mathcal{M}q$ is abelian, then eq commutes with fq and it follows that $pq = q$, that is, $q \leq p$. If $p = 0$, then $e \vee f = (1-e) \vee f = e \vee (1-f) = (1-e) \vee (1-f) = 1$ and, using the parallelogram law ([L], 4.4), we deduce that $e = 1 - (1-e) = (1-e) \vee f - (1-e) \sim f - (1-e) \wedge f = f$ and $1 - e = e \vee f - e \sim f - e \wedge f = f$; hence e and $1 - e$ are equivalent abelian projections in $\mathcal{M}$, which means that in this case $\mathcal{M}$ is of type I_2.

On the other hand, we have $s(e,f)^2 = e + f - ef - fe$, $c(e,f)^2 = e \vee f - e - f + ef + fe$, hence

$$s(e,f)^2 + c(e,f)^2 = e \vee f. \tag{3}$$

We have $ec(e,f)^2 = efe = c(e,f)^2 e$, hence $c(e,f)$ commutes with e and f:

$$s(e,f) \in \mathcal{Z}(\mathcal{M}), c(e,f) \in \mathcal{Z}(\mathcal{M}). \tag{4}$$

Since $e \vee f \in \mathcal{Z}(\mathcal{M})$, we have

$$\mathbf{z}_{\mathcal{M}}(e) \vee \mathbf{z}_{\mathcal{M}}(f) = e \vee f. \tag{5}$$

Also,

$$\mathbf{s}_{\mathscr{M}}(c(e,f)) \leq \mathbf{z}_{\mathscr{M}}(e)\mathbf{z}_{\mathscr{M}}(f). \tag{6}$$

Indeed, let $q \in \mathscr{Z}(\mathscr{M})$ be a projection such that $qe = e$. Then $e \vee f - f \geq q(e \vee f - f) \sim q(e - e \wedge f) = e - e \wedge f \sim e \vee f - f$, hence $q(e \vee f - f) = e \vee f - f$, since by (2) $\mathscr{M}$ is finite. Therefore, $qc(e,f)^2 = q(e \vee f - f - e + ef + fe) = e \vee f - f - e + ef + fe = c(e,f)^2$, so that $qc(e,f) = c(e,f)$. Thus, $\mathbf{s}_{\mathscr{M}}(c(e,f)) \leq \mathbf{z}_{\mathscr{M}}(e)$ and, similarly, $\mathbf{s}_{\mathscr{M}}(c(e,f)) \leq \mathbf{z}_{\mathscr{M}}(f)$.

We now prove that

$$\begin{gathered} \textit{if } \mathbf{s}_{\mathscr{M}}(c(e,f)) = e \vee f \textit{ and } e \vee f - (e+f) = uc(e,f) \\ \textit{is the polar decomposition, then} \\ u = u^*,\, u^2 = e \vee f,\, ueu = f,\, ufu = e. \end{gathered} \tag{7}$$

Indeed, $e \vee f - (e+f)$ is self-adjoint, hence $u = u^*$. Also, $u^2 = u^*u = \mathbf{s}_{\mathscr{M}}(c(e,f)) = e \vee f$. Since $e \vee f = \mathbf{s}_{\mathscr{M}}(c(e,f)) = \mathbf{s}_{\mathscr{M}}(e \vee f - e - f)$, we have $\mathbf{s}_{\mathscr{M}}(efe) = e, \mathbf{s}_{\mathscr{M}}(fef) = f$, for if $q \in \mathscr{M}$ is a projection with $q \leq e, efeq = 0$, then $q \leq e \vee f, eq = e, fq = 0$, so that $(e \vee f - e - f)q = 0$ and hence $q = 0$. We have $e \vee f - ueu - ufu = u(e \vee f - e - f)u = uuc(e,f)u = e \vee f - e - f$, hence $ueu + ufu = e + f$. Finally, $ueu \geq uec(e,f)^2eu = (e \vee f - e - f)e(e \vee f - e - f) = fef$, hence $ueu \geq \mathbf{s}_{\mathscr{M}}(fef) = f$. Similarly, $ufu \geq e$. Thus $ueu = f$ and $ufu = e$.

Note that

$$\|s(e,f)\| = \|e - f\|, \ \|c(e,f)\| = \|ef\|. \tag{8}$$

The first identity is obvious. Since $ec(e,f)^2e = efe$, we have $\|ec(e,f)\| = \|ef\|$. Using (6) and the fact that any induction by a projection with central support equal to 1 is a *-isomorphism ([L], 3.14), it follows that $\|c(e,f)\| = \|e\, c(e,f)\| = \|ef\|$.

For $\lambda > 0$ we show that

$$\begin{gathered} \textit{if for every nonzero projections } e', f' \in \mathscr{M}, e' \leq e, f' \leq f, \\ \textit{we have } \|e'f\| \geq \lambda \textit{ and } \|ef'\| \geq \lambda, \textit{ then } c(e,f) \geq \lambda(e \vee f). \end{gathered} \tag{9}$$

We first prove that $c(e,f) \geq \lambda \mathbf{z}_{\mathscr{M}}(e)$. Otherwise, using the Gelfand representation, we find $0 < \mu < \lambda$ and a nonzero projection $q \in \mathscr{Z}(\mathscr{M})$ such that $qc(e,f) \leq \mu q \mathbf{z}_{\mathscr{M}}(e) \neq 0$. Then $qe \neq 0$ and $qe \leq e$, hence $c(qe,f) = |qe \vee f - qe - f| = q|e \vee f - e - f| + (1-q)|f - f| = qc(e,f) \leq \mu q \mathbf{z}_{\mathscr{M}}(e)$. Thus, by (8) and the assumption, we get $\lambda \leq \|qef\| = \|c(qe,f)\| \leq \|\mu q \mathbf{z}_{\mathscr{M}}(e)\| \leq \mu$, a contradiction. Therefore, $c(e,f) \geq \lambda \mathbf{z}_{\mathscr{M}}(e)$ and, similarly, $c(e,f) \geq \lambda \mathbf{z}_{\mathscr{M}}(f)$, so the desired inequality follows using (5).

In connection with the first equation in (8) we note that

$$\|e - f\| < 1 \Rightarrow e = \mathbf{l}_{\mathscr{M}}(ef) \sim \mathbf{r}_{\mathscr{M}}(ef) = f. \tag{10}$$

Indeed, if $q \in \mathscr{M}$ is a projection such that $q \leq e$ and $qef = 0$, then $qe = q, qf = 0$, hence $\|q\| = \|q(e-f)\| < 1$ and $q = 0$. Thus, $e = \mathbf{l}_{\mathscr{M}}(ef)$ and, similarly, $f = \mathbf{r}_{\mathscr{M}}(ef)$.

Finally, in connection with the second equation in (8), we note that

$$\|ef\| < 1 \Rightarrow f' = e \vee f - e \sim f, \quad e \vee f = e + f', \quad \|f - f'\| = \|ef\|. \tag{11}$$

Indeed, $\|e \wedge f\| = \|(e \wedge f)ef\| < 1$, hence $e \wedge f = 0$, so that $f' = e \vee f - e \sim f - e \wedge f = f$. Then using (8), we obtain $\|f - f'\| = \|e \vee f - e - f\| = \|c(e,f)\| = \|ef\|$.

17.16. *End of the proof of Theorem* 17.9. To prove that $p(\sigma)$ satisfies condition 17.9.(1) we may assume that $p(\sigma) = 0$.

Let $e \in \mathcal{M}$ be a nonzero projection and suppose that

$$\lambda = \inf\{\|f\sigma(f)\|; f \in Proj(\mathcal{M}), 0 \neq f \leq e\} > 0. \tag{1}$$

Choose $\varepsilon > 0$ with $(\lambda + 1)\varepsilon < \lambda$ and a nonzero projection $f \in \mathcal{M}$ such that $\|f\sigma(f)\| \leq \lambda + \varepsilon$. For every nonzero projection $f' \leq f$ we have $\|f'\sigma(f')\| \geq \lambda$, hence $\|f\sigma(f')\| \geq \lambda$, $\|f'\sigma(f)\| \geq \lambda$. From 17.15.(9), we deduce that $c(f, \sigma(f)) \geq \lambda(f \vee \sigma(f))$. Using 17.15.(8), it follows from the choice of f that $c(f, \sigma(f)) \leq (\lambda + \varepsilon)(f \vee \sigma(f))$.

On the other hand, let u be the partial isometry appearing in the polar decomposition of $f \vee \sigma(f) - f - \sigma(f)$. Using 17.15.(7), we obtain $u = u^*, u^2 = f \vee \sigma(f), ufu = \sigma(f), u\sigma(f)u = f$, and $\|\lambda u - uc(f, \sigma(f))\| \leq \|\lambda(f \vee \sigma(f)) - c(f, \sigma(f))\| \leq \varepsilon$, that is,

$$\|\lambda u - (f \vee \sigma(f) - f - \sigma(f))\| \leq \varepsilon. \tag{2}$$

The equation $\tau(x) = u\sigma(x)u^*(x \in f\mathcal{M}f)$ defines a *-automorphism $\tau \in Aut(f\mathcal{M}f)$. Since $p(\sigma) = 0$, we obtain using 17.2.(2), $\tau \notin Int(f\mathcal{M}f)$. By the last statement in Theorem 17.9, which has already been proved (17.11–17.14), there exists a nonzero projection $h \in \mathcal{M}, h \leq f$, such that $\|h\tau(h)\| \leq \varepsilon$.

Then we have $\|hu\sigma(h)\| = \|hu\sigma(h)u^*\| = \|h\tau(h)\| \leq \varepsilon$. Using (2) we infer that $\|h(f \vee \sigma(f) - f - \sigma(f))\sigma(h)\| \leq \varepsilon(\lambda + 1)$. Since $h \leq f$, we have $h(f \vee \sigma(f) - f - \sigma(f))\sigma(h) = -h\sigma(h)$, hence $\|h\sigma(h)\| \leq \varepsilon(\lambda + 1) < \lambda$, contradicting (1).

We conclude that $\lambda = 0$, and this completes the proof of Theorem 17.9. □

17.17. Let $\mathcal{M}$ be a finite W^*-algebra and μ a faithful normal trace on $\mathcal{M}$ with $\mu(1) = 1$. For $x \in \mathcal{M}$ we consider the norms $\|x\|_1 = \mu(|x|)$ and $\|x\|_2 = \mu(x^*x)^{1/2}$. It is easy to see that $\|x\|_1 \leq \|x\|_2$, $\|x\|_2^2 \leq \|x\|\|x\|_1 (x \in \mathcal{M})$.

Recall that the closed unit ball $\mathcal{M}_1$ of $\mathcal{M}$ endowed with s-topology is a complete metrizable space; in fact the metric $\|x - y\|_2$ defines the s-topology on $\mathcal{M}_1$ (Strătilă & Zsidó, 1977, 2005, 8.12).

A family of mutually orthogonal nonzero projections in $\mathcal{M}$ with sum equal to 1 will be called, briefly, a *partition of unity* in $\mathcal{M}$.

Theorem 17.9 is the main technical instrument in proving the following important result:

Theorem (V. Rokhlin, A. Connes). *Let $\mathcal{M}$ be a finite W^*-algebra, μ a faithful normal trace on $\mathcal{M}$ with $\mu(1) = 1$, and $\sigma \in Aut(\mathcal{M})$ an aperiodic *-automorphism such that $\mu \circ \sigma = \mu$.*

For each $n \in \mathbb{Z}, n \geq 1$, and $\varepsilon > 0$, there exists a partition of unity $\{e_1, \ldots, e_n\}$ in $\mathcal{M}$ such that

$$\|\sigma(e_1) - e_2\|_2 \leq \varepsilon, \ldots, \|\sigma(e_{n-1}) - e_n\|_2 \leq \varepsilon, \|\sigma(e_n) - e_1\|_2 \leq \varepsilon.$$

The proof is given in Sections 17.20–17.23, where we use Lemmas 17.18, 17.19. In the present section, we divide the proof into three distinct cases.

There exists a partition of unity $\{p_k\}_{k>0}$ in $\mathcal{Z}(\mathcal{M})$ such that, if $k > 1$, we have $(\sigma|\mathcal{Z}(\mathcal{M})p_k)^k = \iota$ and $(\sigma|\mathcal{Z}(\mathcal{M})p)^j \neq \iota$ for every central projection $0 \neq p \leq p_k$ and every $1 \leq j < k$; then we have $(\sigma|\mathscr{Z}(\mathscr{M})p)^j \neq \iota$ for every central projection $0 \neq p \leq p_0$ and every $j > 1$. Thus, in proving Theorem 17.17, we can consider separately the cases $p_1 = 1$, $p_k = 1$ with $k > 2$, and $p_0 = 1$, that is,

(I) *σ acts identically on $\mathscr{Z}(\mathscr{M})$;*

(II) *there exists $k \geq 2$ such that $(\sigma|\mathscr{Z}(\mathscr{M}))^k = \iota$ but $(\sigma|\mathscr{Z}(\mathscr{M})p)^j \neq \iota$ for every central projection $p \neq 0$ and every $1 \leq j < k$;* and

(III) *$\sigma|\mathscr{Z}(\mathscr{M})$ is aperiodic.*

17.18 Lemma. *Let $\mathscr{M}$ be a W*-algebra, let $n \in \mathbb{Z}, n > 1$, and choose $\varepsilon > 0$ so that $n!\varepsilon < 1$. If $\{f_1, \dots, f_n\} \subset \mathscr{M}$ is a family of projections with $\|f_jf_k\| \leq \varepsilon$ for $j \neq k$, then there exists a family $\{e_1, \dots, e_2\} \subset \mathscr{M}$ of mutually orthogonal projections such that $\sum_{k=1}^{n} e_k = \bigvee_{k=1}^{n} f_k$ and $e_k \sim f_k$, $\|e_k - f_k\| \leq n!\varepsilon$ for $1 \leq k \leq n$.*

Proof. For $n = 1$ the lemma is obvious. Assume the lemma is true for $n - 1$ projections. Then there exists a family $\{e_1, \dots, e_{n-1}\} \subset \mathscr{M}$ of mutually orthogonal projections such that $e = \sum_{k=1}^{n-1} e_k = \bigvee_{k=1}^{n-1} f_k$ and $e_k \sim f_k$, $\|e_k - f_k\| \leq (n-1)!\varepsilon$ for $1 \leq k \leq n - 1$. Then $\|e_kf_n - f_kf_n\| \leq (n-1)!\varepsilon (1 \leq k \leq n-1)$, hence

$$\|ef_n\| \leq \sum_{k=1}^{n-1} \|e_kf_n - f_kf_n\| + \sum_{k=1}^{n-1} \|f_kf_n\| \leq (n-1)(n-1)!\varepsilon + (n-1)!\varepsilon,$$

that is, $\|ef_n\| \leq n!\varepsilon < 1$. Using 17.15.(11) we obtain a projection $e_n \in \mathscr{M}$ such that the family $\{e_1, \dots, e_{n-1}, e_n\}$ satisfies the requirements of the lemma. □

17.19 Lemma. *Let $\mathscr{M} \subset \mathscr{B}(\mathscr{H})$ be a countably decomposable von Neumann algebra, $\xi \in \mathscr{H}$ and $e, f \in \mathscr{M}$ projections. Put $\varepsilon = \|e\xi - f\xi\|$.*

If $fe = wa$ is the polar decomposition of fe, that is, $a = (efe)^{1/2}$, $w^*w = \mathbf{s}(efe)$, then

$$\|(w-e)\xi\|4\varepsilon, \|(w-f)\xi\| \leq 3\varepsilon, \|(w-e)^*\xi\| \leq 3\varepsilon, \|(w-f)^*\xi\| \leq 4\varepsilon. \tag{1}$$

If $e \sim f$, then there exists $u \in \mathscr{M}$ such that $u^*u = e, uu^* = f$ and

$$\|(u-f)\xi\| \leq 6\varepsilon, \|(u-f)^*\xi\| \leq 7\varepsilon. \tag{2}$$

Proof. We have $a^2 = efe \leq e, a^2 \leq a \leq e$, so that

$$\|(w - fe)\xi\| = \|(w(e-a)\xi\| \leq \|(e-a^2)\xi\| = \|e(e-f)e\xi\| \leq \|(e-fe)\xi\|.$$
$$\|(e-fe)\xi\| \leq \|(e-f)\xi\| + \|f(f-e)\xi\| \leq 2\varepsilon,$$

hence $\|(w-e)\xi\| \leq \|(w-fe)\xi\| + \|(e-fe)\xi\| \leq 4\varepsilon$ and $\|(w-f)\xi\| \leq \|(w-fe)\xi\| + \|(fe-f)\xi\| \leq 3\varepsilon$. This proves the first two inequalities in (1). The last two inequalities in (1) then follow since $ef = w^*(waw^*)$ is the polar decomposition of ef.

Assume now that $e \sim f$ and let $\delta > 0$. There exists a projection $p \in \mathscr{Z}(\mathscr{M})$ such that the projections $e_1 = pe, f_1 = pf$ are finite while the projections $e_0 = (1-p)e, f_0 = (1-p)f$ are properly infinite.

Since e_0, f_0 are properly infinite, we can write $e_0 = \sum_{k=1}^{\infty} e_0^k, f_0 = \sum_{k=1}^{\infty} f_0^k$ with $e_0^k \sim e_0 \sim f_0 \sim f_0^k$. Let $e_0' = \sum_{k=1}^{n} e_0^k, f_0' = \sum_{k=1}^{n} f_0^k$ with n so large that

$$\|(e_0 - e_0')\xi\| < \delta, \ \|(f_0 - f_0')\xi\| < \delta.$$

Clearly, $e_0 - e_0' \sim e_0 \sim f_0 \sim f_0 - f_0'$. We define

$$e^1 = e_1 + e_0', f^1 = f_1 + f_0' \text{ and } e^2 = \mathbf{r}(f^1 e^1), f^2 = \mathbf{l}(f^1 e^1).$$

Then $e^1 f^1 e^1 \le e^2 \le e^1, f^1 e^1 f^1 \le f^2 \le f^1$, so that

$$\begin{aligned} \|(e^1 - e^2)\xi\| &\le \|(e^1 - e^1 f^1 e^1)\xi\| \le 2\|(e^1 - f^1)\xi\| \\ &\le 2(\|(e - f)\xi\| + \|(e_0 - e_0')\xi\| + \|(f_0 - f_0')\xi\| \le 2\varepsilon + 4\delta, \end{aligned} \tag{3}$$

and hence $\|(e^1 - e^2)\xi\| \le 2\varepsilon + 4\delta, \|(f^1 - f^2)\xi\| \le 2\varepsilon + 4\delta$. Also, we have $e^2 \sim f^2$ more precisely $e^2 = w^* w, f^2 = w w^*$, where w is the partial isometry appearing in the polar decomposition of $f^1 e^1$. Using (1) and (3) we deduce that $\|(w - f^1)\xi\| \le 3(\varepsilon + 2\delta), \|(w - f^1)^*\xi\| \le 4(\varepsilon + 2\delta)$.

On the other hand, we have $e_1 \sim f_1$ and $e_1 \ge pe^2 \sim pf^2 \le f_1$, hence ([L], E.4.9) $e_1 - pe^2 \sim f_1 - pf^2$. Also, we have $e_0 - (1-p)e^2 \ge e_0 - e_0', f_0 - (1-p)f^2 \ge f_0 - f_0'$; hence $e_0 - (1-p)e^2$ and $f_0 - (1-p)f^2$ are countably decomposable properly infinite projections having equal central supports so that ([L], 4.13) they are equivalent: $e_0 - (1-p)e^2 \sim f_0 - (1-p)f^2$. It follows that $e - e^2 \sim f - f^2$, and there exists $v \in \mathcal{M}$ with $v^* v = e - e^2, vv^* = f - f^2$.

Let $u = w + v$. Clearly, $u^* u = e, uu^* = f$. Then

$$\begin{aligned} \|v\xi\| &= \|vv^* v\xi\| \le \|v^* v\xi\| = \|(e - e^2)\xi\| \le \|(e^1 - e^2)\xi\| + \|(e_0 - e_0')\xi\| \le 2\varepsilon + 5\delta, \\ \|v^*\xi\| &= \|v^* vv^*\xi\| \le \|vv^*\xi\| = \|(f - f^2)\xi\| \le \|(f^1 - f^2)\xi\| + \|(f_0 - f_0')\xi\| \le 2\varepsilon + 5\delta, \\ \|(w - f)\xi\| &\le \|(w - f^1)\xi\| + \|(f_0 - f_0')\xi\| \le 3\varepsilon + 7\delta, \\ \|(w - f)^*\xi\| &\le \|(w - f^1)^*\xi\| + \|(f_0 - f_0')\xi\| \le 4\varepsilon + 9\delta, \end{aligned}$$

so (2) follows on choosing $\delta < \varepsilon/9$. □

17.20 Lemma. *Let $\mathcal{M}$ be a finite W^*-algebra, $\sigma \in Aut(\mathcal{M})$ an aperiodic *-automorphism acting identically on $\mathcal{Z}(\mathcal{M})$, and μ a faithful normal σ-invariant trace on $\mathcal{M}$ with $\mu(1) = 1$.*

For each $n \in \mathbb{Z}, n > 2$, and $\delta > 0$, there exists a unitary element $v \in \mathcal{M}$ and a family $\{f_1, \dots, f_n\} \subset \mathcal{M}$ of mutually orthogonal nonzero projections such that

$$\|v - 1\|_1 \le \delta\mu(f_1 + \dots + f_n), \tag{1}$$

$$(Ad(v)\circ\sigma)(f_j) = f_{j+1} \quad (j = 1, \dots, n), \tag{2}$$

where $f_{n+1} = f_1$.

Proof. Let $\gamma = \delta/12(n+1), m = nr$ with $2m^{-1/2} \le \gamma/2$ and $0 < \varepsilon < 1/m!$ with $2mm|\varepsilon \le \gamma/2$.

Since σ is aperiodic, using Theorem 17.9 we find projections $p_1 \ge p_2 \ge \dots \ge p_m \ne 0$ in $\mathcal{M}$ such that $\|p_k \sigma^k(p_k)\| \le \varepsilon\ (1 \le k \le m)$. Let $p = p_m$. Since $p \le p_k$, it follows that $\|p\sigma^k(p)\| \le \varepsilon\ (1 \le k \le m)$, so that $\|\sigma^i(p)\sigma^j(p)\| \le \varepsilon$ for $i \ne j, 1 \le i, j \le m$.

Let $e = \bigvee_{k=1}^{m} \sigma^k(p)$. Using Lemma 17.18, we find a family $\{e_1, \dots, e_m\} \subset \mathcal{M}$ of mutually orthogonal projections such that $e = \sum_{k=1}^{m} e_k$ and $e_k \sim \sigma^k(p)$, $\|e_k - \sigma^k(p)\| \le m!\varepsilon \le \gamma/4m (1 \le k \le m)$. It follows that

$$\|\sigma(e_k) - e_{k+1}\| \le \gamma/2m \quad (k = 1, \dots, m-1). \tag{3}$$

Since σ acts identically on $\mathcal{Z}(\mathcal{M})$, the canonical central trace $\natural : \mathcal{M} \to \mathcal{Z}(\mathcal{M})$ is σ-invariant. By ([L], 7.12) it follows that $\sigma^k(p) \sim p (1 \le k \le m)$, hence

$$e_1 \sim e_2 \sim \dots \sim e_m. \tag{4}$$

Define

$$f_j = \sum_{s=0}^{r-1} e_{ns+j} \quad (1 \le j \le n). \tag{5}$$

Clearly, $\sum_{j=1}^{n} f_j = e$ and $f_1 \sim f_2 \sim \dots \sim f_n$. □

Let $f = e \vee \sigma(e) = e \vee \sigma^{m+1}(p)$ and $\mathcal{N} = f\mathcal{M}f$. Then $e_k, \sigma(e_k) \in \mathcal{N} (1 \le k \le m)$ and $f_j, \sigma(f_j) \in \mathcal{N} (1 \le j \le n)$. Then $\mu' = (\mu|\mathcal{N})/\mu(f)$ is a faithful normal trace on $\mathcal{N}$ with $\mu'(f) = 1$ and for $x \in \mathcal{N} \subset \mathcal{M}$ we have

$$\|x\|'_2 = \mu'(x^*x)^{1/2} = \mu(f)^{-1/2}\mu(x^*x)^{1/2} = \mu(f)^{-1/2}\|x\|_2. \tag{6}$$

Using (3), (4), and (6), we obtain

$$\|\sigma(f_j) - f_{j+1}\|'_2 \le r\gamma/2m < \gamma \quad (1 \le j \le n-1).$$

On the other hand, from (4) it follows that $\mu'(e_k) \le 1/m$, hence $\mu'(\sigma(e_k)) \le 1/m$ (since $e_k \sim \sigma(e_k)$) for $1 \le k \le m$. Thus, $\|e_1\|'_2 \le m^{-1/2}$, $\|\sigma(e_{nr})\|'_2 \le m^{-1/2}$ and, using (5), (3), and (6), we deduce that

$$\|\sigma(f_n) - f_1\|'_2 \le (r-1)\gamma/2m + 2m^{-1/2} < \gamma.$$

For $1 \le j \le n$, we have $\sigma(f_j) \sim f_j \sim f_{j+1}$ so, by Lemma 17.19, there exists $w_j \in \mathcal{N}$ such that

$$w_i^* w_j = \sigma(f_j), w_j w_j^* = f_{j+1} \text{ and } \|w_j - f_{j+1}\|'_2 \le 6\gamma.$$

Since $f - \sigma(e) \sim f - e$, there exists $w_0 \in \mathcal{N}$ such that

$$w_0^* w_0 = f - \sigma(e), w_0 w_0^* = f - e.$$

Then $w = w_0 + w_1 + \dots + w_n \in \mathcal{N}$ is unitary and

$$w\sigma(f_j)w^* = f_{j+1} \quad (1 \le j \le n). \tag{7}$$

Since $\mu'(f-e) \le \mu'(\sigma^{m+1}(p)) = \mu'(p) \le 1/m$, we obtain

$$\begin{aligned}\|w-f\|_2' &= \|w_0-(f-e)\|_2' + \sum_{j=1}^{n} \|w_j - f_{j+1}\|_2' \\ &\le \|w_0\|_2' + \|f-e\|_2' + \sum_{j=1}^{n} \|w_j - f_{j+1}\|_2' \\ &\le 2m^{-1/2} + 6n\gamma = 6(n+1)\gamma = \delta/2.\end{aligned}$$

Finally, $v = w + (1-f) \in \mathcal{M}$ is unitary, satisfies condition (2) and $\|v-1\|_1 = \mu(|v-1|) = \mu(|w-f|) = \mu(f)\mu'(|w-f|) < 2\mu(f)\|w-f\|_2' \le 2\mu(f)\delta/2 = \delta\mu(f_1 + \ldots + f_n)$. □

17.21. *Proof of Theorem* 17.17 *in case* 17.17.(I). We assume that σ acts identically on $\mathcal{Z}(\mathcal{M})$. Let $n \in \mathbb{Z}, n \ge 1$, and $\delta = \varepsilon/4 > 0$ be fixed.

Let $\mathcal{E}$ be the set of all $n+1$-tuples $(e_1, \ldots, e_n; u)$, where $\{e_1, \ldots, e_n\} \subset \mathcal{M}$ consists of mutually orthogonal equivalent projections and $u \in \mathcal{M}$ is a unitary element such that $\|u-1\|_1 < \delta\mu(e_1 + \ldots + e_n)$ and $u\sigma(e_j)u^* = e_{j+1}$ for $1 \le j \le n$.

We define an order relation "$\le$" on $\mathcal{E}$ writing $(e_1, \ldots, e_n; u) \le (e_1', \ldots, e_n'; u')$ if and only if $e_j \le e_j' (1 \le j \le n)$, and $\|u-u'\|_1 \le \delta\mu((e_1'-e_1) + \ldots + (e_n'-e_n))$.

The set $\mathcal{E}$ is then inductively ordered. Indeed, let $\mathcal{F} \subset \mathcal{E}$ be a totally ordered set. Since the mapping

$$\mathcal{F} \ni (e_1, \ldots, e_n; u) \mapsto \mu(e_1 + \ldots + e_n) \in [0, 1]$$

is an order isomorphism of $\mathcal{F}$ onto a subset of [0, 1], we may assume that $\mathcal{F} = \{(e_1^k, \ldots, e_n^k; u_k)\}_{k>1}$ is an increasing sequence. Then, for each $1 \le j \le n$, the increasing sequence of projections $\{e_j^k\}_{k>1}$ is s-convergent to the projection $e_j = \bigvee_{k\ge1} e_j^k$. Therefore, the projections $\{e_1, \ldots, e_n\}$ are mutually orthogonal and equivalent, and $e_j^k \le e_j$ for all j and k. On the other hand, we have

$$\|u_k - u_{k+1}\|_1 \le \delta\mu\left(\sum_{j=1}^{n}(e_j^{k+1} - e_j^k)\right),$$

hence $\|u_k - u_{k+1}\|_1 \le \delta < +\infty$. By the remarks made at the beginning of Section 17.17, it follows that there exists a unitary operator $u \in \mathcal{M}$ such that $\|u_k - u\|_1 \to 0$ and $u_k \xrightarrow{s} u$. It is now easy to check that $(e_1, \ldots, e_n; u) \in \mathcal{E}$ is an upper bound for $\mathcal{F}$.

By Zorn's lemma, there exists a maximal element $(e_1, \ldots, e_n; u)$ in $\mathcal{E}$. Assume that

$$f = 1 - (e_1 + \ldots + e_n) \ne 0.$$

Let $\mathcal{N} = f\mathcal{M}f$, $\mu' = (\mu|\mathcal{N})/\mu(f)$, and $\sigma' = (\mathrm{Ad}(u)\circ\sigma)|\mathcal{N}$. Then $\sigma' \in Aut(\mathcal{N})$ is aperiodic, acts identically on $\mathcal{Z}(\mathcal{N})$ and $\mu'\circ\sigma' = \mu'$. By Lemma 17.20, there exists an $n+1$-tuple $(f_1, \ldots, f_n; v)$, where $\{f_1, \ldots, f_n\} \subset \mathcal{N}$ is a set of mutually orthogonal and equivalent projections and $v \in \mathcal{N}$ is a unitary element such that $\|v-f\|_1' \le \delta\mu'(f_1 + \ldots + f_n) \ne 0$ and $v\sigma'(f_j)v^* = f_{j+1}$ for $1 \le j \le n$. We now define $e_j' = e_j + f_j, (1 \le j \le n)$, and $u' = v + (1-f)u$. Then $\{e_1', \ldots, e_n'\} \subset \mathcal{M}$ consists of mutually orthogonal and equivalent projections, $u \in \mathscr{M}$ is a unitary element, $f'\sigma(e'_j)u'^* = e'_{j+1}$ for $1 \leq j \leq n$, and we have

$$\|(v + (1 - f)) - 1\|_1 = \|v - f\|_1 = \mu(f)\|v - f\|'_1 \leq \delta\mu(f)\mu'\left(\sum_j f_j\right) = \delta\mu\left(\sum_j f_j\right),$$

$$\|u' - u\|_1 \leq \|u\|\|(v + (1 - f)) - 1\|_1 \leq \delta\mu\left(\sum_j f_j\right) = \delta\mu\left(\sum_j (e'_j - e_j)\right),$$

$$\|u' - 1\|_1 \leq \|u' - u\|_1 + \|u - 1\|_1 \leq \delta\mu\left(\sum_j e'_j\right),$$

hence $(e'_1, \dots, e'_n; u') \in \mathscr{E}$ is strictly greater than $\{e_1, \dots, e_n; u\|$, a contradiction.

Thus, if $(e_1, \dots, e_n; u)$ is a maximal element in $\mathscr{E}$, then $(e_1, \dots, e_n)$ is a partition of unity in $\mathscr{M}$, $u \in U(\mathscr{M})$, $\|u - 1\|_1 < \delta$, and $u\sigma(e_j)u^* = e_{j+1} (1 \leq j \leq n)$. Therefore,

$$\begin{aligned}\|\sigma(e_j) - e_{j+1}\|_2^2 &= \|\sigma(e_j) - u\sigma(e_j)u^*\|_2^2 \\ &\leq 2\|\sigma(e_j) - u\sigma(e_j)u^*\|_1 \leq 4\|u - 1\|_1 \leq 4\delta = \varepsilon.\end{aligned}$$

17.22. *Proof of Theorem* 17.17 *in case* 17.17.(II). We assume that there exists $k \geq 2$ such that $(\sigma|\mathscr{Z}(\mathscr{M}))^k = \iota$, but $(\sigma|\mathscr{Z}(\mathscr{M})p)^j \neq \iota$ for any central projection $p \neq 0$ and j such that $1 \leq j < k$. If $p \in \mathscr{Z}(\mathscr{M})$ is a maximal projection such that $p, \sigma(p), \dots, \sigma^{k-1}(p)$ are mutually orthogonal, then $p + \sigma(p) + \dots + \sigma^{k-1}(p) = 1$. Thus, there exists a partition of unity $\{p_1, \dots, p_k\}$ in $\mathscr{Z}(\mathscr{M})$ such that

$$\sigma(p_i) = p_{i+1} \quad (1 \leq i \leq k),$$

where $p_{k+1} = p_i$.

Let $n \in \mathbb{Z}, n \geq 1$, and $\delta = \varepsilon/k > 0$ be fixed. The *-automorphism $\sigma' = \sigma^k|\mathscr{M}p_1 \in Aut(\mathscr{M}p_1)$ is aperiodic, acts identically on $\mathscr{Z}(\mathscr{M}p_1)$, and preserves the faithful normal trace $\mu' = (\mu|\mathscr{M}p_1)/\mu(p_1) = k\mu|\mathscr{M}p_1$. By Section 17.21, there exists a partition of unity $\{f_0, \dots, f_{n-1}\}$ in $\mathscr{M}p_1$ such that (with $f_n = f_0$)

$$\|\sigma'(f_s) - f_{s+1}\|'_2 \leq \delta \quad (0 \leq s \leq n - 1).$$

The family $\{\sigma^i(f_s); 1 \leq i \leq k, 0 \leq s \leq n - 1\}$ is a partition of unity $\{h_1, \dots, h_{sk}\}$ in $\mathscr{M}$, where

$$h_{sk+i} = \sigma^i(f_s) \quad (1 \leq i \leq k, 0 \leq s \leq n - 1).$$

For $1 \leq r \leq nk$ we have (with $h_{nk+1} = h_1$)

$$\|\sigma(h_r) - h_{r+1}\|_2 \leq k^{-1/2}\delta < \delta.$$

Indeed, if $r = sk + i$ with $0 \leq s \leq n - 1$ and $1 \leq i < k$, then $r + 1 = sk + (i + 1)$, hence $\sigma(h_r) = h_{r+1}$; if $r = sk + k$, then $r + 1 = (s + 1)k + 1$, hence

$$\begin{aligned}\|\sigma(h_r) - h_{r+1}\|_2 &= \|\sigma(\sigma^k(f_s)) - \sigma(f_{s+1})\|_2 = \|\sigma^k(f_s) - f_{s+1}\|_2 \\ &= k^{-1/2}\|\sigma'(f_s) - f_{s+1}\|'_2 \leq k^{-1/2}\delta.\end{aligned}$$

Finally, the projections

$$e_j = h_j + h_{j+n} + \ldots + h_{j+(k-1)n} \quad (1 \le j \le n)$$

constitute a partition of unity in $\mathscr{M}$ and we have (with $e_{n+1} = e_1$)

$$\|\sigma(e_j) - e_{j+1}\|_2 \le k\delta = \varepsilon \quad (1 \le j \le n).$$

17.23. *Proof of Theorem* 17.17 *in case* 17.17. (III). In this case we may assume that $\mathscr{M}$ is abelian. So, in this section we give essentially the proof of the classical Rokhlin theorem.

Let $n \in \mathbb{Z}, n \ge 1$, and $\varepsilon > 0$. Choose $r \in \mathbb{Z}, r > 1$, such that $1/r \le \varepsilon/16$, and write $m = nr$.

Let $p \in \mathscr{M}$ be a maximal projection such that

$$p, \sigma(p), \ldots, \sigma^{m-1}(p) \text{ are mutually orthogonal and define} \tag{1}$$

$$p_k = \sigma^{-m+1}(p)\sigma^k(p) \quad (1 \le k \le m). \tag{2}$$

Then $p_1, \ldots, p_m$ are mutually orthogonal projections and the maximal choice of p with property (1) implies that

$$\sigma^{-m+1}(p) = p_1 + \ldots + p_m; \tag{3}$$

indeed, if $q = \sigma^{-1}(\sigma^{-m+1}(p) - \sum_{k=1}^m p_k) \ne 0$, then $p + q$ contradicts the maximality of p.

Consider the projections

$$\begin{array}{llll} \sigma^{-1}(p_2) & & & \\ \sigma^{-1}(p_3) & \sigma^{-2}(p_3) & & \\ \ldots\ldots\ldots\ldots & & & \\ \sigma^{-1}(p_m) & \sigma^{-2}(p_m)\ldots & \sigma^{-m+1}(p_m), & \end{array} \tag{4}$$

$$\sigma^{-1}(p) \quad \sigma^{-2}(p)\ldots \quad \sigma^{-m+1}(p) \quad p. \tag{5}$$

All the projections appearing in (4) and (5) are mutually orthogonal. Indeed, all the projections in (5) are mutually orthogonal, by (1). Moreover, the projections appearing in the same column of (4) are mutually orthogonal, by (2), and two projections appearing in different columns of (4) are orthogonal, by (3) and (1); hence all the projections in (4) are mutually orthogonal. Finally, using (2) and (1) it follows that $\sigma^{-i}(p_j)\sigma^{-k}(p) = 0$ for all $1 \le i < j \le m, 0 \le k \le m-1$.

Let q be the sum of all the projections in (4) and (5). We claim that $q = 1$. To prove this it is sufficient to show that q is σ-invariant since, as σ is aperiodic, the assumption $1 - q \ne 0$ would then contradict the choice of p maximal with property (1). Furthermore, to show that $\sigma^{-1}(q) = q$ it is sufficient to show just that $\sigma^{-1}(q) \le q$, since μ is a σ-invariant faithful state on $\mathscr{M}$. Looking at (4) and (5) we see that the only projections from the sum defining $\sigma^{-1}(q)$ which do not appear explicitly in the sum defining q are $\sigma^{-m}(p), \sigma^{-2}(p_2), \ldots, \sigma^{-m}(p_m)$. By (3) we have $\sigma^{-m}(p) = \sigma^{-1}(p_1 + p_2 + \ldots + p_m) \le \sigma^{-1}(p_1) \vee q$ and from (2) it follows that $\sigma^{-k}(p_k) \le p \le q$ for all $1 \le k \le m$. Hence $\sigma^{-1}(q) \le q$, and consequently $q = 1$.

We now define the projection

$$e = \sum_{s=0}^{r-1} \sigma^{-ns}(p) + \sum_{k=n+1}^{m} \sum_{s=0}^{s(k)} \sigma^{-sn-1}(p_k), \tag{6}$$

where $s(k)$ is the greatest $s \in \mathbb{Z}, s \geq 0$, such that $sn + 1 < k - n$.

Looking at (4) and (5) we see that

$$e, \sigma^{-1}(e), \ldots, \sigma^{-n+1}(e) \text{ are mutually orthogonal.} \tag{7}$$

We have

$$\|e - \sigma^{-n}(e)\|_1 \leq \varepsilon/4. \tag{8}$$

Indeed, if we compute $\sigma^{-n}(e)$ using (6) and compare with (6), we obtain

$$\begin{aligned}\mu(|e - \sigma^{-n}(e)|) &\leq \mu(p + \sigma^{-nr}(p) + \sum_{k=n+1}^{m} \sigma^{-1}(p_k) + \sigma^{-(s(k)+1)n-1}(p_k) \\ &\leq 2\mu(p) + 2\sum_{k=1}^{m} \mu(p_k) \leq 4\mu(p) \leq 4/m \leq 4/r \leq \varepsilon/4,\end{aligned}$$

since μ is σ-invariant and we have (3).

We also have

$$\|1 - \sum_{i=0}^{n-1} \sigma^{-i}(e)\|_1 \leq \varepsilon/4. \tag{9}$$

Indeed, the sum $q = \sum_{i=0}^{n-1} \sigma^{-i}(e)$ contains all the projections in (5) and the projections in (4) except for at most n projections from each row of (4). Since the sum of the projections in (4) and (5) is 1, we obtain

$$\mu(1 - q) \leq \sum_{k=2}^{m} n\mu(p_k) \leq n\mu\left(\sum_{k=1}^{m} p_k\right) = n\mu(p) \leq n/m < \varepsilon/4,$$

using (3) again, and the σ-invariance of μ.

From (7), (8), and (9), it follows that the projections

$$e_j = \sigma^{-n+j}(e) \; (1 \leq j \leq n-1) \text{ and } e_n = e + \left(1 - \sum_{i=0}^{n-1} \sigma^{-i}(e)\right)$$

satisfy the requirements of Theorem 17.17. □

17.24. Finally, we note a useful consequence of Theorem 17.9.

Corollary. *Let $\sigma : G \to Aut(\mathcal{M})$ be a properly outer action of the finite group G on the countably decomposable infinite W*-algebra $\mathcal{M}$. Then the centralizer $\mathcal{M}^\sigma$ is infinite.*

Proof. Let $\varepsilon > 0$ be such that $\varepsilon(card\ G)!(card\ G)^2 < 1$.

Since every $\sigma_g(e \neq g \in G)$ is properly outer, a familiar maximality argument based on Theorem 17.9 shows that there exists a projection $p \in \mathcal{M}$ such that $|\sigma_g(p)\sigma_h(p)\| < \varepsilon$ for $g, h \in G(g \neq h)$, and $\bigvee_{g\in G} \sigma_g(p) = 1$. Since $\mathcal{M}$ is properly infinite and G is finite, it follows that some $\sigma_g(p)$ is properly infinite and we may therefore assume that p is properly infinite. Thus, there exists a partial isometry $v \in p\mathcal{M}p$ with $v^*v = p$ and $vv^* \neq p$. Let $u = \sum_{g\in G} \sigma_g(v) \in \mathcal{M}^\sigma$. From the choice of $\varepsilon > 0$ it is easy to check that $\mathbf{r}(u) = 1$ while $\mathbf{l}(u) \neq 1$. Hence $\mathcal{M}^\sigma$ is infinite.

Note that a particular case of this result (16.17) has been used (17.14) in proving Theorem 17.9. □

17.25. Here we describe the Ocneanu's Theorem on cocycle conjugacy of outer actions of countable discrete amenable groups.

Let $\mathcal{M}$ be a type II_1 factor and G be a countable discrete group. An *outer action* of G on $\mathcal{M}$ is an injective group homomorphism

$$\widetilde{\alpha} : G \to Out(\mathcal{M}) = Aut(\mathcal{M}) / Int(\mathcal{M}).$$

Equivalently, chosing a lifting $\alpha = \left(\alpha_g\right)_{g\in G}$ to $Aut(\mathcal{M})$ (i.e., each α_g is a noninner automorphism of $\mathcal{M}$ whose class in $Out(\mathcal{M})$ is $\widetilde{\alpha}_g, g \in G$), an outer action is determined by the condition that there exist unitaries $\left(u_{s,t}\right)_{s,t\in G}$ in $\mathcal{M}$ such that

$$\alpha_s \circ \alpha_t = Ad\left(u_{s,t}\right) \circ \alpha_{st}, \quad (s, t \in G).$$

In this case, the condition that α determines a group homomorphism into $Out(\mathcal{M})$ is that, for all $g, h, k \in G$, we have

$$u_{g,h} u_{gh,k} = \lambda_{g,h,k} \alpha_g\left(u_{h,k}\right) u_{g,hk},$$

where $\lambda_{g,h,k} \in \mathbb{T} = \{z \in \mathbb{C}; |z| = 1\}$. One may always assume, by requiring that $\alpha_e = \mathrm{id}$, that

$$u_{e,g} = u_{g,e} = 1, \quad (g \in G).$$

The scalars $\left(\lambda_{g,h,k}\right)_{g,h,k\in G}$ define a 3-cocycle for the group G. The class of this cocycle in $H^3(G, \mathbb{T})$ depends only on $\widetilde{\alpha}$ and is denoted by $Ob\left(\widetilde{\alpha}\right)$.

The following result is a particular case of the Theorem of Ocneanu (1985).

Theorem. *Let $\mathcal{R}$ be the hyperfinite type II_1 factor, let G be a countable discrete amenable group. Let $\widetilde{\alpha}, \widetilde{\beta} : G \to Out(\mathcal{R})$ be two outer actions of G on $\mathcal{R}$. Then $\widetilde{\alpha}$ is conjugated to $\widetilde{\beta}$ (i.e., there exists $\sigma \in Aut(\mathcal{R})$ such that $\widetilde{\alpha}_g = \sigma\widetilde{\beta}_g\sigma_{-1}$ in $Out(\mathcal{R})$ for all $g \in G$) if and only if*

$$Ob\left(\widetilde{\alpha}\right) = Ob\left(\widetilde{\beta}\right).$$

Moreover, every element in $H^3(G,\mathbb{T})$ *may be realized as the invariant* $Ob(\widetilde{\alpha})$ *for some outer action* $\widetilde{\alpha} : G \to Out(\mathcal{R})$.

Corollary. *Let G be a countable discrete amenable group and let* $\mathcal{R}$ *be the hyperfinite type* II_1 *factor. Any two free actions of G on* $\mathcal{R}$ *are outer conjugate.*

Here an *action* $\alpha : G \to Aut(\mathcal{M})$ is called *free* if $\alpha_g \notin Int(\mathcal{M})$ for all $g \in G, g \neq e$. And two actions $\alpha, \beta : G \to Aut(\mathcal{M})$ are called *outer conjugate* if there is a unitary cocycle u for α, that is, unitaries $u_g \in M$, $(g \in G)$, with

$$u_{gh} = u_g\alpha_g(u_h), (g, h \in G),$$

and $\theta \in Aut(\mathcal{M})$ such that

$$\beta_g = \theta \circ Ad(u_g) \circ \alpha_g \circ \theta^{-1}, \quad (g \in G).$$

The Theorem of Ocneanu generalizes previous results of Connes (1975e) for $G = \mathbb{Z}$ and of Jones (1980) for finite groups.

There is a more general form of the theorem which also works for the hyperfinite type II_∞ factor, and a relative form of the Theorem for the McDuff factors. A converse to the Ocneanu's Theorem was formulated and proved by Jones (1983).

The theorem was extended by Kawahigashi et al. (1992) to the type III case. In the most general form of the theorem, one finds a classification also in the case that $\{g \in G; \alpha_g \in Int(\mathcal{R})\}$ is nontrivial. See also Sutherland and Takesaki (1985, 1989).

17.26. Notes. Propositions 17.4–17.7 are due to Kallman (1969, 1970). The results in Section 17.8 are from Choda (1971, 1974) and Tam (1971). The main results, Theorems 17.9 and 17.17, are due to Connes (1975e).

For our exposition, we have used Choda (1974), Combes and Delaroche (1973–1974), Connes (1975e), and Kallman (1969).

An interesting extension of Theorem 17.17 for several commuting *-automorphisms has been obtained by Ocneanu (1981). The notion of a full group was introduced in the commutative case by Dye (1959, 1963) and in the general case by Haga and Takeda (1972) and Connes (1973a).

CHAPTER IV

Crossed Products

18 Hopf–von Neumann Algebras

In this section, we consider a category of objects called Hopf–von Neumann algebras, which in a certain sense generalize locally compact groups, and also their actions on W^*-algebras. The main interest of these objects consists in their giving a natural framework for the duality theory of locally compact groups.

18.1. We first introduce certain notation and conventions that will be used frequently in what follows.

Let $\mathscr{H}, \mathscr{K}$ be Hilbert spaces and $\mathscr{M} \subset \mathscr{B}(\mathscr{H}), \mathscr{N} \subset \mathscr{B}(\mathscr{K})$ von Neumann algebras. There exists a unique unitary operator

$$\sim : \mathscr{H} \bar{\otimes} \mathscr{K} \to \mathscr{K} \bar{\otimes} \mathscr{H}$$

such that $\sim (\xi \otimes \eta) = \eta \otimes \xi$ for all $\xi \in \mathscr{H}, \eta \in \mathscr{K}$. The mapping

$$\tilde{\cdot} : \mathscr{M} \bar{\otimes} \mathscr{N} \ni X \mapsto \widetilde{X} = \sim \circ X \circ \sim \in \mathscr{N} \bar{\otimes} \mathscr{M}$$

is then a $*$-isomorphism, uniquely determined, such that $(x \bar{\otimes} y)^{\sim} = y \bar{\otimes} x$ for $x \in \mathscr{M}, y \in \mathscr{N}$.

The value $\varphi(x)$ of a linear form $\varphi \in \mathscr{M}^*$ on a vector $x \in \mathscr{M}$ will also be denoted by $\langle x, \varphi \rangle$.

Let G be a locally compact group. The elements of the Hilbert space $\mathscr{H} \bar{\otimes} \mathscr{L}^2(G) = \mathscr{L}^2(G, \mathscr{H})$ will be identified with vector-valued functions on G. Also, the linear operators T on $\mathscr{L}^2(G, \mathscr{H})$ will usually be defined by specifying the elements $(T\xi)(g)(\xi \in \mathscr{L}^2(G, \mathscr{H}), g \in G)$. The formal versions of these procedures are standard and well known.

The identity mapping on $\mathscr{M}$ will be denoted by $\iota_{\mathscr{M}}$, or by ι_k, where k indicates a position in tensor products. The same convention will be used for the unit element $1_{\mathscr{M}}$.

18.2. A *Hopf–von Neumann algebra* is a pair $(\mathscr{A}, \delta_{\mathscr{A}})$, where $\mathscr{A}$ is a W^*-algebra and $\delta_{\mathscr{A}} : \mathscr{A} \to \mathscr{A} \bar{\otimes} \mathscr{A}$ is an injective unital normal $*$-homomorphism, called *comultiplication*, which is *coassociative*, that is,

$$(\iota_{\mathscr{A}} \bar{\otimes} \delta_{\mathscr{A}}) \circ \delta_{\mathscr{A}} = (\delta_{\mathscr{A}} \bar{\otimes} \iota_{\mathscr{A}}) \circ \delta_{\mathscr{A}}. \tag{1}$$

The Hopf–von Neumann algebra $(\mathscr{A}, \delta_{\mathscr{A}})$ is said to be commutative if the algebra $\mathscr{A}$ is commutative and is called *cocommutative* if

$$\tilde{\cdot} \circ \delta_{\mathscr{A}} = \delta_{\mathscr{A}}. \tag{2}$$

A *coinvolutive Hopf–von Neumann algebra* is a triple $(\mathscr{A}, \delta_{\mathscr{A}}, j_{\mathscr{A}})$, where $(\mathscr{A}, \delta_{\mathscr{A}})$ is a Hopf–von Neumann algebra and $j : \mathscr{A} \to \mathscr{A}$ is an involutive *-antiautomorphism (i.e., $j_{\mathscr{A}} \circ j_{\mathscr{A}} = \iota_{\mathscr{A}}$), called *coinvolution*, such that

$$(j_{\mathscr{A}} \bar{\otimes} j_{\mathscr{A}}) \circ \delta_{\mathscr{A}} = \tilde{\cdot} \circ \delta_{\mathscr{A}} \circ j_{\mathscr{A}}. \tag{3}$$

An *action* of the Hopf–von Neumann algebra $(\mathscr{A}, \delta_{\mathscr{A}})$ on the W^*-algebra $\mathscr{M}$ is an injective unital normal *-homomorphism $\delta : \mathscr{M} \to \mathscr{M} \bar{\otimes} \mathscr{A}$ such that

$$(\iota_{\mathscr{M}} \bar{\otimes} \delta_{\mathscr{A}}) \circ \delta = (\delta \bar{\otimes} \iota_{\mathscr{A}}) \circ \delta. \tag{4}$$

In this case, we also say that $\mathscr{M}$ is an *$\mathscr{A}$-comodule via δ*.

Note that $\delta_{\mathscr{A}}$ is an action of $(\mathscr{A}, \delta_{\mathscr{A}})$ on $\mathscr{A}$.

The *centralizer* of the action $\delta : \mathscr{M} \to \mathscr{M} \bar{\otimes} \mathscr{A}$ is the set

$$\mathscr{M}^{\delta} = \{x \in \mathscr{M}; \delta(x) = x \bar{\otimes} 1_{\mathscr{A}}\}.$$

Clearly, $\mathscr{M}^{\delta}$ is a unital W^*-subalgebra of $\mathscr{M}$.

Let $\mathscr{M}_1, \mathscr{M}_2$ be $\mathscr{A}$-comodules via the actions $\delta_1 : \mathscr{A} \to \mathscr{M}_1 \bar{\otimes} \mathscr{A}, \delta_2 : \mathscr{A} \to \mathscr{M}_2 \bar{\otimes} \mathscr{A}$, respectively. We shall say that a normal completely positive linear mapping $\sigma : \mathscr{M}_1 \to \mathscr{M}_2$ *intertwines the actions* δ_1, δ_2 or that σ is an *$\mathscr{A}$-comodule mapping* if

$$\delta_2 \circ \sigma = (\sigma \bar{\otimes} \iota_{\mathscr{A}}) \circ \delta_1. \tag{5}$$

In particular, if there exists a *-isomorphism $\sigma : \mathscr{M}_1 \to \mathscr{M}_2$ intertwining δ_1, δ_2 then we say that the actions δ_1, δ_2 are *isomorphic*.

If $\delta : \mathscr{M} \to \mathscr{M} \bar{\otimes} \mathscr{A}$ is an action of $\mathscr{A}$ on $\mathscr{M}$ and $\mathscr{N} \subset \mathscr{M}$ is a unital W^*-subalgebra such that $\delta(\mathscr{N}) \subset \mathscr{N} \bar{\otimes} \mathscr{A}$, then $\delta|\mathscr{N} : \mathscr{N} \to \mathscr{N} \bar{\otimes} \mathscr{A}$ is an action of $\mathscr{A}$ on $\mathscr{N}$ and the canonical injection $\mathscr{N} \hookrightarrow \mathscr{M}$ is an $\mathscr{A}$-comodule mapping. In this case, we shall say that $\mathscr{N}$ is an *$\mathscr{A}$-subcomodule* of $\mathscr{M}$.

Let $\mathscr{A}_1, \mathscr{A}_2$ be Hopf–von Neumann algebras and $\delta_1 : \mathscr{M} \to \mathscr{M} \bar{\otimes} \mathscr{A}_1, \delta_2 : \mathscr{M} \to \mathscr{M} \bar{\otimes} \mathscr{A}_2$ actions of $\mathscr{A}_1, \mathscr{A}_2$ on the same W^*-algebra $\mathscr{M}$. We shall say that δ_1 and δ_2 *commute* if

$$(\delta_1 \bar{\otimes} \iota_{\mathscr{A}_2}) \circ \delta_2 = (\iota_{\mathscr{M}} \bar{\otimes} \tilde{\cdot}) \circ (\delta_2 \bar{\otimes} \iota_{\mathscr{A}_1}) \circ \delta_1. \tag{6}$$

Finally, we remark that if $\delta : \mathscr{M} \to \mathscr{M} \bar{\otimes} \mathscr{A}$ is an action of the Hopf–von Neumann algebra $(\mathscr{A}, \delta_{\mathscr{A}})$ on the W^*-algebra $\mathscr{M}$ and $\mathscr{N}$ is an arbitrary W^*-algebra, then

$$\iota_{\mathscr{N}} \bar{\otimes} \delta : \mathscr{N} \bar{\otimes} \mathscr{M} \to \mathscr{N} \bar{\otimes} \mathscr{M} \bar{\otimes} \mathscr{A}$$

is an action of $(\mathscr{A}, \delta_{\mathscr{A}})$ on $\mathscr{N} \bar{\otimes} \mathscr{M}$ and we have

$$(\mathscr{N} \bar{\otimes} \mathscr{M})^{\iota_{\mathscr{N}} \bar{\otimes} \delta} = \mathscr{N} \bar{\otimes} \mathscr{M}^{\delta}. \tag{7}$$

Indeed, the inclusion "$\supset$" is obvious. Conversely, let $x \in \mathscr{N} \bar{\otimes} \mathscr{M}$ be such that $(\iota_{\mathscr{N}} \bar{\otimes} \delta)(x) = x \bar{\otimes} 1_{\mathscr{A}}$. For $\psi \in \mathscr{N}_*, \varphi \in \mathscr{M}_*, k \in \mathscr{A}_*$, we have

$$\begin{aligned}\langle \delta(E^{\psi}_{\mathscr{M}}(x)), \varphi \bar{\otimes} k\rangle &= \langle E^{\psi}_{\mathscr{M}}(x), (\varphi \bar{\otimes} k) \circ \delta\rangle = \langle x, \psi \bar{\otimes} ((\varphi \bar{\otimes} k) \circ \delta)\rangle \\ &= \langle x, (\psi \bar{\otimes} \varphi \bar{\otimes} k) \circ (\iota_{\mathscr{N}} \bar{\otimes} \delta)\rangle = \langle x \bar{\otimes} 1_{\mathscr{A}}, \psi \bar{\otimes} \varphi \bar{\otimes} k\rangle \\ &= \langle x, \psi \bar{\otimes} \varphi\rangle\langle 1_{\mathscr{A}}, k\rangle = \langle E^{\psi}_{\mathscr{M}}(x) \bar{\otimes} 1_{\mathscr{A}}, \varphi \bar{\otimes} k\rangle,\end{aligned}$$

hence $E^{\psi}_{\mathscr{M}}(x) \in \mathscr{M}^{\delta}$. By Proposition 9.8 it follows that $x \in \mathscr{N} \bar{\otimes} \mathscr{M}^{\delta}$.

18.3. Let $(\mathscr{A}, \delta_{\mathscr{A}})$ be a Hopf–von Neumann algebra. It is easy to check that the predual $\mathscr{A}_*$ with its Banach space structure and multiplication $\mathscr{A}_* \times \mathscr{A}_* \ni (h, k) \mapsto h \cdot k \in \mathscr{A}_*$ defined by

$$\langle a, h \cdot k\rangle = \langle \delta_{\mathscr{A}}(a), h \bar{\otimes} k\rangle \quad (a \in \mathscr{A}) \tag{1}$$

is a *Banach algebra*. Note that

$$(h \cdot k)^* = h^* \cdot k^*(h, k \in \mathscr{A}_*). \tag{2}$$

The Banach algebra $\mathscr{A}_*$ is commutative if and only if the Hopf–von Neumann algebra $(\mathscr{A}, \delta_{\mathscr{A}})$ is cocommutative.

If $(\mathscr{A}, \delta_{\mathscr{A}}, j_{\mathscr{A}})$ is a coinvolutive Hopf–von Neumann algebra, then the Banach algebra $\mathscr{A}_*$ with the involution $\mathscr{A}_* \ni k \mapsto k^0 = k^* \circ j_{\mathscr{A}} \in \mathscr{A}_*$, that is,

$$\langle a, k^0\rangle = \overline{\langle j_{\mathscr{A}}(a^*), k\rangle} \quad (a \in \mathscr{A}), \tag{3}$$

is an *involutive Banach algebra*. Note that

$$(k^0)^* = (k^*)^0 \quad (k \in \mathscr{A}). \tag{4}$$

Let $\delta : \mathscr{M} \to \mathscr{M} \bar{\otimes} \mathscr{A}$ be an action of the Hopf–von Neumann algebra $(\mathscr{A}, \delta_{\mathscr{A}})$ on the W^*-algebra $\mathscr{M}$. For every $k \in \mathscr{A}_*$, we consider the Fubini mapping (9.8) $E^k_{\mathscr{M}} : \mathscr{M} \bar{\otimes} \mathscr{A} \to \mathscr{M}$ and define

$$k \cdot x = E^k_{\mathscr{M}}(\delta(x)) \quad (x \in \mathscr{M}),$$

that is,

$$\langle k \cdot x, \varphi\rangle = \langle \delta(x), \varphi \bar{\otimes} k\rangle \quad (x \in \mathscr{M}, k \in \mathscr{A}_*, \varphi \in \mathscr{M}_*). \tag{5}$$

It is easy to check that

$$\textit{the mapping } (k, x) \mapsto k \cdot x \textit{ is bilinear;} \tag{6}$$

$$\textit{the mappings } x \mapsto k \cdot x \textit{ are w-continuous } (k \in \mathscr{A}_*); \tag{7}$$

$$\|k \cdot x\| \le \|k\|\|x\| \quad (x \in \mathscr{M}, k \in \mathscr{A}_*); \tag{8}$$

$$(k \cdot x)^* = k^* \cdot x^* \quad (x \in \mathscr{M}, k \in \mathscr{A}_*); \tag{9}$$

$$x \in \mathscr{M}^+, k \in \mathscr{A}_*^+ \Rightarrow k \cdot x \in \mathscr{M}^+; \tag{10}$$

$$h \cdot (k \cdot x) = (h \cdot k) \cdot x \quad (x \in \mathscr{M}, h, k \in \mathscr{A}_*); \tag{11}$$

In particular, the mapping $\mathscr{A}_* \times \mathscr{M} \ni (k,x) \mapsto k \cdot x$ determines the structure of a *left Banach $\mathscr{A}_*$-module* on $\mathscr{M}$.

On the other hand, the mapping $\mathscr{M}_* \times \mathscr{A}_* \ni (\varphi, k) \mapsto \varphi \cdot k \in \mathscr{M}_*$, defined by

$$\langle x, \varphi \cdot k \rangle = \langle \delta(x), \varphi \bar{\otimes} k \rangle = \langle k \cdot x, \varphi \rangle \quad (x \in \mathscr{M}), \tag{12}$$

determines the structure of a *right Banach $\mathscr{A}_*$-module* on $\mathscr{M}_*$; definition (1) is just a particular case of this definition. Note that

$$\|\varphi \cdot k\| \leq \|\varphi\| \|k\| \quad (\varphi \in \mathscr{M}_*, k \in \mathscr{A}_*), \tag{13}$$
$$(\varphi \cdot k)^* = \varphi^* \cdot k^* \quad (\varphi \in \mathscr{M}_*, k \in \mathscr{A}_*). \tag{14}$$

Since δ is injective and $\mathscr{M} = (\mathscr{M}_*)^*$, it follows using the Hahn–Banach theorem that

$$\textit{the set } \{\varphi \cdot k; \varphi \in \mathscr{M}_*, k \in \mathscr{A}_*\} \textit{ is total in } \mathscr{M}_*. \tag{15}$$

If $\mathscr{N}$ and $\mathscr{M}$ are two $\mathscr{A}$-comodules and $\sigma : \mathscr{N} \to \mathscr{M}$ is an $\mathscr{A}$-comodule mapping, then

$$\sigma(k \cdot y) = k \cdot \sigma(y) \quad (y \in \mathscr{N}, k \in \mathscr{A}_*). \tag{16}$$

In particular, if $\mathscr{N}$ is an $\mathscr{A}$-subcomodule of $\mathscr{M}$, then

$$y \in \mathscr{N} \subset \mathscr{M}, k \in \mathscr{A}_* \Rightarrow k \cdot y \in \mathscr{N}. \tag{17}$$

18.4. Let G be a locally compact group with neutral element $e \in G$, left Haar measure $\mathrm{d}g = \mathrm{d}^\ell g$ and modular function $\Delta = \Delta_G$. Recall that $\Delta : G \to \mathbb{R}^+ \backslash \{0\}$ is a continuous group homomorphism and we have for $k \in \mathscr{L}^1(G), t \in G$

$$\int k(tg)\,\mathrm{d}g = k(g)\,\mathrm{d}g, \tag{1}$$
$$\int k(gt)\,\mathrm{d}g = \Delta(t)^{-1} \int k(g)\,\mathrm{d}g, \tag{2}$$
$$\int k(g^{-1})\,\mathrm{d}g = \int k(g)\Delta(g)^{-1}\,\mathrm{d}g. \tag{3}$$

In particular, $\mathrm{d}^r g = \Delta(g)^{-1}\,\mathrm{d}g$ is the right Haar measure.

Let $\mathscr{K}(G)$ be the set of all continuous functions on G of compact support.

The set $\mathscr{K}(G)$ endowed with the scalar product from $\mathscr{L}^2(G)$ and with the operations of multiplication and complex conjugation :

$$(\xi\eta)(s) = \xi(s)\eta(s) \quad \bar{\xi}(s) = \overline{\xi(s)} \tag{4}$$

is a commutative Hilbert algebra. It is easy to check that the associated maximal Hilbert algebra is $\mathscr{L}^\infty(G) \cap \mathscr{L}^2(G) \subset \mathscr{L}^2(G)$ with the same operations. The associated modular operator ∇_G and canonical conjugation K_G are given by

$$\nabla_G \xi = \xi \quad K_G \xi = \xi. \tag{5}$$

The associated left and right von Neumann algebras both coincide with the von Neumann algebra $\mathscr{L}^\infty(G)$ acting by multiplication on $\mathscr{L}^2(G)$:

$$(f\xi)(s) = f(s)\xi(s). \tag{6}$$

By the commutation theorem ([L], 10.4.(2)), it follows that $\mathscr{L}^\infty(G)$ is maximal commutative in $\mathscr{B}(\mathscr{L}^2(G))$:

$$\mathscr{L}^\infty(G)' = \mathscr{L}^\infty(G). \tag{7}$$

The natural weight on $\mathscr{L}^\infty(G)$ associated with this Hilbert algebra ([L], 10.16) is denoted by μ_G and is called *the Haar weight* on $\mathscr{L}^\infty(G)$. We have

$$\mu_G(f) = \int f(g)\,\mathrm{d}g \quad (f \in \mathscr{L}^\infty(G)^+). \tag{8}$$

On the other hand, the set $\mathscr{K}(G)$ endowed with the scalar product from $\mathscr{L}^2(G)$ and with the operations of convolution and involution

$$(\xi * \eta)(s) = \int \xi(g)\eta(g^{-1}s)\,\mathrm{d}g, \quad \xi^*(s) = \Delta(s)^{-1}\overline{\xi(s^{-1})} \tag{9}$$

is a left Hilbert algebra. The associated modular operator Δ_G and canonical conjugation J_G are given by

$$(\Delta_G\xi)(s) = \Delta(s)\xi(s) \quad (J_G\xi)(s) = \Delta(s)^{-1/2}\overline{\xi(s^{-1})}. \tag{10}$$

The associated left von Neumann algebra, denoted by $\mathfrak{L}(G)$, is generated by the left regular representation $\boldsymbol{\lambda} : G \ni g \mapsto \lambda(g) \in \mathscr{B}(\mathscr{L}^2(G))$,

$$(\boldsymbol{\lambda}(g)\xi)(s) = \xi(g^{-1}s) \tag{11}$$

while the associated right von Neumann algebra, denoted by $\mathfrak{R}(G)$, is generated by the right regular representation $\boldsymbol{\rho} : G \ni g \mapsto \rho(g) \in \mathscr{B}(\mathscr{L}^2(G))$,

$$(\boldsymbol{\rho}(g)\xi)(s) = \Delta(g)^{1/2}\xi(sg). \tag{12}$$

Thus,

$$\mathfrak{L}(G) = \mathscr{R}\{\boldsymbol{\lambda}(g); g \in G\} \quad \mathfrak{R}(G) = \mathscr{R}\{\boldsymbol{\rho}(g); g \in G\} \tag{13}$$

and, by the commutation theorem ([L], 10.4.(2)), we have

$$\mathfrak{L}(G)' = \mathfrak{R}(G). \tag{14}$$

The natural weight on $\mathfrak{L}(G)$ associated with this left Hilbert algebra ([L], 10.16) is denoted by ω_G and is called *the Plancherel weight* on $\mathfrak{L}(G)$. The properties of the Plancherel weight will be studied later (18.17).

Using the commutation relations (7) and (14) as well as the fact that the only translation invariant functions on G are the constant functions, we obtain

$$\mathscr{B}(\mathscr{L}^2(G)) = \mathscr{R}\{\mathscr{L}^\infty(G), \mathfrak{L}(G)\} = \mathscr{R}\{\mathscr{L}^\infty(G), \mathfrak{R}(G)\}. \tag{15}$$

We define a unitary operator W_G on $\mathscr{L}^2(G)\bar{\otimes}\mathscr{L}^2(G)$ by

$$(W_G\zeta)(s,t) = \zeta(s,st) \tag{16}$$

($\zeta \in \mathscr{L}^2(G\times G); s,t \in G$). Consider also the unitary operator $V_G = \widetilde{W}_G^* = \sim \circ W_G^* \circ \sim$,

$$(V_G\zeta)(s,t) = \zeta(t^{-1}s,t). \tag{17}$$

Using the commutation relations, we get

$$W_G \in \mathscr{L}^\infty(G) \,\bar{\otimes}\, \mathfrak{L}(G), \quad V_G \in \mathfrak{L}(G) \,\bar{\otimes}\, \mathscr{L}^\infty(G). \tag{18}$$

18.5. We now consider the first example of a coinvolutive Hopf–von Neumann algebra associated with a locally compact group G, namely the triple

$$\mathbf{G} = (\mathscr{L}^\infty(G), \pi_G, k_G)$$

consisting of the von Neumann algebra $\mathscr{L}^\infty(G) \subset \mathscr{B}(\mathscr{L}^2(G))$ with comultiplication $\pi_G : \mathscr{L}^\infty(G) \to \mathscr{L}^\infty(G) \,\bar{\otimes}\, \mathscr{L}^\infty(G)$ defined by

$$\pi_G(f) = V_G^*(f \,\bar{\otimes}\, 1_G)V_G. \tag{1}$$

that is, $\pi_G(f)$ is the multiplication operator on $\mathscr{L}^2(G\times G)$ given by the function

$$(\pi_G(f))(s,t) = f(ts), \tag{2}$$

and coinvolution $k_G : \mathscr{L}^\infty(G) \to \mathscr{L}^\infty(G)$ defined by

$$k_G(f) = J_G\bar{f}J_G, \tag{3}$$

that is, $k_G(f)$ is the multiplication operator on $\mathscr{L}^2(G)$ given by the function

$$(k_G(f))(s) = f(s^{-1}). \tag{4}$$

It is easy to check that requirements 18.2.(1) and 18.2.(3) are satisfied.

The predual $\mathscr{L}^\infty(G)_*$ of the von Neumann algebra $\mathscr{L}^\infty(G) \subset \mathscr{B}(\mathscr{L}^2(G))$ is identified with the Banach space $\mathscr{L}^1(G)$ in the usual way:

$$\langle f,k\rangle = \int f(g)k(g)\,\mathrm{d}g \quad (f \in \mathscr{L}^\infty(G), k \in \mathscr{L}^1(G)). \tag{5}$$

Indeed, since $\mathscr{L}^\infty(G) \subset \mathscr{B}(\mathscr{L}^2(G))$ is in standard form, every element $k \in \mathscr{L}^\infty(G)_*$ is of the form $\omega_{\xi,\eta}$ with $\xi, \eta \in \mathscr{L}^2(G)$, hence k will correspond to the function $\xi\bar{\eta} \in \mathscr{L}^1(G)$. By definitions 18.3.(1) and 18.3.(3), $\mathscr{L}^1(G)$ becomes an involutive Banach algebra. It is easy to check that for $h, k \in \mathscr{L}^1(G)$ we have

$$h \cdot k = k * h, \tag{6}$$

$$k^0 = k^*, \tag{7}$$

the convolution "$*$" and the involution "$\sharp$" being defined by 18.4.(9).

The definition of π_G can be extended to an action of $\mathbf{G}$ on $\mathscr{B}(\mathscr{L}^2(G))$, still denoted by

$$\pi_G : \mathscr{B}(\mathscr{L}^2(G)) \to \mathscr{B}(\mathscr{L}^2(G)) \mathbin{\bar{\otimes}} \mathscr{L}^\infty(G),$$

namely

$$\pi_G(x) = V_G^*(x \mathbin{\bar{\otimes}} 1_G)V_G \quad (x \in \mathscr{B}(\mathscr{L}^2(G))). \tag{8}$$

Note that

$$\mathscr{B}(\mathscr{L}^2(G))^{\pi_G} = \mathfrak{R}(G). \tag{9}$$

Indeed, if $x \in \mathscr{B}(\mathscr{L}^2(G))$ and $\pi_G(x) = x \mathbin{\bar{\otimes}} 1_G$, then $V_G(x \mathbin{\bar{\otimes}} 1_G) = (x \mathbin{\bar{\otimes}} 1_G)V_G$. By applying this to a vector of the form $\xi \otimes \eta$ with $\xi, \eta \in \mathscr{K}(G)$ and then taking the scalar product with another vector of the same form, it follows that $x\boldsymbol{\lambda}(t) = \boldsymbol{\lambda}(t)x (t \in G)$. Hence $x \in \mathfrak{L}(G)' = \mathfrak{R}(G)$.

More generally, for any W^*-algebra $\mathscr{M}$, the mapping

$$\iota_{\mathscr{M}} \mathbin{\bar{\otimes}} \pi_G : \mathscr{M} \mathbin{\bar{\otimes}} \mathscr{B}(\mathscr{L}^2(G)) \to \mathscr{M} \mathbin{\bar{\otimes}} \mathscr{B}(\mathscr{L}^2(G)) \mathbin{\bar{\otimes}} \mathscr{L}^\infty(G)$$

determines an action of $\mathbf{G}$ on $\mathscr{M} \mathbin{\bar{\otimes}} \mathscr{B}(\mathscr{L}^2(G))$,

$$(\iota_{\mathscr{M}} \mathbin{\bar{\otimes}} \pi_G)(X) = (1_{\mathscr{M}} \mathbin{\bar{\otimes}} V_G^*)(X \mathbin{\bar{\otimes}} 1_G)(1_{\mathscr{M}} \mathbin{\bar{\otimes}} V_G) \tag{10}$$
$$(X \in \mathscr{M} \mathbin{\bar{\otimes}} \mathscr{B}(\mathscr{L}^2(G)))$$

and, by (9) and 18.2.(7), we have

$$\mathscr{M} \mathbin{\bar{\otimes}} \mathscr{B}(\mathscr{L}^2(G))^{\iota_{\mathscr{M}} \bar{\otimes} \pi_G} = \mathscr{M} \mathbin{\bar{\otimes}} \mathscr{R}(G). \tag{11}$$

18.6. In this section, we show that *the actions of the Hopf–von Neumann algebra* $\mathbf{G}$ *on the* W^**-algebra* $\mathscr{M}$ *actually correspond to the continuous actions of the locally compact group G on* $\mathscr{M}$.

Consider first a continuous action $\sigma : G \ni g \mapsto \sigma_g \in Aut(\mathscr{M})$ of G on $\mathscr{M}$. For each $x \in \mathscr{M}$, the function $G \in g \mapsto \sigma_g^{-1}(x) \in \mathscr{M}$ is w-continuous and bounded by $\|x\|$, hence (Lemma 1/13.1) it defines a unique element $\pi_\sigma(x) \in \mathscr{M} \mathbin{\bar{\otimes}} \mathscr{L}^\infty(G)$ such that

$$\langle \pi_\sigma(x), \varphi \mathbin{\bar{\otimes}} k \rangle = \int \varphi(\sigma_g^{-1}(x))k(g)\,\mathrm{d}g \tag{1}$$

for $\varphi \in \mathcal{M}_*$ and $k \in \mathcal{L}^1(G)$. If $\mathcal{M} \subset \mathcal{B}(\mathcal{H})$ and $\mathcal{M} \bar{\otimes} \mathcal{L}^\infty(G) \subset \mathcal{B}(\mathcal{L}^2(G, \mathcal{H}))$ are realized as von Neumann algebras, then

$$(\pi_\sigma(x)\xi)(g) = \sigma_g^{-1}(x)\xi(g) \quad (\xi \in \mathcal{L}^2(G, \mathcal{H}), g \in G). \tag{2}$$

Using (1) or (2), it is easy to check that

$$\pi_\sigma : \mathcal{M} \to \mathcal{M} \bar{\otimes} \mathcal{L}^\infty(G)$$

is an injective unital normal *-homomorphism. We show that π_σ is an action of $\mathbf{G}$ on $\mathcal{M}$, that is,

$$(\iota_\mathcal{M} \bar{\otimes} \pi_G) \circ \pi_\sigma = (\pi_\sigma \bar{\otimes} \iota_G) \circ \pi_\sigma. \tag{3}$$

Indeed, for $x \in \mathcal{M}, \varphi \in \mathcal{M}_*$, and $h, k \in \mathcal{L}^1(G)$ we have

$$\begin{aligned}
&\langle (\iota_\mathcal{M} \bar{\otimes} \pi_G)(\pi_\sigma(x)), \varphi \bar{\otimes} h \bar{\otimes} k \rangle = \langle \pi_\sigma(x), \varphi \bar{\otimes} ((h \bar{\otimes} k) \circ \pi_G) \\
&\quad = \langle \pi_\sigma(x), \varphi \bar{\otimes} (h \cdot k) \rangle = \langle \pi_\sigma(x), \varphi \bar{\otimes} (k * h) \rangle \\
&\quad = \int \varphi(\sigma_g^{-1}(x))(k * h)(g)\, \mathrm{d}g = \int\int \varphi(\sigma_g^{-1}(x))k(t)h(t^{-1}g)\, \mathrm{d}t\, \mathrm{d}g
\end{aligned}$$

and

$$\begin{aligned}
&\langle (\pi_\sigma \bar{\otimes} \iota_G)(\pi_\sigma(x)), \varphi \bar{\otimes} h \bar{\otimes} k \rangle = \langle \pi_\sigma(x), ((\varphi \bar{\otimes} h) \circ \pi_\sigma) \bar{\otimes} k \rangle \\
&\quad = \int \langle \pi_\sigma(\sigma_t^{-1}(x)), \varphi \bar{\otimes} h \rangle k(t)\, \mathrm{d}t = \int\int \varphi(\sigma_s^{-1}(\sigma_t^{-1}(x)))h(s)k(t)\, \mathrm{d}s\, \mathrm{d}t \\
&\quad = \int\int \varphi(\sigma_{ts}^{-1}(x))h(s)k(t)\, \mathrm{d}s\, \mathrm{d}t = \int\int \varphi(\sigma_g^{-1}(x))k(t)h(t^{-1}g)\, \mathrm{d}g\, \mathrm{d}t,
\end{aligned}$$

and Fubini's theorem insures that the two integrals are equal.

We now show that

$$\mathcal{M}^{\pi_\sigma} = \mathcal{M}^\sigma, \tag{4}$$

that is, for $x \in \mathcal{M}$ we have

$$\pi_\sigma(x) = x \bar{\otimes} 1_G \Leftrightarrow \sigma_g(x) = x \text{ for all } g \in G.$$

Indeed, if $\sigma_g(x) = x$ for all $g \in G$, then for $\varphi \in \mathcal{M}_*$ and $k \in \mathcal{L}^1(G)$ we have $\langle \pi_\sigma(x), \varphi \bar{\otimes} k \rangle = \int \varphi(x)k(g)\, \mathrm{d}g = \langle x, \varphi \rangle \langle 1_G, k \rangle = \langle x \bar{\otimes} 1_G, \varphi \bar{\otimes} k \rangle$, hence $\pi_\sigma(x) = x \bar{\otimes} 1_G$. Conversely, if $\pi_\sigma(x) = x \bar{\otimes} 1_G$, then for $\varphi \in \mathcal{M}_*$ and $k \in \mathcal{L}^1(G)$ we have

$$\int \varphi(\sigma_g^{-1}(x))k(g)\, \mathrm{d}g = \int \varphi(x)k(g)\, \mathrm{d}g.$$

It follows that the continuous function $G \ni g \mapsto \varphi(\sigma_g^{-1}(x))$ coincides almost everywhere with the constant function $\varphi(x)$. Consequently, $\varphi(\sigma_g^{-1}(x)) = \varphi(x)$ for all $g \in G$ and $\varphi \in \mathcal{M}_*$, that is, $x \in \mathcal{M}^\sigma$.

Note that

$$\pi_\sigma(\sigma_g(x)) = (1_{\mathscr{M}} \bar{\otimes} \boldsymbol{\lambda}(g))\pi_\sigma(x)(1_{\mathscr{M}} \bar{\otimes} \boldsymbol{\lambda}(g))^* \quad (x \in \mathscr{M}, g \in G). \tag{5}$$

Indeed, for $k \in \mathscr{L}^1(G) = \mathscr{L}^\infty(G)_*$, the element $k \circ \mathrm{Ad}(\boldsymbol{\lambda}(g)) \in \mathscr{L}^\infty(G)_*$, considered as an $\mathscr{L}^1$-function on G, can be written

$$(k \circ \mathrm{Ad}(\boldsymbol{\lambda}(g)))(t) = k(gt) \quad (t \in G).$$

Consequently, for $\varphi \in \mathscr{M}_*, k \in \mathscr{L}^1(G)$, we have

$$\begin{aligned}\langle (1 \bar{\otimes} \boldsymbol{\lambda}(g))\pi_\sigma(x)(1 \bar{\otimes} \boldsymbol{\lambda}(g))^*, \varphi \bar{\otimes} k\rangle &= \langle \pi_\sigma(x), \varphi \bar{\otimes} (k \circ \mathrm{Ad}(\boldsymbol{\lambda}(g)))\\ &= \int \varphi(\sigma_{t^{-1}}(x))k(gt)\,\mathrm{d}t = \int \varphi(\sigma_{t^{-1}g}(x))k(t)\,\mathrm{d}t\\ &= \int \varphi(\sigma_t^{-1}(\sigma_g(x)))k(t)\,\mathrm{d}t = \langle \pi_\sigma(\sigma_g(x)), \varphi \bar{\otimes} k\rangle.\end{aligned}$$

Recall that for every $k \in \mathscr{L}^1(G)$ and every $x \in \mathscr{M}$, we have defined an element (13.2)

$$\sigma_k(x) = \int \sigma_g(x)k(g)\,\mathrm{d}g$$

and also an element (18.3.(5))

$$k \cdot x = E^k_{\mathscr{M}}(\pi_\sigma(x)).$$

We have

$$k \cdot x = \sigma_{\overline{k^\sharp}}(x) \quad (x \in \mathscr{M}, k \in \mathscr{L}^1(G)). \tag{6}$$

Indeed, for every $\varphi \in \mathscr{M}_*$ we obtain

$$\begin{aligned}\langle k \cdot x, \varphi\rangle = \langle \pi_\sigma(x), \varphi \bar{\otimes} k\rangle &= \int \varphi(\sigma_g^{-1}(x))k(g)\,\mathrm{d}g\\ &= \left\langle \int \sigma_{g^{-1}}(x)k(g)\,\mathrm{d}g, \varphi \right\rangle = \left\langle \int \sigma_g(x)k(g^{-1})\Delta(g^{-1})\,\mathrm{d}g, \varphi \right\rangle\\ &= \left\langle \int \sigma_g(x)\overline{k^\sharp(g)}\,\mathrm{d}g, \varphi\right\rangle = \langle \sigma_{\overline{k^\sharp}}(x), \varphi \rangle.\end{aligned}$$

Note also that, writing $k^t(g) = k(gt)$, we have

$$\sigma_t(k \cdot x) = \Delta(t)(k^t \cdot x) \quad (t \in G), \tag{7}$$

the verification being similar to the above computation.

Using 15.1.(2), we infer from (6) that for any norm-bounded approximate unit $\{k_i\}_{i\in I}$ of $\mathscr{L}^1(G)$ and any $x \in \mathscr{M}$, we have

$$k_i \cdot x \overset{w}{\to} x. \tag{8}$$

Moreover, using (7), from (8) we also obtain

$$\Delta(t)(k_i^t \cdot x) \overset{w}{\to} \sigma_t(x) \quad (t \in G). \tag{9}$$

Conversely, we have the following:

Proposition. *Let $\pi : \mathscr{M} \to \mathscr{M} \bar{\otimes} \mathscr{L}^\infty(G)$ be an action of G on $\mathscr{M}$. There exists a continuous action $\sigma : G \to Aut(\mathscr{M})$ of G on $\mathscr{M}$, uniquely determined such that $\pi = \pi_\sigma$.*

Proof. (I) Consider first the action

$$\pi = \pi_G : \mathscr{B}(\mathscr{L}^2(G)) \to \mathscr{B}(\mathscr{L}^2(G)) \bar{\otimes} \mathscr{L}^\infty(G)$$

of **G** on $\mathscr{B}(\mathscr{L}^2(G))$. There exists a continuous action

$$\sigma = \mathrm{Ad}(\boldsymbol{\lambda}) : G \to Aut(\mathscr{B}(\mathscr{L}^2(G)))$$

of G on $\mathscr{B}(\mathscr{L}^2(G))$, defined by the left regular representation. Since $\mathscr{B}(\mathscr{L}^2(G)) = \mathscr{R}\{\mathscr{L}^\infty(G), \mathfrak{R}(G)\}$, in order to establish that $\pi = \pi_\sigma$ it is sufficient to show that $\pi(x) = \pi_\sigma(x)$ separately for $x \in \mathfrak{R}(G)$ and for $x \in \mathscr{L}^\infty(G)$. If $x \in \mathfrak{R}(G)$, then it is clear that $\pi(x) = x \bar{\otimes} 1 = \pi_\sigma(x)$. Also, for $x = f \in \mathscr{L}^\infty(G)$ it is easy to check that $\pi_\sigma(x) = \pi(x)$ is the multiplication operator on $\mathscr{L}^2(G \times G)$ given by the function $(s,t) \mapsto f(ts)$. Hence $\pi = \pi_\sigma$.

(II) Consider next the action

$$\pi = \iota_{\mathscr{M}} \bar{\otimes} \pi_G : \mathscr{M} \bar{\otimes} \mathscr{B}(\mathscr{L}^2(G)) \to \mathscr{M} \bar{\otimes} \mathscr{B}(\mathscr{L}^2(G)) \bar{\otimes} \mathscr{L}^\infty(G)$$

of **G** on $\mathscr{M} \bar{\otimes} \mathscr{B}(\mathscr{L}^2(G))$ and the continuous action

$$\sigma = \iota_{\mathscr{M}} \bar{\otimes} \mathrm{Ad}(\boldsymbol{\lambda}) : G \to Aut(\mathscr{M} \bar{\otimes} \mathscr{B}(\mathscr{L}^2(G)))$$

of G on $\mathscr{M} \bar{\otimes} \mathscr{B}(\mathscr{L}^2(G))$, that is,

$$\sigma_g(X) = (1_{\mathscr{M}} \bar{\otimes} \boldsymbol{\lambda}(g))X(1_{\mathscr{M}} \bar{\otimes} \boldsymbol{\lambda}(g))^* \quad (g \in G, X \in \mathscr{M} \bar{\otimes} \mathscr{B}(\mathscr{L}^2(G))).$$

Using the result proved in step (I), it is easy to see that in the present situation also, we have $\pi = \pi_\sigma$.

(III) It is clear that if the action π of **G** on $\mathscr{M}$ is isomorphic to an action of the type π_σ, then π is also of the form π_a.

(IV) Now let $\sigma : G \to Aut(\mathscr{M})$ be a continuous action of G on $\mathscr{M}$ and let $\mathscr{N} \subset \mathscr{M}$ be a unital W^*-subalgebra such that $\pi_\sigma(\mathscr{N}) \subset \mathscr{N} \bar{\otimes} \mathscr{L}^\infty(G)$. Then $\pi = \pi_\sigma|\mathscr{N}$ is an action of **G** on $\mathscr{N}$ and $\mathscr{N}$ with the **G**-comodule structure defined by π is a **G**-subcomodule of the **G**-comodule $\mathscr{M}$ defined by

π_σ. Consequently, for $y \in \mathscr{N} \subset \mathscr{M}$ we have (see 18.3.(17))

$$k \cdot y \in \mathscr{N} \quad (k \in \mathscr{L}^1(G))$$

and if $\{k_i\}_{i\in I}$ is a norm-bounded approximate unit of $\mathscr{L}^1(G)$, then, according to (9),

$$\mathscr{N} \ni \Delta(t)(k_i^t \cdot y) \xrightarrow{w} \sigma_t(y) \quad (t \in G).$$

It follows that $\sigma_t(\mathscr{N}) = \mathscr{N}(t \in G)$, and $\pi = \pi_\sigma|\mathscr{N} = \pi_{\sigma|\mathscr{N}}$.

(V) Finally, let $\pi : \mathscr{M} \to \mathscr{N} \bar{\otimes} \mathscr{L}^\infty(G)$ be any action of G on $\mathscr{M}$. By the coassociativity condition $(\iota_{\mathscr{M}} \bar{\otimes} \pi_G) \circ \pi = (\pi \bar{\otimes} \iota_G) \circ \pi$ it follows that $\pi(\mathscr{M})$ is a $\mathbf{G}$-subcomodule of the $\mathbf{G}$-comodule $\mathscr{M} \bar{\otimes} \mathscr{B}(\mathscr{L}^2(G))$, which is isomorphic to the $\mathbf{G}$-comodule $\mathscr{M}$. Thus, the existence assertion of the statement follows from (II), (III), and (IV).

(V) The uniqueness assertion follows easily using (9). □

18.7. We now consider the second example of a coinvolutive Hopf–von Neumann algebra associated with a locally compact group G, namely the triple

$$\hat{\mathbf{G}} = (\mathfrak{L}(G), \delta_G, j_G)$$

consisting of the von Neumann algebra $\mathfrak{L}(G) \subset \mathscr{B}(\mathscr{L}^2(G))$ with comultiptication $\delta_G : \mathfrak{L}(G) \to \mathfrak{L}(G) \bar{\otimes} \mathfrak{L}(G)$ defined by

$$\delta_G(x) = W_G^*(x \bar{\otimes} 1_G)W_G, \tag{1}$$

so that, in particular,

$$\delta_G(\boldsymbol{\lambda}(g)) = \boldsymbol{\lambda}(g) \bar{\otimes} \boldsymbol{\lambda}(g), \tag{2}$$

and coinvolution $j_G : \mathfrak{L}(G) \to \mathfrak{L}(G)$ defined by

$$j_G(x) = K_G x^* K_G, \tag{3}$$

so that, in particular,

$$j_G(\boldsymbol{\lambda}(g)) = \boldsymbol{\lambda}(g^{-1}). \tag{4}$$

It is easy to check that 18.2.(1) and 18.2.(3) are satisfied.

Let $\mathscr{A}(G) = \mathfrak{L}(G)_*$ be the predual of $\mathfrak{L}(G)$ endowed with the corresponding involutive Banach algebra structure (18.3.(1) and 18.3.(3)). For every $k \in \mathscr{A}(G)$, we consider the function $k(\cdot)$ defined by

$$k(g) = \langle \boldsymbol{\lambda}(g), k \rangle \quad (g \in G). \tag{5}$$

Then $k(\cdot)$ is a continuous function in $\mathscr{L}^\infty(G)$ and

$$\|k(\cdot)\|_\infty \leq \|k\|_{\mathscr{A}(G)}.$$

Actually, *the mapping* $\mathscr{A}(G) \ni k \mapsto k(\cdot) \in \mathscr{L}^\infty(G)$ *is an injective *-homomorphism*. Indeed, this mapping is clearly linear and injective and, for $h, k \in \mathscr{A}(G)$ and $g \in G$, we have

$$\begin{aligned} (h \cdot k)(g) &= \langle \boldsymbol{\lambda}(g), h \cdot k \rangle = \langle \delta_G(\boldsymbol{\lambda}(g)), h \bar{\otimes} k \rangle \\ &= \langle \boldsymbol{\lambda}(g) \bar{\otimes} \boldsymbol{\lambda}(g), h \bar{\otimes} k \rangle = h(g)k(g), \end{aligned} \tag{6}$$

$$\begin{aligned} k^0(g) &= \langle \boldsymbol{\lambda}(g), k^0 \rangle = \langle j_G(\boldsymbol{\lambda}(g)), k^* \rangle = \overline{\langle \boldsymbol{\lambda}(g^{-1})^*, k \rangle} \\ &= \overline{\langle \boldsymbol{\lambda}(g), k \rangle} = \overline{k(g)}. \end{aligned} \tag{7}$$

Since $\mathfrak{L}(G) \subset \mathscr{B}(\mathscr{L}^2(G))$ is in standard form, every $k \in \mathscr{A}(G)$ is of form $\omega_{\xi,\eta}$ with $\xi, \eta \in \mathscr{L}^2(G)$. In this case, it is easy to check that $k(\cdot) = \eta \, * \, \check{\xi}$, where $\check{\xi}(g) = \xi(g^{-1})$ $(g \in G)$.

The involutive Banach algebra $\mathscr{A}(G)$ is called *the Eymard algebra*.

The definition of δ_G can be extended with the aid of equality (1) to an action of $\hat{\mathbf{G}}$ on $\mathscr{B}(\mathscr{L}^2(G))$ still denoted by

$$\delta_G : \mathscr{B}(\mathscr{L}^2(G)) \to \mathscr{B}(\mathscr{L}^2(G)) \bar{\otimes} \mathfrak{L}(G). \tag{8}$$

Note that

$$\mathscr{B}(\mathscr{L}^2(G))^{\delta_G} = \mathscr{L}^\infty(G), \tag{9}$$

the proof being similar to the proof of 18.5.(9).

More generally, for any W^*-algebra $\mathscr{M}$, the mapping

$$\iota_{\mathscr{M}} \bar{\otimes} \delta_G : \mathscr{M} \bar{\otimes} \mathscr{B}(\mathscr{L}^2(G)) \to \mathscr{M} \bar{\otimes} \mathscr{B}(\mathscr{L}^2(G)) \bar{\otimes} \mathfrak{L}(G)$$

determines an action of $\hat{\mathbf{G}}$ on $\mathscr{M} \bar{\otimes} \mathscr{B}(\mathscr{L}^2(G))$,

$$\begin{aligned} (\iota_{\mathscr{M}} \bar{\otimes} \delta_G)(X) = (1_{\mathscr{M}} \bar{\otimes} W_G^*)(X \bar{\otimes} 1_G)(1_{\mathscr{M}} \bar{\otimes} W_G) \\ (X \in \mathscr{M} \bar{\otimes} \mathscr{B}(\mathscr{L}^2(G))), \end{aligned} \tag{10}$$

and, by (9) and 18.2.(7), we have

$$(\mathscr{M} \bar{\otimes} \mathscr{B}(\mathscr{L}^2(G)))^{\iota_{\mathscr{M}} \bar{\otimes} \delta_G} = \mathscr{M} \bar{\otimes} \mathscr{L}^\infty(G). \tag{11}$$

Also, there exists another action of $\hat{\mathbf{G}}$ on $\mathscr{B}(\mathscr{L}^2(G))$,

$$\delta_G^* : \mathscr{B}(\mathscr{L}^2(G)) \to \mathscr{B}(\mathscr{L}^2(G)) \bar{\otimes} \mathfrak{L}(G).$$

defined by

$$\delta_G^*(x) = W_G(x \otimes 1_G)W_G^* \quad (x \in \mathscr{B}(\mathscr{L}^2(G))), \tag{12}$$

with centralizer

$$\mathscr{B}(\mathscr{L}^2(G))^{\delta_G^*} = \mathscr{L}^\infty(G) \tag{13}$$

and, correspondingly, an action $\iota_{\mathscr{M}} \bar{\otimes} \delta_G^*$ of $\hat{\mathbf{G}}$ on $\mathscr{M} \bar{\otimes} \mathscr{B}(\mathscr{L}^2(G)))$ with centralizer

$$(\mathscr{M} \bar{\otimes} \mathscr{B}(\mathscr{L}^2(G)))^{\iota_{\mathscr{M}} \bar{\otimes} \delta_G^*} = \mathscr{M} \bar{\otimes} \mathscr{L}^\infty(G). \tag{14}$$

18.8. Assume now that the locally compact group G is abelian and consider the dual locally compact abelian group $\hat{G}$ of all continuous characters on G, endowed with the compact-open topology. We choose the Haar measures $\mathrm{d}g$ on G and $\mathrm{d}\gamma$ on $\hat{G}$ such that the inversion theorem for the Fourier transform holds.

By the Pontryagin duality theorem, G can be identified with the dual group $G^{\wedge\wedge}$ of G. For $g \in G = G^{\wedge\wedge}$ and $\gamma \in \hat{G}$, we shall denote by $\langle g, \gamma\rangle$ the value of γ at g (or the value of g at γ).

By the Plancherel theorem, the restriction of the Fourier transform to $\mathscr{L}^1(G) \cap \mathscr{L}^2(G)$ can be extended to a unitary operator

$$F : \mathscr{L}^2(G) \to \mathscr{L}^2(\hat{G})$$

called the *Fourier–Plancherel isomorphism*, whose inverse is obtained similarly from the inverse Fourier transform.

Recall that if we denote by $\mathscr{B}(G)$ the linear span of continuous positive definite functions on G, then $\mathscr{D}(G) = \mathscr{K}(G) \cap \mathscr{B}(G)$ is a dense translation invariant linear subspace of $\mathscr{L}^2(G)$ and for every $\xi \in \mathscr{D}(G)$ we have

$$(F\xi)(\gamma) = \hat{\xi}(\gamma) = \int_G \xi(g)\langle g, \gamma\rangle \,\mathrm{d}g \quad (\gamma \in \hat{G}). \tag{1}$$

Also, $F(\mathscr{D}(G))$ is a dense linear subspace of $\mathscr{L}^2(\hat{G})$, invariant under multiplication by characters, and for every $\eta \in F(\mathscr{D}(G))$ we have

$$(F^{-1}\eta)(g) = \hat{\eta}(g) = \int_{\hat{G}} \eta(\gamma)\langle g, \gamma\rangle \,\mathrm{d}\gamma \quad (g \in G). \tag{2}$$

The characters $g \in G = G^{\wedge\wedge}$ on $\hat{G}$ are $\mathscr{L}^\infty$-functions. For the sake of clarity, the unitary operator given by the multiplication with the function $g \in \mathscr{L}^\infty(\hat{G})$ on $\mathscr{L}^2(\hat{G})$ will be denoted by $\mathbf{m}(g)$:

$$(\mathbf{m}(g)\eta)(\gamma) = \langle g, \gamma\rangle \eta(\gamma) \quad (\eta \in \mathscr{L}^2(\hat{G}), \gamma \in \hat{G}). \tag{3}$$

Since the trigonometric polynomials (i.e., the linear combinations of characters) are dense in $\mathscr{K}(\hat{G})$ with respect to the topology of uniform convergence on compact subsets, it follows that the von Neumann algebra $\mathscr{L}^\infty(\hat{G}) \subset \mathscr{B}(\mathscr{L}^2(\hat{G}))$ is generated by the operators $\mathbf{m}(g)(g \in G)$.

We recall that the von Neumann algebra $\mathfrak{L}(G) \subset \mathscr{B}(\mathscr{L}^2(G))$ is generated by the left translation operators $\boldsymbol{\lambda}(g), (g \in G)$.

Proposition. *The Fourier–Plancherel isomorphism establishes an isomorphism between the coinvolutive Hopf–von Neumann algebras* $(\mathfrak{L}(G), \delta_G, j_G)$ *and* $(\mathscr{L}^\infty(\hat{G}), \pi_{\hat{G}}, \pi_{\hat{G}})$.

Proof. We define a $*$-isomorphism $\Phi : \mathscr{B}(\mathscr{L}^2(G)) \to \mathscr{B}(\mathscr{L}^2(\hat{G}))$ by

$$\Phi(x) = F \circ x \circ F^{-1} \quad (x \in \mathscr{B}(\mathscr{L}^2(G))) \tag{4}$$

and show that $\Phi(\mathfrak{L}(G)) = \mathscr{L}^\infty(\hat{G})$, $(\Phi \bar{\otimes} \Phi) \circ \delta_G = \pi_{\hat{G}} \circ \Phi$, $\Phi \circ j_G = k_{\hat{G}} \circ \phi$.

For $g \in G, \xi \in \mathscr{D}(G), \gamma \in \hat{G}$ we have

$$\begin{aligned}(\mathbf{m}(g)F\xi)(\gamma) &= \langle g, \gamma\rangle (F\xi)(\gamma) = \langle g, \gamma\rangle \int \xi(t)\langle t, \gamma\rangle \,\mathrm{d}t \\ &= \int \xi(t)\langle gt, \gamma\rangle \,\mathrm{d}t = \int \xi(g^{-1}s)\langle s, \gamma\rangle \,\mathrm{d}s \\ &= \int (\boldsymbol{\lambda}(g)\xi)(s)\langle s, \gamma\rangle \mathrm{d}s = (F\boldsymbol{\lambda}(g)\xi)(\gamma),\end{aligned}$$

that is, $\mathbf{m}(g)F = F\boldsymbol{\lambda}(g)$, and hence

$$\Phi(\boldsymbol{\lambda}(g)) = \mathbf{m}(g) \quad (g \in G), \tag{5}$$

so that $\Phi(\mathfrak{L}(G)) = \mathscr{L}^\infty(\hat{G})$.

For $g \in G$, the operator $\pi_{\hat{G}}(\mathbf{m}(g))$ is (by 18.5.(2)) the multiplication operator by the function

$$\hat{G} \times \hat{G} \ni (\alpha, \beta) \mapsto \langle g, \alpha\beta\rangle = \langle g, \alpha\rangle\langle g, \beta\rangle$$

on $\mathscr{L}^2(\hat{G} \times \hat{G})$, that is, $\pi_{\hat{G}}(\mathbf{m}(g)) = \mathbf{m}(g) \bar{\otimes} \mathbf{m}(g)$. According to (5) and 18.7.(2) we get

$$(\Phi \bar{\otimes} \Phi)(\delta_G(\boldsymbol{\lambda}(g))) = \Phi(\boldsymbol{\lambda}(g)) \bar{\otimes} \Phi(\boldsymbol{\lambda}(g)) = \mathbf{m}(g) \bar{\otimes} \mathbf{m}(g) = \pi_{\hat{G}}(\Phi(\boldsymbol{\lambda}(g))),$$

so that $(\Phi \bar{\otimes} \Phi) \circ \delta_G = \pi_{\hat{G}} \circ \Phi$.

Finally, using (5), 18.5.(4) and 18.7.(4), it is easy to see that

$$\Phi(j_G(\boldsymbol{\lambda}(g))) = \Phi(\boldsymbol{\lambda}(g^{-1})) = \mathbf{m}(g^{-1}) = k_{\hat{G}}(\mathbf{m}(g)) = k_{\hat{G}}(\Phi(\boldsymbol{\lambda}(g))),$$

so that $\Phi \circ j_G = k_{\hat{G}} \circ \Phi$. □

Thus, if G is abelian, the notation $\hat{\mathbf{G}}$ introduced with two different meanings in Sections 18.5 and 18.7, designates the same coinvolutive Hopf–von Neumann algebra.

Also, according to Proposition 18.6, every action $\delta : \mathscr{M} \to \mathscr{M} \bar{\otimes} \mathfrak{L}(G)$ of $\hat{\mathbf{G}} = (\mathfrak{L}(G), \delta_G, j_G)$ on a W^*-algebra $\mathscr{M}$ corresponds to a continuous action $\sigma : \hat{G} \to Aut(\mathscr{M})$ of $\hat{G}$ on $\mathscr{M}$, uniquely determined, such that $\pi_\sigma = (\iota_{\mathscr{M}} \bar{\otimes} \Phi) \circ \delta$.

18.9. We continue to assume that the locally compact group G is abelian. For $\gamma \in \hat{G}$, we denote by $\mathbf{m}(\gamma) \in \mathscr{L}^\infty(G) \subset \mathscr{B}(\mathscr{L}^2(G))$, the unitary operator of multiplication with the function $\gamma \in \mathscr{L}^\infty(G)$ on $\mathscr{L}^2(G)$. Then

$$\hat{G} \ni \gamma \mapsto \mathbf{m}(\gamma) \in \mathscr{B}(\mathscr{L}^2(G))$$

is an *so*-continuous unitary representation of $\hat{G}$ on $\mathscr{L}^2(G)$.

It is easy to check that

$$\boldsymbol{\lambda}(g)\mathbf{m}(\gamma) = \overline{\langle g, \gamma \rangle}\mathbf{m}(\gamma)\boldsymbol{\lambda}(g) \quad (g \in G, \gamma \in \hat{G}). \tag{1}$$

We define a continuous action $\theta : \hat{G} \to Aut(\mathscr{B}(\mathscr{L}^2(G)))$ by

$$\theta_\gamma(x) = \mathbf{m}(\gamma)^* x \mathbf{m}(\gamma) \quad (x \in \mathscr{B}(\mathscr{L}^2(G)), y \in \hat{G}) \tag{2}$$

and the corresponding action $\pi_\theta : \mathscr{B}(\mathscr{L}^2(G)) \to \mathscr{B}(\mathscr{L}^2(G)) \bar{\otimes} \mathscr{L}^\infty(\hat{G})$ of $\hat{\mathbf{G}} = (\mathscr{L}^\infty(\hat{G}), \pi_{\hat{G}}, k_{\hat{G}})$ on $\mathscr{B}(\mathscr{L}^2(G))$.

It is clear that $\theta_\gamma(f) = f$ for every $f \in \mathscr{L}^\infty(G)$ $(\gamma \in \hat{G})$, hence

$$\pi_\theta(f) = f \bar{\otimes} 1_G \quad (f \in \mathscr{L}^\infty(G)). \tag{3}$$

On the other hand, using the definition 18.6.(1) of π_θ and the commutation relations (1), it is easy to check that

$$\pi_\theta(\boldsymbol{\lambda}(g)) = \boldsymbol{\lambda}(g) \bar{\otimes} \mathbf{m}(g) \quad (g \in G). \tag{4}$$

Let ι_G be the identity mapping on $\mathscr{B}(\mathscr{L}^2(G))$ and $\Phi : \mathscr{B}(\mathscr{L}^2(G)) \to \mathscr{B}(\mathscr{L}^2(\hat{G}))$ the Fourier–Plancherel isomorphism. Using (3), (4), 18.7.(2), and 18.8.(4), we obtain

$$\pi_\theta(\iota_G(x)) = (\iota_G \bar{\otimes} \Phi)(\delta_G(x)),$$

valid for $x = f \in \mathscr{L}^\infty(G)$ and for $x = \boldsymbol{\lambda}(g) \in \mathfrak{L}(G)$ and hence (18.4.(15)) for all $x \in \mathscr{B}(\mathscr{L}^2(G))$.

We thus get the following:

Proposition. *Let G be a locally compact abelian group and $\mathscr{M}$ a W^*-algebra. The action*

$$\iota_{\mathscr{M}} \bar{\otimes} \delta_G = \mathscr{M} \bar{\otimes} \mathscr{B}(\mathscr{L}^2(G)) \to \mathscr{M} \bar{\otimes} \mathscr{B}(\mathscr{L}^2(G)) \bar{\otimes} \mathfrak{L}(G)$$

of $\hat{\mathbf{G}} = (\mathfrak{L}(G), \delta_G, j_G)$ on $\mathscr{M}$ corresponds, via the Fourier–Plancherel isomorphism to the continuous action

$$\theta : \hat{G} \to Aut(\mathscr{M} \bar{\otimes} \mathscr{B}(\mathscr{L}^2(G)))$$

of $\hat{\mathbf{G}}$ on $\mathscr{M} \bar{\otimes} \mathscr{B}(\mathscr{L}^2(G))$ defined by

$$\theta_\gamma(X) = (1_{\mathscr{M}} \bar{\otimes} \mathbf{m}(\gamma))^* X (1_{\mathscr{M}} \bar{\otimes} \mathbf{m}(\gamma)) \quad (X \in \mathscr{M} \bar{\otimes} \mathscr{B}(\mathscr{L}^2(G)), \gamma \in \hat{G}).$$

Let $\sigma : G \to Aut(\mathscr{M})$ be a continuous action of the locally compact group G on the W^*-algebra $\mathscr{M}$, $\pi_\sigma : \mathscr{M} \to \mathscr{M} \bar{\otimes} \mathscr{L}^\infty(G)$ the corresponding action of $\mathbf{G}$ on $\mathscr{M}$ and φ a normal weight on $\mathscr{M}$.

Recall that φ is called σ-invariant, if $\varphi \circ \sigma_g = \varphi$ $(g \in G)$. This condition is equivalent to the following condition:

$$\langle \pi_\sigma(x), \varphi \bar{\otimes} k \rangle = \langle x \bar{\otimes} 1_G, \varphi \bar{\otimes} k \rangle \quad (x \in \mathscr{M}^+, k \in \mathscr{L}^1(G)^+).$$

In what follows, we consider a convenient notion of invariance of weights with respect to the action of a coinvolutive Hopf–von Neumann algebra.

18.10. Let $\delta : \mathscr{M} \to \mathscr{M} \bar{\otimes} \mathscr{A}$ be an action of the coinvolutive Hopf–von Neumann algebra $(\mathscr{A}, \delta_{\mathscr{A}}, j_{\mathscr{A}})$ on the W^*-algebra $\mathscr{M}$ and let φ be a normal semifinite weight on $\mathscr{M}$.

We shall say that φ is *δ-invariant* if

$$\langle \delta(x), \varphi \bar{\otimes} k \rangle = \langle x \bar{\otimes} 1_{\mathscr{A}}, \varphi \bar{\otimes} k \rangle \quad (x \in \mathscr{M}^+, k \in \mathscr{A}_*^+). \tag{1}$$

Lemma. *Let φ be a δ-invariant normal semifinite weight on the $\mathscr{A}$-comodule $\mathscr{M}$. Then*

$$\mathbf{s}(\varphi) \in \mathscr{M}^{\delta}, \tag{2}$$

for every $x \in \mathfrak{N}_{\varphi}$ and $k \in \mathscr{A}_*$ we have $k \cdot x \in \mathfrak{N}_{\varphi}$ and

$$\|(k \cdot x)_{\varphi}\|_{\varphi} \leq \|k\| \|x_{\varphi}\|_{\varphi}, \tag{3}$$

and for every $x, y \in \mathfrak{N}_{\varphi}$ and $k \in \mathscr{A}_*^+$ we have $(y^* \bar{\otimes} 1_{\mathscr{A}})\delta(x) \in \mathfrak{M}_{\varphi \bar{\otimes} k}, y^*(k \cdot x) \in \mathfrak{M}_{\varphi}$ and

$$\langle (y^* \bar{\otimes} 1_{\mathscr{A}})\delta(x), \varphi \bar{\otimes} k \rangle = \langle y^*(k \cdot x), \varphi \rangle. \tag{4}$$

Proof. Let $x \in \mathfrak{N}_{\varphi}, k \in \mathscr{A}_*$ with polar decomposition $k = |k|(v\cdot), v \in \mathscr{M}$, and let $f \in \mathscr{M}_*^+, f \leq \varphi$. We have

$$\begin{aligned}
\langle (k \cdot x)^*(k \cdot x), f \rangle &= \langle ((k \cdot x)^* \bar{\otimes} 1_{\mathscr{A}})\delta_{\mathscr{A}}(x), f \bar{\otimes} k \rangle \\
&= \langle ((k \cdot x)^* \bar{\otimes} v)\delta(x), f \bar{\otimes} |k| \rangle \\
&\leq \langle (k \cdot x)^*(k \cdot x), f \rangle^{1/2} \|k\|^{1/2} \langle \delta(x^*x), f \bar{\otimes} |k| \rangle^{1/2} \\
&\leq \langle (k \cdot x)^*(k \cdot x), f \rangle^{1/2} \|k\|^{1/2} \langle \delta(x^*x), \varphi \bar{\otimes} |k| \rangle^{1/2} \\
&= \langle (k \cdot x)^*(k \cdot x), f \rangle^{1/2} \|k\|^{1/2} \langle x^*x, \varphi \rangle^{1/2} \|k\|^{1/2}
\end{aligned}$$

hence

$$\langle (k \cdot x)^*(k \cdot x), f \rangle^{1/2} \leq \|k\| \|x_{\varphi}\|_{\varphi}.$$

Since $f \in \mathscr{M}_*^+, f \leq \varphi$ was arbitrary, (3) follows.

Consider now $x, y \in \mathfrak{N}_{\varphi}$ and $k \in \mathscr{A}_*^+$. From (3) it follows that $y^*(k \cdot x) \in \mathfrak{M}_{\varphi}$. Since $\langle \delta(x)^*\delta(x), \varphi \bar{\otimes} k \rangle = \langle x^*x, \varphi \rangle \langle 1_{\mathscr{A}}, k \rangle < +\infty$, we have $\delta(x) \in \mathfrak{N}_{\varphi \bar{\otimes} k}$, hence $(y^* \bar{\otimes} 1_{\mathscr{A}})\delta(x) \in \mathfrak{M}_{\varphi \bar{\otimes} k}$. Let

$$(y^* \bar{\otimes} 1_{\mathscr{A}})\delta(x) = a_1 - a_2 + \mathrm{i}a_3 - \mathrm{i}a_4 \text{ with } a_j \in \mathfrak{M}_{\varphi \bar{\otimes} k}, a_j \geq 0 \quad (1 \leq j \leq 4). \tag{5}$$

Using 9.8.(5) we get

$$\langle E_{\mathscr{M}}^k(a_j), \varphi \rangle = \langle a_j, \varphi \bar{\otimes} k \rangle < +\infty \quad (1 \leq j \leq 4), \tag{6}$$

hence, $0 \leq E^k_{\mathscr{M}}(a_j) \in \mathfrak{M}_\varphi$. Thus,

$$y^*(k \cdot x) = E^k_{\mathscr{M}}((y^* \bar{\otimes} 1_{\mathscr{A}})\delta(x)) = E^k_{\mathscr{M}}(a_1) - E^k_{\mathscr{M}}(a_2) + \mathrm{i}E^k_{\mathscr{M}}(a_3) - \mathrm{i}E^k_{\mathscr{M}}(a_4)$$

and, using (5) and (6), we obtain (4).

Finally, we prove assertion (2). Clearly, $(1_{\mathscr{M}} - \mathbf{s}(\varphi))_\varphi = 0$ and, by (3), this implies that $(k \cdot (1_{\mathscr{M}} - \mathbf{s}(\varphi))_\varphi = 0$, hence $(k \cdot (1_{\mathscr{M}} - \mathbf{s}(\varphi)))\mathbf{s}(\varphi) = 0$ for all $k \in \mathscr{A}_*$. Consequently, for $k \in \mathscr{A}_*$ and $f \in \mathscr{M}_*$, we have

$$\langle \delta(1_{\mathscr{M}} - \mathbf{s}(\varphi))(\mathbf{s}(\varphi) \bar{\otimes} 1_{\mathscr{A}}), f \bar{\otimes} k\rangle = \langle (k \cdot (1_{\mathscr{M}} - \mathbf{s}(\varphi)))\mathbf{s}(\varphi), f\rangle = 0,$$

that is, $\delta(1_{\mathscr{M}} - \mathbf{s}(\varphi))(\mathbf{s}(\varphi) \bar{\otimes} 1_{\mathscr{A}}) = 0$. Thus,

$$\delta(\mathbf{s}(\varphi)) \leq \mathbf{s}(\varphi) \bar{\otimes} 1_{\mathscr{A}}.$$

On the other hand, let $e = (\mathbf{s}(\varphi) \bar{\otimes} 1_{\mathscr{A}}) - \delta(\mathbf{s}(\varphi))$. Since $\langle \delta(1_{\mathscr{M}} - \mathbf{s}(\varphi)), \varphi \bar{\otimes} k\rangle = \langle (1_{\mathscr{M}} - \mathbf{s}(\varphi)) \bar{\otimes} 1_{\mathscr{A}}, \varphi \bar{\otimes} k\rangle = 0$ $(k \in \mathscr{A}_*^+)$, we have $\langle e, \varphi \bar{\otimes} k\rangle = 0$, so $e(\mathbf{s}(\varphi) \bar{\otimes} \mathbf{s}(k)) = 0$ for all $k \in \mathscr{A}_*^+$ and hence $e = e(\mathbf{s}(\varphi) \bar{\otimes} 1_{\mathscr{A}})e = 0$. □

From (3) it follows that the mappings $E^k_{\mathscr{M}} : x \mapsto k \cdot x$ define bounded linear operators $\pi^\delta_\varphi(k) \in \mathscr{B}(\mathscr{H}_\varphi)$,

$$\pi^\delta_\varphi(k)x_\varphi = (k \cdot x)_\varphi \quad (x \in \mathfrak{N}_\varphi, k \in \mathscr{A}_*) \tag{7}$$

$$\|\pi^\delta_\varphi(k)\| \leq \|k\| \quad (k \in \mathscr{A}_*) \tag{8}$$

and the mapping

$$\pi^\delta_\varphi : \mathscr{A}_* \to \mathscr{B}(\mathscr{H}_\varphi)$$

is a contractive representation of the Banach algebra $\mathscr{A}_*$.

We shall say that the weight φ is $(\delta, j_{\mathscr{A}})$-invariant, if φ is δ-invariant and

$$\langle (y^* \bar{\otimes} 1_{\mathscr{A}})\delta(x), \varphi \bar{\otimes} k\rangle = \langle (\delta(y^*)(x \bar{\otimes} 1_{\mathscr{A}}), \varphi \bar{\otimes} (k \cdot j_{\mathscr{A}})\rangle \tag{9}$$

for all $x, y \in \mathfrak{N}_\varphi$ and $k \in A_*^+$.

Using (4), we see that both sides of (9) are well defined and that (9) is equivalent to the equation:

$$((k \cdot x)_\varphi | y_\varphi)_\varphi = (x_\varphi | (k^0 \cdot y)_\varphi)_\varphi \quad (x, y \in \mathfrak{N}_\varphi, k \in \mathscr{A}_*), \tag{10}$$

that is,

$$\pi^\delta_\varphi(k)^* = \pi^\delta_\varphi(k^0) \quad (k \in \mathscr{A}_*).$$

Consequently, if φ is $(\delta, j_{\mathscr{A}})$*-invariant, then* π^δ_φ *is a contractive *-representation of the involutive Banach algebra* $\mathscr{A}_*$.

18.11. Consider again an action $\delta : \mathscr{M} \to \mathscr{M} \bar{\otimes} \mathscr{A}$ of the coinvolutive Hopf–von Neumann algebra $(\mathscr{A}, \delta_{\mathscr{A}}, j_{\mathscr{A}})$ on the W^*-algebra $\mathscr{M}$ and a normal semifinite weight φ on $\mathscr{M}$.

If there exists $\lambda > 0$, such that

$$x \in \mathfrak{N}_\varphi, k \in \mathscr{A}_* \to k \cdot x \in \mathfrak{N}_\varphi, \|(k \cdot x)_\varphi\|_\varphi \leq \lambda \|k\| \|x_\varphi\|_\varphi, \tag{1}$$

$$x, y \in \mathfrak{N}_\varphi, k \in \mathscr{A}_* \to ((k \cdot x)_\varphi | y_\varphi)_\varphi = (x_\varphi (k^0 \cdot y)_\varphi)_\varphi, \tag{2}$$

then we shall say that π_φ^δ *is a bounded *-representation of the involutive Banach algebra* $\mathscr{A}_*$.

In this case, we can indeed define by 18.10.(7) a bounded *-representation $\pi_\varphi^\delta : \mathscr{A}_* \to \mathscr{B}(\mathscr{H}_\varphi)$ with $\|\pi_\varphi^\delta\| \leq \lambda$.

Also, with the same arguments as in the proof of 18.10.(2), we can show that

$$\delta(\mathbf{s}(\varphi)\mathscr{M}\mathbf{s}(\varphi)) \subset \mathbf{s}(\varphi)\mathscr{M}\mathbf{s}(\varphi) \bar{\otimes} \mathscr{A}. \tag{3}$$

Lemma. *If* π_φ^δ *is a bounded *-representation of the involutive Banach algebra* $\mathscr{A}_*$, *then the action* δ *commutes with the modular automorphism group* $\{\sigma_t^\varphi\}_{t \in \mathbb{R}}$ *of* φ:

$$\delta \circ \sigma_t^\varphi = (\sigma_t^\varphi \bar{\otimes} \iota_{\mathscr{A}}) \circ \delta \quad (i \in \mathbb{R}).$$

Proof. Taking into account (3) we see that, without loss of generality but disregarding the assumption $\delta(1_{\mathscr{M}}) = 1_{\mathscr{M}} \bar{\otimes} 1_{\mathscr{A}}$, we may assume that φ is an n.s.f. weight on $\mathscr{M}$.

By assumption, it follows that for every $k \in \mathscr{A}_*$ and every $x_\varphi \in \mathfrak{A}_\varphi$ we have $S_\varphi \pi_\varphi^\delta(k) x_\varphi = S_\varphi (k \cdot x)_\varphi = ((k \cdot x)^*)_\varphi = (k^* \cdot x^*)_\varphi = \pi_\varphi^\delta(k^*) S_\varphi x_\varphi$. Since $S_\varphi = \overline{S_\varphi | \mathfrak{A}_\varphi}$, we obtain

$$\xi \in D(S_\varphi) \Rightarrow \pi_\varphi^\delta(k)\xi \in D(S_\varphi), \quad S_\varphi \pi_\varphi^\delta(k)\xi = \pi_\varphi^\delta(k^*) S_\varphi \xi.$$

Let $\eta \in D(\Delta_\varphi)$ and $\xi \in D(S_\varphi)$. Then $\eta \in D(S_\varphi), S_\varphi \eta \in D(S_\varphi^*)$ and $(S_\varphi \pi_\varphi^\delta(k)\eta | S_\varphi \xi)_\varphi = (\pi_\varphi^\delta(k^*) S_\varphi \eta | S_\varphi \xi)_\varphi = (S_\varphi \eta | \pi_\varphi^\delta(k^{*0}) S_\varphi \xi)_\varphi = (S_\varphi \eta | S_\varphi \pi_\varphi^\delta(k^0)\xi)_\varphi = (\pi_\varphi^\delta(k^0)\xi | S_\varphi^* S_\varphi \eta)_\varphi = (\xi | \pi_\varphi^\delta(k) \Delta_\varphi \eta)_\varphi$. Therefore, $S_\varphi \pi_\varphi^\delta(k)\eta \in D(S_\varphi^*)$ and $S_\varphi^* S_\varphi \pi_\varphi^\delta(k)\eta = \pi_\varphi^\delta(k)\Delta_\varphi \eta$. Thus,

$$\eta \in D(\Delta_\varphi) \Rightarrow \pi_\varphi^\delta(k)\eta \in D(\Delta_\varphi), \quad \Delta_\varphi \pi_\varphi^\delta(k) = \pi_\varphi^\delta(k)\Delta_\varphi \eta,$$

that is, $\pi_\varphi^\delta(k)$ commutes with Δ_φ.

It follows that for $x \in \mathfrak{N}_\varphi, k \in \mathscr{A}_*, t \in \mathbb{R}$, we have $(\sigma_t^\varphi(k \cdot x))_\varphi = \Delta_\varphi^{it}(k \cdot x)_\varphi = \Delta_\varphi^{it} \pi_\varphi^\delta(k) x_\varphi = \pi_\varphi^\delta(k) \Delta_\varphi^{it} x_\varphi = (k \cdot \sigma_t^\varphi(x))_\varphi$. Consequently, $\sigma_t^\varphi(k \cdot x) = k \cdot \sigma_t^\varphi(x)$ for all $x \in \mathscr{M}, k \in \mathscr{A}_*, t \in \mathbb{R}$ and, for every $f \in \mathscr{M}_*$, we deduce that

$$\begin{aligned} \langle \delta(\sigma_t^\varphi(x)), f \bar{\otimes} k \rangle &= \langle k \cdot \sigma_t^\varphi(x), f \rangle = \langle \sigma_t^\varphi(k \cdot x), f \rangle = \langle k \cdot x, f \circ \sigma_t^\varphi \rangle \\ &= \langle \delta(x), (f \circ \sigma_t^\varphi) \bar{\otimes} k \rangle = \langle (\sigma_t^\varphi \bar{\otimes} 1_{\mathscr{A}})(\delta(x)), f \bar{\otimes} k \rangle, \end{aligned}$$

thus proving the lemma. □

18.12 Theorem. *Let $\delta : \mathscr{M} \to \mathscr{M} \bar{\otimes} \mathscr{A}$ be an action of the convolutive Hopf–von Neumann algebra $(\mathscr{A}, \delta_{\mathscr{A}}, j_{\mathscr{A}})$ on the W^*-algebra $\mathscr{M}$ and φ a normal semifinite weight on $\mathscr{M}$. The following statements are equivalent:*

(i) *φ is $(\delta, j_{\mathscr{A}})$-invariant.*
(ii) *π^{δ}_{φ} is a bounded *-representation of the involutive Banach algebra $\mathscr{A}_*$.*
(iii) *π^{δ}_{φ} is a contractive *-representation of the involutive Banach algebra $\mathscr{A}_*$.*
(iv) *The following conditions are satisfied:*
   (a) *$\mathbf{s}(\varphi) \in \mathscr{M}^{\delta}$;*
   (b) *$\delta(\sigma^{\varphi}_t(x)) = (\sigma^{\varphi}_t \bar{\otimes} \iota_{\mathscr{A}})(\delta(x))$ for all $x \in \mathbf{s}(\varphi)\mathscr{M}\mathbf{s}(\varphi), t \in \mathbb{R}$;*
   (c) *there exists a σ^{φ}-invariant *-subalgebra $\mathscr{B}$ of $\mathfrak{M}_{\varphi}$, w-dense in $\mathbf{s}(\varphi)\mathscr{M}\mathbf{s}(\varphi)$, such that*

$$\langle \delta(x), \varphi \bar{\otimes} k \rangle = \langle x \bar{\otimes} 1_{\mathscr{A}}, \varphi \bar{\otimes} k \rangle \quad (x \in \mathscr{B} \cap \mathscr{M}^+, k \in \mathscr{A}^+_*);$$

   (d) *there exist a $\|\cdot\|_{\varphi}$-dense subset $\mathscr{D}$ of $\mathfrak{N}_{\varphi}$ and a norm-dense subset $\mathscr{F}$ of $\mathscr{A}^+_*$ such that for every $x, y \in \mathscr{D}$ and every $k \in \mathscr{F}$ we have $\delta(x), \delta(y) \in \mathfrak{N}_{\varphi \bar{\otimes} k}$ and*

$$\langle (y^* \bar{\otimes} 1_{\mathscr{A}})\delta(x), \varphi \bar{\otimes} k \rangle = \langle \delta(y^*)(x \bar{\otimes} 1_{\mathscr{A}}), \varphi \bar{\otimes} k \rangle.$$

Proof. (i) $\Rightarrow$ (iii). By Section 18.10.

(iii) $\Rightarrow$ (ii). Obvious.

(ii) $\Rightarrow$ (i). We have to show just that φ is δ-invariant, that is,

$$(\varphi \bar{\otimes} k) \circ \delta = \langle 1_{\mathscr{A}}, k \rangle \varphi \quad (k \in \mathscr{A}^+_*), \tag{1}$$

since then condition 18.10.(9) will follow from 18.11.(2) using 18.10.(4).

Let $k \in \mathscr{A}^+_*$. From (ii) it follows that $(k \cdot (1_{\mathscr{M}} - \mathbf{s}(\varphi)))\mathbf{s}(\varphi) = 0$ and then, using 9.8.(5), we deduce that $\langle \delta(1_{\mathscr{M}} - \mathbf{s}(\varphi)), \varphi \bar{\otimes} k \rangle = \langle k \cdot (1_{\mathscr{M}} - \mathbf{s}(\varphi)), \varphi \rangle = 0$, that is, $\mathbf{s}((\varphi \bar{\otimes} k) \circ \delta) \leq \mathbf{s}(\varphi)$, and therefore, $\mathbf{s}((\varphi \bar{\otimes} k) \circ \delta) = \mathbf{s}((\varphi \bar{\otimes} k) \circ \delta | \mathbf{s}(\varphi)\mathscr{M}\mathbf{s}(\varphi))$. On the other hand, it is clear that $\mathbf{s}(\langle 1_{\mathscr{A}}, k \rangle \varphi) = \mathbf{s}(\langle 1_{\mathscr{A}}, k \rangle \varphi | \mathbf{s}(\varphi)\mathscr{M}\mathbf{s}(\varphi))$. Thus, we may assume that φ is an n.s.f. weight on $\mathscr{M}$.

Then $\mathscr{B} = \mathfrak{T}^2_{\varphi}$ is a w-dense σ^{φ}-invariant *-subalgebra of $\mathscr{M}$ and $\mathscr{B} = lin(\mathscr{B} \cap \mathscr{M}^+) \subset \mathfrak{M}_{\varphi}$.

Let $x \in \mathscr{B} \cap \mathscr{M}^+$ and $k \in \mathscr{A}^+_*$. Then $k \cdot x \in \mathfrak{M}_{\varphi} \cap \mathscr{M}^+$. Indeed, for every $y \in \mathfrak{N}_{\bar{\varphi}}$, we have (18.11.(2)) $\langle y^*(k \cdot x), \varphi \rangle = \langle (k^0 \cdot y)^* x, \varphi \rangle$, so that the assertion follows using 2.13.(3). Consequently, there exist $a, b \in \mathfrak{N}_{\varphi} \cap \mathscr{M}^+$ such that

$$x = a^2, \quad k \cdot x = b^2.$$

By Proposition 2.16, there exists a net $\{v_j\} \in \mathfrak{T}_{\varphi}$ such that

$$\sigma^{\varphi}_{\alpha}(v_j) \xrightarrow{s^*} 1_{\mathscr{M}} \text{ for all } \alpha \in \mathbb{C}.$$

Using Lemma 18.11, we obtain (here $y \in \mathcal{M}$ is identified with $\pi_\varphi(y) \in \mathcal{B}(\mathcal{H}_\varphi)$):

$$\begin{aligned}
\langle x, (\varphi \bar{\otimes} k) \circ \delta \rangle &= \langle \delta(x), \varphi \bar{\otimes} k \rangle = \langle k \cdot x, \varphi \rangle = \langle b^2, \varphi \rangle = (b_\varphi | b_\varphi)_\varphi \\
&= (b_\varphi | S_\varphi b_\varphi)_\varphi = (b_\varphi | J_\varphi \Delta_\varphi^{1/2} b_\varphi)_\varphi = \lim_j (b_\varphi | J_\varphi \sigma^\varphi_{-i/2}(v_j^*) \Delta_\varphi^{1/2} b_\varphi)_\varphi \\
&= \lim_j (b_\varphi | J_\varphi \Delta_\varphi^{1/2} v_j^* b_\varphi)_\varphi = \lim_j (b_\varphi | S_\varphi (v_j^* b_\varphi)_\varphi)_\varphi = \lim_j (b_\varphi | (b v_j)_\varphi)_\varphi \\
&= \lim_j ((k \cdot x)_\varphi | (v_j)_\varphi)_\varphi = \lim_j (x_\varphi | (k^0 \cdot v_j)_\varphi)_\varphi = \lim_j (a_\varphi | (a(k^0 \cdot v_j))_\varphi)_\varphi \\
&= \lim_j (a_\varphi | S_\varphi ((k^0 \cdot v_j) a)_\varphi)_\varphi = \lim_j (a_\varphi | J_\varphi \Delta_\varphi^{1/2} ((k^0 \cdot v_j) a)_\varphi)_\varphi \\
&= \lim_j (a_\varphi | J_\varphi \sigma^\varphi_{-i/2}((k^0 \cdot v_j) \Delta_\varphi^{1/2} a_\varphi)_\varphi = \lim_j (a_\varphi | J_\varphi (k^0 \cdot \sigma^\varphi_{-i/2}(v_j)) \Delta_\varphi^{1/2} a_\varphi)_\varphi \\
&= (a_\varphi | J_\varphi (k^0 \cdot 1_{\mathcal{M}}) \Delta_\varphi^{1/2} a_\varphi)_\varphi = (a_\varphi | S_\varphi a_\varphi)_\varphi \langle 1_{\mathcal{A}}, k \rangle = (a_\varphi | a_\varphi)_\varphi \langle 1_{\mathcal{A}}, k \rangle \\
&= \langle a^2, \varphi \rangle \langle 1_{\mathcal{A}}, k \rangle = \langle x, \langle 1_{\mathcal{A}}, k \rangle \varphi \rangle.
\end{aligned}$$

Consequently, the two weights appearing in (1) are equal on $\mathcal{B}$. In particular, $(\varphi \bar{\otimes} k) \circ \delta$ is semifinite.

Again using Lemma 18.11, we see that $(\varphi \bar{\otimes} k) \circ \delta$ commutes with $\langle 1_{\mathcal{A}}, k \rangle_\varphi$, hence (1) follows from Theorem 6.2.

If the equivalent conditions (i), (ii), (iii) are satisfied, then statement (iv) results as follows: (a) by 18.10.(2), (b) by Lemma 18.11, (c) is clear with $\mathcal{B} = \mathfrak{M}_\varphi$, and (d) is clear with $\mathcal{D} = \mathfrak{N}_\varphi$ and $\mathcal{F} = \mathcal{A}_*^+$.

(iv) $\Rightarrow$ (iii). From condition (a) it follows that we may assume that φ is an n.s.f. weight. Using conditions (b) and (c) and Theorem 6.2, we obtain (1), hence φ is δ-invariant. Then π_φ^δ is defined and contractive and, using condition (d), we obtain $\pi_\varphi^\delta(k)^* = \pi_\varphi^\delta(k^0)$, first for $k \in \mathcal{F}$ and then, by passing to the limit, for every $k \in \mathcal{A}_*^+$. □

Note that in Sections 18.10–18.12, we have not used all the conditions, which define a coinvolutive Hopf–von Neumann algebra and its action on a W^*-algebra, but only the following: *$\mathcal{M}$ and $\mathcal{A}$ are W^*-algebras, $\delta : \mathcal{M} \to \mathcal{M} \bar{\otimes} \mathcal{A}$ is a unital normal *-homomorphism and $j_{\mathcal{A}} : \mathcal{A} \to \mathcal{A}$ is a *-antiautomorphism with $j_{\mathcal{A}} \circ j_{\mathcal{A}} = \iota_{\mathcal{A}}$.* The comultiplication $\delta_{\mathcal{A}}$ of $\mathcal{A}$ appeared only in considering the multiplicative structure on A_*, but this structure has not been used.

18.13 Corollary. *Let $\delta : \mathcal{M} \to \mathcal{M} \bar{\otimes} \mathcal{A}$ be an action of the coinvolutive Hopf–von Neumann algebra $(\mathcal{A}, \delta_{\mathcal{A}}, j_{\mathcal{A}})$ on the W^*-algebra $\mathcal{M}$ and let φ, ψ be $(\delta, j_{\mathcal{A}})$-invariant n.s.f. weights on $\mathcal{H}$. Then*

$$[D\psi : D\varphi]_t \in \mathcal{M}^\delta \quad (t \in \mathbb{R}). \tag{1}$$

Proof. The mapping $\delta_2 = \iota_2 \bar{\otimes} \delta$ is an action of $\mathcal{A}$ on the W^*-algebra $Mat_2(\mathcal{M}) = Mat_2(\mathbb{C}) \bar{\otimes} \mathcal{M}$. Since φ and ψ are $(\delta, j_{\mathcal{A}})$-invariant, the balanced weight $\theta = \theta(\varphi, \psi)$ is $(\delta_2, j_{\mathcal{A}})$-iovariant. By Theorem 18.12, it follows that δ_2 commutes with σ_t^θ $(t \in \mathbb{R})$.

Let $u_t = [D\psi : D\varphi]_t$ $(t \in \mathbb{R})$. Then

$$\begin{aligned}\begin{pmatrix} 0 & 0 \\ \delta(u_t) & 0 \end{pmatrix} &= \delta_2 \begin{pmatrix} 0 & 0 \\ u_t & 0 \end{pmatrix} = \delta_2\left(\sigma_t^\theta\left(\begin{pmatrix} 0 & 0 \\ 1_{\mathscr{M}} & 0 \end{pmatrix}\right)\right) \\ &= (\sigma_t^\theta \bar{\otimes} \iota_{\mathscr{A}})\left(\delta_2\left(\begin{pmatrix} 0 & 0 \\ 1_{\mathscr{M}} & 0 \end{pmatrix}\right)\right) = (\sigma_t^\theta \bar{\otimes} \iota_{\mathscr{A}})\left(\begin{pmatrix} 0 & 0 \\ 1_{\mathscr{M}} & 0 \end{pmatrix} \bar{\otimes} 1_{\mathscr{A}}\right) \\ &= \begin{pmatrix} 0 & 0 \\ u_t \bar{\otimes} 1_{\mathscr{M}} & 0 \end{pmatrix}\end{aligned}$$

hence $\delta(u_t) = u_t \bar{\otimes} 1_{\mathscr{A}}$, that is, $u_t \in \mathscr{M}^\delta$. □

18.14 Proposition. *Let $(\mathscr{A}, \delta, j)$ be a coinvolutive Hopf–von Neumann algebra. If there exists a nonzero δ-invariant normal semifinite weight on $\mathscr{A}$, then this weight is faithful, $\mathscr{A}^\delta = \mathbb{C} \cdot 1_{\mathscr{A}}$ and any two nonzero (δ, j)-invariant normal semifinite weights ω, τ on $\mathscr{A}$ are proportional: $\tau = \lambda\omega (\lambda > 0)$.*

Proof. Let ω be a nonzero δ-invariant normal semifinite weight on $\mathscr{A}$. Let $e \in \mathscr{A}^\delta$ be a projection. Then (18.2.(3)) $\delta(j(e)) = \widetilde{\ }(j \bar{\otimes} j)(\delta(e)) = 1_{\mathscr{A}} \bar{\otimes} j(e)$, hence for every $x \in \mathfrak{M}_\omega \cap \mathscr{A}^+$ and $k \in \mathscr{A}_*^+$ we obtain

$$\begin{aligned}\langle j(e)xj(e), \omega\rangle\langle 1_{\mathscr{A}}, k\rangle &= \langle \delta(j(e)xj(e)), \omega \bar{\otimes} k\rangle \\ &= \langle (1_{\mathscr{A}} \bar{\otimes} j(e))\delta(x)(1_{\mathscr{A}} \bar{\otimes} j(e)), \omega \bar{\otimes} k\rangle \\ &= \langle (\delta(x), \omega \bar{\otimes} k(j(e) \cdot j(e))\rangle \\ &= \langle x, \omega\rangle\langle j(e), k\rangle. \qquad (1)\end{aligned}$$

Since $\omega \neq 0$, we have $\mathbf{s}(\omega) \neq 0$, so that there exists $k \in \mathscr{A}_*^+, k \neq 0$, with $\langle j(1_{\mathscr{A}} - \mathbf{s}(\omega)), k\rangle = 0$. By 18.10.(2), we have $\mathbf{s}(\omega) \in \mathscr{A}^\delta$ so that, replacing e by $1_{\mathscr{A}} - \mathbf{s}(\omega)$ in (1), we get

$$\langle j(1_{\mathscr{A}} - \mathbf{s}(\omega))xj(1_{\mathscr{M}} - \mathbf{s}(\omega)), \omega\rangle = 0 \quad (x \in \mathfrak{M}_\omega \cap \mathscr{A}^+).$$

It follows that $j(1_{\mathscr{A}} - \mathbf{s}(\omega))\mathbf{s}(\omega) = 0$, that is,

$$\mathbf{s}(\omega) = j(\mathbf{s}(\omega))\mathbf{s}(\omega). \qquad (2)$$

Assume now that $\mathbf{s}(\omega) \neq 1_{\mathscr{A}}$. Then there exists $k \in \mathscr{A}_*^+, k \neq 0$, with $\langle j(\mathbf{s}(\omega)), k\rangle = 0$ and, again using (1), we obtain

$$\langle j(\mathbf{s}(\omega))xj(\mathbf{s}(\omega)), \omega\rangle = 0 \quad (x \in \mathfrak{M}_\omega \cap \mathscr{A}^+),$$

that is, $j(\mathbf{s}(\omega))\mathbf{s}(\omega) = 0$. This, together with (2), implies that $\omega = 0$, a contradiction. Hence, ω is faithful.

If $e \in \mathscr{A}^\delta, e \neq 1_{\mathscr{A}}$, there exists $k \in \mathscr{A}_*^+, k \neq 0$, with $\langle j(e), k\rangle = 0$ and using (1) we obtain $\langle j(e)xj(e), \omega\rangle = 0$ for all $x \in \mathfrak{M}_\omega \cap \mathscr{A}^+$, that is, $j(e) = j(e)\mathbf{s}(\omega) = 0$, hence $e = 0$. Consequently, $\mathscr{A}^\delta = \mathbb{C} \cdot 1_{\mathscr{A}}$.

Consider now two nonzero (δ, j)-invariant normal semifinite weights ω and τ on $\mathscr{A}$. Then ω and τ are faithful and, by Corollary 18.13, we have $[D\tau : D\omega]_t \in \mathscr{A}^\delta = \mathbb{C} \cdot 1_{\mathscr{A}}$ $(t \in \mathbb{R})$. It follows that there exists $\lambda > 0$ with $[D\tau : D\omega]_t = \lambda^{it}$ $(t \in \mathbb{R})$, and hence $\tau = \lambda\omega$. □

18.15. A *Kac algebra* is a quadruple $(\mathscr{A},\delta,j,\omega)$, where $(\mathscr{A},\delta,j)$ is a coinvolutive Hopf–von Neumann algebra and ω is a (δ,j)-invariant n.s.f. weight on $\mathscr{A}$ such that $\sigma_t^{\omega\circ j} = \sigma_t^{\omega}$ $(t \in \mathbb{R})$, that is,

$$\sigma_t^{\omega} \circ j = j \circ \sigma_{-t}^{\omega} \quad (t \in \mathbb{R}). \tag{1}$$

By Proposition 18.14, the invariant weight ω appearing in the definition of a Kac algebra is unique up to a positive multiplicative constant. The weight ω will be called the *left Haar weight* on the Kac algebra $\mathscr{A}$.

There exists a canonical way to associate with every Kac algebra $(\mathscr{A},\delta,j,\omega)$ a *dual Kac algebra* $(\hat{\mathscr{A}},\hat{\delta},\hat{j},\hat{\omega})$, a procedure that we now describe without proof.

The von Neumann algebra $\hat{\mathscr{A}}$ is defined by

$$\hat{\mathscr{A}} = \mathscr{R}\{\pi_{\omega}^{\delta}(\mathscr{A}_*)\} \subset \mathscr{B}(\mathscr{H}_{\omega}).$$

Using the δ-invariance of ω, one shows that the mapping $x \bar{\otimes} y \mapsto \delta(x)(1 \bar{\otimes} y)$ defines an isometry on $\mathscr{H}_{\omega} \bar{\otimes} \mathscr{H}_{\omega}$ and, moreover, the adjoint W of this isometry is again an isometry, that is, $W \in \mathscr{B}(\mathscr{H}_{\omega} \bar{\otimes} \mathscr{H}_{\omega})$ is a unitary operator. We have $W \in \hat{\mathscr{A}} \bar{\otimes} \mathscr{A}$ and

$$\delta(x) = W^*(x \bar{\otimes} 1)W \quad (x \in \mathscr{A}).$$

One then considers the unitary operator $\hat{W} = \sim \circ W^* \circ \sim \in \mathscr{A} \bar{\otimes} \hat{\mathscr{A}}$ and one defines by

$$\hat{\delta}(x) = \hat{W}^*(x \bar{\otimes} 1)\hat{W} \quad (x \in \hat{\mathscr{A}})$$

a coassociative comultiplication $\hat{\delta}$ on $\hat{\mathscr{A}}$.

On the other hand, the equation

$$\hat{j}(x) = J_{\omega}x^*J_{\omega} \quad (x \in \hat{\mathscr{A}})$$

defines a coinvolution $\hat{j}$ on $\hat{\mathscr{A}}$, which is compatible with $\hat{\delta}$.

In order to define the weight $\hat{\omega}$, one constructs a $*$-subalgebra $\mathfrak{D}$ of $\mathscr{A}$, which is contained in $\mathfrak{T}_{\omega}^2$ and which, regarded as a subset of $\mathscr{A}_*$ via Proposition 2.13, is an involutive subalgebra of the involutive Banach algebra $\mathscr{A}_*$; moreover, $\overline{S_{\omega}|\mathfrak{D}} = S_{\omega}$. With the involutive algebra structure inherited from $\mathscr{A}_*$ and the scalar product of $\mathscr{H}_{\omega}$. $\mathfrak{D}$ becomes a left Hilbert algebra whose associated von Neumann algebra is equal to $\hat{\mathscr{A}}$. The natural weight associated with the left Hilbert algebra $\mathfrak{D}$ is a $(\hat{\delta},\hat{j})$-invariant n.s.f. weight $\hat{\omega}$ on $\hat{\mathscr{A}}$.

Of course, $\mathfrak{D}$, with the $*$-algebra structure inherited from $\mathscr{A}$ and the scalar product of $\mathscr{H}_{\omega}$, is also a left Hilbert algebra, $\mathscr{A}$ is the associated von Neumann algebra and ω is the corresponding natural weight.

This structure of a "Tomita bi-algebra" on $\mathfrak{D}$ is at the heart of the duality theory for Kac algebras, the fundamental result being that $(\hat{\hat{\mathscr{A}}},\hat{\hat{\delta}},\hat{\hat{j}},\hat{\hat{\omega}})$ *is isomorphic to* $(\mathscr{A},\delta,j,\omega)$.

We now return to the examples considered above.

18.16. Let $\sigma : G \to Aut(\mathscr{M})$ be a continuous action of the locally compact group G on the W^*-algebra $\mathscr{M}$ and $\pi_{\sigma} : \mathscr{M} \to \mathscr{M} \bar{\otimes} \mathscr{L}^{\infty}(G)$ the corresponding action of $\mathbf{G} = (\mathscr{L}^{\infty}(G), \pi_G, k_G)$

on $\mathscr{M}$. For any normal semifinite weight φ on $\mathscr{M}$ it is easy to check that

$$\begin{aligned}\varphi \text{ is } \pi_\sigma\text{-invariant} &\Rightarrow \varphi \text{ is } \sigma\text{-invariant}\\ &\Rightarrow \varphi \text{ is } (\pi_\sigma, k_G)\text{-invariant}\\ &\Rightarrow \varphi \text{ is } \pi_\sigma\text{-invariant}\end{aligned} \tag{1}$$

hence all meaningful notions of invariance coincide.

In particular, since the left Haar measure dg is invariant under left translation and $\pi_G = \pi_{\mathrm{Ad}(\boldsymbol{\lambda})}$, it follows that the left Haar weight μ_G on $\mathscr{L}^\infty(G)$ is (π_G, k_G)-invariant and $\mathbf{G} = (\mathscr{L}^\infty(G), \pi_G, k_G, \mu_G)$ *is a Kac algebra.*

With the notation introduced in 18.10.(7), 13.2, 18.4.(9), it is easy to check that

$$\pi_{\mu_G}^{\pi_G}(k) = \overline{\lambda(k^*)} \in \mathfrak{L}(G) \subset \mathscr{B}(\mathscr{L}^2(G)) \quad (k \in \mathscr{L}^1(G) = \mathscr{L}^\infty(G)_*). \tag{2}$$

18.17. Consider now the coinvolutive Hopf–von Neumann algebra $\hat{\mathbf{G}} = (\mathfrak{L}(G), \delta_G, j_G)$ associated with the locally compact group G and the Plancherel weight ω_G on $\mathfrak{L}(G)$.

Recall (18.4) that ω_G is the natural weight associated with the left Hilbert algebra $\mathfrak{A} = \mathscr{K}(G) \subset \mathscr{L}^2(G)$ with the operations of convolution and involution.

We shall use the notation $\mathfrak{A}', \mathfrak{A}'', L_\xi, R_\eta$, and so on, associated with the left Hilbert algebra $\mathfrak{A} = \mathscr{K}(G)$ (see, 2.12 and [L], 10.1–10.4). Note that $\mathfrak{A}'' = \mathfrak{A}_{\omega_G}$.

By definition, for $\xi \in \mathfrak{A} = \mathscr{K}(G)$ we have

$$L_\xi\eta = \xi * \eta = \boldsymbol{\lambda}(\xi)\eta \quad (\eta \in \mathscr{L}^2(G)). \tag{1}$$

Consequently, for $\eta \in \mathscr{L}^2(G)$, the operator R_η^0 is defined by $R_\eta^0\xi = L_\xi\eta = \xi * \eta$ $(\xi \in \mathfrak{A} = \mathscr{K}(G))$. If R_η is bounded, it follows that

$$R_\eta\xi = \xi * \eta \quad (\xi \in \mathscr{L}^2(G)). \tag{2}$$

Indeed, if $\xi_n \in \mathscr{K}(G)$ and $\xi_n \to \xi$ in $\mathscr{L}^2(G)$, then $R_\eta\xi_n \to R_\eta\xi$ in $\mathscr{L}^2(G)$ and $\xi_n * \eta \to \xi * \eta$ uniformly. In particular, (2) holds whenever $\eta \in \mathfrak{A}'$. Consequently, for $\xi \in \mathscr{L}^2(G)$, the operator L_ξ^0 is defined by $L_\xi^0\eta = R_\eta\xi = \xi * \eta$ $(\eta \in \mathfrak{A}')$. If L_ξ is bounded, we obtain similarly

$$L_\xi\eta = \xi * \eta \quad (\eta \in \mathscr{L}^2(G)). \tag{3}$$

It follows that

$$\xi \in \mathscr{L}^1(G) \cap \mathscr{L}^2(G) \Rightarrow L_\xi \textit{ is bounded and } L_\xi = \boldsymbol{\lambda}(\xi). \tag{4}$$

Using 2.12.(2), we see that for every $\xi, \eta \in \mathscr{L}^1(G) \cap \mathscr{L}^2(G)$ we have $\lambda(\xi), \lambda(\eta) \in \mathfrak{N}_{\omega_G}$ and $\omega(\boldsymbol{\lambda}(\eta)^*\boldsymbol{\lambda}(\xi)) = (\xi|\eta)$, that is,

$$\omega_G(\boldsymbol{\lambda}(\eta^\sharp * \xi)) = \int \xi(g)\overline{\eta(g)}\,\mathrm{d}g = (\eta^\sharp * \xi)(e) \quad (\xi, \eta \in \mathscr{L}^1(G) \cap \mathscr{L}^2(G)). \tag{5}$$

Recall that the convolution of two $\mathscr{L}^2$-functions is equal almost everywhere to a continuous function.

Moreover, for every function $f \in \mathscr{L}^1(G)$, we have $\lambda(f^\sharp * f) \in \mathfrak{L}(G)^+$ and

$$\omega_G(\lambda(f^\sharp * f)) = \int |f(g)|^2 \,\mathrm{d}g = "(f^\sharp * f)(e)" \quad (f \in \mathscr{L}^1(G)) \tag{6}$$

where the last equation is just formal. Indeed, if $f \in \mathscr{L}^1(G) \cap \mathscr{L}^2(G)$, then (6) follows from (5). If $\omega_G(\boldsymbol{\lambda}(f^\sharp * f)) < +\infty$, then $\boldsymbol{\lambda}(f) \in \mathfrak{N}_{\omega_G}$ and so there exists $\xi \in \mathscr{L}^2(G)$ with $L_\xi = \boldsymbol{\lambda}(f)$, that is, $\xi * \eta = f * \eta$ for all $\eta \in \mathscr{L}^2(G)$, which means that $f = \xi \in \mathscr{L}^2(G)$.

In computations involving ω_G, we shall often use the following:

Proposition. *For $x \in \mathfrak{L}(G)^+$, we have $\omega_G(x) < +\infty$ if and only if there exists a continuous function $f \in \mathscr{L}^2(G)$ with $x = L_f$; in this case,*

$$\omega_G(L_f) = f(e). \tag{7}$$

In particular, if $f \in \mathscr{L}^1(G)$ and $\boldsymbol{\lambda}(f) \geq 0$, then $\omega_G(\boldsymbol{\lambda}(f)) < +\infty$ if and only if $f \in \mathscr{L}^1(G) \cap \mathscr{L}^2(G)$; in this case f is (equal almost everywhere to) a continuous function and

$$\omega_G(\boldsymbol{\lambda}(f)) = f(e). \tag{8}$$

Proof. Let $f \in \mathscr{L}^2(G)$ be a continuous function such that the operator L_f is bounded and positive. Then $\varphi = J_G f \in \mathscr{L}^2(G)$ is also a continuous function and the operator $R_\varphi = J_G L_f J_G$ (see 2.12.(4)) is bounded and positive:

$$0 \leq R_\varphi \leq \lambda.$$

Since, for every $\zeta \in \mathscr{K}(G)$ we have $(R_\varphi \zeta | \zeta) = (\zeta * \varphi | \zeta) = (\varphi | \zeta^\sharp * \zeta) = \int \varphi(g)\overline{(\zeta^\sharp * \zeta)(g)}\,\mathrm{d}g$, it follows that the continuous function φ is positive definite and

$$\int \varphi(g)(\zeta^\sharp * \zeta)(g)\,\mathrm{d}g \leq \lambda \|\zeta\|_2^2 \quad (\zeta \in \mathscr{K}(G)).$$

Let $u_\varphi : G \to \mathscr{B}(\mathscr{H}_\varphi)$ be the *so*-continuous cyclic unitary representation of G, with cyclic vector $\xi_\varphi \in \mathscr{H}_\varphi$, associated with φ (see 13.4.5). The corresponding $*$-representation of the involutive Banach algebra $\mathscr{L}^1(G)$ is just the *GNS*-representation defined by the positive form $\varphi \in \mathscr{L}^\infty(G) = \mathscr{L}^1(G)^*$ and will also be denoted by u_φ. For $\zeta \in \mathscr{K}(G)$, we have

$$\|u_\varphi(\zeta)\xi_\varphi\|^2 = \langle \zeta^\sharp * \zeta, \varphi \rangle = \int \varphi(g)(\zeta^\sharp * \zeta)(g)\,\mathrm{d}g \leq \lambda \|\zeta\|_2^2.$$

Thus, there exists a bounded linear operator $T : \mathscr{L}^2(G) \to \mathscr{H}_\varphi$, $\|T\| \leq \lambda$, such that $T\zeta = u_\varphi(\zeta)\xi_\varphi$ $(\zeta \in \mathscr{K}(G))$. The range of T is dense in $\mathscr{H}_\varphi$ since $\xi_\varphi \in \mathscr{K}_\varphi$ is a cyclic vector and

$$T\boldsymbol{\lambda}(g) = u_\varphi(g)T \quad (g \in G).$$

If $T = V|T|$ is the polar decomposition of T, then $V : \mathscr{L}^2(G) \to \mathscr{H}_\varphi$ is a coisometry, that is, $VV^* = 1$, and

$$V\boldsymbol{\lambda}(g) = u_\varphi(g)V \quad (g \in G).$$

Let $\eta = \overline{V^*\xi_\varphi} \in \mathscr{L}^2(G)$. For every $\zeta \in \mathscr{K}(G)$, we have

$$\overline{\overline{\zeta} * \eta} = \zeta * \bar{\eta} = \boldsymbol{\lambda}(\zeta)V^*\xi_\varphi = V^*u_\varphi(\zeta)\xi_\varphi = V^*T\zeta = |T|\zeta,$$

hence the operator R_η is bounded and positive. On the other hand,

$$\varphi(g) = (u_\varphi(g)\xi_\varphi|\xi_\varphi) = (\boldsymbol{\lambda}(g)\bar{\eta}|\bar{\eta}) = \int \bar{\eta}(g^{-1}t)\eta(t)\,\mathrm{d}t \quad (g \in G)$$

that is, $\varphi = \eta * \eta^\flat$, where $\eta^\flat(s) = \overline{\eta(s^{-1})}(s \in G)$.

It follows that $\xi = J_G\eta \in \mathscr{L}^2(G)$ has the property that the operator $L_\xi = J_G R_\eta J_G$ is bounded and positive and $f = J_G\varphi = J_G(\eta * \eta^\flat) = \xi^\sharp * \xi$. Consequently, $\omega_G(L_f) = \omega_G(L_\xi^* L_\xi) = \|\xi\|_2^2 = (\xi^\sharp * \xi)(e) = f(e) < +\infty$.

Consider now $x \in \mathfrak{L}(G)^+$ with $\omega_G(x) < +\infty$. By the definition of the natural weight ω_G, it follows that there exists $\xi \in \mathscr{L}^2(G)$ such that $x^{1/2} = L_\xi$ and $\omega_G(x) = \|\xi\|^2$. Then $f = \xi^\sharp * \xi \in \mathscr{L}^2(G)$ is a continuous function with $x = L_\xi^* L_\xi = L_f$ and $\omega_G(x) = f(e)$. In particular, if $x = \boldsymbol{\lambda}(h)$ with $h \in \mathscr{L}^1(G)$, then $\lambda(h) = L_f$, hence $h = f$ almost everywhere.

Thus, the proposition is completely proved. □

We have

$$k \cdot \boldsymbol{\lambda}(f) = \boldsymbol{\lambda}(k(\cdot)f) \quad (f \in \mathscr{L}^1(G), k \in \mathscr{A}(G)). \tag{9}$$

Indeed, for every $h \in \mathscr{A}(G)$, we get $\langle k \cdot \boldsymbol{\lambda}(f), h\rangle = \langle \delta_G(\boldsymbol{\lambda}(f)), h \bar{\otimes} k\rangle = \langle \int f(g)\boldsymbol{\lambda}(g) \bar{\otimes} \boldsymbol{\lambda}(g)\,\mathrm{d}g, h \bar{\otimes} k\rangle = \langle \int f(g)\langle \boldsymbol{\lambda}(g), k\rangle \boldsymbol{\lambda}(g)\,\mathrm{d}g, h\rangle = \langle \boldsymbol{\lambda}(k(\cdot)f), h\rangle$.

From (9) it follows that

$$\pi_{\omega_G}^{\delta_G}(k) = k(\cdot) \in \mathscr{L}^\infty(G) \subset \mathscr{B}(\mathscr{L}^2(G)) \quad (k \in \mathscr{A}(G) = \mathfrak{L}(G)_*). \tag{10}$$

Indeed, for every $\xi \in \mathscr{K}(G)$, we have $\pi_{\omega_G}^{\delta_G}(k) = \pi_{\omega_G}^{\delta_G}(k)(\boldsymbol{\lambda}(\xi))_{\omega_G} = (k \cdot \boldsymbol{\lambda}(\xi))_{\omega_G} = (\boldsymbol{\lambda}(k(\cdot)\xi))_{\omega_G} = k(\cdot)\xi$.

We are now able to show that

$$\textit{the n.s.f. weight } \omega_G \textit{ on } \mathfrak{L}(G) \textit{ is } (\delta_G, j_G)\textit{-invariant} \tag{11}$$

by checking conditions (iv), (a)–(d), of Theorem 18.12.

(a) $\mathbf{s}(\omega_G) = 1_G \in \mathfrak{L}(G)^{\delta_G}$.

(b) The modular operator associated with ω_G is the multiplication operator defined by the modular function Δ_G; hence it is affiliated to $\mathscr{L}^\infty(G)$, which is the centralizer $\mathscr{B}(\mathscr{L}^2(G))^{\delta_G}$ of the action δ_G (18.7.(9)). Consequently, $\sigma_t^{\omega_G} = \Delta_G^{\mathrm{i}t} \cdot \Delta_G^{-\mathrm{i}t}$ commutes with δ_G.

(c) For $\xi \in \mathscr{K}(G)$ with $\boldsymbol{\lambda}(\xi) \geq 0$ and $k \in \mathscr{A}(G)^+$, we have $\boldsymbol{\lambda}(k(\cdot)\xi) = k \cdot \boldsymbol{\lambda}(\xi) \geq 0$ and, using (8), we get $\langle \delta_G(\boldsymbol{\lambda}(\xi)), \omega_G \bar{\otimes} k\rangle = \langle \boldsymbol{\lambda}(k(\cdot)\xi), \omega_G\rangle = k(e)\xi(e) = \langle \boldsymbol{\lambda}(\xi) \bar{\otimes} 1_G, \omega_G \bar{\otimes} k\rangle$.

(d) This condition amounts to showing that the $\pi_{\omega_G}^{\delta_G}$ is a *-representation, and this follows obviously from (10) as $k^0(\cdot) = \overline{k(\cdot)}$.

We also have

$$\omega_G \circ j_G = \omega_G. \tag{12}$$

Indeed, it is easy to see that $\sigma_t^{\omega_G}(\boldsymbol{\lambda}(g)) = \Delta_G^{it}\boldsymbol{\lambda}(g)\Delta_G^{-it} = \Delta_G(g)^{it}\boldsymbol{\lambda}(g)(g \in G, t \in \mathbb{R})$, so $j_G(\sigma_t^{\omega_G}(\boldsymbol{\lambda}(g))) = \Delta_G(g)^{it} j_G(\boldsymbol{\lambda}(g)) = \Delta_G(g^{-1})^{-it}\boldsymbol{\lambda}(g^{-1}) = \sigma_{-t}^{\omega_G}(\boldsymbol{\lambda}(g^{-1})) = \sigma_{-t}^{\omega_G}(j_G(\boldsymbol{\lambda}(g))\ (g \in G, t \in \mathbb{R})$, and hence the two weights appearing in (12) commute. On the other hand, for $\xi \in \mathscr{K}(G)$ with $\boldsymbol{\lambda}(\xi) \geq 0$, we have $0 \leq j_G(\boldsymbol{\lambda}(\xi)) = \int \xi(g)\boldsymbol{\lambda}(g^{-1})\,\mathrm{d}g = \int \xi(g^{-1})\Delta_G(g)^{-1}\boldsymbol{\lambda}(g)\,\mathrm{d}g = \boldsymbol{\lambda}(\eta)$, where $\eta(g) = \xi(g^{-1})\Delta_G(g)^{-1}, (g \in G)$, and using (8) we get $\langle \boldsymbol{\lambda}(\xi), \omega_G \circ j_G \rangle = \langle \boldsymbol{\lambda}(\eta), \omega_G \rangle = \eta(e) = \xi(e) = \langle \boldsymbol{\lambda}(\xi), \omega_G \rangle$. Consequently, (12) follows using Theorem 6.2.

From (11) and (12), it follows that $\hat{\mathbf{G}} = (\mathfrak{L}(G), \delta_G, j_G, \omega_G)$ *is a Kac algebra*. Actually, $\mathbf{G}$ and $\hat{\mathbf{G}}$ *are dual Kac algebras*.

Finally, we show that

$$\omega_G \circ \mathrm{Ad}(\boldsymbol{\lambda}(g)) = \Delta_G(g)\omega_G; g \in G. \tag{13}$$

Indeed, the two weights appearing in this equation commute because $\mathrm{Ad}(\boldsymbol{\lambda}(g))$ commutes with $\sigma_t^{\omega_G}(t \in \mathbb{R})$. For $\xi \in \mathscr{K}(G)$ with $\boldsymbol{\lambda}(\xi) \geq 0$, we have $0 \leq (\mathrm{Ad}(\boldsymbol{\lambda}(g)))(\boldsymbol{\lambda}(\xi)) = \boldsymbol{\lambda}(g)\boldsymbol{\lambda}(\xi)\boldsymbol{\lambda}(g^{-1}) = \boldsymbol{\lambda}(\eta)$, where $\eta(s) = \Delta_G(g)\xi(gsg^{-1})(s \in G)$, and using (8) we get $\langle \boldsymbol{\lambda}(\xi), \omega \circ \mathrm{Ad}(\boldsymbol{\lambda}(g)) \rangle = \langle \boldsymbol{\lambda}(\eta), \omega \rangle = \eta(e) = \Delta_G(g)\xi(e) = \langle \boldsymbol{\lambda}(\xi), \Delta_G(g)\omega_G \rangle$. Thus, (13) follows by Theorem 6.2.

18.18. The invariance property of a weight with respect to an action can be extended to a similar property of the tensor product of the weight with a normal positive form, namely:

Proposition. *Let* $\delta : \mathscr{M} \to \mathscr{M} \bar{\otimes} \mathscr{A}$ *be an action of the coinvolutive Hopf–von Neumann algebra* $(\mathscr{A}, \delta_{\mathscr{A}}, j_{\mathscr{A}})$ *on the* W^**-algebra* $\mathscr{M}$ *and let* φ *be a* $(\delta, j_{\mathscr{A}})$*-invariant n.s.f. weight on* $\mathscr{M}$. *For any* W^**-algebra* $\mathscr{N}$ *and any* $f \in (\mathscr{A} \bar{\otimes} \mathscr{N})_*^+$ *we have*

$$\langle X, (\varphi \bar{\otimes} f) \circ (\delta \bar{\otimes} \iota_{\mathscr{N}}) \rangle = \langle X \bar{\otimes} 1_{\mathscr{A}}, (\varphi \bar{\otimes} f) \circ (\iota_{\mathscr{M}} \bar{\otimes} \tilde{\cdot}_{\mathscr{N},\mathscr{A}}) \rangle \tag{1}$$
$$(X \in (\mathscr{M} \bar{\otimes} \mathscr{N})^+).$$

Proof. We consider the normal weight $\psi = (\varphi \bar{\otimes} f) \cdot (\delta \bar{\otimes} \iota_{\mathscr{N}})$ on $\mathscr{M} \bar{\otimes} \mathscr{N}$ and the normal positive form $h = f(1_{\mathscr{A}} \bar{\otimes} \cdot)$ on $\mathscr{N}$, that is, $\langle y, h \rangle = \langle 1_{\mathscr{A}} \bar{\otimes} y, f \rangle (y \in \mathscr{N})$. Using the δ-invariance of φ, we obtain, for $x \in \mathfrak{N}_{\varphi} \subset \mathscr{M}$ and $y \in \mathscr{N}$

$$\begin{aligned}
\langle x^*x \bar{\otimes} y^*y, \psi \rangle &= \langle \delta(x^*x) \bar{\otimes} y^*y, \varphi \bar{\otimes} f \rangle = \langle \delta(x^*x), (\varphi \bar{\otimes} f)(\cdot \bar{\otimes} y^*y) \rangle \\
&= \langle \delta(x^*x), \varphi \bar{\otimes} (f(\cdot \bar{\otimes} y^*y)) \rangle = \langle x^*x \bar{\otimes} 1_{\mathscr{A}}, \varphi \bar{\otimes} (f(\cdot \bar{\otimes} y^*y)) \rangle \\
&= \langle x^*x, \varphi \rangle \langle 1_{\mathscr{A}} \bar{\otimes} y^*y, f \rangle = \langle x^*x, \varphi \rangle \langle y^*y, h \rangle = \langle x^*x \bar{\otimes} y^*y, \varphi \bar{\otimes} h \rangle.
\end{aligned}$$

In particular, ψ is semifinite. On the other hand, for every $t \in \mathbb{R}, \sigma_t^{\varphi}$ commutes with δ (Lemma 18.11), hence

$$\begin{aligned}
\psi \circ (\sigma_t^{\varphi} \bar{\otimes} \iota_{\mathscr{N}}) &= (\varphi \bar{\otimes} f) \circ ((\delta \circ \sigma_t^{\varphi}) \bar{\otimes} \iota_{\mathscr{N}}) \\
&= (\varphi \bar{\otimes} f) \circ (\sigma_t^{\varphi} \bar{\otimes} \iota_{\mathscr{A}} \bar{\otimes} \iota_{\mathscr{N}}) \circ (\delta \bar{\otimes} \iota_{\mathscr{N}}) = ((\varphi \circ \sigma_t^{\varphi}) \bar{\otimes} f) \circ (\delta \bar{\otimes} \iota_{\mathscr{N}}) = \psi.
\end{aligned}$$

If $\mathscr{N}$ is a type I factor, then, using Proposition 8.10, we infer that $\psi = \varphi \bar{\otimes} h$, which is equivalent to (1).

In the general case, we can represent $\mathscr{N}$ as a von Neumann algebra $\mathscr{N} \subset \mathscr{B}(\mathscr{H})$ and extend f to a normal positive form on $\mathscr{A} \bar{\otimes} \mathscr{B}(\mathscr{H})$ so that the general case of (1) follows by restriction from the case $\mathscr{N} = \mathscr{B}(\mathscr{H})$, which has already been proved. □

Note that (1) for $\mathscr{N} = \mathbb{C} \cdot 1_{\mathscr{N}}$ just expresses the δ-invariance of φ.

18.19. We now show that the action δ of a Kac algebra on a W^*-algebra $\mathscr{M}$ defines a canonical $\mathscr{M}^\delta$-valued weight on $\mathscr{M}$.

Proposition. *Let $\delta : \mathscr{M} \to \mathscr{M} \bar{\otimes} \mathscr{A}$ be an action of the Kac algebra $(\mathscr{A}, \delta_{\mathscr{A}}, j_{\mathscr{A}}, \omega_{\mathscr{A}})$ on the W^*-algebra $\mathscr{M}$. The formula*

$$P_\delta(x) = E_{\mathscr{M}}^{\omega_{\mathscr{A}} \circ j_{\mathscr{A}}}(\delta(x)) \quad (x \in \mathscr{M}^+) \tag{1}$$

defines a normal faithful $\mathscr{M}^\delta$-valued weight on $\mathscr{M}$.

Proof. Let $x \in \mathscr{M}^+$. (1) defines an element $\mathfrak{m} = P_\delta(x) \in \overline{\mathscr{M}}^+$. We shall show that $\mathfrak{m} = \overline{(\mathscr{M}^\delta)^+}$. By the uniqueness of the spectral decomposition of $\mathfrak{m}$ (11.3.(2)), it is sufficient to show that $\delta(\mathfrak{m}) = \mathfrak{m} \bar{\otimes} 1_{\mathscr{A}}$, that is,

$$\langle \delta(\mathfrak{m}), f\rangle = \langle \mathfrak{m} \bar{\otimes} 1_{\mathscr{A}}, f\rangle \text{ for all } f \in (\mathscr{M} \bar{\otimes} \mathscr{A})_*^+. \tag{2}$$

Let $f \in (\mathscr{M} \bar{\otimes} \mathscr{A})_*^+$. We have $\langle \delta(\mathfrak{m}), f\rangle = \langle \mathfrak{m}, f \circ \delta\rangle = \langle \delta(x), (f \circ \delta) \bar{\otimes} (\omega_{\mathscr{A}} \circ j_{\mathscr{A}})\rangle = \langle x, (f \bar{\otimes} (\omega_{\mathscr{A}} \circ j_{\mathscr{A}})) \circ (\delta \bar{\otimes} \iota_{\mathscr{A}}) \circ \delta\rangle = \langle x, (f \bar{\otimes} (\omega_{\mathscr{A}} \circ j_{\mathscr{A}})) \circ (\iota_{\mathscr{M}} \bar{\otimes} \delta_{\mathscr{A}}) \circ \delta\rangle = \langle \delta(x), (f \bar{\otimes} (\omega_{\mathscr{A}} \circ j_{\mathscr{A}})) \circ (\iota_{\mathscr{M}} \bar{\otimes} \delta_{\mathscr{A}})\rangle$ and $\langle \mathfrak{m} \bar{\otimes} 1_{\mathscr{A}}, f\rangle = \langle \mathfrak{m}, f(\cdot \bar{\otimes} 1_{\mathscr{A}})\rangle = \langle \delta(x), (f(\cdot \bar{\otimes} 1_{\mathscr{A}})) \bar{\otimes} (\omega_{\mathscr{A}} \circ j_{\mathscr{A}})\rangle$. Consequently, (2) is equivalent to the following equality of weights on $\mathscr{M} \bar{\otimes} \mathscr{A}$:

$$(f \bar{\otimes} (\omega_{\mathscr{A}} \circ j_{\mathscr{A}})) \circ (\iota_{\mathscr{M}} \bar{\otimes} \delta_{\mathscr{A}}) = (f(\cdot \bar{\otimes} 1_{\mathscr{A}}) \bar{\otimes} (\omega_{\mathscr{A}} \circ j_{\mathscr{A}}). \tag{3}$$

As in the last part of the proof of Proposition 18.18, we see that it is sufficient to check (3) only when $\mathscr{M}$ is a type I factor. In this case, there exists an involutive *-automorphism $j : \mathscr{M} \to \mathscr{M}, j \circ j = \iota_{\mathscr{M}}$. Let $h = f \circ (j \bar{\otimes} j_{\mathscr{A}}) \circ \tilde{\ }_{\mathscr{A},\mathscr{M}} \in (\mathscr{A} \bar{\otimes} \mathscr{M})_*^+$.

By composing the weights in (3) with the *-antiautomorphism $\tilde{\ }_{\mathscr{A},\mathscr{M}} \circ (j_{\mathscr{A}} \bar{\otimes} j) : \mathscr{A} \bar{\otimes} \mathscr{M} \to \mathscr{M} \bar{\otimes} \mathscr{A}$ and using 18.2.(3), we see that (3) is equivalent to the following equality:

$$(\omega_{\mathscr{A}} \bar{\otimes} h) \circ (\delta_{\mathscr{A}} \bar{\otimes} \iota_{\mathscr{M}}) = [(\omega_{\mathscr{A}} \bar{\otimes} h) \circ (\iota_{\mathscr{A}} \bar{\otimes} \tilde{\ }_{\mathscr{M},\mathscr{A}})](\cdot \bar{\otimes} 1_{\mathscr{A}})$$

of normal weights on $\mathscr{A} \bar{\otimes} \mathscr{M}$, which is an obvious consequence of Proposition 18.18.

Thus, $P_\delta(x) \in \overline{(\mathscr{M}^\delta)}^+$ for every $x \in \mathscr{M}^+$. Taking into account the properties of the Fubini mappings (9.8 and 12.18), it is now easy to check that $P_\delta : \mathscr{M}^+ \to \overline{(\mathscr{M}^\delta)}^+$ is a normal faithful operator-valued weight. □

Note that if $\mathscr{N} \subset \mathscr{M}$ is an $\mathscr{A}$-subcomodule of the $\mathscr{A}$-comodule $\mathscr{M}$ via δ, then for the action $\delta|\mathscr{N}$ of $\mathscr{A}$ on $\mathscr{N}$ we have

$$P_{\delta|\mathscr{N}} = P_\delta|\mathscr{N} \tag{4}$$

as $\delta(\mathscr{N}) \subset \mathscr{N} \bar{\otimes} \mathscr{A}$ and, clearly, $E_{\mathscr{N}}^{\omega_{\mathscr{A}} \circ j_{\mathscr{A}}} = E_{\mathscr{M}}^{\omega_{\mathscr{A}} \circ j_{\mathscr{A}}}|\mathscr{N} \bar{\otimes} \mathscr{A}$.

Also, if $\delta : \mathcal{M} \to \mathcal{M} \bar{\otimes} \mathcal{A}$ is an action of $\mathcal{A}$ on $\mathcal{M}$ and $\mathcal{N}$ is any other W^*-algebra, then for the action $\iota_{\mathcal{N}} \bar{\otimes} \delta : \mathcal{N} \bar{\otimes} \mathcal{M} \to \mathcal{N} \bar{\otimes} \mathcal{M} \bar{\otimes} \mathcal{A}$ of $\mathcal{A}$ on $\mathcal{N} \bar{\otimes} \mathcal{M}$ we have

$$P_{\iota_{\mathcal{N}} \bar{\otimes} \delta} = \iota_{\mathcal{N}} \bar{\otimes} P_\delta \tag{5}$$

as $E^{\omega_{\mathcal{A}} \circ j_{\mathcal{A}}}_{\mathcal{N} \bar{\otimes} \mathcal{M}} = \iota_{\mathcal{N}} \bar{\otimes} E^{\omega_{\mathcal{A}} \circ j_{\mathcal{A}}}_{\mathcal{M}}$.

If $\delta = \delta_{\mathcal{A}}$ is the action of the Kac algebra $\mathcal{A}$ on itself, then $\mathcal{A}^\delta = \mathbb{C} \cdot 1_{\mathcal{A}}$ (18.14) and P_σ coincides with the weight $\omega_{\mathcal{A}} \circ j_{\mathcal{A}}$.

The action $\delta : \mathcal{M} \to \mathcal{M} \bar{\otimes} \mathcal{A}$ is called *integrable* if the operator-valued weight P_δ is semifinite.

18.20. Consider, in particular, a continuous action $\sigma : G \to Aut(\mathcal{M})$ of the locally compact group G on the W^*-algebra $\mathcal{M}$ and the corresponding action $\pi_\sigma : \mathcal{M} \to \mathcal{M} \bar{\otimes} \mathcal{L}^\infty(G)$ of the Kac algebra $\mathbf{G} = (\mathcal{L}^\infty(G), \pi_G, k_G, \mu_G)$ on $\mathcal{M}$. We recall that $\mathcal{M}^\sigma = \mathcal{M}^{\pi_\sigma}$ (18.6.(4)) and put $P_\sigma = P_{\pi_\sigma}$. We shall show that *in this case the operator-valued weight* $P_\sigma : \mathcal{M}^+ \to \overline{(\mathcal{M}^\delta)}^+$ is given by

$$P_\sigma(x) = \int \sigma_g(x)\, \mathrm{d}g \quad (x \in \mathcal{M}^+). \tag{1}$$

Let $x \in \mathcal{M}^+, \varphi \in \mathcal{M}^+_*$ and $h \in \mathcal{L}^1(G) = \mathcal{L}^\infty(G)_*$. Then the element $h \circ k_G \in \mathcal{L}^\infty(G)_*$, regarded as an $\mathcal{L}^1$-function, has the expression $(h \circ k_G)(g) = h(g^{-1})\Delta_G(g^{-1})$ $(g \in G)$ see (18.5.(5)). Thus,

$$\begin{aligned} \langle k_G(E^\varphi_{\mathcal{L}^\infty(G)}(\pi_\sigma(x))), h\rangle &= \langle \pi_\sigma(x), \varphi \bar{\otimes} (h \circ k_G)\rangle \\ &= \int \varphi(\sigma_g^{-1}(x))h(g^{-1})\Delta_G(g^{-1})\, \mathrm{d}g \\ &= \int \varphi(\sigma_g(x))h(g)\, \mathrm{d}g \end{aligned}$$

hence the element $k_G(E^\varphi_{\mathcal{L}^\infty(G)}(\pi_\sigma(x))) \in \mathcal{L}^\infty(G)$ is the function $G \ni g \mapsto \varphi(\sigma_g(x))$.

Consequently

$$\begin{aligned} \langle P_\sigma(x), \varphi\rangle = \langle \pi_\sigma(x), \varphi \bar{\otimes} (\mu_G \circ k_G)\rangle &= \langle k_G(E^\varphi_{\mathcal{L}^\infty(G)}(\pi_\sigma(x))), \mu_G\rangle \\ = \int \varphi(\sigma_g(x))\, \mathrm{d}g &= \left\langle \int \sigma_g(x)\, \mathrm{d}g, \varphi \right\rangle, \end{aligned}$$

proving (1).

Equation (1) justifies the notion of an integrable action (18.19).

Note that if G is compact, then P_σ is finite, that is, P_σ is a normal faithful conditional expectation of $\mathcal{M}$ onto $\mathcal{M}^\sigma$.

Using (1) and 18.4.(2), we see that

$$P_\sigma(\sigma_t(x)) = \Delta_G(t)^{-1} P_\sigma(x) \quad (x \in \mathcal{M}^+, t \in G). \tag{2}$$

In particular, if G is unimodular, for example, abelian or compact, then $P_\sigma \circ \sigma_t = P_\sigma (t \in G)$.

18.21. Consider now the continuous action

$$\sigma \bar{\otimes} \mathrm{Ad}(\boldsymbol{\rho}) : G \to Aut(\mathscr{M} \bar{\otimes} \mathscr{B}(\mathscr{L}^2(G)))$$

where $\sigma : G \to Aut(\mathscr{M})$ is any continuous action of G on $\mathscr{M}$ and $\boldsymbol{\rho} : G \to \mathscr{B}(\mathscr{L}^2(G))$ is the right regular representation of G.

We shall assume $\mathscr{M} \subset \mathscr{B}(\mathscr{H})$ realized as a von Neumann algebra and hence $\mathscr{M} \bar{\otimes} \mathscr{B}(\mathscr{L}^2(G)) \subset \mathscr{B}(\mathscr{L}^2(G, \mathscr{H}))$. Moreover, we may assume (2.24) that there exists an *so*-continuous unitary representation $G \ni g \mapsto v(g) \in \mathscr{B}(\mathscr{H})$ such that $\sigma_g = \mathrm{Ad}(v(g))(g \in G)$. On the other hand, we put $u(g) = 1_{\mathscr{M}} \bar{\otimes} \boldsymbol{\lambda}(g) \in \mathscr{M} \bar{\otimes} \mathscr{B}(\mathscr{L}^2(G))(g \in G)$, and recall (18.6.(5)) that $\pi_\sigma(\sigma_g(x)) = u(g)\pi_\sigma(x)u(g)^*(x \in \mathscr{M}, g \in G)$.

Every compactly supported *w*-continuous function $f : G \to \mathscr{M}$ defines an element

$$T_f^\sigma = \int \pi_\sigma(f(t))u(t)\,\mathrm{d}t \in \mathscr{M} \bar{\otimes} \mathscr{B}(\mathscr{L}^2(G)). \tag{1}$$

On the other hand, every compactly supported s^*-continuous function $G \times G \ni (s, r) \mapsto X(s, r) \in \mathscr{M}$ defines an element $X \in \mathscr{M} \bar{\otimes} \mathscr{B}(\mathscr{L}^2(G))$ uniquely determined such that

$$(X\xi|\eta) = \int\int (X(s, r)\xi(r)|\eta(s))\mathrm{d}r\,\mathrm{d}s \quad (\xi, \eta \in \mathscr{L}^2(G, \mathscr{H})). \tag{2}$$

It is easy to check that if the operator X is positive, then $X(g, g) \geq 0$ for all $g \in G$.

We shall show that *if the operator X just defined is positive, then*

$$P_{\sigma \bar{\otimes} \mathrm{Ad}(\boldsymbol{\rho})}(X) = T_f^\sigma, \tag{3}$$

where the compactly supported w-continuous function $f : G \to \mathscr{M}$ is defined by

$$f(t) = \int \sigma_{\mathrm{tr}}(X(tr, r))\Delta_G(r)\,\mathrm{d}r \quad (t \in G). \tag{4}$$

Indeed, for $\xi \in \mathscr{L}^2(G, \mathscr{H})$ and $g, s \in G$ we have

$$\begin{aligned}
&(((\sigma_g \bar{\otimes} \mathrm{Ad}(\boldsymbol{\rho}(g))(X))\xi)(s) \\
&= ((v(g) \bar{\otimes} \boldsymbol{\rho}(g))X(v(g) \bar{\otimes} \boldsymbol{\rho}(g))^*\xi)(s) \\
&= \Delta_G(g)^{1/2}v(g)(X(v(g) \bar{\otimes} \boldsymbol{\rho}(g))^*\xi)(sg) \\
&= \Delta_G(g)^{1/2}v(g)\int X(sg, r)((v(g)^* \bar{\otimes} \boldsymbol{\rho}(g^{-1}))\xi)(r)\,\mathrm{d}r \\
&= v(g)\int X(sg, r)v(g)^*\xi(rg^{-1})\,\mathrm{d}r \\
&= \Delta_G(g)\int v(g)X(sg, rg)v(g)^*\xi(r)\,\mathrm{d}r \\
&= \Delta_G(g)\int \sigma_g(X(sg, rg))\xi(r)\,\mathrm{d}r
\end{aligned}$$

and, using 18.20.(1), we obtain

$$(P_{\sigma \bar{\otimes} \mathrm{Ad}(\rho)}(X)\xi|\xi) = \int\int\int \Delta_G(g)(\sigma_g(X(sg, rg))\xi(r)|\xi(s))\, \mathrm{d}r\, \mathrm{d}s\, \mathrm{d}g.$$

On the other hand, for the function f defined by (4) we have

$$\begin{aligned}
(T_f^\sigma \xi|\xi) &= \int (\pi_\sigma(f(t))u(t)\xi|\xi)\, \mathrm{d}t \\
&= \int\int ((\pi_\sigma(f(t))u(t)\xi)(s)|\xi(s))\, \mathrm{d}s\, \mathrm{d}t \\
&= \int\int (\sigma_s^{-1}(f(t))(u(t)\xi)(s)|\xi(s))\, \mathrm{d}s\, \mathrm{d}t \\
&= \int\int (\sigma_s^{-1}(f(t))\xi(t^{-1}s)|\xi(s))\, \mathrm{d}s\, \mathrm{d}t \\
&= \int\int\int \Delta_G(r)(\sigma_{s^{-1}tr}(X(tr, r))\xi(t^{-1}s)|\xi(s))\, \mathrm{d}r\, \mathrm{d}s\, \mathrm{d}t \\
(\text{via } r = t^{-1}sg) &= \int\int\int \Delta_G(t^{-1}sg)(\sigma_g(X(sg, t^{-1}sg))\xi(t^{-1}s)|\xi(s))\, \mathrm{d}g\, \mathrm{d}s\, \mathrm{d}t \\
(\text{via } t = sr^{-1}) &= \int\int\int \Delta_G(g)(\sigma_g(X(sg, rg))\xi(r)|\xi(s))\, \mathrm{d}g\, \mathrm{d}s\, \mathrm{d}r.
\end{aligned}$$

Consequently, the element $P_{\sigma \bar{\otimes} \mathrm{Ad}(\rho)}(X)$ is bounded and (3) holds.

18.22. Finally, we consider the action $\delta = \iota_{\mathscr{M}} \bar{\otimes} \delta_G$ of the Kac algebra $\hat{\mathbf{G}} = (\mathfrak{L}(G), \delta_G, j_G, \omega_G)$ on $\mathscr{M} \bar{\otimes} \mathscr{B}(\mathscr{L}^2(G))$, where $\mathscr{M} \subset \mathscr{B}(\mathscr{H})$ is a von Neumann algebra, and compute the values of the operator-valued weight P_δ on two types of elements.

Recall that $\omega_G \circ j_G = \omega_G$ (18.17.(12)) and $u(g) = 1_{\mathscr{M}} \bar{\otimes} \boldsymbol{\lambda}(g) \in \mathscr{M} \bar{\otimes} \mathscr{B}(\mathscr{L}^2(G))$ $(g \in G)$.

Let $f : G \to \mathscr{M}$ be a compactly supported w-continuous function.

We first consider the operator $\Lambda_f = \int (f(g) \bar{\otimes} \boldsymbol{\lambda}(g))\, \mathrm{d}g \in \mathscr{M} \bar{\otimes} \mathfrak{L}(G)$ and show that if $\Lambda_f \geq 0$, then

$$E_{\mathscr{M}}^{\omega_G}(\Lambda_f) = f(e). \tag{1}$$

Indeed, for $\varphi \in \mathscr{M}_*^+$ and $k \in \mathscr{A}(G)$ we have

$$\begin{aligned}
\langle E_{\mathfrak{L}(G)}^{\varphi}(\Lambda_f), k\rangle &= \langle \Lambda_f, \varphi \bar{\otimes} k\rangle = \int \varphi(f(g))k(g)\, \mathrm{d}g \\
&= \left\langle \int \varphi(f(g))\boldsymbol{\lambda}(g)\, \mathrm{d}g, k \right\rangle = \langle \boldsymbol{\lambda}(\varphi \circ f), k\rangle
\end{aligned}$$

hence $E_{\mathfrak{L}(G)}^{\varphi}(\Lambda_f) = \boldsymbol{\lambda}(\varphi \circ f) \geq 0$ and, using 18.17.(8), we obtain

$$\langle E_{\mathscr{M}}^{\omega_G}(\Lambda_f), \varphi\rangle = \langle E_{\mathfrak{L}(G)}^{\varphi}(\Lambda_f), \omega_G\rangle = \langle \boldsymbol{\lambda}(\varphi \circ f), \omega_G\rangle = \langle f(e), \varphi\rangle$$

which proves (1).

Next, *we consider a continuous action* $\sigma : G \to Aut(\mathscr{M})$, *we define the operator (18.21.(1))* $T_f^\sigma = \int \pi_\sigma(f(g))u(g)\,\mathrm{d}g \in \mathscr{M} \bar{\otimes} \mathscr{B}(\mathscr{L}^2(G))$ *and show that*

$$P_{\iota_{\mathscr{M}}\bar{\otimes}\delta_G} = \pi_\sigma(f(e)). \tag{2}$$

Indeed, we have $\pi_\sigma(\mathscr{M}) \subset \mathscr{M} \bar{\otimes} \mathscr{L}^\infty(G) = (\mathscr{M} \bar{\otimes} \mathscr{B}(\mathscr{L}^2(G))^\delta$ and $\delta(u(g)) = u(g) \bar{\otimes} \boldsymbol{\lambda}(g)$, hence $\delta(T_f^\sigma) = \int ((\pi_\sigma(f(g))u(g)) \bar{\otimes} \boldsymbol{\lambda}(g)\,\mathrm{d}g$ so that, using (1), we obtain $P_\delta(T_f^\sigma) = E_{\mathscr{M}}^{\omega_G}(\delta(T_f^\sigma)) = \pi_\sigma(f(e))$.

Given the compactly supported w-continuous functions $f, f_1, f_2 : G \to \mathscr{M}$, we define the functions $f^\sharp, f_1 * f_2 : G \to \mathscr{M}$ by

$$f^\sharp(g) = \Delta_g(g)^{-1} f(g^{-1})^*, (f_1 * f_2)(g) = \int f_1(t) f_2(t^{-1}g)\,\mathrm{d}t; \quad (g \in G).$$

It is easy to check that

$$(\Lambda_f)^* = \Lambda_{f^\sharp} \quad \Lambda_{f_1}\Lambda_{f_2} = \Lambda_{f_1 * f_2} \tag{3}$$

$$(T_f^\sigma)^* = T_{f^\sharp}^\sigma \quad T_{f_1}^\sigma T_{f_2}^\sigma = T_{f_1 * f_2}^\sigma \tag{4}$$

and from (1) and (2), using the polarization relation, we infer that

$$E_{\mathscr{M}}^{\omega_G}(\Lambda_{f_1}^* \Lambda_{f_2}) = (f_1^\sharp * f_2)(e) \tag{5}$$

$$P_{\iota_{\mathscr{M}}\bar{\otimes}\delta_G}((T_{f_1}^\sigma)^*(T_{f_2}^\sigma)) = \pi_\sigma((f_1^\sharp * f_2)(e)) \tag{6}$$

Consider now a compactly supported w-continuous function $G \times G \ni (s, r) \mapsto X(s, r) \in \mathscr{M}$, *the corresponding operator* $X \in \mathscr{M} \bar{\otimes} \mathscr{B}(\mathscr{L}^2(G))$ *uniquely determined by equalities 18.21.(2) and the operator* $F_X \in \mathscr{M} \bar{\otimes} \mathscr{L}^\infty(G)$ *defined by the compactly supported w-continuous function*

$$F_X(g) = \Delta_G(g) X(g, g) \quad (g \in G). \tag{7}$$

If the operator X is positive, then

$$P_{\iota_{\mathscr{M}}\bar{\otimes}\delta_G}(X) = F_X. \tag{8}$$

Indeed, we first compute the element $(\iota_{\mathscr{M}} \bar{\otimes} \delta_G)(X) \in \mathscr{M} \bar{\otimes} \mathscr{B}(\mathscr{L}^2(G)) \bar{\otimes} \mathfrak{L}(G) \subset \mathscr{B}(\mathscr{L}^2(G \times G, \mathscr{H}))$. For $\zeta \in \mathscr{L}^2(G \times G, \mathscr{H})$, we have

$$\begin{aligned}
((\iota_{\mathscr{M}} \bar{\otimes} \delta_G)(X)\zeta|\zeta) &= ((1_{\mathscr{M}} \bar{\otimes} W_G^*)(X \bar{\otimes} 1_G)(1_{\mathscr{M}} \bar{\otimes} W_G)\zeta|\zeta) \\
&= ((X \bar{\otimes} 1_G)(1_{\mathscr{M}} \bar{\otimes} W_G)\zeta|(1_{\mathscr{M}} \bar{\otimes} W_G)\zeta) \\
&= \int\int\int (X(s, r)(((1_{\mathscr{M}} \bar{\otimes} W_G\zeta)(r, t))|((1_{\mathscr{M}} \bar{\otimes} W_G)\zeta)(s, t))\,\mathrm{d}r\,\mathrm{d}s\,\mathrm{d}t \\
&= \int\int\int (X(s, r)\zeta(r, rt)|\zeta(s, st))\,\mathrm{d}r\,\mathrm{d}s\,\mathrm{d}t \\
&= \int\int\int (X(s, r)\zeta(r, rs^{-1}t)|\zeta(s, t))\,\mathrm{d}r\,\mathrm{d}s\,\mathrm{d}t \\
&= \int\int\int (X(s, r^{-1}s)\zeta(r^{-1}s, r^{-1}t)|\zeta(s, t))\Delta_G(r^{-1}s)\,\mathrm{d}r\,\mathrm{d}s\,\mathrm{d}t.
\end{aligned}$$

Let $\varphi = \omega_\xi \in \mathscr{M} \mathbin{\bar{\otimes}} \mathscr{B}(\mathscr{L}^2(G))^+_*$ with $\xi \in \mathscr{K}(G, \mathscr{H})$ and $k \in \mathscr{A}(G)$. There exists $\eta \in \mathscr{L}^2(G)$ such that $k = \omega_\eta$, that is,

$$k(r) = \langle \boldsymbol{\lambda}(r), \omega_\eta \rangle = (\boldsymbol{\lambda}(r)\eta | \eta) = \int \eta(r^{-1}t)\overline{\eta(t)}\, \mathrm{d}t.$$

Then $\varphi \mathbin{\bar{\otimes}} k = \omega_\zeta$ with $\zeta = \xi \mathbin{\bar{\otimes}} \eta$, that is, $\zeta(s,t) = \xi(s)\eta(t)$, and by the above computation it follows that

$$\begin{aligned} \langle E^{\varphi}_{\mathfrak{L}(G)}(\delta(X)), k\rangle &= \langle (\iota_{\mathscr{M}} \mathbin{\bar{\otimes}} \delta_G)(X), \varphi \mathbin{\bar{\otimes}} k\rangle \\ &= \int k(r) \left(\int \Delta_G(r^{-1}s)(X(s, r^{-1}s)\xi(r^{-1}s)|\xi(s))\, \mathrm{d}s \right) \mathrm{d}r = \langle \boldsymbol{\lambda}(f), k\rangle \end{aligned}$$

where $f \in \mathscr{K}(G)$ is defined by

$$f(r) = \int \Delta_G(r^{-1}s)(X(s, r^{-1}s)\xi(r^{-1}s)|\xi(s))\, \mathrm{d}s \quad (r \in G). \tag{9}$$

Consequently, $0 \leq E^{\varphi}_{\mathfrak{L}(G)}(\delta(X)) = \boldsymbol{\lambda}(f)$ and, using 18.17.(8), it follows that

$$\begin{aligned} (P_\delta(X)\xi|\xi) &= \langle E^{\omega_G}_{\mathscr{M} \otimes \mathscr{B}(\mathscr{L}^2(G))}(\delta(X)), \varphi\rangle = \langle \delta(X), \varphi \mathbin{\bar{\otimes}} \omega_G\rangle \\ &= \langle E^{\varphi}_{\mathfrak{L}(G)}(\delta(X)), \omega_G\rangle = \langle \boldsymbol{\lambda}(f), \omega_G\rangle = f(e) \\ &= \int \Delta_G(s)(X(s,s)\xi(s)|\xi(s))\, \mathrm{d}r = (F_X\xi|\xi). \end{aligned}$$

Therefore, the operator $P_\delta(X)$ is bounded and $P_\delta(X) = F_X$.

18.23. Notes. The extension of the Pontryagin and Tannaka duality theories (see Hewitt & Ross, 1963, 1970) to a duality in the category of Kac algebras has been developed in a long series of papers by several different mathematicians: Segal (1953); Stinespring (1959); Kac (1963); Eymard (1964); Tatsuuma (1967); Ernest (1966); Takesaki (1972b); Vainerman and Kac (1973, 1974); Enock and Schwartz (1975); Kirchberg (1975), and so on. The actions of Hopf–von Neumann algebras on W^*-algebras have been considered in Strătilă, Voiculescu, and Zsidó (1976, 1977). The notions of $(\delta, j_{\mathscr{A}})$-invariance, Theorem 18.12, and Proposition 18.14, and also the computations given in Sections 18.21 and 18.22, are from Strătilă et al. (1976, 1977). Proposition 18.19 was proved in several special cases in Haagerup (1978); Landstad (1979); Strătilă et al. (1976, 1977) and the general case was asserted and used in Enock (1977), but the complete proof of it in the general case has only appeared in Zsidó (1978).

For our exposition we have used: Haagerup (1978); Strătilă et al. (1976, 1977); Takesaki (1972b); and Zsidó (1978).

19 Crossed Products

In this section, we introduce the crossed product of a W^*-algebra by the continuous action of a locally compact group as a particular case of crossed products by actions of Kac algebras. The main properties of crossed products, including the duality theory, are described.

19.1. Let $\delta : \mathcal{M} \to \mathcal{M} \bar{\otimes} \mathcal{A}$ be an action of the Kac algebra $(\mathcal{A}, \delta_{\mathcal{A}}, j_{\mathcal{A}}, \omega_{\mathcal{A}})$ on the W^*-algebra $\mathcal{M}$. Consider the W^*-algebra $\mathcal{A}$ realized as a von Neumann algebra $\mathcal{A} \subset \mathcal{B}(\mathcal{H}_{\mathcal{A}})$ in the standard representation associated with the weight $\omega_{\mathcal{A}}$. As we have seen in Section 18.15, one can construct a dual Kac algebra $(\hat{\mathcal{A}}, \hat{\delta}_{\mathcal{A}}, \hat{j}_{\mathcal{A}}, \hat{\omega}_{\mathcal{A}})$, where $\hat{\mathcal{A}} \subset \mathcal{B}(\mathcal{H}_{\mathcal{A}})$ is a von Neumann algebra also acting on $\mathcal{H}_{\mathcal{A}}$.

We define *the crossed product of the W^*-algebra $\mathcal{M}$ by the action δ of the Kac algebra $\mathcal{A}$ on $\mathcal{M}$* to be the W^*-algebra

$$\mathcal{R}(\mathcal{M}, \delta) \subset \mathcal{M} \bar{\otimes} \mathcal{B}(\mathcal{H}_{\mathcal{A}})$$

generated by $\delta(\mathcal{M})$ and $1_{\mathcal{M}} \bar{\otimes} \hat{\mathcal{A}}$.

In particular, let G be a locally compact group and let $\mathbf{G} = (\mathcal{L}^\infty(G), \pi_G, k_G, \mu_G)$ and $\hat{\mathbf{G}} = (\mathfrak{L}(G), \delta_G, j_G, \omega_G)$ be the two associated Kac algebras, which are dual to one another (18.16, 18.17).

If $\sigma : G \to Aut(\mathcal{M})$ is a continuous action of G on the W^*-algebra $\mathcal{M}$, then $\pi_\sigma : \mathcal{M} \to \mathcal{M} \bar{\otimes} \mathcal{L}^\infty(G)$ is an action of $\mathbf{G}$ on $\mathcal{M}$ (18.6) and, according to the general definition above, *the crossed product of the W^*-algebra $\mathcal{M}$ by the continuous action σ of G on $\mathcal{M}$* is the W^*-algebra

$$\mathcal{R}(\mathcal{M}, \sigma) \subset \mathcal{M} \bar{\otimes} \mathcal{B}(\mathcal{L}^2(G))$$

generated by $\pi_\sigma(\mathcal{M})$ and $1_{\mathcal{M}} \bar{\otimes} \mathfrak{L}(G)$, that is,

$$\mathcal{R}(\mathcal{M}, \sigma) = \mathcal{R}\{\pi_\sigma(x), 1_{\mathcal{M}} \bar{\otimes} \boldsymbol{\lambda}(g); x \in \mathcal{M}, g \in G\}. \tag{1}$$

Recall (18.6.(5)) that

$$\pi_\sigma(\sigma_g(x)) = (1_{\mathcal{M}} \bar{\otimes} \boldsymbol{\lambda}(g))\pi_\sigma(x)(1_{\mathcal{M}} \bar{\otimes} \boldsymbol{\lambda}(g))^* \quad (x \in \mathcal{M}, g \in G), \tag{2}$$

which constitutes another proof of the fact (2.24) that there exists a realization $\mathcal{M} \subset \mathcal{B}(\mathcal{H})$ of $\mathcal{M}$ as a von Neumann algebra and an *so*-continuous unitary representation $G \ni g \mapsto v(g) \in \mathcal{B}(\mathcal{H})$ with

$$\sigma_g(x) = v(g)xv(g)^* \quad (x \in \mathcal{M}, g \in G). \tag{3}$$

If $\delta : \mathcal{M} \mapsto \mathcal{M} \bar{\otimes} \mathfrak{L}(G)$ is an action of the Kac algebra $\hat{\mathbf{G}}$ on $\mathcal{M}$, then the corresponding crossed product is the W^*-algebra

$$\mathcal{R}(\mathcal{M}, \delta) \subset \mathcal{M} \bar{\otimes} \mathcal{B}(\mathcal{L}^2(G))$$

generated by $\delta(\mathcal{M})$ and $1_{\mathcal{M}} \bar{\otimes} \mathcal{L}^\infty(G)$, that is,

$$\mathcal{R}(\mathcal{M}, \delta) = \mathcal{M}\{\delta(x), 1_{\mathcal{M}} \bar{\otimes} f; x \in \mathcal{M}, f \in \mathcal{L}^\infty(G)\}. \tag{4}$$

In what follows we are interested only in crossed products by continuous actions of groups. However, the appearance of certain important duality phenomena necessitates the consideration also of crossed products by actions of Kac algebras of type $\hat{\mathbf{G}}$.

Throughout this section, G will denote a locally compact group.

19.2 Lemma. *Let $\sigma : G \to Aut(\mathscr{M})$ be a continuous action of G on the W^*-algebra $\mathscr{M}$. Then*

$$\mathscr{M} \bar{\otimes} \mathscr{L}^\infty(G) = \mathscr{R}\{\pi_\sigma(\mathscr{M}), 1_\mathscr{M} \bar{\otimes} \mathscr{B}(\mathscr{L}^\infty(G))\} \tag{1}$$

$$\begin{aligned} \mathscr{M} \bar{\otimes} \mathscr{B}(\mathscr{L}^2(G)) &= \mathscr{R}\{\pi_\sigma(\mathscr{M}), 1_\mathscr{M} \bar{\otimes} \mathscr{B}(\mathscr{L}^2(G))\} \\ &= \mathscr{R}\{\mathscr{R}(\mathscr{M},\sigma), 1_\mathscr{M} \bar{\otimes} \mathscr{L}^\infty(G)\}. \end{aligned} \tag{2}$$

$$(\mathscr{M} \bar{\otimes} \mathscr{L}^\infty(G))^{\sigma \bar{\otimes} \mathrm{Ad}(\boldsymbol{\rho})} = \pi_\sigma(\mathscr{M}). \tag{3}$$

Proof. We may assume $\mathscr{M} \subset \mathscr{B}(\mathscr{H})$ realized as a von Neumann algebra such that there exists an so-continuous unitary representation $v : G \to \mathscr{B}(\mathscr{H})$ with $\sigma_g = \mathrm{Ad}(v(g))(g \in G)$. Let V be the unitary operator on $\mathscr{H} \bar{\otimes} \mathscr{L}^2(G)$ defined by the bounded so-continuous function $g \mapsto v(g)$, that is,

$$(V\xi)(g) = v(g)\xi(g) \quad (\xi \in \mathscr{L}^2(G,\mathscr{H}), g \in G).$$

It is easy to check that the $*$-automorphism $\theta = \mathrm{Ad}(V^*) = V^* \cdot V$ leaves invariant the von Neumann algebra $\mathscr{M} \bar{\otimes} \mathscr{L}^\infty(G) \subset \mathscr{B}(\mathscr{H} \bar{\otimes} \mathscr{L}^2(G))$; in fact

$$\theta(x \bar{\otimes} 1_G) = \pi_\sigma(x) \quad (x \in \mathscr{M}) \tag{4}$$

$$\theta(1_\mathscr{M} \bar{\otimes} f) = 1_\mathscr{M} \bar{\otimes} f \quad (f \in \mathscr{L}^\infty(G)). \tag{5}$$

Consequently, $\mathscr{M} \bar{\otimes} \mathscr{L}^\infty(G) = \theta(\mathscr{M} \bar{\otimes} \mathscr{L}^\infty(G)) = \theta(\mathscr{R}\{\mathscr{M} \bar{\otimes} 1_G, 1_\mathscr{M} \bar{\otimes} \mathscr{L}^\infty(G)\}) = \mathscr{R}\{\pi_\sigma(\mathscr{M}), 1_\mathscr{M} \bar{\otimes} \mathscr{L}^\infty(G)\}$, proving (1). Equation (2) follows from (1) because $\mathscr{B}(\mathscr{L}^2(G)) = \mathscr{R}\{\mathfrak{L}(G), \mathscr{L}^\infty(G)\}$ and, by definition, $\mathscr{R}(\mathscr{M},\sigma) = \mathscr{R}\{\pi_\sigma(\mathscr{M}), 1_\mathscr{M} \bar{\otimes} \mathfrak{L}(G)\}$.

On the other hand, it is easy to check that

$$(\sigma_g \bar{\otimes} \mathrm{Ad}(\boldsymbol{\rho}(g))) \circ \theta = \theta \circ (\iota_\mathscr{M} \bar{\otimes} \mathrm{Ad}(\boldsymbol{\rho}(g))) \quad (g \in G) \tag{6}$$

as $*$-automorphisms on $\mathscr{M} \bar{\otimes} \mathscr{L}^\infty(G)$. Consequently, for $X \in \mathscr{M} \bar{\otimes} \mathscr{L}^\infty(G)$, we have $X \in (\mathscr{M} \bar{\otimes} \mathscr{L}^\infty(G))^{\sigma \bar{\otimes} \mathrm{Ad}(\boldsymbol{\rho})} \Leftrightarrow (\sigma_g \bar{\otimes} \mathrm{Ad}(\boldsymbol{\rho}(g)))X = X (g \in G) \Leftrightarrow (\iota_\mathscr{M} \bar{\otimes} \mathrm{Ad}(\boldsymbol{\rho}(g)))\theta^{-1}X = \theta^{-1}X, (g \in G) \Leftrightarrow \theta^{-1}X \in \mathscr{M} \bar{\otimes} 1_G \Leftrightarrow X \in \theta(\mathscr{M} \bar{\otimes} 1_G) = \pi_\sigma(\mathscr{M})$, since $(\mathscr{M} \bar{\otimes} \mathscr{L}^\infty(G))^{\iota \bar{\otimes} \mathrm{Ad}(\boldsymbol{\rho})} = \mathscr{M} \bar{\otimes} 1_G$. This proves (3). □

Similarly, one can show that for $X \in \mathscr{M} \bar{\otimes} \mathscr{L}^\infty(G)$ we have

$$X \in \pi_\sigma(\mathscr{M}) \Leftrightarrow (\pi_\sigma \bar{\otimes} \iota_G)(X) = (\iota_\mathscr{M} \bar{\otimes} \pi_G)(X) \tag{7}$$

and that if $\mathscr{N} \subset \mathscr{M}$ is a unital W^*-subalgebra, then for $x \in \mathscr{M}$ we have

$$\pi_\sigma(x) \in \mathscr{N} \bar{\otimes} \mathscr{L}^\infty(G) \Rightarrow x \in \mathscr{N}. \tag{8}$$

19.3. Let $\mathscr{R}(\mathscr{M},\sigma) \subset \mathscr{M} \bar{\otimes} \mathscr{B}(\mathscr{L}^2(G))$ be the crossed product of the W^*-algebra $\mathscr{M}$ by the continuous action $\sigma : G \to Aut(\mathscr{M})$ of G on $\mathscr{M}$.

Consider also the action $\delta = \iota_{\mathcal{M}} \bar{\otimes} \delta_G$ of $\hat{\mathbf{G}}$ on $\mathcal{M} \bar{\otimes} \mathcal{B}(\mathcal{L}^2(G))$. Since, by 18.7.(11), its centralizer is $\mathcal{M} \bar{\otimes} \mathcal{L}^\infty(G)$ and since $\pi_\sigma(\mathcal{M}) \subset \mathcal{M} \bar{\otimes} \mathcal{L}^\infty(G)$, we have

$$\delta(\pi_\sigma(x)) = \pi_\sigma(x) \bar{\otimes} 1_G \in \mathcal{R}(\mathcal{M}, \sigma) \bar{\otimes} \mathfrak{L}(G) \quad (x \in \mathcal{M}). \tag{1}$$

On the other hand (18.7.(2)),

$$\delta(1_{\mathcal{M}} \bar{\otimes} \boldsymbol{\lambda}(g)) = 1_{\mathcal{M}} \bar{\otimes} \boldsymbol{\lambda}(g) \bar{\otimes} \boldsymbol{\lambda}(g) \in \mathcal{R}(\mathcal{M}, \sigma) \bar{\otimes} \mathfrak{L}(G) \quad (g \in G). \tag{2}$$

Thus (19.1.(1)), $\delta(\mathcal{R}(\mathcal{M}, \sigma)) \subset \mathcal{R}(\mathcal{M}, \sigma) \bar{\otimes} \mathfrak{L}(G)$ and hence $\mathcal{R}(\mathcal{M}, \sigma)$ is a $\hat{\mathbf{G}}$-subcomodule of the $\hat{\mathbf{G}}$-comodule $\mathcal{M} \bar{\otimes} \mathcal{B}(\mathcal{L}^2(G))$ via $\delta = \iota_{\mathcal{M}} \bar{\otimes} \delta_G$. The restriction of the action $\iota_{\mathcal{M}} \bar{\otimes} \delta_G$ to $\mathcal{R}(\mathcal{M}, \sigma)$ is denoted by

$$\hat{\sigma} : \mathcal{R}(\mathcal{M}, \sigma) \to \mathcal{R}(\mathcal{M}, \sigma) \bar{\otimes} \mathfrak{L}(G)$$

and is called *the dual action of* $\hat{\mathbf{G}}$ *on* $\mathcal{R}(\mathcal{M}, \sigma)$.

If G is commutative, then, according to Proposition 18.9, *the dual action is determined by the continuous action*

$$\hat{\sigma} : \hat{G} \to Aut(\mathcal{R}(\mathcal{M}, \sigma))$$

defined by

$$\hat{\sigma}_\gamma(X) = (1_{\mathcal{M}} \bar{\otimes} \mathbf{m}(\gamma))^* X (1_{\mathcal{M}} \bar{\otimes} \mathbf{m}(\gamma)) \quad (X \in \mathcal{R}(\mathcal{M}, \sigma), \gamma \in \hat{G}).$$

and equalities (1) and (2) can be reformulated as follows (see 18.9.(1)):

$$\hat{\sigma}_\gamma(\pi_\sigma(x)) = \pi_\sigma(x) \quad (x \in \mathcal{M}, \gamma \in \hat{G}) \tag{3}$$

$$\hat{\sigma}_\gamma((1_{\mathcal{M}} \bar{\otimes} \boldsymbol{\lambda}(g)) = \overline{\langle g, \gamma \rangle} (1_{\mathcal{M}} \bar{\otimes} \boldsymbol{\lambda}(g)) \quad (g \in G, \gamma \in \hat{G}). \tag{4}$$

In the general case, we also consider the continuous action $\sigma \bar{\otimes} \mathrm{Ad}(\boldsymbol{\rho})$ of G on $\mathcal{M} \bar{\otimes} \mathcal{B}(\mathcal{L}^2(G))$. Using 19.2.(3), 18.4.(14), and the definition 19.1.(1) of the crossed product, we see that

$$\mathcal{R}(\mathcal{M}, \sigma) \subset (\mathcal{M} \bar{\otimes} \mathcal{B}(\mathcal{L}^2(G))^{\sigma \bar{\otimes} \mathrm{Ad}(\boldsymbol{\rho})}. \tag{5}$$

In the sequel, we shall write $\mathcal{N} = \mathcal{R}(\mathcal{M}, \sigma)$, $\mathcal{P} = \mathcal{M} \bar{\otimes} \mathcal{B}(\mathcal{L}^2(G))$ and $\beta = \sigma \bar{\otimes} \mathrm{Ad}(\boldsymbol{\rho})$; recall also the notation $\delta = \iota_{\mathcal{M}} \bar{\otimes} \delta_G$.

The identity, $\delta_G(\mathrm{Ad}(\boldsymbol{\rho}(t))(x)) = (\mathrm{Ad}(\boldsymbol{\rho}(t)) \bar{\otimes} \iota_G)(\delta_G(x))$ $(t \in G)$, is easily verified for $x = f \in \mathcal{L}^\infty(G)$ and for $x = \boldsymbol{\lambda}(g) \in \mathfrak{L}(G)$, and so it remains valid for any $x \in \mathcal{B}(\mathcal{L}^2(G))$ (see 18.4.(15)). Consequently,

$$\delta \circ \beta_g = (\beta_g \bar{\otimes} \iota_G) \circ \delta \quad (g \in G) \tag{6}$$

that is, the actions δ and β on $\mathcal{P}$ commute. It follows that

$$X \in \mathcal{P}^\delta \Rightarrow \beta_g(X) \in \mathcal{P}^\delta \text{ and } X \in \mathcal{P}^\beta \Rightarrow \delta(X) \in \mathcal{P}^\beta \bar{\otimes} \mathfrak{L}(G)$$

hence, by restriction, we obtain

$$\text{a continuous action } \beta : G \to Aut(\mathscr{P}^{\delta}) \text{ of } G \text{ on } \mathscr{P}^{\delta}, \tag{7}$$

$$\text{an action } \delta : \mathscr{P}^{\beta} \to \mathscr{P}^{\beta} \mathbin{\bar{\otimes}} \mathfrak{L}(G) \text{ of } \hat{\mathbf{G}} \text{ on } \mathscr{P}^{\beta}. \tag{8}$$

Proposition. *For every continuous action $\sigma : G \to Aut(\mathscr{M})$ of G on the W^*-algebra $\mathscr{M}$ we have*

$$\pi_\sigma(\mathscr{M}) = \mathscr{R}(\mathscr{M},\sigma)^{\hat{\sigma}} \tag{9}$$

where $\hat{\sigma}$ is the dual action.

Proof. With the above notation, we have $\pi_\sigma(\mathscr{M}) \subset \mathscr{N}^{\delta}$ (by (1)), $\mathscr{N} \subset \mathscr{P}^{\beta}$ (by (5)) and $\mathscr{P}^{\delta} = \mathscr{M} \mathbin{\bar{\otimes}} \mathscr{L}^\infty(G)$ (by 18.7.(11)). Since β and δ commute, using Lemma 19.2.(3), we obtain $\pi_\sigma(\mathscr{M}) \subset \mathscr{N}^{\delta} \subset (\mathscr{P}^{\beta})^{\delta} = (\mathscr{P}^{\delta})^{\beta} = (\mathscr{M} \mathbin{\bar{\otimes}} \mathscr{L}^\infty(G))^{\beta} = \pi_\sigma(\mathscr{M})$ and hence $\pi_\sigma(\mathscr{M}) = \mathscr{N}^{\delta}$. □

This proof also showed that

$$(\mathscr{P}^{\beta})^{\delta} = \pi_\sigma(\mathscr{M}). \tag{10}$$

Finally, note that the mapping $\chi : \mathscr{L}^\infty(G) \ni f \mapsto 1_{\mathscr{M}} \mathbin{\bar{\otimes}} f \in \mathscr{P}$ is a unital injective normal *-homomorphism and that

$$\beta_g(\chi(f)) = \chi(\mathrm{Ad}(\boldsymbol{\rho}(g))(f)) \quad (f \in \mathscr{L}^\infty(G), g \in G). \tag{11}$$

19.4. Consider now the crossed product $\mathscr{R}(\mathscr{M},\delta) \subset \mathscr{M} \mathbin{\bar{\otimes}} \mathscr{B}(\mathscr{L}^2(G))$ of the W^*-algebra $\mathscr{M}$ by the action $\delta : \mathscr{M} \to \mathscr{M} \mathbin{\bar{\otimes}} \mathfrak{L}(G)$ of $\hat{\mathbf{G}}$ on $\mathscr{M}$.

Consider also the continuous action $\sigma = \iota_{\mathscr{M}} \mathbin{\bar{\otimes}} \mathrm{Ad}(\boldsymbol{\rho})$ of G on $\mathscr{M} \mathbin{\bar{\otimes}} \mathscr{B}(\mathscr{L}^2(G))$. Since, by 18.4.(14), the centralizer of this action is $\mathscr{M} \mathbin{\bar{\otimes}} \mathfrak{L}(G)$ and since $\delta(\mathscr{M}) \subset \mathscr{M} \mathbin{\bar{\otimes}} \mathfrak{L}(G)$, we have

$$\sigma_g(\delta(x)) = \delta(x) \in \mathscr{R}(\mathscr{M},\delta) \quad (x \in \mathscr{M}, g \in G). \tag{1}$$

On the other hand,

$$\sigma_g(1_{\mathscr{M}} \mathbin{\bar{\otimes}} f) = 1_{\mathscr{M}} \mathbin{\bar{\otimes}} (\mathrm{Ad}(\boldsymbol{\rho}(g))(f)) \in \mathscr{R}(\mathscr{M},\delta) \quad (f \in \mathscr{L}^\infty(G), g \in G). \tag{2}$$

Thus (19.1.(4)), $\sigma_g(\mathscr{R}(\mathscr{M},\delta)) = \mathscr{R}(\mathscr{M},\delta)$ $(g \in G)$. The restriction of the continuous action $\sigma = \iota_{\mathscr{M}} \mathbin{\bar{\otimes}} \mathrm{Ad}(\boldsymbol{\rho})$ to $\mathscr{R}(\mathscr{M},\delta)$ is denoted by

$$\hat{\delta} : G \to Aut(\mathscr{R}(\mathscr{M},\delta))$$

and is called *the dual action of G on $\mathscr{R}(\mathscr{M},\delta)$*.

If G is commutative, then the method of defining the dual action given in that section leads to the same result as the procedure of Section 19.3. This can easily be verified using the Fourier–Plancherel isomorphism (18.8.(4); a continuous action of G can be regarded also as an action of the dual object of the group $\hat{G}$.

19.5. Let $\sigma : G \to Aut(\mathscr{M})$ be a continuous action of G on $\mathscr{M}$. We then have a dual action $\hat{\sigma} : \mathscr{R}(\mathscr{M},\sigma) \to \mathscr{R}(\mathscr{M},\sigma) \mathbin{\bar{\otimes}} \mathfrak{L}(G)$ of $\hat{\mathbf{G}}$ on the crossed product $\mathscr{R}(\mathscr{M},\sigma)$ (19.3) and also a second dual action $\hat{\hat{\sigma}} : G \to Aut(\mathscr{R}(\mathscr{R}(\mathscr{M},\sigma),\hat{\sigma}))$ of G on the second crossed product $\mathscr{R}(\mathscr{R}(\mathscr{M},\sigma),\hat{\sigma})$ (19.4).

Theorem. *Let $\sigma : G \to Aut(\mathscr{M})$ be a continuous action of the locally compact group G on the W^*-algebra $\mathscr{M}$. There exists a *-isomorphism*

$$\Phi : \mathscr{M} \bar{\otimes} \mathscr{B}(\mathscr{L}^2(G)) \to \mathscr{R}(\mathscr{R}(\mathscr{M},\sigma),\hat{\sigma})$$

such that

$$\hat{\hat{\sigma}}_g \circ \Phi = \Phi \circ (\sigma_g \bar{\otimes} \mathrm{Ad}(\boldsymbol{\rho}(g))) \quad (g \in G).$$

Briefly, we shall write

$$(\mathscr{R}(\mathscr{R}(\mathscr{M},\sigma),\hat{\sigma})); \hat{\hat{\sigma}}) \approx (\mathscr{M} \bar{\otimes} \mathscr{B}(\mathscr{L}^2(G)); \sigma \bar{\otimes} \mathrm{Ad}(\boldsymbol{\rho})). \tag{1}$$

Proof. We shall use the notation introduced in Section 19.3, that is, $\mathscr{N} = \mathscr{R}(\mathscr{M},\sigma)$, $\mathscr{P} = \mathscr{M} \bar{\otimes} \mathscr{B}(\mathscr{L}^2(G))$, $\beta = \sigma \bar{\otimes} \mathrm{Ad}(\boldsymbol{\rho})$ and $\delta = \iota_{\mathscr{M}} \bar{\otimes} \delta_G$; recall that $\hat{\sigma} = \delta|\mathscr{N}$.

For each $X \in \mathscr{P}$, we consider the element $\Psi(X) \in \mathscr{P} \bar{\otimes} \mathscr{L}^\infty(G)$ defined by the function $g \mapsto \beta_g(X)$, that is,

$$\langle \Psi(X), \varphi \bar{\otimes} k\rangle = \int \varphi(\beta_g(X))k(g)\,\mathrm{d}g \quad (\varphi \in \mathscr{P}_*, k \in \mathscr{L}^1(G)). \tag{2}$$

It is easy to check that $\Psi : \mathscr{P} \ni X \mapsto \Psi(X) \in \mathscr{P} \bar{\otimes} \mathscr{B}(\mathscr{L}^2(G)))$ is a unital injective normal *-homomorphism. We define another unital injective normal *-homomorphism $\Phi : \mathscr{P} \to \mathscr{P} \bar{\otimes} \mathscr{B}(\mathscr{L}^2(G))$ by

$$\Phi(X) = (1_{\mathscr{M}} \bar{\otimes} W_G^*)\Psi(X)(1_{\mathscr{M}} \bar{\otimes} W_G) \quad (X \in \mathscr{P}). \tag{3}$$

If $X \in \mathscr{N} \subset \mathscr{P}^\beta$ (19.3.(5)), then from (2) it follows that $\Psi(X) = X \bar{\otimes} 1_G$ and from (3) and 18.7.(10) we obtain

$$\Phi(X) = \delta(X) = \hat{\sigma}(X) \quad (X \in \mathscr{N}). \tag{4}$$

Since $\hat{\sigma}(\mathscr{N}) \subset \mathscr{R}(\mathscr{N},\hat{\sigma})^{\hat{\hat{\sigma}}}$ (19.4.(1)) and $\mathscr{N} \subset \mathscr{P}^\beta$ (19.3.(5)), it follows that

$$\hat{\hat{\sigma}}_g(\Phi(X)) = \Phi(X) = \Phi(\beta_g(X)) \quad (X \in \mathscr{N}, g \in G). \tag{5}$$

On the other hand, if $f \in \mathscr{L}^\infty(G)$, then it is easy to check that $\Psi(1_{\mathscr{M}} \bar{\otimes} f)$ is the element $F \in \mathscr{M} \bar{\otimes} \mathscr{L}^\infty(G_s) \bar{\otimes} \mathscr{L}^\infty(G_t) \subset \mathscr{P} \bar{\otimes} \mathscr{B}(\mathscr{L}^2(G_t))$ defined by the function $F : G \times G \ni (s,t) \mapsto f(st) \cdot 1_{\mathscr{M}} \in \mathscr{M}$. For every $\xi \in \mathscr{L}^2(G \times G, \mathscr{H})$ we have

$$\begin{aligned}((1_{\mathscr{M}} \bar{\otimes} W_G^*)F(1_{\mathscr{M}} \bar{\otimes} W_G)\xi)(s,t) &= (F(1_{\mathscr{M}} \bar{\otimes} W_G)\xi)(s,s^{-1}t)\\ = F(s,s^{-1}t)((1_{\mathscr{M}} \bar{\otimes} W_G)\xi)(s,s^{-1}t) &= f(t)\xi(s,t) = ((1_{\mathscr{M}} \bar{\otimes} 1_G \bar{\otimes} f)\xi)(s,t)\end{aligned}$$

hence

$$\Phi(1_{\mathscr{M}} \bar{\otimes} f) = 1_{\mathscr{P}} \bar{\otimes} f \quad (f \in \mathscr{L}^\infty(G)). \tag{6}$$

Using 19.4.(2), we obtain

$$\hat{\hat{\sigma}}_g(\Phi(1_{\mathscr{M}} \bar{\otimes} f)) = 1_{\mathscr{P}} \bar{\otimes} (\mathrm{Ad}(\boldsymbol{\rho}(g))(f)) = \Phi(\beta_g(1_{\mathscr{M}} \bar{\otimes} f)) \quad (f \in \mathscr{L}^\infty(G), g \in G). \tag{7}$$

Thus, using Lemma 19.2.(2), the definition of $\mathscr{R}(\mathscr{N}, \hat{\sigma})$ (19.1.(4)) and (4) and (6), we get

$$\Phi(P) = \Phi(\mathscr{R}\{\mathscr{N}, 1_{\mathscr{M}} \bar{\otimes} \mathscr{L}^\infty(G)\}) = \mathscr{R}\{\hat{\sigma}(\mathscr{N}), 1_{\mathscr{P}} \bar{\otimes} \mathscr{L}^\infty(G)\} = \mathscr{R}(\mathscr{N}, \hat{\sigma}),$$

and from (5) and (7) it follows that $\hat{\hat{\sigma}}_g \circ \Phi = \Phi \circ \beta_g$. □

The above Theorem is usually called "the Takesaki duality theorem."

19.6. In this section, we summarize the essential facts concerning crossed products by continuous group actions.

Thus, let $\sigma : G \to Aut(\mathscr{M})$ be a continuous action of G on the W^*-algebra $\mathscr{M}$ and $\mathscr{N} = \mathscr{R}(\mathscr{M}, \sigma)$. Using the $*$-isomorphism $\pi_\sigma : \mathscr{M} \to \pi_\sigma(\mathscr{M}) \subset \mathscr{N}$, we shall identify $\mathscr{M}$ with a unital W^*-subalgebra of $\mathscr{N}$, hence $\mathscr{M} \subset \mathscr{N}$.

There exist an s-continuous unitary representation

$$G \ni f \mapsto u(g) = 1 \bar{\otimes} \boldsymbol{\lambda}(g) \in \mathscr{N} \tag{1}$$

and a dual action of $\hat{\mathbf{G}}$ on $\mathscr{N}$ (19.3)

$$\delta = \hat{\sigma} : \mathscr{N} \to \mathscr{N} \bar{\otimes} \mathfrak{L}(G), \tag{2}$$

such that (19.3.(2))

$$\delta(u(g)) = u(g) \bar{\otimes} \boldsymbol{\lambda}(g) \quad (g \in G), \tag{3}$$

which characterize the W^*-subalgebra $\mathscr{M}$ of $\mathscr{N}$ (Proposition 19.3)

$$\mathscr{M} = \mathscr{N}^\delta, \tag{4}$$

and the continuous action σ of G on $\mathscr{M}$ (19.1.(2)),

$$\sigma_g(x) = u(g)xu(g)^* \quad (x \in \mathscr{M}, g \in G). \tag{5}$$

In particular,

$$\mathscr{N} = \mathscr{R}\{\mathscr{N}^\delta, u(G)\}. \tag{6}$$

If G is commutative, then the dual action is a continuous action of the dual group $\hat{G}$

$$\theta = \hat{\sigma} : \hat{G} \to Aut(\mathscr{N}) \tag{2'}$$

such that (19.3.(4))

$$\theta_\gamma(u(g)) = \overline{\langle g, \gamma \rangle} u(g) \; (g \in G, \gamma \in \hat{G}) \tag{3'}$$

$$\mathscr{M} = \mathscr{N}^\delta \tag{4'}$$

$$\mathscr{N} = \mathscr{R}\{\mathscr{N}^\delta, u(G)\}. \tag{5'}$$

We shall show that, conversely, the existence of a unitary representation (1) and an action (2) (resp. (2')), which satisfy the commutation relations (3) (resp. (3')) implies the fact that $\mathscr{N}$ is a crossed product $\mathscr{R}(\mathscr{M}, \sigma)$, where $\mathscr{M}$ and σ are determined by (4) (resp. (4')) and (5), in particular the generation relation (6) (resp. (6')) holds.

To this end, we need some preliminary results that are of independent interest and will also be used in other situations.

19.7 Proposition. *Let $\mathscr{N}$ be a W^*-algebra with the property that there exist an s-continuous unitary representation $u : G \to \mathscr{N}$ and an action $\delta : \mathscr{N} \to \mathscr{N} \bar{\otimes} \mathfrak{L}(G)$ of $\hat{\mathbf{G}}$ on $\mathscr{N}$ such that*

$$\delta(u(g)) = u(g) \bar{\otimes} \boldsymbol{\lambda}(g) \quad (g \in G). \tag{1}$$

Then the faithful normal operator-valued weight $P_\delta : \mathscr{N}^+ \to \overline{(\mathscr{N}^\delta)}^+$,

$$P_\delta(x) = E_{\mathscr{N}}^{\omega_G}(\delta(x)) \quad (x \in \mathscr{N}^+) \tag{2}$$

is semifinite and

$$P_\delta(u(g)xu(g)^*) = \Delta_G(g)u(g)P_\delta(x)u(g)^* \quad (x \in \mathscr{N}^+, g \in G). \tag{3}$$

For every $k \in \mathscr{K}(G)$ with $u(k) \geq 0$ we have

$$P_\delta(u(k)) = k(e) \cdot 1_G. \tag{4}$$

For all $h, k \in \mathscr{K}(G)$ we have $u(h), u(k) \in \mathfrak{N}_{P_\delta}$ and

$$P_\delta(u(h)^*u(k)) = (h^* \; * \; k)(e) \cdot 1_G. \tag{5}$$

Proof. Recall (18.17.(12)) that $\omega_G \circ j_G = \omega_G$. Thus, using the definition (2) (or 18.19.(1)) of P_δ, (1) and 18.17.(8), for $k \in \mathscr{K}(G)$ with $u(k) \geq 0$ and $\psi \in \mathscr{N}_*^+$ we have

$$\begin{aligned}
\langle P_\delta(u(k)), \psi \rangle &= \langle \delta(u(k)), \psi \bar{\otimes} \omega_G \rangle = \left\langle \delta\left(\int k(g)u(g)\,\mathrm{d}g \right), \psi \bar{\otimes} \omega_G \right\rangle \\
&= \left\langle \int k(g)(u(g) \bar{\otimes} \boldsymbol{\lambda}(g))\,\mathrm{d}g, \psi \bar{\otimes} \omega_G \right\rangle = \left\langle \int k(g)\psi(u(g))\boldsymbol{\lambda}(g)\,\mathrm{d}g, \omega_G \right\rangle \\
&= \langle \boldsymbol{\lambda}(k(\cdot)\psi(u(\cdot))), \omega_G \rangle = \langle k(e)\psi(u(e)) = (k(e) \cdot 1_{\mathscr{N}}, \psi \rangle,
\end{aligned}$$

which proves (4).

For every $k \in \mathcal{K}(G)$, we have $k^\sharp * k \in \mathcal{K}(G)$ and $u(k^\sharp * k) = u(k)^* u(k) \geq 0$, hence we infer from (4) that $u(k) \in \mathfrak{N}_{P_\delta}$ and $P_\delta(u(k)^* u(k)) = (k^\sharp * k)(e) \cdot 1_{\mathcal{N}}$. Thus (5) follows on applying the polarization relation.

If $\{k_i\}$ is a net in $\mathcal{K}(G)$ such that $u(k_i) \xrightarrow{s} 1_{\mathcal{N}}$, then for every $x \in \mathcal{N}$ we have $xu(k_i) \in \mathfrak{N}_{P_\delta}$ and $xu(k_i) \xrightarrow{s} s$. Hence P_δ is semifinite.

Finally, using 18.17.(13), we obtain for $x \in \mathcal{N}^+, g \in G$ and $\psi \in \mathcal{N}_*^+$

$$\begin{aligned}
\langle P_\delta(u(g)xu(g)^*), \psi\rangle &= \langle \delta(u(g)xu(g)^*), \psi \mathbin{\bar{\otimes}} \omega_G\rangle \\
&= \langle (u(g) \mathbin{\bar{\otimes}} \boldsymbol{\lambda}(g))\delta(x)(u(g) \mathbin{\bar{\otimes}} \boldsymbol{\lambda}(g))^*, \psi \mathbin{\bar{\otimes}} \omega_G\rangle \\
&= \langle \delta(x), (\psi \circ \mathrm{Ad}(u(g))) \mathbin{\bar{\otimes}} (\omega_G \circ \mathrm{Ad}(\boldsymbol{\lambda}(g)))\rangle \\
&= \Delta_G(g)\langle \delta(x), \psi \circ \mathrm{Ad}(u(g))) \mathbin{\bar{\otimes}} \omega_G\rangle \\
&= \Delta_G(g)\langle P_\delta(x), \psi \circ \mathrm{Ad}(u(g))\rangle \\
&= \Delta_G(g)\langle u(g)P_\delta(x)u(g)^*, \psi\rangle,
\end{aligned}$$

which proves (3). □

If G is commutative, then it is more convenient to state the above proposition as follows:

Let $\mathcal{N}$ be a W-algebra with the property that there exist an s-continuous unitary representation $u : G \to \mathcal{N}$ and a continuous action $\theta : \hat{G} \to Aut(\mathcal{N})$ of $\hat{G}$ on $\mathcal{N}$ such that*

$$\theta_\gamma(u(g)) = \overline{\langle g, \gamma\rangle} u(g) \quad (g \in G, \gamma \in \hat{G}). \tag{1'}$$

Then the faithful normal operator-valued weight $P_\theta : \mathcal{N}^+ \to (\overline{\mathcal{N}^\theta})^+$,

$$P_\theta(x) = \int \theta_\gamma(x)\,\mathrm{d}\gamma \quad (x \in \mathcal{N}^+) \tag{2'}$$

is semifinite and

$$P_\theta(u(g)xu(g)^*) = u(g)P_\theta(x)u(g)^* \quad (x \in \mathcal{N}^+, g \in G). \tag{3'}$$

Also, identities similar to (4) and (5) hold.

In this case, the main part of the proof reduces to an application of the Fourier inversion theorem. Indeed, for any positive definite function $k \in \mathcal{K}(G)$ and every $\psi \in \mathcal{N}_*^+$ we have

$$\begin{aligned}
\langle P_\theta(u(k)), \psi\rangle &= \left\langle \int \theta\gamma\left(\int k(g)u(g)\,\mathrm{d}g\right)\mathrm{d}\gamma, \psi\right\rangle \\
&= \int \langle e, \gamma\rangle \left(\int \overline{\langle g, \gamma\rangle} k(g)\psi(u(g))\,\mathrm{d}g\right)\mathrm{d}\gamma = k(e)\psi(u(e)) = \langle k(e) \cdot 1_{\mathcal{N}}, \psi\rangle.
\end{aligned}$$

19.8. In particular, for every continuous action $\sigma : G \to Aut(\mathcal{M})$, the dual action $\hat{\sigma} : \mathcal{R}(\mathcal{M}, \sigma) \to \mathcal{R}(\mathcal{M}, \sigma) \mathbin{\bar{\otimes}} \mathfrak{L}(G)$ defines an n.s.f. operator-valued weight

$$P_{\hat{\sigma}} : \mathcal{R}(\mathcal{M}, \sigma)^+ \to \overline{\pi_\sigma(\mathcal{M})}^+.$$

Thus, if φ is an n.s.f. weight on $\mathscr{M}$, then, according to Proposition 11.6,

$$\hat{\varphi} = \varphi \circ \pi_\sigma^{-1} \circ P_{\hat{\sigma}}$$

is an n.s.f. weight on $\mathscr{R}(\mathscr{M}, \sigma)$, called *the dual weight of* φ.

The theorem in this section will characterize the dual weight of a given φ and also the set of all dual weights on $\mathscr{R}(\mathscr{M}, \sigma)$. In order to obtain such characterizations, it is necessary to know the values of $\hat{\varphi}$ on a wide dass of elements of $\mathscr{R}(\mathscr{M}, \sigma)$. For this reason, we begin by defining and studying this class. As usual, we put $u(g) = 1_{\mathscr{M}} \bar{\otimes} \boldsymbol{\lambda}(g)$ $(g \in G)$.

Every compactly supported w-continuous function $f : G \to \mathscr{M}$ defines an element

$$T_f^\sigma = \int \pi_\sigma(f(g))u(g)\, \mathrm{d}g \in \mathscr{R}\{\pi_\sigma(\mathscr{M}), u(G)\} = \mathscr{R}(\mathscr{M}, \sigma).$$

We recall (18.21, 18.22) that the mapping $f \mapsto T_f^\sigma$ is linear and

$$(T_f^\sigma)^* = T_{f^\sharp}^\sigma, \text{ where } f^\sharp(g) = \Delta_G(g)^{-1} f(g^{-1})^* \quad (g \in G) \tag{1}$$

$$T_{f_1}^\sigma T_{f_2}^\sigma = T_{f_1 * f_2}^\sigma, \text{ where } (f_1 * f_2)(g) = \int f_1(t) f_2(t^{-1}g)\, \mathrm{d}t, \quad (g \in G). \tag{2}$$

We also recall (18.3) that the action $\hat{\sigma}$ of $\hat{\mathbf{G}}$ on $\mathscr{R}(\mathscr{M}, \sigma)$ defines a right $\mathscr{A}(G)$-module structure on $\mathscr{R}(\mathscr{M}, \sigma)$. In this connection, it is easy to check that for every $k \in \mathscr{A}(G)$ we have

$$k \cdot T_f^\sigma = T_{kf}^\sigma, \text{ where } (kf)(g) = k(g) f(g), \ (g \in G). \tag{3}$$

Finally, if $f \in \mathscr{K}(G)$ and if we put $(fx)(g) = f(g)x$, $(x \in \mathscr{M}, g \in G)$, then

$$T_{fx}^\sigma = \pi_\sigma(x) u(f). \tag{4}$$

Consequently,

$$\mathscr{B} = \{T_f^\sigma ; f : G \to \mathscr{H} \ w\text{-continuous with compact support}\}$$

is an s^*-dense $*$-subalgebra of $\mathscr{R}(\mathscr{M}, \sigma)$.

On the other hand, from the definition (19.3) of the dual action $\hat{\sigma}$ and 18.22.(2) it follows that if $T_f^\sigma \geq 0$ (in particular if f is of the form $f^\sharp * f$),then

$$P_{\hat{\sigma}}(T_f^\sigma) = \pi_\sigma(f(e)). \tag{5}$$

Theorem. *Let $\sigma : G \to Aut(\mathscr{M})$ be a continuous action of G on the W^*-algebra $\mathscr{M}$.*

For any n.s.f. weight φ on $\mathscr{M}$, the dual weight $\hat{\varphi} = \varphi \circ \pi_\sigma^{-1} \circ P_{\hat{\sigma}}$ is the unique n.s.f. weight $\mathscr{R}(\mathscr{M}, \sigma)$ with the properties:

(i) *if $f : G \to \mathscr{M}$ is a compactly supported w-continuous function and $T_f^\sigma \geq 0$, then*

$$\hat{\varphi}(T_f^\sigma) = \varphi(f(e)); \tag{6}$$

(ii) *for all* $x \in \mathcal{M}, g \in G, t \in \mathbb{R}$ *we have*

$$\sigma_t^{\hat{\varphi}}(\pi_\sigma(x)) = \pi_\sigma(\sigma_t^\varphi(x)) \tag{7}$$

$$\sigma_t^{\hat{\varphi}}(1_{\mathcal{M}} \bar{\otimes} \boldsymbol{\lambda}(g)) = \Delta_G(g)^{\mathrm{i}t}(1_{\mathcal{M}} \bar{\otimes} \boldsymbol{\lambda}(g))\pi_\sigma([D(\varphi \circ \sigma_g) : D\varphi]_t). \tag{8}$$

Moreover, the dual weights are exactly those weights that are invariant with respect to the dual action, that is, the mapping $\varphi \mapsto \hat{\varphi}$ *establishes a bijection between the sets* $\{\varphi; \varphi$ *n.s.f. weight on* $\mathcal{M}\}$ *and* $\{\Psi : \Psi\ (\hat{\sigma}, j_G)$*-invariant n.s.f. weight on* $\mathcal{R}(\mathcal{M}, \sigma)\}$.

Proof. Let φ be an n.s.f. weight on $\mathcal{M}$. (6) follows obviously from (5). On the other hand, using Theorem 11.9, we obtain (7) and

$$[D(\varphi \circ \sigma_g)\hat{}: D\hat{\varphi}]_t = \pi_\sigma([D(\varphi \circ \sigma_g) : D\varphi]_t) \quad (g \in G, t \in \mathbb{R}),$$

so that, by 3.10,

$$\sigma_t^{\hat{\varphi}}(u(g)) = \sigma_t^{\hat{\varphi},(\varphi \circ \sigma_g)\hat{}}(u(g))\pi_\sigma([D(\varphi \circ \sigma_g) : D\varphi]_t) \quad (g \in G, t \in \mathbb{R}),$$

and (8) will follow once we show that

$$\sigma_t^{\hat{\varphi},(\varphi \circ \sigma_g)\hat{}}(u(g)) = \Delta_G(g)^{\mathrm{i}t}u(g) \quad (g \in G, t \in \mathbb{R}), \tag{9}$$

By 3.15.(1), (9) is equivalent to

$$(\varphi \circ \sigma_g)\hat{}(X) = \Delta_G(g)^{-1}\hat{\varphi}(u(g)Xu(g)^*) \quad (X \in \mathcal{R}(\mathcal{M}, \sigma)^+, g \in G) \tag{10}$$

and, since $(\varphi \circ \sigma_g)\hat{}(X) = (\varphi \circ \pi_\sigma^{-1})(u(g)P_{\hat{\sigma}}(X)u(g)^*)$, $\Delta_G(g)^{-1}\hat{\varphi}(u(g)Xu(g)^*) = (\varphi \circ \pi_\sigma^{-1})(P_{\hat{\sigma}}(u(g)Xu(g)^*))$, (10) follows using 19.7.(3).

Equations (7) and (8) determine the modular automorphism group of $\hat{\varphi}$ uniquely and permit an immediate verification of the fact that the s^*-dense $*$-subalgebra $\mathcal{B}$ of $\mathcal{R}(\mathcal{M}, \sigma)$ is $\sigma^{\hat{\varphi}}$-invariant. On the other hand, (6) determines the values $\hat{\varphi}(X^*X)$ for $X \in \mathcal{B}$. Consequently, according to Theorem 6.2, conditions (i) and (ii) determine the dual weight $\hat{\varphi}$ uniquely.

We now show that the dual weight $\hat{\varphi}$ is $(\hat{\sigma}, j_G)$-invariant by checking conditions (a)–(d) from Theorem 18.12. Since $\hat{\varphi}$ is faithful, condition (a) is trivially satisfied. The commutation condition (b), that is, $\hat{\sigma} \circ \sigma_t^{\hat{\varphi}} = (\sigma_t^{\hat{\varphi}} \bar{\otimes} \iota_G) \circ \hat{\sigma}$, follows immediately using equalities (7), (8), and 19.6.(3), 19.6.(4). For $0 \leq T_f^\sigma \in \mathcal{B}$ and $k \in \mathcal{A}(G)^+$, we have $\langle \hat{\sigma}(T_f^\sigma), \hat{\varphi} \bar{\otimes} k \rangle = \langle k \cdot T_f^\sigma, \hat{\varphi} \rangle = \langle T_{kf}^\sigma, \hat{\varphi} \rangle = k(e)\varphi(f(e)) = \langle 1_G, k \rangle \langle T_f^\sigma, \hat{\varphi} \rangle = \langle T_f^\sigma \bar{\otimes} 1_G, \hat{\varphi} \bar{\otimes} k \rangle$, hence condition (c) is satisfied. Consider now $X = T_{f_1}^\sigma \in \mathcal{B}, Y = T_{f_2}^\sigma \in \mathcal{B}$, and $k \in \mathcal{A}(G)$. Using (3) and 18.7.(7), we see that $k \cdot X \in \mathcal{B}$ is determined by the function $f_3(g) = k(g)f_1(g)$, and $k^0 \cdot Y \in \mathcal{B}$ is determined by the function $f_4(g) = \overline{k(g)}f_2(g)$. From (1), (2), and (6), it follows that $\hat{\varphi}((T_f^\sigma)^*(T_f^\sigma)) = (f^\sharp * f)(e)$; this equality, together with the usual polarization relation, justifies the following computations:

$$\langle (Y^* \bar{\otimes} 1_G)\hat{\sigma}(X), \hat{\varphi} \bar{\otimes} k \rangle = ((k \cdot X)_{\hat{\varphi}} | Y_{\hat{\varphi}})_{\hat{\varphi}} = \hat{\varphi}(Y^*(k \cdot X)) = (f_2^\sharp * f_3)(e),$$

$$\langle \hat{\sigma}(Y^*)(X \bar{\otimes} 1_G), \hat{\varphi} \bar{\otimes} k \circ j_G \rangle = (X_{\hat{\varphi}} | (k^0 \cdot Y)_{\hat{\varphi}})_{\hat{\varphi}} = \hat{\varphi}((k^0 \cdot Y)^* X) = (f_4^\sharp * f_1)(e).$$

Since $(f_2^\sharp * f_3)(e) = (f_4^\sharp * f_1)(e)$, condition (d) is also satisfied.

Finally, let Ψ be a $(\hat{\sigma}, j_G)$-invariant n.s.f. weight on $\mathcal{R}(\mathcal{M},\sigma)$ and φ any n.s.f. weight on $\mathcal{M}$. By Corollary 18.13, $[D\Psi : D\hat{\varphi}]_t \in \mathcal{R}(\mathcal{M},\sigma)^{\hat{\sigma}} = \pi_\sigma(\mathcal{M})$, hence, by Theorem 5.1, there exists an n.s.f. weight ψ on $\mathcal{M}$ such that $\pi_\sigma([D\psi : D\varphi]_t) = [D\Psi : D\hat{\varphi}]_t$ $(t \in \mathbb{R})$. Then, using Theorem 11.9, we deduce that $[D\hat{\psi} : D\hat{\varphi}]_t = \pi_\sigma([D\psi : D\varphi]_t) = [D\Psi : D\hat{\varphi}]_t$ $(t \in \mathbb{R})$, and hence, by Corollary 3.6, $\Psi = \hat{\psi}$.

If $\hat{\psi} = \hat{\varphi}$, then $\pi_\sigma([D\psi : D\varphi]_t) = [D\hat{\psi} : D\hat{\varphi}]_t = 1$ $(t \in \mathbb{R})$, and hence $\psi = \varphi$.

The proof of the theorem is complete. □

19.9. The following important theorem characterizes those W^*-algebras that are crossed products by a continuous action of G.

Theorem (Landstad). *Let $\mathcal{N}$ be a W^*-algebra with the property that there exist an s-continuous unitary representation $u : G \to \mathcal{N}$ and an action $\delta : \mathcal{N} \to \mathcal{N} \bar{\otimes} \mathfrak{L}(G)$ of $\hat{\mathbf{G}}$ on $\mathcal{N}$ such that*

$$\delta(u(g)) = u(g) \bar{\otimes} \boldsymbol{\lambda}(g) \quad (g \in G).$$

Then $\mathcal{N}$ is generated by $\mathcal{N}^\delta$ and $u(G)$,

$$\mathcal{N} = \mathcal{R}\{\mathcal{N}^\delta, u(G)\}, \tag{1}$$

and $u(g)\mathcal{N}^\delta u(g)^* = \mathcal{N}^\delta$ for every $g \in G$.

*Consider the W^*-algebra $\mathcal{M} = \mathcal{N}^\delta$ and the continuous action $\sigma : G \to Aut(\mathcal{M})$ of G on $\mathcal{M}$ defined by $\sigma_g = \mathrm{Ad}(u(g))|\mathcal{M}, (g \in G)$. There exists a *-isomorphism*

$$\Phi : \mathcal{N} \to \mathcal{R}(\mathcal{M},\sigma)$$

such that

$$\Phi(x) = \pi_\sigma(x) \quad (x \in \mathcal{M}) \tag{2}$$
$$\Phi(u(g)) = 1_{\mathcal{M}} \bar{\otimes} \boldsymbol{\lambda}(g) \quad (g \in G) \tag{3}$$
$$(\Phi \bar{\otimes} \iota_G) \circ \delta = \hat{\sigma} \circ \Phi. \tag{4}$$

Briefly,

$$(\mathcal{N},\delta) \approx (\mathcal{R}(\mathcal{M},\sigma),\hat{\sigma}). \tag{5}$$

If G is abelian, then it is more convenient to state the theorem as follows:

Let $\mathcal{N}$ be a W^*-algebra with the property that there exist an s-continuous unitary representation $u : G \to \mathcal{N}$ and a continuous action $\theta : \hat{G} \to Aut(\mathcal{N})$ of $\hat{G}$ on $\mathcal{N}$ such that

$$\theta_\gamma(u(g)) = \overline{\langle g,\gamma\rangle} u(g) \quad (g \in G, \gamma \in \hat{G}). \tag{6}$$

Then $\mathcal{N}$ is generated by $\mathcal{N}^\delta$ and $u(G)$,

$$\mathcal{N} = \mathcal{R}\{\mathcal{N}^\delta, u(G)\}, \tag{1'}$$

and $u(g)\mathcal{N}^\delta u(g)^* = \mathcal{N}^\delta$ for every $g \in G$.

Consider the W-algebra* $\mathscr{M} = \mathscr{N}^{\delta}$ *and the continuous action* $\sigma : G \to Aut(\mathscr{M})$ *of G on* $\mathscr{M}$ *defined by* $\sigma_g = \mathrm{Ad}(u(g))|\mathscr{M}(g \in G)$*. There exists a *-isomorphism*

$$\Phi : \mathscr{N} \to \mathscr{R}(\mathscr{M}, \sigma)$$

such that

$$\Phi(x) = \pi_\sigma(x) \quad (x \in \mathscr{M}) \tag{2'}$$
$$\Phi(u(g)) = 1_{\mathscr{M}} \bar{\otimes} \boldsymbol{\lambda}(g) \quad (g \in G) \tag{3'}$$
$$\Phi \circ \theta_g = \hat{\sigma}_g \circ \Phi \quad (g \in G). \tag{4'}$$

Briefly,

$$(\mathscr{N}, \theta) \equiv (\mathscr{R}(\mathscr{M}, \sigma), \hat{\sigma}). \tag{5'}$$

Of course, if G is abelian, then the above two statements are equivalent (see 18.6, 18.8, and 18.9). The proof of the theorem is contained in Sections 19.10–19.12.

19.10. The main result contained in Theorem 19.9 is the assertion that $\mathscr{N} = \mathscr{R}\{\mathscr{N}^{\delta}, u(G)\}$, which will be proved in the next section. *In this section, we assume that* $\mathscr{N} = \mathscr{R}\{\mathscr{N}^{\delta}, u(G)\}$*; we prove the other assertions in Theorem 19.9.* We shall use the notation of the statement of Theorem 19.9 and assume $\mathscr{N} \subset \mathscr{B}(\mathscr{H})$ realized as a von Neumann algebra.

Assume first that G is abelian. Then, for every $x \in \mathscr{M} = \mathscr{N}^{\theta}, g \in G$ and $\gamma \in \hat{G}$, we have $\theta_\gamma(u(g)xu(g)^*) = \theta_\gamma(u(g))\theta_\gamma(x)\theta_\gamma(u(g^{-1})) = \overline{\langle g, \gamma\rangle}\ \overline{\langle g^{-1}, \gamma\rangle} u(g)xu(g^{-1}) = u(g)xu(g)^*$, hence $u(g)xu(g)^* \in \mathscr{N}^{\theta} = \mathscr{M}$.

Consider now the general case. For every $x \in \mathscr{M} = \mathscr{N}^{\delta}$ and $g \in G$, we have $\delta(u(g)xu(g)^*) = \delta(u(g))\delta(x)\delta(u(g)^*) = (u(g) \bar{\otimes} \boldsymbol{\lambda}(g))(x \bar{\otimes} 1_G)(u(g)^* \bar{\otimes} \boldsymbol{\lambda}(g)^*) = (u(g)xu(g)^*) \bar{\otimes} 1_G$, hence $u(g)xu(g)^* \in \mathscr{N}^{\delta} = \mathscr{M}$.

Let $U \in \mathscr{N} \bar{\otimes} \mathscr{L}^{\infty}(G) \subset \mathscr{B}(\mathscr{L}^2(G, \mathscr{H}))$ be the unitary operator defined by the bounded s-continuous function $G \ni g \mapsto u(g)$, that is, $(U\xi)(g) = u(g)\xi(g)(\xi \in \mathscr{L}^2(G, \mathscr{H}), g \in G)$. We define a unital injective normal *-homomorphism

$$\Phi : \mathscr{N} \to \mathscr{N} \bar{\otimes} \mathscr{B}(\mathscr{L}^2(G))$$

by the equality $\Phi(y) = U^*\delta(y)U(y \in \mathscr{N})$.

For $x \in \mathscr{M} = \mathscr{N}^{\delta}$, we have $\delta(x) = x \bar{\otimes} 1_G$, hence

$$(\Phi(x)\xi)(g) = (U^*(x \bar{\otimes} 1_G)U\xi)(g) = u(g)^*xu(g)\xi(g) = \sigma_g^{-1}(x)\xi(g) = (\pi_\sigma(x)\xi)(g);$$

this proves 19.9.(2).

For $s \in G$, we have $\delta(u(s)) = u(s) \bar{\otimes} \boldsymbol{\lambda}(s)$, hence

$$\begin{aligned}(\Phi(u(s))\xi)(g) &= (U^*(u(s) \bar{\otimes} \boldsymbol{\lambda}(s))U\xi)(g) \\ &= u(g)^*u(s)(U\xi)(s^{-1}g) = u(g)^*u(s)u(s^{-1}g)\xi(s^{-1}g) = ((1 \bar{\otimes} \boldsymbol{\lambda}(s))\xi)(g);\end{aligned}$$

this proves 19.9.(3).

Thus, since $\mathcal{N} = \mathcal{R}\{\mathcal{N}^\delta, u(G)\}$, we get $\Phi(\mathcal{N}) = \Phi(\mathcal{R}\{\mathcal{M}, u(G)\}) = \mathcal{R}\{\pi_\sigma(\mathcal{M}), 1 \bar{\otimes} \mathfrak{L}(G)\} = \mathcal{R}(\mathcal{M}, \sigma)$.

Finally, 19.9.(4) is obvious when applied to elements $x \in \mathcal{M}$ or $u(g)(g \in G)$, and so remains valid also when applied to an arbitrary element of $\mathcal{N} = \mathcal{R}\{\mathcal{M}, u(G)\}$.

Note that in the commutative case this part of the proof is more complicated, as in this case δ does not appear explicitly in the statement, but has to be defined, starting from θ, as the Fourier–Plancherel transform of the action $\pi_\theta : \mathcal{N} \to \mathcal{N} \bar{\otimes} \mathcal{L}^\infty(G)$.

19.11. Assume $\mathcal{N} \subset \mathcal{B}(\mathcal{H})$ realized as a von Neumann algebra. To show that $\mathcal{N} = \mathcal{R}\{\mathcal{M}, u(G)\}$, it is necessary and sufficient to prove that $\delta(\mathcal{N}) \subset \mathcal{R}\{\mathcal{M} \bar{\otimes} 1_G, \delta(u(G))\} \subset \mathcal{B}(\mathcal{H} \bar{\otimes} \mathcal{L}^2(G))$ or, equivalently, to prove the reverse inclusion for the commutants, that is, $(\mathcal{M} \bar{\otimes} 1_G)' \cap \delta(u(G))' \subset \delta(\mathcal{N})'$. Since there exist nets $\{k_i\} \subset \mathcal{K}(G)$ with $u(k_i) \xrightarrow{s^*} 1_{\mathcal{N}}$, it is sufficient to prove that

$$(\mathcal{M} \bar{\otimes} 1_G)' \cap \delta(u(G))' \subset \{\delta(u(k)^* x u(k)); k \in \mathcal{K}(G), x \in \mathcal{N}^+\}' \tag{1}$$

Before starting the proof of (1), we need some preparation. Consider the n.s.f. operator-valued weights

$$E = E_{\mathcal{N}}^{\omega_G} : (\mathcal{N} \bar{\otimes} \mathfrak{L}(G))^+ \to \bar{\mathcal{N}}^+, P = P_\delta = E \circ \delta : \mathcal{N}^+ \to \bar{\mathcal{M}}^+$$

and the unitary operator U on $\mathcal{H} \bar{\otimes} \mathcal{L}^2(G) = \mathcal{L}^2(G, \mathcal{H})$ defined by the function $G \ni g \mapsto u(g)$, i.e. $(U\zeta)(g) = u(g)\zeta(g)(\zeta \in \mathcal{L}^2(G, \mathcal{H}), g \in G)$.

We show that for $X \in \mathcal{N} \bar{\otimes} \mathfrak{L}(G), h, k \in \mathcal{K}(G), \xi \in \mathcal{K}, g \in G$,

$$E(\delta(u(h))X\delta(u(k)u(g)^*))u(g)\xi = \Delta_G(g)[\delta(u(h))XU(\xi \bar{\otimes} k)](g). \tag{2}$$

Indeed, it is easy to check that $\delta(u(\mathcal{K}(G))) \subset \mathfrak{N}_E$ (see 19.7.(5)), hence $\delta(u(h))X\delta(u(k)u(g)^*) \in \mathfrak{M}_E$ and $E(\delta(u(h))X\delta(u(k)u(g)^*))$ depends w-continuously on X (see Proposition 1.14). Consequently, it is sufficient to check (2) for X in a w-dense subset of $\mathcal{N} \bar{\otimes} \mathfrak{L}(G)$. So, we can assume that $X = x \bar{\otimes} \boldsymbol{\lambda}(f)$ with $x \in \mathcal{N}$ and $f \in \mathcal{K}(G)$. Then

$$\begin{aligned}
&E(\delta(u(h))(x \bar{\otimes} \boldsymbol{\lambda}(f))\delta(u(k)u(g)^*))u(g)\xi \\
&= E(\delta(u(h))(x \bar{\otimes} \boldsymbol{\lambda}(f))\delta(u(k))(1 \bar{\otimes} \lambda(g)^*))\xi \\
&= E\left(\left(\int h(s)(u(s) \bar{\otimes} \boldsymbol{\lambda}(s))\,\mathrm{d}s\right)\left(x \bar{\otimes} \int f(r)\boldsymbol{\lambda}(r)\,\mathrm{d}r\right)\right. \\
&\qquad \times \left.\left(\int k(t)(u(t) \bar{\otimes} \boldsymbol{\lambda}(t))\,\mathrm{d}t\right)(1 \bar{\otimes} \boldsymbol{\lambda}(g)^*)\right)\xi \\
&= E\left(\int\int\int h(s)f(r)k(t)(u(s)xu(t) \bar{\otimes} \boldsymbol{\lambda}(srtg^{-1}))\,\mathrm{d}s\,\mathrm{d}r\,\mathrm{d}t\right)\xi \\
&\qquad \text{(via } t \mapsto r^{-1}s^{-1}tg) \\
&= \Delta_G(g)E\left(\int\left(\left(\int\int h(s)f(r)u(s)xk(r^{-1}s^{-1}tg)u(r^{-1}s^{-1}tg)\mathrm{d}s\mathrm{d}r\right) \bar{\otimes} \boldsymbol{\lambda}(t)\right)\mathrm{d}t\right)\xi \\
&\qquad \text{(using 18.22.(1))}
\end{aligned}$$

$$= \Delta_G(g)\left(\int\int h(s)f(r)u(s)xk(r^{-1}s^{-1}g)u(r^{-1}s^{-1}g)\xi \,\mathrm{d}s\,\mathrm{d}r\right)\xi$$
$$= \Delta_G(g)\int\int h(s)f(r)u(s)xk(r^{-1}s^{-1}g)u(r^{-1}s^{-1}g)\xi \,\mathrm{d}s\,\mathrm{d}r$$
$$= \Delta_G(g)\int\int h(s)f(r)u(s)x[U(\xi \bar{\otimes} k)](r^{-1}s^{-1}g)\,\mathrm{d}s\,\mathrm{d}r$$
$$= \Delta_G(g)\int h(s)u(s)x\left(\int f(r)[(1_{\mathcal{N}} \bar{\otimes} \boldsymbol{\lambda}(r))U(\xi \bar{\otimes} k)](s^{-1}g)\,\mathrm{d}r\right)\mathrm{d}s$$
$$= \Delta_G(g)\int h(s)u(s)x[(1_{\mathcal{N}} \bar{\otimes} \boldsymbol{\lambda}(f))U(\xi \bar{\otimes} k)](s^{-1}g)\,\mathrm{d}s$$
$$= \Delta_G(g)\int h(s)u(s)[(x \bar{\otimes} \boldsymbol{\lambda}(f))U(\xi \bar{\otimes} k)](s^{-1}g)\,\mathrm{d}s$$
$$= \Delta_G(g)\int h(s)[(u(s) \bar{\otimes} \boldsymbol{\lambda}(s))(x \bar{\otimes} \boldsymbol{\lambda}(f))U(\xi \bar{\otimes} k)](g)\,\mathrm{d}s$$
$$= \Delta_G(g)\left[\int h(s)\delta(u(s))(x \bar{\otimes} \boldsymbol{\lambda}(f))U(\xi \bar{\otimes} k)\,\mathrm{d}s\right](g)$$
$$= \Delta_G(g)[\delta(u(h))(x \bar{\otimes} \boldsymbol{\lambda}(f))U(\xi \bar{\otimes} k)](g).$$

We show that for $x \in \mathcal{N}, h, k \in \mathcal{N}(G), \xi \in \mathcal{H}, g \in G$, we have

$$P(u(h)^*xu(k)u(g)^*)u(g)\xi = \Delta_G(g)[\delta(u(h)^*x)U(\xi \bar{\otimes} k)](g) \tag{3}$$
$$u(g)^*P(u(g)u(h)^*xu(k))\xi = [\delta(u(h)^*x)U(\xi \bar{\otimes} k)](g^{-1}) \tag{4}$$

Indeed, (3) follows from (2) replacing X by $\delta(x)$ and h by $h^\sharp$, since $u(h)^* = u(h^\sharp)$ and $P = E \circ \delta$, while (4) follows from (3) using 19.7.(3) and replacing g by g^{-1}.

For $f \in \mathcal{K}(G)$, we consider the operator

$$R_f = \int \Delta_G(g)^{-1/2} f(g^{-1})(1_{\mathcal{N}} \bar{\otimes} \boldsymbol{\rho}(g))\,\mathrm{d}g \in (\mathcal{N} \bar{\otimes} \mathfrak{L}(G))'. \tag{5}$$

For $\zeta \in \mathcal{L}^2(G, \mathcal{H})$, we have

$$(R_f\zeta)(s) = \int f(g^{-1})\zeta(sg)\,\mathrm{d}g \quad (s \in G). \tag{6}$$

It is easy to check that for $f, k \in \mathcal{K}(G)$ an $\xi \in \mathcal{H}$ we have

$$R_f U(\xi \bar{\otimes} k) = \delta(u(k))(\xi \bar{\otimes} f). \tag{7}$$

We now prove inclusion (1). Let $X \in (\mathcal{M} \bar{\otimes} 1_G)' \cap \delta(u(G))'$ and $h, k \in \mathcal{K}(G), x \in \mathcal{N}$. Consider also $\xi, \eta \in \mathcal{H}$ and $\varphi, \psi \in \mathcal{K}(G)$. Using (3)–(7), we obtain

$$(X\delta(u(h)^*xu(k))(\xi \bar{\otimes} \varphi)|(\eta \bar{\otimes} \psi))$$
$$= (X\delta(u(h)^*x)\delta(u(k))(\xi \bar{\otimes} \varphi)|(\eta \bar{\otimes} \psi))$$
$$= (X\delta(u(h)^*x)R_\varphi U(\xi \bar{\otimes} k)|(\eta \bar{\otimes} \psi))$$

$$
\begin{aligned}
&= \int ([R_\varphi \delta(u(h)^* x) U(\xi \bar{\otimes} k)](s) | [X^*(\eta \bar{\otimes} \psi)](s))\, ds \\
&= \int\int \varphi(g^{-1})([\delta(u(h)^* x) U(\xi \bar{\otimes} k)](sg) | [X^*(\eta \bar{\otimes} \psi)](s))\, ds\, dg \\
&\qquad\qquad (\text{via } g \mapsto s^{-1} g) \\
&= \int\int \varphi(g^{-1}s)([\delta(u(h)^* x) U(\xi \bar{\otimes} k)](g) | [X^*(\eta \bar{\otimes} \psi)](s))\, ds\, dg \\
&= \int\int \Delta_G(g^{-1})\varphi(g^{-1}s)(P(u(h)^* x u(k) u(g)^*) u(g)\xi | [X^*(\eta \bar{\otimes} \psi)](s))\, ds\, dg \\
&= \int\int \Delta_G(g^{-1})([(P(u(h)^* x u(k) u(g)^*) u(g) \bar{\otimes} \boldsymbol{\lambda}(g))(\xi \bar{\otimes} \varphi)](s) \\
&\qquad\qquad | [X^*(\eta \bar{\otimes} \psi)](s))\, ds\, dg \\
&= \int \Delta_G(g^{-1})(\delta(P(u(h)^* x u(k) u(g)^*) u(g))(\xi \bar{\otimes} \varphi) | X^*(\eta \bar{\otimes} \psi))\, dg \\
&\qquad\qquad (\text{since } X \in (\mathscr{M} \bar{\otimes} 1_G)' \cap \delta(u(G))') \\
&= \int \Delta_G(g^{-1})(X(\xi \bar{\otimes} \varphi) | \delta(u(g)^* P(u(g) u(k)^* x^* u(h)))(\eta \bar{\otimes} \psi))\, dg \\
&= \int\int \Delta_G(g^{-1})([X(\xi \bar{\otimes} \varphi)](s) | [(u(g)^* P(u(g) u(k)^* x^* u(h)) \bar{\otimes} \boldsymbol{\lambda}(g)^*) \\
&\qquad\qquad \times (\eta \bar{\otimes} \psi)](s))\, ds\, dg \\
&= \int\int \Delta_G(g^{-1})([X(\xi \bar{\otimes} \varphi)](s) | \psi(gs) u(g)^* P(u(g) u(k)^* x^* u(h))\eta)\, ds\, dg \\
&= \int\int \Delta_G(g^{-1})([X(\xi \bar{\otimes} \varphi)](s) | \psi(gs)[\delta(u(k)^* x^*) U(\eta \bar{\otimes} h)](g^{-1}))\, ds\, dg \\
&\qquad\qquad (\text{via } g \mapsto (sg)^{-1} = g^{-1}s^{-1}) \\
&= \int\int ([X(\xi \bar{\otimes} \varphi)](s) | \psi(g^{-1})[\delta(u(k)^* x^*) U(\eta \bar{\otimes} h)](sg))\, ds\, dg \\
&= \int ([X(\xi \bar{\otimes} \varphi)](s)\varphi)](s) | [R_\psi \delta(u(k)^* x^*) U(\eta \bar{\otimes} h)](s))\, ds \\
&= (X(\xi \bar{\otimes} \varphi) | \delta(u(k)^* x^*) R_\psi U(\eta \bar{\otimes} h)) \\
&= (X(\xi \bar{\otimes} \varphi) | \delta(u(k)^* x^*) \delta(u(h))(\eta \bar{\otimes} \psi)) \\
&= (\delta(u(h)^* x u(k)) X(\xi \bar{\otimes} \varphi) | (\eta \bar{\otimes} \psi)),
\end{aligned}
$$

hence $X\delta(u(h)^* x u(k)) = \delta(u(h)^* x u(k))X$.

This completes the proof of Theorem 19.9. □

19.12. From the first half of the sequence of equations in the final part of Section 19.11, it follows that for $x \in \mathscr{N}$ and $h, k \in \mathscr{K}(G)$ we have $\delta(u(h)^* x u(k)) = \int \Delta_G(g^{-1})\delta(P(u(h)^* x u(k) u(g)^*) u(g))\, dg = \delta\left(\int P(u(h)^* x u(k) u(g)) u(g)^*\, dg\right)$, hence

$$
u(h)^* x u(k) = \int P(u(h)^* x u(k) u(g)) u(g^{-1})\, dg. \tag{1}
$$

This relation is also sufficient to conclude the proof of Theorem 19.9. Moreover, it contains the main idea of the proof, by showing that for a wide class of elements $a \in \mathscr{N}$ we have

$$a = \int P(au(g))u(g^{-1})\,\mathrm{d}g \tag{2}$$

with $P(au(g)) \in \mathscr{M} = \mathscr{N}^{\delta}$ and $u(g^{-1}) \in u(G)$.

If G is abelian, then (1) is easily reduced to the Fourier inversion theorem. Indeed, let $a = u(h)^* x u(k)$ and consider the function $f : \hat{G} \to \mathscr{N}$ defined by $f(\gamma) = \theta_\gamma(a)(\gamma \in \hat{G})$. Recall (19.7) that in this case we have $P = P_\theta = \int \theta_\gamma(\cdot)\mathrm{d}\gamma$.

Since $a \in \mathfrak{M}_P$ (see 19.7), it follows that the function f is integrable. Then

$$\begin{aligned} f(\gamma) &= \left(\int h^{\sharp}(s)\theta_\gamma(u(s))\,\mathrm{d}s\right)\theta_\gamma(x)\left(\int k(t)\theta_\gamma(u(t))\,\mathrm{d}t\right) \\ &= \int\int \overline{\langle st, \gamma\rangle} h^{\sharp}(s)k(t)u(s)\theta_\gamma(x)u(t)\,\mathrm{d}s\,\mathrm{d}t \\ &\text{(via } t \mapsto s^{-1}t^{-1}) \\ &= \int \langle t, \gamma\rangle \left(\int h^{\sharp}(s)k(s^{-1}t^{-1})u(s)\theta_\gamma(x)u(s^{-1}t^{-1})\,\mathrm{d}s\right)\mathrm{d}t \end{aligned}$$

and

$$\begin{aligned} &\int\int \|h^{\sharp}(s)k(s^{-1}t^{-1})u(s)\theta_\gamma(x)u(s^{-1}t^{-1})\|\,\mathrm{d}s\,\mathrm{d}t \\ &\quad \le \|x\| \int\int |h^{\sharp}(s)k(s^{-1}t^{-1})|\,\mathrm{d}s\,\mathrm{d}t < +\infty \end{aligned}$$

hence f is the Fourier transform of an integrable function. Consequently, we can apply the Fourier inversion theorem to get

$$\begin{aligned} \int P(au(g))u(g^{-1})\,\mathrm{d}g &= \int\left(\int \theta_\gamma(a)\theta_\gamma(u(g))\mathrm{d}\gamma\right)u(g^{-1})\,\mathrm{d}g \\ &= \int\left(\int \overline{\langle g, \gamma\rangle} f(\gamma)\,\mathrm{d}\gamma\right)u(g)u(g^{-1})\,\mathrm{d}g = \int \hat{f}(g)\,\mathrm{d}g = f(\varepsilon) = \theta_\varepsilon(a) = a \end{aligned}$$

where $\varepsilon \in \hat{G}$ denotes the neutral element of $\hat{G}$.

19.13. As a first application of Theorem 19.9, we show that in 19.3.(5) we actually have an equality.

Corollary (Takesaki & Digernes). *Let $\sigma : G \to \mathrm{Aut}(\mathscr{M})$ be a continuous action of G on the W^*-algebra $\mathscr{M}$. Then*

$$\mathscr{R}(\mathscr{M}, \sigma) = (\mathscr{M} \mathbin{\bar{\otimes}} \mathscr{B}(\mathscr{L}^2(G)))^{\sigma \mathbin{\bar{\otimes}} \mathrm{Ad}(\boldsymbol{\rho})}.$$

Proof. We shall use the notation $\mathscr{N} = \mathscr{R}(\mathscr{M}, \sigma), \mathscr{P} = \mathscr{M} \mathbin{\bar{\otimes}} \mathscr{B}(\mathscr{L}^2(G)), \beta = \sigma \mathbin{\bar{\otimes}} \mathrm{Ad}(\boldsymbol{\rho}), \delta = \iota_{\mathscr{M}} \mathbin{\bar{\otimes}} \delta_G$ as in Section 19.3 and $u(g) = 1_{\mathscr{M}} \mathbin{\bar{\otimes}} \boldsymbol{\lambda}(g) \in \mathscr{N} \subset \mathscr{P}^\beta (g \in G)$. As we have seen in Section 19.3, $g \mapsto u(g)$ is an s-continuous unitary representation of G in $\mathscr{P}^\beta$ and $\delta = \mathscr{P}^\beta \to \mathscr{P}^\beta \bar{\otimes} \mathfrak{L}(G)$ is an action of $\hat{\mathbf{G}}$ on $\mathscr{P}^\beta$ such that $\delta(u(g) = u(g) \bar{\otimes} \boldsymbol{\lambda}(g)$ $(g \in G)$, and $(\mathscr{P}^\beta)^\delta = \pi_\sigma(\mathscr{M})$. By Theorem 19.9 it follows that $\mathscr{P}^\beta = \mathscr{R}\{\pi_\sigma(\mathscr{M}), u(G)\} = \mathscr{N}$. □

19.14. The previous corollary enables us to compute the commutant of the crossed product.

Corollary (Takesaki & Digernes). *Let $\sigma : G \to Aut(\mathscr{M})$ be a continuous action of G on the von Neumann algebra $\mathscr{M} \subset \mathscr{B}(\mathscr{H})$ and assume that there exists an so-continuous unitary representation $v : G \to \mathscr{B}(\mathscr{H})$ such that $\sigma_g = \mathrm{Ad}(v(g))|\mathscr{M}$ $(g \in G)$. Then the commutant of the von Neumann algebra $\mathscr{R}(\mathscr{M}, \sigma) \subset \mathscr{B}(\mathscr{H} \bar{\otimes} \mathscr{L}^2(G))$ is given by*

$$\mathscr{R}(\mathscr{M}, \sigma)' = \mathscr{R}\{x' \bar{\otimes} 1_G, v(g) \bar{\otimes} \boldsymbol{\rho}(g); x' \in \mathscr{M}', g \in G\}. \tag{1}$$

Proof. By the von Neumann double commutant theorem, (1) is equivalent to

$$\mathscr{R}(\mathscr{M}, \sigma) = \mathscr{R}\{x' \bar{\otimes} 1_G, v(g) \bar{\otimes} \boldsymbol{\rho}(g); x' \in \mathscr{M}', g \in G\}',$$

which is an obvious consequence of Corollary 19.13. □

The *so*-continuous unitary representation $v : G \to \mathscr{B}(\mathscr{H})$ in the statement of the corollary also defines a continuous action $\sigma' : G \to Aut(\mathscr{M}')$ where $\sigma'_g = \mathrm{Ad}(v(g))|\mathscr{M}'$ $(g \in G)$. We show that

$$\mathscr{R}(\mathscr{M}, \sigma)' \textit{ is spatially isomorphic to } \mathscr{R}(\mathscr{M}', \sigma'). \tag{2}$$

To this end, recall that *the regular representations $\boldsymbol{\lambda}$ and $\boldsymbol{\rho}$ of G are unitarily equivalent*. Indeed, the operator $U \in \mathscr{B}(\mathscr{L}^2(G))$ defined by $(U\xi)(s) = \Delta_G(s)^{-1/2}\xi(s^{-1})(\xi \in \mathscr{L}^2(G), s \in G)$, is unitary and for $g \in G$ we have

$$\begin{aligned}(U\boldsymbol{\lambda}(g)\xi)(s) &= \Delta_G(s)^{-1/2}\xi(g^{-1}s^{-1}) = \Delta_G(g)^{1/2}\Delta_G(sg)^{-1/2}\xi((sg)^{-1}) \\ &= (\boldsymbol{\rho}(g)U\xi)(s),\end{aligned}$$

that is, $U^*\boldsymbol{\rho}(g)U = \boldsymbol{\lambda}(g)$.

On the other hand, let $V \in \mathscr{B}(\mathscr{H} \bar{\otimes} \mathscr{L}^2(G))$ be the unitary operator defined by the functions $s \mapsto v(s)$, that is, $(V\zeta)(s) = v(s)\zeta(s)(\zeta \in \mathscr{L}^2(G, \mathscr{H}), s \in G)$.

Then $W = (1_{\mathscr{H}} \bar{\otimes} U)V \in \mathscr{B}(\mathscr{H} \bar{\otimes} \mathscr{L}^2(G))$ is a unitary operator and, as easily verified, we have

$$W^*(x' \bar{\otimes} 1_G)W = \pi_{\sigma'}(x') \quad (x' \in \mathscr{M}'), \tag{3}$$

$$W^*(v(g) \bar{\otimes} \boldsymbol{\rho}(g))W = 1_{\mathscr{H}} \bar{\otimes} \boldsymbol{\lambda}(g) \quad (g \in G), \tag{4}$$

proving (2). □

19.15. Landstad's Theorem 19.9 enables us to prove a result for actions of $\hat{\mathbf{G}}$ on $\mathscr{M}$ similar to Proposition 19.3.

Proposition. *For every action, $\delta : \mathscr{M} \to \mathscr{M} \bar{\otimes} \mathfrak{L}(G)$ of $\hat{\mathbf{G}}$ on the W^*-algebra $\mathscr{M}$ we have*

$$\delta(\mathscr{M}) = \mathscr{R}(\mathscr{M}, \delta)^{\hat{\delta}} \tag{1}$$

where $\hat{\delta}$ is the dual action.

Proof. Let $\mathcal{M} \subset \mathcal{B}(\mathcal{H})$ and $\mathcal{M} \bar{\otimes} \mathcal{B}(\mathcal{L}^2(G)) \subset \mathcal{B}(\mathcal{K} \bar{\otimes} \mathcal{L}^2(G))$ be realized as von Neumann algebras. We consider the action (18.7.(12))

$$\iota_{\mathcal{H}} \bar{\otimes} \delta_G^* : \mathcal{B}(\mathcal{H}) \bar{\otimes} \mathcal{B}(\mathcal{L}^2(G)) \to \mathcal{B}(\mathcal{H}) \bar{\otimes} \mathcal{B}(\mathcal{L}^2(G)) \bar{\otimes} \mathfrak{L}(G).$$

of $\hat{\mathbf{G}}$. on $\mathcal{B}(\mathcal{H} \bar{\otimes} \mathcal{L}^2(G))$ and show that

$$(\iota_{\mathcal{H}} \bar{\otimes} \delta_G^*)(\delta(\mathcal{M})') \subset \delta(\mathcal{M})' \bar{\otimes} \mathfrak{L}(G).$$

Indeed, if $X \in \delta(\mathcal{M})' \subset \mathcal{B}(\mathcal{H} \bar{\otimes} \mathcal{L}^2(G)))$ and $x \in \mathcal{M}$, then

$$\begin{aligned}
&((\iota_{\mathcal{H}} \bar{\otimes} \delta_G^*)(X))(\delta(x) \bar{\otimes} 1_G) \\
&= (1_{\mathcal{H}} \bar{\otimes} W_G)(X \bar{\otimes} 1_G)(1_{\mathcal{H}} \bar{\otimes} W_G^*)(\delta(x) \bar{\otimes} 1_G)(1_{\mathcal{H}} \bar{\otimes} W_G)(1_{\mathcal{H}} \bar{\otimes} W_G^*) \\
&= (1_{\mathcal{H}} \bar{\otimes} W_G)(X \bar{\otimes} 1_G)((\iota_{\mathcal{H}} \bar{\otimes} \delta_G)(\delta(x))(1_{\mathcal{H}} \bar{\otimes} W_G^*) \\
&= (1_{\mathcal{H}} \bar{\otimes} W_G)(X \bar{\otimes} 1_G)((\delta \bar{\otimes} \iota_G)(\delta(x)))(1_{\mathcal{H}} \bar{\otimes} W_G^*) \\
&= (1_{\mathcal{H}} \bar{\otimes} W_G)((\delta \bar{\otimes} \iota_G)(\delta(x)))(X \bar{\otimes} 1_G)(1_{\mathcal{H}} \bar{\otimes} W_G^*) \\
&= (1_{\mathcal{H}} \bar{\otimes} W_G)((\iota_{\mathcal{H}} \bar{\otimes} \delta_G)(\delta(x)))(X \bar{\otimes} 1_G)(1_{\mathcal{H}} \bar{\otimes} W_G^*) \\
&= (1_{\mathcal{H}} \bar{\otimes} W_G)(1_{\mathcal{H}} \bar{\otimes} W_G^*)(\delta(x) \bar{\otimes} 1_G)(1_{\mathcal{H}} \bar{\otimes} W_G)(X \bar{\otimes} 1_G)(1_{\mathcal{H}} \bar{\otimes} W_G^*) \\
&= (\delta(x) \bar{\otimes} 1_G)((\iota_{\mathcal{H}} \bar{\otimes} \delta_G^*)(X)).
\end{aligned}$$

Consequently, $\iota_{\mathcal{H}} \bar{\otimes} \delta_G^*$ restricts to an action $\delta' : \delta(\mathcal{M})' \to \delta(\mathcal{M})' \bar{\otimes} \mathfrak{L}(G)$ of $\hat{\mathbf{G}}$ on $\delta(\mathcal{M})'$. On the other hand, there exists an s-continuous unitary representation $u' : G \ni g \mapsto u'(g) = 1_{\mathcal{H}} \bar{\otimes} \boldsymbol{\rho}(g) \in \delta(\mathcal{M})'$ and it is easy to check that $\delta'(u'(g)) = u'(g) \bar{\otimes} \boldsymbol{\lambda}(g)$ $(g \in G)$. By Theorem 19.9, it follows that

$$\delta(\mathcal{M})' = \mathcal{R}\{(\delta(\mathcal{M})')^{\delta'}, u'(G)\}. \tag{2}$$

Recall (19.4) that the dual action $\hat{\delta} : G \to Aut(\mathcal{R}(\mathcal{M}, \delta))$ is $\hat{\delta} = (\iota_{\mathcal{M}} \bar{\otimes} \mathrm{Ad}(\rho)) \,|\mathcal{R}(\mathcal{M}, \delta)$ and that $\delta(\mathcal{M}) \subset \mathcal{B}(\mathcal{M}, \delta)^{\hat{\delta}}$. To prove the reverse inclusion, we note that

$$\mathcal{R}(\mathcal{M}, \delta)^{\hat{\delta}} \subset (\mathcal{M} \bar{\otimes} \mathcal{B}(\mathcal{L}^2(G)))^{\iota \bar{\otimes} \mathrm{Ad}(\rho)} = \mathcal{M} \bar{\otimes} \mathfrak{L}(G) \tag{3}$$

and that for every $X \in \mathcal{R}(\mathcal{M}, \delta)$ we have

$$(\iota_{\mathcal{M}} \bar{\otimes} \delta_G)(X) = (\delta \bar{\otimes} \iota_G)(X). \tag{4}$$

Indeed, for $X = \delta(x)(x \in \mathcal{M})$, (4) follows from the fact that δ is an action of $\mathbf{G}$ on $\mathcal{M}$ (18.2.(4)), while for $X = 1_{\mathcal{M}} \bar{\otimes} f(f \in \mathcal{L}^\infty(G))$, (4) is immediate (18.7.(9)). Since $\mathcal{R}(\mathcal{M}, \delta) = \mathcal{R}\{\delta(\mathcal{M}), 1_{\mathcal{M}} \bar{\otimes} \mathcal{L}^\infty(G)\}$, (4) is valid for every $X \in \mathcal{R}(\mathcal{M}, \delta)$.

Consider now $X \in \mathcal{R}(\mathcal{M}, \delta)^{\hat{\delta}}$. Then $X \in \mathcal{M} \bar{\otimes} \mathfrak{L}(G)$ and X satisfies (4). Since $X \in \mathcal{M} \bar{\otimes} \mathfrak{L}(G)$, it is clear that $Xu'(g) = u'(g)X$ for all $g \in G$. Let $Y \in (\delta(\mathcal{M})')^{\delta'}$, that is, $Y \in \delta(\mathcal{M})'$ and

$(1_{\mathscr{H}} \bar{\otimes} W_G)(Y \bar{\otimes} 1_G)(1_{\mathscr{H}} \bar{\otimes} W_G^*) = Y \bar{\otimes} 1_G$, Then

$$\begin{aligned}
XY \bar{\otimes} 1_G &= (X \bar{\otimes} 1_G)(Y \bar{\otimes} 1_G) \\
&= (1_{\mathscr{H}} \bar{\otimes} W_G)(1_{\mathscr{H}} \bar{\otimes} W_G^*)(X \bar{\otimes} 1_G)(1_{\mathscr{H}} \bar{\otimes} W_G)(Y \bar{\otimes} 1_G)(1_{\mathscr{H}} \bar{\otimes} W_G^*) \\
&= (1_{\mathscr{H}} \bar{\otimes} W_G)((\iota_{\mathscr{M}} \bar{\otimes} \delta_G)(X))(Y \bar{\otimes} 1_G)(1_{\mathscr{H}} \bar{\otimes} W_G^*) \\
&= (1_{\mathscr{H}} \bar{\otimes} W_G)((\delta \bar{\otimes} \iota_G)(X))(Y \bar{\otimes} 1_G)(1_{\mathscr{H}} \bar{\otimes} W_G^*) \\
&= (1_{\mathscr{H}} \bar{\otimes} W_G)(Y \bar{\otimes} 1_G)((\delta \bar{\otimes} \iota_G)(X))(1_{\mathscr{H}} \bar{\otimes} W_G^*) \\
&= (1_{\mathscr{H}} \bar{\otimes} W_G)(Y \bar{\otimes} 1_G)((\iota_{\mathscr{M}} \bar{\otimes} \delta_G)(X))(1_{\mathscr{H}} \bar{\otimes} W_G^*) \\
&= (1_{\mathscr{H}} \bar{\otimes} W_G)(Y \bar{\otimes} 1_G)(1_{\mathscr{H}} \bar{\otimes} W_G^*)(X \bar{\otimes} 1_G)(1_{\mathscr{H}} \bar{\otimes} W_G)(1_{\mathscr{H}} \bar{\otimes} W_G^*) \\
&= (Y \bar{\otimes} 1_G)(X \bar{\otimes} 1_G) = YX \bar{\otimes} 1_G,
\end{aligned}$$

hence $XY = YX$. From (2) it follows that $X \in \delta(\mathscr{M})'' = \delta(\mathscr{M})$. □

For $X \in \mathscr{M} \bar{\otimes} \mathfrak{L}(G)$, we have also obtained

$$X \in \delta(\mathscr{M}) \Leftrightarrow (\iota_{\mathscr{M}} \bar{\otimes} \delta_G)(X) = (\delta \bar{\otimes} \iota_G)(X). \tag{5}$$

19.16. We now prove a result similar to Proposition 19.7.

Proposition. *Let $\sigma : G \to Aut(\mathscr{N})$ be a continuous action of G on the W^*-algebra $\mathscr{N}$. If there exists a unital injective normal *-homomorphism $\chi : \mathscr{L}^\infty(G) \to \mathscr{N}$ such that*

$$\sigma_g(\chi(f)) = \chi(\mathrm{Ad}(\boldsymbol{\rho}(g))(f)) \quad (f \in \mathscr{L}^\infty(G), g \in G) \tag{1}$$

then the faithful normal operator-valued weight $P_\sigma : \mathscr{N}^+ \to \overline{(\mathscr{N}^\sigma)^+}$,

$$P_\sigma(x) = \int \sigma_g(x)\, \mathrm{d}g \quad (x \in \mathscr{N}^+) \tag{2}$$

is semifinite. For $f \in \mathscr{L}^\infty(G)^+$, we have

$$P_\sigma(\chi(f)) = \left(\int f(g)\, \mathrm{d}g\right) \cdot 1_{\mathscr{N}}. \tag{3}$$

For $x \in \mathscr{N}^+$ and $s \in G$, we have

$$P_\sigma(\sigma_s(x)) = \Delta_G(s)^{-1} P_\sigma(x). \tag{4}$$

Proof. For $f \in \mathscr{L}^\infty(G)^+$ and $k \in \mathscr{L}^\infty(G)_*^+ \subset \mathscr{L}^1(G)$, we have

$$\int \langle \mathrm{Ad}(\boldsymbol{\rho}(g))(f), k\rangle\, \mathrm{d}g = \int\int f(sg)k(s)\, \mathrm{d}s\, \mathrm{d}g = \left(\int f(g)\, \mathrm{d}g\right)\langle 1_G, k\rangle.$$

Thus, for every $\psi \in \mathcal{N}_*^+$ we obtain

$$\int \langle \sigma_g(\chi(f)), \psi \rangle \, dg = \int \langle \chi(\mathrm{Ad}(\boldsymbol{\rho}(g))(f))\psi \rangle \, dg$$

$$= \int \langle \mathrm{Ad}(\boldsymbol{\rho}(g))(f), \psi \circ \chi \rangle \, dg = \left(\int f(g) \, dg \right) (1_{\mathcal{N}}, \psi)$$

which proves (3) and the semifiniteness of P_σ. (4) has already been proved (18.20.(2)). □

19.17. In particular, for every action $\delta : \mathcal{M} \to \mathcal{M} \bar{\otimes} \mathfrak{L}(G)$ of $\hat{\mathbf{G}}$ on $\mathcal{M}$, the dual action $\hat{\delta} : G \to Aut(\mathcal{R}(\mathcal{M}, \delta))$ defines an n.s.f. operator-valued weight

$$P_{\hat{\delta}} : \mathcal{R}(\mathcal{M}, \sigma)^+ \to \overline{\delta(\mathcal{M})^+}.$$

In this case, the *-homomorphism $\chi : \mathcal{L}^\infty(G) \to \mathcal{R}(\mathcal{M}, \delta)$ is defined by $\chi(f) = 1_{\mathcal{M}} \bar{\otimes} f$ ($f \in \mathcal{L}^\infty(G)$).

Thus, if φ is an n.s.f. weight on $\mathcal{M}$, then, according to Proposition 11.6,

$$\hat{\varphi} = \varphi \circ \delta^{-1} \circ P_{\hat{\delta}}$$

is an n.s.f. weight on $\mathcal{R}(\mathcal{M}, \delta)$, called *the dual weight of* φ.

From Theorem 11.9, it follows that

$$\sigma_t^{\hat{\varphi}}(\delta(X)) = \delta(\sigma_t^{\varphi}(x)) \quad (x \in \mathcal{M}, t \in \mathbb{R}) \tag{1}$$

$$[D\hat{\psi} : D\hat{\varphi}]_t = \delta([D\psi : D\varphi]_t) \quad (t \in \mathbb{R}) \tag{2}$$

for any n.s.f. weights φ and ψ on $\mathcal{M}$.

According to 19.16.(4), *the dual weights are relatively invariant with respect to the dual action*, more precisely,

$$\hat{\varphi}(\hat{\delta}_g(X)) = \Delta_G(g)^{-1} \hat{\varphi}(X) \quad (X \in \mathcal{R}(\mathcal{M}, \delta)^+, g \in G). \tag{3}$$

Conversely, let Ψ be an n.s.f. weight on $\mathcal{R}(\mathcal{M}, \delta)$, which is relatively invariant with respect to the dual action. Then it follows that $[D\Psi : D\hat{\varphi}]_t \in \mathcal{R}(\mathcal{M}, \delta)^{\hat{\delta}} = \delta(\mathcal{M})$ ($t \in \mathbb{R}$), and by Theorem 5.1 there exists an n.s.f. weight ψ on $\mathcal{M}$ such that $[D\Psi : D\hat{\varphi}]_t = \delta([D\psi : D\varphi]_t) = [D\hat{\psi} : D\hat{\varphi}]_t$ ($t \in \mathbb{R}$), hence $\Psi = \hat{\psi}$. Thus, *the dual weights on* $\mathcal{R}(\mathcal{M}, \delta)$ *are exactly those which are relatively invariant with respect to the dual action.*

If G is commutative, then the notions introduced in Sections 19.16 and 19.17 agree with the corresponding notions introduced in Sections 19.7 and 19.8.

19.18. Consider again a continuous action $\sigma : G \to Aut(\mathcal{M})$ of G on $\mathcal{M}$ and an n.s.f. weight φ on $\mathcal{M}$.

On the crossed product $\mathcal{R}(\mathcal{M}, \sigma)$, we have the dual weight (19.8) $\hat{\varphi} = \varphi \circ \pi_\sigma^{-1} \circ P_{\hat{\sigma}}$. By Theorem 19.5, we have

$$(\mathcal{R}(\mathcal{R}(\mathcal{M}, \sigma), \hat{\sigma}); \hat{\hat{\sigma}}) \equiv (\mathcal{M} \bar{\otimes} \mathcal{B}(\mathcal{L}^2(G)); \sigma \bar{\otimes} \mathrm{Ad}(\boldsymbol{\rho})) \tag{1}$$

and hence on $\mathcal{M} \bar{\otimes} \mathcal{B}(\mathcal{L}^2(G))$ we obtain the dual weight $\hat{\hat{\varphi}}$ of the weight $\hat{\varphi}$ (19.17)

$$\hat{\hat{\varphi}} = \hat{\varphi} \circ P_{\sigma \bar{\otimes} \mathrm{Ad}(\rho)} = \varphi \circ \pi_{\sigma}^{-1} \circ P_{\hat{\sigma}} \circ P_{\sigma \bar{\otimes} \mathrm{Ad}(\rho)}. \tag{2}$$

On the other hand, on $\mathcal{M} \bar{\otimes} \mathcal{B}(\mathcal{L}^2(G))$ we can consider the weight $\varphi \bar{\otimes} tr$ where tr is the canonical trace on $\mathcal{B}(\mathcal{L}^2(G))$.

It is then natural to try to compute the Connes cocycle

$$[D\hat{\hat{\varphi}} : D(\varphi \bar{\otimes} tr)]_t \quad (t \in \mathbb{R})$$

that is, to give a more explicit expression of the second dual weight $\hat{\hat{\varphi}}$ using the *-isomorphism (1).

To this end, we consider for each $t \in \mathbb{R}$ the unitary operator $U_t \in \mathcal{M} \bar{\otimes} \mathcal{L}^\infty(G)$ defined by the function

$$G \ni g \mapsto U_t(g) = [D(\varphi \circ \sigma_g) : D\varphi]_t \in \mathcal{M}$$

and recall that the modular function $\Delta = \Delta_G$ can be also regarded as nonsingular positive self-adjoint operator on $\mathcal{L}^2(G)$, acting by multiplication.

Theorem. *Let $\sigma : G \to Aut(\mathcal{M})$ be a continuous action of G on the W^*-algebra $\mathcal{M}$. For every n.s.f. weight φ on $\mathcal{M}$, we have*

$$[D\hat{\hat{\varphi}} : D(\varphi \bar{\otimes} tr)]_t = (1_{\mathcal{M}} \bar{\otimes} \Delta_G^{it})U_t \quad (t \in \mathbb{R}).$$

The following obvious consequence is particularly useful:

Corollary. *Let $\sigma : G \to Aut(\mathcal{M})$ be a continuous action of the unimodular locally compact group G on the W^*-algebra $\mathcal{M}$. For every σ-invariant n.s.f. weight φ on $\mathcal{M}$ we have $\hat{\hat{\varphi}} = \varphi \bar{\otimes} tr$.*

The proof of the theorem is contained in Sections 19.19–19.23.

19.19. We first reformulate Theorem 19.18. Since $\Delta = \Delta_G$ is a nonsingular positive self-adjoint operator on $\mathcal{L}^2(G)$, we can consider the n.s.f. weight tr_Δ on $\mathcal{B}(\mathcal{L}^2(G))$ given by the Pedersen–Takesaki construction (4.4).

Lemma. *The mapping $\mathbb{R} \ni t \mapsto U_t \in \mathcal{M} \bar{\otimes} \mathcal{L}^\infty(G) \subset \mathcal{M} \bar{\otimes} \mathcal{B}(\mathcal{L}^2(G))$ is a unitary cocycle with respect to the modular automorphism group of the n.s.f. weight $\varphi \bar{\otimes} tr_\Delta$ on $\mathcal{M} \bar{\otimes} \mathcal{B}(\mathcal{L}^2(G))$.*

Proof. Recall that the unitary operator $U_t \in \mathcal{M} \bar{\otimes} \mathcal{L}^\infty(G)$ is determined by

$$\langle U_t, \psi \bar{\otimes} k \rangle = \int \psi(U_t(g))k(g)\,\mathrm{d}g \quad (\psi \in \mathcal{M}_*, k \in \mathcal{L}^1(G)),$$

where $U_t(g) = [D(\varphi \circ \sigma_g) : D\varphi]_t$ $(g \in G)$.

Since, for every $g \in G$, the function $t \mapsto U_t(g)$ is s^*-continuous, using the Lebesgue dominated convergence theorem we see also that the function $t \mapsto U_t$ is s^*-continuous.

Let $\theta = \varphi \bar{\otimes} tr_\Delta$. Then $\sigma_t^\theta = \sigma_t^\varphi \bar{\otimes} \mathrm{Ad}(\Delta^{it})$ and hence

$$\sigma_t^\theta | \mathcal{M} \bar{\otimes} \mathcal{L}^\infty(G) = \sigma_t^\varphi \bar{\otimes} \iota_G \quad (t \in \mathbb{R}).$$

Let $s, t \in \mathbb{R}$. Since $U_s \in \mathcal{M} \bar{\otimes} \mathcal{L}^\infty(G)$ is defined by the function $g \mapsto U_s(g)$, the element $\sigma_t^\theta(U_s) \in \mathcal{M} \bar{\otimes} \mathcal{L}^\infty(G)$ is defined by the function $g \mapsto \sigma_t^\varphi(U_s(g))$ and the element $U_t\sigma_t^\theta(U_s) \in \mathcal{M} \bar{\otimes} \mathcal{L}^\infty(G)$ is defined by the function $g \mapsto U_t(g)\sigma_t^\varphi(U_s(g)) = U_{t+s}(g)$. Hence $U_{t+s} = U_t\sigma_t^\theta(U_s)$. □

Thus, by Theorem 5.1, there exists a unique n.s.f. weight on $\mathcal{M} \bar{\otimes} \mathcal{B}(\mathcal{L}^2(G))$, denoted by $(\varphi \bar{\otimes} tr_\Delta)_U$, such that

$$[D((\varphi \bar{\otimes} tr_\Delta)_U) : D(\varphi \bar{\otimes} tr_\Delta)]_t = U_t \quad (t \in \mathbb{R}).$$

Then

$$[D((\varphi \bar{\otimes} tr_\Delta)_U) : D(\varphi \bar{\otimes} tr)]_t = (1 \bar{\otimes} \Delta^{it})U_t \quad (t \in \mathbb{R}).$$

Consequently, Theorem 19.18 asserts that

$$\hat{\hat{\varphi}} = (\varphi \bar{\otimes} tr_\Delta)_U \tag{1}$$

on $\mathcal{M} \bar{\otimes} \mathcal{B}(\mathcal{L}^2(G))$.

The proof of (1) consists in computing the values of the two weights on elements $X \in (\mathcal{M} \bar{\otimes} \mathcal{B}(\mathcal{L}^2(G)))^+$ defined by compactly supported s^*-continuous functions $G \times G \mapsto \mathcal{M}$ (18.21.(2)) and showing that the two weights commute; then (1) will follow from Theorem 6.2.

19.20. Since the values of the weight $(\varphi \bar{\otimes} tr_\Delta)_U$ are not directly computable, we shall first express this weight in another form, which makes possible such computations.

Without loss of generality, we may assume $\mathcal{M} \subset \mathcal{B}(\mathcal{H})$ realized as a von Neumann algebra such that there is an *so*-continuous unitary representation $v : G \to \mathcal{B}(\mathcal{H})$ with $\sigma_g = \mathrm{Ad}(v(g))|\mathcal{M}(g \in G)$. Then the function $g \mapsto v(g)$ defines a unitary operator $V \in \mathcal{M} \bar{\otimes} \mathcal{L}^\infty(G)$ and hence a $*$-automorphism $\mathfrak{S} = \mathrm{Ad}(V)$ of $\mathcal{M} \bar{\otimes} \mathcal{L}^\infty(G)$ (compare to the proof of Lemma 19.2). It is easy to check that if an element $X \in \mathcal{M} \bar{\otimes} \mathcal{L}^\infty(G)$ is defined by a bounded *w*-continuous function $G \ni g \mapsto X(g) \in \mathcal{M}$, then the element $\mathfrak{S}(X) \in \mathcal{M} \bar{\otimes} \mathcal{L}^\infty(G)$ is defined by the function $G \ni g \mapsto \sigma_g(X(g))$:

$$[\mathfrak{S}(X)](g) = \sigma_g(X(g)) \quad (g \in G). \tag{1}$$

Thus, the $*$-automorphism $\mathfrak{S}$ of $\mathcal{M} \bar{\otimes} \mathcal{L}^\infty(G)$ depends only on the continuous action $\sigma : G \to Aut(\mathcal{M})$ and we can consider the n.s.f. weight $(\varphi \bar{\otimes} \mu) \circ \mathfrak{S}$ on $\mathcal{M} \bar{\otimes} \mathcal{L}^\infty(G)$, where $\mu = \mu_G$.

On the other hand, the action $\delta = \iota_\mathcal{M} \bar{\otimes} \delta_G$ of $\hat{\mathbf{G}}$ on $\mathcal{M} \bar{\otimes} \mathcal{B}(\mathcal{L}^2(G))$ defines an n.s.f. operator-valued weight P_δ on $\mathcal{M} \bar{\otimes} \mathcal{B}(\mathcal{L}^2(G))$ with values in $(\mathcal{M} \bar{\otimes} \mathcal{B}(\mathcal{L}^2(G)))^\delta = \mathcal{M} \bar{\otimes} \mathcal{L}^\infty(G)$, and we can consider the n.s.f. weight $((\varphi \bar{\otimes} \mu) \circ \mathfrak{S}) \circ P_\delta$ on $\mathcal{M} \bar{\otimes} \mathcal{B}(\mathcal{L}^2(G))$.

Lemma. *We have*

$$(\varphi \bar{\otimes} tr_\Delta)_U = ((\varphi \bar{\otimes} \mu) \circ \mathfrak{S}) \circ P_\delta. \tag{2}$$

Proof. From the proof of Lemma 19.19, it also follows that the mapping $t \mapsto U_t \in \mathcal{M} \bar{\otimes} \mathcal{L}^\infty(G)$ is a unitary cocycle with respect to the modular automorphism group of the n.s.f. weight $\varphi \bar{\otimes} \mu$ on $\mathcal{M} \bar{\otimes} \mathcal{L}^\infty(G)$, hence there exists a unique n.s.f. weight on $\mathcal{M} \bar{\otimes} \mathcal{L}^\infty(G)$, denoted by $(\varphi \bar{\otimes} \mu)_U$, such that $[D(\varphi \bar{\otimes} \mu)_U : D(\varphi \bar{\otimes} \mu)]_t = U_t$ $(t \in \mathbb{R})$. Actually,

$$(\varphi \bar{\otimes} \mu)_U = (\varphi \bar{\otimes} \mu) \circ \mathfrak{S}. \tag{3}$$

Indeed, for the element $X \in \mathscr{M} \bar{\otimes} \mathscr{L}^\infty(G)$ defined by the bounded w-continuous function $G \ni g \mapsto X(g) \in \mathscr{M}$ we have

$$\begin{aligned}
[\sigma_t^{(\varphi\bar{\otimes}\mu)\circ\mathfrak{S}}(X)](g) &= [(\mathfrak{S}^{-1} \circ (\sigma_t^\varphi \bar{\otimes} \iota_G) \circ \mathfrak{S})(X)](g) = (\sigma_g^{-1} \circ \sigma_t^\varphi \circ \sigma_g)(X(g)) \\
&= \sigma_t^{\varphi\circ\sigma_g}(X(g)) = U_t(g)\sigma_t^\varphi(X(g))U_t(g)^* \\
&= U_t(g)[(\sigma_t^\varphi \bar{\otimes} \iota_G)(X)](g)U_t(g)^* = U_t\sigma_t^{\varphi\bar{\otimes}\mu}(X)U^*](g),
\end{aligned}$$

hence $\sigma_t^{(\varphi\bar{\otimes}\mu)\circ\mathfrak{S}}(X) = U_t\sigma_t^{\varphi\bar{\otimes}\mu}(X)U_t^*$ $(t \in \mathbb{R})$. Then, we similarly verify the corresponding *KMS* condition and hence (3) follows from the uniqueness part of Theorem 3.1.

Furthermore, since P_δ is an n.s.f. $(\mathscr{M} \bar{\otimes} \mathscr{L}^\infty(G))$-valued weight on $\mathscr{M} \bar{\otimes} \mathscr{B}(\mathscr{L}^2(G))$ and $U_t \in \mathscr{M} \bar{\otimes} \mathscr{L}^\infty(G)$ $(t \in \mathbb{R})$, we infer from (3) and Theorem 11.9.(3) that

$$(\varphi \bar{\otimes} \mu)_U \circ P_\delta = ((\varphi \bar{\otimes} \mu) \circ P_\delta)_U. \tag{4}$$

On the other hand, the action δ_G of $\hat{\mathbf{G}}$ on $\mathscr{B}(\mathscr{L}^2(G))$ defines an $\mathscr{L}^\infty(G)$-valued n.s.f. weight $P = P_{\delta_G}$ on $\mathscr{B}(\mathscr{L}^2(G))$. Since $\delta = \iota_{\mathscr{M}} \bar{\otimes} \delta_G$ we have (18.19.(5)) $P_\delta = \iota_{\mathscr{M}} \bar{\otimes} P$. Consequently,

$$(\varphi \bar{\otimes} \mu) \circ P_\delta = \varphi \bar{\otimes} (\mu \circ P). \tag{5}$$

Thus, (2) will follow from (3), (4), and (5) once we have proved that on $\mathscr{B}(\mathscr{L}^2(G))$:

$$tr_\Delta = \mu \circ P. \tag{6}$$

To prove (6), we consider an operator $X \in \mathscr{B}(\mathscr{L}^2(G))^+$ defined as in 18.21.(2) by a compactly supported continuous function $X : G \times G \ni (s, r) \mapsto X(s, r) \in \mathbb{C}$. By 18.22.(7) and 18.22.(8), we have $[P(X)](g) = \int \Delta(g)X(g, g)\, dg$ $(g \in G)$, hence $(\mu \circ P)(X) = \int \Delta(g)X(g, g)\, dg$. On the other hand, the operator ΔX is defined similarly by the function $(\Delta X)(s, r) = \Delta(s)X(s, r)(s, r \in G)$, and using Mercer's formula (Dunford & Schwartz, 1958, 1963, Ch. IX, §8, Ex. 49 (c)) we obtain $tr_\Delta(X) = \int \Delta(g)X(g, g)\, dg$. Therefore, the n.s.f. weights $\mu \circ P$ and tr_Δ are equal on the linear subspace $\mathscr{B}$ of $\mathscr{B}(\mathscr{L}^2(G))$ spanned by the elements $X \in \mathscr{B}(\mathscr{L}^2(G))$, which are defined as above by compactly supported continuous functions $G \times G \to \mathbb{C}$. It is clear that $\mathscr{B}$ is an s^*-dense $*$-subalgebra of $\mathscr{B}(\mathscr{L}^2(G))$. Since $\sigma_t^{tr_\Delta}(X) = \Delta^{it}X\Delta^{-it}(X \in \mathscr{B}(\mathscr{L}^2(G)), t \in \mathbb{R})$, it follows that $\mathscr{B}$ is also σ^{tr_Δ}-invariant. Moreover, since $\Delta^{it} \in \mathscr{L}^\infty(G)$, P is an $\mathscr{L}^\infty(G)$-valued weight on $\mathscr{B}(\mathscr{L}^2(G))$ and $\mathscr{L}^\infty(G)$ is abelian, we have $P(\Delta^{it}X\Delta^{-it}) = \Delta^{it}P(X)\Delta^{-it} = P(X)$ $(X \in \mathscr{B}(\mathscr{L}^2(G))^+)$. Consequently, the weight $\mu \circ P$ is σ^{tr_Δ}-invariant, that is the two weights appearing in (6) commute. Thus, (6) follows from Theorem 6.2. □

19.21 Lemma. *Let $X \in (\mathscr{M} \bar{\otimes} \mathscr{B}(\mathscr{L}^2(G)))^+$ be defined as in 18.21.(2) by the compactly supported s^*-continuous function $G \times G \ni (s, r) \mapsto X(s, r) \in \mathscr{M}$. Then*

$$\hat{\hat{\varphi}}(X) = \int \varphi(\sigma_g(X(g, g)))\Delta(g)\, dg = (\varphi \bar{\otimes} tr_\Delta)_U(X). \tag{1}$$

Proof. By 18.21.(3), we have $P_{\sigma\bar{\otimes}\mathrm{Ad}(\rho)}(X) = T_f^\sigma$, where $f(s) = \int \sigma_{sr}(X(sr, r)\, \Delta(r)\, dr$ $(s \in G)$. Recall that the dual action $\hat{\sigma}$ is the restriction of the action $\iota_{\mathscr{M}} \bar{\otimes} \delta_G$ so that, by 18.19.(4) and 18.22.(2), we have $P_{\hat{\sigma}}(T_f^\sigma) = \pi_\sigma(f(e)) = \pi_\sigma\left(\int \sigma_r(X(r, r))\Delta(r)\, dr\right)$. Finally, using the definition 19.18.(2) of $\hat{\hat{\varphi}}$ and Corollary 2.10, we infer that $\hat{\hat{\varphi}}(X) = \int \varphi(\sigma_r(X(r, r)))\Delta(r)\, dr$.

On the other hand, using 19.20.(2) and 18.22.(2), we obtain

$$
\begin{aligned}
(\varphi \bar{\otimes} tr_\Delta)_U(X) = (\varphi \bar{\otimes} \mu)(\mathfrak{S}(P_\delta(X))) &= \int \varphi([\mathfrak{S}(P_\delta(X))](g))dg \\
&= \int \varphi(\sigma_g([P_\delta(X)](g))dg = \int \varphi(\sigma_g(\Delta(g)X(g,g)))dg \\
&= \int \varphi(\sigma_g(X(g,g)))\Delta(g)dg.
\end{aligned}
$$

□

19.22. Our next objective is to show that the weights $(\varphi \bar{\otimes} tr_\Delta)_U$ and $\hat{\varphi}$ commute. We shall put

$$
\begin{aligned}
\alpha_t &= \sigma_t^{(\varphi \bar{\otimes} tr_\Delta)_U} \in Aut(\mathscr{M} \bar{\otimes} \mathscr{B}(\mathscr{L}^2(G))) \quad (t \in \mathbb{R}) \\
\beta_g &= \sigma_g \bar{\otimes} \mathrm{Ad}(\boldsymbol{\rho}(g)) \in Aut(\mathscr{M} \bar{\otimes} \mathscr{B}(\mathscr{L}^2(G))) \quad (g \in G) \\
\widetilde{U}_t(g) &= \Delta(g)^{it} U_t(g) \quad (t \in \mathbb{R}, g \in G).
\end{aligned}
$$

Lemma. *With the above notation we have*

$$\alpha_t | \mathscr{R}(\mathscr{M}, \sigma) = \sigma_t^{\hat{\varphi}} \quad (t \in \mathbb{R}) \tag{1}$$

$$P_\beta \circ \alpha_t = \alpha_t \circ P_\beta \quad (t \in \mathbb{R}). \tag{2}$$

Corollary. *The weiglts $\hat{\hat{\varphi}}$ and $(\varphi \bar{\otimes} tr_\Delta)_U$ commute.*

Proof of the Corollary. Recall that P_β is an $\mathscr{R}(\mathscr{M}, \sigma)$-valued n.s.f. weight on $\mathscr{M} \bar{\otimes} \mathscr{B}(\mathscr{L}^2(G))$ and $\hat{\hat{\varphi}} = \hat{\varphi} \circ P_\beta$. Using (1) and (2), we get $\hat{\hat{\varphi}} \circ \alpha_t = \hat{\varphi} \circ (P_\beta \circ \alpha_t) = (\hat{\varphi} \circ \sigma_t^{\hat{\varphi}}) \circ P_\beta = \hat{\varphi} \circ P_\beta = \hat{\hat{\varphi}}$ $(t \in \mathbb{R})$, hence $\hat{\hat{\varphi}}$ is α-invariant, that is, $\hat{\hat{\varphi}}$ and $(\varphi \bar{\otimes} tr_\Delta)_U$ commute.

Proof of the Lemma. Note that

$$\alpha_t(X) = \widetilde{U}_t((\sigma_t^\varphi \bar{\otimes} \iota_G)(X))\widetilde{U}_t^* \quad (X \in \mathscr{M} \bar{\otimes} \mathscr{B}(\mathscr{L}^2(G)), t \in \mathbb{R}). \tag{3}$$

Thus, for $f \in \mathscr{L}^\infty(G)$ and $t \in \mathbb{R}$ we have

$$\alpha_t(1_\mathscr{M} \bar{\otimes} f) = \widetilde{U}_t(1_\mathscr{M} \bar{\otimes} f)\widetilde{U}_t^* = 1_\mathscr{M} \bar{\otimes} f, \tag{4}$$

since $\widetilde{U}_t \in \mathscr{M} \bar{\otimes} \mathscr{L}^\infty(G)$ and $\mathscr{L}^\infty(G)$ is commutative. Then, for $x \in \mathscr{M}, t \in \mathbb{R}$ and $g \in G$, we have

$$
\begin{aligned}
[\alpha_t(\pi_\sigma(x))](g) &= [\widetilde{U}_t((\sigma_t^\varphi \bar{\otimes} \iota_G)(\pi_\sigma(x)))\widetilde{U}_t^*](g) \\
&= \widetilde{U}_t(g)\sigma_t^\varphi([\pi_\sigma(x)](g))\widetilde{U}_t(g)^* \\
&= U_t(g)\sigma_t^\varphi(\sigma_g^{-1}(x))U_t(g)^* \\
&= \sigma_t^{\varphi \circ \sigma_g}(\sigma_g^{-1}(x)) \\
&= \sigma_g^{-1}(\sigma_t^\varphi(x)) \\
&= [\pi_\sigma(\sigma_t^\varphi(x))](g) \\
&= [\sigma_t^{\hat{\varphi}}(\pi_\sigma(x))](g),
\end{aligned}
$$

by 19.8.(7), hence

$$\alpha_t(\pi_\sigma((x)) = \sigma_t^{\hat{\varphi}}(\pi_\sigma(x)), \tag{5}$$

Finally, let $g \in G, t \in \mathbb{R}$. We have

$$\begin{aligned}\alpha_t(1_{\mathscr{M}} \bar{\otimes} \boldsymbol{\lambda}(g)) &= \widetilde{U}_t(1_{\mathscr{M}} \bar{\otimes} \boldsymbol{\lambda}(g))\widetilde{U}_t^* \\ &= (1_{\mathscr{M}} \bar{\otimes} \boldsymbol{\lambda}(g))((\mathrm{Ad}(1_{\mathscr{M}} \bar{\otimes} \boldsymbol{\lambda}(g)^*))(\widetilde{U}_t))\widetilde{U}_t^*.\end{aligned} \tag{6}$$

For $r \in G$, we have

$$\begin{aligned}&[(\mathrm{Ad}(1_{\mathscr{M}} \bar{\otimes} \boldsymbol{\lambda}(g)^*))(\widetilde{U}_t)\widetilde{U}_t^*](r) = \widetilde{U}_t(gr)\widetilde{U}_t(r)^* \\ &= \Delta(g)^{it}U_t(gr)U_t(r)^* = \Delta(g)[\pi_\sigma([D(\varphi \circ \sigma_g) : D\varphi]_t))(r)\end{aligned} \tag{7}$$

The last equality is justified as follows: By Corollary 3.5, we have

$$[D(\varphi \circ \sigma_{gr}) : D(\varphi \circ \sigma_r)]_t[D(\varphi \circ \sigma_r) : D\varphi]_t = [D(\varphi \circ \sigma_{gr}) : D\varphi]_t$$

that is,

$$U_t(gr)U_t(r)^* = [D(\varphi \circ \sigma_{gr}) : D(\varphi \circ \sigma_r)]_t$$

and, using Corollary 3.8, we obtain

$$\begin{aligned}U_t(gr)U_t(r)^* &= [D((\varphi \circ \sigma_g) \circ \sigma_r) : D(\varphi \circ \sigma_r)]_t \\ &= \sigma_r^{-1}([D(\varphi \circ \sigma_g) : D\varphi]_t) = [\pi_\sigma([D(\varphi \circ \sigma_g) : D\varphi]_t)(r).\end{aligned}$$

From (6) and (7) we infer that

$$\alpha_t(1_{\mathscr{M}} \bar{\otimes} \boldsymbol{\lambda}(g)) = \Delta(g)^{it}(1_{\mathscr{M}} \bar{\otimes} \boldsymbol{\lambda}(g))\pi_\sigma([D(\varphi \circ \sigma_g) : D\varphi]_t)$$

and hence (19.8.(8))

$$\alpha_t(1_{\mathscr{M}} \bar{\otimes} \boldsymbol{\lambda}(g)) = \sigma_t^{\hat{\varphi}}(1_{\mathscr{M}} \bar{\otimes} \boldsymbol{\lambda}(g)). \tag{8}$$

Since $\mathscr{R}(\mathscr{M},\sigma) = \mathscr{R}\{\pi_\sigma(\mathscr{M}), 1_{\mathscr{M}} \bar{\otimes} \mathfrak{L}(G)\}$, (1) follows from (5) and (8). Since $\mathscr{R}(\mathscr{M},\sigma) \subset (\mathscr{M} \bar{\otimes} \mathscr{B}(\mathscr{L}^2(G)))^\beta$, (5) and (8) also show that

$$(\beta_g \circ \alpha_t)(X) = (\alpha_t \circ \beta_g)(X) \tag{9}$$

for $X \in \mathscr{R}(\mathscr{M},\sigma)$, and using (4) we get (9) for $X \in 1_{\mathscr{M}} \bar{\otimes} \mathscr{L}^\infty(G)$. Thus, (9) is valid for all $X \in \mathscr{M} \bar{\otimes} \mathscr{B}(\mathscr{L}^2(G))$ (see Lemma 19.2.(2)), and, since $P_\beta = \int \beta_g(\cdot)\,\mathrm{d}g$ (18.20), this proves (2). □

19.23. We shall construct an α-invariant s^*-dense $*$-subalgebra

$$\mathscr{B} \subset \mathfrak{N}_{(\varphi \bar{\otimes} tr_\Delta)_U} \subset \mathscr{M} \bar{\otimes} \mathscr{B}(\mathscr{L}^2(G))$$

such that the weights $(\varphi \bar{\otimes} tr_\Delta)_U$ and $\hat{\varphi}$ are equal on $\mathscr{B}$.

To this end, we consider the sets

$$\begin{aligned}
\mathfrak{X} = \{X \in \mathcal{M} \mathbin{\bar{\otimes}} \mathcal{B}(\mathcal{L}^2(G)); X \text{ is defined (18.21.(2)) by a compactly}\\
\text{supported } s^*\text{-continuous function } G \times G \ni (s,r) \mapsto X(s,r) \in \mathcal{M}\},\\
\mathfrak{S} = \{S \in \mathcal{B}(\mathcal{L}^2(G)); S \text{ is defined (18.21.(2)) by a compactly}\\
\text{supported continuous function } G \times G \ni (s,r) \mapsto S(s,r) \in \mathbb{C}\},\\
\mathfrak{W} = \{W : G \times G \to \mathcal{M}; W \text{ is an } s^*\text{-continuous function with } W(s,r)\\
\text{unitary for all } s, r \in G\}.
\end{aligned}$$

For $S \in \mathfrak{S}$, $W \in \mathfrak{W}$, $x \in \mathfrak{N}_\varphi \subset \mathcal{M}$, we consider the operator

$$Y = Y(S, W, x) \in \mathfrak{X} \subset \mathcal{M} \mathbin{\bar{\otimes}} \mathcal{B}(\mathcal{L}^2(G))$$

defined by the function

$$Y(s,r) = S(s,r)W(s,r)\sigma_r^{-1}(x) \quad (s, r \in G). \tag{1}$$

Then Y^* is defined by the function

$$Y^*(s,r) = Y(r,s)^* = \overline{S(r,s)}\sigma_s^{-1}(x^*)W(r,s)^* \quad (s, r \in G), \tag{2}$$

and $X = Y^*Y$ is defined by the function

$$\begin{aligned}
X(s,r) &= \int Y^*(s,h)Y(h,r)\,\mathrm{d}h\\
&= \int \overline{S(h,s)}S(h,r)\sigma_s^{-1}(x^*)W(h,s)^*W(h,r)\sigma_r^{-1}(x)\,\mathrm{d}h \quad (s, r \in G).
\end{aligned}$$

In particular,

$$X(g,g) = \left(\int |S(h,g)|^2\,\mathrm{d}h\right)\sigma_g^{-1}(x^*x) \quad (g \in G)$$

and, using Lemma 19.21, we get

$$(\varphi \mathbin{\bar{\otimes}} tr_\Delta)_U(X) = \int \varphi(\sigma_g(X(g,g)))\Delta(g)\,\mathrm{d}g = \varphi(x^*x)\int\int |S(h,g)|^2\,\mathrm{d}h\,\mathrm{d}g < +\infty.$$

Consequently,

$$Y(S, W, x) \in \mathfrak{N}_{(\varphi \mathbin{\bar{\otimes}} tr_\Delta)_U}.$$

Let $\mathcal{B}$ be the *-subalgebra of $\mathcal{M}$ generated by the elements of form $Y(S, W, x)$ with $S \in \mathfrak{S}$, $W \in \mathfrak{W}$, and $x \in \mathfrak{N}_\varphi$. Then

$$\mathcal{B} \subset \mathfrak{N}_{(\varphi \mathbin{\bar{\otimes}} tr_\Delta)_U}.$$

Also, as $\mathscr{B} \subset \mathfrak{X}$ we have, by Lemma 19.21,

$$\hat{\hat{\varphi}}(X^*X) = (\varphi \bar{\otimes} tr_\Delta)_U(X^*X) \quad (X \in \mathscr{B}). \tag{3}$$

If $W_0 \in \mathfrak{W}$ is defined by $W_0(s,r) = 1_{\mathscr{M}}(s,r \in G)$, then for every $S \in \mathfrak{S}$ and every $x \in \mathfrak{N}_\varphi$, we have

$$Y(S, W_0, x) = (1_{\mathscr{M}} \bar{\otimes} S)\pi_\sigma(x).$$

Using Lemma 19.2.(2), we infer that the *-subalgebra $\mathscr{B}$ is s^*-dense in $\mathscr{M}$.

It remains to be shown that $\mathscr{B}$ is α_t-invariant. Let $S \in \mathfrak{S}$, $W \in \mathfrak{W}$, $x \in \mathfrak{N}_\varphi$ and let the operator $Y = Y(S, W, x) \in \mathfrak{X}$ be defined by the function (1). Recall that the *-automorphism α_t is determined by 19.22.(3). It follows that the operator $\alpha_t(Y) \in \mathfrak{X}$ is determined by the function

$$\begin{aligned}
[\alpha_t(Y)](s,r) &= S(s,r)\widetilde{U}_t(s)\sigma_t^\varphi(W(s,r))\sigma_t^\varphi(\sigma_r^{-1}(x))\widetilde{U}_t(r)^* \\
&= S(s,r)\widetilde{U}_t(s)\sigma_t^\varphi(W(s,r))\widetilde{U}_t(r)^*\widetilde{U}_t(r)\sigma_t^\varphi(\sigma_r^{-1}(x))\widetilde{U}_t(r)^* \\
&= S(s,r)\widetilde{U}_t(s)\sigma_t^\varphi(W(s,r))\widetilde{U}_t(r)^*\sigma_r^{-1}(\sigma_t^\varphi(x)).
\end{aligned}$$

Hence

$$\alpha_t(Y(S, W, x)) = Y(S, W_t, \sigma_t^\varphi(x))$$

where $W_t \in \mathfrak{W}$ is the function $W_t(s,r) = \widetilde{U}_t(s)\sigma_t^\varphi(W(s,r))\widetilde{U}_t(r)^*(s, r \in G)$. Consequently, $\mathscr{B}$ is indeed α_t-invariant.

Thus, the weights $\hat{\hat{\varphi}}$ and $(\varphi \bar{\otimes} tr_\Delta)_U$ commute and satisfy (2), so that they are equal by Theorem 6.2.

The proof of Theorem 19.18 is complete. □

19.24. Notes. The crossed product construction for operator algebras appeared already in the work of Murray and von Neumann (1936, 1937, 1943, I) for actions of discrete groups on commutative W^*-algebras. It was precisely this construction that produced the first concrete examples of factors of types II_1, II_∞ and III. Subsequently, the construction was developed, for discrete groups, in Ching (1969); Diximier (1969); Golodec (1971); Nakamura and Takeda (1958, 1960); Suzuki (1959); Turumaru (1958); Zeller-Meier (1968), and so on, and Connes (1973a) discovered its crucial importance for the structure theory of type III factors. In the general case of locally compact groups, the construction appears in Doplicher, Kastler, Robinson. and the first systematic study of it is due to Takesaki (1973b). The definition of the crossed product of a W^*-algebra by the action of a Kac algebra was introduced in Strătilă, Voiculescu, and Zsidó (1976).

Takesaki (1973b) proved the most important special cases of the main results presented in this section (19.5, 19.8, 19.13, 19.14, 19.18), and so extended the duality principle given in Connes (1973a) to arbitrary W^*-algebras of type III. In the general case, for Theorem 19.5 see Landstad (1977); Strătilă et al. (1976, 1977), for Corollaries 19.13 and 19.14 see Digernes (1975), for Theorem 19.8 see Digernes (1975); Haagerup (1978); Sauvageot (1977); Strătilă et al. (1976, 1977); Zsidó (1977); Theorem 19.18 is proved in Digernes (1975) for abelian groups and in Strătilă et al. (1976, 1977); Strătilă (1977) for arbitrary locally compact groups. The characterization of crossed product W^*-algebras by Theorem 19.9 is due to Landstad (1979).

The theory of crossed products of W^*-algebras by actions of group duals was developed in Landstad (1977); Strătilă et al. (1976, 1977). The key point in this theory is the saturation property (19.15.(5)) introduced in the general case and proved for amenable groups in Strătilă et al. (1976); for arbitrary locally compact groups, Proposition 19.15 is due to Landstad (1977) (see (Strătilă et al., 1976, 1977, II, pp. 83–85). A different approach, working for arbitrary compact groups, is due to Roberts (1976).

For our exposition, we have used Haagerup (1978); Landstad (1979) and Strătilă et al. (1976, 1977).

We should mention also the survey article Nakagami and Takesaki (1979) and the following related references concerning extensions of the commutation theorem for crossed products Day (1957); van Heeswijck (1979); Rousseau (1979, 1980); Rousseau and Van Daele (1979), crossed products with cocycles (Sutherland, 1977, 1980; Zeller-Meier, 1968), crossed products by actions of Kac algebras Enock (1977); Enock and Schwartz (1975) and C^*-crossed products Doplicher et al. (1979); Landstad, Olesen, and Pedersen (1978); Olesen, Pedersen, and Størmer, (1977); Pedersen (1973–1977); Takai (1975); Pimsner and Voiculescu (1982); Rieffel (1981).

20 Comparison of cocycles

In this section, we develop a comparison theory for cocycles with respect to a given continuous group action on a W^*-algebra, characterize the square-integrable cocycles as subcocycles of the dominant cocycle and, as an application, study the dominant actions. Moreover, this theory will give rise, via the Coones cocycle theorem (3.1 and 5.1), (§23) to a corresponding comparison theory for weights on W^*-algebras.

20.1. Let G be a locally compact group and $\sigma : G \to Aut(\mathscr{M})$ a continuous action of G on the W^*-algebra $\mathscr{M}$. Recall (5.1) that a σ-cocycle (of degree 1) is an s^*-continuous function $\sigma : G \ni s \mapsto a(s) \in \mathscr{M}$ with the properties:

$$a(st) = a(s)\sigma_s(a(t)) \text{ and } a(s^{-1}) = \sigma_s^{-1}(a(s)^*) \quad (s, t \in G) \tag{1}$$

and that the set of all σ-cocycles is denoted by $Z_\sigma = Z_\sigma(G; \mathscr{M})$. We denote by $e \in G$ the neutral element of G.

Proposition. *Let $a \in Z_\sigma(G; \mathscr{M})$. Then the elements $a(s) \in \mathscr{M}$ are partial isometries*

$$a(s)a(s)^* = a(e), \quad a(s)^*a(s) = \sigma_s(a(e)) \quad (s \in G) \tag{2}$$

in particular $a(e)$ is a projection and

$$a(e)a(s) = a(s) = a(s)\sigma_s(a(e)) \quad (s \in G). \tag{3}$$

The equation

$$({}_a\sigma)_s(x) = a(s)\sigma_s(x)a(s)^* \quad (x \in \mathscr{M}_{a(e)}, s \in G) \tag{4}$$

defines a continuous action ${}_a\sigma : G \to Aut(\mathcal{M}_{a(e)})$ *whose centralizer is denoted by*

$$\mathcal{M}^a = (\mathcal{M}_{a(e)})^{{}_a\sigma}. \tag{5}$$

For every projection $p \in \mathcal{M}^a$, *the function*

$$a^p; G \ni s \mapsto pa(s) \in \mathcal{M} \tag{6}$$

is a σ*-cocycle* $a^p \in Z_\sigma(G; \mathcal{M})$.

Proof. We have $a(e) = \sigma_e^{-1}(a(e)^*) = a(e)^*$ and $a(e) = a(ee) = a(e)\sigma_e(a(e)) = a(e)^2$, hence $a(e)$ is a projection. Since $a(s)^* = \sigma_s(a(s^{-1}))$, we obtain $a(s)a(s)^* = a(s)\sigma_s(a(s^{-1})) = a(ss^{-1}) = a(e)$ and

$$a(s)^*a(s) = \sigma_s(a(s^{-1}))a(s) = \sigma_s(a(s^{-1})\sigma_{s^{-1}}(a(s))) = \sigma_s(a(s^{-1}s)) = \sigma_s(a(e)),$$

thus proving (2) and (3).

Using (1), (2), and (3) it is easy to see that (4) define a continuous action. For instance, for $s, t \in G$ and $x, y \in a(e)\mathcal{M}a(e)$, we have $({}_a\sigma)_s(xy) = a(s)\sigma_s(xy)a(s)^* = a(s)\sigma_s(x)\sigma_s(a(e))\sigma_s(y)a(s)^* = a(s)\sigma_s(x)a(s)^*a(s)\sigma_s(y)a(s)^* = ({}_a\sigma)_s(x)({}_a\sigma)_s(y)$ and $({}_a\sigma)_{st}(x) = a(st)\sigma_{st}(x)a(st)^* = a(s)\sigma_s(a(t))\sigma_s(\sigma_t(x))\sigma_s(a(t)^*)a(s)^* = a(s)\sigma_s(a(t)\sigma_t(x)a(t)^*)a(s)^* = ({}_a\sigma)_s(({}_a\sigma)_t(x))$.

Consider now a projection $p \in \mathcal{M}^a$. Then $p \le a(e)$ and $a(r)\sigma_r(p)a(r)^* = p$ for all $r \in G$, hence $a^p(st) = pa(st) = pa(s)\sigma_s(a(t)) = pa(s)\sigma_s(p)a(s)^*a(s)\sigma_s(a(t)) = pa(s)\sigma_s(pa(e)a(t)) = a^p(s)\sigma_s(a^p(t))$ and, since $\sigma_r(p) = a(r)^*pa(r)$, $a^p(s^{-1}) = pa(s^{-1}) = a(e)^*pa(s^{-1}) = (a(s^{-1})\sigma_{s^{-1}}(a(s)))^*pa(s^{-1}) = \sigma_s^{-1}(a(s)^*)a(s^{-1})^*\, pa(s^{-1}) = \sigma_s^{-1}(a(s)^*)\sigma_s^{-1}(p) = \sigma_s^{-1}((pa(s))^*) = \sigma_s^{-1}(a^p(s)^*)$. □

If the W^*-algebra $\mathcal{M}^a$ is properly infinite, then we say that *the cocycle* $a \in Z_\sigma(G; \mathcal{M})$ *is of infinite multiplicity*.

A cocycle of the form a^p is called a *subcocycle* of a.

20.2. Let $\mathcal{F}_2$ be the type I_2 factor with the system of matrix units $\{e_{ij}\}_{1\le i,j\le 2}$. We shall identify $\mathcal{M} \bar{\otimes} \mathcal{F}_2$ with $Mat_2(\mathcal{M})$ in the usual way. Let $\iota : G \to Aut(\mathcal{F}_2)$ be the trivial action and $\sigma : G \to Aut(\mathcal{M})$ a continuous action. Then $\sigma \bar{\otimes} \iota : G \to Aut(\mathcal{M} \bar{\otimes} \mathcal{F}_2)$ is a continuous action.

Let $a, b \in Z_\sigma(G; \mathcal{M})$. We define a function $c = c(a, b) : G \to \mathcal{M} \bar{\otimes} \mathcal{F}_2$ by

$$c(s) = a(s) \bar{\otimes} e_{11} + b(s) \bar{\otimes} e_{22} = \begin{pmatrix} a(s) & 0 \\ 0 & b(s) \end{pmatrix} \quad (s \in G) \tag{1}$$

and the set $\mathcal{I}(a, b)$ by

$$\mathcal{I}(a, b) = \{x \in a(e)\mathcal{M}a(e); xb(s) = a(s)\sigma_s(x) \text{ for all } s \in G\}.$$

Proposition. *Let* $\sigma : G \to Aut(\mathcal{M})$ *be a continuous action and* $a, b \in Z_\sigma(G; \mathcal{M})$. *Then* $c = c(a, b) \in Z_{\sigma\bar{\otimes}\iota}(G; \mathcal{M} \bar{\otimes} \mathcal{F}_2)$,

$$a(e) \bar{\otimes} e_{11} \in (\mathcal{M} \bar{\otimes} \mathcal{F}_2)^c, b(e) \bar{\otimes} e_{22} \in (\mathcal{M} \bar{\otimes} \mathcal{F}_2)^c \tag{2}$$

and the following statements are equivalent:

(i) $a(e) \bar{\otimes} e_{11} \prec b(e) \bar{\otimes} e_{22}$ *in the* W^**-algebra* $(\mathcal{M} \bar{\otimes} \mathcal{F}_2)^c$*;*
(ii) *there exists* $u \in \mathcal{M}$ *such that* $u^*u = a(e), uu^* \leq b(e)$ *and* $u = b(s)\sigma_s(u)a(s)^*$ *for all* $s \in G$*;*
(iii) *there exist* $v \in \mathcal{M}$ *and a projection* $q \in \mathcal{M}^b$ *such that* $a(s) = v^*b^q(s)\sigma_s(v)$ *and* $b^q(s) = va(s)\sigma_s(v^*)$ *for all* $s \in G$*.*

On the other hand, we have

$$\mathcal{I}(a,a) = \mathcal{M}^a \tag{3}$$

$$\mathcal{I}(a,b)^* = \mathcal{I}(b,a) \tag{4}$$

$$\mathcal{I}(a,w)\mathcal{I}(w,b) \subset \mathcal{I}(a,b) \text{ for all } w \in Z_\sigma(G;\mathcal{M}) \tag{5}$$

$$x \in \mathcal{I}(a,b) \text{ with polar decomposition } x = v|x| \Rightarrow |x| \in \mathcal{I}(b,b), \quad v \in \mathcal{I}(a,b) \tag{6}$$

$$(\mathcal{M} \bar{\otimes} \mathcal{F}_2)^c = \begin{pmatrix} \mathcal{I}(a,a) & \mathcal{I}(a,b) \\ \mathcal{I}(b,a) & \mathcal{I}(b,b) \end{pmatrix} \tag{7}$$

$$[(\mathcal{M} \bar{\otimes} \mathcal{F}_2)^c]_{a(e) \bar{\otimes} e_{11}} = \mathcal{M}^a, \quad [(\mathcal{M} \bar{\otimes} \mathcal{F}_2)^c]_{b(e) \bar{\otimes} e_{22}} = \mathcal{M}^b \tag{8}$$

$$\mathcal{I}(a,b) = \{0\} \Leftrightarrow \mathbf{z}(a(e) \bar{\otimes} e_{11}) \perp \mathbf{z}(b(e) \bar{\otimes} e_{22}) \text{ in } (\mathcal{M} \bar{\otimes} \mathcal{F}_2)^c \tag{9}$$

$$\mathbf{z}(a(e) \bar{\otimes} e_{11}) = \mathbf{z}(b(e) \bar{\otimes} e_{22}) = \mathbf{z} \in (\mathcal{M} \bar{\otimes} \mathcal{F}_2)^c \Rightarrow \mathbf{z} = a(e) \bar{\otimes} e_{11} + b(e) \bar{\otimes} e_{22}. \tag{10}$$

Proof. By definition (1) it is clear that $c \in Z_{\sigma \bar{\otimes} \iota}(G;\mathcal{M})$ and $a(e) \bar{\otimes} e_{11} \leq c(e)$. For $s \in G$, we have $c(s)[(\sigma \bar{\otimes} \iota)_s(a(e) \bar{\otimes} e_{11})]c(s)^* = \begin{pmatrix} a(s) & 0 \\ 0 & b(s) \end{pmatrix} \begin{pmatrix} \sigma_s(a(e)) & 0 \\ 0 & 0 \end{pmatrix} \begin{pmatrix} a(s)^* & 0 \\ 0 & b(s)^* \end{pmatrix} \begin{pmatrix} a(s)\sigma_s(a(e))a(s)^* & 0 \\ 0 & 0 \end{pmatrix}$ $= \begin{pmatrix} a(e) & 0 \\ 0 & 0 \end{pmatrix} = a(e) \bar{\otimes} e_{11}$, which proves (2).

We now prove the equivalence of statements (i), (ii), and (iii).

(i) $\Leftrightarrow$ (ii). Let $x \in \mathcal{M} \bar{\otimes} \mathcal{F}_2$ with $x^*x = a(e) \bar{\otimes} e_{11}, xx^* \leq b(e) \bar{\otimes} e_{22}$. If $x = [x_{ij}]$ with $x_{ij} \in \mathcal{M}$, then it follows that $x_{11}^*x_{11} + x_{21}^*x_{21} = a(e), x_{12} = x_{22} = 0$ and $x_{11} = 0, x_{21}x_{21}^* \leq b(e)$, that is $x = \begin{pmatrix} 0 & 0 \\ u & 0 \end{pmatrix}$ with $u \in \mathcal{M}, u^*u = a(e), uu^* \leq b(e)$. Hence $x \in c(e)(\mathcal{M} \bar{\otimes} \mathcal{F}_2)c(e)$. If we require that $x \in (\mathcal{M} \bar{\otimes} \mathcal{F}_2)^c$, that is, $c(s)[(\sigma \bar{\otimes} \iota)_s(x)]c(s)^* = x$, then it also follows that $u = b(s)\sigma_s(u)a(s)^*$ $(s \in G)$. Thus, (i) $\Rightarrow$ (ii) and, similarly, (ii) $\Rightarrow$ (i).

(ii) $\Rightarrow$ (iii). Let $v = u$ and $q = uu^* \leq b(e)$. We have $q = uu^* = b(s)\sigma_s(u)a(s)^* a(s)\sigma_s(u^*)b(s^*) = b(s)\sigma_s(ua(e)u^*)b(s)^* = b(s)\sigma_s(q)b(s)^*$, hence $q \in \mathcal{M}^b$ and then, successively, $ua(s) = b(s)\sigma_s(u)a(s)^*a(s) = b(s)\sigma_s(u)\sigma_s(a(e)) = b(s)\sigma_s(u), u^*b(s)\sigma_s(u) = u^*ua(s) = a(e)a(s) = a(s), a(s) = u^*uu^*b(s)\sigma_s(u) = v^*qb(s)\sigma_s(v) = v^*b^q(s)\sigma_s(v), b^q(s) = qb(s) = b(s)\sigma_s(q)b(s)^*b(s) = b(s)\sigma_s(u)\sigma_s(u^*)\sigma_s(b(e)) = b(s)\sigma_s(u)\sigma_s(u^*) = ua(s)\sigma_s(u^*) = va(s)\sigma_s(v^*)$.

(iii) $\Rightarrow$ (ii). Let $u = qva(e)$. Then $\mathbf{s}(u^*u) \leq a(e), \mathbf{s}(uu^*) \leq q$ and $a(s) = u^*b^q(s)\sigma_a(u), b^q(s) = ua(s)\sigma_s(u^*)$ $(s \in G)$. Then $a(e) = u^*b^q(e)u = u^*ua(e)u^*u = (u^*u)^2$, hence $u^*u = a(e)$. Also, $q = b^q(e) = ua(e)u^* = uu^*quu^* = (uu^*)^2$, hence $uu^* = q \leq b(e)$. Finally, we obtain $b^q(s)\sigma_s(u) = ua(s)\sigma_s(u^*u) = ua(s)$ and $u = ua(e) = ua(s)a(s)^* = b^q(s)\sigma_s(u)a(s)^* = qb(s)\sigma_s(u)a(s)^* = b(e)qb(s)\sigma_s(u)a(s)^* = b(s)b(s)^*qb(s)\sigma_s(u)(a(s)^* = b(s)\sigma_s(q)\sigma_s(u)a(s)^* = b(s)\sigma_s(u)a(s)^*$.

Assertions (3), (4), (5), (7), and (8) can be checked without any difficulty.

If $x \in \mathcal{I}(a,b)$, then, by (3), (4), and (5), we have $|x|^2 = x^*x \in \mathcal{I}(b,b) = \mathcal{M}^b$ and hence $|x| \in \mathcal{M}^b$. Since $x \in a(e)\mathcal{M}b(e)$ and $x = v|x|$ is the polar decomposition of x, it follows that $v \in a(e)\mathcal{M}b(e)$

and $vb(s)\sigma_s(|x|) = v|x|b(s) = xb(s) = a(s)\sigma_s(x) = a(s)\sigma_s(v)\sigma_s(|x|)$. Thus, with $\mathcal{M} \subset \mathcal{B}(\mathcal{H})$ realized as a von Neumann algebra, we have $vb(s)\xi = a(s)\sigma_s(v)\xi$ for all $\xi \in \overline{\sigma_s(|x|)\mathcal{H}}$. If $\xi \perp \sigma_s(|x|)\mathcal{H}$ and $\eta \in \mathcal{H}$, then $(vb(s)\xi|\eta) = (b(s)\xi|v^*\eta) = 0$ (as $v^*\mathcal{H} = \overline{|x|\mathcal{H}}$) and, for every $\zeta \in \mathcal{H}$, we have $(b(s)\xi||x|\zeta) = (\xi|b(s)^*|x|\zeta) = (\xi|\sigma_s(|x|)b(s)^*\zeta) = 0$. Consequently, $vb(s) = a(s)\sigma_s(v)$, that is $v \in \mathcal{I}(a,b)$, thus proving (6).

If there exists $0 \neq x \in \mathcal{I}(a,b)$, then

$$\begin{pmatrix} a(e) & x \\ x^* & b(e) \end{pmatrix} \in (\mathcal{M} \bar{\otimes} \mathcal{F}_2)^c \text{ and}$$

$$\begin{pmatrix} a(e) & 0 \\ 0 & 0 \end{pmatrix}\begin{pmatrix} a(e) & x \\ x^* & b(e) \end{pmatrix}\begin{pmatrix} 0 & 0 \\ 0 & b(e) \end{pmatrix} = \begin{pmatrix} 0 & x \\ 0 & 0 \end{pmatrix} \neq 0$$

that is, $(a(e) \bar{\otimes} e_{11})(\mathcal{M} \bar{\otimes} \mathcal{F}_2)^c(b(e) \bar{\otimes} e_{22}) \neq 0$ and consequently ([L], 4.5) the central supports of $a(e) \bar{\otimes} e_{11}$ and $b(e) \bar{\otimes} e_{22}$ in $(\mathcal{M} \bar{\otimes} \mathcal{F}_2)^c$ are not mutually orthogonal. Conversely, if there exists $0 \neq x = [x_{ij}] \in (\mathcal{M} \bar{\otimes} \mathcal{F}_2)^c$ with $(a(e) \bar{\otimes} e_{11})x(b(e) \bar{\otimes} e_{22}) \neq 0$, then $0 \neq a(e)x_{12}b(e) \in \mathcal{I}(a,b)$. We have thus proved (9).

Finally, if $a(e) \bar{\otimes} e_{11}$ and $b(e) \bar{\otimes} e_{22}$ have the same central support z in $(\mathcal{M} \bar{\otimes} \mathcal{F}_2)^c$, then $z \geq a(e) \bar{\otimes} e_{11} + b(e) \bar{\otimes} e_{22} = c(e)$. Since $c(e)$ is the unit element in $(\mathcal{M} \bar{\otimes} \mathcal{F}_2)^c$, it follows that $z = c(e)$, which proves (10). □

The cocycle $c = c(a,b)$ is called *the balanced cocycle associated with the cocycles a and b*.

If the equivalent conditions (i), (ii), and (iii) are satisfied, then we say that the cocycle $a \in Z_\sigma$ is *dominated* by the cocycle $b \in Z_\sigma$ and write $a \precsim b$.

If $a(e) \bar{\otimes} e_{11} \sim b(e) \bar{\otimes} e_{22}$ in $(\mathcal{M} \bar{\otimes} \mathcal{F}_2)^c$ or, equivalently, if $q = b(e)$ in (iii), then we say that a and b are *equivalent* and write $a \approx b$. Note that $a \precsim b$ if and only if a is equivalent to a cocycle of the form b^q with $q \in Proj(\mathcal{M}^b)$.

From ([L], 4.7), it follows that

$$a \approx b \Leftrightarrow a \precsim b \text{ and } b \precsim a. \tag{11}$$

From (8), it follows that

$$a \approx b \Rightarrow \textit{ the } W^*\textit{-algebras } \mathcal{M}^a \textit{ and } \mathcal{M}^b \textit{ are } *\textit{-isomorphic.} \tag{12}$$

If $\mathcal{I}(a,b) = \{0\}$, then we say that the cocycles $a, b \in Z_\sigma$ are *disjoint* and write $a \between b$.

If $a(e) \bar{\otimes} e_{11}$ and $b(e) \bar{\otimes} e_{22}$ have the same central support in $(\mathcal{M} \bar{\otimes} \mathcal{F}_2)^c$, then we say that a and b are *quasi-equivalent* and write $a \sim b$. Clearly, $a \approx b \Rightarrow a \sim b$. On the other hand,

$$\textit{if } \mathcal{M} \textit{ is countably decomposable and the cocycles } a, b \in Z_\sigma(G;\mathcal{M}) \textit{ are of infinite multiplicity, then } a \approx b \Leftrightarrow a \sim b. \tag{13}$$

Indeed, by assumption and assertion (8), the projections $a(e) \bar{\otimes} e_{11}$ and $b(e) \bar{\otimes} e_{22}$ are properly infinite in the countably decomposable W^*-algebra $(\mathcal{M} \bar{\otimes} \mathcal{F}_2)^c$ and hence ([L], 4.13) they are equivalent if and only if they have same central support.

Note that

$$\begin{array}{l}\textit{if } a,b\in Z_\sigma(G;\mathscr{M}) \textit{ and } a\approx b, \textit{ then the continuous actions}\\ {}_a\sigma : G\to Aut(\mathscr{M}_{a(e)}), {}_b\sigma : G\to Aut(\mathscr{M}_{b(e)}) \textit{ are inner conjugate,}\\ \textit{that is, there exists } u\in\mathscr{M} \textit{ with } u^*u=a(e), uu^*=b(e) \textit{ and}\\ ({}_a\sigma)_s(x)=u^*[({}_b\sigma)_s(uxu^*)]u \quad (x\in a(e)\mathscr{M}a(e)).\end{array} \tag{14}$$

Indeed, if $a \approx b$, then, by the above proposition, there exists $u \in \mathscr{M}$ with $u^*u = a(e), uu^* = b(e)$, and $a(s) = u^*b(s)\sigma_s(u)$ $(s \in G)$. Consequently, for every $x \in a(e)\mathscr{M}a(e)$, we have $({}_a\sigma)_s(x) = a(s)\sigma_s(x)a(s)^* = u^*b(s)\sigma_s(u)\sigma_s(x)\sigma_s(u^*)b(s)^*u = u^*[({}_b\sigma)_s(uxu^*)]u$.

20.3. Let $\sigma : G \to Aut(\mathscr{M})$ be a continuous action of G on $\mathscr{M}$. A cocycle $a \in Z_\sigma(G;\mathscr{M})$ with $a(e) = 1_{\mathscr{M}}$ is called a *unitary cocycle*. In this case all the elements $a(s)$ $(s \in G)$, belong to the unitary group $U(\mathscr{M})$ of $\mathscr{M}$. We shall denote by $Z_\sigma(G; U(\mathscr{M}))$ the set of all unitary σ-cocycles.

The function $1 : G \to \mathscr{M}$ defined by $1(s) = 1_{\mathscr{M}}$ $(s \in G)$, is a unitary cocycle, $1 \in Z_\sigma(G; U(\mathscr{M}))$. Clearly, ${}_1\sigma = \sigma$ and $\mathscr{M}^1 = \mathscr{M}^\sigma$.

A cocycle $a \in Z_\sigma(G,\mathscr{M})$ is called *trivial* if $a \approx 1$. In this case, a is a unitary cocycle and there exists a unitary element $v \in \mathscr{M}$ such that $a(s) = v^*\sigma_s(v)$, $(s \in G)$.

Note that

$$\begin{array}{l}\textit{for all } u\in Z_\sigma(G;U(\mathscr{M})) \textit{ and } b\in Z_{({}_u\sigma)}(G;\mathscr{M}) \textit{ the equation}\\ a(s)=b(s)u(s)\ (s\in G) \textit{ defines an element } a\in Z_\sigma(G;\mathscr{M})\\ \textit{whose centralizer is equal to the centralizer of } b\in Z_{({}_u\sigma)}(G;\mathscr{M}).\end{array} \tag{1}$$

Indeed, for $s, t \in G$ we have

$$\begin{aligned} a(st) &= b(st)u(st) = b(s)u(s)\sigma_s(b(t))u(s)^*u(s)\sigma_s(u(t)) \\ &= b(s)u(s)\sigma_s(b(t)u(t)) = a(s)\sigma_s(a(t)) \end{aligned}$$

and

$$\begin{aligned} a(s^{-1}) &= b(s^{-1})u(s^{-1}) = \sigma_s^{-1}(u(s)^*b(s)^*u(s))\sigma_s^{-1}(u(s)^*) \\ &= \sigma_s^{-1}(u(s)^*b(s)^*) = \sigma_s^{-1}(a(s)^*), \end{aligned}$$

hence $a \in Z_\sigma(G;\mathscr{M})$ and, clearly, ${}_u\sigma = {}_b({}_a\sigma)$.

On the other hand, we now show that

$$\begin{array}{l}\textit{if } \mathscr{M} \textit{ is properly infinite, then there is a unitary}\\ \textit{cocycle } u\in Z_\sigma(G;\mathscr{M}) \textit{ of infinite multiplicity.}\end{array} \tag{2}$$

Indeed, since $\mathscr{M}$ is properly infinite, there is a family of partial isometrics $\{u_n\}_{n\in I} \subset \mathscr{M}$ with $u_nu_m^* = 0$ for $n \neq m$, $u_nu_n^* = 1$ and $\sum_n u_n^*u_n = 1$. We define

$$u(s) = \sum_n u_n^*\sigma_s(u_n) \quad (s \in G).$$

Then, $u(s) \in U(\mathscr{M})$ since

$$u(s)u(s)^* = \left(\sum_n u_n^*\sigma_s(u_n)\right)\left(\sum_m \sigma_s(u_m^*)u_m\right) = \sum_n u_n^*u_n = 1$$

and

$$u(s)^*u(s) = \left(\sum_m \sigma_s(u_m^*)u_m\right)\left(\sum_n u_n^*\sigma_s(u_n)\right) = \sigma_s\left(\sum_n u_n^*u_n\right) = 1.$$

Also, $u \in Z_\sigma(G; U(\mathscr{M}))$ since

$$u(s)\sigma_s(u(t)) = \left(\sum_n u_n^*\sigma_s(u_n)\right)\left(\sum_m \sigma_s(u_m^*)\sigma_{st}(u_m)\right) = \sum_n u_n^*\sigma_{st}(u_n) = u(st).$$

Moreover, we have $u_i^*u_j \in \mathscr{M}^u$ since

$$u(s)\sigma_s(u_i^*u_j)u(s)^* = u_i^*\sigma_s(u_i)\sigma_s(u_i^*u_j)\sigma_s(u_j^*)u_j = u_i^*u_j$$

and hence $\mathscr{M}^u$ is properly infinite.

Finally, we note that if $\iota : G \to Aut(\mathscr{M})$ is the trivial action then, clearly

$$Z_\iota(G; U(\mathscr{M})) = \{u : G \to \mathscr{M} \text{ } s\text{-continuous unitary representations}\}. \tag{3}$$

20.4. Let $\sigma : G \to Aut(\mathscr{M})$ and $\tau : G \to Aut(\mathscr{N})$ be continuous actions of G on the W^*-algebras $\mathscr{M}$ and $\mathscr{N}$, respectively. If $a \in Z_\sigma(G; \mathscr{M})$ and $b \in Z_\tau(G; \mathscr{N})$, then the equation

$$(a \mathbin{\bar\otimes} b)(s) = a(s) \mathbin{\bar\otimes} b(s) \quad (s \in G)$$

defines a cocycle $a \mathbin{\bar\otimes} b \in Z_{\sigma\bar\otimes\tau}(G; \mathscr{M} \mathbin{\bar\otimes} \mathscr{N})$, called the *tensor product* of a and b.

Next we take $\mathscr{M}$ countably decomposable, $\mathscr{N} = \mathscr{F}_\infty =$ the countably decomposable type I_∞ factor and $\tau = \iota =$ the trivial action. Then, for $a, b \in Z_\sigma(G; \mathscr{M})$ we have

$$a \sim b \Leftrightarrow a \mathbin{\bar\otimes} 1 \approx b \mathbin{\bar\otimes} 1. \tag{1}$$

Indeed, if $c \in Z_{\sigma\bar\otimes\iota}(G; \mathscr{M} \mathbin{\bar\otimes} \mathscr{F}_2)$ is the balanced cocycle associated with a and b, then $c \mathbin{\bar\otimes} 1 \in Z_{\sigma\bar\otimes\iota\bar\otimes\iota}(G; \mathscr{M} \mathbin{\bar\otimes} \mathscr{F}_\infty \mathbin{\bar\otimes} \mathscr{F}_2)$ is the balanced cocycle associated with $a \mathbin{\bar\otimes} 1$ and $b \mathbin{\bar\otimes} 1, (\mathscr{M} \mathbin{\bar\otimes} \mathscr{F}_\infty \mathbin{\bar\otimes} \mathscr{F}_2)^{c \bar\otimes 1} = (\mathscr{M} \mathbin{\bar\otimes} \mathscr{F}_2)^c \mathbin{\bar\otimes} \mathscr{F}_\infty$ and

$$(a \mathbin{\bar\otimes} 1)(e) \mathbin{\bar\otimes} e_{11} = (a(e) \mathbin{\bar\otimes} e_{11}) \mathbin{\bar\otimes} 1, (b \mathbin{\bar\otimes} 1)(e) \mathbin{\bar\otimes} e_{22} = (b(e) \mathbin{\bar\otimes} e_{22}) \mathbin{\bar\otimes} 1.$$

In general, if p is a projection of the W^*-algebra $\mathscr{P}$, then the central support $\mathbf{z}(p \mathbin{\bar\otimes} 1)$ of $p \mathbin{\bar\otimes} 1$ in $\mathscr{P} \mathbin{\bar\otimes} \mathscr{F}_\infty$ is $\mathbf{z}(p) \mathbin{\bar\otimes} 1$, where $\mathbf{z}(p)$ is the central support of p in $\mathscr{P}$. If $\mathscr{P}$ is countably decomposable, then for $p, q \in \mathscr{P}$ we have $p \mathbin{\bar\otimes} 1 \sim q \mathbin{\bar\otimes} 1 \Leftrightarrow \mathbf{z}(p \mathbin{\bar\otimes} 1) = \mathbf{z}(q \mathbin{\bar\otimes} 1) \Leftrightarrow \mathbf{z}(p) = \mathbf{z}(q)$, since $\mathscr{P} \mathbin{\bar\otimes} \mathscr{F}_\infty$ is properly infinite and countably decomposable ([L], 4.13). In view of the definitions of Section 20.2, these considerations prove (1).

Now, let $\boldsymbol{\rho}$ be the right regular representation of G regarded as a cocycle $\boldsymbol{\rho} \in Z_\iota(G; \mathscr{B}(\mathscr{L}^2(G)))$. For every $u \in Z_\sigma(G, U(\mathscr{M}))$, we have

$$u \,\bar{\otimes}\, \boldsymbol{\rho} \approx 1 \,\bar{\otimes}\, \boldsymbol{\rho} \text{ in } Z_{\sigma \,\bar{\otimes}\, \iota}(G; \mathscr{H} \,\bar{\otimes}\, \mathscr{B}(\mathscr{L}^2(G))). \tag{2}$$

Indeed, let $\mathscr{M} \subset \mathscr{B}(\mathscr{H})$ and $\mathscr{M} \,\bar{\otimes}\, \mathscr{B}(\mathscr{L}^2(G)) \subset \mathscr{B}(\mathscr{L}^2(G; \mathscr{H}))$ be realized as von Neumann algebras. Consider the unitary operator $U \in \mathscr{M} \,\bar{\otimes}\, \mathscr{L}^\infty(G)$ defined by the function $s \mapsto u(s^{-1})$, that is, $(U\xi)(s) = u(s^{-1})\xi(s)$ ($\xi \in \mathscr{L}^2(G, \mathscr{H}), s \in G$). For $s, t \in G$, and $\xi \in \mathscr{L}^2(G, \mathscr{H})$ we have

$$U(1 \,\bar{\otimes}\, \boldsymbol{\rho})(t)\xi](s) = u(s^{-1})[(1 \,\bar{\otimes}\, \boldsymbol{\rho}(t))\xi](s) = \Delta(t)^{1/2} u(s^{-1})\xi(st)$$

and, since $[((\sigma \,\bar{\otimes}\, \iota)_t U)\xi](r) = [((\sigma_t \,\bar{\otimes}\, \iota)U)\xi](r) = \sigma_t(u(r^{-1}))\xi(r)$,

$$\begin{aligned}
&[((u(t) \,\bar{\otimes}\, \boldsymbol{\rho}(t))((\sigma_t \,\bar{\otimes}\, \iota)U)\xi](s) = \Delta(t)^{1/2} u(t)[((\sigma_t \,\bar{\otimes}\, \iota)U)\xi](st) \\
&\quad = \Delta(t)^{1/2} u(t)\sigma_t(u((st)^{-1}))\xi(st) = \Delta(t)^{1/2} u(t)\sigma_t(\sigma_{st}^{-1}(u(st)^*))\xi(st) \\
&\quad = \Delta(t)^{1/2} u(t)\sigma_s^{-1}((u(s)\sigma_s(u(t)))^*)\xi(st) \\
&\quad = \Delta(t)^{1/2} u(t)u(t)^*\sigma_s^{-1}(u(s)^*)\xi(st) = \Delta(t)^{1/2} u(s^{-1})\xi(st),
\end{aligned}$$

we have $U(1 \,\bar{\otimes}\, \boldsymbol{\rho})(t) = ((u \,\bar{\otimes}\, \boldsymbol{\rho})(t))((\sigma_s \,\bar{\otimes}\, \iota)U)$ ($t \in G$) proving (2).

Since, by Corollary 19.13, we have $\mathscr{R}(\mathscr{M}, \sigma) = (\mathscr{M} \,\bar{\otimes}\, \mathscr{B}(\mathscr{L}^2(G))^{\sigma \,\bar{\otimes}\, \mathrm{Ad}(\boldsymbol{\rho})} = (\mathscr{M} \,\bar{\otimes}\, \mathscr{B}(\mathscr{L}^2(G)))^{1 \,\bar{\otimes}\, \boldsymbol{\rho}}$, and since the regular representations $\boldsymbol{\lambda}$ and $\boldsymbol{\rho}$ are equivalent, that is, $\boldsymbol{\lambda} \approx \boldsymbol{\rho}$, it follows from (2) that for every unitary cocycle $u \in Z_\sigma(G; U(\mathscr{M}))$ we have

$$\mathscr{R}(\mathscr{M}, \sigma) \approx (\mathscr{M} \,\bar{\otimes}\, \mathscr{B}(\mathscr{L}^2(G)))^{u \,\bar{\otimes}\, \boldsymbol{\rho}} \approx (\mathscr{M} \,\bar{\otimes}\, \mathscr{B}(\mathscr{L}^2(G)))^{u \,\bar{\otimes}\, \boldsymbol{\lambda}}. \tag{3}$$

20.5. Let $\sigma : G \to Aut(\mathscr{M})$ be a continuous action of G on $\mathscr{M}$. A unitary cocycle $a \in Z_\sigma(G; U(\mathscr{M}))$ is called a *dominant cocycle* if it is of infinite multiplicity and

$$a \,\bar{\otimes}\, 1 \approx a \,\bar{\otimes}\, \boldsymbol{\rho} \text{ in } Z_{\alpha \,\bar{\otimes}\, \iota}(\mathscr{M} \,\bar{\otimes}\, \mathscr{B}(\mathscr{L}^2(G))). \tag{1}$$

According to statement 20.4.(2), (1) is equivalent to the condition

$$a \,\bar{\otimes}\, 1 \approx 1 \,\bar{\otimes}\, \boldsymbol{\rho} \text{ in } Z_{\sigma \,\bar{\otimes}\, \iota}(\mathscr{M} \,\bar{\otimes}\, \mathscr{B}(\mathscr{L}^2(G))). \tag{2}$$

and this implies that the *-automorphisms $({}_a\sigma) \,\bar{\otimes}\, \iota$ and $\sigma \,\bar{\otimes}\, \mathrm{Ad}(\boldsymbol{\rho})$ of $\mathscr{M} \,\bar{\otimes}\, \mathscr{B}(\mathscr{L}^2(G)))$ are inner conjugate (10.2.(14)).

If the locally compact group G is separable, then

$$a \in Z_\sigma(G; U(\mathscr{M})) \textit{ dominant cocycle } \Rightarrow \mathscr{M}^\sigma \approx \mathscr{R}(\mathscr{M}, \sigma). \tag{3}$$

Indeed, according to (1) and 20.4.(3), we have $\mathscr{R}(\mathscr{M}, \sigma) \approx (\mathscr{M} \,\bar{\otimes}\, \mathscr{B}(\mathscr{L}^2(G)))^{a \bar{\otimes} \rho} \approx (\mathscr{M} \,\bar{\otimes}\, \mathscr{B}(\mathscr{L}^2(G)))^{a \bar{\otimes} 1} \approx \mathscr{M}^a \,\bar{\otimes}\, \mathscr{B}(\mathscr{L}^2(G)))$. Since G is separable, there exists $n \in \mathbb{N} \cup \{\infty\}$ such that $\mathscr{B}(\mathscr{L}^2(G)) \approx \mathscr{F}_n$ = the countably decomposable type I_n factor and, since $\mathscr{M}^a$ is properly infinite, we conclude that $\mathscr{M}^a \,\bar{\otimes}\, \mathscr{F}_n \approx \mathscr{M}^a$.

Clearly, the existence of a dominant cocycle $a \in Z_\sigma(G; \mathscr{M})$ implies that $\mathscr{M}$ is properly infinite.

Theorem. *Let $\sigma : G \to Aut(\mathscr{M})$ be a continuous action of the separable locally compact group G on the properly infinite W^*-algebra $\mathscr{M}$. Then there exists a dominant cocycle $a \in Z_\sigma(G; U(\mathscr{M}))$.*

If, moreover, $\mathscr{M}$ is countably decomposable, then any two dominant cocycles $a, b \in Z_\sigma(G; \mathscr{M})$ are equivalent, that is, $a \approx b$.

Proof. We first prove the uniqueness assertion. If $a, b \in Z_\sigma(G, U(\mathscr{M}))$ are dominant, then $a \bar{\otimes} 1 \approx 1 \bar{\otimes} \boldsymbol{\rho} = b \bar{\otimes} 1$. Since $\mathscr{M}$ and $\mathscr{B}(\mathscr{L}^2(G))$ are countably decomposable, we infer, using 20.4.(1), that $a \sim b$ and, since a and b are of infinite multiplicity, it follows by 20.2.(13) that $a \approx b$.

We now prove the existence assertion. By 20.3.(2), there exists a unitary cocycle $u \in Z_\sigma(G; U(\mathscr{M}))$ of infinite multiplicity, that is, the centralizer $\mathscr{M}^{(_u\sigma)}$ is properly infinite. If there exists a dominant cocycle $b \in Z_{(_u\sigma)}(G; U(\mathscr{M}))$, then by 20.3.(1), the equation $a(s) = b(s)u(s)$ $(s \in G)$, defines a cocycle $a \in Z_\sigma(G; \mathscr{M})$ of infinite multiplicity. Since $b \in Z_{(_u\sigma)}(G; U(\mathscr{M}))$ is dominant, we have $b \bar{\otimes} \boldsymbol{\rho} \approx b \bar{\otimes} 1$ in $Z_{(_u\sigma\bar{\otimes}\iota)}(G; \mathscr{M} \bar{\otimes} \mathscr{B}(\mathscr{L}^2(G)))$, that is, there exists $V \in U(\mathscr{M} \bar{\otimes} \mathscr{B}(\mathscr{L}^2(G)))$ such that $V = (b(s) \bar{\otimes} 1)(u(s) \bar{\otimes} 1)((\sigma_s \bar{\otimes} \iota)(V))(u(s)^* \bar{\otimes} 1)(b(s)^* \bar{\otimes} \boldsymbol{\rho}(s)^*)$ or $V = (a(s) \bar{\otimes} 1)(\sigma_s \bar{\otimes} \iota)(V))(a(s)^* \bar{\otimes} \boldsymbol{\rho}(s)^*)$ $(s \in G)$, and this means that $a \bar{\otimes} \boldsymbol{\rho} \approx a \bar{\otimes} 1$ in $Z_{\sigma\bar{\otimes}\iota}(G; \mathscr{M} \bar{\otimes} \mathscr{B}(\mathscr{L}^2(G)))$; hence $a \in Z_\sigma(G; \mathscr{M})$ is dominant.

Thus, in order to prove the existence of a dominant cocycle $a \in Z_\sigma(G; U(\mathscr{M}))$, we may assume that the centralizer $\mathscr{M}^\sigma$ is properly infinite. Then, according to Corollary 9.16, we have $(\mathscr{M}, \sigma) \approx (\mathscr{M} \bar{\otimes} \mathscr{F}_n \bar{\otimes} \mathscr{F}_\infty, \sigma \bar{\otimes} \iota \bar{\otimes} \iota)$ for all $n \in \mathbb{N} \cup \{\infty\}$.

Since G is separable, there exists $n \in \mathbb{N} \cup \{\infty\}$ with $\mathscr{B}(\mathscr{L}^2(G)) \approx \mathscr{F}_n$, so that $(\mathscr{M}, \sigma) \approx (\mathscr{M} \bar{\otimes} \mathscr{B}(\mathscr{L}^2(G)) \bar{\otimes} \mathscr{F}_\infty, \sigma \bar{\otimes} \iota \bar{\otimes} \iota)$. Using the identification, we can define a cocycle $a \in Z_\sigma(G; U(\mathscr{M})), a \approx 1 \bar{\otimes} \boldsymbol{\rho} \bar{\otimes} 1$. Then $\mathscr{M}^a \approx \mathscr{M}^\sigma \bar{\otimes} \mathfrak{L}(G)\, \mathscr{F}_\infty$, hence a is of infinite multiplicity. On the other hand, according to 20.4.(2), we have $\boldsymbol{\rho} \bar{\otimes} \boldsymbol{\rho} \approx 1 \bar{\otimes} \boldsymbol{\rho}$, hence $a \bar{\otimes} \boldsymbol{\rho} \approx a \bar{\otimes} 1$, that is, a is a dominant cocycle. □

20.6. Recall (18.19 and 18.20) that a continuous action $\sigma : G \to Aut(\mathscr{M})$ is called integrable if the faithful normal operator-valued weight $P_\sigma = \int \sigma_a(\cdot)\, ds : \mathscr{M}^+ \to \overline{(\mathscr{M}^\sigma)}^+$ is semifinite. A cocycle $a \in Z_\sigma(G; \mathscr{M})$ is called *square-integrable*, if the continuous action $_a\sigma : G \to Aut(\mathscr{M}_{a(e)})$ is integrable. From Section 20.2, it follows that

$$\textit{if } a, b \in Z_\sigma(G; \mathscr{M}), b \lesssim a \textit{ and } a \textit{ is square-integrable, then } b \textit{ is also square-integrable.} \tag{1}$$

It follows from Proposition 19.16 that the action $\mathrm{Ad}(\rho) : G \to Aut(\mathscr{B}(\mathscr{L}^2(G)))$ is integrable, and hence that

$$\textit{the cocycle } \boldsymbol{\rho} \in Z_\iota(G; \mathscr{B}(\mathscr{L}^2(G))) \textit{ is square-integrable.} \tag{2}$$

It is easy to check that if $\sigma : G \to Aut(\mathscr{M})$ is an integrable action and $\tau : G \to Aut(\mathscr{N})$ is an arbitrary continuous action, then the tensor product action $\sigma \bar{\otimes} \tau : G \to Aut(\mathscr{M} \bar{\otimes} \mathscr{N})$ is integrable.

Conversely, if $\sigma : G \to Aut(\mathscr{M})$ is a continuous action, $\iota : G \to Aut(\mathscr{N})$ is the trivial action and the action $\sigma \bar{\otimes} \iota$ is integrable, then also the action σ is integrable. Indeed, for $X \in (\mathscr{M} \bar{\otimes} \mathscr{N})^+, \psi \in \mathscr{N}_*^+, \varphi \in \mathscr{M}_*^+$, and $s \in G$, we have $\langle \sigma_s(E^\psi_{\mathscr{M}}(X)), \varphi \rangle = \langle E^\psi_{\mathscr{M}}(X), \varphi \circ \sigma_s \rangle = \langle X, (\varphi \circ \sigma_s) \bar{\otimes} \psi \rangle = \langle (\sigma_s \bar{\otimes} \iota)(X), \varphi \bar{\otimes} \psi \rangle = \langle (E^\psi_{\mathscr{M}}((\sigma_s \bar{\otimes} \iota)(X)), \varphi \rangle$, hence

$$P_\sigma(E^\psi_{\mathscr{M}}(X)) = E^\psi_{\mathscr{M}}(P_{\sigma \bar{\otimes} \iota}(X)).$$

Thus, if $X_i \in (\mathscr{M} \bar{\otimes} \mathscr{N})^+, \|P_{\sigma \bar{\otimes} \iota}(X_i)\| < +\infty, X_i \xrightarrow{s} 1$ and $\psi_j \in \mathscr{N}_*^+, \mathrm{s}(\psi_j) \xrightarrow{s} 1$, then $E^{\psi_j}_{\mathscr{M}}(X_i) \in \mathscr{M}^+, \|P_\sigma(E^{\psi_j}_{\mathscr{M}}(X_i))\| < +\infty$ and $E^{\psi_j}_{\mathscr{M}}(X_i) \xrightarrow{s} 1$, that is, P_σ is semifinite.

We now show that

$$\textit{every dominant cocycle } a \in Z_\sigma(G; U(\mathcal{M})) \textit{ is square-integrable.} \tag{3}$$

Indeed, by 20.5.(2) we have $a \bar{\otimes} 1 \approx 1 \bar{\otimes} \boldsymbol{\rho}$ and, by (2), $\boldsymbol{\rho}$ is square-integrable, hence (3) follows from (1) using the above remarks.

If the group G is compact, then every cocycle $a \in Z_\sigma(G; \mathcal{M})$ is square-integrable. In the general case, the following result holds:

Theorem. *Let $\sigma : G \to Aut(\mathcal{M})$ be a continuous action of the separable locally compact group G on the countably decomposable properly infinite W^*-algebra $\mathcal{M}$. Let $a \in Z_\sigma(G; U(\mathcal{M}))$ be a dominant cocycle. Then a cocycle $b \in Z_\sigma(G; \mathcal{M})$ is square-integrable if and only if $b \lesssim a$.*

If $b \lesssim a$, then b is square-integrable, as follows from the previous considerations. The converse will be proved in Section 20.9 using the preliminary results given in Sections 20.7–20.8.

In particular, consider the trivial action $\iota : G \to Aut(\mathcal{B}(\mathcal{H}))$, where $\mathcal{H}$ is any Hilbert space. In this case, a cocycle $a \in Z_\sigma(G; \mathcal{B}(\mathcal{H}))$ is just an *so*-continuous unitary representation $a : G \to \mathcal{B}(\mathcal{K})$ of G on a closed linear subspace $\mathcal{K}$ of $\mathcal{H}$. The cocycle $a \in Z_\iota(G; \mathcal{B}(\mathcal{H}))$ is square-integrable in the sense defined above if and only if the *so*-continuous unitary representation $a : G \to \mathcal{B}(\mathcal{K})$ is square-integrable in the usual sense, that is, $\int |(a(s)\xi|\xi)|^2 \, ds < +\infty$ for a dense subset of vectors $\xi \in \mathcal{K}$. If $\mathcal{H}$ is separable and infinite dimensional, then the dominant cocycle is a multiple $\boldsymbol{\rho} \bar{\otimes} 1$ of the right regular representation $\boldsymbol{\rho}$. It is well known that an irreducible representation of G is square-integrable if and only if it is equivalent to a subrepresentation of the regular representation. Thus, the above Theorem is also an extension of this result.

20.7. With the assumptions of Theorem 20.6, the next result improves 20.3.(2):

Lemma. *For every cocycle $b \in Z_\sigma(G; \mathcal{M})$, there exists a unitary cocycle of infinite multiplicity $\check{b} \in Z_\sigma(G; U(\mathcal{M}))$ such that $b \lesssim \check{b}$. If b is square-integrable, then $\check{b}$ can also be chosen square-integrable.*

Proof. Let $p = \mathbf{z}(b(e))$ be the central support of the projection $b(e) \in \mathcal{M}$. Then, for every $s \in G$ we have $p = \mathbf{z}(b(s)b(s)^*) = \mathbf{z}(b(s)^*b(s)) = \mathbf{z}(\sigma_s(b(e))) = \sigma_s(p)$, hence $p \in \mathcal{M}^\sigma$. Consequently, $(\mathcal{M}, \sigma) \approx (\mathcal{M}p, \sigma) \oplus (\mathcal{M}(1-p), \sigma)$. By Theorem 20.5, there exists a dominant cocycle in $Z_\sigma(G; \mathcal{M}(1-p))$, which is square-integrable (20.6.(3)).

Thus, to prove the lemma, we may assume that $\mathbf{z}(b(e)) = 1$. In this case, there exists a sequence of partial isometries $\{u_n\} \subset \mathcal{M}$ such that $u_n^* u_m = 0$ for $n \neq m$, $u_n^* u_n = b(e)$ and $\sum_n u_n u_n^* = 1$. We define

$$\check{b}(s) = \sum_n u_n b(s) \sigma_s(u_n^*); \quad s \in G.$$

Then $\check{b}(s) \in U(\mathcal{M})$, since $\check{b}(e) = \sum_n u_n b(e) u_n^* = \sum_n u_n u_n^* = 1, \check{b} \in Z_\sigma(G; U(\mathcal{M}))$, since $\check{b}(s)\sigma_s(\check{b}(t)) = \sum_{n,m} u_n b(s) \sigma_s(u_n^*) \sigma_s(u_m) \sigma_s(b(t)) \sigma_{st}(u_m^*) = \sum_n u_n b(s) \sigma_s(b(e)) \sigma_s(b(t)) \sigma_{st}(u_n^*) = \sum_n u_n b(st) \sigma_{st}(u_n^*) = \check{b}(st)$, $u_i u_j^* \in \mathcal{M}^{\check{b}}$, as $\check{b}(s)\sigma_s(u_i u_j^*) \check{b}(s) = u_i b(s) \sigma_s(u_i^*) \sigma_s(u_i u_j^*) \sigma_s(u_j) b(s)^* u_j^* = u_i b(s) \sigma_s(b(e)) b(s)^* u_j^* = u_i b(e) u_j^* = u_i u_j^*$, so that $\check{b}$ is of infinite multiplicity, $q_k = u_k u_k^* \in \mathcal{M}^{\check{b}}$, and finally $b \approx \check{b}^{q_k} \approx \check{b}$, as $\check{b}^{q_k}(s) = q_k \check{b}(s) = u_k b(s) \sigma_s(u_k^*)$.

If b is square-integrable, then there exists a net $\{x_\lambda\}$ in $b(e)\mathscr{M}b(e), x_\lambda \xrightarrow{s} b(e)$ with $\|P_{(b^\sigma)}(x_\lambda)\| < +\infty$. Let $x_{\lambda,n} = u_n x_\lambda u_n^*$. Then $\check{b}(s)\sigma_s(x_{\lambda,n})\check{b}(s)^* = u_n b(s)\sigma_s(x_\lambda)b(s)^* u_n^*$, hence $\|P_{\check{b}^\sigma}(x_{\lambda,n})\| < +\infty$ and $x_{\lambda,n} \xrightarrow{s} u_n b(e)u_n^* = u_n u_n^*$. Since $\sum_n u_n u_n^* = 1$, it follows that $\check{b}$ is square-integrable. □

20.8. We continue with the assumptions of Theorem 20.6. The next lemma contains the main technical part of the proof of this Theorem.

Lemma. *If the cocycle $b \in Z_\sigma(G; \mathscr{M})$ is square-integrable, then, with respect to the pair $(\mathscr{M} \bar{\otimes} \mathscr{B}(\mathscr{L}^2(G)), \sigma \bar{\otimes} \iota)$ we have*

$$b(e) \bar{\otimes} 1 = \vee\{\mathbf{s}(X^*X); X \in \mathscr{I}(b \bar{\otimes} \boldsymbol{\rho}, b \bar{\otimes} 1)\}.$$

Proof. Indeed, let $P = \vee\{\mathbf{s}(X^*X); X \in \mathscr{I}(b \bar{\otimes} \boldsymbol{\rho}, b \bar{\otimes} 1)\}$. Since for any $X \in \mathscr{I}(b \bar{\otimes} \boldsymbol{\rho}, b \bar{\otimes} 1)$, we have $X^*X \in \mathscr{I}(b \bar{\otimes} 1, b \bar{\otimes} 1)$ (see 20.2.(4) and 20.2.(5)) so that $\mathbf{s}(X^*X) \in (\mathscr{M} \bar{\otimes} \mathscr{B}(\mathscr{L}^2(G)))^{b \bar{\otimes} 1} = \mathscr{M}^b \bar{\otimes} \mathscr{B}(\mathscr{L}^2(G))$, it follows that $P \in \mathscr{M}^b \bar{\otimes} \mathscr{B}(\mathscr{L}^2(G))$. On the other hand, for each $U \in U(\mathscr{M}^b \bar{\otimes} \mathscr{B}(\mathscr{L}^2(G)))$ we have $\mathscr{I}(b \bar{\otimes} \boldsymbol{\rho}, b \bar{\otimes} 1)U = \mathscr{I}(b \bar{\otimes} \boldsymbol{\rho}, b \bar{\otimes} 1)$ (see 20.2.(3) and 20.2.(5)) and hence $X \in \mathscr{I}(b \bar{\otimes} \boldsymbol{\rho}, b \bar{\otimes} 1) \Rightarrow XU \in \mathscr{I}(b \bar{\otimes} \boldsymbol{\rho}, b \bar{\otimes} 1) \Rightarrow P \geq \mathbf{s}(U^*X^*XU) = U^*\mathbf{s}(X^*X)U$, so that $P = U^*PU$. Consequently, $P \in \mathscr{Z}(\mathscr{M}^b \bar{\otimes} \mathscr{B}(\mathscr{L}^2(G))) = \mathscr{Z}(\mathscr{M}^b) \bar{\otimes} 1$. Thus, it remains to show that

$$0 \neq q \in \mathscr{Z}(\mathscr{M}^b) \Rightarrow \mathscr{I}(b \bar{\otimes} \boldsymbol{\rho}, b \bar{\otimes} 1)(q \bar{\otimes} 1) \neq 0. \tag{1}$$

Let $0 \neq q \in \mathscr{Z}(\mathscr{M}^b)$. Since b is square-integrable, there exists $x \in \mathscr{M}$ such that $x = xq \neq 0$ and $\|P_{(b^\sigma)}(x^*x)\| < +\infty$. Let $f \in \mathscr{K}(G), f \neq 0$. Consider $\mathscr{M} \subset \mathscr{B}(\mathscr{H})$ and $\mathscr{M} \bar{\otimes} \mathscr{B}(\mathscr{L}^2(G)) \subset \mathscr{B}(\mathscr{L}^2(G, \mathscr{H}))$ realized as von Neumann algebras. For $\xi \in \mathscr{L}^2(G, \mathscr{H})$ and $s \in G$ we define

$$[X\xi](s) = \Delta(s)^{-1/2}{}_b\sigma_s^{-1}(x) \int f(t)\xi(t)\,\mathrm{d}t. \tag{2}$$

Then,

$$\begin{aligned}
\|X\xi\|_2^2 &= \int \|\Delta(s)^{-1/2}{}_b\sigma_s^{-1}(x) \int f(t)\xi(t)\,\mathrm{d}t\|^2 \mathrm{d}s \\
&\leq \int\int \|\Delta(s)^{-1/2} f(t){}_b\sigma_s^{-1}(x)\xi(t)\|^2\,\mathrm{d}t\,\mathrm{d}s \\
&= \int |f(t)|^2 \left(\int \Delta(s)^{-1}\|{}_b\sigma_s^{-1}(x)\xi(t)\|^2\,\mathrm{d}s\right)\mathrm{d}t \\
&= \int |f(t)|^2 \left(\int \|{}_b\sigma_s(x)\xi(t)\|^2\,\mathrm{d}s\right)\mathrm{d}t \\
&= \int |f(t)|^2 \left(\int ({}_b\sigma_s(x^*x)\xi(t)|\xi(t))\,\mathrm{d}s\right)\mathrm{d}t \\
&= \int |f(t)|^2 \|P_{(b^\sigma)}(x^*x)^{1/2}\xi(t)\|^2\,\mathrm{d}t \\
&\leq \|P_{(b^\sigma)}(x^*x)\|\|f\|_\infty\|\xi\|_2,
\end{aligned}$$

so that (2) defines an element $X \in \mathscr{B}(\mathscr{L}^2(G,\mathscr{H}))$. Using the von Neumann double commutant theorem, it is easy to check that $X \in \mathscr{M} \bar{\otimes} \mathscr{B}(\mathscr{L}^2(G))$. Then,

$$\begin{aligned}[X(q \bar{\otimes} 1)\xi](s) &= \Delta(s)^{-1/2}{}_b\sigma_s^{-1}(x) \int f(t)[(q \bar{\otimes} 1)\xi](t)\,\mathrm{d}t \\ &= \Delta(s)^{-1/2}{}_b\sigma_s^{-1}(x)q \int f(t)\xi(t)\,\mathrm{d}t = \Delta(s)^{-1/2}{}_b\sigma_s^{-1}(xq) \int f(t)\xi(t)\,\mathrm{d}t \\ &= \Delta(s)^{-1/2}{}_b\sigma_s^{-1}(x) \int f(t)\xi(t)\,\mathrm{d}t = [X\xi](s),\end{aligned}$$

and $X(q \bar{\otimes} 1) = X \neq 0$. Finally, we show that $X \in \mathscr{I}(b \bar{\otimes} \boldsymbol{\rho}, b \bar{\otimes} 1)$. We have

$$\begin{aligned}[X(b(r) \bar{\otimes} 1)\xi](s) &= \Delta(s)^{-1/2}{}_b\sigma_s(x) \int f(t)b(r)\xi(t)\,\mathrm{d}t \\ &= \Delta(s)^{-1/2}{}_b\sigma_s(x)b(r) \int f(t)\xi(t)\,\mathrm{d}t\end{aligned}$$

and

$$\begin{aligned}[(b(r) \bar{\otimes} \boldsymbol{\rho}(r))((\sigma_r \bar{\otimes} \iota)(X))\xi](s) &= \Delta(r)^{1/2}b(r)[((\sigma_r \bar{\otimes} \iota)(X))\xi](sr) \\ &= \Delta(r)^{1/2}b(r)\Delta(sr)^{-1/2}\sigma_r({}_b\sigma_{sr}^{-1}(x)) \int f(t)\xi(t)\,\mathrm{d}t \\ &= \Delta(s)^{-1/2}b(r)\sigma_r({}_b\sigma_{sr}^{-1}(x))b(r)^*b(r) \int f(t)\xi(t)\,\mathrm{d}t \\ &= \Delta(s)^{-1/2}{}_b\sigma_s^{-1}({}_b\sigma_{sr}^{-1}(x))b(r) \int f(t)\xi(t)\,\mathrm{d}t \\ &= \Delta(s)^{-1/2}{}_b\sigma_s(x)b(r)^*b(r) \int f(t)\xi(t)\,\mathrm{d}t.\end{aligned}$$

This proves (1) and also the lemma. □

20.9. *Proof of Theorem 20.6.* Let $b \in Z_\sigma(G;\mathscr{M})$ be square-integrable. To show that $b \lesssim a$, where $a \in Z_\sigma(G;U(\mathscr{M}))$ is the dominant cocycle, we may assume, by Lemma 20.7, that b is unitary and of infinite multiplicity. Consider $b \bar{\otimes} \boldsymbol{\rho}, b \bar{\otimes} 1 \in Z_{\sigma \bar{\otimes} \iota}(G;U(\mathscr{M} \bar{\otimes} \mathscr{B}(\mathscr{L}^2(G))))$ and the balanced cocycle $c = c(b \bar{\otimes} \boldsymbol{\rho}, b \bar{\otimes} 1) \in Z_{\sigma \bar{\otimes} \iota \bar{\otimes} \iota}(G;U(\mathscr{M} \bar{\otimes} \mathscr{B}(\mathscr{L}^2(G)) \bar{\otimes} \mathscr{F}_2))$.

By Lemma 20.8, it follows that in the W^*-algebra $(\mathscr{M} \bar{\otimes} \mathscr{B}(\mathscr{L}^2(G)) \bar{\otimes} \mathscr{F}_2)^c$ we have ()) Indeed, we have

$$\mathbf{z}(1 \bar{\otimes} 1 \bar{\otimes} e_{11}) \geq \mathbf{z}(1 \bar{\otimes} 1 \bar{\otimes} e_{22}). \tag{1}$$

Indeed, we have

$$(\mathscr{M} \bar{\otimes} \mathscr{B}(\mathscr{L}^2(G)) \bar{\otimes} \mathscr{F}_2)^c = \begin{pmatrix} \mathscr{I}(b \bar{\otimes} \boldsymbol{\rho}, b \bar{\otimes} \boldsymbol{\rho}) & \mathscr{I}(b \bar{\otimes} \boldsymbol{\rho}, b \bar{\otimes} 1) \\ \mathscr{I}(b \bar{\otimes} 1, b \bar{\otimes} \boldsymbol{\rho}) & \mathscr{I}(b \bar{\otimes} 1, b \bar{\otimes} 1) \end{pmatrix}$$

and

$$\begin{pmatrix} 0 & 0 \\ x^* & 0 \end{pmatrix}\begin{pmatrix} 1 & 0 \\ 0 & 0 \end{pmatrix}\begin{pmatrix} 0 & x \\ 0 & 0 \end{pmatrix} = \begin{pmatrix} 0 & 0 \\ 0 & x^*x \end{pmatrix}.$$

Lemma 20.8 shows that the support of x^*x converges to 1 when x runs over $\mathscr{I}(b \bar{\otimes} \boldsymbol{\rho}, b \bar{\otimes} 1)$ [Recall that for any projection f in a W^*-algebra $\mathscr{N}$ we have $\mathbf{z}(f) = \vee\{\mathbf{s}(y^*fy); y \in \mathscr{N}\}$].

On the other hand, the projection

$$1 \bar{\otimes} 1 \bar{\otimes} e_{11} \text{ is properly infinite in } (\mathscr{M} \bar{\otimes} \mathscr{B}(\mathscr{L}^2(G)) \bar{\otimes} \mathscr{F}_2)^c, \tag{2}$$

since the corresponding reduced algebra is *-isomorphic to $\mathscr{M}^b$ which, by assumption, is properly infinite.

The assumptions of Theorem 20.6 also insure that all the W^*-algebras involved in our argument are countably decomposable. Therefore, we infer from (1) and (2) that $1 \bar{\otimes} 1 \bar{\otimes} e_{22} \prec 1 \bar{\otimes} 1 \bar{\otimes} e_{11}$ in $(\mathscr{M} \bar{\otimes} \mathscr{B}(\mathscr{L}^2(G)) \bar{\otimes} \mathscr{F}_2)^c$, that is, $b \bar{\otimes} 1 \precsim b \bar{\otimes} \boldsymbol{\rho}$. Thus, by 20.4.(2) and 20.5.(2), we have $b \bar{\otimes} 1 \precsim b \bar{\otimes} \boldsymbol{\rho} \approx 1 \bar{\otimes} \boldsymbol{\rho} \approx a \bar{\otimes} 1$. Since a and b are both of infinite multiplicity, we infer using 20.4.(1) and 20.2.(13) that $b \precsim a$. □

20.10 Corollary. *Let $\sigma : G \to Aut(\mathscr{M})$ be an integrable continuous action of the separable locally compact group G on the countably decomposable properly infinite W^*-algebra $\mathscr{M}$. Then the centralizer $\mathscr{M}^\sigma$ is a reduced algebra of the crossed product $\mathscr{R}(\mathscr{M}, \sigma)$.*

Proof. By assumption, the trivial cocycle, $1 \in Z_\sigma(G; U(\mathscr{M}))$ is square-integrable and hence, by Theorem 20.6, $1 \precsim a$, where $a \in Z_\sigma(G; U(\mathscr{M}))$ denotes the dominant cocycle. Thus, there exists $u \in \mathscr{M}$ with $u^*u = 1, uu^* = p \in \mathscr{M}^\sigma$, and $u^*a(s)\sigma_s(u) = 1$ for all $s \in G$. Then, $x \in \mathscr{M}^\sigma \Leftrightarrow \sigma_s(x) = x$ for all $s \in G \Leftrightarrow a(s)\sigma_s(u)\sigma_s(x)\sigma_s(u^*)a(s)^* = uxu^*$ for all $s \in G \Leftrightarrow {}_a\sigma_s(uxu^*) = uxu^*$ for all $s \in G \Leftrightarrow uxu^* \in \mathscr{M}^a$, hence $\mathscr{M}^\sigma \approx p\mathscr{M}^a p$. Since, by 20.5.(4), $\mathscr{M}^a \approx \mathscr{R}(\mathscr{M}, \sigma)$, this proves the Corollary. □

20.11 Proposition. *Let $\sigma : G \to Aut(\mathscr{M})$ be a continuous action of the separable locally compact group G on the W^*-algebra $\mathscr{M}$. We assume that the centralizer $\mathscr{M}^\sigma$ is properly infinite. Then the following statements are equivalent:*

(i) *the trivial cocycle $1 \in Z_\sigma(G; U(\mathscr{M}))$ is a dominant cocycle;*

(ii) *the actions $\sigma \bar{\otimes} \mathrm{Ad}(\boldsymbol{\rho})$ and $\sigma \bar{\otimes} \iota$ are inner conjugate, that is, there exists $U \in U(\mathscr{M} \bar{\otimes} \mathscr{B}(\mathscr{L}^2(G)))$ such that*

$$(\sigma \bar{\otimes} \mathrm{Ad}(\boldsymbol{\rho}))_s = \mathrm{Ad}(U) \circ (\sigma \bar{\otimes} \iota)_s \circ \mathrm{Ad}(U^*) \quad (s \in G);$$

(iii) *the actions $\sigma \bar{\otimes} \mathrm{Ad}(\boldsymbol{\rho})$ and $\sigma \bar{\otimes} \iota$ are conjugate, that is, there exists $\Phi \in Aut(\mathscr{M} \bar{\otimes} \mathscr{B}(\mathscr{L}^2(G)))$ such that*

$$(\sigma \bar{\otimes} \mathrm{Ad}(\boldsymbol{\rho}))_s = \Phi \circ (\sigma \bar{\otimes} \iota)_s \circ \Phi^{-1} \quad (s \in G).$$

Proof. (i) ⇒ (ii). By assumption, we have $1 \bar{\otimes} \boldsymbol{\rho} \approx 1 \bar{\otimes} 1$ in $Z_{\sigma \bar{\otimes} \iota}(G; \mathscr{M} \bar{\otimes} \mathscr{B}(\mathscr{L}^2(G)))$, that is, there exists $U \in U(\mathscr{M} \bar{\otimes} \mathscr{B}(\mathscr{L}^2(G)))$ such that $1 \bar{\otimes} \boldsymbol{\rho}(s) = U(\sigma_s \bar{\otimes} \iota)(U^*)$ $(s \in G)$. Then, for every

$X \in \mathcal{M} \bar{\otimes} \mathcal{B}(\mathcal{L}^2(G))$ and $s \in G$ we have

$$\begin{aligned}(\sigma_s \bar{\otimes} \mathrm{Ad}(\boldsymbol{\rho}(s))(X) &= (1 \bar{\otimes} \boldsymbol{\rho}(s))((\sigma_s \bar{\otimes} \iota)(X))(1 \bar{\otimes} \boldsymbol{\rho}(s)^*) \\ &= U((\sigma_s \bar{\otimes} \iota)(U^*XU))U^*.\end{aligned}$$

(ii) ⇒ (iii). Obvious.

(iii) ⇒ (i). Since $\mathcal{M}^\sigma$ is properly infinite and G is separable, we have by Corollary 9.16 $(\mathcal{M}, \sigma) \approx (\mathcal{M} \bar{\otimes} \mathcal{B}(\mathcal{L}^2(G)), \sigma \bar{\otimes} \iota)$. Using assumption (iii), we obtain $(\mathcal{M}, \sigma) \approx (\mathcal{M} \bar{\otimes} \mathcal{B}(\mathcal{L}^2(G)), \sigma \bar{\otimes} \mathrm{Ad}(\boldsymbol{\rho}))$. Consequently, we have to show that $1 \bar{\otimes} 1 \bar{\otimes} \boldsymbol{\rho} \approx 1 \bar{\otimes} 1 \bar{\otimes} 1$ in $Z_{\sigma \bar{\otimes} \mathrm{Ad}(\boldsymbol{\rho}) \bar{\otimes} \iota}(G; \mathcal{M} \bar{\otimes} \mathcal{B}(\mathcal{L}^2(G)) \bar{\otimes} \mathcal{B}(\mathcal{L}^2(G)))$ or, equivalently, that $1 \bar{\otimes} \boldsymbol{\rho} \bar{\otimes} \boldsymbol{\rho} \approx 1 \bar{\otimes} \boldsymbol{\rho} \bar{\otimes} 1$ in $Z_{\sigma \bar{\otimes} \iota \bar{\otimes} \iota}(G; \mathcal{M} \bar{\otimes} \mathcal{B}(\mathcal{L}^2(G)) \bar{\otimes} \mathcal{B}(\mathcal{L}^2(G)))$, which results from 20.4.(2). □

The continuous action $\sigma : G \to Aut(\mathcal{H})$ will be called *dominant* if the centralizer $\mathcal{M}^\sigma$ is properly infinite and the actions $\sigma \bar{\otimes} \mathrm{Ad}(\boldsymbol{\rho})$ and $\sigma \bar{\otimes} \iota$ are inner conjugate. In this case, we have

$$(\mathcal{M}, \sigma) \approx (\mathcal{M} \bar{\otimes} \mathcal{B}(\mathcal{L}^2(G)), \sigma \bar{\otimes} \mathrm{Ad}(\boldsymbol{\rho})). \tag{1}$$

The previous proposition gives us equivalent conditions for the action σ to be dominant when G is separable, but the implications (i) ⇒ (ii) ⇒ (iii) are valid in general, without the assumption of separability on G.

20.12. In this section, we show that the dominant actions are actually dual actions on crossed products. Although such a result is valid in general, we restrict ourselves here to the abelian case, where a more detailed analysis is possible.

Proposition. *Let $\sigma : G \to Aut(\mathcal{M})$ be a continuous action of the separable locally compact abelian group G on the W^*-algebra $\mathcal{M}$. We assume that the centralizer $\mathcal{M}^\sigma$ is properly infinite. Then the following statements are equivalent:*

(i) *σ is a dominant action;*

(ii) *there exists an s-continuous unitary representation $u : \hat{G} \to U(\mathcal{M})$ such that $\sigma_g(u(\gamma)) = \overline{\langle g, \gamma \rangle} u(\gamma)$ for all $g \in G, \gamma \in \hat{G}$;*

(ii') *there exists a Borel function $u : \hat{G} \to U(\mathcal{M})$ such that $\sigma_g(u(\gamma)) = \overline{\langle g, \gamma \rangle} u(\gamma)$ for all $g \in G, \gamma \in \hat{G}$;*

(iii) *there exists a continuous action $\theta : \hat{G} \to Aut(\mathcal{M}^\sigma)$ such that $(\mathcal{M}, \sigma) \approx (\mathcal{R}(M^\sigma, \theta), \hat{\theta})$.*

If the predual $\mathcal{M}_*$ is separable, then another equivalent statement is:

(ii") *for every $\gamma \in \hat{G}$ there exists $u(\gamma) \in U(\mathcal{M})$ such that $\sigma_g(u(\gamma)) = \overline{\langle g, \gamma \rangle} u(\gamma)$ for all $g \in G$.*

Proof. (i) ⇒ (ii). Since the action σ is dominant, we may assume (20.11.(1)) that $(\mathcal{M}, \sigma) = (\mathcal{M} \bar{\otimes} \mathcal{B}(\mathcal{L}^2(G)), \sigma \bar{\otimes} \mathrm{Ad}(\boldsymbol{\rho}))$. In this case, denoting by $\mathbf{m}(\gamma)$ the multiplication operator by the character $\gamma \in \hat{G}$ and defining $u(\gamma) = 1 \bar{\otimes} \mathbf{m}(\gamma)^* (\gamma \in \hat{G})$, we obtain statement (ii).

It is clear that (ii) ⇒ (ii') ⇒ (ii").

We show that (ii") ⇒ (ii') if $\mathcal{M}_*$ is separable. Indeed, $\hat{G} \times U(\mathcal{M})$ is then a polish space and the projection map $\hat{G} \times U(\mathcal{M}) \to \hat{G}$ is continuous. The set $E = \{(\gamma, u) \in \hat{G} \times U(\mathcal{M}); \sigma_g(u) = \overline{\langle g, \gamma \rangle} u$ for all $g \in G\} \subset \hat{G} \times U(\mathcal{M})$ is closed and, by (ii"), its projection on $\hat{G}$ covers the whole of $\hat{G}$. It follows that there exists a Borel section $u : \hat{G} \to U(\mathcal{M})$ with $(\gamma, u(\gamma)) \in E$.

(ii') ⇒ (i). The Borel function $u : \hat{G} \to U(\mathscr{M})$ defines a unitary element $U \in \mathscr{M} \bar{\otimes} \mathscr{L}^\infty(G)$, uniquely determined, such that $\langle U, \varphi \bar{\otimes} k\rangle = \int \langle u(\gamma), \varphi\rangle k(\gamma)\, \mathrm{d}\gamma$ for all $\varphi \in \mathscr{M}_*, k \in \mathscr{L}^1(G)$. Then the element $(\sigma_g \bar{\otimes} \iota)(U) \in \mathscr{M} \bar{\otimes} \mathscr{L}^\infty(\hat{G})$ is defined by the function $\gamma \mapsto \sigma_g(u(\gamma))$, hence $U^*((\sigma_g \bar{\otimes} \iota)(U))$ is defined by the function $\gamma \mapsto u(\gamma)^*\sigma_g(u(\gamma)) = \overline{\langle g, \gamma\rangle}$. Thus, considering the element $g \in G$ as a character on $\hat{G}$, we have $U^*((\sigma_g \bar{\otimes} \iota)(U)) = 1 \bar{\otimes} \mathbf{m}(g)^* \in \mathscr{M} \bar{\otimes} \mathscr{L}^\infty(\hat{G})\ (g \in G)$, or, identifying $\mathscr{L}^\infty(\hat{G})$ with $\mathfrak{L}(G) = \mathfrak{R}(G)$ by the Fourier–Plancherel isomorphism, $U^*((\sigma_g \bar{\otimes} \iota)(U)) = 1 \bar{\otimes} \boldsymbol{\rho}(g) \in \mathscr{M} \bar{\otimes} \mathscr{B}(\mathscr{L}^2(G))\ (g \in G)$. It follows that the cocycle $1 \in Z_\sigma(G; U(\mathscr{M}))$ is dominant, hence σ is a dominant action.

(ii) ⇒ (iii). This follows from Landstad's Theorem (19.9). □

Moreover, it follows from Landstad's theorem that the action θ is defined by the representation u as follows:

$$\theta_\gamma = \mathrm{Ad}(u(\gamma))|\mathscr{M}^\sigma \quad (\gamma \in \hat{G}). \tag{1}$$

Note that the equivalences (i) ⇔ (ii) ⇔ (iii) are valid without the separability assumption on G.

Using the Takesaki duality theorem (19.5), we deduce from (iii) that if σ is a dominant action, then

$$\mathscr{M}^\sigma \approx \mathscr{M}^\sigma \bar{\otimes} \mathscr{B}(\mathscr{L}^2(G)) \approx \mathscr{R}(\mathscr{M}, \sigma). \tag{2}$$

20.13. The definition (19.1) of crossed products is invariant under *-isomorphisms. More precisely, if $\sigma : G \to Aut(\mathscr{M}), \tau : G \to Aut(\mathscr{N})$ are continuous actions of the locally compact group G on the W^*-algebras $\mathscr{M}, \mathscr{N}$ and there exists a *-isomorphism $\Phi : \mathscr{M} \to \mathscr{N}$ such that $\Phi \circ \sigma_g = \tau_g \circ \Phi\ (g \in G)$, then the mapping $\Psi = \Phi \bar{\otimes} \iota : \mathscr{M} \bar{\otimes} \mathscr{B}(\mathscr{L}^2(G)) \to \mathscr{N} \bar{\otimes} \mathscr{B}(\mathscr{L}^2(G))$ is a *-isomorphism and it is easy to check that $\Psi(\pi_\sigma(x)) = \pi_\tau(\Phi(x))\ (x \in \mathscr{M})$, and that $\Psi(1 \bar{\otimes} \boldsymbol{\lambda}(g)) = 1 \bar{\otimes} \boldsymbol{\lambda}(g)\ (g \in G)$, hence $\Psi(\mathscr{R}(\mathscr{M}, \sigma)) = \mathscr{R}(\mathscr{N}, \tau)$. Moreover, since the restriction of Ψ to $1_{\mathscr{M}} \bar{\otimes} \mathscr{B}(\mathscr{L}^2(G))$ is the identity mapping, it follows that Ψ intertwines the dual actions, that is, $(\Psi \bar{\otimes} \iota) \circ \hat{\sigma} = \hat{\tau} \circ \Psi$. We abbreviate these remarks by writing:

$$(\mathscr{M}, \sigma) \approx (\mathscr{N}, \tau) \Rightarrow (\mathscr{R}(\mathscr{M}, \sigma), \pi_\sigma, \boldsymbol{\lambda}, \hat{\sigma}) \approx (\mathscr{R}(\mathscr{N}, \tau), \pi_\tau, \boldsymbol{\lambda}, \hat{\tau}) \tag{1}$$

In particular, if $\sigma : G \to Aut(\mathscr{M})$ is a continuous action of G on $\mathscr{M}$, then every *-automorphism $\Phi \in Aut(\mathscr{M})$ with $\Phi \circ \sigma_g = \sigma_g \circ \Phi\ (g \in G)$ "extends" to a *-automorphism $\Psi \in Aut(\mathscr{R}(\mathscr{M}, \sigma))$, uniquely determined, such that

$$\Psi(\pi_\sigma(x)(1 \bar{\otimes} \boldsymbol{\lambda}(g)) = \pi_\tau(\Phi(x))(1 \bar{\otimes} \boldsymbol{\lambda}(g)) \quad (x \in \mathscr{M}, g \in G). \tag{2}$$

On the other hand, suppose that σ and τ are both actions of G on the same von Neumann algebra $\mathscr{M} \subset \mathscr{B}(\mathscr{H})$ and $\sigma \sim \tau$, that is (15.11) there exists a unitary cocycle $u \in Z_\sigma(G; U(\mathscr{M}))$ such that $\tau = {}_u\sigma$. Consider the unitary operator $U \in \mathscr{M} \bar{\otimes} \mathscr{L}^\infty(G)$ defined by the function $G \ni s \mapsto u(s^{-1})$, that is, $(U\xi)(s) = u(s^{-1})\xi(s)\ (\xi \in \mathscr{L}^2(G, \mathscr{H}), s \in G)$. It is easy to check that

$$\begin{aligned} U\pi_\sigma(x)U^* &= \pi_\tau(X) \quad (x \in \mathscr{M}) \\ U(1 \bar{\otimes} \boldsymbol{\lambda}(g))U^* &= \pi_\tau(u(g)^*)(1 \bar{\otimes} \boldsymbol{\lambda}(g)) \quad (g \in G). \end{aligned} \tag{3}$$

Consequently, $\mathrm{Ad}(U)$ establishes a (spatial) *-isomorphism between $\mathcal{R}(\mathcal{M},\sigma)$ and $\mathcal{R}(\mathcal{M},\tau)$, sending $\pi_\sigma(x)$ into $\pi_\tau(x)$ for all $x \in \mathcal{M}$. Moreover, since $U \in \mathcal{M} \bar{\otimes} \mathcal{L}^\infty(G)$ and $\mathcal{L}^\infty(G)$ is commutative, $\mathrm{Ad}(U)$ intertwines the dual actions $\hat{\sigma}$ and $\hat{\tau}$.

More generally, we shall write $(\mathcal{M},\sigma) \sim (\mathcal{N},\tau)$, if there exists

$$u \in Z_\sigma(G; U(\mathcal{M}))$$

such that $(\mathcal{M}, {}_u\sigma) \approx (\mathcal{N},\tau)$. Combining the above two remarks, we conclude that

$$(\mathcal{M},\sigma) \sim (\mathcal{N},\tau) \Rightarrow (\mathcal{R}(\mathcal{M},\sigma), \pi_\sigma, \hat{\sigma}) \approx (\mathcal{R}(\mathcal{N},\tau), \pi_\tau, \hat{\tau}). \tag{4}$$

Furthermore, if φ, ψ are n.s.f. weights on $\mathcal{M}, \mathcal{N}$ respectively and if $(\mathcal{M},\sigma,\varphi) \approx (\mathcal{N},\tau,\psi)$, then $(\mathcal{R}(\mathcal{M},\sigma), \hat{\varphi}) \approx (\mathcal{R}(\mathcal{N},\tau), \hat{\psi})$ by the same *-isomorphism Ψ as in (1). Also, if $\sigma \sim {}_u\sigma = \tau$ are both actions of G on $\mathcal{M}$ and φ is an n.s.f. weight on $\mathcal{M}$, then the *-isomorphism $\mathrm{Ad}(U)$ appearing in (3) transports the dual weight $\hat{\varphi}$ on $\mathcal{R}(\mathcal{M},\sigma)$ into the dual weight $\hat{\varphi}$ on $\mathcal{R}(\mathcal{M},\tau)$. Therefore,

$$(\mathcal{M},\sigma,\varphi) \sim (\mathcal{N},\tau,\psi) \Rightarrow (\mathcal{R}(\mathcal{M},\sigma), \pi_\sigma, \hat{\sigma}, \hat{\varphi}) \approx (\mathcal{R}(\mathcal{N},\tau), \pi_\tau, \hat{\tau}, \hat{\psi}). \tag{5}$$

20.14. If $\mathcal{M}$ is a properly infinite W^*-algebra, then for any continuous action $\sigma : G \to Aut(\mathcal{M})$ of a locally compact group G on $\mathcal{M}$ we have

$$(\mathcal{M},\sigma) \sim (\mathcal{M} \bar{\otimes} \mathcal{F}, \sigma \bar{\otimes} \iota) \tag{1}$$

where $\mathcal{F}$ is any countably decomposable type I factor.

Indeed, by 20.3.(2), there exists a unitary cocycle $u \in Z_\sigma(G; U(\mathcal{M}))$ of infinite multiplicity. Then the continuous action $\tau = {}_u\sigma$ has a properly infinite centralizer $\mathcal{M}^\tau$ so that, using Corollary 9.16, we get $(\mathcal{M},\sigma) \sim (\mathcal{M},\tau) \approx (\mathcal{M} \bar{\otimes} \mathcal{F}, \tau \bar{\otimes} \iota) \sim (\mathcal{M} \bar{\otimes} \mathcal{F}, \sigma \bar{\otimes} \iota)$.

20.15. Notes. The results in this section are due to Connes and Takesaki [1977]. The s^*-continuity requirement in the definition of a σ-cocycle (20.1) can be replaced, without modifying this notion, by some measurability condition (see Connes [1973a], proof of 1.2.5 and 1.2.10).

For our exposition, we have used Connes and Takesaki [1977].

21 Abelian Groups

In this section, we undertake a more detailed study of crossed products by actions of abelian groups, which includes the connection between the Connes invariant and the dual action on the center of the crossed product, as well as a certain "Galois correspondence" between subgroups and invariant subalgebras of the crossed product.

21.1 Theorem (A. Connes, M. Takesaki). *Let $\sigma : G \to Aut(\mathcal{M})$ be a continuous action of the locally compact abelian group G on the W^*-algebra $\mathcal{M}$. Then the Connes invariant $\Gamma(\sigma)$ is the kernel of the restriction of the dual action $\hat{\sigma}$ to the center of the crossed product $\mathcal{R}(\mathcal{M},\sigma)$:*

$$\Gamma(\sigma) = Ker(\hat{\sigma}) : \hat{G} \to Aut(\mathcal{Z}(\mathcal{R}(\mathcal{M},\sigma))).$$

The proof will be given in Section 21.4, using some auxiliary results of independent interest which are presented in Sections 21.2 and 21.3.

21.2 Lemma. *Let $\sigma : G \to Aut(\mathscr{Z})$ be a continuous action of the locally compact group G on the commutative W^*-algebra $\mathscr{Z}$. If $g_0 \in G$ and $\sigma_{g_0} \neq \iota$, then there exist a neighborhood V of g_0 and a projection $0 \neq p \in \mathscr{Z}$ such that $p\sigma_g(p) = 0$ for all $g \in V$.*

Proof. Let $\mathscr{A} = \{x \in \mathscr{Z}; G \ni g \mapsto \sigma_g(x) \in \mathscr{Z}$ is norm-continuous$\}$. Then $\mathscr{A}$ is an s-dense σ-invariant C^*-subalgebra of $\mathscr{Z}$; indeed, if $\{f_i\} \subset \mathscr{L}^1(G)$ is an approximate unit of the Banach algebra $\mathscr{L}^1(G)$ with $supp\, f_i$ compact sets, then for every $x \in \mathscr{Z}$ we have $\mathscr{A} \ni \sigma_{f_i}(x) \xrightarrow{s} x$. Consequently, $\sigma_{g_0}|\mathscr{A} \neq \iota$.

Let $\mathscr{A} \cong \mathscr{C}(\Omega)$ where Ω is the Gelfand spectrum of $\mathscr{A}$. For each $g \in G$, there exists a unique homeomorphism $T_g : \Omega \to \Omega$ such that $[\sigma_g(x)](\omega) = x(T_g\omega)$ $(x \in \mathscr{A}, \omega \in \Omega)$. For each $x \in \mathscr{A}$, the function $G \times \Omega \ni (g, \omega) \mapsto [\sigma_g(x)](\omega) = x(T_g(\omega)) \in \mathbb{C}$ is continuous in both variables, hence the function $G \times \Omega \ni (g, \omega) \mapsto T_g(\omega) \in \Omega$ is also continuous in both variables. Since $T_{g_0} \neq \iota$, there exists $\omega_0 \in \Omega$ with $T_{g_0}(\omega_0) \neq \omega_0$. Therefore, there exist a neighborhood V of g_0 and a neighborhood D of ω_0 such that $T_g\omega \neq \omega$ for all $g \in V$ and $\omega \in D$.

Let $x \in \mathscr{A}, 0 \neq x \geq 0$ with $supp\, x \subset D$. Then $x\sigma_g(x) = 0$ for $g \in V$. There exist $\varepsilon > 0$ and a spectral projection p of x with $xp \geq \varepsilon p$. Then $0 \leq p\sigma_g(p) \leq \varepsilon^{-2}x\sigma_g(x) = 0$, hence $p\sigma_g(p) = 0$, for $g \in V$. □

21.3. Let $\sigma : G \to Aut(\mathscr{M})$ be a continuous action of the locally compact abelian group G on the W^*-algebra $\mathscr{M}$. Consider the faithful normal operator-valued weight $P_\sigma : \mathscr{M}^+ \ni x \mapsto \int \sigma_g(x)\, dg \in \overline{(\mathscr{M}^\sigma)}^+$.

For $x \in \mathfrak{M}_{P_\sigma}$ and $\gamma \in \hat{G}$ we define

$$\hat{x}(\gamma) = \int \overline{\langle g, \gamma \rangle} \sigma_g(x)\, dg \in \mathscr{M}. \tag{1}$$

Then,

$$\hat{x}(\gamma) \in \mathscr{M}(\sigma; \{\gamma\}) \tag{2}$$

since for every $s \in G$ we have $\sigma_g(\hat{x}(\gamma)) = \int \overline{\langle g, \gamma \rangle} \sigma_{sg}(x)\, dg = \int \overline{\langle s^{-1}g, \gamma \rangle} \sigma_g(x)\, dg = \langle s, \gamma \rangle \hat{x}(\gamma)$. According to the uniqueness theorem for the Fourier transform, it follows that for $x \in \mathfrak{M}_{P_\sigma}$ we have

$$\hat{x}(\gamma) = 0 \text{ for all } \gamma \in \hat{G} \Leftrightarrow x = 0. \tag{3}$$

Recall that for $f \in \mathscr{L}^1(G)$ we defined the Fourier transform $\hat{f}$ by $\hat{f}(\gamma) = \int \langle g, \gamma \rangle f(g)\, dg$, $(\gamma \in \hat{G})$, and that the Haar measures on G and $\hat{G}$ are chosen so that the Fourier inversion theorem holds.

Lemma. *For every $x \in \mathfrak{M}_{P_\sigma}$ and every $f \in \mathscr{L}^1(G)$ we have $\sigma_f(x) \in \mathfrak{M}_{P_\sigma}$ and*

$$\widehat{\sigma_f(x)}(\gamma) = \hat{f}(\gamma)\hat{x}(\gamma) \quad (\gamma \in \hat{G}) \tag{4}$$

and moreover, if $\hat{f} \in \mathscr{L}^1(\hat{G})$*, then*

$$\sigma_f(x) = \int \langle g, \gamma \rangle \sigma_f(x)\hat{}(\gamma)\, \mathrm{d}\gamma. \tag{5}$$

Proof. We may assume $x \geq 0$ and $f \geq 0$. Then $\sigma_f(x) \geq 0$ and, using the Fubini–Tonelli theorem, we obtain

$$\begin{aligned} P_\sigma(\sigma_f(x)) &= \int \sigma_g(\sigma_f(x))\, \mathrm{d}g = \int \sigma_g \left(\int f(s)\sigma_s(x)\, \mathrm{d}s \right) \mathrm{d}g \\ &= \int \int f(s)\sigma_{gs}(x)\, \mathrm{d}s\, \mathrm{d}g = \int f(s) \left(\int \sigma_{gs}(x)\, \mathrm{d}g \right) \mathrm{d}s \\ &= \left(\int f(s)\, \mathrm{d}s \right) \left(\int \sigma_g(x)\, \mathrm{d}g \right) = \|f\|_1 P_\sigma(x) \in \mathscr{M}^+, \end{aligned}$$

hence $\sigma_f(x) \in \mathfrak{M}_{P_\sigma}$. Then, for $\gamma \in \hat{G}$ we have

$$\begin{aligned} \sigma_f(x)\hat{}(\gamma) &= \int \overline{\langle g, \gamma \rangle} \sigma_g(\sigma_f(x))\, \mathrm{d}g = \int \int \overline{\langle g, \gamma \rangle} f(s)\sigma_{gs}(x)\, \mathrm{d}s\, \mathrm{d}g \\ &= \int \langle s, \gamma \rangle f(s) \left(\int \overline{\langle gs, \gamma \rangle} \sigma_{gs}(x)\, \mathrm{d}g \right) \mathrm{d}s \\ &= \left(\int \langle s, \gamma \rangle f(s)\, \mathrm{d}s \right) \left(\int \overline{\langle g, \gamma \rangle} \sigma_g(x)\, \mathrm{d}g \right) = \hat{f}(\gamma)\hat{x}(\gamma). \end{aligned}$$

Finally, if $\hat{f} \in \mathscr{L}^1(\hat{G})$, (5) follows by the Fourier inversion theorem. □

Corollary. *If the action* $\sigma : G \to Aut(\mathscr{M})$ *is integrable, then*

$$\mathscr{M} = \mathscr{R}\{\mathscr{M}(\sigma; \{\gamma\}); \gamma \in \hat{G}\} \tag{6}$$

and, for every open set $V \subset \hat{G}$*, we have*

$$\mathscr{M}(\sigma; V) \neq 0 \Leftrightarrow \textit{ there exists } \gamma \in V \textit{ such that } \mathscr{M}(\sigma; \{\gamma\}) \neq \{0\}. \tag{7}$$

Proof. Let $\{f_i\}$ be a norm-bounded approximate unit of the Banach algebra $\mathscr{L}^1(G)$ with *supp* f_i compact sets. For each $x \in \mathfrak{M}_{P_\sigma}$ we have $\sigma_{f_i}(x) \xrightarrow{w} x$ and for each $\sigma_{f_i}(x)$ (5) holds. Since the action σ is integrable, that is, $\mathfrak{M}_{P_\sigma}$ is w-dense in $\mathscr{M}$, (6) follows using (2) and (5).

We now prove assertion (7). The implication ($\Leftarrow$) is obvious. Conversely, assume that $\mathscr{M}(\sigma; \{\gamma\}) = \{0\}$ for every $\gamma \in V$. By (2) it follows that $\hat{x}(\gamma) = 0$ for all $x \in \mathfrak{M}_{P_\sigma}$ and $\gamma \in V$. Furthermore, using (4) it follows that if $x \in \mathfrak{M}_{P_\sigma}$ and $f \in \mathscr{L}^1(G)$, *supp* $\hat{f} \subset V$, then $\sigma_f(x)\hat{}(\gamma) = 0$ for all $\gamma \in \hat{G}$, hence $\sigma_f(x) = 0$. Since $\mathfrak{M}_{P_\sigma}$ is w-dense in $\mathscr{M}$, it follows that $\sigma_f(x) = 0$ for all $x \in \mathscr{M}$ and all $f \in \mathscr{L}^1(G)$ with *supp* $\hat{f} \subset V$, hence (14.3.(7)) $\mathscr{M}(\sigma; V) = \{0\}$. □

Note that if the action σ is integrable, then

$$x = \int \langle g, \gamma\rangle \hat{x}(\gamma)\,\mathrm{d}\gamma \tag{8}$$

for x in a w-dense *-subalgebra of $\mathcal{M}$.

21.4. *Proof of Theorem 21.1.* Consider the actions $\sigma \bar{\otimes} \iota$ and $\sigma \bar{\otimes} \mathrm{Ad}(\rho)$ of G on $\mathcal{M} \bar{\otimes} \mathcal{B}(\mathcal{L}^2(G))$. By 16.16.(2) and 16.3, we have

$$\Gamma(\sigma) = \Gamma(\sigma \bar{\otimes} \iota) = \Gamma(\sigma \bar{\otimes} \mathrm{Ad}(\boldsymbol{\rho})). \tag{1}$$

On the other hand, putting $\mathcal{B} = \mathcal{B}(\mathcal{L}^2(G))$, we have (19.13 and 19.3)

$$\begin{aligned}
&\mathcal{R}(\mathcal{M}, \sigma) = (\mathcal{M} \bar{\otimes} \mathcal{B})^{\sigma \bar{\otimes} \mathrm{Ad}\,\rho}, \hat{\sigma}_\gamma = \mathrm{Ad}(1 \bar{\otimes} \mathbf{m}(\gamma)^*),\\
&\mathcal{R}(\mathcal{M} \bar{\otimes} \mathcal{B}, \sigma \bar{\otimes} \mathrm{Ad}\,\boldsymbol{\rho}) = (\mathcal{M} \bar{\otimes} \mathcal{B} \bar{\otimes} \mathcal{B})^{\sigma \bar{\otimes} \mathrm{Ad}\,\rho \bar{\otimes} \mathrm{Ad}\,\rho},\\
&(\sigma \bar{\otimes} \mathrm{Ad}\,\boldsymbol{\rho})\hat{}_\gamma = \mathrm{Ad}(1 \bar{\otimes} 1 \bar{\otimes} \mathbf{m}(\gamma)^*)
\end{aligned}$$

so that

$$\begin{aligned}
\mathcal{Z}(\mathcal{R}(\mathcal{M}, \sigma)) &\approx \mathcal{Z}((\mathcal{M} \bar{\otimes} \mathcal{B} \bar{\otimes} \mathcal{B})^{\sigma \bar{\otimes} \mathrm{Ad}\,\rho \bar{\otimes} \iota}), \hat{\sigma}_\gamma \approx \mathrm{Ad}(1 \bar{\otimes} \mathbf{m}(\gamma)^* \bar{\otimes} 1),\\
\mathcal{Z}(\mathcal{R}(\mathcal{M} \bar{\otimes} \mathcal{B}, \sigma \bar{\otimes} \mathrm{Ad}\,\boldsymbol{\rho})) &\approx \mathcal{Z}((\mathcal{M} \bar{\otimes} \mathcal{B} \bar{\otimes} \mathcal{B}))^{\sigma \bar{\otimes} \mathrm{Ad}\,\rho \bar{\otimes} \mathrm{Ad}\,\rho}),\\
(\sigma \bar{\otimes} \mathrm{Ad}\,\boldsymbol{\rho})\hat{}_\gamma &\approx \mathrm{Ad}(1 \bar{\otimes} \mathbf{m}(\gamma)^* \bar{\otimes} 1),
\end{aligned}$$

the operator $1 \bar{\otimes} W \in \mathcal{M} \bar{\otimes} \mathcal{B} \bar{\otimes} \mathcal{B}$, where $(W\xi)(s,t) = \xi(s, ts)$, has the properties $W^*(\boldsymbol{\rho}(g) \bar{\otimes} 1)W = \boldsymbol{\rho}(g) \bar{\otimes} \boldsymbol{\rho}(g)$, $W^*(\mathbf{m}(\gamma) \bar{\otimes} 1)W = \mathbf{m}(\gamma) \bar{\otimes} 1$, and hence implements a *-isomorphism

$$(\mathcal{Z}(\mathcal{R}(\mathcal{M}, \sigma)), \hat{\sigma}) \approx (\mathcal{Z}(\mathcal{R}(\mathcal{M} \bar{\otimes} \mathcal{B}(\mathcal{L}^2(G)), (\sigma \bar{\otimes} \mathrm{Ad}\boldsymbol{\rho})\hat{}\,). \tag{2}$$

From (1) and (2), it follows that in proving Theorem 21.1 we may replace $(\mathcal{M}, \sigma)$ by $(\mathcal{M} \bar{\otimes} \mathcal{B}(\mathcal{L}^2(G)), \sigma \bar{\otimes} \mathrm{Ad}\,\boldsymbol{\rho})$. Similarly, we see that we may replace $(\mathcal{M}, \sigma)$ by $(\mathcal{M} \bar{\otimes} \mathcal{B}(\mathcal{L}^2(G)) \bar{\otimes} \mathcal{F}$, $\sigma \bar{\otimes} \mathrm{Ad}\,\boldsymbol{\rho} \bar{\otimes} \iota)$, where $\mathcal{F}$ is any infinite factor. In this case, $\sigma \bar{\otimes} \mathrm{Ad}\,\boldsymbol{\rho} \bar{\otimes} \iota$, is a dominant action (20.12).

Consequently, in proving the theorem we shall assume, as we may, that $\sigma : G \to Aut(\mathcal{M})$ is a dominant action. Then, by Proposition 20.12, there exists a continuous action $\theta : \hat{G} \to Aut(\mathcal{M}^\sigma)$ such that $(\mathcal{M}, \sigma) \approx (\mathcal{R}(\mathcal{M}^\sigma, \theta), \hat{\theta})$, and there exists an s-continuous unitary representation $u : \hat{G} \to U(\mathcal{M})$ such that

$$\sigma_g(u(\gamma)) = \overline{\langle g, \gamma\rangle} u(\gamma) \text{ and } \theta_\gamma = \mathrm{Ad}(u(\gamma))|\mathcal{M}^\sigma \quad (g \in G, \gamma \in \hat{G}). \tag{3}$$

Thus, we have the identifications

$$(\mathcal{R}(\mathcal{M}, \sigma), \hat{\sigma}) \approx (\mathcal{R}(\mathcal{R}(\mathcal{M}^\sigma, \theta), \hat{\theta})\hat{\hat{\theta}}) \approx (\mathcal{M}^\sigma \bar{\otimes} \mathcal{B}(\mathcal{L}^2(G)), \theta \bar{\otimes} \mathrm{Ad}\,\boldsymbol{\rho}), \tag{4}$$

the last *-isomorphism being given by the Takesaki duality theorem (19.5). It follows that

$$(\mathcal{Z}(\mathcal{R}(\mathcal{M}, \sigma)), \hat{\sigma}) \approx (\mathcal{Z}(\mathcal{M}^\sigma \bar{\otimes} \mathcal{B}(\mathcal{L}^2(G))), \theta \bar{\otimes} \mathrm{Ad}\,\boldsymbol{\rho}) \approx (\mathcal{Z}(\mathcal{M}^\sigma), \theta).$$

Thus, all we have to prove is that

$$\Gamma(a) = Ker(\theta : \hat{G} \to Aut(\mathscr{Z}(\mathscr{M}^{\sigma}))). \tag{5}$$

Let $e \in \mathscr{Z}(\mathscr{M}^{\sigma})$ be a nonzero projection. From (3), it follows that $u(\gamma) \in \mathscr{M}(\sigma; \{\gamma\}) \cap U(\mathscr{M})$, hence $\mathscr{M}(\sigma; \{\gamma\}) = \mathscr{M}^{\sigma} u(\gamma)$. Therefore,

$$\begin{aligned}\mathscr{M}_e(\sigma^e; \{\gamma\}) &= \mathscr{M}(\sigma; \{\gamma\}) \cap \mathscr{M}_e = e\mathscr{M}(\sigma; \{\gamma\})e = e\mathscr{M}^{\sigma} u(\gamma)e \\ &= e\mathscr{M}^{\sigma} u(\gamma)eu(\gamma)^* u(\gamma) = e\mathscr{M}^{\sigma}\theta_{\gamma}(e)u(\gamma) = e\theta_{\gamma}(e)\mathscr{M}^{\sigma} u(\gamma),\end{aligned}$$

so that

$$\mathscr{M}_e(\sigma^e; \{\gamma\}) = \{0\} \Leftrightarrow e\theta_{\gamma}(e) = 0. \tag{6}$$

Thus, $\gamma \in Ker(\theta : \hat{G} \to Aut(\mathscr{Z}(\mathscr{M}^{\sigma})) \Rightarrow \theta_{\gamma}(e) = e$ for all $e \in \mathscr{Z}(\mathscr{M}^{\sigma}) \Rightarrow e\theta_{\gamma}(e) \neq 0$ for all $0 \neq e \in \mathscr{Z}(\mathscr{M}^{\sigma}) \Rightarrow \mathscr{M}_e(\sigma^e; \{\gamma\}) \neq \{0\}$ for all projections $0 \neq e \in \mathscr{Z}(\mathscr{M}^{\sigma}) \Rightarrow \gamma \in Sp\, \sigma^e$ for all projections $0 \neq e \in \mathscr{Z}(\mathscr{M}^{\sigma}) \Rightarrow \gamma \in \Gamma(\sigma)$. Conversely, using Lemmas 21.2 and 21.3 and denoting by $V(\gamma)$ the family of open neighborhoods of $\gamma \in \hat{G}$, we get $\gamma \notin Ker(\theta : \hat{G} \to Aut(\mathscr{Z}(\mathscr{M}^{\sigma}))) \Rightarrow$ there exist $V \in V(\gamma)$ and $0 \neq e \in Proj(\mathscr{Z}(\mathscr{M}^{\sigma}))$ such that $e\theta_{\beta}(e) = 0$ for all $\beta \in V \Rightarrow$ there exist $V \in V(\gamma)$ and $0 \neq e \in Proj(\mathscr{Z}(\mathscr{M}^{\sigma}))$ such that $\mathscr{M}_e(\sigma^e; \{\beta\}) = \{0\}$ for all $\beta \in V \Rightarrow$ there exist $V \in V(\gamma)$ and $0 \neq e \in Proj(\mathscr{Z}(\mathscr{M}^{\sigma}))$ such that $\mathscr{M}_{\sigma}(\sigma^e; V) = \{0\} \Rightarrow$ there exists $0 = e \in Proj(\mathscr{Z}(\mathscr{M}^{\sigma}))$ such that $\gamma \notin Sp\, \sigma^e \Rightarrow \gamma \notin \Gamma(\sigma)$.

The proof of Theorem 21.1 is complete. □

Note that in the above proof the notion of a dominant action appears only in order to simplify the exposition, so that the reference to Section 20.12 is not necessary. Indeed, all facts obtained using this reference follow obviously for $(\mathscr{M} \bar{\otimes} \mathscr{B}(\mathscr{L}^2(G)), \sigma \bar{\otimes} \mathrm{Ad}(\boldsymbol{\rho}))$ from the Takesaki duality theorem. Moreover, since (in contrast to Proposition 20.12) the Takesaki duality theorem does not require restrictive conditions on $\mathscr{M}$ or G, the statement of Theorem 21.1 is also free of any such conditions.

21.5. Let $\sigma : G \to Aut(\mathscr{M})$ be a continuous action of the locally compact abelian group G on the W^*-algebra $\mathscr{M}$ and let $\hat{\sigma} : G \to Aut(\mathscr{R}(\mathscr{M}, \sigma))$ be the dual action. Recall (19.3.(9)) that

$$\mathscr{R}(\mathscr{M}, \sigma)^{\hat{\sigma}} = \pi_{\sigma}(\mathscr{M}). \tag{1}$$

With the help of Landstad's theorem (19.9), this result can be extended as follows: Consider a closed subgroup H of G and its "orthogonal" subgroup

$$H^{\perp} = \{\gamma \in \hat{G}; \langle h, \gamma \rangle = 1 \text{ for all } h \in H\}.$$

It is well known (Hewitt & Ross, 1963, 1970; Rudin, 1962; Strătilă, 1967, 1968) that the dual group $\hat{H}$ can be canonically identified with the quotient group $\hat{G}/H^{\perp}$; namely, denoting by $\tilde{\gamma} \in \hat{G}/H^{\perp}$ the image of $\gamma \in \hat{G}$, we have $\langle h, \tilde{\gamma} \rangle = \langle h, \gamma \rangle$ $(h \in H, \gamma \in \hat{G})$.

Proposition. *In the above situation, we have*

$$\mathscr{R}(\mathscr{M}, \sigma)^{\hat{\sigma}/H^{\perp}} = \mathscr{R}\{\pi_{\sigma}(\mathscr{M}), 1 \bar{\otimes} \boldsymbol{\lambda}(H)\} \approx \mathscr{R}(\mathscr{M}, \sigma|H). \tag{2}$$

Proof. Let $\mathcal{N} = \mathcal{R}(\mathcal{M},\sigma)^{\hat{\sigma}/H^{\perp}}$. We have an s-continuous unitary representation $u : H \ni h \mapsto u(h) = 1 \bar{\otimes} \boldsymbol{\lambda}(h) \in \mathcal{N}$ and a continuous action $\theta : \hat{H} \ni \tilde{\gamma} \mapsto \theta_{\tilde{\gamma}} = \hat{\sigma}_{\gamma}|\mathcal{N} \in Aut(\mathcal{N})$ such that $\theta_{\tilde{\gamma}}(u(h)) = \overline{\langle h, \tilde{\gamma}\rangle} u(h)$ for all $h \in H, \tilde{\gamma} \in \hat{H}$. From (1), it follows that $\mathcal{N}^{\theta} = \pi_{\sigma}(\mathcal{M})$ and by 19.1.(2) we have $\pi_{\sigma}(\sigma_h(x)) = u(h)\pi_{\sigma}(x)u(h)^*$ $(x \in \mathcal{M}, h \in H)$. Thus, the proposition follows using Landstad's theorem. □

Clearly, (1) follows from (2) by taking as H the trivial subgroup, consisting of the neutral element of G.

Recall that $\mathfrak{L}(G) = \mathcal{R}(\mathcal{M},\sigma)$, where $\mathcal{M} = \mathbb{C} \cdot 1_G$ and σ is the trivial action of G, while the dual action $\hat{\sigma}$ is implemented by the multiplication operators defined by the characters: $\hat{\sigma}_{\gamma}(x) = \mathbf{m}(\gamma)^* x \mathbf{m}(\gamma) (x \in \mathfrak{L}(G), \gamma \in \hat{G})$. Thus, from the above proposition it follows that for any closed subgroup H of G we have

$$\{x \in \mathfrak{L}(G); x\mathbf{m}(\gamma) = \mathbf{m}(\gamma)x, (\forall)\gamma \in H^{\perp}\} = \mathcal{R}\{\boldsymbol{\lambda}(h); h \in H\} \approx \mathfrak{L}(H). \quad (3)$$

21.6. An important problem concerning crossed products is to obtain information about the center of a crossed product.

Let $\sigma : G \to Aut(\mathcal{M})$ be a continuous action of the locally compact abelian group G on the W^*-algebra $\mathcal{M}$.

We show that

$$\mathcal{Z}(\mathcal{R}(\mathcal{M},\sigma) \subset \mathcal{Z}(\mathcal{M}^{\sigma}) \bar{\otimes} \mathcal{R}\{\boldsymbol{\lambda}(g); g \in \Gamma(\sigma)^{\perp}\}. \quad (1)$$

Indeed, let $X \in \mathcal{Z}(\mathcal{R}(\mathcal{M},\sigma))$. Since X commutes with $1 \bar{\otimes} \boldsymbol{\lambda}(g)$ $(g \in G)$, it follows that $X \in \mathcal{M} \bar{\otimes} \mathfrak{L}(G)$. Since $X \in \mathcal{R}(\mathcal{M},\sigma) = (\mathcal{M} \bar{\otimes} \mathcal{B}(\mathcal{L}^2(G)))^{\sigma \bar{\otimes} \operatorname{Ad} \rho}$ (see 19.13), it follows that $X \in \mathcal{M}^{\sigma} \bar{\otimes} \mathfrak{L}(G)$. If $x \in \mathcal{M}^{\sigma}$, then X commutes with $\pi_{\sigma}(x) = x \bar{\otimes} 1$, hence $X \in \mathcal{Z}(\mathcal{M}^{\sigma}) \bar{\otimes} \mathfrak{L}(G)$. By Theorem 21.1, for every $\gamma \in \Gamma(\sigma)$ we have $\hat{\sigma}_{\gamma}(X) = X$, that is,

$$(1 \bar{\otimes} \mathbf{m}(\gamma))X = X(1 \bar{\otimes} \mathbf{m}(\gamma)) \quad (\gamma \in \Gamma(\sigma)).$$

Let $\varphi \in \mathcal{M}_*$ and $a = E_G^{\varphi}(X)$. Since $X \in \mathcal{M} \bar{\otimes} \mathfrak{L}(G)$, it follows from the previous identity that $a \in \mathfrak{L}(G)$ and $\mathbf{m}(\gamma)a = a\mathbf{m}(\gamma)$ for all $\gamma \in \Gamma(\sigma)$. Using 21.5.(3), we further deduce that $E_G^{\varphi}(X) = a \in \mathcal{R}\{\boldsymbol{\lambda}(g); g \in \Gamma(\sigma)^{\perp}\}$. Since $\varphi \in \mathcal{M}_*$ was arbitrary, we conclude that $X \in \mathcal{Z}(\mathcal{M}^{\sigma}) \bar{\otimes} \mathcal{R}\{\boldsymbol{\lambda}(g); g \in \Gamma(\sigma)^{\perp}\}$.

On the other hand, we have

$$\mathcal{Z}(\mathcal{M})^{\sigma} \bar{\otimes} 1_G \subset \mathcal{Z}(\mathcal{R}(\mathcal{M},\sigma)). \quad (2)$$

Indeed, let $z \in \mathcal{Z}(\mathcal{M})^{\sigma}$. Then $z \bar{\otimes} 1$ obviously commutes with every $1 \bar{\otimes} \boldsymbol{\lambda}(g)$ $(g \in G)$, and for any $x \in \mathcal{M}$ we have $(z \bar{\otimes} 1)\pi_{\sigma}(x) = \pi_{\sigma}(z)\pi_{\sigma}(x) = \pi_{\sigma}(zx) = \pi_{\sigma}(xz) = \pi_{\sigma}(x)\pi_{\sigma}(z) = \pi_{\sigma}(x)(z \bar{\otimes} 1)$.

We now show that

$$\Gamma(\sigma) = \hat{G} \Leftrightarrow \mathcal{Z}(\mathcal{R}(\mathcal{M},\sigma)) = \mathcal{Z}(\mathcal{M})^{\sigma} \bar{\otimes} 1_G. \quad (3)$$

Indeed, the implication $(\Leftarrow)$ follows immediately from Theorem 21.1, as $\hat{\sigma}_{\gamma} = \operatorname{Ad}(1 \bar{\otimes} \mathbf{m}(\gamma)^*)$ $(\gamma \in \hat{G})$. Conversely, if $\Gamma(\sigma) = \hat{G}$, then from (1) it follows that $\mathcal{Z}(\mathcal{R}(\mathcal{M},\sigma)) \subset \mathcal{Z}(\mathcal{M}^{\sigma}) \bar{\otimes} 1$. Thus, if $X \in \mathscr{Z}(\mathscr{R}(\mathscr{M},\sigma))$, there exists $z \in \mathscr{M}^\sigma$ such that $X = z \bar{\otimes} 1 = \pi_\sigma(z)$, and for every $x \in \mathscr{M}$ we have $\pi_\sigma(zx) = X\pi_\sigma(x) = \pi_\sigma(x)X = \pi_\sigma(xz)$, that is, $z \in \mathscr{Z}(\mathscr{M}) \cap \mathscr{M}^\sigma = \mathscr{Z}(\mathscr{M})^\sigma$. Therefore, $\mathscr{Z}(\mathscr{R}(\mathscr{M},\sigma)) \subset \mathscr{Z}(\mathscr{M})^\sigma \bar{\otimes} 1$ and the reverse inclusion follows by (2).

From (3) and Theorem 21.1 it follows that

$$\mathscr{R}(\mathscr{M},\sigma) \textit{ is a factor } \Leftrightarrow \Gamma(\sigma) = \hat{G} \textit{ and } \mathscr{Z}(\mathscr{M})^\sigma = \mathbb{C}\cdot 1_{\mathscr{M}}. \tag{4}$$

Thus, *if the action σ is ergodic on $\mathscr{Z}(\mathscr{M})$, for instance if $\mathscr{M}$ is a factor, or if $\mathscr{M}^\sigma$ is a factor, then*

$$\mathscr{R}(\mathscr{M},\sigma) \textit{ is a factor} \Leftrightarrow \Gamma(\sigma) = \hat{G}. \tag{5}$$

In particular, we obtain the following:

Corollary. *Let $\sigma : G \to Aut(\mathscr{M})$ be a continuous action of the locally compact abelian group G on the W^*-algebra $\mathscr{M}$. If the centralizer $\mathscr{M}^\sigma$ is a factor and $\sigma_g \neq \iota$ for every $g \neq e$ (= the neutral element of G), then $\mathscr{R}(\mathscr{M},\sigma)$ is a factor.*

Proof. Since $\mathscr{M}^\sigma$ is a factor, (5) holds and $\Gamma(\sigma) = Sp\ \sigma$ (16.1.(3)). Let $g \in \Gamma(\sigma)^\perp = (Sp\ \sigma)^\perp$. By Proposition 14.6, the spectrum of the operator $\sigma_g \in \mathscr{B}(\mathscr{M})$ is equal to $\{\overline{\langle g,\gamma\rangle; \gamma \in Sp\ \sigma}\} = \{1\}$ and hence (14.12.(3)) $\sigma_g = \iota$. By hypothesis it follows that $g = e$. Thus, $\Gamma(\sigma)^\perp = \{e\}$ and $\Gamma(\sigma) = \Gamma(\sigma)^{\perp\perp} = \hat{G}$, that is, $\mathscr{R}(\mathscr{M},\sigma)$ is a factor. □

As a further application, we give another proof of statement 16.5.(1). Indeed, let $g \in Int\ \sigma$ and let $u(g) \in U(\mathscr{M}^\sigma)$ be such that $\sigma_g = \mathrm{Ad}(u(g))$. Then $u(g) \in U(\mathscr{Z}(\mathscr{M}^\sigma))$ and using Corollary 19.13 it is easy to see that

$$u(g) \bar{\otimes} \boldsymbol{\lambda}(g) \in \mathscr{Z}(\mathscr{R}(\mathscr{M},\sigma)). \tag{1}$$

Consequently, if $\gamma \in \Gamma(\sigma)$, then by Theorem 21.1, we have $u(g) \bar{\otimes} \boldsymbol{\lambda}(g) = \hat{\sigma}(u(g) \bar{\otimes} \boldsymbol{\lambda}(g)) = (1 \bar{\otimes} \mathbf{m}(\gamma)^*)(u(g) \bar{\otimes} \boldsymbol{\lambda}(g))(1 \bar{\otimes} \mathbf{m}(\gamma)) = \overline{\langle g,\gamma\rangle}(u(g) \bar{\otimes} \boldsymbol{\lambda}(g))$, so that $\langle g,\gamma\rangle = 1$. Hence $g \in \Gamma(\sigma)^\perp$, thus proving the inclusion $Int\ \sigma \subset \Gamma(\sigma)^\perp$. Recall that if $\mathscr{M}$ is a factor and $Sp\ \sigma/\Gamma(\sigma)$ is compact, then, by Theorem 16.5, $Int\ \sigma = \Gamma(\sigma)^\perp$.

21.7. The Coones invariant $\Gamma(\sigma)$ also gives information about the comparison of σ-cocycles:

Corollary. *Let $\sigma : G \to Aut(\mathscr{M})$ be a continuous action of the separable locally compact abelian group G on the W^*-algebra $\mathscr{M}$ with separable predual $\mathscr{M}_*$. If $\Gamma(\sigma) = \hat{G}$, then every square-integrable unitary cocycle of infinite multiplicity $a \in Z_\sigma(G; U(\mathscr{M}))$ is a dominant cocycle.*

Proof. By Theorems 20.5 and 20.6, there exist a dominant cocycle $u \in Z_\sigma(G; U(\mathscr{M}))$ and a projection $p \in \mathscr{M}^u$ such that $a \approx u^p$; then the actions ${}_a\sigma$ and $({}_u\sigma)^p$ are conjugate. Since ${}_u\sigma$ is a dominant action, proving the corollary amounts to showing that

$$\begin{gathered}\textit{if } \sigma \textit{ is a dominant action with } \Gamma(\sigma) = \hat{G}, \textit{ then for every}\\ \textit{properly infinite projection } p \in \mathscr{M}^\sigma \textit{ with } p \sim 1 \textit{ in } \mathscr{M}\\ \textit{the action } \sigma^p \textit{ is also dominant.}\end{gathered} \tag{1}$$

If σ is a dominant action, then, by Proposition 20.12, there exists a continuous action $\theta : \hat{G} \to Aut(\mathscr{M}^\sigma)$ such that $(\mathscr{M},\sigma) \approx (\mathscr{R}(\mathscr{M}^\sigma,\theta),\hat{\theta})$. According to the Takesaki duality theorem (19.5), it follows that $(\mathscr{Z}(\mathscr{R}(\mathscr{M},\sigma)),\hat{\sigma}) = (\mathscr{Z}(\mathscr{M}^\sigma),\theta)$. By Theorem 21.1, the assumption $\Gamma(\sigma) = \hat{G}$ means that θ acts identically on $\mathscr{Z}(\mathscr{M}^\sigma)$.

Then, for each $\gamma \in \hat{G}$, the properly infinite projections p and $\theta_\gamma(p)$ of $\mathscr{M}^\sigma$ have the same central support in $\mathscr{M}^\sigma$, and $\mathscr{M}^\sigma$ is countably decomposable, since $\mathscr{M}_*$ is separable. Consequently, for every $\gamma \in \hat{G}$ we have $p \sim \theta_\gamma(p)$ in $\mathscr{M}^\sigma$, that is, there exists $w(\gamma) \in \mathscr{M}^\sigma$ such that $w(\gamma)^* w(\gamma) = \theta_\gamma(p)$ and $w(\gamma)w(\gamma)^* = p$.

On the other hand, since the action σ is dominant, it also follows from Propsition 20.12 that there exists an s-continuous unitary representation $u : \hat{G} \to \mathscr{M}$ such that $\sigma_g(u(\gamma)) = \overline{(g,\gamma)}u(\gamma)$ and $\theta_\gamma = \mathrm{Ad}(u(\gamma))|\mathscr{M}^\sigma$ for all $g \in G, \gamma \in \hat{G}$.

Let $v(\gamma) = w(\gamma)u(\gamma)p \in p\mathscr{M}p, (\gamma \in \hat{G})$. Then $v(\gamma) \in U(p\mathscr{M}p)$ as $v(\gamma)^*v(\gamma) = pu(\gamma)^*w(\gamma)^* w(\gamma)u(\gamma)p = pu(\gamma)^*\theta_\gamma(p)u(\gamma)p = p\theta_\gamma^{-1}(\theta_\gamma(p))p = p$ and $v(\gamma)v(\gamma)^* = w(\gamma)u(\gamma)pu(\gamma)^*w(\gamma)^* = w(\gamma)\theta_\gamma(p)w(\gamma)^* = w(\gamma)w(\gamma)^* = p$. Also, $\sigma_g^p(v(\gamma)) = \sigma_g(w(\gamma))\sigma_g(u(\gamma))\sigma_g(p) = w(\gamma)\overline{\langle(g,\gamma)\rangle}u(\gamma)p = \overline{\langle g,\gamma\rangle}v(\gamma)$. Since $\mathscr{M}_*$ is separable, it follows from Proposition 20.12 that σ^p is dominant. □

Thus, under the conditions of the corollary, all square-integrable unitary cocycles of infinite multiplicity are equivalent.

21.8. Proposition 21.5 allows us to establish a certain "Galois correspondence" between the closed subgroups of G and the $\hat{\sigma}$-invariant unital W^*-subalgebras of $\mathscr{R}(\mathscr{M},\sigma)$, which contain $\pi_\sigma(\mathscr{M})$, in the case when $\mathscr{M}$ is a factor.

Theorem. *Let $\sigma : G \to Aut(\mathscr{M})$ be a continuous action of the locally compact abelian group G on the factor $\mathscr{M}$. If $\mathscr{N}$ is a $\hat{\sigma}$-invariant unital W^*-subalgebra of $\mathscr{R}(\mathscr{M},\sigma)$ containing $\pi_\sigma(\mathscr{M})$, then there exists a closed subgroup H of G, uniquely determined, such that*

$$\mathscr{N} = \mathscr{R}(\mathscr{M},\sigma)^{\hat{\sigma}|H^\perp} = \mathscr{R}\{\pi_\sigma(\mathscr{M}), 1 \overline{\otimes} \boldsymbol{\lambda}(H)\} \approx \mathscr{R}(\mathscr{M},\sigma|H) \tag{1}$$

namely,

$$H = \{\gamma \in \hat{G}; \hat{\sigma}_\gamma|\mathscr{N} = \iota\}^\perp = \{g \in G; 1 \overline{\otimes} \boldsymbol{\lambda}(g) \in \mathscr{N}\}. \tag{2}$$

Thus, the mappings $H \mapsto \mathscr{N}$ and $\mathscr{N} \to H$ defined by (1) and (2) are reciprocal order-preserving bijections between the closed subgroups H of G and the $\hat{\sigma}$-invariant unital W^*-subalgebras $\mathscr{N} \supset \pi_\sigma(\mathscr{M})$ of $\mathscr{R}(\mathscr{M},\sigma)$.

Proof. Let $H = \{\gamma \in \hat{G}; \hat{\sigma}_\gamma|\mathscr{N} = \iota\}^\perp$. It is clear that $\mathscr{N} \subset \mathscr{R}(\mathscr{M},\sigma)^{\hat{\sigma}|H^\perp}$ and, by Proposition 21.5, we have $\mathscr{R}(\mathscr{M},\sigma)^{\hat{\sigma}|H^\perp} \approx \mathscr{R}(\mathscr{M},\sigma|H)$ so that, identifying $\mathscr{R}(\mathscr{M},\sigma)^{\hat{\sigma}|H^\perp}$ with $\mathscr{R}(\mathscr{M},\sigma|H)$ via this *-isomorphism, the subalgebra $\mathscr{N} \subset \mathscr{R}(\mathscr{M},\sigma|H)$ satisfies the conditions $\mathscr{N} \supset \pi_{\sigma|H}(\mathscr{M}), (\sigma|H)\hat{}_\chi(\mathscr{N}) = \mathscr{N}(\chi \in \hat{H})$, and the subgroup $\{\chi \in \hat{H}; (\sigma|H)\hat{}_\chi|\mathscr{N} = \iota\}$ reduces to the neutral element of $\hat{H}$.

Thus, it is sufficient to prove the theorem in the case when $\hat{\sigma}_\gamma|\mathscr{N} \neq \iota$ for all $\gamma \in \hat{G}, \gamma \neq \varepsilon$, where ε stands for the neutral element of G.

In this case, we identify $\mathscr{M}$ with $\pi_\sigma(\mathscr{M})$ and put $\mathscr{R} = \mathscr{R}(\mathscr{M},\sigma)$, so that $\mathscr{M} \subset \mathscr{N} \subset \mathscr{R}$. We have $\mathscr{R}^{\hat{\sigma}} = \mathscr{N}^{\hat{\sigma}} = \mathscr{M}$ which, by assumption, is a factor. From Corollary 21.6, it follows that $\tilde{\mathscr{N}} = \mathscr{R}(\mathscr{N},\hat{\sigma})$ is a factor. Putting $\tilde{\mathscr{R}} = \mathscr{R}(\mathscr{R},\hat{\sigma})$ and $\tilde{\mathscr{M}} = \mathscr{R}(\mathscr{M},\hat{\sigma})$, we have $\tilde{\mathscr{M}} \subset \tilde{\mathscr{N}} \subset \tilde{\mathscr{R}}$. If we can show that $\tilde{\mathscr{N}} = \tilde{\mathscr{R}}$, then it will follow that $\mathscr{N} = \tilde{\mathscr{N}}^{\hat{\hat{\sigma}}} = \tilde{\mathscr{R}}^{\hat{\hat{\sigma}}} = \mathscr{R} = \mathscr{R}(\mathscr{M},\sigma)$, thus proving the theorem.

We have $\tilde{\mathscr{R}} = \mathscr{R}(\mathscr{R}(\mathscr{M},\sigma),\hat{\sigma})$ and hence, by the Takesaki duality theorem (19.5), there exists a *-isomorphism $\tilde{\mathscr{R}} \approx \mathscr{M} \bar{\otimes} \mathscr{B}(\mathscr{L}^2(G))$ such that $(\mathscr{R}(\mathscr{M},\sigma),\hat{\sigma}) \approx ((\mathscr{M} \bar{\otimes} \mathscr{B}(\mathscr{L}^2(G)))^{\sigma\bar{\otimes}\mathrm{Ad}\,\rho}, \iota \bar{\otimes} \mathrm{Ad}(\mathbf{m}))$. The image of $\mathscr{M} \equiv \pi_\sigma(\mathscr{M}) = \mathscr{R}(\mathscr{M},\sigma)^{\hat{\sigma}}$ by this *-isomorphism is just $\pi_\sigma(\mathscr{M}) \subset \mathscr{M} \bar{\otimes} \mathscr{B}(\mathscr{L}^2(G))$ and hence the image of $\tilde{\mathscr{M}} = \mathscr{R}(\mathscr{M},\hat{\sigma})$ is (19.2.(1)) $\tilde{\mathscr{M}} \approx \mathscr{R}\{\pi_\sigma(\mathscr{M}), 1 \bar{\otimes} \mathbf{m}(\hat{G})\} = \mathscr{M} \bar{\otimes} \mathscr{L}^\infty(G)$. Therefore, we have $\mathscr{M} \bar{\otimes} \mathscr{L}^\infty(G) \subset \tilde{\mathscr{N}} \subset \mathscr{M} \bar{\otimes} \mathscr{B}(\mathscr{L}^2(G))$ or, passing to the commutants, $\mathscr{M}' \bar{\otimes} 1_G \subset \tilde{\mathscr{N}}' \subset \mathscr{M}' \bar{\otimes} \mathscr{L}^\infty(G)$, where $\tilde{\mathscr{N}}'$ is a factor. By Corollary 10.18, it follows that $\tilde{\mathscr{N}}' = \mathscr{M}' \bar{\otimes} 1_G$, that is, $\tilde{\mathscr{N}} = \mathscr{M} \bar{\otimes} \mathscr{B}(\mathscr{L}^2(G)) = \tilde{\mathscr{R}}$.

We have thus proved, in the general case, that (1) holds with $H = \{\gamma \in \hat{G}; \hat{\sigma}_\gamma|\mathscr{N} = \iota\}^\perp$. It is then immediate that $1 \bar{\otimes} \boldsymbol{\lambda}(h) \in \mathscr{N}$ for every $h \in H$. Conversely, let $g \in G$ be such that $1 \bar{\otimes} \boldsymbol{\lambda}(g) \in \mathscr{N}$. Then, for $\gamma \in H^\perp$ we have $1 \bar{\otimes} \boldsymbol{\lambda}(g) = \hat{\sigma}_\gamma(1 \bar{\otimes} \boldsymbol{\lambda}(g)) = \overline{\langle g,\gamma\rangle}(1 \bar{\otimes} \boldsymbol{\lambda}(g))$, hence $\langle g,\gamma\rangle = 1$. Thus, $g \in H^{\perp\perp} = H$, which proves (2). □

21.9 Corollary. *Let $\sigma : G \to Aut(\mathscr{M})$ be a continuous action of the commutative locally compact group G on the W^*-algebra $\mathscr{M}$. If*

(a) *the action σ is dominant and $\mathscr{M}^\sigma$ is a factor, or if*
(b) *the action σ is integrable, $\Gamma(\sigma) = \hat{G}$ and $\mathscr{M}$ is a factor,*

then the mappings $H \mapsto \mathscr{N}_H = \mathscr{M}^{\sigma|H}$ and $\mathscr{N} \mapsto H_{\mathscr{N}} = \{g \in G; \sigma_g|\mathscr{N} = \iota\}$ are reciprocal order-reversing bijections between the closed subgroups H of G and the σ-invariant unital W^*-subalgebras $\mathscr{N} \supset \mathscr{M}^\sigma$ of $\mathscr{M}$.

Proof. (a) If the action σ is dominant, then (20.12) there exists a continuous action $\theta : \hat{G} \to Aut(\mathscr{M})$ such that $(\mathscr{M},\sigma) \approx (\mathscr{R}(\mathscr{M}^\sigma,\theta),\hat{\theta})$. Since $\mathscr{M}^\sigma$ is assumed to be a factor, the corollary follows immediately from Theorem 21.8.

(b) Let $\mathscr{F}$ be the countably decomposable infinite factor of type I. It is easy to see that if $(\mathscr{M},\sigma)$ satisfies condition (b), then $(\mathscr{M} \bar{\otimes} \mathscr{F}, \sigma \bar{\otimes} \iota)$ also satisfies condition (b) and that if the conclusion of the corollary is true for $(\mathscr{M} \bar{\otimes} \mathscr{F}, \sigma \otimes \iota)$, it remains true for $(\mathscr{M},\sigma)$.

Consequently, we may assume that the centralizer $\mathscr{M}^\sigma$ is properly infinite. Then, by assumption (b) and Corollary 21.7 it follows that σ is a dominant action. Moreover, $\mathscr{M}^\sigma \approx \mathscr{M}^\sigma \bar{\otimes} \mathscr{B}(\mathscr{L}^2(G)) \approx \mathscr{R}(\mathscr{M},\sigma)$ (see 20.12.(2)) is a factor because $\Gamma(\sigma) = \hat{G}$ and $\mathscr{M}$ is a factor (see 21.6.(5)). We can thus restrict attention to case (a), which has been already considered. □

21.10. Let G be a locally compact abelian group and $H \subset G$ a closed subgroup. For $g \in G$, we shall denote by $\tilde{g} \in G/H$ its image under the canonical quotient mapping. It is known (Rudin, 1962; Strătilă, 1967, 1968) that the Haar measures on G, H and G/H can be chosen so that

$$\int_G f(g)\,\mathrm{d}g = \int_{G/H}\left(\int_H f(gh)\,\mathrm{d}h\right)\mathrm{d}\tilde{g} \quad (f \in \mathscr{L}^1(G)). \tag{1}$$

More precisely, for every $f \in \mathscr{L}^1(G)$ there exists a negligible set $\Omega \subset G/H$ such that for every $g \in G$ with $\tilde{g} \in \Omega$, the function $H \ni h \mapsto f(gh)$ belongs to $\mathscr{L}^1(H)$, the function $G/H \ni \tilde{g} \mapsto \int_H f(gh)\,\mathrm{d}h$ belongs to $\mathscr{L}^1(G/H)$, and (1) holds.

If $\sigma : G \to Aut(\mathcal{M})$ is a continuous action of G on the W^*-algebra $\mathcal{M}$, then we can also define a continuous action $\tilde{\sigma} : G/H \to Aut(\mathcal{M}^{\sigma|H})$ by

$$\tilde{\sigma}_{\tilde{g}} = \sigma_g|\mathcal{M}^{\sigma|H} \quad (g \in G). \tag{2}$$

Using (1), it is easy to check that if $x \in \mathcal{M}^+$ and if $\|P_\sigma(x)\| < +\infty$, then $P_{\sigma|H}(x) \in \overline{(\mathcal{M}^{\sigma|H})}^+$, $\|P_{\sigma|H}(x)\| < +\infty$ and

$$P_{\tilde{\sigma}}(P_{\sigma|H}(x)) = P_\sigma(x). \tag{3}$$

It follows that *if the action σ is integrable, then the actions $\sigma|H$ and $\tilde{\sigma}$ are also integrable.*

21.11. As an application of Corollary 21.9, consider the factor $\mathcal{M} = \mathcal{B}(\mathcal{L}^2(G))$ and the continuous action $\sigma : G \times \hat{G} \to Aut(\mathcal{M})$ defined by

$$\sigma_{g,\gamma}(x) = \boldsymbol{\lambda}(g)\mathbf{m}(\gamma)^* x \mathbf{m}(\gamma)\boldsymbol{\lambda}(g)^* \quad (x \in \mathcal{M}, g \in G, \gamma \in \hat{G}).$$

Since $\mathcal{B}(\mathcal{L}^2(G)) = \mathcal{R}\{\mathfrak{L}(G), \mathcal{L}^\infty(G)\} = \mathcal{R}\{\lambda(G), \mathbf{m}(\hat{G})\}$, we have

$$\mathcal{M}^\sigma = \mathbb{C} \cdot 1_G. \tag{1}$$

Consequently, $\Gamma(\sigma) = Sp\ \sigma$, and it is easy to see that

$$\Gamma(\sigma) = (G \times \hat{G})\hat{} = \hat{G} \times G. \tag{2}$$

Finally, the action σ is integrable. Indeed, for $\xi, \eta \in \mathcal{L}^2(G)$ the operator $\xi \bar{\otimes} \bar{\eta} \in \mathcal{B}(\mathcal{L}^2(G))$ (see 4.23) belongs to $\mathfrak{M}_{P_\sigma}$ and

$$P_\sigma(\xi \bar{\otimes} \bar{\eta}) = (\xi|\eta) \cdot 1_G. \tag{3}$$

By Corollary 21.9, it follows that *there exists a decreasing bijection between the σ-invariant von Neumann subalgebras of $\mathcal{B}(\mathcal{L}^2(G))$ and the closed subgroups of $G \times \hat{G}$.*

Let $H \subset G \times \hat{G}$ be a closed subgroup. Then $H^\perp = \{(\gamma, g) \in \hat{G} \times G; \langle h, \gamma\rangle\langle g, \chi\rangle = 1$, for all $(h, \chi) \in H\}$. Using the commutation relations 18.9.(1), for $g, h \in G$ and $\gamma, \chi \in G$ we obtain

$$\sigma_{(h,\chi)}(\boldsymbol{\lambda}(g)\mathbf{m}(\gamma)) = \overline{\langle h, \gamma\rangle}\,\overline{\langle g, \chi\rangle}\boldsymbol{\lambda}(g)\mathbf{m}(\gamma), \tag{4}$$

so that

$$\boldsymbol{\lambda}(g)\mathbf{m}(\gamma) \in \mathcal{M}^{\sigma|H} \Leftrightarrow (\gamma, g) \in H^\perp. \tag{5}$$

The action σ of $G \times \hat{G}$ on $\mathcal{M}$ also defines a natural action $\tilde{\sigma}$ of $(G \times \hat{G})/H$ on $\mathcal{M}^{\sigma|H}$ whose centralizer is $(\mathcal{M}^{\sigma|H})^{\tilde{\sigma}} = \mathcal{M}^\sigma = \mathbb{C} \cdot 1_G$. Since the action σ is integrable, so is the action $\tilde{\sigma}$ (21.10) and, by 21.3.(6), it follows that

$$\mathcal{M}^{\sigma|H} = \mathcal{R}\{\mathcal{M}^{\sigma|H}(\tilde{\sigma}; \{(\gamma, g)\}); \quad (\gamma, g) \in H^\perp\}. \tag{6}$$

Let $(\gamma, g) \in H^{\perp}$ and $x \in \mathscr{M}^{\sigma|H}(\tilde{\sigma}; \{(\gamma, g)\})$. We infer from (4) that

$$\boldsymbol{\lambda}(g)\mathbf{m}(\gamma) \in \mathscr{M}^{\sigma|H}(\tilde{\sigma}; \{(\gamma^{-1}, g^{-1})\}), \text{ hence } x\boldsymbol{\lambda}(g)\mathbf{m}(\gamma) \in (\mathscr{M}^{\sigma|H})^{\tilde{\sigma}} = \mathbb{C} \cdot 1_G.$$

Thus, there exists $\mu \in \mathbb{C}$ with $x = \mu(\boldsymbol{\lambda}(g)\mathbf{m}(\gamma))^*$. We have proved that

$$\mathscr{M}^{\sigma|H}(\tilde{\sigma}; \{(\gamma, g)\}) = \mathbb{C} \cdot (\lambda(g)\mathbf{m}(\gamma))^*. \tag{7}$$

Using (5), (6), and (7), we obtain

$$\mathscr{M}^{\sigma|H} = \mathscr{R}\{\boldsymbol{\lambda}(g)\mathbf{m}(\gamma); (\gamma, g) \in H^{\perp}\}. \tag{8}$$

Equation (8) makes explicit the correspondence between the σ-invariant von Neumann subalgebras of $\mathscr{B}(\mathscr{L}^2(G))$ and the closed subgroups of $G \times \hat{G}$.

On the other hand, by the definition of σ it is clear that

$$\mathscr{M}^{\sigma|H} = \{\boldsymbol{\lambda}(h)\mathbf{m}(\chi); (h, \chi) \in H\}', \tag{9}$$

so that we obtain the following commutation result:

Proposition. *Let G be a locally compact abelian group and $H \subset G \times \hat{G}$ a closed subgroup. Then*

$$\mathscr{R}\{\boldsymbol{\lambda}(h)\mathbf{m}(\chi); (h, \chi) \in H\}' = \mathscr{R}\{\boldsymbol{\lambda}(g)\mathbf{m}(\gamma); (\gamma, g) \in H^{\perp}\}. \tag{10}$$

In particular, let H_1, H_2 be two closed subgroups of G and $H = H_1 \times H_2^{\perp}$. In this case, (10) becomes

$$\mathscr{R}\{\boldsymbol{\lambda}(H_1), \mathbf{m}(H_2^{\perp})\}' = \mathscr{R}\{\boldsymbol{\lambda}(H_2), \mathbf{m}(H_1^{\perp})\}. \tag{11}$$

Note that this contains in particular the commutation relations $\mathscr{L}^\infty(G)' = \mathscr{L}^\infty(G)$ (for $H_1 = H_2 = \{e\}$), $\mathfrak{L}(G)' = \mathfrak{L}(G) (= \mathfrak{R}(G))$ (for $H_1 = H_2 = G$), and $\mathscr{B}(\mathscr{L}^2(G)) = \mathscr{R}\{\mathfrak{L}(G), \mathscr{L}^\infty(G)\}$ (for $H_1 = \{e\}, H_2 = G$).

For every closed subgroup $K \subset G$, the set $\mathbf{m}(K^{\perp}) \subset \mathscr{B}(\mathscr{L}^2(G))$ generates the von Neumann algebra consisting of operators of multiplication by those functions $f \in \mathscr{L}^\infty(G)$, which are constant on the equivalence classes of G with respect to K; with an abuse of notation, we shall denote this von Neumann algebra by $\mathscr{L}^\infty(G/K)$. Then (11) can be written as follows:

$$\mathscr{R}\{\boldsymbol{\lambda}(H_1), \mathscr{L}^\infty(G/H_2)\}' = \mathscr{R}\{\boldsymbol{\lambda}(H_2), \mathscr{L}^\infty(G/H_1)\}. \tag{12}$$

In this form, the result can be extended to arbitrary locally compact groups (see Takesaki, 1969).

21.12. Finally, we mention without proof the following useful result:

Proposition. *Let $\mathscr{M}$ be any W^*-algebra. Then every unitary cocycle $u \in Z_{\iota \bar{\otimes} \mathrm{Ad}\, \boldsymbol{\lambda}}(\mathbb{R}; U(\mathscr{M} \bar{\otimes} \mathscr{L}^\infty(\mathbb{R})))$ is trivial, that is, there exists $v \in U(\mathscr{M} \bar{\otimes} \mathscr{L}^\infty(\mathbb{R}))$ such that $u(t) = v[\iota \bar{\otimes} \mathrm{Ad}(\boldsymbol{\lambda}(t))](v^*)$ $(t \in \mathbb{R})$.*

Note that, if $\mathscr{M} = \mathbb{C}$, the proposition follows easily using the Stone–von Neumann uniqueness theorem (see Loomis, 1952; Mackey, 1949) for the irreducible representation of the Heisenberg commutation relations 18.9.(1), in the case $G = \mathbb{R}$. Indeed, consider more generally a locally compact abelian group G and $u \in Z_{\mathrm{Ad}\, \boldsymbol{\lambda}}(G; \mathscr{L}^\infty(G))$. Then, both $\{\boldsymbol{\lambda}(t), \mathbf{m}(\gamma); t \in G, \gamma \in \hat{G}\}$

and $\{u(t)\boldsymbol{\lambda}(t), \mathbf{m}(\gamma); t \in G, \gamma \in \hat{G}\}$ are irreducible representation of the Heisenberg commutation relations on $\mathscr{L}^2(G)$; hence (Loomis, 1952; Mackey, 1949) there exists a unitary operator $v \in \mathscr{B}(\mathscr{L}^2(G))$ such that $v\boldsymbol{\lambda}(t)v^* = u(t)\boldsymbol{\lambda}(t)$ $(t \in G)$, and $v\mathbf{m}(\gamma)v^* = \mathbf{m}(\gamma)$ $(\gamma \in \hat{G})$. It follows that $v \in U(\mathscr{L}^\infty(G))$ and $u(t) = v[\mathrm{Ad}(\boldsymbol{\lambda}(t))](v^*)$ $(t \in G)$.

For the general case, we refer to Moore (1976).

21.13. Notes. The main results in this section (21.1, 21.8, and 21.11) are due to Connes and Takesaki (1977). As mentioned in Section 21.6, Theorem 21.1 is an important tool for getting information about the center of the crossed product; this result replaces in the general case the previous result (Theorem 16.5) of Connes (1973a). The Galois correspondence type results originated in the works of Dye (1959, 1963); Nakamura and Takeda (1958, 1960). Theorem 21.8 completes the result of Proposition 21.5, which is due to Takesaki (1973b). Proposition 21.11 is an extension (in the commutative case) of the previous results on the same line obtained by Takesaki (1969) as a generalization of the Heisenberg commutation relation. Similar results concerning Galois correspondence and commutation relations in noncommutative settings are contained in Sections 22.5 and 22.9 and in Takesaki (1969). Further information is given in the survey article by Nakagami and Takesaki (1979).

For our exposition, we have used Connes and Takesaki (1977) and Takesaki (1973b).

22 Discrete Groups

In this section, we study crossed products by actions of discrete groups. The results obtained in this case are more precise, and their proofs simpler, than in the general case. Moreover, the first concrete examples of factors appeared as crossed products by discrete groups.

22.1. Let G be a discrete group. Consider the Haar measure on G, which assigns to each point of G the mass 1. Recall that all discrete groups are unimodular.

For $s, t \in G$, we denote by δ_t^s the Kronecker symbol, equal to 1 if $s = t$ and equal to 0 if $s \neq t$. The Dirac functions $\delta_t(s) = \delta_t^s$ $(s, t \in G)$ form an orthonormal basis $\{\delta_t; t \in G\}$ of the Hilbert space $\ell^2(G)$, and we have $(\xi|\delta_t) = \xi(t)$ $(\xi \in \ell^2(G), t \in G)$.

For any von Neumann algebra $\mathscr{M} \subset \mathscr{B}(\mathscr{H})$, we obtain an identification $\mathscr{M} \bar{\otimes} \mathscr{B}(\ell^2(G)) \ni X \mapsto [X_{s,t}] \mapsto Mat_G(\mathscr{M})$ via the matrix representation given by

$$X_{s,t}\xi = [X(\xi \bar{\otimes} \delta_t)](s) \quad (\xi \in \mathscr{H}, s, t \in G).$$

In particular (if $\mathscr{M} = \mathbb{C}$), for $X \in \mathscr{B}(\ell^2(G))$, we have

$$X_{s,t} = [X\delta_t](s) = (X\delta_t|\delta_s) \quad (s, t \in G).$$

Let $\sigma : G \to Aut(\mathscr{M})$ be an action of G on $\mathscr{M}$. Recall that, by definition, $\mathscr{R}(\mathscr{M}, \sigma) = \mathscr{R}\{\pi_\sigma(\mathscr{M}), 1_{\mathscr{M}} \bar{\otimes} \boldsymbol{\lambda}(G)\}$. Thus, if $a : G \ni g \mapsto a(g) \in \mathscr{M}$ is a function such that the family $\{\pi_\sigma(a(g))(1 \bar{\otimes} \boldsymbol{\lambda}(g))\}_{g\in G}$ is w-summable, we obtain an element (compare with 18.21.(1))

$$T_a^\sigma = \sum_{g\in G} \pi_\sigma(a(g))(1 \bar{\otimes} \boldsymbol{\lambda}(g)) \in \mathscr{R}(\mathscr{M}, \sigma).$$

In order to simplify the exposition, we shall in the sequel no longer mention explicitly the w-summability of the family defining T_a^σ; however, this condition will always be satisfied, either by assumption or construction.

It is easy to check that for every $X \in \mathscr{M} \bar{\otimes} \mathscr{B}(\ell^2(G))$, we have

$$(\sigma_g \bar{\otimes} \mathrm{Ad}(\boldsymbol{\rho}(g)))(X)]_{s,t} = \sigma_g(X_{sg,tg}) \quad (s,t,g \in G)$$

hence

$$(\mathscr{M} \bar{\otimes} \mathscr{B}(\ell^2(G)))^{\sigma \bar{\otimes} \mathrm{Ad}\, \boldsymbol{\rho}} = \{X \in \mathscr{M} \bar{\otimes} \mathscr{B}(\ell^2(G)); \sigma_g(X_{sg,tg}) = X_{s,t} \quad (s,t,g \in G)\}.$$

Theorem. *Let $\sigma : G \to Aut(\mathscr{M})$ be an action of the discrete group G on the W^*-algebra $\mathscr{M}$. Then*

$$\mathscr{R}(\mathscr{M},\sigma) = \{T_a^\sigma; a : G \to \mathscr{M}\} = (\mathscr{M} \bar{\otimes} \mathscr{B}(\ell^2(G)))^{\sigma \bar{\otimes} \mathrm{Ad}\, \boldsymbol{\rho}}. \tag{1}$$

More precisely, for $X \in \mathscr{H} \bar{\otimes} \mathscr{B}(\ell^2(G)))$ we have

$$X \in \mathscr{R}(\mathscr{M},\sigma) \Leftrightarrow \sigma_g(X_{sg,tg}) = X_{s,t} \textit{ for all } s,t,g \in G \tag{2}$$

and in this case there exists a unique function $a : G \to \mathscr{M}$ such that $X = T_a^\sigma$, namely

$$X = T_a^\sigma \Leftrightarrow a(g) = \sigma_g(X_{g,e}) \textit{ for all } g \in G, \tag{3}$$

$$X = T_a^\sigma \Leftrightarrow X_{s,t} = \sigma_s^{-1}(a(st^{-1})) \textit{ for all } s,t \in G. \tag{4}$$

In particular,

$$[\pi_\sigma(x)]_{s,t} = \delta_t^s \sigma_s^{-1}(x) \quad (x \in \mathscr{M}, s,t \in G) \tag{5}$$

$$[1 \bar{\otimes} \boldsymbol{\lambda}(g)]_{s,t} = \delta_{gt}^s \quad (s,t,g \in G). \tag{6}$$

Proof. Consider $\mathscr{M} \subset \mathscr{B}(\mathscr{H})$ realized as a von Neumann algebra. For $x \in \mathscr{M}, g \in G, \xi \in \mathscr{H}$, and $s,t \in G$ we have

$$\pi_\sigma(x)]_{s,t}\xi = [\pi_\sigma(x)(\xi \bar{\otimes} \delta_t)](s) = \sigma_s^{-1}(x)[\xi \bar{\otimes} \delta_t](s) = \delta_t^s \sigma_s^{-1}(x)\xi$$

$$[1 \bar{\otimes} \boldsymbol{\lambda}(g)]_{s,t}\xi = [(1 \bar{\otimes} \boldsymbol{\lambda}(g))(\xi \bar{\otimes} \delta_t)](s) = [\xi \bar{\otimes} \delta_t](g^{-1}s) = \delta_t^{g^{-1}s}\xi = \delta_{gt}^s \xi,$$

proving (5) and (6). Then

$$\begin{aligned}[\pi_\sigma(x)(1 \bar{\otimes} \boldsymbol{\lambda}(g))]_{s,t} &= \sum_r [\pi_\sigma(x)]_{s,r}[1 \bar{\otimes} \boldsymbol{\lambda}(g)]_{r,t} \\ &= \sum_r \delta_r^s \delta_{gt}^r \sigma_s^{-1}(x) = \delta_{gt}^s \sigma_s^{-1}(x).\end{aligned}$$

Thus, if $X = T_a$ for some function $a : G \to \mathscr{M}$, then $X_{s,t} = \sum_g \delta_{gt}^s \sigma_s^{-1}(a(g)) = \sigma_s^{-1}(a(st^{-1}))$ $(s,t \in G)$. In particular, $X_{g,e} = \sigma_g^{-1}(a(g))$, hence $a(g) = \sigma_g(X_{g,e})$; and $\sigma_g(X_{sg,tg}) = \sigma_g(\sigma_{sg}^{-1}(a(st^{-1}))) = \sigma_s^{-1}(a(st^{-1})) = X_{s,t}$ $(g,s,t \in G)$, so that

$$\{T_a^\sigma; a : G \to \mathscr{M}\} \subset \mathscr{R}(\mathscr{M},\sigma) \subset (\mathscr{M} \bar{\otimes} \mathscr{B}(\ell^2(G)))^{\sigma \bar{\otimes} \mathrm{Ad}\, \boldsymbol{\rho}}.$$

Conversely, let $X \in \mathscr{M} \bar{\otimes} \mathscr{B}(\ell^2(G))$ be such that $\sigma_g(X_{sg,tg}) = X_{s,t}$ for all $s, t, g \in G$. Then we define a function $a : G \to \mathscr{M}$ by putting $a(g) = \sigma_g(X_{g,e})$ $(g \in G)$, and we have $\sigma_s^{-1}(a(st^{-1})) = \sigma_s^{-1}((\sigma_{st^{-1}}(X_{st^{-1},e})) = \sigma_{t^{-1}}(X_{st^{-1},tt^{-1}}) = X_{s,t}$ $(s, t \in G)$, hence $X = T_a^\sigma$. □

The equality of the extreme terms in (1) has been proved also in the general case (19.13), but the possibility of writing every element $X \in \mathscr{R}(\mathscr{M}, \sigma)$ in the form $X = T_a^\sigma$ occurs only in the case of discrete groups.

Note that if $a(g) = \delta_g^e 1_{\mathscr{M}}$ $(g \in G)$, then $T_a = 1 \in \mathscr{R}(\mathscr{M}, \sigma)$.

22.2. Let $\sigma : G \to Aut(\mathscr{M})$ be an action of the discrete group G on the W^*-algebra $\mathscr{M}$ and let $\hat{\sigma} : \mathscr{R}(\mathscr{M}, \sigma) \to \mathscr{R}(\mathscr{M}, \sigma) \bar{\otimes} \mathfrak{L}(G)$ be the dual action (19.3). Note that in the discrete case the identity $\mathscr{R}(\mathscr{M}, \sigma)^{\hat{\sigma}} = \pi_\sigma(\mathscr{M})$ (19.3.(9)) has a simple proof based on Theorem 22.1.

The dual action $\hat{\sigma}$ defines (19.7, 19.8) an n.s.f. operator-valued weight $P_{\hat{\sigma}} : \mathscr{R}(\mathscr{M}, \sigma)^+ \to \overline{\pi_\sigma(\mathscr{M})}^+$ such that for every finitely supported function $a : G \to \mathscr{M}$ with $T_a^\sigma \geq 0$ we have

$$P_{\hat{\sigma}}(T_a^\sigma) = \pi_\sigma(a(e)). \tag{1}$$

In particular, we have $P_{\hat{\sigma}}(1) = 1$, so that *$P_{\hat{\sigma}}$ is a faithful normal conditional expectation of the crossed product $\mathscr{R}(\mathscr{M}, \sigma)$ onto the image $\pi_\sigma(\mathscr{M})$ of $\mathscr{M}$*, and (1) can be extended by continuity to any function $a : G \to \mathscr{M}$ defining an element of $\mathscr{R}(\mathscr{M}, \sigma)$. Thus using 22.1.(3), we get

$$P_{\hat{\sigma}}(X) = \pi_\sigma(X_{e,e}) \quad (X \in \mathscr{R}(\mathscr{M}, \sigma)). \tag{2}$$

Actually, it is easy to check directly that (1) or (2) define a faithful normal conditional expectation of $\mathscr{R}(\mathscr{M}, \sigma)$ onto $\pi_\sigma(\mathscr{M})$ and that (19.2.(3))

$$P_{\hat{\sigma}}((1 \bar{\otimes} \boldsymbol{\lambda}(g))X(1 \bar{\otimes} \boldsymbol{\lambda}(g))^*) = (1 \bar{\otimes} \boldsymbol{\lambda}(g))P_{\hat{\sigma}}(X)(1 \bar{\otimes} \boldsymbol{\lambda}(g))^* \tag{3}$$

for every $X \in \mathscr{R}(\mathscr{M}, \sigma)$ and $g \in G$. Moreover, from (1) and (2) it follows that

$$P_{\hat{\sigma}}(1 \bar{\otimes} \boldsymbol{\lambda}(g)) = 0 \quad (e \neq g \in G) \tag{4}$$

$$P_{\hat{\sigma}}(\pi_\sigma(x)) = \pi_\sigma(x) \quad (x \in \mathscr{M}) \tag{5}$$

and for every element $X \in \mathscr{R}(\mathscr{M}, \sigma)$ we have (compare with 19.12.(2))

$$\begin{aligned} X &= \sum_g P_{\hat{\sigma}}(X(1 \bar{\otimes} \boldsymbol{\lambda}(g^{-1})))(1 \bar{\otimes} \boldsymbol{\lambda}(g)) \\ &= \sum_g P_{\hat{\sigma}}(X(1 \bar{\otimes} \boldsymbol{\lambda}(g)))(1 \bar{\otimes} \boldsymbol{\lambda}(g^{-1})). \end{aligned} \tag{6}$$

The next proposition gives us a characterization of crossed products by discrete groups, which is different from Landstad's theorem.

Proposition. *Let $\sigma : G \to Aut(\mathscr{M})$ be an action of the discrete group G on the countably decomposable W^*-algebra $\mathscr{M}$. Let $\mathscr{N}$ be a W^*-algebra with the property that there exist:*

- *an injective unital normal *-homomorphism $\pi : \mathscr{M} \to \mathscr{N}$,*
- *a faithful normal conditional expectation $P : \mathscr{N} \to \pi(\mathscr{M})$,*
- *a unitary representation $u : G \to \mathscr{N}$,*

such that

(a) $\mathcal{N} = \mathcal{R}\{\pi(\mathcal{M}), u(G)\}$,
(b) $\pi(\sigma_g(x)) = u(g)\pi(x)u(g)^*$ for all $x \in \mathcal{M}$ *and* $g \in G$,
(c) $P(u(g)) = 0$ for all $e \neq g \in G$.

*Then there exists a *-isomorphism* $\Phi : \mathcal{N} \to \mathcal{R}(\mathcal{M}, \sigma)$ *such that*

$$\Phi(\pi(x)) = \pi_\sigma(x) \quad (x \in \mathcal{M}) \tag{1}$$

$$\Phi(u(g)) = 1 \bar{\otimes} \boldsymbol{\lambda}(g) \quad (g \in G) \tag{2}$$

$$\Phi(P(X)) = P_{\hat{\sigma}}(\Phi(X)) \quad (X \in \mathcal{N}). \tag{3}$$

Proof. Since $\mathcal{M}$ is countably decomposable, there exists a faithful normal state $\varphi \in \mathcal{M}_*$. Then $\psi = \varphi \circ \pi^{-1} \circ P$ is a faithful normal state on $\mathcal{N}$. Let $\pi_\psi : \mathcal{N} \to \mathcal{B}(\mathcal{H}_\psi)$ be the *GNS* representation associated with ψ, with cyclic vector $\xi_\psi \in \mathcal{H}_\psi$. For $x, y \in \mathcal{M}$ and $s, t \in \mathcal{M}$, we have

$$\begin{aligned}
(\pi_\psi(\pi(x)u(s))\xi_\psi \,|\, \pi_\psi(\pi(y)u(t))\xi_\psi) &= (\pi_\psi(u(t)^*\pi(y^*x)u(s))\xi_\psi \,|\, \xi_\psi) \\
= (\pi_\psi(u(t^{-1}s)\pi(\sigma_s^{-1}(y^*x))\xi_\psi \,|\, \xi_\psi) &= \psi(u(t^{-1}s)\pi(\sigma_s^{-1}(y^*x))) \\
= (\varphi \circ \pi^{-1})(P(u(t^{-1}s)\pi(\sigma_s^{-1}(y^*x))) &= \delta_t^s \varphi(\sigma_s^{-1}(y^*x)).
\end{aligned}$$

Consider now $(\mathcal{N}_1, \pi_1, P_1, u_1) = (\mathcal{N}, \pi, P, u)$ and $(\mathcal{N}_2, \pi_2, P_2, u_2) = (R(\mathcal{M}, \sigma), \pi_\sigma, P_{\hat{\sigma}}, 1 \bar{\otimes} \boldsymbol{\lambda})$ and the corresponding states ψ_1 and ψ_2. The previous computation shows that

$$\left\| \sum_{j=1}^n \pi_{\psi_1}(\pi_1(x_j)u_1(s_j))\xi_{\psi_1} \right\|^2 = \left\| \sum_{j=1}^n \pi_{\psi_2}(\pi_2(x_j)u_2(s_j))\xi_{\psi_2} \right\|^2$$

for every $x_1, \dots, x_n \in \mathcal{M}$ and every $s_1, \dots, s_n \in G$. Since $\mathcal{N}_k = \mathcal{R}\{\pi_k(\mathcal{M}), u_k(G)\}$ and $\overline{\pi_{\psi_k}(\mathcal{N}_k)\xi_{\psi_k}} = \mathcal{H}_{\psi_k}$, it follows that there exists a unitary operator $V : \mathcal{H}_{\psi_1} \to \mathcal{H}_{\psi_2}$ such that

$$V\pi_{\psi_1}(\pi_1(x)u_1(s))\xi_{\psi_1} = \pi_{\psi_2}(\pi_2(x)u_2(s))\xi_{\psi_2} \quad (x \in \mathcal{M}, s \in G).$$

It is then easy to check (on vectors of the form $\pi_{\psi_2}(\pi_2(y)u_2(t))\xi_{\psi_2}$ with $y \in \mathcal{M}, t \in G$) that

$$\begin{aligned}
V\pi_{\psi_1}(\pi_1(x))V^* &= \pi_{\psi_2}(\pi_2(x)) \quad (x \in \mathcal{M}) \\
V\pi_{\psi_1}(u_1(g))V^* &= \pi_{\psi_2}(u_2(g)) \quad (g \in G).
\end{aligned}$$

Thus the mapping $\Phi : \mathcal{N} = \mathcal{N}_1 \ni X \mapsto \pi_{\psi_2}^{-1}(V\pi_{\psi_1}(X)V^*) \in \mathcal{N}_2 = \mathcal{R}(\mathcal{M}, \sigma)$ is the required *-isomorphism. □

Note that *if in Proposition 22.2 the action σ is assumed to be properly outer, then condition (c) can be omitted from the statement, since it is automatically satisfied.*

Indeed, from (b) it follows that $\pi(\sigma_g(x))u(g) = u(g)\pi(x)$; hence $\pi(\sigma_g(x))\, P(u(g)) = P(u(g))\pi(x)$, so that either $g = e$, or $P(u(g)) = 0$, since the action σ is properly outer (see 17.6 and 17.4.(2)).

22.3. Let $\sigma : G \to Aut(\mathcal{M})$ be an action of the discrete group G on the W^*-algebra $\mathcal{M}$. By the last remark of Section 17.4, σ is properly outer if and only if

$$a \in \mathcal{M}, g \in G, xa = a\sigma_g(x) \textit{ for all } x \in \mathcal{M} \Leftrightarrow g = e \text{ or } a = 0.$$

We shall say that σ *acts freely on* $\mathcal{Z}(\mathcal{M})$ if each $\sigma_g(e \neq g \in G)$, acts freely on $\mathcal{Z}(\mathcal{M})$ (17.5). By Proposition 17.5, σ acts freely on $\mathcal{Z}(\mathcal{M})$ if and only if

$$a \in \mathcal{M}, g \in G, za = a\sigma_g(z) \text{ for all } z \in \mathcal{Z}(\mathcal{M}) \Leftrightarrow g = e \text{ or } a = 0.$$

Theorem (Relative Commutant Theorem). *Let $\sigma : G \to Aut(\mathcal{M})$ be an action of the discrete group G on the W^*-algebra $\mathcal{M}$. Then*

$$\sigma \textit{ is properly outer } \Leftrightarrow \pi_\sigma(\mathcal{M}') \cap \mathcal{R}(\mathcal{M}, \sigma) = \pi_\sigma(\mathcal{Z}(\mathcal{M})), \tag{1}$$
$$\sigma \textit{ acts freely on } \mathcal{Z}(\mathcal{M}) \Leftrightarrow \pi_\sigma(\mathcal{Z}(\mathcal{M}))' \cap \mathcal{R}(\mathcal{M}, \sigma) = \pi_\sigma(\mathcal{M}). \tag{2}$$

Proof. Let $X = \sum_g \pi_\sigma(a(g))(1 \bar{\otimes} \boldsymbol{\lambda}(g)) \in \mathcal{R}(\mathcal{M}, \sigma)$ and $x \in \mathcal{M}$. Then $\pi_\sigma(x)X = \sum_g \pi_\sigma(xa(g))(1 \bar{\otimes} \boldsymbol{\lambda}(g))$ and $X\pi_\sigma(x) = \sum_g \pi_\sigma(a(g)\sigma_g(x))(1 \bar{\otimes} \boldsymbol{\lambda}(g))$ so that

$$X \in \pi_\sigma(\mathcal{M})' \Leftrightarrow xa(g) = a(g)\sigma_g(x) \text{ for all } x \in \mathcal{M}, g \in G, \tag{3}$$
$$X \in \pi_\sigma(\mathcal{Z}(\mathcal{M}))' \Leftrightarrow za(g) = a(g)\sigma_g(z) \text{ for all } z \in \mathcal{Z}(\mathcal{M}), g \in G, \tag{4}$$

from which (1) and (2) follow immediately. □

Note that if σ acts freely on $\mathcal{Z}(\mathcal{M})$, then both (1) and (2) hold, so that

$$(\pi_\sigma(\mathcal{M})' \cap \mathcal{R}(\mathcal{M}, \sigma))' \cap \mathcal{R}(\mathcal{M}, \sigma) = \pi_\sigma(\mathcal{M}), \tag{5}$$
$$(\pi_\sigma(\mathcal{Z}(\mathcal{M}))' \cap \mathcal{R}(\mathcal{M}, \sigma))' \cap \mathcal{R}(\mathcal{M}, \sigma) = \pi_\sigma(\mathcal{Z}(\mathcal{M})). \tag{6}$$

22.4. Consider a W^*-algebra $\mathcal{N}$ and a unital W^*-subalgebra $\mathcal{M}$ of $\mathcal{N}$ such that $\mathcal{M}' \cap \mathcal{N} \subset \mathcal{M}$ and such that there exists a normal conditional expectation $P : \mathcal{N} \to \mathcal{M}$. By Proposition 10.17, P is uniquely determined and faithful and for the normalizer $\mathcal{N}(P)$ of P we have $\mathcal{N}(P) = \{v \in U(\mathcal{N}); P \circ \mathrm{Ad}(v) = \mathrm{Ad}(v) \circ P\} = \{v \in U(\mathcal{N}); v\mathcal{M}v^* = \mathcal{M}\}$.

Lemma. *For every $v \in \mathcal{N}(P)$, we have $vP(v^*) = P(v)v^* = P(v)P(v^*) = P(v^*)P(v) = v^*P(v) = P(v^*)v = p(\mathrm{Ad}(v)|\mathcal{M})$ (see 17.2).*

Proof. Indeed, we have $vP(v^*)v^* = P(vv^*v^*) = P(v^*)$, hence $vP(v^*) = P(v^*)v$. Similarly, $P(v)v^* = v^*P(v)$. For $x \in \mathcal{M}$, we have $vxv^* \in \mathcal{M}$, hence $P(v)x = P(vx) = P(vxv^*v) = vxv^*P(v)$, so that $v^*P(v) \in \mathcal{M}' \cap \mathcal{N} \subset \mathcal{M}$. Therefore, $v^*P(v) = P(v^*P(v)) = P(v^*)P(v)$. Replacing v here by v^* we get $vP(v^*) = P(v)P(v^*)$. Finally, applying P to the identity $P(v)v^* = v^*P(v)$, we obtain $P(v)P(v^*) = P(v^*)P(v)$. Thus, the first six terms appearing in the statement are all equal to some element $p \in \mathcal{M}' \cap \mathcal{N} = \mathcal{Z}(\mathcal{M})$, which is a projection since $p^* = (P(v)v^*)^* = vP(v^*) = p$ and $p^2 = v^*P(v)vP(v^*) = P(v^*vv)P(v^*) = P(v)P(v^*) = p$.

On the other hand, we have $[\mathrm{Ad}(v)](p) = vpv^* = vv^*P(v)v^* = vP(v^*) = p$. Since $P(v)P(v)^* = p = P(v)^*P(v)$, $P(v)$ is a unitary element in $\mathscr{M}p$ and for every $x \in \mathscr{M}p$ we have $[\mathrm{Ad}(v)](x) = vxv^* = vpxpv^* = vv^*P(v)xP(v^*)vv^* = P(v)xP(v)^*$. Thus, $p \leq p(\mathrm{Ad}(v)|\mathscr{M})$. Let $q = p(\mathrm{Ad}(v)|\mathscr{M}) - p$ and consider a unitary element u in $\mathscr{M}q$ such that $[\mathrm{Ad}(v)](x) = uxu^*$ for all $x \in \mathscr{M}q$. Then $v^*u \in \mathscr{N}, u^*vv^*u = u^*u = q, v^*uu^*v = v^*qv = q$ and $v^*uxu^*v = x$ for all $x \in \mathscr{M}q$. Since $q \in \mathscr{Z}(\mathscr{M})$, it follows that $v^*u \in \mathscr{M}' \cap \mathscr{N} = \mathscr{Z}(\mathscr{M})$, so that there exists a unitary element $z \in \mathscr{Z}(\mathscr{M}q)$ such that $u = vz$; we have $q = uu^* = P(u)P(u^*) = P(v)zz^*P(v^*) = qP(v)P(v^*) = qp = 0$. □

Recall (17.3) that $[G]$ denotes the full group associated with a group $G \subset Aut(\mathscr{M})$.

Proposition. *Let $\mathscr{N}$ be a W^*-algebra and $\mathscr{M} \subset \mathscr{N}$ a unital W^*-subalgebra such that $\mathscr{M}' \cap \mathscr{N} \subset \mathscr{M}$ and such that there exists a normal conditional expectation $P : \mathscr{N} \to \mathscr{M}$. Let $\mathscr{G} \subset \mathcal{N}(P)$ be a subgroup such that $\mathscr{N} = \mathscr{R}\{\mathscr{M}, \mathscr{G}\}$. Then $\{\mathrm{Ad}(v)|\mathscr{M}; v \in \mathcal{N}(P)\} = \{\mathrm{Ad}(w)|\mathscr{M}; w \in \mathscr{G}\}$.*

Proof. Let $v \in \mathcal{N}(P)$. From the previous lemma it follows that

$$p(\mathrm{Ad}(w^*v)|\mathscr{M}) = \mathbf{s}(P(w^*v)) \quad (w \in \mathscr{G}).$$

Thus, in order to prove that $\mathrm{Ad}(v)|\mathscr{M} \in [\mathrm{Ad}(\mathscr{G})|\mathscr{M}]$, it is sufficient to show that for every nonzero projection $q \in \mathscr{Z}(\mathscr{M})$ there exists $w \in \mathscr{G}$ such that $qP(w^*v) \neq 0$. Let $q \in \mathscr{Z}(\mathscr{M})$ be a nonzero projection. Then $0 \neq q = qP(v^*v)$. Since $\mathscr{N} = \mathscr{R}\{\mathscr{M}, \mathscr{G}\}$, there exist $a \in \mathscr{M}, w \in \mathscr{G}$ such that $0 \neq qP(aw^*v) = qaP(w^*v)$, hence $qP(w^*v) \neq 0$.

Conversely, let $\sigma \in [\mathrm{Ad}(\mathscr{G})|\mathscr{M}] \subset Aut(\mathscr{M})$. By Section 17.3, there exists a family $\{(q_i, u_i, w_i)\}_{i\in I}$, where $q_i \in \mathscr{Z}(\mathscr{M})$ are projections, $\sum_i q_i = 1, u_i \in \mathscr{M}, u_i^*u_i = u_iu_i^* = q_i, w_i \in \mathscr{G}, \sigma(q_i) = w_iq_iw_i^*$, and $\sigma(x) = \sum_i w_iu_ixu_i^*w_i^*$ for all $x \in \mathscr{M}$. Then $v = \sum_i w_iu_i \in U(\mathscr{N})$ and $\sigma = \mathrm{Ad}(v)|\mathscr{M}$, hence $v \in \mathscr{N}(P)$. □

Thus, in the setting of the proposition, for every $v \in \mathcal{N}(P)$ there exists a family $\{(q_i, u_i, w_i)\}_{i\in I}$, where the $q_i \in \mathscr{Z}(\mathscr{M})$ are projections, $\sum_i q_i = 1, u_i \in \mathscr{M}, u_i^*u_i = u_iu_i^* = q_i, w_i \in \mathscr{G}, vq_iv^* = w_iq_iw_i^*$, and $vxv^* = \sum_i w_iu_ixu_i^*w_i^*$ for all $x \in \mathscr{M}$. If $x \in \mathscr{M}q_i$, then $vxv^* = w_iu_ixu_i^*w_i^*$, and $v^*w_iu_i \in (\mathscr{M}q_i)' \cap \mathscr{N}$. It follows that $v^*w_iu_i \in \mathscr{Z}(\mathscr{M})q_i$ so that, multiplying u_i by some unitary element of $\mathscr{Z}(\mathscr{M})q_i$ if necessary, we may assume that $v^*w_iu_i = q_i$. Since $\sum_i q_i = 1$, it follows that

$$v = \sum_i w_iu_i \text{ with } w_i \in \mathscr{G}, u_i \in U(\mathscr{Z}(\mathscr{M})q_i), \sum_i q_i = 1, vq_i = w_iu_i. \tag{1}$$

In particular, let $\sigma : G \to Aut(\mathscr{M})$ be a properly outer action of the discrete group G on the W^*-algebra $\mathscr{M}$. Then the conclusion of the previous proposition is valid for $\mathscr{N} = \mathscr{R}(\mathscr{M}, \sigma), \mathscr{M} = \pi_\sigma(\mathscr{M}) \subset \mathscr{N}, P = P_{\hat{\sigma}}, \mathscr{G} = \{1 \bar{\otimes} \boldsymbol{\lambda}(g); g \in G\} \subset \mathcal{N}(P_{\hat{\sigma}})$ (see 22.2 and 22.3), and can be formulated as follows: $[\sigma(G)] = \{\alpha \in Aut(\mathscr{M}); \alpha$ extends to an inner *-automorphism of $\mathscr{R}(\mathscr{M}, \sigma)\}$.

22.5. The next result concerns the structure of certain "intermediate subalgebras" of the crossed product (compare with Theorem 21.8).

Proposition. *Let $\sigma : G \to Aut(\mathscr{M})$ be a (properly) outer action of the discrete group G on the factor $\mathscr{M}$. If $\mathscr{N}$ is a unital W^*-subalgebra of $\mathscr{R}(\mathscr{M}, \sigma)$ containing $\pi_\sigma(\mathscr{M})$ and there exists a normal conditional expectation $P_{\mathscr{N}} : \mathscr{R}(\mathscr{M}, \sigma) \to \mathscr{N}$, then*

$$H = \{g \in G; 1 \bar{\otimes} \boldsymbol{\lambda}(g) \in \mathscr{N}\} \tag{1}$$

is a subgroup of G,

$$\mathcal{N} = \mathcal{R}\{\pi_\sigma(\mathcal{M}), 1 \bar{\otimes} \boldsymbol{\lambda}(H)\} \approx \mathcal{R}(\mathcal{M}, \sigma|H) \tag{2}$$
$$P_{\mathcal{N}}(1 \bar{\otimes} \boldsymbol{\lambda}(g)) = 0 \text{ for all } g \in G \setminus H \tag{3}$$

Proof. We put $u(g) = 1 \bar{\otimes} \boldsymbol{\lambda}(g)$ and identify $\mathcal{M} = \pi_\sigma(\mathcal{M}) \subset \mathcal{N}$.

We first prove (3). For $g \in G$ and $x \in \mathcal{M}$ we have $xP_{\mathcal{N}}(u(g)) = P_{\mathcal{N}}(xu(g)) = P_{\mathcal{N}}(u(g)u(g)^*xu(g)) = P_{\mathcal{N}}(u(g))u(g)^*xu(g)$, hence $P_{\mathcal{N}}(u(g))u(g)^* \in \mathcal{M}' \cap \mathcal{R}(\mathcal{M}, \sigma)$. Since σ is properly outer, it follows that $P_{\mathcal{N}}(u(g))u(g)^* \in \mathcal{Z}(\mathcal{M})$ (22.3). Since $\mathcal{M}$ is a factor, we deduce also that $P_{\mathcal{N}}(u(g)) = \lambda u(g)$ for some $\lambda \in \mathbb{C}$. If $\lambda \neq 0$, then $u(g) \in \mathcal{N}$, that is, $g \in H$. Thus, $P_{\mathcal{N}}(u(g)) = 0$ for all $g \in G \setminus H$.

Consider now $X \in \mathcal{N} \subset \mathcal{R}(\mathcal{M}, \sigma), X = \sum_g a(g)u(g)$ (22.1). Then, using (3), we get $X = P_{\mathcal{N}}(X) = \sum_g a(g)P_{\mathcal{N}}(u(g)) = \sum_g a(g)u(g) \in \mathcal{R}\{\mathcal{M}, u(H)\}$, proving (2). □

22.6. We now study the center of the crossed product.

Theorem. *Let $\sigma : G \to Aut(\mathcal{M})$ be an action of the discrete group G on the W^*-algebra $\mathcal{M}$ and let $X = \sum_g \pi_\sigma(a(g))(1 \bar{\otimes} \boldsymbol{\lambda}(g)) \in \mathcal{R}(\mathcal{M}, \sigma)$ an arbitrary element of the crossed product ($a(g) \in \mathcal{M}$ for all $g \in G$). Then we have $X \in \mathcal{Z}(\mathcal{R}(\mathcal{M}, \sigma))$ if and only if the following conditions are satisfied:*

$$a(g)\sigma_g(x) = xa(g) \quad (x \in \mathcal{M}, g \in G) \tag{1}$$
$$\sigma_s(a(g)) = a(sgs^{-1}) \quad (s, g \in G). \tag{2}$$

In this case

$$a(e) \in \mathcal{Z}(\mathcal{M})^\sigma. \tag{3}$$

Proof. By 22.3.(3), (1) means that $X \in \pi_\sigma(\mathcal{M})'$. On the other hand, it is easy to check that (2) is equivalent to the statement that X commutes with all $1 \bar{\otimes} \boldsymbol{\lambda}(s)$ $(s \in G)$. Finally, (3) follows immediately from (1) and (2), with $g = e$. □

Corollary 1. *If the action $\sigma : G \to Aut(\mathcal{M})$ of the discrete group G on the W^*-algebra $\mathcal{M}$ is properly outer and its restriction to $\mathcal{Z}(\mathcal{M})$ is ergodic, then the crossed product $\mathcal{R}(\mathcal{M}, \sigma)$ is a factor.*

In particular, if $\mathcal{M}$ is a factor, then the ergodicity condition $\mathcal{Z}(\mathcal{M})^\sigma = \mathbb{C} \cdot 1_{\mathcal{M}}$ 1s automatically satisfied.

Also, if G is commutative, we can also deduce the discrete case of Corollary 21.6 from the above Theorem. Indeed, from (2) it follows that $a(g) \in \mathcal{M}^\sigma$ for all $g \in G$ and using (1) we deduce also that $a(g) \in \mathcal{Z}(\mathcal{M}^\sigma)$. Thus, if $\mathcal{M}^\sigma$ is a factor, then $a(g)$ are all scalars. Furthermore, if $a(g) \neq 0$, then it follows from (1) that $\sigma_g = \iota$. Consequently, if $\sigma_g \neq \iota$ for all $g \neq e$, then $a(g) = 0$ for all $g \neq e$.

If $\mathcal{M} = \mathbb{C}$ and $\sigma : G \to Aut(\mathcal{M})$ is the trivial action, then $\mathcal{R}(\mathcal{M}, \sigma) = \mathfrak{L}(G)$. In this case an element $X = \sum_g \alpha(g)\boldsymbol{\lambda}(g) \in \mathfrak{L}(G)$ belongs to $\mathcal{Z}(\mathfrak{L}(G))$ if and only if $\alpha(g) = \alpha(sgs^{-1})$ for all $s, g \in G$. However, the function $\alpha : G \to \mathbb{C}$ is not arbitrary, as $X \in \mathcal{B}(\ell^2(G))$. Thus, $X\delta_e \in \ell^2(G)$, that is,

$$\sum_g |\alpha(g)|^2 = \sum_g |(X\delta_e)(g)| < +\infty. \tag{4}$$

The discrete group G will be called an *ICC-group* if for every $g \in G, g \neq e$, the conjugacy class $\{sgs^{-1}; s \in G\}$ is an infinite set. The above discussion shows

Corollary 2. *Let G be a discrete group. Then $\mathfrak{L}(G)$ is a factor if and only if G is an ICC-group, or $G = \{e\}$.*

It is clear that the free group F_k with k generators ($2 \leq k \leq \infty$) is an ICC-group and hence that $\mathfrak{L}(F_k)$ is a factor.

Also, it is easy to check that the group $S(\infty)$ of all bijections $\gamma : \mathbb{N} \to \mathbb{N}$ with the property that $\gamma(n) = n$ for sufficiently large $n \in \mathbb{N}$, is also an *ICC*-group, so that $\mathfrak{L}(S(\infty))$ is a factor.

In order to extend the result of Corollary 2, let us now consider an arbitrary action $\sigma : G \to Aut(\mathcal{M})$ of the discrete group G on the von Neumann algebra $\mathcal{M} \subset \mathcal{B}(\mathcal{H})$. Let $X = \sum_g \pi_\sigma(a(g))(1 \bar{\otimes} \boldsymbol{\lambda}(g)) \in \mathcal{R}(\mathcal{M}, \sigma) \subset \mathcal{B}(\ell^2(G, \mathcal{H}))$ and $\xi \in \mathcal{H}$. We have $[X(\xi \bar{\otimes} \delta_e))(g) = X_{g,e}\xi = \sigma_g^{-1}(a(g))\xi$, hence

$$\sum_g \|\sigma_g^{-1}(a(g))\xi\|^2 = \|X(\xi \bar{\otimes} \delta_e)\|^2 < +\infty.$$

Furthermore, assuming that $\|\sigma_t(x)\xi\| = \|x\xi\|$ $(x \in \mathcal{M}, t \in G)$, the above inequality becomes

$$\sum_g \|a(g)\xi\|^2 < +\infty,$$

and if $X \in \mathcal{Z}(\mathcal{R}(\mathcal{M}, \sigma))$, then (2) shows that

$$\|a(g)\xi\| = \|a(sgs^{-1})\xi\| \quad (s, g \in G).$$

If G is an *ICC*-group, then we conclude that

$$X \in \mathcal{Z}(\mathcal{R}(\mathcal{M}, \sigma)) \Rightarrow a(g)\xi = 0 \text{ for all } e \neq g \in G.$$

If, moreover, $\xi \in \mathcal{H}$ is a separating vector for $\mathcal{M} \subset \mathcal{B}(\mathcal{H})$, this means that

$$X \in \mathcal{Z}(\mathcal{R}(\mathcal{M}, \sigma)) \Rightarrow a(g) = 0 \text{ for all } e \neq g \in G.$$

Finally, if σ is ergodic on $\mathcal{Z}(\mathcal{M})$, then we infer from (3) that $a(e) \in \mathbb{C} \cdot 1_{\mathcal{M}}$.

We have thus proved that if G is an *ICC*-group acting ergodically on $\mathcal{Z}(\mathcal{M})$ and if there exists a separating vector $\xi \in \mathcal{H}$ for $\mathcal{M}$ such that $\|\sigma_t(x)\xi\| = \|x\xi\|$ $(x \in \mathcal{M}, t \in G)$, then $\mathcal{R}(\mathcal{M}, \sigma)$ is a factor. Using the *GNS*-representation, this result can be formulated as follows:

Corollary 3. *Let $\sigma : G \to Aut(\mathcal{M})$ be an action of the ICC-group G on the W^*-algebra $\mathcal{M}$ whose restriction to $\mathcal{Z}(\mathcal{M})$ is ergodic and such that there exists a σ-invariant faithful normal state on $\mathcal{M}$. Then $\mathcal{R}(\mathcal{M}, \sigma)$ is a factor.*

22.7. We now study the type of the crossed product. Let $\sigma : G \to Aut(\mathcal{M})$ be an action of the discrete group G on the W^*-algebra $\mathcal{M}$. We put $u(g) = 1 \bar{\otimes} \boldsymbol{\lambda}(g)$ $(g \in G)$, and we identify $\mathcal{M} = \pi_\sigma(\mathcal{M}) \subset \mathcal{R}(\mathcal{M}, \sigma)$. Thus, an arbitrary element of $\mathcal{R}(\mathcal{M}, \sigma)$ is of the form $X = \sum_g a(g)u(g)$ with $a(g) \in \mathcal{M}$ $(g \in G)$, and the conditional expectation $P = P_{\hat{\sigma}}$ of $\mathcal{R}(\mathcal{M}, \sigma)$ onto $\mathcal{M}$ is defined by $P(X) = a(e)$.

Theorem 1. *The crossed product $\mathscr{R}(\mathscr{M},\sigma)$ is a finite W^*-algebra if and only if there exists a separating family of σ-invariant finite normal traces on $\mathscr{M}$.*

Proof. Assume that $\mathscr{R}(\mathscr{H},\sigma)$ is finite. Let $a \in \mathscr{M}^+, a \neq 0$. There exists a finite normal trace μ on $\mathscr{R}(\mathscr{M},\sigma)$ with $\mu(a) \neq 0$. Then $\tau = \mu|\mathscr{M}$ is a finite normal trace on $\mathscr{M}$ and $\tau(a) \neq 0$. For $x \in \mathscr{M}, g \in G$, we have $\tau(\sigma_g(x)) = \mu(\sigma_g(x)) = \mu(u(g)xu(g)^*) = \mu(x) = \tau(x)$, hence τ is σ-invariant.

Conversely, let τ be a σ-invariant finite normal trace on $\mathscr{M}$. Then $\mu = \tau \circ P$ is a positive normal linear form on $\mathscr{R}(\mathscr{M},\sigma)$ with $\mathbf{s}(\mu) = \mathbf{s}(\tau)$. For $X = \sum_g a(g)u(g) \in \mathscr{R}(\mathscr{M},\sigma)$, we have $X^*X = \sum_g(\sum_s \sigma_s^{-1}(a(s)^*a(sg)))u(g), XX^* = \sum_g(\sum_s a(gs)\sigma_g(a(s)^*))u(g)$, so that $P(X^*X) = \sum_s \sigma_s^{-1}(a(s)^*a(s)), P(XX^*) = \sum_s a(s)a(s)^*$ and hence $\mu(X^*X) = \sum_s \tau(\sigma_s^{-1}(a(s)^*(a(s)) = \sum_s \tau(a(s)^*(a(s))) = \sum_s \tau(a(s)a(s)^*) = \mu(XX^*)$, that is, μ is a trace. □

Thus, if G is a discrete ICC-group, then $\mathfrak{L}(G)$ is a finite infinite-dimensional factor, that is a factor of type II_1.

Theorem 2. *If there exists a σ-invariant n.s.f. trace on $\mathscr{M}$, then the crossed product $\mathscr{R}(\mathscr{M},\sigma)$ is semifinite.*

Conversely, if the crossed product $\mathscr{R}(\mathscr{M},\sigma)$ is semifinite and σ acts freely on $\mathscr{Z}(\mathscr{M})$, then there exists a σ-invariant n.s.f. trace on $\mathscr{M}$.

Proof. The first assertion is proved with the same arguments as in Theorem 1. If $\mathscr{R}(\mathscr{M},\sigma)$ is semifinite, then there exists an n.s.f. trace μ on $\mathscr{R}(\mathscr{M},\sigma)$. As in the proof of Theorem 1 one can show that $\tau = \mu|\mathscr{M}$ is a σ-invariant faithful normal trace on $\mathscr{M}$. Since μ is semifinite, there exists a net $\{x_i\} \subset \mathscr{R}(\mathscr{M},\sigma)$ such that $\mu(x_i^*x_i) < +\infty$ and $x_i \overset{w}{\to} 1$. Then $a_i = P(x_i) \in \mathscr{M}$ and $a_i \overset{w}{\to} P(1) = 1$. Thus, in order to prove that τ is semifinite, it is sufficient to show that

$$x \in \mathscr{R}(\mathscr{M},\sigma), \mu(x^*x) < +\infty \Rightarrow \mu(P(x)^*P(x)) < +\infty.$$

By Proposition 10.14, there exists $a \in \overline{co}^w\{uxu^*; \sigma \in U(\mathscr{Z}(\mathscr{M}))\} \cap \mathscr{Z}(\mathscr{M})'$. Since σ acts freely on $\mathscr{Z}(\mathscr{M})$, it follows by Theorem 22.3.(2) that $a \in \mathscr{M}$, hence $P(a) = a$. For $u \in U(\mathscr{Z}(\mathscr{M}))$, we have $P(uxu^*) = uP(x)u^* = P(x)$, hence $a = P(a) = P(x)$. Using Proposition 10.14 again, we conclude that $\mu(a^*a) < +\infty$, that is, $\mu(P(x)^*P(x)) < +\infty$. □

22.8. In this section, we consider crossed products of abelian W^*-algebras by actions of countable discrete groups and some concrete examples.

Let Ω be a locally compact Hausdorff topological space with a countable basis of open sets and μ a sigma-finite positive Borel measure on Ω. Consider the Hilbert space $\mathscr{H} = \mathscr{L}^2(\Omega, \mu)$ and the maximal abelian von Neumann algebra $\mathscr{M} = \mathscr{L}^\infty(\Omega, \mu) \subset \mathscr{B}(\mathscr{H})$.

Let G be a countable discrete group of homeomorphisms of Ω; for each $g \in G$ we shall denote by $T_g : \Omega \to \Omega$ the corresponding homeomorphism of Ω. We shall assume that the measure μ is *G-quasi-invariant*, that is, for every $g \in G$ the measures μ and $\mu \circ T_g$ are equivalent in the sense of absolute continuity. In this case, it is clear that the formula $\sigma_g(x) = x \circ T_{g^{-1}}$ $(x \in \mathscr{M}, g \in G)$ defines an action $\sigma : G \to Aut(\mathscr{M})$ of G on $\mathscr{M}$.

We shall say that G *acts (almost) freely* on (Ω, μ) if for every $g \in G, g \neq e$, we have $\mu(\{\omega \in \Omega; T_g\omega = \omega\}) = 0$. In this case, σ also acts freely on $\mathscr{M}$. Indeed, let $g \in G, g \neq e$, and let $E \subset \Omega$ be a μ-measurable set with $\mu(E) > 0$; we have to show that there exists a μ-measurable set $F \subset E$ with $\mu(F) > 0$ such that $F \cap T_gF = \emptyset$ (see 17.5). Since G acts freely on (Ω, μ), we may assume that

$T_g\omega \neq \omega$ for all $\omega \in \Omega$. Then, for every $\omega \in E$ there exists a compact neighborhood $V(\omega)$ of ω such that $V(\omega) \cap T_g V(\omega) = \emptyset$. We have $W(\omega) = V(\omega) \cap E \subset E$ and $W(\omega) \cap T_g W(\omega) = \emptyset$. Since Ω has a countable basis of open sets and $\mu(E) > 0$, there exists $\omega_0 \in E$ with $\mu(W(\omega_0)) > 0$ and we can take $F = W(\omega_0)$.

We shall say that G *acts ergodically* on (Ω, μ) if the only μ-measurable sets $E \subset \Omega$ such that $\mu((E \cup T_g E)\setminus(E \cap T_g E)) = 0$ for all $g \in G$ satisfy either $\mu(E) = 0$ or $\mu(\Omega\setminus E) = 0$. In this case it is easy to check that also σ is ergodic, that is, $\mathcal{M}^\sigma = \mathbb{C} \cdot 1_{\mathcal{M}}$.

Thus, according to Corollary 1/22.6, if *G acts freely and ergodically on (Ω, μ), then $\mathcal{R}(\mathcal{M}, \sigma)$ is a factor*. In the sequel, we assume that these conditions are satisfied.

We shall say that G is *μ-measurable*, if there exists a sigma-finite positive Borel measure ν on Ω, equivalent to μ and G-invariant, that is, $\nu \circ T_g = \nu\ (g \in G)$. Since G acts ergodically on (Ω, μ), the measure ν with the previous properties is unique up to a multiplicative scalar factor. The measure ν defines a σ-invariant n.s.f. trace τ on $\mathcal{M}$ by $\tau(x) = \int x \, \mathrm{d}\nu\ (x \in \mathcal{M}^+)$, and any σ-invariant n.s.f. trace τ on $\mathcal{M}$ is of this form; clearly, the trace τ is finite if and only if the measure ν is finite. Since G acts freely on (Ω, μ), it follows according to Theorem 2/22.7 that *the factor $\mathcal{R}(\mathcal{M}, \sigma)$ is semifinite if and only if G is μ-measurable*; moreover, according to Theorem 1/22.7, *the factor $\mathcal{R}(\mathcal{M}, \sigma)$ is finite if and only if the measure ν is finite*.

Thus, *if G acts freely and ergodically on (Ω, μ) and is not μ-measurable, then $\mathcal{R}(\mathcal{M}, \sigma)$ is a type* III *factor.*

In order to give a concrete example of a type III factor with a separable predual, we first notice the following simple fact: *if the subgroup $G_0 = \{g \in G; \mu \circ T_g = \mu\} \subset G$ acts ergodically on (Ω, μ) and $G_0 \neq G$, then G is not μ-measurable*. Indeed, assume to the contrary, that is, there exists a G-invariant sigma-finite positive Bore measure ν on Ω, equivalent to μ, and consider the Radon–Nikodym derivative $f = \mathrm{d}\mu/\mathrm{d}\nu$. Since for $g \in G_0$ we have

$$f \circ T_g = \mathrm{d}(\mu \circ T_g)/\mathrm{d}\nu = \mathrm{d}\mu/\mathrm{d}\nu = f$$

and since G_0 acts ergodically on (Ω, μ), it follows that the function f is constant and hence the measure μ is G-invariant, contradicting $G_0 \neq G$.

Consider now $\Omega = \mathbb{R}, \mu =$ the Lebesgue measure on $\mathbb{R}$ and the countable group G of homeomorphisms of $\mathbb{R}$ consisting of the transformations $T(\alpha, \beta) : \mathbb{R} \ni \omega \mapsto \alpha\omega + \beta$, where α, β are rational numbers and $\alpha > 0$. Then the measure μ is G-quasi-invariant, G acts freely and $G_0 = \{g \in G; \mu \circ T_g = \mu\} = \{T(\alpha, \beta); \alpha = 1\}\ (\neq G)$ acts ergodically, hence G is not μ-measurable. Thus, the crossed product $\mathcal{R}(\mathcal{M}, \sigma)$ is in this case a concrete example of a type III factors with separable predual. This was the first example of a factor of type III due to Murray and von Neumann. It is actually a factor of type III_1.

If the group G acts freely and ergodically on (Ω, μ), is μ-measurable, and if the G-invariant measure ν equivalent to μ is diffuse, that is, $\nu(\{\omega\}) = 0$ for every $\omega \in \Omega$, then the factor $\mathcal{R}(\mathcal{M}, \sigma)$ is of type II. Indeed, in this case there exists a decreasing sequence $\{E_n\}_{n\geq 1}$ of μ-measurable sets in Ω with $\nu(E_n) > \nu(E_{n+1}), (n \geq 1)$, and $\lim_n \nu(E_n) = 0$. Let τ be the σ-invariant n.s.f. trace on $\mathcal{M}$ defined by ν and let $\tilde{\tau} = \tau \circ P$ be the corresponding n.s.f. trace on $\mathcal{R}(\mathcal{M}, \sigma)$. Then $e_n = \chi_{E_n} \in \mathcal{M} \subset \mathcal{R}(\mathcal{M}, \sigma)$ is a decreasing sequence of non-zero projections in $\mathcal{R}(\mathcal{M}, \sigma)$ and $\tilde{\tau}(e_n) = \nu(E_n) \to 0$ and so the factor $\mathcal{R}(\mathcal{M}, \sigma)$ is of type II.

For instance, consider $\Omega = \{\omega \in \mathbb{C}; |\omega| = 1\}$ the one-dimensional torus, let $\mu =$ the Haar measure on Ω with $\mu(\Omega) = 1$ and let G be any dense countable infinite subgroup of Ω, which acts on Ω by translation, $T_g\omega = g\omega$ ($g \in G, \omega \in \Omega$) (e.g., $G = \{\exp(2\pi it); t \in \mathbb{R}\setminus\mathbb{Q}\}$ or $G = \{\exp(2\pi int_0); n \in \mathbb{Z}\}$ with $t_0 \in \mathbb{R}$ a fixed irrational number). Then the Haar measure μ is finite, diffuse and G-invariant, and G acts freely on (Ω, μ). We shall show that G acts ergodically on (Ω, μ), so that $\mathcal{R}(\mathcal{M}, \sigma)$ is in this case a type II_1 factor. Indeed, let $E \subset \Omega$ be a μ-measurable set such that $\mu((E \cup T_g E)\setminus(E \cap T_g E)) = 0$ for all $g \in G$. Since the functions $\varphi_n(z) = z^n$ ($n \in \mathbb{Z}$) form an orthonormal basis in $\mathcal{L}^2(\Omega, \mu)$, the characteristic function $\chi_E \in \mathcal{L}^2(\Omega, \mu)$ can be written $\chi_E(\omega) = \sum_{n\in\mathbb{Z}} \lambda_n \omega^n$ with $\sum_{n\in\mathbb{Z}} |\lambda_n|^2 < +\infty$. By assumption, we have $\chi_E \circ T_\theta = \chi_E$ in $\mathcal{L}^2(\Omega, \mu)$ for every $g \in G$. Since $(\chi_E \circ T_g)(\omega) = \sum_n \lambda_n g^n \omega^n$, it follows that $\lambda_n = \lambda_n g^n$ for every $g \in G, n \in \mathbb{Z}$. Since G is dense in Ω, we infer that $\lambda_n = 0$ for all $n \neq 0$ hence either $\mu(E) = 0$, or $\mu(\Omega\setminus E) = 0$.

Similarly, if $\Omega = \mathbb{R}$, μ = the Lebesgue measure on $\mathbb{R}$ and G is a dense countable infinite subgroup of Ω, which acts on Ω by translation (e.g., $G = \mathbb{Q}$), then the corresponding crossed product $\mathcal{R}(\mathcal{M}, \sigma)$ is a type II_∞ factor.

22.9. Using Corollary 10.6 and Theorem 1/22.7, we obtain from Proposition 22.5 the following "Galois correspondence" (compare with Theorem 21.8):

Corollary. *Let $\sigma : G \to Aut(\mathcal{M})$ be a (properly) outer action of the discrete group G on the factor $\mathcal{M}$. We assume that there exists a σ-invariant faithful normal finite trace on $\mathcal{M}$. Then the mappings $\mathcal{N} \to H$ and $H \to \mathcal{N}$ defined by*

$$H = \{g \in G; 1 \bar{\otimes} \boldsymbol{\lambda}(g) \in \mathcal{N}\}, \quad \mathcal{N} = \mathcal{R}\{\pi_\sigma(\mathcal{M}), 1 \bar{\otimes} \boldsymbol{\lambda}(H)\}$$

are reciprocal increasing bijections between the subgroups H of G and the unital W^*-subalgebras $\mathcal{N} \supset \pi_\sigma(\mathcal{M})$ of the crossed product $\mathcal{R}(\mathcal{M}, \sigma)$.

If $\mathcal{M}$ is no longer assumed to be a factor, the corresponding result cannot be expressed in terms of subgroups of G, but can be expressed in terms of full subgroups of the full group $[\sigma(G)]$.

More precisely, let $\sigma : G \to Aut(\mathcal{M})$ be a properly outer action of the discrete group G on the W^*-algebra $\mathcal{M}$. We assume that there exists a σ-invariant faithful normal finite trace τ on $\mathcal{M}$. Let $P_{\hat{\sigma}}$ be the faithful normal conditional expectation of the crossed product $\mathcal{R}(\mathcal{M}, \sigma)$ onto $\pi_\sigma(\mathcal{M}) = \mathcal{M}$. By Proposition 22.4, Theorem 22.3.(1), and Proposition 10.17, the full group $[\sigma(G)] \subset Aut(\mathcal{M})$ consists of all *-automorphisms of the form $\mathrm{Ad}(v)|\mathcal{M}$ with $v \in \mathcal{N}(P_{\hat{\sigma}}) = \{v \in U(\mathcal{R}(\mathcal{M}, \sigma)); v\mathcal{M}v^* = \mathcal{M}\}$. If $\mathcal{N} \supset \pi_\sigma(\mathcal{M})$ is a unital W^*-subalgebra of $\mathcal{R}(\mathcal{M}, \sigma)$, then $H_{\mathcal{N}} = \{\mathrm{Ad}(v)|\mathcal{M}; v \in \mathcal{N}(P_{\hat{\sigma}}) \cap \mathcal{N}\}$ is a full subgroup of $[\sigma(G)]$ and $\mathcal{N} = \mathcal{R}\{v \in U(\mathcal{R}(\mathcal{M}, \sigma)); \mathrm{Ad}(v)|\mathcal{M} \in H_{\mathcal{N}}\}$. Conversely, if H is a full subgroup of $[\sigma(G)]$, then $\mathcal{N}_H = \mathcal{R}\{v \in U(\mathcal{R}(\mathcal{M}, \sigma)); \mathrm{Ad}(v)|\mathcal{M} \in H\}$ is a unital W^*-subalgebra of $\mathcal{R}(\mathcal{M}, \sigma)$ containing $\pi_\sigma(\mathcal{M})$ and $H = \{\mathrm{Ad}(v)|\mathcal{M}; v \in \mathcal{N}(P_{\hat{\sigma}}) \cap \mathcal{N}_H\}$. Hence the mappings $\mathcal{N} \mapsto H_{\mathcal{N}}$ and $H \mapsto \mathcal{N}_H$ are reciprocal increasing bijections between the full subgroups H of $[\sigma(G)]$ and the unital W^*-subalgebras $\mathcal{N} \supset \pi_\sigma(\mathcal{M})$ of $\mathcal{R}(\mathcal{M}, \sigma)$.

For the full proofs of these results, we refer to Haga and Takeda (1972); Henle (1979).

22.10. Let T and G be discrete groups and let $\sigma : T \to Aut(G)$ be a group homomorphism. The product set $G \times T$ endowed with the operation $(g, t)(g', t') = (g\sigma_t(g'), tt')$ ($g, g' \in G, t, t' \in T$) is a group, denoted by $G \times_\sigma T$, called *the semidirect product of G by the action σ of T*, with neutral element $(e, 1) \in G \times_\sigma T$ (where $e \in G$ and $1 \in T$ are the corresponding neutral elements of G and T) and inverse given by $(g, t)^{-1} = (\sigma_t^{-1}(g^{-1}), t^{-1})$ ($g \in G, t \in T$). We have a split short exact sequence

$$\{e\} \to G \overset{\alpha}{\to} G \times_\sigma T \underset{\gamma}{\overset{\beta}{\rightleftarrows}} T \to \{1\}$$

where $\alpha(g) = (g, 1), \beta(g, t) = t, \gamma(t) = (e, t)$ $(g \in G, t \in T)$; $\alpha(G)$ is a normal subgroup of $G \times {}_\sigma T$, $\gamma(T)$ is a subgroup of $G \times_\sigma T$, $G \times_\sigma T$ is generated by $\alpha(G) \cup \gamma(T)$ and $(G \times_\sigma T)/\alpha(G)$ is isomorphic to $\gamma(T)$.

The von Neumann algebra $\mathfrak{L}(G \times_\sigma T) \subset \mathscr{B}(\ell^2(G \times_\sigma T))$ is generated by the left translation operators $\boldsymbol{\lambda}(g, t) \in \mathscr{B}(\ell^2(G \times {}_\sigma T))$ $(g \in G, t \in T)$.

On the other hand, the von Neumann algebra $\mathfrak{L}(G) \subset \mathscr{B}(\ell_2(G))$ is generated by the left translation operators $\boldsymbol{\lambda}(g) \in \mathscr{B}(\ell^2(G))$ $(g \in G)$, and the action $\sigma : T \to Aut(G)$ extends to an action $\sigma : T \to Aut(\mathfrak{L}(G))$ defined by

$$\sigma_t\left(\sum_g \alpha(g)\boldsymbol{\lambda}(g)\right) = \sum_g \alpha(g)\boldsymbol{\lambda}(\sigma_t(g)) = \sum_g \alpha(\sigma_t^{-1}(g))\boldsymbol{\lambda}(g) \quad (t \in T);$$

so we can consider the crossed product von Neumann algebra $\mathscr{R}(\mathfrak{L}(G), \sigma) \subset \mathscr{B}(\ell^2(T, \ell^2(G)))$ generated by the operators $\pi_\sigma(\boldsymbol{\lambda}(g))$, $(g \in G)$ and $1 \bar{\otimes} \boldsymbol{\lambda}(t)$ $(t \in T)$, acting on $\ell^2(T, \ell^2(G))$.

It is easy to check that the equation

$$[U\xi](g, t) = [\xi(T)](\sigma_t^{-1}(g)) \quad (\xi \in \ell^2(T, \ell^2(G)), g \in G, t \in T)$$

defines a unitary operator $U : \ell^2(T, \ell^2(G)) \to \ell^2(G \times {}_\sigma T)$ such that

$$U^{-1}\boldsymbol{\lambda}(g, 1)U = \pi_\sigma(\boldsymbol{\lambda}(g)) \quad (g \in G), \tag{1}$$

$$U^{-1}\boldsymbol{\lambda}(e, t)U = 1 \bar{\otimes} \boldsymbol{\lambda}(t)) \quad (t \in T). \tag{2}$$

Indeed, for $\xi \in \ell^2(T, \ell^2(G)), h \in G$ and $s \in T$ we have

$$\begin{aligned}
[(U^{-1}\boldsymbol{\lambda}(g, 1)U\xi)(s)](h) &= [\boldsymbol{\lambda}(g, 1)U\xi](\sigma_s(h), s) \\
&= [U\xi]((g, 1)^{-1}(\sigma_s(h), s)) = [U\xi]((g^{-1}, 1)(\sigma_s(h), s)) \\
&= [U\xi](g^{-1}\sigma_s(h), s) = [\xi(s)](\sigma_s^{-1}(g^{-1}\sigma_s(h))) \\
&= [\xi(s)](\sigma_s^{-1}(g)^{-1}h) = [\boldsymbol{\lambda}(\sigma_s^{-1}(g))\xi(s)](h) \\
&= [\sigma_s^{-1}(\boldsymbol{\lambda}(g))\xi(s)](h) = [(\pi_\sigma(\boldsymbol{\lambda}(g)\xi)(s)](h), \\
[(U^{-1}\boldsymbol{\lambda}(e, t)U\xi)(s)](h) &= [\boldsymbol{\lambda}(e, t)U\xi](\sigma_s(h), s) \\
&= [U\xi]((e, t)^{-1}(\sigma_s(h), s)) = [U\xi]((e, z^{-1})(\sigma_s(h), s)) \\
&= [U\xi](\sigma_{t^{-1}s}, t^{-1}s) = [\xi(t^{-1}s)](\sigma_{t^{-1}s}^{-1}(\sigma_{t^{-1}s}(h))) \\
&= [\xi(t^{-1}s](h) = [((1 \bar{\otimes}\boldsymbol{\lambda}(t))\xi)(s)](h).
\end{aligned}$$

Consequently,

$$U^{-1}\mathfrak{L}(G \times {}_\sigma T)U = \mathscr{R}(\mathfrak{L}(G), \sigma) \tag{3}$$

and hence we have proved the following

Proposition. *Let G, T be discrete groups and $\sigma : T \to Aut(G)$ a group homomorphism. Then the von Neumann algebras $\mathfrak{L}(G \times {}_\sigma T)$ and $\mathscr{R}(\mathfrak{L}(G), \sigma)$ are spatially isomorphic.*

22.11. In particular, let us consider the case $G = F_\infty$, the free group on a countable infinity of generators $\{x_n\}_{n\in\mathbb{Z}} \subset F_\infty, T = \mathbb{Z}, m \in \mathbb{N}$ and ${}_m\sigma \in Aut(F_\infty)$ the unique automorphism of F_∞ such that ${}_m\sigma(x_n) = x_{n+m} (n \in \mathbb{Z})$. We obtain a group homomorphism, also denoted by ${}_m\sigma : \mathbb{Z} \to Aut(F_\infty)$, by putting ${}_m\sigma_k = ({}_m\sigma)^k$ $(k \in \mathbb{Z})$.

Proposition. *The semidirect product* $F_\infty \times_{{}_m\sigma} \mathbb{Z}$ *is isomorphic to the free group* F_{m+1} *on* $m+1$ *generators.*

Proof. We shall put $\sigma = {}_m\sigma$. The neutral element of $F_\infty \times_\sigma \mathbb{Z}$ is $(e, 0)$ and for $(y, i), (z, j) \in F_\infty \times_\sigma \mathbb{Z}$ we have $(y, i)(z, j) = (y\sigma^i(z), i + j), (y, i)^{-1} = (\sigma^{-i}(y^{-1}), -i)$. Let $y_0, y_1, \dots, y_m$ be the generators of the free group F_{m+1}.

We define a homomorphism $\varphi : F_{m+1} \to F_\infty \times_\sigma \mathbb{Z}$ by putting $\varphi(y_k) = (x_k, 1)$ for $k = 0, 1, \dots, m-1$ and $\varphi(y_m) = (e, 1)$. Since $(e, 1)^{-1} = (e, -1), (x_k, 1)(e, -1) = (x_k, 0)$, and $(e, 1)(x_k, 0) = (x_{k+m}, 1)$, it follows that φ is surjective.

Consider also the homomorphism $\psi_1 : F_\infty \to F_{m+1}$ defined by $\psi_1(x_n) = y_m^p y_k y_m^{-p-1}$ $(n, p, k \in \mathbb{Z}, n = mp + k, 0 \le k < m)$, and the homomorphism $\psi_2 : \mathbb{Z} \to F_{m+1}$ defined by $\psi_2(i) = y_m^i$ $(i \in \mathbb{Z})$, and define a mapping $\psi : F_\infty \times_\sigma \mathbb{Z} \to F_{m+1}$ by $\psi(y, i) = \psi_1(y)\psi_2(i)$ $(y \in F_\infty, i \in \mathbb{Z})$. We show that ψ is a group homomorphism. For (y, i) and (z, j) in $F_\infty \times_\sigma \mathbb{Z}$, we have $\psi((y, i)(z, j)) = \psi_1(y)\psi_1(\sigma^i(z))\psi_2(i)\psi_2(j)$ and $\psi(y, i)\psi(z, j) = \psi_1(y)\psi_2(i)\psi_1(z)\psi_2(j)$, and so it is sufficient to show that $\psi_1(\sigma^i(z))\psi_2(i) = \psi_2(i)\psi_1(z)$, that is, $\psi_1(\sigma^i(z)) = y_m^i\psi_1(z)y_m^{-i}$. Moreover, we may assume that $z = x_n$. In this case writing $n = mp + k$ with $p, k \in \mathbb{Z}, 0 \le k < m$, we have $\psi_1(\sigma^i(x_n)) = \psi_1(x_{n+mi}) = \psi_1(x_{m(i+p)+k}) = y_m^{i+p} y_k y_m^{-i-p-1} = y_m^i(y_m^p y_k y_m^{-p-1})y_m^i = y_m^i\psi_1(x_n)y_m^{-i}$.

Finally, we show that $\psi \circ \varphi =$ the identity mapping, so that φ is the required isomorphism. For $k = 0, 1, \dots, m-1$ we have $(\psi \circ \varphi)(y_k) = \psi(x_k, 1) = \psi_1(x_k)\psi_2(1) = y_k y_m^{-1} y_m = y_k$ and for $k = m$ we have $(\psi \circ \varphi)(y_m) = \psi(e, 1) = \psi_1(e)\psi_2(1) = ey_m = y_m$. □

Using Proposition 22.10, we deduce that

Corollary. $\mathcal{R}(\mathfrak{L}(F_\infty), {}_m\sigma)$ *is *-isomorphic to* $\mathfrak{L}(F_{m+1})$.

One can also show (Phillips, 1976) that for every $m, n \ge 2$ there exists $\sigma \in Aut(\mathfrak{L}(F_{m(n-1)+1}))$ such that the *-automorphisms $\sigma, \sigma^2, \dots \sigma^{m-1}$ are all outer, $\sigma^m = \iota$ and $\mathcal{R}(\mathfrak{L}(F_{m(n-1)+1}), \sigma)$ is *-isomorphic to $\mathfrak{L}(F_n)$. Let us mention that it is not yet known whether the factors $\mathfrak{L}(F_2), \mathfrak{L}(F_3), \dots$ are *-isomorphic or not.

22.12. In this section, we give some examples of outer *-automorphisms of the W^*-algebras of the form $\mathfrak{L}(G)$.

Recall that every automorphism σ of the discrete group G has an extension to *-automorphism σ of $\mathfrak{L}(G)$ such that

$$\sigma\left(\sum_g \alpha(g)\boldsymbol{\lambda}(g)\right) = \sum_g \alpha(g)\boldsymbol{\lambda}(\sigma(g)) = \sum_g \alpha(\sigma^{-1}(g))\boldsymbol{\lambda}(g), \quad (\alpha(g) \in \mathbb{C}, g \in G).$$

Proposition. *Let* σ *be an automorphism of the discrete group* G. *For the corresponding *-automorphism* $\sigma \in Aut(\mathfrak{L}(G))$ *we have*

(i) σ *is properly outer if and only if for every* $g \in G$ *the set* $\{\sigma(s)gs^{-1}; s \in G\}$ *is infinite;*
(ii) σ *is ergodic if and only if for every* $g \in G, g \ne e$, *the set* $\{\sigma^n(g); n \in \mathbb{Z}\}$ *is infinite.*

Proof. Assume that for every $g \in G$ the set $\{\sigma(s)gs^{-1}; s \in G\}$ is infinite. Let $a = \sum_g \alpha(g)\boldsymbol{\lambda}(g) \in \mathfrak{L}(G)$ be such that $a\boldsymbol{\lambda}(s) = \sigma(\boldsymbol{\lambda}(s))a = \boldsymbol{\lambda}(\sigma(s))a$, that is, $a = \boldsymbol{\lambda}(\sigma(s^{-1}))a\boldsymbol{\lambda}(s)$ for every $s \in G$. Since $\boldsymbol{\lambda}(\sigma(s^{-1}))a\boldsymbol{\lambda}(s) = \sum_g \alpha(\sigma(s)gs^{-1})\boldsymbol{\lambda}(g)$, it follows that $\alpha(g) = \alpha(\sigma(s)gs^{-1})$ for $s, g \in G$. Since $\sum_g |\alpha(g)|^2 < +\infty$ (22.6.(4)), using the assumption we conclude $\alpha(g) = 0$ for all $g \in G$, that is, $a = 0$. Hence σ is properly outer.

Conversely, assume that σ is properly outer and that there exists $g \in G$ such that the set $S = \{\sigma(s)gs^{-1}; s \in G\}$ is finite. Then $a = \sum_{s\in S} \boldsymbol{\lambda}(s) \in \mathfrak{L}(G)$ and $a = \boldsymbol{\lambda}(\sigma(s))a\boldsymbol{\lambda}(s^{-1})$, that is, $\sigma(\boldsymbol{\lambda}(s))a = a\boldsymbol{\lambda}(s)$ for each $s \in G$, hence $\sigma(x)a = ax$ for all $x \in \mathfrak{L}(G)$, which implies that $a = 0$, a contradiction.

We have proved assertion (i); (ii) can be proved similarly. □

Corollary 1. *If G is a discrete ICC-group and σ is an outer automorphism of G, then the corresponding *-automorphism of the factor $\mathfrak{L}(G)$ is also outer.*

Proof. Assume that $\sigma \in Aut(\mathfrak{L}(G))$ is inner, that is $\sigma = \mathrm{Ad}(a)$ with $a = \sum_g \alpha(g)\boldsymbol{\lambda}(g) \in U(\mathfrak{L}(G))$, $\sum_g |\alpha(g)|^2 < +\infty$. If $g \in G$ and $\alpha(g) \neq 0$, then, as in the proof of the proposition, one shows that the set $\{\sigma(s)gs^{-1}; s \in G\}$ is finite. Since σ is an outer automorphism of G, we infer that there exist $g_1, g_2 \in G, g_1 \neq g_2$, such that the sets $S_1 = \{\sigma(s)g_1s^{-1}; s \in G\}$ and $S_2 = \{\sigma(s)g_2s^{-1}; s \in G\}$ be finite. Then the set $\{sg_1^{-1}g_2s^{-1}; s \in G\} \subset S_1^{-1}S_2$ is finite, hence $g_1 = g_2$, since G is an *ICC*-group. This contradiction proves that σ is an outer *-automorphism of $\mathfrak{L}(G)$. □

Let us mention that there exist countable discrete *ICC*-groups without outer *-automorphisms, for instance (Kallman, 1969) the semidirect product $G = \mathbb{Q} \times_\sigma \mathbb{Q}^*$, where $\mathbb{Q}$ is the additive group of rational numbers, $\mathbb{Q}^*$ is the multiplicative group of nonzero rational numbers and $\sigma : \mathbb{Q}^* \to Aut(\mathbb{Q})$ is defined by $\sigma(r)(q) = rq$, $(r \in \mathbb{Q}^*, q \in \mathbb{Q})$.

By the above results, we get the following examples of outer *-automorphisms:

Corollary 2. *Let F_2 be the free group on two generators and σ the automorphism of F_2 intertwining the two generators. Then $\sigma \in Aut(\mathfrak{L}(F_2))$ is outer.*

Corollary 3. *Let $S(\infty)$ be the group of finite permutations of $\mathbb{Z}$, let π be an infinite permutation of $\mathbb{Z}$ and let σ be the automorphism of $S(\infty)$ defined by $\sigma(s) = \pi \circ s \circ \pi^{-1}$ $(s \in S(\infty))$. Then $\sigma \in Aut(\mathfrak{L}(S(\infty)))$ is outer.*

With the same methods one can show that every countable discrete group can be represented as a group of outer *-automorphisms of the II_1 factor $\mathfrak{L}(S(\infty))$ (Suzuki, 1958; Kallman, 1969). Also, there exist factors of types II_∞ and III enjoying the same property (Saito, 1961; Kallman, 1969).

22.13. We now describe another type of outer *-automorphisms of factors of the form $\mathfrak{L}(G)$.

Let G be a discrete *ICC*-group and γ a character of G, that is, a group homomorphism of G into the one-dimensional torus $\{\omega \in \mathbb{C}; |\omega| = 1\}$; clearly, $\overline{\gamma(g)} = \gamma(g^{-1}) = \gamma(g)^{-1}$ $(g \in G)$. γ being in $\ell^\infty(G)$ defines a multiplication operator $\mathbf{m}(\gamma) \in \mathscr{B}(\ell^2(G))$; we have $\mathbf{m}(\gamma)\mathfrak{L}(G)\mathbf{m}(\gamma)^* = \mathfrak{L}(G)$. Consequently, we obtain a *-automorphism $\sigma_\gamma = \mathrm{Ad}(\mathbf{m}(\gamma))|\mathfrak{L}(G) \in Aut(\mathfrak{L}(G))$, uniquely determined, such that $\sigma_\gamma(\lambda(g)) = \gamma(g)\lambda(g)$ $(g \in G)$.

Proposition. *Let G be a discrete* ICC-*group and $\gamma \neq 1$ a nontrivial character of G. Then the *-automorphism $\sigma_\gamma \in Aut(\mathfrak{L}(G))$ is outer.*

Proof. Assume to the contrary, so that there exists $v = \sum_g \alpha(g)\boldsymbol{\lambda}(g) \in U(\mathfrak{L}(G))$ with $\sigma_\gamma = \mathrm{Ad}(v)$. For every $s \in G$, we have $\gamma(s)\boldsymbol{\lambda}(s) = \sigma_\gamma(\boldsymbol{\lambda}(s)) = v\boldsymbol{\lambda}(s)v^* = \sum_{g,h} \alpha(g)\overline{\alpha(h)}\boldsymbol{\lambda}(gsh^{-1}) =$

$\sum_t(\sum_g \alpha(g)\overline{\alpha(t^{-1}gs)})\boldsymbol{\lambda}(t)$. Let $\xi_s \in \ell^2(G)$ be defined by $\xi_s(g) = \alpha(s^{-1}gs)$ $(s, g \in G)$. The previous computation shows that $(\xi_e|\xi_s) = \sum_g \alpha(g)\overline{\alpha(s^{-1}gs)} = |\gamma(s)| = 1$ and $\|\xi_s\} = \|\xi_e\| = 1$, so that the vectors ξ_s and ξ_e are proportional, in particular $|\alpha(s^{-1}gs)| = |\alpha(g)|$ $(s, g \in G)$. Since $\gamma \neq 1$, we have $\sigma_\gamma \neq \iota$, and there is $g_0 \in G, g_0 \neq e$, with $\alpha(g_0) \neq 0$. Since G is an *ICC*-group, it follows that $\sum_g |\alpha(g)|^2 = +\infty$, a contradiction. □

In particular, if F_∞ is the free group on a countable infinity of generators $\{x_n\}_{n\geq 1} \subset F_\infty$ and $G = \{1, \gamma_1, \gamma_2, \ldots\}$ is an ordered countable infinite subgroup of the one-dimensional torus, then there exists a unique $*$-automorphism $\sigma_G \in Aut(\mathfrak{L}(F_\infty))$ such that $\sigma_G(\lambda(x_n)) = \gamma_n\lambda(x_n)$ $(n \geq 1)$.

One can prove that if G and H are two ordered countable infinite subgroups of the one-dimensional torus, then the $*$-automorphisms $\sigma_G, \sigma_H \in Aut(\mathfrak{L}(F_\infty))$ are outer conjugate if and only if $G = H$ (Phillips, 1976). Thus $Aut(\mathfrak{L}(F_\infty))$ has a continuous infinity of outer conjugacy classes; moreover, there is no "good" classification of these classes (Phillips, 1976).

On the other hand, the set of outer conjugacy classes of $Aut(\mathfrak{L}(S(\infty))$ is countable and completely described by certain simple invariants introduced by Connes (1977a, 1975e).

22.14. In this section we examine, in a slightly more general framework, some interesting consequences of the relative commutant theorem (22.3.(1)).

Proposition. *Let $\sigma : G \to Aut(\mathcal{M})$ be an integrable continuous action of the locally compact unimodular group G on the countably decomposable W^*-algebra $\mathcal{M}$. If $\pi_\sigma(\mathcal{M})' \cap \mathcal{R}(\mathcal{M}, \sigma) = \pi_\sigma(\mathcal{Z}(\mathcal{M}))$, then $(\mathcal{M}^\sigma)' \cap \mathcal{M} = \mathcal{Z}(\mathcal{M})$, in particular $\mathcal{Z}(\mathcal{M}^\sigma) = \mathcal{Z}(\mathcal{M}^\sigma)$.*

Proof. Since $\mathcal{M}$ is countably decomposable, we may assume $\mathcal{M} \subset \mathcal{B}(\mathcal{H})$ realized as a von Neumann algebra with a cyclic and separating vector $\xi_0 \in \mathcal{H}$. Since σ is integrable and G is unimodular, the formula $\varphi(x) = \int(\sigma_g(x)\xi_0|\xi_0)\,\mathrm{d}g$ $(x \in \mathcal{M}^+)$, defines a σ-invariant n.s.f. weight φ on $\mathcal{M}$. By Corollary 2.24, there exists an *so*-continuous unitary representation $u : G \to \mathcal{B}(\mathcal{H}_\varphi)$ such that $\pi_\varphi(\sigma_g(x)) = u(g)\pi_\varphi(x)u(g)^*$ $(x \in \mathcal{M}, g \in G)$, and moreover, by Section 2.25, we have $u(g)a_\varphi = (\sigma_g(a))_\varphi$ and $u(g)J_\varphi = J_\varphi u(g)$ $(g \in G, a \in \mathfrak{N}_\varphi)$.

Since $\int \|\sigma_t^{-1}(a)\xi_0\|^2\,\mathrm{d}t = \varphi(a^*a) = \|a_\varphi\|^2$ $(a \in \mathfrak{N}_\varphi)$, there exists an isometric linear operator $U : \mathcal{H}_\varphi \to \mathcal{L}^2(G, \mathcal{H})$, uniquely determined, such that $[Ua_\varphi](t) = \sigma_t^{-1}(a)\xi_0$ for all $a \in \mathfrak{N}_\varphi, t \in G$. Recall that the crossed product $\mathcal{R}(\mathcal{M}, \sigma) \subset \mathcal{B}(\mathcal{L}^2(G, \mathcal{H}))$ is the von Neumann algebra generated by the operators $\pi_\sigma(x)$ $(x \in \mathcal{M})$ and $1 \bar{\otimes} \boldsymbol{\lambda}(g)$ $(g \in G)$. It is easy to check that $U\pi_\varphi(x) = \pi_\sigma(x)U$, $(x \in \mathcal{M})$, and $Uu(g) = (1 \bar{\otimes} \boldsymbol{\lambda}(g))U$ $(g \in G)$, so that the mapping $\Phi : \mathcal{R}(\mathcal{M}, \sigma) \ni X \mapsto U^*XU \in \mathcal{R}\{\pi_\varphi(\mathcal{M}), u(G)\}$ is a surjective normal $*$-homomorphism. Let Q be the unique central projection in $\mathcal{R}(\mathcal{M}, \sigma)$ such that $Ker\,\Phi = (1 - Q)\,\mathcal{R}(\mathcal{M}, \sigma)$.

It is clear that $\pi_\varphi(\mathcal{M}^\sigma) = \pi_\varphi(\mathcal{M}) \cap u(G)'$ and hence that

$$J_\varphi\pi_\varphi((\mathcal{M}^\sigma)' \cap \mathcal{M})J_\varphi = \mathcal{R}\{\pi_\varphi(\mathcal{M}), u(G)\} \cap \pi_\varphi(\mathcal{M})'.$$

Thus, if $z \in (\mathcal{M}^\sigma)' \cap \mathcal{M}$, there exists $X = XQ \in \mathcal{R}(\mathcal{M}, \sigma)$ such that $\Phi(X) = J_\varphi\pi_\varphi(z)J_\varphi \in \pi_\varphi(\mathcal{M})' = \Phi(\pi_\sigma(\mathcal{M}))'$. It follows that $X \in \pi_\sigma(\mathcal{M})' \cap \mathcal{R}(\mathcal{M}, \sigma) = \pi_\sigma(\mathcal{Z}(\mathcal{M}))$ and therefore

$$\pi_\varphi(z) = J_\varphi\Phi(X)J_\varphi \in J_\varphi\Phi(\pi_\sigma(\mathcal{Z}(\mathcal{M})))J_\varphi = J_\varphi\pi_\varphi(\mathcal{Z}(\mathcal{M}))J_\varphi = \pi_\varphi(\mathcal{Z}(\mathcal{M})),$$

that is, $z \in \mathcal{Z}(\mathcal{M})$. □

In particular,

$$\mathscr{Z}(\mathscr{M}^\sigma) = \mathscr{Z}(\mathscr{M})^\sigma \tag{1}$$

for any properly outer action $\sigma : G \to Aut(\mathscr{M})$ of a finite group G on the countably decomposable W^*-algebra $\mathscr{M}$. This is an extension of 16.17.(3).

22.15. As an application, we prove the following extension of Corollary 16.17.

Corollary. *Let $\sigma : G \to Aut(\mathscr{M})$ be a properly outer action of the finite group G on the countably decomposable W^*-algebra $\mathscr{M}$.*

Then, for $e,f \in Proj(\mathscr{M}^\sigma)$ we have

$$e \sim f \text{ in } \mathscr{M} \Leftrightarrow e \sim f \text{ in } \mathscr{M}^\sigma \tag{1}$$

and every unitary cocycle $u \in Z_\sigma(G; U(\mathscr{M}))$ is trivial, that is, there exists $v \in U(\mathscr{M})$ such that $u(g) = v\sigma_g(v^*)$ $(g \in G)$.

Proof. (1) follows by the same arguments as in the proof of Corollary 16.17, using Corollary 17.24 and 22.14.(1).

Consider now any two unitary cocycles $a, b \in Z_\sigma(G; U(\mathscr{M}))$, the action $\sigma \bar{\otimes} \iota$, of G on the W^*-algebra $\mathscr{M} \bar{\otimes} F_2$ and the balanced cocycle $c = c(a,b) \in Z_{\sigma \bar{\otimes} \iota}(G; U(\mathscr{M} \bar{\otimes} F_2))$ (see 20.2). According to 17.2.(1) and 17.6.(1), the action ${}_c(\sigma \bar{\otimes} \iota)$ of G on $\mathscr{M} \bar{\otimes} F_2$ is also properly outer. On the other hand, the projections $1 \bar{\otimes} e_{11}, 1 \bar{\otimes} e_{22} \in (\mathscr{M} \bar{\otimes} F_2)^c$ are equivalent in $\mathscr{M} \bar{\otimes} F_2$. Hence, using (1) we conclude that $1 \bar{\otimes} e_{11} \sim 1 \bar{\otimes} e_{22}$ in $(\mathscr{M} \bar{\otimes} F_2)^c$, that is, $a \approx b$. □

Thus, if $\tau : G \to Aut(\mathscr{N})$ is another action satisfying the assumptions of the previous corollary and if $(\mathscr{M}, \sigma) \sim (\mathscr{N}, \tau)$, then $(\mathscr{M}, \sigma) \approx (\mathscr{N}, \tau)$.

If, moreover, $\mathscr{M}$ is properly infinite, there exists a projection $p \in \mathscr{M}$ such that the $\sigma_g(p)$ $(g \in G)$ are mutually orthogonal and $\sum_{g \in G} \sigma_a(p) = 1$. Indeed, by the previous remark and 20.14.(1) we have in this case $(\mathscr{M}, \sigma) \approx (\mathscr{M} \bar{\otimes} \mathscr{B}(\ell^2(G)), \sigma \bar{\otimes} \mathrm{Ad}(\boldsymbol{\rho}))$, and the stated property is obvious for the action $\mathrm{Ad}(\boldsymbol{\rho}) : G \to Aut(\mathscr{B}(\ell^2(G)))$.

22.16. Notes. For the classical results concerning crossed products by discrete groups (22.1, 22.3, 22.6, 22.7, 22.8, 22.10) we refer to Ching (1969); Dixmier (1957, 1969); Nakagami and Takesaki (1979); Nakamura and Takeda (1958, 1960); Sakai (1971); Turumaru (1958); Zeller-Meier (1968). The results of Sections 22.2 and 22.4 appeared in Connes (1973a); Haga and Takeda (1972). The material of Section 22.12 is from Dixmier (1957, 1969) and Kallman (1969). Proposition 22.11 is due to Phillips (1976), Proposition 22.13 is due to Paschke (1976) and the Galois correspondence type results 22.5 and 22.9 are due to Haga and Takeda (1972). Extensions of the relative commutant theorem (22.3.(2)) and of the results concerning the type of the crossed products (22.7) have been obtained by Sauvageot (1977). A deep commutative nondiscrete relative commutant theorem, due to Connes and Takesaki (1977), will be given in Section 23.19. A detailed analysis of several relative commutant theorems in general situations, including Proposition 22.14, is due to Paschke (1978) (see also Nakagami & Takesaki, 1979). Corollary 22.15 is due to Connes and Takesaki (1977).

For our exposition, we have used Ching (1969); Connes (1973a); Connes and Takesaki (1977); Dixmier (1957, 1969); Kallman (1969); Nakagami and Takesaki (1979); Paschke (1976, 1978); Phillips (1976) and Sakai (1971).

Important examples of factors arising by the "group measure space construction" (22.8) are the *Pukánszky factors* $\mathscr{P}_\lambda (0 < \lambda < 1)$ (Pukánszky, 1956; Sakai, 1971, p. 192; Connes, 1973a, p. 207), which we now describe. Let F_2 be the free group on two generators and let $\Omega = \prod_{g \in F_2} \Omega_g$, where $\Omega_g = \{0, 1\}$ for each $g \in F_2$. Let $p > 0, q > 0$ be such that $p + q = 1$ and $p = \lambda q$. For each $g \in F_2$, we denote by μ_g the measure on Ω_g defined by $\mu_g(\{0\}) = p, \mu_g(\{1\}) = q$ and then consider the product measure $\mu = \bigotimes_{g \in F_2} \mu_g$ on Ω. Each $g_0 \in F_2$ defines a shift transformation $T_{g_0} : \{\omega_g\}_{g \in F_2} \mapsto \{\omega_{g_0 g}\}_{g \in F_2}$ on Ω and also a transformation $S_{g_0} : \{\omega_g\}_{g \in F_2} \mapsto \{\omega'_g\}_{g \in F_2}$ on Ω such that $\omega'_{g_0} = 1 - \omega_{g_0}$ and $\omega'_g = \omega_g$ for $g \neq g_0$. Let G be the transformation group on Ω generated by the transformations T_g and S_g ($g \in F_2$). Then μ is G-quasi-invariant and G acts freely and ergodically on Ω (cf. *loc. cit.*), so that the corresponding crossed product W^*-algebra is a factor denoted by $\mathscr{P}_\lambda$. Note also that the Powers factors $\mathscr{R}_\lambda$ (A.17) can be obtained by a similar group measure space construction.

In order to obtain factors by the group measure space construction, one has to assume that the action of the transformation group is free and ergodic. Krieger (1970a, 1970b, 1971a, 1971b, 1972, 1974, 1976a) showed how to modify this construction in order to obtain factors even if the action is not free (as long as it is still ergodic). Several variants, extensions and concrete examples of Krieger's construction have appeared in Guichardet (1974); Nielsen (1974–1975); Samuelides and Sauvageot (1975); Strătilă and Voiculescu (1975, 1978).

Zimmer (1977a, 1977b) proved that the factors arising by the classical group measure space construction, or the Krieger construction, are approximately finite dimensional if and only if the action of the group is amenable, as defined in Zimmer (1978) (see also Delaroche, 1979). For instance, the Pukánszky factors are not approximately finite dimensional. Recently, Connes, Feldman, and Weiss (1981) proved that amenable actions of discrete groups are necessarily singly generated.

From the works of Dye (1959, 1963) and Krieger (1968, 1969, 1976b) it follows that the *-isomorphism class of a factor arising by the group measure space construction depends only on the equivalence relation defined by the orbits of the group. A detailed construction and analysis of the von Neumann algebras associated with an ergodic equivalence relation is given by Feldman and Moore (1977).

A factor that is the crossed product of an abelian W^*-algebra by a single *-automorphism is called a *Krieger factor* (cf. Connes, 1975d). There are several abstract characterizations of Krieger factors due to Connes (1974d, 1975d). Two Krieger factors are *-isomorphic if and only if the corresponding transformations are weakly equivalent, that is, if and only if they define isomorphic equivalence relations (Krieger, 1976b).

All Krieger factors are approximately finite dimensional. Connes (1975d) proved that any approximately finite dimensional factor of type III_0 acting on a separable Hilbert space is a Krieger factor and these factors are classified by their flow of weights. On the other hand, every Araki–Woods factor (A.17) can be obtained as a Krieger factor (Krieger, 1970b; Connes, 1977b). Connes [1973a] proved that the class of Krieger factors (and hence the class of approximately finite dimensional factors) is strictly larger than the class of Araki–Woods factors; a simpler proof of this result has been obtained by Connes and Woods (1980a).

Solving an old open problem, Connes (1975a, 1975b) constructed examples or factors of types II_1 and III, which are not *-antiisomorphic to themselves. In particular, not every factor arises as the $\mathfrak{L}(G)$ for some *ICC*-group G.

CHAPTER V

Continuous Decompositions

23 Dominant Weights and Continuous Decompositions

Let $\mathscr{M}$ be a properly infinite W^*-algebra and denote by $W_n(\mathscr{M})$ (resp. $W_{ns}(\mathscr{M})$; resp. $W_{nsf}(\mathscr{M})$) the set of all normal (resp. normal semifinite; resp. n.s.f.) weights on $\mathscr{M}$. In Sections 23.1–23.3, we consider a fixed weight $\omega \in W_{nsf}(\mathscr{M})$ and, using the canonical bijection $W_{ns}(\mathscr{M}) \ni \varphi \mapsto [D\varphi : D\omega] \in Z_{\sigma^\omega}(\mathbb{R}; \mathscr{M})$ given by Theorems 3.1 and 5.1, we apply the results obtained in Section 20 in order to get a comparison theory for weights. In particular, we define the notion of a dominant weight on $\mathscr{M}$, this yields the continuous decomposition $(\mathscr{N}, \theta, \tau)$ of $\mathscr{M}$, the properties of which are analyzed further. The main feature of the continuous decomposition is that it reduces the study of type III W^*-algebras to the study of actions of $\mathbb{R}$ on semifinite W^*-algebras, modulo (outer) conjugation.

23.1. Recall (2.21) that if $\varphi \in W_{ns}(\mathscr{M})$ and $u \in \mathscr{M}$ is a partial isometry with $uu^* \in \mathscr{M}^\varphi$, then the formula $\varphi_u(x) = \varphi(uxu^*)$ $(x \in \mathscr{M}^+)$ defines a weight $\varphi_u \in W_{ns}(\mathscr{M})$ with $\mathbf{s}(\varphi_u) = u^*u$; moreover, if $u^*u \in \mathscr{M}^\varphi$, then $\varphi_u = \varphi_{u^*u} \Leftrightarrow u \in \mathscr{M}^\varphi$. Consider a fixed weight $\omega \in W_{nsf}(\mathscr{M})$.

Proposition. *Let $\varphi, \psi \in W_{ns}(\mathscr{M})$ and $u \in \mathscr{M}$ with $u^*u = \mathbf{s}(\psi)$ and $uu^* \in \mathscr{M}^\varphi$. Consider the W^*-algebra $\mathscr{P} = \mathscr{M} \bar{\otimes} \mathscr{F}_2$ and the balanced weight $\theta = \theta(\varphi, \psi) \in W_{ns}(\mathscr{P})$. The following statements are equivalent:*

(i) $\psi = \varphi_u$;

(ii) $\begin{pmatrix} 0 & u \\ 0 & 0 \end{pmatrix} \in \mathscr{P}^\theta$;

(iii) $[D\psi : D\omega]_t = u^*[D\varphi; D\omega]_t\sigma_t^\omega(u)$ *for all* $t \in \mathbb{R}$.

Proof. (i) $\Leftrightarrow$ (ii). We have $\mathbf{s}(\theta) = \begin{pmatrix} \mathbf{s}(\varphi) & 0 \\ 0 & \mathbf{s}(\psi) \end{pmatrix}$, hence $U = \begin{pmatrix} 0 & u \\ 0 & 0 \end{pmatrix} \in s(\theta)\mathscr{P}\mathbf{s}(\theta)$, since $F = UU^* = \begin{pmatrix} uu^* & 0 \\ 0 & 0 \end{pmatrix}$ and $E = U^*U = \begin{pmatrix} 0 & 0 \\ 0 & \mathbf{s}(\psi) \end{pmatrix}$; it also follows that $E, F \in \mathscr{P}^\theta$. By Proposition 2.21, we have $U \in \mathscr{P}^\theta$ if and only if $\theta_U = \theta_E$. For $X = \begin{pmatrix} x_{11} & x_{12} \\ x_{21} & x_{22} \end{pmatrix} \in \mathscr{P}^+$, we have $UXU^* = \begin{pmatrix} ux_{22}u^* & 0 \\ 0 & 0 \end{pmatrix}$ and $EXE = \begin{pmatrix} 0 & 0 \\ 0 & \mathbf{s}(\psi)x_{22}\mathbf{s}(\psi) \end{pmatrix}$, hence $\theta_U = \theta_E$ if and only if $\psi = \varphi_u$.

(ii) ⇔ (iii). Consider now the W^*-algebra $Q = \mathcal{M} \bar{\otimes} \mathcal{F}_3$ and the weight $\tau \in W_{ns}(Q)$ defined as $\tau([x_{ij}]) = \varphi(x_{11}) + \psi(x_{22}) + \omega(x_{33}), ([x_{ij}] \in Q^+)$. Then by Proposition 3.3, we have

$$\sigma_t^\tau \begin{pmatrix} 0 & 0 & \mathbf{s}(\varphi) \\ 0 & 0 & 0 \\ 0 & 0 & 0 \end{pmatrix} = \begin{pmatrix} 0 & 0 & [D\varphi : D\omega]_t \\ 0 & 0 & 0 \\ 0 & 0 & 0 \end{pmatrix};$$

$$\sigma_t^\tau \begin{pmatrix} 0 & 0 & 0 \\ 0 & 0 & \mathbf{s}(\psi) \\ 0 & 0 & 0 \end{pmatrix} = \begin{pmatrix} 0 & 0 & 0 \\ 0 & 0 & [D\psi : D\omega]_t \\ 0 & 0 & 0 \end{pmatrix};$$

$$\sigma_t^\tau \begin{pmatrix} 0 & 0 & 0 \\ 0 & 0 & 0 \\ 0 & 0 & u \end{pmatrix} = \begin{pmatrix} 0 & 0 & 0 \\ 0 & 0 & 0 \\ 0 & 0 & \sigma_t^\omega(u) \end{pmatrix};$$

$$\sigma_t^\tau \begin{pmatrix} 0 & u & 0 \\ 0 & 0 & 0 \\ 0 & 0 & 0 \end{pmatrix} = \begin{pmatrix} \sigma_t^\theta \begin{pmatrix} 0 & u \\ 0 & 0 \end{pmatrix} & \begin{matrix} 0 & 0 \\ & 0 \end{matrix} \\ 0 & \begin{matrix} 0 & 0 \end{matrix} \end{pmatrix}$$

so that

$$\begin{aligned} &\begin{pmatrix} \sigma_t^\theta \begin{pmatrix} 0 & u \\ 0 & 0 \end{pmatrix} & \begin{matrix} 0 \\ 0 \end{matrix} \\ 0 & \begin{matrix} 0 & 0 \end{matrix} \end{pmatrix} \\ &= \sigma_t^\tau \left(\begin{pmatrix} 0 & 0 & \mathbf{s}(\varphi) \\ 0 & 0 & 0 \\ 0 & 0 & 0 \end{pmatrix} \begin{pmatrix} 0 & 0 & 0 \\ 0 & 0 & 0 \\ 0 & 0 & u \end{pmatrix} \begin{pmatrix} 0 & 0 & 0 \\ 0 & 0 & \mathbf{s}(\psi) \\ 0 & 0 & 0 \end{pmatrix} \right) \\ &= \begin{pmatrix} 0 & [D\varphi : D\omega]_t \sigma_t^\omega(u) [D\psi : D\omega]_t^* & 0 \\ 0 & 0 & 0 \\ 0 & 0 & 0 \end{pmatrix} \end{aligned}$$

hence $U \in \mathcal{P}^\theta$ if and only if $[D\psi : D\omega]_t = u^*[D\varphi : D\omega]_t \sigma_t^\omega(u)$ for every $t \in \mathbb{R}$.

If $a = [D\varphi : D\omega], b = (D\psi : D\omega] \in Z_{\sigma^\omega}(\mathbb{R}, \mathcal{M})$ and $c = \begin{pmatrix} a & 0 \\ 0 & b \end{pmatrix} \in Z_{\sigma^\omega \bar{\otimes} \iota}(\mathbb{R}, \mathcal{P})$, then we have $\mathcal{M}^a = \mathcal{M}^\varphi, \mathcal{M}^b = \mathcal{M}^\varphi$, and $\mathcal{P}^c = \mathcal{P}^\theta$, so that the equivalence (ii) ⇔ (iii) also follows from Proposition 20.2. □

By the previous proposition we have that for $\varphi, \psi \in W_{ns}(\mathcal{M})$ the following conditions are equivalent:

(i) there exists $u \in \mathcal{M}$ with $u^*u = \mathbf{s}(\psi), uu^* \in \mathcal{M}^\varphi$ and $\psi = \varphi_u$;

(ii) there exists $u \in \mathcal{M}$ with $u^*u = \mathbf{s}(\psi)$ and $\begin{pmatrix} 0 & u \\ 0 & 0 \end{pmatrix} \in \mathcal{P}^\theta$;

(iii) $[D\psi : D\omega] \lesssim [D\varphi : D\omega]$ in $Z_{\sigma^\omega}(\mathbb{R}; \mathcal{M})$.

If these conditions are satisfied, we write $\psi \lesssim \varphi$ and we say that the weight ψ is *dominated* by the weight φ. If in condition (i) we have also $uu^* = \mathbf{s}(\varphi)$ or, equivalently, in condition (iii) we have

$[D\psi : D\omega] \approx [D\varphi : D\omega]$, then we write $\psi \approx \varphi$ and say that the weights ψ and φ are *equivalent*. For the relations "$\lesssim$" and "$\approx$," we can thus apply all the results obtained in Section 20. In particular, if $\psi \lesssim \varphi$ and $\varphi \lesssim \psi$, then $\psi \approx \varphi$.

23.2 Proposition. *Let $\varphi \in W_{ns}(\mathcal{M})$ and $u, v \in \mathcal{M}$ be partial isometries such that $uu^*, vv^* \in \mathcal{M}^{\varphi}$. Then $\varphi_u \lesssim \varphi_v$ if and only if $uu^* \prec vv^*$ in $\mathcal{M}^{\varphi}$.*

Proof. Put $e = uu^*, f = vv^*$. We have $u \in \mathcal{M}^{\varphi_e}, uu^* = e = \mathbf{s}(\varphi_e), u^*u = \mathbf{s}(\varphi_u)$ and $\varphi_u = (\varphi_e)_u$, hence $\varphi_u \approx \varphi_e$ and, similarly, $\varphi_v \approx \varphi_f$. Thus, we have to prove that $\varphi_e \lesssim \varphi_f$ if and only if $e \prec f$ in $\mathcal{M}^{\varphi}$.

If $e \prec f$ in $\mathcal{M}^{\varphi}$, there exists $w \in \mathcal{M}^{\varphi}$ with $w^*w = e$ and $ww^* = f$; hence $\varphi_e = \varphi_w \lesssim \varphi_f$, by Proposition 2.21.

Conversely, if $\varphi_e \lesssim \varphi_f$, then there exists $w \in \mathcal{M}$ with $ww^* \in \mathcal{M}^{\varphi_f}, ww^* \leq f, w^*w = e$ and $\varphi_e = (\varphi_f)_w = \varphi_w$. By Proposition 2.21, this implies that $w \in \mathcal{M}^{\varphi}$, and so $e \prec f$ in $\mathcal{M}^{\varphi}$. □

23.3. Recall that a weight $\varphi \in W_{ns}(\mathcal{M})$ is said to be of infinite multiplicity if its centralizer $\mathcal{M}^{\varphi}$ is properly infinite. Since $\mathcal{M}^{\varphi} = \mathcal{M}^{[D\varphi:D\omega]}$, the weight φ is of infinite multiplicity if and only if the cocycle $[D\varphi : D\omega] \in Z_{\sigma^{\omega}}(\mathbb{R}; \mathcal{M})$ is of infinite multiplicity.

Proposition. *Let $\mathcal{M}$ be a W^*-algebra with separable predual and $\varphi \in W_{nsf}(\mathcal{M})$ a weight of infinite multiplicity. Then the following statements are equivalent:*

(i) *$\varphi \approx \lambda\varphi$ for every $\lambda > 0$;*
(ii) $\sigma^{\varphi} : \mathbb{R} \to Aut(\mathcal{M})$ is a dominant action;
(iii) $[D\varphi : D\omega] \in Z_{\sigma^{\omega}}(\mathbb{R}; U(\mathcal{M}))$ is a dominant cocycle.

Proof. (i) ⇔ (ii). We have $\varphi \approx \lambda\varphi$ for each $\lambda > 0$ if and only if for every $s \in \mathbb{R}$ there exists $u(s) \in U(\mathcal{M})$ with $e^s\varphi = \varphi_{u(s)}$, that is, (3.6) if and only if for every $s \in \mathbb{R}$ there exists $u(s) \in U(\mathcal{M})$ with $[D(e^s\varphi) : D\varphi]_t = [D(\varphi_{u(s)}) : D\varphi]_t$ $(t \in \mathbb{R})$, or (3.7) if and only if for every $s \in \mathbb{R}$ there exists $u(s) \in U(\mathcal{M})$ with $\sigma_t^{\varphi}(u(s)) = e^{ist}u(s)$ $(t \in \mathbb{R})$. By Proposition 20.12, this last condition means that the action $\sigma^{\varphi} : \mathbb{R} \to Aut(\mathcal{M})$ is dominant.

(ii) ⇔ (iii). The action σ^{φ} is dominant if and only if the trivial cocycle $1 \in Z_{\sigma^{\varphi}}(\mathbb{R}; U(\mathcal{M}))$ is dominant (20.11) and, since $\sigma^{\varphi} = {}_{[D\varphi:D\omega]}\sigma^{\omega}$, this is equivalent to the fact that the cocycle $[D\varphi : D\omega] \in Z_{\sigma^{\omega}}(\mathbb{R}; U(\mathcal{M}))$ is dominant (see the proof of Theorem 20.5). □

If the predual $\mathcal{M}_*$ is not assumed to be separable, then the previous equivalence still holds if we replace condition (i) by the condition that there exists an s-continuous function $\lambda \mapsto u(\lambda) \in U(\mathcal{M})$ such that $\lambda\varphi = \varphi_{u(\lambda)}$.

A weight $\varphi \in W_{nsf}(\mathcal{M})$ of infinite multiplicity is called a *dominant weight* if $\sigma^{\varphi} : \mathbb{R} \to Aut(\mathcal{M})$ is a dominant action. We shall see later (23.16) that if $\mathcal{M}$ is a type III W^*-algebra with separable predual, then any weight $\varphi \in W_{nsf}(\mathcal{M})$ such that $\varphi \approx \lambda\varphi$ for every $\lambda > 0$ is automatically of infinite multiplicity and hence dominant.

23.4. A weight $\varphi \in W_{ns}(\mathcal{M})$ is called *integrable* if the action $\sigma^{\varphi} : \mathbb{R} \to Aut(\mathcal{M}_{s(\varphi)})$ is integrable, that is, (20.6) if the cocycle $[D\varphi : D\omega_0] \in Z_{\sigma^{\omega_0}}(\mathbb{R}; \mathcal{M})$ is square integrable, where ω_0 is any fixed n.s.f. weight on $\mathcal{M}$. By Theorems 20.5 and 20.6, we have the following result:

Theorem. *Let $\mathcal{M}$ be a properly infinite W^*-algebra. Then there exists a dominant weight $\varphi \in W_{nsf}(\mathcal{M})$. If, moreover, $\mathcal{M}$ is countably decomposable, then all dominant weights on $\mathcal{M}$ are equivalent and a weight $\psi \in W_{ns}(\mathcal{M})$ is integrable if and only if $\psi \lesssim \varphi$.*

There exists also a simpler and more natural proof of the existence of a dominant weight, which we present in the following.

By Stone's theorem, there exists a unique nonsingular positive self-adjoint operator A on the Hilbert space $\mathcal{L}^2(\mathbb{R})$ such that $A^{it} = \boldsymbol{\rho}(t)$ $(t \in \mathbb{R})$, where we denote by $\boldsymbol{\rho}$ the regular representation of $\mathbb{R}$ on $\mathcal{L}^2(\mathbb{R})$. Then the n.s.f. weight $\omega = tr_A$ on $\mathcal{F} = \mathcal{B}(\mathcal{L}^2(\mathbb{R}))$ has the following property:

$$\lambda\omega \approx \omega \quad (\lambda > 0).$$

Indeed, let $\mathbf{m}(s) \in \mathcal{B}(\mathcal{L}^2(\mathbb{R}))$ be the unitary operator defined by $[\mathbf{m}(s)\xi](t) = e^{-ist}\xi(t)$ $(\xi \in \mathcal{L}^2(\mathbb{R}), s, t \in \mathbb{R})$. We have $[D\omega_{\mathbf{m}(s)} : D\,tr]_t = [D\omega_{\mathbf{m}(s)} : D\omega]_t[D\omega : D\,tr]_t = \mathbf{m}(s)^*\sigma_t^\omega(\mathbf{m}(s))\boldsymbol{\rho}(t) = \mathbf{m}(-s)\boldsymbol{\rho}(t)\mathbf{m}(s)\boldsymbol{\rho}(-t)\boldsymbol{\rho}(t) = e^{-ist}\boldsymbol{\rho}(t) = [D(e^{-s}\omega) : D\omega]_t[D\omega : D\,tr]_t = [D(e^{-s}\omega) : D\,tr]_t$, hence $e^{-s}\omega = \omega_{\mathbf{m}(s)} \approx \omega$, for all $s \in \mathbb{R}$.

Consider now any n.s.f. weight φ on $\mathcal{M}$ of infinite multiplicity. Then $\varphi \bar{\otimes} \omega$ is a dominant weight on $\mathcal{M} \bar{\otimes} \mathcal{F}$ since $\mathcal{M}^{\varphi \bar{\otimes} \omega} \supset \mathcal{M}^\varphi \bar{\otimes} 1$ is properly infinite and $\lambda(\varphi \bar{\otimes} \omega) = \varphi \bar{\otimes} (\lambda\omega) \approx \varphi \bar{\otimes} \omega$ for every $\lambda > 0$. Since $\mathcal{M}$ is properly infinite, we have $\mathcal{M} \approx \mathcal{M} \bar{\otimes} \mathcal{F}$ and hence we can say that $\varphi \bar{\otimes} \omega$ is a dominant weight on $\mathcal{M}$.

If $\mathcal{M}$ is countably decomposable, then the uniqueness modulo equivalence of the dominant weight shows that for any two n.s.f. weights φ, ψ on $\mathcal{M}$ there exists a unitary operator $U \in U(\mathcal{M} \bar{\otimes} \mathcal{F})$ such that $\psi \bar{\otimes} \omega = (\varphi \bar{\otimes} \omega) \circ \mathrm{Ad}(U)$.

Also, since $\sigma^{\varphi \bar{\otimes} \omega} = \sigma^\varphi \bar{\otimes} \sigma^\omega = \sigma^\varphi \bar{\otimes} \mathrm{Ad}(\boldsymbol{\rho})$, it follows that the centralizer of the dominant weight $\varphi \bar{\otimes} \omega$ on $\mathcal{M} \bar{\otimes} \mathcal{B}(\mathcal{L}^2(\mathbb{R}))$ is equal to the crossed product $\mathcal{R}(\mathcal{M}, \sigma^\varphi)$ (see Corollary 19.13). Therefore, for any dominant weight φ on $\mathcal{M}$, the centralizer $\mathcal{M}^\varphi$ is *-isomorphic to the crossed product $\mathcal{R}(\mathcal{M}, \sigma^\varphi)$ (see also 20.5.(3)).

On the other hand, consider also the unique n.s.f. weight ω' on $\mathcal{F} = \mathcal{B}(\mathcal{L}^2(\mathbb{R}))$ such that $[D\omega' : D\,tr]_s = \mathbf{m}(s)$ $(s \in \mathbb{R})$. Since $\mathbf{m}(s)\boldsymbol{\rho}(t) = e^{ist}\boldsymbol{\rho}(t)\mathbf{m}(s)$, it follows that the modular automorphism groups $\sigma_t^\omega = \mathrm{Ad}(\boldsymbol{\rho}(t))$ and $\sigma_s^{\omega'} = \mathrm{Ad}(\mathbf{m}(s))$ commute, hence $\sigma_t^{\varphi \bar{\otimes} \omega}$ and $\sigma_t^{\varphi \bar{\otimes} \omega'}$ also commute. However, the weights $\varphi \bar{\otimes} \omega$ and $\varphi \bar{\otimes} \omega'$ do not commute. Indeed, if they did commute, then $[D(\varphi \bar{\otimes} \omega') : D(\varphi \bar{\otimes} \omega)] = 1 \bar{\otimes} [D\omega' : D\omega]$ would be a one-parameter group of unitary operators (see Theorem 4.10); however $[D\omega' : D\omega]_t = [D\omega' : D\,tr]_t[D\,tr : D\omega]_t = \mathbf{m}(t)\boldsymbol{\rho}(-t)$ is not a one-parameter group of unitary operators (see also 4.15).

It follows that if φ is a dominant weight on the countably decomposable W^*-algebra $\mathcal{M}$, then there exists an n.s.f. weight ψ on $\mathcal{M}$ which *anticommutes* with φ, that is, ψ does not commute with φ, but the corresponding modular automorphism groups commute, $\sigma_t^\varphi\sigma_s^\psi = \sigma_s^\psi\sigma_t^\varphi$ $(s, t \in \mathbb{R})$. We shall see later (23.16, 23.17) that this property characterizes the dominant weights on countably decomposable type III factors.

23.5. Consider again the arguments given in the last part of Section 12.4. Let φ be any n.s.f. weight on the W^*-algebra $\mathcal{M}$. On the crossed product $\mathcal{R}(\mathcal{M}, \sigma^\varphi)$, we have a dual weight $\hat{\varphi}$ (19.8) and a dual action $\alpha = (\sigma^\varphi)^\wedge$ (19.3). Namely, the dual action α is defined by $\alpha_s = \mathrm{Ad}(1 \bar{\otimes} \mathbf{m}(s))|\mathcal{R}(\mathcal{M}, \sigma^\varphi)$ $(s \in \mathbb{R})$, and the modular automorphism group associated with the dual weight $\hat{\varphi}$ is characterized by $\sigma_t^{\hat{\varphi}}(\pi_{\sigma^\varphi}(x)) = \pi_{\sigma^\varphi}(\sigma_t^\varphi(x)) = [\mathrm{Ad}(1 \bar{\otimes} \boldsymbol{\lambda}(t))](\pi_{\sigma^\varphi}(x))$ $(x \in \mathcal{M})$, and $\sigma_t^{\hat{\varphi}}(1 \bar{\otimes} \boldsymbol{\lambda}(r)) = 1 \bar{\otimes} \boldsymbol{\lambda}(r) = [\mathrm{Ad}(1 \bar{\otimes} \boldsymbol{\lambda}(t))](1 \bar{\otimes} \boldsymbol{\lambda}(r))$ $(r \in \mathbb{R})$, so that $\sigma_t^{\hat{\varphi}} = \mathrm{Ad}(1 \bar{\otimes} \boldsymbol{\lambda}(t)|\mathcal{R}(\mathcal{M}, \sigma^\varphi)$ $(t \in \mathbb{R})$. Thus, if A is the unique nonsingular positive self-adjoint operator in $\mathcal{L}^2(\mathbb{R})$ such that $\boldsymbol{\lambda}(t) = A^{-it}$ $(t \in \mathbb{R})$, then $\mu = \hat{\varphi}_{1 \bar{\otimes} A}$ is an n.s.f. trace on $\mathcal{R}(\mathcal{M}, \sigma^\varphi)$ (since $\sigma^\mu = \iota$) and we have $[D\hat{\varphi} : D\mu]_t = 1 \bar{\otimes} \boldsymbol{\lambda}(t)$ $(t \in \mathbb{R})$.

Since $\alpha_s((1 \bar{\otimes} A)^{it}) = \alpha_s(1 \bar{\otimes} \boldsymbol{\lambda}(-t)) = 1 \bar{\otimes} \mathbf{m}(s)\boldsymbol{\lambda}(-t)\mathbf{m}(-s) = e^{ist}(1 \bar{\otimes} \boldsymbol{\lambda}(-t)) = e^{ist}(1 \bar{\otimes} A)^{it}$ $(s, t \in \mathbb{R})$, it follows that $\alpha_s(1 \bar{\otimes} A) = e^s(1 \bar{\otimes} A)$. Since the dual weight $\hat{\varphi}$ is invariant with respect to the dual action α (19.8), we infer that

$$\mu \circ \alpha_s = e^{-s}\mu \ (s \in \mathbb{R}). \tag{1}$$

Thus, *the crossed product $\mathscr{R}(\mathscr{M}, \sigma^\varphi)$ is a semifinite W^*-algebra and there exists an n.s.f. trace μ on $\mathscr{R}(\mathscr{M}, \sigma^\varphi)$ which satisfies (1), where $\alpha = (\sigma^\varphi)^\wedge$ is the dual action.*

23.6 Theorem (Connes & Takesaki). *Let $\mathscr{M}$ be a properly infinite W^*-algebra and φ a dominant weight on $\mathscr{M}$. Then the centralizer $\mathscr{M}^\varphi$ is a semifinite W^*-algebra and there exist a continuous action $\theta : \mathbb{R} \mapsto Aut(\mathscr{M}^\varphi)$ and an n.s.f trace τ on $\mathscr{M}^\varphi$ such that $(\mathscr{M}, \sigma^\varphi) \approx (\mathscr{R}(\mathscr{M}^\varphi, \theta), \hat{\theta})$ and $\tau \circ \theta_s = e^{-s}\tau$ $(s \in \mathbb{R})$.*

Proof. Since φ is a dominant weight, we infer by Proposition 20.12 that there exists a continuous action $\theta'' : \mathbb{R} \to Aut(\mathscr{M}^\varphi)$ such that

$$(\mathscr{M}, \sigma^\varphi) \approx (\mathscr{R}(\mathscr{M}^\varphi, \theta''), \hat{\theta}'') \tag{1}$$

By the Takesaki duality theorem (19.5), we have

$$(\mathscr{R}(\mathscr{M}^\varphi, \sigma^\varphi), (\sigma^\varphi)^\wedge) \approx (\mathscr{M}^\varphi \bar{\otimes} \mathscr{B}(\mathscr{L}^2(\mathbb{R})), \theta'' \bar{\otimes} \mathrm{Ad}(\rho)). \tag{2}$$

Since $\mathscr{M}^\varphi$ is properly infinite, by Corollary 9.16 and the result in Section 20.14 we deduce that there exists a continuous action $\theta' : \mathbb{R} \to Aut(\mathscr{M}^\varphi), \theta' \sim \theta''$, whose centralizer $(\mathscr{M}^\varphi)^{\theta'}$ is properly infinite. Then, using Corollary 9.16 again we get $(\mathscr{M}^\varphi, \theta') \approx (\mathscr{M}^\varphi \bar{\otimes} \mathscr{B}(\mathscr{L}^2(\mathbb{R})), \theta' \bar{\otimes} \iota)$. Since $\theta' \bar{\otimes} \iota \sim \theta'' \bar{\otimes} \iota \sim \theta'' \bar{\otimes} \mathrm{Ad}(\boldsymbol{\rho})$, it follows that there exists a continuous action $\theta : \mathbb{R} \to Aut(\mathscr{M}^\varphi), \theta \sim \theta' \sim \theta''$, such that

$$(\mathscr{M}^\varphi, \theta) \approx (\mathscr{M}^\varphi \bar{\otimes} \mathscr{B}(\mathscr{L}^2(\mathbb{R})), \theta'' \bar{\otimes} \mathrm{Ad}(\boldsymbol{\rho})). \tag{3}$$

From (2) and (3), it follows that

$$(\mathscr{R}(\mathscr{M}, \sigma^\varphi), (\sigma^\varphi)^\wedge) \approx (\mathscr{M}^\varphi, \theta). \tag{4}$$

and, since $\theta \sim \theta''$, we infer from (1) and 20.14 that

$$(\mathscr{M}, \sigma^\varphi) \approx (\mathscr{R}(\mathscr{M}^\varphi, \theta), \hat{\theta}). \tag{5}$$

Finally, by (4) and 23.5, there exists an n.s.f. trace τ on $\mathscr{M}^\varphi$ such that

$$\tau \circ \theta_s = e^{-s}\tau \quad (s \in \mathbb{R}). \tag{6}$$

□

23.7. Any triple $(\mathscr{N}, \theta, \tau)$ consisting of a properly infinite semifinite W^*-algebra $\mathscr{N}$, a continuous action $\theta : \mathbb{R} \to Aut(\mathscr{N})$ and an n.s.f. trace τ on $\mathscr{N}$ such that $\tau \circ \theta_s = e^{-s}\tau$ $(s \in \mathbb{R})$, is called a *continuous decomposition.*

By Theorem 23.6, every properly infinite W^*-algebra $\mathcal{M}$ has a continuous decomposition $(\mathcal{N},\theta,\tau)$ such that $\mathcal{M}=\mathcal{R}(\mathcal{N},\theta)$; in this case, $(\mathcal{N},\theta,\tau)$ will be called a *continuous decomposition of* $\mathcal{M}$.

Proposition. *Let $(\mathcal{N},\theta,\tau)$ be any continuous decomposition and let $\mathcal{M}=\mathcal{R}(\mathcal{N},\theta)$ with the dual action $\hat{\theta}:\mathbb{R}\to Aut(\mathcal{M})$ and the dual weight $\varphi=\hat{\tau}\in W_{nsf}(\mathcal{M})$. Then φ is a dominant weight on $\mathcal{M}$, $\sigma^{\varphi}=\hat{\theta}$ and $\mathcal{M}^{\varphi}=\pi_{\theta}(\mathcal{N})$.*

Proof. Indeed, we have $\sigma_t^{\varphi}(\pi_\theta(x))=\pi_\theta(\sigma_t^{\tau}(x))=\pi_\theta(x)$ $(x\in\mathcal{N})$, and $\sigma_t^{\varphi}(1\bar{\otimes}\boldsymbol{\lambda}(s))=(1\bar{\otimes}\boldsymbol{\lambda}(s))\pi_\theta([D(\tau\circ\theta_s):D\tau]_t)=e^{-\mathrm{i}st}(1\bar{\otimes}\boldsymbol{\lambda}(s))$ $(s\in\mathbb{R})$, hence $\sigma_t^{\varphi}=\mathrm{Ad}(1\bar{\otimes}\mathbf{m}(t))=\hat{\theta}_t$ $(t\in\mathbb{R})$. Then $\mathcal{M}^{\varphi}=\mathcal{M}^{\hat{\theta}}=\pi_\theta(\mathcal{N})$ (see 19.3.(9)) and the fact that φ is dominant now follows by using Proposition 20.12. □

Note also that if $(\mathcal{N},\theta,\tau)$ is a continuous decomposition, $\mathcal{M}=\mathcal{R}(\mathcal{N},\theta)$ and $\varphi=\hat{\tau}$, then $\mathbb{R}\ni s\mapsto u(s)=1\bar{\otimes}\boldsymbol{\lambda}(s)\in\mathcal{M}$ is an s-continuous unitary representation and, identifying $\mathcal{N}\equiv\pi_\theta(\mathcal{N})\subset\mathcal{M}$, we have $\mathcal{N}=\mathcal{M}^{\varphi}$, $\mathcal{M}=\mathcal{R}\{\mathcal{N},u(\mathbb{R})\}$, $\theta_s=\mathrm{Ad}(u(s))|\mathcal{N}$ $(s\in\mathbb{R})$; also $\sigma^{\varphi}=\hat{\theta}$ is determined by $\sigma_t^{\varphi}|\mathcal{N}=\iota$ and $\sigma_t^{\varphi}(u(s))=e^{-\mathrm{i}st}u(s)$ $(s,t\in\mathbb{R})$, hence $\varphi_{u(s)}=e^{-s}\varphi$ $(s\in\mathbb{R})$.

23.8. With regard to the uniqueness of the continuous decomposition of a W^*-algebra, we first prove the following result:

Proposition. *Let $(\mathcal{N}_1,\theta_1,\tau_1)$ and $(\mathcal{N}_2,\theta_2,\tau_2)$ be two continuous decompositions with $\mathcal{N}_1,\mathcal{N}_2$ countably decomposable W^*-algebras. Then $\mathcal{R}(\mathcal{N}_1,\theta_1)\approx\mathcal{R}(\mathcal{N}_2,\theta_2)$ if and only if $(\mathcal{N}_1,\theta_1)\sim(\mathcal{N}_2,\theta_2)$.*

Proof. The "if" part is clear by statement 20.13.(4). Conversely, assume that $\mathcal{R}(\mathcal{N}_1,\theta_1)=\mathcal{R}(\mathcal{N}_2,\theta_2)=\mathcal{M}$. By assumption, $\mathcal{M}$ is countably decomposable, hence the dominant weights (23.7) $\varphi_1=\hat{\tau}_1,\varphi_2=\hat{\tau}_2$ on $\mathcal{M}$ are equivalent, that is, $(\mathcal{R}(\mathcal{N}_1,\theta_1),\varphi_1)\approx(\mathcal{R}(\mathcal{N}_2,\theta_2),\varphi_2)$. Since $\sigma^{\varphi_1}=\hat{\theta}_1,\sigma^{\varphi_2}=\hat{\theta}_2$ (23.7), we infer that $(\mathcal{R}(\mathcal{N}_1,\theta_1),\hat{\theta}_1)\approx(\mathcal{R}(\mathcal{N}_2,\theta_2),\hat{\theta}_2)$. Using the Takesaki duality theorem, we also obtain $(\mathcal{N}_1\bar{\otimes}\mathcal{B}(\mathcal{L}^2(\mathbb{R})),\theta_1\bar{\otimes}\mathrm{Ad}(\boldsymbol{\rho}))\approx(\mathcal{N}_2\bar{\otimes}\mathcal{B}(\mathcal{L}^2(\mathbb{R})),\theta_2\bar{\otimes}\mathrm{Ad}(\boldsymbol{\rho}))$. On the other hand, for each $j=1,2$, there exists a continuous action $\theta_j':\mathbb{R}\to Aut(\mathcal{N}_j)$, $\theta_j'\sim\theta_j$, such that $(\mathcal{N}_j\bar{\otimes}\mathcal{B}(\mathcal{L}^2(\mathbb{R})),\theta_j\bar{\otimes}\mathrm{Ad}(\boldsymbol{\rho}))\approx(\mathcal{N}_j,\theta_j')$. It follows that $(\mathcal{N}_1,\theta_1')\approx(\mathcal{N}_2,\theta_2')$ and hence $(\mathcal{N}_1,\theta_1)\sim(\mathcal{N}_2,\theta_2)$. □

Actually, we shall see (23.12) that $\mathcal{R}(\mathcal{N}_1,\theta_1)\approx\mathcal{R}(\mathcal{N}_2,\theta_2)$ if and only if $(\mathcal{N}_1,\theta_1)\approx(\mathcal{N}_2,\theta_2)$. So far as the determination of the trace τ in a continuous decomposition $(\mathcal{N},\theta,\tau)$ is concerned we recall (4.11) that any other n.s.f. trace on $\mathcal{N}$ is of the form τ_A with A a nonsingular positive self-adjoint operator affiliated to $\mathcal{Z}(\mathcal{N})$; it is easy to check that τ_A satisfies the same condition as τ with respect to $(\mathcal{N},\theta)$ if and only if A is affiliated to $\mathcal{Z}(\mathcal{N})^{\theta}$.

23.9. In this section, we give some auxiliary results, which, in particular, will enable us to complete the previous uniqueness result.

Proposition. *Let $\mathcal{N}$ be a semifinite W^*-algebra, $\theta\in Aut(\mathcal{N})$ and τ an n.s.f. trace on $\mathcal{N}$ such that $\tau\circ\theta\le\lambda\tau$ for some λ such that $0<\lambda<1$. Then the action $\mathbb{Z}\ni n\mapsto\theta^n\in Aut(\mathcal{N})$ is integrable, there exists an s-continuous unitary representation $u:\mathbb{T}\mapsto U(\mathcal{N})$ such that $\theta^n(u(\omega))=\omega^n u(\omega)$ $(\omega\in\mathbb{T},n\in\mathbb{Z})$, and there exists a projection $e\in\mathcal{N}$ such that $\sum_{n\in\mathbb{Z}}\theta^n(e)=1$.*

Proof. The family $\{f\in Proj(\mathcal{N});\theta^n(f)\ (n\in\mathbb{Z})$ are mutually orthogonal$\}$ is inductively ordered and hence, by Zorn's lemma, it has a maximal element which we denote by $e\in\mathcal{N}$. Let $e_0=\sum_{n\in\mathbb{Z}}\theta^n(e)$. If $f_0=1-e_0\neq0$, there exists a projection $p\in\mathcal{N},\theta\neq p\le f_0$, with $\tau(p)<+\infty$.

Putting $q = \bigvee_{n=0}^{\infty} \theta^n(p)$ we have $\tau(q) \leq \sum_{n=0}^{\infty} \tau(\theta^n(p)) \leq \sum_{n=0}^{\infty} \lambda^n \tau(p) = (1-\lambda)^{-1}\tau(p) < +\infty$ and $\theta(q) \leq q, \tau(\theta(q)) \leq \lambda\tau(q) < \tau(q)$; hence $f = q - \theta(q) \neq 0$. However the existence of the projection $e+f$ contradicts the maximality of e, since $(e+f)(\theta^n(e)+\theta^n(f)) = e\theta^n(e)+f\theta^n(e)+e\theta^n(f)+f\theta^n(f) = 0$; indeed, we have $e\theta^n(e) = 0$ by the choice of e, then $f\theta^n(e) = 0 = e\theta^n(f)$ since $f \leq 1 - e_0$ and finally $f\theta^n(f) = 0$ as $f = q - \theta(q)$. Hence $e_0 = 1$.

We have $\sum_{n\in\mathbb{Z}} \theta^n(a) \in \mathcal{N}^+$ for any finite partial sum a of the series $\sum_{n\in\mathbb{Z}} \theta^n(e)$. Since these finite partial sums converge to $e_0 = 1$ with respect to the s-topology, it follows that the action $n \mapsto \theta^n$ is indeed integrable.

Recall that $\mathbb{T} = \{\omega \in \mathbb{C}; |\omega| = 1\}$ is the dual group of $\mathbb{Z}$. By defining $u(\omega) = \sum_{n\in\mathbb{Z}} \omega^{-m}\theta^m(e) \in U(\mathcal{N})$ $(\omega \in \mathbb{T})$, we obtain an s-continuous unitary representation $u : \mathbb{T} \to U(\mathcal{N})$ and it is easy to check that $\theta^n(u(\omega)) = \omega^n u(\omega)$ $(\omega \in \mathbb{T}, n \in \mathbb{Z})$. □

Using Landstad's theorem (19.9) and Proposition 20.12, we deduce

Corollary. *Let $\mathcal{N}$ be a semifinite W^*-algebra, $\theta \in Aut(\mathcal{N})$ and τ an n.s.f. trace on $\mathcal{N}$ such that $\tau \circ \theta \leq \lambda t$ for some λ such that $0 < \lambda < 1$. Then there exists a continuous action $u : \mathbb{T} \to Aut(\mathcal{N}^\theta)$ such that $(\mathcal{N}, \theta) \approx (\mathcal{R}(\mathcal{N}^\theta, \sigma), \hat{\sigma})$.*

In particular, if $\mathcal{N}^\theta$ is properly infinite, then $\theta : \mathbb{Z} \ni n \mapsto \theta^n \in Aut(\mathcal{N})$ is a dominant action.

23.10. For continuous decompositions, we obtain the following similar result:

Proposition. *Let $\mathcal{N}$ be a semifinite W^*-algebra, $\theta : \mathbb{R} \to Aut(\mathcal{N})$ a continuous action and τ an n.s.f. trace on $\mathcal{N}$ such that $\tau \circ \theta_s = e^{-s}\tau$ $(s \in \mathbb{R})$. Then θ is an integrable action, there exists a θ-invariant n.s.f. weight φ on $\mathcal{N}$ and there exists an s-continuous unitary representation $v : \mathbb{R} \to U(\mathcal{N})$ such that $\theta_s(v(t)) = e^{ist}v(t)$ $(s, t \in \mathbb{R})$.*

Proof. The *-automorphism $\theta_1 \in Aut(\mathcal{N})$ satisfies $\tau \circ \theta_1 = e^{-1}\tau$. By Proposition 23.9, it follows that $n \mapsto \theta_1^n = \theta_n$ is an integrable action of $\mathbb{Z}$ on $\mathcal{N}$. If $a \in \mathcal{N}^+$ is θ_1-integrable, that is, $\sum_{n\in\mathbb{Z}} \theta_n(a) \in \mathcal{N}^+$, then

$$\int_{-\infty}^{+\infty} \theta_t(a)\, \mathrm{d}t = \sum_n \int_n^{n+1} \theta_t(a)\, \mathrm{d}t = \sum_n \theta_n \left(\int_0^1 \theta_t(a)\, \mathrm{d}t \right)$$
$$= \int_0^1 \theta_t \left(\sum_n \theta_n(a) \right) \mathrm{d}t \in \mathcal{N}^+.$$

Hence θ is an integrable action.

Thus, there exists a θ-invariant n.s.f. operator-valued weight $P_\theta : \mathcal{N}^+ \to \overline{(\mathcal{N}^\theta)}^+$ (see 18.19, 20.6). If ψ is any n.s.f. weight on $\mathcal{N}^\theta$, then $\varphi = \psi \circ P_\theta$ is a θ-invariant n.s.f. weight on $\mathcal{N}$.

Let A be the unique nonsingular positive self-adjoint operator affiliated to $\mathcal{N}$ such that $\varphi = \tau_A$. Since $\tau \circ \theta_s = e^{-s}\tau$ and $\varphi \circ \theta_s = \varphi$, it follows that $\theta_s(A) = e^{-s}A$ $(s \in \mathbb{R})$. Then $\mathbb{R} \ni t \mapsto v(t) = A^{-it} \in \mathcal{N}$ is an s-continuous unitary representation and $\theta_s(v(t)) = e^{ist}v(t)$ $(s, t \in \mathbb{R})$. □

Using Landstad's theorem (19.9) and Proposition 20.12, we deduce

Corollary. *Let $\mathcal{N}$ be a semifinite W^*-algebra, $\theta : \mathbb{R} \to Aut(\mathcal{N})$ a continuous action and τ an n.s.f. trace on $\mathcal{M}$ such that $\tau \circ \theta_s = e^{-s}\tau$ $(s \in \mathbb{R})$. Then there exists a continuous action $\sigma : \mathbb{R} \to Aut(\mathcal{N}^\theta)$ such that $(\mathcal{N}, \theta) \approx (\mathcal{R}(\mathcal{N}^\theta, \sigma), \hat{\sigma})$.*

In particular, if $\mathcal{N}^\theta$ is properly infinite, then $\theta : \mathbb{R} \to Aut(\mathcal{N})$ is a dominant action.

23.11. Let $(\mathcal{N}, \theta, \tau)$ be a continuous decomposition. In this section, we study the type of the crossed product W^*-algebra $\mathcal{R}(\mathcal{N}, \theta)$. We shall use the notation, conventions and results of Section 23.7.

Proposition 1. *We have $\mathcal{Z}(\mathcal{R}(\mathcal{N}, \theta)) = \mathcal{Z}(\mathcal{N})^\theta$. In particular, $\mathcal{R}(\mathcal{N}, \theta)$ is a factor if and only if θ acts ergodically on $\mathcal{Z}(\mathcal{N})$.*

Proof. The inclusion $\mathcal{Z}(\mathcal{N})^\theta \subset \mathcal{Z}(\mathcal{R}(\mathcal{N}, \theta))$ is obvious (21.6.(2)). Conversely, if $z \in \mathcal{Z}(\mathcal{M})$, then $z \in \mathcal{M}^\varphi = \mathcal{N}, z$ commutes with each element of $\mathcal{N}$, that is, $z \in \mathcal{Z}(\mathcal{N})$, and z commutes with $u(s)$ $(s \in \mathbb{R})$, hence $z \in \mathcal{Z}(\mathcal{N})^\theta$. □

Proposition 2. *The following statements are equivalent:*

(i) *$\mathcal{R}(\mathcal{N}, \theta)$ is semifinite;*
(ii) *$\mathcal{N}^\theta$ is semifinite;*
(iii) *$(\mathcal{N}, \theta) \approx (\mathcal{N}^\theta \bar{\otimes} \mathcal{L}^\infty(\mathbb{R}), \iota \bar{\otimes} \mathrm{Ad}(\boldsymbol{\lambda}))$;*
(iv) *there exists an s-continuous unitary representation $\mathbb{R} \ni t \mapsto v(t) \in \mathcal{Z}(\mathcal{N})$ such that $\theta_s(v(t)) = e^{ist}v(t)$ $(s, t \in \mathbb{R})$.*

Proof. (i) ⇔ (ii). Indeed, using Corollary 23.10 and the Takesaki duality theorem (19.5), we get $\mathcal{R}(\mathcal{N}, \theta) \approx \mathcal{N}^\theta \bar{\otimes} \mathcal{B}(\mathcal{L}^2(\mathbb{R}))$.

(iii) ⇔ (iv). This follows immediately using Landstad's theorem (19.9).

(i) ⇒ (iv). If $\mathcal{M}$ is semifinite, then the modular automorphism group σ^φ is inner, that is, there exists an s-continuous unitary representation $\mathbb{R} \ni t \mapsto v(t) \in \mathcal{M}$, such that $\sigma_t^\varphi = \mathrm{Ad}(v(t))$ $(t \in \mathbb{R})$. Then $v(t) \in \mathcal{M}^\varphi = \mathcal{N}$ and for $x \in \mathcal{N} = \mathcal{M}^\varphi$ we have $v(t)xv(t)^* = \sigma_t^\varphi(x) = x$, hence $v(t) \in \mathcal{Z}(\mathcal{N})$ $(t \in \mathbb{R})$. Then $e^{-ist}u(s) = \sigma_t^\varphi(u(s)) = v(t)u(s)v(t)^*$, hence $\theta_s(v(t)) = u(s)v(t)u(s)^* = e^{ist}v(t)$ $(s, t \in \mathbb{R})$.

(iv) ⇒ (i). If condition (iv) is satisfied, then for $x \in \mathcal{N} = \mathcal{M}^\varphi$ we have $\sigma_t^\varphi(x) = x = v(t)xv(t)^*$, and for $s \in \mathbb{R}$ we have $\sigma_t^\varphi(u(s)) = e^{-ist}u(s) = v(t)u(s)v(t)^*$, hence $\sigma_t^\varphi = \mathrm{Ad}(v(t))$ $(t \in \mathbb{R})$, since $\mathcal{M} = \mathcal{R}\{\mathcal{N}, u(\mathbb{R})\}$. Thus, σ^φ is inner and $\mathcal{M}$ is semifinite. Actually, it is also clear that (iii) ⇒ (i). □

Since for every projection $p \in \mathcal{Z}(\mathcal{M}) = \mathcal{Z}(\mathcal{N})^\theta$, the triple $(\mathcal{N}p, \theta|\mathcal{N}p, \tau|\mathcal{N}p)$ is also a continuous decomposition and $\mathcal{M}p = \mathcal{R}(\mathcal{N}p, \theta|\mathcal{N}p)$, we get the following

Corollary. *The following statements are equivalent:*

(i) *$\mathcal{R}(\mathcal{N}, \theta)$ is of type III;*
(ii) *for every nonzero projection $p \in \mathcal{Z}(\mathcal{N})^\theta$, $\mathcal{N}^\theta p$ is not semifinite;*
(iii) *for every nonzero projection $p \in \mathcal{Z}(\mathcal{N})^\theta$,*

$$(\mathcal{N}p, \theta|\mathcal{N}p) \not\approx (\mathcal{N}^\theta p \bar{\otimes} \mathcal{L}^\infty(\mathbb{R}), \iota \bar{\otimes} \mathrm{Ad}(\boldsymbol{\lambda}));$$

(iv) *for every nonzero projection $p \in \mathcal{Z}(\mathcal{N})^\theta$, there is no s-continuous unitary representation $v : \mathbb{R} \to \mathcal{Z}(\mathcal{N})p$ such that $\theta_s(v(t)) = e^{ist}v(t)$ $(s, t \in \mathbb{R})$.*

Proposition 3. *If $\mathcal{R}(\mathcal{N}, \theta)$ is of type* III, *then $\mathcal{N}$ is of type* II_∞.

Proof. Assume that $\mathcal{R}(\mathcal{N}, \theta)$ is of type III, but $\mathcal{N}$ is not of type II_∞. In accordance with Proposition 1, we can then assume that $\mathcal{N}$ is homogeneous of type I, hence $\mathcal{N} = \mathcal{Z} \bar{\otimes} F$, where $\mathcal{Z} \approx \mathcal{Z}(\mathcal{N})$ and $\mathcal{F}$ is an infinite dimensional factor of type I.

Any n.s.f. trace on $\mathcal{N}$ is then of the form $(\mu \bar{\otimes} tr)_A$ with μ an n.s.f. trace on $\mathcal{Z}$, tr the canonical trace on $\mathcal{F}$ and A a nonsingular positive self-adjoint operator affiliated to $\mathcal{Z}(\mathcal{N}) = \mathcal{Z} \bar{\otimes} 1_F$. Hence $\tau = \mu \bar{\otimes} tr$ for some n.s.f. trace μ on $\mathcal{Z}$.

We identify $\mathcal{Z}$ with $\mathcal{Z}(\mathcal{N}) = \mathcal{Z} \bar{\otimes} 1_{\mathcal{A}}$ and consider $\alpha_\tau = (\theta_t|\mathcal{Z}) \bar{\otimes} \iota \in Aut(\mathcal{N})$ $(t \in \mathbb{R})$. Then $\alpha_{-t} \circ \theta_t \in Aut(\mathcal{N})$ acts identically on $\mathcal{Z}(\mathcal{N})$ and hence ([L], 8.11) is inner, that is, there exists $w(t) \in U(\mathcal{N})$ such that $\alpha_{-t}(\theta_t(x)) = w(t)xw(t)^*$ for every $x \in \mathcal{N}$. Consequently, we have

$$\theta_t(x) = \alpha_t(w(t))\alpha_t(x)\alpha_t(w(t))^* \quad (x \in \mathcal{N}, t \in \mathbb{R}).$$

Consider now $a \in \mathcal{F} \subset \mathcal{N}, a \geq 0$, with $tr(a) < +\infty$. For $z \in \mathcal{Z} \subset \mathcal{N}$ and $t \in \mathbb{R}$, we have

$$\begin{aligned} e^{-t}\mu(z)tr(b) &= e^{-t}\tau(zb) = (\tau \circ \theta_t)(zb) = \tau(\alpha_t(w(t))\alpha_t(zb)\alpha_t(w(t))^*) \\ &= \tau(\alpha_t(zb)) = \tau(\theta_t(z)b) = \mu(\theta_t(z))tr(b). \end{aligned}$$

This shows that $\mu \circ \theta_t = e^{-t}\mu$.

Using the equivalence (i) $\Leftrightarrow$ (iv) in Proposition 2 and the same arguments as in the proof of Proposition 23.10, we can now conclude that $\mathcal{R}(\mathcal{N}, \theta)$ is semifinite, a contradiction. □

23.12 Theorem. *Let $(\mathcal{N}, \theta, \tau)$ be a continuous decomposition with $\mathcal{N}$ a countably decomposable W^*-algebra. Then every unitary cocycle $u \in Z_\theta(\mathbb{R}; U(\mathcal{N}))$ is trivial, that is, there exists $v \in U(\mathcal{N})$ such that $u(t) = v^*\theta_t(v)$ $(t \in \mathbb{R})$.*

Proof. By Corollary 23.10, there exists a continuous action $\sigma : \mathbb{R} \to Aut(\mathcal{N}^\theta)$ such that $(\mathcal{N}, \theta) \approx (\mathcal{R}(\mathcal{N}^\theta, \sigma), \hat{\sigma})$. We may assume that $(\mathcal{N}, \theta) = \mathcal{R}(\mathcal{N}^\theta, \sigma), \hat{\sigma})$.

Let $p \in \mathcal{Z}(\mathcal{N}^\theta)$ be the unique projection such that $\mathcal{N}^\theta p$ is semifinite and $\mathcal{N}^\theta(1-p)$ is purely infinite. Then $\sigma_t(p) = p$ $(t \in \mathbb{R})$, hence $p \in \mathcal{Z}(\mathcal{R}(\mathcal{N}^\theta, \sigma)) = \mathcal{Z}(\mathcal{N})$. It is now easy to see that we can divide the proof into two cases, assuming first that $\mathcal{N}^\theta$ is semifinite and then that $\mathcal{N}^\theta$ is purely infinite.

If $\mathcal{N}^\theta$ is semifinite, then, by Proposition 2/23.11, we have $(\mathcal{N}, \theta) \approx (\mathcal{N} \bar{\otimes} \mathcal{L}^\infty(\mathbb{R}), \iota \bar{\otimes} \mathrm{Ad}(\boldsymbol{\lambda}))$ and the required conclusion follows from Proposition 21.12.

If $\mathcal{N}^\theta$ is purely infinite, then, by Corollary 23.10, θ is a dominant action, hence the trivial cocycle $1 \in Z_\theta(\mathbb{R}; U(\mathcal{N}))$ is a dominant cocycle. Let $u \in Z_\theta(\mathbb{R}; U(\mathcal{N}))$ and $\theta' = {}_u\theta$. Then for $x \in \mathcal{N}^+$ and $s \in \mathbb{R}$, we have $\tau(\theta'_s(x)) = \tau(u(s)\theta_s(x)u(s)^*) = \tau(\theta_s(x)) = e^{-s}\tau(x)$, hence also $(\mathcal{N}, \theta', \tau)$ is a continuous decomposition. Since $\theta' \sim \theta$, using Corollary 23.10 and the Takesaki duality theorem (19.5) we get $\mathcal{N}^{\theta'} \bar{\otimes} \mathcal{B}(\mathcal{L}^2(\mathbb{R})) \approx \mathcal{R}(\mathcal{N}, \theta') \approx \mathcal{R}(\mathcal{N}, \theta) \approx \mathcal{N}^\theta \bar{\otimes} \mathcal{B}(\mathcal{L}^2(\mathbb{R}))$, so that $\mathcal{N}^{\theta'}$ is also purely infinite. It follows that the cocycle $1 \in Z_{\theta'}(\mathbb{R}; U(\mathcal{N}))$ is dominant, hence the cocycle $u \in Z_\theta(\mathbb{R}; U(\mathcal{N}))$ is dominant. By Theorem 20.5, we conclude that $u \approx 1$, that is, there exists $v \in U(\mathcal{N})$ such that $u(t) = v^*\theta_t(v)$ $(\tau \in \mathbb{R})$. □

Using the previous Theorem and Proposition 23.8 we obtain:

Corollary. *Let $(\mathcal{N}_1, \theta_1, \tau_1)$ and $(\mathcal{N}_2, \theta_2, \tau_2)$ be two continuous decompositions with $\mathcal{N}_1, \mathcal{N}_2$ countably decomposable W^*-algebras. Then $\mathcal{R}(\mathcal{N}_1, \theta_1) \approx \mathcal{R}(\mathcal{N}_2, \theta_2)$ if and only if $(\mathcal{N}_1, \theta_1) \approx (\mathcal{N}_2, \theta_2)$.*

23.13. In this section, we give a discrete variant of Theorem 23.12. For its proof, we require the following result (compare with 23.9).

Lemma. *Let θ be a *-automorphism of the abelian W^*-algebra $\mathscr{Z}$ such that there exists a nonsingular positive self-adjoint operator A affiliated to $\mathscr{Z}$ with $\theta(A) \leq \lambda A$ for some $0 < \lambda < 1$. Then there exists a projection $e \in \mathscr{Z}$ such that $\sum_{n\in\mathbb{Z}} \theta^n(e) = 1$ and there exists an s-continuous unitary representation $u : \mathbb{T} \to U(\mathscr{Z})$ such that $\theta^n(u(\omega)) = \omega^n u(\omega)$ $(\omega \in \mathbb{T}, n \in \mathbb{Z})$.*

Proof. In order to prove the first assertion, it is sufficient to show that every θ-invariant nonzero projection $p \in \mathscr{Z}$ majorizes a nonzero projection $e \in \mathscr{Z}$ with the property that $\theta^i(e)\theta^j(e) = 0$ for all $i, j \in \mathbb{Z}, i \neq j$. Without loss of generality, we may assume that $p = 1$.

There exists $n \in \mathbb{Z}$ such that the spectral projection e of A corresponding to the interval determined by λ^n and λ^{n+1} is nonzero. If $n \geq 0$, then $e = \chi_{(\lambda^{n+1},\lambda^n]}(A) \neq 0$ and for every $k \geq 1$ we have $\theta^{-k}(A) \geq \lambda^{-k}A$, so that $\theta^{-k}(e) = \chi_{(\lambda^{n+1},\lambda^n]}(\theta^{-k}(A)) \leq \chi_{(0,\lambda^n]}(\theta^{-k}(A)) \leq \chi_{(0,\lambda^n]}(\lambda^{-k}A) = \chi_{(0,\lambda^{n+k}]}(A)$, and hence $e\theta^{-k}(e) = 0$, as $\lambda^{n+k} \leq \lambda^{n+1}$; therefore $\theta^i(e)\theta^j(e) = 0$ for every $i, j \in \mathbb{Z}, i \neq j$. If $n < 0$, a similar argument leads to the same conclusion.

The other assertion in the statement follows now as in the proof of Proposition 23.9. □

Theorem. *Let $\mathscr{N}$ be a countably decomposable semifinite W^*-algebra, $\theta \in Aut(\mathscr{N})$ and τ an n.s.f. trace on $\mathscr{N}$ such that $\tau \circ \theta \leq \lambda\tau$ for some $0 < \lambda < 1$. Then for every $u \in U(\mathscr{N})$ there exists $v \in U(\mathscr{N})$ such that $u = v^*\theta(v)$.*

Proof. By Corollary 23.9, there exists a continuous action $\sigma : \mathbb{T} \to Aut(\mathscr{N}^\theta)$ such that $(\mathscr{N}, \theta) \approx (\mathscr{R}(\mathscr{N}^\theta, \sigma), \hat{\sigma})$. As in the proof of Theorem 23.12, we can consider separately the cases when $\mathscr{N}^\theta$ is semifinite and purely infinite.

Assume that $\mathscr{N}^\theta$ is semifinite and let μ_0 be an n.s.f. trace on $\mathscr{N}^\theta$. Since $\mathbb{T}$ is a compact group, the faithful normal operator-valued weight $P_\sigma : (\mathscr{N}^\theta)^+ \to \overline{((\mathscr{N}^\theta)^\sigma)}^+$ is finite and σ-invariant, hence $\mu = \mu_0 \circ P_\sigma$ is a σ-invariant n.s.f. trace on $\mathscr{N}^\theta$. The dual weight $\varphi = \hat{\mu}$ on $\mathscr{N} = \mathscr{R}(\mathscr{N}^\theta, \sigma)$ is invariant under the dual action $\theta = \hat{\sigma}$. Since μ is a σ-invariant trace, we see using Theorem 19.8 that $\varphi = \hat{\mu}$ is a θ-invariant n.s.f. trace on $\mathscr{N}$. Therefore, there exists a nonsingular positive self-adjoint operator A affiliated to $\mathscr{Z}(\mathscr{N})$ such that $\varphi = \tau_A$. Since $\varphi \circ \theta = \varphi$ and $\tau \circ \theta \leq \lambda\theta$, it follows that $\theta(A) \leq \lambda A$. By the above lemma there exists an s-continuous unitary representation $u : \mathbb{T} \to \mathscr{Z}(\mathscr{N})$ such that $\theta^n(u(\omega)) = \omega^n u(\omega)$ for all $\omega \in \mathbb{T}, n \in \mathbb{Z}$. Using Landstad's theorem (19.9), we infer that $(\mathscr{N}, \theta) \approx (\mathscr{N}^\theta \bar{\otimes} \ell^\infty(\mathbb{Z}), \iota \bar{\otimes} \mathrm{Ad}(\boldsymbol{\lambda}))$. Consider now a unitary element $u \in \mathscr{N}$; since $\mathscr{N} = \mathscr{N}^\theta \bar{\otimes} \ell^\infty(\mathbb{Z}) \equiv \ell^\infty(\mathbb{Z}, \mathscr{N}^\theta)$, u is a sequence $\{u_n\}_{n\in\mathbb{Z}}$ of unitary elements in $\mathscr{N}^\theta$. We define a unitary element $v = \{v_n\}_{n\in\mathbb{Z}}$ in $\mathscr{N}$ by putting $v_0 = 1, v_{n+1} = v_n u_n$ for $n \geq 1$ and $v_n = v_{n+1}u_n^*$ for $n < 0$. Then $\theta(v) = \{w_n\}_{n\in\mathbb{Z}}$ where $w_n = v_{n+1}$ $(n \in \mathbb{Z})$, hence $v^*\theta(v) = \{v_n^* v_{n+1}\}_{n\in\mathbb{Z}} = \{u_n\}_{n\in\mathbb{Z}} = u$.

If $\mathscr{N}^\theta$ is purely infinite, then, using Corollary 23.9 and the correspondence between θ-cocycles and unitary elements $u \in U(\mathscr{N})$ (see 16.15.(2)), the proof is similar to the corresponding part of the proof of Theorem 23.12. □

The above proof also shows that $\mathscr{N}^\theta$ is finite if and only if $\mathscr{N}$ is finite.

23.14. A continuous action $\theta : G \to Aut(\mathscr{N})$ of a locally compact group G on the W^*-algebra $\mathscr{N}$ will be called *stable* if every unitary cocycle $u \in Z_\theta(G; U(\mathscr{M}))$ is trivial, that is, if there exists $v \in U(\mathscr{N})$ such that $u(t) = v^*\theta_t(v)$ $(t \in G)$; in this case, $\mathrm{Ad}(v)$ establishes a *-isomorphism $(\mathscr{N}, \theta) \approx (\mathscr{N}, {}_u\theta)$ so that the actions θ and ${}_u\theta$ are inner conjugate.

In particular, a *-automorphism $\theta \in Aut(\mathscr{N})$ will be called *stable* if the action $\theta : \mathbb{Z} \ni n \mapsto \theta^n \in Aut(\mathscr{N})$ is stable. This means that for every $u \in U(\mathscr{N})$ there exists $v \in U(\mathscr{N})$ with $v^*\theta(v) = u$; in this case, $\mathrm{Ad}(v)$ establishes a *-isomorphism $(\mathscr{N}, \theta) \approx (\mathscr{N}, \mathrm{Ad}(u) \circ \theta)$, hence the *-automorphisms θ and $\mathrm{Ad}(u) \circ \theta$ are inner conjugate.

Notable examples of stable actions are given in Theorems 23.12, 23.13, Proposition 21.12, and Corollary 22.14. Another important example is given in the next corollary.

Corollary. *Let $\mathcal{N}$ be a type II_∞ factor, τ an n.s.f. trace on $\mathcal{N}$ and $\theta \in Aut(\mathcal{N})$. Then θ is stable if and only if $\tau \circ \theta \neq \tau$.*

Proof. If $\tau \circ \theta \neq \tau$, then $\tau \circ \theta = \lambda\tau$ for some $\lambda \neq 1$ and, replacing θ by θ^{-1} if necessary, we may assume that $0 < \lambda < 1$. By Theorem 23.13, it follows that θ is stable.

Assume now that $\tau \circ \theta = \tau$. There exists a projection $e \in \mathcal{N}$ with $0 \neq \tau(e) < +\infty$. Then e is finite and, since $\tau(\theta(e)) = \tau(e)$, we have $\theta(e) \sim e$. Thus, there exists a unitary operator $u \in U(\mathcal{N})$ such that $\theta(e) = ueu^*$. Then $(\mathrm{Ad}(u) \circ \theta)(e) = e$ and hence the positive normal form $\tau(e\cdot)$ on $\mathcal{N}$ is $(\mathrm{Ad}(u) \circ \theta)$-invariant. It follows that $\mathrm{Ad}(u) \circ \theta$ is not integrable. On the other hand, the existence of a dominant cocycle shows that there exists $u' \in U(\mathcal{N})$ such that $\mathrm{Ad}(u') \circ \theta$ is integrable. It is then obvious that $\mathrm{Ad}(u) \circ \theta$ and $\mathrm{Ad}(u') \circ \theta$ are not conjugate, hence θ is not stable. □

23.15. Let $\mathcal{M}$ be a properly infinite W^*-algebra and φ a normal semifinite weight on $\mathcal{M}$. There exists a sequence $\{w_n\} \subset \mathcal{M}$ of isometries such that $\sum_n w_n w_n^* = 1$. Then $u_n = (w_n \mathbf{s}(\varphi))^*$ is a partial isometry with $u_n u_n^* = \mathbf{s}(\varphi)$ and the projections $u_n^* u_n$ are mutually orthogonal. We define $\varphi_n = \varphi_{u_n}$. Then $\varphi_n \approx \varphi$ and φ_n have mutually orthogonal supports. Thus, we can consider the normal semifinite weight $\sum_n \varphi_n$ on $\mathcal{M}$; we have

$$\left(\mathcal{M}, \sum_n \varphi_n\right) \approx (\mathcal{M} \bar{\otimes} \mathcal{F}_\infty, \varphi \bar{\otimes} tr).$$

Indeed, if $\{e_{ij}\}$ is a system of matrix units for $\mathcal{F}_\infty$, then the equation $\pi(x) = \sum_{ij} w_i^* x w_j \bar{\otimes} e_{ij}$ $(x \in \mathcal{M})$, defines a $*$-isomorphism $\pi : (\mathcal{M}, \sum_n \varphi_n) \to (\mathcal{M} \bar{\otimes} \mathcal{F}_\infty, \varphi \bar{\otimes} tr)$.

The weight $\check{\varphi} = \sum_n \varphi_n \in W_{ns}(\mathcal{M})$ is of infinite multiplicity, $\varphi \precsim \check{\varphi}$ and $(\mathcal{M}, \check{\varphi}) \approx (\mathcal{M} \bar{\otimes} \mathcal{F}_\infty, \varphi \bar{\otimes} tr)$. If φ is of infinite multiplicity, then $(\mathcal{M}, \check{\varphi}) \approx (\mathcal{M}, \varphi)$ by Corollary 9.18.

By construction, if $\varphi, \psi \in W_{ns}(\mathcal{M})$ and $\varphi \approx \psi_n$ then $\check{\varphi} \approx \check{\psi}$.

Similarly, for every sequence $\{\psi_n\} \subset W_{ns}(\mathcal{M})$, we can construct a sequence $\{\varphi_n\} \subset W_{ns}(\mathcal{M})$ with $\mathbf{s}(\varphi_n)$ mutually orthogonal and $\varphi_n \approx \psi_n$ such that the equivalence class of the weight $\sum_n \varphi_n$ depends only on the equivalence classes of the weights ψ_n. This equivalence class, or some representative of it, will be denoted by $\sum_n^{\oplus} \psi_n$.

23.16. In the next Theorem, we consider some properties related to the dominance property of a weight. Let ω be the unique n.s.f. weight on $\mathcal{F}_\infty = \mathcal{B}(\mathcal{L}^2(\mathbb{R}))$ such that $[D\omega : D\,tr]_t = \rho(t)$ $(t \in \mathbb{R})$.

Theorem. *Let $\mathcal{M}$ be a properly infinite W^*-algebra with separable predual and φ an n.s.f. weight on $\mathcal{M}$. Consider the statements:*

(i) *φ is a dominant weight;*
(ii) *there exists an n.s.f. weight ψ on $\mathcal{M}$ of infinite multiplicity such that $(\mathcal{M}, \varphi) \approx (\mathcal{M} \bar{\otimes} \mathcal{F}_\infty, \psi \bar{\otimes} \omega)$;*
(iii) *φ is the dual weight of the trace appearing in some continuous decomposition of $\mathcal{M}$;*
(iv) *there exists an n.s.f. weight ψ on $\mathcal{M}$ which anticommutes with φ;*
(v) *$\varphi \approx \lambda\varphi$ for every $\lambda > 0$.*

Then, (v) $\Leftarrow$ (i) $\Leftrightarrow$ (ii) $\Leftrightarrow$ (iii) $\Rightarrow$ (iv). If $\mathcal{M}$ is a factor, then (iv) $\Rightarrow$ (v). If $\mathcal{M}$ is of type III, then (i) $\Leftrightarrow$ (v). Therefore, for factors of type III with separable preduals, (i)–(v) are equivalent.

Proof. The equivalence (i) $\Leftrightarrow$ (ii) follows from Section 23.4 and the equivalence (i) $\Leftrightarrow$ (iii) from Sections 23.6 and 23.7; the implication (i) $\Rightarrow$ (v) is obvious and the implication (i) $\Rightarrow$ (iv) has been proved in Section 23.4.

Assume now that $\mathcal{M}$ is a factor and that there exists an n.s.f. weight ψ on $\mathcal{M}$ which anticommutes with φ. Let $u(t) = [D\psi : D\varphi]_t$ $(t \in \mathbb{R})$. Then, for $x \in \mathcal{M}$ and $s, t \in \mathbb{R}$, we have $u(t)\sigma^{\varphi}_{s+t}(x)u(t)^* = u(t)\sigma^{\varphi}_t(\sigma^{\varphi}_s(x))u(t)^* = \sigma^{\psi}_t(\sigma^{\varphi}_s(x)) = \sigma^{\varphi}_s(\sigma^{\psi}_t(x)) = \sigma^{\varphi}_s(u(t)\sigma^{\varphi}_t(x)u(t)^*) = \sigma^{\varphi}_s(u(t))\sigma^{\varphi}_{s+t}(x)\sigma^{\varphi}_s(u(t))^*$, that is, $\sigma^{\varphi}_s(u(t))u(t)^* \in \mathcal{Z}(\mathcal{M}) = \mathbb{C}\cdot 1_{\mathcal{M}}$. Thus, there exists a continuous function $f : \mathbb{R}\times\mathbb{R} \to \mathbb{T}$ such that $\sigma^{\varphi}_s(u(t)) = f(s,t)u(t)$ $(s, t \in \mathbb{R})$. Since $\sigma^{\varphi}_{s_1+s_2} = \sigma^{\varphi}_{s_1}\sigma^{\varphi}_{s_2}$, we have $f(s_1+s_2, t) = f(s_1, t)f(s_2, t)$ and, since $u(t_1+t_2) = u(t_1)\sigma^{\varphi}_{t_1}(u(t_2))$, we have $f(s, t_1+t_2) = f(s, t_1)f(s, t_2)$. It follows that there exists $r \in \mathbb{R}$ such that $f(s,t) = e^{irst}(s, t \in \mathbb{R})$. Since φ does not commute with ψ, there exists $t \in \mathbb{R}$ with $u(t) \notin \mathcal{M}^{\varphi}$, hence $r \neq 0$. We have $\sigma^{\varphi}_s(u(t)) = (e^{rt})^{is}u(t)$ $(s, t \in \mathbb{R})$, and $\{e^{rt}; t \in \mathbb{R}\} = \{\lambda \in \mathbb{R}; \lambda > 0\}$. Hence for every $\lambda > 0$, there exists $v(\lambda) \in U(\mathcal{M})$ such that $\sigma^{\varphi}_s(u(\lambda)) = \lambda^{is}u(\lambda)$ $(s \in \mathbb{R})$, so that $\varphi \approx \varphi_{v(\lambda)} = \lambda\varphi$. Thus, (iv) $\Rightarrow$ (v) whenever $\mathcal{M}$ is a factor.

Finally, assume that $\mathcal{M}$ is a type III W^*-algebra and $\varphi \approx \lambda\varphi$ for every $\lambda > 0$. Then the weight $\check{\varphi}$ (23.15) is of infinite multiplicity and $\check{\varphi} \approx \lambda\check{\varphi}$ for every $\lambda > 0$, hence $\check{\varphi}$ is dominant (23.3). Since $\varphi \lesssim \check{\varphi}$, φ is integrable.

Let ψ be any dominant weight on $\mathcal{M}$ and $(\mathcal{M}^{\psi}, \theta, \tau)$ the corresponding continuous decomposition of $\mathcal{M}$ (23.6). Since φ is integrable we have $\varphi \lesssim \psi$, so that we may assume $\varphi = \psi_f$ for some projection $f \in \mathcal{M}^{\psi}$. We shall show that the projection f is properly infinite in $\mathcal{M}^{\psi}$, that is, $\tau(fp) = +\infty$ for every nonzero projection $p \in \mathcal{Z}(\mathcal{M}^{\psi})$.

Recall (23.7) that there exists an s-continuous unitary representation $u : \mathbb{R} \to U(\mathcal{M})$ such that $\theta_s = \mathrm{Ad}(u(s))|\mathcal{M}^{\psi}$ and $\psi_{u(s)} = e^{-s}\psi$ $(s \in \mathbb{R})$. Thus, $e^{-s}\varphi = e^{-s}\psi_f = (e^{-s}\psi)_f = (\psi_{u(s)})_f = \psi_{u(s)f}$ $(s \in \mathbb{R})$. Since $\varphi \approx e^{-s}\varphi$, it follows that $\psi_f = \psi_{u(s)f}$, hence $f \sim u(s)fu(s)^*$ in $\mathcal{M}^{\psi}$ (23.2), that is, $\theta_s(f) \sim f$ in $\mathcal{M}^{\psi}$ $(s \in \mathbb{R})$. Thus, for every projection $q \in \mathcal{Z}(\mathcal{M}^{\psi})$, we have $\tau(f\theta_s(q)) = \tau(\theta_s(fq)) = e^{-s}\tau(fq)$ and therefore $\tau(f\theta_s(z)) = e^{-s}\tau(fz)$ for all $z \in \mathcal{Z}(\mathcal{M}^{\psi})^+$ and all $s \in \mathbb{R}$. Let p be the least upper bound of the projections $q \in \mathcal{Z}(\mathcal{M}^{\psi})$ with $\tau(fq) < +\infty$. The above arguments show that $\theta_s(p) = p$ for all $s \in \mathbb{R}$ and that $\tau(fp\cdot)$ is an n.s.f. trace on $\mathcal{Z}(\mathcal{M}^{\psi})p$. Using Proposition 1/23.11, we obtain $p \in \mathcal{Z}(\mathcal{M}^{\psi})^{\theta} = \mathcal{Z}(\mathcal{M})$. On the other hand, using the results of Section 23.10 applied to the triple $(\mathcal{Z}(\mathcal{M}^{\psi})p, \theta|\mathcal{Z}(\mathcal{M}^{\psi})p, \tau(fp\cdot))$ and the equivalence (i) $\Leftrightarrow$ (iv) in Proposition 23.11, we conclude that $\mathcal{M}p$ is semifinite. Since $\mathcal{M}$ is of type III, it follows that $p = 0$. □

23.17. During the proof of Theorem 23.16, we have also shown that if φ and ψ are two anticommuting n.s.f. weights on the factor $\mathcal{M}$, then there exists a unique real number $r = r(\varphi, \psi) \in \mathbb{R}, r \neq 0$, such that

$$\sigma^{\varphi}_s[D\psi : D\varphi]_t) = e^{irst}[D\psi : D\varphi]_t \quad (s, t \in \mathbb{R}). \tag{1}$$

If the above identity holds with $r = 0$, then φ and ψ commute.

The next Theorem shows that the number $r(\varphi, \psi)$ completely determines the equivalence class of the pair (φ, ψ).

Theorem. *Let $\mathcal{M}$ be a factor with separable predual. Let (φ, ψ) and (φ', ψ') be two anticommuting pairs of dominant weights on $\mathcal{M}$. Then $r(\varphi, \psi) = r(\varphi', \psi')$ if and only if there exists $u \in U(\mathcal{M})$ such that $\varphi' = \varphi_u$ and $\psi' = \psi_u$.*

Proof. Assume that $\varphi' = \varphi_u$ and $\psi' = \psi_u$ with $u \in U(\mathscr{M})$. Then $[D\varphi' : D\varphi]_t = u^*\sigma_t^\varphi(u)$ and $[D\psi' : D\psi]_t = u^*\sigma_t^\psi(u)$, hence $[D\psi' : D\varphi']_t = u^*[D\psi : D\varphi]_t u$ and $\sigma_t^{\varphi_u}(x) = u^*\sigma_t^\varphi(uxu^*)u$ $(x \in \mathscr{M}, t \in \mathbb{R})$. It follows that $r(\varphi,\psi) = r(\varphi',\psi')$.

Conversely, assume that $r(\varphi,\psi) = r(\varphi',\psi')$. By the uniqueness modulo equivalence of the dominant weight, it follows that there exists $u \in U(\mathscr{M})$ with $\varphi' = \varphi_u$. We have $r(\varphi_u,\psi') = r(\varphi,\psi) = r(\varphi_u,\psi_u)$. Thus, we are led to showing that if φ,ψ,ψ' are three dominant weights on $\mathscr{M}$ such that (φ,ψ) and (φ,ψ') both anticommute and $r(\varphi,\psi) = r(\varphi,\psi') = r$, then there exists $u \in U(\mathscr{M}^\varphi)$ such that $\psi' = \psi_u$.

Let $(\mathscr{N},\theta,\tau)$ be a continuous decomposition of $\mathscr{M}$ and let $u : \mathbb{R} \to U(\mathscr{M})$ be the s-continuous unitary representation of $\mathbb{R}$ on $\mathscr{M}$ such that $\theta_s = \mathrm{Ad}(u(s))|\mathscr{N}$. We may assume that $\varphi = \hat{\tau}$ (see 23.6, 23.7). Put $v(t) = [D\psi : D\varphi]_t$ and $v'(t) = [D\psi' : D\varphi]_t$ $(t \in \mathbb{R})$. We have $\sigma_s^\varphi(v(t)) = e^{\mathrm{i}rst}v(t)$ and $\sigma_s^\varphi(v'(t)) = e^{\mathrm{i}rst}v'(t)$, hence $v(s+t) = e^{\mathrm{i}rst}v(s)v(t)$ and $v'(s+t) = e^{\mathrm{i}rst}v'(s)v'(t)$ $(s,t \in \mathbb{R})$. Continuous functions $a : \mathbb{R} \to U(\mathscr{M}), a' : \mathbb{R} \to U(\mathscr{M})$ are defined by $a(s) = e^{-\mathrm{i}rs^2/2}v(s)u(rs)^*, a'(s) = e^{-\mathrm{i}rs^2/2}v'(s)u(rs)^*$ $(s \in \mathbb{R})$; we have $a(s+t) = a(s)\theta_{rs}(a(t)), a'(s+t) = a'(s)\theta_{rs}(a'(t))$ $(s,t \in \mathbb{R})$. Using Theorem 23.12, we can find an element $u \in U(\mathscr{N})$ such that $a(s) = ua'(s)\theta_{rs}(u^*)$ $(s \in \mathbb{R})$. It follows that for every $s \in \mathbb{R}$ we have $v(s) = e^{\mathrm{i}rs^2/2}a(s)u(rs) = e^{\mathrm{i}rs^2/2}ua'(s)\theta_{rs}(u^*)u(rs) = e^{\mathrm{i}rs^2/2}ua'(s)u(rs)u^*u(rs)^*u(rs) = uv'(s)u^*$, that is, $\psi' = \psi_u$ with $u \in \mathscr{N} = \mathscr{M}^\varphi$. □

Note that if $\mathscr{M}$ is a type III factor, then the dominance condition imposed on the weights in the statement of the theorem follows automatically from the anticommutation condition, by Theorem 23.16.

23.18. The next result asserts the possibility of approximation by integrable weights in the sense of the metric d introduced in Section 6.10.

Theorem. *Let φ be an n.s.f. weight of infinite multiplicity on the W^*-algebra $\mathscr{M}$. For every $\varepsilon > 0$, there exists an integrable n.s.f. weight ψ of infinite multiplicity, commuting with φ, such that $d(\varphi,\psi) \le \varepsilon$.*

Proof. By assumption, $\mathscr{M}^\varphi$ is properly infinite, so that there exists an infinite type I W^*-subfactor $\mathscr{F} \subset \mathscr{B}(\mathscr{L}^2(\mathbb{R}))$ of $\mathscr{M}^\varphi$. On the other hand, there exists an operator $a \in \mathscr{F}$ with absolutely continuous spectrum such that $1-\varepsilon \le a \le 1+\varepsilon$ and $\{a\}' \cap \mathscr{F}$ is properly infinite. Since a has absolutely continuous spectrum and since the quasi-equivalence class of an so-continuous unitary representation of $\mathbb{R}$ is completely determined by the equivalence class of its associated spectral measure, it follows that the unitary representation $\mathbb{R} \ni t \mapsto a^{\mathrm{i}t} \in \mathscr{B}(\mathscr{L}^2(\mathbb{R}))$ is quasi-equivalent to a subrepresentation of the regular representation $\boldsymbol{\rho}$ of $\mathbb{R}$, so that the weight tr_a is quasi-equivalent to a subweight of the dominant weight on $\mathscr{B}(\mathscr{L}^2(\mathbb{R}))$ and hence it is integrable. Thus, there exists an increasing sequence $\{e_n\} \subset \mathscr{F} \subset \mathscr{M}^\varphi \subset \mathscr{M}$ of projections such that $e_n \uparrow 1$ and $\int a^{\mathrm{i}t}e_n a^{-\mathrm{i}t}\,\mathrm{d}t < +\infty$. Then $\psi = \varphi_a$ is an n.s.f. weight on $\mathscr{M}$ with $[D\psi : D\varphi]_t = a^{\mathrm{i}t}$ and $\sigma_t^\psi = \mathrm{Ad}(a^{\mathrm{i}t}) \circ \sigma_t^\varphi$ $(t \in \mathbb{R})$, so that $\int \sigma_t^\psi(e_n)\,\mathrm{d}t = \int a^{\mathrm{i}t}e_n a^{-\mathrm{i}t}\,\mathrm{d}t < +\infty$ and ψ is integrable. Moreover, ψ is of infinite multiplicity, since $\{a\}' \cap \mathscr{F} \subset \mathscr{M}^\psi$. Finally, as $1-\varepsilon \le a \le 1+\varepsilon$, it follows that $d(\varphi,\psi) \le \varepsilon$. □

23.19 Theorem (Relative Commutant Theorem). *Let $\mathscr{M}$ be a properly infinite W^*-algebra with separable predual. For every n.s.f. weight φ on $\mathscr{M}$, we have*

$$\mathscr{R}(\mathscr{M},\sigma^\varphi)' \cap (\mathscr{M} \bar{\otimes} \mathscr{B}(\mathscr{L}^2(\mathbb{R}))) = \mathscr{Z}(\mathscr{R}(\mathscr{M},\sigma^\varphi)). \tag{1}$$

For every integrable n.s.f. weight φ on $\mathcal{M}$, we have

$$(\mathcal{M}^{\varphi})' \cap \mathcal{M} = \mathcal{Z}(\mathcal{M}^{\varphi}). \tag{2}$$

Proof. Since any integrable n.s.f. weight is equivalent to a subweight of the dominant weight (23.4), in proving (2) we may assume that φ is a dominant weight.

Let ω be the unique n.s.f. weight on $\mathcal{B}(\mathcal{L}^2(\mathbb{R}))$ such that $[D\omega : D\, tr]_t = \boldsymbol{\rho}(t)$ $(t \in \mathbb{R})$. Then, for any n.s.f. weight φ on $\mathcal{M}$, $\varphi \bar{\otimes} \omega$ is a dominant weight on $\mathcal{M} \bar{\otimes} \mathcal{B}(\mathcal{L}^2(\mathbb{R})) \approx \mathcal{M}$, and

$$(\mathcal{M} \bar{\otimes} \mathcal{B}(\mathcal{L}^2(\mathbb{R})))^{\varphi \bar{\otimes} \omega} = (\mathcal{M} \bar{\otimes} \mathcal{B}(\mathcal{L}^2(\mathbb{R})))^{\sigma^{\varphi} \bar{\otimes} \mathrm{Ad}(\boldsymbol{\rho})} = \mathcal{R}(\mathcal{M}, \sigma^{\varphi})$$

by Corollary 19.13. Since any two dominant weights are equivalent (23.4), we see that, in order to prove the entire theorem, it is sufficient to prove (1) just for a particular choice of φ.

Thus, let φ be a faithful normal state on $\mathcal{M}$ and $\sigma = \sigma^{\varphi}$ be its modular automorphism group. We may assume $\mathcal{M} \subset \mathcal{B}(\mathcal{H})$ realized as a von Neumann algebra with a cyclic and separating vector $\xi_0 \in \mathcal{H}$ such that $\varphi = \omega_{\xi_0} | \mathcal{M}$.

Since $\mathcal{R}(\mathcal{M}, \sigma) = \mathcal{R}\{\pi_{\sigma}(\mathcal{M}), 1_{\mathcal{M}} \bar{\otimes} \mathfrak{L}(\mathbb{R})\}$, we have $\mathcal{R}(\mathcal{M}, \sigma)' \cap (\mathcal{M} \bar{\otimes} \mathcal{B}(\mathcal{L}^2(\mathbb{R})) = \pi_{\sigma}(\mathcal{M})' \cap (\mathcal{M} \bar{\otimes} \mathfrak{L}(\mathbb{R})') = \pi_{\sigma}(\mathcal{M})' \cap (\mathcal{M} \bar{\otimes} \mathfrak{L}(\mathbb{R}))$ (see 18.4.(14)). Also (see 21.6.(1))) $\mathcal{Z}(\mathcal{R}(\mathcal{M}, \sigma)) \subset \mathcal{M}^{\varphi} \bar{\otimes} \mathfrak{L}(\mathbb{R}) \subset \mathcal{R}(\mathcal{M}, \sigma)$. Therefore, in order to prove (1), we have to show that

$$\pi_{\sigma}(\mathcal{M})' \cap (\mathcal{M} \bar{\otimes} \mathfrak{L}(\mathbb{R})) \subset \mathcal{M}^{\varphi} \bar{\otimes} \mathfrak{L}(\mathbb{R}). \tag{3}$$

Let $F : \mathcal{L}^2(\mathbb{R}) \to \mathcal{L}^2(\mathbb{R})$ be the unitary operator defined by the Fourier–Plancherel transform (18.8), that is,

$$[F\xi](t) = \int e^{-\mathrm{i}st} \xi(s)\, \mathrm{d}s \quad (\xi \in \mathcal{L}^2(\mathbb{R}), t \in \mathbb{R}),$$

$$[F^* \xi](s) = \int e^{\mathrm{i}st} \xi(t)\, \mathrm{d}t \quad (\xi \in \mathcal{L}^2(\mathbb{R}), s \in \mathbb{R}),$$

where the Haar measures $\mathrm{d}s$ on $\mathbb{R}$ and $\mathrm{d}t$ on $\hat{\mathbb{R}} \equiv \mathbb{R}$ are chosen so that the Fourier inversion formula holds. Also, denote by Φ the *-automorphism of $\mathcal{B}(\mathcal{L}^2(\mathbb{R}))$ implemented by F, that is, $\Phi(x) = FxF^*$ $(x \in \mathcal{B}(\mathcal{L}^2(\mathbb{R})))$, and put $\tilde{F} = 1_{\mathcal{M}} \bar{\otimes} F$, $\tilde{\Phi} = \iota_{\mathcal{M}} \bar{\otimes} \Phi$. Then (3) becomes (see Proposition 18.8):

$$\pi_{\sigma}(\mathcal{M})' \cap \tilde{\Phi}^{-1}(\mathcal{M} \bar{\otimes} \mathcal{L}^{\infty}(\mathbb{R})) = \tilde{\Phi}^{-1}(\mathcal{M}^{\varphi} \bar{\otimes} \mathcal{L}^{\infty}(\mathbb{R})). \tag{4}$$

Since the predual of $\mathcal{M}$ is separable, $\mathcal{M} \bar{\otimes} \mathcal{L}^{\infty}(\mathbb{R})$ can be identified with the W^*-algebra $\mathcal{L}^{\infty}(\mathbb{R}, \mathcal{M})$ of all essentially norm-bounded w-measurable functions $a(\cdot) : \mathbb{R} \to \mathcal{M}$; namely, the element $a \in \mathcal{M} \bar{\otimes} \mathcal{L}^{\infty}(\mathbb{R})$ corresponding to the function $a(\cdot) \in \mathcal{L}^{\infty}(\mathbb{R}, \mathcal{M})$ acts on $\mathcal{H} \bar{\otimes} \mathcal{L}^2(\mathbb{R}) = \mathcal{L}^2(\mathbb{R}, \mathcal{H})$ as follows (see Sakai, 1971, 3.2.2):

$$[a\xi](t) = a(t)\xi(t) \quad (\xi \in \mathcal{L}^2(\mathbb{R}, \mathcal{H}), t \in \mathbb{R}).$$

Let $\mathcal{C}(\mathbb{R}, \mathcal{M})$ be the *-subalgebra of $\mathcal{L}^{\infty}(\mathbb{R}, \mathcal{M})$ consisting of all norm-bounded w-continuous functions $\mathbb{R} \to \mathcal{M}$. We show that $\pi_{\sigma}(\mathcal{M})' \cap \tilde{\Phi}^{-1}(\mathcal{C}(\mathbb{R}, \mathcal{M}))$ is w-dense in $\pi_{\sigma}(\mathcal{M})' \cap \tilde{\Phi}^{-1}(\mathcal{L}^{\infty}(\mathbb{R}, \mathcal{M}))$. To this end, consider the continuous action $\theta : \mathbb{R} \to Aut(\mathcal{M} \bar{\otimes} \mathcal{B}(\mathcal{L}^2(\mathbb{R})))$ defined by $\theta_s = \mathrm{Ad}(1_{\mathcal{M}} \bar{\otimes} \mathbf{m}(s)^*)$, where $[\mathbf{m}(s)\xi](t) = e^{-\mathrm{i}st}\xi(t)$ $(s, t \in \mathbb{R}, \xi \in \mathcal{L}^2(\mathbb{R}))$. Recall (19.3) that the dual action $\hat{\sigma} : \mathbb{R} \to Aut(\mathscr{R}(\mathscr{M}, \sigma))$ is defined by $\hat{\sigma}_s = \theta_s | \mathscr{R}(\mathscr{M}, \sigma)$ so that (19.3.(3)) $\theta_s(\pi_\sigma(\mathscr{M})) = \pi_\sigma(\mathscr{M})$ and hence $\theta_s(\pi_\sigma(\mathscr{M})') = \pi_\sigma(\mathscr{M})'$ $(s \in \mathbb{R})$. Now let $a \in \mathscr{L}^\infty(\mathbb{R}, \mathscr{M})$ be such that $\tilde{\Phi}^{-1}(a) \in \pi_\sigma(\mathscr{M})'$. A simple computation shows that $\pi_\sigma(\mathscr{M})' \ni \theta_s(\tilde{\Phi}^{-1}(a)) = \tilde{\Phi}^{-1}(a_s)$, where $a_s(t) = a(t + s)$ $(s, t \in \mathbb{R})$. Then, for any continuous function $f : \mathbb{R} \to \mathbb{C}$ with compact support, it follows that $\pi_\sigma(\mathscr{M})' \ni \theta_f(\tilde{\Phi}^{-1}(a)) = \tilde{\Phi}^{-1}(a_f)$, where $a_f(t) = \int f(s)a(t+s)\,\mathrm{d}s = \int f(s-t)a(s)\,\mathrm{d}s$ $(t \in \mathbb{R})$. It is easy to see that $a_f \in \mathscr{C}(\mathbb{R}, \mathscr{M})$, hence $\tilde{\Phi}^{-1}(a_f) \in \pi_\sigma(\mathscr{M})' \cap \tilde{\Phi}^{-1}(\mathscr{C}(\mathbb{R}, \mathscr{M}))$. On the other hand, if $\{f_i\}$ is a norm-bounded approximate unit of $\mathscr{L}^1(\mathbb{R})$, then (15.1.(2)) $\tilde{\Phi}^{-1}(a_{f_i}) = \theta_{f_i}(\tilde{\Phi}^{-1}(a)) \overset{w}{\to} \Phi^{-1}(a)$. Consequently, in order to prove (4), it is sufficient to show that

$$\pi_\sigma(\mathscr{M})' \cap \tilde{\Phi}^{-1}(\mathscr{C}(\mathbb{R}, \mathscr{M})) \subset \tilde{\Phi}^{-1}(\mathscr{C}(\mathbb{R}, \mathscr{M}^\varphi))$$

or, equivalently, that

$$\tilde{\Phi}(\pi_\sigma(\mathscr{M})') \cap \mathscr{C}(\mathbb{R}, \mathscr{M}) \subset \mathscr{C}(\mathbb{R}, \mathscr{M}^\varphi). \tag{5}$$

Let $a \in \mathscr{C}(\mathbb{R}, \mathscr{M})$, $x \in \mathscr{M}$, and $\eta, \zeta \in \mathscr{D}(\mathbb{R})$, where $\mathscr{D}(\mathbb{R})$ stands for the space of all complex valued C^∞-functions with compact support on $\mathbb{R}$. Then, as easily verified,

$$\begin{aligned}&(\tilde{F}\pi_\sigma(x)\tilde{F}^*(\xi_0 \bar{\otimes} \eta) | a^*(\xi_0 \bar{\otimes} \zeta))\\ &\quad = \int \left(\int \int e^{\mathrm{i}t(r-s)} \eta(r)\overline{\zeta(s)} \varphi(a(s)\sigma_{-t}(x))\,\mathrm{d}r\,\mathrm{d}s \right) \mathrm{d}t\end{aligned}$$

where for r and s the order of integration is irrelevant. Therefore, if a commutes with $\tilde{F}\pi_\sigma(x)\tilde{F}^*$, we obtain

$$\begin{aligned}&\overline{\int \left(\int \int e^{\mathrm{i}t(r-s)} \eta(r)\zeta(s)\varphi(a(s)\sigma_{-t}(x))\,\mathrm{d}r\,\mathrm{d}s \right) \mathrm{d}t}\\ &\quad = \overline{(\tilde{F}\pi_\sigma(x)\tilde{F}^*(\zeta_0 \otimes \eta) | a^*(\xi_0 \otimes \bar{\zeta}))} = (\tilde{F}\pi_\sigma(x^*)\tilde{F}^*(\xi_0 \otimes \bar{\zeta}) | a(\xi_0 \otimes \eta))\\ &\quad = \int \left(\int \int e^{\mathrm{i}t(r-s)} \overline{\eta(r)\zeta(s)} \varphi(a^*(r)\sigma_{-t}(x^*))\,\mathrm{d}r\,\mathrm{d}s \right) \mathrm{d}t\\ &\quad = \overline{\int \left(\int \int e^{\mathrm{i}t(r-s)} \eta(r)\zeta(s)\varphi(\sigma_{-t}(x)a(r))\,\mathrm{d}r\,\mathrm{d}s \right) \mathrm{d}t}.\end{aligned}$$

Since $\sigma = \sigma^\varphi$ satisfies the *KMS*-condition (2.12), for each $r \in \mathbb{R}$ there exists a function $G(\cdot, r)$ defined, continuous and bounded on the strip $\{\alpha \in \mathbb{C}; 0 \le \operatorname{Im} \alpha \le 1\}$, analytic in the interior of the strip and such that $G(t, r) = \varphi(\sigma_{-t}(x)a(r))$, $G(t + \mathrm{i}, r) = \varphi(a(r)\sigma_{-t}(x))$ for all $t \in \mathbb{R}$. Then,

$$\begin{aligned}&\int \left(\int \int e^{\mathrm{i}t(r-s)} \eta(r)\zeta(s)\varphi(\sigma_{-t}(x)a(r))\,\mathrm{d}r\,\mathrm{d}s \right) \mathrm{d}t\\ &\quad = \int \left(\int \int e^{\mathrm{i}t(r-s)} \eta(r)\zeta(s)G(t, r))\,\mathrm{d}r\,\mathrm{d}s \right) \mathrm{d}t\\ &\quad = \int \left(\int e^{\mathrm{i}tr} \eta(r)[F\zeta](t)G(t, r))\,\mathrm{d}r \right) \mathrm{d}t\\ &\quad = \int \left(\int e^{\mathrm{i}tr} [F\zeta](t)G(t, r))\,\mathrm{d}t \right) \eta(r)\,\mathrm{d}r = \cdot/.\end{aligned}$$

by Fubini's theorem. Since the function $\alpha \mapsto e^{\mathrm{i}\alpha r}[F\zeta](\alpha)G(\alpha, r))$ is analytic on the strip $\{\alpha \in \mathbb{C}; 0 < \operatorname{Im}\alpha < 1\}$ and decays exponentially along horizontal lines, using the theorems of Cauchy and Fubini we can continue the earlier computation as follows:

$$
\begin{aligned}
\cdot/\cdot &= \int \left(\int e^{\mathrm{i}(t+\mathrm{i})r}[F\zeta](t+\mathrm{i})G(t+\mathrm{i}, r)\,\mathrm{d}t \right) \eta(r)\,\mathrm{d}r \\
&= \int \left(\int e^{\mathrm{i}(t+\mathrm{i})r}\eta(r)[F\zeta](t+\mathrm{i})G(t+\mathrm{i}, r)\,\mathrm{d}r \right) \mathrm{d}t \\
&= \int \left(\int \int e^{\mathrm{i}(t+\mathrm{i})(r-s)}\eta(r)\zeta(s)G(t+\mathrm{i}, r)\,\mathrm{d}r\,\mathrm{d}s \right) \mathrm{d}t \\
&= \int \left(\int \int e^{\mathrm{i}(t+\mathrm{i})(r-s)}\eta(r)\zeta(s)\varphi(a(r)\sigma_{-t}(x))\,\mathrm{d}r\,\mathrm{d}s \right) \mathrm{d}t.
\end{aligned}
$$

Thus, if a commutes with $\tilde{F}\pi_\sigma(x)\tilde{F}^*$, then

$$
H : \mathbb{R} \times \mathbb{R} \ni (s, t) \mapsto \varphi(a(s)\sigma_{-t}(x)) \in \mathbb{C}
$$

is a bounded continuous function such that for every $\eta, \zeta \in \mathscr{D}(\mathbb{R})$ we have

$$
\begin{aligned}
&\int \left(\int \int e^{\mathrm{i}t(r-s)}\eta(r)\zeta(s)H(s,t)\,\mathrm{d}r\,\mathrm{d}s \right) \mathrm{d}t \\
&\qquad = \int \left(\int \int e^{\mathrm{i}(t+\mathrm{i})(r-s)}\eta(r)\zeta(s)H(r,t)\,\mathrm{d}r\,\mathrm{d}s \right) \mathrm{d}t.
\end{aligned} \tag{6}
$$

With these conditions, we shall show in the last part of the proof that

$$
H(s,t) = H(s,0) \quad (s, t \in \mathbb{R}). \tag{7}
$$

In our case, this means that $\varphi(a(s)\sigma_{-t}(x)) = \varphi(a(s)x)$ and hence that $\varphi(\sigma_t(a(s))x) = \varphi(a(s)x)$ $(s, t \in \mathbb{R})$.

If $a \in \tilde{\Phi}(\pi_\sigma(\mathscr{M})') \cap \mathscr{C}(\mathbb{R}, \mathscr{M})$, then it follows that $\sigma_t(a(s)) = a(s)$ $(s, t \in \mathbb{R})$, that is, $a \in \mathscr{C}(\mathbb{R}, \mathscr{M}^\varphi)$, which proves the theorem.

Suppose now that a bounded continuous function $H : \mathbb{R}^2 \to \mathbb{C}$ satisfies (6) for all $\eta, \zeta \in \mathscr{D}(\mathbb{R})$. Then

$$
\begin{aligned}
&\int \left(\int \int e^{\mathrm{i}t(r-s)}\xi(r,s)H(s,t)\,\mathrm{d}r\,\mathrm{d}s \right) \mathrm{d}t \\
&\qquad = \int \left(\int \int e^{\mathrm{i}(t+\mathrm{i})(r-s)}\xi(r,s)H(r,t)\,\mathrm{d}r\,\mathrm{d}s \right) \mathrm{d}t.
\end{aligned}
$$

for all $\xi \in \mathscr{D}(\mathbb{R}^2)$, since both sides of this equation define distributions on $\mathscr{D}(\mathbb{R}^2)$ which, by (6), agree on $\mathscr{D}(\mathbb{R}) \bar{\otimes} \mathscr{D}(\mathbb{R})$. Equivalently, this means that

$$
\begin{aligned}
&\int \left(\int \int e^{\mathrm{i}tr}\xi(s+r,s)H(s,t)\,\mathrm{d}r\,\mathrm{d}s \right) \mathrm{d}t \\
&\qquad = \int \left(\int \int e^{\mathrm{i}tr}e^{-r}\xi(s+r,s)H(s+r,t)\,\mathrm{d}r\,\mathrm{d}s \right) \mathrm{d}t
\end{aligned}
$$

or, since the functions $(r,s) \mapsto \xi(s+r,s)$ exhaust all of $\mathscr{D}(\mathbb{R}^2)$,

$$\int \left(\int \int e^{itr} \xi(r,s) H(s,t) \, dr \, ds \right) dt \tag{8}$$
$$= \int \left(\int \int e^{itr} e^{-r} \xi(r, s-r) H(s,t) \, dr \, ds \right) dt$$

for all $\xi \in \mathscr{D}(\mathbb{R}^2)$. The expression

$$\langle \xi, \widetilde{H} \rangle = \int \left(\int \int e^{itr} \xi(r,s) H(s,t) \, dr \, ds \right) dt \quad (\xi \in \mathscr{D}(\mathbb{R}^2))$$

defines a distribution on $\mathscr{D}(\mathbb{R}^2)$. Being the partial Fourier transform of the bounded continuous function H, the distribution $\widetilde{H}$ is tempered and hence extends by continuity to the Schwartz space $\mathscr{S}(\mathbb{R}^2)$. On the other hand, consider the linear operator T defined on $\mathscr{D}(\mathbb{R}^2)$ by

$$[T\xi](r,s) = e^{-r} \xi(r, s-r) \quad (\xi \in \mathscr{D}(\mathbb{R}^2)).$$

Then (8) means that

$$\langle (I-T)\xi, \widetilde{H} \rangle = 0 \quad (\xi \in \mathscr{D}(\mathbb{R}^2)). \tag{9}$$

For a given $\xi \in \mathscr{D}(\mathbb{R}^2)$ with *supp* $\xi \cap (\{0\} \times \mathbb{R}) = \emptyset$, we define a sequence $\{\xi_n\} \subset \mathscr{D}(\mathbb{R}^2)$ by

$$\xi_n(r,s) = \begin{cases} \displaystyle\sum_{k=0}^{n} e^{-kr} \xi(r, s-kr) \text{ if } r \geq 0 \\ \displaystyle -\sum_{k=0}^{n} e^{(k+1)r} \xi(r, s+(k+1)r) \text{ if } r \leq 0. \end{cases}$$

Then

$$[(I-T)\xi_n](r,s) = \xi(r,s) - e^{-(n+1)r} \xi(r, s-(n+1)r) \quad (r \geq 0)$$
$$[(I-T)\xi_n](r,s) = \xi(r,s) - e^{-(n+1)r} \xi(r, s+(n+1)r) \quad (r \leq 0)$$

and it follows that $\lim_n (I-T)\xi_n = \xi$ in the topology of the Schwartz space $\mathscr{S}(\mathbb{R}^2)$, so that $\langle \xi, \widetilde{H} \rangle = \lim_n \langle (I-T)\xi_n, \widetilde{H} \rangle = 0$. This means that *supp* $\widetilde{H} \subset \{0\} \times \mathbb{R}$.

Consequently. for every $\eta \in \mathscr{D}(\mathbb{R})$ with *supp* $\eta \notin 0$ we get

$$0 = \int \left(\int e^{itr} \eta(r) H(s,t) \, dr \right) dt = [F\eta](-t) H(s,t) \, dt.$$

Thus, for each fixed $s \in \mathbb{R}$, the partial Fourier transform of H is a distribution with support equal to $\{0\}$ and is therefore a finite linear combination of derivatives of the Dirac delta function, which in turn implies that $H(s,\cdot)$ is a polynomial. Since H is bounded, it follows that $H(s,\cdot)$ is a constant function, which proves (7).

23.20. Notes. The material of this section is due to Connes (1973a); Connes and Takesaki (1973a), and Takesaki (1973b). They give (Connes & Takesaki, 1977) an explicit construction of the continuous decomposition of a factor of type III arising from the group measure space construction. Also, there are concrete descriptions of the integrable and dominant weights and the continuous decomposition of von Neumann algebras associated with foliations (Connes, 1979a, b).

For our exposition we have used Connes (1973a); Connes and Takesaki (1977); Takesaki (1973b, 1976).

24 The Flow of Weights

In this section, we introduce the (smooth) flow of weights on countably decomposable properly infinite W^*-algebras and we prove a "spectral multiplicity theorem" for integrable weights.

24.1. Let $\mathscr{M}$ be a countably decomposable properly infinite W^*-algebras, $\omega \in W_{nsf}(\mathscr{M})$ a dominant weight on $\mathscr{M}$ and $(\mathscr{N}, \theta, \tau)$ the corresponding continuous decomposition of $\mathscr{M}$, that is, $\mathscr{N} = \mathscr{M}^{\omega}, \theta : \mathbb{R} \to Aut(\mathscr{M}^{\omega})$ is a continuous action and τ is an n.s.f. trace on $\mathscr{M}^{\omega}$ with $\tau \circ \theta_s = e^{-s}\tau$, $(s \in \mathbb{R})$, such that

$$(\mathscr{M}, \sigma^{\omega}, \omega) \approx (\mathscr{R}(\mathscr{M}^{\omega}, \theta), \hat{\theta}, \hat{\tau}). \tag{1}$$

Recall (23.7) that there exists an s-continuous unitary representation $u : \mathbb{R} \to U(\mathscr{M})$ such that $\theta_s = \mathrm{Ad}(u(s))|\mathscr{M}^{\omega}$ and $\omega_{u(s)} = e^{-s}\omega$, that is, $\sigma_t^{\omega}(u(s)) = e^{-ist}u(s)$ $(s, t \in \mathbb{R})$.

By Corollary 23.12, the pair $(\mathscr{Z}(\mathscr{M}^{\omega}), \theta|\mathscr{Z}(\mathscr{M}^{\omega}))$ is uniquely determined modulo *-isomorphisms and will be called *the smooth flow of weights on* $\mathscr{M}$.

Sometimes it is more convenient to regard the continuous action of the additive group $\mathbb{R}$ on $\mathscr{Z}(\mathscr{M}^{\omega})$ as a continuous action $F = F^{\mathscr{M}}$ of the multiplicative group $\mathbb{R}_*^+ = \{\lambda \in \mathbb{R}; \lambda > 0\}$ defined by

$$F_{\lambda} = F_{\lambda}^{\mathscr{M}} = \theta_{-\ln(\lambda)}|\mathscr{Z}(\mathscr{M}^{\omega}) \quad (\lambda \in \mathbb{R}_*^+). \tag{2}$$

Thus, the pair $(\mathscr{Z}(\mathscr{M}^{\omega}), F : \mathbb{R}_*^+ \to Aut(\mathscr{Z}(\mathscr{M}^{\omega}))$ is the smooth flow of weights on $\mathscr{M}$. The justification of this terminology will follow from the contents of this section, which we briefly present in the following:

Let $W_{int}(\mathscr{M})$ be the set of all integrable normal semifinite weights on $\mathscr{M}$ and $W_{int}^{\infty}(\mathscr{M})$ the subset of $W_{int}(\mathscr{M})$ consisting of weights of infinite multiplicity.

For every $\varphi \in W_{ns}(\mathscr{M})$ one defines a canonical surjective mapping

$$c_{\varphi} : W_{ns}(\mathscr{M}) \ni \psi \mapsto c_{\varphi}(\psi) \in Proj(\mathscr{Z}(\mathscr{M}^{\varphi}))$$

whose properties are studied in Sections 24.2 and 24.3.

In particular, the mapping

$$c_{\omega} : W_{int}(\mathscr{M}) \ni \varphi \mapsto c_{\omega}(\varphi) \in Proj(\mathscr{Z}(\mathscr{M}^{\omega}))$$

has the properties $c_\omega(\lambda\varphi) = F_\lambda(c_\omega(\varphi)), \psi \precsim \varphi \Rightarrow c_\omega(\psi) \le c_\omega(\varphi) \Rightarrow \check{\psi} \precsim \check{\varphi}, c_\omega\left(\sum_n^\oplus \varphi_n\right) = \vee_n c_\omega(\varphi_n)$ for all $\varphi, \psi, \varphi_n \in W_{int}(\mathscr{M}), \lambda > 0$. Thus,

$$W_{int}^\infty(\mathscr{M})/\approx$$

can be identified with $Proj(\mathscr{Z}(\mathscr{M}^\omega))$ in such a way that the smooth flow of weights F_λ corresponds to the mapping $\varphi \mapsto \lambda\varphi$ $(\lambda > 0)$.

For each $\varphi \in W_{int}(\mathscr{M})$, there exists a *-isomorphism

$$p_\varphi : \mathscr{Z}(\mathscr{M}^\varphi) \to \mathscr{Z}(\mathscr{M}^\omega)c_\omega(\varphi),$$

uniquely determined, such that $p_\varphi(c_\varphi(\psi)) = c_\omega(\psi)c_\omega(\varphi)$ for all $\psi \in W_{ns}(\mathscr{M})$. Also, there exist an n.s.f. operator-valued weight $P_\varphi : \mathscr{M}^+_{\mathbf{s}(\varphi)} \to \overline{(\mathscr{M}^\varphi)}^+$ and a unique n.s.f. trace τ_φ on $\mathscr{M}^\varphi$ such that $\varphi|\mathscr{M}^+_{s(\varphi)} = \tau_\varphi \circ P_\varphi$. In particular, $\tau_{\lambda\varphi} = \lambda\tau_\varphi(\lambda > 0)$.

Finally, there exists a unique mapping

$$\nu : W_{int}(\mathscr{M}) \ni \varphi \mapsto \nu_\varphi \in W_n(\mathscr{Z}(\mathscr{M}^\omega))$$

such that $\nu_\varphi(c_\omega(\psi)) = \tau_\varphi(c_\varphi(\psi))$ $(\psi \in W_{int}(\mathscr{M}))$. For all $\varphi, \psi, \varphi_1, \dots, \varphi_m \in W_{int}(\mathscr{M})$ and $\lambda > 0$, we have $\nu_{\lambda\varphi} = \lambda\nu_\varphi \circ F_\lambda^{-1}, \psi \le \varphi \Leftrightarrow \nu_\psi \le \nu_\varphi, \nu_{\sum_n^\oplus \varphi_n} = \sum_n \nu_{\varphi_n}$ and, if $\mathscr{M}$ is of type III, then the mapping ν is also surjective. One thus obtains a "spectral multiplicity" theory for integrable weights.

24.2. Let $\varphi, \psi \in W_{ns}(\mathscr{M})$ and $\theta = \theta(\varphi, \psi)$ be the balanced weight on $\mathscr{P} = \mathscr{M} \bar{\otimes} \mathscr{F}_2$ with $\mathbf{s}(\theta) = \begin{pmatrix} \mathbf{s}(\varphi) & 0 \\ 0 & \mathbf{s}(\psi) \end{pmatrix}$. We define a projection $c_\varphi(\psi) \in \mathscr{Z}(\mathscr{M}^\varphi)$ by

$$\mathbf{z}_{\mathscr{P}^\theta}\left(\begin{pmatrix} 0 & 0 \\ 0 & \mathbf{s}(\psi) \end{pmatrix}\right)\begin{pmatrix} \mathbf{s}(\varphi) & 0 \\ 0 & 0 \end{pmatrix} = \begin{pmatrix} c_\varphi(\psi) & 0 \\ 0 & 0 \end{pmatrix}.$$

It is easy to see that the left-hand side necessarily has the form of the right-hand side and that $c_\varphi(\psi) \in \mathscr{Z}(\mathscr{M}^\varphi)$. The next proposition contains some computation rules for $c_\varphi(\psi)$.

Proposition. *Let $\varphi, \varphi_n, \psi, \psi_n, \psi' \in W_{ns}(\mathscr{M}), \lambda > 0, \sigma \in Aut(\mathscr{M})$, and let $w \in \mathscr{M}$ be a partial isometry with $ww^* \in \mathscr{M}^\varphi$. Then*

$$c_\varphi(\varphi_w) = \mathbf{z}_{\mathscr{M}^\varphi}(ww^*) \tag{1}$$

$$c_{\varphi_w}(\psi) = w^* c_\varphi(\psi) w \tag{2}$$

$$c_{\lambda\varphi}(\psi) = c_\varphi(\lambda^{-1}\psi) \tag{3}$$

$$c_{\varphi\circ\sigma}(\psi \circ \sigma) = \sigma^{-1}(c_\varphi(\psi)) \tag{4}$$

$$\psi' \precsim \psi \Rightarrow c_\varphi(\psi') \le c_\varphi(\psi) \tag{5}$$

$$c_{\sum_n^\oplus \varphi_n}(\psi) = \sum_n c_{\varphi_n}(\psi) \tag{6}$$

$$c_\varphi\left(\sum_n^\oplus \psi_n\right) = \bigvee_n c_\varphi(\psi_n) \tag{7}$$

$$c_\varphi(\check{\psi}) = c_\varphi(\psi). \tag{8}$$

Proof. (1) Let $\theta = \theta(\varphi, \varphi_w)$. We have $\mathbf{s}(\varphi_w) = w^*w$ and $\begin{pmatrix} 0 & w \\ 0 & 0 \end{pmatrix} \in \mathscr{P}^\theta$ by Proposition 23.1, hence the projections $\begin{pmatrix} 0 & 0 \\ 0 & \mathbf{s}(\varphi_w) \end{pmatrix}$ and $\begin{pmatrix} ww^* & 0 \\ 0 & 0 \end{pmatrix}$ are equivalent in $\mathscr{P}^\theta$ and therefore have the same central support in $\mathscr{P}^\theta$. Since $\mathbf{s}(\varphi)$ is the unit element in $\mathscr{M}^\varphi$, (1) follows from the definition of $c_\varphi(\varphi_w)$.

(2) Let $\theta = \theta(\varphi, \psi), \theta' = (\varphi_w, \psi)$ and $W = \begin{pmatrix} w & 0 \\ 0 & 1 \end{pmatrix} \in \mathscr{P}$. Then $WW^* \in \mathscr{P}^\theta$ and $\theta' = \theta_W$, hence the mapping $\mathscr{P}^{\theta'} \ni X \mapsto WXW^* \in \mathscr{P}^{\theta_{WW^*}} \subset \mathscr{P}^\theta$ is a *-isomorphism. Since $\mathbf{s}(\varphi_w) = w^*w$ and $\mathbf{s}(\varphi)w = w$, we have

$$\begin{aligned}
\begin{pmatrix} c_{\varphi_w}(\psi) & 0 \\ 0 & 0 \end{pmatrix} &= \mathbf{z}_{\mathscr{P}^{\theta'}}\left(\begin{pmatrix} 0 & 0 \\ 0 & \mathbf{s}(\psi) \end{pmatrix}\right)\begin{pmatrix} \mathbf{s}(\varphi_w) & 0 \\ 0 & 0 \end{pmatrix} \\
&= W^*\mathbf{z}_{\mathscr{P}^\theta}\left(W\begin{pmatrix} 0 & 0 \\ 0 & \mathbf{s}(\psi) \end{pmatrix}W^*\right)W\begin{pmatrix} w^*w & 0 \\ 0 & 0 \end{pmatrix} \\
&= \begin{pmatrix} w^* & 0 \\ 0 & 1 \end{pmatrix}\mathbf{z}_{\mathscr{P}^\theta}\left(\begin{pmatrix} 0 & 0 \\ 0 & \mathbf{s}(\psi) \end{pmatrix}\right)\begin{pmatrix} w & 0 \\ 0 & 0 \end{pmatrix} \\
&= \begin{pmatrix} w^* & 0 \\ 0 & 1 \end{pmatrix}\mathbf{z}_{\mathscr{P}^\theta}\left(\begin{pmatrix} 0 & 0 \\ 0 & \mathbf{s}(\psi) \end{pmatrix}\right)\begin{pmatrix} \mathbf{s}(\varphi) & 0 \\ 0 & 0 \end{pmatrix}\begin{pmatrix} w & 0 \\ 0 & 0 \end{pmatrix} \\
&= \begin{pmatrix} w^* & 0 \\ 0 & 1 \end{pmatrix}\begin{pmatrix} c_\varphi(\psi) & 0 \\ 0 & 0 \end{pmatrix}\begin{pmatrix} w & 0 \\ 0 & 0 \end{pmatrix} = \begin{pmatrix} w^*c_\varphi(\psi)w & 0 \\ 0 & 0 \end{pmatrix},
\end{aligned}$$

which proves (2).

(3) Let $\theta = \theta(\lambda\varphi, \psi)$ and $\theta' = \theta(\varphi, \lambda^{-1}\psi)$. We have $\mathbf{s}(\lambda\varphi) = \mathbf{s}(\varphi), \mathbf{s}(\lambda^{-1}\psi) = \mathbf{s}(\psi)$ and $\theta = \lambda\theta'$, hence $\mathscr{P}^\theta = \mathscr{P}^{\theta'}$ and (3) now follows easily using the definition of $c_\varphi(\psi)$.

(4) Let $\theta = \theta(\varphi, \psi), \theta' = \theta(\varphi \circ \sigma, \psi \circ \sigma)$. We have $\mathbf{s}(\varphi \circ \sigma) = \sigma^{-1}(\mathbf{s}(\varphi)), \mathbf{s}(\psi \circ \sigma) = \sigma^{-1}(\mathbf{s}(\psi)$ and $\theta' = \theta \circ (\sigma \bar{\otimes} \iota)$, hence $\mathscr{P}^{\theta'} = (\sigma \bar{\otimes} \iota)^{-1}(\mathscr{P}^\theta)$ and (4) follows again by definition.

(5) Let $\psi' = \psi_v$ with a partial isometry $v \in \mathscr{M}, vv^* \in \mathscr{M}^\psi$. Let $\theta = \theta(\varphi, \psi), \theta' = \theta(\varphi, \psi')$. For $V = \begin{pmatrix} 1 & 0 \\ 0 & v \end{pmatrix} \in \mathscr{P}$ we have $VV^* \in \mathscr{P}^\theta$ and $\theta' = \theta_V$, hence the mapping $\mathscr{P}^{\theta'} \ni X \mapsto VXV^* \in \mathscr{P}^{\theta_{VV^*}} \subset \mathscr{P}^\theta$ is a *-isomorphism. Since $\mathbf{s}(\psi') = v^*v$ and $vv^* \le \mathbf{s}(\psi)$, we have

$$\begin{aligned}
\begin{pmatrix} c_\varphi(\psi') & 0 \\ 0 & 0 \end{pmatrix} &= \mathbf{z}_{\mathscr{P}^{\theta'}}\left(\begin{pmatrix} 0 & 0 \\ 0 & v^*v \end{pmatrix}\right)\begin{pmatrix} \mathbf{s}(\varphi) & 0 \\ 0 & 0 \end{pmatrix} \\
&= V^*\mathbf{z}_{\mathscr{P}^\theta}\left(V\begin{pmatrix} 0 & 0 \\ 0 & v^*v \end{pmatrix}V^*\right)V\begin{pmatrix} \mathbf{s}(\varphi) & 0 \\ 0 & 0 \end{pmatrix} \\
&= \begin{pmatrix} 1 & 0 \\ 0 & v^* \end{pmatrix}\mathbf{z}_{\mathscr{P}^\theta}\left(\begin{pmatrix} 0 & 0 \\ 0 & vv^* \end{pmatrix}\right)\begin{pmatrix} \mathbf{s}(\varphi) & 0 \\ 0 & 0 \end{pmatrix}\begin{pmatrix} 1 & 0 \\ 0 & v \end{pmatrix} \\
&\le \begin{pmatrix} 1 & 0 \\ 0 & v^* \end{pmatrix}\mathbf{z}_{\mathscr{P}^\theta}\left(\begin{pmatrix} 0 & 0 \\ 0 & \mathbf{s}(\psi) \end{pmatrix}\right)\begin{pmatrix} \mathbf{s}(\varphi) & 0 \\ 0 & 0 \end{pmatrix}\begin{pmatrix} 1 & 0 \\ 0 & v \end{pmatrix} \\
&= \begin{pmatrix} 1 & 0 \\ 0 & v^* \end{pmatrix}\begin{pmatrix} c_\varphi(\psi) & 0 \\ 0 & 0 \end{pmatrix}\begin{pmatrix} 1 & 0 \\ 0 & v \end{pmatrix} = \begin{pmatrix} c_\varphi(\psi) & 0 \\ 0 & 0 \end{pmatrix},
\end{aligned}$$

proving (5).

(6) Assume that the projections $e_n = \mathbf{s}(\varphi_n)$ ($n \in \mathbb{N}$) are mutually orthogonal and $\varphi = \sum_n \varphi_n$. Then $e_n \in \mathcal{M}^\varphi, \varphi_n = \varphi_{e_n}$ and $\sum_n e_n = \mathbf{s}(\varphi)$. Using (2), we obtain $\sum_n c_{\varphi_n}(\psi) = \sum_n e_n c_\varphi(\psi) e_n = (\sum_n e_n) c_\varphi(\psi) = c_\varphi(\psi)$.

(7) Assume that the projections $e_n = \mathbf{s}(\psi_n)$ ($n \in \mathbb{N}$) are mutually orthogonal and that $\psi = \sum_n \psi_n$, so that $e_n \in \mathcal{M}^\psi, \psi_n = \psi_{e_n}$ and $\sum_n e_n = \mathbf{s}(\psi)$. Let $\theta_n = \theta(\varphi, \psi_n)$, $\theta = \theta(\varphi, \psi)$. Then $\theta_n = \theta_{E_n}$, where $E_n = \begin{pmatrix} 1 & 0 \\ 0 & e_n \end{pmatrix} = \mathbf{s}(\theta_n)$, and $\theta = \sum_n \theta_n$. It is clear that $\begin{pmatrix} 0 & 0 \\ 0 & e_n \end{pmatrix}$ has the same central projection in $\mathscr{P}^{\theta_n}$ as in $\mathscr{P}^\theta$. It follows that

$$\begin{aligned}
\begin{pmatrix} c_\varphi(\psi) & 0 \\ 0 & 0 \end{pmatrix} &= \mathbf{z}_{\mathscr{P}^\theta}\left(\begin{pmatrix} 0 & 0 \\ 0 & \mathbf{s}(\psi) \end{pmatrix}\right)\begin{pmatrix} \mathbf{s}(\varphi) & 0 \\ 0 & 0 \end{pmatrix} \\
&= \bigvee_n \mathbf{z}_{\mathscr{P}^\theta}\left(\begin{pmatrix} 0 & 0 \\ 0 & e_n \end{pmatrix}\right)\begin{pmatrix} \mathbf{s}(\psi) & 0 \\ 0 & 0 \end{pmatrix} \\
&= \bigvee_n \begin{pmatrix} c_\varphi(\psi_n) & 0 \\ 0 & 0 \end{pmatrix} = \begin{pmatrix} \bigvee_n c_\varphi(\psi_n) & 0 \\ 0 & 0 \end{pmatrix}.
\end{aligned}$$

proving (7).

(8) This follows easily using (5), (7), and the definition (23.15) of the weight $\check{\varphi}$. □

24.3 Proposition. *For every $\varphi \in W_{ns}(\mathcal{M})$, the mapping $\psi \mapsto c_\varphi(\psi)$ defines an order isomorphism between the equivalence classes of weights $\psi \in W_{ns}(\mathcal{M}), \psi \precsim \check{\varphi}$ of infinite multiplicity, and the projections in $\mathcal{Z}(\mathcal{M}^\varphi)$.*

Proof. Let $\check{\varphi} = \sum_n \varphi_n$ with $\varphi_n = \varphi_{u_n} \in W_{ns}(\mathcal{M})$, where $u_n = (w_n \mathbf{s}(\varphi))^*$ and $\{w_n\}$ is a sequence of isometries in $\mathcal{M}$ such that $\sum_n w_n w_n^* = 1$ (see 23.15). Using Proposition 24.2, for $\psi \in W_{ns}(\mathcal{M})$ we obtain $c_{\check{\varphi}}(\psi) = \sum_n c_{\varphi_n}(\psi) = \sum_n u_n^* c_\varphi(\psi) u_n = \sum_n w_n c_\varphi(\psi) w_n^*$. Recall (23.15) that the mapping $\pi : \mathcal{M} \ni x \mapsto \sum_{ij} w_i^* x w_j \bar{\otimes} e_{ij} \in \mathcal{M} \bar{\otimes} \mathcal{F}$, where $\{e_{ij}\}$ is a system of matrix units of the countably decomposable infinite type I factor $\mathcal{F}$, establishes a *-isomorphism $\pi : (\mathcal{M}, \check{\varphi}) \approx (\mathcal{M} \bar{\otimes} \mathcal{F}, \varphi \bar{\otimes} tr)$. We have $\pi(\mathcal{Z}(\mathcal{M}^{\check{\varphi}})) = \mathcal{Z}((\mathcal{M} \bar{\otimes} \mathcal{F})^{\varphi \bar{\otimes} tr}) = \mathcal{Z}(\mathcal{M}^\varphi) \bar{\otimes} 1$, and the previous equation shows that $\pi(c_{\check{\varphi}}(\psi)) = c_\varphi(\psi) \bar{\otimes} 1$. Thus, we may assume that φ is of infinite multiplicity, that is, $\varphi \approx \check{\varphi}$.

If $e \in \mathcal{Z}(\mathcal{M}^\varphi)$ is a nonzero projection, then $\varphi_e \precsim \varphi$ is of infinite multiplicity, since $\mathcal{M}^{\varphi_e} = \mathcal{M}^\varphi e$ and $\mathcal{M}^\varphi$ is properly infinite. Since $c_\varphi(\varphi_e) = e c_\varphi(\varphi) e = e\mathbf{s}(\varphi) e$, it follows that the mapping c_φ is surjective.

Let $\psi \precsim \varphi, \psi' \precsim \varphi$. There exist partial isometries $w, w' \in \mathcal{M}$ with $e = ww^* \in \mathcal{M}^\varphi, e' = w'w'^* \in \mathcal{M}^\varphi$, such that $\psi = \varphi_w, \psi' = \varphi_{w'}$ and then $c_\varphi(\psi) = \mathbf{z}_{\mathcal{M}^\varphi}(e), c_\varphi(\psi') = \mathbf{z}_{\mathcal{M}^\varphi}(e')$. If ψ and ψ' are both of infinite multiplicity and $c_\varphi(\psi') \le c_\varphi(\psi)$, then the projections $e, e' \in \mathcal{M}^\varphi$ are properly infinite and $\mathbf{z}_{\mathcal{M}^\varphi}(e') \le \mathbf{z}_{\mathcal{M}^\varphi}(e)$; hence $e' \prec e$ in $\mathcal{M}^\varphi$, so that $\psi' \precsim \psi$ by Proposition 23.2. □

24.4 Corollary. *The mapping $\varphi \mapsto c_\omega(\varphi)$ establishes an order isomorphism between the equivalence classes of integrable weights of infinite multiplicity on $\mathcal{M}$ and the projections in $\mathcal{Z}(\mathcal{M}^\omega)$; we have*

$$c_\omega(\lambda\varphi) = F_\lambda(c_\omega(\varphi)); \varphi \in W_{int}(\mathcal{M}), \lambda > 0. \tag{1}$$

Proof. We use the notation of Section 24.1. The first part of the corollary follows clearly from Proposition 24.3 and Theorem 23.4. Since $c_\omega(\varphi) = c_\omega(\check{\varphi})$, in order to prove (1) we may assume that $\varphi \in W^\infty_{int}(\mathcal{M})$. Then there exists a projection $q \in \mathcal{Z}(\mathcal{M}^\omega)$ with $\varphi \approx \omega_q$. For $s \in \mathbb{R}$, we have

$$e^{-s}\varphi \approx e^{-s}\omega_q = (e^{-s}\omega)_q = \omega_{u(s)q} = \omega_{u(s)qu(s)^*u(s)} = \omega_{\theta_s(q)u(s)} \approx \omega_{\theta_s(q)} \tag{2}$$

hence $c_\omega(\varphi) = q$ and $c_\omega(e^{-s}\varphi) = \theta_s(q)$, that is,

$$c_\omega(e^{-s}\varphi) = \theta_s(c_\omega(\varphi)), \tag{3}$$

proving (1), as $F_\lambda = \theta_{-ln(\lambda)}(\lambda > 0)$. □

24.5 Proposition. *For each $\varphi \in W_{int}(\mathcal{M})$, there exists a unique *-isomorphism $p_\varphi : \mathcal{Z}(\mathcal{M}^\varphi) \to \mathcal{Z}(\mathcal{M}^\omega)c_\omega(\varphi)$ such that*

$$p_\varphi(c_\varphi(\psi)) = c_\omega(\psi)c_\omega(\varphi) \quad (\psi \in W_{ns}(\mathcal{M})). \tag{1}$$

Proof. Since $\varphi \in W_{int}(\mathcal{M})$, we have $\varphi \lesssim \omega$, hence there exists a partial isometry $w \in \mathcal{M}$ with $ww^* \in \mathcal{M}^\omega$ such that $\varphi = \omega_w$. We thus obtain a *-isomorphism $\mathcal{M}^\varphi \ni x \mapsto wxw^* \in (\mathcal{M}^\omega)_{ww^*}$ and, by restriction, a *-isomorphism $\mathcal{Z}(\mathcal{M}^\varphi) \ni x \mapsto wxw^* \in \mathcal{Z}(\mathcal{M}^\omega)ww^*$. On the other hand, we have $c_\omega(\varphi) = \mathbf{z}_{\mathcal{M}^\omega}(ww^*)$, hence the mapping $\mathcal{Z}(\mathcal{M}^\omega)c_\omega(\varphi) \ni z \mapsto zww^* \in \mathcal{Z}(\mathcal{M}^\omega)ww^*$ is a *-isomorphism. By composing these two mappings, we obtain a *-isomorphism $p_\varphi : \mathcal{Z}(\mathcal{M}^\varphi) \to \mathcal{Z}(\mathcal{M}^\omega)c_\omega(\varphi)$.

If $e \in \mathcal{Z}(\mathcal{M}^\varphi)$ is a projection, then $\varphi_e = \omega_{we}$, hence $c_\omega(\varphi_e) = \mathbf{z}_{\mathcal{M}^\omega}(wew^*) \in \mathcal{Z}(\mathcal{M}^\omega)$ and, clearly, $c_\omega(\varphi_e) \le c_\omega(\varphi)$. We have $c_\omega(\varphi_e)ww^* = \mathbf{z}_{\mathcal{M}^\omega}(wew^*)ww^* = wew^*$ (since $e \in \mathcal{Z}(\mathcal{M}^\varphi)$, hence $wew^* \in \mathcal{Z}(\mathcal{M}^\omega)_{ww^*}$) and therefore $c_\omega(\varphi_e) = p_\varphi(e)$.

Assume now that $e = c_\varphi(\psi)$ with $\psi \in W_{ns}(\mathcal{M})$. Since $\varphi = \omega_w$, we have $e = c_{\omega_w}(\psi) = w^*c_\omega(\psi)w$, hence $p_\varphi(e)ww^* = ww^*c_\omega(\varphi)ww^* = c_\omega(\psi)ww^* = c_\omega(\psi)\mathbf{z}_{\mathcal{M}^\omega}(ww^*)ww^* = c_\omega(\psi)c_\omega(\varphi)ww^*$, and therefore $p_\varphi(c_\varphi(\psi)) = c_\omega(\psi)c_\omega(\varphi)$.

The uniqueness of p_φ follows obviously from the surjectivity of c_φ. □

24.6. Let $\varphi \in W_{int}(\mathcal{M})$. Then the continuous action $\sigma^\varphi : \mathbb{R} \to Aut(\mathcal{M}_{\mathbf{s}(\varphi)})$ is integrable, so that it defines a σ^φ-invariant n.s.f. operator-valued weight $P_\varphi : \mathcal{M}^+_{\mathbf{s}(\varphi)} \to (\overline{\mathcal{M}^\varphi})^+$.

Proposition. *For each $\varphi \in W_{int}(\mathcal{M})$, there exists a unique n.s.f. trace τ_φ on $\mathcal{M}^\varphi$ such that*

$$\varphi|\mathcal{M}^+_{\mathbf{s}(\varphi)} = \tau_\varphi \circ P_\varphi. \tag{1}$$

Proof. If $\varphi = \omega$, we can take $\tau_\omega = \tau$ from the continuous decomposition $(\mathcal{M}^\omega, \theta, \tau)$ of $\mathcal{M}$ constructed in Theorem 23.6 (see also Theorem 23.16).

If φ is just integrable, then there exists a projection $e \in \mathcal{M}^\omega$ such that $\varphi \approx \omega_e$ and we may assume that $\varphi = \omega_e$. Then $\mathbf{s}(\varphi) = e, \mathcal{M}^\varphi = e\mathcal{M}^\omega e, \sigma^\varphi_t = \sigma^\omega_t|e\mathcal{M}e, c_\omega(\varphi) = \mathbf{z}_{\mathcal{M}^\omega}(e)$ and the *-isomorphism

$$p_\varphi : \mathcal{Z}(e\mathcal{M}^\omega e) \to \mathcal{Z}(\mathcal{M}^\omega)\mathbf{z}_{\mathcal{M}^\omega}(e)$$

is the inverse of the mapping $z \mapsto ze$. The operator-valued weight $P_\varphi : (e\mathcal{M}e)^+ \to \overline{(e\mathcal{M}^\omega e)}^+ \subset \overline{(\mathcal{M}^\omega)}^+$ is defined by $P_\varphi(exe) = \int \sigma_t^\varphi(exe)\,\mathrm{d}t = e\left(\int \sigma_t^\omega(x)\,\mathrm{d}t\right)e = eP_\omega(x)e = P_\omega(exe)$ $(x \in \mathcal{M}^+)$, hence $P_\varphi = P_\omega | e\mathcal{M}^+e$. Then

$$\tau_\varphi = \tau_\omega | e\mathcal{M}^\omega e \tag{2}$$

is an n.s.f. trace on $\mathcal{M}^\varphi = e\mathcal{M}^\omega e$ and for every $x \in e\mathcal{M}^+e$ we have

$$(\tau_\varphi \circ P_\varphi)(x) = \tau_\omega(P_\omega(x)) = \omega(x) = \varphi(x).$$

The uniqueness of τ_φ is obvious. □

Note that

$$\tau_{\lambda\varphi} = \lambda\tau_\varphi \quad (\lambda > 0). \tag{3}$$

24.7. We now prove the "spectral multiplicity theorem" for integrable weights.

Theorem. *For every $\varphi \in W_{int}(\mathcal{M})$, there exists a unique normal weight ν_φ on $\mathcal{Z}(\mathcal{M}^\omega)$ such that*

$$\nu_\varphi(c_\omega(\psi)) = \tau_\varphi(c_\varphi(\psi)) \quad (\psi \in W_{int}(\mathcal{M})). \tag{1}$$

The mapping $\varphi \mapsto \nu_\varphi$ establishes a bijective correspondence between the equivalence classes of weights $\varphi \in W_{int}(\mathcal{M})$ and a certain set of normal weights ν_φ on $\mathcal{Z}(\mathcal{M}^\omega)$; for $\varphi, \psi, \varphi_1, \dots, \varphi_m \in W_{int}(\mathcal{M})$ and $\lambda > 0$ we have

$$\nu_{\lambda\varphi} = \lambda\nu_\varphi \circ F_\lambda^{-1} \tag{2}$$

$$\psi \precsim \varphi \Leftrightarrow \nu_\psi \le \nu_\varphi \tag{3}$$

$$\nu_{\sum_n^\oplus \varphi_n} = \sum_n \nu_{\varphi_n}. \tag{4}$$

If $\mathcal{M}$ is of type III, then every normal weight on $\mathcal{Z}(\mathcal{M}^\omega)$ is of the form ν_φ for some $\varphi \in W_{int}(\mathcal{M})$, hence the mapping $\varphi \mapsto \nu_\varphi$ establishes an order isomorphism $\nu : W_{int}(\mathcal{M})/\approx \to W_n(\mathcal{Z}(\mathcal{M}^\omega))$.

Proof. Using Propositions 24.5, 24.6, we define ν_φ by

$$\nu_\varphi(z) = \tau_\varphi(p_\varphi^{-1}(zc_\omega(\varphi))) \quad (z \in \mathcal{Z}(\mathcal{M}^\omega)^+) \tag{5}$$

and then, for $\psi \in W_{int}(\mathcal{M})$ we have $\nu_\varphi(c_\omega(\psi)) = \tau_\varphi(p_\varphi^{-1}(c_\omega(\psi)c_\omega(\varphi))) = \tau_\varphi(p_\varphi^{-1}(p_\varphi(c_\omega(\psi))) = \tau_\varphi(c_\omega(\psi))$, which proves the existence of ν_φ. The uniqueness of ν_φ is obvious.

If $\varphi \approx \varphi'$, then $(\mathcal{M}^\varphi, c_\varphi, \tau_\varphi) \approx (\mathcal{M}^{\varphi'}, c_{\varphi'}, \tau_{\varphi'})$, hence $\nu_\varphi = \nu_{\varphi'}$. Thus, in proving (3) and (4), we can consider just weights of the form $\varphi = \omega_e$ with $e \in Proj(\mathcal{M}^\omega)$. In this case, it follows from (5) and 24.6.(2) that

$$\nu_\varphi(z) = \tau_\omega(ze) \quad (z \in \mathcal{Z}(\mathcal{M}^\omega)^+). \tag{6}$$

Note that $\nu_\omega = \tau_\omega|\mathcal{Z}(\mathcal{M}^\omega)^+ = +\infty$, as $\mathcal{M}^\omega$ is properly infinite, that is, it has no nonzero finite central projection.

Let $e, f \in Proj(\mathcal{M}^\omega)$ and $\varphi = \omega_e, \psi = \omega_f$. If $\nu_\psi \leq \nu_\varphi$, then, using (6), we obtain $\tau_\omega(zf) \leq \tau_\omega(ze)$ for all $z \in \mathcal{Z}(\mathcal{M}^\omega)^+$. By ([L], E.7.13) we infer that $f \prec e$ in $\mathcal{M}^\omega$ and, by Proposition 23.2, we deduce $\psi = \omega_f \lesssim \omega_e = \varphi$. Conversely, if $\psi \lesssim \varphi$, then $f \prec e$ in $\mathcal{M}^\omega$, hence $\tau_\omega(zf) \leq \tau_\omega(ze)$ for every $z \in \mathcal{Z}(\mathcal{M}^\omega)^+$, so that $\nu_\psi \leq \nu_\varphi$. We have thus proved (3).

If e and f are orthogonal, then $\varphi + \psi = \omega_{e+f}$, hence $\nu_{\varphi+\psi} = \tau_\omega(z(e+f)) = \tau_\omega(ze) + \tau_\omega(zf) = \nu_\varphi(z) + \nu_\psi(z)$ for $z \in \mathcal{Z}(\mathcal{M}^\omega)^+$, and $\nu_{\varphi+\psi} = \nu_\varphi + \nu_\psi$ which proves (4).

Using 24.2.(3) and 24.6.(3), we obtain $\nu_{\lambda\varphi}(c_\varphi(\psi)) = \tau_{\lambda\varphi}(c_{\lambda\varphi}(\psi)) = \lambda\tau_\varphi(c_\varphi(\lambda^{-1}\psi)) = \lambda\nu_\varphi(c_\omega(\lambda^{-1}\psi)) = \lambda\nu_\varphi(F_\lambda^{-1}(c_\omega(\psi)))$ and this proves equality (2).

If $\mathcal{M}$ is of type III, then, by Proposition 3/23.11, $\mathcal{M}^\omega$ is of type II$_\infty$ and so, by Corollary 12.14, every normal weight on $\mathcal{Z}(\mathcal{M}^\omega)$ is of the form $\tau_\omega(e\cdot) = \nu_{\omega_e}$ for some projection $e \in \mathcal{M}^\omega$; this proves the last assertion of the theorem. □

24.8. If the W^*-algebra $\mathcal{M}$ is semifinite, then, by Proposition 2/23.11, the continuous decomposition $(\mathcal{N}, \theta)$ of $\mathcal{M}$ is $*$-isomorphic to $(\mathcal{N}^\theta \bar{\otimes} \mathcal{L}^\infty(\mathbb{R}), \iota \bar{\otimes} \mathrm{Ad}(\boldsymbol{\lambda}))$; hence the smooth flow of weights on $\mathcal{M}$ is the pair $(\mathcal{Z}(\mathcal{N}^\theta) \bar{\otimes} \mathcal{L}^\infty(\mathbb{R}), \iota \bar{\otimes} \mathrm{Ad}(\boldsymbol{\lambda}))$, or $(\mathcal{Z}(\mathcal{N}^\theta) \bar{\otimes} \mathcal{L}^\infty(\mathbb{R}_*^+), \iota \bar{\otimes} \mathrm{Ad}(\boldsymbol{\lambda}))$. If we represent the abelian W^*-algebra $\mathcal{Z}(\mathcal{N}^\theta)$ in the form $\mathcal{L}^\infty(\Omega)$ (see Dixmier, 1957, 1969; Sakai, 1970; Strătilă & Zsidó, 1977, 2005, 9.37), then the smooth flow of weights on $\mathcal{M}$ becomes a continuous action $F : \mathbb{R}_*^+ \to Aut(\mathcal{L}^\infty(\Omega \times \mathbb{R}_*^+))$, defined by the point transformations $F_\lambda(\xi, \mu) = (\xi, \lambda^{-1}\mu)$ $(\xi \in \Omega;\ \lambda, \mu \in \mathbb{R}_*^+)$.

In particular, if $\mathcal{M}$ is a semifinite factor, then $\mathcal{N}^\theta$ is a factor (by Proposition 1/23.11), hence the smooth flow of weights is the continuous action $F : \mathbb{R}_*^+ \to Aut(\mathcal{L}^\infty(\mathbb{R}_*^+))$ defined by $F_\lambda(\mu) = \lambda^{-1}\mu$, $(\lambda, \mu \in \mathbb{R}_*^+)$.

By Proposition 1/23.11, $\mathcal{M}$ is a factor if and only if the smooth flow of weights on $\mathcal{M}$ is ergodic. For type III factors, there are descriptions of the smooth flow of weights based on their "discrete decomposition" (see 30.8, 30.10).

24.9. The smooth flow of weights is just the continuous part of a more general, but rather artificial, object called the *global flow of weights* on $\mathcal{M}$, which we describe in the following.

There exists a pair $(Q_\mathcal{M}, q_\mathcal{M})$ consisting of an abelian W*-algebra $Q_\mathcal{M}$ and a mapping $q_\mathcal{M}$ of $W_{ns}(\mathcal{M})$ onto the set of countably decomposable projections of $Q_\mathcal{M}$, such that for all $\varphi, \psi, \varphi_n \in W_{ns}(\mathcal{M})$ we hare

$$q_\mathcal{M}(\varphi) = q_\mathcal{M}(\check{\varphi}) \tag{1}$$

$$q_\mathcal{M}(\psi) \leq q_\mathcal{M}(\varphi) \Leftrightarrow \check{\psi} \lesssim \check{\varphi} \tag{2}$$

$$q_\mathcal{M}\left(\sum_n^{\oplus} \varphi_n\right) = \bigvee_n (q_\mathcal{M}(\varphi_n)); \tag{3}$$

and there exist $*$-isomorphisms

$$q_\varphi : \mathcal{Z}(\mathcal{M}^\varphi) \to Q_\mathcal{M} q_\mathcal{M}(\varphi) \quad (\varphi \in W_{ns}(\mathcal{M})), \tag{4}$$

uniquely determined, such that

$$q_\varphi(e) = q_{\mathcal{M}}(\varphi_e) \quad (e \in Proj(\mathcal{Z}(\mathcal{M}^\varphi))). \tag{5}$$

Indeed, let $\{e_{\varphi,\psi}\}_{\varphi,\psi \in W_{ns}(\mathcal{M})}$ be a system of matrix units in $\mathcal{B}(\ell^2(W_{ns}(\mathcal{M})))$ put $\mathcal{R} = \mathcal{M} \bar{\otimes} \mathcal{B}(\ell^2(W_{ns}(\mathcal{M})))$ and consider the n.s.f. weight Φ on $\mathcal{R}$ defined by $\Phi(\sum x_{\varphi,\psi} \bar{\otimes} e_{\varphi,\psi}) = \sum_\varphi \varphi(x_{\varphi,\varphi})$. Then the conditions of the earlier statement are satisfied with $Q_{\mathcal{M}} = \mathcal{Z}(\mathcal{R}^\Phi)$ and $q_{\mathcal{M}}(\varphi) = \mathbf{z}_{\mathcal{R}^\Phi}(\mathbf{s}(\varphi) \bar{\otimes} e_{\varphi,\varphi}) = \sum_\psi c_\psi(\varphi) \bar{\otimes} e_{\psi,\psi}$ $(\varphi \in W_{ns}(\mathcal{M}))$. The verification is similar to that in the case of the smooth flow of weights.

The *global flow of weights* on $\mathcal{M}$ is an action $\mathfrak{F}^{\mathcal{M}} : \mathbb{R}^+_* \to Aut(Q_{\mathcal{M}})$, uniquely determined, such that $\mathfrak{F}^{\mathcal{M}}(q_{\mathcal{M}}(\varphi)) = q_{\mathcal{M}}(\lambda\varphi)$ $(\varphi \in W_{ns}(\mathcal{M}), \lambda \in \mathbb{R}^+_*)$.

Proposition. *The weight $\varphi \in W_{ns}(\mathcal{M})$ is integrable if and only if the mapping*

$$\mathbb{R}^+_* \ni \lambda \to \mathfrak{F}^{\mathcal{M}}_\lambda(q_{\mathcal{M}}(\varphi)) \in Q_{\mathcal{M}} \tag{6}$$

is s-continuous.

Proof. Assume that φ is integrable. Since $\check{\varphi}$ is integrable and $q_{\mathcal{M}}(\check{\varphi}) = q_{\mathcal{M}}(\varphi)$, we may assume that φ is of infinite multiplicity. In this case, there exists by Corollary 24.4, a unique projection $e \in \mathcal{Z}(\mathcal{M}^\omega)$ such that $\varphi \approx \omega_e$. Using (5) and 24.4.(2), we get

$$\mathfrak{F}^{\mathcal{M}}_\lambda(q_{\mathcal{M}}(\varphi)) = q_\omega(F_\lambda(e)) \quad (\lambda \in \mathbb{R}^+_*), \tag{7}$$

so that the mapping (6) is s-continuous.

Conversely, assume that the mapping (6) is s-continuous. Let $e = q_{\mathcal{M}}(\varphi) \in Q_{\mathcal{M}}$ and $f = \bigvee_{\lambda>0} \mathfrak{F}^{\mathcal{M}}_\lambda(e) \in Q_{\mathcal{M}}$. By assumption, if $\{\lambda_n\}$ is a sequence of nonzero positive rational numbers converging to $\lambda > 0$, then $\mathfrak{F}^{\mathcal{M}}_{\lambda_n}(e) \xrightarrow{s} \mathfrak{F}^{\mathcal{M}}_\lambda(e)$. It follows that f is a countably decomposable projection. Since f is clearly $\mathfrak{F}^{\mathcal{M}}$-invariant, we infer that $f = q_{\mathcal{M}}(\omega)$, where ω is the fixed dominant weight on $\mathcal{M}$. Finally, since $q_{\mathcal{M}}(\varphi) = e \le f = q_{\mathcal{M}}(\omega)$, it follows that $\varphi \lesssim \check{\varphi} \lesssim \omega$, and so φ is indeed integrable. □

Corollary. *The projection $q_{\mathcal{M}}(\omega) \in Q_{\mathcal{M}}$ is the largest countably decomposable projection $q \in Q_{\mathcal{M}}$ such that the mapping $\mathbb{R}^+_* \ni \lambda \mapsto \mathfrak{F}^{\mathcal{M}}_\lambda(q) \in Q_{\mathcal{M}}$ is s-continuous and $(Q_{\mathcal{M}} q_{\mathcal{M}}(\omega), \mathfrak{F}^{\mathcal{M}} | Q_{\mathcal{M}} q_{\mathcal{M}}(\omega)) \approx (\mathcal{Z}(\mathcal{M}^\omega), F^{\mathcal{M}})$.*

More precisely, by (7) we have

$$\mathfrak{F}^{\mathcal{M}}_\lambda(q_{\mathcal{M}}(\varphi)) = q_\omega(F^{\mathcal{M}}_\lambda(c_\omega(\varphi))) \quad (\varphi \in W_{int}(\mathcal{M}), \lambda \in \mathbb{R}^+_*) \tag{8}$$

so that

$$q_\omega : (\mathcal{Z}(\mathcal{M}^\omega), F^{\mathcal{M}}) \approx (Q_{\mathcal{M}} q_{\mathcal{M}}(\omega), \mathfrak{F}^{\mathcal{M}} | Q_{\mathcal{M}} q_{\mathcal{M}}(\omega)). \tag{9}$$

In what follows, we shall always understand by *the flow of weights on $\mathcal{M}$ the smooth* flow of weights on $\mathcal{M}$.

24.10. Notes. The material of this section is due to Connes and Takesaki (1977). For type III factors arising from the group measure space construction, they computed (Connes & Takesaki, 1977) the smooth flow of weights using Mackey's procedure based on the theory of virtual groups (Mackey, 1966; Ramsay, 1971). For further results, and connections with the theory of measure groupoids, we refer to Connes (1978b); Hahn (1978a, 1978b); Ramsay (1971); Samuelides (1978); Sauvageot (1978). The flow of weights has a natural interpretation for von Neumann algebras associated with foliations (Connes, 1979a, 1979b). Connes and Woods (1980b, 1985) obtained a characterization of Araki–Woods factors in terms of the flow of weights. Previously, another characterization of Araki–Woods factors had been obtained by Størmer (1971, 1972, 1975).

For our exposition, we have used (Connes & Takesaki, 1977).

25 The Fundamental Homomorphism

In this section, we introduce the fundamental homomorphism $mod : Aut(\mathscr{M}) \to Aut(F^{\mathscr{M}})$, prove its continuity and relate it to a classical invariant, called the fundamental group.

25.1. Let $\mathscr{M}$ be a countably decomposable properly infinite W^*-algebra with (smooth) flow of weights $(\mathscr{P}_{\mathscr{M}}, F_{\mathscr{M}})$. Abstractly, $\mathscr{P}_{\mathscr{M}}$ represents the equivalence classes of integrable normal semifinite weights of infinite multiplicity on $\mathscr{M}$ and the action of $F_\lambda^{\mathscr{M}}$ is multiplication by $\lambda > 0$, that is, $\mathscr{P}_{\mathscr{M}} = W_{int}^{\infty}(\mathscr{M})/ \approx \,, F_\lambda^{\mathscr{M}} \varphi = \lambda\varphi$. Concretely, one considers a dominant weight ω on $\mathscr{M}$ and the corresponding continuous decomposition $(\mathscr{M}^\omega, \theta, \tau)$, and then (24.1) $\mathscr{P}_{\mathscr{M}} = \mathscr{Z}(\mathscr{M}^\omega), F_\lambda^{\mathscr{M}} = \theta_{-\ln(\lambda)}|\mathscr{Z}(\mathscr{M}^\omega)$ $(\lambda > 0)$.

Consider the groups $Aut(\mathscr{M})$ and $Aut(F^{\mathscr{M}}) = \{\mathfrak{G} \in Aut(\mathscr{P}_{\mathscr{M}}); \mathfrak{G} \circ F_\lambda^{\mathscr{M}} = F_\lambda^{\mathscr{M}} \circ \mathfrak{G}$ for all $\lambda > 0\} \subset Aut(\mathscr{P}_{\mathscr{M}})$ endowed with the u-topology (2.23).

Let $\sigma \in Aut(\mathscr{M})$. Then $\omega \circ \sigma^{-1}$ is also a dominant weight on $\mathscr{M}$. By Proposition 24.5, there exists a *-isomorphism

$$p_\sigma : \mathscr{Z}(\mathscr{M}^{\omega\circ\sigma^{-1}}) \to \mathscr{Z}(\mathscr{M}^\omega), \tag{1}$$

uniquely determined, such that

$$p_\sigma(c_{\omega\circ\sigma^{-1}}(\varphi)) = c_\omega(\varphi) \quad (\varphi \in W_{int}(\mathscr{M})). \tag{2}$$

On the other hand, we have $\mathscr{M}^{\omega\circ\sigma^{-1}} = \sigma(\mathscr{M}^\omega)$, and we obtain a *-isomorphism $\sigma : \mathscr{Z}(\mathscr{M}^\omega) \to \mathscr{Z}(\mathscr{M}^{\omega\circ\sigma^{-1}})$. We thus obtain a *-automorphism

$$mod(\sigma) = p_\sigma \circ \sigma : \mathscr{Z}(\mathscr{M}^\omega) \to \mathscr{Z}(\mathscr{M}^\omega), \tag{3}$$

uniquely determined, such that (see 24.2.(4))

$$(mod(\sigma))(c_\omega(\varphi)) = c_\omega(\varphi \circ \sigma^{-1}) \quad (\varphi \in W_{int}(\mathscr{M})). \tag{4}$$

Since for every $\varphi \in W_{int}(\mathscr{M})$ and $\lambda > 0$, we have $[mod(\sigma)](F_\lambda^{\mathscr{M}}(c_\omega(\varphi))) = [mod(\sigma)](c_\omega(\lambda\varphi)) = c_\omega(\lambda\varphi \circ \sigma^{-1}) = F_\lambda^{\mathscr{M}}(c_\omega(\varphi \circ \sigma^{-1})) = F_\lambda^{\mathscr{M}}([mod(\sigma)](c_\omega(\varphi)))$, it follows that $mod(\sigma) \in Aut(F^{\mathscr{M}})$.

We thus obtain a group homomorphism $mod : Aut(\mathcal{M}) \to Aut(F^{\mathcal{M}})$, called *the fundamental homomorphism associated with* $\mathcal{M}$.

Note that for every $\sigma \in Aut(\mathcal{M})$ the weight $\omega \circ \sigma^{-1}$ is also dominant, hence $\omega \circ \sigma^{-1} \approx \omega$, and we have

$$u \in U(\mathcal{M}), \omega \circ \sigma^{-1} = \omega \circ \mathrm{Ad}(u) \Rightarrow mod(\sigma) = (\mathrm{Ad}(u) \circ \sigma)|\mathcal{Z}(\mathcal{M}^{\omega}). \tag{5}$$

Indeed, using 23.2.(2) and 23.2.(4), for $\varphi \in W_{int}(\mathcal{M})$ we obtain $(\mathrm{Ad}(u) \circ \sigma)(c_{\omega}(\varphi)) = uc_{\omega \circ \sigma^{-1}}(\varphi \circ \sigma^{-1})u^* = c_{\omega}(\varphi \circ \sigma^{-1})$. In particular,

$$\omega \circ \sigma = \omega \Leftrightarrow mod(\sigma) = \sigma|\mathcal{Z}(\mathcal{M}^{\omega}), \tag{6}$$

$$mod(\sigma) = \iota \Leftrightarrow \left[\begin{array}{l} \textit{there exists } u \in U(\mathcal{M}) \textit{ such that} \\ \omega \circ \mathrm{Ad}(u) \circ \sigma = \omega, (\mathrm{Ad}(u) \circ \sigma)|\mathcal{Z}(\mathcal{M}^{\omega}) = \iota. \end{array} \right. \tag{7}$$

Thus, $mod(\sigma) = \iota$ for all $\sigma \in Int(\mathcal{M})$. Since $Int(\mathcal{M})$ is a normal subgroup of $Aut(\mathcal{M})$, it follows that the fundamental homomorphism can be factored to a homomorphism

$$mod : Out(\mathcal{M}) = Aut(\mathcal{M})/Int(\mathcal{M}) \to Aut(F^{\mathcal{M}}).$$

Note also that from (4), 24.2.(4), and 24.5.(1), we obtain

$$p_{\varphi \circ \sigma} = mod(\sigma^{-1}) \circ p_{\varphi} \circ \sigma \quad (\varphi \in W_{int}(\mathcal{M}), \sigma \in Aut(\mathcal{M})). \tag{8}$$

25.2 Theorem. *For every properly infinite W^*-algebra $\mathcal{M}$ with a separable predual, the fundamental homomorphism $mod : Aut(\mathcal{M}) \to Aut(F^{\mathcal{M}})$ is u-continuous.*

The proof will be given in Section 25.5, after some preparation.

Since $mod(\sigma) = \iota$ for every $\sigma \in \overline{Int(\mathcal{M})}$, we infer that

Corollary. *If $\mathcal{M}$ is a properly infinite W^*-algebra with separable predual, then $mod(\sigma) = \iota$ for every $\sigma \in \overline{Int(\mathcal{M})} \subset Aut(\mathcal{M})$.*

25.3. Let $\mathcal{M}$ be a W^*-algebra with separable predual. Then (2.23) $U(\mathcal{M})$ is a polish group, that is, there exists a bounded complete metric d which defines the two-sided uniform structure associated with the topology on $U(\mathcal{M})$.

Let $\sigma : G \to Aut(\mathcal{M})$ be a continuous action of the separable locally compact group G on $\mathcal{M}$. There exists an increasing sequence $\{K_n\}$ of compact subsets of G with $G = \bigcup_n \mathring{K}_n$. Then the set $\mathscr{C}(G, U(\mathcal{M}))$ of all continuous functions $G \to U(\mathcal{M})$, endowed with the topology of uniform convergence on compact subsets, is also a polish topological space, since the equation

$$\delta(u, v) = \sum_n 2^{-n} \sup_{g \in K_n} d(u(g), v(g)) \quad (u, v \in \mathscr{C}(G, U(\mathcal{M}))$$

defines a complete metric δ on $\mathscr{C}(G, U(\mathcal{M}))$ compatible with the topology.

Since the set of all unitary σ-cocycles $Z_{\sigma}(G; U(\mathcal{M}))$ is closed in $\mathscr{C}(G, U(\mathcal{M}))$ it follows that $Z_{\sigma}(G; U(\mathcal{M}))$ *endowed with the topology of uniform convergence on compact subsets is a polish topological space.*

We show that *the mapping* $\partial = \partial_\sigma : U(\mathcal{M}) \to Z_\sigma(G; U(\mathcal{M}))$ *defined by* $(\partial u)(g) = u^*\sigma_g(u)$ $(u \in U(\mathcal{M}), g \in G)$ *is continuous*. Indeed, let $u_n \to u$ in $U(\mathcal{M})$, let $K \subset G$ be a compact set and $\mathcal{L} \subset \mathcal{M}_*$ a $\sigma(\mathcal{M}_*, \mathcal{M})$-compact set. Then (13.5) the set $K\mathcal{L} = \{\varphi \circ \sigma_g; \varphi \in \mathcal{L}, g \in K\} \subset \mathcal{M}_*$ is $\sigma(\mathcal{M}_*, \mathcal{M})$-compact. Since $u_n \to u$ with respect to the Mackey topology τ_ω, we have $\varphi(\sigma_g(u_n)) \to \varphi(\sigma_g(u))$ uniformly with respect to $g \in K$ and $\varphi \in \mathcal{L}$. Since $\mathcal{L}$ was arbitrary, it follows that $\sigma_g(u_n) \to \sigma_g(u)$ in $U(\mathcal{M})$, uniformly for $g \in K$, hence $(\partial u_n)(g) = u_n^*\sigma_g(u_n) \to u^*\sigma_g(u) = (\partial u)(g)$ uniformly for $g \in K$. Consequently, $\partial u_n \to \partial u$ in $Z_\sigma(G; U(\mathcal{M}))$.

Since for $u, v \in U(\mathcal{M})$ we have $\partial u = \partial v \Leftrightarrow uv^* \in U(\mathcal{M}^\sigma)$ and since $U(\mathcal{M}^\sigma)$ is a closed subgroup of $U(\mathcal{M})$, it follows that the homogeneous space $U(\mathcal{M})/U(\mathcal{M}^\sigma)$ with the quotient topology is a polish space and that ∂ can be factored to an injective mapping $\underline{\partial}_\sigma : U(\mathcal{M})/U(\mathcal{M}^\sigma) \to Z_\sigma(G; U(\mathcal{M}))$. According to classical results in the theory of Borel spaces, it follows that $\partial_\sigma(U(\mathcal{M}))$ is *a Borel set in* $Z_\sigma(G; U(\mathcal{M}))$ *and* $\underline{\partial}_\sigma : U(\mathcal{M})/U(\mathcal{M}^\sigma) \to \partial_\sigma(U(\mathcal{M}))$ *is a Borel isomorphism.*

25.4. Let $\mathcal{M}$ be a W^*-algebra with separable predual and ω a fixed n.s.f. weight on $\mathcal{M}$. By Theorem 5.1, the mapping

$$D_\omega : W_{nsf}(\mathcal{M}) \ni \varphi \mapsto [D\varphi : D\omega] \in Z_{\sigma^\omega}(\mathbb{R}; U(\mathcal{M})) \tag{1}$$

is a bijection. We shall consider on $W_{nsf}(\mathcal{M})$ the unique topology such that D_ω be a homeomorphism.

The topology introduced on $W_{nsf}(\mathcal{M})$ does not depend on the fixed n.s.f. weight ω on $\mathcal{M}$, as can be easily seen using Corollary 3.5.

We show that *the mapping*

$$Aut(\mathcal{M}) \ni \sigma \mapsto \omega \circ \sigma^{-1} \in W_{nsf}(\mathcal{M}) \tag{2}$$

is continuous. Indeed, let f be a faithful normal state on $\mathcal{M}$ and $\sigma_n \to \sigma$ in $Aut(\mathcal{M})$. Since $\|f \circ \sigma_n^{-1} - f \circ \sigma^{-1}\| \to 0$, by Proposition 7.18 it follows that $[D(f \circ \sigma_n^{-1}) : D(f \circ \sigma^{-1})]_t \to 1$ uniformly on compact subsets. Also, $\sigma_n([D\omega : Df]_t) \to \sigma([D\omega : Df]_t)$ uniformly on compact subsets (see the last remark in Section 2.23). Therefore,

$$[D(\omega \circ \sigma_n^{-1}) : D(\omega \circ \sigma^{-1})]_t = \sigma_n([D\omega : Df]_t)[D(f \circ \sigma_n^{-1}) : D(f \circ \sigma^{-1})]_t\sigma([Df : D\omega]_t)$$

converges to 1 uniformly on compact subsets, that is, $\omega \circ \sigma_n^{-1} \to \omega \circ \sigma^{-1}$ in $W_{nsf}(\mathcal{M})$.

Clearly, $W_{nsf}(\mathcal{M})$ is a polish space. Consider now

$$W(\omega) = \{\varphi \in W_{nsf}(\mathcal{M}); \varphi \approx \omega\}. \tag{3}$$

We show that $W(\omega)$ *is a Borel set in* $W_{nsf}(\mathcal{M})$ *and that there exists a Borel mapping*

$$u : W(\omega) \ni \varphi \mapsto u(\varphi) \in U(\mathcal{M}) \tag{4}$$

such that

$$\varphi = \omega \circ \mathrm{Ad}(u(\varphi)) \quad (\varphi \in W(\omega)). \tag{5}$$

Indeed, if we identify $W_{nsf}(\mathcal{M})$ with $Z_{\sigma^\omega}(\mathbb{R}; U(\mathcal{M}))$ by the mapping D_ω, then the range of the mapping ∂ considered in Section 25.3 is just the set $W(\omega)$ which is, therefore, a Borel set. Moreover, from Section 25.3 it follows that the mapping $\partial : U(\mathcal{M})/U(\mathcal{M}^{\omega}) \to W(\omega)$ is a Borel isomorphism, so that the existence of the mapping u is a consequence of the classical result concerning the existence of a Borel section for the canonical mapping $U(\mathcal{M}) \to U(\mathcal{M})/U(\mathcal{M}^{\omega})$.

25.5. Proof of Theorem 25.2. Let ω be a dominant weight on $\mathcal{M}$. The mapping

$$\mathfrak{m} : Aut(\mathcal{M}) \ni \sigma \mapsto \mathrm{Ad}(u(\omega \circ \sigma^{-1})) \circ \sigma \in Aut(\mathcal{M}) \tag{1}$$

is a Borel mapping since the mapping $Aut(\mathcal{M}) \ni \sigma \mapsto \omega \circ \sigma^{-1} \in W(\omega)$ is continuous (25.4.(2)), the mapping $W(\omega) \ni \varphi \mapsto u(\varphi) \in U(\mathcal{M})$ is a Borel mapping (25.4.(4)), and the mapping $U(\mathcal{M}) \ni u \mapsto \mathrm{Ad}(u) \in Aut(\mathcal{M})$ is continuous (2.23). Also, by 25.1.(5) we have

$$mod(\sigma) = \mathfrak{m}(\sigma)|\mathcal{Z}(\mathcal{M}^{\sigma}) \quad (\sigma \in Aut(\mathcal{M})). \tag{2}$$

It follows that mod is a Borel homomorphism between the polish groups $Aut(\mathcal{M})$ and $Aut(F^{\mathcal{M}})$, and so is continuous.

25.6. Let $\mathcal{M}$ be an infinite semifinite factor with separable predual. We recall (24.8) that $(\mathcal{P}_{\mathcal{M}}, F^{\mathcal{M}}) \approx (\mathcal{L}^{\infty}(\mathbb{R}_*^+), \mathrm{Ad}(\boldsymbol{\lambda}))$. It is then easy to check that the mapping $\mathbb{R}_*^+ \ni \lambda \mapsto F_{\lambda}^{\mathcal{M}} \in Aut(F^{\mathcal{M}})$ is a surjective group isomorphism.

Let τ be an n.s.f. trace on $\mathcal{M}, \sigma \in Aut(\mathcal{M})$ and $\lambda \in \mathbb{R}_*^+$. Then

$$mod(\sigma) = F_{\lambda}^{\mathcal{M}} \Leftrightarrow \tau \circ \sigma^{-1} = \lambda\tau. \tag{1}$$

Indeed, let $\tau \circ \sigma^{-1} = \lambda\tau$ and $mod(\sigma) = F_{\mu}^{\mathcal{M}}$ for some $\lambda, \mu \in \mathbb{R}_*^+$. Then we have $\varphi \circ \sigma^{-1} \approx \mu\varphi$ for every $\varphi \in W_{int}^{\infty}(\mathcal{M})$. Let $\varepsilon > 0$. By Section 23.18, there exists $a \in \mathcal{M}^+$ with $1 - \varepsilon \leq a \leq 1 + \varepsilon$ such that $\varphi = \tau_a \in W_{int}^{\infty}(\mathcal{M})$. Then, there exists $u \in U(\mathcal{M})$ such that $\mu\varphi = \varphi \circ \sigma^{-1} \circ \mathrm{Ad}(u)$. Thus, for $x \in \mathcal{M}^+$ we obtain $\mu\tau_a(x) = \mu\varphi(x) = \varphi(\sigma^{-1}(uxu^*)) = (\tau \circ \sigma^{-1})(\sigma(a)^{1/2}uxu^*\sigma(a)^{1/2}) = \lambda\tau(uu^*\sigma(a)^{1/2}uxu^*\sigma(a)^{1/2}uu^*) = \lambda\tau_{u^*\sigma(a)u}(x)$, hence $\mu a = \lambda u^*\sigma(a)u$. It follows that $(1 - \varepsilon)\lambda \leq (1 + \varepsilon)\mu$ and $(1 - \varepsilon)\mu \leq (1 + \varepsilon)\lambda$. Since $\varepsilon > 0$ was arbitrary, we conclude that $\mu = \lambda$.

Clearly, the earlier result is interesting only for type II_{∞}, factors since if $\mathcal{M}$ is a type I_{∞} factor, $Aut(\mathcal{M}) = Int(\mathcal{M})$, and $mod(\sigma) = \iota, \tau \circ \sigma = \tau$ for all $\sigma \in Aut(\mathcal{M})$.

25.7. Consider now a type II_1 factor $\mathcal{N}$ and let ν be the unique faithful finite normal trace on $\mathcal{N}$ with $\nu(1) = 1$. Consider also the type II_{∞} factor $\mathcal{M} = \mathcal{N} \bar{\otimes} \mathcal{F}_{\infty}$ and the n.s.f. trace $\tau = \nu \bar{\otimes} tr$ on $\mathcal{M}$, where tr is the canonical trace on the countably decomposable type I_{∞} factor $\mathcal{F}_{\infty}$.

Let $0 < \lambda \leq 1$ be fixed. If e and e_1 are projections in $\mathcal{N}$ with $\nu(e) = \nu(e_1) = \lambda$, then $e \sim e_1$ in $\mathcal{N}$ and so the reduced algebras $e\mathcal{N}e$ and $e_1\mathcal{N}e_1$ are *-isomorphic. Thus, the isomorphism class of $e\mathcal{N}e$ does not depend on the projection $e \in \mathcal{N}$ with $\nu(e) = \lambda$ and will be denoted by $\mathcal{N}^{\lambda}$. We show that

$$\mathcal{N}^{\lambda} \approx \mathcal{N} \Leftrightarrow \textit{there exists } \sigma \in Aut(\mathcal{M}) \textit{ with } \tau \circ \sigma = \lambda\sigma. \tag{1}$$

Indeed, let $e \in \mathcal{N}$ be a projection with $\nu(e) = \lambda$ and let $\pi : \mathcal{N} \to e\mathcal{N}e \subset \mathcal{N}$ be a *-isomorphism. Since $e \bar{\otimes} 1_{\mathcal{F}}$ and $1_{\mathcal{N}} \bar{\otimes} 1_{\mathcal{F}} = 1_{\mathcal{M}}$ are properly infinite projections in the countably decomposable factor $\mathcal{M}$, we have $e \bar{\otimes} 1_{\mathcal{F}} \sim 1_{\mathcal{M}}$, hence there exists $w \in \mathcal{M}$ such that $w^*w = e \bar{\otimes} 1_{\mathcal{F}}$ and $ww^* = 1_{\mathcal{M}}$. Then the mapping $x \mapsto wxw^*$ defines a *-isomorphism $\rho : e\mathcal{N}e \bar{\otimes} \mathcal{F}_{\infty} \to \mathcal{N} \bar{\otimes} \mathcal{F}_{\infty}$.

We thus obtain a $*$-automorphism $\sigma = \rho \circ (\pi \bar{\otimes} \iota) \in Aut(\mathcal{M})$. If $f \in \mathcal{F}_\infty$ is any minimal projection, then $\tau(1_{\mathcal{N}} \bar{\otimes} f) = 1$ and $(\tau \circ \sigma)(1_{\mathcal{N}} \bar{\otimes} f) = \tau(w(e \bar{\otimes} f)w^*) = \tau(e \bar{\otimes} f) = \nu(e) = \lambda$, hence $\tau \circ \sigma = \lambda\tau$.

Conversely, let $\sigma \in Aut(\mathcal{M})$ with $\tau \circ \sigma = \lambda\tau$, let $e \in \mathcal{N}$ be a projection with $\nu(e) = \lambda$ and let $f \in \mathcal{F}_\infty$ be any minimal projection. Then $\tau(\sigma(1_{\mathcal{N}} \bar{\otimes} f)) = \lambda\tau(1_{\mathcal{N}} \bar{\otimes} f) = \tau(e \bar{\otimes} f)$, hence $e \bar{\otimes} f \sim \sigma(1_{\mathcal{N}} \bar{\otimes} f)$ in $\mathcal{M}$. It follows that $e\mathcal{N}e \approx (e \bar{\otimes} f)(\mathcal{N} \bar{\otimes} \mathcal{F}_\infty)(e \bar{\otimes} f) \approx \sigma(1_{\mathcal{N}} \bar{\otimes} f)(\mathcal{N} \bar{\otimes} \mathcal{F}_\infty)\sigma(1_{\mathcal{N}} \bar{\otimes} f) \approx (1_{\mathcal{N}} \bar{\otimes} f)(\mathcal{N} \bar{\otimes} \mathcal{F}_\infty)(1_{\mathcal{N}} \bar{\otimes} f) \approx \mathcal{N}$, and so $\mathcal{N}^\lambda \approx \mathcal{N}$.

For $\lambda \geq 1$, we shall write $\mathcal{N}^\lambda \approx \mathcal{N}$ if $\mathcal{N}^{1/\lambda} \approx \mathcal{N}$. The assertion (1) extends obviously to every $\lambda > 0$. It follows that the set

$$G(\mathcal{N}) = \{\lambda > 0; (\mathcal{N}^\lambda \approx \mathcal{N})\}$$

is a subgroup of $\mathbb{R}_*^+$, called *the fundamental group* of the type II_1 factor $\mathcal{N}$.

For the type II_∞ factor $\mathcal{M}$, we shall define *the fundamental group* $G(\mathcal{M})$ by

$$G(\mathcal{M}) = \{\lambda > 0; \text{there exists } \sigma \in Aut(\mathcal{M}) \text{ with } \tau \circ \sigma = \lambda\tau\}$$

where τ is any n.s.f. trace on $\mathcal{M}$.

From the considerations of this section and Section 25.6, it follows that the fundamental homomorphism, more precisely its range, is an extension of the fundamental group to properly infinite factors. For type III factors, the fundamental homomorphism will be studied in more detail later (see 30.9, 30.10).

25.8. Notes. The material or this section is due to Connes and Takesaki (1977). The fundamental group of a type II_1 factor is a classical invariant introduced by Murray and von Neumann (1936, 1937, 1943, IV).

For our exposition, we have used Connes and Takesaki (1977); Murray and von Neumann (1936, 1937, 1943, IV).

26 The Extension of the Modular Automorphism Group

In this section, we extend the modular automorphism groups $\{\sigma_t^\varphi\}_{t\in\mathbb{R}}$ to families $\{\sigma_c^\varphi\}_c$ indexed by unitary $F^{\mathcal{M}}$-cocycles; similar extensions are given for Connes cocycles.

26.1. Let $\mathcal{M}$ be a countably decomposable properly infinite W^*-algebra and $(\mathcal{P}_{\mathcal{M}}, F^{\mathcal{M}})$ the flow of weights on $\mathcal{M}$.

Consider the set $Z(F^{\mathcal{M}}) = Z_{F^{\mathcal{M}}}(\mathbb{R}_*^+; U(\mathcal{P}_{\mathcal{M}}))$ of all unitary $F^{\mathcal{M}}$-cocycles, the mapping $\partial : U(\mathcal{P}_{\mathcal{M}}) \to Z(F^{\mathcal{M}})$ defined by $(\partial v)(\lambda) = v^* F_\lambda^{\mathcal{M}}(v)$ $(v \in U(\mathcal{P}_{\mathcal{M}}), \lambda \in \mathbb{R}_*^+)$, and put $B(F^{\mathcal{M}}) = \partial(U(\mathcal{P}_{\mathcal{M}}))$. It is easy to check that, with respect to pointwise multiplications, $Z(F^{\mathcal{M}})$ is an abelian group and $B(F^{\mathcal{M}})$ is a subgroup of $Z(F^{\mathcal{M}})$, so that we can consider the quotient group $H(F^{\mathcal{M}}) = Z(F^{\mathcal{M}})/B(F^{\mathcal{M}})$.

We define an injective group homomorphism $\mathbb{R} \ni t \mapsto \bar{t} \in Z(F^{\mathcal{M}})$ by $\bar{t}(\lambda) = \lambda^{-it}$, $(t \in \mathbb{R}, \lambda \in \mathbb{R}_*^+)$.

Let ω be a dominant weight on $\mathcal{M}$ and $(\mathcal{N}, \theta, \tau)$ a continuous decomposition of $\mathcal{M}$ with $\mathcal{N} = \mathcal{M}^\omega$ and $(\mathcal{M}, \sigma^\omega, \omega) \approx (\mathcal{R}(\mathcal{N}, \theta), \hat{\theta}, \hat{\tau})$. The $*$-isomorphism $\mathcal{M} \approx \mathcal{R}(\mathcal{M}^\omega, \theta)$ shows that there exists an s-continuous unitary representation $u : \mathbb{R} \to U(\mathcal{M})$ such that $\mathcal{M} = \mathcal{R}\{\mathcal{M}^\omega, u(\mathbb{R})\}, \theta_s = \mathrm{Ad}(u(s))|\mathcal{M}^\omega$ and $\omega \circ \mathrm{Ad}(u(s)) = e^{-s}\omega$, that is, $\sigma_t^\omega(u(s)) = e^{-ist}u(s)$ $(s, t \in \mathbb{R})$ (see 23.7).

We shall identify $\mathbb{R}$ with $\mathbb{R}_*^+$ via the mapping $s \mapsto e^{-s}$. Then the flow of weights is the pair $(\mathscr{Z}(\mathscr{M}^\omega), \theta|\mathscr{Z}(\mathscr{M}^\omega)), Z(F^{\mathscr{M}})$ consists of all continuous functions $c : \mathbb{R} \to U(\mathscr{Z}(\mathscr{M}^\omega))$ with the property $c(s+t) = c(s)\theta_s(c(t))$ $(s, t \in \mathbb{R})$, for $v \in U(\mathscr{Z}(\mathscr{M}^\omega))$ we have $(\partial v)(s) = v^*\theta_s(v)$ $(s \in \mathbb{R})$, and for $t \in \mathbb{R}$ we have $\bar{t}(s) = e^{-ist}$ $(s \in \mathbb{R})$.

26.2. Assume that the W^*-algebra $\mathscr{M}$ is realized as a von Neumann algebra $\mathscr{M} \subset \mathscr{B}(\mathscr{H})$. Note that $\mathscr{M} \approx \mathscr{R}(\mathscr{M}^\omega, \theta) \subset \mathscr{B}(\mathscr{L}^2(\mathbb{R}, \mathscr{H}))$ is also a realization of $\mathscr{M}$ as a von Neumann algebra.

Let $c \in Z(F^{\mathscr{M}})$. We define a unitary operator $U_c \in \mathscr{Z}(\mathscr{M}^\omega) \bar{\otimes} \mathscr{L}^\infty(\mathbb{R}) \subset \mathscr{B}(\mathscr{L}^2(\mathbb{R}, \mathscr{H}))$ by

$$(U_c\xi)(s) = \theta_s^{-1}(c(s))\xi(s) = c(-s)^{-1}\xi(s) \quad (\xi \in \mathscr{L}^2(\mathbb{R}, \mathscr{H}), s \in \mathbb{R}).$$

For $a \in \mathscr{M}^\omega$, we have $\pi_\theta(a) \in \mathscr{M}^\omega \bar{\otimes} \mathscr{L}^\infty(\mathbb{R})$, hence $U_c\pi_\theta(a)U_c^* = \pi_\theta(a)$. For $u(s) = 1 \bar{\otimes} \boldsymbol{\lambda}(s)$ $(s \in \mathbb{R})$, we have

$$\begin{aligned}(U_c(1 \bar{\otimes} \boldsymbol{\lambda}(s))U_c^*\xi)(r) &= c(-r)^{-1}((1 \bar{\otimes} \boldsymbol{\lambda}(s))U_c^*\xi)(r) = c(-r)^{-1}(U_c^*\xi)(r-s)\\ &= c(-r)^{-1}c(s-r)\xi(r-s) = c(-r)^{-1}c(-r)\theta_r^{-1}(c(s))\xi(r-s)\\ &= \theta_r^{-1}(c(s))((1 \bar{\otimes} \boldsymbol{\lambda}(s))\xi)(r) = (\pi_\theta(c(s))(1 \bar{\otimes} \boldsymbol{\lambda}(s))\xi)(r)\ (\xi \in \mathscr{L}^2(\mathbb{R}, \mathscr{H}), r \in \mathbb{R}).\end{aligned}$$

It follows that U_c determines a unique $*$-automorphism

$$\sigma_c^\omega = \mathrm{Ad}(U_c)|\mathscr{M} \in Aut(\mathscr{M}) \tag{1}$$

unique with the properties

$$\sigma_c^\omega(a) = a \quad (a \in \mathscr{M}^\omega) \tag{2}$$
$$\sigma_c^\omega(u(s)) = c(s)u(s) \quad (s \in \mathbb{R}). \tag{3}$$

If $c_1, c_2 \in Z(F^{\mathscr{M}})$, then the unitary operators $U_{c_1}, U_{c_2} \in \mathscr{Z}(\mathscr{M}^\omega) \bar{\otimes} \mathscr{L}^\infty(\mathbb{R})$ commute and $U_{c_1}U_{c_2} = U_{c_1c_2}$, so that

$$\sigma_{c_1c_2}^\omega = \sigma_{c_1}^\omega\sigma_{c_2}^\omega. \tag{4}$$

Hence the mapping $Z(F^{\mathscr{M}}) \ni c \mapsto \sigma_c^\omega \in Aut(\mathscr{M})$ is a group homomorphism.

For $t \in \mathbb{R}$, the $*$-automorphism $\sigma_{\bar{t}}^\omega \in Aut(\mathscr{M})$ is uniquely determined by $\sigma_{\bar{t}}^\omega(a) = a$ $(a \in \mathscr{M}^\omega)$ and $\sigma_{\bar{t}}^\omega(u(s)) = e^{-ist}u(s)$ $(s \in \mathbb{R})$, hence

$$\sigma_{\bar{t}}^\omega = \sigma_t^\omega \quad (t \in \mathbb{R}). \tag{5}$$

Since ω is the dual weight corresponding to the trace τ, we have $\omega = \tau \circ P_\omega$ where $P_\omega : \mathscr{M}^+ \to \overline{(\mathscr{M}^\omega)^+}$ is the n.s.f. operator-valued weight defined by $P_\omega(x) = \int \sigma_t^\omega(x)\,\mathrm{d}t$ $(x \in \mathscr{M}^+)$. Since σ_c^ω commutes with σ_t^ω $(t \in \mathbb{R})$, and $\sigma_c^\omega|\mathscr{M}^\omega = \iota$, it follows that $P_\omega \circ \sigma_c^\omega = \sigma_c^\omega \circ P_\omega = P_\omega$ and hence

$$\omega \circ \sigma_c^\omega = \omega \quad (c \in Z(F^{\mathscr{M}})). \tag{6}$$

If $x \in \mathcal{M}(\sigma^\omega, \{s\})$, that is, $x \in \mathcal{M}$ and $\sigma_t^\omega(x) = e^{-ist}x$ $(t \in \mathbb{R})$, then $u(s)^*x \in \mathcal{M}^\omega$, so that $u(s)^*x = \sigma_c^\omega(u(s)^*x) = u(s)^*c(s)^*x$. Therefore,

$$x \in \mathcal{M}(\sigma^\omega, \{s\}) \Rightarrow \sigma_c^\omega(x) = c(s)x \textit{ for all } c \in Z(F^{\mathcal{M}}). \tag{7}$$

This condition also determines uniquely the *-automorphism σ_c^ω.

Let $\alpha \in Aut(\mathcal{M})$ be such that $\omega \circ \alpha = \omega$ and let $c \in Z(F^{\mathcal{M}})$. Then $\alpha(\mathcal{Z}(\mathcal{M}^\omega)) = \mathcal{Z}(\mathcal{M}^\omega)$ and the function $\alpha(c) : \mathbb{R} \ni s \mapsto \alpha(c(s)) \in \mathcal{Z}(\mathcal{M}^\omega)$ defines a unique element $\alpha(c) \in Z(F^{\mathcal{M}})$. If $x \in \mathcal{M}(\sigma^\omega, \{s\})$, then $\alpha(x) \in \mathcal{M}(\sigma^\omega, \{s\})$ and so $\sigma_{\alpha(c)}^\omega(\alpha(x)) = \alpha(c(s))\alpha(x) = \alpha(c(s)x) = \alpha(\sigma_c^\omega(x))$.

Consequently,

$$\alpha \in Aut(\mathcal{M}), \omega \circ \alpha = \omega \Rightarrow \alpha \circ \sigma_c^\omega = \sigma_{u(c)}^\omega \circ \alpha \textit{ for all } c \in Z(F^{\mathcal{M}}). \tag{8}$$

We show that for $c \in Z(F^{\mathcal{M}})$ we have

$$\sigma_c^\omega \in Int(\mathcal{M}) \Leftrightarrow c \in B(F^{\mathcal{M}}). \tag{9}$$

Indeed, if $c \in B(F^{\mathcal{M}})$, then there exists a unitary element $v \in \mathcal{Z}(\mathcal{M}^\omega)$ such that $c(s) = v^*\theta_s(v)$ $(s \in \mathbb{R})$. We have $[\mathrm{Ad}(v^*)](a) = a$ for $a \in \mathcal{M}^\omega$ and $[\mathrm{Ad}(v^*)](u(s)) = v^*u(s)vu(s)^*u(s) = v^*\theta_s(v)u(s) = c(s)u(s)$ for $s \in \mathbb{R}$, hence $\sigma_c^\omega = \mathrm{Ad}(v^*) \in Int(\mathcal{M})$. Conversely, let $\sigma_c^\omega = \mathrm{Ad}(v^*)$ for some $v \in U(\mathcal{M})$. Using (2), we infer that $v \in (\mathcal{M}^\omega)' \cap \mathcal{M}$; hence $v \in \mathcal{Z}(\mathcal{M}^\omega)$ by the relative commutant theorem (23.19) and from (3) it follows that $c = \partial v \in B(F^{\mathcal{M}})$.

Thus, the mapping $c \mapsto \sigma_c^\omega$ can be factored to a group homomorphism

$$\delta_\omega : H(F^{\mathcal{M}}) \to Out(\mathcal{M}), \tag{10}$$

where $\delta_\omega(c/B(F^{\mathcal{M}})) = \sigma_c^\omega/Int(\mathcal{M})$ $(c \in Z(F^{\mathcal{M}}))$. We shall see later that the mapping δ_ω does not depend on the choice of the dominant weight ω on $\mathcal{M}$.

Finally, we notice the following important result

$$\{\sigma_c^\omega; c \in Z(F^{\mathcal{M}})\} = \{\sigma \in Aut(\mathcal{M}); \sigma|\mathcal{M}^\omega = \iota\}. \tag{11}$$

Indeed, the inclusion "$\subset$" follows from (2). Conversely, let $\sigma \in Aut(\mathcal{M})$ be such that $\sigma|\mathcal{M}^\omega = \iota$, and define $c(s) = \sigma(u(s))u(s)^* \in U(\mathcal{M})$ $(s \in \mathbb{R})$. For $x \in \mathcal{M}^\omega$ and $s \in \mathbb{R}$, we have $c(s)xc(s)^* = \sigma(u(s))u(s)^*xu(s)\sigma(u(s)^*) = \sigma(u(s))\theta_s^{-1}(x)\sigma(u(s)^*) = \sigma(u(s)\theta_s^{-1}(x)u(s)^*) = \sigma(\theta_s(\theta_s^{-1}(x))) = \sigma(x) = x$, hence $c(s) \in (\mathcal{M}^\omega)' \cap \mathcal{M} = \mathcal{Z}(\mathcal{M}^\omega)$ by the relative commutant theorem (23.19). For $s, t \in \mathbb{R}$, we have $c(s+t) = \sigma(u(s+t))u(s+t)^* = \sigma(u(s))\sigma(u(t))u(t)^*u(s)^* = \sigma(u(s))u(s)^*u(s)\sigma(u(t))u(t)^*u(s)^* = \sigma(u(s))u(s)^*\theta_s(\sigma(u(t))u(t)^*) = c(s)\theta_s(c(t))$, hence $c \in Z(F^{\mathcal{M}})$. Since $\sigma(u(s)) = c(s)u(s)^*$ $(s \in \mathbb{R})$, and $\sigma(a) = a$ $(a \in \mathcal{M}^\omega)$, it follows that $\sigma = \sigma_c^\omega$.

26.3. We now extend the previous construction to integrable n.s.f. weights on $\mathcal{M}$. In formulating the uniqueness condition, we shall use the notation $c_\omega(\varphi)$ and p_φ of Sections 24.2, 24.6.

Theorem. *Let $\mathcal{M}$ be a properly infinite W^*-algebra with separable predual and ω a dominant weight on $\mathcal{M}$.*

*Let φ be an integrable n.s.f. weight on $\mathcal{M}$. For every $c \in Z(F^{\mathcal{M}})$, there exists a unique *-automorphism $\sigma_c^\varphi \in Aut(\mathcal{M})$ such that*

$$x \in \mathcal{M}(\sigma^\varphi; \{s\}) \Rightarrow \sigma_c^\varphi(x) = p_\varphi^{-1}(c(s)c_\omega(\varphi))x \quad (s \in \mathbb{R}). \tag{1}$$

The mapping $Z(F^{\mathcal{M}}) \ni c \mapsto \sigma_c^\varphi \in Aut(\mathcal{M})$ is a group homomorphism and

$$\varphi \circ \sigma_c^\varphi = \varphi \tag{2}$$

$$\sigma_{\tilde{t}}^\varphi = \sigma_t^\varphi. \tag{3}$$

Let φ, ψ be integrable n.s.f. weights on $\mathcal{M}$ and $\theta(\varphi, \psi)$ the corresponding balanced weight on $Mat_2(\mathcal{M})$. For every $c \in Z(F^{\mathcal{M}})$, there exists a unique unitary element $[D\psi : D\varphi]_c \in \mathcal{M}$ such that

$$\sigma_c^{\theta(\varphi,\psi)}\left(\begin{pmatrix} 0 & 0 \\ 1 & 0 \end{pmatrix}\right)\begin{pmatrix} 0 & 0 \\ [D\psi : D\varphi]_c & 0 \end{pmatrix} \tag{4}$$

and we have

$$\sigma_c^\psi = \mathrm{Ad}([D\psi : D\varphi]_c) \circ \sigma_c^\varphi \tag{5}$$

$$[D\psi : D\varphi]_{c_1c_2} = [D\psi : D\varphi]_{c_1}\sigma_{c_1}^\varphi([D\psi : D\varphi]_{c_2}) \tag{6}$$

$$[D\psi, : D\varphi]_{\tilde{t}} = [D\psi : D\varphi]_t. \tag{7}$$

Proof. Since φ is integrable, we have $\varphi \lesssim \omega$ and, as φ is faithful, there exists $v \in \mathcal{M}$ such that $v^*v = \mathbf{s}(\varphi) = 1, e = vv^* \in \mathcal{M}^\omega$, and $\varphi = \omega_v$. The mapping $\Phi_v : \mathcal{M} \ni x \mapsto vxv^* \in e\mathcal{M}e$ is a *-isomorphism such that $\omega_e \circ \Phi_v = \varphi$. It follows that $(\sigma_t^\omega | e\mathcal{M}e) \circ \Phi_v = \Phi_v \circ \sigma_t^\varphi$ $(t \in \mathbb{R})$, so that $\Phi_v(\mathcal{M}^\varphi) = e\mathcal{M}^\omega e$ and $\Phi_v(\mathcal{Z}(\mathcal{M}^\varphi)) = \mathcal{Z}(\mathcal{M}^\omega)e$. Also, we have $c_\omega(\varphi) = \mathbf{z}_{\mathcal{M}^\omega}(e) = c_\omega(\omega_e)$ (see 24.2) and $\Phi_v \circ p_\varphi^{-1} = p_{\omega_e}^{-1} : \mathcal{Z}(\mathcal{M}^\omega)c_\omega(\varphi) \to \mathcal{Z}(\mathcal{M}^\omega)e$ is just the mapping $z \to ze$.

We define $\sigma_c^\varphi = \Phi_v^{-1} \circ (\sigma_c^\omega | e\mathcal{M}e) \circ \Phi_v$, that is, $\sigma_c^\varphi(x) = v^*\sigma_c^\omega(vxv^*)v$ $(x \in \mathcal{M})$.

If $x \in \mathcal{M}(\sigma^\varphi, \{s\})$, then $vxv^* = \Phi_v(x) \in e\mathcal{M}(\sigma^\omega, \{s\})e$ and $\sigma_c^\varphi(x) = v^*\sigma_c^\omega(vxv^*)v = v^*c(s)vxv^*v = v^*c(s)c_\omega(\varphi)evx = [\Phi_v^{-1}(\Phi_v \circ p_\varphi^{-1})(c(s)c_\omega(\varphi))]x = p_\varphi^{-1}(c(s)c_\omega(\varphi))x$, thus proving (1). Since φ is integrable, it follows by Corollary 21.3 that $\mathcal{M}$ is generated by the union $\bigcup_s \mathcal{M}(\sigma^\varphi; \{s\})$, hence condition (1) completely determines the *-automorphism σ_c^φ.

The fact that the mapping $c \to \sigma_c^\varphi$ is a group homomorphism, and equations (2) and (3), now follow easily from the previous considerations and from the similar properties of ω, already proved in Section 26.2.

If the weights φ and ψ are integrable, then, as easily verified, also the weight $\theta(\varphi, \psi)$ on $\bar{\mathcal{M}} = \mathcal{M} \otimes \mathcal{F}_2$ is also integrable. Furthermore, $\bar{\omega} = \omega \otimes tr$ is a dominant weight on $\bar{\mathcal{M}}, \overline{\mathcal{M}^\omega} = \mathcal{M}^\omega \otimes \mathcal{F}_2$ and $(\mathcal{M}^\omega \otimes \mathcal{F}_2, \theta \otimes \iota, \tau \otimes tr)$ is a continuous decomposition of $\bar{\mathcal{M}}$. It follows that the flow of weights on $\bar{\mathcal{M}}$ is isomorphic to the flow of weights on $\mathcal{M}$. Using the uniqueness assertion based on (1), for $x \in \mathcal{M}$ we obtain

$$\sigma_c^{\theta(\varphi,\psi)}\left(\begin{pmatrix} x & 0 \\ 0 & 0 \end{pmatrix}\right) = \begin{pmatrix} \sigma_c^\varphi(x) & 0 \\ 0 & 0 \end{pmatrix}, \sigma_c^{\theta(\varphi,\psi)}\left(\begin{pmatrix} 0 & 0 \\ 0 & x \end{pmatrix}\right) = \begin{pmatrix} 0 & 0 \\ 0 & \sigma_c^\psi(x) \end{pmatrix}.$$

Equations (4), (5), and (6) can now be proved with the same arguments as in the proof of Theorem 3.1, while (7) follows immediately from (3) and (4). □

Besides the previous basic properties, we also have the following computation rules for the extensions σ_c^φ and $[D\psi : D\varphi]_c$:

$$\sigma_c^{\psi\circ\alpha} = \alpha^{-1} \circ \sigma^\varphi_{[mod(\alpha)](c)} \circ \alpha \tag{8}$$

$$[D(\psi\circ\alpha) : D(\varphi\circ\alpha)]_c = \alpha^{-1}([D\psi : D\varphi]_{[mod(\alpha)](c)} \tag{9}$$

$$[D\psi_u : D\varphi]_c = u^*[D\psi : D\varphi]_c \sigma_c^\psi(u) \tag{10}$$

$$[D\psi : D\varphi]_c = [D\varphi : D\psi]_c^* \tag{11}$$

$$[D\chi : D\varphi]_c = [D\chi : D\psi]_c [D\psi : D\varphi]_c \tag{12}$$

where φ, ψ, χ are integrable n.s.f. weights on $\mathscr{M}, \alpha \in Aut(\mathscr{M}), u \in U(\mathscr{M})$ and $c \in Z(F^{\mathscr{M}})$. (8) follows using (1) and 25.1.(8), (9) follow from (4) and (8), and (10), (11), (12) are easily checked using the definitions and the special case $c = \bar{\imath}$ considered in Section 3.

It is possible (see Connes & Takesaki, 1977, IV.2.2) to show that 26.2.(11) can also be extended to integrable n.s.f. weights φ on $\mathscr{M}$:

$$\{\sigma_c^\varphi ; c \in Z(F^{\mathscr{M}})\} = \{\sigma \in Aut(\mathscr{M}); \sigma | \mathscr{M}^\varphi = \imath\}. \tag{13}$$

Note that the condition $\mathbf{s}(\varphi) = \mathbf{1}$ was necessary only in order to obtain σ_c^φ as a *-automorphism of $\mathscr{M}$ itself. Without assuming this condition we can still define, by the same method, $\sigma_c^\varphi \in Aut(\mathscr{M}_{\mathbf{s}(\varphi)})$.

26.4. From 26.3.(5) and 26.2.(9), it follows that for every integrable n.s.f. weight φ on $\mathscr{M}$ and $c \in Z(F^{\mathscr{M}})$ we have

$$\sigma_c^\varphi \in Int(\mathscr{M}) \Rightarrow c \in B(F^{\mathscr{M}}). \tag{1}$$

Thus, the mapping $c/B(F^{\mathscr{M}}) \to \sigma_c^\varphi / Int(\mathscr{M})$ $(c \in Z(F^{\mathscr{M}}))$ defines an injective group homomorphism

$$\delta_{\mathscr{M}} : H(F^{\mathscr{M}}) \to Out(\mathscr{M})$$

which, by 26.3.(5), does not depend on the integrable n.s.f. weight φ on $\mathscr{M}$. In particular, $\delta_{\mathscr{M}} = \delta_\omega$ (26.2.(10)).

Since $\omega \circ \sigma_c^\omega = \omega$ and hence $mod(\sigma_c^\omega) = \sigma_c^\omega | \mathscr{Z}(\mathscr{M}^\omega)$ (see 25.1.(6)), it follows that the composition of $\delta_{\mathscr{M}}$ with $mod : Out(\mathscr{M}) \to Aut(F^{\mathscr{M}})$ is the trivial homomorphism:

$$mod(\delta_{\mathscr{M}}(c)) = \imath \quad (c \in H(F^{\mathscr{M}})). \tag{2}$$

Note that $\delta_{\mathscr{M}}(H(F^{\mathscr{M}}))$ is a normal subgroup of $Out(\mathscr{M})$, in fact for every $c \in H(F^{\mathscr{M}})$ and $\alpha \in Aut(\mathscr{M})$ we have

$$\alpha \delta_{\mathscr{M}}(c) \alpha^{-1} = \delta_{\mathscr{M}}([mod(\alpha)](c)). \tag{3}$$

Indeed, $\omega \circ \alpha$ is still a dominant weight, so that there exists $v \in U(\mathscr{M})$ with $\omega \circ \alpha \circ \mathrm{Ad}(v) = \omega$ and we may assume that $\omega \circ \alpha = \omega$; in this case, using 26.2.(8) and 25.1.(6), we obtain $\alpha \circ \sigma_c^\omega \circ \sigma^{-1} = \sigma^\omega_{[mod(v)](c)}$.

As an application, in the next theorem we compute the group $Out(\mathcal{M})$ in terms of the continuous decomposition $(\mathcal{N}, \theta, \tau)$ of $\mathcal{M}$. To this end, we consider the subgroups $Aut_{\theta,\tau}(\mathcal{N}) = \{\beta \in Aut(\mathcal{N}); \tau \circ \beta = \tau, \beta \circ \theta_s = \theta_s \circ \beta \text{ for all } s \in \mathbb{R}\} \subset Aut(\mathcal{N})$ and $Out_{\theta,\tau}(\mathcal{N}) = \{\beta/Int(\mathcal{N}); \beta \in Aut_{\theta,\tau}(\mathcal{N})\} \subset Out(\mathcal{N})$.

Theorem. *Let $\mathcal{M}$ be a countably decomposable properly infinite W^*-algebra and $(\mathcal{N}, \theta, \tau)$ a continuous decomposition of $\mathcal{M}$. There exists a homomorphism $\bar{\gamma}$ of $Out(\mathcal{M})$ onto $Out_{\theta,\tau}(\mathcal{N})$ such that the following sequence is exact:*

$$\{1\} \to H(F^{\mathcal{M}}) \overset{\delta_{\mathcal{M}}}{\to} Out(\mathcal{M}) \overset{\bar{\gamma}}{\to} Out_{\theta,\tau}(\mathcal{N}) \to \{1\}. \tag{4}$$

Proof. We shall use the notation introduced in Section 26.1, in particular $\mathcal{N} = \mathcal{M}^{\omega}$. Let $Aut_{\omega}(\mathcal{M}) = \{\sigma \in Aut(\mathcal{M}); \omega \circ \sigma = \omega\}$. For every $\sigma \in Aut(\mathcal{M}), \omega \circ \sigma$ is also a dominant weight and so there exists $u \in U(\mathcal{M})$ such that $\omega \circ \sigma \circ \mathrm{Ad}(u) = \omega$. It follows that $Out(\mathcal{M}) = Aut_{\omega}(\mathcal{M})/Int(\mathcal{M})$. If $\sigma \in Aut_{\omega}(\mathcal{M})$, then $\sigma(\mathcal{M}^{\omega}) = \mathcal{M}^{\omega}$, so that we can define a homomorphism $\gamma : Aut_{\omega}(\mathcal{M}) \to Out(\mathcal{N})$ by putting

$$\gamma(\sigma) = (\sigma|\mathcal{N})/Int(\mathcal{N}) \quad (\sigma \in Aut_{\omega}(\mathcal{M})). \tag{5}$$

Since $Aut_{\omega}(\mathcal{M}) \cap Int(\mathcal{M}) = \{\mathrm{Ad}(u); u \in U(\mathcal{M}^{\omega})\}$, it follows that γ can be factored to a homomorphism $\bar{\gamma} : Out(\mathcal{M}) \to Out(\mathcal{N})$.

If $\sigma = \sigma_c^{\omega}$ with $c \in Z(F^{\mathcal{M}})$, then $\sigma|\mathcal{N} = \iota$ (26.2.(2)), and $\gamma(\sigma) = \iota$. Conversely, if $\sigma \in Aut_{\omega}(\mathcal{M})$ and $\gamma(\sigma) = \iota$, that is, $\sigma|\mathcal{N} = \mathrm{Ad}(v^*)$ with $v \in U(\mathcal{N})$, then $(\sigma \circ \mathrm{Ad}(v))|\mathcal{N} = \iota$ and so there exists $c \in Z(F^{\mathcal{M}})$ such that $\sigma \circ \mathrm{Ad}(v) = \sigma_c^{\omega}$ (26.2.(11)). Thus, $Ker(\bar{\gamma}) = Range(\delta_{\mathcal{M}})$.

Let $\sigma \in Aut_{\omega}(\mathcal{M})$ and $\beta = \sigma|\mathcal{N} \in Aut(\mathcal{N})$. For $s \in \mathbb{R}$, we have $\sigma_t^{\omega}(\sigma(u(s))) = \sigma(\sigma_t^{\omega}(u(s))) = e^{-ist}\sigma(u(s))$, hence $a(s) = \sigma(u(s))u(s)^* \in U(\mathcal{N})$ $(s \in \mathbb{R})$. As in the proof of 26.2.(11) one sees that the function $s \to a(s)$ defines a cocycle $a \in Z_{\theta}(\mathbb{R}; U(\mathcal{N}))$. By Theorem 23.12, there exists $v \in U(\mathcal{N})$ such that $a(s) = v^*\theta_s(v)$ $(s \in \mathbb{R})$. It follows that $\sigma(u(s)) = v^*\theta_s(v)u(s)$, or $(\sigma \circ \mathrm{Ad}(v))(u(s)) = u(s)$, that is, $\beta \circ \mathrm{Ad}(v) = (\sigma \circ \mathrm{Ad}(v))|\mathcal{N}$ commutes with θ_s $(s \in \mathbb{R})$. Let $P_{\omega} : \mathcal{M}^+ \to \bar{\mathcal{N}}^+$ be the n.s.f. operator-valued weight associated with ω (24.6). Then $\omega = \tau \circ P_{\omega}$ and, as $\omega \circ \sigma = \omega$, we get $(\tau \circ \beta) \circ P_{\omega} = \tau \circ (\sigma|\mathcal{N}) \circ P_{\omega} = \tau \circ P_{\omega} \circ \sigma = \omega \circ \sigma = \omega = \tau \circ P_{\omega}$, hence $\tau \circ \beta = \tau$ and $\tau \circ (\beta \circ \mathrm{Ad}(v)) = \tau$. Thus, $Range\ (\bar{\gamma}) \subset Out_{\theta,\tau}(\mathcal{N})$.

Finally, let $\beta \in Aut_{\theta,\tau}(\mathcal{N})$. Since $(\mathcal{M}, \sigma^{\omega}, \omega) \approx (\mathcal{R}(\mathcal{N}, \theta), \hat{\theta}, \hat{\tau})$, it follows that there exists $\sigma \in Aut_{\omega}(\mathcal{M})$ with $\sigma|\mathcal{N} = \beta$ (20.13.(2)). Thus, $Out_{\theta,\tau}(\mathcal{N}) \subset Range(\bar{\gamma})$. □

If $\mathcal{N}$ has separable predual, then the subgroup $Out_{\theta,\tau}(\mathcal{N})$ also has the following description:

$$Out_{\theta,\tau}(\mathcal{N}) = \{\beta \in Out(\mathcal{N}); \tau \circ \beta = \tau, \mathfrak{o}_{\mathcal{N}}(\theta_s)\beta = \beta \circ \mathfrak{o}_{\mathcal{N}}(\theta_s)\ (s \in \mathbb{R})\} \tag{6}$$

where $\mathfrak{o}_{\mathcal{N}} : Aut(\mathcal{N}) \to Out(\mathcal{N})$ is the canonical quotient mapping.

Instead of a complete proof, which involves some nontrivial technical complications (see Connes & Takesaki, 1977, IV.3.2), we give just a sketch of the proof here.

By familiar arguments it follows from the theory of standard Borel spaces that there exists a Borel mapping $v : Int(\mathcal{N}) \to U(\mathcal{N})$ such that $\alpha = \mathrm{Ad}(v(\alpha))$ for every $\alpha \in Int(\mathcal{N})$.

If $\beta \in Aut(\mathcal{N})$ and $\mathfrak{o}_{\mathcal{N}}(\beta)$ commutes with $\mathfrak{o}_{\mathcal{N}}(\theta_s)$, then $\beta \circ \theta_s \circ \beta^{-1} \circ \theta_s^{-1} \in Int(\mathcal{N})$ and $b(s) = v(\beta \circ \theta_s \circ \beta^{-1} \circ \theta_s^{-1}) \in U(\mathcal{N})$ $(s \in \mathbb{R})$. Then $\mathrm{Ad}(b(s)) \circ \theta_s = \beta \circ \theta_s \circ \beta^{-1}$, so that $\mathrm{Ad}(b(s)\theta_s(b(t))) = \mathrm{Ad}(b(s+t))$ and so $c(s,t) = b(s)^*b(s+t)\theta_s(b(t)^*) \in \mathcal{Z}(\mathcal{N})$ $(s, t \in \mathbb{R})$. By direct computation, one checks that $c(r,s)c(r+s,t) = \theta_s(c(s,t))c(r,s+t), (r,s,t \in \mathbb{R})$, and by cohomological arguments, it follows that there exists a Borel function $d : \mathbb{R} \to U(\mathscr{Z}(\mathscr{N}))$ such that $c(s,t) = d(s)^* d(s+t)\theta_s(d(t)^*)$ for almost all $s,t \in \mathbb{R}$. Then $bd : \mathbb{R} \to U(\mathscr{N})$ is a Borel function which, modified on a negligible set, gives rise to a cocycle $a \in Z_\theta(\mathbb{R}; U(\mathscr{N}))$ such that $\mathrm{Ad}(a(s)) \circ \theta_s = \beta \circ \theta_s \circ \beta^{-1}$ $(s \in \mathbb{R})$. Using Theorem 23.12, we infer that $a(s) = w^*\theta_s(w)$ $(s \in \mathbb{R})$ for some $w \in U(\mathscr{N})$. Then $\mathrm{Ad}(w) \circ \beta$ commutes with θ_s $(s \in \mathbb{R})$. We thus obtain the inclusion "$\supset$" in (6), the other inclusion being obvious.

26.5. In order to explain the nature of the extensions σ_c^φ and $[D\psi : D\varphi]_c$ defined in Section 26.3, we shall consider the case of an infinite semifinite factor $\mathscr{M}$ with separable predual. Let μ be an n.s.f. trace on $\mathscr{M}$.

By Theorem 4.10, every n.s.f. weight φ on $\mathscr{M}$ is of the form $\varphi = \mu_A$ for some nonsingular positive self-adjoint operator A affiliated to $\mathscr{M}$. In particular, for the dominant weight ω on $\mathscr{M}$, there exists a nonsingular positive self-adjoint operator D affiliated to $\mathscr{M}$ such that $\omega = \mu_D$. The weight $\varphi = \mu_A$ is integrable if and only if A is unitarily equivalent to the restriction of D to some invariant subspace (see 23.4).

The structure of the operator D can be better analyzed if we consider the $*$-isomorphism $(\mathscr{M}, \mu) \approx (\mathscr{M} \bar{\otimes} \mathscr{F}, \mu \bar{\otimes} tr)$ (see 9.18) where $\mathscr{F} = \mathscr{B}(\mathscr{L}^2(\mathbb{R}_*^+))$. Let D_0 be the operator defined in $\mathscr{L}^2(\mathbb{R}_*^+)$ by $(D_0\xi)(\lambda) = \lambda\xi(\lambda)$ $(\xi \in Dom(D_0) \subset \mathscr{L}^2(\mathbb{R}_*^+), \lambda \in \mathbb{R}_*^+)$. As in Section 23.4, it is easy to see that $\mu \bar{\otimes} tr_{D_0}$ is a dominant weight, so that the operator $D = 1 \bar{\otimes} D_0$ has absolutely continuous spectrum.

Thus, $\varphi = \mu_A$ is integrable if and only if A has absolutely continuous spectrum. In this case, the functional calculus associated with A can be extended to functions $f \in \mathscr{L}^\infty(\mathbb{R}_*^+)$. Indeed, this fact is obvious for D_0 : $(f(D_0)\xi)(\lambda) = f(\lambda)\xi(\lambda)$ $(\xi \in Dom(f(D_0)) \subset \mathscr{L}^2(\mathbb{R}_*^+), \lambda \in \mathbb{R}_*^+)$. Note that the commutant of D_0 in $\mathscr{F}$ is just the set $\{f(D_0); f \in \mathscr{L}^\infty(\mathbb{R}_*^+)\}$.

The centralizer $\mathscr{M}^\varphi$ of $\varphi = \mu_A$ is the commutant of A in $\mathscr{M}$. Consequently,

$$\mathscr{Z}(\mathscr{M}^\varphi) = \{f(A); f \in \mathscr{L}^\infty(\mathbb{R}_*^+)\}. \tag{1}$$

By the construction (24.5) of the $*$-isomorphism $p_\varphi : \mathscr{Z}(\mathscr{M}^\varphi) \to \mathscr{Z}(\mathscr{M}^\omega)c_\omega(\varphi)$ it follows that

$$p_\varphi(f(A)) = f(D)c_\omega(\varphi) \quad (f \in \mathscr{L}^\infty(\mathbb{R}_*^+)). \tag{2}$$

Recall (24.8) that the flow of weights on $\mathscr{M}$ is the pair consisting of $\mathscr{Z}(\mathscr{M}^\omega)$, identified with $\mathscr{L}^\infty(\mathbb{R}_*^+)$ via the mapping

$$\mathscr{L}^\infty(\mathbb{R}_*^+) \ni f \mapsto f(D) \in \mathscr{Z}(\mathscr{M}^\omega), \tag{3}$$

and the continuous action $F^\mathscr{M}$ of $\mathbb{R}_*^+$ on $\mathscr{L}^\infty(\mathbb{R}_*^+)$ defined by

$$(F_\lambda^\mathscr{M} f)(\alpha) = f(\lambda^{-1}\alpha) \quad (f \in \mathscr{L}^\infty(\mathbb{R}_*^+, \lambda, \alpha \in \mathbb{R}_*^+). \tag{4}$$

As mentioned in Section 21.12, every unitary cocycle $c \in Z(F^\mathscr{M})$ is trivial, that is, there exists a unitary element $f \in \mathscr{L}^\infty(\mathbb{R}_*^+)$, uniquely determined up to a scalar multiple, such that $c(\lambda) = fF_\lambda^\mathscr{M}(f^*)$ $(\lambda \in \mathbb{R}_*^+)$.

If $\varphi = \mu_A, \psi = \mu_B$ *are integrable n.s.f. weights on* $\mathcal{M}$ *and* $c \in Z(F^{\mathcal{M}})$ *is defined by* $c(\lambda) = fF^{\mathcal{M}}_\lambda(f^*)$ $(\lambda \in \mathbb{R}^+_*)$, *then*

$$\sigma^\varphi_c = \mathrm{Ad}(f(A)) \tag{5}$$

$$[D\psi : D\varphi]_c = f(B)f(A)^*. \tag{6}$$

Indeed, let $\lambda > 0$ and let $w \in \mathcal{M}(\sigma^\varphi, \{-\ln(\lambda)\})$ be a partial isometry. Then w^*w and ww^* belong to $\mathcal{M}^\varphi$ so that they commute with A. We have $A^{it}wA^{-it} = \sigma^\varphi_t(w) = \lambda^{it}w$, hence $w^*A^{it}w = w^*w(\lambda A)^{it} = w^*w(\lambda A)^{it}w^*w$ $(t \in \mathbb{R})$. For $h \in \mathcal{L}^\infty(\mathbb{R}^+_*)$, it follows that $w^*h(A)w = w^*wh(\lambda A)w^*w = w^*wh(\lambda A)$, whence

$$h(A)w = wh(\lambda A). \tag{7}$$

Using (7), (4), (2), (3), and 26.3.(1), we get $[\mathrm{Ad}(f(A))](w) = f(A)wf(A)^* = wf(\lambda A)f(A)^* = f(A)f(\lambda^{-1}A))^*w = [c(\lambda)](A)w = p^{-1}_\varphi([c(\lambda)](D)c_\omega(\varphi))w = p^{-1}_\varphi(c(\lambda)c_\omega(\varphi))w = \sigma^\varphi_c(w)$. Thus, the $*$-automorphisms σ^φ_c and $\mathrm{Ad}(f(A))$ coincide on every partial isometry $w \in \mathcal{M}(\sigma^\varphi, \{-\ln(\lambda)\})$. Since both these $*$-automorphisms act identically on $\mathcal{M}^\varphi$, it follows that they are equal, which proves (5).

The balanced weight $\theta(\varphi, \psi)$ is obtained from the trace $\theta(\mu, \mu)$ with the help of the operator $\begin{pmatrix} A & 0 \\ 0 & B \end{pmatrix}$ so that, using (5) and 26.3.(4), we obtain

$$\begin{pmatrix} 0 & 0 \\ [D\psi : D\varphi]_c & 0 \end{pmatrix} = \sigma^{\theta(\varphi,\psi)}_c\left(\begin{pmatrix} 0 & 0 \\ 1 & 0 \end{pmatrix}\right)$$

$$= \begin{pmatrix} f(A) & 0 \\ 0 & f(B) \end{pmatrix}\begin{pmatrix} 0 & 0 \\ 1 & 0 \end{pmatrix}\begin{pmatrix} f(A)^* & 0 \\ 0 & f(B)^* \end{pmatrix} = \begin{pmatrix} 0 & 0 \\ f(B)f(A)^* & 0 \end{pmatrix}$$

so that (6) is also proved.

Thus, the extension of the modular automorphism group to a family of $*$-automorphism indexed by $Z(F^{\mathcal{M}})$ is actually a generalization of the usual functional calculus.

26.6 Proposition. *Let* $\mathcal{M}$ *be a properly infinite* W^**-algebra with separable predual and let* $c \in Z(F^{\mathcal{M}})$. *Let* φ *be an integrable n.s.f. weight on* $\mathcal{M}$ *and* A *be a nonsingular positive self-adjoint operator affiliated to* $\mathcal{M}^\varphi$ *such that the weight* φ_A *is integrable. Then* $c(A) = [D\varphi_A : D\varphi]_c$ *belongs to the center of* $\{A\}' \cap \mathcal{M}^\varphi$.

Proof. Let $\psi = \theta(\varphi, \varphi_A)$ be the balanced weight on $\mathcal{P} = Mat_2(\mathcal{M})$ and let $u = \begin{pmatrix} 0 & 0 \\ 1 & 0 \end{pmatrix} \in \mathcal{P}$. Then (26.3.(4)) $\begin{pmatrix} 0 & 0 \\ c(A) & 0 \end{pmatrix} = \sigma^\psi_c(u)$. We have $\sigma^\psi_t(u^*)u \in \mathcal{P}^\psi$, so that $\sigma^\psi_t(u^*)u = \sigma^\psi_c(\sigma^\psi_t(u^*)u) = \sigma^\psi_t(\sigma^\psi_c(u^*))\sigma^\psi_c(u)$ and hence $\begin{pmatrix} c(A) & 0 \\ 0 & 0 \end{pmatrix} = u^*\sigma^\psi_c(u) \in \mathcal{P}^\psi$, that is, $c(A) \in \mathcal{M}^\varphi$. For every $x \in \{A\}' \cap \mathcal{M}^\varphi \subset \mathcal{M}^\varphi \cap \mathcal{M}^{\varphi_A}$, we have $x = \sigma^{\varphi_A}_c(x) = c(A)\sigma^\varphi_c(x)c(A)^* = c(A)xc(A)^*$ and therefore $c(A) \in (\{A\}' \cap \mathcal{M}^\varphi)' \cap \mathcal{M}^\varphi = \mathcal{Z}(\{A\}' \cap \mathcal{M}^\varphi)$. □

Using Theorem 26.3.(6), we infer that

$$(c_1c_2)(A) = c_1(A)c_2(A) \quad (c_1, c_2 \in Z(F^{\mathscr{M}})). \tag{1}$$

26.7. Notes. The material of this section is due to Connes and Takesaki (1977). In our exposition, we have defined the extended modular automorphisms and the extended Connes cocycles only for integrable weights. In fact, it is possible to define these objects for arbitrary n.s.f. weigh by restricting the class of cocycles $c \in Z(F^{\mathscr{M}})$ to those which are twice continuously differentiable in norm (Connes & Takesaki, 1977, IV.2.6). For factors arising by the group measure space construction, the extended modular automorphisms are explicitly computed in (Connes & Takesaki, 1977).

CHAPTER VI

Discrete Decompositions

27 The Connes Invariant $T(\mathscr{M})$

In this section, we introduce the invariant $T(\mathscr{M})$ as the group of inner periods of modular automorphism groups on $\mathscr{M}$.

27.1. Let $\mathscr{M}$ be a W^*-algebra and $\mathfrak{o}_{\mathscr{M}} : Aut(\mathscr{M}) \to Out(\mathscr{M})$ the canonical quotient mapping. Recall (3.2) that the modular homomorphism $\delta_{\mathscr{M}} : \mathbb{R} \to Out(\mathscr{M})$ defined by $\delta_{\mathscr{M}}(t) = \mathfrak{o}_{\mathscr{M}}(\sigma_t^{\varphi})$ $(t \in \mathbb{R})$ does not depend on the choice of the n.s.f. weight φ on $\mathscr{M}$ and that $\delta_{\mathscr{M}}(\mathbb{R})$ is contained in the center of the group $Out(\mathscr{M})$. Therefore, the kernel of the modular homomorphism is a subgroup of the additive group $\mathbb{R}$ and an algebraic invariant of $\mathscr{M}$, which we denote by

$$T(\mathscr{M}) = \{t \in \mathbb{R}; \delta_{\mathscr{M}}(t) = \iota\}.$$

Theorem. *Let $\mathscr{M}$ be a W^*-algebra and $t \in \mathbb{R}$. The following statements are equivalent:*

(i) $t \in T(\mathscr{M})$;
(ii) *for every* $\varphi \in W_{nsf}(\mathscr{M})$ *we have* $\sigma_t^{\varphi} \in Int(\mathscr{M})$;
(iii) *for every* $\varphi \in W_{nsf}(\mathscr{M})$ *there exists a unitary element* $u \in \mathscr{Z}(\mathscr{M}^{\varphi})$ *such that* $\sigma_t^{\varphi} = \mathrm{Ad}(u)$;
(iv) *for every* $\varphi \in W_{nsf}(\mathscr{M})$ *there exists a (strictly semifinite)* $\psi \in W_{nsf}(\mathscr{M})$ *commuting with* φ *such that* $\sigma_t^{\psi} = \iota$;
(v) *there exists* $\varphi \in W_{nsf}(\mathscr{M})$ *such that* $\sigma_t^{\varphi} \in Int(\mathscr{M})$.

If $\mathscr{M}$ is countably decomposable and $t \in T(\mathscr{M})$, then

(vi) *there exists a faithful normal state* φ *on* $\mathscr{M}$ *with* $\sigma_t^{\varphi} = \iota$.

Proof. It is clear that (vi) $\Rightarrow$ (iv) $\Rightarrow$ (v) $\Rightarrow$ (i) $\Rightarrow$ (ii).

(ii) $\Rightarrow$ (iii). If $\sigma_t^{\varphi} = \mathrm{Ad}(u)$ with $u \in U(\mathscr{M})$, then $\varphi(uxu^*) = \varphi(\sigma_t^{\varphi}(x)) = \varphi(x)$ for $x \in \mathscr{M}^+$, whence $u \in \mathscr{M}^{\varphi}$. For $x \in \mathscr{M}^{\varphi}$, we have $uxu^* = \sigma_t^{\varphi}(x) = x$, so $u \in \mathscr{Z}(\mathscr{M}^{\varphi})$.

(iii) $\Rightarrow$ (iv). By assumption, we have $\sigma_t^{\varphi} = \mathrm{Ad}(u)$ for some unitary $u \in \mathscr{Z}(\mathscr{M}^{\varphi})$. There exists a nonsingular positive self-adjoint operator A affiliated to $\mathscr{Z}(\mathscr{M}^{\varphi})$ such that $A^{-it} = u$. Then the n.s.f. weight $\psi = \varphi_A$ commutes with φ and $\sigma_t^{\psi} = \mathrm{Ad}(A^{it}) \circ \sigma_t^{\varphi} = \iota$. On the other hand, we recall (10.9) that if $\sigma_t^{\psi} = \iota$, then the weight ψ is strictly semifinite.

(iii) $\Rightarrow$ (vi). Assume that $\mathscr{M}$ is countably decomposable. By (iii) there exists an n.ss.f. weight ψ on $\mathscr{M}$ such that $\sigma_t^{\psi} = \iota$. Since ψ is strictly semifinite and $\mathscr{M}$ is countably decomposable, it follows by Theorem 10.9 that there exists a sequence $\{\psi_n\}_{n\geq 1} \subset \mathscr{M}_*^+$ with supports $\mathbf{s}(\psi_n) \in \mathscr{M}^{\psi}$, mutually orthogonal, such that $\psi = \sum_n \psi_n$. Then $\varphi = \sum_n (2^n \|\psi_n\|)^{-1} \psi_n$ is a faithful normal state with $\sigma_t^\varphi = \iota$. □

27.2 Proposition. *Let $\mathcal{M}$ be a W^*-algebra. If $\mathcal{M}$ is semifinite, then $T(\mathcal{M}) = \mathbb{R}$. If $T(\mathcal{M}) = \mathbb{R}$ and $\mathcal{M}$ has separable predual, then $\mathcal{M}$ is semifinite.*

Proof. If μ is an n.s.f. trace on $\mathcal{M}$, $\sigma_t^\mu = \iota$, for all $t \in \mathbb{R}$, and $T(\mathcal{M}) = \mathbb{R}$. Conversely, if $T(\mathcal{M}) = \mathbb{R}$ and φ is any n.s.f. weight on $\mathcal{M}$, then $\sigma_t^\varphi \in Int(\mathcal{M})$ for all $t \in \mathbb{R}$. If $\mathcal{M}$ has separable predual, then, by Theorem 15.16, there exists an s-continuous unitary representation $\mathbb{R} \ni t \to u(t) \in \mathcal{M}$ such that $\sigma_t^\varphi = \mathrm{Ad}(u(t))$ $(t \in \mathbb{R})$. Also, there exists a nonsingular positive self-adjoint operator A affiliated to $\mathcal{M}^\varphi$ such that $u(t) = A^{-it}$ and hence $\sigma_t^{\varphi_A} = \iota$, for all $t \in \mathbb{R}$, that is, $\mu = \varphi_A$ is an n.s.f. trace on $\mathcal{M}$. □

Thus, for any nonsemifinite W^*-algebra $\mathcal{M}$ with separable predual the center of the group $Out(\mathcal{M})$ is nontrivial.

27.3 Proposition. *For any two W^*-algebras $\mathcal{M}, \mathcal{N}$ we have*

$$T(\mathcal{M} \bar{\otimes} \mathcal{N}) = T(\mathcal{M}) \cap T(\mathcal{N}), \tag{1}$$

$$T(\mathcal{M} \oplus \mathcal{N}) = T(\mathcal{M}) \cap T(\mathcal{N}), \tag{2}$$

if $e \in \mathcal{M}$ is a projection with central support equal to 1,

$$T(\mathcal{M}_e) = T(\mathcal{M}), \tag{3}$$

and if $\mathcal{M} \subset \mathcal{B}(\mathcal{H})$ is realized as a von Neumann algebra

$$T(\mathcal{M}') = T(\mathcal{M}). \tag{4}$$

Proof. Let φ and ψ be n.s.f. weights on $\mathcal{M}$ and $\mathcal{N}$, respectively. Then $\varphi \bar{\otimes} \psi$ is an n.s.f. weight on $\mathcal{M} \bar{\otimes} \mathcal{N}$ and, by Proposition 17.6.(3), $\sigma_t^{\varphi \bar{\otimes} \psi} = \sigma_t^\varphi \bar{\otimes} \sigma_t^\psi \in Int(\mathcal{M} \bar{\otimes} \mathcal{N})$ if and only if $\sigma_t^\varphi \in Int(\mathcal{M})$ and $\sigma_t^\psi \in Int(\mathcal{N})$, which proves (1). Also, $\varphi \oplus \psi$ is an n.s.f. weight on $\mathcal{M} \oplus \mathcal{N}$ and $\sigma_t^{\varphi \oplus \psi} = \sigma_t^\varphi \oplus \sigma_t^\psi \in Int(\mathcal{M} \oplus \mathcal{N})$ if and only if $\sigma_t^\varphi \in Int(\mathcal{M})$ and $\sigma_t^\psi \in Int(\mathcal{N})$, from which (2) follows.

In view of (2), to prove (3) we may consider separately the cases e finite and e properly infinite. If e is finite, then $\mathcal{M}_e$ and $\mathcal{M}$ are both semifinite, whence $T(\mathcal{M}_e) = \mathbb{R} = T(\mathcal{M})$. If e is properly infinite and $\mathcal{M}$ is countably decomposable, then $e \sim 1$ in $\mathcal{M}$, whence $\mathcal{M}_e \approx \mathcal{M}$ and $T(\mathcal{M}_e) = T(\mathcal{M})$. If $\mathcal{M}$ is not countably decomposable, then we can decompose $\mathcal{M}$ into a direct sum of uniform W^*-algebras (see [L], 8.4) and, using (2), (1), Proposition 27.2, and the previous remarks, we conclude again that $T(\mathcal{M}_e) = T(\mathcal{M})$.

Equation (4) is obvious if $\mathcal{M} \subset \mathcal{B}(\mathcal{H})$ is in standard form, since then $\mathcal{M}'$ is *-antiisomorphic to $\mathcal{M}$. In general, $\mathcal{M}$ is *-isomorphic to some standard form of itself via an amplification and a faithful induction ([L], E.8.8). These operations change the commutant $\mathcal{M}'$ either into $\mathcal{M}' \bar{\otimes} \mathcal{B}(\mathcal{K})$ (for some Hilbert space $\mathcal{K}$) or into $\mathcal{M}'_{e'}$ (for some projection $e' \in \mathcal{M}'$ with central support equal to 1) without modifying $T(\mathcal{M}')$. □

27.4. In this section, we compute the invariant $T(\mathcal{M})$ for crossed products by properly outer actions of discrete groups. More generally, let $\mathcal{M}$ be a W^*-algebra, $\mathcal{N} \subset \mathcal{M}$ a semifinite unital

W^*-subalgebra with $\mathscr{Z}(\mathscr{N}) = \mathscr{N}' \cap \mathscr{M}, P : \mathscr{M} \to \mathscr{N}$ a faithful normal conditional expectation of $\mathscr{M}$ onto $\mathscr{N}$ and $\mathscr{G} \subset \mathcal{N}(P)$ a subgroup of the normalizer $\mathcal{N}(P)$ of P such that $\mathscr{M} = \mathscr{R}\{\mathscr{N}, \mathscr{G}\}$ (see 10.17).

By Theorem 4.10, for every n.s.f. trace τ on $\mathscr{M}$ and $u \in \mathcal{N}(P)$ there exists a unique nonsingular positive self-adjoint operator $A_{\tau,u}$ affiliated to $\mathscr{Z}(\mathscr{N})$ such that

$$\tau(u \cdot u^*) = \tau_{A_{\tau,u}}. \tag{1}$$

Note that

$$\sigma_t^{\tau \circ P} = uA_{\tau,u}^{it} \quad (u \in \mathcal{N}(P), t \in \mathbb{R}). \tag{2}$$

Indeed, using Corollary 3.7, Theorem 11.9, and Corollary 4.8, we get

$$\begin{aligned} u^* \sigma_t^{\tau \circ P}(u) &= [D(\tau \circ P)_u : D(\tau \circ P)]_t = [D(\tau(u \cdot u^*) \circ P) : D(\tau \circ P)]_t \\ &= [D(\tau(u \cdot u^*)) : D\tau]_t = [D(\tau_{A_{\tau,u}}) : D\tau]_t = A_{\tau,u}^{it}. \end{aligned}$$

Theorem. *In the previous setting, the following statements concerning $t \in \mathbb{R}$ are equivalent:*

(i) *$t \in T(\mathscr{M})$;*
(ii) *for every n.s.f. trace τ on $\mathscr{N}$ there exists a unitary element $v \in \mathscr{Z}(\mathscr{N})$ such that $A_{\tau,u}^{it} = u^*vuv^*$ for all $u \in \mathscr{G}$;*
(iii) *there exists an n.s.f. trace τ on $\mathscr{N}$ such that $A_{\tau,u}^{it} = 1$ for all $u \in \mathscr{G}$.*

Proof. (i) ⇒ (ii). Since $t \in T(\mathscr{M})$, given the weight $\varphi = \tau \circ P$ on $\mathscr{M}$ there exists $v \in U(\mathscr{M})$ such that $\sigma_t^\varphi = \mathrm{Ad}(v)$. For $y \in \mathscr{N}$, we have (11.9) $vyv^* = \sigma_t^\varphi(y) = \sigma_t^\tau(y) = y$, hence $v \in \mathscr{N}' \cap \mathscr{M} = \mathscr{Z}(\mathscr{N})$. By (2), for $u \in \mathscr{G}$ we have $vuv^* = \sigma_t^\varphi(u) = uA_{\tau,u}^{it}$, hence $A_{\tau,u}^{it} = v^*vuv^*$.

(ii) ⇒ (iii). Let μ be any n.s.f. trace on $\mathscr{N}$ and let $v \in \mathscr{Z}(\mathscr{N})$ be such that $u^*vuv^* = A_{\tau,u}^{it} = [D(\mu(u \circ u^*)) : D\mu]_t$ for all $u \in \mathscr{G}$. There exists an invertible element $b \in \mathscr{Z}(\mathscr{N})^+$ such that $b^{it} = v^*$. Then $\tau = \mu_b$ is an n.s.f. trace on $\mathscr{N}$ with $[D\tau : D\mu]_t = b^{it} = v^*$, so that for $u \in \mathscr{G}$ we obtain (3.8) $[D(\tau(u \cdot u^*)) : D(\mu(u \cdot u^*))]_t = u^*[D\tau : D\mu]_t u = u^*v^*u$ and $A_{\tau,u}^{it} = [D(\tau(u \cdot u^*)) : D\tau]_t = (u^*v^*u)(u^*vuv^*)v = 1$.

(iii) ⇒ (i). Let $\varphi = \tau \circ P$ with τ as in (iii). For $y \in \mathscr{N}$ we have $\sigma_t^\varphi(y) = \sigma_t^\tau(y) = y$ and for $u \in \mathscr{G}$ we have $\sigma_t^\varphi(u) = uA_{\tau,u}^{it} = u$, whence $\sigma_t^\varphi = \iota$ since $\mathscr{M} = \mathscr{R}\{\mathscr{N}, \mathscr{G}\}$. □

Proposition. *In the above setting, consider the following statements concerning $t \in \mathbb{R}$ and $\psi \in W_{nsf}(\mathscr{N})$:*

(i) *$t \in T(\mathscr{M})$;*
(ii) *$\sigma_t^\psi \in [\{\mathrm{Ad}(w)|\mathscr{N}; w \in \mathscr{G}\}]$;*
(iii) *there exists $u \in \mathscr{G}$ such that $\sigma_t^\psi = \mathrm{Ad}(u)|\mathscr{N}$.*

Then (i) ⇒ (ii) and, if $\mathscr{G}$ is abelian and $\psi \circ \mathrm{Ad}(w) = \psi$ for all $w \in \mathscr{G}$, (iii) ⇒ (i).

Proof. Assume that $t \in T(\mathscr{M})$ and consider the n.s.f. weight $\varphi = \psi \circ P$ on $\mathscr{M}$. Then there exists $u \in U(\mathscr{M})$ such that $\sigma_t^\varphi = \mathrm{Ad}(u)$. For $y \in \mathscr{N}$, we have $uyu^* = \sigma_t^\varphi(y) = \sigma_t^\psi(y) \in \mathscr{N}$, hence $y \in \mathcal{N}(P)$ and $\sigma_t^\psi = \sigma_t^\varphi|\mathscr{N} = \mathrm{Ad}(u)|\mathscr{N} \in [\{\mathrm{Ad}(w)|\mathscr{N}; w \in \mathscr{G}\}]$, by Proposition 22.4.

Assume now that $\mathscr{G}$ is abelian, that $\psi \circ \mathrm{Ad}(w) = \psi$ for all $w \in \mathscr{G}$ and $\sigma_t^\psi = \mathrm{Ad}(u)|\mathscr{N}$ for some $u \in \mathscr{G}$. Let $\varphi = \psi \circ P \in W_{nsf}(\mathscr{M})$. For every $w \in \mathscr{G}$, we have (3.7, 11.9) $w^*\sigma_t^\varphi(w) = [D(\varphi \circ \mathrm{Ad}(w)) : D\varphi]_t = [D(\psi \circ \mathrm{Ad}(w)) : D\psi]_t = 1$, hence $\sigma_t^\varphi(w) = w = uwu^* = [\mathrm{Ad}(u)](w)$ since $\mathscr{G}$ is abelian. For $y \in \mathscr{N}$, we have $\sigma_t^\varphi(y) = \sigma_t^\psi(y) = [\mathrm{Ad}(u)](y)$. Since $\mathscr{M} = \mathscr{R}\{\mathscr{N}, \mathscr{G}\}$, it follows that $\sigma_t^\varphi = \mathrm{Ad}(u)$, so that $t \in T(\mathscr{M})$. □

27.5. The previous results are valid if, for instance, $\mathscr{M} = \mathscr{R}(\mathscr{N}, \sigma)$ is the crossed product of the semifinite W^*-algebra $\mathscr{N}$ by a properly outer action $\sigma : G \to Aut(\mathscr{N})$ of the discrete group G, since then $\mathscr{M}$ is generated by $\mathscr{N} \equiv \pi_\sigma(\mathscr{N}) \subset \mathscr{M}$ and $\mathscr{G} = \{1 \bar{\otimes} \boldsymbol{\lambda}(g); g \in G\}, \mathscr{N}' \cap \mathscr{M} = \mathscr{Z}(\mathscr{N})$ (see 22.3), and $P_{\hat{\sigma}} : \mathscr{M} \to \mathscr{N}$ is a faithful normal conditional expectation such that $\mathscr{G} \subset \mathcal{N}(P_{\hat{\sigma}})$ (see 22.2.(3)).

In particular, let $G \ni g \mapsto T_g$ be an action of the discrete countable group G as homeomorphisms of the locally compact Hausdorff topological space Ω with a countable basis of open sets and let μ be a G-quasi-invariant sigma-finite positive Borel measure on Ω; we assume that G acts freely on (Ω, μ) (see 22.8). We then obtain a free action $\sigma : G \to Aut(\mathscr{N})$ of G on the abelian W^*-algebra $\mathscr{N} = \mathscr{L}^\infty(\Omega, \mu)$ and a corresponding crossed product W^*-algebra $\mathscr{M} = \mathscr{R}(\mathscr{N}, \sigma)$ (see 22.8). For any n.s.f. trace τ on $\mathscr{N}$, there is a Borel measure ν on Ω, equivalent to μ, such that $\tau(x) = \int x\,\mathrm{d}\nu$, $(x \in \mathscr{N}^+)$. For every $g \in G$, there exists a unique nonsingular positive self-adjoint operator $A_{\tau\cdot g}$ in $\mathscr{H} = \mathscr{L}^2(\Omega, \mu)$, affiliated to $\mathscr{N} = \mathscr{L}^\infty(\Omega, \mu)$, such that $\tau \circ \sigma_g = \tau_{A_{\tau,g}}$. Since for $x \in \mathscr{L}^\infty(\Omega, \mu)^+$ and $g \in G$, we have $\tau(\sigma_g(x)) = \int x(T_{g^{-1}}(\omega))\,\mathrm{d}\nu(\omega) = \int x(\omega)\,\mathrm{d}(\nu \circ T_g)(\omega) = \int x(\omega)[\mathrm{d}(\nu \circ T_g)/\mathrm{d}\nu))(\omega)\mathrm{d}\nu(\omega)$, it follows that $A_{\tau,g}$ is the multiplication operator defined by the Radon–Nikodym derivative $A_{\tau,g} = \mathrm{d}(\nu \circ T_g)/\mathrm{d}\nu$. Note that a function $f : \Omega \to \mathbb{R}^+$ has the property $f^{it} = 1$ if and only if $f(\omega) \in \{e^{2n\pi/t}; n \in \mathbb{Z}\}$. We thus infer from Theorem 27.4 the following

Corollary. *In the previous setting, a real number t belongs to $T(\mathscr{M})$ if and only if there exists a positive measure ν on Ω, equivalent to μ, such that $[\mathrm{d}(\nu \circ T_g)/\mathrm{d}\nu](\omega) \in \{e^{2n\pi/t}; n \in \mathbb{Z}\}$ for all $g \in G, \omega \in \Omega$.*

27.6 Proposition. *Let $\mathscr{M}$ be a countably decomposable W^*-algebra, φ a faithful normal state on $\mathscr{M}$ and $t \in \mathbb{R}$. Then $t \in T(\mathscr{M})$ if and only if there exists a unitary element $u \in \mathscr{M}$ such that*

$$\Delta_\varphi^{it} = \pi_\varphi(u)J_\varphi\pi_\varphi(u)J_\varphi. \tag{1}$$

Proof. If $t \in T(\mathscr{M})$, there exists $u \in U(\mathscr{M}^\varphi)$ such that $\sigma_t^\varphi = \mathrm{Ad}(u)$ and for $x \in \mathscr{M}$ we have $\Delta_\varphi^{it}\pi_\varphi(x)\xi_\varphi = \Delta_\varphi^{it}\pi_\varphi(x)\Delta_\varphi^{-it}\xi_\varphi = \pi_\varphi(\sigma_t^\varphi(x))\xi_\varphi = \pi_\varphi(u)\pi_\varphi(x)\pi_\varphi(u)^*\xi_\varphi = \pi_\varphi(u)\pi_\varphi(x)S_\varphi\pi_\varphi(u)\xi_\varphi = \pi_\varphi(u)\pi_\varphi(x)J_\varphi\Delta_\varphi^{1/2}\pi_\varphi(u)\xi_\varphi = \pi_\varphi(u)\pi_\varphi(x)J_\varphi\pi_\varphi(u)\xi_\varphi = \pi_\varphi(u)\pi_\varphi(x)J_\varphi\pi_\varphi(u)J_\varphi\xi_\varphi = \pi_\varphi(u)J_\varphi\pi_\varphi(u)J_\varphi\pi_\varphi(x)\xi_\varphi$, since $u \in \mathscr{M}^\varphi$ and $J_\varphi\pi_\varphi(u)J_\varphi \in \pi_\varphi(\mathscr{M})'$; this proves (1).

Conversely, if there exists $u \in U(\mathscr{M}^\varphi)$ satisfying (1), then for every $x \in \mathscr{M}$ we have $\pi_\varphi(\sigma_t^\varphi(x))\xi_\varphi = \Delta_\varphi^{it}\pi_\varphi(x)\xi_\varphi = \pi_\varphi(u)J_\varphi\pi_\varphi(u)J_\varphi\pi_\varphi(x)\xi_\varphi = \ldots = \pi_\varphi(uxu^*)\xi_\varphi$, hence $\sigma_t^\varphi = \mathrm{Ad}(u)$ and $t \in T(\mathscr{M})$. □

If $\mathscr{M}$ has separable predual, the $U(\mathscr{M}^\varphi)$ and $U(\mathscr{B}(\mathscr{H}_\varphi))$ are polish topological groups and the mappings $t \mapsto \Delta_\varphi^{it}$ and $u \mapsto \pi_\varphi(u)J_\varphi\pi_\varphi(u)J_\varphi$ are continuous. Consequently,

if $\mathscr{M}$ has separable predual, then $T(\mathscr{M})$ is an analytic subgroup of $\mathbb{R}$. (2)

In particular, $T(\mathscr{M})$ is in this case Lebesgue measurable. If $T(\mathscr{M})$ has positive measure, then $T(\mathscr{M})$ is open and hence $T(\mathscr{M}) = \mathbb{R}$. Therefore,

$$\text{if } \mathscr{M} \text{ is a type III W}^*\text{-algebra with separable predual, then } T(\mathscr{M}) \text{ has zero Lebesgue measure.} \tag{3}$$

On the other hand, we mention that every subgroup of $\mathbb{R}$ is of the form $T(\mathscr{M})$ for some countably decomposable factor $\mathscr{M}$ (Connes, 1973a, 1.5.8).

27.7. A W^*-algebra $\mathscr{M}$ is called *normal* if for every unital W^*-subalgebra $\mathscr{N} \subset \mathscr{M}$ the condition $\mathscr{N} = (\mathscr{N}' \cap \mathscr{M})' \cap \mathscr{M}$ holds. By the von Neumann double commutant theorem ([L], 3.2), any type I factor is normal. As an application of the invariant $T(\mathscr{M})$, we prove the following result.

Proposition. *No type III factor with separable predual is normal.*

Proof. Let $\mathscr{M} \subset \mathscr{B}(\mathscr{H})$ be a type III factor acting on the separable Hilbert space $\mathscr{H}$. Then $\mathscr{N} = \mathscr{M}' \cap \mathscr{B}(\mathscr{H})$ is a type III factor in standard form (see [L]. 8.13) and $\mathscr{M} \approx \mathscr{N}' \bar{\otimes} \mathscr{F}$, where $\mathscr{F}$ is the countably decomposable infinite type I factor. Since $T(\mathscr{N})$ has zero Lebesgue measure (27.6.(3)), there is a $t \in \mathbb{R}$ such that $nt \notin T(\mathscr{N})$ for every $n \in \mathbb{Z}$. Let ψ be an n.s.f. weight on $\mathscr{N}$ and $\sigma = \sigma_t^{\psi} \in Aut(\mathscr{N})$. Then, for every $n \in \mathbb{Z}$, σ^n is an outer $*$-automorphism of $\mathscr{N}$. On the other hand, as $\mathscr{N}$ is in standard form, there is $v \in U(\mathscr{B}(\mathscr{H}))$ such that $\sigma = \mathrm{Ad}(v)$. Let $\mathscr{P} = \mathscr{R}\{\mathscr{N} \bar{\otimes} 1, v \bar{\otimes} \boldsymbol{\rho}(1)\} \subset \mathscr{B}(\mathscr{H} \bar{\otimes} \ell^2(\mathbb{Z}))$, where $\boldsymbol{\rho}$ is the right regular representation of $\mathbb{Z}$, and $Q = \mathscr{P}' \subset (\mathscr{N} \bar{\otimes} 1)' = \mathscr{N}' \bar{\otimes} \mathscr{F} = \mathscr{M}$. We have $\mathscr{N} \bar{\otimes} 1 \neq \mathscr{P}$ since $v \bar{\otimes} \boldsymbol{\rho}(1) \notin \mathscr{N} \bar{\otimes} 1$, hence $Q \neq \mathscr{M}$. On the other hand, we have $Q' \cap \mathscr{M} = \mathscr{P}'' \cap (\mathscr{N} \bar{\otimes} 1)' = \mathscr{P} \cap (\mathscr{N} \bar{\otimes} 1)' = \mathbb{C} \cdot 1$; the last equality follows as in the relative commutant theorem (22.3.(1)), since the action $n \mapsto \sigma^n$ of $\mathbb{Z}$ on $\mathscr{N}$ is properly outer (see also 19.14.(1)). Thus, $(Q' \cap \mathscr{M})' \cap \mathscr{M} = \mathscr{M} \neq Q$. □

It is a classical result (Fuglede & Kadison, 1951) that no type II factor is normal. Thus the only normal factors with separable predual are those of type I.

27.8. Notes. The invariant $T(\mathscr{M})$ was introduced by Connes (1973a), who obtained the results presented in this section. Moreover, Connes (Connes, 1973a, 1.3.7) computed the invariant $T(\mathscr{M})$ for infinite tensor product factors, and obtained an explicit expression of $T(\mathscr{M})$ in terms of the eigenvalue list for Araki–Woods factors (Connes, 1973a, 1.3.9). In particular, for the Powers factors $\mathscr{R}_\lambda$ $(0 < \lambda < 1)$, we have $T(\mathscr{R}_\lambda) = \{nt; n \in \mathbb{Z}\}$ where $t = -2\pi/\ln(\lambda)$, from which it follows that $\{\mathscr{R}_\lambda\}_{0<\lambda<1}$ is a continuous family of approximately finite dimensional (or injective) nonisomorphic type III factors (Powers, 1967, 1970). Since $T(\mathscr{R}_\lambda \bar{\otimes} \mathfrak{L}(F_2)) = T(\mathscr{R}_\lambda)$, we see that $\{\mathscr{R}_\lambda \bar{\otimes} \mathfrak{L}(F_2)\}_{0<\lambda<1}$ is a continuous family of noninjective nonisomorphic type III factors (see also Sakai (1971)). For an arbitrary factor $\mathscr{M}$, Araki and Woods (1968) introduced the invariant $\rho(\mathscr{M})$ as the set of all $\lambda \in [0, 1]$ such that $\mathscr{M} \bar{\otimes} \mathscr{R}_\lambda \approx \mathscr{R}_\lambda$. For $\lambda \in (0, 1)$ and $t = -2\pi/\ln(\lambda)$, Connes (1973a, 3.6.2) proved that $\lambda \in \rho(\mathscr{M}) \Rightarrow t \in T(\mathscr{M})$ for any factor $\mathscr{M}$ and that $\lambda \in \rho(\mathscr{M}) \Leftrightarrow t \in T(\mathscr{M})$ whenever $\mathscr{M}$ is an Araki–Woods factor; note that $T(\mathfrak{L}(F_2)) = \mathbb{R}$ but $\rho(\mathfrak{L}(F_2)) = \emptyset$ (Connes, 1973a, 3.6.5). For every subgroup $\mathbb{G}$ of $\mathbb{R}$, there exists a countably decomposable factor $\mathscr{M}$ such that $T(\mathscr{M}) = \mathbb{G}$; moreover, there exists a countably decomposable factor $\mathscr{M}$ of type III such that $T(\mathscr{M}) = \mathbb{R}$ (compare with 27.2 and 27.6.(3)); if the group $\mathbb{G}$ is countable, then there exists a factor $\mathscr{M}$ with separable predual such that $T(\mathscr{M}) = \mathbb{G}$ (Connes, 1973a, 1.5.8).

For our exposition, we have used Connes (1973a).

28 The Connes Invariant $S(\mathscr{M})$

In this section, we introduce the invariant $S(\mathscr{M})$ as the intersection of the spectra of all modular operators.

28.1. We shall identify the dual group $\hat{\mathbb{R}}$ of the additive group $\mathbb{R}$ with the multiplicative group $\mathbb{R}_*^+$ so that

$$\langle t, \lambda \rangle = \lambda^{it} \quad (t \in \mathbb{R}, \lambda \in \mathbb{R}_*^+).$$

Then the Fourier transform of a function $f \in \mathscr{L}^1(\mathbb{R})$ is given by

$$\hat{f}(\lambda) = \int_{-\infty}^{+\infty} f(t)\lambda^{it}\, dt \quad (\lambda \in \mathbb{R}_*^+)$$

and for any positive nonsingular self-adjoint operator Δ we have

$$\hat{f}(\Delta) = \int_{-\infty}^{+\infty} f(t)\Delta^{it}\, dt; \tag{1}$$

recall that if $\sigma : \mathbb{R} \to Aut(\mathscr{M})$ is a continuous action of $\mathbb{R}$ on the W^*-algebra $\mathscr{M}$, then

$$\sigma_f(x) = \int_{-\infty}^{+\infty} f(t)\sigma_t(x)\, dt \quad (x \in \mathscr{M}). \tag{2}$$

Note that for every n.s.f. weight φ on $\mathscr{M}$ we have

$$a \in \mathfrak{N}_\varphi, f \in \mathscr{L}^1(\mathbb{R}) \Rightarrow \sigma_f^\varphi(a) \in \mathfrak{N}_\varphi, \quad (\sigma_f^\varphi(a))_\varphi = \hat{f}(\Delta_\varphi)a_\varphi. \tag{3}$$

For the proof, we may assume that $\int_{-\infty}^{+\infty} f(t)\, dt = 1$. Then $\sigma_t^\varphi(a)$ belongs to the w-closed convex hull of the set $\{\sigma_t^\varphi(a); t \in \mathbb{R}\}$, which is contained in the w-closed convex set $\{x \in \mathscr{M}; \varphi(x^*x) \leq \varphi(a^*a)\}$ (see 5.9), whence $\sigma_f^\varphi(a) \in \mathfrak{N}_\varphi$. Then, for every $b \in \mathfrak{N}_\varphi$ we have

$$\begin{aligned}((\sigma_t^\varphi(a))_\varphi, b_\varphi)_\varphi = \varphi(b^*\sigma_f^\varphi(a)) &= \int_{-\infty}^{+\infty} f(t)\varphi(b^*\sigma_t^\varphi(a))\, dt \\ &= \int_{-\infty}^{+\infty} f(t)(\Delta_\varphi^{it} a_\varphi | b_\varphi)_\varphi\, dt = (\hat{f}(\Delta_\varphi)a_\varphi | b_\varphi)_\varphi,\end{aligned}$$

and $(\sigma_t^\varphi(a))_\varphi = \hat{f}(\Delta_\varphi)a_\varphi$.

Proposition. *For every $\varphi \in W_{nsf}(\mathbb{R})$, we have*

$$Sp\, \sigma^\varphi = Sp(\Delta_\varphi) \cap \mathbb{R}_*^+. \tag{4}$$

Proof. Using (3), for every $f \in \mathcal{L}^1(\mathbb{R})$ we have $\sigma_f^\varphi = 0 \Leftrightarrow \sigma_f^\varphi(a) = 0$ for all $a \in \mathfrak{N}_\varphi \Leftrightarrow \hat{f}(\Delta_\varphi)a_\varphi = 0$ for all $a \in \mathfrak{N}_\varphi \Leftrightarrow \hat{f}(\Delta_\varphi) = 0 \Leftrightarrow \hat{f}|Sp(\Delta_\varphi) \cap \mathbb{R}_*^+ = 0$. Thus, if $\lambda \in Sp(\Delta_\varphi) \cap \mathbb{R}_*^+$, then $\hat{f}(\lambda) = 0$ for every $f \in \mathcal{L}^1(\mathbb{R})$ with $\sigma_f^\varphi = 0$, that is, $\lambda \in Sp\, \sigma^\varphi$. Conversely, if $\lambda \in \mathbb{R}_*^+$ but λ does not belong to the closed subset $Sp(\Delta_\varphi) \cap \mathbb{R}_*^+$ of $\mathbb{R}_*^+$, then there exists a function $f \in \mathcal{L}^1(\mathbb{R})$ such that $\hat{f}|Sp(\Delta_\varphi) \cap \mathbb{R}_*^+ = 0$, that is, $\sigma_f^\varphi = 0$, but $\hat{f}(\lambda) \neq 0$, and so $\lambda \notin Sp\, \sigma^\varphi$. □

In particular, it follows that $1 \in Sp(\Delta_\varphi)$ for each $\varphi \in W_{\text{nsf}}(\mathcal{M})$; if φ is finite, this is obvious since $\Delta_\varphi \xi_\varphi = \xi_\varphi$.

Note that $Sp(\Delta_\varphi)$ is symmetrical about 1 since $\Delta_\varphi = J_\varphi \Delta_\varphi^{-1} J_\varphi$.

28.2. In what follows we restrict our considerations to the case of factors. For each factor $\mathcal{M}$, the set

$$S(\mathcal{M}) = \bigcap \{Sp(\Delta_\varphi); \varphi \in W_{\text{nsf}}(\mathcal{M})\}$$

is a closed subset of $\mathbb{R}_*^+ = [0, +\infty)$ and an algebraic invariant of $\mathcal{M}$.

Proposition. *A factor $\mathcal{M}$ is semifinite if and only if $0 \notin S(\mathcal{M})$; in this case, $S(\mathcal{M}) = \{1\}$.*

A factor $\mathcal{M}$ is finite if and only if there exists a faithful normal state φ on $\mathcal{M}$ such that $0 \notin Sp(\Delta_\varphi)$.

Proof. If $\mathcal{M}$ is semifinite and τ is an n.s.f. trace on $\mathcal{M}$, $\Delta_\tau = 1$, $Sp(\Delta_\tau) = \{1\}$ and $S(\mathcal{M}) = \{1\}$. Conversely, assume that $0 \notin S(\mathcal{M})$, so that there is a $\varphi \in W_{\text{nsf}}(\mathcal{M})$ such that $0 \notin Sp(\Delta_\varphi)$. Then there is $\lambda \geq 1$ such that $Sp(\Delta_\varphi) \subset [\lambda^{-1}, \lambda]$. Since $\pi_\varphi(\sigma_t^\varphi(x)) = \Delta_\varphi^{it} \pi_\varphi(x) \Delta_\varphi^{-it}$ ($x \in \mathcal{M}, t \in \mathbb{R}$), it follows in accordance with Theorem 15.8 that there exists an s-continuous unitary representation $\mathbb{R} \ni t \mapsto u(t) \in \mathcal{M}$ such that $\sigma_t^\varphi = \text{Ad}(u(t))$ ($t \in \mathbb{R}$), thus $\mathcal{M}$ is semifinite by Proposition 27.2.

Assume now that there exists a faithful normal state φ on $\mathcal{M}$ such that Δ_φ is bounded. We show that the $*$-operation is s-continuous on the closed unit ball of $\mathcal{M}$, so that $\mathcal{M}$ is finite ([L], 7.23). If $\{x_i\} \subset \mathcal{M}$ is a net such that $\|x_i\| \leq 1$ and $x_i \xrightarrow{s} x \in \mathcal{M}$, then $\varphi((x_i - x)(x_i - x)^*) = \|(x_i^*)_\varphi - x_\varphi^*\|_\varphi^2 = \|J_\varphi \Delta_\varphi^{1/2}(x_i)_\varphi - J_\varphi \Delta_\varphi^{1/2} x_\varphi\|_\varphi^2 \leq \|\Delta_\varphi^{1/2}\| \|(x_i)_\varphi - x_\varphi\|_\varphi = \|\Delta_\varphi^{1/2}\| \varphi((x_i - x)^*(x_i - x)) \to 0$, hence $x_i^* \xrightarrow{s} x^*$. □

28.3 Theorem. *Let $\mathcal{M}$ be a factor. For every $\varphi \in W_{nsf}(\mathcal{M})$, we have*

$$S(\mathcal{M}) \cap \mathbb{R}_*^+ = \Gamma(\sigma^\varphi) \tag{1}$$

$$S(\mathcal{M}) \cap \mathbb{R}_*^+ = \bigcap \{Sp(\Delta_{\varphi_e}); 0 \neq e \in Proj(\mathcal{M}^\varphi)\} \tag{2}$$

$$S(\mathcal{M}) \cap \mathbb{R}_*^+ = \bigcap \{Sp(\Delta_{\varphi_e}); 0 \neq e \in Proj(\mathcal{Z}(\mathcal{M}^\varphi))\} \tag{3}$$

thus $S(\mathcal{M}) \cap \mathbb{R}_*^+$ is a closed subgroup of $\mathbb{R}_*^+$ and

$$(S(\mathcal{M}) \cap \mathbb{R}_*^+) \cdot Sp(\Delta_\varphi) \subset Sp(\Delta_\varphi). \tag{4}$$

Also, we have

$$S(\mathcal{M}) = \bigcap \{Sp(\Delta_\psi); \psi \in W_{nsf}(\mathcal{M}) \text{ strictly semifinite}\}. \tag{5}$$

If $\mathscr{M}$ is countably decomposable, then the identity

$$S(\mathscr{M}) = \bigcap\{Sp(\Delta_\psi); \psi \text{ faithful normal state on } \mathscr{M}\} \tag{6}$$

holds, except when $\mathscr{M}$ is both infinite and semifinite; in this case $S(\mathscr{M}) = \{1\}$, while the right hand side of (6) is equal to $\{0, 1\}$.

Proof. By Theorem 16.6, $\Gamma(\sigma^\varphi)$ is the intersection of the spectra $Sp\ \sigma$ of all continuous actions $\sigma : \mathbb{R} \to Aut(\mathscr{M})$ with $\sigma \sim \sigma^\varphi$. By Theorems 3.1 and 5.1, we have $\sigma \sim \sigma^\varphi$ if and only if $\sigma = \sigma^\psi$ for some $\psi \in W_{\rm nsf}(\mathscr{M})$. Thus, (1) follows from the definition of $S(\mathscr{M})$ and Proposition 28.1. The fact that $S(\mathscr{M}) \cap \mathbb{R}_*^+$ is a closed subgroup of $\mathbb{R}_*^+$ and the proof of (4) follows from (1) and Proposition 16.1. Since, for every $0 \neq e \in Proj(\mathscr{M}^\varphi)$, σ^{φ_e} is just the restriction of σ^φ to $e\mathscr{M}e$, (2) and (3) follow from the definition of $\Gamma(\sigma^\varphi)$ (16.1).

We now prove (5). The semifinite case being trivial, we shall assume that $\mathscr{M}$ is of type III. There exists a strictly semifinite $\varphi \in W_{\rm nsf}(\mathscr{M})$. For every $0 \neq e \in Proj(\mathscr{M}^\varphi)$, $\psi = \varphi_e \bar{\otimes}\ tr$ is a strictly semifinite normal faithful weight on $\mathscr{M}_e \bar{\otimes} \mathscr{F}_\infty \approx \mathscr{M}$ and $Sp(\Delta_\psi) = Sp(\Delta_{\varphi_e} \bar{\otimes} 1) = Sp(\Delta_{\varphi_e})$. Hence (5) follows from (2).

Finally, assuming $\mathscr{M}$ countably decomposable, we prove (6). If $\mathscr{M}$ is of type III, φ is a faithful normal state on $\mathscr{M}$ and $0 \neq e \in Proj(\mathscr{M}^\varphi)$, then φ_e is a faithful normal state on $\mathscr{M}_e \approx \mathscr{M}$, so that there exists a faithful normal state ψ on $\mathscr{M}$ with $Sp(\Delta_\psi) = Sp(\Delta_{\varphi_e})$ and we can apply (2). If $\mathscr{M}$ is finite, then, by Proposition 28.2, both sides of (6) are reduced to $\{1\}$. If $\mathscr{M}$ is semifinite and infinite, then, again by Proposition 28.2, $S(\mathscr{M}) = \{1\}$, while the right hand side of (6) contains the set $\{0, 1\}$. Let $\lambda > 0, \lambda \neq 1$, and $t \in \mathbb{R} = T(\mathscr{M})$ with $\lambda^{it} \neq 1$. By Theorem 27.1. (vi), there exists a faithful normal state φ on $\mathscr{M}$ with $\sigma_t^\varphi = \iota$, that is, $\Delta_\varphi^{it} \in \mathscr{Z}(\mathscr{M}) = \mathbb{C} \cdot 1$, whence $\lambda \notin Sp(\Delta_\varphi)$ and $\lambda \notin S(\mathscr{M})$. □

From (3) it follows that for $\varphi \in W_{nsf}(\mathscr{M})$ we have

$$\mathscr{M}^\varphi \text{ is a factor } \Rightarrow S(\mathscr{M}) = Sp(\Delta_\varphi). \tag{7}$$

Also, using (1) and Proposition 16.10, we see that

$$S(\mathscr{M}) = \{0, 1\} \Rightarrow \mathscr{Z}(\mathscr{M}^\varphi) \text{ has no minimal projections.} \tag{8}$$

28.4 Proposition. *Let $\mathscr{M}$ be a factor. If $\mathscr{N}$ is a semifinite factor,*

$$S(\mathscr{M} \bar{\otimes} \mathscr{N}) = S(\mathscr{M}); \tag{1}$$

if $e \in \mathscr{M}$ is a nonzero projection,

$$S(\mathscr{M}_e) = S(\mathscr{M}); \tag{2}$$

and if $\mathscr{M} \subset \mathscr{B}(\mathscr{H})$ is realized as a von Neumann algebra,

$$S(\mathscr{M}') = S(\mathscr{M}). \tag{3}$$

Proof. (1) If φ is an n.s.f. weight on $\mathscr{M}$ and τ is an n.s.f. trace on $\mathscr{N}$, then $\varphi \bar{\otimes} \tau$ is an n.s.f. weight on $\mathscr{M} \bar{\otimes} \mathscr{N}$ and $\mathscr{Z}((\mathscr{M} \bar{\otimes} \mathscr{N})^{\varphi \bar{\otimes} \tau}) = \mathscr{Z}(\mathscr{M}^{\varphi}) \bar{\otimes} 1$; thus (1) follows from 28.3.(3) and 10.21.(4).

(2) It is clear that $\mathscr{M}_e$ is semifinite if and only if $\mathscr{M}$ is semifinite, and so (2) follows from 28.3.(1) using 16.2.(3).

(3) Using (1) and (2), the proof is similar to the proof of 27.3.(4). □

28.5. In this section, we compute the invariant $S(\mathscr{M})$ for crossed products by properly outer actions of discrete groups. More generally, let $\mathscr{M}$ be a W^*-algebra, $\mathscr{N} \subset \mathscr{M}$ a semifinite unital W^*-subalgebra with $\mathscr{Z}(\mathscr{N}) = \mathscr{N}' \cap \mathscr{M}, P : \mathscr{M} \to \mathscr{N}$ a faithful normal conditional expectation of $\mathscr{M}$ onto $\mathscr{N}$ and $\mathscr{G} \subset \mathcal{N}(P)$ a subgroup of the normalizer $\mathcal{N}(P)$ of P such that $\mathscr{M} = \mathscr{R}\{\mathscr{N}, \mathscr{G}\}$. Let τ be a fixed n.s.f. trace on $\mathscr{M}$. For every $u \in \mathcal{N}(P)$, there exists a unique positive nonsingular self-adjoint operator A_u affiliated to $\mathscr{Z}(\mathscr{N})$ such that $\tau(u \cdot u^*) = \tau_{A_u}$.

Theorem. *In the above setting if $\lambda > 0$ we have $\lambda \in S(\mathscr{M})$ if and only if for $\varepsilon > 0$ and e a nonzero projection in $\mathscr{Z}(\mathscr{N})$ there exist a nonzero projection $p \in \mathscr{Z}(\mathscr{N})$ and $u \in \mathscr{G}$ such that $p \leq e, upu^* \leq e$ and $Sp(A_u p|\mathscr{N}p) \subset (\lambda - \varepsilon, \lambda + \varepsilon)$.*

Proof. Let $\varphi = \tau \circ P \in W_{nsf}(\mathscr{M})$.

We shall first prove that for every nonzero projection $e \in \mathscr{Z}(\mathscr{N})$ and closed set $F \subset \mathbb{R}^+_*$ we have

$$\bigvee\{\mathbf{r}(x); x \in \mathscr{M}(\sigma^{\varphi}; F) \cap e\mathscr{M}e\} = \bigvee\{e(u^*eu)\chi_F(A_u); u \in \mathscr{G}\}. \tag{1}$$

Indeed, let $d = \bigvee\{e(u^*eu)\chi_F(A_u); u \in \mathscr{G}\} \in \mathscr{Z}(\mathscr{N})$. For $f \in \mathscr{L}^1(\mathbb{R})$ and $u \in \mathscr{G}$ we have $\sigma^{\varphi}_f(eu\chi_F(A_u)e) = e\sigma^{\varphi}_f(u)\chi_F(A_u)e$ since both e and $\chi_F(A_u)$ belong to $\mathscr{N} \subset \mathscr{M}^{\varphi}$. Then, using 27.4.(2), we obtain $\sigma^{\varphi}_f(u) = \int f(t)\sigma^{\varphi}_t(u)\,\mathrm{d}t = \int f(t)uA^{it}_u\,\mathrm{d}t = u\int f(t)A^{it}_u\,\mathrm{d}t = u\hat{f}(A_u)$. Thus, if $f \in \mathscr{L}^1(\mathbb{R})$ and $\hat{f}$ vanishes on a neighborhood of $F, \sigma^{\varphi}_f(eu\chi_F(A_u)e) = 0$, that is, $eu\chi_F(A_u)e \in \mathscr{M}(\sigma^{\varphi}; F) \cap e\mathscr{M}e$. Since $\mathbf{r}(eu\chi_F(A_u)e) = \mathbf{s}(e\chi_F(A_u)u^*eu\chi_F(A_u)e) = e(u^*eu)\chi_F(A_u)$, it follows that

$$d \leq \bigvee\{\mathbf{r}(x); x \in \mathscr{M}(\sigma^{\varphi}; F) \cap e\mathscr{M}e\}.$$

It remains to be shown that for every $x \in \mathscr{M}(\sigma^{\varphi}; F) \cap e\mathscr{M}e$ we have $xd = x$. Let $f \in \mathscr{L}^1(\mathbb{R})$ be such that $\hat{f}$ vanishes on a neighborhood of F and let $u \in \mathscr{G}$. We have $P(u^*\sigma^{\varphi}_f(x)) = P(u^*\int f(t)\sigma^{\varphi}_t(x)\,\mathrm{d}t) = \int f(t)P(u^*\sigma^{\varphi}_t(x))\,\mathrm{d}t = \int f(t)P(\sigma^{\varphi}_t(\sigma^{\varphi}_{-t}(u^*)x))\,\mathrm{d}t = \int f(t)\sigma^{\varphi}_t(P(\sigma^{\varphi}_{-t}(u^*)x))\,\mathrm{d}t = \int f(t)P(\sigma^{\varphi}_{-t}(u)^*x)\,\mathrm{d}t = \int f(t)P(A^{it}_u u^*x)\,\mathrm{d}t = \int f(t)A^{it}_u P(u^*x)\,\mathrm{d}t = \hat{f}(A_u)P(u^*x)$. If $x \in \mathscr{M}(\sigma^{\varphi}; F) \cap e\mathscr{M}e$, then $\sigma^{\varphi}_f(x) = 0$, and $\hat{f}(A_u)P(u^*x) = 0$. For every $\gamma \notin F$, there exists $f \in \mathscr{L}^1(\mathbb{R})$ such that $\hat{f}(\gamma) \neq 0$ and $\hat{f}$ vanishes on a neighborhood of F. Since $P(u^*x) \in \mathscr{N}$ and $A_u \in \mathscr{Z}(\mathscr{N})$, it follows that $\mathbf{r}(P(u^*x)) \leq \chi_F(A_u)$. We have $ex = xe = x$, hence $P(u^*x)e = P(u^*xe) = P(u^*x), P(u^*x)u^*eu = u^*euP(u^*x) = P(u^*euu^*x) = P(u^*x)$. Thus, $\mathbf{r}(P(u^*x)) \leq e(u^*eu)\chi_F(A)$. Consequently, for every $u \in \mathscr{G}$ we have $P(u^*x)d = P(u^*x), P(u^*(xd - x)) = 0$, hence $P(y^*(xd - x)) = 0$ for all $y \in \mathscr{R}\{\mathscr{N}, \mathscr{G}\} = \mathscr{M}$, in particular $P((xd - x)^*(xd - x)) = 0$, that is, $xd = x$.

Note that $\mathscr{N} \subset \mathscr{M}^{\varphi}$ and $\mathscr{Z}(\mathscr{M}^{\varphi}) \subset \mathscr{N}' \cap \mathscr{M} = \mathscr{Z}(\mathscr{N})$. By Theorem 28.3 and Proposition 28.1, it follows that

$$S(\mathscr{M}) \cap \mathbb{R}^+_* = \bigcap\{Sp\,\sigma^{\varphi_e}; 0 \neq e \in Proj(\mathscr{Z}(\mathscr{N}))\}. \tag{2}$$

Consider now $0 \neq \lambda \in S(\mathscr{M}), \varepsilon > 0, 0 \neq e \in Proj(\mathscr{Z}(\mathscr{N}))$ and $F = \left[\lambda - \frac{\varepsilon}{2}, \lambda + \frac{\varepsilon}{2}\right]$. From (2), we infer that $\lambda \in Sp\,\sigma^{\varphi_e}$, hence $\bigvee\{\mathbf{r}(x); x \in \mathscr{M}(\sigma^{\varphi}; F) \cap e\mathscr{M}e\} \neq 0$ and from (1), it follows that there exists $u \in \mathscr{G}$ such that $p = e(u^*eu)\chi_F(A_u) \neq 0$. Then $p \in \mathscr{Z}(\mathscr{N}), p \leq e, upu^* \leq e$, and $Sp(A_u p|\mathscr{N}p) \subset F \subset (\lambda - \varepsilon, \lambda + \varepsilon)$.

Conversely, let $0 \neq e \in Proj(\mathscr{Z}(\mathscr{N})), \varepsilon > 0$ and $F = [\lambda-\varepsilon, \lambda+\varepsilon]$. By assumption, we can find $0 \neq p \in Proj(\mathscr{Z}(\mathscr{N})), p \leq e$, and $u \in \mathscr{G}$ with $upu^* \leq e$ and $Sp(A_u|\mathscr{N}p) \subset F$; thus $\bigvee\{e(u^*eu)\chi_F(A_u); u \in \mathscr{G}\} \neq 0$. From (1), it follows that $\mathscr{M}(\sigma^{\varphi}; F) \cap e\mathscr{M}e \neq 0$, hence $\lambda \in Sp\,\sigma^{\varphi_e}$. Using (2), we conclude that $\lambda \in S(\mathscr{M})$. □

28.6. For instance, Theorem 28.5 holds if $\mathscr{M} = \mathscr{R}(\mathscr{N}, \sigma)$ is the crossed product of a semifinite W^*-algebra $\mathscr{N}$ by a properly outer action $\sigma : G \to Aut(\mathscr{N})$ of a discrete group G which is ergodic on $\mathscr{Z}(\mathscr{N})$ (see 27.5 and Corollary 1/22.6).

In particular, let $G \ni g \to T_g$ be a free ergodic action of G on the measure space (Ω, μ) with μ a G-quasi-invariant measure (see 22.8, 27.5). We then obtain a free ergodic action $\sigma : G \to Aut(\mathscr{N})$ of G on the abelian W^*-algebra $\mathscr{N} = \mathscr{L}^\infty(\Omega, \mu)$ and the crossed product $\mathscr{M} = \mathscr{R}(\mathscr{N}, \sigma)$ is a factor.

For the dynamical system (Ω, μ, G), one defines (cf. Araki & Woods, 1968; Krieger, 1970b) an invariant called "ratio set" as being the set $\mathbf{r}(G)$ of all $\lambda \geq 0$ with the following property: for every $\varepsilon > 0$ and every μ-measurable set $A \subset \Omega$ with $\mu(A) > 0$, there exist a μ-measurable set $B \subset A$ with $\mu(B) > 0$ and $g \in G$ such that $T_g(B) \subset A$ and

$$|[\mathrm{d}(\mu \circ T_g)/\mathrm{d}\mu](\omega) - \lambda| < \varepsilon \text{ for all } \omega \in B.$$

It is easy to see that $0 \notin \mathbf{r}(G)$ if and only if there exists a G-invariant sigma-finite positive measure ν on Ω which is equivalent to μ (see Krieger, 1970b), that is, if and only if the crossed product $\mathscr{M} = \mathscr{R}(\mathscr{N}, \sigma)$ is semifinite (22.8).

If τ is the n.s.f. trace on $\mathscr{N}$ defined by the measure μ, then the unique positive nonsingular self-adjoint operator A_g affiliated to $\mathscr{N}$ such that $\tau \circ \sigma_g = \tau_{A_g}$ is the multiplication operator defined by the function $A_g = \mathrm{d}(\mu \circ T_g)/\mathrm{d}\mu$ $(g \in G)$ (see 27.5).

Thus, we infer from Theorem 28.5 the following:

Corollary. *In the previous situation, we have $S(\mathscr{M}) = \mathbf{r}(G)$.*

28.7. In this section, we give a spatial characterization of the invariant $S(\mathscr{M})$.

Let Δ be a positive self-adjoint operator in the Hilbert space $\mathscr{H}$ and $D \subset Dom(\Delta)$ a vector subspace such that $\Delta = \overline{\Delta|D}$. For $\lambda \in \mathbb{C}$, it is easy to check that

$$\lambda \in Sp(\Delta) \Leftrightarrow \text{ there exist } \xi_n \in D, \|\xi_n\| = 1, \|(\lambda - D)\xi_n\| \to 0, \tag{1}$$

$$\lambda \notin Sp(\Delta) \Rightarrow \|(\lambda - \Delta)^{-1}\| = \mathrm{dist}(\lambda, Sp(\Delta))^{-1}; \tag{2}$$

for $\varepsilon > 0$, it follows that

$$\mathrm{dist}(\lambda, Sp(\Delta)) < \varepsilon \Leftrightarrow \text{ there exists } \xi \in D, \|\xi\| > 1, \|(\lambda - \Delta)\xi\| < \varepsilon. \tag{3}$$

Lemma. *Let $\mathscr{M} \subset \mathscr{B}(\mathscr{H})$ be a von Neumann algebra with a cyclic and separating vector $\xi \in \mathscr{H}$, $\lambda \geq 0$, and $\varepsilon > 0$. Consider the statements:*

$$\operatorname{dist}(\lambda^{1/2}, Sp(\Delta_\xi^{1/2})) < \varepsilon \tag{$A(\varepsilon)$}$$

there exist $x \in \mathscr{M}$ and $x' \in \mathscr{M}'$ such that
$$\|\lambda^{1/2}x\xi - x'\xi\| < \varepsilon, \|x^*\xi - \lambda^{1/2}x'^*\xi\| < \varepsilon, \|x\xi\| > 1. \tag{$B(\varepsilon)$}$$

The $A(\varepsilon) \Rightarrow B(\varepsilon) \Rightarrow A(2\varepsilon)$.

Proof. Put $\Delta = \Delta_\xi, J = J_\xi$. Since $\Delta^{1/2} = \overline{\Delta^{1/2}|\mathscr{M}\xi}$, from statement $A(\varepsilon)$ it follows that there is an $x \in \mathscr{M}$ such that $\|x\xi\| > 1$ and $\|(\lambda^{1/2} - \Delta^{1/2})x\xi\| < \varepsilon$. Then $x' = Jx^*J \in \mathscr{M}'$ and $\|\lambda^{1/2}x\xi - x'\xi\| = \|\lambda^{1/2}x\xi - Jx^*\xi\| = \|\lambda^{1/2}x\xi - \Delta^{1/2}x\xi\| < \varepsilon, \|x^*\xi - \lambda^{1/2}x'^*\xi\| = \|J\Delta^{1/2}x\xi - \lambda^{1/2}Jx\xi\| = \|\Delta^{1/2}x\xi - \lambda^{1/2}x\xi\| < \varepsilon$. Hence $A(\varepsilon) \Rightarrow B(\varepsilon)$.

Conversely, assume that $B(\varepsilon)$ holds and consider $\eta = x\xi, \zeta = Jx'^*\xi = \Delta^{-1/2}x'\xi$. Then $\eta, \zeta \in Dom(\Delta^{1/2})$, $\|\eta\| > 1$ and $\|\lambda^{1/2}\eta - \Delta^{1/2}\zeta\| < \varepsilon, \|\Delta^{1/2}\eta - \lambda^{1/2}\zeta\| < \varepsilon$. Since $\|\Delta^{1/2}(\lambda^{1/2} + \Delta^{1/2})^{-1}\| \leq 1, \|\lambda^{1/2}(\lambda^{1/2} + \Delta^{1/2})^{-1}\| \leq 1$, it follows that $\|\lambda(\lambda^{1/2} + \Delta^{1/2})^{-1}\eta - \lambda^{1/2}\Delta^{1/2}(\lambda^{1/2} + \Delta^{1/2})^{-1}\zeta\| < \varepsilon, \|\Delta(\lambda^{1/2} + \Delta^{1/2})^{-1}\eta - \lambda^{1/2}\Delta^{1/2}(\lambda^{1/2} + \Delta^{1/2})^{-1}\zeta\| < \varepsilon$; hence $\|(\lambda - \Delta)(\lambda^{1/2} + \Delta^{1/2})^{-1}\eta\| < 2\varepsilon$. Thus $\|\eta\| > 1$ and $\|(\lambda^{1/2} - \Delta^{1/2})\eta\| < 2\varepsilon$, that is, $\operatorname{dist}(\lambda^{1/2}, Sp(\Delta^{1/2})) < 2\varepsilon$. Therefore, $B(\varepsilon) \Rightarrow A(2\varepsilon)$. □

Theorem. *Let $\mathscr{M} \subset \mathscr{B}(\mathscr{H})$ be a factor and let $\lambda \geq 0$. Then $\lambda \in S(\mathscr{M})$ if and only if for every $0 \neq \xi \in \mathscr{H}$ and $\varepsilon > 0$ there exist $x \in \mathscr{M}$ and $x' \in \mathscr{M}'$ such that*

$$\|x\xi\| > 1, \|x\xi - x'\xi\| < \varepsilon, \|x^*\xi - \lambda x'^*\xi\| < \varepsilon. \tag{4}$$

Proof. Assume that $\lambda \in S(\mathscr{M})$. Let $\varepsilon > 0, 0 \neq \xi \in \mathscr{H}$ and $e = p_\xi \in \mathscr{M}, e' = p'_\xi \in \mathscr{M}'$ be the cyclic projections onto $\overline{\mathscr{M}'\xi}, \overline{\mathscr{M}\xi}$, respectively. Then $e'_e = ee'$ is a projection in $\mathscr{M}'_e \subset \mathscr{B}(e\mathscr{H})$ with central support equal to the identity of $\mathscr{M}'_e$, so that the canonical induction

$$I : \mathscr{M}_e \to \mathscr{N} = \mathscr{M}_{ee'} \subset \mathscr{B}(ee'\mathscr{H})$$

is a *-isomorphism and $\xi \in ee'\mathscr{H}$ is a cyclic and separating vector for $\mathscr{N}$. Let Δ be the modular operator associated with the factor $\mathscr{N} \subset \mathscr{B}(ee'\mathscr{H})$ and the vector $\xi \in ee'\mathscr{H}$. By 28.4.(2), we have $\lambda \in S(\mathscr{M}) = S(\mathscr{M}_e) = S(\mathscr{N})$, hence $\lambda \in Sp(\Delta)$.

If $\lambda = 0$, the previous lemma shows that there exists $y \in \mathscr{N}$ such that $\|y\xi\| > 1$ and $\|y^*\xi| < \varepsilon$. Since $y\xi \in ee'\mathscr{H} \subset \overline{\mathscr{M}'\xi}$, there exists $x' \in \mathscr{M}'$ such that $\|y\xi - x'\xi\| < \varepsilon$. Then, with $x = I^{-1}(y)$ we have $x\xi = y\xi, x^*\xi = y^*\xi$. The pair (x, x') satisfies (4).

If $\lambda \neq 0$, the previous lemma shows that there exist $y \in \mathscr{N}, y' \in \mathscr{N}'$ such that $\|y\xi\| > 1, \|y\xi - \lambda^{-1/2}y'\xi\| < \varepsilon$, and $\|y^*\xi - \lambda^{1/2}x'^*\xi\| < \varepsilon$. Then there exist $x \in \mathscr{M}, x' \in \mathscr{M}'$ such that $x\xi = y\xi, x^*\xi = y^*\xi, \lambda^{1/2}x'\xi = y'\xi, \lambda^{1/2}x'^*\xi = y'^*\xi$, and the pair (x, x') satisfies (4).

Conversely, assume that the condition in the statement is satisfied. If $\mathscr{M}$ contains a finite projection, then it also contains a cyclic finite projection. Otherwise, every projection is purely infinite. Consequently, there exists a cyclic projection $0 \neq e \in \mathscr{M}$ which is not simultaneously infinite and semifinite. Thus, it follows from 28.4.(2) and 28.3.(6) that we have to prove that $\lambda \in Sp(\Delta_\varphi)$ for every faithful normal state φ on $\mathscr{M}_e$.

Let φ be any faithful normal state on $\mathscr{M}_e$. Since e is cyclic, there exists $\xi \in \mathscr{H}$ such that $p_\xi = e$ and $\varphi = \omega_\xi | e\mathscr{M}e$. Put $e' = p'_\xi \in \mathscr{M}', \mathscr{N} = \mathscr{M}_{ee'}, \mathscr{N}' = \mathscr{M}'_{ee'}$ and denote by $I : \mathscr{M}_e \to \mathscr{N}, I' : \mathscr{M}'_{e'} \to \mathscr{N}'$ the canonical inductions. Note that $\xi \in ee'\mathscr{H}$ is a cyclic and separating vector for $\mathscr{N}$.

By assumption, there exist $x \in \mathscr{M}, x' \in \mathscr{M}'$ satisfying (4). Let $x_1 = exe \in \mathscr{M}_e, x'_1 = e'x'e' \in \mathscr{M}'_{e'}$. We have $\|x_1\xi\| \geq \|ex'\xi\| - \|ex\xi - ex'\xi\| \geq \|x'\xi\| - \|x\xi - x'\xi\| \geq \|x\xi\| - 2\|x\xi - x'\xi\| \geq 1 - 2\varepsilon, \|x_1\xi - x'_1\xi\| = \|ee'(x\xi - x'\xi)\| < \varepsilon, \|x_1^*\xi - \lambda x_1'^*\xi\| = \|ee'(x^*\xi - \lambda x'^*\xi)\| < \varepsilon$. It follows that $y = (1-2\varepsilon)^{-1}I(x_1) \in \mathscr{N}, y' = \lambda^{1/2}(1-2\varepsilon)^{-1}I'(x'_1) \in \mathscr{N}'$ and $\|y\xi\| > 1, \|\lambda^{1/2}y\xi - y'\xi\| < \lambda^{1/2}(1-2\varepsilon)^{-1}\varepsilon, \|y^*\xi - \lambda^{1/2}y'^*\xi\| < (1-2\varepsilon)^{-1}\varepsilon$. As $\varepsilon > 0$ was arbitrary, using the previous lemma we conclude that $\lambda \in Sp(\Delta_\varphi)$. □

28.8. We shall say that the W^*-algebra $\mathscr{M}$ has property L_λ $(0 \leq \lambda \leq 1/2)$ if for every faithful normal state φ on $\mathscr{M}$ and $\varepsilon > 0$ there exists a partial isometry $u \in \mathscr{M}$ such that $u^2 = 0, uu^* + u^*u = 1$, and $\|\lambda\varphi(u\cdot) - (1-\lambda)\varphi(\cdot u)\| \leq \varepsilon$. Clearly, property L_λ is an algebraic invariant of $\mathscr{M}$.

As an application of the spatial characterization of $S(\mathscr{M})$, we shall prove the following result.

Proposition. *If $\mathscr{M}$ is a factor with property L_λ, then $\lambda/(1-\lambda) \in S(\mathscr{M})$ $(0 < \lambda < 1/2)$.*

For the proof, we need the following:

Lemma. *Let $\mathscr{M} \subset \mathscr{B}(\mathscr{H})$ be a factor in standard form, let $\varepsilon > 0$ and let $\xi_1, \xi_2 \in \mathscr{H}$ with $\|(\omega_{\xi_1} - \omega_{\xi_2})|\mathscr{M}\| < \varepsilon$. Then there exists a unitary element $u' \in \mathscr{M}'$ such that $\|u'\xi_1 - \xi_2\|^2 < 4\varepsilon$.*

Proof. Assume first that $\omega_{\xi_1} = \omega_{\xi_2}$ on $\mathscr{M}$. Then there exists a partial isometry $w' \in \mathscr{M}'$ such that $w'x\xi_1 = x\xi_2$ $(x \in \mathscr{M})$ and $w'([\mathscr{M}\xi_1]^\perp) = 0$, whence $w'^*w' = p'_{\xi_1}, w'w'^* = p'_{\xi_2}$. Using the comparison theorem in $\mathscr{M}'$ ([L], 4.6), it follows, for instance, that $1 - p'_{\xi_1} < 1 - p'_{\xi_2}$, and there exists a partial isometry $v' \in \mathscr{M}'$ such that $v'^*v' = 1 - p'_{\xi_1}, v'v'^* \leq 1 - p'_{\xi_2}$. Then $w' + v' \in \mathscr{M}'$ is an isometry with $(w' + v')\xi_1 = \xi_2$. Since every isometry is the s-limit of a sequence of unitary elements (see Dixmier & Marechal, 1971, Lemma 2), it follows that there exists $u' \in U(\mathscr{M}')$ with $\|u'\xi_1 - \xi_2\|^2 < \varepsilon$.

In the general case, consider $\varphi_1 = \omega_{\xi_1}, \varphi_2 = \omega_{\xi_2}$ with $\|\varphi_1 - \varphi_2\| < \varepsilon$ and $\psi = \varphi_1 + (\varphi_1 - \varphi_2)^- = \varphi_2 + (\varphi_1 - \varphi_2)^+$. Since $\mathscr{M}$ is in standard form, there exists $\eta \in \mathscr{H}$ such that $\psi = \omega_\eta$ ([L], 10.25) and since $\varphi_1 \leq \psi$, there exists $a' \in \mathscr{M}'$, $0 \leq a' \leq 1$ such that $\varphi_1 = \omega_{a'\eta}$. Then $\|a'\eta - \eta\|^2 = ((1-a')^2\eta|\eta) \leq ((1-a')(1+a')\eta|\eta) = \|\eta\|^2 - \|a'\eta\|^2 = (\psi - \varphi_1)(1) = \|\psi - \varphi_1\| \leq \|\varphi_1 - \varphi_2\| < \varepsilon$, and there exists $\eta_1 \in \mathscr{H}$ with $\varphi_1 = \omega_{\eta_1}$ and $\|\eta_1 - \eta\|^2 < \varepsilon$. By the first part of the proof it follows that there is a $u'_1 \in U(\mathscr{M}')$ with $\|u'_1\xi_1 - \eta\|^2 < 2\varepsilon$. Similarly, there is a $u'_2 \in U(\mathscr{M}')$ with $\|u'_2\xi_2 - \eta\|^2 < 2\varepsilon$. Hence $u' = (u'_2)^{-1}u'_1 \in U(\mathscr{M}')$ and $\|u'\xi_1 - \xi_2\|^2 < 4\varepsilon$. □

Proof of the Proposition. We may assume that $\mathscr{M} \subset \mathscr{B}(\mathscr{H})$ is in standard form. Let $\mathscr{H}_1$ be a four-dimensional Hilbert space with orthonormal basis $\{\varepsilon_{ij}\}_{1\leq i,j\leq 2}$, let $\mathscr{F}$ be the type I_2 factor generated in $\mathscr{B}(\mathscr{H}_1)$ by the system of matrix units $\{e_{ij}\}_{1\leq i,j\leq 2}$, where $e_{ij}\varepsilon_{kl} = \delta_{jk}\varepsilon_{il}$, and let $\mathscr{F}'$ be the type I_2 factor generated in $\mathscr{B}(\mathscr{H}_1)$ by the system of matrix units $\{e'_{ij}\}_{1\leq i,j\leq 2}$, where $e'_{ij}\varepsilon_{kl} = \delta_{il}\varepsilon_{kj}$ and $\eta_1 = \lambda^{1/2}\varepsilon_{11} + (1-\lambda)^{1/2}\varepsilon_{22} \in \mathscr{H}_1$. Then $\mathscr{F}'$ is the commutant of $\mathscr{F}$ and η_1 is a cyclic and separating vector for $\mathscr{F}$.

Consider now $\xi \in \mathscr{H}, \|\xi\| = 1, \varphi = \omega_\xi|\mathscr{M}$, and $0 < \varepsilon < \lambda/2$. By assumption, there is a partial isometry $u \in \mathscr{M}$ such that $u^2 = 0, uu^* + u^*u = 1$ and

$$|\lambda\varphi(ux) - (1-\lambda)\varphi(xu)| \leq \varepsilon\|x\| \quad (x \in \mathscr{M}) \tag{1}$$

$$|\lambda\varphi(xu^*) - (1-\lambda)\varphi(u^*x)| \leq \varepsilon\|x\| \quad (x \in \mathscr{M}). \tag{2}$$

Let $e = u^*u \in \mathcal{M}$. Then $1 - e = uu^*$ and from (1) with $x = u^*$ we deduce that $|\lambda\varphi(1-e) - (1-\lambda)\varphi(e)| \le \varepsilon, |\lambda - \varphi(e)| \le \varepsilon$, and, as $\varepsilon < \lambda/2$, we get

$$\varphi(e) < \lambda/2. \tag{3}$$

On the other hand, since $(u^*)^2 = 0$, from (2) with $x = uyuu^*$ we deduce that $|(1-\lambda)\varphi(ey(1-e))| \le \varepsilon\|y\|$ and, as $\lambda < 1/2, 1-\lambda > 1/2$, we obtain $|\varphi(ey(1-e))| \le 2\varepsilon\|y\|, |\varphi((1-e)ye)| \le 2\varepsilon\|y\|$, so that $|\varphi(y) - \varphi(eye) - \varphi((1-e)y(1-e))| \le 4\varepsilon\|y\|$ $(y \in \mathcal{M})$. Using this inequality and also (1) with $x = eyu^*$ and with $x = u^*y(1-e)$, we conclude that

$$\|\varphi(\cdot) - \lambda\varphi(e\cdot e) - \lambda\varphi(u\cdot u^*) - (1-\lambda)\varphi((1-e)\cdot(1-e)) - (1-\lambda)\varphi(u^*\cdot u)\| \le 6\varepsilon. \tag{4}$$

Let $\mathcal{N}$ be the subalgebra of $\mathcal{M}$ generated by u and u^*. Then $\{E_{11} = e = u^*u, E_{12} = u, E_{21} = u^*, E_{22} = 1-e = uu^*\}$ is a system of matrix units in $\mathcal{N}$, so that $\mathcal{N}$ is a type I_2 factor. Every element $x \in \mathcal{M}$ can be uniquely written in the form (9.15.(3))

$$x = x_{11}e + x_{12}u + x_{21}u^* + x_{22}(1-e) \tag{5}$$

with $x_{ij} \in \mathcal{N}' \cap \mathcal{M}$. Since $\mathcal{M}$ is a factor, it follows that $\mathcal{N}' \cap \mathcal{M}$ is also a factor.

Let $\mathcal{N}_1 = \mathcal{N}$ and $\pi_1 : \mathcal{N}_1 \to \mathcal{F}$ be the *-isomorphism defined by $\pi_1(E_{ij}) = e_{ij}$ $(1 \le i,j \le 2)$. Let $\mathcal{N}_2 = \mathcal{N}' \cap \mathcal{M}$ and $\pi_2 : \mathcal{N}_2 \to \mathcal{B}(\mathcal{H}_2)$ be the faithful standard representation of $\mathcal{N}_2$ associated with $\varphi|\mathcal{N}_2 = \omega_\xi|\mathcal{N}_2$ with the cyclic and separating vector $\eta_2 \in \mathcal{H}_2$, that is, $\varphi(x) = (\pi_2(x)\eta_2|\eta_2)$ $(x \in \mathcal{N}_2)$. Using (5), we define a faithful standard representation $\pi = \pi_1 \bar{\otimes} \pi_2 : \mathcal{M} \to \mathcal{F} \bar{\otimes} \pi_2(\mathcal{N}_2) \subset \mathcal{B}(\mathcal{H}_1 \bar{\otimes} \mathcal{H}_2)$, that is, $\pi(x) = \sum_{ij} e_{ij} \bar{\otimes} \pi_2(x_{ij})$ $(x \in \mathcal{M})$. Thus, for $x \in \mathcal{M}$ we have

$$(\pi(x)(\eta_1 \bar{\otimes} \eta_2)|\eta_1 \bar{\otimes} \eta_2) = \sum_{ij}(e_{ij}\eta_1|\eta_1)\varphi(x_{ij}) = \lambda\varphi(x_{11}) + (1-\lambda)\varphi(x_{22}) \tag{6}$$

Since $x_{11} = exe + uxu^*, x_{22} = (1-e)x(1-e) + u^*xu$, from (6) and (4) it follows that

$$\|(\pi(\cdot)(\eta_1 \otimes \eta_2)|\eta_1 \otimes \eta_2) - \varphi(\cdot)\| \le 6\varepsilon. \tag{7}$$

Since any two standard forms are spatially isomorphic, from the previous lemma and (7) it follows that there exists a unitary operator $V : \mathcal{H} \to \mathcal{H}_1 \bar{\otimes} \mathcal{H}_2$ such that $\pi(x) = VxV^*$ $(x \in \mathcal{M})$ and

$$\|V^*(\eta_1 \otimes \eta_2) - \xi\|^2 \le 25\varepsilon. \tag{8}$$

Let $t = \lambda^{1/2}(1-\lambda)^{1/2}$ and $u' = t^{-1}V^*(e'_{12} \bar{\otimes} 1)V \in \mathcal{M}'$. By (3), we have $\|u\xi\|^2 = \varphi(u^*u) = \varphi(e) \ge \lambda/2$, hence

$$\|u\xi\| \ge \lambda^{1/2}/2. \tag{9}$$

From (8), we obtain $\|u\xi - u'\xi\| \le \|uV^*(\eta_1 \otimes \eta_2) - u'V^*(\eta_1 \otimes \eta_2)\| + 5\varepsilon^{1/2}(\|u\| + \|u'\|), \|u^*\xi - t^2u'^*\xi\| \le \|u^*V^*(\eta_1 \otimes \eta_2) - t^2u'^*V^*(\eta_1 \otimes \eta_2)\| + 5\varepsilon^{1/2}(\|u\| + t^2\|u'\|)$. Since $e_{12}\eta_1 = t^{-1}e'_{12}\eta_1$,

$e_{21}\eta_1 = te'_{21}\eta_1$, and $VuV^* = \pi(u) = e_{12} \bar{\otimes} 1$, $Vu'V^* = t^{-1}(e'_{12} \bar{\otimes} 1)$, it follows that

$$\|u\xi - u'\xi\| \leq 5\varepsilon^{1/2}(1 + t^{-1}), \ \|u^*\xi - t^2 u'^*\xi\| \leq 5\varepsilon^{1/2}(1 + t). \tag{10}$$

From Theorem 28.7, (9), and (10), we conclude that $\lambda/(1 - \lambda) = t^2 \in S(\mathcal{M})$. □

28.9. Recall that if $S \subset \mathbb{R}_*^+ \equiv \hat{\mathbb{R}}$ is a closed subgroup, then its annihilator in $\mathbb{R}$ is the set $T = \{t \in \mathbb{R}; \lambda^{it} = 1 \text{ for all } \lambda \in S\}$; T is a closed subgroup of $\mathbb{R}$ and $S = \{\lambda \in \mathbb{R}_*^+; \lambda^{it} = 1 \text{ for all } t \in T\}$.

The invariants $T(\mathcal{M})$ and $S(\mathcal{M})$ are connected by the following partial duality result.

Theorem. *If $\mathcal{M}$ is a factor and $S(\mathcal{M}) \neq \{0, 1\}$, then $T(\mathcal{M})$ is the annihilator of $S(\mathcal{M}) \cap \mathbb{R}_*^+$ in $\mathbb{R}$.*

Proof. If $S(\mathcal{M}) = \{1\}$, then $\mathcal{M}$ is semifinite (28.2) and hence $T(\mathcal{M}) = \mathbb{R}$ (27.2.).

Let φ be an n.s.f. weight on $\mathcal{M}$. If (28.3.(1)) $\Gamma(\sigma^\varphi) = S(\mathcal{M}) \cap \mathbb{R}_*^+ \neq \{1\}$, then the quotient group $\hat{\mathbb{R}}/\Gamma(\sigma^\varphi)$ is compact and so, by Theorem 16.5, the annihilator of $S(\mathcal{M}) \cap \mathbb{R}_*^+ = \Gamma(\sigma^\varphi)$ is $Int(\sigma^\varphi) = T(\mathcal{M})$. □

We mention that for every $t \in \mathbb{R}$ there exists a factor $\mathcal{M}$ with separable predual such that $T(\mathcal{M}) = \{nt; n \in \mathbb{Z}\}$ and $S(\mathcal{M}) = \{0, 1\}$ (see Connes, 1973a, 3.4.4).

28.10. A general duality result can be obtained by considering the Bohr compactification $\mathbb{B}$ of $\mathbb{R}_*^+$ (Rudin (1962)), that is, $\mathbb{B}$ is the dual group of the discrete additive group $\mathbb{R}$ which we shall denote by $\mathbb{R}_d$. The identity mapping $\iota : \mathbb{R}_d \to \mathbb{R}$ is a continuous group isomorphism and the dual mapping $\beta : \mathbb{R}_*^+ \to \mathbb{B}$ is an injective continuous group homomorphism with dense range.

Proposition. *For any factor $\mathcal{M}$, the invariant $T(\mathcal{M}) \subset \mathbb{R}_d$ is the annihilator of the set*

$$\bigcap \{\overline{\beta(Sp(\Delta_\varphi) \cap \mathbb{R}_*^+)}; \varphi \in W_{nsf}(\mathcal{M})\} \subset \mathbb{B}.$$

Proof. Using Propositions 28.1 and 14.9, for every $\varphi \in W_{\text{nsf}}(\mathcal{M})$ we have

$$\overline{\beta(Sp(\Delta_\varphi) \cap \mathbb{R}_*^+)} = Sp(\sigma^\varphi \circ \iota).$$

Let $\omega \in W_{\text{nsf}}(\mathcal{M})$ be fixed. By Theorem 16.5, $T(\mathcal{M}) \subset \mathbb{R}_d$ is the annihilator of $\Gamma(\sigma^\omega \circ \iota) \subseteq \mathbb{B}$. It is therefore sufficient to show that

$$\Gamma(\sigma^\omega \circ \iota) = \bigcap \{Sp(\sigma^\varphi \circ \iota); \varphi \in W_{nsf}(\mathcal{M})\}. \tag{1}$$

For every $\varphi \in W_{\text{nsf}}(\mathcal{M})$ we have $\sigma^\varphi \sim \sigma^\omega$ (3.1), hence $\sigma^\varphi \circ \iota \sim \sigma^\omega \circ \iota$ and therefore $\Gamma(\sigma^\omega \circ \iota) = \Gamma(\sigma^\varphi \circ \iota) \subset Sp(\sigma^\varphi \circ \iota)$ (16.3). We have thus proved the inclusion "$\subset$" in (1).

By the definition of the invariant Γ (see 16.1), we have

$$\Gamma(\sigma^\omega \circ \iota) = \bigcap \{Sp(\sigma^{\omega_e} \circ \iota); 0 \neq e \in Proj(\mathcal{M}^\omega)\}.$$

Let $0 \neq e \in Proj(\mathcal{M}^\omega)$. By Theorems 16.6 and 5.1, we can find an n.s.f. weight φ on $\mathcal{M}$ such that $Sp\ \sigma^\varphi \subset Sp\ \sigma^{\omega_e}$ and by Proposition 14.9 it follows that $Sp(\sigma^\varphi \circ \iota) \subset Sp(\sigma^{\omega_e} \circ \iota)$. This proves the inclusion "$\supset$" in (1). □

28.11. Let $\mathcal{M}$ be a type III factor. Since $S(\mathcal{M}) \cap \mathbb{R}_*^+$ is a closed subgroup of $\mathbb{R}_*^+$ and $S(\mathcal{M})$ is a closed subset of $[0, +\infty)$, we have either $S(\mathcal{M}) = \{0, 1\}$, or $S(\mathcal{M}) = [0, +\infty)$, or $S(\mathcal{M}) = \{0\} \cup \{\lambda^n; n \in \mathbb{Z}\}$ for some $0 < \lambda < 1$.

If $S(\mathcal{M}) = \{0, 1\}$, we say that $\mathcal{M}$ is of *type* III_0.

If $S(\mathcal{M}) = [0, +\infty)$, we say that $\mathcal{M}$ is of *type* III_1. In this case, we have $T(\mathcal{M}) = \{0\}$, by Theorem 28.9.

If $S(\mathcal{M}) = \{0\} \cup \{\lambda^n; n \in \mathbb{Z}\}$, we say that $\mathcal{M}$ is of *type* III_λ, $(0 < \lambda < 1)$. In this case, by Theorem 28.9, we have $T(\mathcal{M}) = \{2\pi k/\ln(\lambda); k \in \mathbb{Z}\}$; note that $-2\pi/\ln(\lambda)$ is the smallest positive member of $T(\mathcal{M})$.

Recall (28.2) that $\mathcal{M}$ is semifinite if and only if $0 \notin S(\mathcal{M})$ and that in this case $S(\mathcal{M}) = \{1\}$.

In the next sections, we shall study in more detail the type III_λ factors for $0 \leq \lambda < 1$. For these factors there exists, besides the continuous decomposition (§23), a discrete decomposition.

28.12. Notes. The invariant $S(\mathcal{M})$ was introduced by Connes (1973a). Previously, similar (but different) invariants had been considered by Araki and Woods (1968), Golodec (1972), and Krieger (1970b). Actually, the starting point for Connes' work (Connes, 1973a) was the confrontation between the classification by Araki and Woods of ITPFI-factors and the Tomita–Takesaki modular theory of W^*-algebras. The group property of the invariant $S(\mathcal{M})$ and 28.3.(4) was first proved by Connes and van Daele (1973) by a different method. The other results presented in this section are due to Connes (1973a).

For an arbitrary factor $\mathcal{M}$, Araki and Woods (1968) introduced the invariant $r_\infty(\mathcal{M})$ as the set of all $\lambda \geq 0$ such that $\mathcal{M} \bar{\otimes} \mathcal{R}_\lambda \approx \mathcal{M}$ and Connes (1973a, 3.6.1) proved that $\lambda \in r_\infty(\mathcal{M}) \Rightarrow \mathcal{M}$ has property $L_{1/(1+\lambda)} \Rightarrow \lambda \in S(\mathcal{M})$. If $\mathcal{M}$ is an Araki–Woods factor, then $S(\mathcal{M}) = r_\infty(\mathcal{M})$ coincides with the "asymptotic ratio set" (Araki & Woods, 1968) which is expressed only in terms of the eigenvalue list of $\mathcal{M}$ (Araki & Woods, 1968; Connes, 1973a, 3.6.1). Property $L_\lambda (0 < \lambda < 1/2)$ was introduced by Powers (1967, 1970). Then Araki (1971) considered a stronger property, called L'_λ, and showed that a W^*-algebra $\mathcal{M}$ has this property if and only if $\lambda/(1-\lambda) \in r_\infty(\mathcal{M})$. Connes (1973a, 3.7.9) proved that the Pukánszky factor $P_{\lambda/(1-\lambda)}$ has property L_λ without having property L'_λ. For some classes of factors, including those with $\{0, 1\} \neq S(\mathcal{M}) \neq [0, +\infty)$, Connes (1973a, 3.7.2. 3.7.7) proved that property L_λ is equivalent with $\lambda/(1-\lambda) \in S(\mathcal{M})$ $(0 < \lambda < 1/2)$. Using Corollary 28.6 and the results of Krieger (1971a, 1971b), Connes (1973a, 3.3.5) proved that for every subgroup $\mathbb{G} \neq \{1\}$ of $\mathbb{R}_*^+$ there exists a factor $\mathcal{M}$ with separable predual such that $S(\mathcal{M}) = \{0, 1\}$ and $S(\mathcal{M} \otimes \mathcal{M}) = \{0\} \cup \mathbb{G}$. In particular, using the duality between $S(\mathcal{M} \bar{\otimes} \mathcal{M})$ and $T(\mathcal{M} \bar{\otimes} \mathcal{M}) = T(\mathcal{M})$ (28.9), it follows that for every $t \in \mathbb{R}$ there exists a factor $\mathcal{M}$ with separable predual such that $S(\mathcal{M}) = \{0, 1\}$ and $T(\mathcal{M}) = \{nt; n \in \mathbb{Z}\}$ (Connes, 1973a, 3.4.4). This shows on the one hand the wealth of the class of factors $\mathcal{M}$ with $S(\mathcal{M}) = \{0, 1\}$ (see also Araki & Woods, 1968) and on the other that the assumptions of Theorems 16.5 and 28.9 are essential.

For our exposition we have used Connes (1973a).

29 Factors of Type III_λ $(0 \leq \lambda < 1)$

In this section, we give canonical constructions for factors of type III_λ $(0 \leq \lambda < 1)$ and study some important classes of weights on these factors. All W^*-algebras that appear will be countably decomposable, either by assumption or by construction.

29.1. We first give a construction which leads to factors of type III_λ with $0 < \lambda < 1$.

Let $0 < \lambda < 1$ be fixed and let $t = -2\pi/\ln(\lambda)$. An n.s.f. weight φ on a factor of type III_λ such that $\varphi(1) = +\infty$ and $\sigma_t^\varphi = \iota$ will be called a *λ-trace*. Note that every λ-trace is strictly semifinite.

Proposition. *Let $\mathcal{N}$ be a factor of type II_∞, τ an n.s.f. trace on $\mathcal{N}$ and let $\theta \in Aut(\mathcal{N})$ be such that $\tau \circ \theta = \lambda\tau$. Then the action $\theta : \mathbb{Z} \ni n \mapsto \theta^n \in Aut(\mathcal{N})$ is properly outer, the crossed product $\mathcal{M} = \mathcal{R}(\mathcal{N}, \theta)$ is a factor of type III_λ, and the dual weight $\varphi = \hat{\tau}$ is a λ-trace on $\mathcal{M}$. Moreover, identifying $\mathcal{N} \equiv \pi_\theta(\mathcal{N}) \subseteq \mathcal{M}$ and putting $u = 1 \bar{\otimes} \boldsymbol{\lambda}(1) \in \mathcal{M}$, we have*

$$\mathcal{N} = \mathcal{M}^\varphi, \varphi|\mathcal{N} = \tau \tag{1}$$

$$u \in \mathcal{M}(\sigma^\varphi, \{\lambda\}), \quad \varphi \circ \mathrm{Ad}(u) = \lambda\varphi. \tag{2}$$

Proof. Since $\tau \circ \theta^n = \lambda^n \tau \neq \tau$, the $*$-automorphism θ^n $(n \neq 0)$ is properly outer, hence the action $\theta : \mathbb{Z} \to Aut(\mathcal{N})$ is properly outer. By Corollary 1/22.6 and Theorem 22.3, $\mathcal{M}$ is a factor and $\mathcal{N}' \cap \mathcal{M} = \mathcal{Z}(\mathcal{N})$.

Let $P : \mathcal{M} \to \mathcal{N}$ be the faithful normal conditional expectation associated with the crossed product (22.2) and let $\mathcal{G} = \{1 \bar{\otimes} \boldsymbol{\lambda}(n); n \in \mathbb{Z}\} = \{u^n; n \in \mathbb{Z}\} \subset \mathcal{N}(P)$. Since $\tau(u^n \cdot u^{-n}) = \tau \circ \theta^n = \lambda^n \tau$, we infer from Theorem 28.5 that a positive number $\mu > 0$ belongs to the Connes invariant $S(\mathcal{M})$ if and only if for each $\varepsilon > 0$ there exists $n \in \mathbb{Z}$ such that $Sp(\lambda^{-n} \cdot 1) \subset (\mu - \varepsilon, \mu + \varepsilon)$, hence $S(\mathcal{M}) = \{0\} \cup \{\lambda^n; n \in \mathbb{Z}\}$, that is, $\mathcal{M}$ is a factor of type III_λ.

The dual weight $\varphi = \hat{\tau} = \tau \circ P$ is an n.s.f. weight on $\mathcal{M}$ and for every $x \in \mathcal{M}^+$ we have $(\varphi \circ \mathrm{Ad}(u))(x) = \varphi(uxu^*) = \tau(P(uxu^*)) = \tau(uP(x)u^*) = (\tau \circ \theta)(P(x)) = \lambda\tau(P(x)) = \lambda\varphi(x)$, whence $\varphi \circ \mathrm{Ad}(u) = \lambda\varphi$. Then, for $s \in \mathbb{R}$ we have $\lambda^{\mathrm{i}s} = [D(\lambda\varphi) : D\varphi]_s = [D(\varphi_u) : D\varphi]_s = u^* \sigma_s^\varphi(u)$, that is, $\sigma_s^\varphi(u) = \lambda^{\mathrm{i}s} u = \langle s, \lambda \rangle u$, so that $u \in \mathcal{M}(\sigma^\varphi, \{\lambda\})$.

On the other hand, we have $\sigma_t^\varphi|\mathcal{N} = \iota$ and $\sigma_t^\varphi(u) = \lambda^{\mathrm{i}t} u = u$, hence $\sigma_t^\varphi = \iota$, since $\mathcal{M} = \mathcal{R}\{\mathcal{N}, u\}$. Also, $\varphi|\mathcal{N} = (\tau \circ P)|\mathcal{N} = \tau$, in particular $\varphi(1) = \tau(1) = +\infty$, so that φ is a λ-trace.

Finally, since $\varphi = \tau \circ P$, from Theorem 11.9 we infer that $\mathcal{N} \subset \mathcal{M}^\varphi$. The mapping $E : \mathcal{M} \to \mathcal{M}^\varphi$ defined by $E(x) = t^{-1} \int_0^t \sigma_s^\varphi(x) \mathrm{d}s$ $(x \in \mathcal{M})$ is a faithful normal conditional expectation of $\mathcal{M}$ onto $\mathcal{M}^\varphi$, for $x \in \mathcal{N} \subset \mathcal{M}^\varphi$ we have $E(x) = x$ and $E(u^n) = (t^{-1} \int_0^t \lambda^{\mathrm{i}ns} \mathrm{d}t) u^n = 0$ for $0 \neq n \in \mathbb{Z}$. Since every element of $\mathcal{M}$ is of the form $\sum_n a_n u^n$ with $a_n \in \mathcal{N}$ (see 22.1), it follows that $E = P$ and hence $\mathcal{M}^\varphi = \mathcal{N}$. □

Recall (19.3) that on $\mathcal{M}$ we also have the dual action $\hat{\theta} : \mathbb{T} = \hat{\mathbb{Z}} \to Aut(\mathcal{M})$. Consider the surjective continuous group homomorphism $\alpha : \mathbb{R} \ni s \mapsto e^{2\pi\mathrm{i}s/t} = \overline{\lambda^{-\mathrm{i}s}} \in \mathbb{T}$. Since $\mathcal{M} = \mathcal{R}\{\mathcal{N}, u\}, \hat{\theta}_{\alpha(s)}|\mathcal{N} = \iota = \sigma_s^\varphi|\mathcal{N}, \sigma_s^\varphi(u) = \lambda^{\mathrm{i}s} u$, and (19.3.(4)) $\hat{\theta}_{\alpha(s)}(u) = \overline{\langle 1, \alpha(s) \rangle} u = \lambda^{\mathrm{i}s} u$, it follows that

$$\hat{\theta}_{\alpha(s)} = \sigma_s^\varphi \quad (s \in \mathbb{R}). \tag{3}$$

Thus, the identity $E = P$ established in the above proof is nothing but the equality of $P_{\hat{\theta}}$ and P. that is, just the definition of P (22.2). Also, some other parts of the previous proof are consequences of (3) and of general properties of the dual action.

A triple $(\mathcal{N}, \theta, \tau)$ consisting of a type II_∞ factor $\mathcal{N}$, a $*$-automorphism $\theta \in Aut(\mathcal{N})$, and an n.s.f. trace τ on $\mathcal{N}$ such that $\tau \circ \theta = \lambda\tau$ will be called a *discrete decomposition of type* III_λ $(0 < \lambda < 1)$.

29.2. Before giving a similar construction for factors of type III_0 we shall study some properties of diffuse abelian W^*-algebras.

An abelian W^*-algebra is called *diffuse* if it has no minimal projections.

Lemma. *Let $\mathcal{Z}$ be a diffuse abelian W^*-algebra and $\theta \in Aut(\mathcal{Z})$ an ergodic *-automorphism. For each projection $0 \neq e \in \mathcal{Z}$, there exists a sequence of projections $\{e_n\}_{n\ge1} \subset \mathcal{Z}$, uniquely determined, such that*

$$e = \sum_{n\ge1} e_n, \theta^n(e_n) \le e \text{ and } e\theta^k(e_n) = 0 \text{ for } k = 1, \dots, n-1. \tag{1}$$

Moreover, we have

$$e = \sum_{n\ge1} \theta^n(e_n). \tag{2}$$

Proof. Since θ is ergodic and $\mathcal{Z}$ is diffuse, it follows that θ is conservative, that is,

$$p \in Proj(\mathcal{Z}), p\theta^n(p) = 0 \text{ for all } n \ge 1 \Rightarrow p = 0. \tag{3}$$

Indeed, if $p\theta^n(p) = 0$ for all $n \ge 1$, then $\theta^i(q)\theta^j(q) = 0$ for all $i \neq j$ in $\mathbb{Z}$ and for every projection $0 \neq q \le p$ in $\mathcal{Z}$, hence $\sum_{k\in\mathbb{Z}} \theta^k(q) = 1$ by the ergodicity of θ. It follows that $0 \neq q \le p \Rightarrow q = p$, that is, p is a minimal projection, contradicting the fact that $\mathcal{Z}$ is diffuse.

Let e_1 be the largest projection of $\mathcal{Z}$ such that $e_1 \le e$ and $\theta(e_1) \le e$. Then, if $e_1, \dots, e_{n-1}$ have been already chosen, we define e_n inductively to be the largest projections of $\mathcal{Z}$ such that $e_n \le e - (e_1 + \dots + e_{n-1}), \theta^n(e_n) \le e$. □

By this definition, we have $\theta^n(e_n) \le e$.

Let $1 \le k < n, p = \theta^k(e_n)e$, and $q = \theta^{-k}(p) = e_n\theta^{-k}(e)$. We have $qe_k = 0, q \le e - (e_1 + \dots + e_{k-1})$ and $\theta^n(q) = p \le e$, hence $q = 0$ by the definition of e_k. Thus, $\theta^k(e_n)e = \theta^k(q) = 0$.

Let $p \le e$ be such that $pe_n = 0$ for every $n \ge 1$ and $q_n = p\theta^{-n}(e)$. We have $q_n \le p \le e - (e_1 + \dots + e_{n-1})$ and $\theta^k(q_n) = \theta^n(p)e \le e$, whence $q_n \le e_n$. Since $q_n \le p$, we have $q_n \le pe_n = 0$, so that $\theta^n(p)e = 0$. In particular, $p\theta^n(p) = 0$ for every $n \ge 1$, hence $p = 0$, by (3). Thus, $e = \sum_{n\ge1} e_n$.

To prove the uniqueness assertion, let $e = \sum_{n\ge1} f_n$ with $\theta^n(f_n) \le e$ and $e\theta^k(f_n) = 0$ for $1 \le k < n$. Let $n \ge 1$ be fixed. For $k < n$ we have $\theta^k(e_kf_n) \le e$, hence $e_kf_n = 0$. Thus, $f_n \le e - (e_1 + \dots + e_{n-1})$ and $\theta^n(f_n) \le e$, so $f_n \le e_n$. Since $\sum_{n\ge1} f_n = e = \sum_{n\ge1} e_n$, it follows that $f_n = e_n$.

It remains to prove (2). If $n < m$, then $\theta^n(e_n)\theta^m(e_m) = \theta^n(e_n\theta^{m-n}(e_m)) \le \theta^n(e\theta^{m-n}(e_m)) = 0$, hence $\sum_{n\ge1} \theta^n(e_n)$ is a projection $\le e$. Consider now a projection $p \le e$ with $p\theta^k(e_k) = 0$ for every $k \ge 1$. Since $p \le e$ and $\theta^n(e_k)e = 0$ for $k > n$ it follows that $p\theta^n(e_k) = 0$ for every $k \ge n$, hence

$$p\theta^n(e - (e_1 + \dots + e_{n-1}) = 0. \tag{4}$$

We show by induction that

$$p\theta^n(e) = 0 \text{ for every } n \ge 1. \tag{5}$$

For $n = 1$ this is just (4). We assume that (5) has already been proved for $1, \dots, n-1$ and prove it for n. Let $1 \le j < n$. By the induction hypothesis, we have $p\theta^{n-j}(e) = 0$, hence $\theta^{-n}(p)\theta^{-j}(e) = 0$. As

$\theta^j(e_j) \leq e$, we have $e_j \leq \theta^{-j}(e)$. Thus, $\theta^{-n}(p)e_j = 0$ and hence $p\theta^n(e_j) = 0$ for every $j = 1, \ldots, n-1$. Using (4) we now obtain (5), and from (5) it follows, in particular, that $p\theta^n(p) = 0$ for every $n \geq 1$, that is, $p = 0$ by (3). □

In particular, for every $n \geq 1$ and every nonzero projection $e \in \mathscr{Z}$, there exists a nonzero projection $p \in \mathscr{Z}, p \leq e$, such that $p\theta^n(p) = 0$. Thus, if θ is an ergodic *-automorphism of a diffuse abelian W^*-algebra $\mathscr{Z}$, then the action $k \mapsto \theta^k$ of $\mathbb{Z}$ on $\mathscr{Z}$ is free.

29.3. We now give the construction that leads to factors of type III_0.

An n.s.f. weight φ on a W^*-algebra such that 1 is an isolated point in $Sp(\Delta_\varphi)$, that is, (28.1.(4)) in $Sp\ \sigma^\varphi$ will be called a *lacunary weight*. It is easy to see that every lacunary weight is strictly semifinite.

Proposition. *Let $\mathscr{N}$ be a type II_∞ W^*-algebra with diffuse center, τ an n.s.f. trace on $\mathscr{N}, \theta \in Aut(\mathscr{N})$ a *-automorphism acting ergodically on $\mathscr{Z}(\mathscr{N})$ and $0 < \lambda < 1$ such that $\tau \circ \theta \leq \lambda\tau$. Then the action $\theta : \mathbb{Z} \ni n \mapsto \theta^n \in Aut(\mathscr{N})$ is free on the center of $\mathscr{N}$, the crossed product $\mathscr{M} = \mathscr{R}(\mathscr{N}, \theta)$ is a factor of type III_0 and the dual weight $\varphi = \hat{\tau}$ is a lacunary weight of infinite multiplicity on $\mathscr{M}$. Moreover, identifying $\mathscr{N}$ and $\pi_\theta(\mathscr{N}) \subset \mathscr{M}$, putting $u = 1 \bar{\otimes} \boldsymbol{\lambda}(1) \in \mathscr{M}$ and considering the faithful normal conditional expectation $P : \mathscr{M} \to \mathscr{N}$ associated with the crossed product, we have*

$$\mathscr{N} = \mathscr{M}^\varphi, \varphi|\mathscr{N} = \tau \tag{1}$$

$$u \in \mathscr{M}(\sigma^\varphi; (0,1)) \tag{2}$$

$$v \in \mathscr{N}(P) \cap \mathscr{M}(\sigma^\varphi; (0,1)), \mathscr{M} = \mathscr{R}\{\mathscr{N}, v\} \Rightarrow u^*v \in \mathscr{M}^\varphi \tag{3}$$

$$x - P(x) \in \mathscr{M}(\sigma^\varphi; \mathbb{R}_*^+ \setminus (\lambda, \lambda^{-1})) \quad (x \in \mathscr{M}). \tag{4}$$

Proof. Since $\mathscr{Z}(\mathscr{N})$ is diffuse and θ is ergodic on $\mathscr{Z}(\mathscr{N})$, the action $\theta : \mathbb{Z} \to Aut(\mathscr{N})$ is ergodic and free on $\mathscr{Z}(\mathscr{N})$ (29.2) and hence (17.5) properly outer on $\mathscr{N}$. Using 22.6 and 22.3, we conclude that $\mathscr{M}$ is a factor and $\mathscr{Z}(\mathscr{N})' \cap \mathscr{M} = \mathscr{N}, \mathscr{N}' \cap \mathscr{M} = \mathscr{Z}(\mathscr{N})$.

We now compute the invariant $S(\mathscr{M})$ using Theorem 28.5. Let A_n be the unique nonsingular positive self-adjoint operator affiliated to $\mathscr{Z}(\mathscr{N})$ such that $\tau \circ \theta^n = \tau_{A_n}$ $(n \in \mathbb{Z})$. Since $\tau \circ \theta \leq \lambda\tau$, it follows that

$$n > 0 \Rightarrow A_n \leq \lambda^n < 1, \tag{5}$$

$$n < 0 \Rightarrow A_n \geq \lambda^n > 1. \tag{6}$$

Let $\mu \in (0,1)$. There exists $m \geq 1$ such that $\lambda^m < \mu < 1$. Since $\mathscr{Z}(\mathscr{N})$ is diffuse and θ is ergodic on $\mathscr{Z}(\mathscr{N})$, it follows from Lemma 29.2 that there exists a nonzero projection $e \in \mathscr{Z}(\mathscr{N})$ such that $e\theta^n(e) = 0$ for all $n = 1, \ldots, m$. Let $\varepsilon > 0$ be such that $\lambda^m < \mu - \varepsilon < \mu + \varepsilon < 1$. Assume that $\mu \in S(\mathscr{M})$. Then (28.5) there exist $n \in \mathbb{Z}$ and a nonzero projection $p \in \mathscr{Z}(\mathscr{N})$ such that $p \leq e, \theta^n(p) \leq e$ and $Sp(A_np|\mathscr{N}p) \subset (\mu - \varepsilon, \mu + \varepsilon)$. If $n > 0$, then $A_n \leq \lambda^n$, that is, $Sp(A_np|\mathscr{N}p) \subset [0, \lambda^n]$, so that $n \leq m$; it follows that $\theta^n(p) \leq e\theta^n(e) = 0$, that is, $p = 0$, a contradiction. If $n \leq 0$, then $A_n \geq 1$, hence $Sp(A_np|\mathscr{N}p) \subset [1, +\infty)$, which is again a contradiction, since $\mu + \varepsilon < 1$.

As $S(\mathscr{M}) \cap \mathbb{R}_*^+$ is a closed subgroup of $\mathbb{R}_*^+$, it follows that $S(\mathscr{M}) \subset \{0, 1\}$. Consequently, in order to show that $S(\mathscr{M}) = \{0, 1\}$, it is enough to show that $\mathscr{M}$ is not semifinite (see 28.2). By Theorem 2/22.7, this amounts to showing that there exist no θ-invariant n.s.f. traces on $\mathscr{N}$.

Otherwise, let μ be a θ-invariant n.s.f. trace on $\mathcal{N}$ and let A be the unique positive nonsingular self-adjoint operator affiliated to $\mathcal{Z}(\mathcal{N})$ such that $\tau = \mu_A$. Then $\mu_{\theta^{-1}(A)} = \tau \circ \theta \leq \lambda\tau = \mu_{\lambda A}$, hence $\theta^{-1}(A) \leq \lambda A$ and $\theta^{-k}(A) \leq \lambda^k A, \theta^k(A) \geq \lambda^{-k} A$, for all $k > 0$. There exists $n \in \mathbb{Z}$ such that $e = \chi_{(\lambda^{n+1},\lambda^n]}(A) \neq 0$ and for every $k > 0$ we have

$$\theta^k(e) = \chi_{(\lambda^{n+1},\lambda^n]}(\theta^k(A)) \leq \chi_{(0,\lambda^n]}(\theta^k(A)) \leq \chi_{(0,\lambda^n]}(\lambda^{-k}A) = \chi_{(0,\lambda^{n+k}]}(A),$$

hence $e\theta^k(e) = 0$. As $\mathcal{Z}(\mathcal{N})$ is diffuse and θ is ergodic on $\mathcal{Z}(\mathcal{N})$, it follows that $e = 0$ (29.2.(3)), a contradiction.

Hence $\mathcal{M}$ is a factor of type III$_0$.

By the general theory of crossed products, we know that $u \in \mathcal{N}(P)$ and $\theta^n = \operatorname{Ad}(u^n)|\mathcal{N}$ $(n \in \mathbb{Z})$. Since $\varphi = \hat{\tau} = \tau \circ P$, it follows (see 27.4.(2)) that $\sigma_t^\varphi(u^n) = u^n A_n^{it}$ $(n \in \mathbb{Z}, t \in \mathbb{R})$. Thus, for $f \in \mathcal{L}^1(\mathbb{R})$, we obtain $\sigma_f^\varphi(u^n) = \int f(t)\sigma_t^\varphi(u^n)\,dt = u^n\left(\int f(t)A_n^{it}\,dt\right) = u^n\hat{f}(A_n)$ and hence $Sp_{\sigma^\varphi}(u^n) = Sp(A_n) \cap \mathbb{R}_*^+$. Using (5) and (6), we infer that

$$n > 0 \Rightarrow Sp_{\sigma^\varphi}(u^n) \subset (0, \lambda^n] \subset (0, \lambda] \subset (0, 1)$$
$$n < 0 \Rightarrow Sp_{\sigma^\varphi}(u^n) \subset [\lambda^n, +\infty) \subset [\lambda^{-1}, +\infty) \subset [1, +\infty).$$

In particular, this proves (2).

Since $\varphi = \tau \circ P$, we have $\varphi|\mathcal{N} = \tau$ and $\mathcal{N} \subset \mathcal{M}^\varphi = \mathcal{M}(\sigma^\varphi; \{1\})$ (11.9). Let $a \in \mathcal{N}$ and $0 \neq n \in \mathbb{Z}$. Using the previous remarks and 15.3.(1), it follows that $au^n \in \mathcal{M}(\sigma^\varphi; \mathbb{R}_*^+\setminus(\lambda, \lambda^{-1}))$. Since $P(au^n) = aP(u^n) = 0$ (22.2.(4)), we also have $au^n - P(au^n) \in \mathcal{M}(\sigma^\varphi; \mathbb{R}_*^+\setminus(\lambda, \lambda^{-1}))$. Since the set $\{au^n; a \in \mathcal{N}, n \in \mathbb{Z}\}$ is w-total in $\mathcal{M}$ (22.1), it follows (14.5.1) that $(Sp\ \sigma^\varphi) \cap (\lambda, \lambda^{-1}) = \{1\}$ and $x - P(x) \in \mathcal{M}(\sigma^\varphi; \mathbb{R}_*^+\setminus(\lambda, \lambda^{-1}))$ for all $x \in \mathcal{M}$. If $x \in \mathcal{M}^\varphi$, then $x - P(x) \in \mathcal{M}^\varphi \cap \mathcal{M}(\sigma^\varphi; \mathbb{R}_*^+\setminus(\lambda, \lambda^{-1})) = \mathcal{M}(\sigma^\varphi; \emptyset) = \{0\}$, hence $x = P(x) \in \mathcal{N}$. Thus, $\mathcal{M}^\varphi = \mathcal{N}$ and $\varphi|\mathcal{M}^\varphi = \tau$ is semifinite.

Therefore, φ is a lacunary weight of infinite multiplicity on $\mathcal{M}$. Also, we have proved assertions (1) and (4).

To prove (3) we consider $u \in \mathcal{N}(P)$ and $\mathcal{G} = \{v^n; n \in \mathbb{Z}\} \subset \mathcal{N}(P)$. By assumption $\mathcal{M} = \mathcal{R}\{\mathcal{N}, \mathcal{G}\}$, and so (22.4.(1)) there exist a family of mutually orthogonal projections $\{q_k\}_{k\in\mathbb{Z}} \subset \mathcal{Z}(\mathcal{N})$ with $\sum_k q_k = 1$ and a family $\{w_k\}_{k\in\mathbb{Z}} \subset \mathcal{N}$ with $w_k^* w_k = w_k w_k^* = q_k$ $(k \in \mathbb{Z})$ such that $u = \sum_k v^k w_k$ and $uq_k = v^k w_k$ $(k \in \mathbb{Z})$. Since $v \in \mathcal{M}(\sigma^\varphi; (0, 1))$, it follows that $v^k \in \mathcal{M}(\sigma^\varphi; [1, +\infty))$ for all $k \leq 0$ and, as $u \in \mathcal{M}(\sigma^\varphi; (0, 1)), q_k \in \mathcal{M}^\varphi, w_k \in \mathcal{M}^\varphi$, we obtain $\mathcal{M}(\sigma^\varphi; (0, 1)) \ni uq_k = v^k w_k \in \mathcal{M}(\sigma^\varphi; [1, +\infty))$, so that $uq_k = v^k w_k = 0$ and $w_k = q_k = 0$ for all $k \leq 0$. Thus, there exist a family of mutually orthogonal projections $\{q_n\}_{n\geq 1} \subset \mathcal{Z}(\mathcal{N})$ with $\sum_{n\geq 1} q_n = 1$ and a family $\{w_n\}_{n\geq 1} \subset \mathcal{N}$ with $w_n^* w_n = w_n w_n^* = q_n$ such that

$$u = \sum_{n\geq 1} v^n w_n \text{ and } uq_n = v^n w_n \text{ for all } n \geq 1. \tag{7}$$

Similarly, there exist a family of mutually orthogonal projections $\{q_n'\}_{n\geq 1} \subset \mathcal{Z}(\mathcal{N})$ with $\sum_{n\geq 1} q_n' = 1$ and a family $\{w_n'\}_{n\geq 1} \subset \mathcal{N}$ with $w_n'^* w_n' = w_n' w_n'^* = q_n'$ such that

$$v = \sum_{n\geq 1} u^n w_n' \text{ and } vq_n' = u^n w_n' \text{ for all } n \geq 1. \tag{8}$$

Let $m > 1$ be fixed. We show that

$$P(u^* v^n q'_m) = 0 \text{ for all } n \geq 1. \tag{9}$$

Indeed, $vq'_m = u^m w'_m$, $u^* vq'_m = u^{m-1} w'_m$ and hence $P(u^* vq'_m) = P(u^{m-1} w'_m) = P(u^{m-1}) w'_m = 0$, which proves equality (9) for $n = 1$. If $n > 1$, then, by the same argument as above, we can find a family $\{w''_j\}_{j\geq 1} \subset \mathcal{N}$ such that $v^{n-1} = \sum_{j\geq 1} u_j w''_j$ and we have $u^* vq'_m = (u^* v^{n-1}) vq'_m = \sum_{j\geq 1} u^* u^j w''_j u^m w'_m = \sum_{j\geq 1} u^{j-1} u^m (u^{-m} w''_j u^m) w'_m = \sum_{i\geq 1} u^i a_i$ with $a_i \in \mathcal{N}$, hence

$$P(u^* v^n q'_m) = \sum_{i\geq m} P(u^i) a_i = 0.$$

From (7) and (9), it follows that $q'_m = P(u^* u q'_m) = P(u^* (\sum_{n\geq 1} v^n w_n) q'_m) = \sum_{n\geq 1} P(u^* v^n q'_m) w_n = 0$. Since $m > 1$ was arbitrary, we conclude from (8) that $v = uw'_1$, hence $u^* v = w'_1 \in \mathcal{N} = \mathcal{M}^{\varphi}$. □

A triple $(\mathcal{N}, \theta, \tau)$ consisting of a type II_∞ W^*-algebra with a diffuse center, a $*$-automorphism $\theta \in Aut(\mathcal{N})$ which acts ergodically on $\mathcal{Z}(\mathcal{N})$, and an n.s.f. trace τ on $\mathcal{N}$ such that $\tau \circ \theta \leq \lambda \tau$ for some $0 < \lambda < 1$ will be called *a discrete decomposition of type* III_0.

29.4. Let $(\mathcal{N}, \theta, \tau)$ be a discrete decomposition of type III_0 and $e \in \mathcal{Z}(\mathcal{N})$ a nonzero projection. By Lemma 29.2, there exists a sequence of projections $\{e_n\}_{n\geq 1} \subset \mathcal{Z}(\mathcal{N})$, uniquely determined, such that $e = \sum_n e_n$, $\theta^n(e_n) \leq e$ and $e\theta^k(e_n) = 0$ for $k = 1, \ldots, n-1$; moreover, we have $e = \sum_n \theta^n(e_n)$. Then a $*$-automorphism $\theta_e \in Aut(\mathcal{N} e)$ is defined by

$$\theta_e(x) = \sum_n \theta^n(xe_n) \quad (x \in \mathcal{N} e).$$

Also, $\tau_e = \tau | \mathcal{N} e$ is an n.s.f. trace on $\mathcal{N} e$.

The triple $(\mathcal{N}, \theta, \tau)$ determines a crossed product W^*-algebra $\mathcal{M} = \mathcal{R}(\mathcal{N}, \theta)$, a canonical embedding $\pi = \pi_\theta : \mathcal{N} \to \mathcal{M}$, a conditional expectation $P : \mathcal{M} \to \pi(\mathcal{N})$, a unitary element $u = 1 \bar{\otimes} \boldsymbol{\lambda}(1) \in \mathcal{M}$, and a dual weight $\varphi = \hat{\tau}$ on $\mathcal{M}$. Similarly, the triple $(\mathcal{N} e, \theta_e, \tau_e)$ determines the objects $\mathcal{M}_0 = \mathcal{R}(\mathcal{N} e, \theta_e), \pi_0 : \mathcal{N} e \to \mathcal{M}_0, P_0 : \mathcal{M}_0 \to \pi_0(\mathcal{N} e), u_0 \in \mathcal{M}_0$, and $\varphi_0 = \hat{\tau}_e$.

Proposition. *Let* $(\mathcal{N}, \theta, \tau)$ *be a discrete decomposition of type* III_0 *and* $e \in \mathcal{Z}(\mathcal{N})$ *a nonzero projection. Then* $(\mathcal{N} e, \theta_e, \tau_e)$ *is also a discrete decomposition of type* III_0 *and there exists a* $*$*-isomorphism*

$$\Phi : (\mathcal{M}_0, \pi_0, P_0, \varphi_e) \to (\pi(e)\mathcal{M}\pi(e), \pi | \mathcal{N} e, P | \pi(e)\mathcal{M}\pi(e), \varphi_{\pi(e)}) \tag{1}$$

such that

$$\Phi(u_0) = \sum_{n\geq 1} u^n \pi(e_n). \tag{2}$$

Proof. Indeed, $\mathcal{N} e$ is a type II_∞ W^*-algebra with diffuse center, $\theta_e \in Aut(\mathcal{N} e)$, τ_e is an n.s.f. trace on $\mathcal{N} e$ and, if $\tau \circ \theta \leq \lambda \tau (0 < \lambda < 1)$, then $\tau_e \circ \theta_e \leq \lambda \tau_e$ since for $x \in (\mathcal{N} e)^+$ we have $\tau_e(\theta_e(x)) = \tau(\sum_{n\geq 1} \theta^n(xe_n)) \leq \sum_{n\geq 1} \lambda^n \tau(xe_n) \leq \lambda \tau_e(x)$. Also, we show that θ_e acts ergodically on $\mathcal{Z}(\mathcal{N} e) = \mathcal{Z}(\mathcal{N})e$. Let $p \in \mathcal{Z}(\mathcal{N})e$ be a projection such that $\theta_e(p) = p$, that is, $\sum_n \theta^n(pe_n) = \sum_n pe_n$. For

$j < m, k < n$ and $(j,m) \neq (k,n)$, we have

$$\theta^j(e_m)\theta^k(e_n) = 0,$$

since, if $j < k$ then $e\theta^{k-j}(e_n) = 0$, hence $\theta^j(e_m)\theta^k(e_n) = \theta^j(e_m\theta^{k-j}(e_n)) \le \theta^j(e\theta^{k-j}(e_n)) = 0$ and if $j = k$ then $m \neq n$, hence $\theta^j(e_m)\theta^k(e_n) = \theta^k(e_m e_n) = 0$. Consequently, we obtain a projection

$$q = \sum_{n=1}^{\infty}\sum_{k=0}^{n-1} \theta^k(pe_n) \in \mathscr{Z}(\mathscr{N}).$$

Since $e_n \le e$ and $e\theta^k(e_n) = 0$ for $1 \le k \le n-1$, it follows that $eq = \sum_{n=1}^{\infty} pe_n = p$. On the other hand, $\theta(q) = \sum_{n=1}^{\infty}\sum_{k=1}^{n}\theta^k(pe_n)$, hence $\theta(q) - q = \sum_{n=1}^{\infty}(\theta^n(pe_n) - pe_n) = 0$. Since θ is ergodic on $\mathscr{Z}(\mathscr{N})$, it follows that either $q = 0$, that is, $p = 0$, or $q = 1$, that is, $p = e$. Hence $(\mathscr{N}e, \theta_e, \tau_e)$ is a discrete decomposition of type III$_0$.

Consider the W^*-algebra $\mathscr{N}e$ and $\theta_e \in Aut(\mathscr{N}e)$. In order to establish the desired *-isomorphism, we shall use Proposition 22.2. We have a W^*-algebra $\mathscr{M}_1 = \pi(e)\mathscr{M}\pi(e)$, an injective unital normal *-homomorphism

$$\pi_1 = \pi|\mathscr{N}e : \mathscr{N}e \to \mathscr{M}_1,$$

a faithful normal conditional expectation $P_1 = P|\mathscr{M}_1 : \mathscr{M}_1 \to \pi_1(\mathscr{N}e)$ and a unitary element $u_1 = \sum_{n\ge 1} u^n\pi(e_n) \in \mathscr{M}_1$ such that

$$\pi_1(\theta_e(x)) = u_1\pi_1(x)u_1^* \quad (x \in \mathscr{N}e) \tag{3}$$

$$\mathscr{M}_1 = \mathscr{R}\{\pi_1(\mathscr{N}e), u_1\}. \tag{4}$$

Indeed, u_1 is unitary:

$$\begin{aligned} u_1u_1^* &= \sum_{n,m} u^n\pi(e_n)\pi(e_m)u^{-m} = \sum_n u^n\pi(e_n)u^{-n} = \sum_n \pi(\theta^n(e_n)) \\ &= \pi\left(\sum_n \theta^n(e_n)\right) = \pi(e) \\ u_1^*u_1 &= \sum_{n,m}\pi(e_n)u^{-n}u^m\pi(e_m) = \sum_{n,m} u^{-n}\pi(\theta^n(e_n))u^m\pi(e_m) \\ &= \sum_{n,m} u^{-n}u^m\pi(\theta^{n-m}(e_n)e_m) = \sum_n \pi(e_n) = \pi(e), \end{aligned}$$

since $\theta^{n-m}(e_n)e_m = \theta^{-m}(\theta^n(e_n)\theta^m(e_m)) = 0$ for $n \neq m$. For $x \in \mathscr{N}e$, we have

$$\begin{aligned} u_1\pi_1(x)u_1^* &= \sum_{n,m} u^n\pi(e_n)\pi(x)\pi(e_m)u^{-m} = \sum_{n,m} u^n\pi(xe_ne_m)u^{-m} \\ &= \sum_n u^n\pi(xe_n)u^{-n} = \pi\left(\sum_n \theta^n(xe_n)\right) = \pi_1(\theta_e(x)). \end{aligned}$$

This proves (3). Let $\mathscr{M}_2 = \mathscr{R}\{\pi_1(\mathscr{N}e), u_1\}$. It is clear that $\mathscr{M}_2 \subset \mathscr{M}_1$ and, since $\mathscr{M} = \mathscr{R}\{\pi(\mathscr{N}), u\}$ and $\mathscr{M}_1 = \pi(e)\mathscr{M}\pi(e)$, we have

$$\mathscr{M}_1 = \mathscr{R}\{\pi(e)\pi(\mathscr{N})\pi(e), \pi(e)u^n\pi(e)\} \quad (n \in \mathbb{Z}).$$

As $\pi(e)\pi(\mathscr{N})\pi(e) = \pi_1(\mathscr{N}e)$ and $\pi \circ \theta = \mathrm{Ad}(u) \circ \pi$, to prove (4) (i.e., $\mathscr{M}_1 = \mathscr{M}_2$) it is sufficient to show that for every projection $p \in \mathscr{Z}(\mathscr{N})$ we have

$$p \leq e, \theta^n(p) \leq e \Rightarrow u^n\pi(p) \in \mathscr{M}_2 \quad (n \geq 1).$$

Since $e\theta^n(pe_k) \leq e\theta^n(e_k) = 0$ for $k > n$, it follows that $pe_k = 0$ for $k > n$, and we may assume that

$$p \leq e_{k_1}, \text{ with } k_1 \leq n;$$

then $\theta^{k_1}(p) \leq e, \theta^{n-k_1}(\theta^{k_1}(p)) \leq e$ and, by the same argument, we may assume also that

$$\theta^{k_1}(p) \leq e_{k_2}, \text{ with } k_1 + k_2 \leq n;$$

by continuing this procedure inductively, we may assume that

$$\theta^{k_1+k_2}(p) \leq e_{k_3} \text{ with } k_1 + k_2 + k_3 \leq n$$
$$\cdots\cdots\cdots\cdots$$
$$\theta^{k_1+k_2+\ldots+k_{j-1}}(p) \leq e_{k_j} \text{ with } k_1 + k_2 + \ldots + k_j = n.$$

In this case, we have

$$\begin{aligned} u_1\pi_1(p) &= \sum_n u^n\pi(e_n)\pi(p) = u_1^{k_1}\pi(p) = \pi(\theta^{k_1}(p))u^{k_1}, \\ u_1^2\pi_1(p) &= u_1u^{k_1}\pi(e_{k_1}) = \sum_n u^n\pi(e_n)\pi(\theta^{k_1}(p))u^{k_1} = u^{k_2}\pi(\theta^{k_1}(p))u^{k_1} \\ &= \pi(\theta^{k_1+k_2}(p))u^{k_1+k_2}, \end{aligned}$$

and, by induction,

$$u_1^j\pi_1(p) = \pi(\theta^n(p))u^n = u^n\pi(p),$$

so that $u^n\pi(p) = u_1^j\pi_1(p) \in \mathscr{M}_2$. Thus, (4) is proved.

Since $(\mathscr{N}e, \theta_e, \tau_e)$ is a discrete decomposition of type III_0, the action $\theta_e : \mathbb{Z} \to Aut(\mathscr{N}e)$ is properly outer (29.3). Using Proposition 22.2 and the last remark in Section 22.2, we deduce that there exists a *-isomorphism

$$\Phi : (\mathscr{M}_0, \pi_0, P_0, u_0) \to (\mathscr{M}_1, \pi_1, P_1, u_1).$$

Consider now $x \in \mathscr{M}_0^+$ and $y \in (\mathscr{N}e)^+$ with $P_0(x) = \pi_0(y)$. Then $\varphi_0(x) = \tau(y)$. On the other hand, $P_1(\Phi(x)) = \Phi(P_0(x)) = \Phi(\pi_0(y)) = \pi_1(y)$, so that $P(\Phi(x)) = \pi(y)$, and $\varphi(\Phi(x)) = \tau(y)$. Thus, $\varphi_0 = \varphi_{n(e)} \circ \Phi$.

Since $\varphi_0 = \varphi_{n(e)} \circ \Phi$, we also have

$$\Phi \circ \sigma_s^{\varphi_0} = \sigma_s^{\varphi} \circ \Phi \quad (s \in \mathbb{R}). \tag{5}$$

□

Note that, as the factor $\mathcal{M}$ is of type III (and countably decomposable), we have, for every nonzero projection $e \in \mathcal{Z}(\mathcal{N}), \mathcal{M} \approx \mathcal{M}_{\pi(e)}$. Thus, the discrete type III$_0$ decompositions $(\mathcal{N}, \theta, \tau)$ and $(\mathcal{N}e, \theta_e, \tau_e)$ give rise to *-isomorphic crossed products which are factors of type III$_0$.

In the constructions of factors described in Sections 29.1 and 29.3, the weights dual to the traces appearing in the corresponding discrete decompositions are lacunary. Thus, in order to obtain a discrete decomposition corresponding to any given factor of type III$_\lambda$ $(0 \le \lambda < 1)$, it is first necessary to study the lacunary weights on such factors.

29.5. Consider first a factor $\mathcal{M}$ of type III$_\lambda$, for $0 < \lambda < 1$, and put $t = -2\pi/\ln(\lambda)$. Then $t \in T(\mathcal{M})$ (see 28.11) and so (27.1.(vi)) there exists a faithful normal state ψ on $\mathcal{M}$ such that $\sigma_t^{\psi} = \iota$. Also $\varphi = \psi \bar{\otimes}\, tr$ is an n.s.f. weight on $\mathcal{M} \approx \mathcal{M} \bar{\otimes} \mathcal{F}_\infty$ with $\varphi(1) = +\infty$ and $\sigma_t^{\varphi} = \iota$, that is, a λ-trace.

On the other hand, since t is the smallest positive number in $T(\mathcal{M})$ (28.11), for every n.s.f. weight φ on $\mathcal{M}$ with $\sigma_t = \iota$ we have $\sigma_s^{\varphi} = \iota$ if and only if there exists $k \in \mathbb{Z}$ with $s = kt$; using 16.4.(2), 28.1.(4), and 28.3.(1), we infer that $Sp\, \sigma^{\varphi} = \{\lambda^n; n \in \mathbb{Z}\} = \Gamma(\sigma^{\varphi})$ and $Sp(\Delta_\varphi) = S(\mathcal{M})$, in particular $Sp(\Delta_\varphi) \cap (\lambda, \lambda^{-1}) = \{1\}$. Hence every n.s.f. weight φ on $\mathcal{M}$ with $\sigma_t^{\varphi} = \iota$ is a lacunary weight and $\mathcal{M}^{\varphi}$ is a factor (see 16.4.(5)).

Consider now two λ-traces φ and ψ on $\mathcal{M}$. Since $\sigma_t^{\varphi} = \sigma_t^{\psi} (= \iota)$ and $\mathcal{M}$ is a factor, it follows that $[D\psi : D\varphi]_t = \alpha \cdot 1_{\mathcal{M}}$. Let $\mu > 0$ be such that $\mu^{it} = \alpha$. Then $\mathcal{M} \bar{\otimes} \mathcal{F}_2$ is a factor of type III$_\lambda$ (28.4) and the balanced weight $\nu = \theta(\mu\varphi, \psi)$ is a λ-trace on $\mathcal{M} \bar{\otimes} \mathcal{F}_2$, since

$$\sigma_t^{\nu}\begin{pmatrix} x & 0 \\ 0 & y \end{pmatrix} = \begin{pmatrix} \sigma_t^{\varphi}(x) & 0 \\ 0 & \sigma_t^{\psi}(y) \end{pmatrix} = \begin{pmatrix} x & 0 \\ 0 & y \end{pmatrix} \quad (x, y \in \mathcal{M})$$

$$\sigma_t^{\nu}\begin{pmatrix} 0 & 0 \\ 1 & 0 \end{pmatrix} = \begin{pmatrix} 0 & 0 \\ [D\psi : D\mu\varphi]_t & 0 \end{pmatrix} = \begin{pmatrix} 0 & 0 \\ \alpha\mu^{-it} & 0 \end{pmatrix} = \begin{pmatrix} 0 & 0 \\ 1 & 0 \end{pmatrix};$$

hence $\sigma_t^{\nu} = \iota$ and $\nu(1) = +\infty$. It follows that $(\mathcal{M} \bar{\otimes} \mathcal{F}_2)^{\nu}$ is a factor and that the restriction of ν to $\mathcal{M} \bar{\otimes} \mathcal{F}_2$ is an n.s.f. trace on $\mathcal{M} \bar{\otimes} \mathcal{F}_2$ (see 10.9). Since $\nu\begin{pmatrix} 1 & 0 \\ 0 & 0 \end{pmatrix} = \nu\begin{pmatrix} 0 & 0 \\ 0 & 1 \end{pmatrix} = +\infty$, it follows that the projections $\begin{pmatrix} 1 & 0 \\ 0 & 0 \end{pmatrix}$ and $\begin{pmatrix} 0 & 0 \\ 0 & 1 \end{pmatrix}$ are equivalent in $(\mathcal{M} \bar{\otimes} \mathcal{F}_2)^{\nu}$. Using Proposition 23.1, we conclude that there exists a unitary element $v \in \mathcal{M}$ such that $\psi = \mu(\varphi \circ \mathrm{Ad}(v))$.

We have thus proved the following.

Proposition. *Let $\mathcal{M}$ be a factor of type* III$_\lambda$ $(0 < \lambda < 1)$ *and $t = -2\pi/\ln(\lambda)$. There exists a λ-trace on $\mathcal{M}$. For any two λ-traces φ, ψ on $\mathcal{M}$, there exist a unitary element $v \in \mathcal{M}$ and $\mu > 0$ such that $\psi = \mu(\varphi \circ \mathrm{Ad}(v))$; in this case $\mu^{it} = [D\psi : D\varphi]_t$.* □

In particular, for every λ-trace φ on $\mathcal{M}$, there exists a unitary element $u \in \mathcal{M}$ such that $\lambda\varphi = \varphi \circ \mathrm{Ad}(u)$.

29.6. For type III$_0$ factors, we first prove the existence of lacunary weights of infinite multiplicity.

Proposition. *Let $\mathcal{M}$ be a factor of type* III$_0$*, φ a faithful normal strictly semifinite weight on $\mathcal{M}$ and $p \in \mathcal{M}^{\varphi}$ a nonzero projection. There exist a nonzero projection $e \in \mathcal{M}^{\varphi}, e \le p$, an element*

$a \in \mathcal{M}^\varphi \cap e\mathcal{M}e, a \geq 0$, which is invertible in $e\mathcal{M}e$, and a faithful normal lacunary state ψ on $e\mathcal{M}e$ such that $a \in (e\mathcal{M}e)^\psi$ and $\varphi_e = \psi(a\cdot)$.

Proof. Since φ is strictly semifinite, $\varphi|\mathcal{M}^\varphi$ is an n.s.f. trace on $\mathcal{M}^\varphi$ and we may hence assume that $\varphi(p) < +\infty$. We have $\Gamma(\sigma^\varphi p) = \Gamma(\sigma^\varphi) = S(\mathcal{M}) \cap \mathbb{R}_*^+$ so that, by Corollary 2/16.2, there exist a nonzero projection $e \in \mathcal{M}^\varphi p = \mathcal{M}^\varphi \cap p\mathcal{M}p$ and $\varepsilon > 0$ such that

$$(Sp\, \sigma^{\varphi_e}) \cap \exp([-2\varepsilon, -\varepsilon] \cup [\varepsilon, 2\varepsilon]) = \emptyset.$$

Using Proposition 15.12, we obtain an element $h \in \mathcal{M}^\varphi \cap e\mathcal{M}e, -\varepsilon/2 \leq h \leq \epsilon/2$, such that for the action $\sigma : \mathbb{R} \to Aut(e\mathcal{M}e)$ defined by $\sigma_s(x) = e^{-ish}\sigma_s^{\varphi_e}(x)e^{ish}$ $(x \in e\mathcal{M}e, s \in \mathbb{R})$ we have

$$(Sp\, \sigma) \cap \exp(-\varepsilon, \varepsilon) = \{1\}.$$

Then the element $a = e^h \in \mathcal{M}^\varphi \cap e\mathcal{M}e$ is positive and invertible in $e\mathcal{M}e$ and $\psi = \varphi_e(a^{-1}\cdot)$ is a faithful normal state on $e\mathcal{M}e$ with $\sigma^\psi = \sigma$, so that ψ is lacunary. It is clear that $a \in (e\mathcal{M}e)^\varphi$ and $\varphi_e = \psi(a\cdot)$. □

Corollary. *On every factor $\mathcal{M}$ of type III_0, there exists a lacunary weight of infinite multiplicity.*

Proof. Since every nonzero projection in $\mathcal{M}$ is equivalent to 1, the previous proposition shows that there exists a faithful normal lacunary weight ψ on $\mathcal{M}$. Then $\varphi = \psi \bar{\otimes} tr$ is a lacunary weight of infinite multiplicity on $\mathcal{M} \bar{\otimes} \mathcal{F}_\infty \approx \mathcal{M}$. □

29.7 Proposition. *Let φ_1, φ_2 be lacunary weights of infinite multiplicity on the type III_0 factor $\mathcal{M}$. Given $\varepsilon > 0$ there exist nonzero projections $e_1 \in \mathcal{Z}(\mathcal{M}^{\varphi_1}), e_2 \in \mathcal{Z}(\mathcal{M}^{\varphi_2})$ and a partial isometry $v \in \mathcal{M}$ with $v^*v = e_1, vv^* = e_2$, such that the mapping $x \mapsto vxv^*$ defines a *-isomorphism*

$$\Phi : (e_1\mathcal{M}e_1, \mathcal{M}^{\varphi_1} \cap e_1\mathcal{M}e_1) \to (e_2\mathcal{M}e_2, \mathcal{M}^{\varphi_2} \cap e_2\mathcal{M}e_2)$$

with the property that for every $x \in e_1\mathcal{M}e_1$ we have

$$Sp_{\sigma^{\varphi_2}}(\Phi(x)) \subset Sp_{\sigma^{\varphi_1}}(x) \cdot \exp[-\varepsilon, \varepsilon],$$
$$Sp_{\sigma^{\varphi_1}}(x) \subset Sp_{\sigma^{\varphi_2}}(\Phi(x)) \cdot \exp[-\varepsilon, \varepsilon].$$

Proof. Put $\sigma^{\varphi_1} = \sigma_1, \sigma^{\varphi_2} = \sigma_2$ and consider the balanced weight $\varphi = \theta(\varphi_1, \varphi_2)$ on $\mathcal{P} = Mat_2(\mathcal{M})$ and the *-isomorphisms $I_1 : \mathcal{M} \to \mathcal{P}, I_2 : \mathcal{M} \to \mathcal{P}$ defined by $I_1(x) = \begin{pmatrix} x & 0 \\ 0 & 0 \end{pmatrix}, I_2(x) = \begin{pmatrix} 0 & 0 \\ 0 & x \end{pmatrix}$, $(x \in \mathcal{M})$.

With $\sigma = \sigma^\varphi$, we have

$$\sigma \circ I_k = I_k \circ \sigma_k \quad (k = 1, 2).$$

Since φ_1 and φ_2 are of infinite multiplicity, the projections $\begin{pmatrix}1 & 0\\ 0 & 0\end{pmatrix}, \begin{pmatrix}0 & 0\\ 0 & 1\end{pmatrix}$ are properly infinite in $\mathscr{P}^\sigma$, in particular $\mathscr{P}^\sigma$ is properly infinite. Let

$$p_1 = \mathbf{z}_{\mathscr{P}^\sigma}\left(\begin{pmatrix}1 & 0\\ 0 & 0\end{pmatrix}\right) \in \mathscr{Z}(\mathscr{P}^\sigma), p_2 = \mathbf{z}_{\mathscr{P}^\sigma}\left(\begin{pmatrix}0 & 0\\ 0 & 1\end{pmatrix}\right) \in \mathscr{Z}(\mathscr{P}^\sigma).$$

We have

$$Sp\ \sigma^{p_k} = Sp\ \sigma_k \quad (k = 1, 2),$$

and 1 is an isolated point in $Sp\ \sigma^{p_k}(k = 1, 2)$. By Corollary 15.17, there exists a nonzero partial isometry $w \in \mathscr{P}$ such that $p_1 \ge q_1 = w^*w \in \mathscr{Z}(\mathscr{P}^\sigma), p_2 \ge q_2 = ww^* \in \mathscr{Z}(\mathscr{P}^\sigma)$ and

$$(Sp_\sigma(w))(Sp_\sigma(w))^{-1} \subset \exp[-\varepsilon, \varepsilon].$$

We define

$$r_1 = \begin{pmatrix}1 & 0\\ 0 & 0\end{pmatrix} q_1, \quad r_2 = \begin{pmatrix}0 & 0\\ 0 & 1\end{pmatrix} q_2.$$

Each r_k is a properly infinite projection in $\mathscr{P}^\sigma$ and $\mathbf{z}_{\mathscr{P}^\sigma}(r_k) = q_k p_k = q_k$; hence $r_k \sim q_k$ in $\mathscr{P}^\sigma$, so that there exists a partial isometry $w_k \in \mathscr{P}^\sigma$ with $w_k^* w_k = r_k$ and $w_k w_k^* = q_k$ $(k = 1, 2)$. Then $u = w_2^* w w_1 \in \mathscr{P}, u^*u = r_1, uu^* = r_2$ and, since $w_1, w_2 \in \mathscr{P}^\sigma$,

$$(Sp_\sigma(u))(Sp_\sigma(u))^{-1} \subset \exp[-\varepsilon, \varepsilon].$$

On the other hand, it is easy to see that $r_k \in I_k(\mathscr{Z}(\mathscr{M}^{\sigma_k}))$, so that there exists $e_k \in \mathscr{Z}(\mathscr{M}^{\sigma_k})$ such that $r_k = I_k(e_k)$ $(k = 1, 2)$. Since $u^*u = r_1, uu^* = r_2$, it follows that u is of the form $u = \begin{pmatrix}0 & 0\\ 0 & v\end{pmatrix}$ with $v \in \mathscr{M}$ and $v^*v = e_1, vv^* = e_2$. For $x \in e_1\mathscr{M}e_1$, we have $Sp_{\sigma_1}(x) = Sp_\sigma(I_1(x))$ and $Sp_{\sigma_2}(vxv^*) = Sp_\sigma(I_2(vxv^*)) = Sp_\sigma\left(\begin{pmatrix}0 & 0\\ 0 & vxv^*\end{pmatrix}\right) = Sp_\sigma\left(\begin{pmatrix}0 & 0\\ v & 0\end{pmatrix}\begin{pmatrix}x & 0\\ 0 & 0\end{pmatrix}\right.$ $\left.\times \begin{pmatrix}0 & v^*\\ 0 & 0\end{pmatrix}\right) = Sp_\sigma(uI_1(x)u^*)$.

The assertions of the proposition are now easily verified. □

29.8 Proposition. *Let φ be a faithful normal lacunary weight on the W^*-algebra $\mathscr{M}$. Then every maximal abelian *-subalgebra $\mathscr{A}$ of $\mathscr{M}^\varphi$ is maximal abelian in $\mathscr{M}$.*

Proof. Put $\sigma = \sigma^\varphi$. By assumption, there exists $\varepsilon > 0$ such that $(Sp\ \sigma) \cap \exp[-\varepsilon, \varepsilon] = \{1\}$. Since $\mathscr{A} \subset \mathscr{M}^\sigma, \mathscr{A}' \cap \mathscr{M}$ is a σ-invariant vector subspace of $\mathscr{M}$.

Assume that there exists $x \in \mathscr{A}' \cap \mathscr{M}, x \notin \mathscr{M}^\varphi$. In view of Lemma 15.1, we may further assume that

$$(Sp_\sigma(x))(Sp_\sigma(x))^{-1} \subset \exp[-\varepsilon, \varepsilon].$$

If $1 \in Sp_\sigma(x)$, then $Sp_\sigma(x) \subset (Sp\ \sigma) \cap \exp[-\varepsilon, \varepsilon] = \{1\}$, and $x \in \mathscr{M}^\varphi$, which is not possible. Consequently, $1 \notin Sp_\sigma(x)$. Since $Sp_\sigma(x) \cap \exp[-\varepsilon, \varepsilon) \subset (Sp\sigma) \cap \exp[-\varepsilon, \varepsilon] = \{1\}$, it follows that we have either $Sp_\sigma(x) \subset \exp(\varepsilon, +\infty)$, or $Sp_\sigma(x) \subset \exp(-\infty, -\varepsilon)$. By replacing if necessary x by x^*, we may assume that

$$Sp_\sigma(x) \subset \exp(-\infty, -\varepsilon).$$

Let $x = u|x|$ be the polar decomposition of x, with $u^*u = \mathbf{r}(x), uu^* = \mathbf{l}(x)$. Since $(Sp_\sigma(x))(Sp_\sigma(x))^{-1} \cap (Sp\ \sigma) = \{1\}$, using Proposition 15.17.(2) we get $Sp_\sigma(u) = Sp_\sigma(x) \subset \exp(-\infty, -\varepsilon), u \in \mathscr{A}' \cap \mathscr{M}$, and $u^*u, uu^* \in \mathscr{A}' \cap \mathscr{M}^\varphi = \mathscr{A}$. It follows that $(u^*u)u = u(u^*u) = u$ and $u(uu^*) = (uu^*)u = u$, so that $u^*u = u^*uuu^* = uu^*$. Since $Sp_\sigma(u)$ is a compact subset of $\exp(-\infty, -\varepsilon)$, there is a function $f \in \mathscr{L}^1(\mathbb{R})$ such that $u = \sigma_f(u), 0 \leq \hat{f} \leq 1$, and $supp\ \hat{f} \subset \exp(-\infty, -\varepsilon/2)$. Then $\sup\{\gamma^{1/2}\hat{f}(\gamma); \gamma \in \mathbb{R}_*^+\} \leq \exp(-\varepsilon/4)$ and (28.1.(3)) $\Delta_\varphi^{1/2}\hat{f}(\Delta_\varphi)u_\varphi = \Delta_\varphi^{1/2}(\sigma_f(u))_\varphi = \Delta_\varphi^{1/2}u_\varphi$, whence $\varphi(uu^*) = \|(u^*)_\varphi\|_\varphi = \|\Delta_\varphi^{1/2}u_\varphi\|_\varphi \leq \exp(-\varepsilon/4)\|u_\varphi\|_\varphi = \exp(-\varepsilon/4)\varphi(u^*u)$, contradicting $u^*u = uu^*$. □

Note that if $\{e_i\}$ is a family of projections of the W^*-algebra $\mathscr{M}$ with $\sum_i e_i = 1$ and if, for each i, $\mathscr{A}_i$ is a maximal abelian *-subalgebra in $e_i\mathscr{M}e_i$, then $\mathscr{A} = \overline{(\sum_i \mathscr{A}_i)}^w$ is a maximal abelian *-subalgebra in $\mathscr{M}$. Indeed, if $x \in \mathscr{A}' \cap \mathscr{M}$, then x commutes with each e_i, so that $x = \sum_i e_ixe_i$ and $e_ixe_i \in \mathscr{A}_i' \cap e_i\mathscr{M}e_i = \mathscr{A}_i$.

29.9 Theorem. *Let φ be a faithful normal stricly semifinite weight on the factor $\mathscr{M}$ of type* III_λ $(0 \leq \lambda < 1)$. *There exists a maximal abelian *-subalgebra of $\mathscr{M}$ contained in $\mathscr{M}^\varphi$. In particular,* $(\mathscr{M}^\varphi)' \cap \mathscr{M} = \mathscr{Z}(\mathscr{M}^\varphi)$.

Proof. Consider first the case $\lambda > 0$ and put $t = -2\pi/\ln(\lambda)$. Since φ is strictly semifinite, there exists a family of mutually orthogonal nonzero projections $\{e_i\} \subset \mathscr{M}^\varphi$, with $\varphi(e_i) < +\infty$ and $\sum_i e_i = 1$. For each i, $\varphi_i = \varphi_{e_i}$ is a faithful normal positive form on the W^*-algebra $\mathscr{M}_i = e\mathscr{M}e_i$ and $e_i \sim 1$ in $\mathscr{M}$, so that $\mathscr{M}_i \approx \mathscr{M}$ and $T(\mathscr{M}_i) = T(\mathscr{M}) \ni t$. By Theorem 27.1, we see that there exists a positive element $a_i \in \mathscr{M}_i^{\varphi_i} = \mathscr{M}^\varphi \cap \mathscr{M}_i$, invertible in $\mathscr{M}_i$, such that for $\psi_i = \varphi_i(a_i\cdot)$ we have $\sigma_t^{\psi_i} = \iota$. Thus, ψ_i is a lacunary faithful normal positive form on $\mathscr{M}_i$. Let $\mathscr{A}_i$ be any maximal abelian *-subalgebra of $\mathscr{M}_i^{\varphi_i}$ containing a_i. Then $\mathscr{A}_i$ is maximal abelian in $\mathscr{M}_i$, by Proposition 29.8, and for every $x \in \mathscr{A}_i$ we have $xa_i^{it} = a_i^{it}x$ and $\sigma_t^{\psi_i}(x) = x$, hence $x \in \mathscr{M}_i^{\varphi_i} \subset \mathscr{M}^\varphi$. Using the last remark in Section 29.8, we conclude that $\mathscr{A} = \overline{(\sum_i \mathscr{A}_i)}^w \subset \mathscr{M}^\varphi$ is a maximal abelian *-subalgebra of $\mathscr{M}$.

If $\lambda = 0$, the desired conclusion can be obtained by the same arguments, using Proposition 29.6.

Finally, if $\mathscr{A}$ is a maximal abelian *-subalgebra of $\mathscr{M}$ contained in $\mathscr{M}^\varphi$ and $x \in (\mathscr{M}^\varphi)' \cap \mathscr{M}$, then $x \in \mathscr{A}' \cap \mathscr{M} = \mathscr{A} \subset \mathscr{M}^\varphi$, hence $(\mathscr{M}^\varphi)' \cap \mathscr{M} = \mathscr{Z}(\mathscr{M}^\varphi)$. □

29.10 Corollary. *For every faithful normal strictly semifinite weight φ on a factor $\mathscr{M}$ of type* III_λ $(0 \leq \lambda < 1)$, *there exists a unique faithful normal conditional expectation $P_\varphi : \mathscr{M} \to \mathscr{M}^\varphi$.*

Proof. This follows from Theorem 29.9, Corollary 10.9, and Proposition 10.17. □

29.11 Corollary. *Let φ, ψ be n.s.f. weights on the factor $\mathscr{M}$ of type* III_λ $(0 \leq \lambda < 1)$. *If φ is strictly semifinite, then we have $\mathscr{M}^\varphi \subset \mathscr{M}^\psi$ if and only if there exists a nonsingular positive self-adjoint operator A affiliated to $\mathscr{Z}(\mathscr{M}^\varphi)$, such that $\psi = \varphi_A$.*

Proof. Let $u_s = [D\psi : D\varphi]_s$ $(s \in \mathbb{R})$. If $\mathcal{M}^\varphi \subset \mathcal{M}^\psi$, then $x \in \mathcal{M}^\varphi \Rightarrow x \in \mathcal{M}^\psi \Rightarrow x = \sigma_s^\psi(x) = u_s\sigma_s^\varphi(x)u_s^* = u_sxu_s^*$, and $u_s \in (\mathcal{M}^\varphi)' \cap \mathcal{M} = \mathcal{Z}(\mathcal{M}^\varphi)$ $(s \in \mathbb{R})$, by Theorem 29.9, that is, $\psi = \varphi_A$ with A affiliated to $\mathcal{Z}(\mathcal{M}^\varphi)$. Conversely, if this condition holds, then $u_s \in \mathcal{Z}(\mathcal{M}^\varphi)$ $(s \in \mathbb{R})$, and hence $x \in \mathcal{M}^\varphi \Rightarrow \sigma_s^\psi(x) = u_s\sigma_s^\varphi(x)u_s^* = u_sxu_s^* = x$ $(s \in \mathbb{R}) \Rightarrow x \in \mathcal{M}^\psi$. □

29.12 Corollary. *Let φ be a faithful normal strictly semifinite weight on a factor $\mathcal{M}$ of type* III$_\lambda$ $(0 < \lambda < 1)$, *and let $t = -2\pi/\ln(\lambda)$. The following statements are equivalent:*

(i) $\sigma_t^\varphi = \iota$;
(ii) $Sp(\Delta_\varphi) = S(\mathcal{M})$, *that is,* $Sp\,\sigma^\varphi = \Gamma(\sigma^\varphi)$;
(iii) $\mathcal{M}^\varphi$ *is a factor;*
(iv) $(\mathcal{M}^\varphi)' \cap \mathcal{M} = \mathbb{C} \cdot 1_{\mathcal{M}}$;
(v) *every n.s.f. weight ψ on $\mathcal{M}$ with $\mathcal{M}^\varphi \subset \mathcal{M}^\psi$ is proportional to φ;*
(vi) *$\mathcal{M}^\varphi$ is maximal among the semiftnite unital W^*-subalgebras $\mathcal{N} \subset \mathcal{M}$ with the property that there exists a faithful normal conditional expectation $P : \mathcal{M} \to \mathcal{N}$.*

Proof. The implication (i) ⇒ (ii) has already been proved in Section 29.5, the equivalence (ii) ⇔ (iii) follows from 16.4.(5), the equivalence (iii) ⇔ (iv) follows from Theorem 29.9, and the implication (iv) ⇒ (v) follows easily from Theorem 29.9 and Corollary 29.11.

(v) ⇒ (vi). Let $\mathcal{M}^\varphi \subset \mathcal{N}$ with $\mathcal{N} \subset \mathcal{M}$ a semifinite unital W^*-subalgebra, let $P : \mathcal{M} \to \mathcal{N}$ be a faithful normal conditional expectation and let τ be an n.s.f. trace on $\mathcal{N}$. Then $\psi = \tau \circ P$ is an n.s.f. weight on $\mathcal{M}$ with $\mathcal{M}^\varphi \subset \mathcal{N} \subset \mathcal{M}^\psi$, so that ψ is proportional to φ and therefore $\mathcal{M}^\varphi = \mathcal{N} = \mathcal{M}^\psi$.

(vi) ⇒ (i). Since $t \in T(\mathcal{M})$, there exists a nonsingular positive self-adjoint operator A affiliated to $\mathcal{Z}(\mathcal{M}^\varphi)$ such that $\psi = \varphi_A$ satisfies $\sigma_i^\psi = \iota$. Then the weight φ is strictly semifinite, that is, there exists a faithful normal conditional expectation $P : \mathcal{M} \to \mathcal{M}^\psi$ and $\mathcal{M}^\psi$ is semifinite (10.9). By Corollary 29.11, we know that $\mathcal{M}^\varphi \subset \mathcal{M}^\psi$; assumption (vi) implies that $\mathcal{M}^\varphi = \mathcal{M}^\psi$. Since the weight ψ satisfies condition (i), it also satisfies condition (v), so that φ is proportional to ψ and $\sigma_t^\varphi = \sigma_t^\psi = \iota$. □

29.13. Notes. The constructions and properties of factors of types III$_\lambda$ $(0 \leq \lambda < 1)$ given in this section are due to Connes (1972, 1973a) (see also Araki, 1973; Takesaki, 1973, $(\lambda > 0)$). An important feature of factors concerns the existence of almost periodic states. A faithful normal state φ on a W^*-algebra $\mathcal{M}$ is called *almost periodic* if the modular operator Δ_φ is diagonalizable, that is, if the set of eigenvalues of Δ_φ is total in $\mathcal{H}_\varphi$ (Connes, 1973a, 3.7.1; see Connes (1972, 1974d) for equivalent definitions). Clearly, any λ-trace on a type III$_\lambda$ factor is (almost) periodic. Using results of Krieger (1970b), Connes (1973a, 5.3.8; 1974d, 1.5) proved the existence of almost periodic faithful normal states on every factor of type III$_0$. Connes (1974d) also introduced the notion of a *full factor* ($\mathcal{M}$ is a full W^*-algebra if $Int(\mathcal{M})$ is closed in $Aut(\mathcal{M})$) and showed that there exist full factors of type III$_1$ with faithful normal almost periodic states as well as factors of type III$_1$ with no almost periodic state or weight. Moreover, Connes (1974d) exhibited a nonsmooth uncountable family of type III$_1$ factors with separable preduals. While every factor of type III$_\lambda$ with $0 \leq \lambda < 1$ arises from an essentially unique discrete decomposition (see §30), Connes (1974d) showed that there exist factors of type III$_1$ with separable preduals which are isomorphic to no crossed product of a semifinite von Neumann algebra by a discrete abelian group.

Connes (1979b) has shown that every type of factor can be obtained as von Neumann algebras associated with a foliation.

For our exposition, we have used Connes (1973a).

30 The Discrete Decomposition of Factors of Type III$_\lambda$ $(0 \le \lambda < 1)$

In this section, we show that every factor of type III$_\lambda$ $(0 \le \lambda < 1)$ has a certain type of discrete decomposition which is essentially unique; we relate the discrete to the continuous decomposition. As in the preceding section, all W^*-algebras are countably decomposable, either by assumption or by construction.

30.1 Theorem. *Let* $0 < \lambda < 1$ *and* $\mathscr{M}$ *be factors of type* III$_\lambda$. *There exists a discrete decomposition* $(\mathscr{N}, \theta, \tau)$ *of type* III$_\lambda$ *such that* $\mathscr{M} \approx \mathscr{R}(\mathscr{N}, \theta)$. *If* $(\mathscr{N}_1, \theta_1, \tau_1)$ *and* $(\mathscr{N}_2, \theta_2, \tau_2)$ *are two discrete decompositions of type* III$_\lambda$ *such that* $\mathscr{R}(\mathscr{N}_1, \theta_1) \approx \mathscr{R}(\mathscr{N}_2, \theta_2)$, *then* $(\mathscr{N}_1, \theta_1) \approx (\mathscr{N}_2, \theta_2)$.

Proof. Let $t = -2\pi / \ln(\lambda)$. By Proposition 29.5, there exist a λ-trace φ on $\mathscr{M}$ and a unitary element $u \in \mathscr{M}$ such that $\lambda\varphi = \varphi \circ \mathrm{Ad}(u)$. Let $\mathscr{N} = \mathscr{M}^\varphi$ and let $P = P_\varphi : \mathscr{M} \to \mathscr{N}$ be the faithful normal conditional expectation defined by $\varphi : P_\varphi(x) = t^{-1} \int_0^t \sigma_s^\varphi(x)\, \mathrm{d}s\ (x \in \mathscr{M})$.

Since $\lambda\varphi = \varphi \circ \mathrm{Ad}(u)$, we have (3.7) $u^*\sigma_s^\varphi(u) = [D(\varphi \circ \mathrm{Ad}(u)) : D\varphi]_s = \lambda^{is}$, whence $\sigma_s^\varphi(u) = \lambda^{is}u\ (s \in \mathbb{R})$, that is, $u \in \mathscr{M}(\sigma^\varphi; \{\lambda\})$ and $u^n \in \mathscr{M}(\sigma^\varphi; \{\lambda^n\})$ for all $n \in \mathbb{Z}$. Since $\sigma_t^\varphi = \iota$, it follows by Corollary 29.12 that $Sp\, \sigma^\varphi = \{\lambda^n; n \in \mathbb{Z}\}$, so that there exists an open covering $\{V_n\}_{n\in\mathbb{Z}}$ of $\mathbb{R}_*^+$ such that $V_n \cap (Sp\, \sigma^\varphi) = \{\lambda^n\}$ for each $n \in \mathbb{Z}$. Using Proposition 14.3.(4), we infer that $\mathscr{M} = \mathscr{R}\{\bigcup_{n\in\mathbb{Z}} \mathscr{M}(\sigma^\varphi; \{\lambda^n\})\}$. Let $x \in \mathscr{M}(\sigma^\varphi; \{\lambda^n\})$. As $u^{-n} \in \mathscr{M}(\sigma^\varphi; \{\lambda^{-n}\})$, it follows that $xu^{-n} \in \mathscr{M}^\varphi = \mathscr{N}$ (15.3.(2)). Thus $\mathscr{M}(\sigma^\varphi; \{\lambda^n\}) = u^n\mathscr{N}$ and $\mathscr{M} = \mathscr{R}\{\mathscr{N}, u\}$. Also, for every $n \neq 0$ we have $P(u^n) = \left(t^{-1} \int_0^t \lambda^{ins}\, \mathrm{d}s\right) u^n = 0$. Finally, since $u \in \mathscr{M}(\sigma^\varphi; \{\lambda\})$ and $u^* \in \mathscr{M}(\sigma^\varphi; \{\lambda^{-1}\})$, for $x \in \mathscr{N} = \mathscr{M}^\varphi$, we have $uxu^* \in \mathscr{M}^\varphi = \mathscr{N}$ (15.3.(2)). We thus obtain a $*$-automorphism $\theta = \mathrm{Ad}(u)|\mathscr{N} \in Aut(\mathscr{N})$ and conclude, using Proposition 22.2, that $\mathscr{M} \approx \mathscr{R}(\mathscr{N}, \theta)$.

By Corollary 29.12, $\mathscr{N} = \mathscr{M}^\varphi$ is a factor. Since φ is strictly semifinite, its restriction $\tau = \varphi|\mathscr{N}$ is an n.s.f. trace on $\mathscr{N}$ with $\tau(1) = \varphi(1) = +\infty$ and $\tau \circ \theta = (\varphi \circ \mathrm{Ad}(u))|\mathscr{N} = \lambda\varphi|\mathscr{N} = \lambda\tau$. Thus, $\mathscr{N}$ is a factor of type II$_\infty$ and $(\mathscr{N}, \theta, \tau)$ is a discrete decomposition of type III$_\lambda$.

Consider now two discrete decompositions $(\mathscr{N}_1, \theta_1, \tau_1), (\mathscr{N}_2, \theta_2, \tau_2)$ of type III$_\lambda$ with $\mathscr{R}(\mathscr{N}_1, \theta_1) = \mathscr{R}(\mathscr{N}_2, \theta_2) = \mathscr{M}$; let u_1, u_2 be the unitary elements corresponding to $1 \bar{\otimes} \boldsymbol{\lambda}(1)$ in the two crossed products and, identifying $\mathscr{N}_1$ and $\mathscr{N}_2$ with the corresponding W^*-subalgebras of $\mathscr{M}$, denote by $P_1 : \mathscr{M} \to \mathscr{N}_1, P_2 : \mathscr{M} \to \mathscr{N}_2$ the canonical faithful normal conditional expectations. We know (29.1) that the dual weights $\varphi_1 = \tau_1 \circ P_1, \varphi_2 = \tau_2 \circ P_2$ are λ-traces on $\mathscr{M}$ and $\mathscr{N}_1 = \mathscr{M}^{\varphi_1}, \mathscr{N}_2 = \mathscr{M}^{\varphi_2}$. Using Proposition 29.5 and modifying, if necessary, the trace τ_2, we obtain a unitary element $u \in \mathscr{M}$ such that $\varphi_2 = \varphi_1 \circ \mathrm{Ad}(u)$ and, in particular, $\mathscr{N}_2 = [\mathrm{Ad}(u)](\mathscr{N}_1)$. Since $\lambda\varphi_1 = \varphi_1 \circ \mathrm{Ad}(u_1), \lambda\varphi_2 = \varphi_2 \circ \mathrm{Ad}(u_2)$ (see 29.5), it follows that $\lambda\varphi_2 = \varphi_2 \circ \mathrm{Ad}(u^*u_1u)$. Thus, u_2 and u^*u_1u belong to the spectral subspace $\mathscr{M}(\sigma^{\varphi_1}; \{\lambda\})$ and hence $v = uu_1^*u^*u_2 \in \mathscr{M}^{\varphi_1} = \mathscr{N}_2$. Since $\theta_1 = \mathrm{Ad}(u_1)|\mathscr{N}_1$ and $\theta_2 = \mathrm{Ad}(u_2)|\mathscr{N}_2$, it follows that $[(\mathrm{Ad}(u))\theta_1^{-1}(\mathrm{Ad}(u))^{-1}]\theta_2 = \mathrm{Ad}(v) \in Int(\mathscr{N}_2)$. Finally, using Theorem 23.13, we conclude that $(\mathscr{N}_1, \theta_1) \approx (\mathscr{N}_2, \theta_2)$. □

The discrete decomposition $(\mathscr{N}, \theta, \tau)$ of type III$_\lambda$ such that $\mathscr{R}(\mathscr{N}, \theta) \approx \mathscr{M}$ will be called *the discrete decomposition of the type* III$_\lambda$ *factor* $\mathscr{M}$ $(0 < \lambda < 1)$. The previous proof and Proposition 29.1 show that the λ-traces on $\mathscr{M}$ are the weights dual to the traces appearing in the discrete decomposition of $\mathscr{M}$.

The previous theorem can also be proved, using Landstad's theorem and the Takesaki duality theorem as in the proof of Theorem 23.6, by factorizing the action $\sigma^\varphi : \mathbb{R} \to Aut(\mathscr{M})$ with $\sigma_t^\varphi = \iota$ to get an action $\sigma : \mathbb{T} \to Aut(\mathscr{M})$.

30.2. For a factor $\mathscr{M}$ of type III$_0$, it is natural to start with a lacunary weight φ of infinite multiplicity on $\mathscr{M}$ and to take $\mathscr{N} = \mathscr{M}^\varphi, \tau = \varphi|\mathscr{N}$, but the procedure for finding a unitary element $u \in \mathscr{M}$ which implements a convenient $*$-automorphism $\theta = \mathrm{Ad}(u)|\mathscr{N}$ is considerably more complicated. We consider separately this part of the proof in the following.

Lemma. *Let $\sigma : \mathbb{R} \to Aut(\mathscr{M})$ be a continuous action of $\mathbb{R}$ on the factor $\mathscr{M}$. Assume that $\mathscr{M}^\sigma$ is properly infinite, that 1 is an isolated point in $Sp\,\sigma \subset \mathbb{R}_*^+$ and that σ is not outer conjugate with the trivial action. Then there exists a unitary element $u \in \mathscr{M}$ such that $\mathscr{M} = \mathscr{R}\{\mathscr{M}^\sigma, u\}, u\mathscr{M}^\sigma u^* = \mathscr{M}^\sigma$, and $Sp_\sigma(u) \subset (1, +\infty)$.*

Proof. For the proof it is more convenient to use the natural identification of the dual group $\hat{\mathbb{R}}$ of $\mathbb{R}$ with $\mathbb{R}$ itself, rather than $\mathbb{R}_*^+$, as in the statement. Thus, by assumption, there exists $\varepsilon > 0$ such that $(Sp\,\sigma) \cap [-\varepsilon, \varepsilon] = \{0\}$.

Let $\mathscr{V}$ be the set of all partial isometries $v \in \mathscr{M}$ with the properties

$$v^*v \in \mathscr{Z}(\mathscr{M}^\sigma),\ vv^* \in \mathscr{Z}(\mathscr{M}^\sigma),\ v\mathscr{M}^\sigma v^* \subset \mathscr{M}^\sigma,\ v^*\mathscr{M}^\sigma v \subset \mathscr{M}^\sigma.$$

It is easy to check that

$$\begin{gathered} v_1, v_2 \in \mathscr{V} \Rightarrow v_1 v_2 \in \mathscr{V}, \\ v_1, v_2 \in \mathscr{V},\ v_1^*v_1 \perp v_2^*v_2,\ v_1v_1^* \perp v_2^*v_2 \Rightarrow v_1 + v_2 \in \mathscr{V}. \end{gathered}$$

Putting $e = v^*v, e' = vv^*, e_1 = v_1^*v_1, e_1' = v_1v_1^*, e_2 = v_2^*v_2, e_2' = v_2v_2^*$, we have

$$\begin{aligned} v = v_1v_2 &\Rightarrow e = v_2^*e_1v_2, \quad e' = v_1e_2'v_1^* \\ v = v_1 + v_2 &\Rightarrow e = e_1 + e_2, \quad e' = e_1' \neq e_2', \end{aligned}$$

and we shall write

$$v_1 \prec v_2 \Leftrightarrow e_1 \le e_2 \quad \text{and} \quad v_1 = v_2 e_1;$$

in this case we also have $v_1 = e_1'v_2$. Note that

$$v_1 \prec v_2 \Leftrightarrow v_1^* \prec v_2^*,\quad vv_1 \prec vv_2,\quad v_1v \prec v_2v.$$

We shall consider the sets

$$\begin{aligned} \mathscr{V}_0 &= \{v \in \mathscr{V}; Sp_\sigma(v) \subset (\varepsilon, +\infty), Sp_\sigma(v) - Sp_\sigma(v) \subset [-\varepsilon/2, \varepsilon/2]\} \\ \mathscr{V}_1 &= \{v \in \mathscr{V}_0; v_1, v_2 \in \mathscr{V}_0, v_1v_2 \prec v \Rightarrow v_1v_2 = 0\} \end{aligned}$$

and establish five properties for them.

(I) *For each $0 \neq v \in \mathscr{V}_0$ there exist $v_1, \dots, v_p \in \mathscr{V}_1$ such that $0 \neq v_1 \dots v_p \prec v$.*

We prove this property by induction over n, assuming at step n that $Sp_\sigma(v) \subset (0, n\varepsilon)$.

For $n = 1$, we have $\{v \in \mathscr{V}_0; Sp_\sigma(v) \subset (0, \varepsilon)\} = \emptyset$. Suppose that the property holds for all elements of $\mathscr{V}_0$ with spectrum contained in $(0, n\varepsilon)$. Let $0 \neq v \in \mathscr{V}_0$ be such that $Sp_\sigma(v) \subset (0, (n+1)\varepsilon)$. If $v \in \mathscr{V}_1$, then the property is satisfied with $p = 1$ and $v_1 = v$, so that we may assume that $v \notin \mathscr{V}_1$. Then there exist $w_1, w_2 \in \mathscr{V}_0$ with $w_1w_2 \prec v$ but $w_1w_2 \neq 0$. Let $e = w_1^*w_1$ and $w_3 = ew_2$. We have $w_3 \in \mathscr{V}_0$

since $e \in \mathcal{M}^\sigma$ and $w_1w_2 = w_1ew_2 = w_1w_3, w_3w_3^* = ew_2w_2^*e \leq e = w_1^*w_1, w_3 = w_1^*w_1w_2 = w_1^*fv$, where $f = (w_1w_2)(w_1w_2)^*$. Since $f \in \mathcal{M}^\sigma$ and $w_1 \in \mathcal{V}_0$, we have $Sp_\sigma(w_3) \subset Sp_\sigma(v) - Sp_\sigma(w_1) \subset (0,(n+1)\varepsilon) - (\varepsilon,+\infty) \subset (0,n\varepsilon)$. Since $w_3 \in \mathcal{V}_0$ and $Sp_\sigma(w_3) \subset (0,n\varepsilon)$, by the induction hypothesis there exist $v_1,\dots,v_p \in \mathcal{V}_1$ such that $0 \neq v_1 \dots v_p \prec w_3$. We have $\mathcal{Z}(\mathcal{M}^\sigma) \ni f' = v_1 \dots v_pv_p^* \dots v_1^* \leq e$ and $0 \neq w_1(v_1 \dots v_p) \prec w_1w_3 = w_1w_2 \prec v$, that is, $w_1f'v_1 \dots v_p = f''v$ with $f'' \in \mathcal{Z}(\mathcal{M}^\sigma), f'' \leq vv^*$, hence $w_4 = w_1f' = f''v(v_1 \dots v_p)^*$. Since $f'' \in \mathcal{M}^\sigma$ and $v_1,\dots,v_p \in \mathcal{V}_0$, we have $Sp_\sigma(w_4) \subset Sp_\sigma(v) - \sum_{j=1}^p Sp_\sigma(v_j) \subset (0,n\varepsilon)$ and since $f' \in \mathcal{M}^\sigma$ and $w_1 \in \mathcal{V}_0$, we have $w_4 \in \mathcal{V}_0$. Again by the induction hypothesis, we can find $v_{p+1},\dots,v_{p+q} \in \mathcal{V}_1$ such that $0 \neq v_{p+1} \dots v_{p+q} \prec w_4 = w_1f'$. Since $\mathbf{r}(v_{p+1} \dots v_{p+q}) \leq f' = \mathbf{l}(v_1 \dots v_p)$, we have $0 \neq (v_{p+1} \dots v_{p+q})(v_1 \dots v_p) \prec w_1w_3 = w_1w_2 \prec v$.

(II) *If* $u \in \mathcal{V}, Sp_\sigma(u) \subset (\varepsilon,+\infty)$, *and* $Sp_\sigma(u) - Sp_\sigma(u) \subset [-\varepsilon,\varepsilon]$, *then there exists* $0 \neq w \in \mathcal{V}_0$ *such that* $w \prec u$.

By Lemma 15.1, there exists $h \in \mathcal{L}^1(\mathbb{R})$ with *supp* $\hat{h}$ − *supp* $\hat{h} \subset [-\varepsilon/2,\varepsilon/2]$ such that $(\sigma_h(u))^*u \neq 0$. By Proposition 15.17, there exist $v \in \mathcal{V}$ and $a \in \mathcal{M}^\sigma$ such that $va = \sigma_h(u)$ and $Sp_\sigma(v) \subset Sp_\sigma(\sigma_h(u))$. Since $a^*v^*u = (\sigma_h(u))^*u \neq 0$, we have $v^*u \neq 0$ and, since $Sp_\sigma(v) \subset (supp\ \hat{h}) \cap Sp_\sigma(u)$, it follows that $v \in \mathcal{V}_0$. Then, $Sp_\sigma(v^*u) \subset (Sp_\sigma(u) - Sp_\sigma(u)) \cap (Sp\ \sigma) \subset [-\varepsilon,\varepsilon] \cap (Sp\ \sigma) = \{0\}$, hence $v^*u \in \mathcal{M}^\sigma$. Let $w = vv^*u$. We have $w \neq 0$ since $v^*w = v^*u \neq 0$, then $w \in \mathcal{V}_0$ since $v \in \mathcal{V}_0, v^*u \in \mathcal{M}^\sigma$ and $u, v \in \mathcal{V}$, and finally $w \prec u$ since $vv^* \in \mathcal{Z}(\mathcal{M}^\sigma)$.

(III) If $v_1, v_2 \in \mathcal{V}_1$, then $v_1^*v_2, v_1v_2^* \in \mathcal{M}^\sigma$.

We show that $v_1^*v_2 \in \mathcal{M}^\sigma$. We have $Sp_\sigma(v_1^*v_2) - Sp_\sigma(v_1^*v_2) \subset Sp_\sigma(v_2) - Sp_\sigma(v_1) + Sp_\sigma(v_1) - Sp_\sigma(v_2) \subset [-\varepsilon/2,\varepsilon/2] + [-\varepsilon/2,\varepsilon/2] = [-\varepsilon,\varepsilon]$. Since $(Sp\ \sigma) \cap [-\varepsilon,\varepsilon] = \{0\}$, it follows that either $Sp_\sigma(v_1^*v_2) = \{0\}$, that is, $v_1^*v_2 \in \mathcal{M}^\sigma$, or $Sp_\sigma(v_1^*v_2) \subset (\varepsilon,+\infty)$, or $Sp(v_2^*v_1) \subset (\varepsilon,+\infty)$. Thus, we have to show that the inclusion $Sp(v_1^*v_2) \subset (\varepsilon,+\infty)$ leads to a contradiction.

Consider $u = v_1^*v_2 \in \mathcal{V}$ with $Sp_\sigma(u) \subset (\varepsilon,+\infty)$ and $Sp_\sigma(u) - Sp_\sigma(u) \subset [-\varepsilon,\varepsilon]$. By property (II), there exists $0 \neq w \in \mathcal{V}_0$ with $w \prec u$. We have $v_1w \prec v_1u = v_1v_1^*v_2 \prec v_2$ and $ww^* \leq uu^* \leq v_1^*v_1$, hence $v_1w \neq 0$, contradicting $v_2 \in \mathcal{V}_1$.

Similarly, in order to show that $u = v_1v_2^* \in \mathcal{M}^\sigma$, it is sufficient to show that the inclusion $Sp_\sigma(u) \subset (\varepsilon,+\infty)$ leads to a contradiction. As above, we can find using property (II) $0 \neq w \in \mathcal{V}_0, w \prec u$, and we have $wv_2 \prec uv_2 = v_1v_2^*v_2 \prec v_1$ and $w^*w \leq u^*u \leq v_2v_2^*$, so that $wv_2 \neq 0$, contradicting $v_1 \in \mathcal{V}_1$.

(IV) $\mathcal{M} = \mathcal{R}\{\mathcal{M}^\sigma,\mathcal{V}_1\}$.

Let $x \in \mathcal{M}$. In order to show that $x \in \mathcal{R}\{\mathcal{M}^\sigma,\mathcal{V}_1\}$, we may assume that $Sp_\sigma(x) - Sp_\sigma(x) \subset [-\varepsilon/2,\varepsilon/2]$ (see Lemma 15.1). Since, by assumption, $(Sp\ \sigma) \cap [-\varepsilon,\varepsilon] = \{0\}$, we may assume further that $Sp_\sigma(x) \subset (\varepsilon,+\infty)$. Using Proposition 15.17, we can write $x = va$ with $v \in \mathcal{V}_0,\ a \in \mathcal{M}^\sigma$. Thus it remains to be shown that $\mathcal{V}_0 \subset \mathcal{R}\{\mathcal{M}^\sigma, V_1\}$.

Let $v \in \mathcal{V}_0$ and $e = v^*v$. From property (I), it follows that there exists a projection $f \in \mathcal{Z}(\mathcal{M}^\sigma), 0 \neq f \leq e$ with $vf \in \mathcal{R}\{\mathcal{M}^\sigma,\mathcal{V}_1\}$. By a standard maximality argument, we infer that $v = ve \in \mathcal{R}\{\mathcal{M}^\sigma,\mathcal{V}_1\}$.

(V) *If* $\{v_j\} \subset \mathcal{V}_1$ *is a maximal family with the property that* $v_jv_k^* = v_j^*v_k = 0\ (k \neq j)$, *then* $\sum_j v_j$ *is unitary.*

Let $e_j = v_j^*v_j, e_j' = v_jv_j^*, e = \sum_j e_j, e' = \sum_j e_j'$; note that $e_j, e_j', e, e' \in \mathcal{Z}(\mathcal{M}^\sigma)$.

Assume that $1 - e \neq 0$. Using the assumption and Theorem 16.6, we see that the action σ^{1-e} is not trivial, so that there exists $y \in (1-e)\mathcal{M}(1-e)$ with $Sp_\sigma(y) \neq \{0\}$. Since $(Sp\ \sigma) \cap [-\varepsilon,\varepsilon] = \{0\}$, we infer using Lemma 15.1 that there exists $x \in (1-e)\mathcal{M}(1-e)$ with $Sp_\sigma(x) \subset (\varepsilon,+\infty)$ and $Sp_\sigma(x) - Sp_\sigma(x) \subset [-\varepsilon/2,\varepsilon/2]$. Furthermore, using Proposition 15.17, we deduce that there exists $0 \neq v \in \mathcal{V}_0$ with $v^*v \leq 1-e, vv^* \leq 1-e$ and, finally, using property (I), we find an element $0 \neq w \in \mathcal{V}_1$ with $w^*w \leq 1-e$. It is clear that $v_jw^* = 0$ for all j. By property (III), it follows that $w^*v_j \in \mathcal{M}^\sigma$; then $\mathbf{r}(w^*v_j) \leq e_j, \mathbf{l}(w^*v_j) \leq 1-e \leq 1-e_j$, and, since $e_j \in \mathcal{Z}(\mathcal{M}^\sigma)$, we deduce that $w^*v_j = 0$. The existence of w contradicts the maximality of the family $\{v_j\}$. Hence $e = 1$.

Assuming that $1 - e' \neq 0$, we obtain similarly an element $0 \neq w \in \mathscr{V}_1$ with $ww^* \le 1 - e'$; we deduce that $v_j^* w = 0 = wv_j^*$ for all j, which contradicts the maximality of the family $\{v_j\}$. Hence $e' = 1$.

We are now in a position to finish the proof of the lemma. Let $\{v_j\} \subset \mathscr{V}_1$ be a maximal family with the property that $v_j v_k^* = v_j^* v_k = 0$, $(j \neq k)$, and let $u = \sum_j v_j$. By (V) we know that $u \in \mathscr{M}$ is unitary. Since $v_j \in \mathscr{V}$, we have $u\mathscr{M}^\sigma u^* = \mathscr{M}^\sigma$. Since $Sp_\sigma(v_j) \subset [\varepsilon, +\infty)$, it follows that $Sp_\sigma(u) \subset [\varepsilon, +\infty) \subset (0, +\infty)$. By (III), for every $v \in \mathscr{V}_1$ we have $v^* v_j \in \mathscr{M}^\sigma$, hence $v^* u \in \mathscr{M}^\sigma$, that is, $v \in u\mathscr{M}^\sigma$ and by (IV) this implies that $\mathscr{M} = \mathscr{R}\{\mathscr{M}^\sigma, u\}$. □

Note that $u^* \in \mathscr{M}$ is unitary, $\mathscr{M} = \mathscr{R}\{\mathscr{M}^\sigma, u\}, u^*\mathscr{M}^\sigma u = \mathscr{M}^\sigma$, and $Sp_\sigma(u^*) \subset (0, 1)$.

30.3 Theorem. *Let $\mathscr{M}$ be a factor of type* III$_0$. *There exists a discrete decomposition $(\mathscr{N}, \theta, \tau)$ of type* III$_0$ *such that $\mathscr{M} \approx \mathscr{R}(\mathscr{N}, \theta)$. If $(\mathscr{N}_1, \theta_1, \tau_1)$ and $(\mathscr{N}_2, \theta_2, \tau_2)$ are two discrete decompositions of type* III$_0$ *such that $\mathscr{R}(\mathscr{N}_1, \theta_1) \approx \mathscr{R}(\mathscr{N}_2, \theta_2)$, there exist nonzero projections $e_1 \in \mathscr{Z}(\mathscr{N}_1)$ and $e_2 \in \mathscr{Z}(\mathscr{N}_2)$ such $(\mathscr{N}_1 e_1, \theta_{1e_1}) \approx (\mathscr{N}_2 e_2, \theta_{2e_2})$.*

Proof. By Corollary 29.6, there exists a lacunary weight φ on $\mathscr{M}$ of infinite multiplicity. Since $\mathscr{M}$ is of type III, the action $\sigma = \sigma^\varphi : \mathbb{R} \to Aut(\mathscr{M})$ is not outer conjugate to the trivial action. By Lemma 30.2, there exists a unitary element $u \in \mathscr{M}$ such that $u\mathscr{M}^\varphi u^* = \mathscr{M}^\varphi$, $\mathscr{M} = \mathscr{R}\{\mathscr{M}^\varphi, u\}$, and $Sp_\sigma(u) \subset (0, 1)$. We define $\mathscr{N} = \mathscr{M}^\varphi, \theta = \mathrm{Ad}(u)|\mathscr{N}$, and $\tau = \varphi|\mathscr{N}$.

Since φ is strictly semifinite, there exists a faithful normal conditional expectation $P : \mathscr{M} \to \mathscr{N}$, τ is an n.s.f. trace on $\mathscr{N}$ and $\varphi = \tau \circ P$. Let A be the unique nonsingular positive self-adjoint operator affiliated to $\mathscr{Z}(\mathscr{N})$ such that $\tau \circ \theta = \tau_A$. Then, by 27.4.(2), we have $\sigma_s^\varphi(u) = uA^{is}$ $(s \in \mathbb{R})$, and for every $f \in \mathscr{L}^1(\mathbb{R})$ we obtain (27.1.(1)) $\sigma_f^\varphi(u) = u\hat{f}(A)$. It follows that $Sp(A) = Sp_\sigma(u) \subset (0, 1)$, so that there is $0 < \lambda < 1$, such that $A \le \lambda$, and $\tau \circ \theta \le \lambda\tau$.

It follows that the action $\theta : \mathbb{Z} \to Aut(\mathscr{N})$ is properly outer, that is, $p(\theta^n) = 0$ for every $n \neq 0$. Indeed, let $n > 0$ and $e \in \mathscr{Z}(\mathscr{N})$ be a projection such that $\theta^n(e) = e$ and $\theta^n|e\mathscr{N}e \in Int(e\mathscr{N}e)$. Then, for every $x \in (e\mathscr{N}e)^+$, we have $\tau(x) = \tau(\theta^n(x)) \le \lambda^n\tau(x)$, which is not possible, since $\lambda < 1$ and $\tau|e\mathscr{N}e$ is an n.s.f. trace on $e\mathscr{N}e$.

Using Proposition 22.2 and the last remark in Section 22.2, we conclude that $\mathscr{M} \approx \mathscr{R}(\mathscr{N}, \theta)$.

The W^*-algebra $\mathscr{N} = \mathscr{M}^\varphi$ is semifinite and properly infinite, since φ is strictly semifinite and of infinite multiplicity.

The *-automorphism θ acts ergodically on $\mathscr{Z}(\mathscr{N})$ since, if $z \in \mathscr{Z}(\mathscr{N})$ and $uz = zu$, then $z \in \mathscr{Z}(\mathscr{R}\{\mathscr{N}, u\}) = \mathscr{Z}(\mathscr{M})$, and $\mathscr{M}$ is a factor.

The center $\mathscr{Z}(\mathscr{N})$ of $\mathscr{N}$ is diffuse since otherwise, by Proposition 16.10 and Theorem 5.1, we could find an n.s.f. weight ψ on $\mathscr{M}$ with $Sp\, \sigma^\psi = \Gamma(\sigma^\psi) = \Gamma(\sigma^\varphi) = \{1\}$ (as $\mathscr{M}$ is of type III$_0$), that is, σ^ψ would be the trivial action, which contradicts the fact that $\mathscr{M}$ is of type III.

We now prove that $\mathscr{N}$ is of type II$_\infty$. Since the splitting of $\mathscr{N}$ into types I$_\infty$ and II$_\infty$ is given by a θ-invariant projection in $\mathscr{Z}(\mathscr{N})$ and since θ acts ergodically on $\mathscr{Z}(\mathscr{N})$, it is sufficient to assume that $\mathscr{N}$ is of type I$_\infty$ and reach a contradiction. If $\mathscr{N}$ is of type I$_\infty$, then $\mathscr{N} = \mathscr{Z}(\mathscr{N}) \bar{\otimes} \mathscr{F}_\infty$ where, as usually, $\mathscr{F}_\infty$ stands for the countably decomposable type I$_\infty$ factor. Consider the *-automorphism $\theta' \in Aut(\mathscr{N})$ defined by $\theta' = (\theta|\mathscr{Z}(\mathscr{N})) \bar{\otimes} \iota$. Then $\theta^{-1}\theta'$ acts identically on $\mathscr{Z}(\mathscr{N})$ and hence is inner ([L], 8.11), that is, $\theta' = \theta \circ \mathrm{Ad}(v)$ for some $v \in U(\mathscr{N})$. It follows that $\tau \circ \theta' \le \lambda\tau$ and $\mathscr{M} \approx \mathscr{R}(\mathscr{N}, \theta')$. By Corollary 23.9, there exists a continuous action $\alpha : \mathbb{T} \to Aut(\mathscr{N}^{\theta'})$ such that $(\mathscr{N}, \theta') \approx (\mathscr{R}(\mathscr{N}^{\theta'}, \alpha), \hat{\alpha})$ and, by the Takesaki duality theorem (19.5) we infer that $\mathscr{M} \approx \mathscr{R}(\mathscr{N}, \theta') \approx \mathscr{N}^{\theta'} \bar{\otimes} \mathscr{F}_\infty = \mathscr{Z}(\mathscr{N})^\theta \bar{\otimes} \mathscr{F}_\infty \bar{\otimes} \mathscr{F}_\infty \approx \mathscr{F}_\infty$, which is impossible, since $\mathscr{M}$ is of type III.

Thus, $(\mathcal{N}, \theta, \tau)$ is a discrete decomposition of type III_0.

Consider now two discrete decompositions of type III_0 $(\mathcal{N}_1, \theta_1, \tau_1)$ and $(\mathcal{N}_2, \theta_2, \tau_2)$ with $\mathcal{R}(\mathcal{N}_1, \theta_1) = \mathcal{R}(\mathcal{N}_2, \theta_2) = \mathcal{M}$ and, identifying $\mathcal{N}_1$ and $\mathcal{N}_2$ with the corresponding W^*-subalgebras of $\mathcal{M}$, let $P_1 : \mathcal{M} \to \mathcal{N}_1$ and $P_2 : \mathcal{M} \to \mathcal{N}_2$ be the canonical faithful normal conditional expectations. By Proposition 29.3, the dual weights $\varphi_1 = \tau_1 \circ P_1$ and $\varphi_2 = \tau_2 \circ P_2$ are lacunary, of infinite multiplicity on $\mathcal{M}$, and $\mathcal{N}_1 = \mathcal{M}^{\varphi_1}, \mathcal{N}_2 = \mathcal{M}^{\varphi_2}$. If $\tau_1 \circ \theta_1 \leq \lambda_1 \tau_1$ and $\tau_2 \circ \theta_2 \leq \lambda_2 \tau_2$ with $\lambda_1, \lambda_2 \in (0, 1)$, then we can find $\varepsilon > 0$ such that $\lambda_1 e^{\varepsilon} < 1$ and $\lambda_2 e^{\varepsilon} < 1$. By Proposition 29.7, there exist nonzero projections $e_1 \in \mathcal{Z}(\mathcal{N}_1), e_2 \in \mathcal{Z}(\mathcal{N}_2)$, and a *-isomorphism

$$\Phi : (e_1 \mathcal{M} e_1, \mathcal{N}_1 e_1) \to (e_2 \mathcal{M} e_2, \mathcal{N}_2 e_2) \tag{1}$$

such that, putting $\sigma_1 = \sigma^{\varphi_1}, \sigma_2 = \sigma^{\varphi_2}$, we have

$$Sp_{\sigma_2}(\Phi(x)) \subset Sp_{\sigma_1}(x) \cdot \exp[-\varepsilon, \varepsilon] \quad (x \in e_1 \mathcal{M} e_1). \tag{2}$$

By Proposition 29.4, we have $e_1 \mathcal{M} e_1 \approx \mathcal{R}(\mathcal{N}_1 e_1, \theta_{1e_1})$, the canonical image of $\mathcal{N}_1 e_1$ in the crossed product corresponding to the W^*-subalgebra $\mathcal{N}_1 e_1$ of $e_1 \mathcal{M} e_1$, and there is a unitary element $u_1 \in e_1 \mathcal{M} e_1$ such that

$$e_1 \mathcal{M} e_1 = \mathcal{R}\{\mathcal{N}_1 e_1, u_1\}, \tag{3}$$

$$\theta_{1e_1} = \mathrm{Ad}(u_1) | \mathcal{N}_1 u_1, \tag{4}$$

$$Sp_{\sigma_1}(u_1) \subset (0, \lambda_1). \tag{5}$$

Then $v_2 = \Phi(u_1) \in e_2 \mathcal{M} e_2$ is a unitary element and from (3) and (1) it follows that

$$e_2 \mathcal{M} e_2 = \mathcal{R}\{\mathcal{N}_2 e_2, v_2\}, \tag{6}$$

Using (4) and (1), we obtain

$$v_2 (\mathcal{N}_2 e_2) v_2^* = \mathcal{N}_2 e_2, \tag{7}$$

and from (5), (2) and the choice of ε, it follows that

$$Sp_{\sigma_2}(v_2) \subset (0, 1). \tag{8}$$

Since, by Proposition 29.4, $(\mathcal{N}_2 e_2, \theta_{2e_2}, \tau_{2e_2})$ is a discrete decomposition of type III_0 and $e_2 \mathcal{M} e_2 \approx \mathcal{R}(\mathcal{N}_2 e_2, \theta_{2e_2})$, using statement 29.3.(3), we infer from (6), (7), and (8) the existence of a unitary element $w_2 \in \mathcal{N}_2 e_2$ such that the unitary element $u_2 = w_2 v_2 \in e_2 \mathcal{M} e_2$ satisfies the equality $\theta_{2e_2} = \mathrm{Ad}(u_2) | \mathcal{N}_2 e_2$.

We have thus obtained a *-isomorphism $\Phi : (\mathcal{N}_1 e_1, \theta_{1e_1}) \to (\mathcal{N}_2 e_2, \theta'_2)$, where $\theta'_2 = \mathrm{Ad}(v_2) | \mathcal{N}_2 e_2$ and $(\theta'_2)^{-1} \theta_{2e_2} = \mathrm{Ad}(w) \in Int(\mathcal{N}_2 e_2)$. Using Theorem 23.13, we conclude that $(\mathcal{N}_1 e_1, \theta_{1e_1}) \approx (\mathcal{N}_2 e_2, \theta_{2e_2})$. □

A discrete decomposition $(\mathcal{N}, \theta, \tau)$ of type III_0 such that $\mathcal{R}(\mathcal{N}, \theta) \approx \mathcal{M}$ will be called a *discrete decomposition of the type* III_0 *factor* $\mathcal{M}$. For every nonzero projection $e \in \mathcal{Z}(\mathcal{N}), (\mathcal{N} e, \theta_e, \tau_e)$ is also a discrete decomposition of $\mathcal{M}$. The previous proof and Proposition 29.3 show that the lacunary weights of infinite multiplicity on $\mathcal{M}$ are those dual to the traces appearing in some discrete decomposition of $\mathcal{M}$.

30.4. In this section, we recapitulate the construction which leads to the discrete decomposition of type III factors $(0 \leq \lambda < 1)$ and establish the notation for the following sections, where we shall relate the continuous decomposition to the discrete decomposition. The results that follow have already been proved for type III$_0$ and can be easily checked, with similar arguments, also for type III$_\lambda$ with $\lambda \neq 0$.

Let $\mathscr{M}$ be a factor of type III$_\lambda$ $(0 \leq \lambda < 1)$. We choose a lacunary weight φ of infinite multiplicity on $\mathscr{M}$ (29.5, 29.6). Then

$$\mathscr{N} = \mathscr{M}^\varphi \tag{1}$$

is a type II$_\infty$ W^*-algebra (30.3). If $\lambda = 0$, then the center $\mathscr{Z}(\mathscr{N})$ of $\mathscr{N}$ is diffuse (30.3). If $\lambda > 0$, then we can choose φ so that $\mathscr{N}$ is a factor (30.1). In both cases, we have (29.9)

$$\mathscr{N}' \cap \mathscr{M} = \mathscr{Z}(\mathscr{N}). \tag{2}$$

There exists unique faithful normal conditional expectation (10.9, 10.17)

$$P : \mathscr{M} \to \mathscr{N}.$$

Also (10.9)

$$\tau = \varphi|\mathscr{N}$$

is an n.s.f. trace on $\mathscr{N}$ such that

$$\varphi = \tau \circ P. \tag{3}$$

There exists a unitary element $u \in \mathscr{M}$ such that (30.2)

$$\mathscr{M} = \mathscr{R}\{\mathscr{N}, u\}, \tag{4}$$

$$u\mathscr{N}u^* = \mathscr{N}, \tag{5}$$

$$Sp_{\sigma^\varphi}(u) \subset (0, 1). \tag{6}$$

Let $\lambda_0 = \sup Sp_{\sigma^\varphi}(u) < 1$. Then (30.3)

$$\theta = \mathrm{Ad}(u)|\mathscr{N} \in Aut(\mathscr{N}), \tag{7}$$

$$\tau \circ \theta \leq \lambda_0 \tau. \tag{8}$$

For each $n \in \mathbb{Z}$, there exists a unique nonsingular positive self-adjoint operator A_n, affiliated to $\mathscr{Z}(\mathscr{N})$, such that

$$\tau \circ \theta^n = \tau_{A_n}. \tag{9}$$

Put $A = A_1$. It is easy to check that

$$A_{m+n} = A_m \theta^{-m}(A_n), \tag{10}$$

$$n > 0 \Rightarrow A_n \leq \lambda_0^n < 1, \tag{11}$$

$$n < 0 \Rightarrow A_n \geq \lambda_0^n > 1, \tag{12}$$

$$\varphi \circ (\mathrm{Ad}(u))^n = \varphi_{A_n}, \tag{13}$$

$$\sigma_s^\varphi(u^n) = u^n A_n^{\mathrm{i}s}, \tag{14}$$

$$Sp(A) = Sp_{\sigma^\varphi}(u). \tag{15}$$

If $v \in \mathscr{M}$ is another unitary element satisfying conditions (4), (5), (6), and $\theta' = \mathrm{Ad}(v)|\mathscr{N}$, then (29.3.(3)) $u^*v \in \mathscr{N}$, hence $\theta^{-1}\theta' \in Int(\mathscr{N})$ and $\tau \circ \theta^n = \tau \circ \theta'^n$; thus, the corresponding operators A_n remain unchanged.

Recall that if H and K are positive self-adjoint operators, then we write $H < K$ if $H \leq K$ and $\mathbf{s}((K - H)\mathbf{s}(K)) = \mathbf{s}(K)$.

30.5. Let $\mathscr{M}$ be a factor of type III_λ $(0 \leq \lambda < 1)$ and let φ be a lacunary weight of infinite multiplicity on $\mathscr{M}$. We use the notation introduced in Section 30.4.

Theorem. *Let*

$$a \in \mathscr{N}^+ \textit{ with } A\mathbf{s}(a) \leq a < 1 \textit{ and } \psi = \varphi_a = \tau_a \circ P,$$
$$a' \in \mathscr{N}^+ \textit{ with } A\mathbf{s}(a') \leq a' < 1 \textit{ and } \psi' = \varphi_{a'} = \tau_{a'} \circ P.$$

If $v \in \mathscr{M}, vv^* = \mathbf{s}(\psi), v^*v = \mathbf{s}(\psi')$, and $\psi' = \psi_v$, then $v \in \mathscr{N}$. In particular, if $\varphi_a \approx \varphi_{a'}$ in $\mathscr{M}$, then $\tau_a \approx \tau_{a'}$ in $\mathscr{N}$.

Proof. Let $B = a + A(1 - \mathbf{s}(a)), B' = a' + A(1 - \mathbf{s}(a'))$. Then B and B' are nonsingular positive self-adjoint operators affiliated to $\mathscr{N}$ and $A \leq B < 1, A \leq B' < 1$. Since $\psi' = \psi_v$, we have (23.1) $[D\psi' : D\varphi]_t = v^*[D\psi : D\varphi]_t\sigma_t^\varphi(v)$ $(t \in \mathbb{R})$. Since $[D\psi' : D\varphi]_t = a'^{\mathrm{i}t}, [D\psi : D\varphi]_t = a^{\mathrm{i}t}$, and $\mathbf{s}(a') = \mathbf{s}(\psi') = v^*v, \mathbf{s}(a) = \mathbf{s}(\psi) = vv^*$, it follows that $a'^{\mathrm{i}t} = v^*a^{\mathrm{i}t}\sigma_t^\varphi(v)$ and $vB'^{\mathrm{i}t} = v\mathbf{s}(a')B'^{\mathrm{i}t} = va'^{\mathrm{i}t} = vv^*a^{\mathrm{i}t}\sigma_t^\varphi(v) = \mathbf{s}(a)a^{\mathrm{i}t}\sigma_t^\varphi(v) = B^{\mathrm{i}t}\mathbf{s}(a)\sigma_t^\varphi(v) = B^{\mathrm{i}t}\sigma_t^\varphi(\mathbf{s}(a)v) = B^{\mathrm{i}t}\sigma_t^\varphi(v)$, that is,

$$vB'^{\mathrm{i}t} = B^{\mathrm{i}t}\sigma_t^\varphi(v) \quad (t \in \mathbb{R}). \tag{1}$$

Since $\mathscr{M} = \mathscr{R}(\mathscr{N}, \theta)$, we can write $v = \sum_k x_k u^k$ with $x_k \in \mathscr{N}$; we have $B^{\mathrm{i}t}\sigma_t^\varphi(v) = B^{\mathrm{i}t}(\sum_k x_k\sigma_t^\varphi(u^k)) = B^{\mathrm{i}t}(\sum_k x_k u^k A_k^{\mathrm{i}t}) = \sum_k B^{\mathrm{i}t}x_k\theta^k(A_k)^{\mathrm{i}t}u^k$ and $vB'^{\mathrm{i}t} = \sum_k x_k u^k B'^{\mathrm{i}t} = \sum x_k\theta^k(B')^{\mathrm{i}t}u^k$. Consequently, it follows from (1) that $B^{\mathrm{i}t}x_k\theta^k(A_k)^{\mathrm{i}t} = x_k\theta^k(B')^{\mathrm{i}t}$, that is,

$$B^{\mathrm{i}t}x_k = x_k\theta^k(B'A_k^{-1})^{\mathrm{i}t} \quad (k \in \mathbb{Z}, t \in \mathbb{R}). \tag{2}$$

We choose and fix $k \neq 0$ and show that $x_k = 0$. Note that in general if u_1 and u_2 are unitary elements and $x = w|x|$ is the polar decomposition of some operator x such that $u_1x = xu_2$, then $u_1w = wu_2, u_2(w^*w) = (w^*w)u_2$ and $u_1(ww^*) = (ww^*)u_1$. Consequently, we can assume that the operator x_k appearing in (2) is a partial isometry such that $x_k^*x_k$ commutes with $\theta^k(B'A_k^{-1})^{\mathrm{i}t}$ and $x_kx_k^*$ commutes with $B^{\mathrm{i}t}$. If $k > 0$, then

$$B^{\mathrm{i}t}x_kx_k^* = x_k\theta^k(B'A_k^{-1})^{\mathrm{i}t}x_k \quad (t \in \mathbb{R}) \tag{3}$$

and if $k < 0$, then

$$x_k^*x_k\theta^k(B')^{\mathrm{i}t} = x_k^*(B\theta^k(A_k))^{\mathrm{i}t} \quad (t \in \mathbb{R}) \tag{4}$$

since $B^{it}\theta^k(A_k)^{it}x_k = x_k\theta^k(B')^{it}$. On the other hand, if $k > 0$, then $0 \le B < 1$ and $A_k \le A \le B'$, so that

$$0 \le B \le 1 \text{ and } \theta^k(B'A_k^{-1}) \ge 1; \tag{5}$$

if $k < 0$, then $0 \le B' \le 1$ and $1 = A_0 = A_{-k+k} = A_{-k}\theta^k(A_k)$, i.e. $\theta^k(A_k) = A_{-k}^{-1}, \theta^k(A_k)^{-1} = A_{-k} \le A \le B$, and hence

$$0 \le \theta^k(B') \le 1 \text{ and } B\theta^k(A_k) \ge 1. \tag{6}$$

Note that, in general, if H and K are nonsingular positive self-adjoint operators such that $0 \le H \le 1$ and $K \ge 1$, then the functions

$$\{z \in \mathbb{C}; \mathrm{Im}(z) \le 0\} \ni z \mapsto H^{iz}$$
$$\{z \in \mathbb{C}; \mathrm{Im}(z) \ge 0\} \ni z \mapsto K^{iz}$$

are analytic and bounded, hence an identity of the form $xH^{it} = yK^{it}$ $(t \in \mathbb{R})$ leads to a bounded entire analytic function which is necessarily constant (by Liouville's theorem), so that $xH^{it} = 0$ and $x = 0$.

Therefore, it follows from (3) and (5) that $x_k = 0$ for all $k > 0$, and it follows from (4) and (6) that $x_k = 0$ for all $k < 0$, so that $v = x_0 \in \mathcal{N}$. □

Corollary. *Let $a \in \mathcal{N}$ with $\mathbf{As}(a) \le a < 1$. Then*

$$\mathcal{M}^{\varphi_a} = \mathcal{N}^{\tau_a} = \{x \in \mathcal{N}; xa = ax\}.$$

In particular, every subweight of φ_a is of the form φ_b for some $b \in \mathcal{N}$ with $\mathbf{As}(b) \le b < 1$.

Proof. Put $\psi = \varphi_a, \mu = \tau_a$. Let $v \in \mathcal{M}^\psi$ be unitary. We have $vv^* = v^*v = \mathbf{s}(\psi)$ and $\psi = \psi_v$ by Proposition 2.21. According to the previous theorem, it follows that $v \in \mathcal{N}$. Hence $\mathcal{M}^\psi \subset \mathcal{N}$.

Since $\psi = \mu \circ P$, for $x \in \mathcal{N}$ we have (11.9) $\sigma_t^\psi(x) = \sigma_t^\mu(x) = a^{it}xa^{-it}$ $(t \in \mathbb{R})$. Thus, $\mathcal{M}^\psi = \mathcal{N}^\mu = \{x \in \mathcal{N}; xa = ax\}$.

A subweight of ψ is of the form ψ_e for some projection $e \in \mathcal{M}$ and is hence of the form φ_b with $b = ae \in \mathcal{N}, \mathbf{As}(b) \le b < 1$. □

30.6. Recall that the operators A_n are affiliated to $\mathcal{Z}(\mathcal{N})$ and that we have $A_n \le \lambda_0 A_{n-1}$.

We now show that *for every positive self-adjoint operator D affiliated to $\mathcal{N}$ there exists a sequence of mutually orthogonal projections $\{e_n\}_{n\in\mathbb{Z}} \subset \{D\}' \cap \mathcal{N}$, uniquely determined, such that*

$$\sum_n e_n = \mathbf{s}(D), \tag{1}$$

$$e_nA_n \le e_nD < e_nA_{n-1}. \tag{2}$$

Indeed, let $e_n = \mathbf{s}((A_{n-1} - D)^+) - \mathbf{s}((A_n - D)^+)$, $(n \in \mathbb{Z})$. Then e_n are clearly mutually orthogonal projections and

$$\sum_n e_n = \lim_{n\to-\infty} \mathbf{s}((A_n - D)^+) - \lim_{n\to+\infty} \mathbf{s}((A_n - D)^+).$$

If $e \in \mathcal{N}$ is a projection such that $e \le \mathbf{s}((A_n - D)^+)$ for all $n > 0$, then $e(A_n - D) \ge 0$ and $eD \le eA_n \le \lambda_0^n e \to 0$, hence $e \le 1 - \mathbf{s}(D)$. Consequently, $\lim_{n\to+\infty} \mathbf{s}((A_n - D)^+) = 1 - \mathbf{s}(D)$. On the other hand, if $e \ge \mathbf{s}((A_n - D)^+)$ for every $n < 0$, then $1 - e \le 1 - \mathbf{s}((A_n - D)^+) = 1 - \mathbf{s}(A_n - D) + \mathbf{s}((A_n - D)^-)$, and $(1 - e)(A_n - D) \le 0$, or $(1 - e)D \ge (1 - e)A_n \ge \lambda_0^n(1 - e) \to +\infty$, and hence $1 - e = 0, e = 1$. Consequently, $\lim_{n\to-\infty} \mathbf{s}((A_n - D^+) = 1$. Hence

$$\sum_n e_n = \mathbf{s}(D).$$

Since $e_n \le \mathbf{s}((A_{n-1} - D)^+)$, we have $e_n(A_{n-1} - D) \ge 0$, that is, $e_nD \le e_nA_{n-1}$. If $e \le e_n$ and $e(A_{n-1} - D) = 0$, then $e\mathbf{s}(A_{n-1} - D) = 0$, so that $e\mathbf{s}((A_{n-1} - D)^+) = 0$ and $e = 0$. Therefore,

$$e_nD < e_nA_{n-1}.$$

Since $e_n \le 1 - \mathbf{s}((A_n - D)^+)$, we have $e_n(A_n - D) \le 0$, hence

$$e_nA_n \le e_nD.$$

Thus, *if* $D \neq 0$, *then there exist* $n \in \mathbb{Z}$ *and a nonzero projection* $e \in \{D\}' \cap \mathcal{N}$, *such that*

$$eA_n \le eD < eA_{n-1}. \tag{3}$$

Consider now a normal semifinite weight μ on $\mathcal{N}$. There exists a unique positive self-adjoint operator D_μ affiliated to $\mathcal{N}$, such that

$$\mu = \tau_{D_\mu}. \tag{4}$$

For each $k \in \mathbb{Z}$, we have a normal semifinite weight $\mu_k = \mu \circ \theta^k$ on $\mathcal{N}$ and a normal semifinite weight $\psi_k = \mu_k \circ P$ on $\mathcal{M}$. Note that $\mu = \mu_0$ and write $\psi = \psi_0 = \mu \circ P$. Then

$$\psi = \varphi_{D_\mu}. \tag{5}$$

For every $k \in \mathbb{Z}$, we have $\psi_k \approx \psi$, more precisely

$$\psi_k = \psi \circ \mathrm{Ad}(u^k) \quad (k \in \mathbb{Z}). \tag{6}$$

Indeed, for every $x \in \mathcal{M}^+$, we have $\psi_k(x) = \mu_k(P(x)) = \mu(\theta^k(P(x))) = \mu(u^kP(x)u^{-k}) = \mu(P(u^kxu^{-k})) = (\psi \circ \mathrm{Ad}(u^k))(x)$.

Also,

$$D_{\mu_k} = A_k\theta^{-k}(D_\mu) \quad (k \in \mathbb{Z}). \tag{7}$$

Indeed, for every $x \in \mathcal{N}^+$, we have

$$\mu_k(x) = \mu(\theta^k(x)) = \tau(D_\mu\theta^k(x)) = \tau(\theta^k(\theta^{-k}(D_\mu)x)) = \tau(A_k\theta^{-k}(D_\mu)x).$$

Finally, for every $k \in \mathbb{Z}$, we have

$$\mathbf{s}(D_\mu)A_n \le D_\mu < \mathbf{s}(D_\mu)A_{n-1} \Rightarrow \mathbf{s}(D_{\mu_k})A_{k+n} \le D_{\mu_k} < \mathbf{s}(D_{\mu_k})A_{k+n-1}. \tag{8}$$

Indeed, the desired conclusion can be obtained by applying θ^{-k} to the left hand side, multiplying by A_k and using the fact that

$$A_k\theta^{-k}(A_m) = A_{k+m}.$$

By (3), we can show inductively that the inequalities in (8) are satisfied, starting with an arbitrary nonzero normal semifinite weight μ and then considering a convenient subweight of μ.

Also, the previous arguments show that if μ is a normal semifinite weight on $\mathcal{N}$ such that

$$\mathbf{s}(D_\mu)A_n \le D_\mu < \mathbf{s}(D_\mu)A_{n-1}$$

for some $n \in \mathbb{Z}$, then, for any $m \in \mathbb{Z}, \nu = \mu_{m-n}$ is a normal semifinite weight on $\mathcal{N}, \nu \circ P \approx \mu \circ P$ on $\mathcal{M}$ and

$$\mathbf{s}(D_\nu)A_m \le D_\nu < \mathbf{s}(D_\nu)A_{m-1}.$$

30.7. Let $\mathcal{M}$ be a factor of type III$_\lambda$ $(0 \le \lambda < 1)$ and φ a lacunary weight of infinite multiplicity on $\mathcal{M}$. We use the notation introduced in Section 30.4.

Theorem. *For every normal semifinite weight ψ on $\mathcal{M}$, there exists $a \in \mathcal{N}$ with $A\mathbf{s}(a) \le a < 1$ such that $\psi \approx \varphi_a = \tau_a \circ P$.*

Proof. (I) We assume first that the weight ψ is of infinite multiplicity and show that there exists a nonzero positive self-adjoint operator D, affiliated to $\mathcal{N}$, such that $\varphi_D \precsim \psi$.

Since 1 is an isolated point in $S(\mathcal{M}) \cap \mathbb{R}_*^+$, we can assume using Theorem 28.3.(2) that $(Sp\ \sigma^\psi) \cap \exp((-2\varepsilon, -\varepsilon) \cup (\varepsilon, 2\varepsilon)) = \emptyset$ for some $\varepsilon > 0$. By Proposition 15.12, it follows that there exists an invertible positive element $c \in \mathscr{Z}(\mathcal{M}^\psi)$ such that the weight ψ_c is lacunary. Then $\mathcal{M}^\psi \subset \mathcal{M}^{\psi_c}$, and so ψ_c is also of infinite multiplicity.

By Proposition 29.7, there exist nonzero projections $e \in \mathscr{Z}(\mathcal{M}^\varphi), f \in \mathscr{Z}(\mathcal{M}^{\psi_c})$ and $v \in \mathcal{M}$ such that $e = v^*v, f = vv^*$ and

$$v(\mathcal{M}^{\varphi_e})v^* = \mathcal{M}^{(\psi_c)_f} = \mathcal{M}^{\psi_{cf}}.$$

We have $c^{-1} \in \mathcal{M}^{\psi_c}$ and $\psi = (\psi_c)_{c^{-1}}$; as $f \in \mathscr{Z}(\mathcal{M}^{\psi_c}), f$ commutes with c^{-1} and so, $\sigma_t^\psi(f) = c^{-it}\sigma_t^{\psi_c}(f)c^{it} = f (t \in \mathbb{R})$, that is, $f \in \mathcal{M}^\psi$. It is therefore meaningful to consider the weight ψ_f and we have $\psi_f = ((\psi_c)_{c^{-1}})_f = (\psi_c)_{c^{-1}f}$, that is,

$$\psi_f = (\psi_c)_b,$$

where $b = c^{-1}f$. Since $vv^* = f \in \mathcal{M}^{\psi_f}$, we may also consider the weight

$$(\psi_f)_v = \psi_v \precsim \psi,$$

and we have

$$\psi_{fv} = ((\psi_c)_b)_v = (\psi_c)_{bv} = ((\psi_c)_v)_{v^*bv}.$$

Since

$$\mathcal{M}^{(\psi_c)_v} = v^*(\mathcal{M}^{\psi_{cf}})v = \mathcal{M}^{\varphi_e},$$

there exists by Corollary 29.11 a positive self-adjoint operator H, affiliated to $\mathcal{Z}(\mathcal{M}^{\varphi_e})$, such that

$$(\varphi_c)_v = (\varphi_e)_H.$$

It follows that the positive self-adjoint operator $D = eHv^*bv$ is affiliated to $\mathcal{N} = \mathcal{M}^{\varphi}$ and $\varphi_D = \psi_v \lesssim \psi$.

(II) Using the results of Section 30.6, it follows from (I) that if ψ is of infinite multiplicity, then there is an $a \in \mathcal{N}$ with $A\mathbf{s}(a) \leq a < 1$ such that $\varphi_a \lesssim \psi$.

(III) We now prove the same result, but without assuming that ψ is of infinite multiplicity.

We have $\psi \lesssim \check{\psi}$ with $\check{\psi}$ of infinite multiplicity (23.15). Thus, there exists a projection $f \in \mathcal{M}^{\check{\psi}}$ such that $\psi \approx (\check{\psi})_f$ and, by (II), there exists $a \in \mathcal{N}$ with $A\mathbf{s}(a) \leq a < 1$ such that $\varphi_a \lesssim \check{\psi}$; consequently, $\varphi_a \approx (\check{\psi})_e$ for some projection $e \in \mathcal{M}^{\check{\psi}}$. By the comparison theorem ([L], 4.6), there exists a projection $p \in \mathcal{Z}(\mathcal{M}^{\check{\psi}})$ such that $pe \prec pf$ and $(1-p)f \prec (1-p)e$ in $\mathcal{M}^{\check{\psi}}$.

If $p \neq 0$, there exists $0 \neq v \in \mathcal{M}^{\check{\psi}}$ with $vv^* = pe$ and $v^*v \leq pf$. Using the results of Section 2.21, we obtain $(\check{\psi}_{pe})_v = (\check{\psi}_{pf})_{v^*v} \lesssim \check{\psi}_{pf} \lesssim \psi$ and $(\check{\psi}_{pe})_v \approx \psi_{ep} \lesssim \varphi_a$ so that, by Corollary 30.5, we have $(\check{\psi}_{pe})_v \approx \varphi_b$ for some $b \in \mathcal{N}$ with $A\mathbf{s}(b) \leq b < 1$, and thus $\varphi_b \lesssim \psi$.

If $p = 0$, then $f \prec e$ in $\mathcal{M}^{\check{\psi}}$, and there exists $v \in \mathcal{M}^{\check{\psi}}$ such that $vv^* = f$ and $v^*v \leq e$. By the results of Section 2.21, we have $\psi \approx (\check{\psi})_f \approx ((\check{\psi})_f)_v = ((\check{\psi})_e)_{v^*v} \lesssim \varphi_e$ and, again by Corollary 30.5, $\psi \approx \varphi_b$ for some $b \in \mathcal{N}$ with $A\mathbf{s}(b) \leq b < 1$.

(IV) The above part of the proof shows that if $\{e_n\}$ is a maximal family of mutually orthogonal nonzero projections in $\mathcal{M}$ such that for each n there exists $a_n \in \mathcal{N}$ with $A\mathbf{s}(a_n) \leq a_n < 1$ and $\psi_{e_n} = \varphi_{a_n}$, then $\sum_n e_n = \mathbf{s}(\psi)$. The supports $\mathbf{s}(a_n) = \mathbf{s}(\varphi_{a_n}) = \mathbf{s}(\psi_{e_n}) = e_n$ are mutually orthogonal, hence $a = \sum_n a_n \in \mathcal{N}, A\mathbf{s}(a) \leq a < 1$, and $\psi = \varphi_a$. □

In view of the results presented in Section 30.6, Theorems 30.5 and 30.7 remain still valid when the condition $A\mathbf{s}(a) \leq a < 1$ is replaced by the condition $A_n\mathbf{s}(a) \leq a < A_{n-1}\mathbf{s}(a)$ for some $n \in \mathbb{Z}$.

Corollary. *For every n.s.f. weight ψ on a factor $\mathcal{M}$ of type* III$_0$, *we have $\mathcal{Z}(\mathcal{M}^{\psi})' \cap \mathcal{M} = \mathcal{M}^{\psi}$.*

Proof. Since $\mathcal{M}$ is of type III$_0$, $\mathcal{Z}(\mathcal{N})$ is diffuse and θ acts ergodically on $\mathcal{Z}(\mathcal{N})$ (30.3), hence the action $\theta : \mathbb{Z} \to Aut(\mathcal{Z}(\mathcal{N}))$ is free (29.2), and $\mathcal{Z}(\mathcal{N})' \cap \mathcal{M} = \mathcal{N}$.

By the previous theorem, we may assume that $\psi = \varphi_a$ for some $a \in \mathcal{N}$ with $\mathbf{s}(a) = 1$ and $A \leq a < 1$. In this case, by Theorem 30.5 we have $\mathcal{M}^{\psi} = \{y \in \mathcal{N}; ya = ay\}$, whence $\mathcal{Z}(\mathcal{N}) \subset \mathcal{Z}(\mathcal{M}^{\psi})$. Let $x \in \mathcal{Z}(\mathcal{M}^{\psi})' \cap \mathcal{M}$. Then $x \in \mathcal{Z}(\mathcal{N})' \cap \mathcal{M} = \mathcal{N}$ and, as $a \in \mathcal{Z}(\mathcal{M}^{\psi})$, we have $xa = ax$, so that $x \in \mathcal{M}^{\psi}$. □

30.8. Consider now a fixed dominant weight ν on $\mathcal{N}$ which defines the (smooth) flow of weights $(\mathcal{Z}(\mathcal{N}^{\nu}), F^{\mathcal{N}})$ on $\mathcal{N}$ (24.1).

Recall that for every normal semifinite weight μ on $\mathcal{N}$ we denote by D_{μ} the unique positive self-adjoint operator affiliated to $\mathcal{N}$, such that $\mu = \tau_{D_{\mu}}$.

Let H and K be positive self-adjoint operators affiliated to $\mathscr{Z}(\mathcal{N})$ such that $H < K$. Put

$$\begin{aligned} W(H,K) &= \{\mu \in W^{\infty}_{int}(\mathcal{N}); \mathbf{s}(D_\mu)H \le D_\mu < \mathbf{s}(D_\mu)K\}, \\ [H,K) &= \bigvee \{c_\nu(\mu); \mu \in W(H,K)\} \in Proj(\mathscr{Z}(\mathcal{N}^\nu)). \end{aligned}$$

It is easy to check that

$$\begin{aligned} &\mu' \in W^{\infty}_{int}(\mathcal{N}), \mu' \lesssim \mu \in W(H,K) \Rightarrow \mu' \in W(H,K), \\ &\{\mu_n\} \subset W(H,K), \mu = \sum_n^{\oplus} \mu_n \Rightarrow \mu \in W(H,K) \end{aligned}$$

and, using these remarks, it follows for $\mu \in W^{\infty}_{int}(\mathcal{N})$ that

$$c_\nu(\mu) \le [H,K) \Leftrightarrow \mathbf{s}(D_\mu)H \le D_\mu < \mathbf{s}(D_\mu)K. \tag{1}$$

In particular, we show that, for the projections $[A_n, A_{n-1}) \in \mathscr{Z}(\mathcal{N}^\nu)$,

$$\sum_{n\in\mathbb{Z}} [A_n, A_{n-1}) = 1. \tag{2}$$

Indeed, let $n < m$. If $\mu \in W^{\infty}_{int}(\mathcal{N})$ and $c_\nu(\mu) \le [A_n, A_{n-1}), c_\nu(\mu) \le [A_m, A_{m-1})$, then, by (1), $A_n\mathbf{s}(D_\mu) \le D_\mu < A_{m-1}\mathbf{s}(D_\mu)$, while $A_{m-1} < A_n$ since $n < m$. It follows that $D_\mu = 0, \mu = 0$ and $c_\nu(\mu) = 0$. Hence

$$[A_n, A_{n-1})[A_m, A_{m-1}) = 0.$$

On the other hand, using the first result in Section 30.6, we see that every $\mu \in W^{\infty}_{int}(\mathcal{N}^\nu)$ can be written in the form $\mu = \sum^{\oplus}_{n\in\mathbb{Z}} \mu_n$ with $c_\nu(\mu_n) \le [A_n, A_{n-1})$, so that $c_\nu(\mu) \le \sum_{n\in\mathbb{Z}} [A_n, A_{n-1})$. This proves (2).

The *-automorphism $\theta \in Aut(\mathcal{N})$ defines a *-automorphism (25.1)

$$mod(\theta) \in Aut(F^{\mathcal{N}}) \subset Aut(\mathscr{Z}(\mathcal{N}^\nu)),$$

uniquely determined, such that $[mod(\theta)](c_\nu(\mu)) = c_\nu(\mu \circ \theta^{-1})$ for every $\mu \in W_{ns}(\mathcal{N})$.

Using the definitions of $mod(\theta)$ and $[A_n, A_{n-1})$, and assertion 30.6.(8) with $k = -1$, we get

$$[mod(\theta)]([A_n, A_{n-1})) = [A_{n-1}, A_{n-2}). \tag{3}$$

Furthermore, by (2) and (3), it follows that the mappings

$$\begin{aligned} \Phi_n &: \mathscr{Z}(\mathcal{N}^\nu)^{mod(\theta)} \ni z \mapsto z[A_n, A_{n-1}) \in \mathscr{Z}(\mathcal{N}^\nu)[A_n, A_{n-1}) \\ \Psi_n &: \mathscr{Z}(\mathcal{N}^\nu)[A_n, A_{n-1}) \ni z \mapsto \sum_{n\in\mathbb{Z}} [mod(\theta)]^n(z) \in \mathscr{Z}(\mathcal{N}^\nu)^{mod(\theta)} \end{aligned}$$

are *-isomorphisms, inverse to one another. Thus,

$$\mathscr{Z}(\mathcal{N}^\nu)^{mod(\theta)} \approx \mathscr{Z}(\mathcal{N}^\nu)[A_n, A_{n-1}). \tag{4}$$

Note that for $\mu \in W_{ns}(\mathcal{N})$ and $\psi = \mu \circ P \in W_{ns}(\mathcal{M})$ we have

$$\psi \in W_{int}(\mathcal{M}) \Leftrightarrow \mu \in W_{int}(\mathcal{N}). \tag{5}$$

Indeed, if $\mu \in W_{int}(\mathcal{N})$, there exists a net $\{y_i\} \subset \mathcal{N}^+$ with $\int \sigma_t^\mu(y_t)\mathrm{d}t < +\infty$, such that $y_i \overset{w}{\to} 1$. Since $\sigma_t^\psi | \mathcal{N} = \sigma_t^\mu$ $(t \in \mathbb{R})$, the integrability of ψ follows using the net. Conversely, if $\psi \in W_{int}(\mathcal{M})$, then there exists a net $\{x_i\} \subset \mathcal{M}^+$ with $\int \sigma_t^\psi(x_t)\mathrm{d}t < +\infty$, such that $x_i \overset{w}{\to} 1$. Then $y_i = P(x_i) \in \mathcal{N}^+, y_i \overset{w}{\to} 1$, and (10.5) $\int \sigma_t^\mu(y_i)\mathrm{d}t = \int \sigma_t^\mu(P(x_i))\mathrm{d}t = \int P(\sigma_t^\psi(x_i)\mathrm{d}t = P\left(\int \sigma_t^\psi(x_i)\mathrm{d}t\right) < +\infty$, whence $\mu \in W_{int}(\mathcal{N})$.

The previous remark and Theorems 30.7, 30.5, show that, given an arbitrary $n \in \mathbb{Z}$, every weight $\psi \in W_{int}^\infty(\mathcal{M})$ is equivalent to a weight of the form $\tau_a \circ P$ with $a \in \mathcal{N}, A_n\mathbf{s}(a) \leq a < A_{n-1}\mathbf{s}(a)$, and $\tau_a \in W_{int}^\infty(\mathcal{N})$; moreover the equivalence class of the weight τ_a with these properties depends only on the equivalence class of the weight ψ.

We now consider a dominant weight ω on $\mathcal{M}$ which defines the (smooth) flow of weights $(\mathcal{Z}(\mathcal{M}^\omega), F^{\mathcal{M}})$ on $\mathcal{M}$ (24.1).

Define a mapping

$$I : Proj(\mathcal{Z}(\mathcal{N}^\nu)) \to Proj(\mathcal{Z}(\mathcal{M}^\omega))$$

by $I(c_\nu(\mu)) = c_\omega(\mu \circ P)$ $(\mu \in W_{int}^\infty(\mathcal{N}))$.

This mapping is well defined and increasing, since if $\mu, \mu' \in W_{int}^\infty(\mathcal{N})$ and $c_\nu(\mu) \leq c_\nu(\mu')$, then (24.4) $\mu \lesssim \mu'$ in $\mathcal{N}$, so that $\mu \circ P \lesssim \mu' \circ P$ in $\mathcal{M}$ and hence $c_\omega(\mu \circ P) \leq c_\omega(\mu' \circ P)$. Similarly, one shows that this mapping is completely additive.

For $n \in \mathbb{Z}$, the restriction of the mapping I to $[A_n, A_{n-1})$ is bijective, so that it defines a *-isomorphism

$$I : \mathcal{Z}(\mathcal{N}^\nu)[A_n, A_{n-1}) \to \mathcal{Z}(\mathcal{M}^\omega).$$

Indeed, this follows from the fact that every weight $\psi \in W_{int}^\infty(\mathcal{M})$ can be written in the form $\psi = \mu \circ P$ for some $\mu \in W_{int}(\mathcal{M})$ with $c_\nu(\mu) \leq [A_n, A_{n-1})$; in this case we have (30.5) $\mathcal{M}^\mu = \mathcal{M}^\psi$, hence $\mu \in W_{int}^\infty(\mathcal{M})$, and the equivalence class of μ depends only on that of ψ.

Also, we have

$$I \circ mod(\theta) = I.$$

Indeed, for $\mu \in W_{int}^\infty(\mathcal{N})$ we have (30.6.(6)) $I([mod(\theta)](c_\nu(\mu))) = I(c_\nu(\mu \circ \theta^{-1})) = c_\omega(\mu \circ \theta^{-1} \circ P) = c_\omega(\mu \circ P \circ \mathrm{Ad}(u^{-1})) = c_\omega(\mu \circ P) = I(c_\nu(\mu))$.

It follows that $J(z) = I(\Phi_u(z))$ $(z \in \mathcal{Z}(\mathcal{N}^\nu)^{mod(\theta)})$ defines a *-isomorphism

$$J : \mathcal{Z}(\mathcal{N}^\nu)^{mod(\theta)} \to \mathcal{Z}(\mathcal{M}^\omega),$$

which is independent of $n \in \mathbb{Z}$. Moreover,

$$JF_\lambda^{\mathcal{N}} = F_\lambda^{\mathcal{M}} J \quad (\lambda \in \mathbb{R}_*^+),$$

since for $\mu \in W^{\infty}_{int}(\mathcal{M})$ and $\lambda \in \mathbb{R}^+_*$ we have $I(F^{\mathcal{N}}_\lambda(c_\nu(\mu))) = I(c_\nu(\lambda\mu)) = c_\omega((\lambda\mu)\circ P) = c_\omega(\lambda(\mu\circ P)) = F^{\mathcal{M}}_\lambda c_\omega(\mu\circ P) = F^{\mathcal{M}}_\lambda I(c_\nu(\mu))$.

Thus, we obtain the following result for the smooth flows of weights $(\mathcal{P}_{\mathcal{M}}, F^{\mathcal{M}})$ and $(\mathcal{P}_{\mathcal{N}}, F^{\mathcal{N}})$.

Theorem. $(\mathcal{P}_{\mathcal{M}}, F^{\mathcal{M}}) \approx ((\mathcal{P}_{\mathcal{N}})^{mod(\theta)}, F^{\mathcal{N}})$.

A similar result can be obtained for the global flow of weights (24.9).

30.9. We mention the following result concerning the fundamental homomorphism $mod : Out(\mathcal{M}) \to Aut(F^{\mathcal{M}})$.

Proposition. *Let $\mathcal{M}$ be a factor of type* III$_\lambda$ $(0 \le \lambda < 1)$ *with separable predual and let $\sigma \in Aut(\mathcal{M})$. The following statements are equivalent:*

(i) *$mod(\sigma) = \iota$;*

(ii) *there exist $\sigma' \in Aut(\mathcal{M})$ with $\sigma'/Int(\mathcal{M}) = \sigma/Int(\mathcal{M})$ and an n.s.f. weight ψ on $\mathcal{M}$ such that $\psi\circ\sigma' = \psi$ and $\sigma'|\mathcal{Z}(\mathcal{M}^\psi) = \iota$;*

(iii) *for each $\varepsilon > 1$ with $S(\mathcal{M}) \cap [\varepsilon^{-1}, \varepsilon] = \{1\}$ there exist $\sigma' \in Aut(\mathcal{M})$ with $\sigma'/Int(\mathcal{M}) = \sigma/Int(\mathcal{M})$ and a faithful normal state ψ on $\mathcal{M}$ such that $\psi\circ\sigma' = \psi, \sigma'|\mathcal{Z}(\mathcal{M}^\psi) = \iota$ and $Sp(\Delta_\psi) \cap [\varepsilon^{-1}, \varepsilon] = \{1\}$.*

Proof. (iii) $\Rightarrow$ (ii). Obvious.

(ii) $\Rightarrow$ (i). Assume that there exists an n.s.f. weight ψ on $\mathcal{M}$ such that $\psi\circ\sigma = \psi$ and $\sigma|\mathcal{Z}(\mathcal{M}^\psi) = \iota$. Since $\mathcal{M}$ is properly infinite, modifying σ if necessary by an inner *-automorphism, we may assume that $(\mathcal{M}, \sigma) \approx (\mathcal{M} \bar{\otimes} \mathcal{F}_\infty, \sigma \bar{\otimes} \iota)$ (20.14). Let ω_0 be the n.s.f. weight on $\mathcal{F}_\infty = \mathcal{B}(\mathcal{L}^2(\mathbb{R}))$ with $[D\omega_0 : D\, tr]_t = \boldsymbol{\rho}(t)$ $(t \in \mathbb{R})$, where, as usual, $\boldsymbol{\rho}$ stands for the right regular representation of $\mathbb{R}$. Put $\tilde{\mathcal{M}} = \mathcal{M} \bar{\otimes} \mathcal{F}_\infty \bar{\otimes} \mathcal{F}_\infty, \tilde{\sigma} = \sigma \bar{\otimes} \iota \bar{\otimes} \iota, \tilde{\omega} = \psi \bar{\otimes} \omega_0 \bar{\otimes} tr$. Then $\tilde{\omega}$ is a dominant weight on $\tilde{\mathcal{M}}$ and we have

$$\tilde{\omega}\circ\tilde{\sigma} = \tilde{\omega}. \tag{1}$$

On the other hand,

$$\tilde{\mathcal{M}}^{\tilde{\omega}} = (\mathcal{M} \bar{\otimes} \mathcal{F}_\infty)^{\psi \bar{\otimes} \omega_0} \bar{\otimes} \mathcal{F}_\infty = (\mathcal{M} \bar{\otimes} \mathcal{F}_\infty)^{\sigma\psi \bar{\otimes} \mathrm{Ad}(\rho)} \bar{\otimes} \mathcal{F}_\infty;$$

hence (19.13, 21.6.(1))

$$\tilde{\mathcal{M}}^{\tilde{\omega}} = \mathcal{R}(\mathcal{M}, \sigma^\psi) \bar{\otimes} \mathcal{F}_\infty$$

and

$$\mathcal{Z}(\tilde{\mathcal{M}}^{\tilde{\omega}}) = \mathcal{Z}(\mathcal{R}(\mathcal{M}, \sigma^\psi)) \bar{\otimes} \mathbb{C} \subset \mathcal{Z}(\mathcal{M}^\psi) \bar{\otimes} \mathfrak{L}(\mathbb{R}) \bar{\otimes} \mathbb{C},$$

and therefore

$$\tilde{\sigma}|\mathcal{Z}(\tilde{\mathcal{M}}^{\tilde{\omega}}) = \iota. \tag{2}$$

From (1), (2), and 25.1.(7), it follows that $mod(\sigma) = mod(\tilde{\sigma}) = \iota$.

(i) $\Rightarrow$ (iii). Let ω be a dominant weight on $\mathcal{M}$. Using 25.1.(7) and modifying σ if necessary by an inner $*$-automorphism, we can rephrase the condition $mod(\sigma) = \iota$ as follows:

$$\omega \circ \sigma = \omega, \sigma | \mathcal{Z}(\mathcal{M}^{\omega}) = \iota. \tag{3}$$

As in the first part of the proof of Theorem 30.7, we can find an element $a \in \mathcal{Z}(\mathcal{M}^{\omega})^{+}$ such that $\psi' = \omega_a$ is a lacunary weight, namely

$$(Sp\, \sigma^{\psi'}) \cap [\varepsilon^{-1}, \varepsilon] = \{1\}. \tag{4}$$

Since $\psi' = \omega_a$ with $a \in \mathcal{Z}(\mathcal{M}^{\omega})$, we infer from (3) that

$$\psi' \circ \sigma = \psi'. \tag{5}$$

Let $e = \mathbf{s}(\psi') = \mathbf{s}(a) \in \mathcal{Z}(\mathcal{M}^{\omega})$. For $x \in e\mathcal{M}^{\omega}e$, we have $\sigma_t^{\psi'}(x) = a^{\mathrm{i}t}\sigma_t^{\omega}(x)a^{-\mathrm{i}t} = a^{\mathrm{i}t}xa^{-\mathrm{i}t} = x\ (t \in \mathbb{R})$, whence $e\mathcal{M}^{\omega}e \subset \mathcal{M}^{\psi'}$. Using the relative commutant theorem (23.19), we get $\mathcal{Z}(\mathcal{M}^{\psi'}) \subset (\mathcal{M}^{\psi'})' \cap \mathcal{M} \subset (e\mathcal{M}^{\omega}e)' \cap \mathcal{M} \subset \mathcal{Z}(\mathcal{M}^{\omega})$ and so from (3) it follows that

$$\sigma | \mathcal{Z}(\mathcal{M}^{\psi'}) = \iota. \tag{6}$$

Being lacunary, the weight ψ' is strictly semifinite and hence $\mu = \psi' | \mathcal{M}^{\psi'}$ is an n.s.f. trace on $\mathcal{M}^{\psi'}$. Since $\mu \circ \sigma = \mu$ and $\sigma | \mathcal{Z}(\mathcal{M}^{\psi'}) = \iota$, it follows that every projection $p \in \mathcal{M}^{\psi'}$ is equivalent in $\mathcal{M}^{\psi'}$ to $\sigma(p)$. Let $p \in \mathcal{M}^{\psi'}$ be a projection such that $0 \neq \psi(p) = \mu(p) < +\infty$. Since $p \sim \sigma(p), e-p \sim \sigma(e-p) = e - \sigma(p)$, there exists $v \in \mathcal{M}^{\psi}$ such that $v^*v = vv^* = e$ and $\sigma(p) = v^*pv$. Composing with the inner $*$-automorphism defined by the unitary operator $u = v + (1-e)$, we may assume that $\sigma(p) = p$. Then $\psi'' = \psi'(p \cdot p)/\psi'(p)$ is a faithful normal state on $p\mathcal{M}p$ which satisfies conditions similar to (4), (5), and (6).

Since $\mathcal{M}$ is a countably decomposable type III factor, the projection p is equivalent to 1 in $\mathcal{M}$ and so there exists $w \in \mathcal{M}$ such that $w^*w = 1$ and $ww^* = p$. We then define $\psi = \psi''_w$ and $\sigma' \in Aut(\mathcal{M})$ by $\sigma'(x) = w^*\sigma(wxw^*)w\ (x \in \mathcal{M})$, that is, $\sigma' = \mathrm{Ad}(w^*\sigma(w)) \circ \sigma$. Then ψ and σ' satisfy the requirements of statement (iii). $\square$

30.10. Consider now a factor $\mathcal{M}$ of type III_{λ} $(0 < \lambda < 1)$ with separable predual and let $(\mathcal{N}, \theta, \tau)$ be the discrete decomposition of $\mathcal{M}$ (30.1). By Theorem 30.8, we know that $(\mathcal{P}_{\mathcal{M}}, F^{\mathcal{M}}) \approx ((\mathcal{P}_{\mathcal{N}})^{mod(\theta)}, F^{\mathcal{N}})$. On the other hand, by the results of Sections 24.8 and 25.6, we have $(\mathcal{P}_{\mathcal{N}}, F^{\mathcal{N}}) \approx (\mathcal{L}^{\infty}(\mathbb{R}_*^+), \mathrm{Ad}(\boldsymbol{\lambda}))$, and $mod(\theta) = F^{\mathcal{N}}_{\lambda^{-1}} = \mathrm{Ad}(\boldsymbol{\lambda}(\lambda^{-1}))$, since $\tau \circ \theta = \lambda\tau$. As $\Gamma(\mathcal{M}) = S(\mathcal{M}) \cap \mathbb{R}_*^+ = \{\lambda^n; n \in \mathbb{Z}\}$, it follows that

$$(\mathcal{P}_{\mathcal{M}}, F^{\mathcal{M}}) \approx (\mathcal{L}^{\infty}(\mathbb{R}_*^+/\Gamma(\mathcal{M})), \mathrm{Ad}(\boldsymbol{\lambda})). \tag{1}$$

It is now easy to check that the mapping

$$\mathbb{R}_*^+/\Gamma(\mathcal{M}) \ni \nu/\Gamma(\mathcal{M}) \mapsto F_{\nu}^{\mathcal{M}} \in Aut(F^{\mathcal{M}}) \tag{2}$$

is a continuous surjective group isomorphism. Since $\mathbb{R}_*^+/\Gamma(\mathcal{M})$ is compact, this mapping is also a topological isomorphism.

Let $\sigma \in Aut(\mathcal{M})$. Then $mod(\sigma) \in Aut(F^{\mathcal{M}})$ and there exists $\nu \in \mathbb{R}^+$ such that $mod(\sigma) = F^{\mathcal{M}}_\nu$, that is, $\psi \circ \sigma^{-1} \approx \nu\psi$, for every $\psi \in W^\infty_{int}(\mathcal{M})$. On the other hand, by Proposition 29.5, there exists a λ-trace φ on $\mathcal{M}$ and $\mu \in \mathbb{R}^+$ such that $\varphi \circ \sigma^{-1} \approx \mu\varphi$. We shall show that $\mu\nu^{-1} \in \Gamma(\mathcal{M})$:

$$mod(\sigma) = F^{\mathcal{M}}_\nu, \quad \varphi \circ \sigma^{-1} \approx \mu\varphi \Rightarrow \mu\nu^{-1} \in \Gamma(\mathcal{M}). \tag{3}$$

Indeed, let $\varepsilon > 0$. As in Section 23.18, there exists $a_\varepsilon \in \mathcal{M}^\varphi$ such that $1 - \varepsilon \leq a_\varepsilon \leq 1 + \varepsilon$ and $\psi = \varphi_{a_\varepsilon} \in W^\infty_{int}(\mathcal{M})$. Also, there exist $u, v \in U(\mathcal{M})$ such that $\varphi \circ \sigma^{-1} = \mu\varphi \circ \mathrm{Ad}(u)$ and $\psi \circ \sigma^{-1} = \nu\psi \circ \mathrm{Ad}(v)$. Then, for $x \in \mathcal{M}^+$ we have

$$\begin{aligned}\nu\varphi(a_\varepsilon^{1/2} vxv^* a_\varepsilon^{1/2}) &= \nu\varphi_a(vxv^*) = \nu(\psi \circ \mathrm{Ad}(v))(x) = \psi(\sigma^{-1}(x)) \\ &= \varphi(a_\varepsilon^{1/2}\sigma^{-1}(x)a_\varepsilon^{1/2}) = (\varphi \circ \sigma^{-1})(\sigma(a_\varepsilon^{1/2})x\sigma(a_\varepsilon^{1/2})) \\ &= \mu(\varphi \circ \mathrm{Ad}(u))(\sigma(a_\varepsilon^{1/2})x\sigma(a_\varepsilon^{1/2}))\end{aligned}$$

or $\nu\varphi(vv^* a_\varepsilon^{1/2} vxv^* a_\varepsilon^{1/2} vv^*) = \mu(\varphi \circ \mathrm{Ad}(u))(\sigma(a_\varepsilon)^{1/2} x \sigma(a_\varepsilon)^{1/2})$, that is,

$$\nu(\varphi \circ \mathrm{Ad}(v))_{v^* a_\varepsilon v} = \mu(\varphi \circ \mathrm{Ad}(u))_{\sigma(a_\varepsilon)}.$$

It follows that

$$[D(\varphi \circ \mathrm{Ad}(v))_{v^* a_\varepsilon v} : D(\varphi \circ \mathrm{Ad}(u))_{\sigma(a_\varepsilon)}]_t = \mu^{\mathrm{i}t}\nu^{-\mathrm{i}t}$$

for every $t \in \mathbb{R}$. Using the "chain rule," we deduce that

$$v^* a_\varepsilon^{\mathrm{i}t} v [D(\varphi \circ \mathrm{Ad}(v)) : D(\varphi \circ \mathrm{Ad}(u))]_t \sigma(a_\varepsilon^{-\mathrm{i}t}) = \mu^{\mathrm{i}t}\nu^{-\mathrm{i}t}$$

for all $t \in \mathbb{R}$. For $t = -2\pi/\ln(\lambda)$, we have

$$D(\varphi \circ \mathrm{Ad}(v)) : D(\varphi \circ \mathrm{Ad}(u))]_t = v^* \sigma_t^\varphi(v) \sigma_t^\varphi(u^*) u = 1,$$

so that

$$v^* a_\varepsilon^{\mathrm{i}t} v \sigma(a_\varepsilon^{-\mathrm{i}t}) = \mu^{\mathrm{i}t}\nu^{-\mathrm{i}t}$$

and therefore, letting ε tend to 0, $(\mu\nu^{-1})^{\mathrm{i}t} = 1$; this means that $\mu\nu^{-1} \in \Gamma(\mathcal{M})$.

30.11. For factors of type III$_\lambda$ with $0 < \lambda < 1$, it is possible also to compute the group $Out(\mathcal{M})$ in terms of the discrete decomposition (compare with Theorem 26.4).

Theorem. *Let $\mathcal{M}$ be a factor of type* III$_\lambda$ $(0 < \lambda < 1)$ *with discrete decomposition* $(\mathcal{N}, \theta, \tau)$. *Let* $t = -2\pi/\ln(\lambda)$ *and denote by G the quotient of the commutant of* $\mathfrak{o}(\theta)$ *in* $Out(\mathcal{N})$ *by* $\{\mathfrak{o}_{\mathcal{N}}(\theta)^k; k \in \mathbb{Z}\}$. *There exists a group homomorphism* $\gamma : Out(\mathcal{M}) \to G$ *such that the following sequence is exact:*

$$\{0\} \to \mathbb{Z} \xrightarrow{n \to nt} \mathbb{R} \xrightarrow{\delta_{\mathcal{M}}} Out(\mathcal{M}) \longrightarrow G \longrightarrow \{\imath\}. \tag{1}$$

Proof. Recall that $\mathcal{M} = \mathcal{R}(\mathcal{N}, \theta)$ and let $\pi : \mathcal{N} \to \pi(\mathcal{N}) \subset \mathcal{M}$ denote the canonical embedding, $P : \mathcal{M} \to \pi(\mathcal{N})$ the faithful normal conditional expectation, $\varphi = \tau \circ \pi^{-1} \circ P$ the corresponding λ-trace on $\mathcal{M}$ and $u = 1 \bar{\otimes} \boldsymbol{\lambda}(1)$ the unitary element of $\mathcal{M}$ such that $\pi(\theta(x)) = u\pi(x)u^*$ $(x \in \mathcal{N})$.

Let $\xi \in Out(\mathcal{M})$ and $\sigma_1 \in Aut(\mathcal{M})$ be such that $\mathfrak{o}_{\mathcal{M}}(\sigma_1) = \xi$. Then $\varphi \circ \sigma_1$ is also a λ-trace on $\mathcal{M}$ so that, by Proposition 29.5, there exists $\sigma \in Aut(\mathcal{M})$ such that $\mathfrak{o}_{\mathcal{M}}(\sigma) = \xi$ and $\varphi \circ \sigma$ is proportional to φ. Since $\pi(\mathcal{N}) = \mathcal{M}^{\varphi} = \mathcal{M}^{\varphi\circ\sigma}$, it follows that $\sigma(\pi(\mathcal{N})) = \pi(\mathcal{N})$ and so there exists $\nu \in Aut(\mathcal{N})$ such that $\sigma \circ \pi = \pi \circ \nu$. For $x \in \mathcal{N}$, we have $\pi(\nu\theta\nu^{-1}\theta^{-1}(x)) = \sigma(u)u^*\pi(x)u\sigma(u)^*$ and, since $\sigma_s^{\varphi}(u) = \lambda^{it}u$ and $\sigma_s^{\varphi}(\sigma(u)) = \sigma_s^{\varphi\circ\sigma^{-1}}(\sigma(u)) = \sigma(\sigma_s^{\varphi}(u)) = \lambda^{is}\sigma(u)$ $(s \in \mathbb{R})$, we have $\sigma(u)u^* \in \mathcal{M}^{\varphi} = \pi(\mathcal{N})$. Consequently, $\nu\theta\nu^{-1}\theta^{-1} \in Int(\mathcal{N})$ and so $\mathfrak{o}_{\mathcal{N}}(\nu)$ belongs to the commutant of $\mathfrak{o}_{\mathcal{N}}(\theta)$ in $Out(\mathcal{N})$. In order to define

$$\gamma(\xi) = \text{the image of } \mathfrak{o}_{\mathcal{N}}(\nu) \text{ in } G, \tag{2}$$

we must show that $\gamma(\xi)$ does not depend on $\sigma \in Aut(\mathcal{M})$ with $\mathfrak{o}_{\mathcal{M}}(\sigma) = \xi$ and that $\varphi \circ \sigma$ is proportional to φ.

Accordingly, consider $\sigma, \sigma' \in Aut(\mathcal{M})$ with $\mathfrak{o}_{\mathcal{M}}(\sigma) = \mathfrak{o}_{\mathcal{M}}(\sigma') = \xi$ and $\varphi \circ \sigma, \varphi \circ \sigma'$ both proportional to φ. By the above arguments, there exist $\nu, \nu' \in Aut(\mathcal{N})$ such that $\sigma \circ \pi = \pi \circ \nu, \sigma' \circ \pi = \pi \circ \nu'$, and $\mathfrak{o}_{\mathcal{N}}(\nu), \mathfrak{o}_{\mathcal{N}}(\nu')$ belong to the commutant of $\mathfrak{o}_{\mathcal{N}}(\theta)$ in $Out(\mathcal{N})$. We have $\sigma^{-1}\sigma' \in Int(\mathcal{M})$ and $\sigma^{-1}\sigma'(\pi(\mathcal{N})) = \pi(\mathcal{N})$, so that $\sigma^{-1}\sigma' = \mathrm{Ad}(v)$ for some $v \in U(\mathcal{M})$ with $v\pi(\mathcal{N})v^* = \pi(\mathcal{N})$. Then v belongs to the normalizer (10.17) $\mathcal{N}(P)$ of P. Since $u \in \mathcal{N}(P)$ and $\mathcal{M} = \mathcal{R}\{\pi(\mathcal{N}), u\}$, we infer by Proposition 22.4 that $\nu^{-1}\nu'$ belongs to the full group $[\mathrm{Ad}(u^n)|\mathcal{N}; n \in \mathbb{Z}] \subset Aut(\mathcal{N})$; since $\mathcal{N}$ is a factor it follows that there exists $k \in \mathbb{Z}$ such that $\nu^{-1}\nu'\theta^{-k} \in Int(\mathcal{N})$ and hence $\mathfrak{o}_{\mathcal{N}}(\nu)^{-1}\mathfrak{o}_{\mathcal{N}}(\nu') = \mathfrak{o}_{\mathcal{N}}(\theta)^k$. The mapping $\gamma : Out(\mathcal{M}) \to G$ is thus well defined by (2). Also, by construction, it is a group homomorphism.

We now compute the kernel of γ. Let $\xi \in Out(\mathcal{M})$ be such that $\gamma(\xi) = \iota$. Then there is a $\sigma \in Aut(\mathcal{M})$ with $\mathfrak{o}_{\mathcal{M}}(\sigma) = \xi$ and $\varphi \circ \sigma$ proportional to φ and a $\nu \in Aut(\mathcal{N})$ with $\sigma \circ \pi = \pi \circ \nu$ such that $\mathfrak{o}_{\mathcal{N}}(\nu) \in \{\mathfrak{o}_{\mathcal{N}}(\theta)^k; k \in \mathbb{Z}\}$. Thus there exist $k \in \mathbb{Z}$ and $a \in U(\mathcal{N})$ such that $\nu \circ \theta^k \circ \mathrm{Ad}(a) = \iota$. Replacing σ by $\sigma \circ \mathrm{Ad}(u^k a)$ we may assume that $\nu = \iota$, that is, $\sigma \circ \pi = \pi$. Then for $x \in \mathcal{N}$ we get $\sigma(u)\pi(x)\sigma(u^*) = \sigma(u\pi(x)u^*) = u\pi(x)u^*$, hence $u^*\sigma(u) \in \pi(\mathcal{N})' \cap \mathcal{M} = \mathbb{C} \cdot 1_{\mathcal{M}}$ and so $u^*\sigma(u) = \lambda^{is}$ for some $s \in \mathbb{R}$. Since also $\sigma_s^{\varphi} \circ \pi = \pi$ and $\sigma_s^{\varphi}(u) = \lambda^{is}\, u$, it follows that $\sigma = \sigma_s^{\varphi}$, and so $\xi = \delta_{\mathcal{M}}(s)$. Thus, the sequence (1) is exact at $Out(\mathcal{M})$.

Finally, we compute the range of γ. Let $g \in G$. There exists $\nu \in Aut(\mathcal{N})$ with $\mathfrak{o}_{\mathcal{N}}(\nu)\mathfrak{o}_{\mathcal{N}}(\theta) = \mathfrak{o}_{\mathcal{N}}(\theta)\mathfrak{o}_{\mathcal{N}}(\nu)$ such that g is the image of $\mathfrak{o}_{\mathcal{N}}(\nu)$ in G. Then $\nu \circ \theta = \mathrm{Ad}(a) \circ \theta \circ \nu$ for some $a \in U(\mathcal{N})$. Let $\pi_1 = \pi \circ \nu : \mathcal{N} \to \mathcal{M}$. For $x \in \mathcal{N}$ we have $\pi_1(\theta(x)) = \pi(\nu(\theta(x))) = \pi(a)\pi(\theta(\nu(x)))\pi(a)^* = \pi(a)u\pi(\nu(x))u^*\pi(a)^* = (\pi(a)u)\pi_1(x)(\pi(a)u)^*$. Also, $\pi_t(\mathcal{N}) = \pi(\mathcal{N}), \mathcal{M} = \mathcal{R}\{\pi_1(\mathcal{N}), \pi(a)u\}$, and there exists a faithful normal conditional expectation $P : \mathcal{M} \to \pi_1(\mathcal{N}) = \pi(\mathcal{N})$ such that $P(\pi(a)u) = \pi(a)P(u) = 0$. By Proposition 22.2 we infer that there exists $\sigma \in Aut(\mathcal{M})$ such that $\sigma \circ \pi = \pi_1$ and $\sigma(u) = \pi(a)u$. It follows that $\varphi \circ \sigma$ and φ are proportional and, as $\sigma \circ \pi = \pi \circ \nu$, we get $\gamma(\mathfrak{o}_{\mathcal{M}}(\sigma)) = g$. Thus, the sequence (1) is also exact at G.

The exactness of (1) at $\mathbb{Z}$ and $\mathbb{R}$ is obvious. □

30.12. Finally, we state without proof two important results:

Theorem 1. (U. Haagerup). *Let $\mathcal{M}$ be a factor of type* III_λ $(0 < \lambda < 1)$ *and let φ, ψ be n.s.f. weights on $\mathcal{M}$. If either $\varphi(1) = \psi(1) = 1$ or $\varphi(1) = \psi(1) = +\infty$, then there exists a unitary element $u \in \mathcal{M}$ such $\lambda\psi \le \varphi \circ \mathrm{Ad}(u) \le \lambda^{-1}\psi$.*

This result might suggest that any two faithful normal states on a factor of type III$_1$ are unitarily equivalent. However, if this is the case then $\mathcal{M} = \mathbb{C} \cdot 1_{\mathcal{M}}$ (Haagerup, U. [1979]). Instead, the following holds.

Theorem 2. (A. Connes, E. Størmer). *Let $\mathcal{M}$ be a factor of type* III$_1$. *For any two faithful normal states φ, ψ on $\mathcal{M}$ and $\varepsilon > 0$ there exists a unitary element $u \in \mathcal{M}$ such that $\|\psi - \varphi \circ \mathrm{Ad}(u)\| \le \varepsilon$.*

Note that if $\mathcal{M}$ is a factor of type III$_1$, then the flow of weights $F^{\mathcal{M}}$ is trivial $F^{\mathcal{M}}_\lambda = \iota$ $(\lambda \in \mathbb{R}^+_*)$. It follows that any integrable n.s.f. weight of infinite multiplicity on $\mathcal{M}$ is dominant. Using Theorem 23.18, it is easy to see that for any two n.s.f. weights of infinite multiplicity φ, ψ on $\mathcal{M}$ and for any $\varepsilon > 0$ there exists a unitary element $u \in \mathcal{M}$ such that $d(\psi, \varphi \circ \mathrm{Ad}(u)) < \varepsilon$ and $\mathfrak{M}_\psi = u\mathfrak{M}_\varphi u^*$ (see Connes & Takesaki, 1977; II.4.8–II.4.10).

30.13. We record here some results of Florin Rădulescu (1992a, 1992b, 1995), concerning some examples of nonhyperfinite factors of type III$_1$ and of the type III$_\lambda$, $(\lambda \in (0, 1))$, based on D.-V. Voiculescu's theory of free probability.

Theorem 1. *There exists a type* III$_1$ *factor with core isomorphic to the tensor product $\mathfrak{L}(F_\infty) \overline{\otimes} \mathcal{B}(\mathcal{H})$, where F_∞ is the free group with a countable infinity of generators and $\mathcal{H}$ is a separable infinite dimensional Hilbert space.*

Corollary. *There exists a one-parameter group $(\alpha_t)_{t>0}$ of automorphisms of the tensor product $\mathfrak{L}(F_\infty) \overline{\otimes} \mathcal{B}(\mathcal{H})$ such that α_t scales the canonical semifinite trace on the tensor product $\mathfrak{L}(F_\infty) \overline{\otimes} \mathcal{B}(\mathcal{H})$ by t, for all $t > 0$.*

In particular, this implies that the fundamental group $\mathfrak{L}(F_\infty)$ is $\mathbb{R}_+ \backslash \{0\}$.

Theorem 2. *Let λ be any number in* $(0, 1)$. *Consider the state φ_λ on $Mat_2(\mathbb{C})$ having the matrix unit $\{e_{ij}\}_{i,j=1,2}$, defined by the conditions*

$$\varphi_\lambda(e_{11}) / \varphi_\lambda(e_{22}) = \lambda \quad \text{and} \quad \varphi_\lambda(e_{ij}) = 0, \quad \text{if} \quad i \neq j.$$

Let ψ be the state on $\mathcal{L}^\infty$ $([0, 1])$ *induced by the Lebesgue measure on* $[0, 1]$.

Consider the free product state $\varphi * \psi$ on the algebraic free product $Mat_2(\mathbb{C}) * \mathfrak{L}^\infty$ $([0, 1])$ *and the free product von Neumann algebra $\mathcal{A} = Mat_2(\mathbb{C}) \overline{*} \mathcal{L}^\infty$* $([0, 1])$ *obtained by the GNS construction with respect to $\varphi_\lambda * \psi$.*

Then $\mathcal{A}$ is a type III$_\lambda$ *factor with core isomorphic to $\mathfrak{L}(F_\infty) \overline{\otimes} \mathcal{B}(\mathcal{H})$.*

Later, this was generalized by Dykema (1994a). Further, Shlyakhtenko (1997) introduced a theory of quasi-free factors that contained the above examples. Invariants for such factors were widely investigated. Related references are Houdayer and Vaes (2013) and Houdayer, Shlyakhtenko, and Vaes (2019).

30.14. Notes. The existence and (essential) uniqueness of the discrete decomposition for factors of types III$_\lambda$ $(0 \le \lambda < 1)$, as well as the computation of $Out(\mathcal{M})$ in terms of the discrete decomposition, are due to Connes (1973a). The connection between the discrete and the continuous decompositions was established by Connes and Takesaki (1977). The results mentioned in Section 30.12 are due to Haagerup (1979b); Connes and Størmer (1978).

The structure theory developed by Connes (1973a, 1974b) for factors of types III_λ $(0 \leq \lambda < 1)$, made necessary the study of outer conjugacy classes of *-automorphisms of W^*-algebras. This work was done by Connes (1977a, 1975e); see also Connes (1976b, 1978b) for the approximately finite dimensional factors of types II_1 and II_∞. A similar classification for measure space automorphisms was obtained by Connes and Krieger (1977).

From the work of Araki and Woods (1968) (see also Connes, 1973a, 3.6.3) it is known that the Powers factor $\mathscr{R}_\lambda$ is the only Araki–Woods factor of type III_λ $(0 < \lambda < 1)$. As a consequence of the work of Connes (1977a, 1975e, 1976a), we now know that $\mathscr{R}_\lambda$ is in fact the only injective factor of type III_λ $(0 < \lambda < 1)$. Nevertheless, for each fixed $0 < \lambda < 1$, there exists an uncountable infinity of nonisomorphic (noninjective) factors of type III_λ (Connes, 1973a, 4.4.5).

Connes (1976a, 1975e) proved that any injective factor of type III_0 is a Krieger factor so that, according to the classification theorem of Krieger [1976b], it follows that two injective factors of type III_0 are *-isomorphic if and only if their flows of weights are isomorphic; moreover, any ergodic nontransitive flow appears in this way. A direct proof of this result has been proposed by Connes (1978b, p. 476). There are also uncountable many noninjective nonisomorphic factors of type III_0.

There are factors of type III_1 for which there is nothing resembling a discrete decomposition (Connes, 1974d, 5.5); this fact makes the classification of type III_1 factors more difficult. Araki and Woods (1968) proved the existence of a unique Araki–Woods factor of type III_1. A. Connes conjectured and U. Haagerup (1987) proved the uniqueness of the approximate finite dimensional factor of type III_1 (see also Connes (1985)). There are uncountably many nonisomorphic factors of type III_1 (Connes, 1974d, 4.5).

For details concerning the classification of injective factors we refer to the fundamental contributions of Connes (1977a, 1975e, 1976a) and the survey articles of Connes (1974b, 1976b, 1978b, 1977b).

The classification of noninjective factors is an open problem. A definite negative result in this direction says that the Borel space of isomorphism classes of noninjective type III factors acting on a separable Hilbert space is not standard and not even countably separated (Woods, 1973; Connes, 1973a, 3.6.4; Connes, 1974d, 4.5). So far as type II_1 factors are concerned, besides the uncountable infinity pointed out by D. McDuff and S. Sakai (1971), we have the examples given by Connes (1975b).

For our exposition, we have used Connes (1973a, 1978b); Connes and Størmer (1978); Connes and Takesaki (1977); and Haagerup (1979b).

Appendix

In this section, we review some facts used in the main text concerning positive self-adjoint operators in Hilbert spaces, W^*-algebras and infinite tensor products. Throughout this section, $\mathscr{H}$ will denote a complex Hilbert space.

A.1. Let A be a positive self-adjoint linear operator on $\mathscr{H}$ with domain $D(A) \subset \mathscr{H}$. For each $n \in \mathbb{N}$, we put $e_n(A) = \chi_{[0,n]}(A) = \chi_{[1/(n+1),1]}((1+A)^{-1})$ (see [L], 9.9) and for each $\varepsilon > 0$, we define $A_\varepsilon = A(1+\varepsilon A)^{-1}$; note that $\varepsilon A_\varepsilon + (1+\varepsilon A)^{-1} = 1$. From ([L], 9.9–9.11) we see, using Lebesgue's dominated convergence theorem, that a vector $\xi \in \mathscr{H}$ belongs to $D(A^{1/2})$ if and only if

$$\|A^{1/2}\xi\|^2 = \lim_{n\to\infty}(Ae_n(A)\xi|\xi) = \lim_{\varepsilon\to 0}(A_\varepsilon\xi|\xi) < +\infty;$$

if $\xi \notin D(A^{1/2})$, then we put $\|A^{1/2}\xi\| = +\infty$, so that the above limit gives $\|A^{1/2}\xi\|$ for any vector $\xi \in \mathscr{H}$.

Recall ([L], 9.20) that the restriction of A to $\mathbf{s}(A)\mathscr{H}$ is the analytic generator of the *so*-continuous unitary representation $\mathbb{R} \ni t \mapsto A^{it} \in \mathscr{B}(\mathbf{s}(A)\mathscr{H})$; the operators A^{it} can be also regarded as partial isometries acting on the whole of $\mathscr{H}$.

A.2. Let Ω a be a subset of $\mathbb{C}$. A continuous function $f : \Omega \to \mathbb{C}$ is said to be *operator continuous* if the mapping

$$\{x \in \mathscr{B}(\mathscr{H}); x \text{ normal}, Sp(x) \subset \Omega\} \ni x \mapsto f(x) \in B(H)$$

is continuous with respect to the *so*-topology and the *s*-topology.

Theorem (I. Kaplansky, R. V. Kadison). *Let $\Omega \subset \mathbb{C}$ be such that $(\overline{\overline{\Omega}\setminus\Omega}) \cap \Omega = \emptyset$ and let f : $\Omega \to \mathbb{C}$ be a continuous function such that* $\sup\{|f(\lambda)|(1+|\lambda|)^{-1}; \lambda \in \Omega\} < +\infty$. *Then f is operator continuous.*

Proof. We first prove the theorem for $\Omega = \mathbb{C}$. Note that the functions $\lambda \mapsto \lambda, \lambda \mapsto \bar{\lambda}$ and $\lambda \mapsto (1+|\lambda|^2)^{-1}$ are operator continuous on $\mathbb{C}$. Indeed, if $x_i, x \in \mathscr{B}(\mathscr{H})$ are normal operators such that $x_i \xrightarrow{so} x$ and $\xi \in \mathscr{H}$, then

$$\begin{aligned}\|x_i^*\xi - x^*\xi\|^2 &= \|x_i^*\xi\|^2 - (x_i^*\xi|x^*\xi) - (x^*\xi|x_i^*\xi) + \|x^*\xi\|^2 \\ &= \|x_i^*\xi\|^2 - (\xi|x_ix^*\xi) - (x_ix^*\xi|\xi) + \|x\xi\|^2 \to 0\end{aligned}$$

and

$$\begin{aligned}&\|(1+x_i^*x_i)^{-1}\xi-(1+x^*x)^{-1}\xi\|\\&=\|(1+x_i^*x_i)^{-1}[x_i^*(x_i-x)+(x_i^*-x^*)x](1+x^*x)^{-1}\xi\|\\&\le\|(x_i-x)(1+x^*x)^{-1}\xi\|+\|(x_i^*-x^*)x(1+x^*x)^{-1}\xi\|\to 0.\end{aligned}$$

A similar argument holds for the s-topology.

Denote by $\mathscr{C}$ the set of all operator continuous functions on $\mathbb{C}$ and by $\mathscr{C}_b$ the subset of bounded functions in $\mathscr{C}$. Then $\mathscr{C}$ is a uniformly closed self-adjoint vector subspace of the *-algebra of all continuous complex functions on $\mathbb{C}$ and $\mathscr{C}_b\mathscr{C}\subset\mathscr{C}$. Hence $\mathscr{C}_b$ is a uniformly closed *-subalgebra containing the functions $\lambda\to(1+|\lambda|^2)^{-1}$ and $\lambda\mapsto\lambda(1+|\lambda|^2)^{-1}$. By the Stone–Weierstrass theorem, it follows that $\mathscr{C}_b$ contains the *-subalgebra $\mathscr{C}_0(\mathbb{C})$ of continuous complex functions vanishing at infinity on $\mathbb{C}$. Let $f:\mathbb{C}\to\mathbb{C}$ be a continuous function such that $\sup\{|f(\lambda)|(1+|\lambda|)^{-1};\lambda\in\mathbb{C}\}<+\infty$. Then $\lambda\to(1+|\lambda|^2)^{-1}f(\lambda)$ belongs to $\mathscr{C}_0(\mathbb{C})\subset\mathscr{C}_b$ and since $\mathscr{C}_b\mathscr{C}\subset\mathscr{C}$ it follows that $\lambda\to\bar{\lambda}(1+|\lambda|^2)^{-1}f(\lambda)$ belongs to $\mathscr{C}$ (actually to $\mathscr{C}_b$), so that $\lambda\mapsto\lambda\bar{\lambda}(1+|\lambda|^2)^{-1}f(\lambda)$ belongs to $\mathscr{C}$ and we conclude that f, as a sum of two functions in $\mathscr{C}$, also belongs to $\mathscr{C}$.

Consider now the general case. Let $x_i,x\in\mathscr{B}(\mathscr{H})$ be normal operators with $Sp(x_i)$ and $Sp(x)$ contained in Ω and $x_i\overset{so}{\to}x$. Since $(\overline{\overline{\Omega}\backslash\Omega})\cap\Omega=\emptyset$, there exists a compact neighborhood N of $Sp(x)$ such that $(\overline{\Omega}\backslash\Omega)\cap N=\emptyset$. Then $\Omega\cap N=\overline{\Omega}\cap N$, hence Ω is closed in $\Omega\cup N$ and, by the Tietze–Urysohn extension theorem, f can be extended to a continuous function g on $\Omega\cup N$. Let $h:\mathbb{C}\to\mathbb{C}$ be a continuous function such that $supp\ h\subset N$ and $h(\lambda)=1$ for $\lambda\in Sp(x)$. We obtain a continuous function with compact support by putting $k(\lambda)=g(\lambda)h(\lambda)$ for $\lambda\in\Omega\cup N$ and $k(\lambda)=0$ for $\lambda\notin\Omega\cup N$. Consider also the functions $l:\mathbb{C}\to\mathbb{C},F:\Omega\to\mathbb{C}$ defined by $l(\lambda)=(1+|\lambda|)(1-h(\lambda))$ and $F(\lambda)=f(\lambda)(1+|\lambda|)^{-1}$. By construction, $l(x)=0$ and, by assumption, there exists a $\mu_0\in(0,+\infty)$ such that $|F(\lambda)|\le\mu_0$ for all $\lambda\in\Omega$. Since $f(\lambda)=f(\lambda)h(\lambda)+f(\lambda)(1-h(\lambda))=k(\lambda)+F(\lambda)l(\lambda)$ $(\lambda\in\Omega)$, it follows that

$$f(x)-f(x_i)=[k(x)-k(x_i)]+F(x_i)[l(x)-l(x_i)].$$

By the first part of the proof, we have $k(x_i)\overset{so}{\to}k(x)$ and $l(x_i)\overset{so}{\to}l(x)=0$, while so $\|F(x_i)\|$ remains bounded by μ_0, so that $f(x_i)\overset{so}{\to}f(x)$. The same argument applies for the s-topology. Hence f is operator continuous.

Note that this result holds in particular for $\Omega=[0,+\infty)$. □

A.3. Let A_k,A be positive self-adjoint operators on $\mathscr{H}$. We shall write $A_k\overset{so}{\to}A$ if $(1+A_k)^{-1}\overset{so}{\to}(1+A)^{-1}$. If $A_k\overset{so}{\to}A$, then $A_k^{it}\overset{so}{\to}A^{it}$ for all $t\in\mathbb{R}$. Actually, we have (see Reed & Simon, 1972, VIII.21 and Exercise VIII.21; Araki & Woods, 1973; Kallman, 1971):

$$A_k\overset{so}{\to}A\Leftrightarrow A_k^{it}\overset{so}{\to}A^{it}\text{ for all }t\in\mathbb{R}\tag{1}$$

$$A_k\overset{so}{\to}A\Leftrightarrow A_k^{it}\overset{so}{\to}A^{it}\text{ uniformly for }|t|\le t_0.\tag{2}$$

It follows that

$$A_k\overset{so}{\to}A\Leftrightarrow(1+\varepsilon A_k)^{-1}\overset{so}{\to}(1+\varepsilon A)^{-1}\Leftrightarrow(A_k)_\varepsilon\overset{so}{\to}A_\varepsilon\quad(\varepsilon>0)\tag{3}$$

Proposition. *Let A_k, A be positive self-adjoint operators on $\mathscr{H}$. If there exists a vector subspace $D \subset D(A) \cap \bigcap_k D(A_k)$ such that $\overline{A|D} = A$ and $A_k\xi \to A\xi$ for all $\xi \in D$, then $A_k \overset{so}{\to} A$.*

Proof. Indeed, let $\xi \in D$ and $\eta = (1 + A)\xi$. Then

$$\|(1 + A)^{-1}\eta - (1 + A_k)^{-1}\eta\| = \|(1 + A_k)^{-1}(A_k\xi - A\xi)\| \le \|A_k\xi - A\xi\| \to 0.$$

Since $\overline{A|D} = A$, by ([L], E.9.1), $(1 + A)D$ is dense in $\mathscr{H}$ and $(1 + A_k)^{-1} \overset{so}{\to} (1 + A)^{-1}$. □

A.4. Let A, B be positive self-adjoint operators on $\mathscr{H}$. We shall write $A \le B$ if $(1+A)^{-1} \ge (1+B)^{-1}$. This condition means that ([L], E.2.6) $(1+B)^{-1/2}x = (1+A)^{-1/2}x$ for some $x \in \mathscr{B}(\mathscr{H})$, $\|x\| \le 1$, that is, $D((1 + B)^{1/2}) \subset D((1 + A)^{1/2})$ and $\|(1 + A)^{1/2}\xi\|^2 \le \|(1 + B)^{1/2}\xi\|^2$ for every $\xi \in D((1 + B)^{1/2})$. Using ([L], E.9.29) we infer that

$$A \le B \Leftrightarrow D(B^{1/2}) \subset D(A^{1/2}) \text{ and } \|A^{1/2}\xi\| \le \|B^{1/2}\xi\|, \; (\xi \in D(B^{1/2})). \tag{1}$$

It follows that

$$A \le B \Leftrightarrow (1 + \varepsilon A)^{-1} \ge (1 + \varepsilon B)^{-1} \Leftrightarrow A_\varepsilon \le B_\varepsilon; \; (\varepsilon > 0) \tag{2}$$

$$A \le B \text{ and } B \le A \Leftrightarrow A = B. \tag{3}$$

Also, if $A \le B$, then $A^{1/2} \le B^{1/2}$ and $x + A \le x + B$ for every $x \in \mathscr{B}(\mathscr{H})^+$. Note that there exist characterizations of the relation $A \le B$ in terms of the *so*-continuous unitary representations associated with A and B, which are in fact particular cases of Corollary 3.13 and Proposition 4.5 in the main text.

A.5. Let A_k, A be positive self-adjoint operators on $\mathscr{H}$. We shall write $A_k \uparrow A$ if $(1+A_k)^{-1} \downarrow (1+A)^{-1}$; this means that $\{(1 + A_k)^{-1}\}$ is a decreasing net and $(1 + A_k)^{-1} \overset{so}{\to} (1 + A)^{-1}$. Using (A.2, A.4) we obtain

$$A_k \uparrow A \Leftrightarrow (1 + \varepsilon A_k)^{-1} \downarrow (1 + \varepsilon A)^{-1} \Leftrightarrow (A_k)_\varepsilon \uparrow A_\varepsilon \quad (\varepsilon > 0). \tag{1}$$

Note that

$$A_\varepsilon \uparrow A \text{ for } \varepsilon \downarrow 0 \tag{2}$$

since $0 < \varepsilon' \le \varepsilon \Rightarrow A_\varepsilon \le A_{\varepsilon'} \le A$ and, with $a = (1 + A)^{-1}$, we have

$$(1 + A_\varepsilon)^{-1} = (a + \varepsilon(1 - a))(1 + \varepsilon(1 - a))^{-1} \overset{so}{\to} a = (1 + A)^{-1}$$

by ([L], 2.20). Also,

$$Ae_n(A) \uparrow A \text{ for } n \uparrow \infty \tag{3}$$

since $m \le n \Rightarrow Ae_m(A) \le Ae_n(A)$ and

$$(1 + Ae_n(A))^{-1} = (1 - e_n(A)) + (1 + A)^{-1}e_n(A) \overset{so}{\to} (1 + A)^{-1}.$$

Let B_k, B to other positive self-adjoint operators on $\mathscr{H}$. Using (1) and (A.4) it is easy to check that

$$A_k \uparrow A, B_k \uparrow B, A_k \le B_k \text{ for all } k \Rightarrow A \le B. \tag{4}$$

Proposition. *Let $\{A_k\}$ be an increasing net of positive self-adjoint operators on $\mathscr{H}$. There exists a positive self-adjoint operator A on $\mathscr{H}$ such that $A_k \uparrow A$ if and only if the vector subspace $D = \{\xi \in \mathscr{H}; \lim_k \|A_k^{1/2}\xi\| < +\infty\}$ is dense in $\mathscr{H}$. In this case, $D(A^{1/2}) = D$.*

Proof. Assume that $A_k \uparrow A$. Since $(A_k)_\varepsilon \uparrow A_\varepsilon, (A_k)_\varepsilon \uparrow_\varepsilon A_k$ and $\|A_k^{1/2}\xi\|^2 = \lim_\varepsilon (A_\varepsilon\xi|\xi)$ for every $\xi \in \mathscr{H}$, we have $\xi \in D(A^{1/2}) \Rightarrow ((A_k)_\varepsilon\xi|\xi) \le (A_\varepsilon\xi|\xi) \le \|A_k^{1/2}\xi\|^2 < +\infty$ for all k and $\varepsilon \Rightarrow (A_k\xi|\xi) \le \|A_k^{1/2}\xi\|^2$ for all $k \Rightarrow \xi \in D$ and $\xi \in D \Rightarrow ((A_k)_\varepsilon\xi|\xi) \le (A_k\xi|\xi) \le \mu < +\infty$ for all k and $\varepsilon \Rightarrow (A_\varepsilon\xi|\xi) \le \mu < +\infty \Rightarrow \|A^{1/2}\xi\|^2 \le \mu < +\infty \Rightarrow \xi \in D(A^{1/2})$. Therefore $D = D(A^{1/2})$ is dense in $\mathscr{H}$.

Conversely, assume that D is dense in $\mathscr{H}$. Since the net $\{A_k\}$ is increasing, we infer from ([L], 2.16) that $(1 + A_k)^{-1} \downarrow a$ for some $a \in \mathscr{B}(\mathscr{H}), 0 \le a \le 1$. If $\eta \in \mathscr{H}$ and $a\eta = 0$, then for every $\xi \in D$ we have

$$\begin{aligned}(|\xi|\eta)|^2 &= |(\xi|(1 + A_k)^{1/2}(1 + A_k)^{-1/2}\eta)|^2 \\ &= |((1 + A_k)^{1/2}\xi|(1 + A_k)^{-1/2}\eta)|^2 \\ &\le ((1 + A_k)\xi|\xi)((1 + A_k)^{-1}\eta|\eta)| \to 0,\end{aligned}$$

hence $\eta \perp D$ and $\eta = 0$ since D is dense in $\mathscr{H}$. Thus, a is injective, so that $A = a^{-1} - 1$ is a positive self-adjoint operator on $\mathscr{H}$. Moreover,

$$(1 + A_k)_\varepsilon = (\varepsilon + (1 + A_k)^{-1})^{-1} \uparrow (\varepsilon + a)^{-1} = (1 + A)_\varepsilon,$$

so that $1 + A_k \uparrow 1 + A$ and $A_k \uparrow A$. □

A.6. Let A, B be positive self-adjoint operators on $\mathscr{H}$ and assume that A and B commute ([L], E.9.24). Then

$$D = \bigcup_{n,m} e_n(A)e_m(B)\mathscr{H}$$

is a dense vector subspace of $\mathscr{H}$ and the linear operator AB defined by $(AB)\xi = A(B\xi)$ ($\xi \in D$) is positive and $AB \subset (AB)^*$, so that AB is preclosed, $\overline{AB}$ is positive and $\overline{AB} \subset (AB)^*$. Conversely, let $\eta \in D((AB)^*)$. There exists $\eta^* = (AB)^*\eta \in \mathscr{H}$ such that $(AB\xi|\eta) = (\xi|\eta^*)$ for all $\xi \in D$. With $p_{nm} = e_n(A)e_m(B)$ we have $D \ni p_{nm}\eta \underset{m,n}{\to} \eta$ and, for every $\xi \in D$,

$$\begin{aligned}(\xi|ABp_{nm}\eta) &= (\xi|(AB)^*p_{nm}\eta) = (AB\xi|p_{nm}\eta) \\ &= (ABp_{nm}\xi|\eta) = (p_{nm}\xi|\eta^*) = (\xi|p_{nm}\eta^*),\end{aligned}$$

whence $ABp_{nm}\eta = p_{nm}\eta^* \underset{n,m}{\to} \eta^* = (AB)^*\eta$. Thus, $\eta \in D(\overline{AB})$ and $(\overline{AB})\eta = (AB)^*\eta$. We conclude that $\overline{AB}$ is a positive self-adjoint operator on $\mathscr{H}$. Since $Ae_n(A)Be_m(B)\xi \underset{n,m}{\to} AB\xi$ for all $\xi \in D$, by

Proposition A.3 we obtain

$$Ae_n(A)Be_m(B) \uparrow \overline{AB}. \tag{1}$$

Also, for fixed $n, m \in \mathbb{N}, \xi \in \mathcal{H}$ and $\varepsilon \downarrow 0, \delta \downarrow 0$ we have

$$A_\varepsilon B_\delta e_n(A)e_m(B)\xi = (Ae_n(A))_\varepsilon (Be_m(B))_\delta \xi \to Ae_n(A)Be_m(B)\xi = ABe_n(A)e_m(B)\xi$$

whence

$$A_\varepsilon B_\delta \uparrow \overline{AB}. \tag{2}$$

In particular, $\overline{AB} = \overline{BA}$. Using ([L], 9.20) it is easy to check that

$$(\overline{AB})^{it} = A^{it}B^{it} \quad (t \in \mathbb{R}). \tag{3}$$

A.7. A *positive quadratic form* on $\mathcal{H}$ is a mapping $q : D(q) \to [0, +\infty)$ such that

$$D(q) \text{ is a dense vector subspace of} \mathcal{H} \tag{1}$$

$$q(\lambda\xi) = |\lambda|^2 q(\xi) \text{ for all } \xi \in D(q) \text{ and all } \lambda \in \mathbb{C} \tag{2}$$

$$q(\xi + \eta) + q(\xi - \eta) = 2q(\xi) + 2q(\eta) \text{ for all } \xi, \eta \in D(q). \tag{3}$$

Then

$$q(\xi, \eta) = \frac{1}{4}(q(\xi + \eta) - q(\xi - \eta) + iq(\xi + i\eta) - iq(\xi - i\eta)) \quad (\xi, \eta \in D(q))$$

defines a positive sesquilinear form $q : D(q) \times D(q) \to \mathbb{C}$, uniquely determined, such that

$$q(\xi, \xi) = q(\xi) \quad (\xi \in D(q)).$$

The positive quadratic form q is said to be *closed* if $D(q)$ is complete with respect to the scalar product $(\xi|\eta)_q = (\xi|\eta) + q(\xi, \eta)$, $(\xi, \eta \in D(q))$, that is, if the conditions

$$\{\xi_n\} \subset D(q), \xi \in \mathcal{H}, \|\xi_n - \xi\| \to 0, q(\xi_n - \xi_m) \to 0$$

imply $\xi \in D(q)$ and $q(\xi_n - \xi) \to 0$.

For every positive self-adjoint operator A on $\mathcal{H}$, we obtain a closed positive quadratic form q_A on $\mathcal{H}$ by putting

$$q_A(\xi) = \|A^{1/2}\xi\|^2 \quad (\xi \in D(q_A) = D(A^{1/2})).$$

Conversely,

Proposition. *Let q be a closed positive quadratic form on $\mathcal{H}$. There exists a unique positive self-adjoint operator A on $\mathcal{H}$ such that $q_A = q$.*

Proof. For any fixed $\xi \in \mathscr{H}$, the mapping $D(q) \ni \eta \mapsto \overline{(\xi|\eta)}$ is a bounded linear functional on the Hilbert space $D(q)$ with norm $\leq \|\xi\|$ (since $|\overline{(\xi|\eta)}| \leq \|\xi\|\|\eta\| \leq \|\xi\|\|\eta\|_q$ for $\eta \in D(q)$) and therefore there is a unique vector $T\xi \in D(q)$, $\|T\xi\|_q \leq \|\xi\|$, such that

$$(\xi|\eta) = (T\xi|\eta)_q = (T\xi|\eta) + q(T\xi, \eta) \quad (\eta \in D(q)). \tag{4}$$

We thus obtain a bounded linear operator $T : \mathscr{H} \to D(q)$ with norm ≤ 1 such that

$$q(T\xi, \eta) = (\xi - T\xi|\eta) \quad (\xi \in \mathscr{H}, \eta \in D(q)). \tag{5}$$

From (4) it follows that T is injective and $T(\mathscr{H})$ is dense in the Hilbert space $D(q)$, so that $T(\mathscr{H})$ is dense in $\mathscr{H}$. Therefore, we obtain a densely defined linear operator A, on $\mathscr{H}$ by putting

$$A\zeta = T^{-1}\zeta - \zeta \quad (\zeta \in D(A) = T(\mathscr{H}) \subset D(q))$$

and we have

$$q(\zeta, \eta) = (A\zeta|\eta) \quad (\zeta \in D(A), \eta \in D(q)). \tag{6}$$

It follows that A is positive and, as $(1 + A)D(A) = T^{-1}(T(\mathscr{H})) = \mathscr{H}, A$ is self-adjoint ([L], 9.5). For $\zeta \in D(A) \subset D(A^{1/2})$, we have $\|A^{1/2}\zeta\|^2 = (A\zeta|\zeta) = q(\zeta)$.

Let $\xi \in D(q)$. Since $D(A) = T(\mathscr{H})$ is dense in the Hilbert space $D(q)$, there exists a sequence $\{\xi_n\} \subset D(A)$ such that $\|\xi_n - \xi\| \to 0$ and $\|A^{1/2}\xi_n - A^{1/2}\xi_m\|^2 = q(\xi_n - \xi_m) \to 0$. Since $A^{1/2}$ is closed, it follows that $\xi \in D(A^{1/2})$ and $\|A^{1/2}\xi_n - A^{1/2}\xi\| \to 0$, so that $\|A^{1/2}\xi\|^2 = \lim_n \|A^{1/2}\xi_n\|^2 = \lim_n q(\xi_n) = q(\xi)$.

Conversely, using the fact that $A^{1/2} = \overline{A^{1/2}|D(A)}$ and the closedness of the form q, we see that $D(A^{1/2}) \subset D(q)$.

Hence $q_A = q$. The uniqueness of A, subject to this condition, follows from (6). □

A.8. Let p, q be two positive quadratic forms on $\mathscr{H}$. We say that p is an *extension* of q if $D(q) \subset D(p)$ and $p(\xi) = q(\xi)$ for every $\xi \in D(q)$. The form q is said to be *preclosed* (or closable) if it has a closed extension.

Proposition. *A positive quadratic form q on $\mathscr{H}$ is preclosed if and only if*

$$\{\xi\}_n \subset D(q), \|\xi_n\| \to 0, q(\xi_n - \xi_m) \to 0 \Rightarrow q(\xi_n) \to 0. \tag{1}$$

In this case, there exists a closed extension $\bar{q}$ of q such that any closed extension of q is also an extension of $\bar{q}$.

Proof. Assume that q is preclosed and let p be a closed extension of q. Then from the assumptions in (1) it follows that $q(\xi_n) = p(\xi_n - 0) \to 0$.

Conversely, assume that (1) holds. Let $D(\bar{q})$ be the set of all $\xi \in \mathscr{H}$ such that there exists a sequence $\{\xi_n\} \subset D(q)$ with the properties $\|\xi_n - \xi\| \to 0$ and $q(\xi_n - \xi_m) \to 0$. From (1) it follows that for $\xi \in D(\bar{q})$ the number $\bar{q}(\xi) = \lim_n q(\xi_n)$ does not depend on the sequence $\{\xi_n\} \subset D(q)$ with $\|\xi_n - \xi\| \to 0$ and $q(\xi_n - \xi_m) \to 0$. It is then easy to check that $\bar{q} : D(\bar{q}) \to [0, +\infty)$ is a closed extension of q and that any closed extension of q is also an extension of $\bar{q}$. □

The form $\bar{q}$ is called the *closure* of q.

A.9. Consider again a positive quadratic form $q : D(q) \to [0, +\infty)$ on $\mathscr{H}$. We extend the definition of q to the whole of $\mathscr{H}$ by putting $q(\xi) = +\infty$ for $\xi \in \mathscr{H} \setminus D(q)$.

Proposition. *Let q be a positive quadratic form on $\mathscr{H}$. Then*

$$q \text{ is closed } \Leftrightarrow q \text{ is lower semicontinuous on } \mathscr{H}; \tag{1}$$

$$q \text{ is preclosed } \Leftrightarrow q \text{ is lower semicontinuous on } D(q). \tag{2}$$

Proof. Assume that q is closed and let A be the unique positive self-adjoint operator on $\mathscr{H}$ such that $q = q_A$ (A.7). Then

$$q(\xi) = \sup\{|(A\zeta|\xi)|; \zeta \in D(A), (A\zeta|\zeta) = 1\} \quad (\xi \in \mathscr{H}). \tag{3}$$

Indeed, for every $\zeta \in D(A) \subset D(A^{1/2})$ and $\xi \in D(q)$ we have $|(A\zeta|\xi)|^2 = |(A^{1/2}\zeta|A^{1/2}\xi)|^2 \leq \|A^{1/2}\zeta\|^2 \|A^{1/2}\xi\|^2 = q(\zeta)q(\xi)$, hence

$$|(A\zeta|\xi)|^2 \leq q(\zeta)q(\xi) \quad (\zeta \in D(A), \xi \in \mathscr{H}) \tag{4}$$

and (3) follows replacing ζ here by $\zeta_n = (Ae_n(A)\xi|\xi)^{-1/2} e_n(A)\xi$ and taking the limit as $n \to +\infty$. Now from (3) it is clear that q is (even weakly) lower semicontinuous on $\mathscr{H}$.

If q is preclosed, then its closure $\bar{q}$ is lower semicontinuous on $\mathscr{H}$, and q is lower semicontinuous on $D(q)$.

Conversely, assume that q is lower semicontinuous on $D(q)$ and consider $\{\xi_n\} \subset D(q)$, $\|\xi_n\| \to 0$, $q(\xi_n - \xi_m) \to 0$. Let $\varepsilon > 0$ and choose $n_\varepsilon \in \mathbb{N}$ such that $q(\xi_n - \xi_m) \leq \varepsilon$ for $n, m \geq n_\varepsilon$. Since $\xi_n \to 0$ and q is lower semicontinuous on $D(q)$, it follows that $q(\xi_n) \leq \lim_m \inf q(\xi_n - \xi_m) \leq \varepsilon$ for all $n \geq n_\varepsilon$. Hence $q(\xi_n) \to 0$. By Proposition A.8, it follows that q is preclosed.

Finally, assume that q is preclosed but is not closed. Then there exists $\xi \in D(\bar{q})$, $\xi \notin D(q)$. By the construction of $\bar{q}$ (A.8) there exists a sequence $\{\xi_n\} \subset D(q)$, $\|\xi_n - \xi\| \to 0$, such that $\lim_n q(\xi_n) = \bar{q}(\xi) < +\infty = q(\xi)$. Thus q is not lower n semicontinuous on $\mathscr{H}$. □

A.10. Let $q : D(q) \to [0, +\infty)$ be a lower semicontinuous positive quadratic form on $\mathscr{H}$. By Propositions A.7–A.9, there exists a unique positive self-adjoint operator A on $\mathscr{H}$ such that $q_A = \bar{q}$, that is,

$$\|A^{1/2}\xi\|^2 = \bar{q}(\xi) \quad (\xi \in D(A^{1/2}) = D(\bar{q})). \tag{1}$$

In particular,

$$D(A^{1/2}) \supset D(q) \text{ and } \|A^{1/2}\xi\|^2 = q(\xi) \text{ for all } \xi \in D(q) \tag{2}$$

and, as is easily verified,

$$A^{1/2} = \overline{A^{1/2}|D(q)}. \tag{3}$$

Moreover, A is the greatest positive self-adjoint operator on $\mathscr{H}$ satisfying (2). Indeed, let B be another positive self-adjoint operator in $\mathscr{H}$ such that $D(B^{1/2}) \supset D(q)$ and $\|B^{1/2}\xi\|^2 = q(\xi)$ for all $\xi \in D(q)$. Let $\xi \in D(A^{1/2}) = D(\bar{q})$. There exists a sequence $\{\xi_n\} \subset D(q)$ such that $\|\xi_n - \xi\| \to 0$ and $q(\xi_n) \to \bar{q}(\xi)$. Then

$$\|B^{1/2}\|^2 \leq \lim_n \|B^{1/2}\xi_n\|^2 = \lim_n q(\xi_n) = \bar{q}(\xi) = \|A^{1/2}\xi\|^2.$$

Consequently, $B \leq A$.

Recall (A.9) that if q is lower semicontinuous on the whole $\mathscr{H}$, then $q = \bar{q}$ is closed, so that in this case the best characterization of A is given by (1).

A.11. Let A, B be positive self-adjoint operators in $\mathscr{H}$. Assume that $D = D(A^{1/2}) \cap D(B^{1/2})$ is dense in $\mathscr{H}$. Then $q : \mathscr{H} \ni \xi \to q_A(\xi) + q_B(\xi) \in [0, +\infty]$ is a lower semicontinuous positive quadratic form on $\mathscr{H}$ with $D(q) = D$. By (A.9) it follows that there exists a unique positive self-adjoint operator $A\hat{+}B$ on $\mathscr{H}$ such that

$$\|(A\hat{+}B)^{1/2}\xi\|^2 = \|A^{1/2}\xi\|^2 + \|B^{1/2}\xi\|^2 \quad (\xi \in D((A\hat{+}B)^{1/2}) = D). \tag{1}$$

The operator $A\hat{+}B$ is called the *weak sum* (or *form sum*) of A and B.

A.12. Let $\varphi : \mathbb{R}\backslash\{0\} \to \mathbb{R}$ be the function defined by

$$\varphi(t) = 2(e^{\pi t} - e^{-\pi t})^{-1} \quad (t \in \mathbb{R}\backslash\{0\}).$$

For any bounded continuous function $f : \mathbb{R} \to \mathbb{C}$, we shall write

$$\begin{aligned} PV \int_{-\infty}^{+\infty} \varphi(t)f(t)\,\mathrm{d}t &= \lim_{\varepsilon\to 0} \left[\int_{-\infty}^{-\varepsilon} \varphi(t)f(t)\,\mathrm{d}t + \int_{+\varepsilon}^{+\infty} \varphi(t)f(t)\,\mathrm{d}t \right] \\ &= \lim_{\varepsilon\to 0} \int_{+\varepsilon}^{+\infty} \varphi(t)(f(t) - f(-t))\,\mathrm{d}t \end{aligned}$$

whenever the limit exists. Similar notation will be used for functions f with values in $\mathscr{H}$ or $\mathscr{B}(\mathscr{H})$.

Proposition. *For every nonsingular positive self-adjoint operator A on $\mathscr{H}$, we have*

$$PV \int_{-\infty}^{+\infty} \varphi(t)A^{\mathrm{i}t}\xi\,\mathrm{d}t = \mathrm{i}(A-1)(A+1)^{-1}\xi \quad (\xi \in \mathscr{H}). \tag{1}$$

Proof. Since $(1+A^{1/2})^{-1} + (1+A^{-1/2})^{-1} = 1$, we have $D(A^{1/2}) + D(A^{-1/2}) = \mathscr{H}$; thus it is sufficient to prove (1) only for $\xi \in D(A^{1/2})$ and for $\xi \in D(A^{-1/2})$.

Let $\xi \in D(A^{1/2})$. Then ([L], 9.15) the function $\alpha \mapsto A^{\alpha}A^{1/2}\xi$ is defined, bounded, continuous on the strip $\{\alpha \in \mathbb{C}; -1/2 \leq \operatorname{Re} \alpha \leq 0\}$ and analytic in the interior of the strip. On the other hand, the function $\alpha \mapsto 2(e^{-\pi \mathrm{i}\alpha} + e^{\pi \mathrm{i}\alpha})^{-1}$ has the same properties, except at the point $\alpha = -1/2$ where it has a pole with residue $(\pi \mathrm{i})^{-1}$; moreover, this function tends to zero when $z \to \infty$ on this strip. Therefore, by integrating along the boundary of the strip the function $\alpha \mapsto 2(e^{-\pi i t} + e^{\pi i t})^{-1} A^{\alpha} A^{1/2}$, we obtain

$$\int_{-\infty}^{+\infty} 2(e^{-\pi t} + e^{\pi t})^{-1} A^{it} A^{1/2} \xi \, dt = \xi + PV \int_{-\infty}^{+\infty} -i\varphi(t) A^{it} \xi \, dt.$$

By ([L], Corollary 9.23), the left hand side is equal to $2A(A+1)^{-1}\xi$ and (1) follows.

For $\xi \in D(A^{-1/2})$, note that $(A-1)(A+1)^{-1} = -(A^{-1}-1)(A^{-1}+1)^{-1}$ and apply the first part of the proof. □

A.13. Let $b \in \mathscr{B}(\mathscr{H}), 0 \le b \le 1$, and let A be a nonsingular positive self-adjoint on $\mathscr{H}$. Put

$$a = \int_{-\infty}^{+\infty} 2(e^{-\pi t} + e^{\pi t})^{-1} A^{it} b A^{-it} \, dt \in \mathscr{B}(\mathscr{H}) \tag{1}$$

and note that $0 \le a \le 1$. With the same arguments as in Section A.12, applied to the function $\alpha \mapsto 2(e^{-\pi i \alpha} + e^{\pi i \alpha})^{-1} (bA^{-\alpha} A^{-1/2} \xi | A^{\bar{\alpha}} A^{1/2} \eta)$ on the strip $\{\alpha \in \mathbb{C}; -1/2 \le \operatorname{Re} \alpha \le 0\}$, we obtain

$$PV \int_{-\infty}^{+\infty} \varphi(t) (A^{it} b A^{-it} \xi | \eta) \, dt = i(aA^{-1/2}\xi | A^{1/2}\eta) - i(b\xi | \eta) \tag{2}$$

whenever $\xi \in D(A^{-1/2})$ and $\eta \in D(A^{1/2})$.

Proposition. *If $\xi \in D(\ln(A))$, then $PV \int_{-\infty}^{+\infty} \varphi(t) A^{it} b A^{-it} \xi \, dt$ exists. If, moreover, $\xi \in D(A^{-1/2})$, then $aA^{-1/2}\xi \in D(A^{1/2})$ and*

$$A^{1/2} a A^{-1/2} \xi = b\xi - iPV \int_{-\infty}^{+\infty} \varphi(t) A^{it} b A^{-it} \xi \, dt.$$

Proof. Note that $A^{it} a A^{-it} \xi = A^{it} a \xi + A^{it} a (A^{-it} - 1)$ and that if $\xi \in D(\ln(A))$, then (see Reed & Simon, 1972, Theorem VIII.7) the function

$$t \mapsto \varphi(t) A^{it} a (A^{-it} - 1)\xi = \frac{2t}{e^{\pi t} - e^{-\pi t}} a \left[\frac{A^{-it} - 1}{t} \xi \right]$$

is continuous at $t = 0$ and hence everywhere; as it tends to zero rapidly at infinity it is integrable. This remark, combined with Proposition A.12 proves the first statement.

The second statement is now an obvious consequence of (2) since $A^{1/2}$ is self-adjoint. □

A.14. Let $\mathscr{H}$ be the Hilbert space $\mathscr{L}^2(\mathbb{R})$. Consider the operator $b \in \mathscr{B}(\mathscr{H}), 0 \le b \le 1$, defined as multiplication by the function

$$b(s) = 1 \text{ when } s < 0 \text{ and } b(s) = 0 \text{ when } s \ge 0,$$

the unique nonsingular positive self-adjoint operator A on $\mathscr{H}$ such that ([L], 9.20)

$$[A^{it}\xi](s) = \xi(s - t) \quad (\xi \in \mathscr{L}^2(\mathbb{R}), s, t \in \mathbb{R})$$

and the corresponding operator $a \in \mathscr{B}(\mathscr{H}), 0 \leq a \leq 1$, defined by A.13.(1). Note that if we replace A by A^{-1}, the operator a remains unchanged.

We shall show, by an indirect argument, that

$$\begin{aligned} &\textit{either there exists } \zeta \in D(A^{-1/2}) \textit{ such that } aA^{-1/2}\zeta \notin D(A^{1/2}) \\ &\textit{or there exists } \zeta \in D(A^{1/2}) \textit{ such that } aA^{1/2}\zeta \notin D(A^{-1/2}). \end{aligned} \tag{1}$$

So assume that $\zeta \in D(A^{-1/2}) \Rightarrow aA^{-1/2}\zeta \in D(A^{1/2})$ and $\zeta \in D(A^{1/2}) \Rightarrow aA^{1/2}\zeta \in D(A^{-1/2})$. Then we can define the linear operators T_1, T_2 on $\mathscr{H}$ by

$$\begin{aligned} T_1\zeta &= A^{1/2}aA^{-1/2}\zeta - b\zeta \quad (\zeta \in D(A^{-1/2})) \\ T_2\zeta &= -A^{-1/2}aA^{1/2}\zeta + b\zeta \quad (\zeta \in D(A^{1/2})). \end{aligned}$$

For $\xi \in D(A^{-1/2}) \cap D(A^{1/2})$, we have $\xi \in D(\ln(A))$ (see Reed & Simon, 1972, Theorem VIII.7 and [L], Corollary 9.21) and hence, by Proposition A.13,

$$T_1\xi = -\mathrm{i}PV\int_{-\infty}^{+\infty} \varphi(t)A^{\mathrm{i}t}bA^{-\mathrm{i}t}\xi \, \mathrm{d}t = T_2\xi.$$

As $\mathscr{H} = D(A^{-1/2}) + D(A^{1/2})$, we can define a linear operator T on $\mathscr{H}$ by putting

$$T(\xi_1 + \xi_2) = T_1\xi_1 + T_2\xi_2 \quad (\xi_1 \in D(A^{-1/2}), \xi_2 \in D(A^{1/2})).$$

For $\xi \in D(A^{-1/2}) \cap D(A^{1/2})$ and $\eta \in D(A^{-1/2})$, we have

$$\begin{aligned} (-T\xi|\eta) = (-T_2\xi|\eta) &= (A^{-1/2}aA^{1/2}\xi|\eta) - (b\xi|\eta) \\ &= (\xi|A^{1/2}aA^{-1/2}\eta) - (\xi|b\eta) = (\xi|T_1\eta) = (\xi|T\eta). \end{aligned}$$

Hence $-T \subset T^*$, so that T is closed and hence bounded. Furthermore, for $\zeta \in D(A^{-1/2}) \cap D(\ln(A))$ and $\zeta \in D(A^{1/2}) \cap D(\ln(A))$ we have

$$T\zeta = -\mathrm{i}PV\int_{-\infty}^{+\infty} \varphi(t)A^{\mathrm{i}t}bA^{-\mathrm{i}t}\zeta \, \mathrm{d}t.$$

Since $D(\ln(A)) = D(\ln(A)) \cap D(A^{-1/2}) + D(\ln(A)) \cap D(A^{1/2})$, this identity holds for any $\zeta \in D(\ln(A))$. As T is bounded, it follows that the mapping

$$D(\ln(A)) \ni \zeta \mapsto PV\int_{-\infty}^{+\infty} \varphi(t)A^{\mathrm{i}t}bA^{-\mathrm{i}t}\zeta \, \mathrm{d}t \in \mathscr{H} \tag{2}$$

is continuous.

However, using again Proposition A.13 and Fubini's theorem, for $\zeta \in D(\ln(A))$ and $\eta \in \mathscr{H}$ we get

$$PV\int_{-\infty}^{+\infty} \varphi(t)(A^{\mathrm{i}t}bA^{-\mathrm{i}t}\zeta|\eta) \, \mathrm{d}t = \int_{-\infty}^{+\infty} \psi(s)\zeta(s)\overline{\eta(s)} \, \mathrm{d}s$$

with $\psi(s) = \int_{|s|}^{+\infty} \varphi(t)\,\mathrm{d}t\ (s \in \mathbb{R})$. Since $\varphi(t) \sim t^{-1}$ for small $|t|$, φ is not integrable and therefore ψ is not bounded. In this case, it is easy to see that the mapping (2) cannot be continuous. Indeed, consider the functions $\zeta_n(t) = n^{1/2}\exp(-n^2t^2),\ t \in \mathbb{R}$. Then $\zeta_n \in D(\ln(A))$ (since the ζ_n are "entire analytic vectors" for $t \mapsto A^{it}$), $\|\zeta_n\|$ is independent of n, while $\int_{-\infty}^{+\infty} \psi(s)\zeta_n(s)\overline{\eta(s)}\,\mathrm{d}s \to +\infty$ whenever η is not zero in a neighborhood of $s = 0$.

This contradiction proves (1).

In the next sections, we review some basic facts concerning W^*-algebras.

A.15. Let $\mathscr{A}$ be a C^*-algebra, let $\mathscr{H}_{\mathscr{A}} = \oplus\{\mathscr{H}_\varphi; \varphi \in S(\mathscr{A})\}$ and let

$$\pi_{\mathscr{A}} = \bigotimes_{\varphi\in S(\mathscr{A})} \pi_\varphi : \mathscr{A} \to \mathscr{B}(\mathscr{H}_{\mathscr{A}})$$

where $S(\mathscr{A})$ denotes the set of all positive linear forms φ on $\mathscr{A}$ with $\|\varphi\| = 1$ and $\pi_\varphi : \mathscr{A} \to \mathscr{B}(\mathscr{H}_{\mathscr{A}})$ is the corresponding Gelfand–Naimark–Segal representation ([L], 5.18; Sakai, 1971; Strătilă & Zsidó, 1977, 2005). Then $\pi_{\mathscr{A}}$ is an isometric nondegenerate $*$-representation of $\mathscr{A}$, called the *universal $*$-representation of $\mathscr{A}$*, the w-closure $\mathscr{N}_{\mathscr{A}}$ of $\pi_{\mathscr{A}}(\mathscr{A})$ is a von Neumann subalgebra of $\mathscr{B}(\mathscr{H}_{\mathscr{A}})$ called the *enveloping von Neumann algebra of $\mathscr{A}$* and the following statements hold:

For every $\psi \in \mathscr{A}^*$, there exists a unique wo-continuous form $F_{\mathscr{A}}(\psi)$ on $\mathscr{N}_{\mathscr{A}}$ such that $\psi = F_{\mathscr{A}}(\psi)\circ\pi_{\mathscr{A}}$ and the map $F_{\mathscr{A}}$ is a linear isometry of A^* onto $(\mathscr{N}_{\mathscr{A}})_*$.

The map $\Phi_{\mathscr{A}} : \mathscr{N}_{\mathscr{A}} \to \mathscr{A}^{**}$ defined by $[\Phi_{\mathscr{A}}(y)](\psi) = [F_{\mathscr{A}}(\psi)](y)\ (\psi \in A^*, y \in \mathscr{N}_{\mathscr{A}})$ is a surjective linear isometry and a $(w, \sigma(\mathscr{A}^{**}, \mathscr{A}_*))$-homeomorphism; $\Phi_{\mathscr{A}}\circ\pi_{\mathscr{A}}$ is the canonical embedding of $\mathscr{A}$ into $\mathscr{A}^{**}$.

A.16. A W^*-algebra is a C^*-algebra $\mathscr{M}$ which is isometrically isomorphic to the dual space of some Banach space. In this case, there is a unique norm-closed vector subspace $\mathscr{F}$ of $\mathscr{M}^*$ such that the map $\Phi : \mathscr{M} \to \mathscr{F}^*$ defined by $[\Phi(x)](\psi) = \psi(x)\ (x \in \mathscr{M}, \psi \in \mathscr{F})$, is a surjective linear isometry. We write $\mathscr{M}_* = \mathscr{F}$ and call $\mathscr{M}_*$ the *predual* of $\mathscr{M}$.

Let $\pi_{\mathscr{M}} : \mathscr{M} \to \mathscr{N}_{\mathscr{M}} \subset \mathscr{B}(\mathscr{H}_{\mathscr{M}})$ be the universal $*$-representation of $\mathscr{M}$. There exists a unique central projection $p_{\mathscr{M}}$ in $\mathscr{N}_{\mathscr{M}}$ such that the map $x \mapsto \pi_{\mathscr{M}}(x)p_{\mathscr{M}}$ is a $*$-isomorphism of $\mathscr{M}$ onto the von Neumann algebra $\mathscr{N}_{\mathscr{M}}p_{\mathscr{M}} \subset \mathscr{B}(p_{\mathscr{M}}\mathscr{H}_{\mathscr{M}})$. Moreover, this map is a $(\sigma(\mathscr{M}, \mathscr{M}_*), w)$ homeomorphism and $\mathscr{M}_* = p_{\mathscr{M}}\mathscr{M}^*$.

A linear form $\psi \in \mathscr{M}^*$ satisfies $\psi = p_{\mathscr{M}}\cdot\psi$, that is, $\psi \in \mathscr{M}_*$, if and only if $\psi(\sum_{i\in I} e_i) = \sum_{i\in I}\psi(e_i)$ for every family $\{e_i\}_{i\in I}$ of mutually orthogonal projections in $\mathscr{M}$. In this case, ψ is called a *normal linear form* on $\mathscr{M}$.

A linear functional $\psi \in \mathscr{M}^*$ satisfies $p_{\mathscr{M}}\cdot\psi = 0$ if and only if for every nonzero projection $e \in \mathscr{M}$ there exists a nonzero projection $f \in \mathscr{M}, f \leq e$, with $\psi(f) = 0$; moreover, f can be chosen such that $f\cdot\psi\cdot f = 0$. In this case, ψ is called a *singular linear form* on $\mathscr{M}$.

Thus, W^*-algebras are essentially the same thing as von Neumann algebras. The $\sigma(\mathscr{M}, \mathscr{M}_*)$-topology on $\mathscr{M}$ is also called the w-topology on $\mathscr{M}$. The s-topology (resp. the s^*-topology) on $\mathscr{M}$ is defined by the seminorms $x \mapsto \psi(x^*x)^{1/2}$ (resp. $x \mapsto \psi(x^*x+xx^*)^{1/2}$ where ψ ranges over all normal positive forms on $\mathscr{M}$.

The uniqueness of the predual of a W^*-algebra shows that every $*$-isomorphism between W^*-algebras is w-continuous, and also s-continuous and s^*-continuous.

Every $*$-isomorphism of a W^*-algebra $\mathscr{M}$ onto a von Neumann algebra will be called a *realization* of $\mathscr{M}$ as a von Neumann algebra. A W^*-algebra $\mathscr{M}$ with separable predual has a realization as a von Neumann algebra acting on a separable Hilbert space.

A W^*-*subalgebra* of the W^*-algebra $\mathscr{M}$ is a w-closed *-subalgebra of $\mathscr{M}$. Note that, while every W^*-algebra $\mathscr{M}$ has a unit element $1_{\mathscr{M}}$, the unit element of a W^*-subalgebra $\mathscr{N}$ of $\mathscr{M}$ can be a proper projection $e_{\mathscr{N}}$ in $\mathscr{M}$. If $e_{\mathscr{N}} = 1_{\mathscr{M}}$, then $\mathscr{N}$ is called a *unital* W^*-*subalgebra* of $\mathscr{M}$.

If $\mathscr{A}$ is any C^*-algebra, there exists a unique C^*-algebra structure on the Banach space $\mathscr{A}^{**}$ such that the canonical image of $\mathscr{A}$ in $\mathscr{A}^{**}$ is a C^*-subalgebra of $\mathscr{A}$. Moreover, the C^*-algebra $\mathscr{A}^{**}$ is a W^*-algebra, *-isomorphic to the enveloping von Neumann algebra of $\mathscr{A}$.

Let Φ be a *-isomorphism of the von Neumann algebra $\mathscr{M} \subset \mathscr{B}(\mathscr{H})$ onto the von Neumann algebra $\mathscr{N} \subset \mathscr{B}(\mathscr{H})$. By ([L], 9.25), ϕ extends to a one-to-one correspondence $A \mapsto B = \Phi(A)$ between the positive self-adjoint operators A in $\mathscr{H}$ affiliated to $\mathscr{M}$ and the positive self-adjoint operators B in $\mathscr{K}$ affiliated to $\mathscr{N}$; Φ is unique subject to the condition $\Phi((1+A)^{-1}) = (1+B)^{-1}$. It is therefore meaningful to speak about *positive self-adjoint operators A affiliated to a* W^*-*algebra* $\mathscr{M}$ without explicitly mentioning the specific realization of $\mathscr{M}$ as a von Neumann algebra.

For the proof of the preceding results, we refer to Sakai (1971); Strătilă and Zsidó (1977, 2005); Takesaki (1969–1970).

A.17. For each $k \in \mathbb{N}$, consider a Hilbert space $\mathscr{H}_k$ and a vector $\xi_k \in \mathscr{H}_k$, $\|\xi_k\| = 1$. Then the mappings

$$I_k : \mathscr{H}_1 \bar{\otimes} \ldots \bar{\otimes} \mathscr{H}_k \ni \zeta \mapsto \zeta \otimes \xi_{k+1} \in \mathscr{H}_1 \bar{\otimes} \ldots \bar{\otimes} \mathscr{H}_k \otimes \mathscr{H}_{k+1}$$

are linear and isometric. The completion of the direct limit of the Hilbert spaces $\mathscr{H}_1 \bar{\otimes} \ldots \bar{\otimes} \mathscr{H}_k$ along the mappings I_k is again a Hilbert space, denoted by $\otimes_k(\mathscr{H}_k, \xi_k)$ and called the (incomplete) *infinite tensor product of the Hilbert spaces* $\mathscr{H}_k$ *along the vectors* $\xi_k \in \mathscr{H}_k$. For any sequence of vectors $\eta_k \in \mathscr{H}_k$ such that

$$\sum_k |1 - \|\eta_k\|| < +\infty \text{ and } \sum_k |1 - (\eta_k|\xi_k)| < +\infty \tag{1}$$

there corresponds a "decomposable vector"

$$\otimes_k \eta_k = \lim_k \eta_1 \otimes \ldots \otimes \eta_k \otimes \xi_{k+1} \otimes \xi_{k+2} \otimes \ldots \in \otimes_k(\mathscr{H}_k, \xi_k) \tag{2}$$

which depends linearly on each η_k. If $\otimes_k\zeta_k \in \otimes_k(\mathscr{H}_k, \xi_k)$ is another decomposable vector, then

$$(\otimes_k\eta_k | \otimes_k \zeta_k) = \prod_k (\eta_k|\zeta_k) \tag{3}$$

where the infinite product is absolutely convergent. The set of decomposable vectors is total in $\otimes_k(\mathscr{H}_k, \xi_k)$.

Let $x_k \in \mathscr{B}(\mathscr{H}_k)$ $(k \in \mathbb{N})$. For each fixed $k_0 \in \mathbb{N}$, there exists a unique bounded linear operator $x_1 \otimes \ldots \otimes x_{k_0} \otimes 1$ on $\otimes_k(\mathscr{H}_k, \xi_k)$ such that

$$(x_1 \otimes \ldots \otimes x_{k_0} \otimes 1)(\otimes_k\eta_k) = x_1\eta_1 \otimes \ldots \otimes x_{k_0}\eta_{k_0} \otimes \eta_{k_0+1} \otimes \eta_{k_0+2} \otimes \ldots \tag{4}$$

for each decomposable vector $\otimes_k\eta_k$. In certain special situations, the sequence $\{x_1 \otimes \ldots \otimes x_k \otimes 1\}_{k\in\mathbb{N}}$ is *so*-convergent to a bounded linear operator on $\otimes_k(\mathscr{H}_k, \xi_k)$ which is then denoted by $\otimes_k x_k$.

For each $k \in \mathbb{N}$, consider also a von Neumann algebra $\mathcal{M}_k \subset \mathcal{B}(\mathcal{H}_k)$. Then

$$\otimes_k (\mathcal{M}_k, \xi_k) = \{x_1 \otimes \ldots \otimes x_k \otimes 1; x_1 \in \mathcal{M}_1, \ldots, x_k \in \mathcal{M}_k, k \in \mathbb{N}\}'' \subset \mathcal{B}(\otimes_k(\mathcal{H}_k, \xi_k)) \quad (5)$$

is a von Neumann algebra, called the *infinite tensor product of the von Neumann algebras* $\mathcal{M}_k \subset \mathcal{B}(\mathcal{H}_k)$ *along the vectors* $\xi_k \in \mathcal{H}_k$.

On the other hand, for each $k \in \mathbb{N}$, let $\mathcal{M}_k$ be a W*-algebra and let φ_k be a normal state on $\mathcal{M}_k$. Let $\pi_k : \mathcal{M}_k \to \pi_k(\mathcal{M}_k) \subset \mathcal{B}(\mathcal{H}_k)$ be the normal cyclic *-representation of $\mathcal{M}_k$ associated with φ_k, with the cyclic vector $\xi_k \in \mathcal{H}_k, \|\xi_k\| = 1$, that is, $\varphi_k(x) = (\pi_k(x)\xi_k|\xi_k)$ for every $x \in \mathcal{M}_k$ $(k \in \mathbb{N})$. Then the von Neumann algebra

$$\otimes_k (\mathcal{M}_k, \varphi_k) = \otimes_k(\pi_k(\mathcal{M}_k), \xi_k) \quad (6)$$

is called the *infinite tensor product of the W*-algebras* $\mathcal{M}_k$ *along the normal states* φ_k on $\mathcal{M}_k$. An alternative equivalent definition is given in Sakai (1971), 4.4.

If all the $\mathcal{M}_k$ are factors, then $\otimes_k(\mathcal{M}_k, \xi_k)$ and $\otimes_k(\mathcal{M}_k, \varphi_k)$ are also factors.

Assume that each $\mathcal{M}_k$ is a finite discrete factor, say of type I_{n_k} $(1 < n_k < +\infty)$ and denote by tr_k the canonical trace on $\mathcal{M}_k(tr_k(1) = n_k)$. For any (normal) state φ_k on $\mathcal{M}_k$, there exists a unique positive element $a_k \in \mathcal{M}_k$ such that $\varphi_k = tr_k(a_k\cdot)$. We shall denote by

$$Sp(\varphi_k/\mathcal{M}_k) = \{\lambda_{k,1} \geq \lambda_{k,2} \geq \ldots \lambda_{k,n_k}\} \quad (7)$$

the set of eigenvalues of a_k, each repeated according to its multiplicity; note that $\lambda_{k,1} + \ldots + \lambda_{k,n_k} = 1$. In this case, the factor $\otimes_k(\mathcal{M}_k, \varphi_k)$ is called an *Araki–Woods factor*, or an ITPFI-*factor* (ITPFI = infinite tensor product of factors of type I), and the set $\{\lambda_{k,j}; 1 \leq j \leq n_k, k \in \mathbb{N}\}$ is called the *eigenvalue list* of $\otimes_k(\mathcal{M}_k, \varphi_k)$. The *-isomorphy class of an ITPFI-factor depends only on its eigenvalue list.

Actually, Araki and Woods (1968) (Definition 3.2) introduced an invariant $r_\infty(\mathcal{M}, \Omega)$ for an ITPFI factor $\mathcal{M}$ based on its eigenvalue list Ω, called *asymptotic ratio set*, and Connes (1973a) proved that

$$r_\infty(\mathcal{M}, \Omega) = S(\mathcal{M})$$

and also that this set is equal to the set

$$r_\infty(\mathcal{M}) = \left\{\lambda > 0; \mathcal{M} \text{ is isomorphic to } \mathcal{M}\overline{\otimes}\mathcal{R}_\lambda\right\},$$

where the Powers factor $\mathcal{R}_\lambda$ is defined below.

In particular, assume that all $\mathcal{M}_k$ are factors of type I_2. Then $Sp(\varphi_k/\mathcal{M}_k) = \{q_k, p_k\}$ with $q_k \geq p_k \geq 0$ and $p_k + q_k = 1$, and the *-isomorphism class of $\otimes_k(\mathcal{M}_k, \varphi_k)$ depends only on $\lambda_k = p_k/q_k \in [0, 1]$ $(k \in \mathbb{N})$. We shall denote this factor by

$$\mathcal{R}(\{\lambda_k\}_{k\in\mathbb{N}} = \otimes_k(\mathcal{M}_k, \varphi_k).$$

If $\lambda_k = \lambda \in (0, 1]$ for all $k \in \mathbb{N}$, then

$$\mathcal{R}_\lambda = \otimes_k(\mathcal{M}_k, \varphi_k)$$

is called the *Powers factor* corresponding to $\lambda \in (0, 1]$. For $\lambda = 1, \mathscr{R}_1$ is nothing but the hyperfinite type II_1 factor $\mathscr{R} \approx \mathfrak{L}(S(\infty))$ (see 10.31, 22.6, 6, 22.16).

There are approximately finite dimensional non-ITPFI factors (Connes & Woods, 1980a). There is a characterization of ITPFI factors by their flow of weights (Connes & Woods, 1980b, 1985).

For details concerning the construction of infinite tensor products, we refer to Araki and Woods (1968); Bures (1963, 1968); Connes (1973a); Guichardet (1969); Neumann (1938); Sakai (1971); Størmer (1971b).

A.18. Notes. The material contained in Sections A.12–A.14 is due to van Daele (1976). Proposition A.9 is a result due to T. Kato. The other results in this Appendix are classical.

For our exposition, we have used Araki and Woods (1968, 1973); Bures (1963, 1968); Van Daele (1976); Faris (1975); Guichardet (1969); Kadison (1968); Kallman (1971); Kaplansky (1951); Kato (1966); Neumann (1938); Pedersen and Takesaki (1973); Reed and Simon (1972); Sakai (1971); Størmer (1971b); Strătilă and Zsidó (1977, 2005); Takesaki (1969–1970).

References

Aarnes, J. F. (1968). On the Mackey topology for von Neumann algebras. *Mathematica Scandinavica*, **22**, 87–107.

Abdalla, J. and Szücs, J. (1974). On an ergodic type theorem for von Neumann algebras. *Acta Scientiarum Mathematicorum*, **36**, 167–72.

Akemann, C. A. (1967). The dual space of an operator algebra. *Transactions of the American Mathematical Society*, **126**, 286–302.

Akemann, C. A., Elliott, G. A., Pedersen, G. K. and Tomiyama, J. (1976). Derivations and multipliers of C*-algebras. *American Journal of Mathematics*, **98**, 679–708.

Araki, H. (1969). A classification of factors II. *Publications RIMS, Kyoto University*, **4**, 585–93.

———. (1971). Asymptotic ratio set and property L'_λ. *Publications RIMS,Kyoto University*, **6**, 443–60.

———. (1973). Structure of some von Neumann algebras with isolated discrete modular spectrum. *Publications RIMS, Kyoto University*, **9**, 1–44.

———. (1974). Some properties of the modular conjugation operator of von Neumann algebras and a non-commutative Radon–Nikodym theorem with a chain rule. *Pacific Journal of Mathematics*, **50**, 309–54.

———. (1976). Recent developments in the theory of operator algebras and their significance in theoretical physics, *Symposia Mathematica*, XX, 395–424.

Araki, H. and Woods, E. J. (1968). A classification of factors. *Publications RIMS,Kyoto University*, **3**, 51–130.

———. (1973). Topologies induced by representations of the canonical commutation relations. *Reports on Mathematical Physics*, **4**, 227–54.

Arveson, W. B. (1969). Subalgebras of C*-algebras. *Acta Mathematica*, **123**, 141–244.

———. (1974). On groups of automorphisms of operator algebras. *Journal of Functional Analysis*, **15**, 217–43.

Barnett, L. (1995). Free product von Neumann algebras of type III. *Proceedings of the American Mathematical Society*, **123**, 543–53.

Bisognano, J. J. and Wichmann, E. H. (1975). On the duality for a hermitian scalar field. *Journal of Mathematical Physics*, **16**, 985–1007.

Borchers, H. J. (1966). Energy and momentum as observable in quantum field theory. *Communication in Mathematical Physics*, **2**, 49–54.

———. (1969). On groups of automorphisms with semibounded spectrum. *CNRS Conference on Systems with Infinitely Many Degrees of Freedom*, pp. 125–42.

———. (1973). *Uber Ableitungen von C*-Algebren, Nachricten Akademie Wissenschaften Gottingen*, 1–18.

———. (1974). Characterization of inner *-automorphisms of W*-algebras. *Publications RIMS, Kyoto University*, **10**, 11–50.

Bunce, J. and Paschke, W. L. (1978). Quasi-expectations and amenable von Neumann algebras. *Proceedings of the American Mathematical Society*, **71**, 232–6.

Bures, D. (1963). Certain factors constructed as infinite tensor products. *Compositio Mathematica*, **15**, 169–91.

———. (1968). Tensor products of W*-algebras. *Pacific Journal of Mathematics*, **27**, 13–37.

Ching, W. M. (1969). Non-isomorphic non-hyperfinite factors. *Canadian Journal of Mathematics*, **21**, 1293–308.

Choda, H. (1971). On ergodic and abelian automorphism groups of von Neumann algebras. *Proceedings of Japan Academy*, **47**(Supplement I), 982–5.

———. (1974). On freely acting automorphisms of operator algebras. *Kodai Mathematical Seminar Reports*, **26**, 1–21.

Choda, H. and Echigo, M. (1963). A new algebraic property of certain von Neumann algebras. *Proceedings of Japan Academy*, **39**, 651–6.

———. (1964). Some remarks on von Neumann algebras with property Q. *Memoirs Osaka Gagukei University*, **13**, 13–21.

Choi, M. D. and Effros, E. G. (1976). Separable nuclear C*-algebras and injectivity. *Duke Mathematical Journal*, **43**, 309–22.

———. (1977a). Injectivity and operator spaces. *Journal of Functional Analysis*, **24**, 156–209.

———. (1977b). Nuclear C*-algebras and injectivity; the general case. *Indiana University Mathematical Journal*, **26**, 443–6.

Colojoara, I. and Foias, C. (1968). *Theory of Generalized Spectral Operators*. London: Gordon & Breach.

Combes, F. (1968). Poids sur une C*-algèbre. *Journal de Mathématiques Pureset Appliquées*, **47**, 57–100.

———. (1971a). Poids associés à une algèebre hilbertienne à gauche. *Compositio Mathematica*, **23**, 49–77.

———. (1971b). Poids et espérances conditionnelles dans les algèbres de von Neumann. *Bulletin de la Société Mathématique de France*, **99**, 73–112.

Combes, F. and Delaroche, C. (1973–1974). *Séminaire sur les algèbres d'opérateurs*. Orléans: Université d'Orléans.

———. (1975). Groupe modulaire d'une espérance conditionnelle dans une algèbre de von Neumann, *Bulletin de la Société Mathématique de France*, **103**, 385–426.

Connes, A. (1972). États presque périodiques sur une algèbre de von Neumann. *Comptes Rendus de l'Académie des Sciences de Paris*, **274**, 1402–5.

———. (1973a). Une classification des facteurs de type III. *Annales de l'Ecole Normale Supérieure*, **6**, 133–252.

———. (1973b). Sur le théoreme de Radon Nikodym pour les poids normauxfidéles sémifinis. *Bulletin de la Société Mathématique de France*, **97**, 253–8.

———. (1973c). Almost periodic state and Krieger factors. Unpublished preprint, Marseille.

———. (1974a). Almost periodic states and factors of type III_1. *Journal of Functional Analysis*, **16**, 415–45.

———. (1974b). Structure theory for type III factors. *Proceedings of the International Congress of Mathematicians Vancouver*, **II**, 87–91.

———. (1974c). Caractérisation des espaces vectoriels ordonnés sous-jacentsaux algèbres de von Neumann. *Annales de l'Institut Fourier*, **24**, 121–55.

———. (1974d). Almost periodic states and factors of type III_1. *Journal of Functional Analysis*, **16**, 415–45.

———. (1975a). A factor not antiisomorphic with itself. *Annals of Mathematics*, **101**, 536–54.

———. (1975b). Sur la classification des facteurs de type II. *Comptes Rendusde l'Academie des Sciences de Paris*, **281**, 13–15.

———. (1975c). On the hierarchy of Krieger. *Illinois Mathematical Journal*, **19**, 428–32.

———. (1975d). On hyperfinite factors of type III_0 and Krieger's factors. *Journal of Functional Analysis*, **18**, 318–327.

———. (1975e). Outer conjugacy classes of automorphisms of factors. *Annales de l'Ecole Normale Supérieure*, **4**, 383–419.

———. (1976a). Classification of injective factors. *Annals of Mathematics*, **104**, 73–115.

———. (1976b). Outer conjugacy of automorphisms of factors. *Symposia Mathematica*, **XX**, 149–60.

———. (1977a). Periodic automorphisms of the hyperfinite factor of type II_1. *Acta Scientiarum Mathematicarum*, **39**, 39–66.

———. (1977b). Quelques aspects de la théorie des algèbres d'opérateurs. *RCP 25 Strasbourg*, **24**, 13–87.

———. (1978a). On the cohomology of operator algebras. *Journal of Functional Analysis*, **28**, 248–53.

———. (1978b). On the classification of von Neumann algebras and their automorphisms. *Symposia Mathematica*, **XX**, 435–78.

———. (1979a). Sur la théorie non-commutative de l'intégration. *Lecture Notes in Mathematics No.*, **725**, 19–143.

———. (1979b). The von Neumann algebra of a foliation. *Lecture Notes in Physics*, **40**, 145–51.

———. (1980). On a spatial theory of von Neumann algebras. *Journal of Functional Analysis*, **35**, 153–64.

———. (1985). Factors of type III_1 and property L'_λ. *Journal of Operator Theory*, **14**, 189–211.

Connes, A., Feldman. J. and Weiss, B. (1981). An amenable equivalence relations is generated by a single transformation. *Ergodic Theory and Dynamical Systems*, **4**, 431–50.

Connes, A., Ghez, P., Lima, R., Testard, D. and Woods, E. J. (1973). Review of Crossed Products of von Neumann Algebras (paper by Golodec). Unpublished preprint, CNRS Marseille.

Connes, A., Haagerup, U. and Størmer, E. (1983). Diameters of state spaces of type III factors. *Lecture Notes in Mathematics*, **1132**, 91–116, Springer Verlag.

Connes, A. and Krieger, W. (1977). Measure space automorphisms, the normalizers of their full groups and approximate finiteness. *Journal of Functional Analysis*, **24**, 336–52.

Connes, A. and Størmer, E. (1978). Homogeneity of the state space of factors of type III_1. *Journal of Functional Analysis*, **28**, 187–96.

Connes, A. and Takesaki, M. (1974). Flôt des poids sur les facteurs de type III. *Comptes Rendus de l'Académie des Sciences de Paris*, **278**, 945–8.

———. (1977). The flow of weights on a factor of type III. *Tohoku Mathematical Journal*, **29**, 473–575. Errata 1978: ibidem **30**, 653–5.

Connes, A. and Van Daele, A. (1973). The group property of the invariant S. *Mathematica Scandinavica*, **32**, 187–92.

Connes, A. and Woods, E. J. (1974). Existence des facteurs infinis asymptotiquement abéliens. *Comptes Rendus de l'Académie des Sciences de Paris*, **279**, 189–91.

———. (1980a). A construction of approximately finite-dimensional non-ITPFI factors. *Canadian Mathematical Bulletin*, **23**, 227–30.

———. (1980b). Characterization of ITPFI factors by their flow of weights. Unpublished preprint.

———. (1985). Approximately transitive flows and ITPFI factors. *Ergodic Theory and Dynamical Systems*, **5**, 203–36.

Danilenko, A. I. and Silva, C. E. (2012). Ergodic theory: non-singular transformations. In R. A. Meyers, ed., *Mathematics of Complexity and Dynamical Systems*, 1–3 vols. Berlin: Springer, pp. 329–56.

Day, M. M. (1957). Amenable semigroups. *Illinois Journal of Mathematics*, **1**, 509–44.

Delaroche, C. (1979). Action moyennable d'un groupe localement compact sur une algèbre de von Neumann. *Mathematica Scandinavica*, **45**, 289–304.

Digernes, T. (1975). Duality for weights on covariant systems and its applications. Dissertation, UCLA.

Dinculeanu, N. (1967). *Vector Measures*. Oxford: Pergamon Press.

Dixmier, J. (1949). Les anneaux d'opérateurs de classe finie. *Annales de l'Ecole Normale Superieure*, **66**, 209–61.

———. (1952). Applications ♮ dans les anneaux d'operateurs. *Compositio Mathematica*, **10**, 1–55.

———. (1953). Formes linéaires sur un anneau d'opérateurs. *Bulletin de la Société Mathématique de France*, **81**, 9–39.

———. (1957, 1969). *Les algèbres d'opérateurs dans l'espace hilbertien* (Algèbresde von Neumann), Paris: Gauthier Villars.

———. (1969). Sur les groupes d'automorphismes normiquement continus. *Comptes Rendus de l'Académie des Sciences de Paris*, **269**, 643–4.

Dixmier, J. and Marechal, O. (1971). Vecteurs totalisateurs d'une algèbre de von Neumann. *Communications in Mathematical Physics*, **22**, 44–50.

Doplicher, S., Kastler, D. and Robinson, D. W. Covariance algebras in field theory and statistical mechanics. *Communications in Mathematical Physics*, **3**, 1–28.

Dunford, N. and Schwartz, L. (1958, 1963). *Linear Operators I, II*. Interscience Publishers.

Dye, H. A. (1959, 1963). On groups of measure preserving transformations I, II. *American Journal of Mathematics*, **81**, 119–59, **85**, 551–76.

Dykema, K. (1994a). Factoriality and Connes invariant T(M) for free products of von Neumann algebras. *Journal fur Reine und Angewandte Mathematik*, **450**, 159–80.

———. (1994b). Interpolated free group factors. *Pacific Journal of Mathematics*, **163**, 123–35.

Effors, E. G. (1972). Injectives and tensor products for convex sets and C*-algebras. *Lecture given at the NATO Conference*, University of Swansea, Wales.

Effors, E. G. and Lance, E. C. (1977). Tensor products of operator algebras. *Advancesin Mathematics*, **25**, 1–34.

Elliott, G. A. (1975). On the Radon Nikodym derivative with a chain rule in von Neumann algebras. *Canadian Mathematical Bulletin*, **18**, 661–9.

Elliott, G. A. and Woods, E. J. (1976). The equivalence of various definitions of hyperfiniteness of a properly infinite von Neumann algebra. *Proceedings of the American Mathematical Society*, **60**, 175–8.

Enock, M. (1977). Produit croisé d'une algèbre de von Neumann par une algèbre de Kac. *Journal of Functional Analysis*, **26**, 16–47.

Enock, M. and Schwartz, J. M. (1975). Une dualité dans les algèbres de von Neumann. *Bulletin de la Société Mathématique de France*, Mémoire No. **44**, 1–144.

———. (1980). Produit croisé d'une algèbre de von Neumannpar une algebre de Kac, II. *Publications RIMS, Kyoto University*, **16**, 189–232.

Ernest, J. (1966). Hopf-von Neumann algebras. *Lecture at the Functional Analysis Conference*, Irvine, Academic Press, pp. 195–215.

Eymard, P. (1964). L'algèbre de Fourier d'un groupe localement compact. *Bulletin de la Société Mathématique de France*, **92**, 181–236.

Falcone, A. J. and Takesaki, M. (1999). Operator valued weights without structure theory. *Transactions of the American Mathematical Society*, **351**, 323–41.

———. (2001). Non-commutative flow of weights on a von Neumann algebra. *Journal of Functional Analysis*, **182**, 170–206.

Faris, W. G. (1975). Self-adjoint operators. *Lecture Notes in Mathematics*, **433**.

Feldman, J. and Moore, C. C. (1977). Ergodic equivalence relations, cohomology and von Neumann algebras. *Transactions of the American Mathematical Society*, **234**, 289–361.

Fuglede, B. and Kadison, R. V. (1951). On a conjecture of Murray and von Neumann. *Proceedings of the National Academy of Science USA*, **37**, 420–5.

Ge, L. M. (2001). Problems on von Neumann algebras by R.V.Kadison 1967 with a previously unpublished manuscript by Kadison. *International Workshop on Operator Algebras and Operator Theory* (Linfen 2001) *Acta Mathematica Sinica* (English Series) **19**, 619–24.

Gleason, A. M. (1958). Projective topological spaces. *Illinois Journal of Mathematics*, **2**, 482–9.

Golodec, V. Ya. (1971). Crossed products of von Neumann algebras (in Russian). *Uspekhi Matematicheskih Nauk*, **26**(5), 3–50.

———. (1972). Spectral properties of modular operators (in Russian). *Funkcionalnyi Analiz I Prilozheniya*, **6**(1), 70–72.

———. (1978). Modular operators and asymptotic abelianess in von Neumann algebras (in Russian). *Uspekhi Matematicheskih Nauk*, **33**(1):43–94.

Greenleaf, F. P. (1969). Invariant means on topological groups and their applications, Van Nostrand.

Guichardet, A. (1969). *Tensor products of C∗-algebras*, Aarhus University Lecture Notes Series, No. 12, 13.

———. (1974). *Systèmes dynamiques non-commutatifs*, *Astérisque*, No. 13, 14.

Haagerup, U. (1975a). Normal weights on W*-algebras. *Journal of Functional Analysis*, **19**, 302–318.

———. (1975b). The standard form of von Neumann algebras. *Mathematica Scandinavica*, **37**, 271–83.

———. (1977). An example of a weight with type III centralizer. *Proceedings of the American Mathematical Society*, **62**, 278–80.

———. (1978). On the dual weights for crossed products of von Neumannalgebras, I, II. *Mathematica Scandinavica*, **43**, 99–118, 119–40.

———. (1979a). Operator valued weights in von Neumann algebras, I, II. *Journal of Functional Analysis*, **32**, 175–206, **33**, 339–61.

———. (1979b). L^p spaces associated with an arbitrary von Neumann algebra,in Algèbres d'opérateurs et leurs applications en physique mathématique. *Proceedings Colloques Internation aux CNRS Marseille 274*, CNRS, pp. 175–84.

———. (1979c). A density theorem for left Hilbert algebras, *Lecture Notes in Mathematics No. 725*, 170–9.

———. (1985). A new proof of the equivalence of injectivity and hyper-finiteness for factors on a separable Hilbert space. *Journal of Functional Analysis*, **62**, 160–201.

Haga, Y. and Takeda, Z. (1972). Correspondence between subgroups and subalgebras in a cross product von Neumann algebra. *Tohoku Mathematical Journal*, **24**, 167–90.

Hahn, P. F. (1978a). Haar measure for measure groupoids. *Transactions of the American Mathematical Society*, **242**, 1–33.

———. (1978b). The regular representation for measure groupoids. *Transactions of the American Mathematical Society*, **242**, 35–72.

Hakeda, J. and Tomiyama, J. (1967). On some extension properties of von Neumann algebras. *Tohoku Mathematical Journal*, **19**, 315–23.

Halpern, H. (1969). Module homomorphisms of a von Neumann algebra into its center. *Transactions of the American Mathematical Society*, **140**, 183–94.

Hamachi, T. and Kosaki, H. (1993). Orbital factor map. *Ergodic Theory and Dynamical Systems*, **13**, 515–32.

Hansen, F. (1977). Inner one parameter groups acting on a factor. *Mathematica Scandinavica*, **41**, 113–6.

Hasumi, M. (1958). The extension property of complex Banach spaces. *Tohoku Mathematical Journal*, **10**, 135–42.

Henle, M. (1979). Galois Theory of W*-algebras. Unpublished preprint.

Herman, R. H. and Takesaki, M. (1970). States and automorphism groups of operator algebras. *Communications in Mathematical Physics*, **19**, 142–60.

Hewitt, E. and Ross, K. A. (1963, 1970). *Abstract Harmonic Analysis I, II*. Berlin: Springer Verlag.

Hilsum, M. (1981). Les espaces L^p d'une algèbre de von Neumann. *Journal of Functional Analysis*, **40**, 151–69.

Hormander, L. (1966). *An Introduction to Complex Analysis in Several Variables*. Van Nostrand.

Houdayer, C., Shlyakhtenko, D. and Vaes, S. (2019). Classification of a family of non-almost-periodic free Araki-Woods factors. *Journal of European Mathematical Society*, **21**, 3113–42.

Houdayer, C. and Vaes, S. (2013). Type III factors with unique Cartan decomposition. *Journal de Mathématiques Pures et Appliquées*, **100**, 564–90.

Ikunishi, A. and Nakagami, Y. (1976). On invariants G(σ) and $\Gamma(\sigma)$ for an automorphism group of a von Neumann algebra. *Publications RIMS, Kyoto University*, **12**, 1–30.

Johnson, B. E. (1972). Cohomology of Banach algebras. *Memoirs of the American Mathematical Society*, **127**.

Johnson, B. E., Kadison, R. V. and Ringrose, J. R. (1972). Cohomology of operator algebras, III. *Bulletin de la Société Mathématique de France*, **100**, 73–96.

Jones, V. F. R. and Takesaki, M. (1984). Actions of compact abelian groups on semifinite injective factors. *Acta Mathematica*, **153**, 213–58.

Jones, V. F. R. (1977). Sur la conjugaison des sous-facteurs de type II_1. *Comptes Rendus de l'Académie des Sciences de Paris*, **284**, 597–8.

———. (1980). Actions of finite groups on the hyper finite type II_1 factors. *Memoirs of the American Mathematical Society*, **237**.

———. (1983). A converse to Ocneanu's theorem. *Journal of Operator Theory*, **10**, 61–63.

Kac, G. I. (1963, 1964). Group rings and the duality principle I, II (in Russian). *Trudy Moskovskogo Matematicheskoe Obshchestvo*, **12**, 259–301, **13**, 84–113.

Kadison, R. V. (1965). Transformations of states in operator theory and dynamics. *Topology*, **3**, 117–98.

———. (1966). Derivations of operator algebras. *Annals of Mathematics*, **83**, 280–93.

———. (1968). Strong continuity of operator functions. *Pacific Journal of Mathematics*, **26**, 121–9.

———. (1969). Some analytic methods in the theory of operator algebras. *Lecture Notes in Mathematics*, **140**, 8–29.

———. (1971). On the problem of isomorphism of hyperfinite II_∞ factors.A critique of the article by Golodec. *Soviet Journal of Functional Analysis and its Applications*, **5**, 37–44. Unpublished preprint.

Kadison, R. V. and Ringrose, J. R. (1967). Derivations and automorphisms of operator algebras. *Communications in Mathematical Physics*, **4**, 32–63.

Kallman, R. R. (1969). A generalization of free action. *Duke Mathematical Journal*, **36**, 781–9.

———. (1970). A decomposition theorem for automorphisms of von Neumann algebras. In *Inequalities*, II vols. Cambridge, MA: Academic Press, pp. 33–5.

———. (1971). Groups of inner automorphisms of von Neumann algebras. *Journal of Functional Analysis*, **7**, 43–60.

Kaplansky, I. (1951). A theorem on rings of operators. *Pacific Journal of Mathematics*, **1**, 227–32.

Katayama, Y. (1974). The tensor product of weights. *Proceedings of Japan Academy*, **50**, 430–2.

Katayama, Y., Sutherland, C. E. and Takesaki, M. (1998). The characteristic square of a factor and the cocycle conjugacy of discrete amenable group actions on factors. *Inventiones Mathematicae*, **132**, 331–80.

Kato, T. (1966). *Perturbation Theory for Linear Operators*. Berlin: Springer Verlag.

Katznelson, Y. and Weiss, B. (1991). The classification of non-singular actions, revisited. *Ergodic Theory and Dynamical Systems*, **11**, 333–48.

Kawahigashi, Y., Sutherland, C. E. and Takesaki, M. (1992). The structure of the automorphismgroup of an injective factor and the cocycle conjugation of discrete abelian group actions. *Acta Mathematica*, **169**, 105–30.

Kirchberg, E. (1975).Darstellungen coinvolutiver Hopf-W*-Algebren und ihre Anwendung der nichtabelschen Dualitatstheorie lokalkompakter Gruppen. Dissertation, Berlin.

Kishimoto, A. and Takai, H. (1978). On the invariant $\Gamma(\sigma)$ in C*-dynamical systems. *Tohoku Mathematical Journal*, **30**, 83–94.

Kosaki, H. (1986). Extension of Jones' theory on index to arbitrary factors. *Journal of Functional Analysis*, **66**, 123–40.

———. (1992). A remark on Sakai's quadratic Radon Nikodym theorem. *Proceedings of the American Mathematical Society*, **116**, 783–6.

Kovács, I. and Szücs, J. (1966). Ergodic type theorems in von Neumann algebras. *Acta Scientiarum Mathematicarum*, **27**, 233–46.

Krieger, W. (1968). On the isomorphy problem for ergodic equivalence relations. *Mathematische Zeitschrift*, **103**, 78–84.

———. (1969). On non-singular transformations of a measure space, I, II. *Z. Wahrscheinlichkeitstheorie und verw. Gebiete*, **11**, 83–97, 98–119.

———. (1970a). On constructing non-isomorphic hyperfinite factors of type III. *Journal of Functional Analysis*, **6**, 97–109.

———. (1970b). On the Araki-Woods asymptotic ratio set and non-singular transformations of a measure space. *Lecture Notes in Mathematics*, **160**, 158–77.

———. (1971a). On a class of hyperfinite factors which arise from null-recurrent Markov chains. *Journal of Functional Analysis*, **7**, 27–42.

———. (1971b). On the hyperfinite factors and non-singular transformations of a measure space. Unpublished preprint.

———. (1972, 1974). On the infinite product construction of non-singular transformations of a measure space. *Inventiones Mathematicae*, **15**, 144–63, **26**, 323–8.

———. (1974). On the Dye weak equivalence theorem. *Proceedings of the International School of Mathematical Physics*, University of Camerino, Camerino, pp. 355–65.

———. (1976a). On the construction of factors from ergodic non-singulartransformations. In C*-*Algebras and their Applications to Statistical Mechanics and Quantum Field Theory*. North-Holland, pp. 114–20.

———. (1976b). On ergodic flows and the isomorphism of factors. *Mathematische Annalen*, **223**, 19–70.

Lance, E. C. (1973). On nuclear C*-algebras. *Journal of Functional Analysis*, **12**, 157–76.

Landstad, M. (1977). Duality for dual covariant algebras. *Communications in Mathematical Physics*, **52**, 191–202.

———. (1979). Duality theory for covariant systems. *Transactions of the American Mathematical Society*, **248**, 223–76.

Landstad, M., Olesen, D. and Pedersen, G. K. (1978). Towards a Galois theory for crossed products of C*-algebras. *Mathematica Scandinavica*, **43**, 311–21.

Loomis, L. H. (1952). Note on a theorem of Mackey. *Duke Mathematical Journal*, **19**, 641–5.

Lyubic, Yu. I. (1971). On the spectrum of a representation of an abelian topological group (in Russian). *Doklady Akademii Nauk SSSR*, **200**, 777–80.

Lyubic, Yu. I. and Macaev, V. I. (1962). On operators with separable spectrum (in Russian). *Matematicheskii Sbornik*, **56**, 433–68.

Lyubic, Yu. I., Macaev, V. I. and Feldman, G. M. (1973). On representations with separable spectrum (in Russian). *Funkcionalnyi Analiz I Prilozheniya*, **7**, 52–61.

Mackey, G. W. (1949). A theorem of Stone and von Neumann. *Duke Mathematical Journal*, **16**, 313–26.

Mackey, G. W. (1966). Ergodic theory and virtual groups. *Mathematische Annalen*, **166**, 187–207.

Mokobodsky, G. (1962). Sur des mesures qui définissent des graphes d'applications, *Séminaire Choquet*.

Moore, C. C. (1967). Invariant measures on product spaces. *Proceedings V-th Berkeley Symposium on Mathematics, Statistics and Probability*, **II**, Part II, 447–59.

———. (1976). Group extensions and cohomology for locally compact groups, III, IV. *Transactions of the American Mathematical Society*, **221**, 1–33, 34–58.

Murray, F. J. and von Neumann, J. (1936, 1937, 1943). On rings of operators, I, II, IV. *Annals of Mathematics*, **37**, 116–229, *Transactions of the American Mathematical Society*, **41**, 208–48, *Annals of Mathematics*, **44**, 716–808.

Nakagami, Y. and Takesaki, M. (1979). Duality for crossed products of von Neumannalgebras, *Lecture Notes in Mathematics*, **731**.

Nakamura, M. (1954). On the direct product of finite factors. *Tohoku Mathematical Journal*, **6**, 205–7.

Nakamura, M. and Takeda, Z. (1958). On some elementary properties of the crossed products of von Neumann algebras, *Proceedings Japan Academy*, **34**, 489–94.

———. (1960). On the fundamental theorem of Galois theory. *Proceedings Japan Academy*, **36**, 313–8.

Nielsen, O. A. (1974–1975). Reduced crossed products of von Neumann algebras, I. Unpublished preprint No. 24, Aarhus University.

Ocneanu, A. (1980). Actions des groupes moyennables sur les algèbres de von Neumann. *Comptes Rendus de l'Académie des Sciences de Paris*, **291**, 399–401.

Ocneanu, A. (1981). A Rokhlin type theorem for groups acting on von Neumannalgebras. *Proceedings Vth International Conference on Operator Theory*, Timisoara, Romania, pp. 247–58, Birkhauser Verlag.

Ocneanu, A. (1985). Actions of discrete amenable groups on factors. *Lecture Notes in Mathematics*, **1138**.

Olesen, D. (1974). On norm continuity and compactness of the spectrum. *Mathematica Scandinavica*, **35**, 223–36.

———. (1975). Inner *-automorphisms of simple C*-algebras. *Communications in Mathematical Physics*, **44**, 175–90.

———. (1976). On spectral subspaces and their applications to automorphism groups. *Symposia Mathematica*, **XX**, 253–96.

Olesen, D. and Pedersen, G. K. (1974). Derivations of C*-algebras have semi continuous generator. *Pacific Journal of Mathematics*, **53**, 563–72.

———. (1978, 1980). Applications of the Connes spectrum to C^*-dynamical systems. *Journal of Functional Analysis*, **30**, 179–97, **36**, 18–32.

Olesen, D., Pedersen, G. K. and Stormer, E. (1977). Compact abelian groups of automorphisms of simple C*-algebras. *Inventiones Mathematicae*, **39**, 55–64.

Palmer, W. T. (1974). Characterization of *-homomorphisms and expectations. *Proceedings of the American Mathematical Society*, **46**, 265–72.

Paschke, W. L. (1976). Inner product modules arising from compact automorphism groups of von Neumann algebras. *Transactions of the American Mathematical Society*, **224**, 87–102.

———. (1978). Relative commutant of a von Neumann algebra in its crossed product by a group action. *Mathematische Zeitschrift*, **163**, 5–13.

Pedersen, G. K. (1973–1977). *An Introduction to* C*-*algebra Theory*. Copenhagen: Kobenhavn Universitet.

———. (1978). Groupes localement compacts d'automorphismes d'une C*-algèbre et conditions spectrales. Unpublished manuscript.

———. (1980). Remarks on the Connes spectrum for C*-dynamical systems. *Journal of Operator Theory*, **3**, 143–8.

Pedersen, G. K. and Takesaki, M. (1973). The Radon–Nikodym theorem for von Neumann algebras. *Acta Mathematica*, **130**, 53–88.

Phillips, J. (1976). Automorphisms of full II_1 factors with applications to factors of type III. *Duke Mathematical Journal*, **43**, 375–85.

Pimsner, M. and Voiculescu, D. V. (1980). Exact sequences for K-groups and Ext-groups of certain cross-product C*-algebras. *Journal of Operator Theory*, **4**, 93–118.

———. (1982). K-groups of reduced crossed products by free groups. *Journal of Operator Theory*, **8**, 131–56.

Polya, G. and Szegö, G. (1964). *Aufgaben und Lehrsatze aus der Analysis, I.* Berlin: Springer Verlag.

Popa. S. (1986). A short proof that injectivity implies hyperifniteness for finite von Neumann algebras. *Journal of Operator Theory*, **16**, 261–72.

Powers, R. T. (1967, 1970). Representations of uniformly hyperfinite algebras and their associated von Neumann algebras. *Annals of Mathematics*, **86**, 138–71; *Cargese Lectures in Physics*, **4**, 137–68.

Pukánszky, L. (1956). Some examples of factors. *Publicationes Mathematicae Debrecen*, **4**, 135–56.

Rădulescu, F. (1992a). A one-parameter group of automorphisms of $L(F_\infty)\otimes B(H)$ scaling the trace. *Comptes Rendus de l'Académie des Sciences de Paris*, **314**, 1027–32.

———. (1992b). The fundamental group of the von Neumann algebra of a free group with infinitely many generators is $\mathbb{R}_+\setminus\{0\}$. *Journal of the American Mathematical Society*, **5**, 517–32.

———. (1995). A type III_λ factor with core isomorphic to the von Neumann algebra of a free group tensor B(H). *Recent Advances in Operator Algebras* (Orleans 1992), *Asterisque No.* **232**, 203–9.

Ramsay, A. (1971). Virtual groups and group actions. *Advances in Mathematics*, **6**, 253–322.

Reed, M. C. and Simon, B. (1972). *Methods of Modern Mathematical Physics, I.* Cambridge, MA: Academic Press.

Rieffel, M. A. (1981). C*-algebras associated with irrational rotations. *Pacific Journal of Mathematics*, **93**, 415–29.

Roberts, J. E. (1976). Cross products of von Neumann algebras by group duals. *Symposia Mathematica*, **XX**, 335–64.

Rousseau, R. (1979). The covariance algebra of an extended covariant system, I, II. *Mathematical Proceedings of Cambridge Philosophical Society*, **85**, 271–80, *Simon Stevin*, **53**, 281–95.

———. (1980). Le commutant d'une algèbre de covariance. *Revue Roumainede Mathématiques Pures et Appliquées*, **25**, 445–71.

Rousseau, R. and Van Daele, A. (1979). Crossed products of commutation systems. *Mathematische Annalen*, **239**, 7–20.

Rousseau, R., Van Daele, A. and Van Heeswijck, L. (1977). A necessary and sufficient condition for a von Neumann algebra to be in standard form. *Journal of London Mathematical Society*, **15**, 147–54.

Rudin, W. (1962). *Fourier Analysis on Groups*. Interscience Publishers.

Saito, T. (1961). On a representation of a countable infinite group. *Tohoku Mathematical Journal*, **13**, 268–73.

Sakai, S. (1957). On topological properties of W*-algebras. *Proceedings of Japan Academy*, **33**, 439–44.

———. (1966). Derivations of W*-algebras. *Annals of Mathematics*, **83**, 273–79.

———. (1970). An uncountable family of non-hyper finite type III factors. In C. O. Wilde, ed., *Functional Analysis*. Cambridge, MA: Academic Press, pp. 65–70.

———. (1971). C*-*algebras and* W*-*algebras*, Berlin: Springer Verlag.

Samuelides, M. (1978). Mesure de Haar et W*-couple d'un groupoide mesuré. *Bulletin de la Société Mathématique de France*, **106**, 261–78.

Samuelides, M. and Sauvageot, J. L. (1975). Algèbre de Krieger d'un système dynamique. *Comptes Rendus de l'Académie des Sciences de Paris*, **280**, 709–12.

Sauvageot, J. L. (1977). Sur le type d'un produit croisé d'une algèbre de von Neumann par un groupe localement compact d'automorphismes. *Bulletin de la Société Mathématique de France,* **105**, 349–68.

———. (1978). Image d'un homomorphisme et flôt des poids d'une relation d' équivalence mesuré. *Mathematica Scandinavica*, **42**, 71–100.

Schatten, R. (1950). *A Theory of Cross Spaces*, Princeton, NJ: Princeton University.

Schwartz, J. M. (1977). Relations entre "ring groups" et algèbres de Kac. *Bulletin des Sciences Mathématiques*, **100**, 289–300.

Schwartz, J. T. (1963). Two finite, non hyperfinite, non-isomorphic factors. *Communications in Pure and Applied Mathematics*, **16**, 19–26.

———. (1970). Recent progress of the structure theory of factors. In C. O. Wilde ed., *Functional Analysis*. Cambridge, MA: Academic Press, pp. 37–53.

Segal, I. E. (1953). A non-commutative extension of abstract integration. *Annals of Mathematics*, **57**, 401–57, 595–6.

Shlyakhtenko, D. (1997). Free quasi-free states. *Pacific Journal of Mathematics*, **177**, 329–68.

Skau, C. F. (1980). Geometric aspects of the Tomita-Takesaki theory, I. *Mathematica Scandinavica*, **47**, 311–28.

Stinespring, W. F. (1959). Integration theorems for gages and duality for unimodular groups. *Transactions of the American Mathematical Society*, **90**, 15–56.

Størmer, E. (1970). States and invariant maps of operator algebras. *Journal of Functional Analysis*, **5**, 44–65.

———. (1971, 1972, 1975). Hyperfinite product factors I, II, III. *Arkiv for Mathematik*, **9**, 165–70, *Journal of Functional Analysis*, **10**, 471–81, *American Journal of Mathematics*, **97**, 589–95.

———. (1971a). Automorphisms and invariant states of operator algebras. *Acta Mathematica*, **127**, 1–9.

———. (1971b). On infinite tensor products of von Neumann algebras. *American Journal of Mathematics*, **93**, 810–18.

———. (1972). On projection maps on von Neumann algebras. *Mathematica Scandinavica*, **30**, 46–50.

———. (1973). Automorphisms and equivalence in von Neumann algebras. *Pacific Journal of Mathematics*, **44**, 371–84.

———. (1974a). Positive linear maps of C*-algebras. *Lecture Notes in Physics,* **29**, 85–106.

———. (1974b). Some aspects of ergodic theory in operator algebras. *Proceedings International Congress of Mathematicians*, Vancouver, pp. 111–4.

Strătilă, Ş. V. (1967, 1968). Elements of harmonic analysis (in Romanian). *Studii si Cercetari Matematice*, **19**, 1209–45, 1473–511, **20**, 347–426, 868–926.

———. (1977). Twisted Plancherel theorem for weights: a non-commutative case. *Proceedings Colloque Marseille*, CNRS, Paris 1979, pp. 353–9.

———. (1979). On the tensor product of weights on W*-algebras. *Acta Scientiarum Mathematicarum*, **41**, 385–90.

Strătilă, Ş. V. and Voiculescu, D. V. (1975). Representations of AF-algebras and of the group U(∞). *Lecture Notes in Mathematics*, **486**.

———. (1978). On a class of KMS states for the unitary group U(∞). *Mathematische Annalen*, **235**, 87–110.

Strătilă, Ş. V. and Zsidó, L. (1972). An algebraic reduction theory for W*-algebras I. *Journal of Functional Analysis*, 11, 295–313.

———. (1977, 2005). Operator Algebras. Unpublished manuscript, Institute of Mathematics of the Romanian Academy, 500p.

———. (1979, 2019). *Lectures on von Neumann Algebras*, Editura Academiei and Abacus Press; Cambridge University Press. (Romanian Edition Editura Academiei 1975).

Strătilă, Ş. V., Voiculescu, D. V. and Zsidó, L. (1976). Sur les produits croisés. *Comptes Rendus de l'Académie des Sciences de Paris*, **282**, 779–82.

———. (1976, 1977). On crossed products I, II. *Revue Roumaine de Mathématiques Pures et Appliquées*, **21**, 1411–49, **22**, 83–117.

Sutherland, C. E. (1977). Crossed products, direct integrals and Connes classification of type III factors. *Mathematica Scandinavica*, **40**, 209–14.

———. (1980). Cohomology and extensions of von Neumann algebras I, II. *Publications of the RIMS Kyoto University*, **16**, 105–33, 135–74.

Sutherland, C. E. and Takesaki, M. (1985). Actions of discrete amenable groups and groupoids on von Neumann algebras II. *Publications of the RIMS Kyoto University*, **21**, 1087–120.

———. (1989). Actions of discrete amenable groups on injective factors of type III λ, $\lambda 1$, *Pacific Journal of Mathematics*, **137**, 405–44.

Suzuki, N. (1958). A linear representation of a countably infinite group. *Proceedings of Japan Academy*, **34**, 575–79.

———. (1959). Crossed products of rings of operators. *Tohoku Mathematical Journal*, **11**, 113–24.

Takai, H. (1975). On a duality for crossed products of C*-algebras. *Journal of Functional Analysis*, **19**, 25–39.

Takesaki, M. (1958). On the direct product of W*-factors. *Tohoku Mathematical Journal*, **10**, 116–9.

———. (1969). A generalized commutation relation for the regular representation. *Bulletin de la Société Mathématique de France*, **97**, 289–297.

———. (1969–1970). The Theory of Operator Algebras, Lecture Notes, UCLA.

———. (1970).Tomita's theory of modular Hilbert algebras and its applications. *Lecture Notes in Mathematics*, **128**.

———. (1972a). Conditional expectations in von Neumann algebras. *Journal of Functional Analysis*, **9**, 306–21.

———. (1972b). Duality and von Neumann algebras. *Lecture Notes in Mathematics*, **247**, 666–779.

———. (1973a). The structure of a von Neumann algebra with a homogeneousperiodic state. *Acta Mathematica*, **131**, 79–122.

———. (1973b). Duality for crossed products and the structure of von Neumann algebras of type III. *Acta Mathematica*, **131**, 249–310.

———. (1976). Relative commutant theorem in crossed products and the exact sequence for the automorphism group of a factor of type III. *Symposia Mathematica*, **XX**, 179–186.

Tam, P. K. (1971). On an ergodic abelian M-group. *Proceedings of Japan Academy*, **47**, 456–57.

Tatsuuma, N. (1967). A duality theorem for locally compact groups. *Journal of Mathematics Kyoto University*, **6**, 187–295.

Tomiyama, J. (1957–1959). On the projection of norm one in W*-algebras I, II, III. *Proceedings of Japan Academy*, **33**, 608–12, *Tohoku Mathematical Journal*, **10**, 204–9, **11**, 125–9.

———. (1959). On the product projections of norm one in direct product of operator algebra. *Tohoku Mathematical Journal*, **11**, 305–13.

———. (1969). On the tensor product of von Neumann algebras. *Pacific Journal of Mathematics*, **30**, 263–70.

———. (1971). Tensor Products and Projections of Norm One in von Neumann Algebras, *Lecture Notes*, København Universitet.

———. (1972). On some types of maximal abelian subalgebras. *Journal of Functional Analysis*, **10**, 373–86.

———. (1981). Completely positive maps in tensor products of von Neumann algebras. *Journal of the Mathematical Society of Japan*, **33**, 559–60.

Turumaru, T. (1958). Crossed products of operator algebras. *Tohoku Mathematical Journal*, **10**, 355–65.

Ueda, Y. (2011). On type III_1 factors arising as free products. *Mathematical Research Letters*, **18**, 909–20.

Umegaki, H. (1954, 1956, 1959, 1962). Conditional expectation in an operator algebra I, II, III, IV. *Tohoku Mathematical Journal*, **6**, 177–81, **8**, 86–100, *Kodai Mathematical Seminary Reports*, **11**, 51–74, **14**, 59–85.

Vainerman, L. I. and Kac, G. I. (1973, 1974). Non commutative ring groups and Hopf-vonNeumann algebras (in Russian). *Doklady Akademii Nauk SSSR*, **211**, 1031–1034, *Matematicheskii Sbornik*, **94**, 194–225, 335.

Van Daele, A. (1973). The upper envelope of invariant functionals majorized by an invariant weight. *Pacific Journal of Mathematics*, **46**, 283–302.

———. (1976). A Radon Nikodym theorem for weights on von Neumann algebras. *Pacific Journal of Mathematics*, **61**, 527–42.

———. (1978). A framework to study commutation problems. *Bulletin de la Société Mathématique de France*, **106**, 289–309.

Van Heeswijck, L. (1979). Duality in the theory of crossed products. *Mathematica Scandinavica*, **44**, 313–29.

Voiculescu, D. V. (1977). Amenability and Kac algebras. *Proceedings Colloque Marseille*, CNRS Paris, 1979, pp. 451–57.

———. (1985). Symmetries of some reduced free product C*-algebras, In *Operator Algebras and their Connection with Topology and Ergodic Theory* (Busteni1983), pp. 556–88, *Lecture Notes in Mathematics*, **1132**, Springer Verlag.

———. (1990). Circular and semicircular systems and free product factors. In *Operator Algebras, Unitary Representations, Enveloping Algebras and In-Variant Theory* (Paris 1989), pp. 45–60, *Progress in Mathematics*, **92**, Birkhauser Verlag,

———. (1991). Limit laws for random matrices and free products. *Inventiones Mathematicae*, **104**, 201–20.

von Neumann, J. (1938). On infinite direct products. *Compositio Mathematica*, **6**, 1–77.

———. (1940). On rings of operators III. *Annals of Mathematics*, **41**, 94–161.

Wassermann, S. (1977). Injective W*-algebras. *Mathematical Proceedings of the Cambridge Philosophical Society*, **82**, 39–48.

Woods, E. J. (1973). The classification of factors is not smooth. *Canadian Journal of Mathematics*, **25**, 96–102.

Woronowicz, S. L. (1979). Operator systems and their applications to the Tomita-Takesaki theory. *Journal of Operator Theory*, **2**, 169–209.

Zeller-Meier, G. (1967). Sur les automorphismes des algèbres de Banach. *Comptes Rendus de l'Académie des Sciences de Paris*, **264**, 1131–2.

———. (1968). Produits croisés d'une C*-algèbre par un groupe d'automorphismes. *Journal de Mathématiques Pures et Appliquées*, **47**, 101–239.

Zimmer, R. J. (1976). Extensions of ergodic group actions. *Illinois Journal of Mathematics*, **20**, 373–409.

———. (1977a). On the von Neumann algebra of an ergodic group action. *Proceedings of the American Mathematical Society*, **66**, 289–93.

———. (1977b). Hyperfinite factors and amenable ergodic actions. *Inventiones Mathematicae*, **41**, 23–32.

———. (1978). Amenable ergodic group actions and an application to Poisson boundaries of random walks. *Journal of Functional Analysis*, **27**, 350–72.

Zsidó, L. (1977). Invariance and dual weights. *Proceedings Colloque Marseille*, CNRS Paris, 529–33, 1979.

———. (1978). On the equality of two weights. *Revue Roumaine de Mathématiques Pures et Appliquées*, **23**, 631–46.

Notation Index

Subject Index